Organosilicon Chemistry IV

Edited by N. Auner and J. Weis

Organosilicon Chemistry IV

From Molecules to Materials

Edited by Norbert Auner and Johann Weis

WILEY-VCH

Weinheim · New York · Chichester · Brisbane · Singapore · Toronto

Prof. Dr. N. Auner
Inst. für Anorganische Chemie
der Universität Frankfurt
Marie-Curie-Strasse 11
D-60439 Frankfurt am Main
Germany
Phone: 069/798-29180, -29591
Fax: 069/798-29188

Dr. J. Weis
Wacker-Chemie GmbH
Geschäftsbereich S
Werk Burghausen
Johannes-Hess-Strasse 24
D-84489 Burghausen
Germany

Library of Congress Card No. applied for.

A catalogue record for this book is available from the British Library.

Die Deutsche Bibliothek – CIP Cataloguing-in-Publication Data
A catalogue record for this publication is available from Die Deutsche Bibliothek

Printed on acid-free and chlorine-free paper.

Printing: betz-druck, D-64291 Darmstadt
Bookbinding: Großbuchbinderei J. Schäffer GmbH & Co KG, D-67269 Grünstadt
Printed in the Federal Republic of Germany

Preface

This volume collects the lectures and poster contributions that were presented at the *IV. Münchner Silicontage* which were held in April of 1998. This symposium was — as were its three predecessors in 1992, 1994, and 1996 — again organized jointly by the Gesellschaft Deutscher Chemiker, and by Wacker Chemie, GmbH. The exceptionally large number of participants from both industry and academia, especially students and young scientists, was convincing evidence of the high interest this meeting continues to find, now having developed into a well-established international symposium. This favorable response, and that of organosilicon researchers from all over the world to the "*Organosilicon Chemistry — From Molecules to Materials*" proceedings encouraged us to continue the series with the present edition of Volume IV. As was said about the first three volumes, this is also not considered a text book in the common sense which would lay out the basic knowledge of a discipline but provides, written by researchers and experts in a fascinating field of rapidly growing main group chemistry, accounts and summaries of the latest results. Furthermore, in editing the present volume, we used this occasion to update the authors' contributions to the different topics reviewed at the symposium with references of papers that were published during the past two years, making it relevant up to the publishing date.

In Volume III, the contributions to each of the different categories were accompanied by a short overview and a summary of current research directions and developments in organosilicon chemistry. In order to avoid repetition, and in consideration of the fact that research emphasis changes only slightly over periods of only two to three years, we have omitted a similarly detailed work-up of the material. Nevertheless, a concise summary of current material producing processes and new, clearly recognizable research trends precedes the contributions. This preamble particularly addresses the trend — as seen recently at the "*12th International Symposium on Organosilicon Chemistry*" in Sendai, Japan, 23–18 May 1999 — towards generating silicon-based novel materials and understanding their properties, as well as the fact that the *physics* of the silicon atom and its compounds becomes increasingly important, as for example for optoelectronic applications. In addition, the biotechnological design of new Si–O-containing materials is possibly developing into a future research area [1]. This trend is also acknowledged here: The contribution by Prof. Dr. D. Morse, Marine Science Institute, Santa Barbara, on "*Silicon Biology: Proteins, Genes, and Molecular Mechanisms Controlling Biological Nanofabrication with Silicon*" opens our wide-ranging collection of conference lectures.

During the *I. Münchner Silicontage*, the two pioneers of organosilicon chemistry, Prof. Dr. Richard Müller and Prof. Dr. Eugene Rochow, were awarded the *Wacker Silicon-Preis* on the occasion of the 50th anniversary of the "Direct Synthesis." While a few weeks ago the latter celebrated his 90th birthday to which we congratulate him on behalf of the whole family of "silicon chemists," Prof. Müller has passed away several months ago [2]. We will honor the name of this great pioneer which is irrevocably connected with the "Müller-Rochow Synthesis." As Volume III of these proceedings was dedicated to the memory of the late Prof. Dr. E. Hengge (Graz), it is the express wish of the editors that the present Volume IV shall keep alive the memory of this excellent organosilicon chemist and outstanding personality — Prof. Dr. Richard Müller.

At the *IV. Müncher Silicontage*, Prof. Dr. Robert Corriu, Montpellier, was honored with the *Wacker Silicon-Preis* for his innovative work in all fields of organosilicon chemistry. His contributions treat fundamental molecular questions as well as the synthesis and structure of novel Silicon containing compounds and materials.
In collecting and publishing the papers in this volume, it is our desire to bring out the diversity of silicon chemistry as well as the fascination with this element. As much as we know today, the chemistry — but in particular also the *physics* and the *biology* of silicon and its compounds — have not at all been exhaustively treated yet. The future will certainly bring many beautiful and certainly also surprising results. This book shall essentially stimulate young researchers to immerse themselves in this research: There are still many challenging problems to be solved, in basic research as well as with the development of new materials.

October 1999 *Prof. Dr. Norbert Auner, Dr. Johann Weis*

References:

[1] R. Tacke, *Angew. Chem. Int. Ed. Engl.* **1999**, *38*, 3015–3018.
[2] H. Schmidbaur, *Nachr. Chem. Tech. Lab.* **1999**, *47*, 1261.

Acknowledgment

First of all we thank the authors for their contributions and intense cooperation, which made this overview of current organosilicon chemistry possible. The tremendous work to achieve the attractive layout of this volume was performed by Dr. Barbara Patzke. Dr. Thomas Müller and Georgios Tsantes helped to organize the editorial work. The coworkers of Norbert Auner: Michael Amon, Christian Bauch, Martin Bleuel, Jens Elsner, Andreas Frost, Dr. Bernhard Herrschaft, Sven Holl, Hannelore Inacker-Bovermann, Bärbel Köhler-Abegg, Dr. Keramatollah Mehraban and Dr. Bahman Solouki were very active to read, compare and correct.

We thank all of them for their admirable engagement!

Prof. Dr. Norbert Auner
Johann Wolfgang Goethe-Universität
Frankfurt

Dr. Johann Weis
Wacker-Chemie GmbH
München

Contents

ORGANOSILICON CHEMISTRY: FROM MOLECULES TO MATERIALS

Introduction

Norbert Auner

Johann Wolfgang Goethe-Universität
Frankfurt, Germany

Johann Weis

Wacker–Chemie GmbH
München, Germany

In Volume III of the series *Organosilicon Chemistry: From Molecules To Materials* the editors gave a comprehensive summary about the focus of basic research activities in the wide-spread field of organosilicon chemistry and cited the relevant literature. With respect to the title the introduction of this volume deals mainly with a summary of "how to make silicon-containing compounds and how to transfer them into new materials".

Organosilanes are compounds containing a silicon–carbon bond. Generally, the chemistry of these compounds is similar to the chemistry of organic compounds, but due to the difference in electronegativity (EN(C) 2.5; EN(Si) 1.7), the silicon–carbon bond is polar, with silicon being the electropositive partner. This leads to very basic distinctions: (i) nucleophilic substitution at silicon is more facile than at carbon, (ii) the bond energies of silicon with electronegative elements, such as oxygen and halides, are greater than with carbon, (iii) a silicon–carbon bond stabilizes a carbanion in the α- and a carbocation in the β-position and (iv) the silicon–hydrogen bond is polarized with $Si^{\delta+}$ and $H^{\delta-}$ (compared to $C^{\delta-}$ and $H^{\delta+}$) resulting in an *anti*-Markovnikov addition to olefins and the ability to hydride transfer. Mainly two different routes are used to produce organosilanes, which can not be found in natural sources (Scheme 1): starting from elemental silicon, which is derived from silica or quartz, first chlorosilanes are synthesized by use of chlorine or hydrogen chloride. These compounds can then be reacted with Grignard reagents or alkali alkyls and aryls to give organic substitutions. A very efficient route to silicon–carbon bond formation is through hydrosilylation of an olefin under catalytic conditions.

$$SiO_2 \xrightarrow[-\,CO_2]{C\ /\ \text{electric furnace}} Si \xrightarrow[\Delta]{2\,Cl_2} SiCl_4 \xrightarrow[-\,LiCl]{RLi} RSiCl_3$$

$$Si \xrightarrow[-\,SiCl_4,\ -\,H_2]{3\,HCl,\ \Delta} HSiCl_3 \qquad RSiCl_3 \xrightarrow[-\,LiCl]{RLi} R_2SiCl_2$$

$$RCH_2CH_2SiCl_3 \xleftarrow[cat]{RHC{=}CH_2} HSiCl_3 \xrightarrow[-\,MgCl_2]{+\,RMgCl} RHSiCl_2 \dashleftarrow R_2SiCl_2 \dashrightarrow$$

Scheme 1. From silicon to organosilanes.

For the large scale production of methylchlorosilanes, especially the dimethyldichlorosilane, which is the source for the production of silicones, e.g. polydimethylsiloxane (PDMS), silicon is reacted with methyl chloride under copper catalysis ("Direct Process", "Müller-Rochow Process", see Preface and Scheme 2).

$$2\,MeCl \;+\; Si \xrightarrow[\Delta]{Cu} Me_2SiCl_2 \xrightarrow[-2\,HCl]{2\,H_2O} \left[Me_2Si(OH)_2\right] \longrightarrow \text{silicone polymers}$$

Scheme 2. The Direct Process and PDMS formation.

The chloro substituents of the organochlorosilanes are often used for the introduction of hydrogen, alkoxy, acyloxy, amino and hydroxyl groups by reaction of the silicon–chlorine bond with hydride reducing agents, alcohols, anhydrides, amines, and water. As mentioned, the reaction with water is particularly important in that it provides a basis for silicone manufacture. Dimethyldichlorosilane is difunctional and reacts with water to form cyclic siloxanes and/or straight chain polymers. Because of their high moisture sensitivity chlorosilanes are often replaced by alkoxysilanes in technological processes, e.g. in sol gel processes. Alkoxysilanes $R_nSi(OR)_{4-n}$ undergo most of the reactions of chlorosilanes but are generally more convenient reagents. As they are more resistant to hydrolysis, polymerization reactions to form silicones are easier to control. Furthermore increasing the size or steric bulk of the alkoxy groups decreases reactivity and the byproduct of hydrolysis is alcohol rather than hydrogen chloride.

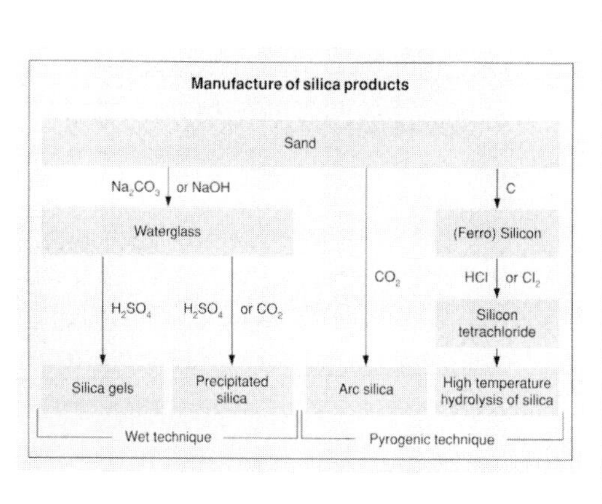

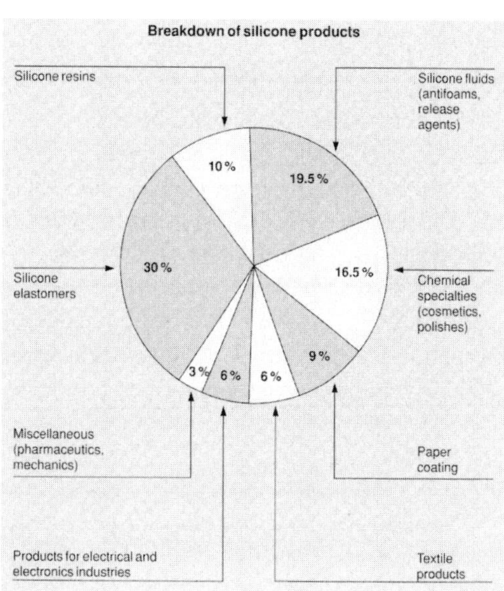

Fig. 1. Manufacture of silica and breakdown of silicone products (printed from A. Tomanek: "Silicon and Industry", published by Wacker Chemie GmbH, Munich, 1991).

The *technology* of silicon deals mainly with the big scale production of hyperpure silicon, silicon carbide, silicates and silica, and industrial silicone products. The surface synthesis and modification of a wide range of substrates may be accomplished with silanes. Deposition techniques include the preparation of silicon for semiconductors with silane, dichlorosilane, and trichlorosilane. Silicon carbide deposition is carried out at high temperatures with methyltrichlorosilane and other suitable precursors. Silicon dioxide may be deposited from acyloxysilanes by thermal processes or by hydrolysis of esters. The manufacture of silica as well as the breakdown of silicone products is summarized in Figure 1.

To understand the wide-spread use of silicones and the enormous number of tons produced per year, one has to consider their outstanding material properties that render them indispensable for many applications: (i) the thermal resistance of silicones is such that their physical and mechanical properties alter only very gradually with changes in temperature. Extensive breakdown of properties occurs at temperatures in the range of 200-300°C. Obviously the high thermostability stems from the strength of the Si–O–Si linkage (E_{Si-O} 444, E_{C-O} 339 kJ/mol). (ii) The resistance to low temperatures manifests itself in low pour points and glass-transition temperatures of around –60 and –120 °C respectively. (iii) Silicones rank among those plastics with the best dielectric properties, e.g., they are well established insulators. Values for dielectric strength, resistivity and dielectric constant do not change over a temperature range of about 20–200°C. (iv) Due to the organic substituents, especially the methyl groups in PDMS, silicones are strongly hydrophobic; silicone fluids spread out spontaneously over surfaces for which they have an affinity, e.g., glass, building materials, and fibers. Therefore they are used in masonry protection and in improving the water repellency of textiles. (v) Closely allied with hydrophobicity is the ability of silicones to act as release agents for tacky substances such as fats, oils and plastics. (vi) With the exception of low-viscosity volatile compounds, silicones are remarkably inert to living organisms. Due to their polymeric nature the passage through biological membranes is hampered; furthermore, the silicon–carbon and silicon–oxygen bonds are rather stable under physiological conditions. These facts mainly explain the use of silicones in medical applications and as materials for prosthetic devices. (vii) Silicones enter the environment mainly in form of fluids and derivatives, e.g. antifoams, waxes, emulsion and cosmetics. There is no evidence for any toxic mutagenic or teratogenic effects on animals or aquatic life and there is no evidence that silicones adversely effect ecosystems. Silicones degrade in contact with *micaceous* clays to form cyclic siloxanes and silanols. UV light and nitrate exposure converts siloxanes to silica which can be absorbed by *siliceous algae*.

Although silicones are still the most important class of technologically produced silicon based materials up to date, the products from hydrolytic condensation reactions of trifunctional organosilicon monomers (e.g. $RSiCl_3$ or $RSi(OR)_3$) are coming more and more into scientific focus; this class of compounds is that of the polyhedral oligosilsesquioxanes (Scheme 3).

$$n\,RSi(OH)_3 \xrightarrow[-x\,H_2O]{} \underset{\text{incompletely-condensed silsesquioxane}}{[RSiO_{3/2}]_n(H_2O)_{3n/2-x}} \xrightarrow[-(3n/2-x)\,H_2O]{} \underset{\text{fully-condensed silsesquioxane}}{[RSiO_{3/2}]_n}$$

Scheme 3. Formation of polyhedral oligosilsesquioxanes.

These reactions are uniquely sensitive to a highly interdependent combination of experimental factors and generally produce a wide varity of products, ranging from small oligomers and discrete clusters to complex mixtures, e.g. the resin materials and “T-gels”. Fully condensed silsesquioxane frameworks containing 6, 8, 10 and 12 Si atoms could be isolated from those experiments (Scheme 4).

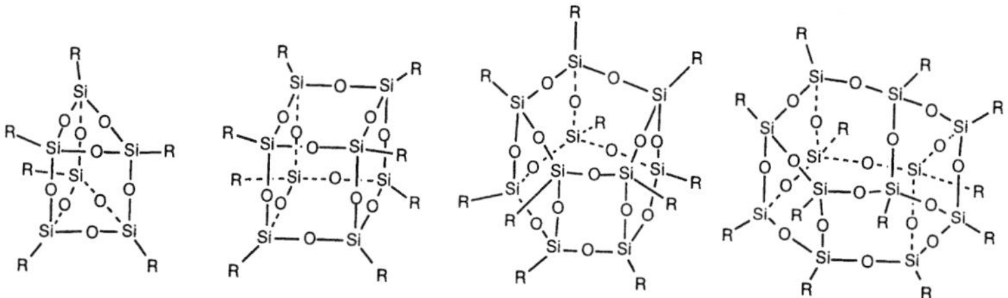

(n = 6) $R_6Si_6O_9$ (n = 8) $R_8Si_8O_{12}$ (n = 10) $R_{10}Si_{10}O_{15}$ (n = 12) D_{2d}-$R_{12}Si_{12}O_{18}$

Scheme 4. Examples of polyhedral silsesquioxanes.

The polyhedral silsesquioxanes have recently attracted wide-spread interest as precursors to hybrid inorganic–organic materials. Incompletely condensed frameworks have been used as comonomers of silsesquioxane-based polymers and as building blocks for network solids, while fully condensed frameworks have been used for very specific applications, including photocurable resins, liquid crystals, electroactive films and building blocks for catalytically active organometallic gels. One of the most promising new applications for silsesquioxanes is the use of fully condensed frameworks possessing one potentially polymerizable pendant group as comonomers in traditional thermoplastic resins.

Another class of materials becoming important for both academia and industry is that of polysilanes, compounds consisting of silicon–silicon bonds in the polymeric chain. Except for the pioneering studies of Kipping, little work was done on polysilanes before 1955. In the late 1950s and 1960s, cyclic and linear (short-chain) silanes were investigated in several laboratories. In 1970 cyclic alkylpolysilanes were found to have unusual behaviour resulting from delocalization of the σ-electrons in the silicon–silicon bonds. Remarkably, also at this time, polysilanes were found to be photochemical precursors to divalent silicon species, the precursors to disilenes, compounds with doubly bonded silicon (for silicon in unusual coordination numbers, including silylenes, Si=E species (E = Si, C, Ge, N, P, O, S), silyl cations etc., see: Introduction of *Organosilicon Chemistry: From Molecules to Materials, Volume III*). Most recently, attention has been focused on the polysilane high polymers, both because of their unique photophysical properties and their growing technological importance, which is due to their wide range of applications including their use as precursors to silicon carbide, as photoresists for microelectronics, and as photoinitiators.

Recently, silicon biotechnology develops as a highly active field in organosilicon chemistry. This is displayed also in the present volume with several contributions. As D. E. Morse pointed out in his lecture (p. 5) “we learn from nature that the earth’s biota produces gigatons of silica annually. The precise control of nanofabrication of these biosilica structures under remarkable conditions in many instances exceeds the present capabilities of human engineering”. From basic understanding “how mother nature produces silicates from silicic acid” the future might show that the transfer of this knowledge into application and possibly into the technology of silicate and resin fabrication, or even into the controlled formation of discrete structures (e.g. silsesquioxanes) with unique physical (e.g. optoelectronic) properties, might heavily influence organosilicon research. Especially this contribution shows that nature transfers molecules into materials in a highly controlled fashion.

Silicon Biotechnology: Proteins, Genes and Molecular Mechanisms Controlling Biosilica Nanofabrication Offer New Routes to Polysiloxane Synthesis

Daniel E. Morse

Marine Biotechnology Center
and Department of Molecular, Cellular and Developmental Biology
University of California
Santa Barbara, CA 93106, USA
Tel.: Int. code + (805)893 3157 and 8982 — Fax: Int. code + (805)893 8062
E-mail: d_morse@lifesci.lscf.ucsb.edu

Keywords: Biotechnology / Biosynthesis / Biosilica / Silicatein / Protein / Gene / DNA / Catalysis

Summary: Analysis of occluded proteins from the biosilica spicules produced by a sponge, and analysis of the cloned DNAs that code for these proteins, reveal that the proteins can act both catalytically and as macromolecular templates in vitro, directing the condensation of polysiloxanes from silicon alkoxides at neutral pH. These proteins (called "silicateins") prove capable of directing the polymerization of silica and silsesquioxanes from the corresponding alkoxides under mild physiological conditions, whereas the anthropogenic synthesis of these materials typically requires acid or base catalysis. Homology between the silicateins and members of a well-characterized family of hydrolytic enzymes suggests a possible mechanism for protein-mediated catalysis of the rate-limiting step in polysiloxane synthesis. Genetic engineering and biotechnology offer the prospects of identifying the structural determinants of the polysiloxane synthesis-directing activities, developing new environmentally benign routes to synthesis of more ordered silsesquioxanes, and discovering the proteins, genes and molecular mechanisms controlling silicon metabolism in the sponge and other living systems.

Biological Silica Synthesis

Biology produces a remarkable diversity of silica structures [1, 2], many of which are strikingly beautiful. The biological synthesis of the wide variety of shells, spines and granules made by diatoms, other unicellular organisms, sponges, mollusks and plants is remarkable in three respects:

(1) The precise control of nanofabrication of these biosilica structures in many instances exceeds the present capabilities of human engineering. We know that this precision is under direct genetic control. Thus, the structures of the siliceous walls made by each of the

tens of thousands of species of diatoms is distinct (Fig. 1), yet essentially the same for each member of a given species. This genetic control tells us that the blueprint for the synthesis and structure is contained in the DNA of each organism that builds with silicon, and that this blueprint is translated via the protein products of the DNA.

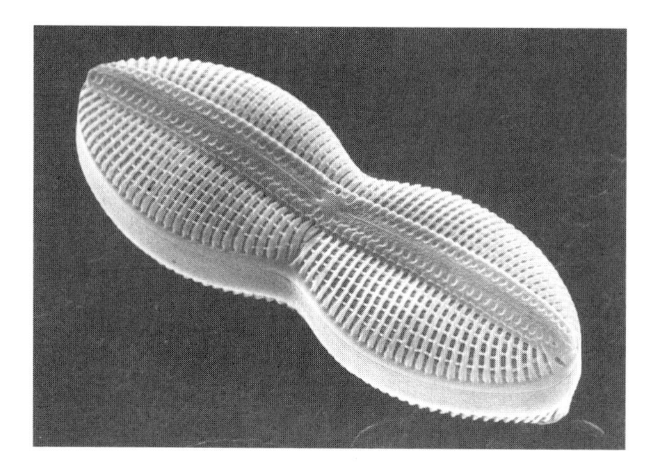

Fig. 1. Silica outer wall of the marine diatom, *Diploneis* (a unicellular alga). Length 50 μm. Reprinted by permission from [3].

(2) Equally remarkable is the scale of this biological synthesis of silica. Earth's biota produce gigatons of silica annually; much of this production occurs in the world's oceans, where it contributes an estimated 3 km^3 to the ocean floor sediments each year. Ancient deposits of diatomite thousands of meters thick are mined today for industrial purposes.

(3) The conditions of synthesis of the biosilicas are also remarkable: unlike the anthropogenic and geological syntheses of siloxanes, which typically require extremes of temperature, pressure or pH, the genetically controlled biosynthesis of silica is accomplished under mild ambient physiological conditions, including the near-zero temperatures of the polar seas in which many diatoms flourish in vast populations. Until recently, little was known of the molecular mechanisms responsible for the remarkable biosynthesis of silica.

Taking advantage of marine organisms that produce relatively large masses of silica, molecular biologists are now using the tools of biotechnology to reveal the molecular mechanisms by which living organisms build with silicon [4–8]. In a major breakthrough, Hildebrand, Volcani and their colleagues recently characterized a protein that transports silicic acid from solution (in the surrounding ocean) into the diatom cell [4]. They accomplished this discovery by cloning and characterizing the DNA that codes for the transporter protein. This is the first silicon transporter to be directly characterized, although we know from physiological experiments that such transporters must exist in many other forms of life. From their analysis of the protein structure encoded by the cloned DNA, Hildebrand et al. deduced that the transporter is somewhat homologous to proteins known to transport other ions into cells; the 12 α-helical domains of the protein are thought to loop back and forth in the cell membrane, forming an ion-specific channel for silicic acid. Dramatic support for their conclusion was obtained by introducing messenger RNA (made from the cloned DNA in vitro) into frog cells in culture; these cells, which never before would take up significant quantities of silicic or germanic acid, suddenly began this specific transport when the messenger RNA was introduced and directed the synthesis of the transporter protein. The transporter protein uses the cell's metabolic energy to pump silicic acid into the cell against a concentration gradient. In cells building silica structures, a similar transporter may pump the intracellular silicic acid (or some conjugate of silicic acid) into the "silica deposition vesicle" (SDV; also known in some systems as the silicalemma), the membrane-enclosed envelope within the cell inside which condensation and nanofabrication of the silica occurs [1]. Kröger and his colleagues recently have characterized the proteins that are closely associated with the silica cell wall of the diatom [5]. From the structures of these proteins (deduced from the sequences of the cloned DNAs), and immunohistochemical localization of the proteins with labeled antibodies, they demonstrated that the proteins most closely associated with the outer surface of the silica contain regularly repeating domains that are rich in hydroxyl groups. Hydroxyls have long been thought to play a role in the organization of biosilicification [9–15]. Hecky et al proposed that regular arrays of protein hydroxyl groups could facilitate silica synthesis by alignment of silicic acid precursors, either by direct condensation or by hydrogen bonding [9], and thermodynamic calculations have shown that such a mechanism would be energetically plausible [12].

Silicateins: Enzyme-like Proteins in Biosilica

We have worked with the readily available marine sponge, *Tethya aurantia*, because 75 % of the dry weight of the organism is composed of silica spicules (glassy needles), permitting us to obtain biochemical quantities of the proteins that are occluded within the biosilica [6–8]. The silica spicules are ca. 2 mm long × 30 μm diameter; dissolution of the biosilica with buffered HF reveals that each spicule contains an occluded, central axial filament that is ca. 2 mm long × 1-2 μm diameter (Fig. 2) consisting primarily (≥91 %) of protein [6]. The purified protein filaments contain no residual silica (as determined by the molybdate assay after hydrolysis). Similar protein filaments have been found occluded within the spicules and more elaborate silica structures made by a wide variety of sponges. They had been suspected to participate in the control of biosilicification, although their mechanism of action was unknown [16]. Microscopic observations in another species suggested that the synthesis of the protein filaments preceded silicification [17].

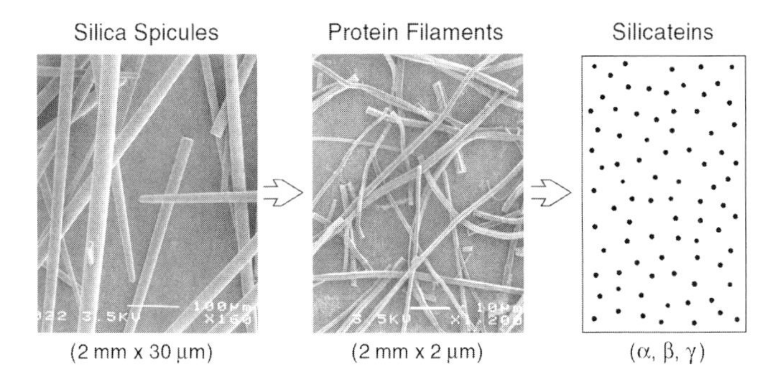

Fig. 2. Relationship of silica spicules, occluded axial protein filaments and the constituent silicatein subunits. Image of spicules and filaments reprinted from [6]; subunits illustrated schematically.

We have now found that in those rare cases (≪1 %) in which the silica spicules made by *Tethya aurantia* exhibit a bifurcation or cruciform structural anomaly, the axial filament within the silica had first bifurcated, supporting the suggestion that the silicification reaction was directed by (and followed the course of) the protein filament.

We found that the macroscopic protein filaments occluded in the cores of the *Tethya* spicules (and comprising only 0.5 % of the weight of the spicules) are composed primarily of three protein subunits that can be dissociated from one another and purified electrophoretically ([6]; cf. schematic illustration in Fig. 2). The subunits are quite similar in amino acid composition and pI, with apparent molecular masses of approximately 29, 28, and 27 kDa. We have named these protein subunits silicatein (for "silica protein") α, β, and γ, respectively. Densitometric analyses reveal these subunits to be present in the relative proportions of approximately $\alpha:\beta:\gamma = 12:6:1$. Small-angle X-ray diffraction of the protein filaments exhibits a sharp peak at $2\theta = 0.51°$ corresponding to a periodicity of ca. 17 nm, suggesting that the macroscopic silicatein filaments are built up from a highly regular, repeating subassembly of the subunits. Earlier work with electron microscopy detected pararacrystallinity in the axial filaments of another sponge species, consistent with this finding [18].

The complete amino acid sequence of silicatein α was determined both by direct sequencing of the protein and its peptide fragments, and from the sequence of the cloned DNA that codes for the protein [6]. Results from the two methods matched perfectly. The major unanticipated finding to emerge from this analysis was the discovery that silicatein α is highly homologous to members of the cathepsin L subfamily of papain-like proteolytic enzymes [6]. The positions of all six cysteines (amino acids with sulfhydryl sidechains) that form intramolecular disulfide bridges in cathepsin L are perfectly conserved in silicatein α, suggesting that the three-dimensional structures of the two proteins are highly similar. Two of the three amino acids of the "catalytic triad" (the catalytic active site) of the cathepsin also are exactly conserved in the silicatein.

Analysis of the amino acid sequence also revealed that silicatein α is made biosynthetically as a "pre-pro-protein", with two peptide fragments successively removed from the amino-terminal end by proteolytic enzymes that cut the protein as it is secreted into the membrane-enclosed silica deposition vesicle and then folded to its final three-dimensional conformation [6]. The "signal peptide" that facilitates recognition and secretion into the SDV is cleaved by a specific "signal peptidase". After folding of the protein within the SDV, the remaining N-terminal "propeptide" is then removed to release the mature silicatein. The sequences of amino acids that specify the sites of these two cleavages are homologous to those found in the precursors of other members of the papain family as well. Computer-assisted analysis of the alignment of the sequence of silicatein α with that of cathepsin L reveals that 45 % of the corresponding amino acids are identical in the two full-length proteins; they differ in length by only three residues. When the mature silicatein α and protease sequences are compared (after removal of the signal and propeptides), the sequence identity is 52 %; correspondence of residues with identical or functionally similar sidechains is 75 %. These findings strongly suggest that silicatein α and the cathepsin L/papain family of proteolytic enzymes evolved from an evolutionarily shared common ancestor. The facts that the pre-pro-peptides of silicatein α and cathepsin L are similar over their entire lengths, and that both proteins are localized within membrane-enclosed vesicles that are essential for their function (the SDV in the case of silicatein; the digestive vesicle known as the lysosome in the case of cathepsin L), further suggest a common evolutionary origin of the two proteins. These findings raised the previously unexpected possibility that the silicateins might possess an enzyme-like catalytic activity, in addition to their previously anticipated "scaffolding" or templating-like activity.

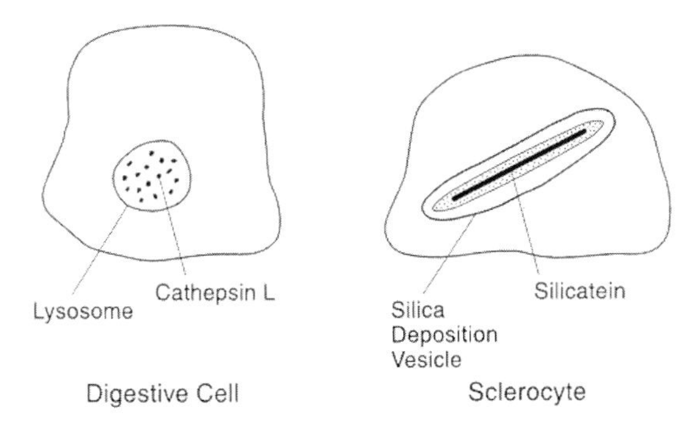

Fig. 3. Similarities between the intracellular membrane-enclosed localizations of the homologous cathepsin L and the silicatein (silicification-directing) proteins. The specialized silica producing cells in the sponge are called sclerocytes. The silica spicule surrounds the occluded silicatein filament.

Silicateins Catalyze Condensation of Silicon Alkoxides at Neutral pH

Because the proteases to which the silicateins are homologous function catalytically as hydrolases (capable of cleaving peptide and ester bonds at neutral pH), we explored the possibility that the silicateins might exhibit similar activity with silicon alkoxide substrates [7]. Hydrolysis of the alkoxide is usually the rate-limiting step in the condensation of these substrates, generally requiring either acid or base catalysis [19]. In contrast, the intact silicatein filaments, the dissociated silicatein subunits, and the purified silicatein α subunit (produced from a recombinant DNA template cloned in bacteria) all accelerate the hydrolysis and condensation of tetraethoxysilane (TEOS) to form silica $(SiO_2)_n$ at neutral pH in vitro [7]. The rate of product formation was quantified by alkaline hydrolysis of the collected silica followed by spectrophotometric determination with the molybdate reagent. Under the reaction conditions we employed (at neutral pH and low temperature), the rate of spontaneous condensation of TEOS in the absence of the silicateins was quite low. The activity of the silicatein filaments and subunits was in each case abolished by thermal denaturation of the protein, indicating that the catalytic activity is dependent on the three-dimensional structure of the native silicatein. Furthermore, the effect is specific for the silicateins, since the related proteolytic enzymes papain and trypsin (an unrelated protein, serum albumin) are without effect. When the macroscopic silicatein filaments are used, the silica product is formed over the surface of the filament, following the contours of the underlying macromolecular topology (Fig 4A, B).

Thus the silicatein filaments exhibit both "scaffolding" (macroscopic structure-directing) and catalytic activities in directing the condensation of the alkoxides in vitro, although the catalysis apparently is limited by passivation of the active surface. Solid-state ^{29}Si NMR confirmed the identity of the silica product, revealing Q^2, Q^3 and Q^4 siloxanes in an amorphous, incompletely polymerized opal-like silica typical of that made in biological systems [7]. Neither silk fibroin nor cellulose fibers — both of which posses high surface densities of hydroxyl groups — promote the condensation of TEOS at neutral pH (e.g., Fig. 4D), thus demonstrating that more than clusters of surface hydroxyls may be required.

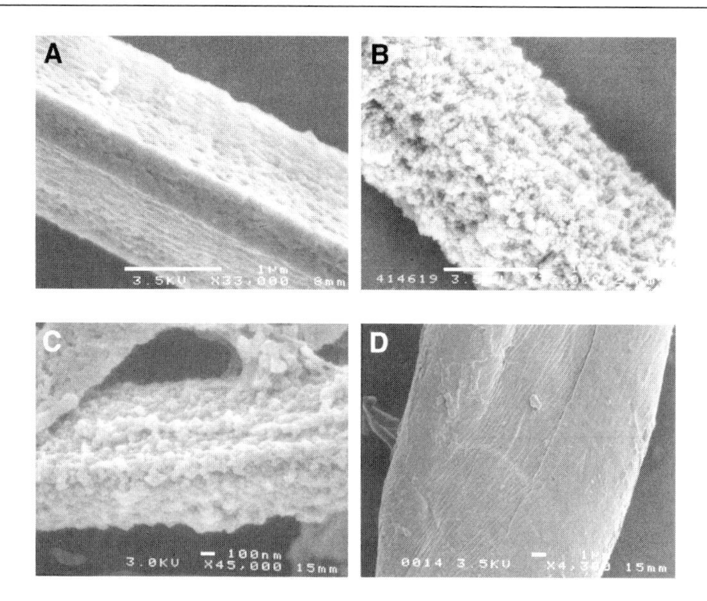

Fig. 4. (A) Purified silicatein filament; (B) silicatein filament covered with silica formed by reaction with TEOS; (C) silicatein filament covered with phenylsilsesquioxane formed by reaction with phenyltriethoxysilane; (D) cellulose filament after reaction with TEOS, showing no silica formed. Scanning electron micrographs reprinted from [7].

When organically substituted silicon triethoxides are provided as substrates, the silicateins also catalyze their condensation to form the corresponding polysilsesquioxanes ($RSiO_{3/2}$) at neutral pH [7]. Reaction of the macroscopic silicatein filaments with phenyl- or methyl-triethoxysilane promotes rapid polymerization with scaffolding of the resulting silsesquioxane polymer network on the silicatein filament (Fig. 4C). In the absence of the silicatein, no condensation was observed at neutral pH under the reaction conditions we employed; polymerization normally would require either an acid or base catalyst [19]. Cross-polarization ^{29}Si NMR confirmed the structure of the phenylsilsesquioxane product, revealing the expected T^3, T^2 and T^1 species, with no Q^4 as predicted for the silsesquioxane [7].

Possible Mechanism of Action

The silicatein α subunit alone proved sufficient to catalyze the acceleration of TEOS condensation at neutral pH [7]. For these experiments, we purified and reconstituted the silicatein α subunit that we produced from a recombinant DNA that we cloned in bacteria — cells which normally would make no silicatein without the introduced recombinant DNA template (Fig. 5). This method allowed us to be sure that the catalysis of silica synthesis we observed was due solely to the silicatein α protein produced from the cloned DNA and purified from the bacteria, since no other proteins from the sponge could be present.

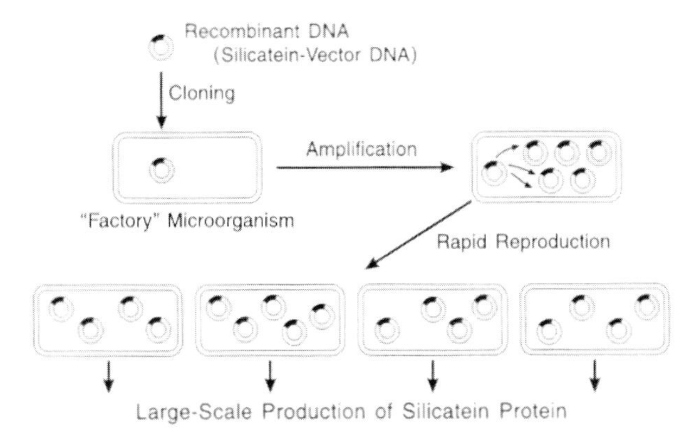

Fig. 5. Amplified production of silicatein protein from recombinant DNA cloned in bacteria. DNA coding for the silicatein α protein was integrated enzymically into a double-stranded circular DNA "vector" molecule to produce a recombinant DNA. When introduced into bacteria, the recombinant DNA is replicated by intracellular enzymes to produce many copies in each cell. These DNA molecules serve as templates, directing the synthesis of the silicatein protein they encode. Rapid reproduction of the bacteria greatly amplifies this synthesis.

The finding that silicatein α is sufficient to catalyze silica synthesis from TEOS at neutral pH is significant for two reasons:

(1) it provides insights into the mechanism of catalysis, since the complete amino acid sequence and three-dimensional structure of the silicatein α are highly homologous to that of cathepsin L [6];
(2) it makes it possible to test the suggested mechanism by producing genetically engineered structural variants of the protein, by a method known as site-directed mutagenesis [20].

Because the three-dimensional conformation and the putative catalytic active site of the silicatein α are both known to be closely related to those of the homologous cathepsin L (see above), because the condensation of the silicon alkoxides promoted by the silicateins and the cleavage of peptides catalyzed by the proteases both proceed through an obligatory hydrolysis reaction, and because both reactions can be accelerated by general acid–base catalysis, it is possible to suggest that the mechanism of action of silicatein α in this process might be formally quite similar to that of its homologous enzyme counterparts. This suggestion is illustrated in Fig. 6, which emphasizes the parallels between the known mechanism of action of the homologous proteases [21] and the possible role of the corresponding serine–histidine pair in silicatein α. Significantly, we found that a serine residue is substituted for cysteine in the putative active site of the silicatein α [6]. We have suggested that the silicatein's putative active-site serine–histidine sidechain couple functions as a general acid–base catalyst of the rate-limiting hydrolysis of the silicon alkoxides, in a manner parallel to the catalysis of peptide-bond hydrolysis by the homologous serine–histidine pair in the proteolytic enzymes (Fig. 6). In the silicatein α-mediated reaction, hydrogen bonding between the imidazole nitrogen of the conserved histidine (at position 165 in the protein chain) and the hydroxyl of the active-site serine (at position 26) is postulated to increase the nucleophilicity of the serine oxygen, potentiating its attack on Si. Nucleophilic attack on the Si would then displace EtOH, with formation of a covalent protein–O–Si intermediate that is formally analogous to the covalent enzyme–substrate intermediate formed by the protease. This intermediate formed between the silicon and silicatein potentially might be further stabilized as a pentavalent Si species via a donor bond from the imidazole N. Addition of water would complete the hydrolysis of the first alkoxide bond (the rate-limiting step), with condensation then initiated by nucleophilic attack of the released Si–OH on the second Si.

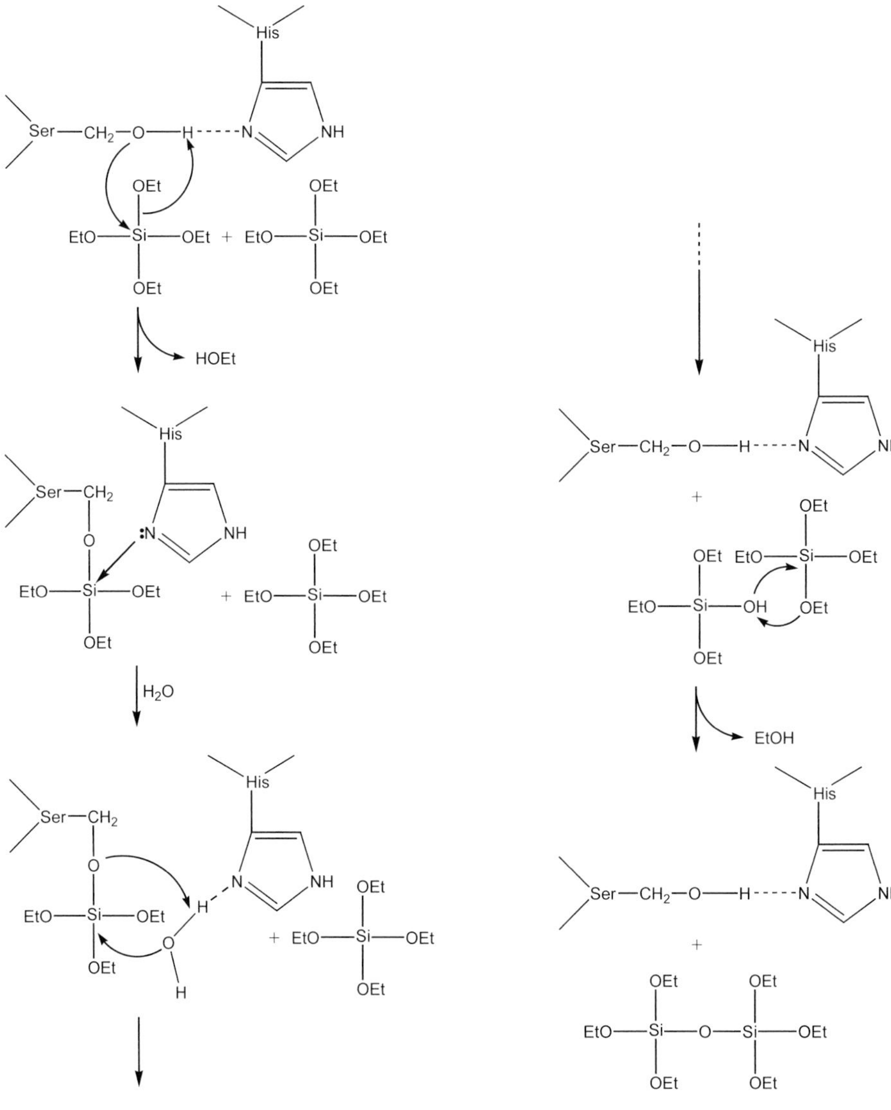

Fig. 6. Suggested reaction mechanism for silicatein-mediated catalysis of polysiloxane synthesis from silicon ethoxides at neutral pH. The mechanism suggested is closely analogous to that known for the structurally similar proteolytic enzymes; explanation in text. Reprinted from [7].

If this mechanism reflects the reaction pathway by which silicatein accelerates the condensation of silicon alkoxides at neutral pH, it is interesting to note that the action of silicatein would be closely parallel to that of the proteases in providing an alternative pathway — through the formation and stabilization of the transitory protein–substrate covalent intermediate — that accomplishes the rate-limiting hydrolysis at neutral pH. This is the condition under which the rate of spontaneous hydrolysis of the silicon alkoxides normally is lowest [19].

We now are using site-directed mutagenesis to produce specifically altered variants of the silicatein molecule, to probe the structural determinants of its catalytic and scaffolding activities. By substituting pre-selected nucleotides into the sequence of the cloned silicatein DNA during its enzymic replication in vitro, we can change the genetically encoded sequence of the protein at any selected site, replacing one amino acid with another. The initial results of these experiments, in which we have replaced the amino acids at the suspected catalytically active site in the silicatein with residues bearing nonfunctional sidechains, so far support the predictions of the reaction mechanism outlined above [20].

While the silicatein-mediated catalysis of silicon alkoxide condensation is interesting and potentially useful, it is unclear whether this reaction plays a significant role in biosilicification. This is largely because the identity of the proximate substrate for the biological synthesis remains unknown. Although it has been shown that the recently identified silicon transporter from diatoms mediates the uptake of dissolved silicic acid from the external environment [4], it is not yet clear whether this is the species that is transported into the SDV for condensation, or if that silicic acid is first conjugated or otherwise modified prior to condensation (cf. [13–15, 22]).

The template-like scaffolding activity of the silicatein filaments (directing the growth of the silica and silsesquioxane layers over the filaments; cf. Fig. 4B, C) may reflect the contribution of the numerous hydroxyl-rich domains of the silicateins to the high formation constants for juxtaposed alignment of silicic acid moieties favoring their polymerization [7]. The role of such hydroxyl-rich polymer domains, previously suggested to play a role in various systems in which biosilicification is thought to be directed either by proteins [9–12] or polysaccharides [13–15], remains to be determined. We are working to resolve the relative contributions of the catalytic and scaffolding mechanisms to the control of silicification (from both free silicic acid and from the alkoxides) through the combined use of solid-state ^{29}Si NMR analyses of the reaction products in conjunction with site-directed mutagenesis of the cloned silicatein cDNA.

Further Genetic and Molecular Engineering

In addition to site-directed mutagenesis, which produces specific substitutions in pre-targeted regions of the gene product under investigation, combinatorial mutagenesis also can be used to identify and optimize determinants of Si–O polymerization and structure controlled by the silicateins (Fig. 7). In this method, mutations in the silicatein (or other silicon metabolism-controlling protein) genes are introduced at random locations in the gene, either by conventional chemical or radiative mutagenesis, or by variations of the nucleotide-substitution/enzymic replication technique described above, and the resulting variants of the protein encoded by these mutants are then screened for alterations of interest. In the example illustrated (Fig. 7), this screening is facilitated by first coupling the silicatein gene to the gene coding for a protein that will be expressed on the outer surface of a bacterial virus (in this case, the tail-fiber protein of the virus).

This "display" method thus places the various mutationally altered silicatein proteins on the exterior surface of the different viruses which express the mutant genes. Samples of each of the individual mutant viruses are robotically isolated in micro-wells, and the precursor of a fluorogenic (or other labeled) silsesquioxane is added. After allowing time for the silicateins to direct polysiloxane synthesis, the micro-wells are robotically scanned with a laser to detect silsesquioxanes with unique spectral signatures indicating the presence of specific structural alterations in the product. In this way, we hope to identify and optimize functional groups on the silicatein surface that act as structural determinants controlling the order and orientation of the polysiloxane product, in the hope of developing modified catalysts capable of greater control over the molecular topology and coherence of the final product than at present afforded by simple acid- or base-catalyzed bulk-condensation routes.

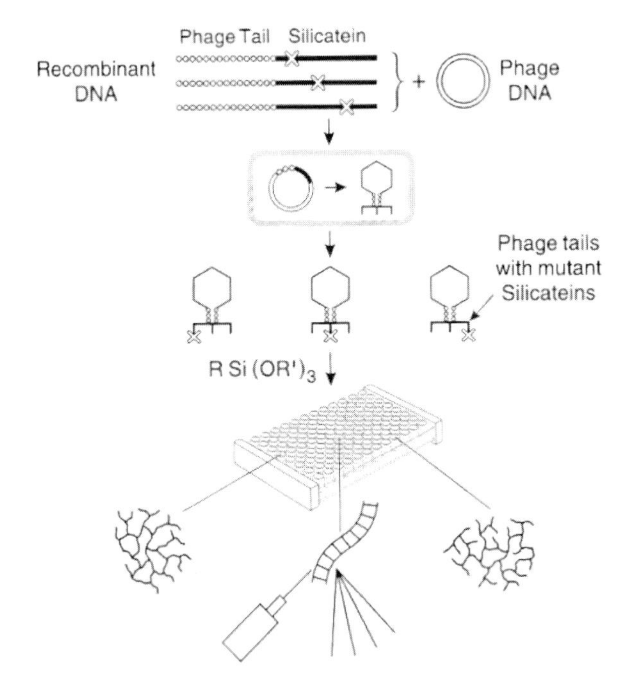

Fig. 7. Combinatorial mutagenesis to produce variants of silicatein capable of directing synthesis of more highly ordered silsesquioxanes. Randomly generated mutations in the silicatein DNA and in the corresponding positions in the silicatein protein are indicated by hollow ×. Exposed on the surface tail-fibers of bacterial viruses (known as bacteriophages, or phages), the protein variants are robotically screened for their ability to direct the synthesis of labeled silsesquioxanes that yield altered fluorescence.

Both site-directed and combinatorial mutagenesis results are analyzed with reference to the detailed three-dimensional structure of the silicatein molecule to identify the precise spatial location of the determinants of the catalytic and scaffolding activities, and to define and optimize the reactive surface controlling the alignment of the substrates and final products. Because silicatein α and β already are known to be highly homologous to cathepsin L and other members of the protease family (molecules whose structures previously have been resolved by X-ray diffraction), good three-dimensional structural approximations of the silicateins can be obtained simply by mapping the sequence of the silicateins on the three-dimensional structures of the homologous proteases. Our final objective is then to use the information obtained from the studies of polysiloxane synthesis with the mutationally altered proteins to design synthetic peptide-based catalysts to test the validity of our conclusions, and to guide the design of synthetic non-peptide-based catalysts and structure-directing scaffolds that are both less expensive and more robust than the natural and genetically engineered proteins. Preliminary results with simple synthetic block-copolymer peptides already demonstrate the feasibility of this approach [23].

Conclusions and Future Prospects

Our findings suggest that the silicateins and the silica deposition vesicle system that control biosilicification in sponges evolved as a functional unit from the ancestral protease/vesicle system now represented by the modern cathepsin L/lysosome complex that mediates the digestion of proteins. Cloning and characterization of the genes controlling silicon metabolism in other species can be expected to help reveal the mechanisms responsible for the apparent involvement of silicon in the formation of bone in higher animals [24–26] and for the essential requirement for silicon for optimal development and growth in many plants and animals [24, 25, 27, 28]. Studies in vit*ro* suggest that the silicateins can play a dynamic role in the control of silicification, manifesting both catalytic and scaffolding activities in vitro. The results summarized here demonstrate that the silicateins can hydrolyze silicon alkoxides to produce silica and organically substituted silsesquioxane polymer networks at neutral pH, and that the silicatein filaments can direct the structures of the resulting products. We do not yet know whether the catalytic activity of the silicateins that we have observed in vitro is physiologically significant in organisms that are alive today, or simply a vestigial legacy of the proteins' evolutionary history. In either case, the activities described here suggest that the molecular mechanisms of action of the silicateins may provide useful insights for the development of environmentally benign new catalysts, and new routes to the synthesis and enhanced structural control of patterned silicon-based materials. Identification of the proximate biological silica precursor (if other than silicic acid) may provide useful alternative feedstocks for chemical synthesis in conjunction with these new catalysts.

Acknowledgment: I thank my close colleagues Katsuhiko Shimizu, Jennifer Cha, Yan Zhou, Sean Christiansen, Galen Stucky and Bradley Chmelka for allowing me to report the results of our collaborative research; Norbert Auner, Mark Brzezinski, Gregg Zank, Dimitri Katsoulis, Udo Pernisz, Paal Klykken, Forrest Stark and Gordon Fearon for stimulating discussions; and Johann Weis, Norbert Auner and *Wacker-Chemie GmbH* for their kind invitation to present this contribution and for their generous hospitality. This work was supported by grants from the *U. S. Army Research Office* Multidisciplinary University Research Initiative Program, the *U. S. Office of Naval Research*, the *U. S. Department of Commerce* — NOAA National Sea Grant College Program, the *California Sea Grant College System*, the Materials Research Science and Engineering Centers Program of the *U. S. National Science Foundation* and a generous donation from the *Dow Corning Corporation*.

References

[1] T. L. Simpson, B. E. Volcani, in: *Silicon and Siliceous Structures in Biological Systems*, Springer-Verlag, New York, **1981**.
[2] M. G. Voronkov, G. I. Zelchan, E. J. Lukevits, *Silicon and Life*, *2nd ed.*, Zinatne Publishing, Riga, **1977**.
[3] G. Hallegraeff, *Plankton: A Microscopic World*, CSIRO/Robert Brown and Assocs., Melbourne, **1988**.
[4] M. Hildebrand, B. E. Volcani, W. Gassman, J. I. Schroeder, *Nature (London)* **1997**, *385*, 688.
[5] N. Kröger, G. Lehmann, R. Rachel, M. Sumper, *Eur. J. Biochem.* **1997**, *250,* 99.
[6] K. Shimizu, J. Cha, G. D. Stucky, D. E. Morse, *Proc. Natl. Acad. Sci. USA* **1998**, *95,* 6234.
[7] J. N. Cha, K. Shimizu, Y. Zhou, S. Christiansen, B. F. Chmelka, G. D. Stucky, D. E. Morse, *Proc. Natl. Acad. Sci. USA* **1998**, submitted.
[8] D. E. Morse, *Trends Biotechnol.* **1998**, in press.
[9] R. E. Hecky, K. Mopper, P. Kilham, E. T. Degens, *Mar. Biol.* **1973**,*19,* 323.
[10] D. M. Swift, A. P. Wheeler, *Phycology* **1992**, *28,* 209.
[11] D. W. Schwab, R. E. Shore, *Nature (London)* **1971**, *232,* 501.
[12] K. D. Lobel, J. K. West, L. L. Hench, *Mar. Biol.* **1996**, *126,* 353.
[13] C. C. Harrison, *Phytochemistry* **1996**, *41,* 37.
[14] C. C. Harrison and N. J. Loton, *J. Chem. Soc., Faraday Trans.* **1995**, *91,* 4287.
[15] C. C. Perry and L. J. Yun, *J. Chem. Soc., Faraday Trans.* **1992**, *88,* 2915.
[16] R. E. Shore, *Biol. Bull.* **1972**, *143,* 689.
[17] G. Imsiecke, R. Steffen, M. Custodio, R. Borojevic, W. E. G. Müller, *In Vitro Cell. Dev. Biol.* **1995**, *31,* 528.
[18] R. Garrone, in: *Phylogenesis of Connective Tissue*, S. Karger, Basel, **1978**, p. 176.
[19] R. K. Iler, in: *The Chemistry of Silica: Solubility, Polymerization, Colloid and Surface Properties, and Biochemistry*, John Wiley, New York, **1979**, p. 98.
[20] Y. Zhou, K. Shimizu, J. N. Cha, G. D. Stucky, D. E. Morse, *Angew. Chem., Int. Ed. Engl.* **1999**, *38*, 779.
[21] A. Lehninger, D. Nelson, M. Cox, in: *Principles of Biochemistry*, Worth Publishers, New York, **1993**, p. 223.
[22] P. Bhattacharya, B. E. Volcani, *Biochem. Biophys. Res. Commun.* **1983**, *114,* 365.
[23] J. Cha, T. J. Deming, G. D. Stucky, D. E. Morse, submitted for publication.
[24] E. M. Carlisle, *Fed. Proc.* **1974**, *33,* 1758.
[25] K. Schwarz, *Fed. Proc.* **1974**, *33,* 1748.
[26] E. M. Carlisle, in: *Silicon and Siliceous Structures in Biological Systems* (Eds.: T. L. Simpson, B. E. Volcani), Springer-Verlag, New York, **1981**, p. 69.
[27] E. Epstein, *Proc. Natl. Acad. Sci. USA* **1994**, *91,* 11.
[28] M. M. Rafi, E. Epstein, R. H. Falk, *J. Plant Physiol.* **1997**, *151,* 497.

Investigations on the Nature of Biomineralized Silica by X-Ray Absorption Spectroscopy and Sorption Measurements

Gallus Schechner, Peter Behrens*

Institut für Anorganische Chemie, Universität Hannover
Callinstr. 9, D-30167 Hannover, Germany
Tel.: Int. code + (511)762 4897 — Fax.: Int. code + (511)762 3006.
E-mail: Peter.Behrens@mbox.acb.uni-hannover.de

Michael Fröba

Institut für Anorganische und Angewandte Chemie
Universität Hamburg, Germany.

Bernd Pillep

Institut für Anorganische Chemie
Ludwig-Maximilians-Universität München, Germany.

Joe Wong

Lawrence Livermore National Laboratory, University of California
PO Box 808, Livermore, USA.

Keywords: Biomineralization / Rice Husks / Silica / XANES / BET Isotherme

Summary: Biomineralized silicas are generally described as amorphous gel-like or opaline-type varieties of hydrated SiO_2. Our X-ray absorption spectroscopic investigation reveals mid-range order in one of these materials (rice husk silicas) at a length scale of 4–10 Å. The degree of ordering depends on the method used to extract the silica from the samples. This is clearly shown by the investigation of three differently treated rice husk silicas. As confirmed by N_2 sorption experiments, these commercially interesting materials exhibit high specific surface areas and broad pore size distributions.

The precipitation of inorganic compounds in biological systems — the phenomenon of biomineralization — has been known for a long time, although until today these processes have not been well understood. In view of the variety of inorganic solids that are produced by living organisms, silica is exceptional in some respects. Siliceous structures of diatoms, radiolaria or many other silica-producing organisms show very complex geometric shapes defined on the length scale of micrometers. Since the last century, biologists have revealed these fascinating structures by microscopic methods. The corresponding morphologies of all other inorganic biogenic materials (e.g. calcite, magnetite) are limited, probably because of their rigid crystalline architecture. The results of investigations of the solid-state properties of biological silica have so far led to a picture within which these substances are identified as highly polymeric silica and are best described as amorphous gel-like or opaline-type varieties of hydrated silica. The amorphous nature may allow the molding of structures as complex as observed. In contrast, the occurrence of crystalline forms like quartz or cristobalite is very doubtful [1–3].

Our current work has so far focused on two silica-containing plants: horse tail (*Equisetum arvense:* complete plants without roots) and rice husks (*Oryz sativa*). The latter can be of commercial interest, because they are cheap and available in huge amounts as agricultural waste [4–6]. Rice husk ash is already used as raw material in some ceramic processes [6].

According to the literature, pure plant silica may be prepared by calcination, i.e. burning out the organic part at high temperatures [3, 4], or by "wet ashing", i.e. boiling in concentrated oxidizing acids like HNO_3 [7]. In our opinion, both methods are likely to strongly affect the properties of the original plant silica. Therefore we tried to employ milder conditions to dissolve the organic part of the biological composite. We tested Caro's acid (H_2SO_5) and Fenton's reagent (H_2O_2 and $FeSO_4$) at room temperature. The duration of the treatments was about four weeks in the case of H_2SO_5 and about three months for Fenton's reagent. The materials obtained were characterized by elemental analysis, N_2 sorption, X-ray diffraction, and solid-state NMR, IR and X-ray absorption spectroscopy. Here we present our XANES (X-ray absorption near-edge structure) and N_2 sorption measurements on rice husk silicas.

Si*K* edge spectra were measured on beamline 3–3 at the Stanford Synchroton Radiation Laboratory with the storage ring SPEAR operating at an electron energy of 3.0 GeV and injection currents of 95 mA. Monochromatization was achieved by a YB_{66}-(400) double-crystal monochromator [8] (estimated resolution: 0.8 eV) having an energy resolution of 0.8 eV at the Si*K* edge (1.838 keV). Spectra processing was performed with the program WINXAS [9].

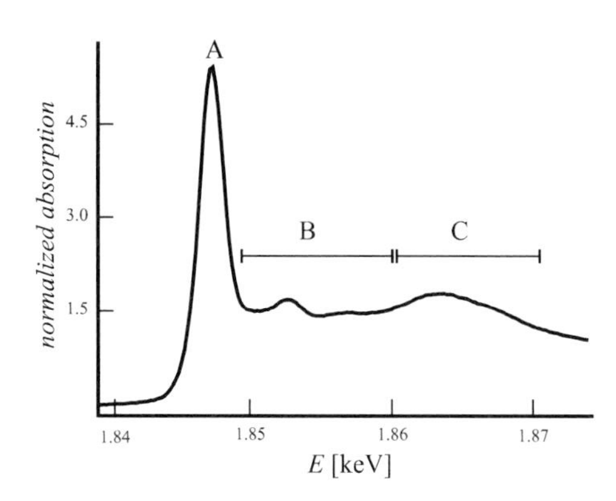

Fig. 1. Si*K* XANES spectrum of cristobalite.

Figure 1 shows a Si*K* XANES spectrum typical of silica compounds. Three prominent features can be distinguished [10]: The feature at an energy value of 1847.1 eV is attributable to a 1s → 3p electronic transition in the Si absorber atoms [11–13]. In region B from 1849 to 1860 eV, features are observed which are caused by multiple scattering events in coordination shells beyond the second shell around the Si atoms [12, 13]. The broad feature C from 1860 to 1870 eV is due to single and multiple scattering processes in the nearest (Si–O) and next-nearest (Si–O–Si) coordination shells and has a very similar appearance for all silicas built from corner-sharing $[SiO_{4/2}]$ tetrahedra. In contrast to A and C, region B is characteristic of a particular compound and can thus be taken as its "fingerprint". For the further analysis of Si*K* XANES spectra, several attempts to calculate the multiple scattering processes in silica materials have been performed [12, 13], but for absorbing elements with low *Z* number such as Si these are very tedious and hence experimental spectra are not well reproduced. Nevertheless, these calculations have revealed that the features in region B can only be reproduced if clusters of a size from ca. 4 to at least 7 Å are taken into account. It can thus be concluded that compounds that exhibit features in region B must possess a sufficient degree of order in a distance range of at least 7 Å to maybe roughly 10 Å around the absorbing Si atom.

In Fig. 2, a zoom into region B of the Si*K* edge spectra of differently treated rice husk silicas and of some reference materials is shown. The spectra of the reference materials show features in line with the statements given above. For cristobalite (as for other crystalline silica materials) strong peaks and distinct valleys are observed, corresponding to a crystallographically well-defined mid-range order at distances of ca. 4–10 Å around the silicon atom. On the contrary, for samples exhibiting no distinct order beyond the first or the first two coordination shells, e.g. fused or precipitated silica (Fig. 2, b, c), the XANES line shapes are essentially featureless in region B.

The spectra of all biological samples exhibit distinct features in the multiple scattering region B, although these are less prominent than those of crystalline compounds, e.g. cristobalite. However, there is clearly a stronger modulation of the absorption than in the spectra of "truly" amorphous samples such as precipitated or fused silica. The term "amorphous" should thus be used with respect to the analytical method employed. Obviously, there is some kind of mid-range ordering at a length-scale of 4–10 Å in the biomineralized silicas.

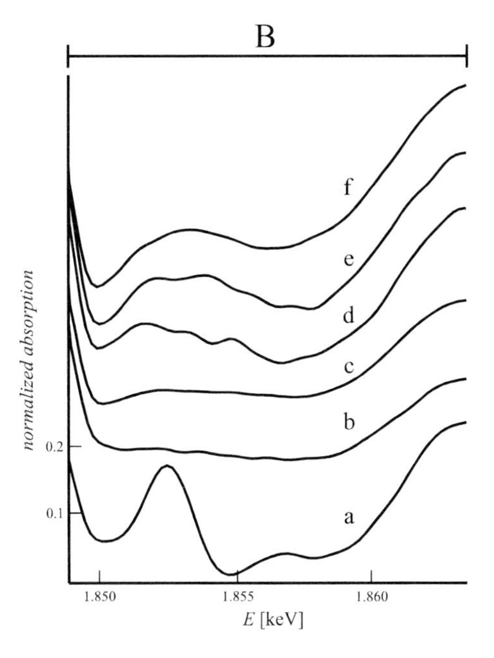

Fig. 2. Zoom into region B of Si*K* XANES spectra of rice husk silicas obtained by different treatments, and of some reference compounds: a, cristobalite; b, fused silica; c, precipitated silica; d, calcined rice husks (600°C in air for 4 h); e, rice husks oxidized by Caro's acid; f; rice husks oxidized by Fenton's reagent.

Another remarkable result are the differences between materials that were treated by different methods in order to extract the silica (Fig. 2, d, e, f): the treatments clearly affect their Si*K* edge XANES spectra, i.e. the materials differ in mid-range order. Infrared spectra (not shown here) support these findings. So, our idea of treating the biological materials as "mildly" as possible is justified. Former analytical investigations on the solid-state properties of biogenic silica may not have considered this aspect sufficiently.

The detection of mid-range order at a length scale of 4–10 Å in biogenic silicas is of special significance since it is generally assumed that the biominerals are templated by biological macromolecules, e.g. peptides, proteins or polymeric hydrocarbons [15]. The repetitive units of these macromolecules have characteristic length scales of 4–10 Å.

In order to interpret our data in a more detailed fashion, suitable reference compounds have to be found. In this context, a comparison with synthetic mesostructured silicas like hexagonal MCM-41 [16, 17] may be of interest, because these materials already show some similarities to biominerals; for example, they can be synthesized under relatively mild conditions and they can be obtained with morphologies reminiscent of biological shapes [18]. Indeed, Si*K* XANES spectra in the multiple scattering region B of mesostructured M41S type silicas exhibit similar features as the biological samples [19]. Si*K* XANES spectroscopy thus may become a suitable tool to evaluate structural relationships between biomineralized silica on the one hand and its biomimetically synthesized counterparts on the other.

In view of possible commercial applications of rice husk silicas, e.g. as catalyst supports, their surface areas and porosities are important properties. Nitrogen sorption measurements of the three differently treated rice husk silicas show that these are porous materials with moderately large surface areas. The surface areas are, in detail: 73 m^2/g for the calcined material, 75 m^2/g in the case of the material oxidized by Fenton's reagent and 51 m^2/g for the rice husks treated with Caro's acid. As the shapes of the ad- and desorption isotherms reveal (Fig. 3), pores ranging from micro- up to macropores are present.

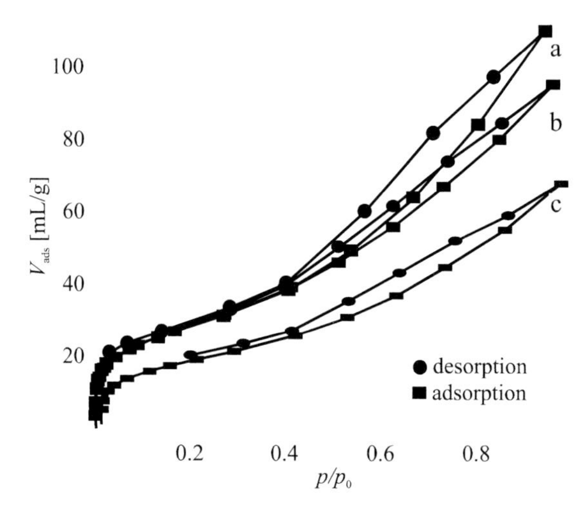

Fig. 3. BET isotherms of rice husk silicas, obtained by different treatments: a, calcined (600°C in air for 4 h); b, oxidized by Fenton's reagent; c, oxidized by Caro's acid.

In order to put our work on broader foundations, we plan to extend our investigations to biosilicas produced by other species like diatoms, radiolaria and sponges. Another aim is the optimization of mild conditions for the preparation of silica from the biological organic–inorganic composites.

Acknowledgment: We are grateful to *SSRL* for allocating beam time and for general support during the experiments. G. Schechner and P. Behrens thank Dr. Bartha from the *Refratechnik GmbH* for a generous gift of rice husks. M. Fröba and J. Wong thank the *Deutscher Akademischer Austauschdienst (DAAD)* and the *National Science Foundation (NSF)* for generous travel support. B. Pillep also thanks the DAAD for a travel grant. The German groups acknowledge financial support from the *Fonds der Chemischen Industrie (FCI).*

References:

[1] S. Mann, *J. Mater. Chem.* **1995**, *7*, 935–946.
[2] T. L. Simpson, B. E. Volcani, "*Silicon and Siliceous Structures in Biological Systems*", Springer, New York, **1981**.
[3] R. K. Iler, "*The Chemistry of Silica: Solubility, Polymerisation, Colloid and Surface Properties, and Biochemistry*", Wiley, New York, **1979**, pp.730–787.
[4] C. Real, M. D. Alcala, J. M. Criado, *J. Am. Ceram. Soc.* **1996**, *79*, 2012–2016.
[5] P. Bartha, *Keram. Z.* **1995**, *47*, 780–785.
[6] A. E. Eminov, B. A. Muslimov, *Keram. Z.* **1996**, *48*, 901–902.
[7] C. C. Perry, E. J. Moss, R. J. P. Williams, *Proc. Royal Soc.* B **1990**, *241*, 47–50.
[8] M. Rowen, Z. U. Rek, J. Wong, T. Tanaka, G. N. George, I. J. Pickering, G. H. Via, G. E. Brown Jr., *Synchrotron Radiation News* **1993**, *6*, 25.
[9] T. Ressler, *J. Phys. IV France* **1997**, *7*, C2–269.
[10] M. Fröba, J. Wong, P. Behrens, P. Sieger, M. Rowen, T. Tanaka, Z. Rek, J. Felsche, *Physica B* **1995**, *208/209*, 65–67.
[11] P. Lagarde, A. M. Flank, G. Tourillon, R. C. Liebermann, J. P. Itie, *J. Phys. I France* **1992**, *2*, 1043.
[12] I. Davoli, E. Paris, S. Stizza, M. Benfatto, M. Fanfoni, A. Gargano, A. Bianconi, F. Seifert, *Phys. Chem. Minerals* **1992**, *19*, 171–175.
[13] J. Chaboy, M. Benfatto, I. Davoli, *Phys. Rev.* **1995**, *B 52*, 10014–10020.
[14] J. Wong, G. N. Georgem I. J. Pickering, Z. U. Rek, M. Rowen, T. Tanaka, G. H. Via, B. DeVries, D. E. W. Vaughan, G. E. Brown Jr., *Solid State Commun.* **1994**, *92*, 559.
[15] R. E. Hecky, K. Mopper, P. Kilham, *Marine Biol.* **1973**, *19*, 323–331.
[16] J. S. Beck , J. C. Vartuli, W. J. Roth, M. E. Leonowicz, C. T. Kresge, K. D. Schnitt, C. T.W. Chu, D. H. Olson, E. W. Sheppard, S. B. McCullen, J. B. Giggins, J. L. Schlenker, *J. Am. Chem. Soc.* **1992**, *114*, 10834.
[17] Q. Huo, D. L. Margolese, U. Ciesla, P. Feng, T. E. Gier, P. Sieger, R. Leon, P. M Petroff, F. Schüth, G. D. Stucky, *Nature (London)* **1994**, *368*, 317–323.
[18] a) G. A. Ozin, H. Yang, I. Sokolov, N. Coombs, *Adv. Mater.* **1997**, *9*, 662–667.
b) W. Zhang, T. R. Pauly, T. J. Pinnavaia, *Chem. Mater.* **1997**, *9*, 2491–2498.
[19] M. Fröba, P. Behrens, J. Wong, G. Engelhardt, Ch. Haggenmüller, G. van de Goor, M. Rowen, T. Tanaka, W. Schwieger, *Mater. Res. Soc. Symp. Proc.* **1995**, *371*, 99–104.

Influence of Substituted Trimethoxysilanes on the Synthesis of Mesostructured Silica Materials

Stephan Altmaier, Konstanze Nusser, Peter Behrens*

Institut für Anorganische Chemie, Universität Hannover
Callinstr. 9, D-30167 Hannover, Germany
Tel.: Int. code + (511)762 3660 — Fax.: Int. code + (511)762 3006
E-mail: Peter.Behrens@mbox.acb.uni-hannover.de

Keywords: Structure-Directed Synthesis / Mesoporous Materials / Mesostructured Materials / Trialkoxysilanes / M41S Materials

Summary: M41S-type mesostructured materials can be obtained in different topologies from silica gels containing amphiphilic surfactants as structure-directing agents. A part of the silica precursor tetraethoxysilane (TEOS) can be replaced by trialkoxysilanes R–Si(OR´)$_3$ such as phenyltrimethoxysilane (PTMOS). This should allow the direct modification of the pore walls during the synthesis process when hydrolyzed PTMOS molecules are able to take part in the condensation process of hydrolyzed TEOS. Series of synthesis experiments, conducted with variation of the concentrations of the silica precursor molecules TEOS and PTMOS as well as of the surfactant C_nTMA$^+$ (an alkyltrimethylammonium cation) establish a synthesis field diagram (SFD) which shows the educt concentration regions from which different product structures form. The effect of the PTMOS molecules on mesostructure formation can be determined by comparing this SFD with another one which was established under the same conditions without using a trialkoxysilane.

M41S, a new class of mesoporous inorganic materials, was first described in 1992 [1, 2]. These solids show sharp pore width distributions with pore sizes in the range from 15 to 100 Å and regular arrangements of the pore systems. Therefore, M41S materials are suitable candidates for catalysts (e.g. for heavy-oil cracking, an application so far reserved for microporous zeolites), for catalyst supports as well as for selective adsorbents of pollutants or for chromatographic separation.

The synthesis of M41S-type materials can be carried out under hydrothermal conditions using alkyltrimethylammonium cations C_nTMA$^+$ as structure-directing agents and tetraethoxysilane (TEOS) as silica source. Similarly to the lyotropic phases formed in surfactant–water mixtures, the surfactants form micellar structures. Their self-assembly is strongly influenced by the presence of silicate anions which form by the hydrolysis of TEOS. After the synthesis, the space in between the surfactant micelles is filled by silica which develops by polycondensation of the silicate anions and which forms walls of a thickness of ca. 8 Å [3, 4].

These mesostructured solids occur in different structural topologies: hexagonal (MCM-41), cubic (MCM-48), lamellar and LMU-1 [5] (or KIT-1 [6]). The topologies are shown in Fig. 1.

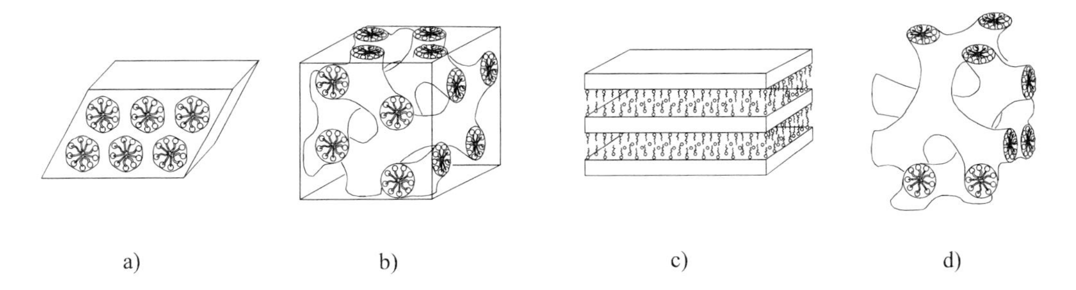

Fig. 1. Mesostructures of M41S type: a) hexagonal MCM-41, b) cubic MCM-48, c) lamellar, d) LMU-1 (or KIT-1) with a disordered three-dimensional channel system.

The organic surfactant molecules can be removed by calcination or extraction of the "as-synthesized" material. The walls of the obtained mesoporous solid consist of partially condensed silica. They carry silanol groups which can be used to modify the pore walls, e.g. by trimethylsilylation with hexamethyldisilazane (hydrophobization) or phenylsilylation with phenyltrimethoxysilane (functionalization).

In contrast to these post-synthetic modifications, it is also possible to functionalize the pore walls directly during the synthesis, as was first shown by Mann and co-workers [7, 8] and Stucky and co-workers [9], who used trialkoxysilanes $R{-}Si(OR')_3$. In our approach, such $R{-}Si(OR')_3$ molecules substitute for part of the TEOS. After hydrolysis, they serve as additional framework components during the hydrothermally induced condensation. An essential condition for this approach is that the trialkoxysilane does not destroy the micellar arrangement of the surfactant, which gives rise to the mesostructure. In mesostructures produced in this way, the R residues should be covalently linked to the silica walls. After the synthesis, the organic surfactant molecules can be removed by extraction so that a modified mesoporous material should remain. For example, when using phenyltrimethoxysilane (PTMOS), phenyl groups may become attached to the walls of the mesopores; these can be utilized for further modifications, e.g. the immobilization of metal complexes.

As a first step, we investigated the influence of the synthesis procedure on the structure of the products obtained. For that purpose, tetraethoxysilane (TEOS), trimethylhexadecylammonium bromide ($C_{16}TMA^+Br^-$) and a substituted trimethoxysilane ($R{-}Si(OR')_3$) were combined in a 0.165 M KOH solution in different sequences. The trimethoxysilanes which were employed are shown in Fig. 2.

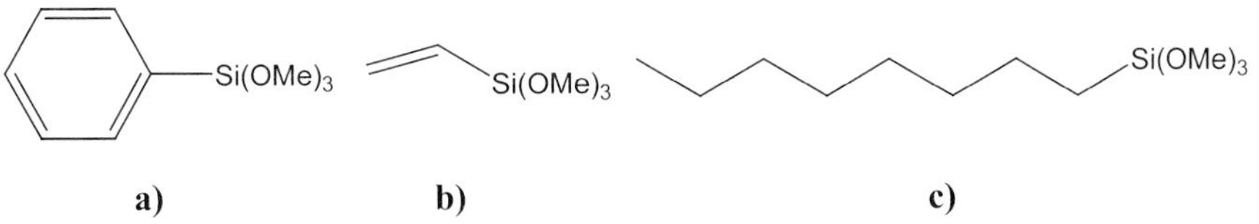

$Si(OMe)_3$ $Si(OMe)_3$ $Si(OMe)_3$

a) **b)** **c)**

Fig. 2. $R{-}Si(OR')_3$ molecules employed in mesostructure synthesis: a) phenyltrimethoxysilane (PTMOS), b) vinyltrimethoxysilane (VTMOS), c) octyltrimethoxysilane (OTMOS).

For all these trialkoxysilanes, the sequence of addition of the three educts to the reaction mixture had no significant effect on the type of mesostructure formed. For example, Fig. 3 shows the X-ray powder diffraction patterns of materials formed by substituting 10 % of TEOS by VTMOS on a molar basis. Independently of the synthesis method (listed in Table 1), an LMU-1 structure is obtained. However, minor variations in the 2Θ values and, correspondingly, in the d values of the first reflection are apparent.

Table 1. Different synthesis procedures used in the preparation of VTMOS/TEOS-based mesostructures[a].

Method	Synthesis procedure
A	After hydrolysis of TEOS, VTMOS is added; after complete hydrolysis of VTMOS, the surfactant is dissolved.
B	Simultaneous addition of TEOS and VTMOS; after complete hydrolysis, the surfactant is dissolved.
C	After hydrolysis of TEOS, VTMOS is added slowly dropwise; after complete hydrolysis, the surfactant is dissolved.
D	After dissolving the surfactant, TEOS and VTMOS are added.

[a] Starting point is in all cases a 0.165 M KOH solution.

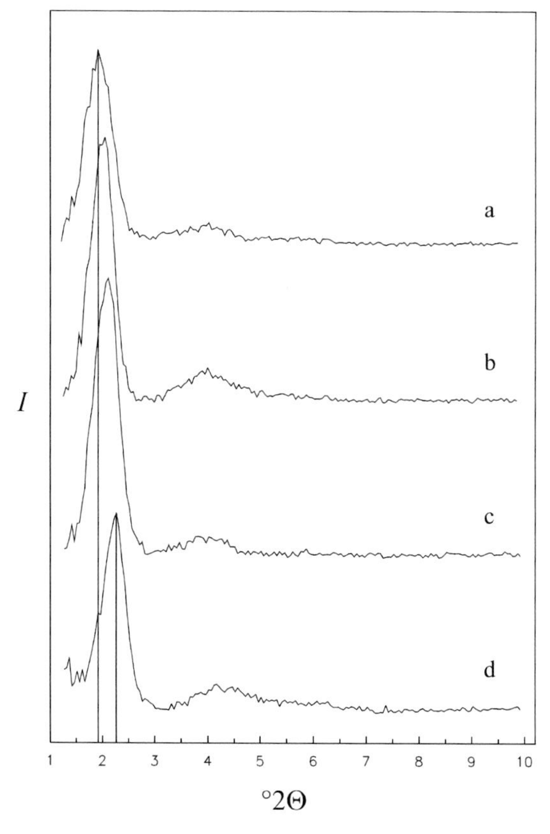

Fig. 3. X-ray diffraction patterns of mesostructures formed by different synthesis methods with identical starting compositions (silica source: 10 mol % VTMOS/90 mol % TEOS; reaction time: 2 d; reaction temperature: 110°C; 0.165 M KOH solution); a) method A; b) method B; c) method C; d) method D. For a description of the synthesis methods see Table 1.

In a further study using method A as described in Table 1, we investigated the products obtained by variation of both the concentrations of $C_{14}N(CH_3)_3^+Br^-$ and of the silica source, which consisted of 10 % PTMOS and 90 % TEOS on a molar basis. The results were used to construct a synthesis field diagram (SFD) where the different mesostructures occupy delineated fields (Fig. 4, left). At a low concentration $c_{TEOS+PTMOS}$ of the silica precursors, no solid product is obtained, giving rise to a "solution" region. When $c_{TEOS+PTMOS}$ is high, a solid material showing no discrete X-ray reflection, but exhibiting mesoporosity, results ("disordered mesoporous silica"). In between is the region where ordered mesostructures occur. An SFD based on synthesis experiments conducted without any trialkoxysilanes, i.e. with pure TEOS as the silica source, has already been published (Fig. 4, right) [3]. A comparison between these two SFDs shows that the PTMOS added to the synthesis mixture has a strong effect on the micellar arrangement and the cooperative self-assembly process. Only the borderline between the solution and the mesostructured region at low silica concentrations remains the same. An astonishing result is the observation that the mesostructured region extends to higher silica concentrations when the trimethoxysilane is applied, indicating that the PTMOS molecules do not impede the mesostructure formation. They do, however, exert an important influence on the type of mesostructure formed. Within the region of mesostructures, the field of hexagonal MCM-41 is clearly enlarged when PTMOS is applied. This occurs largely at the cost of the MCM-48 field, which becomes very small and occurs at low surfactant and medium TEOS concentrations.

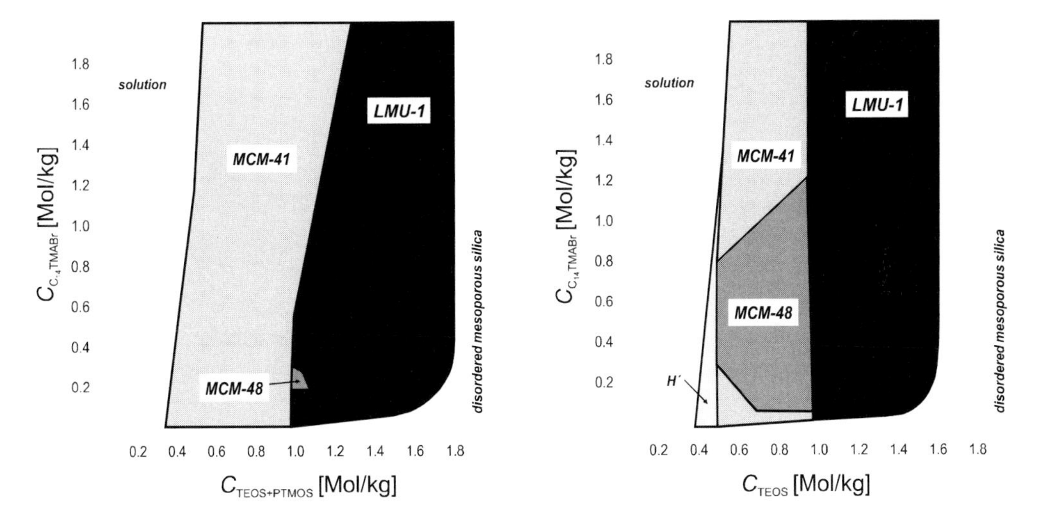

Fig. 4. Synthesis field diagrams (SFDs) for mesostructures obtained with the surfactant $C_{14}N(CH_3)_3^+Br^-$ (reaction time: 2 d; reaction temperature: 110°C; 0.33 M KOH solution). Left: SFD obtained by replacing 10 % of the TEOS by PTMOS; right: standard synthesis system using pure TEOS [3]. H': hexagonal MCM-41 with a significantly smaller lattice constant [3].

Extraction of the materials containing PTMOS is possible with preservation of the mesostructure. After extraction a modified mesoporous phase is isolated which still carries organic phenyl groups coupled to the pore walls. In which way and to what extent the phenyl groups have been covalently attached to the pore walls has to be tested by further characterization.

Acknowledgment: We thank the *Deutsche Forschungsgemeinschaft* (Schwerpunktprogramm "Spezifische Phänomene der Si-Chemie") and the *Fonds der Chemischen Industrie* for support of this work.

References:

[1] C. T. Kresge, M. E. Leonowicz, W. J. Roth, J. C. Vartuli, J. S. Beck, *Nature* (London) **1992**, *359*, 710.

[2] J. S. Beck, J. C.Vartuli, W. J. Roth, M. E. Leonowitcz, C. T. Kresge, K. D. Schmitt, C. T. W. Chu, D. H. Olson, E. W. Sheppard, S. B. McCullen, J. B. Higgins, J. L. Schlenker, *J. Am. Chem. Soc.* **1992**, *114*, 10834.

[3] P. Behrens, G. D. Stucky, *Angew. Chem., Int. Ed. Engl.* **1993**, *32*, 696.

[4] G. D. Stucky, A. Monnier, F. Schüth, Q. Huo, D. Margolese, D. Kumar, M. Krishnamurty, P. Petroff, A. Firouzi, M. Janicke, B. F. Chmelka, *Mol. Cryst. Liq. Cryst.* **1994**, *240*, 187.

[5] P. Behrens, A. Glaue, C. Haggenmüller, G. Schechner, *Solid State Ionics* **1997**, *101–103*, 255–260.

[6] R. Ryoo, J. M. Kim, C. H. Shin, *J. Phys. Chem.* **1996**, *100*, 17713.

[7] S. L. Burkett, S. D. Sims, S. Mann, *J. Chem. Soc., Chem. Commun.* **1997**, 1367.

[8] C. E. Fowler, S. L. Burkett, S. Mann, *J. Chem. Soc., Chem. Commun.* **1997**, 1769.

[9] Q. Huo, D. I. Margolese, G. D. Stucky, *Chem. Mater.* **1996**, *8*, 1147–1160.

Biocatalysis in Preparative Organosilicon Chemistry: Microbial Reduction of *rac*-1-(4-Fluorophenyl)-1-methyl-1-sila-2-cyclohexanone and Microbial Hydrolysis of *rac*-(Si*S*,C*R*/Si*R*,C*S*)-2-Acetoxy-1-(4-fluorophenyl)-1-methyl-1-silacyclohexane

Markus Merget, Reinhold Tacke*

Institut für Anorganische Chemie, Universität Würzburg
Am Hubland, D-97074 Würzburg, Germany
Tel.: Int. code + (931)888 5250 — Fax: Int. code + (931)888 4609
E-mail: r.tacke@mail.uni-wuerzburg.de

Keywords: Biotransformations / Chirality / Diastereoselectivity / Enantioselectivity / Microbial Reduction / Microbial Hydrolysis

Summary: *rac*-1-(4-Fluorophenyl)-1-methyl-1-sila-2-cyclohexanone (*rac*-**1**) and *rac*-(Si*S*,C*R*/Si*R*,C*S*)-2-acetoxy-1-(4-fluorophenyl)-1-methyl-1-silacyclohexane [*rac*-(Si*S*,C*R*/Si*R*,C*S*)-**3a**] were synthesized as substrates for stereoselective microbial transformations. Resting free cells of the yeast *Saccharomyces cerevisiae* (DSM 11285) and growing cells of the yeasts *Trigonopsis variabilis* (DSM 70714) and *Kloeckera corticis* (ATCC 20109) were found to reduce *rac*-**1** diastereoselectively to yield mixtures of the enantiomers (Si*S*,C*R*)- and (Si*R*,C*S*)-1-(4-fluorophenyl)-1-methyl-1-sila-2-cyclohexanol [(Si*S*,C*R*)-**2a** and (Si*R*,C*S*)-**2a**]. In the case of *Kloeckera corticis* (ATCC 20109), diastereoselective reduction of *rac*-**1** gave a quasi-racemic mixture of (Si*S*,C*R*)-**2a** and (Si*R*,C*S*)-**2a** (diastereomeric purity 95 % *de*, yield 95 %). Enantioselective ester hydrolysis (kinetic resolution) of the 2-acetoxy-1-silacyclohexane *rac*-(Si*S*,C*R*/Si*R*,C*S*)-**3a** yielded the optically active 1-sila-2-cyclohexanol (Si*R*,C*S*)-**2a** [enantiomeric purity ≥99 % *ee*; yield 56 % (relative to (Si*R*,C*S*)-**3a**)].

Introduction

Stereoselective biotransformation with growing cells, resting free or immobilized cells, or isolated enzymes has been demonstrated to be a useful preparative method for the synthesis of centrochiral optically active organosilicon compounds [1–3]. In continuation of our own studies in this field, we have investigated stereoselective microbial transformations of *rac*-1-(4-fluorophenyl)-1-methyl-1-sila-2-cyclohexanone (*rac*-**1**) and *rac*-(Si*S*,C*R*/Si*R*,C*S*)-2-acetoxy-1-(4-fluorophenyl)-1-methyl-1-silacyclohexane [*rac*-(Si*S*,C*R*/Si*R*,C*S*)-**3a**]. We report here on (i) the synthesis of *rac*-**1** and *rac*-(Si*S*,C*R*/Si*R*,C*S*)-**3a**, (ii) the diastereoselective microbial reduction of *rac*-**1** [→ (Si*S*,C*R*)-**2a**, (Si*R*,C*S*)-**2a**], and (iii) the enantioselective microbial hydrolysis of *rac*-(Si*S*,C*R*/Si*R*,C*S*)-**3a** [→ (Si*R*,C*S*)-**2a**].

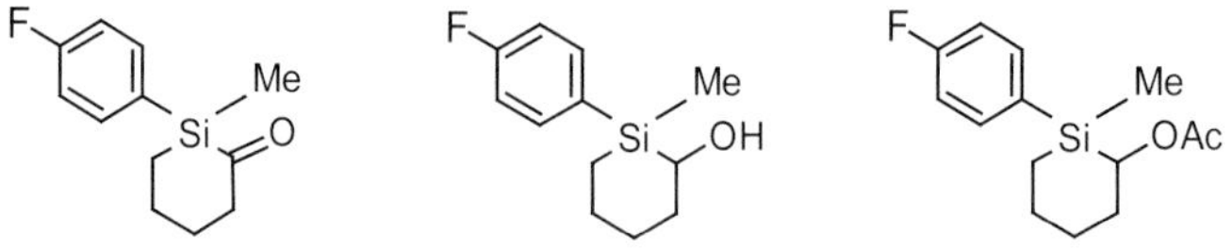

Syntheses

The 1-sila-2-cyclohexanone *rac*-**1** and the 2-acetoxy-1-silacyclohexane *rac*-(Si*S*,C*R*/Si*R*,C*S*)-**3a** were synthesized according to Scheme 1, starting from dichloro(methyl)silane (**4**).

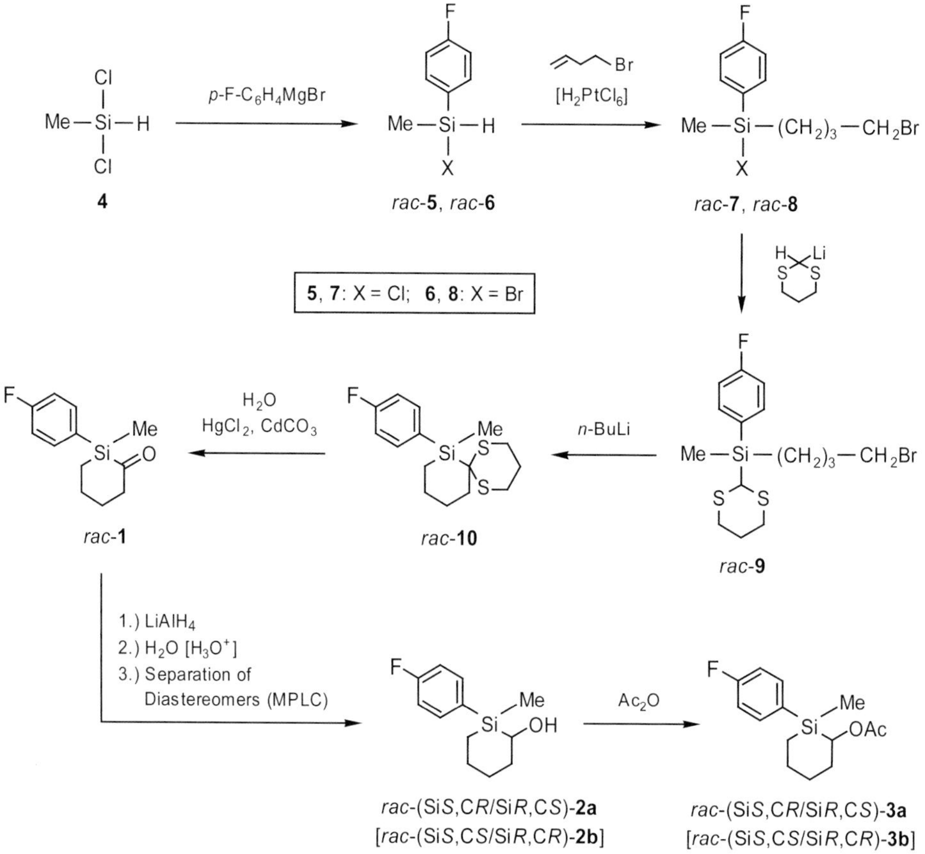

Scheme 1. Synthesis of the substrates *rac*-**1** and *rac*-(Si*S*,C*R*/Si*R*,C*S*)-**3a**.

In the first step, the dichlorosilane **4** was treated with (4-fluorophenyl)magnesium bromide to give a mixture of the chloro(4-fluorophenyl)silane *rac*-**5** and the bromo(4-fluorophenyl)silane *rac*-**6** (mixture separated by distillation only for analytical purposes). Subsequent hydrosilylation of 4-bromo-1-butene with *rac*-**5**/*rac*-**6**, catalyzed by H_2PtCl_6 (isopropanol), yielded a mixture of the chloro(4-bromobutyl)silane *rac*-**7** and the bromo(4-bromobutyl)silane *rac*-**8** (mixture separated by distillation only for analytical purposes). Reaction of *rac*-**7**/*rac*-**8** with (1,3-dithian-2-yl)lithium gave the (1,3-dithian-2-yl)silane *rac*-**9** (not isolated), which upon treatment with *n*-butyllithium yielded the 1-sila-cyclohexane *rac*-**10**. Subsequent hydrolysis of this compound in the presence of $HgCl_2$ and $CdCO_3$ gave the 1-sila-2-cyclohexanone *rac*-**1**. Its reduction with lithium aluminum hydride, followed by work-up with hydrochloric acid, yielded a mixture of the diastereomeric 1-sila-2-cyclohexanols *rac*-(Si*S*,C*R*/Si*R*,C*S*)-**2a** and *rac*-(Si*S*,C*S*/Si*R*,C*R*)-**2b** (molar ratio ca. 7:3). The racemates were separated from each other by medium-pressure liquid chromatography (MPLC) on silica gel [eluent *n*-hexane/diethyl ether (2:1)] and subsequently esterified with acetic anhydride to give the diastereomerically pure 2-acetoxy-1-silacyclohexanes *rac*-(Si*S*,C*R*/Si*R*,C*S*)-**3a** and *rac*-(Si*S*,C*S*/Si*R*,C*R*)-**3b**. The identity of compounds *rac*-**1**, *rac*-**2a**, *rac*-**2b**, *rac*-**3a**, *rac*-**3b**, *rac*-**5**–*rac*-**8**, and *rac*-**10** was established by elemental analyses, NMR-spectroscopic studies, and mass-spectrometric investigations.

Microbial Reductions

The 1-sila-2-cyclohexanone *rac*-**1** was found to be reduced diastereoselectively by resting free cells of *Saccharomyces cerevisiae* (DSM 11285) and by growing cells of *Trigonopsis variabilis* (DSM 70714) and *Kloeckera corticis* (ATCC 20109) to give mixtures of the enantiomeric 1-sila-2-cyclohexanols (Si*S*,C*R*)-**2a** and (Si*R*,C*S*)-**2a** as the major products (Scheme 2).

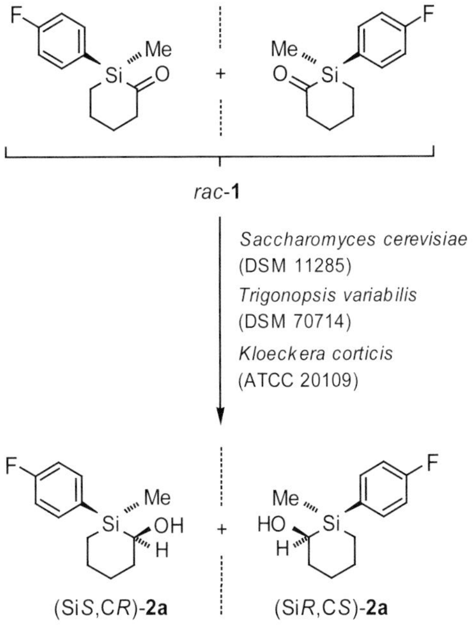

rac-**1**

Saccharomyces cerevisiae (DSM 11285)
Trigonopsis variabilis (DSM 70714)
Kloeckera corticis (ATCC 20109)

(Si*S*,C*R*)-**2a** (Si*R*,C*S*)-**2a**

Scheme 2. Microbial reductions of *rac*-**1**.

The bioconversions with these yeasts were performed on a preparative scale [4]. The yields and diastereomeric purities of the products and the molar ratios (Si*S*,C*R*)-**2a**/(Si*R*,C*S*)-**2a** are listed in Table 1.

Table 1. Data for the microbial reductions of *rac*-**1**.[a]

Microorganism	Yield [%]	*de* [%]	(Si*S*,C*R*)-2a/ (Si*R*,C*S*)-2a	(Si*S*,C*S*)-2b/ (Si*R*,C*R*)-2b
Saccharomyces cerevisiae (DSM 11285)	87	96	57:43	[b]
Trigonopsis variabilis (DSM 70714)	93	60	57:43	22:78
Kloeckera corticis (ATCC 20109)	95	95	50:50	[b]

[a] See [6]. — [b] Not determined.

The best preparative results were obtained with *Kloeckera corticis* (ATCC 20109). Reduction of *rac*-**1** with growing cells of this particular microorganism yielded a 1:1 mixture of (Si*S*,C*R*)-**2a** and (Si*R*,C*S*)-**2a** in 95 % yield; the diastereomeric purity of the product was 95 % *de*.

Microbial Hydrolysis

The yeast *Pichia pijperi* (ATCC 20127) was found to be a highly effective biocatalyst for the enantioselective ester hydrolysis (kinetic resolution) of the 2-acetoxy-1-silacyclohexane *rac*-(Si*S*,C*R*/Si*R*,C*S*)-**3a** (Scheme 3).

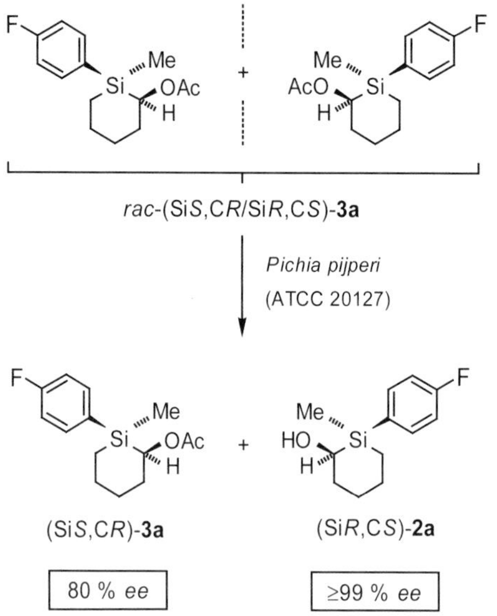

Scheme 3. Microbial hydrolysis of *rac*-(Si*S*,C*R*/Si*R*,C*S*)-**3a**.

Bioconversion of this compound with growing cells of *Pichia pijperi* (ATCC 20127) on a preparative scale [5] yielded a mixture of the 1-sila-2-cyclohexanol (Si*R*,C*S*)-**2a** and the non-converted substrate (Si*S*,C*R*)-**3a** (Scheme 3). After chromatographic separation of these compounds on silica gel [eluent *n*-hexane/diethyl ether (2:1)], the optically active 1-sila-2-cyclohexanol (Si*R*,C*S*)-**2a** was obtained in 56 % yield [relative to (Si*R*,C*S*)-**3a**] as an almost enantiomerically pure product (≥99 % *ee*). The non-converted substrate (Si*S*,C*R*)-**3a** (enantiomeric purity 80 % *ee*) was isolated in 54 % yield [relative to (Si*S*,C*R*)-**3a**].

Combination of the Biotransformations

As demonstrated in Scheme 4, the optically active 1-sila-2-cyclohexanol (Si*R*,C*S*)-**2a** can be prepared by a three-step synthesis, starting from the 1-sila-2-cyclohexanone *rac*-**1**. A highly diastereoselective microbial reduction [*rac*-**1** → (Si*S*,C*R*/Si*R*,C*S*)-**2a** (50:50)], a non-enzymatic (chemical) esterification [(Si*S*,C*R*/Si*R*,C*S*)-**2a** (50:50) → (Si*S*,C*R*/Si*R*,C*S*)-**3a** (50:50)], and a highly enantioselective microbial hydrolysis [(Si*S*,C*R*/Si*R*,C*S*)-**3a** (50:50) → (Si*R*,C*S*)-**2a**] are involved in this synthesis. Following this synthetic strategy, compound (Si*R*,C*S*)-**2a** was prepared in 23 % yield (total yield, relative to *rac*-**1**) as an almost diastereomerically and enantiomerically pure product (95 % *de*, ≥99 % *ee*). These results again emphasize the high preparative potential of biocatalysis for synthetic organosilicon chemistry.

(*S*)-**1** + (*R*)-**1**
Racemate
Kloeckera corticis (ATCC 20109) →
(Si*S*,C*R*)-**2a** + (Si*R*,C*S*)-**2a**
1:1 Mixture
↓ Ac_2O
(Si*S*,C*R*)-**3a** + (Si*R*,C*S*)-**3a**
1:1 Mixture
← *Pichia pijperi* (ATCC 20127)
(Si*S*,C*R*)-**3a** + (Si*R*,C*S*)-**2a**
80 % *ee* | ≥99 % *ee*

Scheme 4. Combination of the biotransformations.

Acknowledgment: Financial support of this work by the *Fonds der Chemischen Industrie* is gratefully acknowledged.

References:

[1] L. Fischer, S. A. Wagner, R. Tacke, *Appl. Microbiol. Biotechnol.* **1995**, *42*, 671.

[2] S. A. Wagner, S. Brakmann, R. Tacke, *"Biotransformation as a Preparative Method for the Synthesis of Optically Active Silanes, Germanes, and Digermanes: Studies on the (R)-Selective Microbial Reduction of $MePh(Me_3C)ElC(O)Me$ (El = Si, Ge), $MePh(Me_3Ge)GeC(O)Me$, and $MePh(Me_3Si)GeC(O)Me$ Using Resting Cells of Saccharomyces cerevisiae (DHW S-3) as Biocatalyst"*, in: *Organosilicon Chemistry II: From Molecules to Materials* (Eds.: N. Auner, J. Weis), VCH, Weinheim, **1996**, pp. 237–242.

[3] R. Tacke, S. A. Wagner, *"Chirality in Bioorganosilicon Chemistry"*, in: *The Chemistry of Organic Silicon Compounds, Part 3* (Eds.: Z. Rappoport, Y. Apeloig), Vol. 2, Wiley, Chichester, **1998**, pp. 2363–2400.

[4] *Saccharomyces cerevisiae:* 100 mg scale, phosphate buffer (pH 6.5), substrate concentration 0.25 g L^{-1}, 30°C. *Trigonopsis variabilis:* 100 mg scale, medium containing a phosphate buffer (pH 6.8), substrate concentration 0.25 g L^{-1}, 30°C. *Kloeckera corticis:* 100 mg or 600 mg scale, medium containing a phosphate buffer (pH 6.8), substrate concentration 0.25 g L^{-1}, 30°C.

[5] *Pichia pijperi:* 100 mg scale, medium containing a phosphate buffer (pH 6.8), substrate concentration 0.25 g L^{-1}, 27°C.

[6] The diastereomeric purities of **2a** were determined by capillary gas chromatography or ^{1}H-NMR studies (quantitative determination of the molar ratio **2a**/**2b**). The enantiomeric purities of **2a** were established, after derivatization with *(S)*-2-methoxy-2-(trifluoromethyl)-2-phenylacetyl chloride [*(S)*-MTPA-Cl], by ^{1}H- or ^{19}F-NMR studies [quantitative determination of the molar ratio (Si*S*,C*R*)-**2a**/(Si*R*,C*S*)-**2a**]. The enantiomeric purities of **3a** (after hydrolytic conversion into **2a**) were determined analogously to those of **2a**. The absolute configurations at the Si*C*H(OH)C carbon atoms of the enantiomers of **2a** were established, after derivatization with *(S)*-MTPA-Cl, by the NMR-spectroscopic correlation method described in [7]. The relative configurations of **2a** and **2b** were assigned by comparison with the diastereoisomers of 1-methyl-1-phenyl-1-sila-2-cyclohexanol (in this context, see [8]).

[7] J. A. Dale, H. S. Mosher, *J. Am. Chem. Soc.* **1973**, *95*, 512.

[8] R. Tacke, *"Recent Results in Bioorganosilicon Chemistry: Novel Sila-Drugs and Microbial Transformations of Organosilicon Compounds"*, in: *Organosilicon and Bioorganosilicon Chemistry: Structure, Bonding, Reactivity and Synthetic Application* (Ed.: H. Sakurai), Ellis Horwood, Chichester, **1985**, pp. 251–262.

Preparation of Silicon- and Germanium-Containing α-Amino Acids

Markus Merget, Stefan Bartoschek, Reiner Willeke, Reinhold Tacke*

Institut für Anorganische Chemie, Universität Würzburg
Am Hubland, D-97074 Würzburg, Germany
Tel.: Int. code + (931)888 5250 — Fax: Int. code + (931)888 4609
E-mail: r.tacke@mail.uni-wuerzburg.de

Keywords: α-Amino Acids / Enantioselective Syntheses / β-Silylalanine / β-Germylalanine / Crystal Structure

Summary: The silicon-containing α-amino acids *(R)*- and *(S)*-β-(trimethylsilyl)alanine [*(R)*-**1a** and *(S)*-**1a**] as well as the germanium-containing α-amino acids *(R)*- and *(S)*-β-(trimethylgermyl)alanine [*(R)*-**1b** and *(S)*-**1b**] were synthesized according to the Schöllkopf approach. In addition, *rac*-β-(trimethylsilyl)alanine (*rac*-**1a**) was studied by single-crystal X-ray diffraction.

Introduction

Synthetic amino acids with unnatural side chains have proven useful for probing the structural requirements for the biological activity of numerous peptides and proteins [1, 2]. In addition, unnatural amino acids are of interest as precursors of drugs and plant-protective agents [3]. In the course of our systematic studies on biologically active organosilicon compounds (for reviews, see [4, 5]), we have started a research program concerning silicon-containing α-amino acids. In the context of our investigations on C/Si/Ge bioisosterism (for recent examples, see [6, 7]), we also became interested in germanium-containing α-amino acids. We report here on the syntheses of the *(R)*- and *(S)*-enantiomers of β-(trimethylsilyl)alanine (**1a**) and β-(trimethylgermyl)alanine (**1b**). In addition, the crystal structure of *rac*-β-(trimethylsilyl)alanine (*rac*-**1a**) is described.

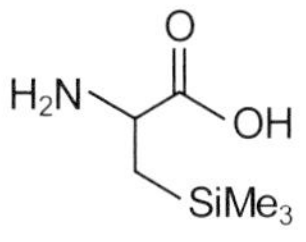

O H$_2$N OH SiMe$_3$ O H$_2$N OH GeMe$_3$

1a 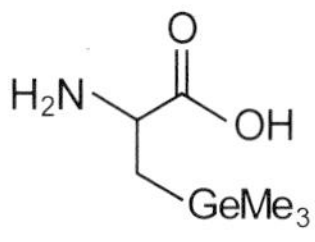   **1b**

Syntheses of *rac*-**1a** [8–10], *(R)*-**1a** [11–13], *(S)*-**1a** [13], *(R)*-β-(dimethylphenylsilyl)alanine [12], and *(R)*-β-(methyldiphenylsilyl)alanine [12] have already been reported. In addition, the ethyl ester and the *N*-Fmoc and *N*-Boc derivatives of *(R)*-**1a**, as well as renin inhibitory peptides bearing *(R)*-**1a**, have been described in the literature [14]. In contrast, germanium-containing α-amino acids have not yet been reported.

Syntheses of the α-Amino Acids *(R)*-1a, *(S)*-1a, *(R)*-1b, and *(S)*-1b

The α-amino acids *(R)*-**1a**, *(S)*-**1a**, *(R)*-**1b**, and *(S)*-**1b** were prepared by three-step syntheses, starting from the bislactim ether *(R)*-**2** (Scheme 1). The strategy of these syntheses is based on the Schöllkopf approach [14, 15].

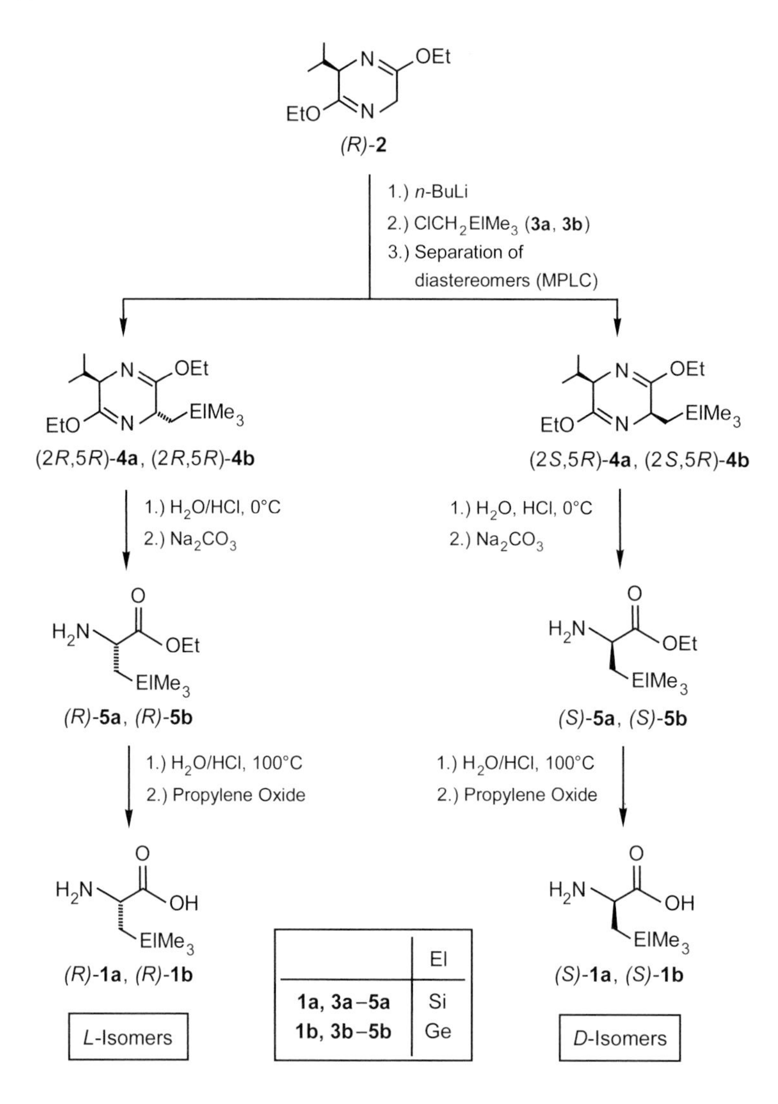

Scheme 1. Syntheses of the α-amino acids *(R)*-**1a**, *(S)*-**1a**, *(R)*-**1b**, and *(S)*-**1b**.

Metalation of the bislactim ether *(R)*-**2** with *n*-butyllithium and subsequent reaction with the (chloromethyl)silane **3a** yielded a mixture of the diastereomeric bislactim ethers (2*R*,5*R*)-**4a** and (2*S*,5*R*)-**4a** (molar ratio 85:15). The analogous reaction with the (chloromethyl)germane **3b** instead of the (chloromethyl)silane **3a** gave a mixture of the corresponding germanium analogues (2*R*,5*R*)-**4b** and (2*S*,5*R*)-**4b** (molar ratio 86:14). The mixtures of the diastereomers of **4a** and **4b** were separated by column chromatography (MPLC) on silica gel, and the diastereomerically pure

(GC studies) bislactim ethers (2*R*,5*R*)-**4a**, (2*S*,5*R*)-**4a**, (2*R*,5*R*)-**4b**, and (2*S*,5*R*)-**4b** were subsequently hydrolyzed with hydrochloric acid at 0°C to yield the respective β-(trimethylsilyl)alanine esters *(R)*-**5a** and *(S)*-**5a** and the β-(trimethylgermyl)alanine esters *(R)*-**5b** and *(S)*-**5b**. As shown by ^{1}H-NMR studies using the chiral solvating agent *(R)*-(–)-1-(9-anthryl)-2,2,2-trifluoroethanol, the enantiomeric purities of these esters were ≥98 % *ee*. Hydrolysis of the esters *(R)*-**5a**, *(S)*-**5a**, *(R)*-**5b**, and *(S)*-**5b** in boiling hydrochloric acid and subsequent treatment of the resulting α-amino acid hydrochlorides with propylene oxide yielded the title compounds *(R)*-**1a**, *(S)*-**1a**, *(R)*-**1b**, and *(S)*-**1b** as colorless crystalline solids.

The identity of the α-amino acids *(R)*-**1a**, *(S)*-**1a**, *(R)*-**1b**, and *(S)*-**1b** and compounds (2*R*,5*R*)-**4a**, (2*S*,5*R*)-**4a**, (2*R*,5*R*)-**4b**, (2*S*,5*R*)-**4b**, *(R)*-**5a**, *(S)*-**5a**, *(R)*-**5b**, and *(S)*-**5b** was established by elemental analyses (C, H, N), solution-state NMR studies (^{1}H, ^{13}C, ^{29}Si), and mass-spectrometric investigations.

Crystal Structure Analysis of the α-Amino Acid *rac*-1a

rac-β-(Trimethylsilyl)alanine (*rac*-**1a**) was structurally characterized by single-crystal X-ray diffraction [$T = -100$°C, $\lambda = 0.71073$ Å; space group $P2_1/c$, $a = 13.736(3)$ Å, $b = 7.0775(14)$ Å, $c = 9.779(2)$ Å, $\beta = 102.67(3)$°, $V = 927.5(3)$ Å^3, $Z = 4$]. Crystals were obtained by crystallization from water (slow evaporation of the solvent at room temperature). Although the quality of the crystal structure analysis is limited ($R1 = 10.1$ %, $wR2 = 19.6$ %), the following conclusion can be drawn: As expected for an α-amino acid, *rac*-**1a** exhibits a zwitterionic structure in the crystal (Fig. 1), the zwitterions being connected by intermolecular N–H···O hydrogen bonds.

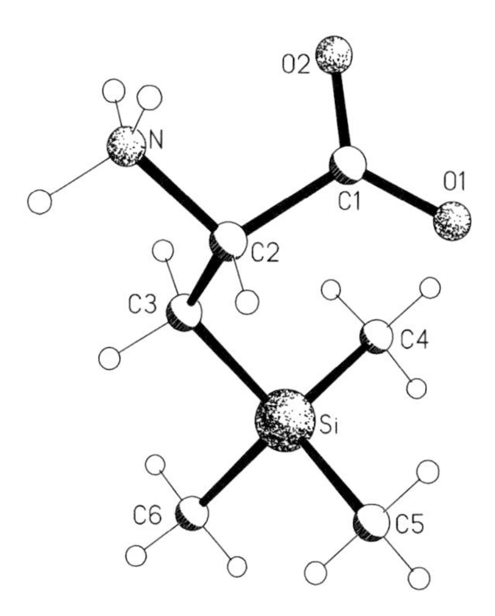

Fig. 1. Structure of *(R)*-β-(trimethylsilyl)alanine in the crystal of *rac*-**1a**.

Acknowledgment: Financial support by *Degussa AG* and the *Fonds der Chemischen Industrie* is gratefully acknowledged. In addition, we would like to thank Prof. Dr. K. Drauz and Dr. C. Weckbecker *(Degussa AG)* for helpful suggestions.

References:

[1] A. Giannis, T. Kolter, *Angew. Chem.* **1993**, *105*, 1303; *Angew. Chem., Int. Ed. Engl.* **1993**, *32*, 1244.

[2] J. Gante, *Angew. Chem.* **1994**, *106*, 1780; *Angew. Chem., Int. Ed. Engl.* **1994**, *33*, 1699.

[3] A. Kleemann, W. Leuchtenberger, B. Hoppe, H. Tanner, *"Amino Acids"*, in: *Ullmann's Encyclopedia of Industrial Chemistry* (Ed.: W. Gerhartz), 5th edn., Vol. A2, VCH, Weinheim, **1985**, pp. 57–97.

[4] R. Tacke, H. Linoh, *"Bioorganosilicon Chemistry"*, in: *The Chemistry of Organic Silicon Compounds, Part 2* (Eds.: S. Patai, Z. Rappoport), Wiley, Chichester, **1989**, pp. 1143–1206.

[5] R. Tacke, S. A. Wagner, *"Chirality in Bioorganosilicon Chemistry"*, in: *The Chemistry of Organic Silicon Compounds, Part 3* (Eds.: Z. Rappoport, Y. Apeloig), Vol. 2, Wiley, Chichester, **1998**, pp. 2363–2400.

[6] R. Tacke, D. Reichel, P. G. Jones, X. Hou, M. Waelbroeck, J. Gross, E. Mutschler, G. Lambrecht, *J. Organomet. Chem.* **1996**, *521*, 305.

[7] R. Tacke, U. Kosub, S. A. Wagner, R. Bertermann, S. Schwarz, S. Merget, K. Günther, *Organometallics* **1998**, *17*, 1687.

[8] L. Birkofer, A. Ritter, *Angew. Chem.* **1956**, *68*, 461.

[9] L. Birkofer, A. Ritter, *Liebigs Ann. Chem.* **1958**, *612*, 22.

[10] T. H. Porter, W. Shive, *J. Med. Chem.* **1968**, *11*, 402.

[11] R. Fitzi, D. Seebach, *Tetrahedron* **1988**, *44*, 5277.

[12] R. D. Walkup, D. C. Cole, B. R. Whittlesey, *J. Org. Chem.* **1995**, *60*, 2630.

[13] H. Yamanaka, T. Fukui, T. Kawamoto, A. Tanaka, *Appl. Microbiol. Biotechnol.* **1996**, *45*, 51.

[14] B. Weidmann, *Chimia* **1992**, *46*, 312.

[15] U. Schöllkopf, *Tetrahedron* **1983**, *39*, 2085.

Fascinating Silicon-Chemistry — Retrospection and Perspectives[†]

Hans Bock

Institut für Anorganische Chemie
Johann Wolfgang Goethe-Universität
Marie-Curie-Str. 11, D-60439 Frankfurt am Main, Germany
Tel: Int. code + (69)798 29181 — Fax: Int. code + (69)798 29188

Keywords: Molecular States / Photoelectron Spectroscopy / EPR Spectroscopy / Radical Anions and Cations / Sterically Overcrowded Molecules / DFT Calculations / Long-Range Interactions

Summary: From silicon and its compounds it has been possible to design new materials, which, from computers to space travel, have helped to shape the technology of our 20th century. Conversely, this technology has stimulated the tremendous development of silicon chemistry together with adequate methods to measure and to calculate the properties of silicon compounds. Within the last 30 years the Frankfurt Group has participated with about 25 of its 100 co-workers and has succeeded, for instance, in contributing:

- proof of positive charge delocalization in polysilanes (Kipping Award, American Chemical Society 1975),
- determination of the tremendous donor effect by β-trialkylsilyl substituents (Chemistry Award, Academy of Sciences, Göttingen 1969),
- oxidation of numerous organosilicon compounds of low first ionization energy to their radical cations using the $AlCl_3/H_2CCl_2$ single-electron transfer system,
- generation of short-lived intermediates in the gas phase such as silylenes, silaethene, silabenzene, and even silaisonitrile $R{-}N^{\delta+}{\equiv}Si^{\delta-}$, a molecule with a singly coordinate, triply bonded silicon center (Klemm Prize, German Chemical Society 1987),
- structural studies of van der Waals interactions in sterically overcrowded organosilicon molecules and density functional theory calculations on long-range Coulomb attractions in seven- and eight-coordinate silicon heterocyclic aggregates (Heyrovsky Medal, Academy of Science, Prague 1996).

Selected experimental results are presented and rationalized within a molecular state approach.

[†] 34th Essay on "Molecular Properties and Models"; shortened version of a plenary lecture at the Wacker Silicon Days, Munich, April 8th 1998.

Introduction: Molecular States of Organosilicon Compounds

At a time when an average of at least one chemical publication appears in every one of the 525 600 minutes of a 365-day year, perfect planning and evaluation of experiments is a necessity [1]. In addition, the real building blocks of a chemist are no longer the over 10^7 molecules now known, but increasingly their numerous molecular and molecular-ion states accessible via various routes of energy transfer [2]. These states are revealed and characterized by spectroscopic band or signal patterns [1–5]. This analytical "earmarking" provides valuable, but often unused, information about the energies of the compound investigated and the symmetries of its various states, as well as the energy-dependent electron distribution over its effective nuclear potentials [4]. Let us, therefore, take a "magnifying glass" (Fig. 1) to examine the structure and energy of selected silicon compounds: within a recommended qualitative molecular state model, structural facets are best covered by a combination of (two-dimensional) topological connectivity between subunits and their three-dimensional arrangement as denoted by symmetry, whereas the energy is specified as usual by the distribution of the electron density over the atomic potentials within the molecular skeleton [4, 5] (Fig. 1).

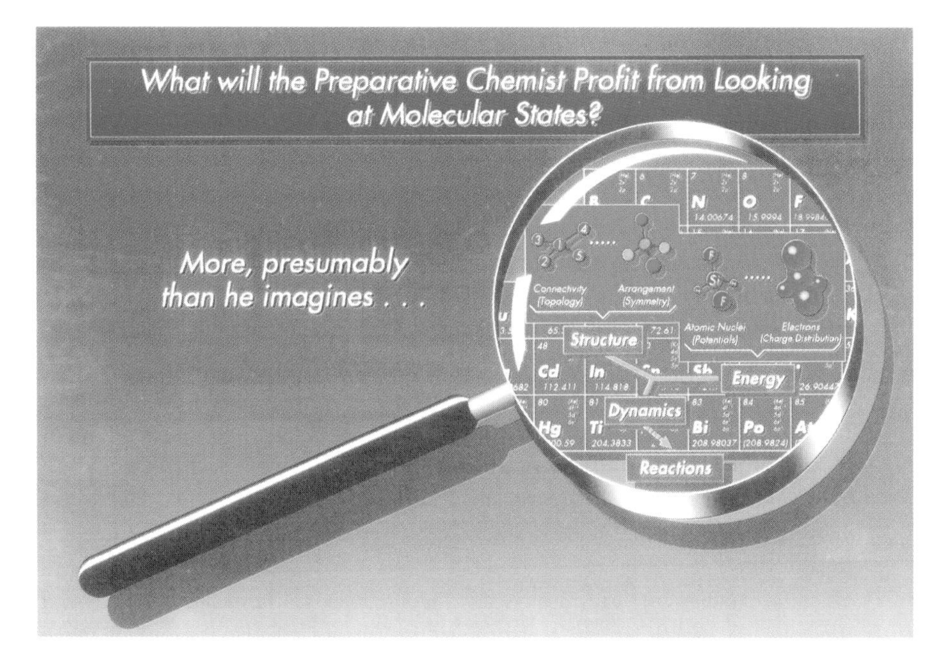

Fig. 1. Qualitative molecular state model based on topology, symmetry, effective nuclear potentials as well as electron distribution, emphasizing the structure ↔ energy relationship as well as molecular dynamics as essential prerequisites for reactions (cf. text).

Molecular states enlarge the portfolio of chemical compounds by specifying in addition their energy [1, 4, 5]. The individual states are arranged along a scale of the molecular total energy (Fig. 2), starting from the ground state of the neutral molecule, denoted Γ(M), in order of increasing energy via its nuclear spin and vibrational states (Fig. 2: NMR and IR) to the UV/Vis electronically excited states denoted $\mathcal{X}_i^J$(M), each with vibrational sublevels. At higher energy, the molecule is ionized to the doublet ground state $\Gamma(M^{\bullet+})$ of its radical cation, for which the spin distribution can be deduced from its ESR multiplet signals, as well as to electronically excited states as reflected in the PES ionization pattern. Still higher energies or oxidation in a stabilizing solvent can generate a dication by expelling another electron, and so forth. On the negative total energy side below the ground state $\Gamma(M)$, the molecular radical anions, dianions, and multianions are located [6].

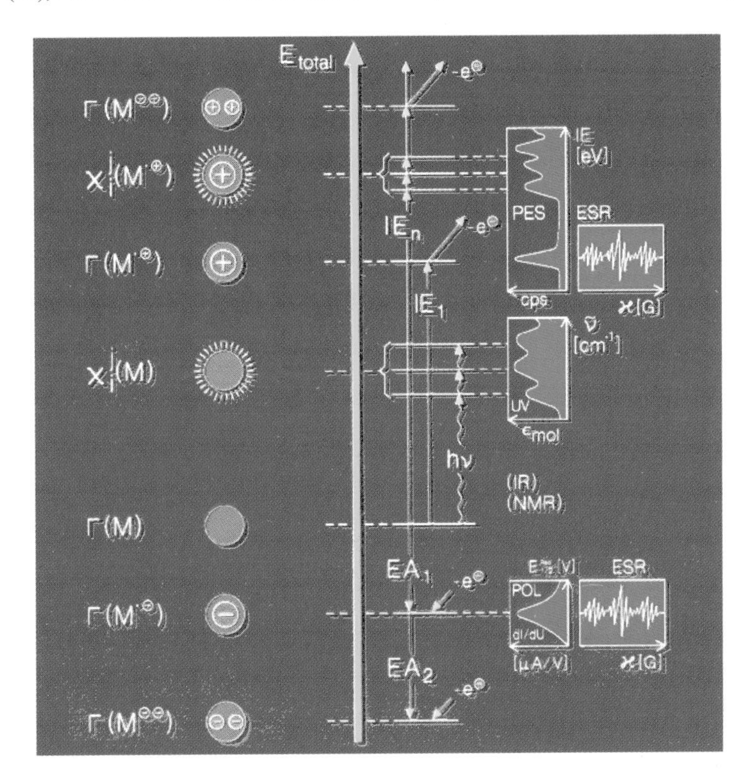

Fig. 2. Molecular states and relevant measurement data: Total energy scale for electronic ground (Γ) and excited (χ_i^J) states of a neutral molecule M and its radical cations $M^{\bullet+}$ or dications M^{2+} generated by ionization or oxidation or its radical anions $M^{\bullet-}$ and dianions M^{2-} resulting from electron insertion. In addition, the measurement methods used by the Frankfurt Group are indicated, especially UV, PES, ESR, and ENDOR spectroscopy as well as polarography (POL) or cyclic voltammetry (CV).

Frequently, molecular state phenomena and their measurement are specified to be either vertical or adiabatic, with a borderline of about 10^{-14} s between them. Vertical processes in time intervals of about 10^{-15} s or less, such as excitation or ionization of molecules, therefore, are accompanied only by changes in the charge distribution. In contrast, adiabatic processes in time intervals about 10^{-14} s or longer are characterized by the onset of molecular dynamics, which changes the molecular structure and allows relaxation or dissociation in molecular states of excessive energy. The schematic time scale for the molecular states and their changes (Fig. 3) indicates that in microwave- or radiowave-excited molecules, for instance, molecular rotations as well as spin resonance phenomena can be observed [7–9].

For molecular state properties, the time scale (Fig. 3) and molecular dynamics are essential. To begin with the vertical ionization of molecules to their various radical cation states in the femtosecond range (Fig. 2): fast electron expulsion leaves the radical cation with the "frozen" structure of the neutral molecule. Due to the lack of structural changes, photoelectron spectra in general can be assigned via Koopmans' theorem, $IE_n^V = \varepsilon_J^{SCF}$, by using the eigenvalues ε_J^{SCF} of uncorrelated quantum chemical calculations. In contrast, ESR spectra recorded in solution with microwave irradiation, exhibit molecular dynamics, at least at higher temperatures, when the rigid matrix becomes a liquid and the necessary activitation energies can be supplied. Molecular dynamics play an essential role in chemical reactions, as is obvious for thermal fragmentations in the gas phase under unimolecular conditions. Microscopic reaction pathways of smaller molecules have been investigated intensively [10–12] and, using femtosecond laser spectroscopy, even transition states have been visualized experimentally [10]. For medium-sized molecules with numerous degrees of freedom, however, and in spite of impressive efforts in both measurements and quantum chemical calculations, the complexity of their microscopic pathways still provides a considerable challenge [11, 12]. In exothermic reactions in solution, the reaction enthalpy stored as vibrational, rotational, and translational energy of the resulting molecules is generally quickly dissipated by frequent collisions, especially with solvent molecules. For redox reactions a network of equilibria is activated, including electron transfer, ion solvation, contact ion formation, and aggregation.

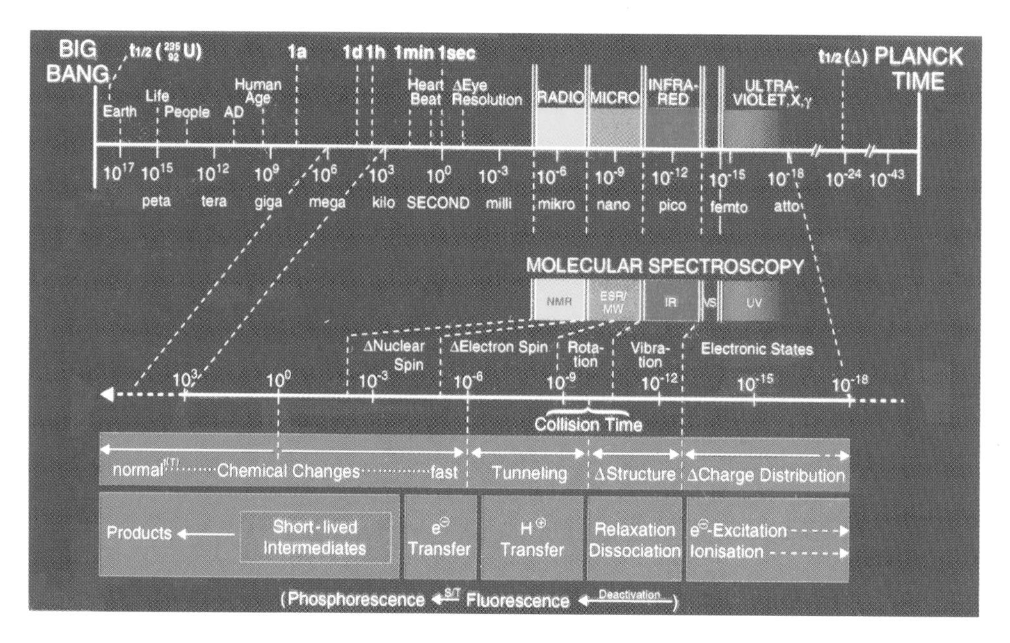

Fig. 3. Timescale for molecular states and their changes (in seconds, the approximate time interval between human heartbeats as well as the resolution of stimuli by the eye; cf. text).

Summing up, chemistry today — including its important organosilicon branch — has developed into a molecular state portfolio, which in addition to the basic synthesis of compounds and materials depends on physical measurements and quantum chemical calculation.

Retrospection: Successful Applications of the Molecular State Approach to Organosilicon Chemistry

For well over 30 years, the Frankfurt Group has investigated numerous compounds of the main group nonmetal elements B, C, N, P, S, and Si by means of extensive preparative efforts, mostly starting from quantum chemical precalculated predictions and using the methods of measurement outlined in the Introduction (Fig. 2). The investigations directed at molecular states of silicon-containing molecules are summarized in a chronological pictogram (Fig. 4) [1].

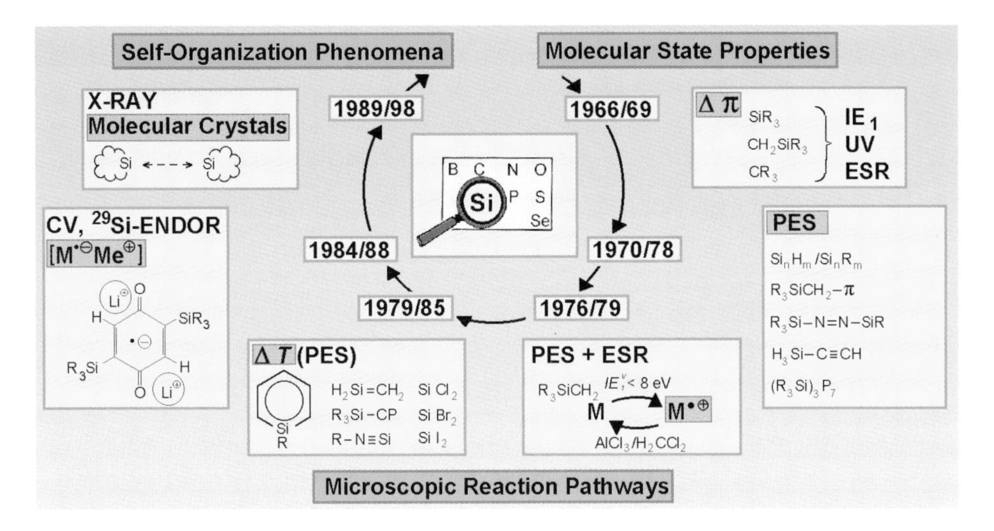

Fig. 4. Representative investigations of molecular state properties, microscopic reaction pathways, and self-organization phenomena of silicon compounds by the Frankfurt Group since 1966 (cf. Summary) [1].

In the early investigations of molecular state properties of silicon compounds we tested the advantageous qualitative molecular state model for Si substituent effects and the important relationship between structure and energy (Fig. 1). The extremely low effective nuclear potential of silicon has been demonstrated by photoelectron spectroscopic comparison of disilane and ethane (Fig. 5), the topology and symmetry of the structural components have been exemplified by the delocalization of positive charge in silicon chains and rings (Fig. 6), and some tremendous substituent perturbations in lone-pair n_x heteromolecules (Fig. 7) and carbon π systems (Fig. 8) [1] could be evaluated for a wide range of compounds [13]. The differing charge (spin) distributions determined from both the ESR spectra and the isolated Birch reduction products of 1,4-bis(trimethylsilyl)- and 1,4-bis(*tert*-butyl)benzene derivatives (Fig. 9) started our redox investigations, which peaked in the generation of numerous organosilicon radical cations from compounds exhibiting low first vertical ionization potentials by means of the powerful and novel single-electron oxidation system $AlCl_3/H_2CCl_2$ (Fig. 10) [1].

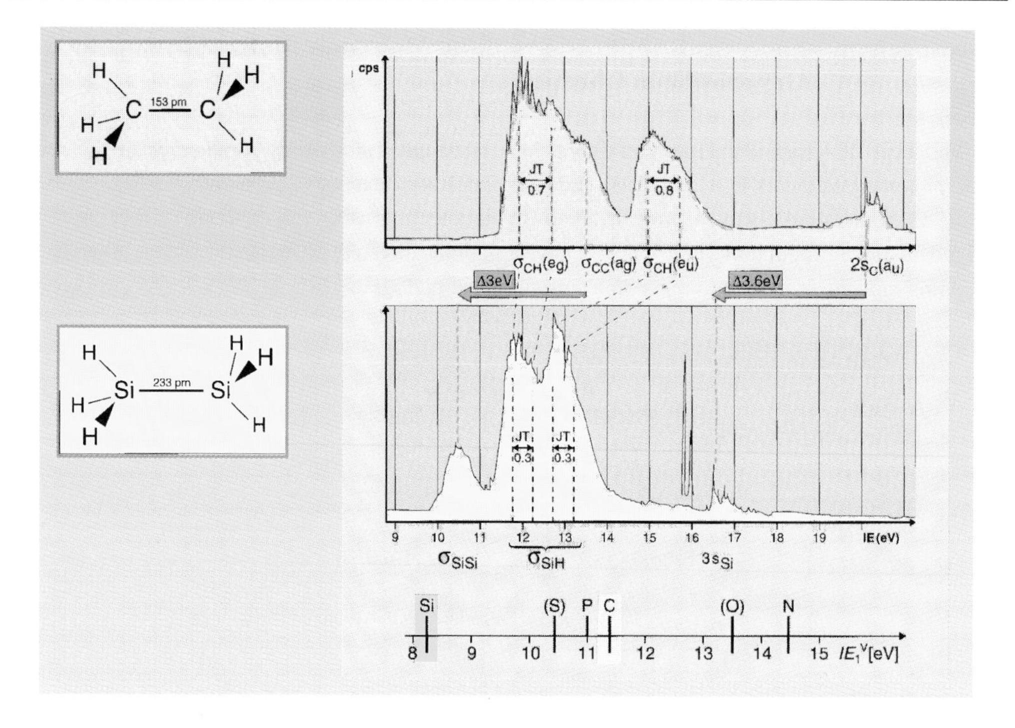

Fig. 5. Photoelectron spectroscopic radical cation state comparison of the iso(valence)electronic molecules disilane and ethane as well as an approximate sequence of effective nuclear potentials for some main group element centers (cf. text) [1].

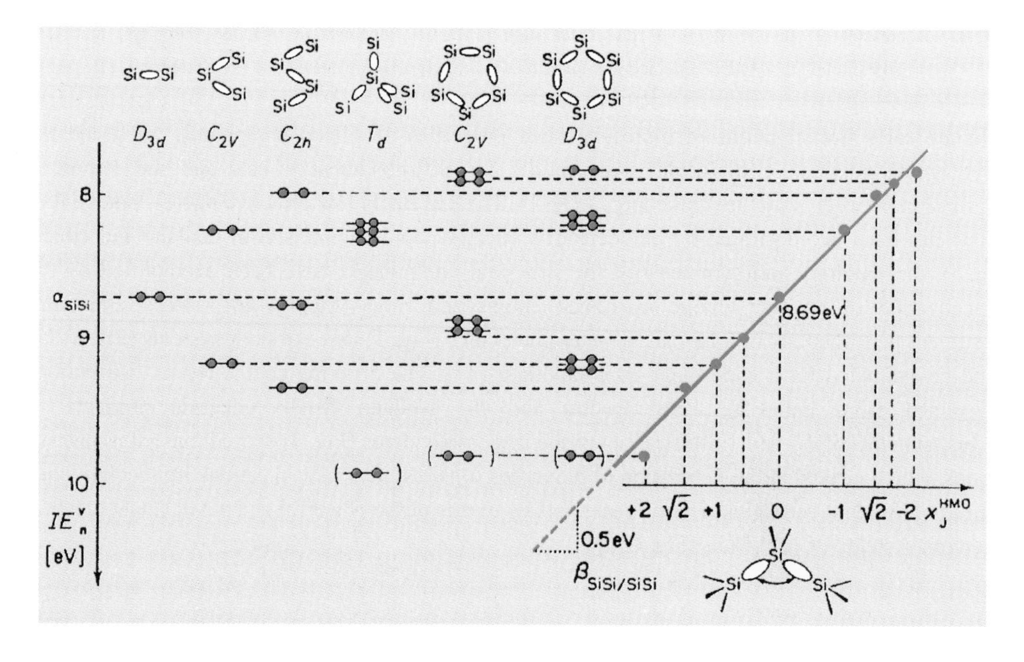

Fig. 6. Si–Si ionization patterns of permethylated polysilanes and their correlation with topological eigenvalues for the respective skeletal symmetries (cf. text) [1].

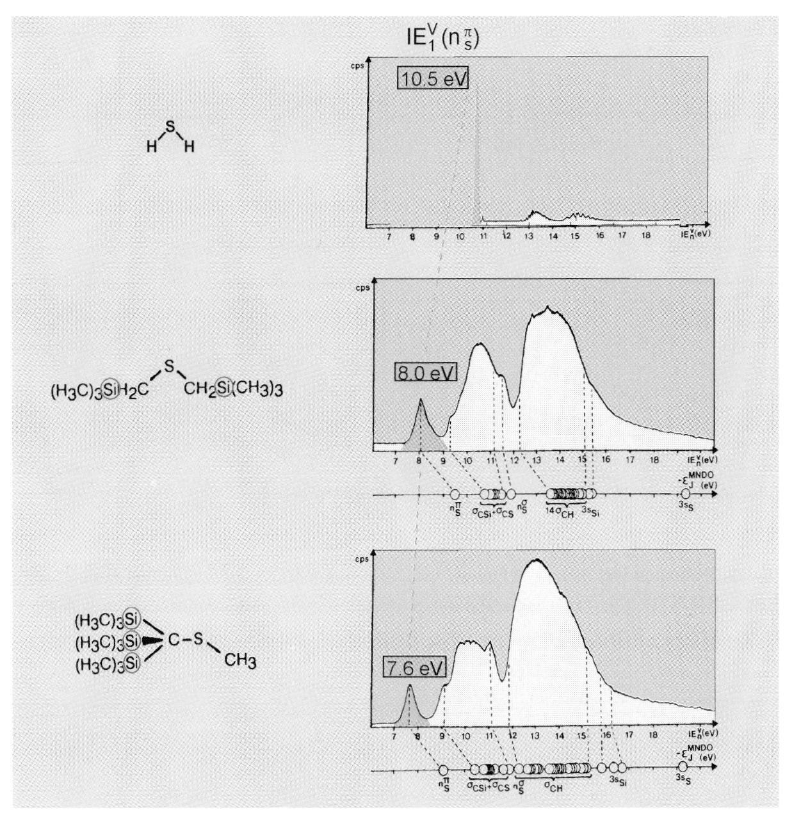

Fig. 7. (He) Photoelectron spectra of H_2S, $(R_3SiCH_2)_2S$ and $(R_3Si)_3C–S–CH_3$ with Koopmans' assignment (see text) [21].

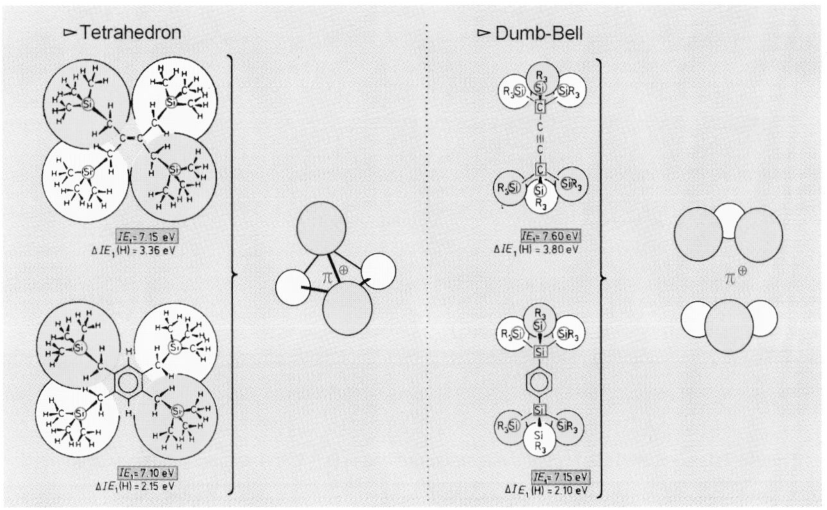

Fig. 8. Vertical first ionization energies of tetrakis(trimethylsilylmethyl)ethene as well as 2,3,4,5-benzene derivatives and of hexakis(trimethylsilyl)-substituted dimethylacetylene as well as 1,4-disilylbenzene derivatives (see text) [1].

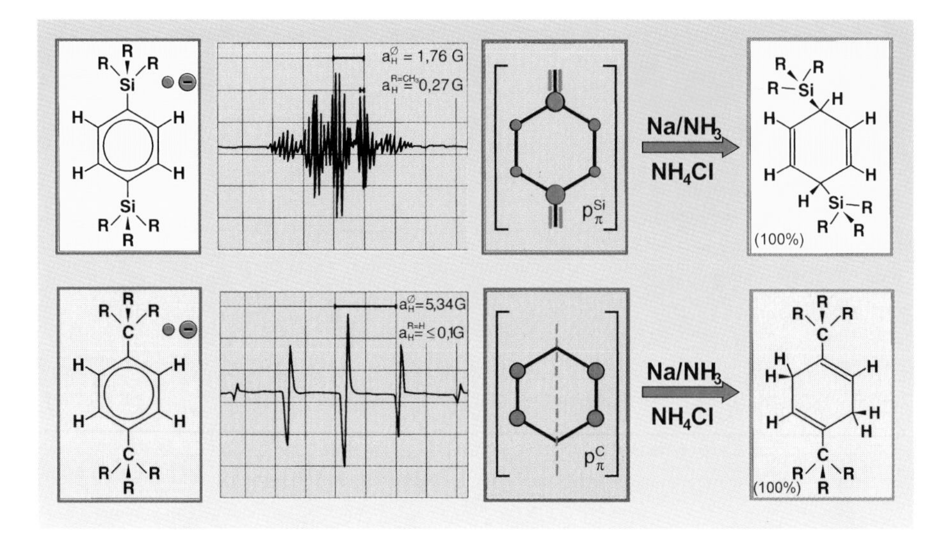

Fig. 9. Spin distribution in the iso(valence)electronic radical anions of 1,4-bis(trimethylsilyl)- and 1,4-dimethyl-substituted (R = H, CH_3) benzenes and the protonation quenching products of their Birch reduction (R = CH_3). Coupling constants of phenyl hydrogen atoms are $a_H^{ø}$ and of substituents a_H^R [1].

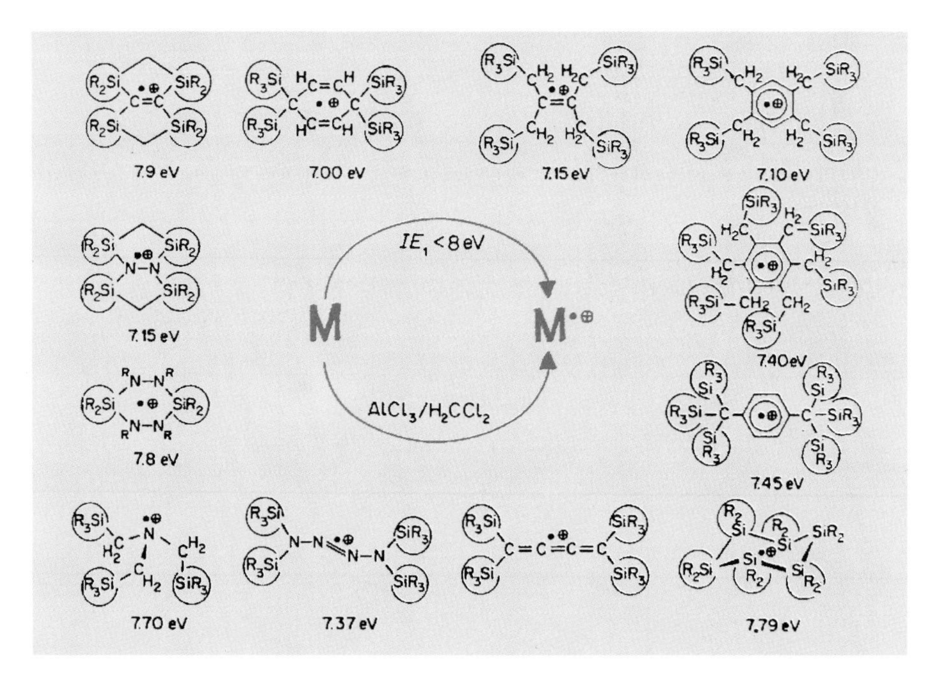

Fig. 10. Representative examples of organosilicon compounds (R = CH_3) exhibit first vertical ionization potentials below 8 eV and, therefore, can be oxidized to their radical cations by the powerful one-electron redox system $AlCl_3/H_2CCl_2$ (see text) [1].

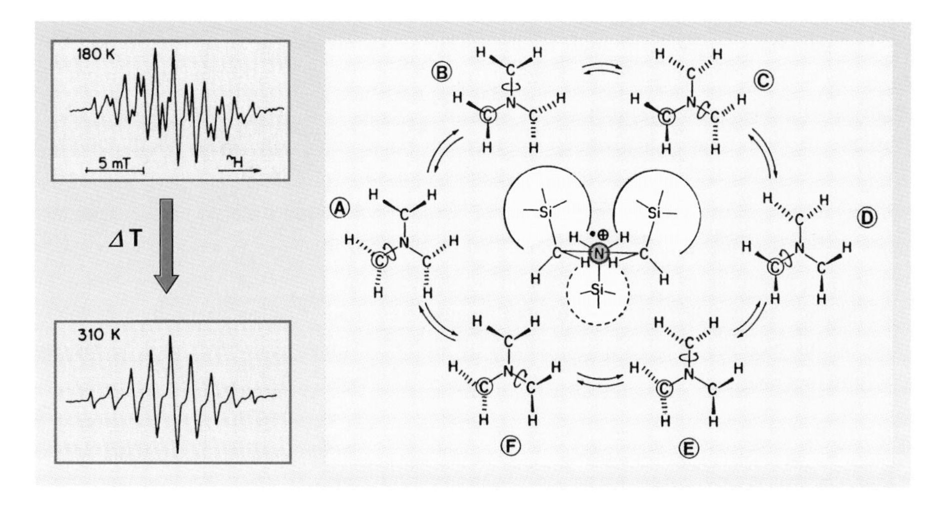

Fig. 11. Temperature-dependent ESR spectra of the tris(trimethylsilylmethyl)amine radical cation (cf. Fig. 10), proving its "cogwheel dynamics" in H_2CCl_2 solution above ~ 260 K (see text) [1].

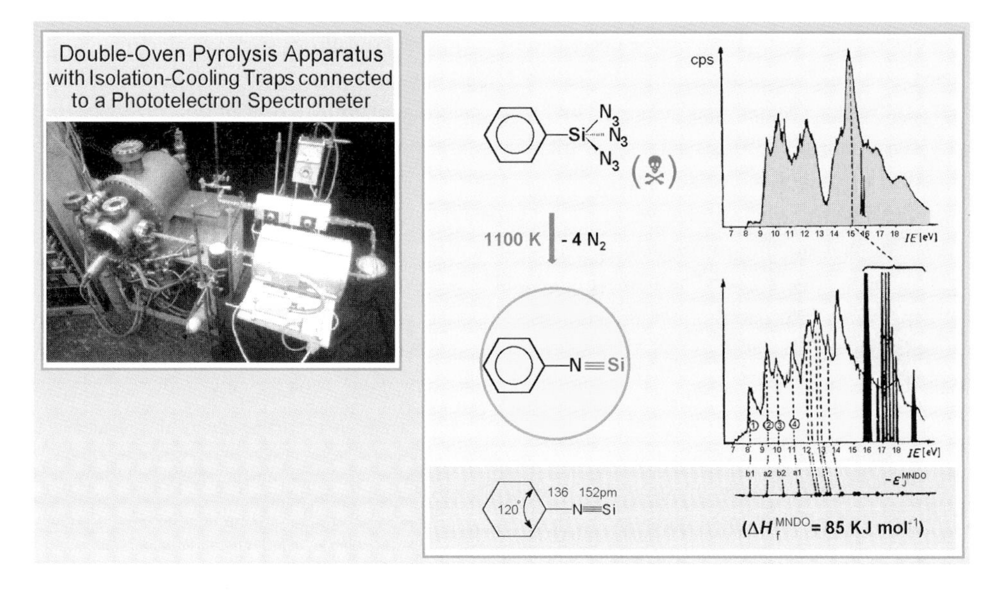

Fig. 12. Generation of phenyl silaisocyanide, the first molecule reported to contain a triple bonded, singly coordinated silicon atom, by controlled explosion of triazido(phenyl)silane and its PE-spectroscopic identification together with the ionization fingerprints of N_2 (black), a most advantageous leaving molecule. Triazido(phenyl)silane is prepared from trichloro(phenyl)silane by azide exchange with trimethylsilyl azide in the presence of $AlCl_3$ [11].

The thermal generation of numerous thermodynamically stable but kinetically unstable silicon organic intermediates such as silylenes, silabenzene, silaethene, or silaisonitrile $R–N^{\delta+}{\equiv}Si^{\delta-}$ with singly coordinated, triply bonded silicon could be accomplished under unimolecular conditions. The molecules were identified by PE spectroscopic radical cation state ionization patterns [2b, 12], and several of them were matrix-isolated in addition. Dynamic phenomena could be observed repeatedly, as in the temperature-dependent cogwheel gearing of the substituent in the tris(trimethylsilylmethyl)amine radical cation (Fig. 11) or the concentration-dependent ESR proof of 2,5-trimethylsilylquinone radical anion contact ion pairs and triples with alkali metal cations (Fig. 14).

All of the above projects, the highlights of which will be presented in subsequent short subsections with essential details, helped to tackle successfully molecular self-recognition and self-organization phenomena of organosilicon compounds in molecular crystals. Both van der Waals attractions in sterically overcrowded aggregates and Coulomb-dominated long-range intramolecular interactions in silicon complexes of high coordination number will be presented in the concluding perspectives.

The Effective Nuclear Potential of Silicon: Radical Cation-State Comparison of the Iso(valence)electronic Molecules H_3SiSiH_3 and H_3CCH_3

Within the qualitative molecular state model recommended (Fig. 1), the potentials at the individual nuclear centers [14, 15] should not be described by ill-defined quantities such as "electronegativity" or "hard and soft", but rather by the rational and advantageous "effective nuclear charges" introduced by J. C. Slater and inherent in most quantum chemical procedures [16]. They are (mostly) easy to parametrize using experimental data and can then be applied to the substituent perturbation of parent molecules [1, 4, 14, 15]. For an illustration, the considerable difference between the effective nuclear charges of Si and C centers, important for the electron-donor effect in β-silyl-substituted organic molecules, is demonstrated by the ionization patterns of the "smallest chain polymers" disilane and ethane (Fig. 5) [1]. A comparison of the iso(valence)electronic molecules reveals surprisingly large differences: The energy of the $H_3SiSiH_3^{\bullet+}$ state with the largest $3s_{Si}$ component at $IE_6^v = 20.1$ eV [2b] is lowered tremendously by $\Delta IE_1^v = 3.6$ eV (Fig. 5). The rather small effective nuclear charge of the silicon centers may even cause changes in the $M^{\bullet+}$-state sequence: The $\sigma_{cc}(a_g)$ ionization band of ethane is shifted by 3.0 eV to become the $\sigma_{SiSi}(a_g)$ ground state of the disilane radical cation at 10.6 eV (Fig. 5).

The low effective nuclear charge of Si centers in silicon organic molecules relative to those of other main group elements such as S, P, C, O, or N (Fig. 5: IE_1^v) has numerous essential consequences, e.g., for the significantly different properties of silanes and alkanes (Fig. 5), for the "band structure" of doped polysilanes [1] or for the breathtaking donor effect of β-trimethylsilyl groups, which allow many $(H_3C)_3SiCH_2$-substituted organic π-systems to be oxidized in solution to stable radical cations $[\pi(CH_2Si(CH_3)_3)_n]^{\bullet+}$ [1, 3] (Fig. 10).

Topology and Symmetry: SiSi Ionization Patterns of Polysilanes

The investigation, which won the Frederic Stanley Kipping Award of the American Chemical Society in 1975, was based on the photoelectron spectra [2b] of both unsubstituted and permethylated (Fig. 6) polysilanes, which exhibit characteristic band patterns in their low-energy region up to 10 eV due to ionizations into radical cation states with a predominant SiSi framework contribution [1, 2b] (Fig. 6). The low vertical ionization energies IE_n^v [eV] of the permethylated polysilanes with differing numbers of dissimilarly connected σ_{SiSi} building blocks in skeletons of different symmetry (Fig. 6) can be correlated with the topological eigenvalues from a linear combination of σ_{SiSi} bond orbitals. The satisfactory linear regression (Fig. 6) is calibrated by the first vertical ionization energy of hexamethyldisilane, $IE_1^v = 8.69$ eV $= \alpha_{SiSi}$; defining the Coulomb term and its gradient defines the interaction parameter $\beta_{SiSi/SiSi} \approx 0.5$ eV. The applicability of the topological LCBO-MO model [1, 14] is further supported by the approximately equally spaced splitting patterns around the Coulomb term α_{SiSi} for all "alternating"[2, 10] polysilanes (Fig. 6), as well as by the almost coincident lowest ring ionizations of the two cyclic organosilanes as expected from their identical eigenvalues.

Altogether, 21 IE_n^v values for polysilanes are incorporated into the simple topologically "isoconjugate" [1] radical cation state model (Fig. 6). From the measurement data, the Coulomb term $\alpha_{SiSi} = 8.69$ eV, and the "resonance" interaction parameter $\beta_{SiSi/SiSi} \approx 0.5$ eV, result. Although smaller than the analogous π interaction in the $M^{\bullet +}$ states of linear polyenes, $\beta_{C=C/C=C} \approx 1.2$ eV [2b, 15], clear-cut experimental evidence is provided that in polysilane subunits of organosilicon radical cations a positive charge will be largely delocalized over the "saturated" Si–Si bonds.

Substituent Perturbation of Parent Molecules: The Powerful Donor Effect of β-Trimethylsilyl Groups

The widespread and useful subdivision of larger molecules into parent systems and substituents and their comparison based on first- and second-order perturbation arguments [1, 14] have proved tremendously valuable for organosilicon compounds as well [1]. Especially important is the extreme electron donor effect of $(H_3C)_3SiCH_2$ substituents, discovered over 30 years ago [17-19] by way of the low first ionization energies of organosilicon molecules containing β-silyl-substituted lone pairs [20] (Fig. 7) or prototype π-systems such as benzene [17], ethene [18], or acetylene [19] (Fig. 8).

The donor substituent perturbation of lone pair centers is transparently demonstrated in sulfides (Fig. 7). The parent molecule H_2S exhibits for its π-type sulfur lone pair perpendicular to the molecular plane an "ionization needle" at 10.5 eV [20]. Two $(H_3C)_3SiCH_2$ substituents broaden the needle considerably, indicating charge delocalization and lower its first vertical ionization potential by 2.5 eV = 241 kJ mol^{-1} to 8.0 eV. In the dimethyl sulfide derivative with three R_3Si substituents at the same carbon center, the total shift reaches almost 3 eV (Fig. 7), an amazing demonstration of the extreme donor power of β-trimethylsilyl substituents. Tremendous substituent perturbations are also observed at other lone-pair heterocenters such as nitrogen in amines [1, 2b], with the ionization energies IE_1^v lowered from 10.85 eV in NH_3 to 7.66 eV in $N(CH_2SiR_3)_3$, a difference of 3.19 eV!

The donor substituent perturbation of prototype π-systems with their π-electrons each formally only at one nuclear potential, as are the lone pairs in heteromolecules, also reaches breathtaking ionization energy differences: on comparing the parent ethene $H_2C{=}CH_2$ with $IE_1^v = 10.51$ eV to $(R_3SiH_2C)_2C{=}C(CH_2SiR_3)_2$ with $IE_1^v = 7.15$ eV, a lowering $IE_1^v = 3.36$ eV results! According to first or second order pertubation estimates, the lowering of the ionization energy depends on the partial charge ($c_{J\mu}^2$) at the respective center and on its effective nuclear charge ($\Delta\alpha$) as well as on the electron delocalization (β_{xy}) [14]. In addition, steric overcrowding can be essential (Fig. 8): for instance, in tetrasubstituted ethene and benzene derivatives with their bulky $(H_3C)_3Si$ groups arranged alternately above and below the molecular plane, the radical cations resulting from electron expulsion are almost tetrahedrally surrounded and, therefore, despite the rather different π-systems, their vertical first ionization energies are almost identical [1, 2b]. An analogous "dumb-bell" shielding is observed in molecules such as $(R_3Si)_3C$-substituted acetylene and $(R_3Si)_3Si$-substituted 1,4-benzene, which are each lipophilically wrapped by 18 methyl groups exhibiting van der Waals radii of 200 pm [7].

Spin and Charge Distribution in Organosilicon (Radical) Anions: the Birch Reduction of 1,4-Bis(trimethylsilyl)benzene

The ESR spectroscopic "fingerprints" of both 1,4-bis(trimethylsilyl)- and 1,4-dialkyl-substituted benzene radical anions (Fig. 9), each generated by contact of their aprotic ($c_{H^+} < 0.1$ ppm) dimethoxyethane solutions with a potassium mirror in a glass apparatus sealed under vacuum, are rather different [21]. The radical anion of 1,4-bis(trimethylsilyl)benzene exhibits a quintet due to four equivalent phenyl hydrogen atoms with a small coupling constant and is further split by the 18 equivalent methyl hydrogen atoms. By contrast, the *p*-xylene radical anion shows a quintet from the phenyl hydrogen atoms with a larger coupling a_H, but no detectable further splitting from the six methyl protons: therefore, its 1,4-substitution centers must be located in a nodal plane with roughly zero spin density ρ_π and, concomitantly, a negligible charge density. On the other hand, in the R_3Si-substituted benzene radical anion, the smaller ring-proton couplings a_H prove a smaller spin density at the ring centers 2, 3, 5 and 6 relative to the large ones at the ring centers 1 and 4 as revealed by the resolved methyl hydrogen multiplets (Fig. 9). The differing ESR signal patterns suggest that Birch reductions using Na/NH_3 followed by proton quenching of the respective dianions using NH_4Cl should take place at different centers each of highest charge density [22] (Fig. 9, all R = CH_3): indeed, exclusively and almost quantitatively, the corresponding cyclohexa-2,5-diene and cyclohexa-1,4-diene derivatives are isolated.

The One-Electron Oxidation of Organosilicon Compounds to their Radical Cations

The small effective nuclear charge of silicon (Fig. 5) and the powerful donor effect of β-trimethylsilyl groups (Fig. 7) lower the PE-spectroscopically determined first ionization energies of organosilicon compounds considerably relative to those of their carbon analogues [1, 2b]. Their expected oxidation to stable radical cations [1, 3, 23] is advantageously achieved by the selective one-electron oxidation system of $AlCl_3$ in water-free CH_2Cl_2 discovered in 1976 [1, 2b] with the chlorocarbenium ion H_2CCl^+, produced by Cl^- abstraction from the solvent, as plausible powerful redox reagent with a cyclovoltammetrically determined oxidation potential of +1.6 V [24]. The correlation of (often irreversible) oxidation potentials of organosilicon molecules with their first ionization energies (Fig. 10) provides an enormously useful prediction: all molecules — including those of other heteroelements such as B, N, P, S, or Si [24] — that exhibit a first ionization energy below 8 eV ought to be oxidized selectively to their radical cations by the oxygen-free $AlCl_3/H_2CCl_2$ system, in a reactions sometimes called "Bock oxidation" and used worldwide.

ESR/ENDOR spectroscopy, and especially the ^{29}Si-ENDOR method developed [25], yield a wealth of information about the ground states of numerous organosilicon radical cations generated so elegantly in $AlCl_3/CH_2Cl_2$ solution (Fig. 10). The recorded signal patterns permit spin distribution to be "read off", thereby establishing structural changes due to "adiabatic" relaxation. Temperature-dependent measurements allow observation of molecular dynamics phenomena on the ESR/ENDOR time scale of approximately 10^{-7} s such as the "cogwheel gearing" in the tris(trimethylsilyl)methylamine radical cation [26] (Fig. 11): Often organosilicon molecules with several β-trimethylsilyl groups (cf. Fig. 8) are sterically overcrowded and therefore rigid with respect to substituent rotations. In contrast, the planar radical cation $[(H_3C)_3SiH_2Cl_3N]^{\bullet+}$ with 120° N-C-N angles between its ligands shows a characteristic ESR line broadening on raising the temperature from 180 to 310 K and its 25 signals are reduced to nine. This phenomenon is due to the rotation of each $(H_3C)_3SiH_2C$ substituent to the other side of the NC_3 molecular plane on the ESR time scale of 10^{-6} to 10^{-8} s. Because each side of the NC_3 plane offers room for only two of the bulky $(H_3C)_3Si$ groups, a "cog-wheel"-like coupled dynamic process (Fig. 11: **A** to **F**) results and the CH_2 protons become equivalent on the ESR time scale [3, 26].

In summary, it has been established that all organosilicon compounds with first vertical ionization energies below 8 eV (Fig. 10) can be oxidized by $AlCl_3/H_2CCl_2$ in aprotic solutions to radical cations. Their ESR/ENDOR signal patterns show the influence of Si substitution on spin distributions and allow detection of structural changes as well as molecular dynamics phenomena (Fig. 11) on the ESR time scale. The temperature-dependent ESR spectra thus reveal the storage of energy in the $3n–6$ degrees of freedom of three-dimensional molecules with n centers.

The Thermal Generation of Phenyl Silaisocyanide in the Gas Phase

Reactions in flow tubes, thermally enforced under unimolecular conditions or heterogeneously catalyzed, are advantageously analyzed by real-time photoelectron spectroscopy [12] (Fig. 12). Numerous short-lived, kinetically unstable intermediates such as the organosilicon molecules dichlorosilylene $SiCl_2$, silaethene $H_2C{=}SiH_2$, or silabenzene C_5SiH_6 [1, 2, 12], can be generated elegantly this way and characterized by their "ionization fingerprints" [12].

Starting from the literature report [27] that $H–N^{\delta+}{\equiv}Si^{\delta-}$ is produced on photolysis of H_3SiN_3 in a 4 K argon matrix, the total energy, the structure, and the ionization pattern of the more stable phenyl derivative have been calculated (Fig. 12). Small scale azide exchange of phenyltrichlorosilane yields triazido(phenyl)silane, the controlled explosion of which in a flash pyrolysis apparatus under approximately unimolecular conditions ($p \approx 10^{-3}$ Pa, $c \approx 10^{-3}$ mM) can be carried out without danger [28]. Real-time PE-spectroscopic analysis of the band intensities [12] revealed that the optimum reaction conditions require a temperature of 1100 K (i.e., red heat) for the complete elimination of four molecules of N_2 (Fig. 12: blackened ionization bands). The identification of the first organosilicon compound with a formal $–N^{\delta+}{\equiv}Si^{\delta-}$ triple bond to silicon and a silicon coordination number of 1 was accomplished by Koopmans' correlation of the total of nine resolved low-energy ionization bands (Fig. 12) with calculated MNDO eigenvalues. The formation of the valence isomer $H_5C_6–Si{\equiv}N$, which is predicted to be 400 kJ mol^{-1} less stable, can be excluded, for example, by comparison of the radical cation state sequence with that of the valence-isoelectronic carbon species $C_6H_5–N^{\delta+}{\equiv}C^{\delta-}$. In the meantime, the PE-spectroscopic results have been confirmed by matrix isolation of $H_5C_6–N^+{\equiv}Si^-$ [29].

Approximate Hyperface Calculation for the Thermal Generation of Silabenzene

Dynamic phenomena are important molecular state properties (Fig. 1), as has been demonstrated by the selected example of the radical cation $(R_3SiCH_2)_3N^{\bullet+}$ (Fig. 11). Along the microscopic pathways of chemical reactions, they often play a decisive role [10]. As concerns medium-sized molecules, however, their $3n$–6 degrees of freedom mostly lead to nearly hopeless situations of multidimensional complexity [8, 11]. Even when some of them are neglected for specific molecular conformations by introducing additional assumptions, number-crunching hypersurface calculations have to be performed. Such a crude simplification of the rather complex reality will be illustrated for the pyrolysis of 1-silacyclohexa-2,5-diene to silabenzene and H_2 [21] (Fig. 13) in a yellow-glowing flow tube at 800 K and under 10^{-4} Pa pressure — that is under nearly unimolecular conditions.

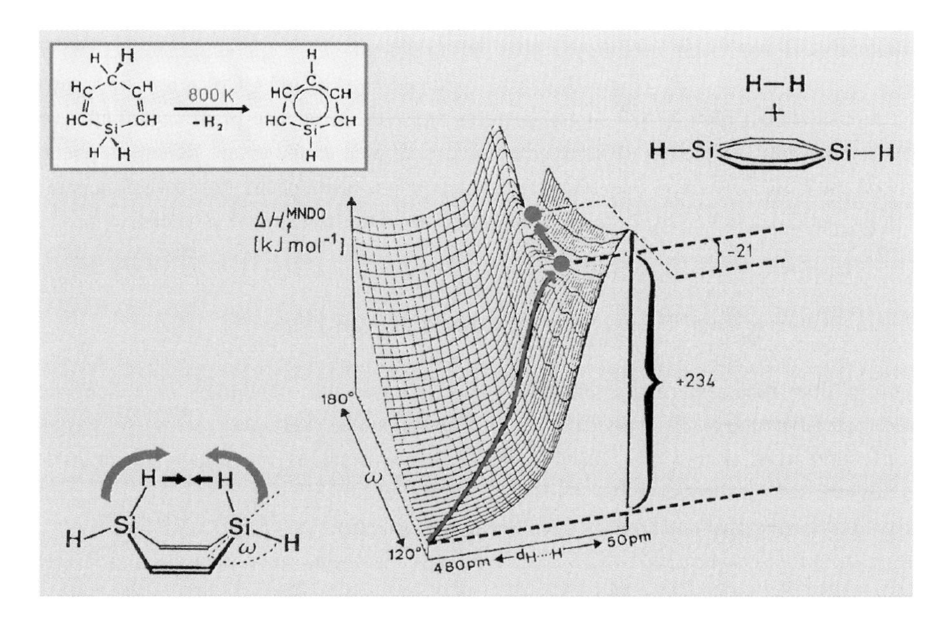

Fig. 13. Approximate energy hypersurface for the thermal intramolecular H_2 elimination from 1,4-disilacyclohexa-2,5-diene to 1,4-disilabenzene in the gasphase (cf. text).

In order to carry out an approximate energy hypersurface calculation, the 36 degrees of freedom of the 14-atom molecule 1-silacyclohexa-2,5-diene must be drastically reduced [8]. Possibilities are: introducing a mirror plane into the siladerivative by incorporating an additional silicon atom at the 4-position, selecting the intramolecular bending of its boat conformation as the dominant vibrational mode characterized by both the angles ω between the C-Si-C triangle and the $(C{=}C)_2$ plane, and using the $d_{H\cdots H}$ distance as an additional "joint" between the two Si centers [1] (Fig. 13). Their variation ($\Delta\omega = 120^\circ \rightarrow 180^\circ$ and $\Delta d_{H\cdots H} = 480 \rightarrow 50$ pm) in over 100 separate, each geometry-optimized, MNDO calculations [1] produces a folded total energy hypersurface, on which the "product valley" for 1,4-disilabenzene + H_2 is reached by crossing the saddlepoint indicated. Additional calculations of selected points for the 1-silacyclohexa-2,5-diene that was actually investigated [1], which exhibits only C_s symmetry, yield values closer to experiment: an activation enthalpy of 175 kJ mol^{-1}, a saddlepoint structure with ω = 140° and $d_{H \cdots H}$ = 120 pm, and an enthalpy difference –120 kJ mol^{-1} for the considerably deeper product valley of the ensemble "C_5SiH_8", which is still situated below the energetically rather favorable products silabenzene and hydrogen. In the meantime, the analogous H_2 elimination of the parent molecule cyclohexa-1,4-diene has been carefully studied by polarized laser spectroscopy [30] and interpreted by large-scale correlated quantum chemical calculations [31].

Reduction Pathway for 2,5-Bis(trimethylsilal)-*p*-benzoquinone through an Network of Electron Transfer, Ion Pair Formation and Solvation Equilibria

The pinpointing of reaction pathways in solution by ESR and ENDOR measurements will be illustrated by the reduction of 2,5-bis(trimethylsilyl)-*p*-benzoquinone without or with added lithium tetraphenylborate (Fig. 14) [1, 13].

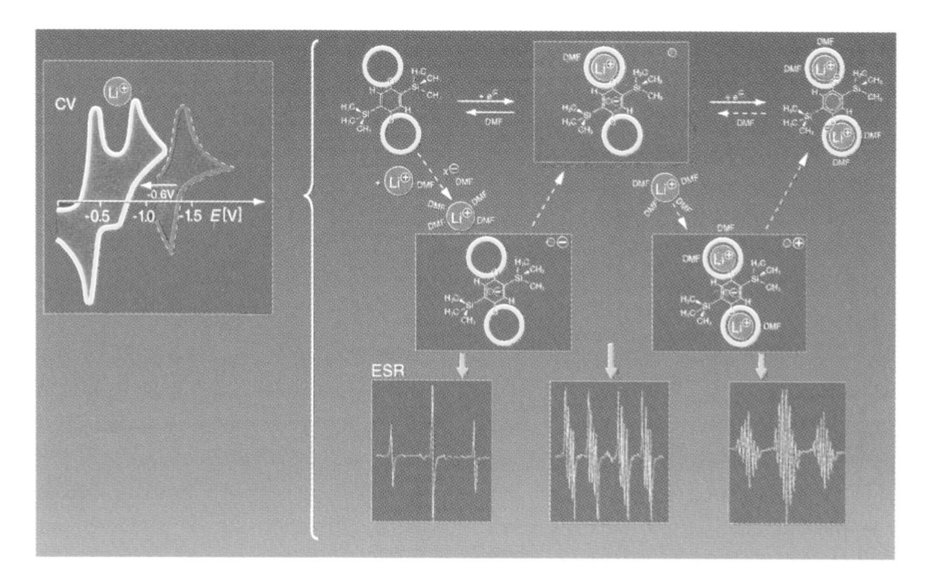

Fig. 14. Electron transfer and contact ion formation on reduction of 2,5-bis(trimethylsilyl)-*p*-benzoquinone in aprotic DMF solution with added $Li^+[B^-(C_6H_5)_4]$. Discovered by the tremendous decrease in the half-wave reduction potential on salt addition, the individual contact ion multiples along the microscopic reaction pathway can be characterized by their ESR multiplet signal "fingerprints" (cf. text).

Preceding cyclovoltammetric studies proved that on reduction in aprotic DMF solution (c_{H^+} < 0.1 ppm), its second and reversible half-wave potential is lowered by 0.6 V(!) on addition of the soluble salt $Li^+[B^-(C_6H_5)_4]$ and simultaneously becomes irreversible [13] (Fig. 14: CV). The presumed microscopic reduction pathway in the presence of excess lithium cation, which, owing to its small ionic radius (r_{Li^+} = 60 pm), possesses a high effective ionic charge, is supported by independent ESR/ENDOR measurements [13] in THF (Fig. 14: ESR): the solvated radical anion $M^{\bullet -}$, with two equivalent quinone hydrogen atoms, can be detected by its (1:2:1) triplet. In the subsequently formed contact ion-pair radical $[M^{\bullet -}Li^+]^{\bullet}$, the Li^+ countercation "docks" at the quinone radical anion. Due to reduced symmetry $C_{2v} \rightarrow C_s$, a doublet of doublets ESR signal pattern results, the lines of which are each split into quartets by the ^{7}Li nuclear spin ($I = 3/2$). Via additional equilibria a triple radical cation $[Li^+M^{\bullet-}Li^+]^{\bullet+}$ finally results, which expectedly displays a triplet of the again-equivalent quinone hydrogen atoms split into septets by the two contact Li^+ ions. In this solvent-shared contact triple ion, the final redox reaction product, i.e., the dilithium salt of 2,5-bis(trimethylsilyl)hydroquinone, is already preformed, and therefore the second electron is inserted irreversibly at a drastically lowered reduction potential (Fig. 14: CV). Other contact ion pairs of 2,5-bis(trimethylsilyl)-*p*-benzosemiquinone investigated contained larger, and therefore more loosely coordinated, countercations [13], for which approximate energy hypersurface calculations [13] suggest the fluctuation between the two equivalent docking sites, which can be proven experimentally by temperature-dependent ESR measurements [13].

Obviously, appropriately selected measurement data for molecular states [1, 4, 5] (Fig. 2) — combined with approximate quantum chemical hypersurface calculations (Fig. 13) [4, 8] — can provide a rationale for both the properties of molecules along their energy scale (Fig. 2) and — on inclusion of molecular dynamics aspects along the time scale (Fig. 3) — some insight into the microscopic pathways of reactions (Fig. 14). The organosilicon example chosen, a consecutive two-electron reduction of 2,5-bis(trimethylsilyl)-*p*-benzoquinone, demonstrates the usefulness of the qualitative molecular state approach in planning experiments and discussing the results of the preparative chemist (Fig. 1). In conclusion it might be pointed out that the partially dynamic contact ion pairs discovered in the investigation could be of interest, both bioinorganically because of their carrier properties and industrially as molecular switches.

Summary and Properties

The "torrent" of novel silicon compounds and the "sea" of their molecular states [2] have allowed us at best to illustrate only some aspects of selected examples. Nevertheless, it must be kept in mind that all chemical species, formally characterized by the electrons at the specific potentials of centers connected in their skeletons, will exhibit different dynamics and hence differing properties (Figs. 2 and 3). For rationalization, the preparative chemist should advantageously compare measurement data for equivalent states of chemically related molecules. This approach — guided by adequate quantum chemical model calculations — will further stimulate and support the chemist's chemical intuition [1–5]. All the molecular state correlations presented here involving silicon compounds, such as the Birch protonation of benzene dianions (Fig. 9) or the one-electron oxidizability of molecules with first ionization energies below 8 eV by the novel $AlCl_3/CH_2Cl_2$ redox system (Fig. 10), proved to be reliable and their predictions valuable [1]. The determination of the differing substituent effects on individual molecular states is useful for both the planning and the evaluation of experiments [1]. Altogether, these and numerous further details of the molecular states of silicon-containing compounds [1–7] contribute to our essential knowledge of silicon chemistry.

A decade ago, band structures of polysilanes, novel photoconducting materials, contact ion pairing, and intramolecular interactions were among the topics of academic and, to some extent, industrial interest [1]. Continuing along these lines, two molecular crystal investigations, on van der Waals and on long-range Coulombic interactions in organosilicon compounds, will illustrate some of the current interests.

The "van der Waals Skin" of Sterically Overcrowded Organosilicon Molecule

The structure of a molecule can change considerably as its energy and thus its electron distribution vary within the time domain of dynamic relaxation [1]. Crystals contain molecules generally in their ground state close to the global minimum of the total energy and with largely "frozen" molecular dynamics. The structures of molecules and molecular ions as determined by an analysis of crystal lattices with negligible packing effects is therefore a suitable starting point for the discussion of essential molecular properties as well as of their quantum chemical calculation [7]. In addition, interactions between molecules become visible in crystals and provide valuable information on molecular recognition and self-organization [32]. Design, synthesis, and structure determination of molecular skeletons which are distorted by steric congestion and/or by charge perturbation has therefore become an additional research project of the Frankfurt Group since 1989 [7]. The over 400 crystals grown and structurally characterized comprise about 60 organosilicon molecules and the study of their structures aims at an improved understanding of the spatial requirements of bulky substituents such as trialkylsilyl groups and their effects on the molecular properties [22, 33].

In molecules with bulky substituents separated by a spacer along a threefold axis, a valuable criterion for steric overcrowding is supplied by the difference in their dihedral angles [22, 33]. The structure of hexakis(trimethylsilyl)silane (Fig. 15) shows that the half-shells of the two bulky $Si(Si(CH_3)_3)_3$ groups are connected by a central SiSi bond of normal (240 pm) length. The shortest non-bonding C···C distances between the two molecular halves are, however, only 352 pm and thus are about 12 % shorter than the sum of the van der Waals radii of two methyl groups $C(H_3)\cdots(H_3)C$ (about 400 pm) [7]. Unexpectedly, the molecular skeleton $Si_3Si–SiSi_3$ of D_3 symmetry exhibits rather different dihedral angles (43° and 77°). Structural correlation (Fig. 15) with other molecules also containing two bulky substituent half-shells and separated along their central C_3 axes by spacers of different lengths, demonstrates that at distances Y···Y below 333 pm different dihedral angles $\omega(X_3Y\text{-}YX_3)$ between the half-shell substituents are observed, and above 414 pm identical ones. The observed torsion $D_{3d} \rightarrow D_3$ within the molecular skeleton, therefore, provides a valuable criterion for steric overcrowding. The drastic cogwheel-meshing of the methyl groups, especially between the two molecular halves, causes extremely short non-bonded C···C distances, down from 400 pm by 48 pm (12 %!) to 352 pm.

Intramolecular van der Waals bonding is of considerable importance for organosilicon compounds such as polysilane materials. The structural investigations of disilane and trisilane provide an interesting facet [7]: correlation of their Si–Si bond lengths versus their Pauling bond orders results in a linear regression. The distance-dependent Pauling bond orders range from 1.00 in hexamethyldisilane with an Si–Si bond length of 235 pm to 0.26 for hexakis(*tert*-butyl)disilane with an extremely elongated Si–Si spacer distance of 270 pm between its bulky $Si(C(CH_3)_3)_3$ half-shells (Fig. 15). To rationalize the sometimes considerably weakened Si–Si bonds — hexakis(*tert*-butyl)disilane does not dissociate into two radicals — it has been proposed [34] that additional attractive van der Waals interactions within the hydrocarbon wrapping skin contribute to the bonding within the respective organosilicon molecules. This assumption is further supported by the extremely short non-bonded $C(H_3)\cdots(H_3)C$ distances, determined to be only 352 pm.

The vertical first ionization energy of $(R_3Si)_3Si–Si(SiR_3)_3$, IE_1^V = 7.7 eV [22], is the smallest one observed so far for a disilane derivative [33] and suggests a perfect delocalization of the cation charge over the σ_{SiSi} skeleton (cf. Fig. 6).

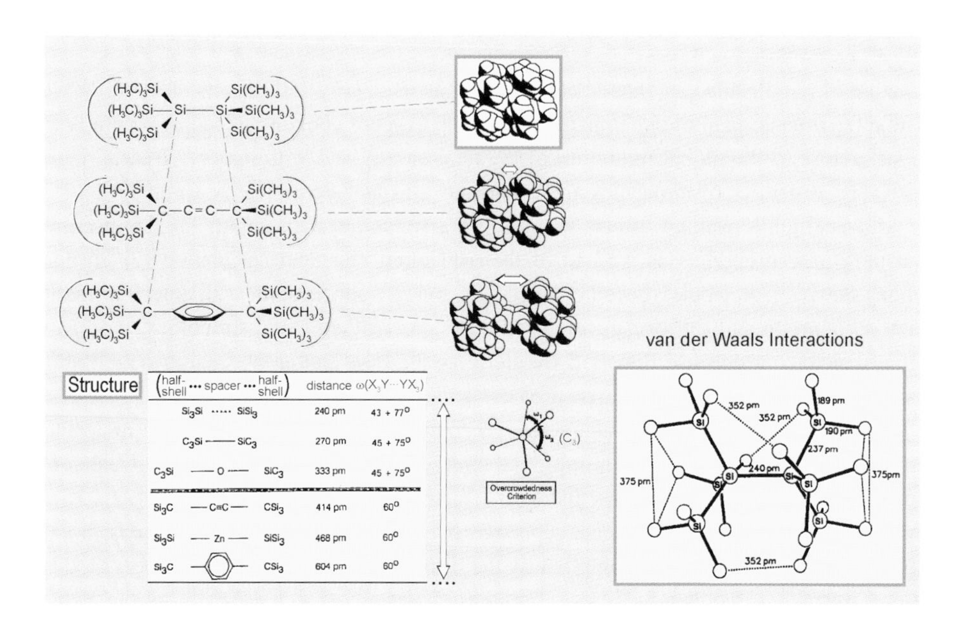

Fig. 15. Organosilicon model compounds with spacers of different lengths between two bulky half-shell substituents with threefold axes: overlap criterion based on deviating dihedral angles and intramolecular (as well as peripheral) van der Waals bonding [22].

Long-Range Interactions in Organosilicon Compounds

Si centers in organosilicon compounds exhibit coordination numbers between 1 and 10 [1] and the resulting distances vary over wide ranges, as exemplified by the histogram for N → Si interactions registered in the Cambridge Structural Database [35] (Fig. 16 with *N* hits within 10 pm ranges), which contains a total of 574 entries for $d_{Si\cdots N}$ distances greater than 200 pm (Fig. 16). The shorter interactions are assigned to so-called "SiN single bonds" such as in the 1-amino-8-silylnaphthalene derivative sitrol (Fig. 16). Compounds with Si–N distances between 220 and 260 pm have repeatedly been defined as intramolecular donor–acceptor complexes. What type of interaction is, however, represented by the numerous distances Si···N exceeding 275 pm?

For density functional theory calculations, the known structures of tris[(2-dimethylaminomethyl)phenyl]silane with a heptacoordinate Si center and of bis[2,6-bis(dimethylaminomethyl)phenyl]silane with an octacoordinate Si center, which complement each other, were selected. For the calculations based on the experimental structures of the relatively large molecules, an NEC SX4 supercomputer had to be used (for structural and computational details cf. [35]): Twisting the nitrogen lone pairs n_N perpendicular to the Si···N interaction axis, and accounting for the changed intramolecular van der Waals interactions, yielded approximate energy differences $\Delta E_{Si\cdots N}$ for the additional long bonds in the individual fragments (Fig. 16). The values of 5, 14, and 2 kJ mol^{-1} for bond lengths 289, 300, and 312 pm indicated a distance-dependent potential minimum. Additional, geometry-optimized and MP4-correlated calculations with large basis sets for the simplest analogous complex imaginable, $H_3N\cdots SiH_4$, not only yielded the expected minimum at about 300 pm Si···N, but in addition reproduced a steeper repulsive gradient at shorter distances and a shallower dissociative one at longer distances (Fig. 16).

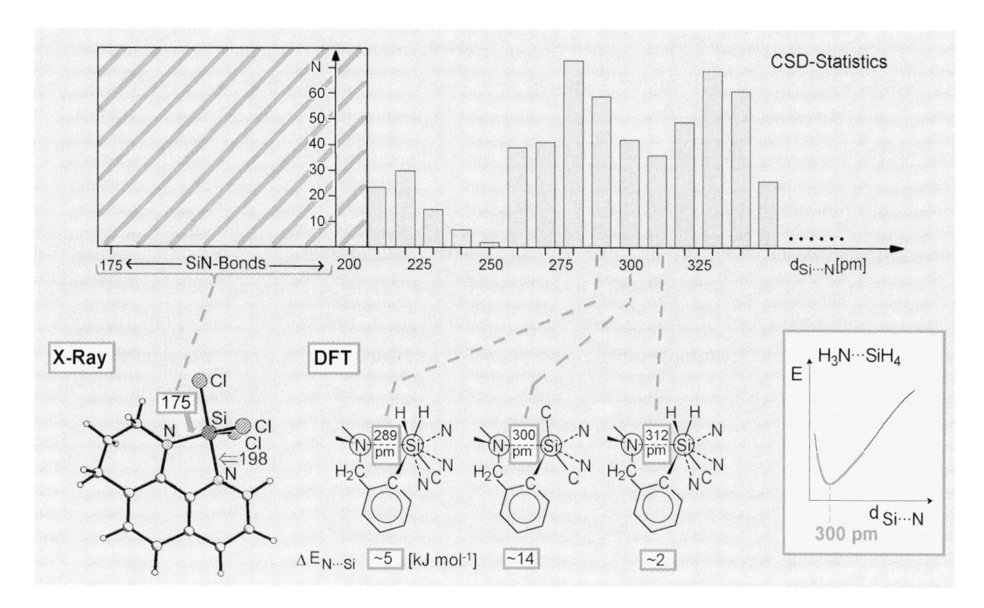

Fig. 16. Observed Si···N bond lengths in hypercoordinate organosilicon compounds (Cambridge Structural Database, Version 5.10) and density functional theory (DFT) calculations of energy differences $\Delta E_{N\cdots Si}$ based on the experimental structural data and correlated potential for the simplest model complex $H_3N\cdots SiH_4$ (cf. text) [35].

For a tentative suggestion concerning the possible origin of the long-range N···Si interactions in bond notation terms, the charge distribution has been calculated by natural bond orbital (NBO) analysis [35]. The pure electrostatic attraction between the resulting charges q_{Si} of +1.6 and q_N of –0.5 at 300 pm distance is estimated to be about 60 kJ mol^{-1}. Accordingly, a Coulomb-dominated donor–acceptor bonding is proposed for these special hypercoordinate N···Si compounds with long-range interactions.

In conclusion, the importance of long-range interactions for the chemist is emphasized from crystal growth [7] to macromolecular structure [32]. Therefore, an extension of the presented DFT calculations to other systems and especially those with Si···O interactions, which differ even more in their effective nuclear charges, will definitely stimulate progress in the polysiloxane and silicate world.

Concluding Remarks

The examples discussed in this chapter represent rather different topics: organosilicon compounds, their measured molecular state properties and quantum chemical efforts to rationalize and to correlate the valuable data. All of the examples presented emphasize that molecular state considerations (Figs. 1–3) are rather useful in the fascinating world of silicon (Fig. 4). In addition, they demonstrate the increasing impact of both the constantly improved and the newly developed techniques of physical measurement and the effect of the ever-faster and numerically more precise quantum chemical calculations, even for larger molecules, which were honored by last year's Nobel prizes. Although both hardware and software are readily available, "intellectual tunneling" between the only seemingly different viewpoints of preparative and theoretical chemistry, which meet in the reality of molecular states, still can and should be improved. It would be most rewarding, could the arguments presented convince the preparative silicon chemist of how profitable it is to take a look at his fast-developing and interesting area of research through the molecular-state magnifying glass (Fig. 1).

Acknowledgments: Thanks are due, above all, to my "theoretical" teachers and friends Edgar Heilbronner and Michael J. S. Dewar, who convinced a preparative inorganic chemist some 20 years ago of the usefulness of simple molecular orbital models and semiempirical calculations. The results of our research group presented here have been obtained by committed and competent co-workers, and are cited in the references. They also reflect cooperation and stimulating discussions with many colleagues world-wide. For generous support of our research, we would like to thank the *Deutsche Forschungsgemeinschaft*, the *Fonds der Chemischen Industrie*, the *State of Hessen, Hoechst AG* and *Wacker Chemie GmbH*, *Degussa AG*, and the *Dow Corning Corporation*. Special thanks go to Dipl.-Chem. V. Krenzel for his graphic efforts.

References:

[1] H. Bock "*Fundamentals of Silicon Chemistry: Molecular States of Silicon-Containing Compounds*", *Angew. Chem.* **1989**, *101*, 1659–1682; *Angew. Chem. Int. Ed. Engl.* **1989**, *28*, 1627–1650.

[2] Comprehensive reviews of silicon chemistry can be found in the two editions of *The Chemistry of Organic Silicon Compounds*, Wiley, Chichester, **1989** (Eds.: Z. Rappoport, Y. Apeloig) and **1989** (Eds. S. Patai, Z. Rappoport), from which the following contributions are emphasized within this context: a) Y. Apeloig, "*Theoretical Aspects of Organosilicon Compounds*", pp. 57–226 (377 references); b) H. Bock, B. Solouki, "*Photoelectron Spectra of Silicon Compounds*", pp. 555–653 (252 references); c) G. Raabe, J. Michl, "*Multiple Bonds to Silicon*", pp. 1015–1142 (406 references); d) R. J. P. Corriu, J. C. Young, "*Hypervalent Silicon Compounds*", pp. 1241–1288 (182 references).

[3] H. Bock, W. Kaim, "*Organosilicon Radical Cations*", *Acc. Chem. Res.* **1982**, *15*, 9–17.

[4] Cf., e.g., H. Bock, "*Molecular States and Molecular Orbitals*", *Angew. Chem.* **1977**, *89*, 631–655; *Angew. Chem. Int. Ed. Engl.* **1977**, *16*, 613–637, and the numerous references cited therein.

[5] Cf., e.g., H. Bock, "*99 Semesters of Chemistry — a Personal Retrospective on the Molecular State Approach by Preparative Chemists*", *Coll. Czech. Chem. Commun.* **1977**, *62*, 1–41, and references cited therein.

[6] H. Bock, K. Gharagozloo-Hubmann, C. Näther, N. Nagel, Z. Havlas, *Angew. Chem.* **1996**, *108*, 720; *Angew. Chem. Int. Ed. Engl.* **1996**, *35*, 631.

[7] H. Bock, K. Ruppert, C. Näther, Z. Havlas, H.-F. Herrmann, C. Arad, I. Göbel, A. John, J. Meuret, S. Nick, A. Rauschenbach, W. Seitz, T. Vaupel, B. Solouki, *Angew. Chem.* **1992**, *104*, 564–595; *Angew. Chem. Int. Ed. Engl.* **1992**, *31*, 550–581.

[8] H. Bock, R. Dammel, B. Roth: "*Molecular State Fingerprints and Semiempirical Hypersurface Calculations: Useful Correlations to Track Short-Lived Molecules in Rings, Clusters and Polymers of the Main Group Elements*" (Ed.: A. H. Cowley). *ACS Symp. Ser.* **1983**, *232*, 139–165.

[9] H. Bock, U. Stein, A. Semkow, *Chem. Ber.* **1980**, *113*, 3208.

[10] Review: A. Zewail "*Femtochemistry*", *J. Phys. Chem.* **1993**, *97*, 12427–12446.

[11] H. Bock, R. Dammel "*The Pyrolysis of Azides in the Gas Phase*", *Angew. Chem.* **1987**, *99*, 518–540; *Angew. Chem. Int. Ed. Engl.* **1987**, *26*, 504–526, and references cited therein.

[12] H. Bock, B. Solouki. "*Photoelectron and Molecular Properties: Real-Time Gas Analysis in Flow Systems*", *Angew. Chem.* **1981**, *93*, 425–442; *Angew. Chem. Int. Ed. Engl.* **1981**, *20*, 4227–444. Cf. also H. Bock, B. Solouki, S. Aygen, M. Bankmann, O. Breuer, J. Dörr, M. Haun, T. Hirabayashi, D. Jaculi, J. Mintzer, S. Mohmand, H. Müller, P. Rosmus, B. Roth, J. Wittmann, H. P. Wolf, "*Optimization of Gas-Phase Reactions Using Real-Time PES-Analysis: Short-Lived Molecules and Heterogeneously Catalyzed Processes*", *J. Mol. Spectrosc.* **1988**, *173*, 31–49.

[13] H. Bock, B. Solouki, P. Rosmus, R. Dammel, P. Hänel, B. Hierholzer, U. Lechner-Knoblauch, H.P. Wolf, "*Recent Investigations on Short-Lived Organosilicon Molecules and Molecular Ions*", in *Organosilicon and Bioorganosilicon Chemistry*, (Ed.: H. Sakurai) Ellis Horwood, Chichester, **1985**, pp. 45–73, and references cited therein. Cf. also H. Bock, "*Reaction of Carbon and Organosilicon Compounds as Viewed by a Preparative Chemist*", *Rev. L'Actualité, Chim.* **1985/3**, 33; *Polyhedron* **1988**, 7, 2429.

[14] Cf., e.g., E. Heilbronner, H. Bock, *The HMO Model and its Application, 3 Vol.,* Verlag Chemie, Weinheim, **1968** and **1978** (in German); Wiley, London, **1976**, (English translation); Hirokawa, Tokyo, **1972**; (Japanese translation); Kirin University Press, **1986,** (Chinese tranlation).

[15] H. Bock, B. G. Ramsey, *Angew. Chem.* **1973**, *85*, 773–792; *Angew. Chem. Int. Ed. Engl.* **1973**, *12*, 734–752 and references cited therein.

[16] Cf., e.g., K. F. Purcell, J. C. Kotz, *Inorganic Chemistry*, W. B. Saunders, Philadelphia, **1977**, pp. 44–60.

[17] H. Bock, H. Alt, *Angew. Chem.* **1967**, 79, 934; *Angew. Chem. Int. Ed. Engl.* **1967**, *6*, 943. See also H. Bock, H. Seidl, M. Fochler, *Chem. Ber.* **1968**, *101*, 2815 or H. Bock, H. Alt, *J. Am. Chem. Soc.* **1970**, *92*, 1569.

[18] H. Bock, H. Seidl, *Angew. Chem.* **1967**, *79*, 1106; *Angew. Chem. Int. Ed. Engl.* **1967**, *6*, 1085. See also H. Bock, H. Seidl, *J. Organomet. Chem.* **1968**, *23*, 87 or H. Bock, H. Seidl, *J. Am. Chem. Soc.* **1968**, *90*, 5694.

[19] H. Bock, H. Seidl, *J. Chem. Soc.* **1968**, 1148. See also H. Bock, H. Alt, *Chem. Ber.* **1970**, *103*, 1784.

[20] H. Bock, J. Meuret, U. Stein, *J. Organomet. Chem.* **1990**, *398*, 65. See also H. Bock, J. Meuret, *J. Organomet. Chem.* **1993**, *459*, 43.

[21] a) F. Gerson, J. Heinzer, H. Bock, H. Alt, H. Seidl, *Helv. Chim. Acta* **1968**, *51*, 707; b) H. Alt, E. R. Franke, H. Bock, *Angew. Chem.* **1969**, *81*, 538; *Angew. Chem.* **1969**, *8*, 525, and references cited therein.

[22] H. Bock, J. Meuret, K. Ruppert, *Angew. Chem.* **1993**, *105*, 413; *Angew. Chem. Int. Ed. Engl.* **1993**, *32*, 414, as well as H. Bock, J. Meuret, K. Ruppert, *J. Organomet. Chem.* **1993**, *445*, 19.

[23] O. Graalmann, M. Hesse, U. Klingebiel, W. Clegg, M. Haase, G. M. Sheldrick, *Angew.*

Chem. **1983**, *95*, 630; *Angew. Chem. Int. Ed. Engl.* **1983**, *22*, 621; *Angew. Chem. Suppl.* **1983**, 874.

[24] H. Bock, U. Lechner-Knoblauch, *J. Organomet. Chem.* **1985**, *294*, 295, and references cited therein.

[25] H. Bock, B. Hierholzer, H. Kurreck, W. Lubitz, *Angew. Chem.* **1983**, *96*, 817; *Angew. Chem. Int. Ed. Engl.* **1983**, *22*, 787; *Angew. Chem. Suppl.* **1983**, 1088–1105; cf. [10].

[26] H. Bock, W. Kaim, M. Kira, H. Osawa, H. Sakurai, *J. Organomet. Chem.* **1979**, *164*, 295.

[27] J. F. Ogilvie, S. Cradock, *J. Chem. Soc. Chem. Commun.* **1966**, 364; cf. additional calculations: J. N. Murrell, H. W. Kroto, M. F. Guest, *J. Chem. Soc., Chem. Commun.* **1977**, 619; R. Preuss, R. J. Buenker, S. D. Peyerimhoff, *J. Mol. Struct.* **1978**, *49*, 171.

[28] H. Bock, R. Dammel, *Angew. Chem.* **1985**, *97*, 128; *Angew. Chem. Int. Ed. Engl.* **1985**, *24*, 111.

[29] J. G. Radzisewski, D. Littmann, V. Balaji, L. Fabry, G. Gross, J. Michl, *Organometallics* **1993**, *12*, 4186. Cf. Also Y. Apeloig, K. Albrecht, *J. Am. Chem. Soc.* **1995**, *117*, 7263.

[30] E. F. Cromwell, D.-J. Lin, M. J. J. Vrakking, A. H. Kung, Y. T. Lee, *J. Chem. Phys.* **1990**, *92*, 3230.

[31] R. J. Rico, M. Page, C. Poubleday Jr., *J. Am. Chem. Soc.* **1992**, *114*, 1131.

[32] J.-M. Lehn, *Supramolecular Chemistry: Concepts and Perspectives*, VCH, Weinheim, **1995**.

[33] H. Bock, J. Meuret, C. Näther, K. Ruppert, in *Organosilicon Chemistry: From Molecules to Materials*, VCH,. Weinheim, **1984**, p. 11.

[34] N. Wiberg, H. Schuster, A. Simon, K. Peters, *Angew. Chem.* **1986**, *98*, 100; *Angew. Chem. Int. Ed. Engl.* **1986**, *25*, 79.

[35] H. Bock, Z. Havlas, V. Krenzel, *Angew. Chem.* **1998**, *110*, 3305; *Angew. Chem. Int. Ed. Engl.* **1998**, *37*, 3165.

Raman Spectra of Molecular SiS_2 and GeS_2 in Solid CH_4

M. Friesen, H. Schnöckel

Institut für Anorganische Chemie der Universität Karlsruhe
Engesserstr. Geb 30.45, 76128 Karlsruhe, Germany
Tel: Int code + (721)608 2981 — Fax: Int code + (721)608 4854
E-mail: hg@achpc9.chemie.uni-karlsruhe.de

Keywords: Matrix-Isolation / Silicon Disulfide / Germanium Disulfide / Raman Spectra

Summary: Molecular SiS_2 and GeS_2 are generated under matrix conditions in solid CH_4 after the reaction of SiS or GeS, respectively, with S atoms formed by photolysis of COS. Raman spectra exhibit the symmetric stretching vibrations from which the force constants *f*(SiS) and *f*(GeS) can be obtained. The experimental data are in line with results from ab initio calculations.

Introduction

From the IR spectra of matrix isolated SiS_2 and GeS_2 symmetry-adapted force constants *f*(SiS)-*f*(SiS/SiS) have been obtained some time ago [1, 2]. In order to get the internal force constants *f*(SiS) and f(GeS), respectively, and the corresponding interaction force constants *f*(MS/MS) Raman spectra of these species have been measured. These experimental results are a good basis for a discussion of bonding by a comparison with the data obtained from ab initio calculations.

Results

The generation of molecular SiS_2 and GeS_2 under matrix conditions is not trivial because these molecules cannot be obtained in the gas phase by vaporization of the solid compounds SiS_2 and GeS, for example, heating of SiS_2 under vacuum condition will result in the formation of $SiS_{(g)}$ and gaseous sulfur species. Therefore we have prepared, for example, molecular SiS_2 in solid methane by a co-condensation of molecular SiS – generated by passing H_2S over silicon heated to about 1500 K – and COS with a large excess of CH_4. After photolysis the absorptions of SiS ($^{28}Si^{32}S$ at 735.4 cm^{-1}) resp. GeS ($^{74}Ge^{32}S$ at 556.5 cm^{-1}) as well as those of COS decrease whereas new bands at 513.9 cm^{-1} and 474.7 cm^{-1}, respectively, appear which can be assigned to the symmetric stretching mode of SiS_2 and GeS_2 (Figs. 1 and 3). This assignment is confirmed for SiS_2, since the isotopomer $^{28}Si^{32}S^{34}S$ (10 % of $^{28}Si^{32}S_2$) could be detected at 506.2 cm^{-1} under high resolution (Fig. 2). This isotopic shift is in line with the calculated one. In the case of GeS_2 the similar isotopomer could not be observed, since the signal to noise ratio did not allow to detect bands of small intensity.

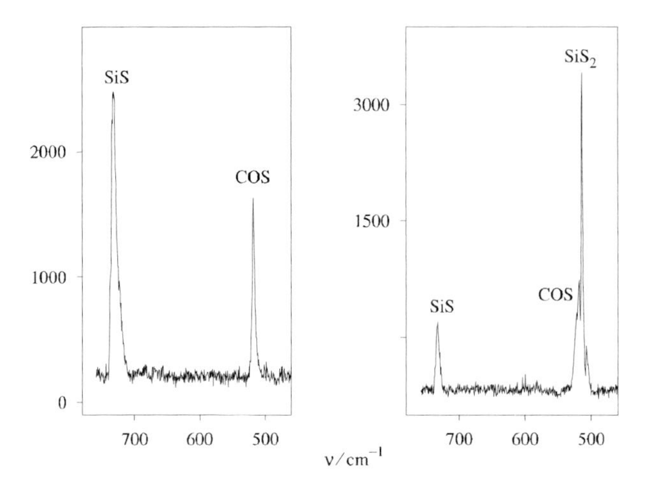

Fig. 1. Raman spectra of SiS and COS in an excess of CH_4 (left side), and after photolysis with a medium-pressure Hg lamp (right side).

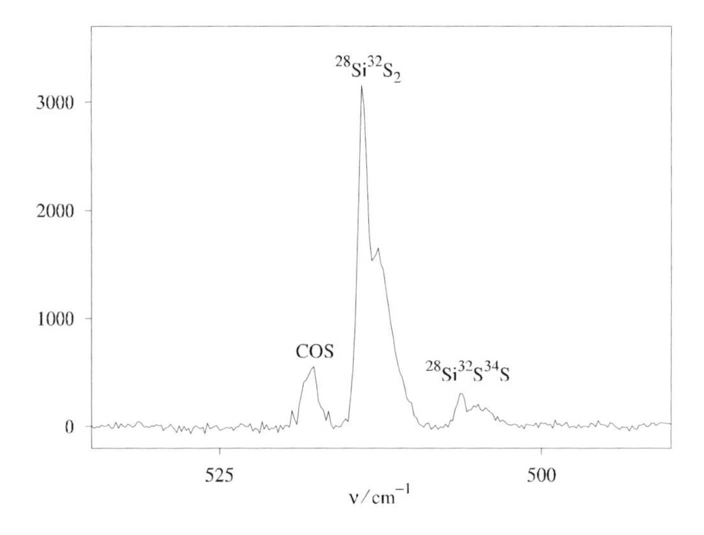

Fig. 2. High-resolution Raman spectra of SiS_2 in solid CH_4.

In order to elucidate the bonding situation for molecular SiS_2 and GeS_2 ab initio calculations have been performed in which electron correlation has been involved (MP2 and DFT) [3, 4]. The results (energy, atomic distances) of the MP2 calculations are summarized in Table 2. Afterwards population analysis and frequency calculations have been performed. Comparing the latter with the experimental data demonstrates that MP2 calculations with TZVP basis sets give the best fit. Therefore only these theoretical results are listed in Table 1.

Table 1. Experimentally observed (exp), corrected (corr) as well as calculated (calc) MP2 frequencies of $^{28}Si^{32}S_2$ and $^{74}Ge^{32}S_2$ [cm^{-1}]

	$^{28}Si^{32}S_2$			$^{74}Ge^{32}S_2$		
	ν(exp)[a]	ν(corr)[b]	ν(calc)[c]	ν(exp)[a]	ν(corr)[b]	ν(calc)[c]
Σ^+_g	513.9	523.8	513.1	474.7	490.6	496.3
Σ^-_u	(918.0)	931.0	931.8	(653.4)	663.9	676.3

[a] CH_4 matrix, (Ar matrix) — [b] Correction procedure is described in the text — [c] MP2 calculation, c.f. text.

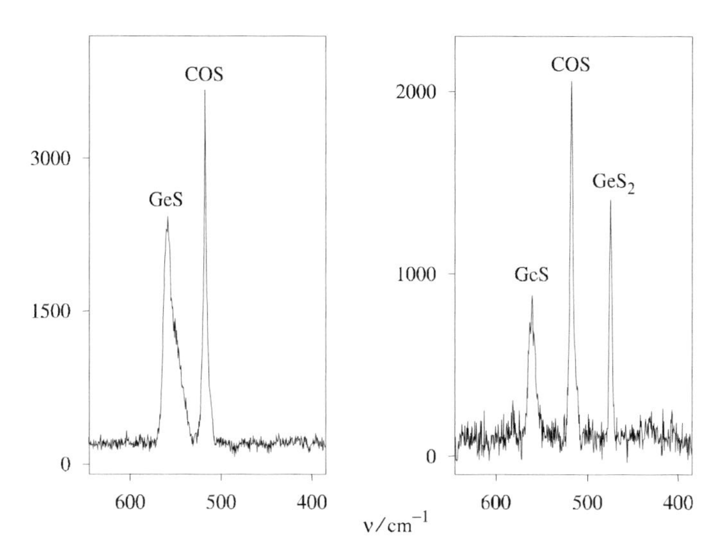

Fig. 3. Raman spectra of SiS and COS in an excess of CH_4 (left side), and after photolysis with a medium-pressure Hg lamp (right side).

Discussion

It is very surprising that the Si–S distances as well as the Ge–S distances for the disulfides are shorter than the distances for the monosulfides, because for the similar carbon compounds the "triple" bond in CS is significantly shorter than the "double" bond in CS_2 (151.9 pm/154.7 pm). This unexpected increase of bond strength from SiS to SiS_2 and from GeS to GeS_2 is now confirmed by similar changes of the force constants, which are strengthened by about 1 % in the Si- and 7 % in the Ge case. These force constant calculations are based on corrected frequencies because the frequencies of matrix-isolated species could not be compared with any data from ab initio calculations since the latter ones correspond to the "harmonic" potential of the gas-phase molecule.

In order to perform a simple correction for the matrix frequencies we used the respective ratio of the harmonic frequencies ω for $^{28}Si^{32}S$ (749.6 cm^{-1}) and for $^{74}Ge^{32}S$ (575.8 cm^{-1}) [5] and the corresponding bands under matrix conditions: SiS: 739.1 cm^{-1} (Ar); 735.4 cm^{-1} (CH_4); GeS: 566.7 cm^{-1} (Ar); 556.5 cm^{-1} (CH_4). The following factors have been obtained — SiS: 1.0142 (Ar); 1.0193 (CH_4); and GeS: 1.0161 (Ar); 1.0235(CH_4) — which we have used to correct the experimental data in order to get the values listed in Tables 1 and 2.

Table 2. Total energies, distances, force constants, interaction force constants and bond energies of SiS_2 and GeS_2.

	E_{MP2} [a. u.]	d(M = S)[a] [pm]	f[b] [mdyn/Å]	f'[b] [mdyn/Å]	BE[c] (M = S) [kJ/mol]
SiS	–686.68933	196.9	5.00	–	615.8
SiS_2	–1084.38203	195.2	5.07	0.10	458.0
GeS	–2473.25739	202.6	4.22	–	547.1
GeS_2	–2870.93314	200.8	4.50	0.04	411.1

a) MP2 calculation — b) C.f. text — c) Refs. [1, 2, 6].

With respect to bond distances and force constants SiS_2 and GeS_2 exhibit stronger bonds than the monosulfides. This behavior, which reflects resistance against small changes from the equilibrium geometry, is in contrast to the corresponding bond energies which are a measure for the complete breaking of the bond into the neutral gaseous atoms. The values for the bond energies are listed in Table 1. In accordance with chemical intuition, the bond energy for a "triple" bond is significantly larger than for the "double"-bonded molecules. The values for the monosulfides are based on the measured thermodynamic data [6]. The corresponding values for SiS_2 and GeS_2 listed in Table 2 have been obtained in the following way. With the help of ab initio calculations the reaction energy for SiS + COS = SiS_2 + CO could be determined to be about –1 kJ/mol (–43.6 kJ/mol for the corresponding formation of of GeS_2). On the basis of these reaction energies and the tabulated formation enthalpies of SiS, COS, CO and GeS the ΔH_f^{298} values of SiS_2 and GeS_2 are 87.9 kJ/mol and 105.9 kJ/mol, respectively. A comparison of the bond energy for the MS and MS_2 molecules clearly shows that breaking the double bonds requires only about 75 % of the energy for the triple bonds. This difference may be caused by different polarity of SiS/GeS in comparison to SiS_2/GeS_2. As population analysis has shown, the positive charge on the Si atom is increased from +0.5 (SiS) to +0.95 (SiS_2) and on the Ge atom from +0.43 (GeS) to +0.71 (GeS_2). Since bond energy is defined as a breaking into the neutral atoms, any ionic contribution will lower this bonding property. The different behavior with respect to the distances and to the force constants is not intelligible so far. It will be discussed later on, when additional experimental data, e.g. for SiO_2 and GeO_2, will be available.

References:

[1] H. Schnöckel, R. Köppe, *J. Am. Chem. Soc.* **1989**, *111*, 4583.

[2] H. Schnöckel, R. Köppe, *J. Mol. Struct.* **1990**, *238*, 429.

[3] R. Ahlrichs, M. Bär, M. Häser, H. Horn, C. Kölmel, *Chem. Phys. Lett.* **1989**, *162*, 165.

[4] M. J. Frisch, G. W. Trucks, H. B. Schlegel, P. M. W. Gill, B. G. Johnson, M. A. Robb, J. R. Cheeseman, T. Keith, G. A. Petersson, J. A. Montgomery, K. Raghavachari, M. A. Al-Laham, V. G. Zakrzweski, J. V. Ortiz, J. B. Foresman, C. Y. Peng, P. Y. Ayala, W. Chen, M. W. Wong, J. L. Andres, E. S. Repogle, R. Gomperts, R. L. Martins, D. J. Fox, J. S. Binkley, D. J. Defrees, J. Baker, J. P. Stewart, M. Head-Gordon, C. Gonzales, J. A. Pople, *Gaussian 94, Revision B.3*, Gaussian Inc., Pittsburgh (PA), **1995.**

[5] K. P. Huber, G. Herzberg, *Molecular Spectra and Molecular Structure, Vol. IV. Constants of Diatomic Molecules*, Van Nostrand Reinhold, New York, **1979.**

[6] M. W. Chase, Jr., C. A. Davies, J. R. Downey, Jr., D. J. Frurip, R. A. McDonald, A. N. Syverud, *Janaf Thermochemical Tables*, 3rd edn., American Chemical Society, American Institute of Physics, National Bureau of Standards, New York, Part I & II, **1985.**

The Quest for Silaketene

Günther Maier*, Hans Peter Reisenauer, Heiko Egenolf

Institut für Organische Chemie
Justus-Liebig-Universität Gießen
Heinrich-Buff-Ring 58, D-35392 Gießen, Germany
Fax: Int. Code + (641)9934309
E-mail: Guenther.Maier@org.Chemie.uni-giessen.de

Keywords: Matrix Isolation / Photoisomerizations / Cocondensation

Summary: Evaporation of silicon atoms and consecutive cocondensation with formaldehyde in an argon matrix was used to generate two CH_2OSi isomers, namely silaoxiranylidene and silaketene. The latter can be considered as a complex of carbon monoxide and silylene rather than as a true (planar) ketene. Structural assignments for the observed species are carried out by comparison of experimental and calculated IR spectra and thus are a good example for the combination of quantum chemical calculations and experimental work. Theoretical aspects of the CH_2OSi energy hypersurface as well as the photochemistry of the observed silaketene are compared with earlier results.

Introduction

During the past three years, we have studied the reactions of thermally generated silicon atoms with low molecular weight reactants. Reaction products were isolated in argon matrices and identified by means of IR spectroscopy, aided by calculated vibrational spectra. The method turned out to be very versatile and successful [1–4]. When the reactions of silicon atoms with unsaturated substrates are compared, a general sequence of products can be outlined as follows: if a heteroatom is participating in the multiple bond, the silicon atom is bound to a lone pair in the primary product. This primary product can be isomerized photochemically to a cyclic silylene, the formal addition product of the Si atom to the π bond, which is the first product in case of substrates with C–C multiple bonds. Consecutive photoisomerizations lead to the formal insertion product of the Si atom into a C–H bond and, finally, to the product(s) of (successive) migration of H atoms to the Si atom.

For the reaction of atomic silicon with formaldehyde, this sequence would lead to the following products:

$$\cdot\ddot{Si}\cdot \xrightarrow[?]{CH_2O} H_2C{-}O{\sim}Si \longrightarrow \text{(cyclic } H_2C{-}O{-}\ddot{Si}\text{)} \longrightarrow H\ddot{Si}{-}C(H){=}O \longrightarrow H_2Si{=}C{=}O$$

Especially the possible final product, silaketene, would be a very interesting molecule from both theoretical and experimental points of view.

The CH_2OSi Energy Hypersurface

Prior to the experiments, the geometries of these and other possible CH_2OSi isomers and their vibrational spectra were calculated with the density functional method B3LYP at the 6-31G** level of theory (Scheme 1).

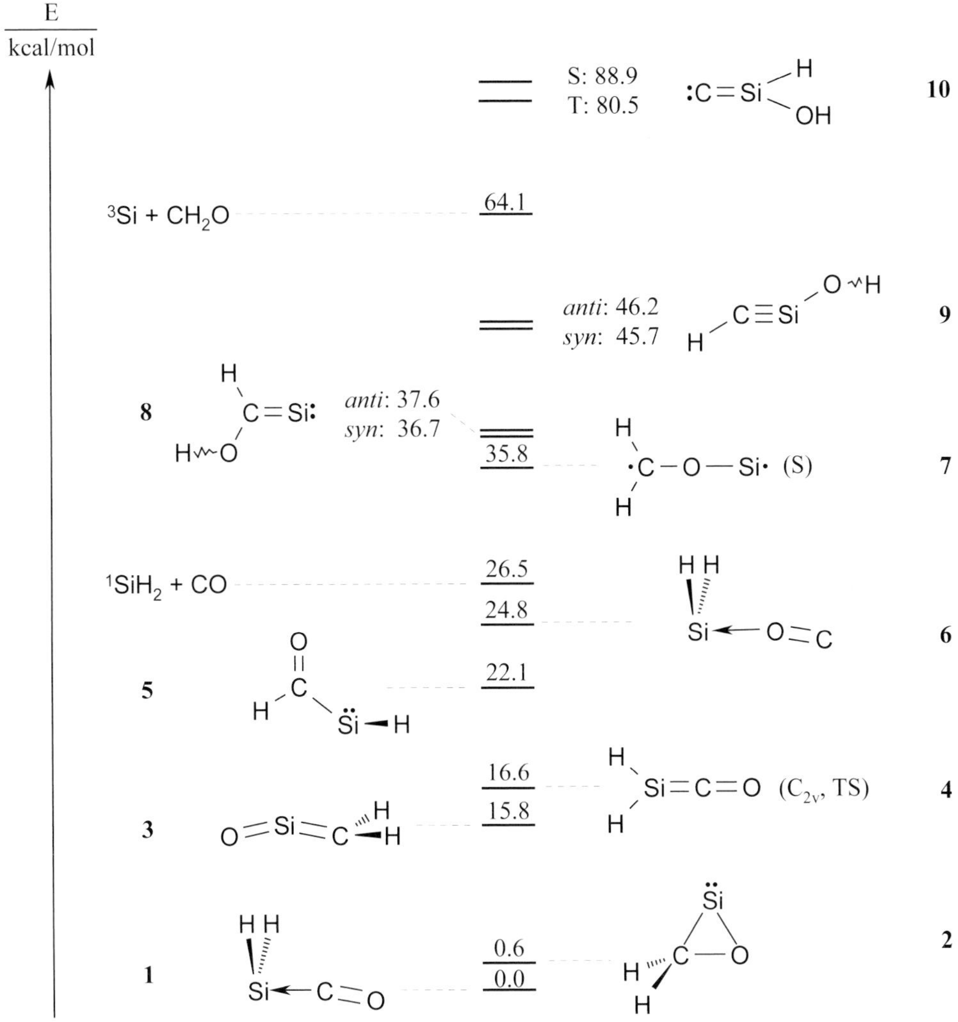

E / kcal/mol

10: S: 88.9, T: 80.5

$^3Si + CH_2O$: 64.1

9: *anti*: 46.2, *syn*: 45.7

8: *anti*: 37.6, *syn*: 36.7

7: 35.8 (S)

$^1SiH_2 + CO$: 26.5

6: 24.8

5: 22.1

4: 16.6 (C_{2v}, TS)

3: 15.8

2: 0.6

1: 0.0

Scheme 1. Relative energies, corrected by zero-point vibrational energies, of several CH_2OSi species and of possible fragments (B3LYP/6-31G**).

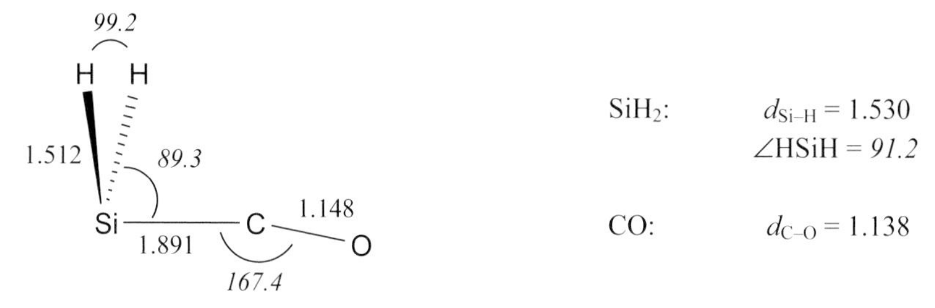

Fig. 1. Calculated geometries of silaketene (**1**), SiH_2, and CO (B3LYP/6-31G**, bond lengths in Å, angles in degrees).

In agreement with earlier calculations [5, 6], the planar form of silaketene (**4**) is only a transition state, the imaginary frequency representing an out-of-plane vibration of the hydrogen atoms. If the C_{2v} symmetry restriction is removed, the optimized geometry of the molecule is bent (Fig. 1). It can be regarded as a silylene molecule complexed by carbon monoxide, which yields a stabilization energy of 26.5 kcal mol^{-1} compared with the separated fragments, SiH_2 and CO.

Considering the oxophilicity of silicon, one might think that complexation of SiH_2 with the oxygen atom of CO may even be stronger. But in this case (**6**), the Si–O bond is calculated to be very long (2.49 Å), resulting in a very low stabilization energy (1.7 kcal mol^{-1}) and smaller deviations of the geometries in the SiH_2 and CO moieties.

CO complexes of substituted silylenes have been claimed to be observed after photolysis of silylene precursors in CO-doped argon or organic matrices by Arrington [7] and by West [8] and coworkers, respectively.

Cocondensation of Atomic Silicon with Formaldehyde

The experimental procedure has been described elsewhere [1]. Paraformaldehyde was depolymerized by heating in vacuo; the glass walls of the bulb containing the gaseous mixture (CH_2O/Ar = 1:80) were covered with a paraffin layer to prevent repolymerization of the formaldehyde.

The matrix IR spectrum after the codeposition of Si atoms and CH_2O shows, apart from the usual byproducts [2] and CH_2O, several new bands. Upon irradiation of the matrix with light of wavelengths >385 nm one band at 1290 cm^{-1} disappears, whereas the other new bands increase. The 1290 cm^{-1} absorption may be assigned to the C=Si=O stretching vibration of the 1-silaketene (**3**), which according to the calculated IR spectrum is indeed characterized by only one strong band at 1295 cm^{-1}. Ketene **3** may be formed via the reaction sequence Si + CH_2O → **7** → **2**→ **3**. The occurrence of **7** as the primary product would be in analogy to the reaction of silicon atoms with HCN, yielding the HCNSi molecule in the first step [3]. **7** has also been reported to be observed in collisional-activation and NRMS experiments with tetramethoxysilane [9, 10].

Experiments with CD_2O support this assignment, for the band is shifted to 1291 cm^{-1} (+1 cm^{-1}), whereas the calculations predict an isotopic shift of +1.5 cm^{-1} for the [D_2]isotopomer of **3**.

The bands arising upon >385 nm irradiation can, on the other hand, definitely be assigned to oxasila-cyclopropenylidene (**2**), the formal addition product of a silicon atom to the π bond of formaldehyde, which shows an intense absorption at 590 cm^{-1} for the ring deformation vibration.

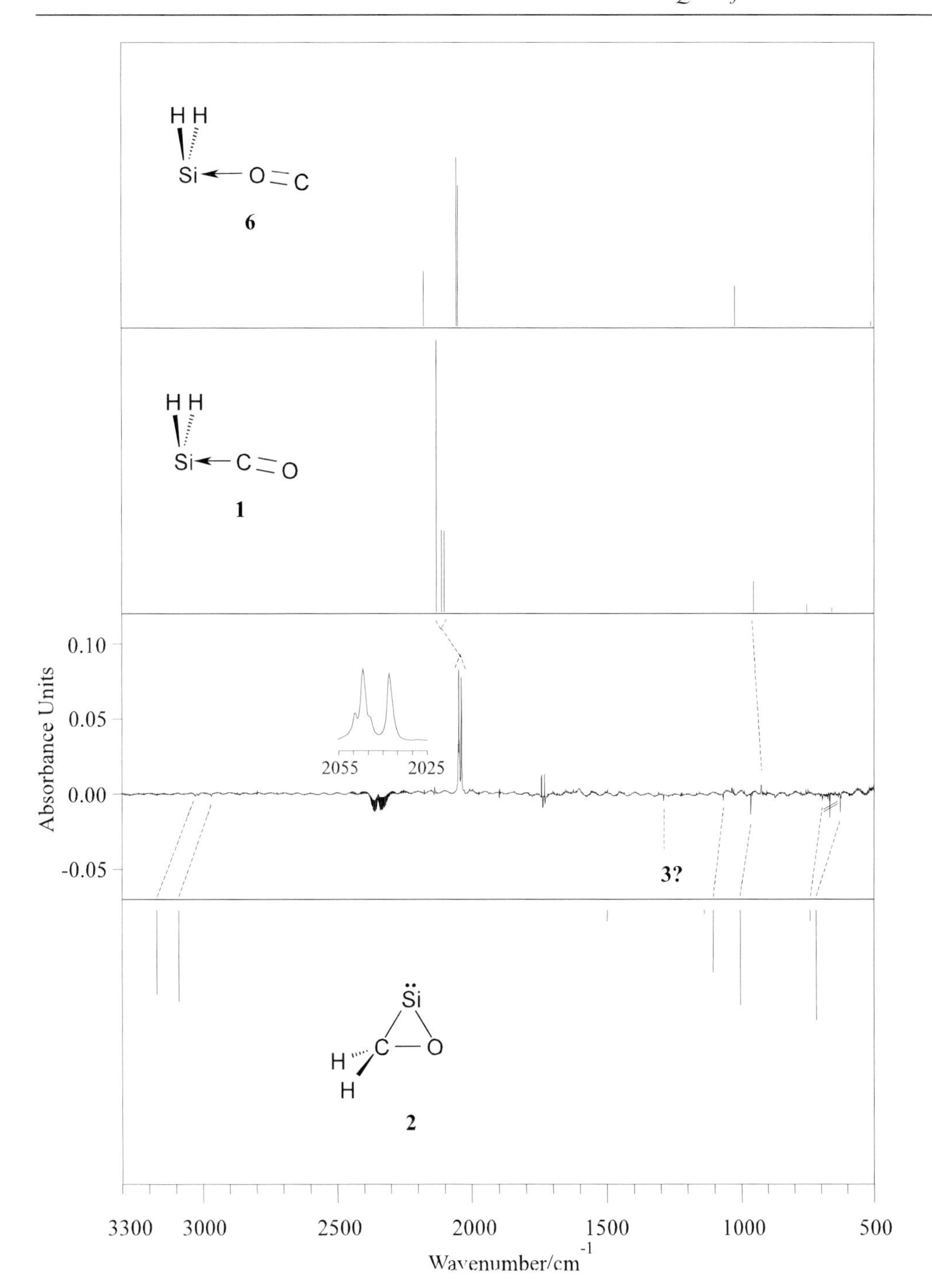

Fig. 2. Difference IR spectrum for the photoreaction **2** → **1** (λ > 385 nm, Ar matrix, 12 K), calculated (B3LYP/ 6-31G**) IR spectra for the CH_2OSi isomers **1**, **2** and **6**.

In this case, isotopic labeling supports the assignment. One might also consider **2** as the primary product of the reaction, for its absorptions are present immediately after the codeposition. But as the bands grow upon irradiation, it is more likely that **2** is produced, at least to some extent, by irradiation with the light emitted from the hot crucible during deposition from another product, maybe **7**.

Upon further irradiation with light of wavelengths >385 nm, the absorptions of **2** decrease, and new bands appear, which can be assigned to silaketene (**1**, Fig. 2). Careful comparisons with calculated spectra and the isotopic labeling experiments clearly rule out the possible presence of the H_2Si–OC complex (**6**). The close accordance of calculated and experimental IR spectra confirms that silaketene is indeed a silylene–CO complex rather than a true planar ketene.

$$\cdot\ddot{Si}\cdot \xrightarrow[\text{Ar, 10 K}]{CH_2O} [H_2\dot{C}-O-Si\cdot]\ (\mathbf{7}) \xrightarrow[\text{fast}]{> 385\ \text{nm}} H_2C(\ddot{Si})O\ (\mathbf{2}) \xrightarrow[\text{slow}]{> 385\ \text{nm}} H_2Si \leftarrow C{=}O\ (\mathbf{1})$$

The bands of silaketene (**1**, intense Si–H and C–O stretching vibrations at 2038, 2047 and 2050 cm^{-1}) disappear upon irradiation with light of wavelength 290 nm. Under these conditions hydrogen is eliminated from **1** and the SiCO molecule (1898 cm^{-1}) [11] is formed [12].

Silaketene — a Precursor for Disilene?

The CO complexes of silylenes with bulky substituents which were observed by West et al. [8] in organic matrices eliminated CO upon annealing of the matrix; the remaining silylenes dimerized to the corresponding disilenes. If this reaction took place in the CH_2OSi system, it would lead to the longsought-after parent disilene molecule.

Annealing of an argon matrix containing silaketene (**1**) to 35 K indeed leads to a decrease of its IR absorptions, and new bands appear at 1720, 1941 and 1971 cm^{-1}. But the correspondence with calculated IR spectra for disilene and its isomer silylsilylene (and with those of other CH_2OSi isomers) is very poor, leaving the origin of these new absorptions unknown.

References:

[1] G. Maier, H. P. Reisenauer, H. Egenolf, in: *Organosilicon Chemistry III: From Molecules to Materials* (Eds.: N. Auner, J. Weis), VCH, Weinheim, **1998**, pp. 31–35.

[2] G. Maier, H. P. Reisenauer, A. Meudt, H. Egenolf, *Chem. Ber./Recueil* **1997**, *130*, 1043–1046.

[3] G. Maier, H. P. Reisenauer, H. Egenolf, J. Glatthaar, *Eur. J. Org. Chem.* **1998**, 1307–1311.

[4] G. Maier, H. P. Reisenauer, H. Egenolf, *Eur. J. Org. Chem.* **1998**, 1313–1317.

[5] G. Trinquier, J.-P. Malrieu, *J. Am. Chem. Soc.* **1987**, *109*, 5303–5315.

[6] T. P. Hamilton, H. F. Schaefer III, *J. Chem. Phys.* **1989**, *90*, 1031–1035.

[7] C. A. Arrington, J. T. Petty, S. E. Payne, W. C. K. Haskins, *J. Am. Chem. Soc.* **1988**, *110*, 6240–6241.

[8] M.-A. Pearsall, R. West, *J. Am. Chem. Soc.* **1988**, *110*, 7228–7229.

[9] J. Hrusák, R. Srinivas, D. K. Böhme, H. Schwarz, *Angew. Chem.* **1991**, *103*, 1396–1398.

[10] R. Srinivas, D. K. Böhme, J. Hrusák, D. Schröder, H. Schwarz, *J. Am. Chem. Soc.* **1992**, *114*, 1939–1942.

[11] R. R. Lembke, R. F. Ferrante, W. Weltner, Jr, *J. Am. Chem. Soc.* **1977**, *99*, 416–423.

[12] Note added in proof: For a detailed discussion of **1** see: G. Maier, H. P. Reisenauer, H. Egenolf, *Organometallics* **1999**, *18*, 2155–2161.

Donor-Substituted Silylenes: From Gas-Phase Reactions to Stable Molecules

J. Heinicke*, A. Oprea, S. Meinel, S. Mantey

Institut für Anorganische Chemie, Universität Greifswald,
Soldmannstr. 16, D-17487 Greifswald, Germany
Tel.: Int. code. + (3834)864337 — Fax: Int. code. + (3834)864319
E-mail: heinicke@mail.uni-greifswald.de

Keywords: Silylene / Disilane / Thermolysis / Cycloaddition / Electronic Stabilization

Summary: Monoaminosilylenes and bis(diethylamino)silylene are formed by thermolysis of amino-substituted disilanes and characterized by trapping with dienes. In the case of $MeSiNMe_2$, cycloaddition reactions were extended to 1,4-diheterodienes, unsaturated ketones and imines allowing an easy synthetic access to functionally substituted unsaturated silicon heterocycles. The syntheses of an isolable, unsymmetric diaminosilylene and of related germylenes and stannylenes are described. The results are presented and discussed in relation to theoretical work on stabilization of donor-substituted silylenes and recent work on isolable diaminosilylenes done by or in cooperation with other groups.

Introduction

Usually silylenes are highly reactive due to the presence of a lone electron pair and a low-lying unoccupied orbital at silicon. In the last years some isolable derivatives could be obtained, however, by steric shielding of these reactive centers or by their inclusion in the chemical bond, e.g. the highly coordinated decamethylsilicocene [1], a tetracoordinated bis(diphosphinomethanido)silylene [2] or silylene complexes with donor and acceptor bonds [3]. Quantum chemical calculations revealed that silylenes are stabilized also by π-interactions between the empty p_z-orbital of silicon and a free electron pair of donor substituents, increasing in the order [4, 5].

HSi–Cl	< HSi–F	< HSi–OH	≈ HSi–SH	< $HSi–NH_2$ [4]	<< $Si(NH_2)_2$ [5]
7.0	9.3	15.0	15.1	22.3	37.2 kcal/mol

As shown recently, a small additional stabilization results from inclusion of the SiN_2 substructure into cyclodelocalized Hückel 6π or 10π systems [5–8]: this has allowed monomeric diaminosilylenes to be isolated. In acyclic hydroxy-, amino- and diaminosilylenes intermolecular donor–acceptor interactions are stronger than the π-bonds and, thus, O- or N-bridging dimers are the most stable form [9]. Since the bridging bonds are easily cleaved, the question is raised of whether the π-stabilization has an influence on the reactivity, the properties and the tendency towards formation of donor-substituted silylenes.

Gas-Phase Reactions

Unsymmetric, single donor-substituted silylenes generated by thermolysis of $MeX_2Si–SiX_2Me$ undergo α-elimination to give $MeSiX_3$ and MeSiX [10]. MeSiCl could be detected directly at 600–700°C by thermolysis–mass spectrometry [11]. The interpretation of the spectra was facilitated by the pattern of the isotope peaks of the $^{35/37}$Cl-containing fragments. The thermolysis of tetraalkoxy- and tetraaminodisilanes was observed by PE spectroscopy. The disilane degradation was detected by the disappearance of the low-energy band attributed to ionization of the Si–Si bond, but a new low-energy band for the ionization of the HOMO of the respective silylenes, as calculated by AM1 and PM3, did not appear. We assume that the lifetime of the unsymmetric silylenes was not sufficient for detection in the equiment used [12]. The symmetric silylenes SiH_2 and $SiCl_2$, however, were observed by this method by Bock et al. [13] and Beauchamp and coworkers [14]. The formation of the alkoxy- and aminomethylsilylenes was lastly confirmed by trapping reactions with 1,3-dienes, mainly 2,3-dimethyl-buta-1,3-diene, which gives [1+4]-cycloaddition products and allows one to distinguish silylenes from other subvalent silicon species [10]. Comparison of cothermolyses of $Me_2Si_2X_4$ (X = Cl, OMe, NMe_2 etc.) and dimethylbutadiene at 400°C for ca. 50 s shows complete degradation of the $Me_2Si_2(OMe)_4$ but only 75 % decay of $Me_2Si_2(NMe_2)_4$ in spite of the higher stability of aminosilylenes. This indicates a kinetic control of the α-elimination by intramolecular donor–acceptor interactions which are stronger for O→Si than for N→Si. The conversion yield in the cycloadditions of MeSiX with X = Cl, OMe, NMe_2 is in each case ca. 80 %. The reactivity under the thermal conditions is too high to allow the observation of a differentiated behavior of the differently donor-substituted silylenes. Cothermolysis of $(Et_2N)_2HSi–SiH(NEt_2)_2$ or $Me_3Si–Si(NEt_2)_3$, respectively, with 2,3-dimethylbutadiene led to the trapping of products $HSiNEt_2$ and $Si(NEt_2)_2$ (Scheme 1).

420°C

+ $HSi(NEt_2)_3$

δ ^{29}Si: +0.5; −28.5

+ $H_2Si(NEt_2)_2$

C_6H_6 550°C

+ Me_3SiNEt_2

Scheme 1. Cothermolysis of disilanes with 2,3-dimethylbutadiene.

Cycloadditions of monoalkoxy- or monoaminosilylenes were observed also by cothermolyses of $Me_2Si_2X_4$ (X = OMe, NMe_2) with various conjugated 1-mono- and 1,4-diheterodienes. Iminoketones, diimines and α,α′-dipyridyl, also benzil or 3,5-di-*tert*-butyl-*o*-benzoquinone in low yields, gave [1+4]-cycloaddition products. Unsaturated ketones or imines furnished double-bond isomers, oxa- or azasilacyclopent-3-enes and -4-enes (Scheme 2). The 4-enes are usually the main products; 3-enes dominated only in the case of *N*-phenylimines. The occurrence of the isomers is in accordance with primary complexation of the silylene at the heteroatom and [1+2]-cycloaddition at the carbonyl group, followed by ring-opening/ring-closure isomerization to five-membered rings. The primary attack at the carbonyl group is supported by competition experiments. MeOSiMe reacts with an equimolar mixture of PhCH=CH–C(Me)=O and 2,3-dimethylbutadiene to give the strongly preferred cycloaddition product to the enone [15]. Cycloadditions of transient diaminosilylenes with hetero- or diheterodienes have not yet been investigated.

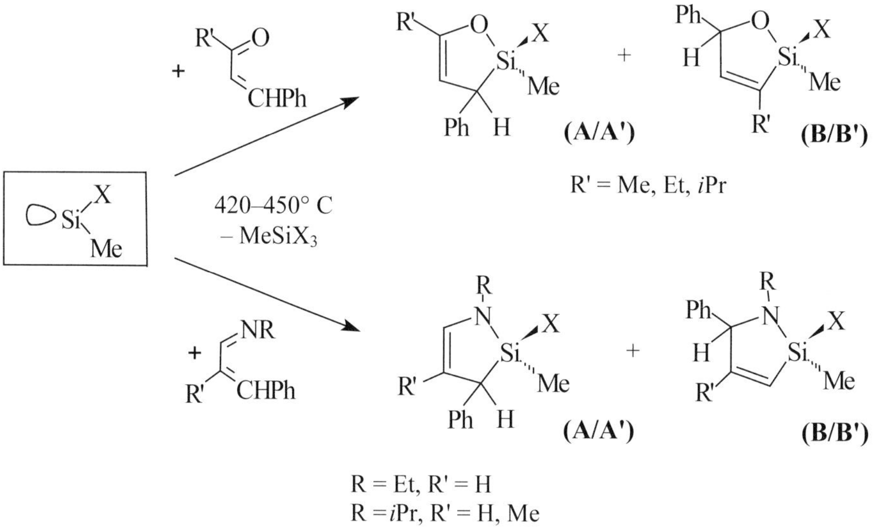

420–450° C
– $MeSiX_3$

(A/A') (B/B')

R' = Me, Et, *i*Pr

(A/A') (B/B')

R = Et, R' = H
R = *i*Pr, R' = H, Me

Scheme 2. Cycloadditions of silylenes to unsaturated ketones and imines.

Isolated Diaminosilylenes, Analogous Germylenes and Stannylenes

The synthesis of isolable monomeric cyclic diaminostannylenes [16] and diaminogermylenes [17] and the above-mentioned calculated stabilization of cyclodelocalized diaminosilylenes encouraged attempts to synthesize analogous diaminosilylenes (Fig. 1). The reduction of the respective diaminodichlorosilanes by Denk, West and others led in 1994 to the discovery of di-*tert*-butyl-1,3,2λ^2-diazasilole, the first isolable twice-coordinate silylene, and to a less stable 4,5-dihydrodiazasilole [6]. In 1995 two benzo-1,3,2λ^2-diazasiloles were reported by Gehrhus, Lappert, Heinicke, Boese and Bläser [8].

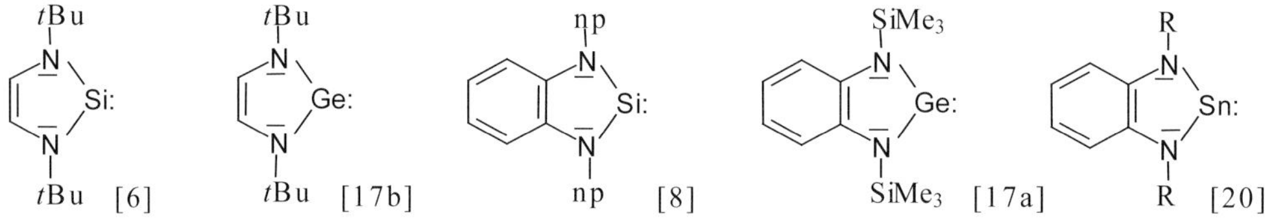

[6] [17b] [8] [17a] [20]

Fig. 1. Diaminosilylenes, germylenes and stannylenes.

The initial problems getting suitable single crystals of the benzodiazasilole prompted us to search for other annulated diazasiloles which might crystallize more easily or exhibit a varied reactivity. Naphtho[2,3]- and differently pyrido-annulated dichlorodiazasiloles were synthesized by cyclization of the respective dilithium diamides with $SiCl_4$. However, the reduction proved to be critical. Only the 1,3,2λ^2-diazasilolo[4,5-*b*]pyridine was available by treatment with potassium in THF [18]. In the other cases the respective silylenes could not be detected, not even in the crude mixtures. For analogy-estimations of whether the putative silylenes would be stable and similar in behavior to the other diazasiloles, we investigated the analogous germanium(II) and tin(II) compounds. The naphtho[2,3]- and [*b*]pyridine-annulated 1,3,2λ^2-diazagermoles and 1,3,2λ^2-diazastannoles were formed in good yields whereas the [*c*]pyridine-annulated derivatives could not be obtained (Scheme 3) [18, 19]. A dilithium reagent formed from *N-tert*-butylpyridine-2-aldimine and lithium in THF reacted with $GeCl_2$·dioxane to give a low yield of the corresponding germylene whereas $SnCl_2$ was reduced to elemental tin.

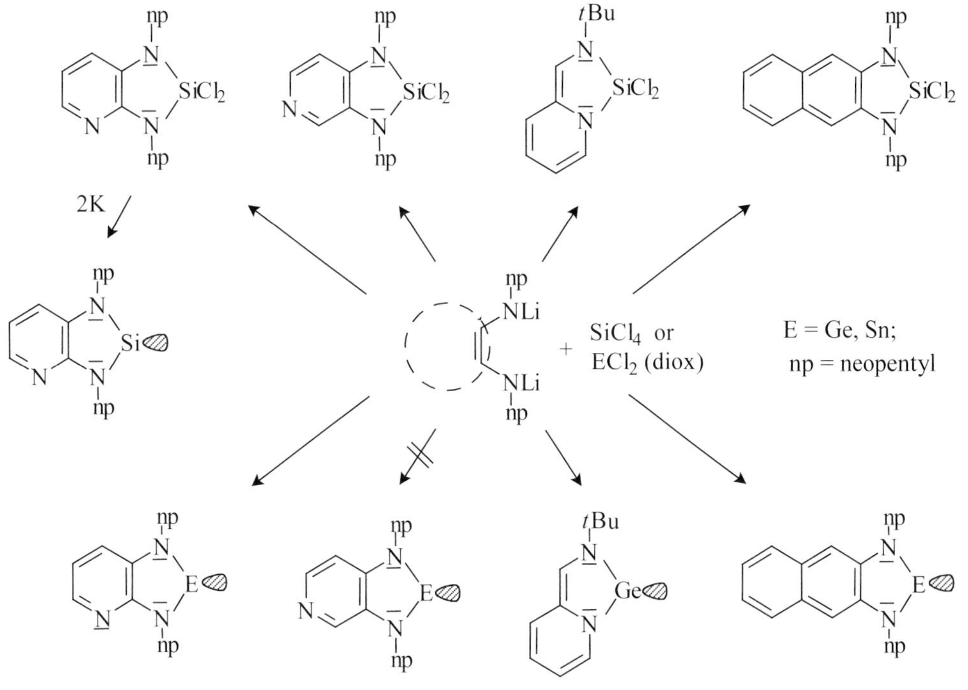

Scheme 3. Synthesis of cyclic diaminosilylenes, -germylenes and stannylenes.

The annulated 1,3,2λ^2-diazasiloles and -germoles are monomeric and easily soluble in THF, ether, benzene and, except the naphthodiazagermole, also in saturated hydrocarbons. The ^{29}Si resonance in [D_8]THF is similar to that in C_6D_6, indicating no significant interactions with Lewis bases. Similarly, addition of tertiary amines or phosphines has no influence on the Si chemical shift. The related stannylenes behave differently. The ^{119}Sn resonance of the pyridodiazastannole(II) appears in [D_8]THF (δ = 100.5) at much higher field than in C_6D_6 (δ = 241.6), indicating coordination by the Lewis-basic solvent. The X-ray structure analysis of both dineopentylbenzodiazasilole(II) [8] and dineopentylbenzodiazastannole(II) [20] shows molecules arranged in pairs with the silicon(II) and the tin(II), respectively, over the benzene ring of the neighboring molecule. The distance of Si(II) to the middle of this benzene ring (350.0 pm) is slightly shorter than the van der Waals distance (370 pm). The tin(II)–benzene distance in the benzodiazastannole(II) is even shorter (323 pm), in spite of the larger radius of tin (van der Waals distance 390 pm), thus indicating dimers with π-coordinate tin(II) [20]. The decreasing solubility in benzene in the order benzo-, pyrido-, naphtho-1,3,2λ^2-diazastannole accounts for an increasing strength of such π-interactions.

The symmetric 1,3,2λ^2-diazasiloles and 1,3,2λ^2-benzodiazasiloles exhibit a surprisingly high thermal stability whereas the unsymmetrical 1,3,2λ^2-diazasilolo[4,5-*b*]pyridine is decomposed to considerable amount during sublimation. The reactivity of the more stable symmetric representatives was investigated mainly by Denk, West et al., Gehrhus and Lappert [6, 8, 21–23]. Various oxidative additions, cycloadditions, complexation and reactions with tin(II) and germanium(II) compounds have been described recently. Differences in the reactivity compared to transient acyclic donor-substituted silylenes may be due to the π-character of the HOMO in the 1,3,2λ^2-diazasiloles or annulated derivatives and the lowering of the orbital energy of the lone electron pair at silicon, attributed to the increased s-character and the angle fixation near 90° in the five-membered rings. The lone electron pairs of nitrogen are delocalized as seen by the strong deshielding of the ^{15}N nuclei of the di-*tert*-butyl-1,3,2λ^2-diazasilole as compared to the respective dichloro-1,3,2λ^4-diazasilole ($\Delta\delta$ ca. 120 ppm) [6] or the strong deshielding in the benzo-1,3,2λ^2-diazasilole. The silicon atoms of the Hückel systems profit from the distributed π-charge density. The ^{29}Si nucleus of the 1,3,2λ^2-diazasilole (δ = 78) is much less deshielded than that of 4,5-dihydro-1,3,2λ^2-diazasilole (δ = 119) [6]. In the benzo- and pyridodiazasiloles the differences in silicon(II) and silicon(IV) resonances are very similar to that in diazasiloles. The s-character of the lone electron pair at silicon, the strongly polar Si–N bond and the back-donation of π-charge density are in accordance with a spherical electron distribution, described by Arduengo, Bock et al. [24] and expressed by an alternative structure suggestion as a diimine–silicon(0) complex. Finally, it should be mentioned that the failure in the attempts to synthesize a 1,3,2λ^2-diazasilolo[4,5-*c*]pyridine and its germanium and tin homologues cannot be attributed to a lower thermodynamic stabilization compared to the respective [*b*]pyridine and benzo compounds — it has about the same value. The calculated charge density shows, however, that a nodal plane crosses the pyridine-N atom in the HOMO of the isolable 1,3,2λ^2-diazasilolo[4,5-*b*]pyridine and generates a nearly symmetric π-charge distribution, similar to that in the benzo-1,3,2λ^2-diazasilole, whereas the π-charge density in the HOMO of the 1,3,2λ^2-diazasilolo[4,5-*b*]pyridine is highly unsymmetric and facilitates consecutive reactions [18]. This shows the importance of the symmetry in the persistence of highly reactive compounds.

Acknowledgment: We thank the *Deutsche Forschungsgemeinschaft* and the *Fonds der Chemischen Industrie* for the support of this research.

References:

[1] P. Jutzi, U. Holtmann, D. Kanne, C. Krüger, R. Blom, R. Gleiter, I. Hyla-Kryspin, *Chem. Ber.* **1989**, *122*, 1629.

[2] H. H. Karsch, U. Keller, S. Gamper, G. Müller, *Angew. Chem., Int. Ed. Engl.* **1990**, *29*, 295.

[3] C. Zybill, H. Handwerker, H. Friedrich, *Adv. Organomet. Chem.* **1994**, *36*, 229 and references cited therein.

[4] B. T. Luke, J. A. Pople, M.-B. Krogh-Jespersen, Y. Apeloig, J. Chandrasekar, P. v. R. Schleyer, *J. Am. Chem. Soc.* **1986**, *108*, 260; T. N. Truong, M. S. Gordon, *J. Am. Chem. Soc.* **1986**, *108*, 1775; L. Nyulaszi, A. Belghazi, S. Kis-Szetsi, T. Veszpremi, J. Heinicke, *J. Mol. Structure (Theochem)* **1994**, *313*, 73.

[5] L. Nyulaszi, T. Karpati, T. Veszpremi, *J. Am. Chem. Soc.* **1994**, *116*, 7239.

[6] M. Denk, R. Hayashi, R. West, *J. Chem. Soc., Chem. Commun.* **1994**, 33; M. Denk, J. C. Green, N. Metzler, M. Wagner, *J. Chem. Soc., Dalton Trans.* **1994**, 2405; R. West, M. Denk, *Pure Appl. Chem.* **1996**, *68*, 785.

[7] C. Heinemann, W. A. Herrmann, W. Thiel, *J. Organomet. Chem.* **1994**, *475*, 73.

[8] B. Gehrhus, M. F. Lappert, J. Heinicke, R. Boese, D. Bläser, *J. Chem. Soc., Chem. Commun.* **1995**, 1931; B. Gehrhus, M. F. Lappert, J. Heinicke, R. Boese, D. Bläser, *J. Organomet. Chem.* **1996**, *521*, 211; P. Blakeman, B. Gehrhus, J. C. Green, J. Heinicke, M. F. Lappert, M. K. Kindermann, T. Veszpremi, *J. Chem. Soc., Dalton Trans.* **1996**, 1475.

[9] M. Karni, Y. Apeloig, *J. Am. Chem. Soc.* **1990**, *112*, 8589; Y. Apeloig, T. Müller, *J. Am. Chem. Soc.* **1995**, *117*, 5363.

[10] W. H. Atwell, D. R. Weyenberg, *Angew. Chem. Int. Ed. Engl.* **1969**, *8*, 469; G. Maier, H.-P. Reisenauer, K. Schöttler, U. Wessolek-Kraus, *J. Organomet. Chem.* **1989**, *366*, 25; J. Heinicke, B. Gehrhus, S. Meinel, *J. Organomet. Chem.* **1994**, *474*, 71.

[11] J. Heinicke, D. Vorwerk, G. Zimmermann, *J. Anal. Appl. Pyrol.*, **1994**, *28*, 93.

[12] J. Heinicke, M. K. Kindermann, P. Rademacher, unpublished results.

[13] H. Bock, B. Solouki, G. Maier, *Angew. Chem.* **1985**, *97*, 205.

[14] G. H. Kruppa, S. K. Shin, J. L. Beauchamp, *J. Phys. Chem.*, **1990**, *94*, 327.

[15] J. Heinicke, B. Gehrhus, *J. Organomet. Chem.* **1992**, *423*, 13; J. Heinicke, B. Gehrhus, *Heteroatom Chem.* **1995**, *6*, 461; B. Gehrhus, J. Heinicke, S. Meinel, *Main Group Met. Chem.* **1998**, *21*, 99.

[16] Review: M. F. Lappert, *Main Group Met. Chem.* **1994**, *17*, 183.

[17] a) J. Pfeiffer, M. Maringgelle, M. Noltemeyer, A. Meller, *Chem. Ber.* **1989**, *122*, 245;
b) W. A. Herrmann, M. Denk, J. Behm, W. Scherer, F.-R. Klingan, H. Bock, B. Solouki, M. Wagner, *Angew. Chem.* **1992**, *104*, 1489.

[18] J. Heinicke, A. Oprea, M. K. Kindermann, T. Karpati, L. Nyulászi, T. Veszprémi, *Chem. Eur. J.* **1998**, *4*, 537.

[19] J. Heinicke, A. Oprea, *Heteroatom Chem.* **1998**, *9*, 439.

[20] H. Braunschweig, B. Gehrhus, P. B. Hitchcock, M. F. Lappert, *Z. Anorg. Allg. Chem.* **1995**, *621*, 1922.

[21] M. Denk, R. K. Hayashi, R. West, *J. Chem. Soc., Chem. Commun.* **1994**, 33; N. Metzler, M. Denk, *J. Chem. Soc., Chem. Commun.* **1996**, 2657.

[22] C. Drost, B. Gehrhus, P. B. Hitchcock, M. F. Lappert, *J. Chem. Soc., Chem. Commun.* **1997**, 1845; B. Gehrhus, P. B. Hitchcock, M. F. Lappert, *Angew. Chem.* **1997**, *109*, 2624; B. Gehrhus, P. B. Hitchcock, M. F. Lappert, H. Maciejewski, *Organometallics*, **1998**, *17*, 5599.

[23] A. Schäfer, W. Saak, M. Weidenbruch, H. Marsmann, G. Henkel, *Chem. Ber./ Receuil* **1997**, *130*, 1733.

[24] A. J. Arduengo III, H. Bock, H. Chen, M. Denk, D. A. Dixon, J. C. Green, W. A. Herrmann, N. L. Jones, M. Wagner, R. West, *J. Am. Chem. Soc.* **1994**, *475*, 73.

Different Reactivities of Divalent Silicon Compounds Towards Metallocene Derivatives of Molybdenum and Tungsten

Stefan H. A. Petri, Dirk Eikenberg and Peter Jutzi*

Fakultät für Chemie, Universität Bielefeld
Universitätsstr. 25, D-33615 Bielefeld, Germany
Tel.: Int. code + (521)106 6181 — Fax.: Int. code + (521)106 6026
E-Mail: peter.jutzi@uni-bielefeld.de

Keywords: Divalent Silicon Compounds / Molybdenum Compounds / Silylene Complex / Tungsten Compounds / X-Ray Crystal Structures

Summary: The thermal and photochemical reactivity of divalent silicon species, decamethylsilicocene (Cp^*_2Si) and 1,3-di-*tert*-butyl-1,3,2-diazasilol-2-ylidene (SiL^N_2), towards Cp_2MH_2 (M = Mo, W) and $Cp_2Mo(PEt_3)$ is investigated. In the case of Cp^*_2Si no conversion of the silicon(II)-species is observed. This is attributed to the high steric demand of the bulky Cp* ligands. In SiL^N_2 the silicon atom is less shielded. So the reactivity of SiL^N_2 in analogous reactions is much higher. The reactions with Cp_2MH_2 lead via silylene insertion to the corresponding metallosilanes. The photolysis of an equimolar amount of SiL^N_2 and $Cp_2Mo(PEt_3)$ results in the formation of the first molybdenum silylene complex with a tricoordinated silicon atom.

Introduction

Only little is known about the reactivity of divalent silicon species towards transition metal compounds. In our group the reactions of decamethylsilicocene (Cp^*_2Si) with ClAu(CO) [1], $Ti(NMe_2)_4$ [2], and $HMn(CO)_5$ [3] have been investigated in detail. Berry and coworkers have reported on the insertion of $SiMe_2$ into a tantalum hydrogen bond [4], and Denk and West have described the reactions of 1,3-di-*tert*-butyl-1,3,2-diazasilol-2-ylidene (SiL^N_2) with transition metal carbonyl compounds [5, 6]. In this contribution we present our results concerning the different reactivities of Cp^*_2Si and SiL^N_2 towards molybdocene and tungstenocene derivatives.

Reactivity of Cp^*_2Si and SiL^N_2 towards Cp_2MoH_2 and Cp_2WH_2

Decamethylsilicocene, Cp^*_2Si, shows no reactivity towards Cp_2MH_2 (M = Mo, W); neither thermal treatment (5 d, 110 °C) of an equimolar amount of Cp^*_2Si and Cp_2MH_2 (Eq. 1) nor photolysis of Cp_2MH_2 in the presence of Cp^*_2Si (Eq. 2) leads to a conversion of the silicon(II)-species.

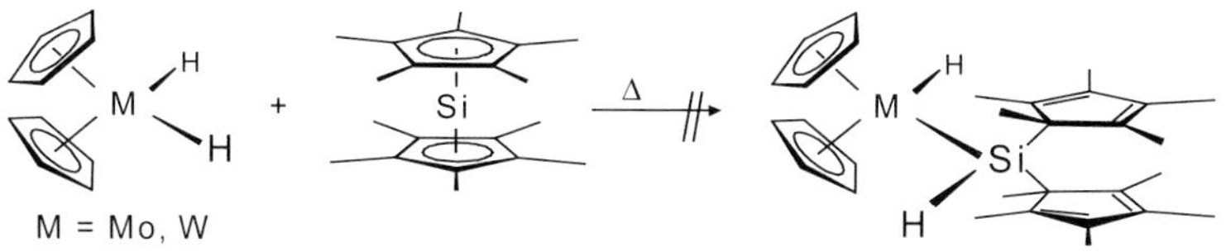

Eq. 1. Thermal treatment of an equimolar amount of Cp^*_2Si and Cp_2MH_2 (M = Mo, W).

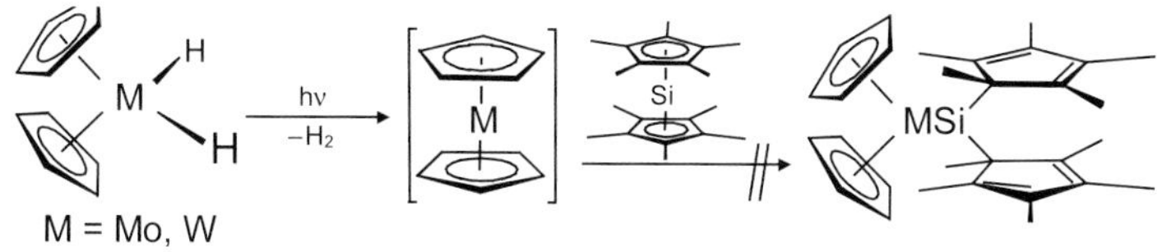

Eq. 2. Photolysis of Cp_2MH_2 (M = Mo, W) in the presence of Cp^*_2Si.

In contrast SiL^N_2 easily reacts with Cp_2MH_2 (Eq. 3) to form the corresponding metallosilanes $Cp_2M(H)SiL^N_2H$ (M = Mo **1**, W **2**). In the case of M = Mo the insertion into the Mo–H bond takes place at room temperature without further activation, while the insertion into the W–H bond needs photolytic conditions or prolonged heating up to 110°C.

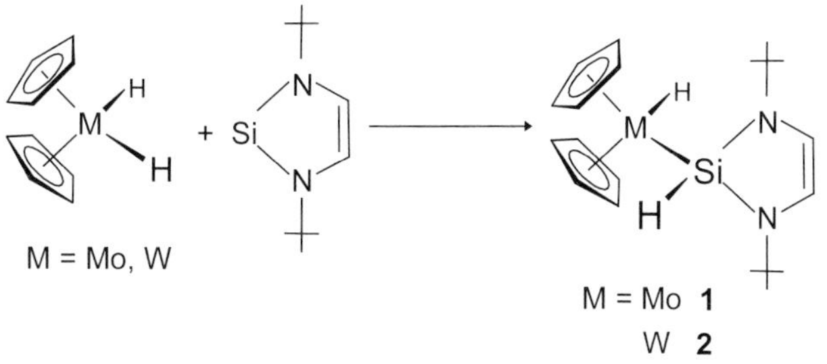

Eq. 3. Insertion of SiL^N_2 into an M–H bond of Cp_2MH_2 (M = Mo, W).

The molybdenum complex **1** is structurally characterized by single-crystal X-ray diffraction analysis. An ORTEP drawing, showing 50 % probability thermal ellipsoids, is given in Fig. 1 including some characteristic bonding parameters. Compound **1** possesses the typical geometry of a bent-sandwich complex with pseudotetrahedral surroundings at the molybdenum. The molybdenum–silicon bond length is within the range of typical $Cp_2Mo(H)SiR_3$ complexes [7–9].

Reactivity of Cp^*_2Si and SiL^N_2 towards $Cp_2Mo(PEt_3)$

$Cp_2Mo(PEt_3)$ is known to be a valuable precursor for molybdocene, Cp_2Mo [10]. But it was shown to be impossible to generate the silylene complex $Cp_2Mo(SiCp^*_2)$ via ligand exchange. Under thermal conditions Cp^*_2Si does not react with $Cp_2Mo(PEt_3)$, and the educts are recovered quantitatively. Photolytic treatment of the mixture only leads to unspecific decomposition products.

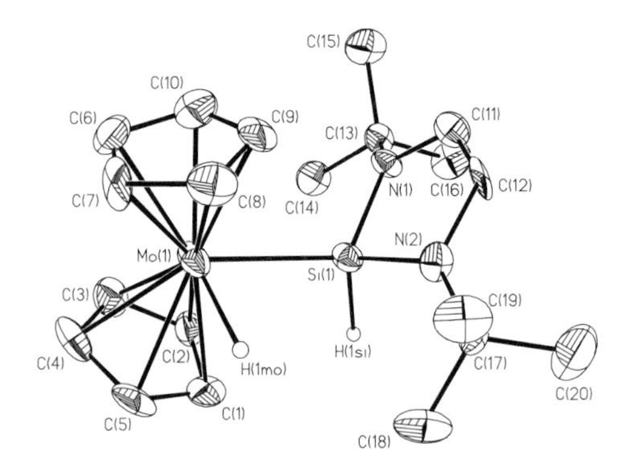

Selected bond lengths and bond angles:

Mo(1)–Si(1):	2.538(3) Å
Mo(1)–H(1mo):	1.84(11) Å
Cp(1)–Mo(1):	1.952 Å
Cp(2)–Mo(1):	1.964 Å
Cp(1)-Mo(1)-Cp(2):	148.5°
H(1mo)-Mo(1)-Si(1):	78(3)°

Cp1 and Cp2 are the centroids of the η^5-Cp.

Fig. 1. Molecular structure and selected bonding parameters of **1**.

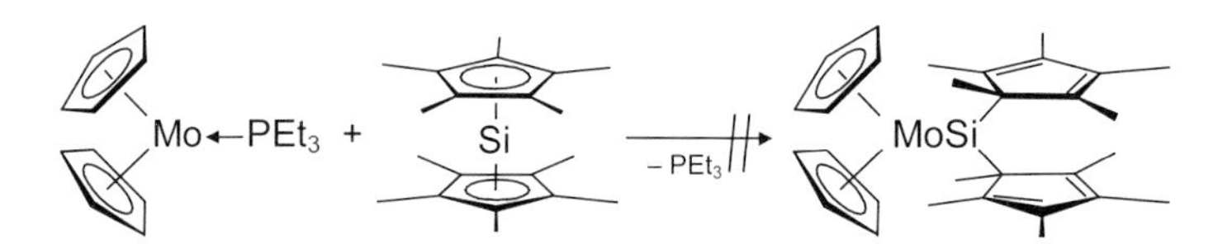

Eq. 4. Reaction of Cp^*_2Si with $Cp_2Mo(PEt_3)$.

The photolysis of an equimolar amount of SiL^N_2 and $Cp_2Mo(PEt_3)$ in hexane results in the formation of the molybdenum–silylene complex $Cp_2Mo(SiL^N_2)$ (**3**) (Eq. 5). Complex **3** is a deep red, extremly air- and moisture-sensitive solid showing a ^{29}Si-NMR signal at 139.9 ppm. This is significantly downfield-shifted in comparison both with the free silylene SiL^N_2 (78.3 ppm [11] and with other silylene complexes $((CO)_2Ni(SiL^N_2)_2$: 97.5 ppm $(CO)_5MoSiPhAr^N$: 97.5 ppm [12]) indicating a distinct silicenium cation character of **3**.

Eq. 5. Reaction of SiL^N_2 with $Cp_2Mo(PEt_3)$.

The silylene complex **3** is characterized by X-ray diffractometry. An ORTEP drawing including some characteristic bonding parameters is depicted in Figure 2. **3** is a typical bent-sandwich complex with trigonal planar surroundings both at the molybdenum and the silicon atom. The molybdenum–silicon distance (2.4125(13) Å) is the shortest ever observed. On the basis of the covalent radii of double-bonded molybdenum and silicon a bond length of only 2.3 Å or less is expected. Theoretical considerations concerning the nature of the molybdenum–silicon bond reveal the interaction to be a dative bond from silicon to molybdenum without any double bond character [8, 13].

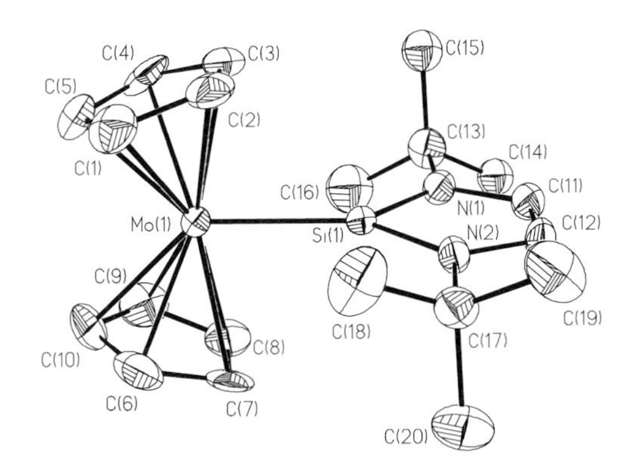

Selected bond lengths and bond angles:

Mo(1)–Si(1):	2.4125(13) Å
Cp(1)–Mo(1):	1.936 Å
Cp(1)–Mo(1):	1.934 Å
Cp(1)-Mo(1)-Cp(2):	148.6°
Sum of angles at Si(1):	360°

Cp1 and Cp2 are the centroids of the η^5-Cp.

Fig. 2. Molecular structure and selected bonding parameters of **3**.

Acknowledgment: This research was generously supported by a grant from the *Deutsche Forschungsgemeinschaft* as part of the Forschergruppe "Nanometerschichtsysteme". Prof. Dr. M. Veith and Dr. A. Rammo (Saarbrücken) are acknowledged for a generous gift of $SiL^N{}_2$.

References:

[1] P. Jutzi, A. Möhrke, *Angew. Chem.* **1990**, *102*, 913.

[2] U. Holtmann, *Dissertation*, Universität Bielefeld, **1988**.

[3] a) P. Jutzi, D. Eikenberg, "*Unprecedented Multistep Reactions of Decamethylsilicocene, $(Me_5C_5)_2Si$:, with CO_2, CS_2, COS, RNCS (R = Me, Ph), with CF_3CCCF_3, and with $HMn(CO)_5$*", in: *Organosilicon Chemistry III: From Molecules to Materials* (Eds.: N. Auner, J. Weis), Wiley–VCH, Weinheim, **1998**, p. 76.
b) D. Eikenberg, *Dissertation*, Universität Bielefeld, **1997**.

[4] D. H. Berry, Q. Jiang, *J. Am. Chem. Soc.* **1987**, *109*, 6210.

[5] M. Denk, R. K. Hayashi, R. West, *J. Chem. Soc., Chem. Commun.* **1994**, 33.

[6] M. Denk, R. West, R. Hayashi, Y Apeloig, R. Pauncz, M. Karni, "*Silylenes, Stable and Unstable*", in: *Organosilicon Chemistry II: From Molecules to Materials* (Eds.: N. Auner, J. Weis), VCH, Weinheim, **1996**, p. 251.

[7] a) P. Jutzi, S. H. A. Petri, "*Novel Synthetic Approach to Molybdenum–Silicon Compounds: Structures and Reactivities*", in: *Organosilicon Chemistry III, From Molecules to Materials* (Eds.: N. Auner, J. Weis), Wiley–VCH, Weinheim, **1998**, p. 275.
b) S. H. A. Petri, B. Neumann, H.-G. Stammler, P.Jutzi, *J. Organomet. Chem.* **1998**, *553*, 317.

[8] S. H. A. Petri, *Dissertation*, Universität Bielefeld, **1998**.

[9] T. S. Koloski, D. C. Pestana, P. J. Carrol, D.H. Berry, *Organometallics* **1994**, *13*, 489.

[10] G. L. Geoffroy, M. G. Bradley, *J. Organomet. Chem.* **1977**, *134*, C27–C31; J.M. Galante, J.W. Bruno, P.N. Hazin, K. Folting, J.C. Huffmann, *Organometallics* **1988**, *7*, 1066–1073.

[11] M. Denk, R. Lennon, R. Hayashi, R. West, A. V. Belyakov, H. P. Verne, A. Haaland, M. Wagner, N. Metzler, *J. Am. Chem. Soc.* **1994**, *116*, 2691.

[12] B. P. S. Chauhan, R. J. P. Corriu, G. F. Lanneau, C. Priou, *Organometallics* **1995**, *14*, 1657.

[13] S. H. A. Petri, B. Neumann, H.-G. Stammler, P. Jutzi, *Organometallics* **1999**, *18*, 2615.

Reactions of Decamethylsilicocene with Inorganic and Organometallic Main Group Element Compounds

Thorsten Kühler, Peter Jutzi*

Fakultät für Chemie, Universität Bielefeld
Universitätsstr. 25, D-33615 Bielefeld, Germany
Tel.: Int. code + (521)106 6181 — Fax.: Int. code + (521)106 6026

Keywords: Decamethylsilicocene / Silylene

Summary: Decamethylsilicocene **1** is reacted with main group element halides and organo-substituted derivatives. Reduction takes place with GaX_3 and InX_3 (X= Cl, Br) to form the corresponding monohalides and $Cp^*{}_2SiX_2$ **4a,b** (Cp*= 1,2,3,4,5-Pentamethylcyclopentadienyl). Cp* transfer occurs with AlX_3 (X= Cl, Br), Me_2AlCl, InCl, ECl_2 (E= Ge, Sn, Pb) and ECl_3 (E= As, Sb) to form $[Cp^*{}_2Al^+AlCl_4^-]$ **2**, Cp*Al(Me)Cl **3**, Cp*In **6**, Cp*ECl (E= Ge **7**, Sn **8**, Pb **9**) and Cp^*ECl_2 (E= As **10**, Sb **11**) respectively. Insertionreactions are observed with "Me_2GaCl" where a mixture of products is formed (**5a,b**).

Recently the chemistry of nucleophilic silylenes has received a lot of attention. Decamethylsilicocene (**1**) can be considered as a hypercoordinated nucleophilic silylene [1]. Here we report on reactions of **1** with Lewis-acidic compounds of elements of groups 13 to 15.

1 reacts with two equivalents of AlX_3 (X = Cl, Br) to yield the compound $Cp^*{}_2Al^+\ AlX_4^-$ (Eq. 1) [2]. This is the more surprising since the oxidation state of +II at the silicon is preserved (formation of $[SiX_2]$).

$$\mathbf{1} + 2\,AlCl_3 \xrightarrow{-[SiCl_2]} [Cp^*{}_2Al]^{\oplus}\ AlCl_4^{\ominus}\ \mathbf{2}$$

Eq. 1

Reaction with organosubstituted derivatives like Me_2AlCl (Eq. 2) leaves the Cp* transfer incomplete, leading to Cp*Al(Me)Cl (**3**) [3] as the only identified product in good yield. So far attempts to isolate or trap the postulated Cp*SiMe have not been successful.

$$\mathbf{1} + Me_2AlCl \xrightarrow{-[Cp^*SiMe]} \tfrac{1}{2}\ [Cp^*Al(Me)(\mu\text{-}Cl)]_2\ \mathbf{3}$$

Eq. 2

1 reacts with $GaCl_3$ and $GaBr_3$ with formation of the correspondig Cp*-substituted silicon dihalides (**4a, 4b**) (Eq. 3). $Cp^*_2SiX_2$ is the only silicon product observed and the yield is almost quantitative. Thus [GaX] is most probably formed. Under the chosen conditions this gallium compound disproportionates to GaX_3 and Ga^0.

1 + GaX_3 ⟶ $Cp^*_2SiX_2$ (**4 a,b**) + [GaX]

X = Cl, Br

Eq. 3

A hint to the mechanism of this process is given by the reaction of **1** with "Me_2GaCl" (Eq. 4). A complicated mixture of products is formed due to ligand scrambling of the gallium species. Figure 1 shows the structure of the main product **5a** (R = Cl). It is formed by the insertion of **1** into the Ga–Cl bond of Cl_2GaMe.

1 + "Me_2GaCl" ⟶ $[Cp^*_2Si(R)–Ga(Me)(\mu\text{-}Cl)]_2$

R = 75% Cl **(5a)**
25% Me **(5b)**

Eq. 4

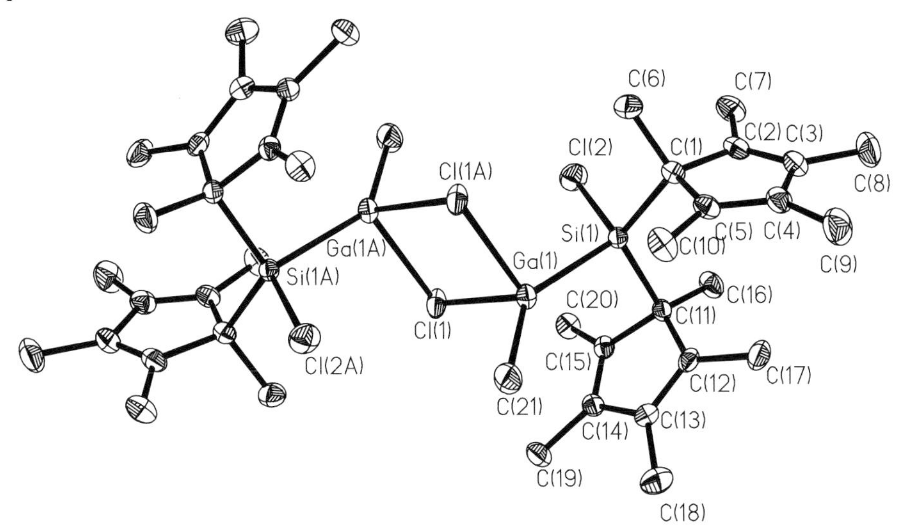

Fig. 1. ORTEP-Plot of **5a**.

1 reacts with species of the type InX_3 (X = Cl, Br) in analogy to the Ga compounds (Eq. 5). Reduction leads to the indium(I)halides, which can be isolated.

1 + InX_3 → $Cp^*_2SiX_2$ (4 a,b) + InX

X = Cl, Br

Eq. 5

1 slowly reacts with indium(I)halides. Cp* transfer from Si to In is observed. The reaction with InCl yields Cp*In (**6**) [4] as the only identified product (Eq. 6).

1 + InCl —(- [Cp*SiCl])→ Cp*In (6)

Eq. 6

With the dichlorides of germanium, tin and lead **1** also reacts as a Cp* transfer reagent (Eq. 7). The corresponding Cp*ECl species **7** [5], **8** [6] and **9** [7] are formed in nearly quantitative yield.

1 + ECl_2 —(- [Cp*SiCl])→ Cp*ECl

E = Ge, Sn, Pb

7, 8, 9

Eq. 7

In the reaction with the corresponding tetravalent compound ECl_4 reduction occurs and $Cp^*_2SiCl_2$ is formed.

The Lewis-acidity of the halides of arsenic and antimony is high enough to allow the nucleophilic attack of **1** (Eq. 8). The reaction leads to the formation of the corresponding Cp^*ECl_2 species **10** [8] and **11** [9] (E = As, Sb).

1 + ECl_3 —(- [Cp*SiCl])→ Cp^*ECl_2

E = As, Sb

10, 11

Eq. 8

The reactivity of decamethylsilicocene to Lewis-acidic main group element compounds can be summarized as follows. **1** reacts as a reducing or a Cp* transfer reagent depending on the element species involved. In all reactions where transfer of one Cp* ligand is observed, a polymeric species of the composition $(Cp^*SiCl)_n$ is formed, probably by polymerization of an intermediate silylene. This can be concluded from the following analytical results. The ^{1}H-NMR spectrum of the reaction mixture shows a broad resonance at ca. 1.8 ppm for the Cp* methyl groups. In the mass spectrum of the residue, signals for high masses are observed, which can be assigned to multiples of Cp*SiCl. The nature of the polymer is a topic of our current research.

Acknowledgments: This work was generously supported by the *Fonds der Chemischen Industrie.*

References:

[1] a) P. Jutzi, D. Kannne, C. Krüger, *Angew. Chem. Int. Ed. engl.* **1986**, *25*, 164.
b) P. Jutzi, U. Holtmann, D. Kannne, C. Krüger, R. Blom, R. Gleiter, I. Hyla-Kryspin, *Chem. Ber.* **1989**, *122*, 1629.

[2] C. Dohmeier, H. Schnöckl, C. Robl, U. Schneider, R. Ahlrichs, *Angew. Chem. Int. Ed. Engl.* **1993**, *32*, 1655.

[3] P. R. Schonberg, R. T. Paine, C. F. Campana, *J. Am. Chem. Soc.* **1997**, *101*, 7726.

[4] O. T. Beachley, Jr., R. Blom, M. R. Churchill, K. Faegri, Jr., J. C. Fettinger, J. C. Pazik, L. Victoriano, *Organometallics* **1989**, *8*, 346.

[5] F. X. Kohl, P. Jutzi, *J. Organomet. Chem.* **1983**, *243*, 31.

[6] S. R. Constanine, G. M. de Lima, P. B. Hitchcock, J. M. Keates, G. A. Lawless, I. Marziani, *Organometallics* **1997**, *16*, 793.

[7] P. Jutzi, R. Dickbreder, H. Nöth, *Chem. Ber.* **1989**, *122*, 865.

[8] P. Jutzi, H. Saleske, D. Nadler, *J. Organomet. Chem.* **1976**, *118*, C8.

[9] P. Jutzi, U. Meyer, S. Opiela, M. M. Olmstead, P. P. Power, *Organometallics* **1990**, *9*, 1459.

Unusual Silylene Reactions with M(II) Compounds (M = Ge, Sn or Pb)

Barbara Gehrhus, Peter B. Hitchcock, Michael F. Lappert

School of Chemistry, Physics and Environmental Science
University of Sussex
Brighton, UK BN1 9QJ
Tel.: Int. code + (1273)678316 — Fax: Int. code + (1273)677196
E-mail: m.f.lappert@sussex.ac.uk

Keywords: Insertion / Silylene / Stannylene

Summary: Treatment of the stable bis(amino)silylene $\overline{Si[(N}CH_2tBu)_2C_6H_4$-1,2] **1** with MX_2 (M = Ge, Sn or Pb and X = $N(SiMe_3)_2$; M = Sn and X = $OC_6H_3tBu_2$-2,6; or M = Ge and X = $OC_6H_2tBu_2$-2,6-Me-4) gave the bis(silyl)metal(II) compounds $M[(\mathbf{1})X]_2$ (M = Sn or Pb) or disilagermoles (M = Ge).

Introduction

Two thermally stable and dicoordinate silylenes have to date been reported: $\overline{Si[N(tBu)CHCHN}tBu]$ [1] and X-ray analysis authenticated $\overline{Si[(N}CH_2tBu)_2C_6H_4$-1,2] **1** [2]. The chemistry of $\overline{Si[N(tBu)CHCHN}tBu]$ was explored in the context of its reactions with $[Ni(CO)_4]$, Me_3SiN_3, Ph_3CN_3 or $B(C_6F_5)_3$ [3]. A wider range of reactions has been studied for **1**. Thus, oxidative addition reactions of **1** with X–Y or E yielded the silicon compounds [**1**(X)Y] (X–Y = Me–I or EtO–H) or the cyclodisilachalcogenides $\overline{\mathbf{1}E\mathbf{1}E}$ (E = S, Se or Te) [4]; and **1** underwent a series of cycloaddition reactions with various C–C, C–O and C–N multi-bonded compounds [5]. Treatment of the silylene **1** with (i) $[NiCl_2(PPh_3)_2]$, (ii) $[Ni(COD)_2]$ or (iii) $[PtCl_2(PPh_3)_2]$ led to a new range of stable and donor-free silylene–transition metal complexes: for (i), $[Ni(\mathbf{1})_3PPh_3]$; for (ii), $[Ni(\mathbf{1})_4]$; or for (iii), $[Pt\{\mathbf{1}(Cl)\}_2(\mathbf{1})_2]$ [6].

Recently we reported on the reaction of **1** with Sn(Ar)X, which afforded the first stable, heteroleptic silylstannylene Sn(Ar)[(**1**)X] [X = Ar = $C_6H_3(NMe_2)_2$-2,6, or X = $N(SiMe_3)_2$] [7]. We now present further results on such new insertion reactions of silylenes, which can be extended to other M(II) compounds (M = Ge, Sn, Pb) [8].

Results and Discussion

Treatment of **1** with MX_2 [X = $N(SiMe_3)_2$ for M = Sn or Pb [8], or X = $OC_6H_3tBu_2$-2,6 for M = Sn) afforded the new dark green or black-green (M = Pb) bis(silyl)M(II) compounds **2**, **3** and **4** (Scheme 1).

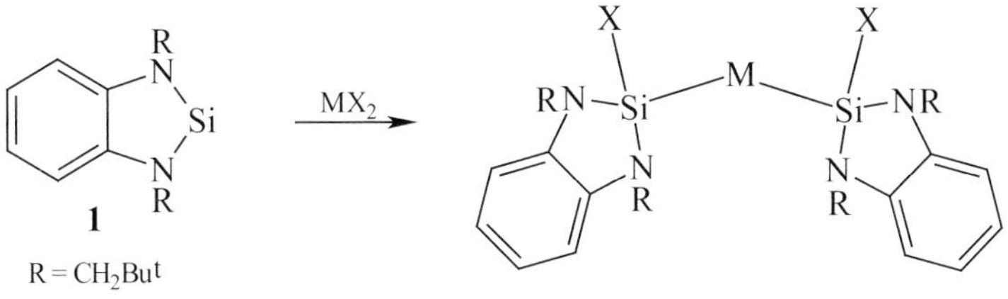

1

R = CH_2Bu^t

2 M = Sn and X = $N(SiMe_3)_2$
3 M = Pb and X = $N(SiMe_3)_2$
4 M = Sn and X = $OC_6H_3Bu^t{}_2$-2,6

Scheme 1.

Each of the compounds **2–4** gave satisfactory microanalysis and EI mass and multinuclear NMR spectra. For **2**, no δ[^{119}Sn] signal has yet been observed; the δ[^{29}Si{^{1}H}] signals were at 105.02, 9.04 and 7.00. For compound **3** it proved difficult to record the heavier nuclei due to decomposition of the sample. The δ[^{119}Sn{^{1}H}] and δ[^{29}Si{^{1}H}] signals for **4** were at –218.1 and 77.05, respectively. Although the value of the ^{119}Sn-NMR spectroscopic shift for **4** points to a higher-coordinated species in solution, the X-ray structure shows its divalent nature. Solid-state NMR spectroscopic investigations will be examined.

The X-ray structures of **2–4** are shown in Fig. 1; selected bond lengths and angles are listed in Table 1.

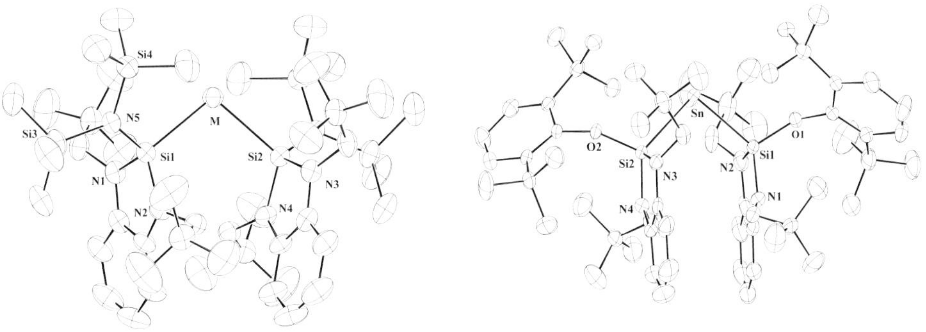

Fig. 1. Molecular structures of crystalline **2** (M = Sn), **3** (M = Pb) and **4**.

Table 1. Some comparative structural data of compounds **2**–**4**.

Property	**Complex**		
	2[a]	**3**[b]	**4**[c]
d(Si–M) [Å]	2.712(1)	2.776(3)	2.679(2)
d(Si1–N1), [Si1–N2] [Å]	1.745(3), [1.754(3)]	1.751(9), [1.755(8)]	1.730(5), [1.735(5)]
d(Si2–N3), [Si2–N4] [Å]			1.741(5), [1.747(5)]
Si1-M-Si2 [deg.]	106.77(5)	105.8(1)	100.87(5)
N1-Si1-N2,[N3-Si2-N4] [deg.]	92.0(2)	92.6(4)	94.1(2), [94.0(3)]

[a] M = Sn, lies on a twofold rotation axis. — [b] M = Pb, lies on a twofold rotation axis. — [c] M = Sn.

A noteworthy feature is the very long Sn–Si bond in compound **2**, particularly in relation to the sum of the Sn and Si covalent radii (2.58 Å).

The reaction of Scheme 1 probably proceeds via the successive intermediates **5** and **6.** A cyclic transition state **7** may be involved in the transformation of **5** to **6**, and a similar one between **6** and the ultimate product **2**, **3** or **4**.

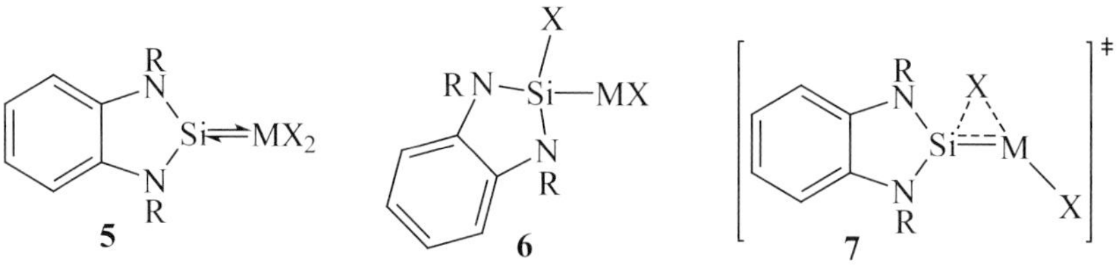

5 **6** **7**

Precedents for compounds related to **5** in Group 14 element chemistry include the crystallographically characterized binuclear M^{II}–M^{II} compounds $[M\{CH(SiMe_3)_2\}_2]_2$ (M = Ge or Sn) [9], $[Sn\{Si(SiMe_3)_3\}_2]_2$ [10], $[Sn\{C_6H\textit{t}Bu\text{-}2\text{-}Me_3\text{-}3,5,6\}_2]_2$ (for magnetic resonance spectral data, see ref. [11b]) [11a], and $X_2SnSnCl_2$ [X = $CH(SiMe_3)C_9H_6N$-8] [12], and the transient heterobinuclear complexes $Ge(C_6H_2Me_3\text{-}2,4,6)_2M'X_2$ [$M'X_2$ = $Si(C_6H_2Me_3\text{-}2,4,6)_2$ [13] or $Sn(C_6H_2\textit{i}Pr_3\text{-}2,4,6)_2$] [14]. (characterized spectroscopically or by trapping experiments). As for **5** → **6**, 1,2-migrations of a ligand X^- from M → M' are well documented: for X = $C_6H_2Me_3$-2,4,6 from Ge → Si [13], for Cl^- from Sn → Sn [12], and $^-C_6F_5$ from B → Si [15].

Treatment of **1** with the corresponding GeX_2 compound [X = $N(SiMe_3)_2$ or $OC_6H_2\textit{t}Bu_2$-2,6-Me-4] led to the colorless germaindole **8** or **9**, Scheme 2, which have been satisfactorily characterized by microanalysis and EI mass and multinuclear NMR spectra as well as their X-ray structures.

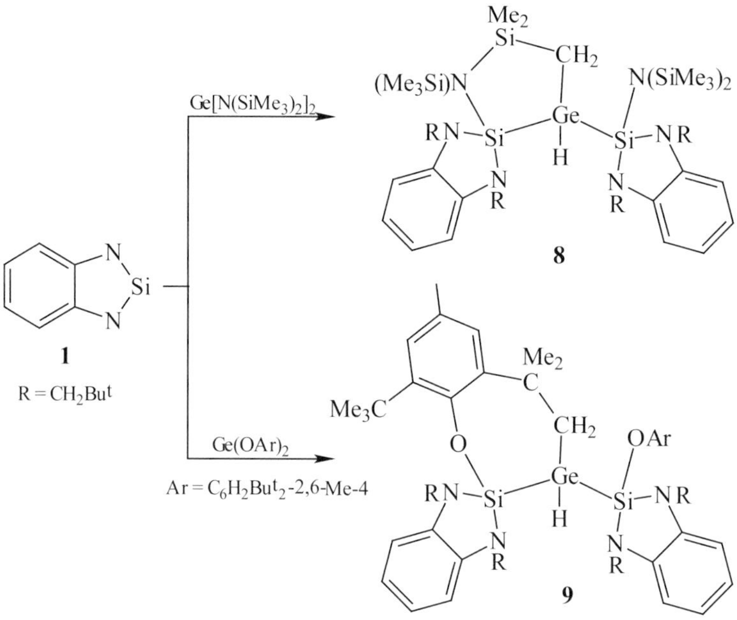

$Ge[N(SiMe_3)_2]_2$

$Ge(OAr)_2$

Ar = $C_6H_2Bu^t_2$-2,6-Me-4

1

R = CH_2Bu^t

8 **9**

Scheme 2.

The synthesis of **8** and **9** is believed to follow the same pathway as for **2**–**4**. Steric acceleration, however, caused the bis(silyl)germylene intermediate **10** (the Ge analogue of **2** or **4**) to undergo cyclometallation by insertion of the labile germylene **10** into a C–H bond of $N(SiMe_3)_2$ or $OC_6H_2tBu_2$-2,6-Me-4, respectively. Precedents for related C–H activations have been described [16].

Acknowlegment: We thank the *EPSRC* for the award of a fellowship to B. G. and for other support.

References:

[1] M. Denk, R. Lennon, R. Hayashi, R. West, A. V. Belyakow, H. P. Verne, A. Haaland, M. Wagner, N. Metzler, *J. Am. Chem. Soc.* **1994**, *116*, 2691.

[2] B. Gehrhus, M. F. Lappert, J. Heinicke, R. Boese, D. Bläser, *J. Chem. Soc., Chem. Commun.* **1995**, 1931.

[3] a) M. Denk, R. Hayashi, R. West, *J. Chem. Soc., Chem. Commun.* **1994**, 33; M. Denk, R. K. Hayashi, R. West, *J. Am. Chem. Soc.* **1994**, *116*, 10813.
b) N. Metzler, M. Denk, *Chem. Commun.* **1996**, 1475.

[4] B. Gehrhus, M. F. Lappert, J. Heinicke, R. Boese, D. Bläser, *J. Organomet. Chem.* **1996**, *521*, 211.

[5] a) B. Gehrhus, P. B. Hitchcock, M. F. Lappert, *Organometallics* **1997**, *16*, 4861.
b) B. Gehrhus, P. B. Hitchcock, M. F. Lappert, *Organometallics* **1998**, *17*, 1378.
c) B. Gehrhus, M. F. Lappert, *Polyhedron* **1998**, *17*, 999.

[6] B. Gehrhus, P.B. Hitchcock, M. F. Lappert, H. Maciejewski, *Organometallics* **1998**, *17*, 5599.

[7] C. Drost, B. Gehrhus, P. B. Hitchcock, M. F. Lappert, *Chem. Commun.* **1997**, 1845.

[8] B. Gehrhus, P. B. Hitchcock, M. F. Lappert, *Angew. Chem., Int. Ed. Engl.* **1997**, *22*, 2514.
see also: A. Schäfer, W. Saak, M. Weidenbruch, H. Marsmann, G. Henkel, *Chem. Ber./Recueil* **1997**, *130*, 1733.

[9] a) P. J. Davidson, D. H. Harris, M. F. Lappert, *J. Chem. Soc., Dalton Trans.* **1976**, 2268.
b) D. E. Goldberg, P. B. Hitchcock, M. F. Lappert, K. M. Thomas, A. J. Thorne, T. Fjeldberg, A. Haaland, B. E. R. Schilling, *J. Chem. Soc., Dalton Trans.* **1986**, 2387.

[10] K. W. Klinkhammer, W. Schwarz, *Angew. Chem.* **1995**, *107*, 1448; *Angew. Chem., Int. Ed. Engl.* **1995**, *34*, 1334.

[11] a) M. Weidenbruch, H. Kilian, K. Peters, H. G. von Schnering, H. Marsmann, *Chem. Ber.* **1995**, *128*, 983.
b) M. A. DellaBona, M. C. Cassani, J. M. Keates, G. A. Lawless, M. F. Lappert, M. Stürmann, M. Weidenbruch, *J. Chem. Soc., Dalton Trans.* **1998**, 1187.

[12] W.-P. Leung, W.-H. Kwok, F. Xue, T. C. W. Mak, *J. Am. Chem. Soc.* **1997**, *119*, 1145.

[13] K. M. Baines, J. A. Cooke, C. E. Dixon, H. W. Liu, M. R. Netherton, *Organometallics* **1994**, *13*, 631.

[14] M.-A. Chaubon, J. Escudié, H. Ranaivonjatovo, J. Satgé, *Chem. Commun.* **1996**, 2621.

[15] N. Metzler, M. Denk, *Chem. Commun.* **1996**, 2657.

[16] a) P. Jutzi, H. Schmidt, B. Neumann, H.-G. Stammler, *Organometallics* **1996**, *15*, 741.
b) C. R. Bennett, D. C. Bradley, *J. Chem. Soc., Chem. Commun.* **1974**, 29.
c) S. M. Hawkins, P. B. Hitchcock, M. F. Lappert, A. K. Rai, *J. Chem. Soc., Chem. Commun.* **1986**, 1689.

Silylene and Disilene Reactions with some 1,3-Dienes and Diynes

Lars Kirmaier, Manfred Weidenbruch*

Fachbereich Chemie, Universität Oldenburg
Carl-von-Ossietzky-Straße 9, D-26111 Oldenburg, Germany
Tel.: Int. code + (441)798 3655 — Fax: Int. code + (441)798 3352

Keywords: Silylene / Disilene / Silaheterocycles / 1,3-Dienes / Diynes

Summary: Di-*tert*-butylsilylene (**2**), generated together with tetra-*tert*-butyldisilene (**3**) by photolysis of hexa-*tert*-butylcyclotrisilane (**1**), reacted with the five-membered heterocyclic compounds 2,5-dimethyltellurophene and *N*-methylpyrrole to furnish the corresponding substituted 1,3-ditellura-2,4-disiletane (**8**) and 2-aza-3-silabicyclo[2.2.0]hex-2-ene (**9**) ring systems. Upon heating, compound **9** rearranged to a 1-aza-2-silacyclohexa-3,5-diene derivative. Further treatment of **9** with **2** yielded the 1-aza-2,5-disilacylohepta-3,6-diene **11**. Reaction of the intermediates **2** and **3** with the triple bonds of hexa-2,4-diyne afforded the rearranged addition products 2-methylene-5-methyl-3,4,7,8-tetrasilabicyclo[4.2.0]oct-1(6)ene (**12**) and 2-methyl-3-propynylsilirene (**13**). Renewed treatment of **13** with **2** afforded the 2,5-disilabicyclo[2.2.0]hexa-1(6),3-diene ring system (**15**). A similar reaction of **1** with 1,4-bis(trimethylsilyl)buta-1,3-diyne led to an unprecedented insertion of **2** into the C–C single bond of the diyne to provide the bis(alkynyl)silane **16**, which was isolated together with the bicyclic compound **17**, in which the methyl groups of **15** are replaced by $SiMe_3$ groups.

Introduction

Although the chemistry of disilenes, i.e., compounds containing an Si=Si double bond, has been under intensive investigation for many years and has meanwhile been documented in a series of review articles [1], [4+2] cycloadducts of the Diels–Alder type with unequivocally confirmed structures were unknown until a short time ago. We recently isolated and unambiguously characterized the Diels–Alder product **4** from the photolysis reaction of hexa-*tert*-butylcyclotrisilane **1**, which decomposes under the prevailing conditions to di-*tert*-butylsilylene **2** and tetra-*tert*-butyldisilene **3** [2], in the presence of cyclopentadiene [3] (Scheme 1).

Furan also undergoes a primary [4+2] cycloaddition with **3**, but this is followed by a [2+1] addition of **2** to the newly formed double bond to furnish the isolated tricyclic compound **5**. Thiophene behaves differently: its reaction with **3** furnishes the disilathiirane **6** as a formal sulfur-abstraction product [3]. This strongly divergent behavior of five-membered ring systems prompted us to investigate the photolyses of **1** also in the presence of a selenophene, a tellurophene, and a pyrrole. We report here on the photolysis of **1** in the presence of these cyclic dienes as well as similar reactions with acyclic diynes.

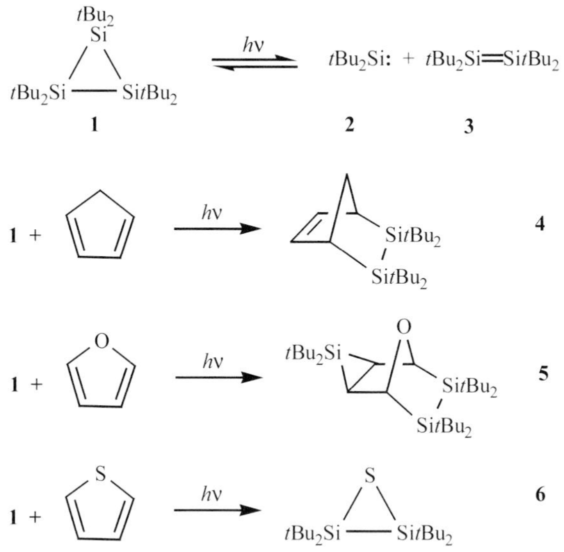

Scheme 1. Photolysis reactions of hexa-*tert*-butylcyclotrisilane **1**.

Results and Discussion

Irradiation of **1** in the presence of selenophene afforded the 2,4-disila-1,3-diseletane **7** in low yield. Similarly, reaction of **1** with the readily accessible [4] 2,5-dimethyltellurophene furnished the 2,4-disila-1,3-ditelluretane **8** in acceptable yield (Scheme 2) [5].

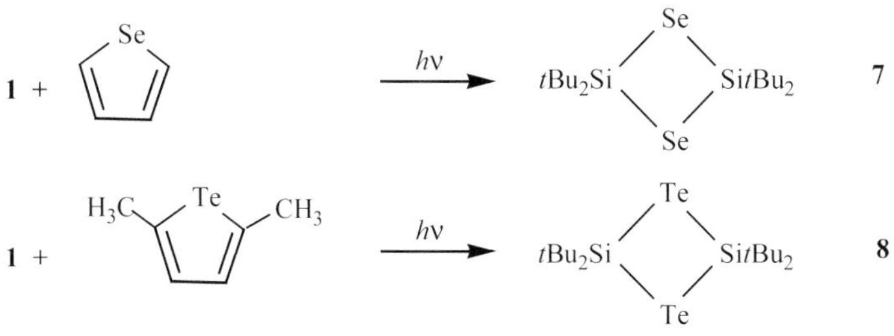

Scheme 2. Photolysis reactions of hexa-*tert*-butylcyclotrisilane **1** in the presence of selenophene or 2,5-dimethyltellurophene, respectively.

Both four-membered ring compounds are completely planar with the smaller endocyclic angles at the chalcogen atoms and the larger angles at the silicon atoms [5]. Surprisingly, both compound **7** and compound **8** represent new ring systems of the element combinations silicon/selenium [6] and silicon/tellurium and thus supplement the known palettes of small rings made up of these elements.

The described photolyses of **1** in the presence of cyclopentadiene, furan, and its heavier homologues impressively demonstrate that the replacement of the CH_2 group by the isoelectronic chalcogen atoms can lead to a completely different reaction behavior. Hence we selected the — also isoelectronic heterocyclic — pyrrole system for further investigations and, in order to avoid possible reactions of the N–H bond, chose *N*-methylpyrrole instead of the parent compound.

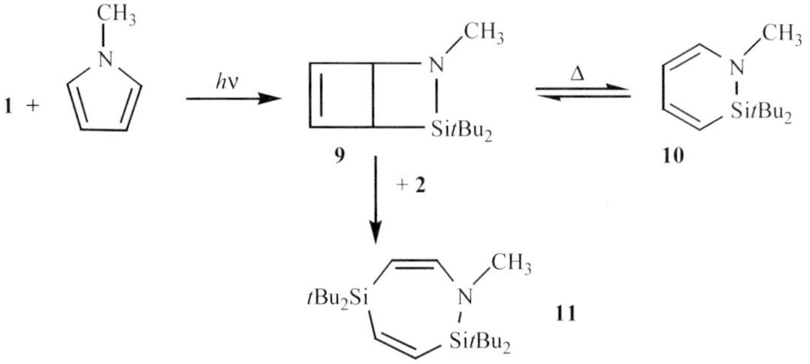

Scheme 3. Irradiation of **1** in the presence of *N*-methylpyrrole.

Irradiation of **1** in the presence of *N*-methylpyrrole furnished the bicyclic product **9** in almost quantitative yield. On longer heating at 100°C the bicyclic product **9** underwent an electrocyclic rearrangement to afford the cyclohexadiene derivative **10**. Renewed attack of **1** (**2**) on **9** finally led to the seven-membered heterocyclic product **11** which was characterized by X-ray crystallography [7]. This example again illustrates that exchange of the heteroatom in these five-membered ring compounds can result in a completely different product spectrum.

These unusual and, for the most part, from a mechanistic point of view as yet understood reactions of **1** with heterocyclic dienes encouraged us to extend our investigations to include corresponding acyclic diynes.

Photolysis of **1** in the presence of hexa-2,4-diyne gave two compounds in high yield, of which one was easily identified as the propynylsilirene **13** (Scheme 4). Elucidation of the constitution of the 2:1 cycloadduct of the disilene **3** and hexa-2,4-diyne was more difficult; finally use of a combination of NMR methods and X-ray crystallography demonstrated the formation of the bicyclic product **12**. The mechanism of formation of **12** presumably begins with the addition of two molecules of disilene **3** to the C≡C triple bond, followed by a thermally allowed sigmatropic 1,5-hydrogen shift. Silyl radicals could be formed by homolysis of the thus-formed cyclobutane ring and then two consecutive cyclization steps would furnish the isolated bicyclic product **12** [8].

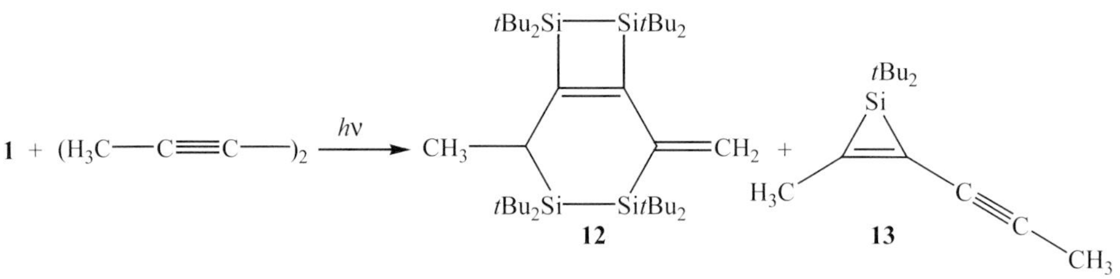

Scheme 4. Photolysis of **1** in the presence of hexa-2,4-diyne.

When the silirene **13** is allowed to react with **1** again, another bicyclic compound (**15**) (Scheme 5) is obtained. A plausible explanation for the formation of this novel ring system involves the addition of a second molecule of **2** to the still-existing triple bond to afford the bis(silirene) **14**, which then undergoes rearrangement through bond cleavage and re-formation to yield the markedly less strained bicyclic product **15** [8].

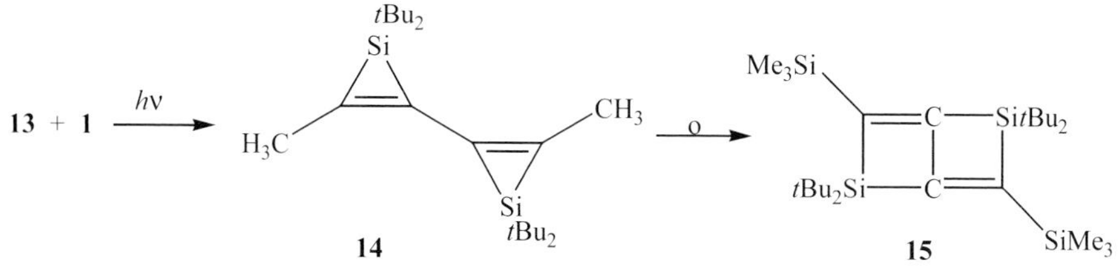

Scheme 5. Reaction of the propynylsilirene **13** with **1** yields the bicyclic compound **15**.

Similarly to the cases with the cyclic 1,3-dienes, slight changes are also sufficient to effect completely different reaction courses with the diynes. For example, when the two methyl groups of hexa-2,4-diyne are replaced by trimethylsilyl groups and this diyne is then subjected to photolysis with **1** under otherwise identical conditions, the bicyclic product **17** — the analogue of **15** — is obtained in one step. More unusual and previously never observed is the insertion of the silylene **2** into the C–C single bond of the diyne, proceeding even at –20°C. It must be assumed that a silirene of the type **13** is formed initially and then rearranges to the strain-free compound **16** [9] (Scheme 6). A similar reaction behavior has been observed with "titanocene" or "zirconocene" which can also effect a cleavage of the central C–C single bond of diacetylenes by way of cyclocumulene intermediates [10].

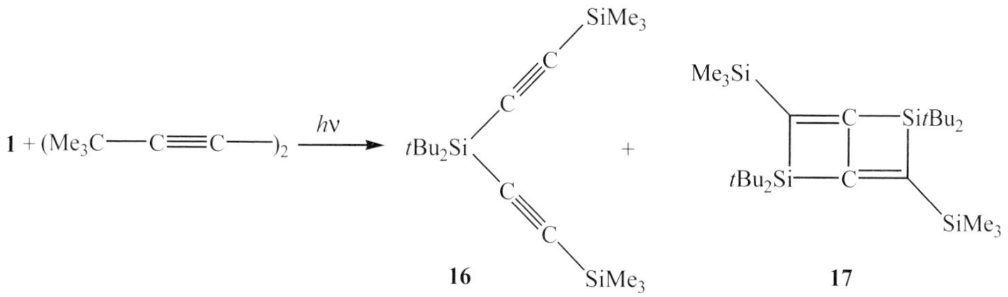

Scheme 6. Photolysis of **1** in the presence of 1,4-bis(trimethylsilyl)buta-1,3-diyne yields the bis(alkynyl)silane **16** and the bicyclic compound **17**.

Acknowledgment: Financial support of our work by the *Deutsche Forschungsgemeinschaft* and the *Fonds der Chemischen Industrie* is gratefully acknowledged.

References:

[1] Reviews:
a) R. West, *Angew. Chem.* **1987**, *99*, 1231; *Angew. Chem., Int. Ed. Engl.* **1987**, *26*, 1202.
b) G. Raabe, J. Michl, in: *The Chemistry of Organic Silicon Compounds Part 2* (Eds.: S. Patai, Z. Rappoport) Wiley, Chichester, **1989**, p. 1015.

c) T. Tsumuraya, S. A. Batcheller, S. Marsamune, *Angew. Chem.* **1991**, *103*, 916; *Angew. Chem., Int. Ed. Engl.* **1991**, *30*, 902.

d) M. Weidenbruch, *Coord. Chem. Rev.* **1994**, *130*, 275.

e) R. Okazaki, R. West, in: *Multiple Bonded Main Group Metals and Metalloids* (Eds.: R. West, F. G. A. Stone) Academic Press, San Diego, USA, **1996**, p. 232.

[2] Review: M. Weidenbruch, *Chem. Rev.* **1995**, *95*, 1479.

[3] E. Kroke, M. Weidenbruch, W. Saak, S. Pohl, H. Marsmann, *Organometallics* **1995**, *14*, 5695.

[4] W. Mack, *Angew. Chem.* **1966**, *78*, 940; *Angew. Chem., Int. Ed. Engl.* **1966**, *5*, 896.

[5] M. Weidenbruch, L. Kirmaier, E. Kroke, W. Saak, *Z. Anorg. Allg. Chem.* **1997**, *623*, 1277.

[6] Very recently, the structure of a similar ring with an Si_2Se_2 skeleton, namely the compound $[\{Cp'(CO)_2)Fe\}_2Me_2Si_2Se_2]$, Cp' = $CH_3C_5H_4$, was reported in a review article: K. Merzweiler, U. Linder, in: *Organosilicon Chemistry II, from Molecules to Materials* (Eds.: N. Auner, J. Weis) Wiley–VCH, Weinheim, **1996**, p. 531.

[7] M. Weidenbruch, L. Kirmaier, H. Marsmann, P. G. Jones, *Organometallics* **1997**, *16*, 3080.

[8] L. Kirmaier, M. Weidenbruch, H. Marsmann, K. Peters, H. G. von Schnering, *Organometallics* **1998**, *17*, 1237.

[9] D. Ostendorf, L. Kirmaier, W. Saak, H. Marsmann, M. Weidenbruch, *Eur. J. Inorg. Chem.*, in press.

[10] Review: A. Ohff, S. Pulst, C. Lefeber, N. Peulecke, P. Arndt, V. V. Burkalov, U. Rosenthal, *Synlett* **1996**, 111; see also S. Pulst, F. G. Kirchbauer, B. Heller, W. Baumann, U. Rosenthal, *Angew. Chem.* **1998**, *110*, 2029; *Angew. Chem., Int. Ed. Engl.* **1998**, *37*, 1925.

Sterically Overloaded Silanes, Silylenes and Disilenes with Supersilyl Substituents tBu_3Si

Nils Wiberg*, Wolfgang Niedermayer, Kurt Polborn[a], Heinrich Nöth[a], Jörg Knizek[a]

Institut für Anorganische Chemie, Ludwig-Maximilians-Universität München
Butenandtstraße 5–13 (Haus D), D-81377 München, Germany
Tel. Int code + (89)2180 7458 — Fax. Int. code + (89)2180 7865
E-mail: niw@cup.uni-muenchen.de

Dieter Fenske[a], Gerhard Baum[a]

Institut für Anorganische Chemie, Universität Karlsruhe
Engesserstrasse Geb. Nr 30.45, D-76131 Karlsruhe, Germany
Tel.Int. code + (721)608 2085

Keywords: Silanes / Structures / Silanides / Silylenes / Disilenes

Summary: Silanes tBu_3SiSiX_3 and $(tBu_3Si)_2SiX_2$ with bulky supersilyl substituents are easily accessible by reaction of halosilanes with tBu_3SiNa. Their decomposition into silylenes and their transformation into their silanides as well as into a stable disilene with phenyl- and supersilyl-substituents are described. The constitutions of tBu_3SiSiI_3, $(tBu_3Si)_2SiCl_2$ and $tBu_3SiPhSi{=}SiPhSitBu_3$ are solved by X-ray structure analyses.

Mono- and Disupersilylated Silanes and Silanides

Reactions of halosilanes (e.g. H_2SiCl_2, SiF_4, $MeHSiCl_2$, $PhHSiCl_2$) with tBu_3SiNa lead according to Scheme 1 to the monosupersilylated compounds **1** (cf. Table 1 and Fig. 1). Compounds **1** react with tBu_3SiNa to form sterically overloaded disupersilylsilanes **2** (cf. Table 2 and Fig. 2). Silanes with 3 supersilyl substituents are – for steric reasons – not accessible. Due to the steric requirement of compounds **2**, the Si-Si-Si bond angles are very large (e.g. $(tBu_3Si)_2SiX_2$: 142.67(1)° (X = H), 142.7(3)° (F), 142.41(3)° (Cl), 141.53(3)° (Br)).

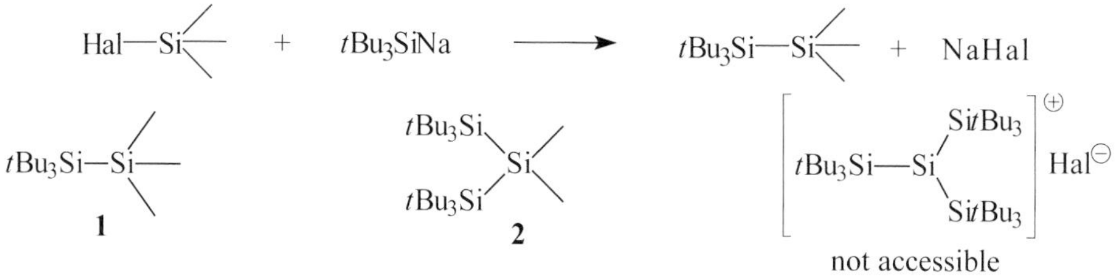

$$\text{Hal—Si} \lt + tBu_3SiNa \longrightarrow tBu_3Si\text{—Si} \lt + \text{NaHal}$$

tBu_3Si—Si (**1**) $(tBu_3Si)_2Si$ (**2**) $[tBu_3Si\text{—}Si(SitBu_3)_2]^{\oplus}\,Hal^{\ominus}$ not accessible

Scheme 1. Syntheses of mono- and disupersilylated silanes.

---

[a] X-ray structure analyses

Table 1. Synthesized compounds of type **1**.

R, R, R for $tBu_3Si–SiR_3$ (1)				
H, H, H	I, I, I	H, H, OTf	Me, H, H	Ph, H, H
F, F, F	OTf, OTf, OTf	H, OTf, OTf	Me, H, Cl	Ph, H, Cl
Cl, Cl, Cl	H, H, Cl	H, Br, Br	Me, Br, Br	Ph, Br, Br
Br, Br, Br	H, H, Br	Cl, Br, Br	Me, Br, Cl	Ph, Br, Cl

Table 2. Synthesized compounds of type **2**.

R, R for $(tBu_3Si)_2SiR_2$ (2)			
H, H	I, I	H, Br	Me, F
F, F	OTf, OTf	H, I	Me, Br
Cl, Cl	H, F	H, OTf	Me, Cl
Br, Br	H, Cl	Me, H	Me, OTf

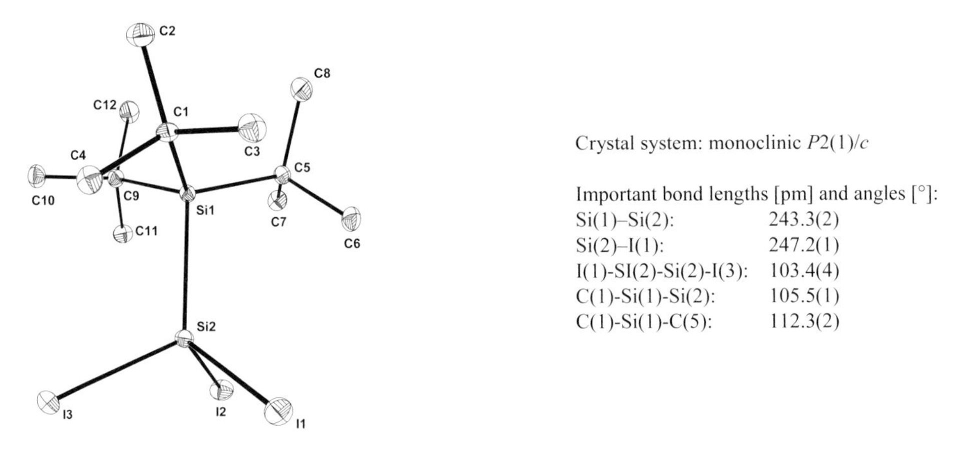

Fig. 1. Structure of tBu_3SiSiI_3 in the crystal. Structure: H. Nöth, J. Knizek.

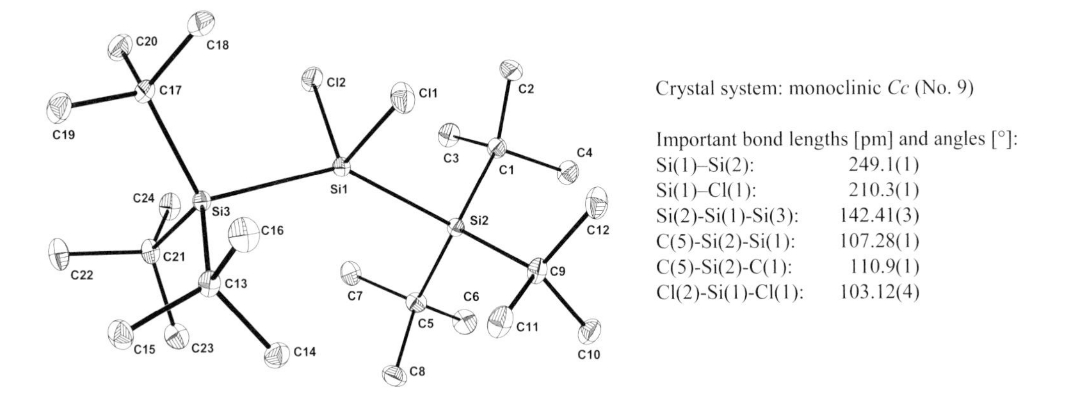

Fig. 2. Structure of $(tBu_3Si)_2SiCl_2$ in the crystal. Structure: D. Fenske, G. Baum.

Reactions of disupersilylated silanes $(tBu_3Si)_2SiXY$ with halogens lead to cleavage of an Si–SitBu_3 bond. (Eq. 2).

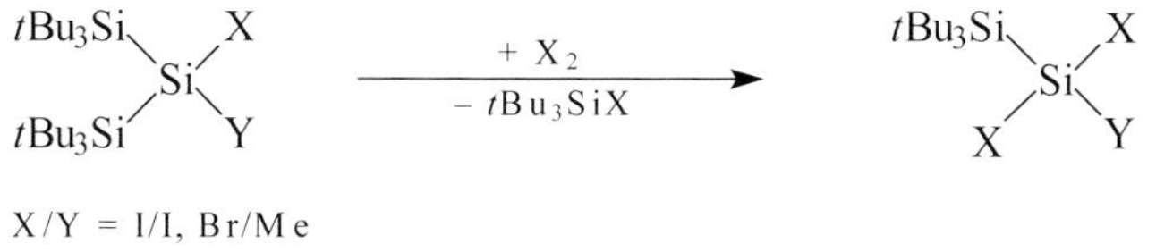

$$(tBu_3Si)_2SiXY \xrightarrow[-\ tBu_3SiX]{+\ X_2} (tBu_3Si)SiX_2Y$$

X/Y = I/I, Br/Me

Eq. 1.

The silanes $(tBu_3Si)_2SiHBr$ and $(tBu_3Si)_2SiMeBr$ react according to Eq. 2 with sodium at room temperature to give the yellow silanides $(tBu_3Si)_2SiHNa$ ($\delta\ ^{29}Si = -174.18$, $^{1}J(Si, H) = 99.7$ Hz) and $(tBu_3Si)_2SiMeNa$ ($\delta\ ^{29}Si = -37.42$).

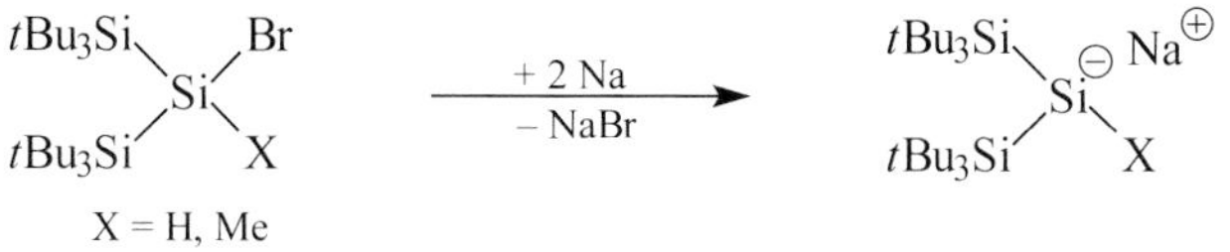

$$(tBu_3Si)_2SiXBr \xrightarrow[-\ NaBr]{+\ 2\ Na} (tBu_3Si)_2SiX^{\ominus}\ Na^{\oplus}$$

X = H, Me

Eq. 2.

Metalation of $(tBu_3Si)_2SiBr_2$ with stoichiometric amounts of lithium naphthalenide at low temperature leads to $(tBu_3Si)_2SiBrLi$ (Scheme 2). Trapping the silanide with methanol gives the silane $(tBu_3Si)_2SiHBr$ which – on the other hand – cannot be obtained by bromination of $(tBu_3Si)_2SiH_2$.

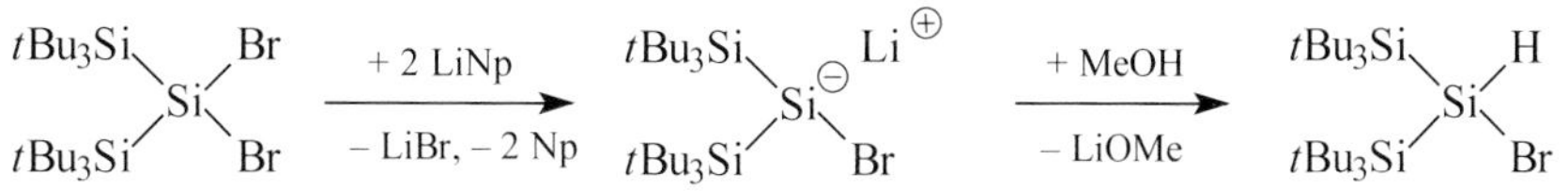

$$(tBu_3Si)_2SiBr_2 \xrightarrow[-\ LiBr,\ -\ 2\ Np]{+\ 2\ LiNp} (tBu_3Si)_2SiBr^{\ominus}\ Li^{\oplus} \xrightarrow[-\ LiOMe]{+\ MeOH} (tBu_3Si)_2SiHBr$$

Scheme 2.

Mono- and Disupersilylated Silylenes

By warming solutions of $(tBu_3Si)_2SiBrLi$ to room temperature, the silanide — as in Scheme 3 — eliminates LiBr and the silylene formed thereby stabilizes itself by insertion of the divalent silicon into a C–H bond of a tBu-group [1].

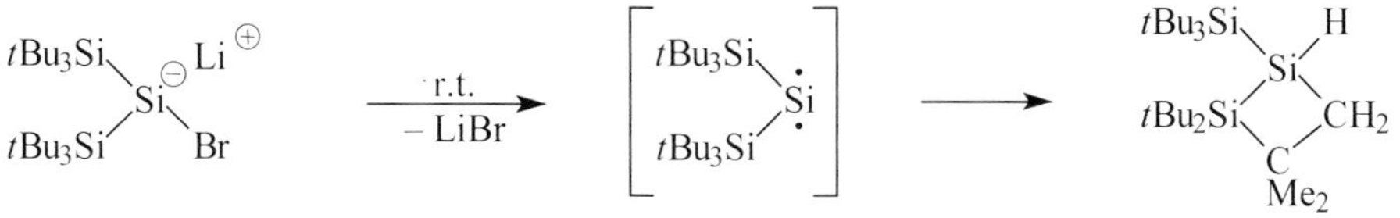

$$(tBu_3Si)_2SiBr^{\ominus}\ Li^{\oplus} \xrightarrow[-\ LiBr]{r.t.} \left[(tBu_3Si)_2Si:\right] \longrightarrow tBu_3Si(H)Si\text{–}CH_2\text{–}CMe_2\text{–}SitBu_2 \text{ (ring)}$$

Scheme 3.

The silanes $(tBu_3Si)_2SiH_2$, $(tBu_3Si)_2SiMeH$ and $(tBu_3Si)_2SiPhH$ decompose at 180°C, 160°C and >–78°C, respectively. They eliminate tBu_3SiH and obviously form silylenes (singlet state?) which easily can be trapped by Et_3SiH with formation of $(tBu_3Si)(Et_3Si)SiH_2$, $(tBu_3Si)(Et_3Si)SiMeH$ and $(tBu_3Si)(Et_3Si)SiPhH$ (Scheme 4). Interestingly, the silylene $(tBu_3Si)_2Si$ (triplet state [2]) cannot be trapped by Et_3SiH.

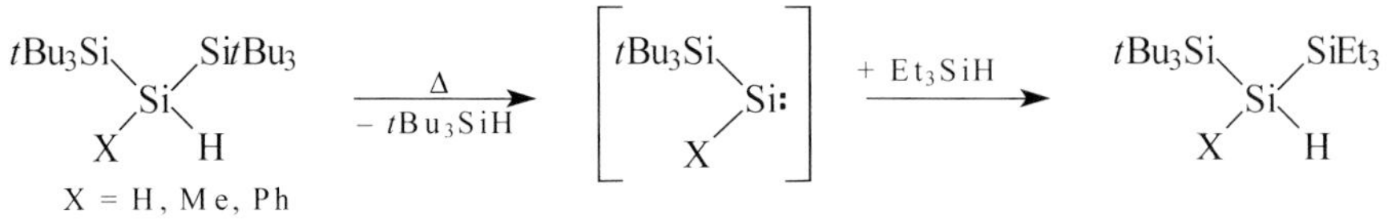

$$(tBu_3Si)_2SiXH \xrightarrow[-\,tBu_3SiH]{\Delta} [(tBu_3Si)XSi:] \xrightarrow{+\,Et_3SiH} (tBu_3Si)(Et_3Si)SiXH$$

X = H, Me, Ph

Scheme 4.

A Supersilylated Disilene

According to Scheme 5 the reaction of $tBu_3SiPhSiHCl$ with half an equivalent of tBu_3SiNa (–78°C) leads to the silylene $tBu_3SiPhSi$: (cf. Scheme 4), which inserts into the Si–H bond of its own precursor with formation of 2 diastereomers of $tBu_3SiPhSiH–SiClPhSitBu_3$. They transform by way of bromination with bromine and dehalogenation with tBu_3SiNa into the disilene $tBu_3SiPhSi=SiPhSitBu_3$.

$$tBu_3SiPhSiClH + HSiPhCl(SitBu_3) \xrightarrow[-\,NaCl,\ -\,tBu_3SiH]{+\,tBu_3SiNa} [tBu_3SiPhSi:] + HSiPhCl(SitBu_3) \longrightarrow$$

$$tBu_3SiPhSiH–SiClPhSitBu_3 \xrightarrow[-\,HBr]{+\,Br_2} tBu_3SiPhSiBr–SiClPhSitBu_3 \xrightarrow[-\,tBu_3SiBr]{+\,tBu_3SiNa} tBu_3SiPhSiNa–SiClPhSitBu_3 \xrightarrow{-\,NaCl} tBu_3SiPhSi=SiPhSitBu_3$$

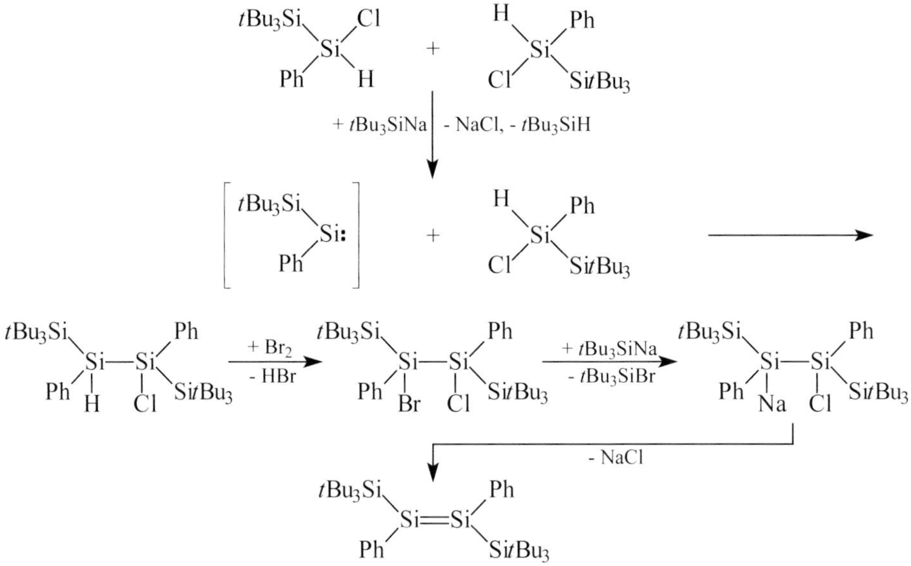

Scheme 5.

The disilene $tBu_3SiPhSi=SiPhSitBu_3$ forms yellow, air-sensitive crystals. The molecule has a planar skeleton >Si=Si<, and the phenyl substituents are nearly orthogonal to the plane mentioned [1, 3].

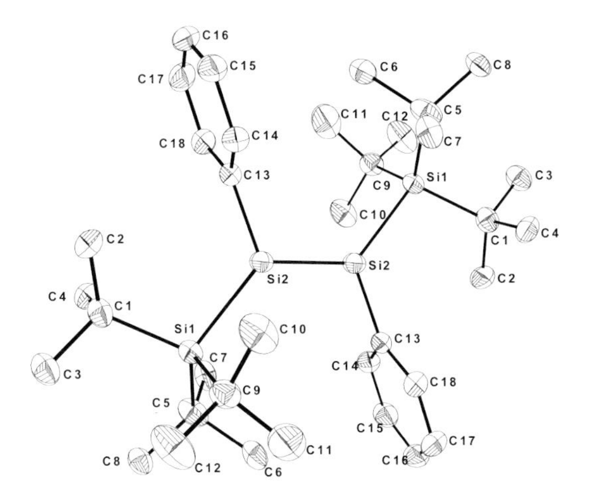

Crystal system: monoclinic, P21/n

Important bond lengths [pm] and angles [°]:

Si(2)-Si(2A):	218.2(2)
Si(1)-Si(2):	240.5(1)
Si(2A)-Si(2)-Si(1):	131.82(6)
Si(2A)-Si(2)-C(13):	113.8(1)
Si(1)-Si(2)-C(13):	113.9(1)
C(1)-Si(1)-C(5):	111.0(2)

Fig. 3. Structure of $t\mathrm{Bu_3SiPhSi{=}SiPhSi}t\mathrm{Bu_3}$ in the crystal. Structure: K. Polborn.

Acknowledgment: The authors wish to thank the *Deutsche Forschungsgemeinschaft* for financial support and the *Wacker AG* for a kind donation of chemicals.

References:

[1] Compare: N. Wiberg, *Coord. Chem. Rev.* **1997**, *163*, 2 17.

[2] Cf.: M. C. Holthausen, W. Koch, Y. Apeloig, *J. Am. Chem. Soc.* **1999**, *121*, 2623.

[3] R. Okazaki, R. West, *Adv. Organomet. Chem.* **1996**, *39*, 231.

A New Conformational Polymorph of Solid Tetramesityldisilene $Mes_2Si{=}SiMes_2$, found by Raman, UV–Vis and Fluorescence Spectroscopy

Larissa A. Leites*, Sergey S. Bukalov

Institute of Organoelement Compounds
Scientific and Technical Center on Raman Spectroscopy
Russian Academy of Sciences
Vavilova ul. 28, Moscow, 117813 Russia
Fax: Int. code + (095)1355085
E-mail: buklei@ineos.ac.ru

Robert West, John E. Mangette and Thomas A. Schmedake

Department of Chemistry, University of Wisconsin
Madison, WI 53706, USA

Keywords: Disilene / Conformational Polymorphism / Raman / UV-Vis

Summary: Three modifications of tetramesityldisilene (**1**) reported to date (orange unsolvated **1a** and two yellow 1:1 solvates with toluene **1b** and THF **1c**) have been found to readily transform to a new modification, yellow unsolvated **1d**. According to the spectral data, the new form **1d** is also a disilene but differs from **1a**–**c** in both crystal and molecular structure. **1d** appears to be the most stable conformational polymorph of **1** but can be converted to the orange **1a** upon illumination in the region 514–457 nm. A quasi-*trans* conformation with two aromatic rings nearly orthogonal and the other two nearly coplanar to the double-bond plane is tentatively proposed for **1d**.

Introduction

Tetramesityldisilene (**1**) was the first Si=Si doubly bonded compound to be synthesized [1] and is undoubtedly the best studied disilene [2, 3]. However, its optical properties, in particular vibrational and electronic spectra, were not studied in detail. We have investigated Raman, pre-resonance Raman, IR, UV–Vis and fluorescence spectra of solid **1**, which exists as several modifications. Three different crystalline forms have been reported: an unsolvated modification **1a**, orange at room temperature, obtained from solution in hexane [4], a yellow, crystalline 1:1 toluene solvate, ($Mes_4Si_2{\cdot}C_7H_8$), **1b** [5], and a similar solvate with one molecule of THF, (Mes_4Si_2.THF), **1c** [6]. The X-ray data of **1a**–**c** [4–6] show that not only crystal but also molecular structures of **1a**–**c** differ in several respects (see Fig. 1), especially in the aromatic ring orientations. Initially our aim was to obtain and to compare vibrational and electronic spectra of the reported modifications **1a**–**c**. However, in the course of experiments we discovered the fourth modification of **1** — yellow unsolvated **1d**. Thus, we had to solve one more problem: to determine the spectra and structure of this newly found form **1d** and to study interconversions of **1a**–**d**.

Results

Interconversions of 1a–d (Fig. 1)

While experimenting with the orange crystals **1a** we found that, when heated or sublimed or ground in a mortar, they readily transform to a yellow powder **1d**, whose UV band at ca. 425 nm confirms it is also a disilene [1–3, 7]. All experiments were carried out in an inert atmosphere or in a high vacuum to prevent sample oxidation, which was checked by the absence in the spectra of the bands corresponding to products of degradation. The same substance **1d** was obtained when solvates **1b** or **1c** were exposed to a high vacuum and heated to remove the solvent. The identity of **1d**, obtained from unsolvated **1a** and from the solvates **1b–c**, was proved by the identity of their Raman and UV–Vis spectra.

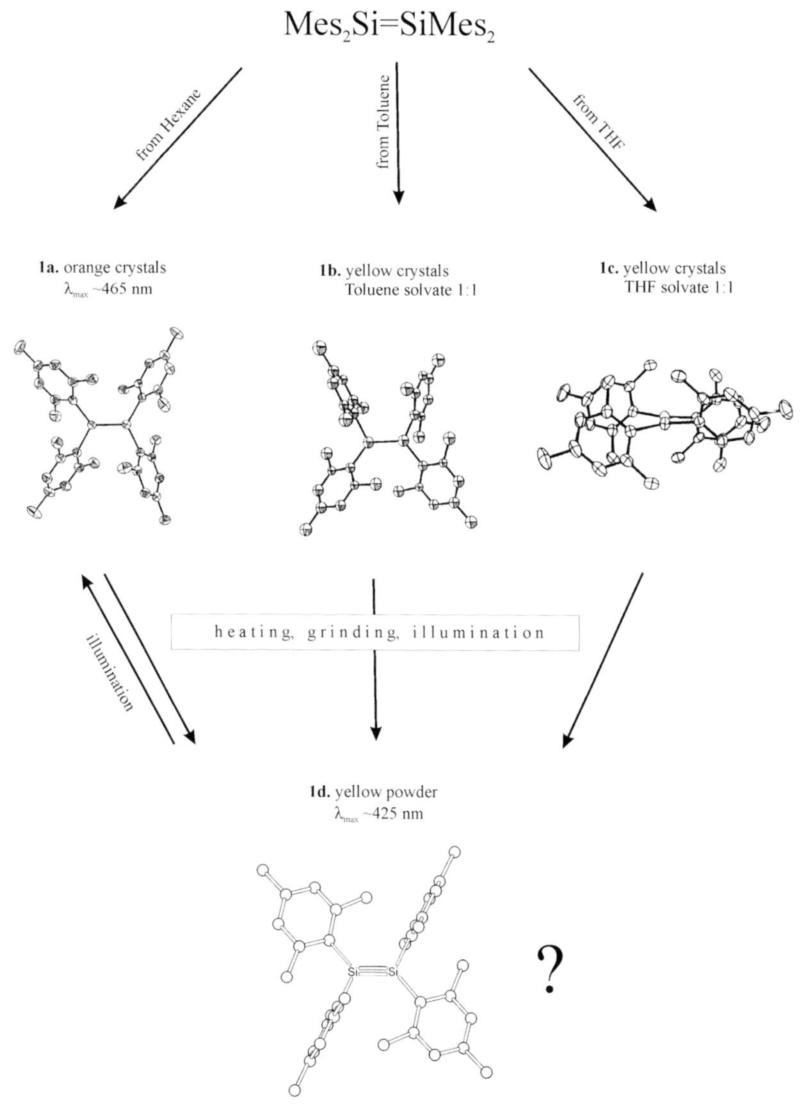

Fig. 1. Interconversions of solid **1**.

Unsolvated forms **1a** and **1d** were found to interconvert upon laser irradiation, if the laser light density overcame a certain threshold. Moreover, the yellow solvates **1b** and **1c** can also, sometimes, be converted to orange **1a** similarly. For instance, long exposure to the 514.5 nm laser beam produced an orange spot in the yellow sample of **1c** sealed in a capillary. The spot persisted when the capillary was removed from light. When exposed to light of the same wavelength but of higher intensity, the crystals of **1b**, sealed in vacuo in a quartz cell, on one occasion suddenly emitted a yellow "cloud", a solid part of which settled on the upper wall of the cell, well separated from the initial crystals. According to the Raman spectrum, this solid film appeared to be a mixture of **1a** and **1d** with **1a** predominant.

Vibrational Spectra of Solid 1a–d

Previous reports on the ν(Si=Si) Raman line and the IR spectra of **1** were reviewed in a recently published paper [8], presenting the Raman and IR spectra of three disilenes including **1a,** as well as normal coordinate analysis (NCA) for several model disilenes. The NCA results show that there is no normal mode in the spectra of disilenes which is well localized in the Si=Si bond. The Si=Si and Si–C stretching coordinates of a $C_2Si{=}SiC_2$ moiety are heavily mixed. Their in-phase combination results in a normal mode with frequency in the range 460–550 cm^{-1} (ν_1) while their out-of-phase combination gives a normal mode near 700 cm^{-1} (ν_2). Particular contributions of the Si=Si and Si–C stretching coordinates to the eigenvectors of ν_1 and ν_2 depend on the particular molecular structure but both are always significant.

For mesityl substituted disilenes, the results of NCA [8] predict that the ν_1 mode, still being an in-phase combination of the Si=Si and Si-C stretching coordinates, also becomes mixed with mesityl group angle deformations. The degree of mixing (the eigenvector elements), but not the ν_1 mode frequency, depends strongly on the mutual orientation of the disilene and mesityl moieties. The more coplanar these fragments are, the more coupled are their vibrations. Therefore mesityl-containing disilenes with different ring orientations are expected to exhibit Raman bands, corresponding to ν_1, with similar frequencies but with different intensities, the intensity being determined by the mode eigenvector. NCA results also show that two of the symmetric Raman active vibrations localized in the mesityl group (the ring breathing and deformation modes, ν_3 and ν_4, respectively) should fall in the same frequency range as ν_1, 500–550 cm^{-1}.

The Raman spectra of **1a-d** (excited by the 514.5 nm line) in the diagnostic region 450-750 cm^{-1} are presented in Fig. 2 and agree well with the NCA predictions [8]. All of the spectra exhibit a triplet in the region 500–550 cm^{-1}, which is a result of an overlap of three lines, corresponding to ν_1, ν_3 and ν_4. An intense line at about 680 cm^{-1} evidently corresponds to ν_2. It is notable that the latter line is absent in the Raman spectrum of $Mes_2Si(O)_2SiMes_2$, the main oxidation product of **1,** indicating that the contribution of the Si=Si stretch to the ν_2 mode for this compound is very important.

However, the precise frequency and intensity values in the Raman spectra of **1a–d** differ slightly but distinctly. Figure 2 demonstrates intensity redistribution in the triplet 500–550 cm^{-1}, as well as different frequency values of ν_2. Also notable are different intensity ratios of the ν_2 line and the neighboring line at 720 cm^{-1}. (The latter value, as well as the intensity of the line at 540 cm^{-1}, is the greatest for **1a,** indicating that the spectrum of **1a** is most enhanced due to pre-resonance.) These differences are in accord with different molecular structures of **1a–d**, allowing identification of each modification by its Raman spectrum.

The reason for the intensity redistributions is not only kinematic (different ring orientations in the molecules of **1a–d** leading to a change in corresponding eigenvector elements) but also electronic. The change in the aromatic ring arrangements affects the degree of π–π conjugation, which in its turn affects the Raman intensity of the ν_1, ν_2 and, possibly, ν_3, since the Raman intensity of appropriate lines is known to be very sensitive to conjugation [9].

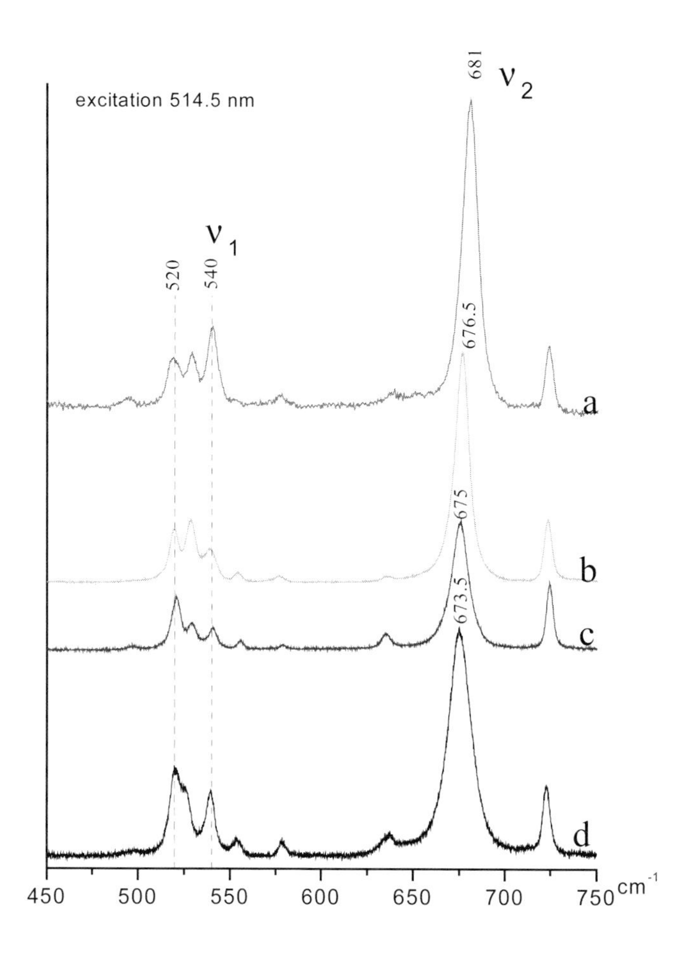

Fig. 2. Raman spectra of **1a–d** in the region of ν_1-ν_4.

There are also differences between the four forms in the low-frequency region of the Raman spectrum where the crystal lattice modes are situated (Fig. 3). These differences agree well with the X-ray data [4–6], which point to different crystal structures of **1a–c**. However, Fig. 3 shows that the crystal structure of **1d** (which is still not characterized by X-ray data) also differs from those of **1a–c**. Thus, the Raman data allow the conclusion that **1d** is really a new modification of **1,** whose crystal and molecular structure is distinct from those of **1a–c**.

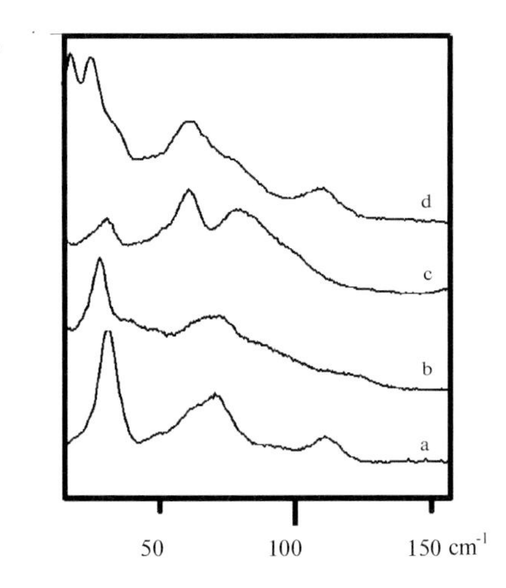

Fig. 3. Raman spectra of **1a–d** in the region of lattice vibrations.

The dependence of the Raman spectra of **1a–d** on the exciting wavelength was determined, using laser lines at 457.9, 514.5, 632.8, 647.1 and 1064 nm. The results obtained for **1a–d** are similar and show evident enhancement of the lines corresponding to ν_1 (540 cm^{-1}) and ν_2 (ca. 680 cm^{-1}) on approaching resonance. This is one more strong piece of evidence of the complex origin of ν_2 with a substantial Si=Si stretch contribution.

Electronic Absorption and Fluorescence Spectra

The UV–Vis absorption spectrum of **1** was investigated in 3-Me-pentane at 77 K and in hexane solutions and was reported to contain a lowest energy band at ca. 420 nm, assigned to the first π–π^* transition localized in a Si=Si chromophore [1–3, 7]. We obtained the band with a similar λ_{max} at ca. 425 nm for the solid sample of yellow unsolvated form **1d** prepared as a thin film deposited on a cold target of the cryostat in vacuo or as a nujol mull (Fig. 4). Our first attempt to obtain the UV–Vis spectrum of the solid **1a,** by subliming the orange crystals in vacuo onto a cold window of the cryostat, resulted in an excellent spectrum, but this was surprisingly that of **1d**. However, different colors of **1a** and **1d** suggested that they should differ also in the electronic absorption. Indeed, the UV–Vis spectrum of real **1a** exhibited a band at ca. 465 nm which is ca. 40 nm red-shifted compared to that of **1d** (Fig. 4). To obtain this result, it appeared necessary to prepare the sample without exposing the compound **1a** to any mechanical or thermal stress. The best way was to put small crystals in nujol, apiezon or silicone grease in a dry box and just to slightly press this suspension between the quartz windows almost without grinding. Otherwise spectra of mixtures of **1a** and **1d** in various proportions are obtained, the corresponding examples being given in Fig. 4.

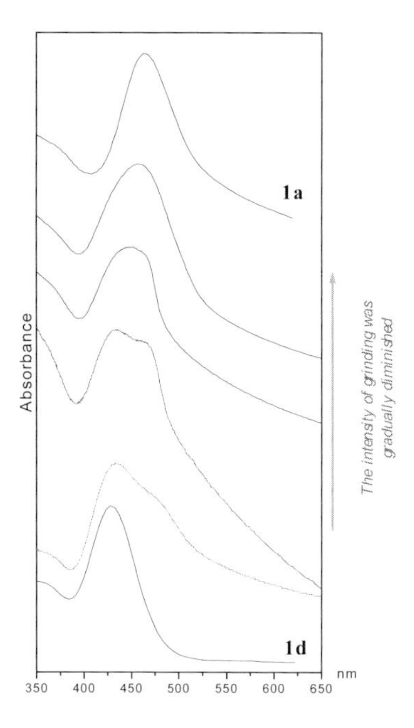

Fig. 4. UV–Vis absorption spectra of solid **1a** and **1d** (nujol or apiezon mulls) and also of their mixtures obtained as a result of grinding of **1a** in a mortar.

The fluorescence emission and excitation spectra for a solution of **1** were reported in [7]. We recorded the fluorescence emission spectra excited by the 457.9 nm laser line for both **1a** and **1d** as solids (Fig. 5). The λ_{max} values of these broad bands differ significantly (ca. 560 and 515 nm, respectively) in good accord with the position of electronic absorption bands. A large Stokes shift points to a substantial geometry change upon excitation. Under illumination, **1d** slowly transformed into **1a** (see Fig. 5, lower curves).

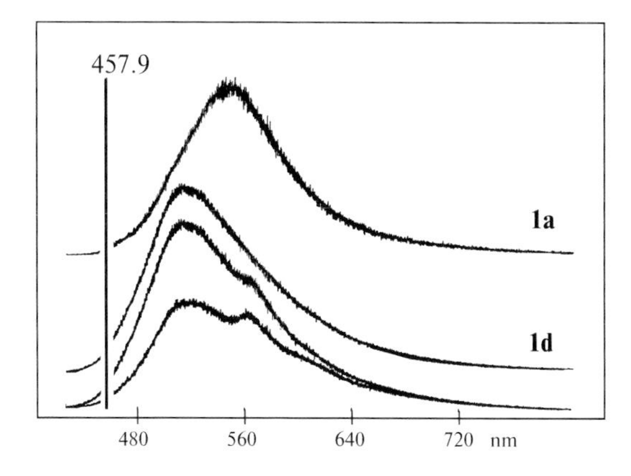

Fig. 5. Fluorescence emission spectra of solid **1a** and **1d**. Two lower curves demonstrate gradual conversion of **1d** into **1a** upon laser illumination.

Discussion

Thus, solid **1** exists as at least four forms differing in both crystalline and molecular structure: two yellow crystalline 1:1 solvates with toluene and THF (**1b** and **1c**), and two solvent-free forms: orange crystals (**1a**) and a yellow powder (**1d**). Bernstein [10] uses a term "conformational polymorphism" for this phenomenon, that is, existence of several forms of a conformationally flexible molecule (with energy difference between the conformers less then 2 kcal/mol) depending on crystallization conditions. Our experience with a step-by-step serendipitous finding of new conformational polymorphs of **1** ([1, 4–6] and this paper) closely resembles the classical story about analogous forms of dimethylbenzylideneaniline, described vividly in [10].

The existence of conformational polymorphs of **1** is evidently a result of interplay between the steric hindrance of the mesityl groups, the tendency of the whole molecule towards planarity which would maximize π–π conjugation, and crystal forces favoring close packing.

The facts that both the solvates as well as the orange form **1a** readily convert to **1d** clearly show that the forms **1a–c** are metastable while the form **1d** is the most stable thermodynamically. **1b–d** can be converted to **1a**, but only upon strong illumination.

1d usually appears as a thin film on a glass surface, and until now we have not succeeded in obtaining a single crystal for the X-ray analysis. However, there are some reasons to speculate about its structure. For disilenes of the type RR'Si=SiR'R, photochemical *cis–trans* isomerization in solution was shown to occur [11], the *trans* isomer being predominant under equilibrium conditions. Of course, symmetrically substituted **1** cannot have real *cis–trans* isomers, but, by analogy, a similar equilibrium with predominance of a conformer, close to the *trans* one, seems likely in solution. This assumption is confirmed by Raman polarization measurements for a solution of **1** in hexane, because the selection rules observed for the conformer predominant in solution are consistent with C_{2h} symmetry, that is, with a quasi-*trans* structure of this conformer [12]. As both the Raman and UV–Vis absorption and fluorescence spectra of solid **1d** are similar to those of **1** in hexane solution [1, 12], we can suggest for **1d** also a quasi-*trans* structure as shown in Fig. 1.

Acknowledgement: The Russian authors acknowledge financial support from the *Russian Foundation for Basic Research* (grant no. 96-03-34079) and from *Dow Corning Corp.* (grant CRDF RC1-602).

References:

[1] R. West, M. J. Fink, J. Michl, *Science* **1981,** *214,* 1343.

[2] R. Okazaki, R. West, *Adv. Organomet. Chem.* **1996,** *39,* 232.

[3] G. Raabe, J. Michl, in: *The Chemistry of Organic Silicon Compounds* (Eds.: S. Patai, Z. Rappoport), Wiley, Chichester, **1989,** ch. 17, p. 1017.

[4] B. D. Shepherd, C. F. Campana, R. West, *Heteroatom Chem.* **1990**, *1,* 1.

[5] M. J. Fink, M. J. Michalczyk, K. J. Haller, R. West, J. Michl, *Organometallics* **1984,** *3,* 793.

[6] M. Wind, D. R. Powell, R. West, *Organometallics* **1996,** *15,* 5772.

[7] R. West, *Pure Appl. Chem.* **1984**, *56*, 163.

[8] L. A. Leites, S. S. Bukalov, 1. A. Garbuzova, R. West, H. Spitzner, *J. Organomet. Chem.* **1997,** *536–537*, 425.

[9] For reviews, see: P. P. Shorygin, *Russ. Chem. Rev.* **1971**, *40,* 367; *Uspekhi Khim.* **1978,** *47,* 1698.

[10] See: J. Bernstein, *J. Phys. D. Appl. Phys.* **1993,** *26,* B66 and refs. cited.

[11] a) M. J. Michalczyk, R. West, J. Michl, M. J. Fink, *J. Am. Chem. Soc.* **1984,** *106,* 821; (b) M. J. Michalczyk, R. West, J. Michl, *Organometallics* **1985,** *4,* 826.

[12] L. A. Leites, S. S. Bukalov, R. West, J. E. Mangette, to be published.

Synthesis and Reactivity of Unsaturated Tin Compounds: $Me_2Sn{=}C(SiMe_3)_2$ and $Me_2Sn{=}N(Si\textit{t}Bu_2Me)$

N. Wiberg, S. Wagner, S.-K. Vasisht

Institut für Anorganische Chemie,
Ludwig-Maximilians-Universität München
Butenandtstraße 5–13 (Haus D), D-81377 München, Germany
Tel. Int code + (89)2180 7458 — Fax. Int. code + (89)2180 7865
E-mail: niw@cup.uni-muenchen.de

Keywords: Stannaethene / Stannanimine / [4+2]-Cycloadditions / Ene reactions / [2+3]-Cycloadditions

Summary: The stannaethene $Me_2Sn{=}C(SiMe_3)_2$ (**1**) is a highly reactive, only short-lived intermediate which can be generated by cycloreversion of the Diels–Alder adduct of $Me_2Sn{=}C(SiMe_3)_2$ with anthracene. Mechanistic aspects of [4+2]-cycloaddition and ene reaction (especially stereoselectivity, regioselectivity and conformational effects) of **1** are studied. With azido-di-*tert*-butylmethylsilane **1** can be trapped as a triazoline. This [2+3]-cycloadduct is a suitable source for the stannanimine $Me_2Sn{=}N(Si\textit{t}Bu_2Me)$ (**2**).

Synthesis of a "Source" for $Me_2Sn{=}C(SiMe_3)_2$ (1)

The labile stannaethene $Me_2Sn{=}C(SiMe_3)_2$ (**1**), which dimerizes at only –100°C, can be regenerated from its Diels–Alder adduct with anthracene (**1**·anthracene) at about 100°C. A suitable synthesis for **1**·anthracene involves the reaction of the bromine compound $Me_2SnBr\text{-}CBr(SiMe_3)_2$ with sodium in the presence of anthracene and benzophenone (activator for sodium) [1].

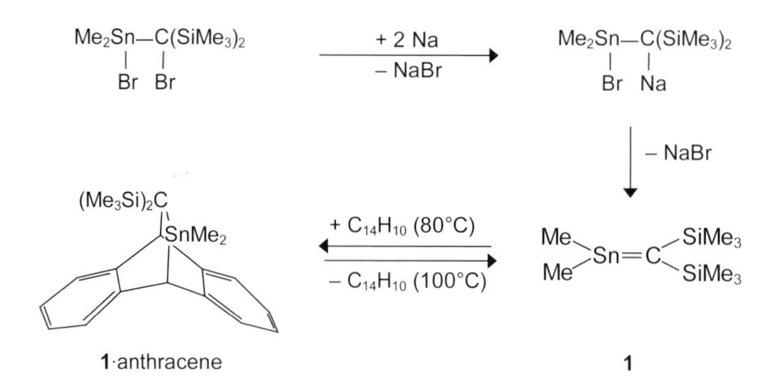

Fig. 1. Synthesis of **1**·anthracene.

Diels–Alder and Ene Reactions of $Me_2Sn{=}C(SiMe_3)_2$ (1)

Thermolysis of **1**·anthracene at 100°C in the presence of butadiene, propene and their organic derivatives leads to [4+2] cycloadducts and ene reaction products.

Diels–Alder as well as ene reactions of **1** proceed stereo- and regioselectively. They are subjected to conformational effects and are accelerated by an electron-donating substituent in the 2-position of the diene or ene (Fig. 2) [2].

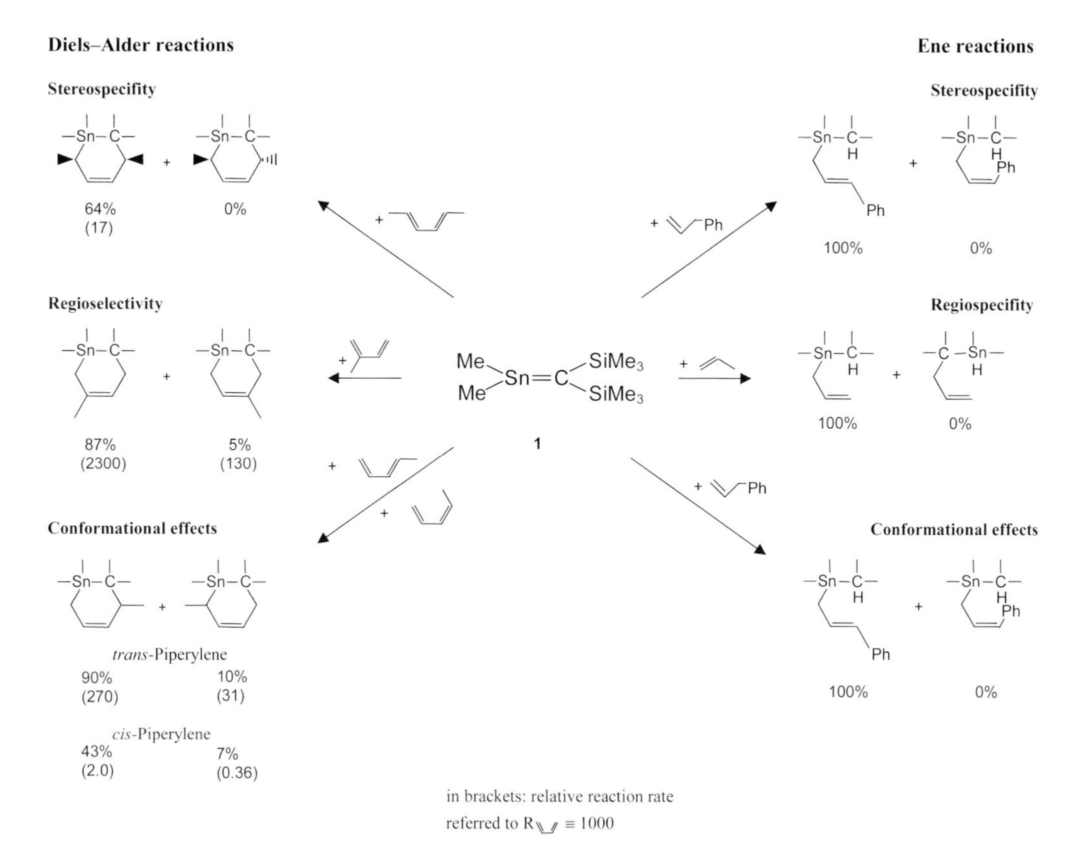

Fig. 2. Diels–Alder and Ene reactions of **1**.

Diels–Alder and ene reactions of **1** occur — like those of silaethene $Me_2Si{=}C(SiMe_3)_2$ [3–9] and germaethene $Me_2Ge{=}C(SiMe_3)_2$[10] — in a concerted way. Thus the unsaturated compounds $Me_2E{=}C(SiMe_3)_2$ (E = Si, Ge, Sn) behave as carbon analogues.

The enophilic behavior of $Me_2E{=}C(SiMe_3)_2$ increases at the cost of dienophilic behavior in the direction germaethene < stannaethene < silaethene [2]. This is concluded from the proportion $R_{butadiene}$ / $R_{propene}$ = $R_{b/p}$ of corrected relative reaction rates R of butadiene and propene with the regarded unsaturated compound: $R_{b/p}$ ($Me_2Si{=}C(SiMe_3)_2$) = 19; $R_{b/p}$ ($Me_2Ge{=}C(SiMe_3)_2$) = 100; $R_{b/p}$ ($Me_2Sn{=}C(SiMe_3)_2$) = 41. That is to say, the reaction of $Me_2E{=}C(SiMe_3)_2$ with a mixture of 3 mol butadiene and 2 mol propene in the case of $Me_2Si{=}C(SiMe_3)_2$ leads to 95 % Diels–Alder and 5 % ene reaction product, in the case of $Me_2Ge{=}C(SiMe_3)_2$ to 99 % Diels–Alder and 1 % ene reaction product, and in case of $Me_2Sn{=}C(SiMe_3)_2$ to 98 % Diels–Alder and 2 % ene reaction product. Over all, [4+2]-cycloaddition is preferred. On the other hand, if ethenes are given a choice between Diels–Alder and ene reaction pathways, they choose exclusively [4+2]-cycloaddition [11, 12]. Thus the carbon analogy decreases in the direction $Me_2Ge{=}C(SiMe_3)_2$ > $Me_2Sn{=}C(SiMe_3)_2$ > $Me_2Si{=}C(SiMe_3)_2$. To explain facts of this case, it is assumed, as for organic Diels–Alder and ene reactions, that the E–C bonding relation in the transition state of the ene reaction is more progressive (stronger) than in the transition state of the Diels–Alder reaction. Therefore the enophilic behavior of $Me_2E{=}C(SiMe_3)_2$ should increase at the cost of the dienophilicity with increasing Lewis acidity of the unsaturated compound [$Me_2Ge{=}C(SiMe_3)_2 < Me_2E{=}C(SiMe_3)_2$ with E = Si, Sn; electronegativities: $EN_{Ge} > EN_{Si} \sim EN_{Sn}$] and with increasing E–C bond strength (SiC > GeC > SnC) [13]. A factor of importance for the dieno- versus enophilic character of the compound $Me_2E{=}C(SiMe_3)_2$ is obviously the bond strength of the new formed E–C bond in the transition state of the Diels–Alder or ene reaction. The strong decrease in bond energy in the direction SiC > SnC is obviously the reason for the experimentally observed path $R_{b/p}(Ge) > R_{b/p}(Sn) > R_{b/p}(Si)$.

Synthesis of "Sources" for $Me_2Sn{=}N(Si\mathit{t}Bu_2Me)$ (2)

The thermolabile stannanimine $Me_2Sn{=}N(Si\mathit{t}Bu_2Me)$ (**2**) can be generated either by heating the [3+2] cycloadduct (**1**·$N_3Si\mathit{t}Bu_2Me$) of stannaethene **1** with $\mathit{t}Bu_2MeSiN_3$ or by heating the [3+2] cycloadduct (**2**·$N_3Si\mathit{t}Bu_2Me$) of stannanimine **2** with $\mathit{t}Bu_2MeSiN_3$ (Fig. 3). A suitable synthesis for **2**·$N_3Si\mathit{t}Bu_2Me$ involves the reaction of the bromine compound $Me_2SnBr–CBr(SiMe_3)_2$ with phenyllithium in the presence of $N_3Si\mathit{t}Bu_2Me$ [14].

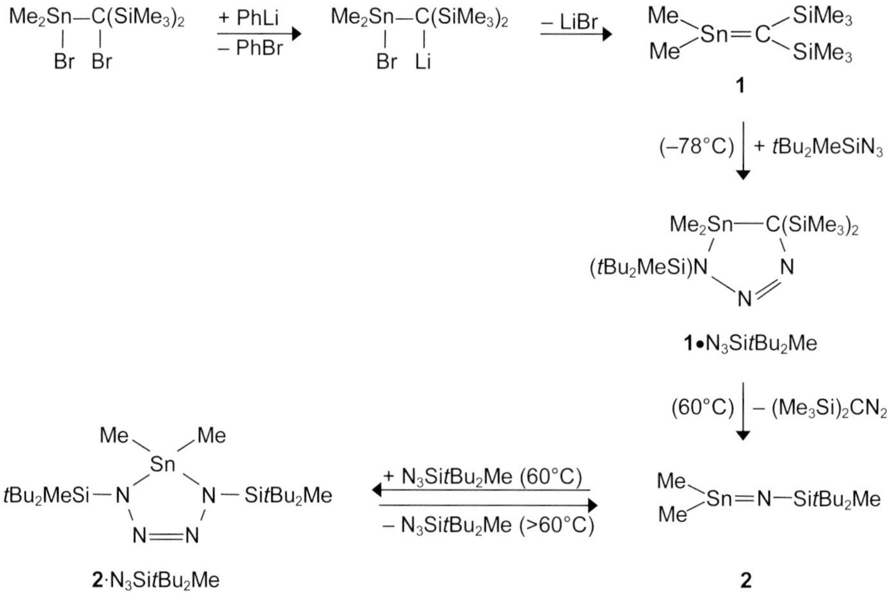

$Me_2Sn(Br)–C(Br)(SiMe_3)_2$ —(+ PhLi, – PhBr)→ $Me_2Sn(Br)–C(Li)(SiMe_3)_2$ —(– LiBr)→ $Me_2Sn{=}C(SiMe_3)_2$ (**1**)

1 —((–78°C) + $\mathit{t}Bu_2MeSiN_3$)→ **1•**$N_3Si\mathit{t}Bu_2Me$ —((60°C) – $(Me_3Si)_2CN_2$)→ $Me_2Sn{=}N{-}Si\mathit{t}Bu_2Me$ (**2**)

2 ⇄ **2**·$N_3Si\mathit{t}Bu_2Me$ (+ $N_3Si\mathit{t}Bu_2Me$ (60°C); – $N_3Si\mathit{t}Bu_2Me$ (>60°C))

Fig. 3. Synthesis of **2**·$N_3Si\mathit{t}Bu_2Me$.

Reactions of $Me_2Sn{=}NSi\textit{t}Bu_2Me$ (2)

Thermolysis of **1**·$N_3Si\textit{t}Bu_2Me$ at 60°C in the presence of 2,3-dimethylbutadiene leads to the ene reaction product of **2** with the ene dimethylbutadiene [14]. In the absence of a trapping reagent only the dimerization product of **2** is observed [14].

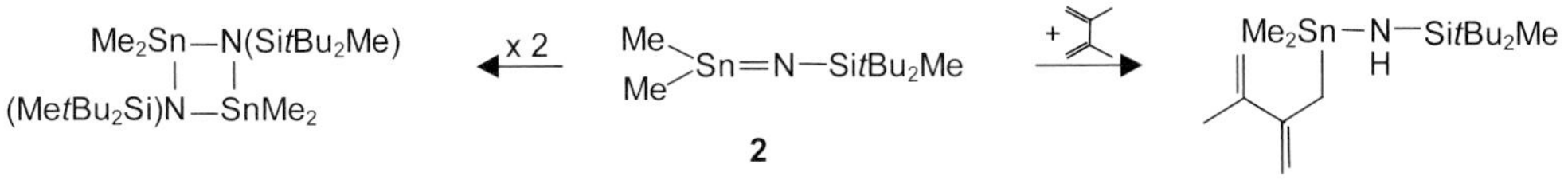

Me_2Sn—N($Sit Bu_2Me$)
($MetBu_2Si$)N—$SnMe_2$
x 2
Me, Me >Sn=N—$SitBu_2Me$
2
Me_2Sn—N(H)—$SitBu_2Me$

Fig. 4. Reactions of **2**.

Acknowledgment: The authors thank the *Deutsche Forschungsgemeinschaft* for financial support.

References:

[1] N. Wiberg, S. Wagner, S.-K. Vasisht, K. Polborn, *J. Can. Chem.* **1999**, in press.
[2] N. Wiberg, S. Wagner, S.-K. Vasisht, *Chem. Eur. J.*, **1998**, *4*, 2571.
[3] N. Wiberg, in: *Organosilicon Chemistry II: From Molecules to Materials* (Eds.: N. Auner, J. Weis), VCH, Weinheim, **1996**, p. 367.
[4] N. Wiberg, G. Preiner, G. Wagner, H. Köpf, *Z. Naturforsch. Teil B* **1987**, *42*, 1055, 1062.
[5] N. Wiberg, K. Schurz, G. Fischer, *Chem. Ber.* **1986**, *119*, 3498.
[6] N. Wiberg, G. Fischer, K. Schurz, *Chem. Ber.* **1987**, *120*, 1605.
[7] N. Wiberg, G. Fischer, S. Wagner, *Chem. Ber.* **1991**, *124*, 769.
[8] N. Wiberg, S. Wagner, G. Fischer, *Chem. Ber.* **1991**, *124*, 1981.
[9] N. Wiberg, S. Wagner, *Z. Naturforsch. Teil B* **1996**, *51*, 629.
[10] N. Wiberg, S. Wagner, *Z. Naturforsch. Teil B* **1996**, *51*, 838.
[11] J. Sauer, R. Sustmann, *Angew. Chem.* **1980**, *92*, 773; *Angew. Chem., Int. Ed. Engl.* **1980**, *19*, 779.
[12] H. M. R. Hoffmann, *Angew. Chem.* **1969**, *81*, 597; *Angew. Chem. Int. Ed. Engl.* **1969**, *8*, 556; W. Oppolzer, V. Snieckus, *Angew. Chem.* **1978**, *90*, 506; *Angew. Chem., Int. Ed. Engl.* **1978**, *17*, 476; G. V. Boyd in *The Chemistry of Double Bonded Functional Groups* (Ed.: S. Patai), John Wiley, London, **1989**, p. 477.
[13] *Gmelin's Handbook of Inorganic Chemistry* (*Organogermanium Compounds*, Part 1), 8th edn., Springer Verlag, Berlin, **1988**, p.26.
[14] N. Wiberg, S.-K. Vasisht, *Angew. Chem.* **1991**, *103*, 105; *Angew. Chem. Int. Ed. Engl.* **1991**, *30*, 93.

The Si_4H_6 Potential Energy Surface

Thomas Müller*

Institut für Anorganische Chemie,
Johann Wolfgang Goethe-Universität Frankfurt
Marie-Curie-Str.11, D-60437 Frankfurt am Main, Germany
Tel: Int. code + (69)798 29166 — Fax: Int. code + (69)798 29188
e-mail: h0443afs@rz.hu-berlin.de

Keywords: Conjugation / Density Functional Calculations / Tetrasilacyclobutene / Tetrasilabutadiene / Trisilacyclopropene

Summary: The Si_4H_6 potential energy surface was explored using density functional methods at the hybrid B3LYP/6-311+G** level. Tetrasila[1.1.0]bicyclobutane **5** was found to be the most stable species. The relative energies of the isomeric compounds are 3.0 (tetrasilacyclobutene, **6**), 10.9 (*trans*-tetrasila[1.1.0]bicyclobutane, **7**), 21.6 (1-silyl-trisilacyclopropene, **8**), 25.7 (2-silyl-trisilacyclopropene, **9**), and 33.3 kcal mol^{-1}, (*s-trans*-tetrasilabutadiene, **10**). The calculations suggest that the thermal isomerization **5** $\rightarrow$ **6** proceeds in one step via a 1,2-H-shift. The alternative multistep reaction via the intermediacy of **10** is energetically by 10.5 kcal mol^{-1} less favored. The conjugation between the Si=Si double bonds in **10** is less pronounced than between the multiple bonds in 1,3-butadiene and 1,4-disila-1,3-butadiene **12**, but stronger than in 2,3-disila-1,3-butadiene **13**.

Introduction

The recent synthesis of tetrasilabutadiene **1** [1] and of tetrasilacyclobutene **2** [2, 3] in the groups of Weidenbruch and Kira, respectively, has renewed the interest in Si_4H_6 isomers, their properties and their interconversions. The bicyclic compound **3** [4] has attracted considerable interest in the past, both theoretically and experimentally, due to the possibility of "bond-stretching" isomers [5,6]. More recently the groups of Kira [7] and Sekiguchi [8] were also succesful in the synthesis of silyl-substituted trisilacyclopropenes **4**.

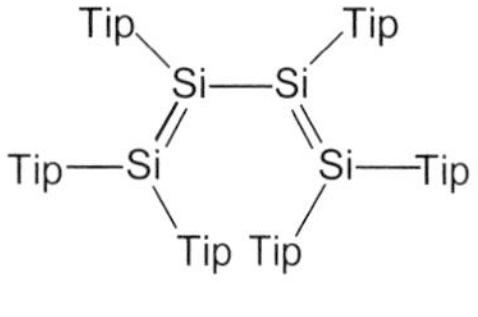

1
Tip = 2,4,6-tri-isopropylphenyl

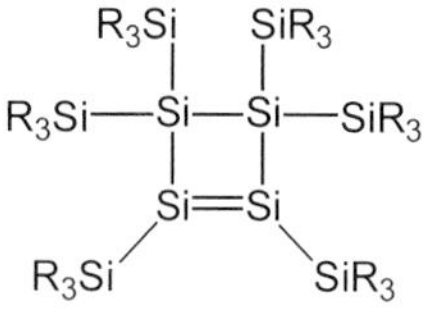

2
SiR_3 = tBuMe$_2$Si

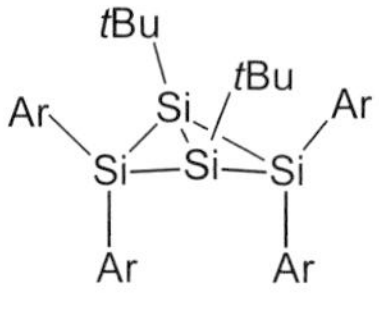

3
Ar = 2,6-diethyl-phenyl

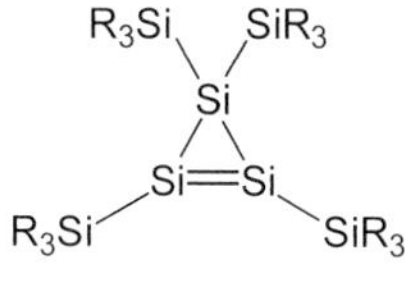

4
SiR_3 = tBu$_2$MeSi
SiR_3 = iPr$_3$Si

This tremendous experimental progress prompted us to study also theoretical the potential energy surface (PES) of Si_4H_6 isomers, their relative energies, properties and monomolecular rearrangements. We applied in this study density functional theory using the hybrid B3LYP/6-311+G**//B3LYP/6-311+G** level [9, 10]. Structures were verified to be minima or saddle points on the PES by subsequent frequency calculations. Unscaled zeropoint energies were added to obtain more reliable relative energies.

Results and Discussion

Relative Energies of Si_4H_6 Isomers.

The most stable Si_4H_6 isomer is the bicyclic structure **5**, however, the energy difference to tetrasila-cyclobutene **6** is small (see Fig. 1).

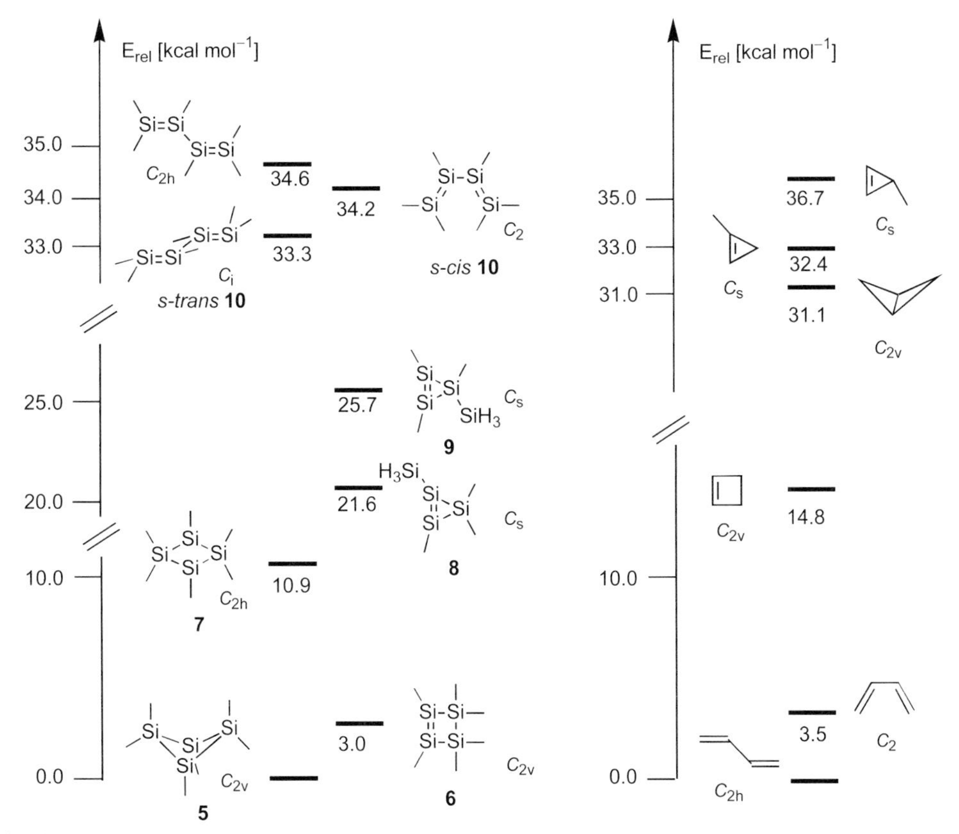

Fig. 1. Relative energies of Si_4H_6 (left) and C_4H_6 isomers (right) at B3LYP/6-311+G**//B3LYP/6-311+G** + ZPVE.

The *trans* isomer **7** with C_{2h} symmetry is by 10.9 kcal mol^{-1} less stable than **5**. The silyl-cyclotrisilenes **8** and **9** are considerably higher in energy than the isomeric four membered ring compound **6** (by 18.6 and 22.7 kcal mol^{-1}, for **8** and **9**, respectively). This is nearly the same energy difference than calculated for 2-methyl cyclopropene and cyclobutene (21.9 kcal mol^{-1}). Tetrasilabutadiene **10** is the least stable compound in this series. Its most stable conformer, the *s-trans* isomer with the molecular point group C_i, is by 30.3 kcal mol^{-1} less stable than **6**. This is in striking contrast to the all-carbon PES, where the *s-trans* 1,3-butadiene is the global minimum and is more stable than cyclobutene by 14.8 kcal mol^{-1}. The most obvious difference between the Si_4H_6 and the C_4H_6 PES is the higher stability of bicyclic and cyclic structures compared to acyclic unsaturated molecules in the silicon case.

Structures of Si_4H_6 Isomers.

Comparison of Figs. 2 and 3 reveals the overall good agreement between the calculated structures for Si_4H_6 and the experimentally observed structures of substituted derivatives, Si_4R_6. Generally, the calculations seems to underestimate the SiSi distances, however all structurally investigated Si_4R_6 derivatives are heavily substituted with bulky groups. The strain created by these groups certainly elongates the SiSi bonds. In addition, the Si=Si and Si–Si bond lengths are increased also by electronic substituent effects of attached silyl groups (i.e. compare the calculated structures of isomers **8** and **9**). The most striking difference between calculation and experiment is found for the tetrasilabicyclobutanes **3** and **5**. The calculation predict for **5** a very long transannular Si–Si distance, while the experimental structure of **3** reveals a transannular distance which is in the range of a normal Si–Si single bond. An isomer of **5** with a short transannular Si–Si bond could not be located at B3LYP/6-311+G**. As already suggested by Schleyer and coworkers [5] the bulky substituents seems to play a decisive role in determing the structure and in particular the short transannular SiSi distance of tetrasilabicyclo[1.1.0]butane **3**.

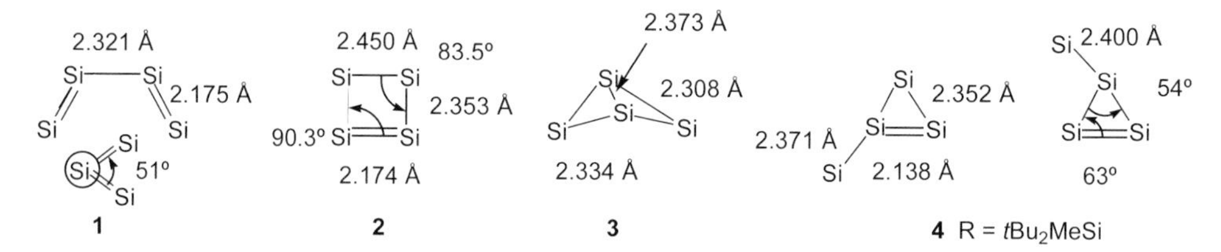

Fig. 2. Experimental geometries of several Si_4R_6 isomers. Only the Si_4-core is shown [1–4, 8].

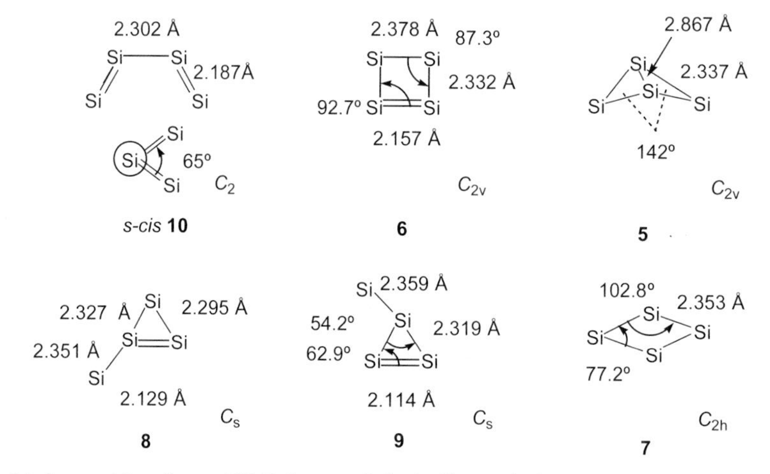

Fig. 3. Calculated geometries of several Si_4H_6 isomers. Only the Si_4-core is shown.(at B3LYP/6-311+G**).

All rotamers of tetrasilabutadiene **10** lie energetically very close. The most stable *s-trans* isomer adopts a trans bent configuration of the Si=Si double bonds, with a pyramidalization angle of 29.9° (=SiH_2 group) and 32.9° (=SiSiH group). However, it is merely by 1.3 kcal mol^{-1} more stable than its planar C_{2h} isomer. Therefore, sterically demanding substituents might chance the energetics in favor of planar Si=Si bonds. Remarkably *s-cis* **10** adopts a strongly twisted structure of C_2 symmetry. The interplanar angle between the two Si=Si bonds is 65°, while for the carbon analogue, *s-cis* butadiene, only 33° are calculated. The twist angle in *s-cis* **10** is even larger than found experimentally for the heavily substituted **1**, suggesting that electronic effects, which are already present in the parent compound, contribute strongly to the observed large twisting in *s-cis* configurated tetrasilabutadienes.

Isomerization Reactions.

Kira and coworkers observed the slow transformation of tetrasilabicyclo[1.1.0]butane **11** into the silacyclobutane **2** at room temperature [2]. Two possible reaction mechanism might be envisaged: (i) a multistep isomerization via the intermediacy of tetrasilabutadiene **10**, or, (ii) the 1,2 shift of the substituent with reorganization of the Si_4–skeleton (see Scheme 1).

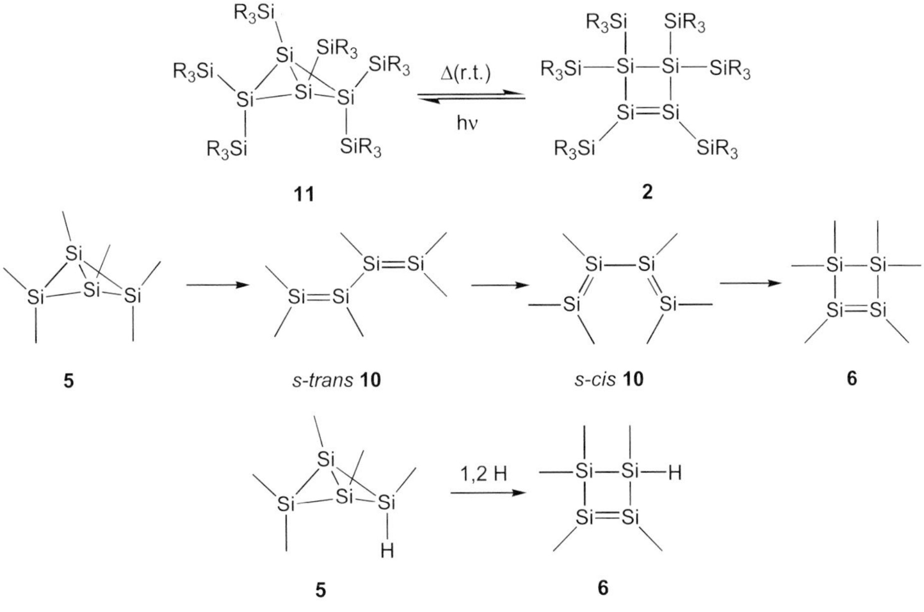

Scheme 1. Isomerization of tetrasilabicyclo[1.1.0]butane **11** to tetrasilacyclobutene **2** and possible mechanisms for the rearrangement.

The model calculations for both suggested reaction pathways are summarized in Fig. 4. In contrast to the experimental results the calculations predict that bicyclobutane **5** is more stable than cyclobutene **6**. Obviously, the silyl substituents in the experiment destabilize **11** compared to **2**. The preferred reaction path for the isomerization **5** → **6** proceeds via a 1,2-H-shift. It involves, however, a significant barrier of 39.8 kcal mol^{-1}. The highest point along the alternative multistep pathway, the transition state for the concerted ring opening of bicyclobutane **5** to form *s-trans* butadiene **10** is

10.5 kcal mol^{-1} higher in energy. This computational result suggests that butadiene **10** is not formed during the isomerization **5** → **6**. This is in qualitative agreement with recent NMR spectroscopic investigations [2], which indicate that the thermal isomerization **11** → **2** occurs via a 1,2-silyl shift. The electrocyclic ringclosure of *s-cis* butadiene **10** is a symmetry allowed, conrotatory process and it involves only a small barrier of 2.6 kcal mol^{-1} (3.0 kcal mol^{-1} at QCISD/6-31G*). Obviously, the bulky Tip substituents stabilize **1** kinetically and prevent the energetically strongly favored cyclisation to the isomeric tetrasilacyclobutene.

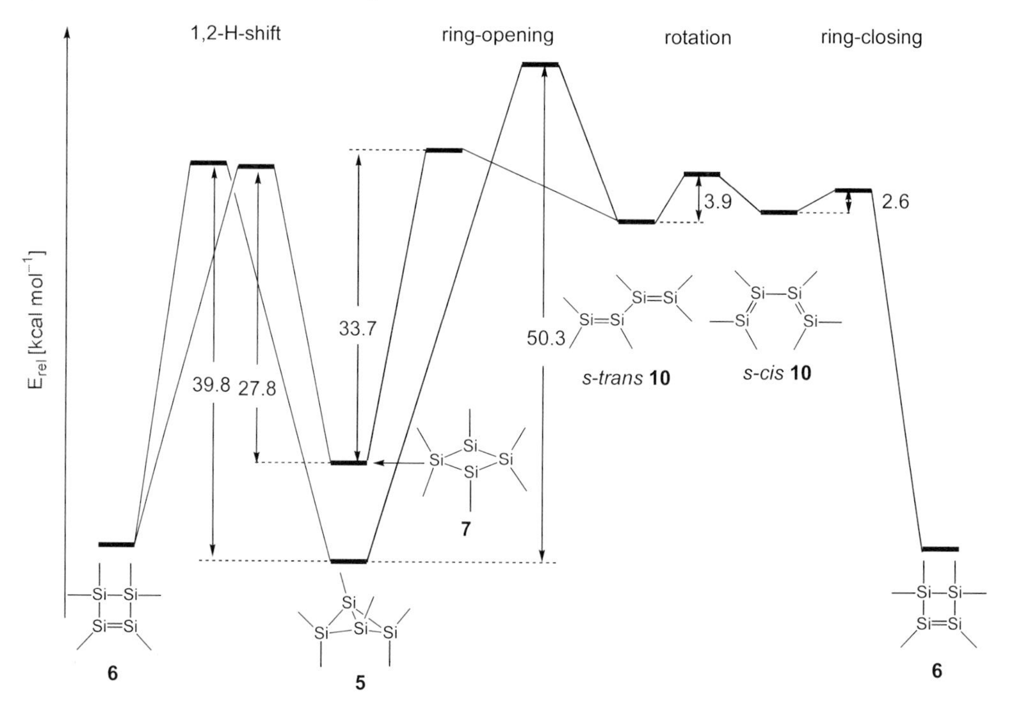

Fig. 4. Calculated reaction channels for the thermal isomerization **5** → **6** (at B3LYP/6-311+G**//B3LYP/6-311+G** + ZPVE).

Conjugation in Tetrasilabutadienes.

Tetrasilabutadiene **1** is the first example of a compound with two Si=Si double bonds, only separated by a Si–Si single bond. The question arises to which extent conjugation between the two Si=Si double bonds in **1** is possible. We applied quantum chemical methodologies to answer this fundamental question. Tetrasilabutadiene **10** was used as model for **1** and was compared with 1,3-butadiene, 1,4-disila-1,3-butadiene **12** and 2,3-disila-1,3-butadiene **13** [11].

The short central Si–Si bond in **1** was put forward as an indication for conjugation between the two Si=Si bonds [1]. Calculations, summarized in Scheme 2, show, that also in **10** the central Si-Si bond is considerably shorter than in disilane. However, the short Si–Si single bond is not a good gauge for conjugation in **10**. An appreciable part of bond shortening can also be found in the TS for the rotation around the central Si–Si in **10**, although there is no conjugation possible between the two Si=Si double bonds. Instead, the observed bond shortening in **1** is mainly due to the sp^2 hybridization of the silicon atoms. Similar results were obtained for 1,3-butadiene and **12**.

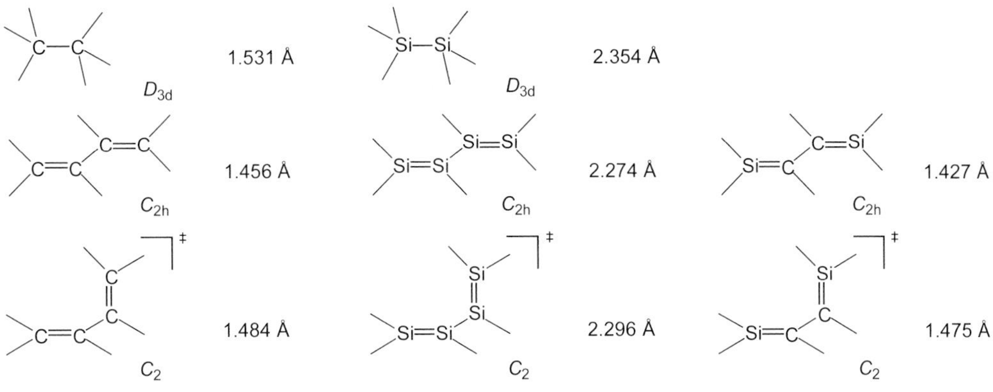

D_{3d} 1.531 Å D_{3d} 2.354 Å

C_{2h} 1.456 Å C_{2h} 2.274 Å C_{2h} 1.427 Å

‡ C_2 1.484 Å ‡ C_2 2.296 Å ‡ C_2 1.475 Å

10 **12**

Scheme 2. Calculated E–E single bond lengths in 1,3-butadiene, **10** and **12** (at B3LYP/6-311+G**).

Conjugation between the Si=Si bonds stabilizes the ground state and raises the barrier for the rotation around the Si–Si bond. The computed barrier in **10** suggests some degree of conjugation in **10,** although it is smaller than in 1,3-butadiene and in **12** (see Scheme 3).

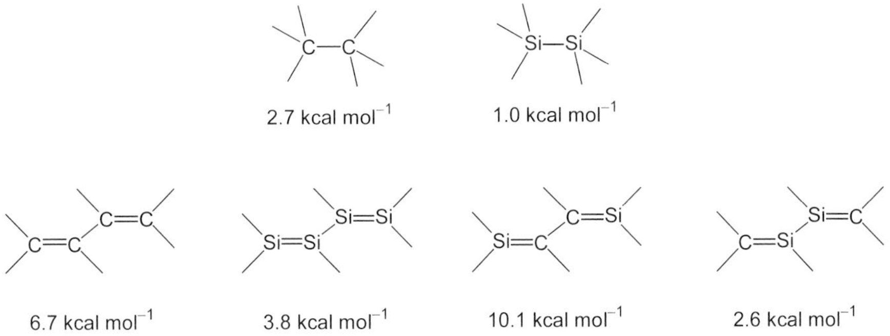

2.7 kcal mol^{-1} 1.0 kcal mol^{-1}

6.7 kcal mol^{-1} 3.8 kcal mol^{-1} 10.1 kcal mol^{-1} 2.6 kcal mol^{-1}

10 **12** **13**

Scheme 3. Calculated rotation barriers around the E-E single bond in 1,3-butadiene, **10**, **12** and **13** (at B3LYP/6-311+G**).

The energetics of isodesmic equations can be used to calculate resonance energies. The computed reaction energies of Eq. 1 are 13.2, 11.0, 13.2, and 6.0 kcal mol^{-1} for 1,3-butadiene, **10**, **12**, and **13**, respectively. This indicates that the Si=Si bonds in **10** are more stable than the isolated Si=Si bond in disilene. The stabilization energy is however smaller than computed for the C=C bonds in 1,3-butadiene and for the Si=C bonds in **12**.

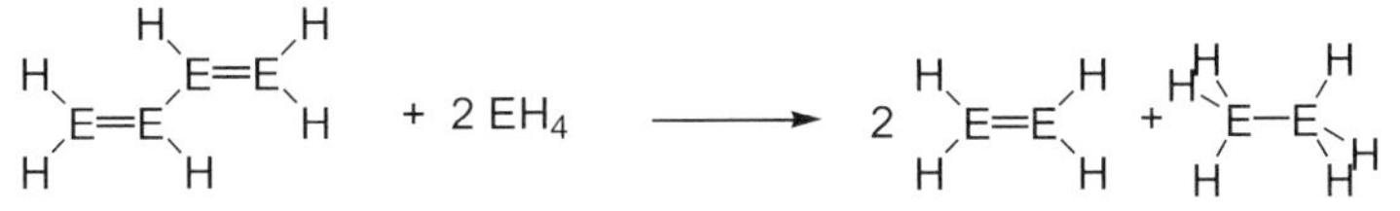

$$H_2E{=}EH{-}EH{=}EH_2 + 2\,EH_4 \longrightarrow 2\,H_2E{=}EH_2 + H_3E{-}EH_3$$

Eq. 1. Isodesmic equation (E = Si, C).

The more balanced, homomolecular homodesmotic Eq. 2 [12] gives similar results, i.e. the resonance energy is in **10** approximatively 80% of that calculated for 1,3–butadiene (5.3 and 4.2 kcal mol^{-1} for 1,3-butadiene and **10**, respectively).

$$H_2E{=}EH{-}EH{=}EH_2 + H_3E{-}EH_2{-}EH_2{-}EH_3 \longrightarrow 2\, H_2E{=}EH{-}EH_2{-}EH_3$$

Eq. 2. Homomolecular homodesmotic equation (E = Si, C).

The analysis of the computational results suggests that there is considerable resonance interaction between the Si=Si bonds in **10**, which is however smaller than between the multiple bonds in 1,3-butadiene and in 1,4-disila-1,3-butadiene **12**.

Acknowledgment: I wish to thank the *Deutschen Forschungsgemeinschaft* for a scholarship and Prof. N. Auner, the *Humboldt Universität* Berlin and the *Johann Wolfgang Goethe Universität* Frankfurt for support. I am grateful to Prof. M. Weidenbruch, *Universität Oldenburg*, for stimulating discussions.

References:

[1] M. Weidenbruch, S. Willms, W. Saak, G. Henkel, *Angew. Chem. Int. Ed.* **1997**, *36*, 2503.

[2] a) M. Kira, T. Iwamoto, C. Kabuto, *J. Am. Chem. Soc* **1996**, *118*, 10303.
b) T. Iwamoto, M. Kira, *Chem. Lett.* **1998**, 277.
c) M. Kira, T. Iwamoto, C. Kabuto *30th Organosilicon Symposium*, London, Ontario, B3, **1997**.

[3] For iodine substituted tetrasilacyclobutenes see: N. Wiberg, H. Auer, H. Nöth, J. Knizek, K. Polborn, *Angew. Chem. Int. Ed.* **1998**, *37*, 2869.

[4] S. Masamune, Y. Kabe, S. Collins, D. J. Williams, R. Jones, *J. Am. Chem. Soc* **1985**, *107*, 5552.

[5] P. v. R. Schleyer, A. F. Sax, J. Kalcher, R. Janoschek, *Angew. Chem. Int. Ed.* **1987**, *26*, 367.

[6] J. A. Boatz, M. S. Gordon, *J. Phys. Chem.* **1988**, *92*, 3037.

[7] M. Kira, C. Kabuto, T. Iwamoto, *J. Am. Chem. Soc*, **1999**, *121*, 886.

[8] M. Ichinohe, T. Matsuno, A. Sekiguchi, *Angew. Chem.* **1999**, *111*, 2331.

[9] All calculations were performed with *Gaussian 94*, *Revisions C.2–E2*, Gaussian, Inc., Pittsburgh (PA), **1995**.

[10] a) A. D. Becke, *Phys. Rev.* **1988**, *A 38*, 3098.
b) A. D. Becke, *J. Chem. Phys.* **1993**, *98*, 5648.
c) B. G. Johnson, P. M. W. Gill, J. A. Pople, *J. Chem. Phys.* **1993**, *98*, 5612.

[11] For a similar treatment of conjugation in digerma-1,3-butadienes see:
a) C. Jouany, G. Trinquier, *Organometallics* **1997**, *16*, 3148.
b) C. Jouany, S. Mathieu, M.-A. Chaubon-Deredempt, G. Trinquier, *J. Am. Chem. Soc.* **1994**, *116*, 3973.

[12] D. B. Chesnut, K. M. Davis, *J. Comp. Chem.*, **1996**, *18*, 584.

Hexaaryltetrasilabuta-1,3-diene: Synthesis and some Reactions

Stefan Willms, Manfred Weidenbruch*

Fachbereich Chemie, Universität Oldenburg
Carl-von-Ossietzky-Straße 9, D-26111 Oldenburg, Germany
Tel.: Int. code + (441)798 3655 — Fax: Int. code + (441)798 3352

Keywords: Disilenes / Tetrasilabutadiene / Oxidation Reactions / Hydrolysis Reactions

Summary: Hexakis(2,4,6-triisopropylphenyl)tetrasilabuta-1,3-diene (**12**), the first molecule containing conjugated Si–Si double bonds, was prepared according to the following reaction sequence. Treatment of tetrakis(2,4,6-triisopropylphenyl)disilene with excess lithium furnished the disilenyllithium compound **10**. In the second step half of **10** reacted with mesityl bromide to give the bromodisilene **11**. Intermolecular LiBr elimination from **10** and **11** afforded the butadiene s-*cis*-**12** in the form of reddish brown crystals. On account of the large 1,4-separation of the silicon atoms, all attempted [4+2] cycloadditions of **12** with several alkenes failed. However, reactions of **12** with water, atmospheric oxygen, and *m*-chloroperbenzoic acid yielded the corresponding hydrolysis and oxidation products including the oxatetrasilacyclopentane ring system **16** analogous to tetrahydrofuran.

Since the isolation of the first molecular compound with an Si–Si double bond in 1981 [1] the chemistry of the disilenes has experienced an almost explosive development, as is reflected in numerous review articles [2]. Addition or cycloaddition reactions to this double bond provide an entry to a series of three- and four-membered ring compounds that are hardly accessible by other routes. Of particular interest among these systems are the disilaoxiranes and disiladioxetanes, both of which open up new questions about the bonding situation in small rings made up of main group elements.

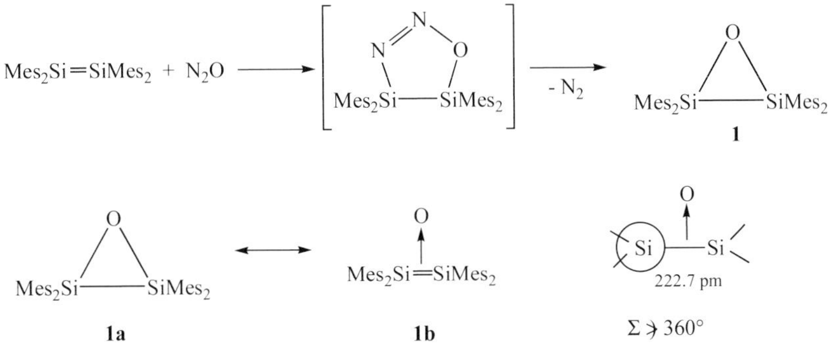

$Mes_2Si{=}SiMes_2$ + N_2O ⟶ [N, N, O, Mes_2Si—$SiMes_2$] ⟶ ($-N_2$) O, Mes_2Si—$SiMes_2$ **1**

O, Mes_2Si—$SiMes_2$ **1a** ⟷ O, $Mes_2Si{=}SiMes_2$ **1b**; O, Si, Si, 222.7 pm, $\Sigma \nrightarrow 360°$

Scheme 1. Tetramesityldisilaoxirane (**1**), easily accessible from tetramesityldisilene and N_2O, exist as a continuum between the classical three-membered ring structure **1a** and that of a π-complex **1b** [3].

For example, tetramesityldisilaoxirane (**1**), easily accessible from tetramesityldisilene and N_2O (Scheme 1), exhibits an Si–Si bond length of 222.7 pm which is about 15 pm shorter than an Si–Si single bond with a comparable substitution pattern and rather approaches the value for an Si–Si double bond. This bond shortening is accompanied by a planar arrangement of the two *ipso*-carbon atoms and the second silicon atom about the silicon atom in question. Both observations allow the assumption that compound **1** and similar molecules exist as a continuum between the classical three-membered ring structure **1a** and that of a π-complex **1b** [3].

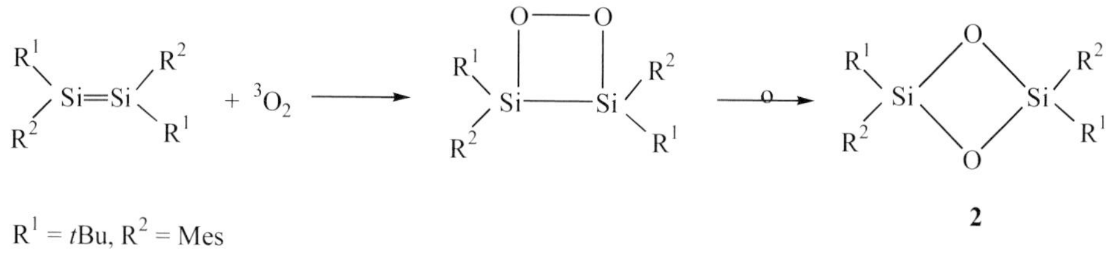

R^1 = *t*Bu, R^2 = Mes

2

Scheme 2. The reaction of disilene with atmospheric $^3O_2$ yields 3,4-disiladioxetane, which rearranges to 2,4-disiladioxetane (**2**) [4].

The addition of atmospheric oxygen to the double bond of disilenes takes an unusual course in which the initial result is a 3,4-disiladioxetane that then undergoes intramolecular rearrangement with retention of the configuration at silicon to furnish the 2,4-disiladioxetane (Scheme 2) [4]. A characteristic feature of compounds of the type **2** is the transannular Si···Si separation of around 240 pm, i.e., in the range of Si–Si single bond lengths [4]. The at-first controversial discussion about possible Si···Si interactions has died down again since recent theoretical calculations and experimental results rather discredit the existence of such Si–Si bonding effects [2]. The unusual bonding situation in, above all, disiliranes such as, for example, compound **1**, prompted us to prepare the corresponding three- and four-membered ring compounds from unsymmetrically substituted disilenes in order not only to determine their molecular structures but also to obtain further information about the bonding situations in the rings from the $^1J(Si,Si)$ coupling constants.

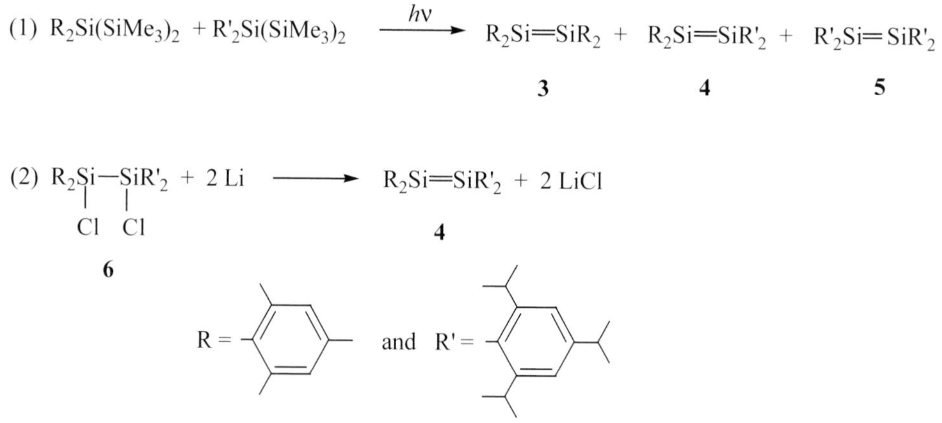

$$(1)\quad R_2Si(SiMe_3)_2 + R'_2Si(SiMe_3)_2 \xrightarrow{h\nu} \underset{\mathbf{3}}{R_2Si{=}SiR_2} + \underset{\mathbf{4}}{R_2Si{=}SiR'_2} + \underset{\mathbf{5}}{R'_2Si{=}SiR'_2}$$

$$(2)\quad \underset{\mathbf{6}}{R_2Si(Cl){-}SiR'_2(Cl)} + 2\,Li \longrightarrow \underset{\mathbf{4}}{R_2Si{=}SiR'_2} + 2\,LiCl$$

R = and R' =

Scheme 3. Syntheses of disilenes with different substitution patterns.

Two routes are available for the synthesis of disilenes with different substitution patterns at the two silicon atoms (Scheme 3): West et al. selected the cophotolysis of two trisilanes with different substitutions at the respective middle silicon atoms which provided the desired disilene **4** smoothly. The disadvantage of this procedure is the simultaneous formation of the symmetrically substituted species **3** and **5** that can hardly be separated from the desired **4** on account of their similar properties [5]. We, on the other hand, have investigated the elimination of halogen from 1,2-dichlorodisilanes which, however, was only successful for compound **6** (→ **4**) (Scheme 3) [6].

	$R_2Si{=}SiR'_2$	R_2Si–SiR'_2 (disilaoxirane)	R_2Si–SiR_2 in:	
d(Si,Si)	215	222.9(1)	248.8(2)	pm
^1J(Si,Si)	159	123	85	Hz

Scheme 4. Reaction of the disilene **4** with atmospheric oxygen or with *m*-chloroperbenzoic acid (*m*-CPBA) yields the 2,4-disiladioxetane **7** or the disilaoxirane **8**, respectively.

The action of atmospheric oxygen on **4** furnished the 2,4-disiladioxetane **7,** which exhibited a short Si···Si separation of 236.7 pm (Scheme 4). However, it was not possible to determine the ^2J(Si,Si) coupling constants. The situation is different with the disilaoxirane **8** which, similar to **1**, has a short Si–Si bond length and angular sums of 360° at each silicon atom. The ^1J(Si,Si) coupling constant of 123 Hz recorded for **8** lies almost exactly between that of the disilene **4** [5, 6] and that of a disilane with a similar substitution pattern [7]. This result provides further support for the assumption that three-membered rings of this sort do indeed have bonds lying between those shown in the limiting structures **8a** and **8b** (Scheme 4).

In the preparation of the disilene **4** from the 1,2-dichlorodisilane **6** and lithium we observed that, in spite of the use of stoichiometric amounts of the two components, the yield of **4** mostly did not exceed 40 % and that half of the dichlorodisilane **6** was always recovered. Close monitoring of the reaction by ^{1}H-NMR spectroscopy showed that at first the dehalogenation of **6** to the disilene **4** occurs on the lithium surface until a maximum of about 40 % of **4** is achieved. Thereafter, the amount of **4** decreases continuously while increasing amounts of 1,3,5-triisopropylbenzene are concomitantly detected in the reaction mixture. Since at the end of the reaction the lithium has been consumed completely, it is reasonable to suggest that, with increasing concentrations of **4** in the reaction mixture, not only does the dehalogenation occur but also a regiospecific cleavage of Si–R' bonds with formation of LiR' takes place and the latter species reacts further with the solvent to form R'H and a disilenyllithium compound (Scheme 5).

$$R_2Si{=}SiR_2 + 2\,Li \longrightarrow R_2Si{=}Si(R)(Li) + LiR\ (\rightarrow RH)$$

9, 10

$$\xrightarrow[-\,LiMes]{+\,MesBr} R_2Si{=}Si(R)(Br) \xrightarrow[-\,LiBr]{+\,\mathbf{10}} R_2Si{=}Si(R){-}Si(R){=}SiR_2$$

11, s-*cis*-12

R = ; Mes =

Scheme 5. Formation of the hexaaryltetrasilabuta-1,3-diene **12**.

In order to confirm this reaction route and at the same time to keep the reaction process as simple as possible, the symmetrically substituted tetraaryldisilene **9** [8] was prepared by the method of West et al. [9] and allowed to react with an excess of lithium. The observed color changes, first to dark green and then to brown–red, suggest the initial transfer of an electron to an aromatic residue [10], followed by extrusion of LiR and formation of the postulated disilenyllithium compound. In a second step of the reaction sequence, **10** was allowed to react with bromomesitylene in anticipation that the poor solubility of the aryllithium compound would favor the halogenation over the competing transarylation. It appears that the bromodisilene **11** is indeed formed and then reacts further with **10** by intermolecular cleavage of lithium bromide to furnish the isolated hexaaryltetrasilabuta-1,3-diene **12** (Scheme 5) [11].

Compound **12** was isolated in a total yield of 60 % in the form of brown–red, highly air-sensitive crystals that are thermally stable up to 267°C. In the solid state and presumably also in solution compound **12** exists in approximately the s-*cis* form with a dihedral angle of 51° between the planes formed by the atoms Si1-Si2-Si2a and Si2-Si2a-Si1a. Similarly to the situation in other systems with conjugated double bonds, the Si–Si double bonds are slightly lengthened to 217.5(2) pm, in consideration of the substituent pattern the central single bond of only 232.1(2) pm is appreciably strained. This result, together with the electronic spectrum, which reveals a bathochromic shift of the band at longest wavelength to 518 nm, i.e., by about 100 nm in comparison with the values for disilenes [2], clearly shows that the conjugation between the two double bonds is present both in the crystal state and in solution [11].

The synthesis of **12** represents an access to a further representative of the compounds with an Si_4R_6 skeleton. The tetrasilabicyclo[1.1.0]butane **13** (Fig. 1) [12] has been known for a considerable time, whereas the tetrasilacyclobutenes **14** were prepared for the first time just recently [13, 14]. Interestingly, ab initio calculations on the parent compound Si_4H_6 show an energy minimum for the compound with the bicyclo[1.1.0]butane skeleton, and that the energy of the cyclobutene derivative of type **14** is only slightly higher. On the other hand the tetrasilabutadienes of type **12** have energy maxima with the s-*trans* form **12b** being slightly favored over the s-*cis* form **12a** [15]. The formation of the various different isomers apparently depends strongly on the respective substitution pattern.

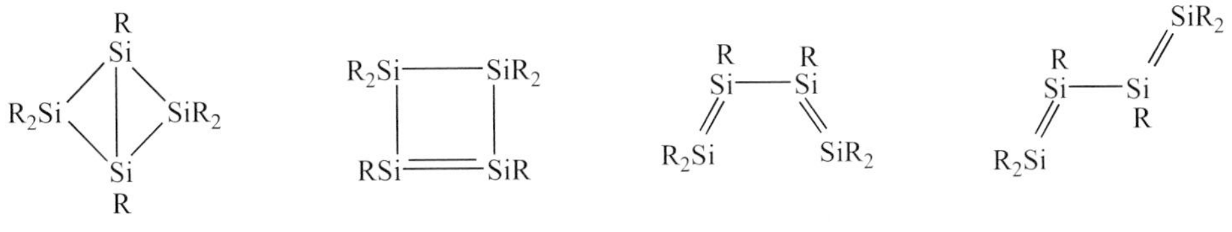

Fig. 1. Some isomers of Si_4R_6.

On account of the large energy difference between **13** and **14** on the one hand and **12** on the other, we have examined whether **12** could be converted into one of these ring systems by thermal or photochemical means. Although irradiation of **12** did not lead to any detectable change, heating at 300°C does cause conversion to a colorless solid which, as yet, has not been identified.

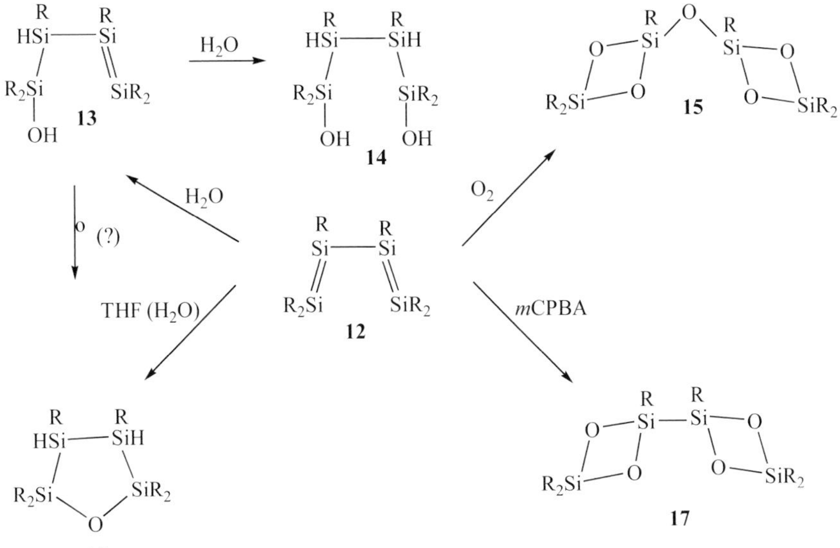

Scheme 6. Oxidation and hydrolysis reactions of **12**.

In view of the preference of the tetrasilabutadiene **12** for the s-*cis* form, it seemed worthwhile to examine its behavior in [4+2] cycloadditions of the Diels–Alder type. Since **12**, like many disilenes, should behave as an electron-rich diene, we attempted to bring it to reaction with various rather electron-poor olefins. No reaction was detected in any case. The effective shielding of the double bonds by the bulky aryl groups and, above all, the 1,4-separation of about 540 pm — which is too large to allow reaction with C–C multiple bonds — appear to be responsible for these failures. Attempted reactions with the disilene, $R_2Si{=}SiR_2$, R = 2,4,6-$i$$Pr_3C_6H_2$, were also unsuccessful, presumably because of the spatial overcrowding in both reaction partners [16].

Reactions of **12** with small molecules, as summarized in Scheme 6, were more successful. Thus reaction with water finally led by way of compound **13** to the tetrasilane-1,4-diol **14**, which showed no tendency to eliminate water. Even more surprising was the successful formation of the oxatetrasilacyclopentane **16**, with its skeletal analogy to tetrahydrofuran, upon heating of **12** in THF containing water. Compound **16** is presumably formed by rearrangement of **13** although no supporting evidence is yet available. Product **15** is formed smoothly on exposure of **12** to atmospheric oxygen. In addition to the presence of the two 2,4-disiladioxetane units (see above), this conversion is of particular interest on account of the insertion of oxygen into the sterically shielded Si–Si bond which proceeds at room temperature with formation of a disiloxane group. In contrast, the Si–Si single bond is not attacked by *m*-chloroperbenzoic acid (*m*CPBA); instead, compound **17** is finally formed by way of the disilaoxirane ring. The structures of products **15**, **16**, and **17** have been confirmed by X-ray crystallography [16].

Acknowledgment: Financial support of our work by the *Deutsche Forschungsgemeinschaft* and the *Fonds der Chemsichen Industrie* is gratefully acknowledged.

References:

[1] R. West, M. J. Fink, J. Michl, *Science* **1981**, *214*, 1343.

[2] a) R. West, *Angew. Chem.* **1987,** *99*, 1231; *Angew. Chem., Int. Ed. Engl.* **1987**, *26*, 1201.
b) G. Raabe, J. Michl, in: *The Chemistry of Organic Silicon Compounds, Part 2* (Eds.: S. Patai, Z. Rappoport), Wiley, Chichester, **1989**, p. 1015.
c) T. Tsumuraya, S. A. Batcheller, S. Masamune, *Angew. Chem.* **1991**, *103*, 916; *Angew. Chem., Int. Ed. Engl.* **1991**, *30*, 902.
d) M. Weidenbruch, *Coord. Chem. Rev.* **1994**, *130*, 275.
e) R. Okazaki, R. West, in: *Multiply Bonded Main Group Metals and Metalloids* (Eds.: R. West, F. G. A. Stone), Academic Press, San Diego, **1996**, p. 232.

[3] H. B. Yokelson, A. J. Millevolte, G. R. Gillette, R. West, *J. Am. Chem. Soc.* **1987**, *109*, 6865.

[4] a) K. L. McKillop, G. R. Gillette, D. R. Powell, R. West, *J. Am. Chem. Soc.* **1992**, *114*, 5203.
b) H. Son, R. P. Tan, D. R. Powell, R. West, *Organometallics* **1994**, *13*, 1390.

[5] R. S. Archibald, Y. van den Winkel, D. R. Powell, R. West, *J. Organomet. Chem.* **1993**, *446*, 67.

[6] M. Weidenbruch, A. Pellmann, Y. Pan, S. Pohl, W. Saak, H. Marsmann, *J. Organomet. Chem.* **1993**, *450*, 67.

[7] M. Weidenbruch, A. Pellmann, S. Pohl, W. Saak, H. Marsmann, *Chem. Ber.* **1995**, *128*, 935.

[8] H. Watanabe, K. Takeuchi, N. Fukawa, M. Kato, M. Goto, Y. Nagai, *Chem. Lett.* **1987**, 1341.

[9] A. J. Millevolte, D. R. Powell, S. G. Johnson, R. West, *Organometallics* **1992**, *11*, 1091.

[10] M. Weidenbruch, K. Kramer, A. Schäfer, J. K. Blum, *Chem. Ber.* **1985**, *118*, 107.

[11] M. Weidenbruch, S. Willms, W. Saak, G. Henkel, *Angew. Chem.* **1997**, *109*, 2612; *Angew. Chem., Int. Ed. Engl.* **1997**, *36*, 2503.

[12] a) S. Masamune, Y. Kabe, S. Collins, D. J. Williams, R. Jones, *J. Am. Chem. Soc.* **1985**, *107*, 5552.
b) R. Jones, D. J. Williams, Y. Kabe, S. Masamune, *Angew. Chem.* **1986**, *98*, 176; *Angew. Chem., Int. Ed. Engl.* **1986**, *25*, 173.

[13] a) M. Kira, T. Iwamoto, C. Kabuto, *J. Am. Chem. Soc.* **1996**, *118*, 10303.
b) T. Iwamoto, M. Kira, *Chem. Lett.* **1998**, 277.
[14] N. Wiberg, H. Auer, H. Nöth, J. Knizek, K. Polborn, *Angew. Chem.* **1998**, *110*, 3030; *Angew. Chem., Int. Ed. Engl.* **1998**, 36, 2869.
[15] T. Müller, "*The Si_4H_6 Potential Energy Surface*", in: *Organosilicon Chemistry IV: From Molecules to Materials* (Eds.: N. Auner, J. Weis), Wiley–VCH, Weinheim, **1999**, p. 110.
[16] S. Willms, M. Weidenbruch, unpublished results.

Products of the Reaction of Tetrasupersilyl-*tetrahedro*-tetrasilane $(tBu_3Si)_4Si_4$ with Iodine

Nils Wiberg*, Harald Auer, Kurt Polborn[a]

Institut für Anorganische Chemie, Ludwig-Maximilians-Universität München
Butenandtstraße 5–13 (Haus D), D-81377 München, Germany
Tel. Int code + (89)2180 7458 — Fax. Int. code + (89) 2180 7865
E-mail: niw@cup.uni-muenchen.de

Michael Veith[a], ***Volker Huch***[a]

Institut für Anorganische Chemie der Universität des Saarlands
Im Stadtwald, D-66123 Saarbrücken, Germany
Tel.: Int. code + (681)302 3415 — Fax: *Int. code* + (681)302 3995

Keywords: Silicon / Supersilyl / Cyclotetrasilene / Cyclotetrasilane / X-Ray Structure Analyses

Summary: Reactions of $(tBu_3Si)_4Si_4$ with I_2 in mole ratios 1:1, 1:3 and 1:4 lead quantitatively to the cyclotetrasilene $(tBu_3Si)_4Si_4I_2$ (**2**), the cyclotetrasilane $(tBu_3Si)_3Si_4I_5$ (**4**) and the cyclotetrasilane $(tBu_3Si)_2Si_4I_6$ (**5**), respectively. X-ray structure analyses of **5** and **6** are presented.

Introduction

In 1993, for the first time, we were able to prepare a molecular silicon compound with an Si_4 tetrahedron: tetrasupersilyl-*tetrahedro*-tetrasilane $R^*_4Si_4$ with supersilyl R^* = tri-*tert*-butylsilyl $SitBu_3$ [1]. As noted, the yellow-orange crystals display high thermo- and photostability and are stable to water and air. The compound cannot be reduced by sodium in benzene in the presence of [18]crown-6 but in solution it reacts with oxidants. These reactions normally lead to "mysterious" mixtures of products. Recently we discovered a reaction of $R^*_4Si_4$ which surprisingly occurs in one direction. It is described hereafter.

Reaction of $R^*_4Si_4$ (1) with Iodine

Reaction of the tetrahedrane $R^*_4Si_4$ (**1**) (which can be prepared easily by debromination of $R^*SiBr_2–SiBr_2R^*$ with sodium in heptane [1]) and one equivalent of iodine in benzene at 0°C quantitatively yields thermally stable and oxygen-insensitive, but water-sensitive red 1,2-diiodo-

[a] X-ray structure analyses

tetrasupersilylcyclotetrasilene (**2**) (Scheme 1(a)), the structure of which has been solved by X-ray structure analysis (cf. [2]). Indeed, after $(tBuMe_2Si)_6Si_4$ [3] it represents the second silicon compound with a *cyclo*-tetrasilene skeleton which could be synthesized. Compared with the latter compound, (i) the reactivity of the Si–Si double bond — for steric reasons — is lower and (ii) the additional silicon—iodine groups allow reactions not possible with $(tBuMe_2Si)_6Si_4$. Accordingly, **2** – in contrast to the latter compound – does not react with dry oxygen under normal conditions. On the other hand water or methanol gives, as in Scheme 1(b) and 1(c) the substitution products **3** (pale yellow) and **4** (orange-red) [2].

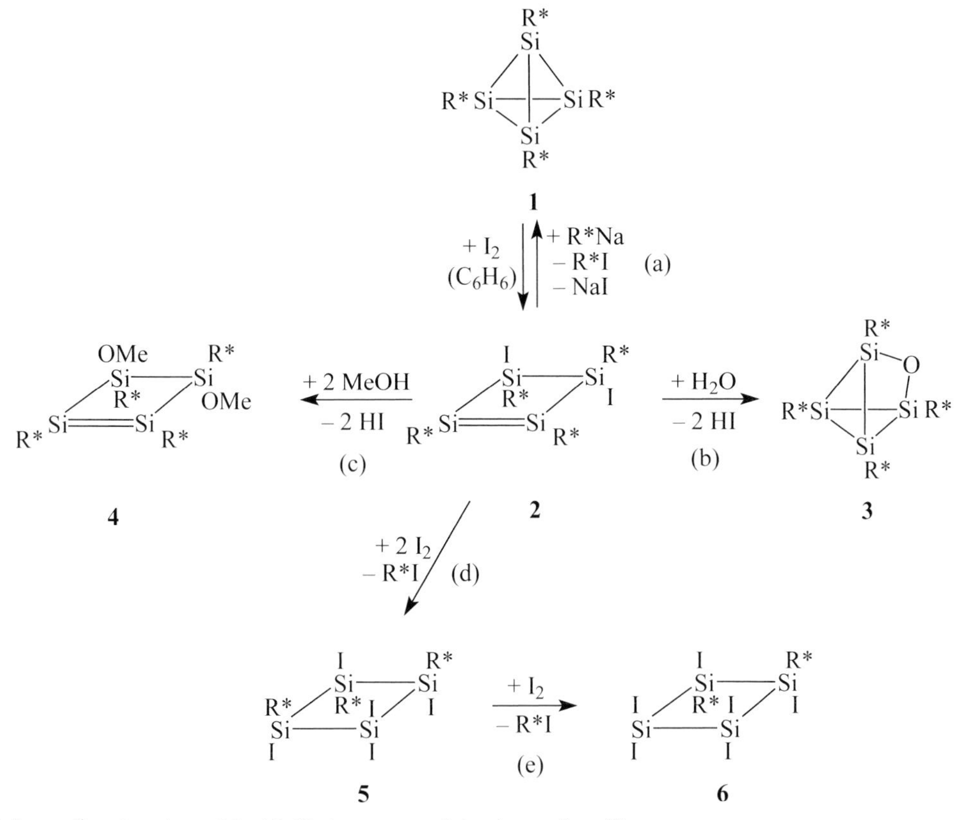

Scheme 1. Reactions of the 1,2-diiodo-tetrasupersilylcyclotetrasilene (**2**).

The cyclotetrasilene **2** not only undergoes acid–base reactions (Scheme 1(b), 1(c)), but also redox reactions. So, reduction of **2** with supersilyl sodium R*Na leads, as in Scheme 1(a), back to the tetrahedron **1**. Over and above that, **2** reacts further with additional iodine at room temperature with elimination of supersilyl iodide R*I and formation of an air- and moisture-stable yellow cyclotetrasilane **5** (Scheme 1(d); Fig.1). The expected intermediate $R^*_4Si_4I_4$ could not be detected and possibly **5** is formed by other intermediates, hitherto not known. Heating a solution of **5** in benzene to 120°C in the presence of iodine leads, with elimination of another supersilyl iodide R*I, quantitatively to the reasonably air-stable pale yellow cyclotetrasilane **6** (Scheme 1(e); Fig. 2).

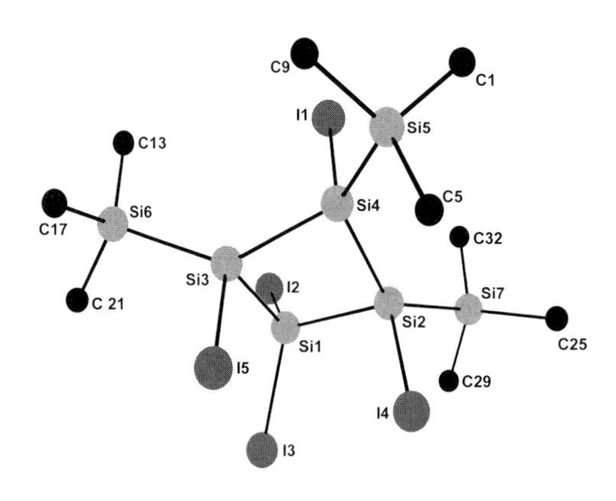

Crystal system: monoclinic *C2/c*
$R_1 = 0.1215$, $wR_2 = 0.2920$
Formula: $C_{36}H_{81}I_5Si_7 \cdot 0.5\ C_6H_6$,
colorless needles from benzene

Important bond lengths [Å] and angles [°]:

Si(1)–Si(2): 2.384(11)	Si(2)-Si(1)-Si(3): 97.2(4)
Si(1)–Si(3): 2.384(10)	Si(1)-Si(2)-Si(4): 84.2(4)
Si(2)–Si(4): 2.507(11)	Si(1)-Si(3)-Si(4): 84.1(3)
Si(3)–Si(4): 2.511(12)	Si(2)-Si(4)-Si(3): 90.9(4)
Si(1)–I(2): 2.466(9)	
Si(1)–I(3): 2.466(9)	Dihedral angles in ring:
Si(2)–I(4): 2.487(9)	14.0 / 14.8 °
Si(3)–I(5): 2.506(9)	
Si(4)–I(1): 2.485(9)	

Structure: M. Veith, V. Huch

Fig. 1. Structure of $R^*_3Si_4I_5$ (**5**).

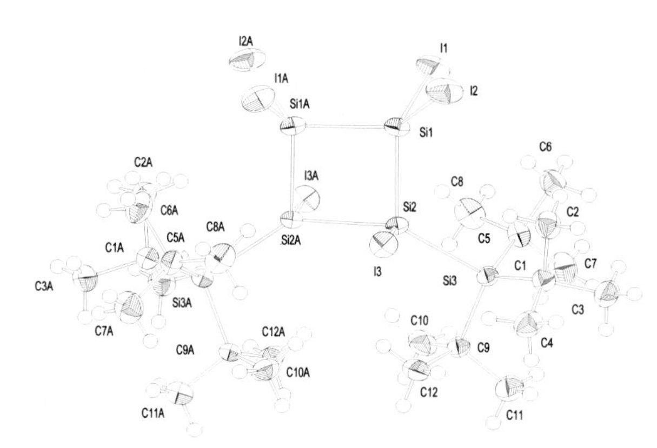

Crystal system: monoclinic *C2/c*
$R_1 = 0.0219$, $wR_2 = 0.0278$
Formula: $C_{24}H_{54}I_6Si_6$,
pale yellow needles from benzene

Important bond lengths [Å] and angles [°]:

Si(1)–Si(2): 2.413(1)	Si(1A)-Si(1)-Si(2): 90.4
Si(1)–Si(1A): 2.362(2)	Si(1)-Si(2)-Si(2A): 88.0
Si(1)–I(1): 2.442(1)	
Si(1)–I(2): 2.457(1)	Dihedral angle in ring:
Si(2)–I(3): 2.512(1)	13.57 °

Structure: K. Polborn

Fig. 2. Structure of $R^*_2Si_4I_6$ (6).

Acknowledgment: The authors thank the *Deutsche Forschungsgemeinschaft* for financial support and *Wacker AG* for the donation of chemicals.

References:

[1] N. Wiberg, C .M. M. Finger, K. Polborn, *Angew. Chem.* **1993**, *105*, 1141; *Angew. Chem., Int. Ed. Engl.* **1993**, *32*, 1054.

[2] N. Wiberg, H. Auer, H. Nöth, K. Polborn, *Angew. Chem.* **1998**, *110*, 3030; *Angew. Chem., Int. Ed. Engl.* **1998**, *37*, 2869.

[3] M. Kira, T. Iwamoto, C. Kabuto, *J. Am. Chem. Soc.* **1996**, *118*, 10303.

Intramolecularely π-Stabilized Silyl Cations

Norbert Auner, Thomas Müller*, Markus Ostermeier*

Institut für Anorganische Chemie
Johann Wolfgang Goethe-Universität Frankfurt
Marie-Curie-Str.11, D-60439 Frankfurt am Main, Germany
Tel: Int. code + (69)798 29180 — Fax: Int. code + (69)798 29188
E-mail: müller@chemie.uni-frankfurt.de

Julia Schuppan, Hans-Uwe Steinberger

Fachinstitut für Anorganische Chemie der Humboldt Universität Berlin
Hessische Str. 1–2, D-10115 Berlin, Germany

Keywords: Silyl Cations / Ab Initio Calculations / Density Functional Calculations / NMR Spectroscopy / Silanorbornenes / Silanorbornadienes

Summary: The synthesis of intramolecularely π-stabilized silylium ions such as 7-sila-norbornadienyl- and 2-silanorbornyl cations was attempted. While benzo-7-sila-norbornadienyl cation **7** could not be detected at room temperature in hydrocarbon solvents, its donor–acceptor complex **8** was identified by NMR spectroscopy. In contrast, persistent 2-silanorbornyl cations could be synthesized and characterized by NMR spectroscopy and modern quantum chemical methods. Silanorbornyl cations **5** and **25** are stable at room temperature and have a bridged structure in which the silicon adopts a [3+2] coordination.

Introduction

The quest for stable trivalent silylium ions has been one of the most difficult practical challenges for organo silicon chemists for the last five decades [1]. Although silylium ions are thermodynamically more stable than their carbon analogues, they are due to their high inherent electrophilicity kinetically extremely unstable. In solution or in the solid state these species interact with a wide variety of both π- and σ-electron donating compounds forming tetra- or higher valent positively charged species. In general there are two strategies to overcome the high kinetic instability of silylium ions. Recently the classical approach of blocking an attacking nucleophil by bulky substituents attached to the silicon center has succesfully been applied by Lambert and Zhao performing the synthesis of trimesitylsilylium **1**, the up-to-date only example of a trivalent silylium ion stable in the condensed phase [2]. The second approach utilizes the intramolecular stabilization of the positively charged silicon by electron donating groups and has succesfully been applied by Corriu and his group [3] and later by Belzner and coworkers [4] in the synthesis of silylcations like **2** and **3**, in which the silicon is five coordinated and the charge is largely transfered to the nitrogen donors.

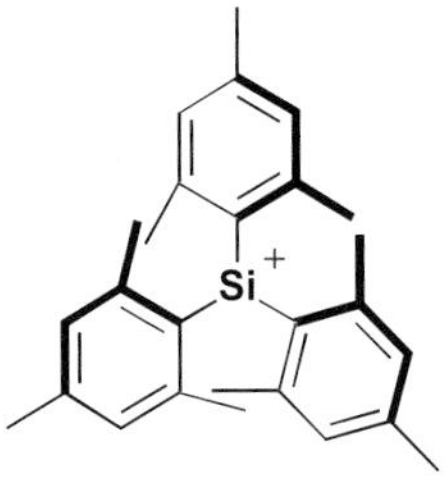

1 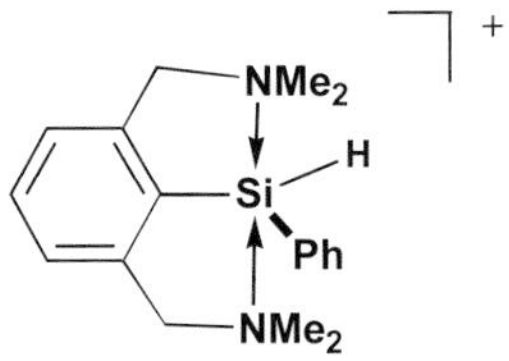 2 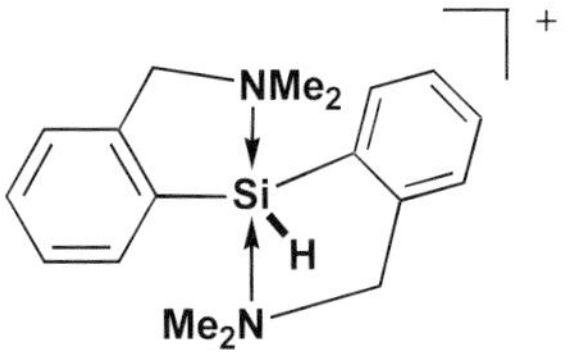 3

In 1995 Schleyer and coworkers reported the results of calculations for the silicon congeners of a number of non-classical carbocations including the 7-silanorbornadienyl cation **4** and the 2-silanorbornyl cation **5** [5]. In both cations the electron deficiency at silicon is strongly reduced by charge transfer from a remote C=C double bond, thus stabilizing the cations both thermodynamically and kinetically. In a joint theoretical-experimental effort our groups and the group of Schleyer investigated the possibilities and the limitations of this approach to stable silyl cations in the condensed phase.

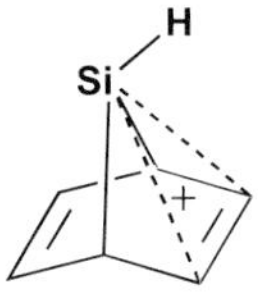

4 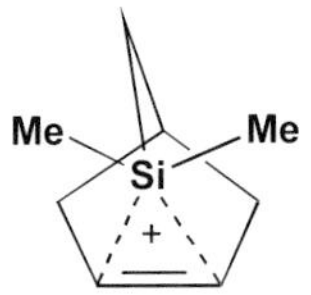 5

Results and Discussion

7-Silanorbornadienyl cations

Recent ab initio calculations by Schleyer and coworkers indicate that **4** is strongly stabilized by interaction of the empty 3p Si orbital with the π-orbital of one allylic double bond [5]. This interaction leads to a pyramidalized silicon center and relatively short Si–C(sp^2) distances and stabilizes the cation **4** by 15.4 kcal mol^{-1} compared to trimethylsilylium (at MP2/6-31G*//MP2/6-31G*). The bonding situation in **4** is best described in terms of three dimensional aromaticity with the 3p(Si) and the C=C double bond obeying the "4n+2 interstitial electron" rule. Due to the reduced positive charge at silicon, cations such as **4** should be also less reactive towards nucleophiles. A facile synthetic approach to this novel typ of silyl cation is outlined in Scheme 1.

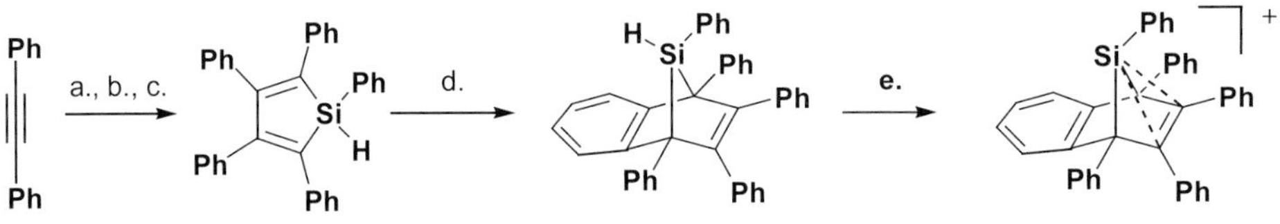

anti-**6** *anti*-**7**

a. Li/Et_2O, RT. — b. $HSiCl_3$/ Et_2O –196°C → RT. — c. PhMgBr/THF, RT. — d. Mg, *ortho*-C_6H_4FBr, THF, 56°C. — e. $[Ph_3C]^+[B(C_6F_5)_4]^-$, benzene, RT.

Scheme 1. Synthesis of benzosilanorbornadienyl cations.

The benzosilanorbornadiene **6** was obtained along this route exclusively as the *anti* stereoisomer in moderate yields and was fully characterized by NMR spectroscopy and by a single-crystal structure analysis (see Fig.1).

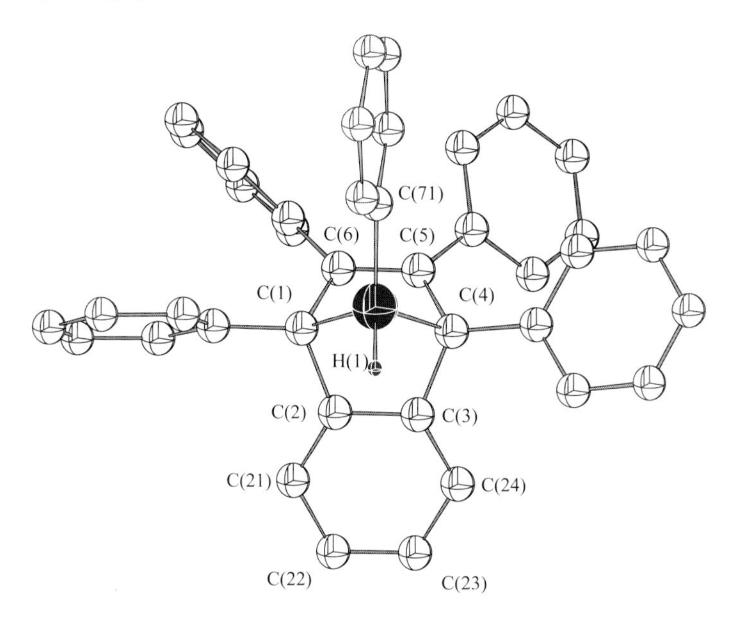

Fig. 1. Molecular structure of benzosilanorbornadiene *anti*-**6** in the crystal. Selected bond length [pm] and angles [°] of **6**. C(1)–C(2) 153.3(4), C(1)–C(6) 154.1(4), C(1)–Si(7) 190.8(3), C(2)–C(21) 137.7(4), C(2)–C(3) 140.2(4), C(21)–C(22) 139.1(4), C(22)–C(23) 137.7(5), C(23)–C(24) 139.8(5), C(24)–C(3) 138.4(4), C(3)–C(4) 153.5(4), C(4)–C(5) 155.4(4), C(4)–Si(7) 190.3(3), C(5)–C(6) 135.7(4), Si(7)–C(71) 186.0(3), C(2)-C(1)-Si(7) 98.6(2), C(6)-C(1)-Si(7) 96.0(2), C(3)-C(2)-C(1) 110.2(3), C(2)-C(3)-C(4) 112.7(2), C(3)-C(4)-C(5) 103.9(3), C(3)-C(4)-Si(7) 97.6(2), C(5)-C(4)-Si(7) 95.35(19), C(6)-C(5)-C(4) 112.3(3), C(5)-C(6)-C(1) 112.0(3), C(71)-Si(7)-C(4) 119.17(16), C(71)-Si(7)-C(1) 119.08(14), C(4)-Si(7)-C(1) 82.97(15).

For the attempted synthesis of 2,3-benzo-1,4,5,6,7-pentaphenyl-7-silanorbornadien-7-ylium (**7**) the well established hydride transfer reaction between silane **6** and a trityl salt with the less nucleophilic tetrakispentafluorphenylborat as counteranion was used. Toluene and benzene served as solvents of low nucleophilicity [6]. Reaction of **6** with *t*rityl-tetrakis(*p*entafluor*p*henyl)*b*orat (TPFPB) in [D_6]benzene at room temperature gives instantaneously a deep brown solution, which separates into two layers. The disappearence of the characteristic SiH proton resonance in the NMR spectrum of the sample and the formation of triphenylmethane proves the occurence of the hydride transfer reaction from the silane **6** to the trityl cation with the intermediate formation of a silylium ion. ^{13}C- and ^{1}H-NMR spectra of the highly viscous oily lower phase show, however, only relatively broad signals, which indicate intensive polymerization of the formed intermediate. Similar results were obtained in toluene at room temperature and at –10°C. While the synthesis of **7** in arenes failed, the use of more nucleophilic solvents was more succesful. Thus, the reaction of **6** with TPFPB in a 1:2 mixture of [D_3]acetonitril and [D_6]benzene gives an orange brown solution, showing one ^{29}Si-NMR resonance signal at $\delta(^{29}Si) = 8.9$. The ^{13}C-NMR spectrum shows a complete set of 17 signals which can be assigned to the carbon skeleton of the 2,3-benzo-1,4,5,6,7-

pentaphenyl-7-silanorbornadiene. The NMR spectroscopic data suggest the formation of the complex *anti*-**8** formed by acetonitril and the cationic species **7**. This finding is strongly supported by the good agreement of the experimentally determined $\delta(^{29}Si)$ for **8** with the calculated value for the model compound **9** ($\delta(^{29}Si) = -11.4$ at GIAO/B3LYP/6 311+G(2df,p) (Si), 6-31G(d) (C,H)// B3LYP/6-31G*) [7–9] (see Fig. 2) taking into account that the substituent effect of +17.7 ppm for the two β-phenyl groups at the bridgehead carbon atoms of the benzosilanorbornadiene skeleton is not considered in the calculations [10].

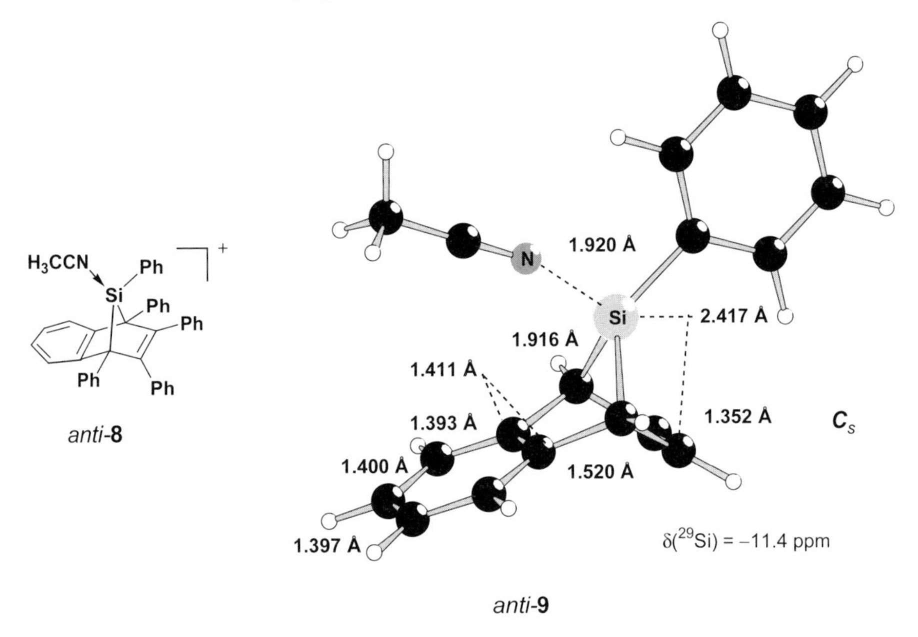

Fig. 2. Acetonitril complex anti-**8** and the calculated structure of the model compound anti-**9**.

The question arises, why **7** is not stable in arene solution and can only be stabilized intermoleculareiy by the relatively strong nucleophil acetonitril? A detailed theoretical analysis [7] of the $C_{10}H_9Si^+$ potential energy surface (PES) at the hybrid density functional B3LYP/6-31G* level [8] of theory reveals that *syn*-**10** (R = H) is less stable than naphtylsilylium **11** and 4,5-benzo-silatropylium **12** by 21.9 and 16.9 kcal mol^{-1}, respectively.

anti-**10** **11** **12** **13**

Furthermore, the calculations indicate that *anti*-**10** (R = H) is not a stable species, but collapses to the π-type complex **13** (R = H), which is by 12.2 kcal mol^{-1} more stable than *syn*-**10** (R = H). The situation is different for cations bearing an aryl substituent at silicon. Thus, both stereoisomers *syn*- and *anti*-**10** (R=Ph) are predicted to be stable compounds on the potential energy surface, the *anti* isomer being by 5.9 kcal mol^{-1} more stable. However, the complex **13** (R = Ph) again is lower by

10.9 kcal mol^{-1} in energy than *syn*-**10** (R = Ph) (for calculated structures, see Fig. 3). Since the barrier for the interconversions similar to *anti*-**10** → **13** was shown to be rather small (2–3 kcal mol^{-1}) [11], this monomolecular rearrangement can easily occur at room temperature. Although complexes such as **13** are stable species in the gas phase and can be detected by CID/MS spectrometry [12], its fate in the condensed phase is still unknown. Its reactivity under the reaction conditions applied can be expected to be rather large, resulting in very fast subsequent reactions and thus preventing a detection in an NMR experiment at room temperature. However, in the presence of nucleophilic solvents such as acetonitril, the cation *anti*-**10** is stabilized by intermolecular donor–acceptor interaction, i.e., the acetonitril complexation calculated for **9** is exothermic by 28.6 kcal mol^{-1} (B3LYP/6-31G*). This complexation energy increases decisively the barrier for the rearrangement to **13** and allows the spectroscopic indentification of the donor–acceptor complexes. Thus, while donor–acceptor complexes of 7-silanorbornadienyl cations are accessible compounds, the results of the density functional calculations suggest that the oppurtunities to obtain donor-free and stable 7-silanorbornadienyl cations are rather small. This is due to fast monomolecular rearrangements into more stable isomers.

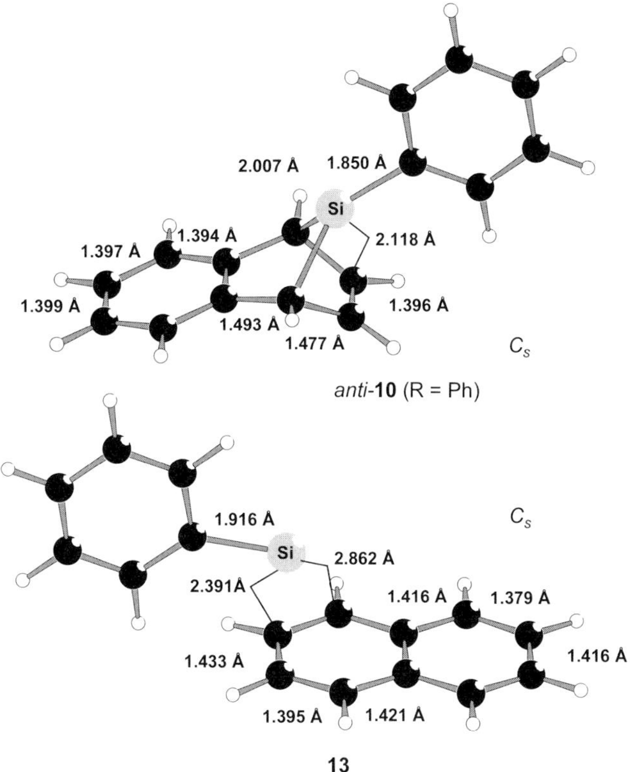

Fig. 3. Calculated structures of *anti*-**10** (R = Ph) and **13**.

2-Silanorbornylcations

During the course of our systematic investigations on the reactivity of small silaheterocycles, we found that 2-silanorbornenes **14** can be easily cleaved by treatment with acids yielding cyclopentenyl compounds **15** (Scheme 2) [13].

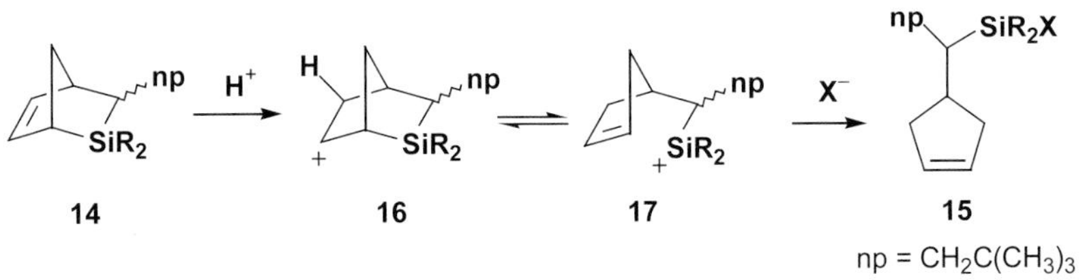

Scheme 2. Allylic cleavage of 2-silanorbornenes by acids.

Originally, we rationalized the formation of **15** with the intermediacy of a β-silyl-substituted carbocation **16** which is in an equilibrium with the silylium ion **17** (see Scheme 2). The reaction course strongly depends on the nature of the substituents R at silicon and on the strength of the acid. Thus, treating the silanorbornenes **14** with R = methyl or methoxy groups with HCl/Et_2O yielded the corresponding cyclopentenyl compounds **15**, exclusively.

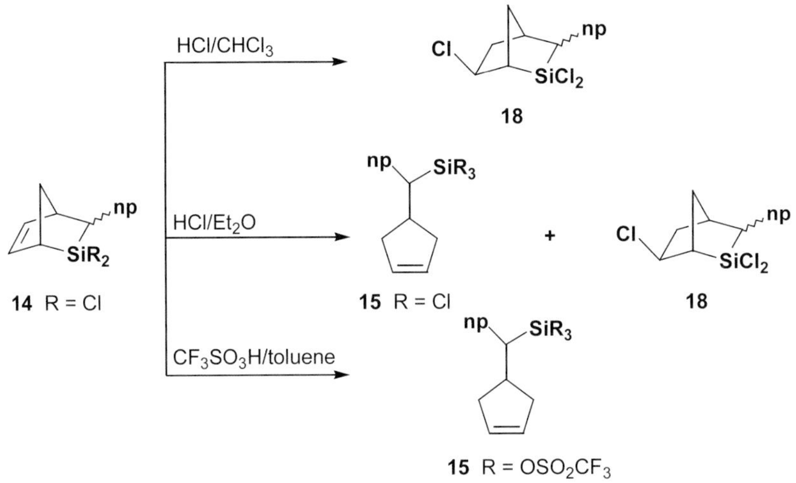

Scheme 3. Reaction of **14** (R = Cl, np = neopentyl) with acids.

However, under comparable conditions 2,2-dichloro-2-silanorbornene **14** (R = Cl) gave a 2:1 mixture of the 1,2 addition product **18** and the product of the allylic cleavage **15** (R = Cl). Using $CHCl_3$ as a solvent, **18** was formed exclusively from the reaction with HCl, while the stronger acid CF_3SO_3H gave the ring-cleaved **15** (R = OSO_2CF_3) in toluene as single product (see Scheme 3). The silanorbornane **18** was formed as a 60:40 mixture of *exo*/*endo* neopentyl stereoisomers. The absence of other stereoisomeric silanorbornanes suggests the regio- and stereoselective addition of HCl across the C=C double bond. The formation of the more stable β-silyl-substituted carbocation in the first step of the HCl addition explains the observed regiochemistry in a straight forward manner. The stereoselective *exo*-attack of the nucleophil on the intermediate cation accounts for the formation of only one possible stereoisomer. The stereoselectivity of the reaction is also demonstrated by deuterium labeling experiments (Scheme 4).

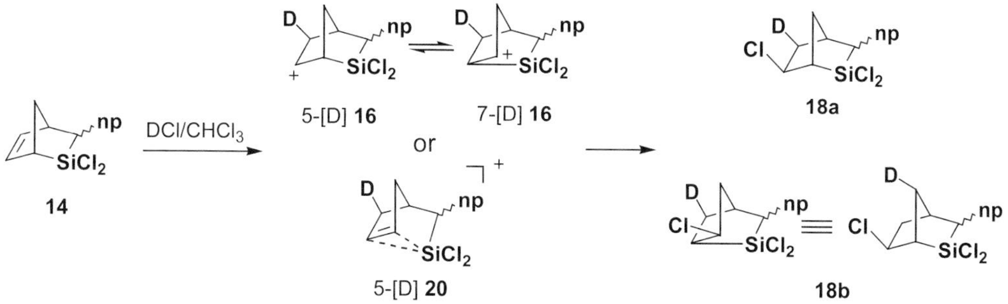

Scheme 4. DCl addition to **14** (R = Cl).

Thus, DCl addition to a mixture of *exo/endo* **14** gave the *exo*-6-chlor isomers **18a,b**, exclusively. The corresponding *endo*-6-chlor isomers **18c,d** could not be detected, indicating the intermediate formation of either a bridged ion **20** or a fast degenerate equilibrium between two β-silyl-substituted carbocations **16**, which both block the endo attack of a incoming nucleophil efficiently [14].

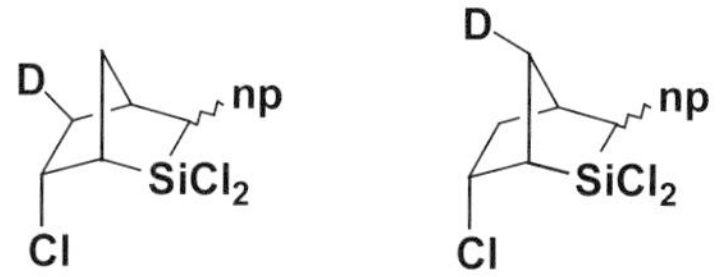

18c **18d**

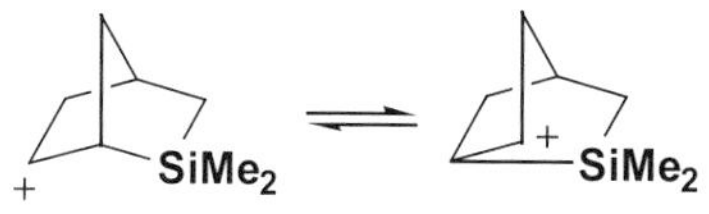

21

A theoretical investigation of the problem by Schleyer and coworkers revealed that the β-silyl-substituted carbocation **21** does not correspond to a local minimum on the ground-state potential energy surface but collapses to the symmetrically bridged ion **5** (see Scheme 5) [15]. Furthermore, **5** is by 22.0 kcal mol^{-1} more stable than the isomeric open silylium ion **22**. Coordination to the solvent, i.e., benzene, does not change the relative energy order, although the association energy for **5** with benzene is substantial (20.8 kcal mol^{-1}, calculated for **23**). The weak complex **24** formed from **5** and benzene is still by 6.3 kcal mol^{-1} more stable than **23** (all calculations at B3LYP/6-31G* +0.89 ZPVE(SCF/6-31G*, see Scheme 5). Thus, theory predicts a bridged compound similar to **5** to be the common intermediate of HCl addition to C-neopentyl substituted silanorbornenes. Furthermore, the intermediate formed is expected to be an isolated cationic species in benzene at room temperature.

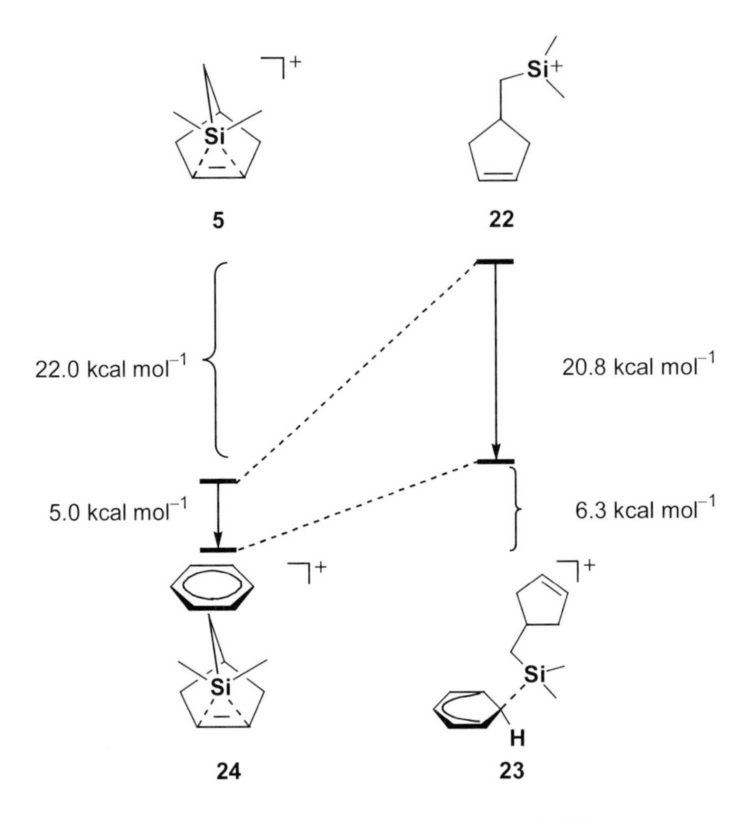

Scheme 5. Relative energies of cations **5**, **22** and their benzene complexes **23**, **24**.

While in the course of the HCl addition silanorbornyl cations were formed along the σ-route (Scheme 6), we choose the π-route for the synthesis and spectroscopic characterization of a persistent silanorbornyl cation (Scheme 6), i.e., the intramolecular addition of a preformed silylium ion to the C=C double bond.

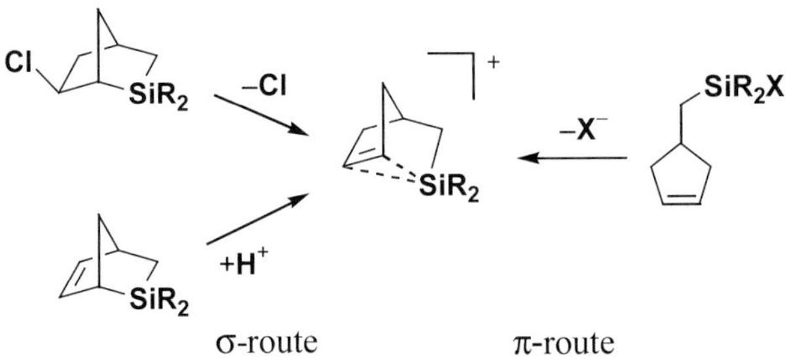

Cl
SiR_2
$-Cl$
SiR_2
$-X^-$
SiR_2X
SiR_2
$+H^+$
σ-route
π-route

Scheme 6. Synthetic approaches to stable silanorbornyl cations.

This is successfully achieved by reaction of methylenecyclopentenyl silanes **24** with TPFPB in arene solvents (Scheme 7).

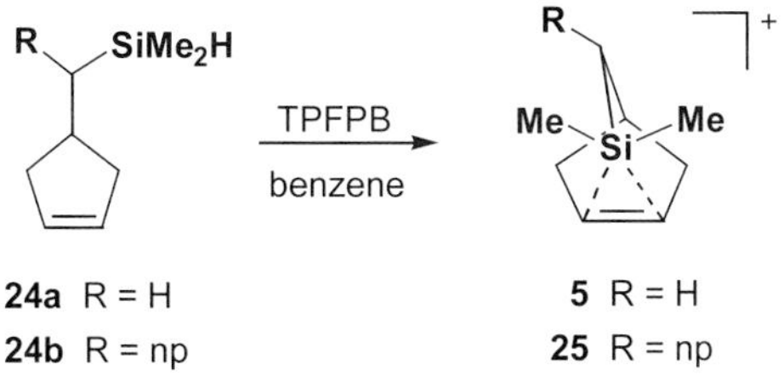

Scheme 7. Reaction of **24** with TPFPB.

Addition of the neat silanes **24** to a solution of TBFPB in [D$_6$]benzene at room temperature gives instantaneously deeply red solutions which separate into two layers. The ^{29}Si-NMR spectra of the highly viscous oily lower phases show in both cases one single peak at $\delta(^{29}Si)$ = 87.4, (**25**) [15], and 87.2 (**5**) [16]. The intramolecular complexation of the positively charged silicon can be shown by the ^{13}C chemical shift of the vinylic methin carbons at $\delta(^{13}C)$ = 150.5, 153.7 (**25**) and 150.6 (**5**), which is by approximately 20 ppm low field shifted compared to the starting silanes. This low field shift indicates substantial charge transfer from the silicon to the C=C bond. The measured direct CH coupling constants for the vinylic carbon hydrogen bond of 174 Hz (**25**) and 170.4 Hz (**5**) can be used to distinguish between a degenerate fast equilibrium **21** and a bridged structure. The observed $^1J_{CH}$ coupling constants are larger than those determined for the starting silanes (by 7.9 or 12.4 Hz for **5** and **25**, respectively). A fast degenerate equilibrium between two β-silyl-substituted carbocations is expected to result in a smaller $^1J_{CH}$ coupling constant. This is due to the averaging of the $^1J_{CH}$ of a sp^2 carbon (typically 171 Hz in trivalent secondary carbocations) and $^1J_{CH}$ of a sp^3 carbon (130 Hz for a CH group in β-position of a carbocation) [17]. These findings are supported by the results of density functional calculations of the $^1J_{CH}$ coupling constants for protonated

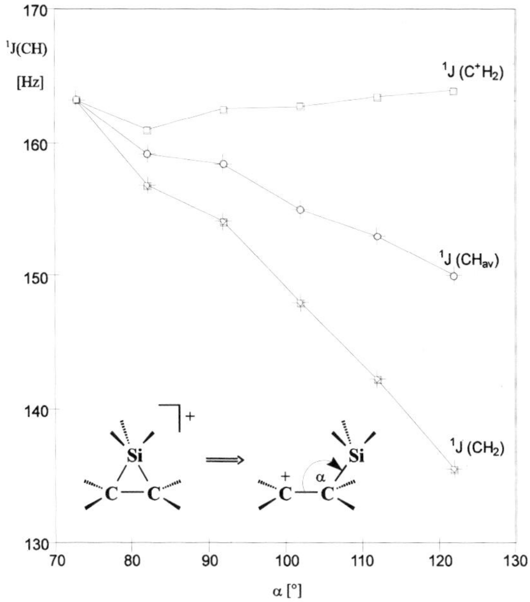

Fig. 4. Calculated $^1J_{CH}$ for protonated silacyclopropane and for β-silyl-substituted ethyl cations at different bond angles α(C$^+$-C-Si) (SOS/DFPT/IGLO/BIII) [18, 19].

silacyclopropane [18] and its isomeric β-silyl-substituted ethyl cations (see Fig. 4), which indicate that the time averaged $^1J_{CH}$ for a fast equilibrium of two β-silyl ethyl cations is markedly smaller than for the symmetrically brigded silacyclopropyl cation and, notably, smaller than that observed for silanorbornyl cations [19]. In general, the agreement of the calculated NMR parameters and the experimentally determined values for **5** gives strong evidence for the static bridged structures of **5** and **25** and argues against a fast equilibrium of two β-silyl-substituted carbocations (see Fig. 5).

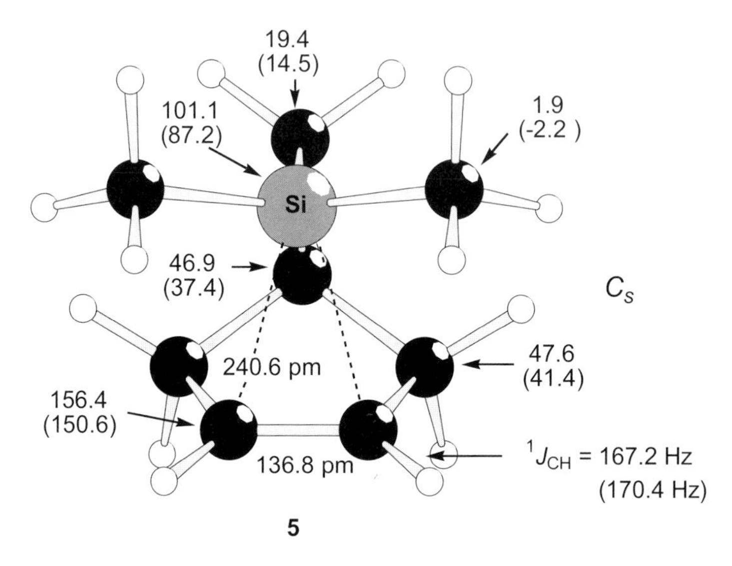

Fig. 5. Calculated structure and NMR parameters of **5** ($\delta(^{29}Si)$ and $\delta(^{13}C)$ at DFPT/IGLO/Basis III//B3LYP/6-31G* using the Perdew Wang exchange correlation functional (TMS: $\sigma(^{29}Si) = 346.7$, $\sigma(^{13}C) = 183.6$), $^1J_{CH}$ at DFPT/IGLO/Basis III//B3LYP/6-31G* using the Perdew exchange functional. Experimental values for **5** in paranthesis [16].

In the absence of moisture cation **25** is stable over a period of weeks, **5**, however, decomposes slowly at room temperature to non-identified products [14–16]. Both cations **5** and **25** show distinct behaviour towards nucleophilic solvents. While their ^{29}Si-NMR and ^{13}C-NMR resonances are not influenced by arene solvents, the addition of the higher nucleophilic acetonitril as cosolvent leads to the formation of silylated nitrilium cations **26** and **27** (Scheme 8). They are characterized by their ^{29}Si and ^{13}C chemical shifts (see Figs. 6 and 7) [14, 16]. The silicon resonances are high field shifted by ca 51–56 ppm and similarely, the signals of the vinylic carbon atoms are shifted by 20 ppm to higher field in the NMR spectra of the nitrilium ions. This clearly indicates the breakdown of the intramolecular stabilization of the positively charged silicon center on the expense of intermolecular interactions between the silyl cations and acetonitril.

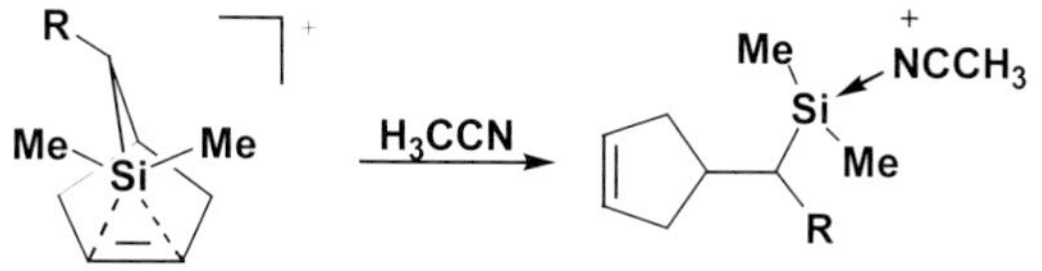

H_3CCN

5 R = H
25 R = np

26 R = H
27 R = np

Scheme 8. Formation of silylated nitrilium cations **26** and **27**.

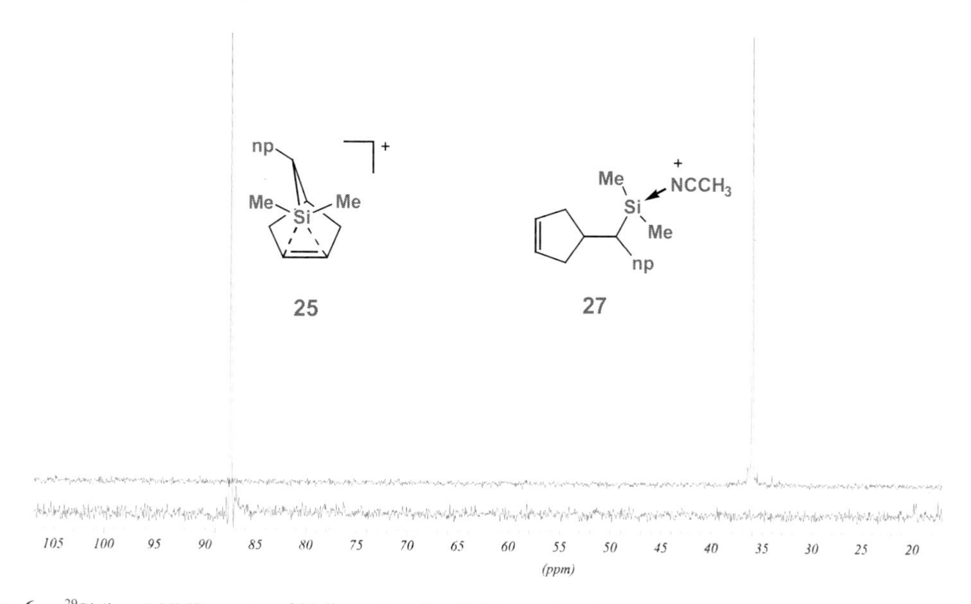

Fig. 6. ^{29}Si (inept) NMR spectra of **25** (lower trace) and **27** (upper trace).

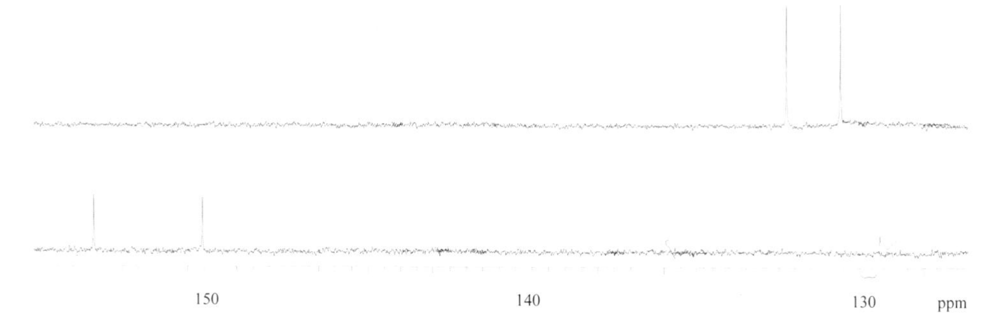

Fig. 7. ^{13}C (dept-135) NMR spectra of **25** (lower trace) and **27** (upper trace).

In summary, the silanorbornyl cations **5** and **25** are stable compounds in arene solvents at ambient temperature, lacking direct coordination to solvent or counterion. They are stabilized by intramolecular π-interaction and can be therefore regarded as an almost symmetrically bridged β-silyl-substituted carbocation with siliconium ion character.

Conclusion

The intramolecular π-stabilization of silyl cations has been successfully verified by the synthesis of 2-silanorbornyl cations and will be extended to other silyl, germyl and stannyl cations in the near future. The synthetic potential of intramolecularly π-stabilized silyl cations in polymerization and hydrosilylation reactions is currently under investigation in our laboratories.

Acknowledgment: Financial support by *Deutsche Forschungsgemeinschaft* (scholarship to T. M.) and the *Dow Corning Cooperation* is gratefully acknowledged. *Chemetall* and *Wacker GmbH* supported this research with generous gifts of alkyllithium compounds and chlorosilanes. Thanks also goes to the *Rechenzentrum der Humboldt Universität, Berlin* for excellent services.

References:

[1] For recent reviews on silyl cations see:
a) C. Maerker, P. v. R. Schleyer, in: *The Chemistry of Organosilicon Compounds, Vol. 2* (Eds.: Z. Rappoport, Y. Apeloig) Wiley, Chichester, **1998**, 513.
b) P. D. Lickiss, in: *The Chemistry of Organosilicon Compounds, Vol. 2* (Eds.: Z. Rappoport, Y. Apeloig) Wiley, Chichester, **1998**, 557.
c) J. B. Lambert, L. Kania, S. Zhang, *Chem. Rev.* **1995**, 1191.

[2] a) J. B. Lambert, Y. Zhao *Angew. Chem.* **1997**, *109*, 389; *Angew. Chem. Int. Ed. Engl.* **1997**, *36*, 400.
b) T. Müller, Y. Zhao, J. B. Lambert, *Organometallics* **1998**, *17*, 278.

[3] a) C. Chuit, R. J. P. Corriu, A. Mehdi, C. Reye, *Angew. Chem.* **1993**, *105*, 1372; *Angew. Chem. Int. Ed. Engl.* **1993**, *32,* 1311.
b) M. Chauhan, C. Chuit, R. J. P. Corriu, C. Reye, *Organometallics* **1993**, *15,* 4326.

[4] J. Belzner, D. Schär, B. O. Kneisel, R. Herbst-Irmer, *Organometallics* **1995**, *14*, 1840.

[5] C. Maerker, J. Kapp, P. v. R. Schleyer, in: *Organosilicon Chemistry: From Molecules to Materials II* (Eds. N. Auner, J. Weis) VCH, Weinheim, **1996**, 329.

[6] a) J. Y. Corey, *J. Am. Chem. Soc.* **1975**, *97,* 3237.
b) H. Mayr, N. Basso, *J. Am. Chem. Soc.* **1992**, *114,* 3060.
c) Y. Apeloig, O. Merin-Aharoni, D. Danovich, A. Ioffe, S. Shaik, *Isr. J. Chem.* **1993**, *33*, 387.

[7] All calculations were performed with *Gaussian 94*, *Revisions C.2–E2*, Gaussian, Inc., Pittsburgh (PA), **1995**.

[8] a) A. D. Becke, *Phys. Rev.* **1988**, *A 38*, 3098.
b) A. D. Becke, *J. Chem. Phys.* **1993**, *98*, 5648.
c) B. G. Johnson, P. M. W. Gill, J. A. Pople, *J. Chem. Phys.* **1993**, *98*, 5612.

[9] a) R. Ditchfield, *Mol. Phys.* **1974**, *27*, 789.
b) K. Wolinski, J. F. Hilton, P. Pulay, *J. Am. Chem. Soc.* **1982**, *104*, 5667.
c) J. R. Cheeseman, G. W. Trucks, T. A. Keith, M. J. Frisch, *J. Chem. Phys.* **1996**, *104*, 5497.

[10] a) H. Sakurai, Y. Nakadaira, T. Koyama, H. Sakaba, *Chem. Lett.* **1983**, 213.
b) E. A. Williams, in: *The Chemistry of Organosilicon Compounds, Vol. 1* (Eds.: S. Patai, Z. Rappoport) Wiley, Chichester, **1989**, 511.

[11] A. Nicolaides, L. Radom, *J. Am. Chem. Soc.* **1996**, *118,*10561.
[12] R. J. Jarek, S. K. Shin, *J. Am. Chem. Soc.* **1997**, *119,* 6376.
[13] N. Auner, H.-U. Steinberger, *Z. Naturforsch., Teil B* **1994**, *49*, 1743.
[14] H.-U. Steinberger *Ph.D. Thesis*, Humboldt Universtät Berlin, **1998**.
[15] H.-U. Steinberger, T. Müller, N. Auner, C. Maerker, P. v. R. Schleyer, *Angew. Chem.* **1997**, *109*, 667; *Angew. Chem. Int. Ed. Engl.* **1997**, *36,* 626.
[16] M. Ostermeier *Diploma Thesis*, Goethe Universität Frankfurt, **1999**.
[17] H.-O. Kalinowski, S. Berger, S. Braun, *^{13}C-NMR-Spektroskopie,* Thieme, Stuttgart, New York, **1984**, p. 459.
[18] Unpublished results C. Maerker, P. v. R. Schleyer, **1997**.
[19] a) V. G. Malkin, O. L. Malkina, M. E. Casida, D. R. Salahub, *J. Am. Chem. Soc.* **1994**, *116*, 5898.
b) V. G. Malkin, O. L. Malkina, L. A. Erikson, D. R. Salahub, in: *Modern Density Functional Theory* (Eds: J. M. Seminario, P. Politzer) Elsevier, Amsterdam, **1995**, 273.
c) V. G. Malkin, O. L. Malkina, D. R. Salahub, *Chem. Phys. Lett.* **1996**, 261, 335.

Silyl Effects in Hypercoordinated Carbocations

Hans-Ullrich Siehl*, Martin Fuß

Abteilung für Organische Chemie I, Universität Ulm
D–89069 Ulm, Germany
Tel.: Int. code + (731)502 2800
E-mail: ullrich.siehl@chemie.uni-ulm.de

Keywords: Silyl Effects / Stable Carbocations / NMR / Ab Initio Calculations

Summary: The 1-(trimethylsilyl)bicyclobutonium ion and the 3-*endo*-(*tert*-butyldimethylsilyl)bicyclobutonium ion were investigated by NMR spectroscopy in superacid solution and by quantum chemical ab initio calculations. The 1-(trimethylsilyl)bicyclobutonium ion undergoes a threefold degenerate methylene rearrangement. The 3-*endo*-(*tert*-butyldimethylsilyl)bicyclobutonium ion is the first static bicyclobutonium ion. The NMR spectra of this carbocation are a direct proof for the hypercoordinated and puckered structure of bicyclobutonium ions.

Introduction

Cyclobutyl cations [1] are intermediates in solvolysis reactions of cyclopropylmethyl, cyclobutyl, and homoallyl halides. Experimental investigations, in particular NMR spectroscopic investigations, of persistent carbocations in solution show that the cyclobutyl ring of the unsubstituted cyclobutyl cation $[C_4H_7]^+$ **1** is puckered and bridged between C_α and C_γ. The puckering and bridging can be explained by interaction of the back lobe of the *endo*-C_γ-H σ-bond orbital with the vacant p-orbital at C_α (Fig. 1). Bridged cyclobutyl cations have a pentacoordinated γ-carbon and are called bicyclobutonium ions.

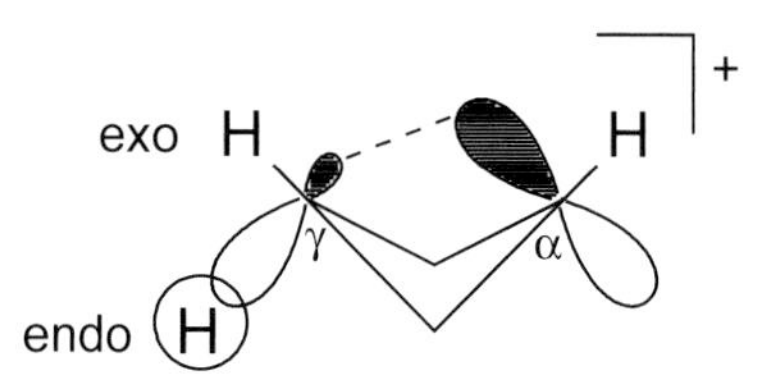

Fig. 1. Interaction of the *endo*- C_γ–H σ-bond orbital of the bicyclobutonium ion **1** with the vacant p-orbital at C_α.

The averaged ^{1}H- and ^{13}C-NMR methylene signals observed for the parent bicyclobutonium ion **1** (Fig. 2, **I**, R = H) are in accord with a fast methylene rearrangement.

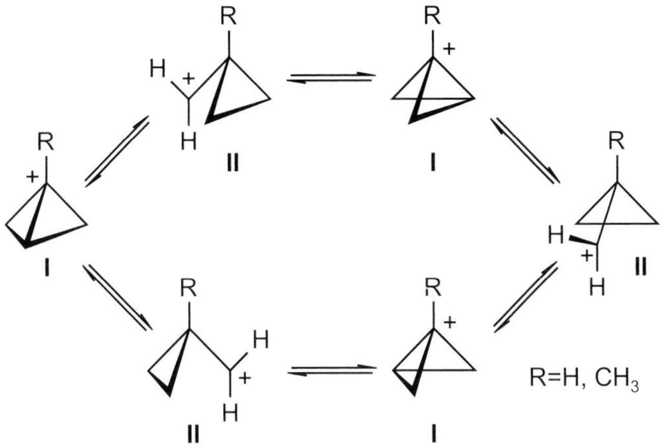

Fig. 2. Threefold degenerate rearrangement of bicyclobutonium ions via cyclopropylmethyl cation structures.

The temperature dependence observed for the ^{13}C-NMR chemical shifts of the parent $[C_4H_7]^+$ cation system indicates that besides **1** smaller amounts of isomeric cyclopropylmethyl cation structures **2** (Fig. 2, **II**, R = H) are involved in the rearrangement process (Fig. 2) which contribute to the observed averaged chemical shifts.

The 1-methylbicyclobutonium ion **3** also undergoes a fast methylene rearrangement (Fig. 2, **I**, R = CH_3). Contrary to the parent cation system $[C_4H_7]^+$ where the isomeric cyclopropylmethyl cation structure **2** (Fig. 2, **II**, R = H) is an energy minimum (MP2/6-31G*), in the $[1\text{-}CH_3\text{-}C_4H_6]^+$ system the (1'-methylcyclopropyl)methyl cation **4** (Fig. 2, **II**, R = CH_3) is a transition state (MP2/6-31G*). (1'-Methylcyclopropyl)methyl cation structures **4** do not contribute to the averaged chemical shifts observed for the threefold degenerate methylene rearrangement of 1-methylbicyclobutonium ions **3**. The different energy profiles for the methylene rearrangement of the parent ($C_4H_7^+$) bicyclobutonium/cyclopropylmethyl cations system **1/2** and that of 1-methylbicyclobutonium ions **3** ($[1\text{-}CH_3\text{-}C_4H_6]^+$) are shown in Fig. 3.

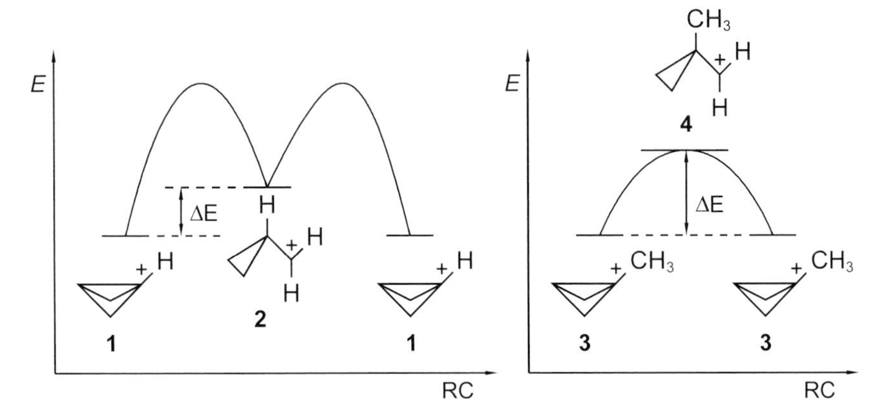

Fig. 3. Qualitative energy profiles, energy (*E*) vs. reaction coordinate (RC) for the rearrangement of parent bicyclobutonium ions **1** and cyclopropylmethyl cations **2**, and of 1-methylbicyclobutonium ions **3** and (1'-methylcyclopropyl)methyl cations **4**.

We have investigated the silyl effect on the structure and stability in various types of carbocations [2]. The introduction of a trialkylsilyl substituent into cyclobutyl/bicyclobutonium cation structures will lead to systems which have a different energy surface as compared to the parent system $[C_4H_7]^+$ (**1**/**2**) or the methyl-substituted system $[C_4H_6CH_3]^+$ (**3**/**4**) [3]. Scheme 1 shows one route (**A** → **B**) for the generation of 1-trialkylsilyl-substituted bicyclobutonium ions **B** and two possible reaction routes (**A** → **B** → **C** and **D** → **C**) to obtain 3-trialkylsilyl-substituted bicyclobutonium ions **C**. We have investigated experimentally route **A** → **B** for **A** (R = R' = Me) which leads to **5**, route **A** → **B** → **C** for **A** (R = *tert*-butyl; R' = Me) which involves a 1,3-hydride shift from the intermediate cation **8** to yield cation **9**, and the reaction pathway **D** → **C** starting from the cyclobutyl progenitor **D** (R = R' = Me) to yield cation **12**.

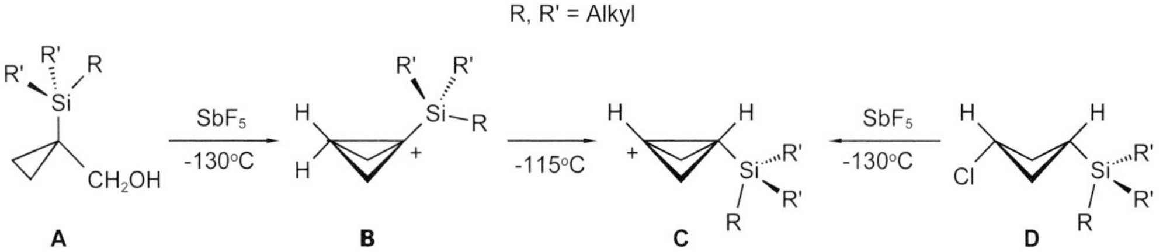

R, R' = Alkyl

A —(SbF_5, -130°C)→ B —(-115°C)→ C ←(SbF_5, -130°C)— D

Scheme 1. Generation and rearrangement pathways for trialkylsilyl ($SiRR'_2$)-substituted bicyclobutonium cations.

1-Silyl-Substituted Bicyclobutonium Ions [3, 4]

Experimental Results

Matrix co-condensation of [1'-(trimethylsilyl)cyclopropyl]methanol (Scheme 1, **A** (R = R' = Me)) with SbF_5 onto a surface of SO_2ClF/SO_2F_2 at –196 °C yields after homogenization at –130 °C a yellow solution. The ^{1}H-NMR spectrum at –128 °C (Fig. 4) shows only two signals at 3.24 ppm (3H) and 4.05 ppm (3H) in addition to the trimethylsilyl signal at 0.38 ppm (9H).

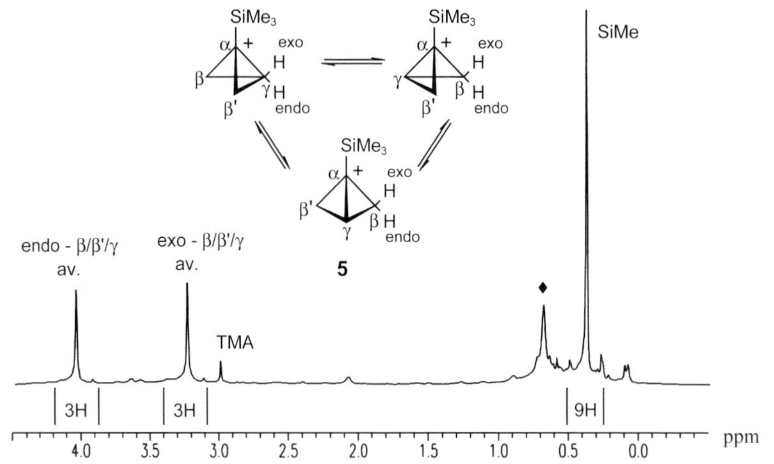

Fig. 4. 400 MHz ^{1}H-NMR spectrum of the 1-(trimethylsilyl)bicyclobutonium ion **5** (♦: $FSiMe_3$) at –128 °C (internal standard TMA $\delta(NMe_4^+)$ = 3.00 ppm).

The ^{13}C-NMR spectrum at –128 °C (Fig. 5) shows a singlet at 137.4 ppm and a doublet of doublets which appears as a pseudo triplet at 48.9 ppm ($^1J_{CH}$ = 177 Hz) in addition to the quartet of the trimethylsilyl group at -5.2 ppm ($^1J_{CH}$ = 119 Hz).

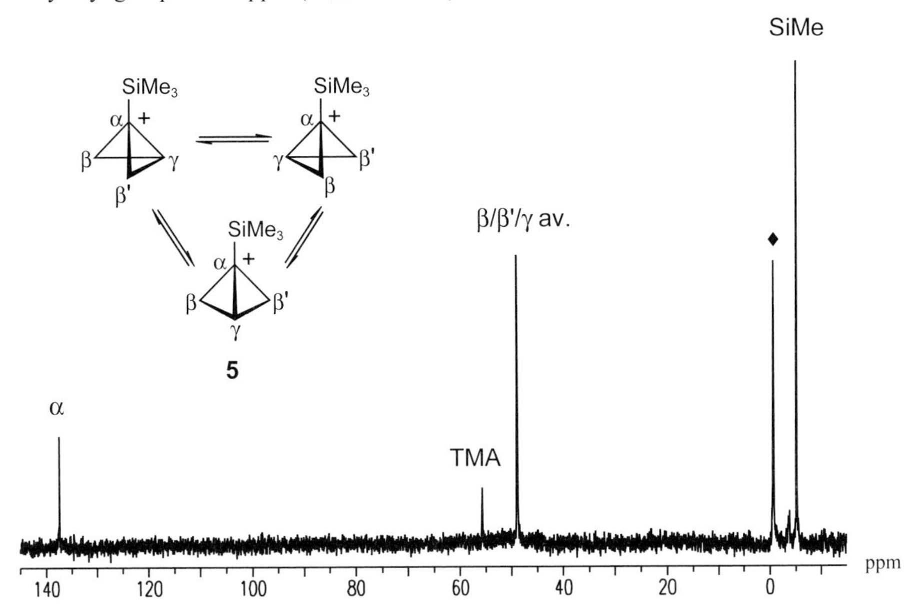

Fig. 5. 100 MHz ^{13}C-NMR spectrum of the 1-(trimethylsilyl)bicyclobutonium ion **5** (♦: $FSiMe_3$) at –128 °C (internal standard TMA $\delta(NMe_4^+)$ = 55.65 ppm).

Selective decoupling of either proton resonance at 3.24 ppm or 4.05 ppm causes the pseudo triplet of the ^{13}C signal at 48.9 ppm to collapse to a doublet. No temperature dependence of the ^1H- and ^{13}C-NMR chemical shifts was observed in the accessible range from –120 °C to –145 °C.

Computational Results [5]

At the MP2/6-31G(d) level of theory the 1-silylcyclobutyl cation $[1\text{-}SiH_3\text{–}C_4H_6]^+$ which serves as a model compound for $[1\text{-}Si(CH_3)_3\text{–}C_4H_6]^+$ **5** has a hypercoordinated puckered 1-silylbicyclobutonium structure **6** (C_α–C_γ distance: 166.0 pm; Fig. 6) which is an energy minimum (number of imaginary frequencies (NImag) = 0). The 1-silylbicyclobutonium ion **6** is about 2.8 kcal/mol lower in energy than the isomeric (1'-silylcyclopropyl)methyl cation **7** (Fig. 6). **7** is characterized as a transition state (NImag = 1) on the flat energy surface.

The ^{13}C-NMR chemical shifts calculated (GIAO-MP2/tzpdz) for the optimized (MP2/6-31G(d)) geometry of the 1-silylbicyclobutonium model cation **6** ($C_{\beta/\beta'}$: 79.6 ppm; C_γ : –14.5 ppm; $C_\beta/C_{\beta'}/C_\gamma$ av.: 48.2 ppm; C_α: 133.9 ppm) are in very good agreement with the experimental values for the 1-trimethylsilyl-substituted bicyclobutonium ion **5** ($C_\beta/C_{\beta'}/$ C_γ av.: 48.9 ppm; C_α: 137.4 ppm), while this is not the case for the chemical shifts calculated for the (1'-silylcyclopropyl)methyl cation **7** (see Fig. 7).

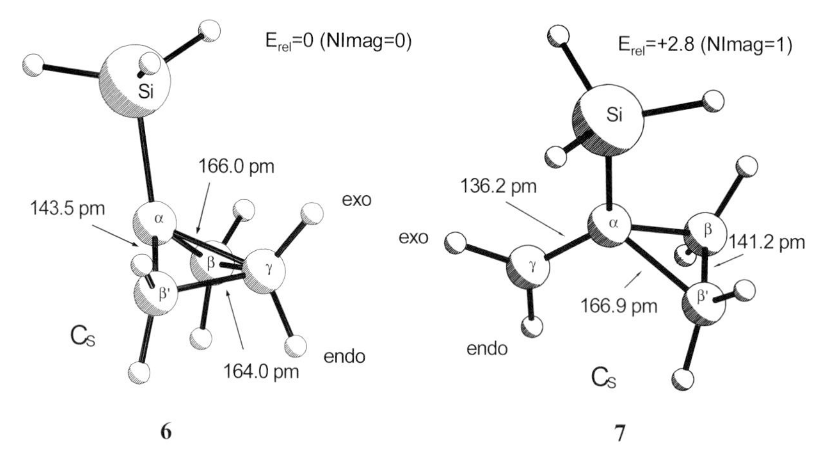

Fig. 6. Ab initio calculated (MP2/6-31G(d)) geometries of the 1-silylbicyclobutonium ion **6** and the (1'-silylcyclopropyl)methyl cation **7**; selected bond lengths in ppm and relative energies in kcal/mol (ZPE included).

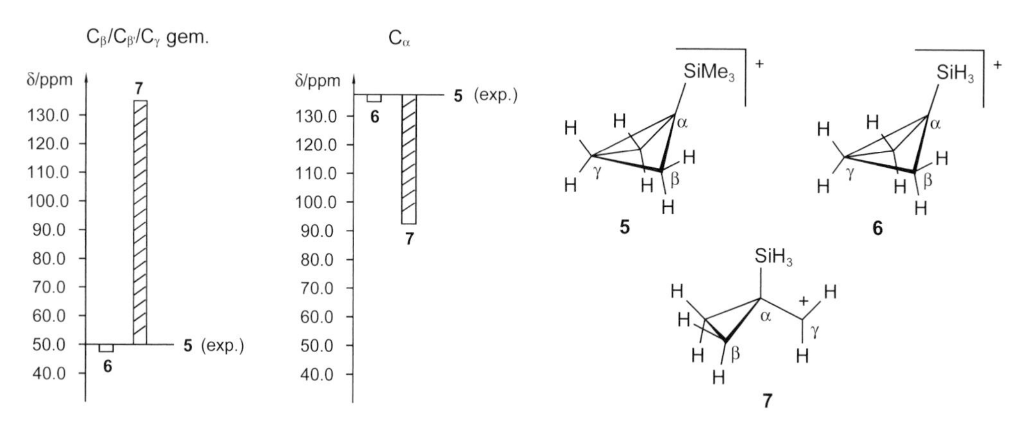

Fig. 7. Comparison of the experimental ^{13}C-NMR chemical shifts of **5** with the data calculated (GIAO-MP2/tzpdz) for **6** and **7**.

3-Silyl-Substituted Bicyclobutonium Ions

Kinetic investigations of γ-silyl-substituted cyclohexyl tosylates have shown that a γ-silyl group is superior to a γ-hydrogen in the stabilization of a positive charge (γ-silyl effect [2a]). We have successfully generated a static 3-trialkylsilylbicyclobutonium ion stabilized by the silyl group in the γ-position.

Experimental Results

Matrix co-condensation of (1'-(*tert*-butyldimethylsilyl)cyclopropyl)methanol (Scheme 1, **A** (R = *tert*-butyl; R' = Me)) with SbF_5 onto a surface of SO_2ClF/SO_2F_2 at –196°C yields after homogenization at –130°C a yellow solution. The initial ^{13}C-NMR spectrum (Fig. 8, lower spectrum), obtained at –130°C shows two sets of signals, one set corresponding to the

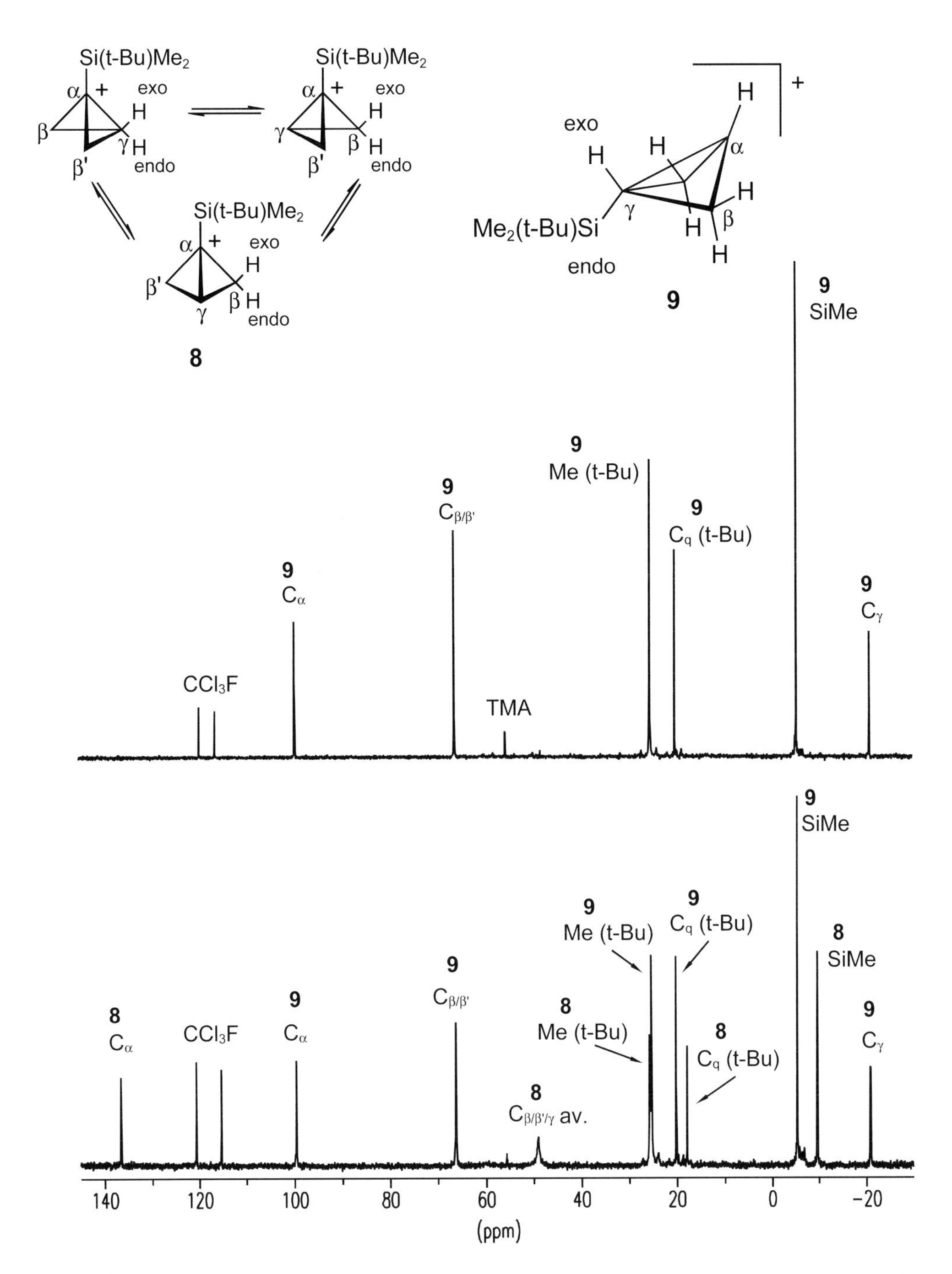

Fig. 8. ^{13}C-NMR spectra; rearrangement of the 1-(*tert*-butyldimethylsilyl)bicyclobutonium ion **8** to the 3-*endo*-(*tert*-butyldimethylsilyl)bicyclobutonium ion **9** at –115°C (internal standard TMA $\delta(NMe_4^+)$ = 55.65 ppm).

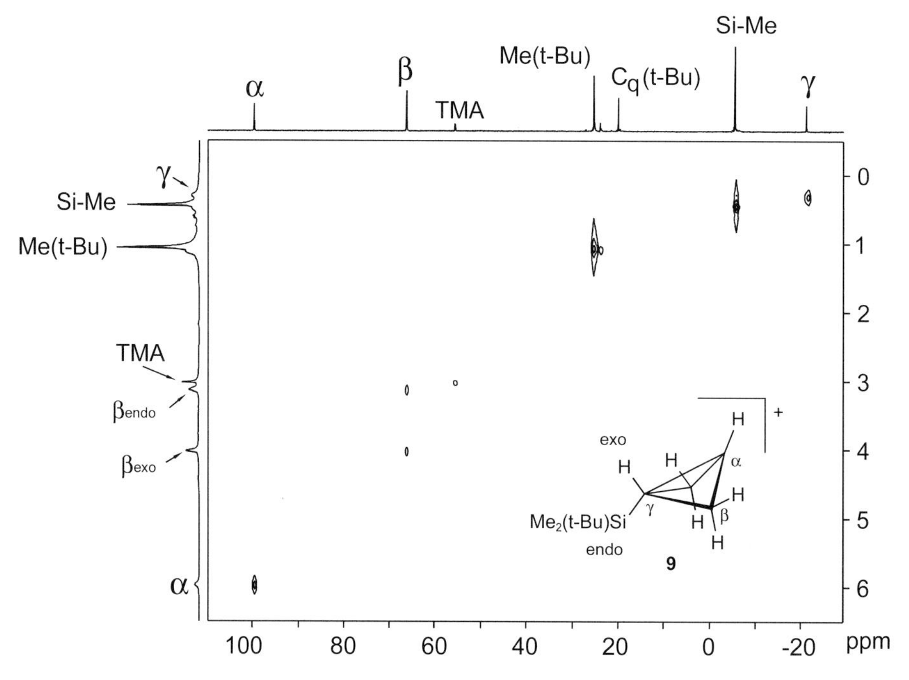

Fig. 9. HC-COSY-NMR spectrum of **9** at –115°C; internal standard TMA $\delta(NMe_4^+)$ = 55.65 ppm (^{13}C) and 3.00 ppm (^1H).

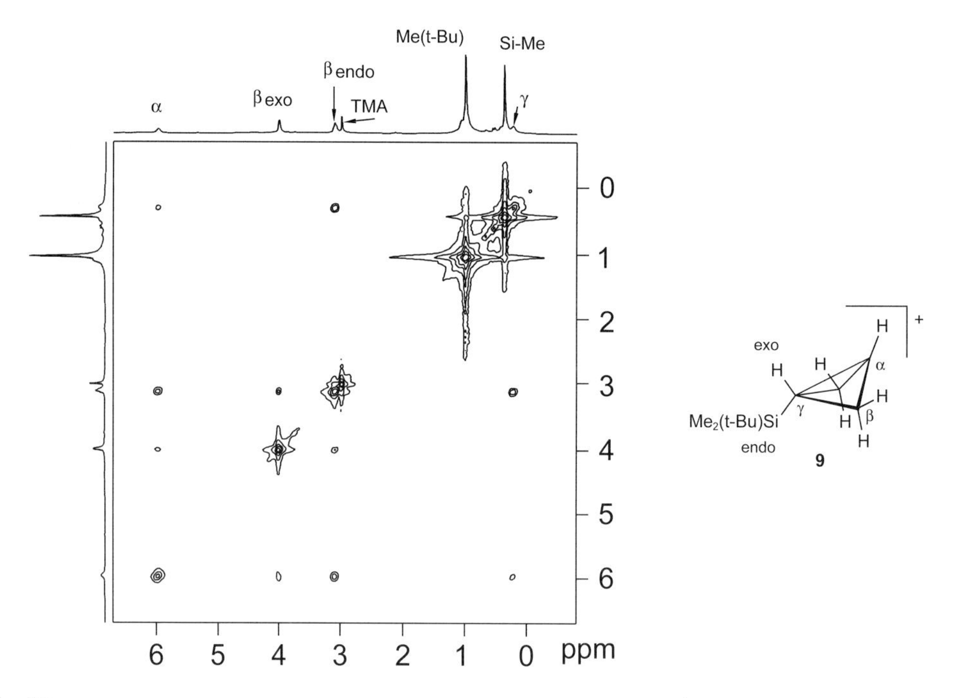

Fig. 10. COSY45-NMR spectrum of **9** at –115°C (internal standard TMA $\delta(NMe_4^+)$ = 3.00 ppm).

1-(*tert*-butyldimethylsilyl)bicyclobutonium ion **8** (136.5 ppm; 49.0 ppm and signals for the Me, *tert*-butyl groups), and another set of signals (99.6 ppm (d/$^{1}J_{CH}$ = 184 Hz); 66.3 ppm (t/$^{1}J_{CH}$ = 171 Hz); –21.0 ppm (d/$^{1}J_{CH}$ = 156 Hz) and signals for the Me, *tert*-butyl groups) corresponding to the 3-*endo*-(*tert*-butyldimethylsilyl)bicyclobutonium ion **9**.

At –115 °C the first set of signals disappears within 10 minutes and only the peaks for cation **9** remain (Fig. 8; upper spectrum). Structural assignment for cation **9** was confirmed by HC-COSY- and COSY45-NMR spectra shown in Figs. 9 and 10, respectively. Experiments with β-CD_2-labeled progenitors and quantum chemical model calculations of transition states for 1,3-hydride shifts indicate that the rearrangement of the 1-silyl-substituted bicyclobutonium ion **8** to the 3-silyl-substituted bicyclobutonium ion **9** occurs most probably by a 1,3-hydride shift from C_γ to C_α across the bridging bond.

3-Silyl-substituted bicyclobutonium ions are also accessible from direct ionization of 3-silyl-substituted cyclobutyl chlorides. Matrix co-condensation of *cis/trans*-3-(trimethylsilyl)cyclobutyl chloride (Scheme 1, **D** (R = R' = Me)) with SbF_5 onto a surface of SO_2ClF/SO_2F_2 at –196°C yields after homogenization at –130°C a yellow solution of carbocation **12**. The ^{13}C-NMR spectrum obtained for cation **12**, except for the alkyl groups at silicon, is very similar to the ^{13}C-NMR spectrum of carbocation **9**. The differences of the signals for the C_α-, $C_\beta/C_{\beta'}$-carbons are smaller than 1 ppm.

Computational Results [5]

The geometries of the model structures 3-*endo*-silylbicyclobutonium ion [3-endo-SiH_3-C_4H_6]$^+$ **10** and 3-*exo*-silylbicyclobutonium ion [3-*exo*-SiH_3-C_4H_6]$^+$ **11** were optimized at the MP2/6-31G(d) level of theory (Fig. 11). The 3-*endo*-silylbicyclobutonium ion **10** is an energy minimum (NImag = 0) and is calculated to be 7.9 kcal/mol lower in energy than the 3-*exo*-silylbicyclobutonium ion **11**, which is characterized by a frequency calculation as a transition state (NImag = 1). The distance between the C_α and the C_γ carbon in cation **10** (164.1 pm) is shorter than the C_α–C_γ distance calculated for the unsubstituted bicyclobutonium ion **1** (165.4 pm). This indicates a stronger bonding interaction between C_α and C_γ for **10** which is due to the stronger stabilizing interaction of the *endo*-silyl group at the C_γ carbon with the formally positively charged carbon C_α as compared to the C_γ–*endo*-H– C_α interaction in **1**.

The ^{13}C-NMR chemical shifts calculated (GIAO-MP2/tzpdz) for the 3-*endo*-silylbicyclobutonium ion **10** are in good agreement with the experimental values for the 3-*endo*-(*tert*-butyldimethylsilyl)bicyclobutonium ion **9** (Fig. 12) and the 3-*endo*-(trimethylsilyl)bicyclobutonium ion **12**. The calculated chemical shifts for C_α and $C_\beta/C_{\beta'}$ of the γ-*endo*-silyl isomer **10** are in better agreement with the experimental data than those calculated for the γ-*exo*-silyl isomer **11**. The assignment is also confirmed by SOS-DFT (Perdew/IGLO-III) calculation of the cross ring $^{3}J_{H_\alpha H_\gamma}$ spin–spin coupling constant, which is 5.5 Hz measured experimentally and 5.9 Hz calculated for the *endo*-silyl isomer **10** but is only 1.2 Hz calculated for the *exo*-silyl isomer **11**.

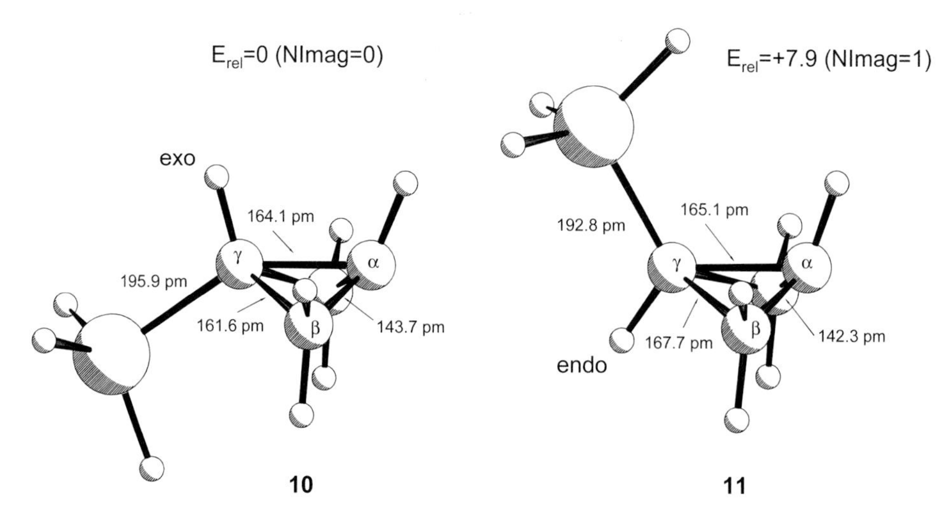

Fig. 11. Ab initio calculated geometries (MP2/6-31G(d)) and relative energies [kcal/mol] (ZPE included) of the 3-*endo*-silylbicyclobutonium ion **10** and the 3-*exo*-silylbicyclobutonium ion **11**.

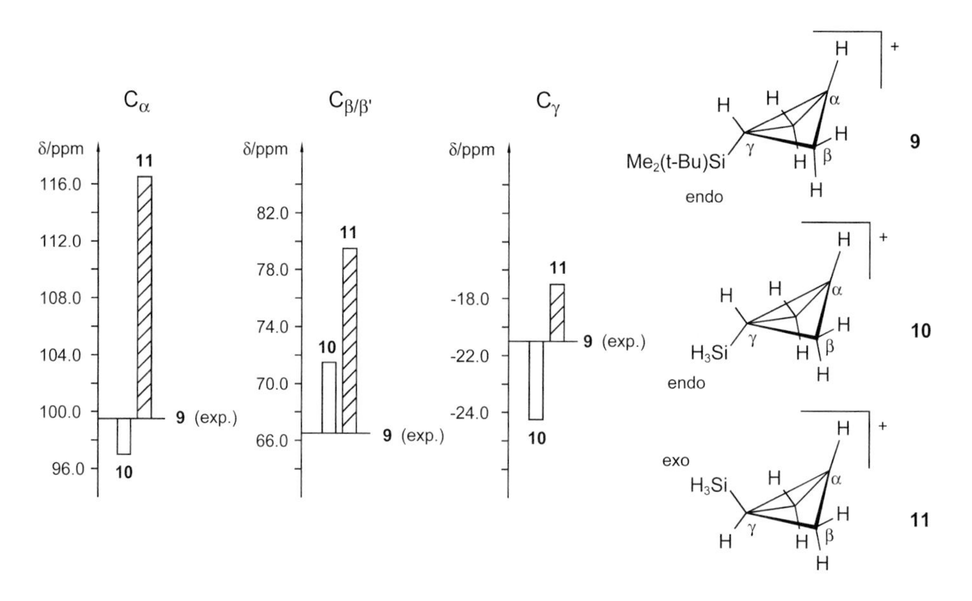

Fig. 12. Comparison of the experimental ^{13}C-NMR chemical shifts of **9** with the data calculated (GIAO-MP2/tzpdz) for **10** and **11**, geometries: MP2/6-31G(d).

Conclusions

The 1-(trimethylsilyl)cyclobutyl cation is obtained by reaction of [1'-(trimethylsilyl)cyclopropyl]-methanol with SbF_5 at –130°C. It has a hypercoordinated puckered 1-(trimethylsilyl)-bicyclobutonium structure **5**. Cation **5** undergoes a fast threefold degenerate methylene rearrangement (compare Fig. 2 with R = $SiMe_3$). Averaged NMR signals are observed for the *exo*-methylene protons (*exo* $H_\beta/H_{\beta'}/H_\gamma$), for the *endo*-methylene protons (*endo* $H_\beta/H_{\beta'}/H_\gamma$), and for the methylene carbon atoms ($C_\beta/ C_{\beta'}/ C_\gamma$) respectively. The [1'-(trimethylsilyl)cyclopropyl]methyl cation does not contribute to the NMR chemical shifts. The reaction of [1'-(*tert*-butyldimethylsilyl)cyclopropyl]methanol with SbF_5 at –130°C leads to the 1-(*tert*-butyldimethylsilyl)bicyclobutonium ion **8**. Like **5**, cation **8** undergoes also a fast methylene rearrangement leading to ^{13}C- and ^{1}H-NMR spectra with averaged methylene signals. At –115°C **8** is converted to the 3-*endo*-(*tert*-butyldimethylsilyl)bicyclobutonium ion **9**. Cation **9** is the first bicyclobutonium ion that is static on the NMR time scale. This is due to the efficient stabilization of the positive charge by the γ-*endo*-trialkylsilyl substituent. The CH-COSY- and COSY45-NMR spectra confirm the structural and stereochemical assignment. The 3-*endo*-(trimethylsilyl)bicyclobutonium ion **12** is generated directly from *cis/trans*-3-(trimethylsilyl)cyclobutyl chloride. The quantum chemical calculations of chemical shifts and spin–spin coupling constants fully support the interpretation of the experimental results.

References:

[1] For reviews see: a) D. Lenoir, H.-U. Siehl, in: *Houben-Weyl Methoden der Organischen Chemie*, Vol. E19c, *Carbokationen* (Ed.: M. Hanack), Thieme, Stuttgart, **1990**, p. 413.
b) G. A. Olah, V. P. Prakash Reddy, G. K. Surya Prakash, *Chem. Rev.* **1992**, *92*, 69.
c) G. A. Olah, J. Sommer, G. K. Surya Prakash, *Superacids*, John Wiley, New York, **1985**, p. 143.

[2] a)H.-U. Siehl, T. Müller in: *The Chemistry of Organic Silicon Compounds*, *Vol. 2* (Eds.: Z. Rappoport, Y. Apeloig), John Wiley, New York, **1998**, Chapter 14.
b) H.-U. Siehl, B. Müller, M. Fuß, Y. Tsuji, in: *Organosilicon Chemistry II: From Molecules to Materials* (Eds.: N. Auner, J. Weis), VCH, Weinheim, **1996**; p. 360.
c) H.-U. Siehl, B. Müller, O. Malkina, in: *Organosilicon Chemistry III: From Molecules to Materials* (Eds.: N. Auner, J. Weis), Wiley–VCH, Weinheim, **1997**, p. 25.

[3] H.-U. Siehl, M. Fuß, J. Gauss, *J. Am. Chem. Soc.* **1995**, *117*, 5983.

[4] Martin Fuß, *Dissertation*, Universität Tübingen, **1997**.

[5] a) Geometry optimizations and frequency calculations have been performed using the *Gaussian 94* program suite: M. J. Frisch, G. W. Trucks, H. B. Schlegel, P. M. W. Gill, B. G. Johnson, M. A. Robb, J. R. Cheeseman, T. Keith, G. A. Petersson, J. A. Montgomery, K. Raghavachari, M. A. Al-Laham, V. G. Zakrzewski, J. V. Ortiz, J. B. Foresman, J. Cioslowski, B. B. Stefanov, A. Nanayakkara, M. Challacombe, C. Y. Peng, P. Y. Ayala, W. Chen, M. W. Wong, J. L. Andres, E. S. Replogle, R. Gomperts, R. L. Martin, D. J. Fox, J. S. Binkley, D. J. Defrees, J. Baker, J. P. Stewart, M. Head-Gordon, C. Gonzalez, J. A. Pople, *Gaussian 94, Revision E.1*, Gaussian, Inc., Pittsburgh (PA), **1995**.
b) GIAO-MP2 calculations of chemical shifts have been performed using the *ACES II* program suite: J. F. Stanton, J. Gauss, J. D. Watts, W. J. Lauderdale, R. J. Bartlett, *ACES II*, University of Florida, Gainesville (Fl), **1993**; for a detailed description of *ACES II* see: J. F. Stanton, J. Gauss, J. D. Watts, W. J. Lauderdale, R. J. Bartlett, *Int. J. Quantum Chem. Symp.* **1992**, *26*, 879.
c) D. Salahub, V. Malkin, O. Malkina, *deMon/MASTER-JMN*, Université de Montreal, Quebec, Canada, **1993/1994**.

Silyl Stabilization in Cyclopropylmethyl Cations: NMR-Spectroscopic and Quantum Chemical Results

Hans-Ullrich Siehl*, Alexander Christian Backes

Abteilung für Organische Chemie I der Universität Ulm
Albert-Einstein-Allee 11, D–89069 Ulm, Germany
Tel./Fax: Int. code + (731) 5022800
E-mail: ullrich.siehl@chemie.uni-ulm.de

Olga Malkina

Computing Center, Slovak Academy of Sciences
SK-84236 Bratislava, Slovakia

Keywords: Carbocations / Silicon / NMR Spectroscopy / DFT Calculations

Summary: The *E*-1-cyclopropyl-2-(triisopropylsilyl)ethyl cation is formed by protonation of *E*-1-(triisopropylsilyl)ethenyl-cyclopropane with FSO_3H/SbF_5 at –135°C and was characterized by ^{1}H- and ^{13}C-NMR spectra in SO_2F_2/SO_2ClF solution at –105°C. Quantum chemical DFT ab initio calculations of NMR chemical shifts and spin–spin coupling constants were performed for the model structures *E*- and *Z*-1-cyclopropyl-2-(trimethylsilyl)ethyl cations. The calculated NMR data of the *E*-1-cyclopropyl-2-(trimethylsilyl)ethyl cation are in good agreement with the experimental data and show that this β-silyl-substituted secondary cyclopropylmethyl carbocation is static on the NMR time scale in the observed temperature range.

Introduction

An important property of trialkylsilyl groups is their ability to stabilize a positive charge in the β–position [1]. The combination of experimental high resolution NMR spectroscopy in solution and quantum chemical ab initio calculations of NMR chemical shifts and spin–spin coupling constants is a powerful tool to obtain detailed information on the structure and stabilization mode in silicon-substituted reactive carbocations.

The interplay of experimental and computational investigations is demonstrated for the first β-silyl substituted cyclopropylmethyl cation, the *E*-1-cyclopropyl-2-(triisopropylsilyl)ethyl cation (**1**). Comparison of experimental NMR data of **1** with calculated NMR data for the analogous $SiMe_3$-substituted model structures *E*-1-cyclopropyl-2-(trimethylsilyl)ethyl cation (**2a**) and *Z*-1-cyclopropyl-2-(trimethylsilyl)ethyl cation (**2b**) allows the detailed interpretation of the observed data and an unequivocal assignment of the *E*-stereochemistry of **1**.

1 **2a** **2b**

Results

Matrix-co-condensation of *E*-1-(triisopropylsilyl)ethenyl-cyclopropan (**3**) with a 2:1 mixture of FSO_3H and SbF_5 onto a surface of SO_2ClF/SO_2F_2 cooled to –196°C yields after homogenization at –135°C a red solution of *E*-1-cyclopropyl-2-(triisopropylsilyl)ethyl cation (**1**).

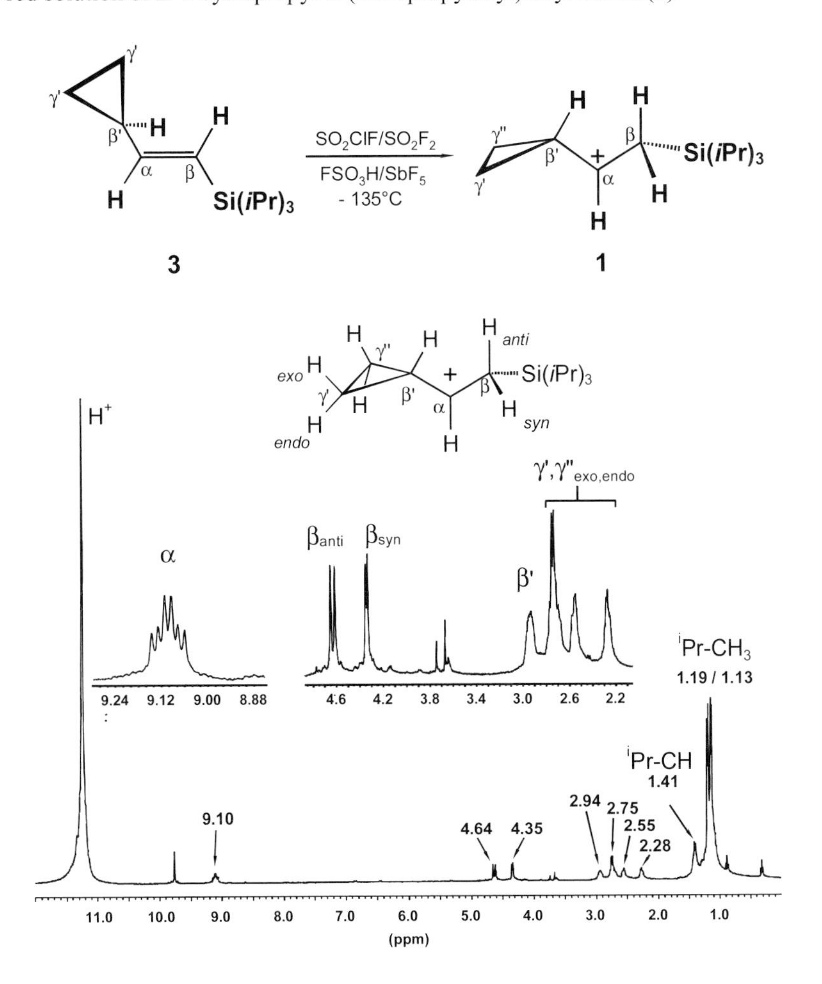

Fig. 1. 400.13 MHz ^{1}H-NMR spectrum of *E*-1-cyclopropyl-2-(triisopropylsilyl)ethyl cation (**1**) at –105°C in SO_2ClF/SO_2F_2 (internal standard TMA, $\delta(NMe_4^+) = 3.00$ ppm).

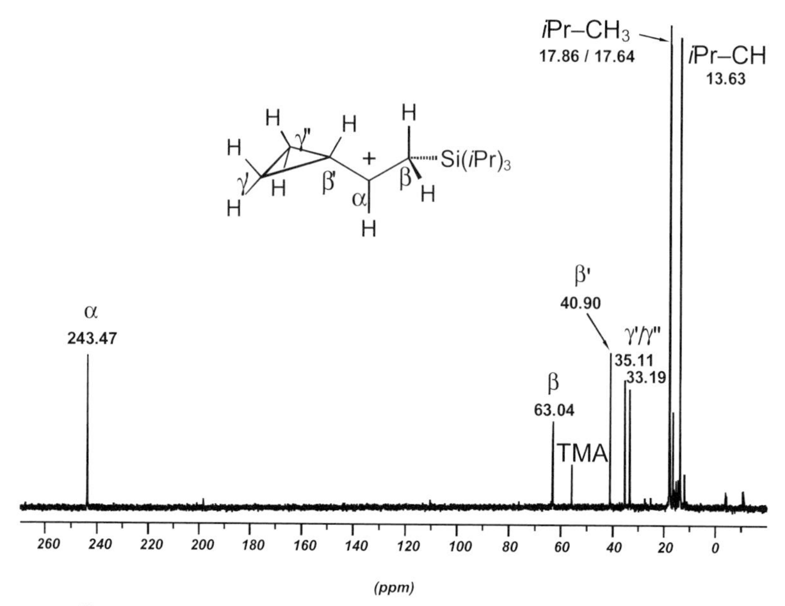

Fig. 2. 100 MHz ^{13}C-NMR spectrum of *E*-1-cyclopropyl-2-(triisopropylsilyl)ethyl cation (**1**) at −105°C in SO_2ClF/SO_2F_2 (internal standard TMA, $\delta(NMe_4^+)$ = 55.65 ppm).

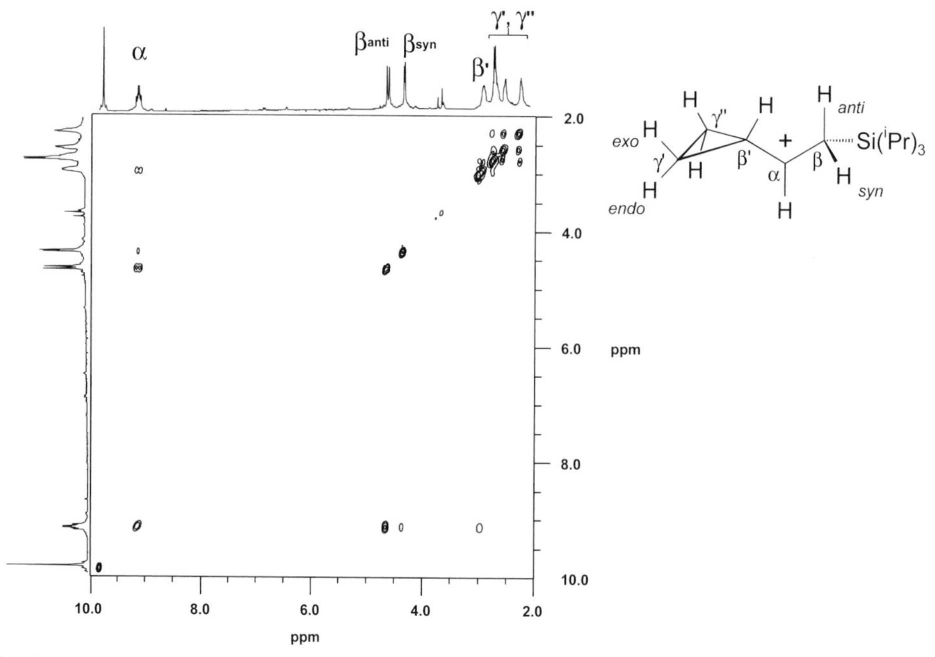

Fig. 3. 400.13 MHz H,H-COSY45 NMR spectrum of *E*-1-cyclopropyl-2-(triisopropylsilyl)ethyl cation (**1**) at −105°C in SO_2ClF/SO_2F_2 (internal standard TMA, $\delta(NMe_4^+)$ = 3.00 ppm).

The 400.13 MHz ^{1}H-NMR spectrum in SO_2ClF/SO_2F_2 at –105°C (Fig. 1) shows the signal of the proton H_α attached to the positively charged carbon atom at 9.10 ppm as a doublet of doublets of doublets. The coupling pattern results from coupling with $H_{\beta(syn)}$ at 4.35 ppm (doublet, $^3J_{HH}$ = 7.0 Hz), coupling with $H_{\beta(anti)}$ at 4.64 ppm (doublet, $^3J_{HH}$ = 14.1 Hz) and a $^3J_{HH}$ doublet coupling of 13.30 Hz with $H_{\beta'}$ (multiplet) at 2.94 ppm. Since the two coupling constants are similar, the doublet of doublets of doublets collapses to a doublet of triplets. The assignment was confirmed by specific H,H-decoupling experiments and a 400.13 MHz H,H-COSY45 NMR experiment (Fig. 3).

The two different $^3J_{HH}$ coupling constants $^3J_{H\alpha H\beta(syn)}$ = 7.0 Hz and $^3J_{H\alpha H\beta(anti)}$ = 14.1 Hz for the magnetically non-equivalent C_β-methylene protons ($H_{\beta(syn)}$, 4.35 ppm; $H_{\beta(anti)}$, 4.64 ppm) are comparable to those known from carbocations of similar structure (i.e. 1-phenyl-2-(triisopropylsilyl)ethyl cation: 5.7 Hz and 14.9 Hz for *syn*- and *anti*-$^3J_{HH}$ respectively) [2]. The proton signals are temperature-independent between –150°C and –80°C. This is in accord with a static structure and suggests the fixed conformation **1-1** (Fig. 4) for carbocation **1**. The suggested stable conformation **1-1** has a parallel orientation of the C_β–Si- bond and the vacant p_z orbital at C_α and thus allows better overlap than in the other conformations **1-2** and **1-3**.

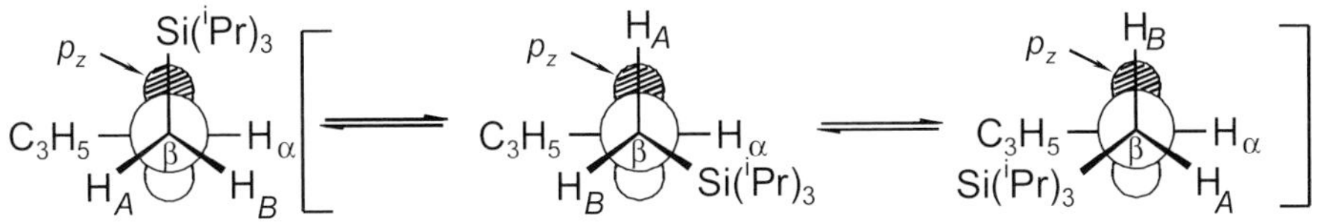

Fig. 4. Newman projections of staggered conformations **1-1**, **1-2** and **1-3** in *E*-1-cyclopropyl-2-(triisopropylsilyl)ethyl cation (**1**); H_A and H_B correspond to $H_{\beta(anti)}$ and $H_{\beta(syn)}$ in text.

The 100 MHz ^{13}C-NMR spectrum in SO_2ClF/SO_2F_2 at –105°C (Fig. 2) shows the C^+ signal at 243.47 ppm (doublet; $^1J_{CH}$ = 159.9 Hz). Assignments were done by ^{1}H-coupled ^{13}C-NMR spectra and a two-dimensional C,H-COSY spectrum. The chemical shift for the C^+ carbon signal is in good agreement with comparable signals in other cyclopropylmethyl carbocations. In the *E*-1-(cyclopropyl)ethyl cation the C^+ signal appears at 252.2 ppm (–80°C in SO_2ClF solution; internal reference $CFCl_3$, δ = 117.9 ppm)[3]. The C^+ carbon signal in the 1-phenyl-2-(triisopropylsilyl)ethyl cation appears at 213.74 ppm (–123°C in SO_2ClF/SO_2F_2 solution; capillary reference TMS in CD_3COCl/SO_2ClF, δ = 0.0 ppm) [2].

Calculations of NMR chemical shifts were performed with the Gaussian 94 program package [4] with the GIAO approach using the B3LYP hybrid method and 6-311G(d,p) basis sets, for MP2/6-31G(d) and B3LYP/6-31G(d) optimized geometries, on model cations *E*-1-cyclopropyl-2-(trimethylsilyl)ethyl (**2a**) and *Z*-1-cyclopropyl-2-(trimethylsilyl)ethyl (**2b**) (Tables 2, 3; Figs. 5, 6).

Table 1. Experimental ^{1}H-NMR chemical shifts [ppm] of *E*-1-cyclopropyl-2-(triisopropylsilyl)ethyl cation (**1**) and calculated ^{1}H-NMR chemical shifts [ppm] of the model cations *E*-1-cyclopropyl-2-(trimethylsilyl)ethyl cation (**2a**) and *Z*-1-cyclopropyl-2-(trimethylsilyl)ethyl cation (**2b**).

		$H_{\beta(syn)}$	$H_{\beta(anti)}$	H_α	$H_{\beta'}$	$H_{\gamma'/\gamma''(exo/endo)}$	Other
1	Expt.[a]	4.35	4.64	9.10	2.94	2.27 – 2.75	*i*Pr-CH 1.41
							*i*Pr-CH_3 1.13 – 1.19
2a	Calc.[b]	4.55	4.68	8.60	2.57	2.23 – 2.86	
	Calc.[c]	4.34	4.42	8.83	2.69	2.44 – 3.01	
2b	Calc.[b]	4.30	3.76	10.15	2.89	2.38 – 3.17	
	Calc.[c]	4.18	3.57	10.41	3.02	2.69 – 3.39	

[a] 400.13 MHz, $T = -105$°C in SO_2ClF/SO_2F_2, internal reference NMe_4^+, $\delta = 3.00$ ppm. — [b] GIAO-B3LYP/6-311G(d,p) at MP2/6-31G(d) geometry, [NImag = 0]. — [c] GIAO-B3LYP/6-311G(d,p) at B3LYP/6-31G(d) geometry, [NImag = 0]. Abs. chemical shifts for TMS: GIAO-B3LYP/6-311G(d,p) at MP2/6-31G(d) geometry (T_d symmetry; NImag = 0): $\delta(^1H) = 32.00$ ppm; GIAO-B3LYP/6-311G(d,p) at B3LYP/6-31G(d) geometry (T_d symmetry; NImag = 0): $\delta(^1H) = 31.92$ ppm.

Table 2. Experimental ^{13}C-NMR chemical shifts [ppm] of *E*-1-cyclopropyl-2-(triisopropylsilyl)ethyl cation (**1**) and calculated ^{13}C-NMR chemical shifts [ppm] of the model cations *E*-1-cyclopropyl-2-(trimethylsilyl)ethyl cation (**2a**) and *Z*-1-cyclopropyl-2-(trimethylsilyl)ethyl cation (**2b**).

		C_β	C_α	$C_{\beta'}$	C_γ	$C_{\gamma'}$	Other
1	Expt.[a]	63.04	243.47	40.90	35.11	33.19	*i*Pr-CH 13.63
							*i*Pr-CH_3 17.86, 17.64
2a	Calc.[b]	80.24	241.47	41.87	36.06	33.27	
	Calc.[c]	76.28	248.84	45.61	40.55	37.12	
2b	Calc.[b]	71.86	243.76	40.59	31.34	53.10	
	Calc.[c]	68.83	252.04	44.49	36.20	56.88	

[a] 100 MHz, $T = -105$°C in SO_2ClF/SO_2F_2, internal reference NMe_4^+, $\delta = 55.65$ ppm. — [b] GIAO-B3LYP/6-311G(d,p) at MP2/6-31G(d) geometry, [NImag = 0]. — [c] GIAO-B3LYP/6-311G(d,p) at B3LYP/6-31G(d) geometry, [NImag = 0]. Abs. chemical shifts for TMS: GIAO-B3LYP/6-311G(d,p) at MP2/6-31G(d) geometry (T_d symmetry; NImag = 0): $\delta(^{13}C) = 184.45$ ppm; GIAO-B3LYP/6-311G(d,p) at B3LYP/6-31G(d) geometry (T_d symmetry; NImag = 0): $\delta(^{13}C) = 183.86$ ppm.

The agreement with the experiment for the ^{1}H-NMR chemical shifts calculated for MP2 and B3LYP optimized geometries is better for *E*-isomer **2a** than for Z-isomer **2b** (Fig. 7). The difference for H_α of **2a** ($\delta_{exp.} = 9.10$ ppm) is $\Delta\delta_{(exp.-calc.)} = -0.5$ and -0.27 ppm for MP2 and B3LYP geometry whereas for **2b** $\Delta\delta = +1.05$ and $+1.31$ ppm (MP2 and B3LYP geometry). The shift difference between $H_{\beta(syn)}$ and $H_{\beta(anti)}$ ($\Delta\delta_{exp.} = 0.29$ ppm) is calculated for **2a** as $\Delta\delta = 0.13$ and 0.08 ppm (MP2 and B3LYP geometry) whereas for the *Z*-isomer **2b** $\Delta\delta = 0.54$ and 0.61 ppm (MP2 and B3LYP geometry). The better agreement of the data calculated for **2a** is evidence for the *E*-configuration of **1**.

The calculated ^{13}C NMR chemical shifts for the *E*-isomer **2a** are also in agreement with the *E*-configuration of **1** (Table 2; Fig. 8). In particular the large difference calculated for the C_γ and $C_{\gamma'}$ carbons in *Z*-isomer **2b** ($\Delta\delta_{(C\gamma' - C\gamma'')} = 21.76$ and 20.68 ppm for MP2 and B3LYP geometries), which is due to the geometric distortions of the cyclopropyl ring caused by large steric hindrance of the silyl substituent (Fig. 6), is not in accord with the experimental shifts ($\Delta\delta_{C(\gamma' - C\gamma'')}$ =1.92 ppm) observed for **1**. The chemical shift difference $\Delta\delta_{C(\gamma' - C\gamma'')}$ = 2.79 and 3.43 ppm (MP2 and B3LYP geometries) calculated for the *E*-isomer **2a** is in reasonably good agreement with the experimental observation. This confirms the assignment of the *E*-configuration to **1**.

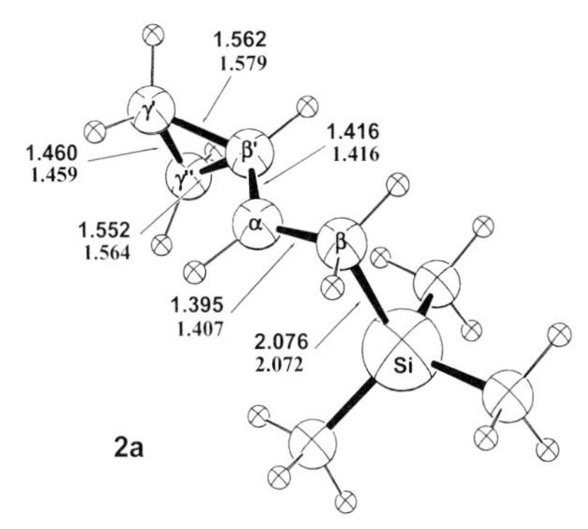

Fig. 5. Calculated geometry of model structure *E*-1-cyclopropyl-2-(trimethylsilyl)ethyl cation (**2a**) with selected bond lengths in Å (upper: MP2/6-31G(d); lower: B3LYP/6-31G(d); angle Si-C_β-C_α 100.6° (MP2), 107.4° (B3LYP); dihedral angle Si-C_β-C_α-$C_{\beta'}$ –90.8° (MP2), –94.9° (B3LYP). Rel. energies: MP2 + ZPE 0 kcal/mol, B3LYP + ZPE 0 kcal/mol.

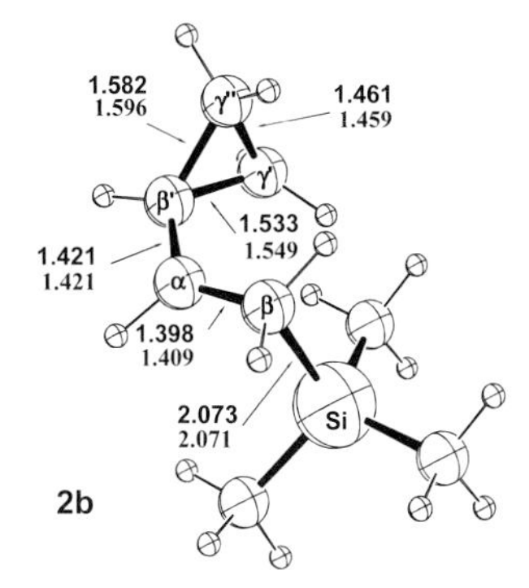

Fig. 6. Calculated geometry of model structure *Z*-1-cyclopropyl-2-(trimethylsilyl)ethyl cation (**2b**) with selected bond lengths in Å (upper: MP2/6-31G(d); lower: B3LYP/6-31G(d)); angle Si-C_β-C_α 101.8° (MP2), 107.6° (B3LYP); dihedral angle Si-C_β-C_α-$C_{\beta'}$ –100.6° (MP2), –102.9° (B3LYP). Rel. energies: MP2 + ZPE 2.81 kcal/mol, B3LYP + ZPE 3.07 kcal/mol.

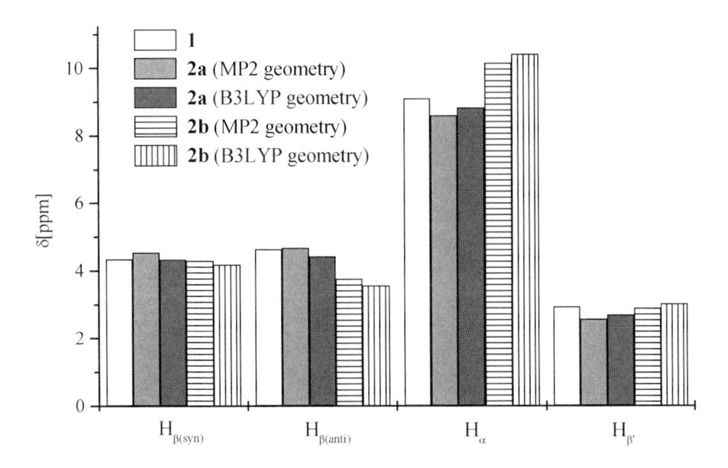

Fig. 7. Experimental ^{1}H-NMR chemical shifts [ppm] of *E*-1-cyclopropyl-2-(triisopropylsilyl)ethyl cation (**1**) and calculated ^{1}H-NMR chemical shifts [ppm] of the model structures *E*-1-cyclopropyl-2-(trimethylsilyl)ethyl cation (**2a**) and *Z*-1-cyclopropyl-2-(trimethylsilyl)ethyl cation (**2b**) (B3LYP/6-311G(d,p) both at MP2/6-31G(d) and B3LYP/6-31G(d) geometry).

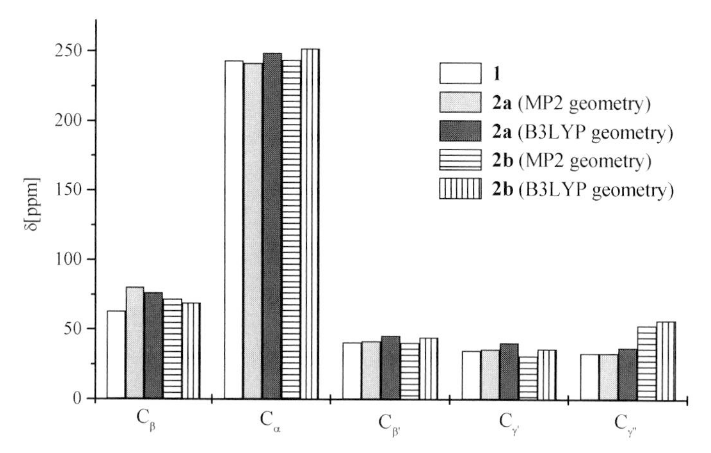

Fig. 8. Experimental ^{13}C-NMR chemical shifts [ppm] of *E*-1-cyclopropyl-2-(triisopropylsilyl)ethyl cation (**1**) and calculated ^{13}C-NMR chemical shifts [ppm] of the model structures *E*-1-cyclopropyl-2-(trimethylsilyl)ethyl cation (**2a**) and *Z*-1-cyclopropyl-2-(trimethylsilyl)ethyl cation (**2b**) (B3LYP/6-311G(d,p) both at MP2/6-31G(d) and B3LYP/6-31G(d) geometry).

Calculations of spin–spin coupling constants $^3J_{H\alpha H\beta'}$ (12.3 Hz (**2a**) and 7.9 Hz (**2b**); exp. 13.3 Hz), $^3J_{H\alpha H\beta(syn)}$ (6.9 Hz (**2a**) and 6.4 Hz (**2b**); exp. 7.0 Hz) and $^3J_{H\alpha H\beta(anti)}$ (14.2 Hz (**2a**) and 15.8 Hz (**2b**); exp. 14.1 Hz) were performed with the SOS-DFT method implemented in the deMon / MASTER-JMN program using the Perdew DFT functional and IGLO-III level basis sets for the MP2/6-31G(d) optimized geometries of **2a** and **2b** [5]. The agreement of the calculated values with the experimental data of **1** is generally better for *E*-isomer **2a** than for *Z*-isomer **2b**. Especially, the $^3J_{H\alpha H\beta'}$ coupling constant calculated for the *cis* orientation of H_α and $H_{\beta'}$ in the *Z*-isomer **2b** (7.9 Hz) is much smaller than the experimental value for **1** (13.3 Hz). The agreement for $^3J_{H\alpha H\beta'}$ in *E*-isomer **2a** (12.3 Hz) with the experimental value of 13.3 Hz in **1** confirms the *trans* arrangement of H_α and $H_{\beta'}$ and thus the *E*-configuration of **1**.

Conclusion

The combination of experimental and calculated determination of NMR chemical shifts and spin–spin coupling constants allows the characterization of the *E*-1-cyclopropyl-2-(triisopropylsilyl)-ethyl cation (**1**) and an unequivocal assignment of its stereochemistry. Due to the stabilizing β-hyperconjugative interaction of both the strained cyclopropyl C–C bonds and the β-C–Si bond with the vacant orbital at C_α the rotation around the C_α–C_β and C_α–$C_{\beta'}$ bonds is frozen under the experimental conditions between –150°C and –80°C.

Acknowledgment: This work was supported by the *Deutsche Forschungsgemeinschaft* and the *Fonds der Chemischen Industrie.*

References:

[1] a) H.-U. Siehl, B. Müller, Y. Tsuji in: *Organosilicon Chemistry II: From Molecules to Materials* (Eds.: N. Auner, J. Weis), VCH, Weinheim, **1996**, p. 361.
b) H.-U. Siehl, *Pure Appl. Chem.* **1995**, *76*, 769.
c) J. B. Lambert, Y. Zhao, *J. Am. Chem. Soc.* **1996**, *118*, 7867.
d) J. B. Lambert, R. W. Emblidge, S. Malany *J. Am.Chem. Soc.* **1993**, *115*, 1317.
e) N. Shimizu, *Rev. Het. Chem.* **1993**, 55.
f) A. K. Nguyen, M. S. Gordon, G. Wang, J. B. Lambert, *Organometallics* **1991**, *10*, 2798.
g) H.-U. Siehl, F. P. Kaufmann, Y. Apeloig, V. Braude, D. Danovich, A. Berndt, N. Stamatis, *Angew. Chem., Int. Ed. Engl.* **1991**, *30*, 1479.

[2] a) B. Müller, Ph. D. thesis, Universität Tübingen, **1995**.
b) H.-U. Siehl, B. Müller, O. Malkina, in: *Organosilicon Chemistry III: From Molecules to Materials* (Eds.: N. Auner, J. Weis), Wiley–VCH, Weinheim, **1998**, p. 25.

[3] C. Falkenberg-Andersen, K. Ranganayakulu, L. R. Schmitz, T. S. Sorensen, *J. Am. Chem. Soc.* **1984**, *106*, 178.

[4] M. J. Frisch, G. W. Trucks, H. B. Schlegel, P. M. W. Gill, B. G. Johnson, M. A. Robb, J. R. Cheeseman, T. Keith, G. A. Petersson, J. A. Montgomery, K. Raghavachari, M. A. Al-Laham, V. G. Zakrzewski, J. V. Ortiz, J. B. Foresman, J. Cioslowski, B. B. Stefanov, A. Nanayakkara, M. Challacombe, C. Y. Peng, P. Y. Ayala, W. Chen, M. W. Wong, J. L. Andres, E. S. Replogle, R. Gomperts, R. L. Martin, D. J. Fox, J. S. Binkley, D. J. Defrees, J. Baker, J. P. Stewart, M. Head-Gordon, C. Gonzalez, J. A. Pople, *Gaussian 94*, *Revision E.1*, Gaussian, Inc., Pittsburgh (PA), **1995**.

[5] D. Salahub, V. Malkin, O. Malkina, *deMon/MASTER-JMN*, Université de Montreal, Quebec, Canada, **1993/1994**.

N,*O*-Dimethyl-*N*-silylhydroxylamine, a Compound with Steeply Pyramidal Nitrogen Coordination

Norbert W. Mitzel*

Anorganisch-chemisches Institut der Technische Universität München
Lichtenbergstr. 4 , D-85747 Garching, Germany
Tel. Int. code + (89)289 13066 — Fax. Int. code + (89) 289 13125
E-mail: N.Mitzel@lrz.tu-muenchen.de

Heinz Oberhammer

Institut für Physikalische und Theoretische Chemie der Universität Tübingen
Auf der Morgenstelle 8 , D-72076 Tübingen, Germany

Keywords: Silylamines / Hydroxylamines / Gas-Phase Electron-Diffraction / Ab Initio Calculations

Summary: *N,O*-Dimethyl-N-silylhydroxylamine, $H_3SiMeNOMe$, has been prepared by reaction of HMeNOMe with H_3SiBr and 2,6-lutidine as an auxiliary base. The solution NMR data indicate aggregation of the compound. $H_3SiMeNOMe$ decomposes at ambient temperature and formation of methylnitrene is probably a first step in the mechanism. The energy of $H_3SiMeNOMe$ relative to its potential rearrangement isomer MeHN–H_2Si–OMe has been estimated by ab initio calculations to be 289 kJ mol^{-1}. The molecular structure of $H_3SiMeNOMe$ has been determined by gas-phase electron diffraction and by ab initio calculations. $H_3SiMeNOMe$ possesses a steeply pyramidal nitrogen atom. Hence it is the first silylamine reported with a typically pyramidal nitrogen coordination for electronic reasons. Results of an NBO analysis are discussed to rationalize the bonding situation.

Introduction

Silylated nitrogen compounds generally have planar coordination at the nitrogen atoms [1] and only a few exceptions have been found so far [2, 3]. The nitrogen coordination can be forced to be pyramidal by incorporation into a small ring cycle as a result of ring strain [4]. However, the pyramidal nitrogen coordination is an inherent phenomenon in the chemistry of strain-free silylhydroxylamine derivatives [5].

The simplest member of this class of compounds, $(H_3Si)_2NOMe$, has been studied in both the gas phase and the solid state to prove this [6]. It shows a slight deviation from planarity at nitrogen with a sum of angles of 351.8° and the corresponding declination of the N–O vector from the NSi_2 plane (N–O/NSi_2) is 33.2°.

The non-planar nitrogen atoms in silylhydroxylamines are the result of substituent effects. Silyl groups lower the inversion barrier of nitrogen centers, whereas alkoxy groups increase it. In a compound with only one silyl group, the nitrogen coordination should thus be more steeply pyramidal. This is why we studied $H_3SiMeNOMe$.

Synthesis

$H_3SiMeNOMe$ was synthesized by the reaction of bromosilane with *N,O*-dimethylhydroxylamine. 2,6-Lutidine served as an auxiliary base. The reaction was carried out at low temperature and in absence of a solvent. The yield was 72 %.

$$\text{HMeNOMe} + \text{H}_3\text{SiBr} + \text{2,6-lutidine} \rightarrow \text{H}_3\text{SiMeNOMe} + \text{2,6-lutidine}\cdot\text{HBr}$$

Decomposition of $H_3SiMeNOMe$

The compound decomposes to give SiH_4 and an insoluble solid residue. A nitrene extrusion, a reaction which is known for *O*-trimethylsilyl-hydroxylamines [7], is a likely decomposition pathway, as traces of H_3SiOMe could be detected by NMR. We suggest the following reaction.

$$\text{H}_3\text{SiMeNOMe} \rightarrow \text{H}_3\text{C-}\underline{\text{N}}| + \text{H}_3\text{SiOMe} \rightarrow \text{H}_3\text{C(H)NSiH}_2\text{OMe}$$

Although $H_3C(H)NSiH_2OMe$ is a likely product and consistent with the appearance of a new triplet in the ^{29}Si NMR spectrum at –35.5 ppm [t m, $^{1}J(SiH)$ = 233.5 Hz] and two new singlets in the proton NMR at 3.42 and 2.91 ppm, we could not further prove its identity. An ab initio calculation (MP2/6-311G**) on $H_3C(H)NSiH_2OMe$ predicted it to be 289 kJ mol^{-1} lower in energy than $H_3SiMeNOMe$ (compare hydrazoic acid, HN_3, $\Delta H_0 = 269$ kJ mol^{-1}).

Spectroscopic Characterization of $H_3SiMeNOMe$

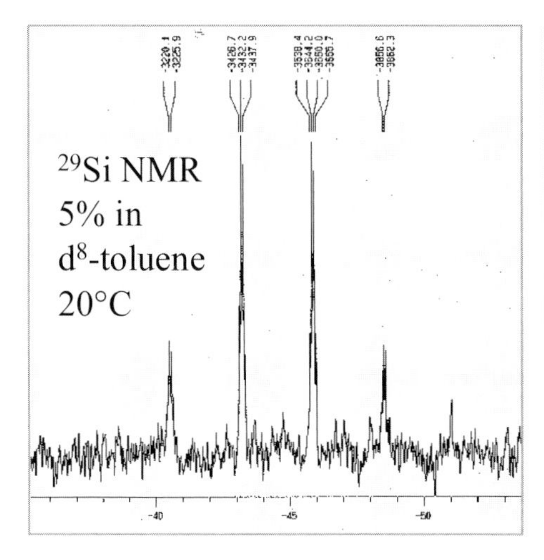

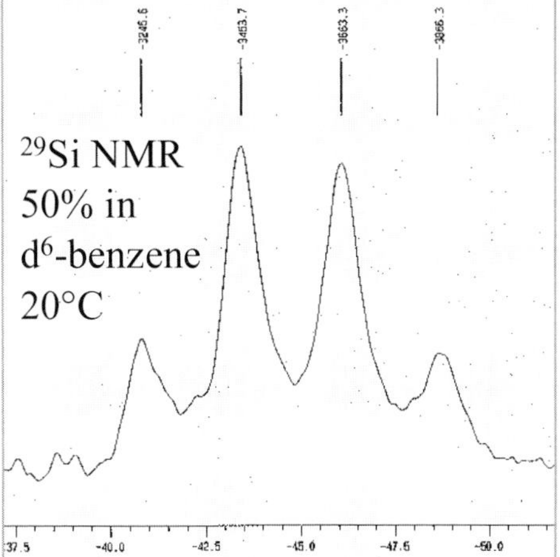

Fig. 1 : ^{29}Si-NMR spectra of $H_3SiMeNOMe$.

Surprisingly broad NMR resonances appear in the proton and ^{29}Si NMR spectrum of $H_3SiMeNOMe$ dissolved in C_6D_6 (Fig.1). Only in concentrations below ca. 5 % do these resonances sharpen to give features as expected for such a compound. However, the broadening of these peaks is also dependent on temperature, and a 5 % solution in $[D_8]$-toluene at –50°C also shows markedly broadened signals. This behavior has not been observed in the spectra of $(H_3Si)_2NOMe$. It can be interpreted as an indication of intermolecular aggregation probably through Si–O contacts, as the linewidth in the ^{15}N-NMR spectrum is not increased.

Gas Phase Structure of $H_3SiMeNOMe$

The molecular structure of $H_3SiMeNOMe$ was determined in the gasphase by means of electron diffraction. The analysis of the data was complicated due to the low symmetry of the molecule. However, we used the SARACEN method for data analysis [8], which is a natural extension of Bartell's "predicate value" method and Schäfer's MOCED method [9]. SARACEN combines electron diffraction intensities and restraints from ab initio calculations, which are assigned uncertainties on the basis of a graded series of calculations and estimating the likely error of the computed parameters. Recently, this method has been successfully applied even in the gas-phase structure determination of relatively large systems of low symmetry [10]. The geometry restraints were computed at the MP2/6-311G** level of theory. Calculated amplitudes of vibration (based on a MP2/6-31G* force field) were used in most cases where free refinement of amplitudes was not possible and in many cases restraints on vibrational amplitudes were applied as well.

Selected geometrical parameter values are listed in Table 1. Fig. 2 shows the radial distribution curve derived by Fourier inversion of the experimental electron diffraction intensities.

Table 1. Molecular parameters for $H_3SiMeNOMe$ (distances in Å, angles in °).

Parameter	**MP2/6-311G**** r_e / $<_e$	**GED** r_a / $<_a$
Si4–N3	1.758	1.742(1)
C5–N3	1.455	1.460(3)
N3–O2	1.444	1.449(4)
O2–C1	1.420	1.425(4)
C1–H6	1.092	1.115(2)
∠O2-N3-Si4	105.4	104.3(4)
∠O2-N3-C5	107.9	106.2(12)
∠N3-O2-C1	107.8	103.2(12)
∠Si4-N3-C5	120.7	121.8(5)
τC1-O2-N3-Si4	120.7	127.9(17)
τC1-O2-N3-C5	–109.1	–102.4(15)

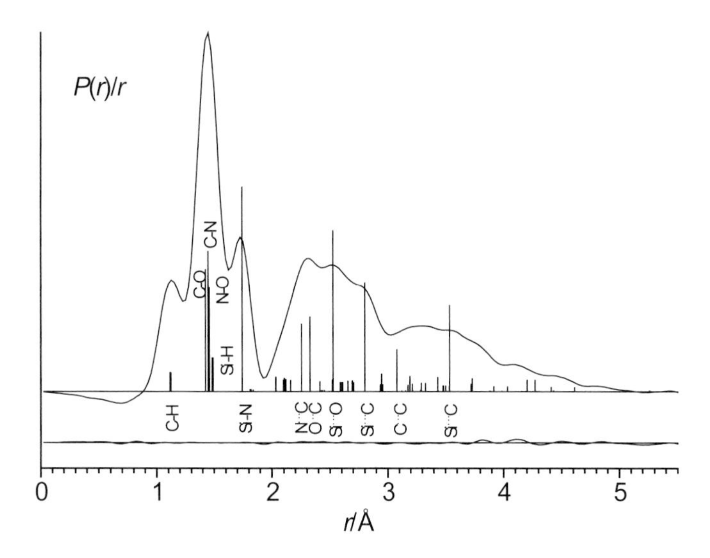

Fig. 2. Radial distribution and difference curve for $H_3SiMeNOMe$.

The steeply pyramidal coordination of the nitrogen atom is obvious from Fig. 3. The sum of angles at nitrogen is 332.3°, which is almost the value for a nitrogen atom with ideal tetrahedral bond angles. The N–O bond is declined by 57.2° from the N3/Si4/C5 plane. Hence, the coordination in $H_3SiMeNOMe$ is substantially more pyramidal than in $(H_3Si)_2NOMe$ (N–O/NSi_2 33.2°, sum of angles at N 351.8°).

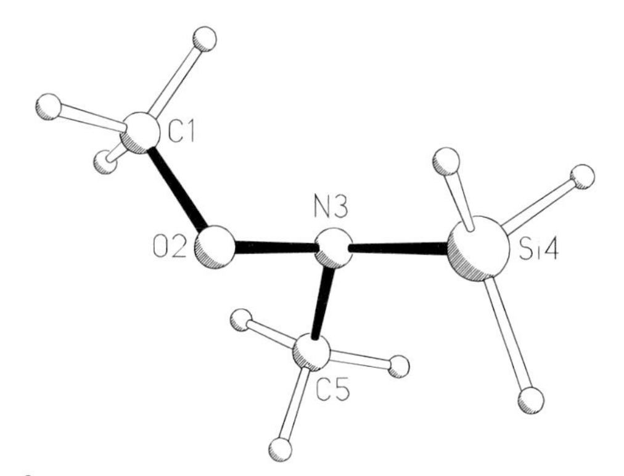

Fig. 3. Gas phase structure of $H_3SiMeNOMe$ as determined by electron diffraction.

The Si–N bond length (r_a = 1.742(1) Å) is only slightly longer than in $(H_3Si)_2NOMe$ (r_a = 1.736(1) Å) or comparable compounds with planar nitrogen atoms, which is surprising as we expected an Si–N bond between a silicon atom and a pyramidal (sp^3) nitrogen atom to be longer than one involving a planar (sp^2) nitrogen atom. Such a case is unprecedented, as Si/N compounds almost always contain (completely) flattened nitrogen atoms.

The O-N-Si angle is 104.3(4)°. This is significantly smaller than the corresponding one in $(H_3Si)_2NOMe$ [110.6(6)°]. It may be speculated whether the slight angle contraction relative to $(H_3Si)_2NOMe$ is due to attractive forces between the negatively charged oxygen and the positively charged silicon atom (β-donor interactions) [11]. The angle O-N-C is 106.2(12)° (predicted 107.9°) and thus also relatively small. The C-O-N angle refined to 103.2(12)°. This is slightly smaller than the calculated one (107.8°) or the one found in $(H_3Si)_2NOMe$ [109.1(4)°].

$(H_3Si)_2NOMe$ is much flatter than the steeply pyramidal $H_3SiMeNOMe$, because the large Si-N-Si angle [131.8(2)°] in the former is about 11° larger than the C-N-Si angle in the latter.

Theoretical Description

A natural bond orbital (NBO) analysis has been carried out for $H_3SiMeNOMe$ to obtain a better description of the bonding situation. The results describe the silicon atom as sp^3 hybridized. In the NBO picture the Si–N bond is made up from an Si-$sp^{3.09}d^{0.08}$ and an N-$sp^{1.94}$ hybrid. This can be used to rationalize the non-elongated Si–N bond, as nitrogen does not use a sp^3 type orbital for the Si–N bond, despite its pyramidal geometry. The sp^2-type hybrids used for bonding of the N atom to Si and C (N–C bond: N-$sp^{2.05}$ and C-$sp^{2.80}$ hybrids) explain the large Si-N-C angle of 121.8(5)° in the gas phase, (predicted ab initio 120.7°). Hence, $H_3SiMeNOMe$ has a steeply pyramidal nitrogen atom, which is substantially distorted away from a pseudo-tetrahedron.

Negative hyperconjugation (p–σ*) is often quoted as the reason for the chemical and structural differences between nitrogen compounds of second row and third row elements. The planar core of $(H_3Si)_3N$ (proved by gas-phase and crystal structures) and the flattened nitrogen coordination in H_3SiNH_2 have been rationalized in terms of p(lp-N)–σ*(Si–H) interactions. In $H_3SiMeNOMe$ the stabilization gained through this interaction is smaller than in H_3SiNH_2, but not as much as could be expected from the steeply pyramidal nitrogen coordination relative to the flat N atom in H_3SiNH_2 (ab initio calculation). Negative hyperconjugation [p(lp-N)–σ*(Si-X)] can thus not solely be responsible for the flattening of the nitrogen coordination spheres in silylamines, as $H_3SiMeNOMe$ also donates electron density from its nitrogen lone pair into the anti-bonding orbitals of the Si–H bonds, with the magnitude of stabilization comparable to that of the clearly flattened H_3SiNH_2.

Acknowledgment: This work was supported by the *Bayerischer Staatsminister für Unterricht, Kultus, Wissenschaft und Kunst* (Bayerischer Habilitationsförderpreis 1996), by the *Deutsche Forschungsgemeinschaft*, and the *Fonds der Chemischen Industrie* and the *Leonhard-Lorenz-Stiftung*. The *Leibniz-Rechenzentrum-München* provided computational resources. We are grateful to Professor H. Schmidbaur for generous support.

References:

[1] a) K. Hedberg, *J. Am. Chem. Soc.* **1955**, *77*, 6491.
b) B. Beagley, A. R. Conrad, *J. Chem. Soc., Faraday Trans.* **1970**, 2740.

[2] H. Bock, *Angew. Chem.* **1989**, *101*, 1659; *Angew. Chem. Int. Ed. Engl.* **1989**, 28, 1627.

[3] K. Ruhlandt-Senge, R. A. Bartlett, M. M. Olmstead, P. P. Power, *Angew. Chem.* **1993**, *105,* 495; *Angew. Chem. Int. Ed. Engl.* **1993**, *32*, 425.

[4] G. Huber, A. Jokisch, H. Schmidbaur, *Eur. J. Inorg. Chem.* **1998**, *1*, 107.

[5] N. W. Mitzel, K. Angermaier, H. Schmidbaur, *Organometallics*, **1994**, *13*, 1762.

[6] N. W. Mitzel, E. Breuning, A. J. Blake, H. E. Robertson, B. A. Smart, D. W. H. Rankin, *J. Am. Chem. Soc.* **1996**, *118*, 2664.

[7] H. Schwarz, B. Steiner, G. Zon, Y. H. Chang, *Z. Naturforsch., Teil B* **1978**, *33*, 129.

[8] a) N. W. Mitzel, B. A. Smart, A. J. Blake, H. E. Robertson, D. W. H. Rankin, *J. Phys. Chem.* **1996**, *100*, 9339.

b) A. J. Blake, P. T. Brain, H. McNab, J. Miller, C. Morrison, S. Parsons, D. W. H. Rankin, H. E. Robertson, B. A. Smart, *J. Phys. Chem.* **1996**, *100*, 12280.

[9] a) *Molecular Structure by Diffraction Methods, Specialist Periodical Reports*, The Chemical Society, London, **1975**, p. 72;

b) V. J. Klimkowski, J. D. Ewbank, C. Van Alsenoy, J. N. Scarsdale, L. Schäfer, *J. Am. Chem. Soc.* **1982**, *104*, 1476–1480.

[10] N. W. Mitzel, H. Schmidbaur, D. W. H. Rankin, B. A. Smart, M. Hoffmann, P. v. R. Schleyer, *Inorg. Chem.* **1996**, *100*, 4360.

[11] N. W. Mitzel, U. Losehand, *Angew. Chem.* **1997**, *109*, 2897; *Angew. Chem. Int. Ed. Engl.* **1997**, *36*, 2807.

Strong β-Donor–Acceptor Bonds in Hydroxylaminosilanes

Norbert W. Mitzel*, Udo Losehand

Anorganisch-chemisches Institut der Technische Universität München
Lichtenbergstr. 4 , D-85747 Garching, Germany
Tel. Int. code + (89)289 13066 — Fax. Int. code + (89) 289 13125
E-mail: N.Mitzel@lrz.tu-muenchen.de

Keywords: β-Donor Bonds / Gas-Phase Electron-Diffraction / Crystal Structure / Ab Initio Calculations / Hydroxylamines

Summary: $ClH_2SiONMe_2$ and its ethyl analogue have been prepared and studied by a variety of spectroscopic methods. $ClH_2SiONMe_2$ adopts an extremely small Si-O-N angle in the solid state corresponding to an unusually strong secondary bond between Si and N atoms (β-donor bond). In the gas phase (electron diffraction) two conformers are present: *anti*, with a strong β-donor bond (weaker than in the crystal), and *gauche*, with a comparatively weak β-donor bond. The gas/solid differences can be explained in terms of increasing dipole moment with stronger β-donor bonds, the differences between the two conformers can be rationalized in terms of negative hyperconjugation of the type p(lp-N) $\rightarrow$ σ*(Si-X), whereby the nature of X and its orientation relative to the Si-O-N plane are of major importance. Electrostatic interaction between Si and N atoms are not suitable to rationalize the experimental facts. The behavior in solution is still not well understood.

Introduction

Only recently were we able to show that O-silylhydroxylamines form intramolecular donor acceptor bonds between Si and N centers, separated by one oxygen atom only [1]. This type of secondary bonds between p-block donor and acceptor atoms in β-positions relative to one another had so far been unprecedented or neglected. For the prediction of molecular geometries of compounds having electropositive and electronegative atoms in the β-position to one another, this interaction turns out to be an important factor and should be considered in structure prediction in addition to the general VSEPR concept [2] and Bartell's two bond radii model [3].

The finding of small angles in Si-O-N units was completely unexpected, as molecules with Si–O linkages almost always adopt substantially widened angles at the oxygen atom, [$O(CH_3)_2$ (111.4°), $O(SiH_3)_2$ (144°), H_3COSiH_3 (120.6°) [4]. However, comparatively small angles at oxygen have been found in the peroxide $Me_3SiOOSiMe_3$ [5].

Weak β-donor interactions could be the reason for the frequently reported high reactivity of hydroxylaminosilanes. These include claims for new cross-linking agents and cold-curing catalysts for silicone polymers [6], and more recently, the catalysis of the alkoholysis of Si–H functions in poly-phenylsilane [7].

Here we report on the compound $ClH_2SiONMe_2$ and its ethyl analogue, which show the strongest β-donor bonds found so far in Si–O–N compounds.

Synthesis

$ClH_2SiONMe_2$ and $ClH_2SiONEt_2$ have been prepared by reacting the corresponding lithiated hydroxylamines with dichlorosilane in diethylether at low temperatures.

$$HONR_2 + nBuLi \rightarrow nBuH + LiONR_2$$
$$H_2SiCl_2 + LiONR_2 \rightarrow LiCl + ClH_2SiONR_2; \qquad R = Me, Et$$

The achievable yields are higher than 75 %. $ClH_2SiONMe_2$ and $ClH_2SiONEt_2$ are extremely air-sensitive, but not pyrophoric. $ClH_2SiONMe_2$ is fairly stable, but its ethyl analogue $ClH_2SiONEt_2$ undergoes decomposition even at temperatures as low as 0°C.

Spectroscopy

It is known from other compounds containing the $SiONEt_2$ linkage, that the appearance of ^1H NMR spectra are temperature dependent. In all known examples of this type there was a quartet observed at elevated temperatures and two quartets of doublets observed at lower temperatures, corresponding to the topomerization of the nitrogen centers (N inversion and N–O rotation, including β-donor secondary bond cleavage). The coalescence temperatures were always in the range of 30–40°C at 400 MHz proton frequency.

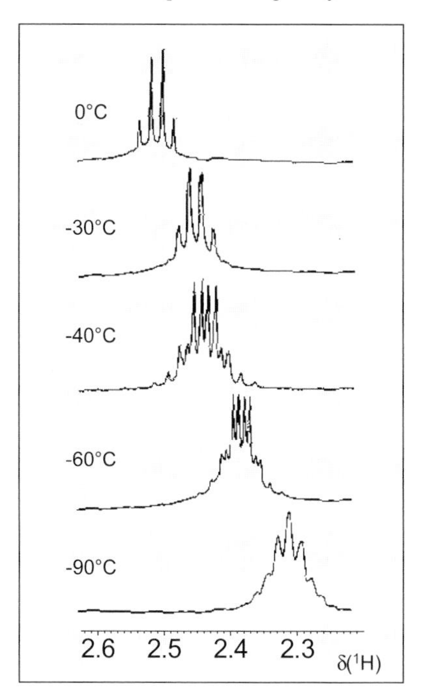

Fig. 1. ^1H-NMR spectra of $ClH_2SiONEt_2$ recorded at different temperatures.

$ClH_2SiONEt_2$ behaves differently (see Fig. 1), but so far we are unable to explain the results. The interpretation is complicated by an additional interconversion process of *gauche* and *anti* conformers, corresponding to different orientations of the Cl–Si bond relative to the Si–O–N plane (see below). The two conformers have about the same energy (in the gas phase) and are separated by a barrier of about 25 kJ/mol.

Crystal Structure of $ClH_2SiONMe_2$

A single crystal of $ClH_2SiONMe_2$ could be grown by in situ methods. The structure, determined by X-ray diffraction, is shown in Fig. 2. The most striking result is the extremely small Si-O-N angle of only 79.6(1)°, corresponding to an Si···N distance of 2.028(1) Å, which is not too far from the sum of the covalent radii of Si and N (1.87 Å). The crystal structure contains the *anti*-conformer of $ClH_2SiONMe_2$ only. It is also important to mention the absence of any intermolecular interaction between the silicon acceptor centers and the nitrogen or oxygen donor functions. This indicates that the coordination sphere of silicon is saturated by the β-donor bond and this interaction (almost) completely involves the nitrogen lone pair of electrons. Si–O and Si–Cl bond lengths are in the established range, whereas the N–O bond is slightly lengthened as compared to $HONMe_2$.

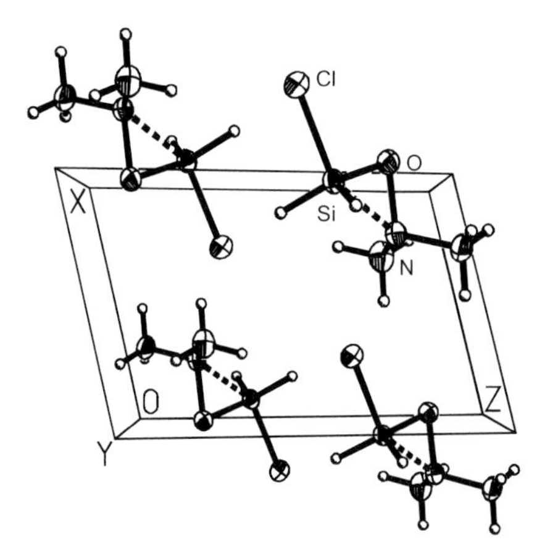

Fig. 2. Unit cell of the crystal structure of $ClH_2SiONMe_2$.

Gas-Phase Structure

The analysis of gas-phase electron diffraction data of $ClH_2SiONMe_2$ was complicated due to the presence of a second conformer (*gauche*), which turned out to be the predominant one. The *anti*/*gauche* ratio was refined to 33:67, which corresponds to zero difference in the energies of the two conformers in the gas phase. We used the SARACEN method for data analysis [8] which is a natural extension of Bartell's "predicate value" method and Schäfer's MOCED method [9]. SARACEN combines electron diffraction intensities and restraints from ab initio calculations, which are assigned uncertainties on the basis of a graded series of calculations and estimating the likely error of the computed parameters. The geometry restraints were computed at the MP2/6-311G** level of theory. Calculated amplitudes of vibration (based on an MP2/6-31G* force field)

were used in most cases where free refinement of amplitudes was not possible and in many cases restraints on vibrational amplitudes were applied as well.
Selected geometrical parameter values are listed in Table 1 together with the crystallographic data and those calculated at the MP2/6-311G** level of theory.

Table 1. Geometrical parameter values for $ClH_2SiONMe_2$ in comparison of *anti* and *gauche* conformers.

		Anti conformer			*Gauche* conformer	
Parameter		**XRD (r)**	**GED (r_a)**	**MP2/6-311G**(r_e)**	**GED (r_a)**	**MP2/6-311G**(r_e)**
r Si–O	[Å]	1.668(1)	1.654(4)	1.678	1.641(3)	1.664
r Si–Cl	[Å]	2.108(1)	2.050(4)	2.063	2.042(2)	2.054
r Si···N	[Å]	2.028(1)	2.160(7)	2.259	2.468(25)	2.477
∠Si-O-N	[°]	79.7(1)	87.1(9)	91.6	104.7(11)	104.5
∠O-Si-Cl	[°]	101.1(1)	105.5(17)	105.4	109.4(9)	112.1

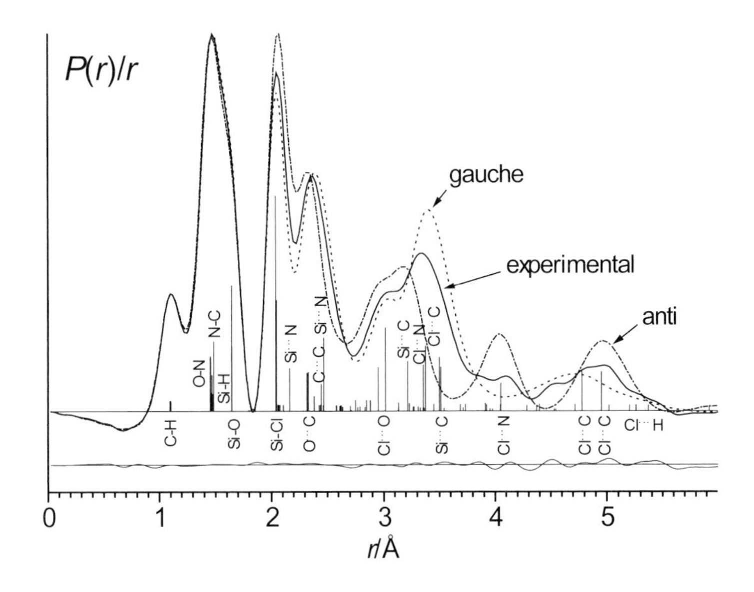

Fig. 3. Radial distribution and difference curve curve for $ClH_2SiONMe_2$.

Figure 3 shows the radial distribution curve derived by Fourier inversion of the experimental electron diffraction intensities. It is obvious that it is not possible to fit the experimental data with one single conformer only.

In the gas phase the Si-O-N angle of the *anti* conformer is substantially wider than in the solid state (Table 1). This can be explained by the molecular dipole moment, which strongly increases with decreasing Si-O-N angle and thus strengthens the lattice forces (Table 2).

Table 2. Dipole moments μ [Debye] and atomic charges q [Mulliken charges in e] for selected compounds, conformers and conformers with fixed SiON angles [°] (calculated at MP2/6-31G*).

Compound	∠SiON	μ	q(Si)	q(N)
$ClH_2SiONMe_2$ *anti* (minimum)	89.8	4.57	1.06	–0.33
$ClH_2SiONMe_2$ *anti* (∠SiON fixed)	80.0	5.51	1.08	–0.37
$ClH_2SiONMe_2$ *anti* (∠SiON fixed)	104.0	3.54	1.05	–0.28
$ClH_2SiONMe_2$ *gauche* (minimum)	104.5	2.60	1.04	–0.28
$ClH_2SiONMe_2$ *gauche* (∠SiON fixed)	90.0	3.04	1.05	–0.33
$H_3SiONMe_2$	102.2	1.63	0.92	–0.29
$H_2Si(ONMe_2)_2$	96.2	1.45	1.24	–0.31

The Si-O-N angle in the *anti* conformer in the gas phase is wider by more than 17° than in the *gauche* conformer (Fig. 4). This is an unusually large difference for conformational isomers. It points the analysis of the nature of bonding towards a highly orientation-dependent effect. Figure 5 shows the interdependence of energy, Si-O-N angle and torsion angle Cl-Si-O-N as a trajectory through the potential hypersurface calculated at MP2/6-31G*. The maximum change in ∠SiON is as large a 30° (torsion of 0 and 180°).

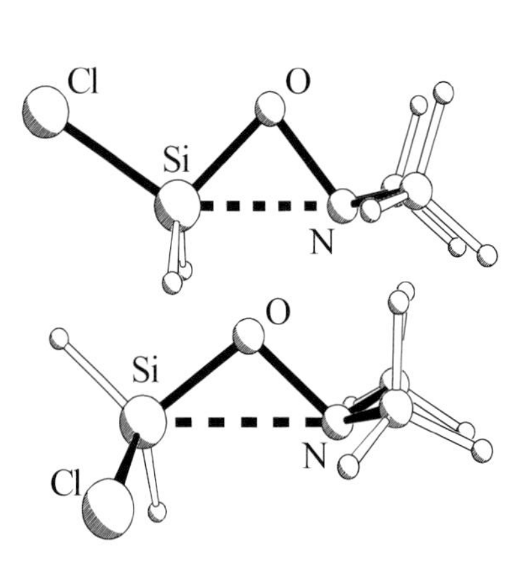

Fig. 4. Molecular structures of the *anti* and *gauche* conformers in the gas phase.

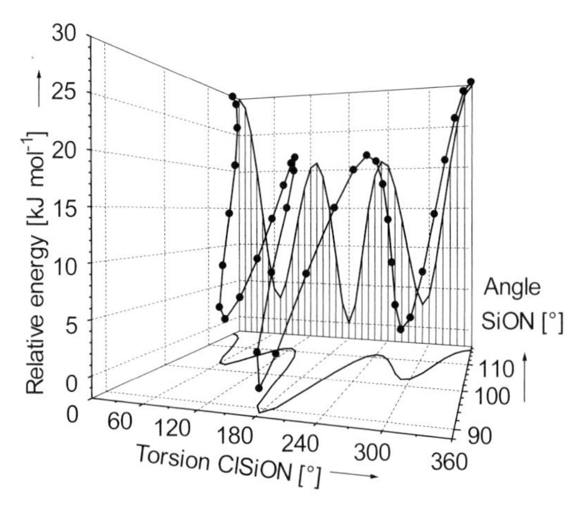

Fig. 5. Interdependence of energy, angle Si-O-N and torsion angle Cl-Si-O-N in $ClH_2SiONMe_2$.

Theoretical Description of the β-Donor Bond

On the first glance one could attribute the small Si-O-N angles in hydroxylaminosilanes to an electrostatic interaction between the positively charged silicon atom and the negatively charged nitrogen atom. Table 2 contains the calculated Mulliken charges of the Si and N atoms in the two conformers of $ClH_2SiONMe_2$ as well as some isomers with artificially fixed angles and other compounds for comparison. It shows that the charges in $ClH_2SiONMe_2$ are not mainly dependent on the type of conformer, but on the Si-O-N angle. Neglecting other effects, there would be no reason for the *anti* and *gauche* conformers to adopt angles as different as 17.6° in the gas phase as observed in the experiment. $H_2Si(ONMe_2)_2$, which has the highest positive charge on Si in Table 2 has not the smallest Si-O-N angle, while $H_3SiONMe_2$ has the smallest charge on Si, but an Si-O-N angle which is even slightly smaller than the *gauche* conformer of $ClH_2SiONMe_2$. Hence, pure electrostatic attraction is not a satisfactory model for the description of the β-donor bond.

We have performed a natural bond orbital (NBO) analysis for several silylhydroxylamines. It turns out that negative hyperconjugation of the type p(lp-N) $\rightarrow$ σ*(Si–X) is an important factor in this description of bonding. In the NBO picture the difference in the strength of β-donor bonds of *anti* and *gauche* conformers of $ClH_2SiONMe_2$ is due to the nature of the silicon substituent in the *anti* position relative to the nitrogen atom, whereas the nature of the silicon substituent in the *gauche* position is not so important. A suitable *anti* substituent X is the prerequisite for the formation of a strong β-donor bond (X: F>Cl>O>H).

This explains why very small Si-O-N angles are observed in *anti*-$ClH_2SiONMe_2$ and *anti*-$H_2Si(ONMe_2)_2$, but larger ones in *gauche*-$ClH_2SiONMe_2$ and *gauche*-$H_3SiONMe_2$. The latter two differ only in one *gauche* substituent (Cl vs. H) and the Si-O-N angles are almost the same.

Although it is only a coarse model, negative hyperconjugation, p(lp-N) $\rightarrow$ $\sigma^*(Si–X_{anti})$, is suitable to rationalize the experimental results of a variety of compounds with β-donor bonds.

Acknowledgment: This work was supported by the *Bayerischer Staatsminister für Unterricht, Kultus, Wissenschaft und Kunst* (Bayerischer Habilitationsförderpreis 1996), by the *Deutsche Forschungsgemeinschaft,* and the *Fonds der Chemischen Industrie*, the *Leonhard-Lorenz-Stiftung* and the *Bayer AG, Leverkusen*, by chemical donations. The *Leibniz-Rechenzentrum-München* provided computational resources. We are grateful to Professor H. Oberhammer for GED data collection, to Mr. J. Riede for establishing the crystallographic data and to Professor H. Schmidbaur for generous support.

References:

[1] a) N. W. Mitzel, A. J. Blake, D. W. H. Rankin, *J. Am. Chem. Soc.* **1997**, *119*, 4143.
b) N. W. Mitzel, U. Losehand, *Angew. Chem., Int. Ed. Engl.* **1997**, *36*, 2807.
c) N. W. Mitzel, *Chem. Eur. J.*, **1998**, 4, 692.

[2] R. J. Gillespie, E. A. Johnson, *Angew. Chem., Int. Ed. Engl.* **1996**, *35*, 495.

[3] L. S. Bartell, *J. Chem. Phys.* **1960**, *32*, 827.

[4] S. Shambayati, J. F. Blake, S. G. Wierschke, W. L. Jorgensen, S. L. Schreiber, *J. Am. Chem. Soc.* **1990**, *112*, 697.

[5] a) H. Oberhammer, J. E. Boggs, *J. Am. Chem. Soc.* **1980**, *102*, 7241.
b) D. Käss, H. Oberhammer, D. Brandes, A. Blaschette, *J. Mol. Struct.* **1977**, *40*, 65.

[6] A collection of patent literature references can be found in: V. G. Voronkov, E. A. Maletina, V. K. Roman, *Heterosiloxanes, Vol. 2: Derivates of Nitrogen and Phosphorus,* Harwood Academic Publishers, Chur, Switzerland, **1991**.

[7] Y. Hamada, S. Mori, *Proceedings of the 29th Organosilicon Symposium, March 1996, Evanston, USA.*

[8] a) N. W. Mitzel, B. A. Smart, A. J. Blake, H. E. Robertson, D. W. H. Rankin, *J. Chem. Phys.* **1996**, *100*, 9339; b) A. J. Blake, P. T. Brain, H. McNab, J. Miller, C. A. Morrison, S. Parsons, D. W. H. Rankin; H. E. Robertson, B. A. Smart, *J. Chem. Phys.* **1996**, *100*, 12280.

[9] K. R. Leopold, M. Canagaratna, J. A. Phillips, *Acc. Chem. Res.* **1997**, *30*, 57.

β-Donor Bonds in Hydroxylamino- and Oximatosilanes

Norbert W. Mitzel*, Udo Losehand

Anorganisch-chemisches Institut der Technische Universität München
Lichtenbergstr. 4 , 85747 Garching, Germany
Tel. Int. code + (89)289 13066 — Fax. Int. code + (89) 289 13125
E-mail: N.Mitzel@lrz.tu-muenchen.de

Keywords: β-Donor Bonds / Hydroxylamines / Oximes

Summary: β-Donor-acceptor bonds between geminally positioned N and Si atoms are an inherent structure phenomenon of O-silylated hydroxylamines and oximes. These interactions lead to considerably compressed Si-O-N angles in hydroxylaminosilanes, culminating in $(Me_2NO)_2SiH_2$. The strength of the β-donor bond can be estimated by temperature-dependent NMR investigation of Et_2NOSiH_3. The β-donor–acceptor bonds in the oximatosilanes $Me_2C{=}NOSiH_3$ and $(Me_2C{=}NO)_2SiH_2$ are much weaker due to the lower basicity of nitrogen atoms involved in a double bond. However, the Si-O-N angles of oximatosilanes are still markedly compressed as compared to related compounds, e.g. isopropoxysilane, $Me_2HCOSiH_3$.

Introduction

The investigation in model systems containing SiON fragments unequivocally shows the formation of Si/N β-donor bonds [1]. Such interactions between geminal atoms are well established in transition metal chemistry, e.g. in the eight-coordinate $Ti(ONMe_2)_4$ [2], but almost unknown in the chemistry of p-block elements. The small angles in Si–O–N units are surprising in the light of numerous established molecular structures of Si–O–X linkages which almost always adopt substantially widened angles at the oxygen atom [3]. In order to examine the electronic means of these weak donor–acceptor interactions in detail it seemed useful to synthesize highly hydrogenated hydroxylamino- and oximatosilanes model compounds to avoid steric and electronic influences caused by large organic substituents. Detailed knowledge of β-donor interactions can be expected to contribute to our general understanding of the molecular geometries of systems involving acceptor and donor atoms in geminal positions to one another. The formation of β-donor interactions leads to an enlargement of the coordination sphere at the silicon center, which should result in a higher reactivity of the acceptor atom. In the present cases an increased reactivity towards nucleophilic substitutions can be expected for the silicon atoms. In this way the present studies might contribute to a better understanding of the high reactivity of hydroxylamino- and oximatosilanes, which have been used as cold curing catalysts in silicon polymer synthesis [4].

We report here on the synthesis and crystal structure determination of Me_2NOSiH_3, $(Me_2NO)_2SiH_2$ [1], $Me_2C{=}NOSiH_3$ and $(Me_2C{=}NO)_2SiH_2$ and on the dynamic behavior of the ethyl analogues $(Et_2NO)_{4-n}SiH_n$.

Synthesis

The best way to prepare the hydroxylamino- and oximatosilanes proved to be the reaction of O-lithiated hydroxylamine or acetone oxime with the corresponding halogenosilane in diethyl or dimethyl ether. All products can be isolated in good yields as colorless liquids. The compounds are extremely sensitive to air moisture, yet not pyrophoric.

Hydroxylamines	Me_2NOH + *n*-BuLi → Me_2NOLi + *n*-BuH
	Me_2NOLi + H_3SiBr → **Me_2NOSiH_3** + LiBr
	2 Me_2NOLi + H_2SiCl_2 → **$(Me_2NO)_2SiH_2$** + 2 LiCl
Oximes	$Me_2C{=}NOH$ + *n*-BuLi → $Me_2C{=}NOLi$ + *n*-BuH
	$Me_2C{=}NOLi$ + H_3SiBr → **$Me_2C{=}NOSiH_3$** + LiBr
	2 $Me_2C{=}NOLi$ + H_2SiCl_2 → **$(Me_2C{=}NO)_2SiH_2$** + 2 LiCl

Spectroscopy

The identity of the compounds was proven by ^{1}H-, ^{13}C-, ^{15}N-, ^{17}O-, and ^{29}Si-NMR spectroscopy, gas-phase IR spectroscopy and mass spectrometry.

The diastereotopic methylene protons of Et_2NOSiH_3 (Fig. 1) show a temperature-dependent signal in the ^{1}H-NMR experiment. The slightly broadened singlet observed at elevated temperatures transforms into a broad singlet at the coalescence temperature and splits into two quartets of doublets at low temperatures (Fig. 2). This is due to the barrier of topomerization in hydroxylaminosilanes, a process which consists of inversion at the nitrogen atom, the rotation about the N–O bond and β-donor bond cleavage. By comparison of the topomerization barriers of silylhydroxylamines with those reported for organic hydroxylamines, the strength of the β-donor bond can be estimated to 12 kJ mol^{-1}.

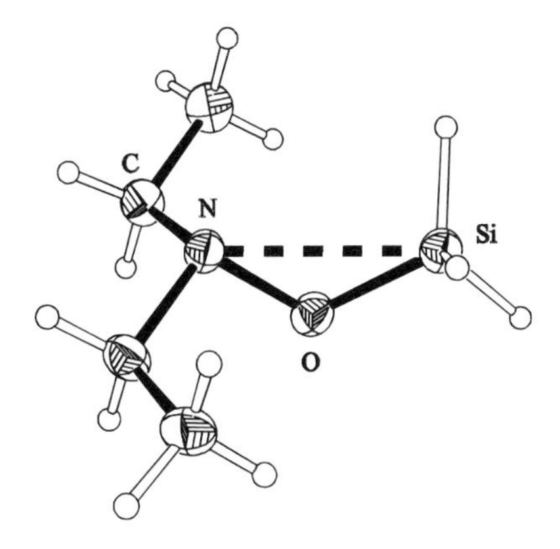

Fig. 1. Molecular structure of Et_2NOSiH_3.

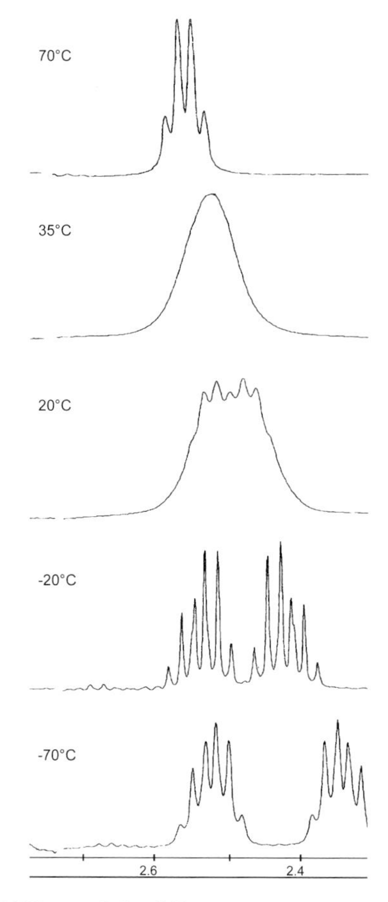

Fig. 2. ^{1}H-NMR spectra of Et_2NOSiH_3 recorded at different temperatures.

Crystal structures

All structural studies were carried out by X-ray diffraction (XRD) on single crystals grown in situ. Selected bond lengths and angles are given in Table 1.

Table 1. Si-O-N angles and Si···N distances.

Compound	Si···N [Å]	Si-O-N/Si-O-C [°]
Me_2NOSiH_3	2.453(1)	102.6(1)
$(Me_2NO)_2SiH_2$	2.300(1), 2.336(1)	94.2(1), 96.2(1)
$Me_2C{=}NOSiH_3$	2.493(2)	106.0(1)
$(Me_2C{=}NO)_2SiH_2$	2.423(1), 2.501(1)	102.5(1), 107.5(1)
$Me_2HCOSiH_3$		118.4(1)

The simplest silylated hydroxylamine, Me_2NOSiH_3 (Fig. 3), shows a small Si-O-N angle of 102.6(1)° and a short Si···N distance of 2.453(1) Å. This is much smaller than in the isoelectronic compound isopropoxysilane, $H_3SiOCHMe_2$, which is not capable of forming a β-donor bond due to the absence of a lone pair of electrons. Comparison of the angles Si-O-C/Si-O-N shows the β-donor–acceptor interaction in Me_2NOSiH_3 to be responsible for a compression of almost 16°.

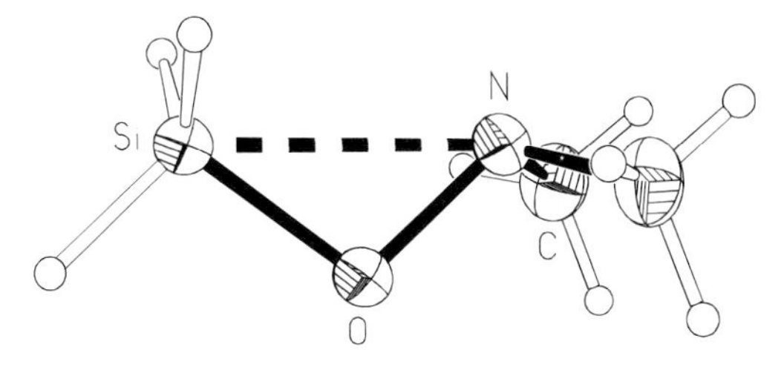

Fig. 3. Molecular structure of Me_2NOSiH_3.

The β-donor bond in $(Me_2NO)_2SiH_2$ (Fig. 4) is even stronger. The compound adopts almost C_{2v} symmetry in the crystal, with two *anti*-oriented NOSiO units. The Si-O-N angle of this compound is as small as 95.1° on average, and the Si···N distances of 2.300(1) and 2.336(1) Å are ca. 0.1 Å shorter than in $H_3SiONMe_2$.

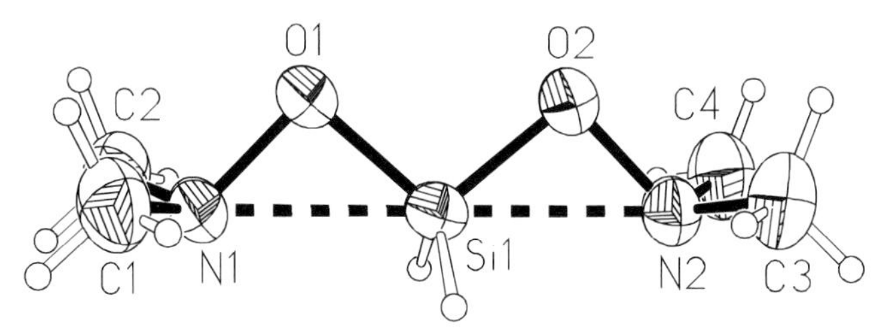

Fig. 4. Molecular structure of $(Me_2NO)_2SiH_2$.

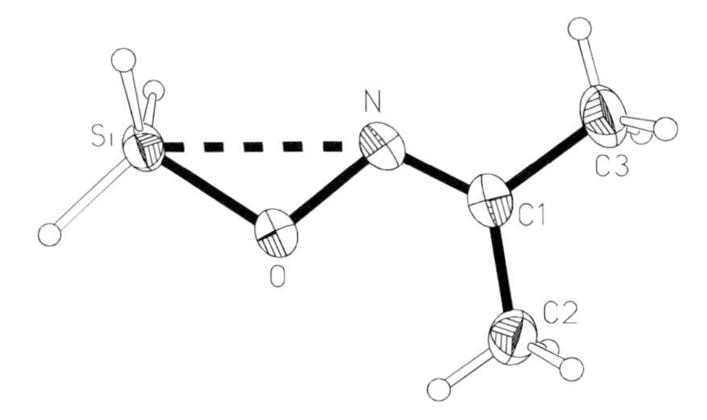

Fig. 5. Molecular structure of $Me_2C{=}NOSiH_3$.

The β-donor bonds in the investigated silylated oximes $Me_2C{=}NOSiH_3$ (Fig. 5) and $(Me_2C{=}NO)_2SiH_2$ are weaker than in the related hydroxylamines. The Si-O-N angle in $Me_2C{=}NOSiH_3$ [106.0(1)°] is 3.4° wider than in Me_2NOSiH_3 and thus the Si···N distance is 0.04 Å longer.

The crystal structure analysis of $(Me_2C{=}NO)_2SiH_2$ (Fig. 6) shows two small Si-O-N angles of 102.5(1) and 107.5(1)°, both substantially wider than in $(Me_2NO)_2SiH_2$ [94.2(1), 96.2(1)°]. The compound crystallizes in a lower symmetry than the related $(Me_2NO)_2SiH_2$ (pseudo-C_{2v}), with one *anti* and one *gauche* arrangement of the NOSiO units. As has been shown for $ClH_2SiONMe_2$ [5], the strength of a β-donor bond is mainly dependent on the orientation of the silicon substituent in the *anti* position relative to the nitrogen donor center. The smaller Si-O-N angle (102.5°) occurs in a SiON linkage which has an oxygen atom in an *anti* position relative to its nitrogen center, whereas the SiON linkage with the weaker β-donor bond (∠Si-O-N 107.5°) has a hydrogen atom as an *anti* substituent. The situation of the weaker β-donor interaction can thus be compared to $Me_2C{=}NOSiH_3$ (∠SiON 106.0°). In the other case a stronger interaction has to be expected, as $(Me_2NO)_2SiH_2$ (*anti* substituent: O) has also a stronger β-donor interaction than the simple Me_2NOSiH_3 (*anti* substituent: H).

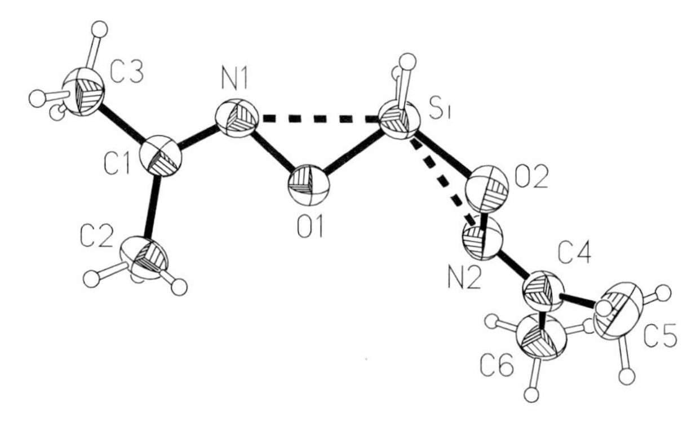

Fig. 6. Molecular structure of $(Me_2C{=}NO)_2SiH_2$.

The bond lengths and angles of all compounds determined by X-ray crystallography fall within less than 1 pm or 2° as compared to the results of ab initio calculations up to the MP2/6-311G** level of theory.

Due to the lower basicity of formally sp^2-hybridized nitrogen atoms in oximatosilanes, the β-donor–acceptor interaction between silicon and nitrogen atoms is weaker than in the related hydroxylaminosilanes. Nevertheless, the Si-O-N angles are still markedly smaller than the Si-O-C angle in isopropoxysilane.

Even stronger β-donor bonds have been established in the chlorine substituted compound $ClH_2SiONMe_2$ (see [5]) and have also been predicted for $FH_2SiONMe_2$ [6].

References:

[1] N. W. Mitzel, U. Losehand, *Angew. Chem., Int. Ed. Engl.* **1997**, *36*, 2807.

[2] N. W. Mitzel, A. J. Blake, S. Parsons, D. W. H. Rankin, *J. Chem. Soc., Dalton Trans.* **1996**, 2089.

[3] S. Shambayati, J. F. Blake, S. G. Wierschke, W. L. Jorgensen, S. L. Schreiber, *J. Am. Chem. Soc.* **1990**, *112*, 697.

[4] V. G. Voronkov, E. A. Maletina, V. K. Roman, *Heterosiloxanes, Vol. 2: Derivates of Nitrogen and Phosphorus,* Harwood Academic Publishers, Chur, Switzerland, **1991**.

[5] N. W. Mitzel, U. Losehand, "*Strong β-Donor Acceptor Bonds in Hydroxylaminosilanes*" in: *Organosilicon Chemistry IV — From Molecules to Materials*, (Eds.: N. Auner, J. Weis), VCH–Wiley, Weinheim, **1999**, pp.164–169.

[6] N. W. Mitzel, A. J. Blake, D. W. H. Rankin, *J. Am. Chem. Soc.* **1997**, *119*, 4143.

Synthesis and Rotational Isomerism of Some Iodosilanes and Iododisilanes

Reinhard Hummeltenberg, Karl Hassler

Institut für Anorganische Chemie
Erzherzog Johann Universität
Stremayrgasse 16, A-8010 Graz, Austria
Tel.: Int. code + (316)873 8206 — FAX: Int. code + (316) 873 8701
E-mail: hassler@anorg.tu-graz.ac.at

Keywords: Ethyliodosilane / Ethyldiiodosilane / 1-Iodo-2-methyldisilane / 1,1-Diiodo-2-methyldisilane / Rotational Isomerism

Summary: The synthesis of the ethylsilanes $CH_3CH_2SiH_2I$ (**1**) and $CH_3CH_2SiHI_2$ (**2**) as well as of the disilanes $CH_3SiH_2SiH_2I$ (**3**) and $CH_3SiH_2SiHI_2$ (**4**), which differ from the former by substitution of a carbon by a silicon atom are described. From temperature-dependent Raman spectroscopic measurements and supported by ab initio calculations, conformational energies and compositions have been determined for the liquid state. The preferred conformations are *gauche* for **1** and *anti* for **2**, **3** and **4**, with ΔH-values of 1.01, 1.17, 1.52 and 1.70 kJ/mol, respectively.

Ab initio calculations of rotational barriers of single bonds between elements of group 14 predict a linear relationship between the bond lengths and the barrier, for instance for the series H_3SiXH_3 with X = C, Si, Ge, Sn and Pb [1]. The barriers decrease from about 6.7 kJ/mol for Si–C to about 3.8 for Si–Si and 1.7 kJ/mol for Si–Pb bonds. For the spectroscopist interested in molecular conformations, barriers that lie well above RT at room temperature (2.48 kJ/mol) support the expectation that the interconversion of rotamers will be slow on the time scale typical for Raman vibrational spectroscopy ($\approx 10^{-13}$ sec) and that the rotamers can be distinguished by their individual vibrational spectra. It is our objective to compare conformational stabilities of carbosilanes (Si–C bonds) with those of disilanes (Si–Si bonds) bearing identical substituents. Fore this purpose, we have prepared the title compounds **1**–**4** and investigated their conformational compositions by variable temperature Raman spectroscopy.

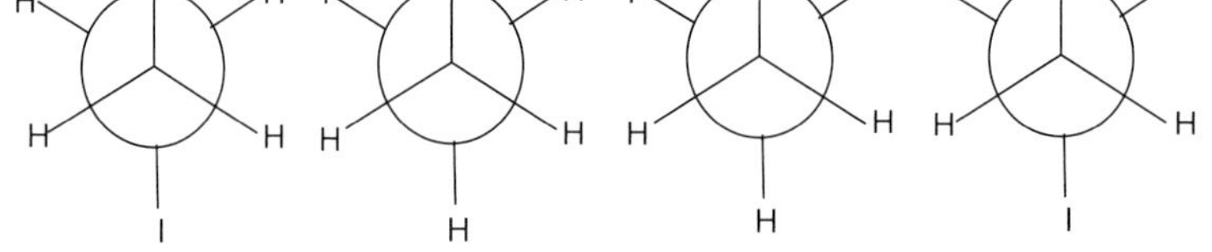

Fig. 1. Newman projections of *anti* and *gauche* $CH_3CH_2SiH_2I$ and $CH_3SiH_2SiH_2I$ (left) and *anti* and *gauche* $CH_3CH_2SiHI_2$ and $CH_3SiH_2SiHI_2$ (right).

As is illustrated in Fig. 1 by use of Newman projections, the compounds **1**–**4** can exist as mixtures of two rotamers, *gauche* and *anti*, at room temperature. The Raman spectra clearly prove that the barriers are considerably larger than at RT and that all four compounds actually comprise mixtures of rotamers. Tables 1 and 2 summarize the Raman spectra in the range of the Si–C, Si–Si and Si–I vibrations from 80 to 700 cm^{-1} together with the assignments which are supported by the ab initio calculations. They have been performed on the SCF/HF level of theory using pseudopotentials of Stevens, Krauss and Basch [2] for the core electrons of the heavy atoms. A valence double-zeta basis with polarization functions on all atoms has been chosen. By numerical differentiation at the equilibrium geometries, harmonic frequencies and harmonic force constants have been calculated. With the use of the program *ASYM40* [3], the force field obtained in cartesian

Table 1. Raman spectra of the liquid ethyliodosilanes **1** and **2** from 90 to 700 cm^{-1}.

$CH_3CH_2SiH_2I$				$CH_3CH_2SiHI_2$			
Ra (l)	**Ab initio**		**Assignment**	**Ra (l)**	**Ab initio**		**Assignment**
	anti	***gauche***			***anti***	***gauche***	
				95vs	119	101	τSiI_2
	112	126	δCSiI	130w	132		γSiI_2
231m	222		δCCSi	160w		155	γSiI_2
272w		261	δCCSi	215m		247	δCCSi
344vs		332	νSiI	245w			
378m	373		νSiI	295vs	284		$\nu_s SiI_2$
	471		ρSiH_2	337s		330	$\nu_s SiI_2$
528mw		523	ρSiH_2	362ms	355		δCCSi
620sh	616	616	νSiC	379mw	373		$\nu_{as} SiI_2$
640ms		638	τSiH_2	407mw		398	$\nu_{as} SiI_2$
680m	657		τSiH_2	636m	611	628	νSiC
				686m	668	666	δCSiH

Table 2. Raman spectra of the liquid methyliododisilanes **3** and **4** from 80 to 700 cm^{-1}.

$CH_3SiH_2SiH_2I$				$CH_3SiH_2SiHI_2$			
Ra (l)	**Ab initio**		**Assignment**	**Ra (l)**	**Ab initio**		**Assignment**
	anti	***gauche***			***anti***	***gauche***	
85s	75	79	δSiSiI	83s	87		δISiI
167m	152		δSiSiC	120mw		111	δISiI
194m		165	δSiSiC	165mw		159	δSiSiC
314vs		310	νSiI	209m	201		δSiSiC
325m	334		νSiI	294vs	287	284	$\nu_s SiI_2$
345vw	349		ρSiH_2	369s	358		$\nu_{as} SiI_2$
379mw		402	ρSiH_2	376ms		370	$\nu_{as} SiI_2$
436w	429	419	νSiSi	443mw	433	431	νSiSi
	481	471	ρSiH_2	460mw	450	451	ρSiH_2
	600	596	τSiH_2	608w	603	592	τSiH_2
655w	654	663	νSiC	663w	650	654	νSiC
685vw	684	636	τSiH_2	693m	680	680	γSiH_2

coordinates has been transformed into a force field defined with symmetry coordinates, which was then used to calculate potential energy distributions by normal coordinate analyses.

The Raman spectra change considerably with temperature; this is caused by the changing conformational composition. One line pair for each compound has been chosen to monitor the intensity ratio with temperature. Fig. 2 illustrates the intensity changes that have been observed for the line pairs 295/337 cm^{-1} of **2** and 209/165 cm^{-1} of **4**.

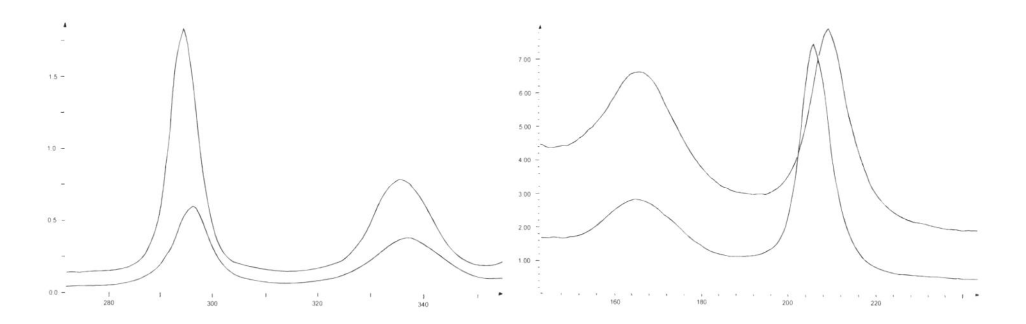

Fig. 2. Observed intensity changes for the line pairs 295/337 cm^{-1} (left) of ethyldiiodosilane (**2**) and 209/165 cm^{-1} (right) of 1,1-diiodo-2-methyldisilane (**4**).

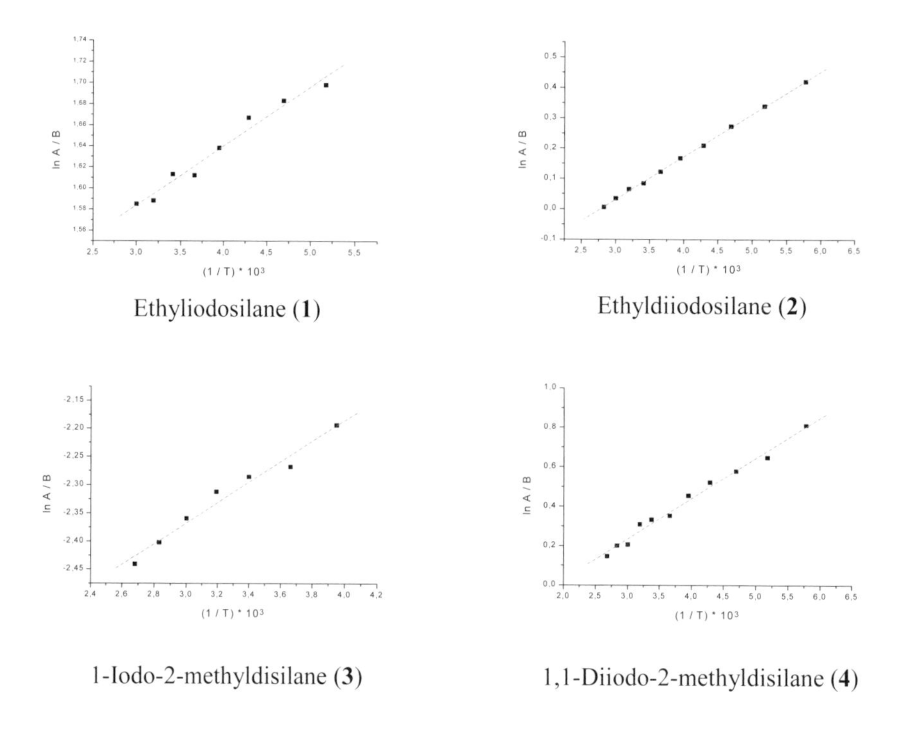

Fig. 3. van't Hoff plots for the line pairs 344/378 cm^{-1} of **1**, 295/337 cm^{-1} of **2**, 325/314 cm^{-1} of **3** and 209/165 cm^{-1} of **4**.

By fitting the observed intensity ratio of a line pair to the equation

$$\ln (I_{gauche}/I_{anti}) = -\Delta H/RT + \text{const.}$$

the van't Hoff plots presented in Fig. 3 have been obtained. From the slopes of the resulting straight lines obtained by least square fits, ΔH-values of 1.01, −1.17, −1.52 and −1.70 kJ/mol ($\Delta H = H_{gauche} - H_{anti}$) have been obtained for **1**, **2**, **3** and **4**, respectively. As one cannot be absolutely sure that the line pairs chosen are "pure" meaning that each line originates from a single conformer, the error limits are estimated to be about ±0.3 kJ/mol, maybe even larger.

A simple model for explaining relative conformational stabilities of organic compounds uses the so-called *gauche* effect stating that the conformer with the larger number of *gauche* interactions between the most polar bonds will be of lower energy [4]. As the polarities of Si–H and Si–I bonds must be very similar because of the identical electronegativities of H and I, the relative stabilities of **1**, **2**, **3** and **4** cannot be explained using this model.

1, **2**, **3** and **4** have been synthesized by protodearylation of $EtMesSiH_2$, $EtPh_2SiH$, $MeH_2SiSiPhH_2$ and $MeH_2SiSiPh_2H$ with liquid hydrogen iodide. The arylated ethylsilanes have been prepared from $EtSiCl_3$ with either MesLi or PhMgBr, followed by a reduction with $LiAlH_4$. Both disilanes have been prepared from $MeCl_2SiSiPh_3$ and trifluoromethanesulfonic acid by selectively replacing one or two phenyl groups with OSO_2CF_3, and a subsequent reduction of the trifluoromethanesulfonyloxydisilanes with $LiAlH_4$.

Table 3 summarizes the ^{29}Si chemical shifts and important coupling constants of all compounds that have been prepared during this work.

Table 3. ^{29}Si chemical shifts [$\delta(^{29}Si)$, ppm against TMS] and coupling constants [Hz] of ethyliodosilanes and iodomethyldisilanes prepared for this work.

	$\delta(^{29}Si)$	$^1J(SiH)$		$\delta(^{29}Si)$	$\delta(^{29}Si^*)$	$^1J(SiH)$	$^1J(Si^*H)$
$EtSiMesCl_2$	20.2		$MeCl_2SiSi^*Ph_3$	29.2	−23.6		
$EtSiMesH_2$	−42.4	190.2	$MeCl_2SiSi^*Ph_2Cl$	22.2	−7.3		
$EtSiIH_2$	−40.4	266.6	$MeCl_2SiSi^*PhCl_2$	18.0	3.0		
$EtSiI_2H$	−52.5	265.9	$MeH_2SiSi^*Ph_2H$	−67.5	−38.8	186.3	189.3
			$MeH_2SiSi^*PhH_2$	−67.7	−60.2	187.3	188.4
			$MeH_2SiSi^*I_2H$	−54.5	−75.5	204.5	239.6

Acknowledgement: Financial support by the *Österreichische Nationalbank* is gratefully acknowledged.

References:

[1] P. v. R. Schleyer, M. Kaupp, F. Hampel, M. Bremer, K. Mislow, *J. Am. Chem. Soc.* **1992**, *114* 6791.

[2] a) W. J. Stevens, H. Basch, M. Krauss, *J. Chem. Phys.* **1984**, *81*, 6026;
b)W. J. Stevens, H. Basch, M. Krauss, P. G. Jasien, *Can. J. Chem.* **1992**, *70*, 612.

[3] L. Hedberg, I. A. Mills, *J. Mol. Spectrosc.* **1993**, *160*, 117.

[4] S. Wolfe, *Acc. Chem. Research* **1972**, *5*, 102.

The Rotational Isomerism of the Disilanes $tBuX_2SiSiX_2tBu$ (X = Cl, Br and I): A Combined Raman Spectroscopic, X-Ray Diffraction and Ab Initio Study

Robert Zink, Harald Siegl, Karl Hassler

Institut für Anorganische Chemie
Technische Universität Graz
Stremayrgasse 16, A-8010 Graz, Austria
E-mail: hassler@anorg.tu-graz.ac.at

Keywords: *tert*-Butyldisilane / Rotational Isomerism / Raman Spectroscopy / Ab Initio Calculations / X-Ray Diffraction

Summary: Ab initio calculations predict the existence of *anti* and *gauche* rotamers for the disilanes $tBuX_2SiSiX_2tBu$ (X = Br and I). For the chloro compound a third backbone conformer with a CSiSiC dihedral angle of 95° (*ortho*) was located on the potential energy surface. Due to the fact that the vibrational spectra of these disilanes are hardly sensitive to the conformation around the Si–Si bond we have not been able to determine energy differences between conformers from variable-temperature Raman spectra. Infrared and Raman spectroscopy suggest that *anti* and twisted conformers are present in liquid and solid $tBuCl_2SiSiCl_2tBu$. For X = Br and I the *anti* conformation is adopted in the solid state, as proven by X-ray diffraction and vibrational spectroscopy.

Introduction

Previous studies of the rotational isomersim of the methylated disilanes MeX_2SiSiX_2Me (X = H, F, Br and I) showed that both *anti* and *gauche* rotamers are present in the liquid state [1]. Energy differences between conformers were obtained from variable-temperature Raman spectra employing van't Hoff plots of the logarithm of intensity ratios versus the inverse temperature. Replacing the methyl ligands against the bulky *tert*-butyl groups leads to the disilanes $tBuX_2SiSiX_2tBu$ that may be expected to exhibit unusual conformational behavior because of the possible occurrence of 1,3- and 1,4-substituent interactions. Very recently we reported the conformational properties of the disilanes $tBuH_2SiSiH_2tBu$ [2] and $tBuF_2SiSiF_2tBu$ [3]. Although calculations predicted the existence of *anti* and *gauche* conformers vibrational spectra established the presence of just a single conformer, *anti*, which is probably due to the barrier of interconversion lying below the lowest torsional vibrational level. In this work we report a combined Raman spectroscopic, X-ray diffraction and ab initio study of the rotational isomerism of the disilanes $tBuX_2SiSiX_2tBu$ (X = Cl, Br and I).

Synthesis

All disilanes were prepared according to the literature method [4].

Ab Initio Calculations

Calculations for the chloro compound tBuCl$_2$SiSiCl$_2t$Bu were undertaken at the 6-31G*/SCF level of theory. Computations for the di-*tert*-butyldisilanes with X = Br and I employed effective core potentials given by Stevens, Basch, Krauss and Jasien [5, 6] for the heavy atoms and a valence double zeta-basis with a single polarization function on all heavy atoms (HF/CEP-31G*/SCF). For X = Br and I calculations predict the existence of *gauche* and *anti* rotamers; however, the 180° *anti* conformations (C_{2h} symmetry) split into enantiomeric pairs with CSiSiC dihedral angles of around ±170° (C_2 symmetry). The calculated torsional energy profile for tBuI$_2$SiSiI$_2t$Bu is presented in Fig. 1.

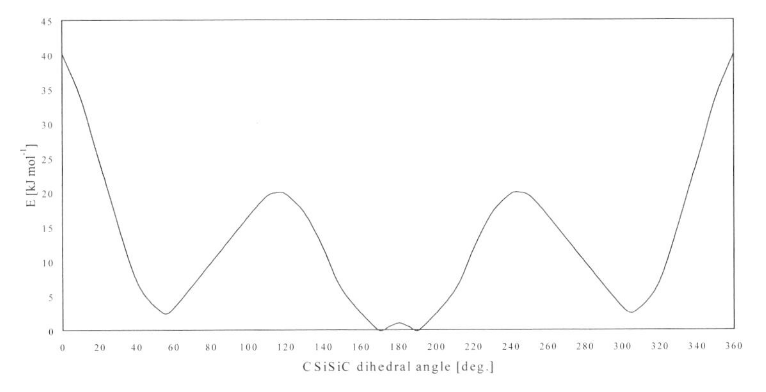

Fig. 1. Torsional energy profile for tBuI$_2$SiSiI$_2t$Bu (CEP-31G*/SCF).

Moreover, for tBuCl$_2$SiSiCl$_2t$Bu three twisted backbone conformers with the CSiSiC dihedral angle assuming values of ±56.7° (*gauche*, $E_{rel.}$ = 8.50 kJ mol^{-1}), ±94.7° (*ortho*, $E_{rel.}$ = 8.42 kJ mol^{-1}) and ±169.6° (*anti*, $E_{rel.}$ = 0 kJ mol^{-1}) are predicted (see Fig. 2).

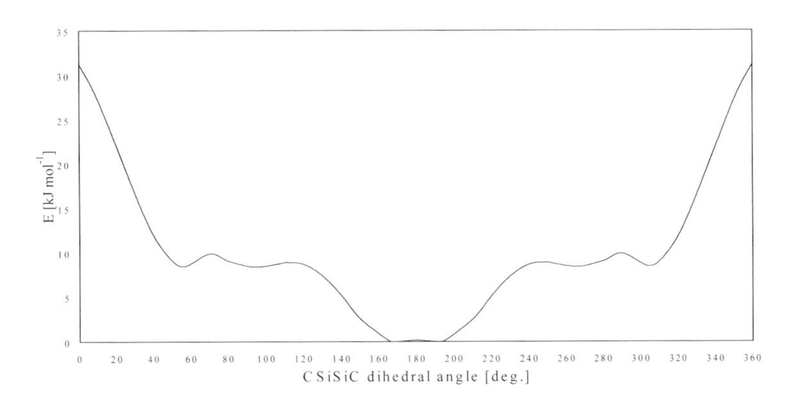

Fig. 2. Torsional energy profile for tBuCl$_2$SiSiCl$_2t$Bu (6-31G*/SCF).

The splitting of the "ordinary" *gauche* minimum with the idealized backbone CSiSiC dihedral angle of 60° into two twisted conformations is presumably due to nonbonded 1,4-substituent interactions (Cl–Si-Si-C-C, C–Si-Si-C-C). Similarly, the twisting of the CSiSiC dihedral angle away from 180° reduces repulsive steric four-bond C–Cl interactions and relieves strain in the SiSiC angle. The existence of three backbone conformers is an unusual phenomenon that has also been reported for n-C_4F_{10} [7, 8], n-Si_4Me_{10} [9] and $CF_3SiMe_2SiMe_2CF_3$ [10].

Rotational Isomerism

Studies of the rotational isomerism of the disilanes MeX_2SiSiX_2Me revealed that Raman-active skeletal vibrations are highly sensitive to the conformation around the Si–Si bond. For example, for $MeBr_2SiSiBr_2Me$ the mode ν_sSiBr_2 is separated by 30 cm^{-1} between *anti* and *gauche* rotamers. Surprisingly, for the disilanes $t$$BuX_2SiSiX_2$$t$Bu (X = Br and I) calculations predict that the Raman spectra of *gauche* and *anti* isomers hardly differ from each other.

Table 1. Calculated and observed wavenumbers of Raman-active modes for $t$$BuBr_2SiSiBr_2$$t$Bu.

Species	Vib. no.	Approx. description	Ab initio Unscaled	Ab initio Scaled by 0.92	Observed Solid	Observed Liquid	PED (>10 %)
A_g(Ra)	ν_1	ρ_1CH_3	1324.6	1219	1200	1192	69(1),17(6),10(9)
	ν_2	ρ_2CH_3	1118.4	1029	1007	1008	80(2),11(5)
	ν_3	ρ_3CH_3	1316.9	1212	1186	1192	52(3),38(5),11(7)
	ν_4	ν_sCC_3	880.1	810	810	809	74(4),18(9),17(1)
	ν_5	$\nu_{as}CC_3$	1021.5	940	939	940	52(5),30(3),18(2)
	ν_6	δ_sCC_3	449.0	413	410	412	60(6),22(10),18(12),14(14)
	ν_7	$\delta_{as}CC_3$	400.7	369	381	381	84(7),13(3)
	ν_8	ρCC_3	260.1	239	244	236	72(8),16(10),10(7)
	ν_9	νSiC	653.3	601	622	621	51(9),18(4),14(11),12(6),11(14)
	ν_{10}	ν_sSiBr_2	227.7	210	223	220	28(10),20(9),18(11),10(14)
	ν_{11}	$\nu SiSi$	572.0	526	537	538	43(11),24(10),18(12),13(8)
	ν_{12}	$\delta SiSiC$	152.1	140	141	141	42(12),22(14),19(13)
	ν_{13}	$\delta SiBr_2$	80.1	74	77	79	73(13),27(12),22(14),10(11)
	ν_{14}	$\gamma SiBr_2$	110.6	102	114	107	35(14),18(12),11(11)
B_g(Ra)	ν_{15}	ρ_1CH_3	1117.3	1028	1007	1008	81(15),11(18)
	ν_{16}	ρ_2CH_3	1038.7	956	–	–	100(16)
	ν_{17}	ρ_3CH_3	1319.8	1214	1186	1192	53(17),39(18),11(19)
	ν_{18}	$\nu_{as}CC_3$	1024.5	943	939	940	51(18),32(17),16(15)
	ν_{19}	$\delta_{as}CC_3$	406.9	374	381	381	78(19),13(21),12(17)
	ν_{20}	ρCC_3	250.3	230	244	236	69(20),19(21),14(19)
	ν_{21}	$\nu_{as}SiBr_2$	502.9	463	470	473	71(21),43(23),20(20)
	ν_{22}	$\tau SiBr_2$	99.9	92	96	92	105(22)
	ν_{23}	$\rho SiBr_2$	140.7	129	131	129	65(23),11(20)

Indeed, variable-temperature Raman spectra do not reveal spectral features due to a single conformer only. Vibrational spectra of solid tBuBr$_2$SiSiBr$_2t$Bu and tBuI$_2$SiSiI$_2t$Bu indicate that the *anti* conformation (point group C_{2h}) is adopted in the solid state since the spectra obey the rule of mutual exclusion. Raman vibrational spectra of both disilanes are summarized in Tables 1 and 2, respectively.

Table 2. Calculated and observed wavenumbers of Raman-active modes for tBuI$_2$SiSiI$_2t$Bu.

Species	Vib. no.	Approx. description	Ab initio Unscaled	Ab initio Scaled by 0.92	Observed Solid	PED (>10 %)
A_g(Ra)	ν_1	ρ_1CH_3	1311.0	1206	1182	38(1),25(3),19(5)
	ν_2	ρ_2CH_3	1118.9	1029	1005	81(2)
	ν_3	ρ_3CH_3	1319.7	1214	1182	35(1),28(3),19(5)
	ν_4	ν_sCC_3	875.5	805	806	80(4),16(9),15(1)
	ν_5	$\nu_{as}CC_3$	1020.2	939	938	53(5),30(3),17(2)
	ν_6	δ_sCC_3	436.6	402	396	54(6),25(12),15(10),12(14)
	ν_7	$\delta_{as}CC_3$	398.4	367	375	82(7),12(3)
	ν_8	ρCC_3	253.0	233	239	75(8),13(11)
	ν_9	νSiC	635.0	584	601	56(9),19(6),15(4),10(11)
	ν_{10}	ν_sSiI_2	204.1	188	198	34(10),20(14),19(9),11(11)
	ν_{11}	$\nu SiSi$	537.9	495	502	51(11),18(12),17(8),15(10)
	ν_{12}	$\delta SiSiC$	123.8	114	118	60(12),19(10)
	ν_{13}	δSiI_2	59.0	54	55	76(13),27(14),20(12)
	ν_{14}	γSiI_2	100.2	92	100	51(14),18(13),11(11)
B_g(Ra)	ν_{15}	ρ_1CH_3	1117.5	1028	1005	82(15),10(18)
	ν_{16}	ρ_2CH_3	1038.6	956	–	100(16)
	ν_{17}	ρ_3CH_3	1318.9	1213	1182	54(17),39(18),11(19)
	ν_{18}	$\nu_{as}CC_3$	1023.5	942	938	52(18),31(17),15(15)
	ν_{19}	$\delta_{as}CC_3$	399.6	368	375	70(19),26(21),10(17),10(23)
	ν_{20}	ρCC_3	239.3	220	230	71(20),25(21),10(19)
	ν_{21}	$\nu_{as}SiI_2$	461.4	425	425	45(21),45(23),29(20),15(19)
	ν_{22}	τSiI_2	82.1	76	80	102(22),19(23)
	ν_{23}	ρSiI_2	122.9	113	118	47(23),10(21)

X-ray crystal diffraction shows that the disilanes tBuBr$_2$SiSiBr$_2t$Bu and tBuI$_2$SiSiI$_2t$Bu crystallize in a triclinic crystal system (space group $P\bar{1}$) and adopt C_i symmetry. Selected theoretical and experimental geometric parameters are given in Table 3. The structure of the iodo compound as determined by X-ray diffraction is depicted in Fig. 3.

Table 3. Selected calculated and experimental geometric parameters for $tBuX_2SiSiX_2tBu$ (X = Br and I).

Parameter	$tBuBr_2SiSiBr_2tBu$ anti (C_2) Ab initio	anti (C_i) X-ray diffraction	Parameter	$tBuI_2SiSiI_2tBu$ anti (C_2) Ab initio	anti (C_i) X-ray diffraction
r(Si–Si)	241.1	235.5	*r*(Si–Si)	243.2	236.5
r(Si–C)	192.5	189.7	*r*(Si–C)	194.1	190.4
r(Si–Br)	225.5/225.6	221.8/223.2	*r*(Si–I)	249.5/249.6	246.1/246.8
∠SiSiC	118.9	119.2	∠SiSiC	119.1	120.0
∠SiSiBr	105.1/108.1	106.1/106.5	∠SiSiI	105.4/108.3	106.7/105.4
∠CSiSiC	168.0	180.0	∠CSiSiC	168.9	180.0

Table 4. Calculated and observed wavenumbers of Raman active modes for $tBuCl_2SiSiCl_2tBu$.

Species	Vib. no.	Approx. description	Ab initio Unscaled	Scaled by 0.92	Observed Solid	Liquid	PED (>10 %)
A_g(Ra)	ν_1	ρ_1CH_3	1347.8	1240	1202	1203	70(1),17(6)
	ν_2	ρ_2CH_3	1132.4	1042	1006	1008	86(2)
	ν_3	ρ_3CH_3	1328.0	1222	1188	1187	58(3),33(5),11(7)
	ν_4	ν_sCC_3	878.7	808	814	815	73(4),20(9),16(1)
	ν_5	$\nu_{as}CC_3$	1024.7	943	939	940	61(5),28(3),12(2)
	ν_6	δ_sCC_3	479.2	441	437	439	55(6),27(10),17(14),10(11)
	ν_7	$\delta_{as}CC_3$	411.8	379	383	383	74(7),12(3)
	ν_8	ρCC_3	245.0	225	233	224	51(8),19(11)
	ν_9	νSiC	667.7	614	632	633	46(9),20(4),16(11),13(14)
	ν_{10}	ν_sSiCl_2	327.7	301	307	306	22(10),21(8),16(7),12(12)
	ν_{11}	$\nu SiSi$	626.5	576	585	589	49(10),29(11),16(12)
	ν_{12}	$\delta SiSiC$	117.4	108	111	109	67(12),48(13)
	ν_{13}	$\delta SiCl_2$	198.6	183	186	185	34(13),32(14),11(12)
	ν_{14}	$\gamma SiCl_2$	138.9	128	130/137	133	42(14),24(11),12(13)
B_g(Ra)	ν_{15}	ρ_1CH_3	1131.1	1041	1006	1008	88(15)
	ν_{16}	ρ_2CH_3	1062.0	977	–	–	100(16)
	ν_{17}	ρ_3CH_3	1330.8	1224	1188	1187	59(17),33(18),11(19)
	ν_{18}	$\nu_{as}CC_3$	1028.6	946	939	940	60(18),29(17),10(15)
	ν_{19}	$\delta_{as}CC_3$	421.2	388	391	393	72(19),11(17)
	ν_{20}	ρCC_3	281.2	259	274	263	52(20),22(19),13(23)
	ν_{21}	$\nu_{as}SiCl_2$	588.0	541	548	552	94(21),28(23)
	ν_{22}	$\tau SiCl_2$	139.5	128	130/137	133	98(22)
	ν_{23}	$\rho SiCl_2$	173.4	160	158	154	69(23),30(20)

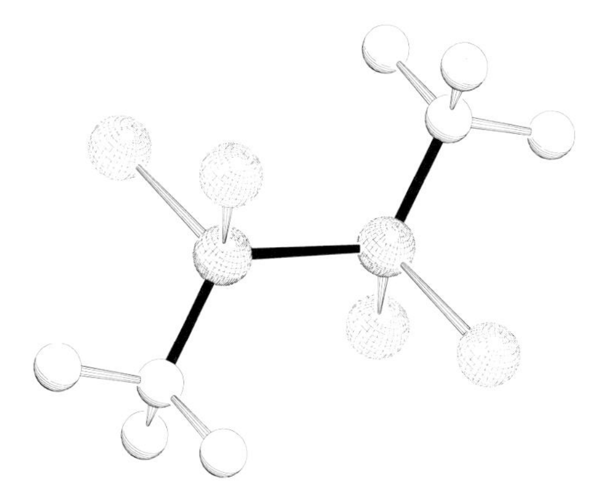

Fig. 3. Structure of tBuI$_2$SiSiI$_2t$Bu as determined by X-ray diffraction.

For the chloro compound tBuCl$_2$SiSiCl$_2t$Bu calculations predict that vibrations of *gauche* and *ortho* rotamers do not differ from each other. However, the modes ν_sSiCl_2 and $\nu_{as}SiCl_2$ are predicted to be separated by approximately 10 cm^{-1} between anti and twisted conformers. The mode ν_sSiCl_2 (*anti*) is assigned to the strong Raman band at 307 cm^{-1}. A weak shoulder at 290 cm^{-1} could be due to twisted (*gauche* + *ortho*) conformations. Further, the Raman band at 550 cm^{-1} ($\nu_{as}SiCl_2$) is very broad and asymmetric. Vibrational spectra of the solid do not follow the rule of mutual exclusion, suggesting that a certain amount of a twisted conformer is present. The X-ray diffraction analysis for tBuCl$_2$SiSiCl$_2t$Bu is being carried out currently. Raman vibrational spectra of the chloro compound are presented in Table 4.

Conclusions

Variable-temperature Raman spectra could not be employed for the determination of energy differences between *anti* and *gauche* rotamers of the disilanes tBuX$_2$SiSiX$_2t$Bu (X = Br and I) due to the insensitivity of skeletal modes to the conformation around the Si–Si bond. For tBuCl$_2$SiSiCl$_2t$Bu Raman spectra suggest that a small amount of the twisted conformations (*gauche* and *ortho*) is present in the liquid and solid state. Matrix-isolation IR spectroscopy might be most appropriate for investigating the rotational isomerism of the disilanes tBuX$_2$SiSiX$_2t$Bu (X = Cl, Br and I).

References:

[1] R. Zink, K. Hassler, M. Ramek, *Vibrational Spectrosc.* **1998**, *18*, 123.

[2] D. Hnyk, R. S. Fender, H. E. Robertson, D. W. H. Rankin, M. Bühl, K. Hassler, K. Schenzel, *J. Mol. Struct.* **1995**, *346*, 215.

[3] B. A. Smart, H. E. Robertson, N. W. Mitzel, D. W. H. Rankin, R. Zink, K. Hassler, *J. Chem. Soc., Dalton Trans.* **1997**, 2475.

[4] B. Reiter, K. Hassler, *J. Organomet. Chem.* **1994**, *21*, 467.

[5] W. J. Stevens, H. Basch, M. Krauss, *J. Chem. Phys.* **1984**, *81*, 6026.

[6] W. J. Stevens, M. Krauss, H. Basch, P. G. Jasien, *Can. J. Chem.* **1992**, *70*, 612.
[7] B. Albinsson, J. Michl, *J. Am. Chem. Soc.* **1995**, *117*, 6378.
[8] B. Albinsson, J. Michl, *J. Phys. Chem.* **1996**, *100*, 3418.
[9] B. Albinsson, H. Teramae, J. W. Downing, J. Michl, *Chem. Eur. J.* **1996**, *2*, 529.
[10] R. Zink, K. Hassler, A. Roth, R. Eujen, to be published.

The Rotational Isomerism of the Disilanes MeX_2SiSiX_2Me (X = H, F, Br and I): A Combined Ab Initio and Raman Spectroscopic Study

Robert Zink, Karl Hassler

Institut für Anorganische Chemie
Technische Universität Graz
Stremayrgasse 16, A-8010 Graz, Austria
E-mail: hassler@anorg.tu-graz.ac.at

Keywords: Rotational Isomerism / Raman Spectroscopy / Ab Initio Calculations / Disilane

Summary: Ab initio calculations and Raman spectroscopy suggest that the methylated disilanes MeX_2SiSiX_2Me (X = H, F, Br and I) exist as mixtures of *anti* and *gauche* rotamers in the liquid state. Energy differences between the conformers have been determined from variable-temperature Raman spectra employing van't Hoff plots of the logarithm of intensity ratios versus the inverse temperature. For X = F, Br and I the *anti* arrangement is preferred energetically. For X = H the *gauche* rotamer is slightly stabilized by 0.43 kJ mol^{-1}.

Introduction

1,2-Disubstituted ethanes have been the subject of a large number of conformational analyses. For example, the rotational isomerism of $ClCH_2CH_2Cl$ was investigated by Mizushima and co-workers more than forty years ago [1]. Just a few experimental studies of the conformational behavior of substituted di- and oligosilanes have been reported so far. This is indeed surprising as backbone conformations are known to exert immense influences on the near-UV spectra and electronic properties of oligo- and polysilanes [2, 3] that are of great industrial interest because of their suitability as photoconductors, nonlinear optical materials or photoresists for microlithographic processes in microelectronics. In this work we report the results of a combined ab initio and Raman spectroscopic study of the conformational properties of the disilanes MeX_2SiSiX_2Me (X = H, F, Br and I) in continuation of the work done on $MeCl_2SiSiCl_2Me$ [4].

Synthesis

MeH_2SiSiH_2Me was obtained by reducing the chloro compound $MeCl_2SiSiCl_2Me$ with $LiAlH_4$ in di-*n*-butyl ether. The disilanes $MeBr_2SiSiBr_2Me$ and MeI_2SiSiI_2Me were synthesized by reacting $MePh_2SiSiPh_2Me$ with the respective hydrogen halide HX (X = Br, I).

The fluoro compound was prepared by reacting $Me(OMe)_2SiSi(OMe)_2Me$ with BF_3 at a pressure of about 25 atm.

Ab Initio Calculations

A fairly simple computational scheme was used for the complete series of disilanes MeX_2SiSiX_2Me (X = H, F, Br and I): geometry optimizations as well as calculations of harmonic frequencies were undertaken at the SCF level of theory employing the effective core potentials of Stevens, Basch, Krauss and Jasien [5, 6] for the heavy atoms and a valence double-zeta basis with a single polarization function on all atoms (CEP-31G**/SCF). For X = H, Br and I calculations predict the existence of two rotamers, *anti* and *gauche*, separated by barriers to interconversion that are large enough to enable the observation of both species on the time scale which is typical for vibrational (IR, Ra) spectroscopy. For X = F only a single conformer, *anti*, was located on the potential energy surface. In contrast, at the 6-31G**/SCF level two rotamers are also predicted for the fluoro compound with the barrier to interconversion lying 0.53 kJ mol^{-1} above the *gauche* structure. The barrier even increases to 0.97 kJ mol^{-1} at the 6-31G**/MP2 level. In agreement with the classical picture of the rotational isomerism of YX_2MMX_2Y systems all *anti* conformers of the disilanes MeX_2SiSiX_2Me are predicted to reflect overall C_{2h} symmetry whereas the *gauche* minima possess C_2 symmetry with the CSiSiC dihedral angle assuming a value of around 70°.

The harmonic ab initio cartesian force fields were then converted into symmetry force fields that served for the calculation of potential energy distributions and the assignment of vibrations to the chosen set of symmetry coordinates.

Rotational Isomerism

It is well known that variable-temperature Raman spectroscopy is a useful technique for the investigation of the rotational isomerism of substituted di- or oligosilanes as the frequencies of skeletal modes are usually highly sensitive to the conformation around Si–Si bonds. Energy differences between rotational isomers can then be determined from variable-temperature spectra by applying van't Hoff's equation:

$$\ln I_a/I_g = -\Delta E/RT + \text{const.}$$

I_a and I_g denote the temperature dependent intensities of vibrational bands due to a single conformer only.

For MeH_2SiSiH_2Me theory predicts that the strongly Raman-active mode νSiSi differs by 15 cm^{-1} between *gauche* and *anti* rotamers. The two Raman bands at 411 cm^{-1} and 421 cm^{-1} are easily assigned to the *gauche* and *anti* rotamer, respectively. Upon solidification the band at 411 cm^{-1} disappears. This unambiguously demonstrates that the *anti* conformation is preferred in the solid state. However, a van't Hoff plot employing the line pair at 415 cm^{-1} yields the surprising result that the *gauche* rotamer is slightly stabilized in the liquid state by 0.43 kJ mol^{-1} (see Fig. 1). Even more striking is the fact that at low temperatures the bandwidth of νSiSi for the *gauche* isomer is nearly twice as large as the bandwidth for the *anti* rotamer. Moreover, as can be seen from Fig. 1, if one used peak heights instead of peak areas for the van't Hoff analysis one would draw the wrong conclusion that the *anti* rotamer is more stable.

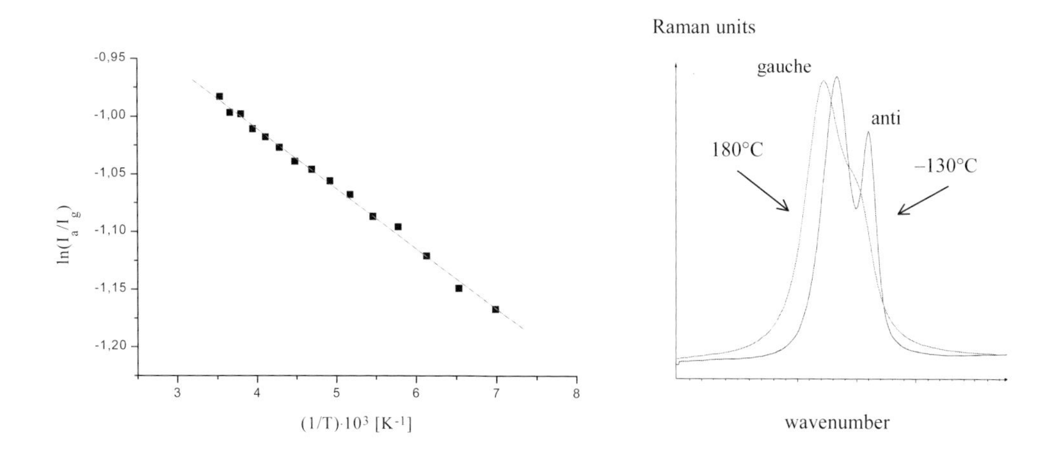

Fig. 1. van't Hoff plot for MeH_2SiSiH_2Me (left) and the employed line pair νSiSi (right).

For MeF_2SiSiF_2Me the stretching vibration νSiC around 720 cm^{-1} splits into a doublet due to the occurrence of rotational isomerism (see Fig. 2). Unfortunately, because of the presence of a third Raman band in the closer vicinity of this line pair we were unable to reliably determine peak areas. Hence peak heights were employed for the van't Hoff plot yielding the result that the *anti* rotamer is more stable by 1.80 kJ mol^{-1} (see Fig. 2).

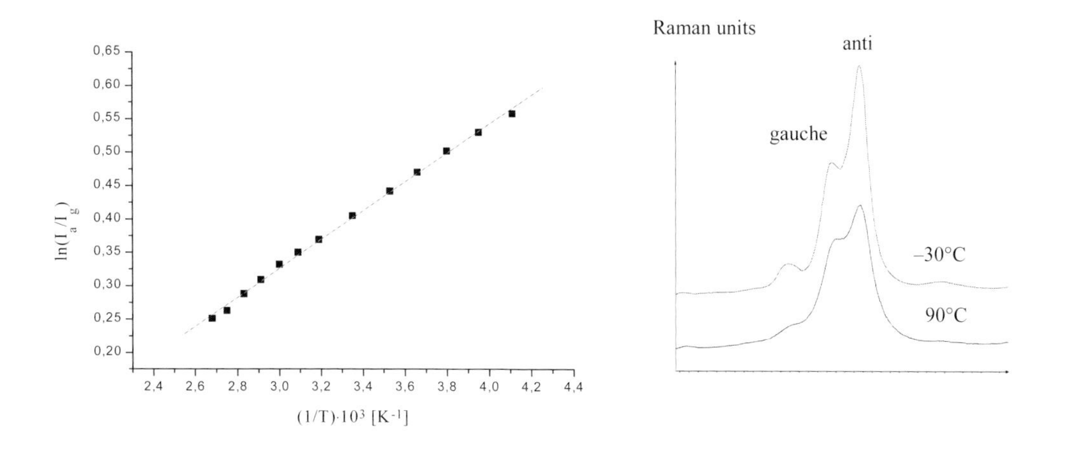

Fig. 2. van't Hoff plot for MeF_2SiSiF_2Me (left) and the employed line pair νSiC (right).

For the disilanes $MeBr_2SiSiBr_2Me$ and MeI_2SiSiI_2Me calculations predict that methyl rocking vibrations and several stretching vibrations (SiC, SiSi and SiX; X = Br and I) are sensitive to the conformation. Table 1 summarizes such vibrations for the bromo compound. The strongly Raman-active modes ν_sSiX_2 (X = Br and I) were found to be most suitable for the determination of energy differences. Variable-temperature Raman spectra unambiguously suggest that for X = Br and I the *anti* rotamers are more stable than the *gauche* isomers. It is also of some note that for the bromo and iodo compounds both *anti* and *gauche* conformers are present in the solid state.

Table 1. Collection of Raman-active vibrations that are sensitive to the conformation of $MeBr_2SiSiBr_2Me$.

Anti			*Gauche*		
Mode	**Calc. [cm^{-1}][a]**	**Obs. [cm^{-1}]**	**Mode**	**Calc. [cm^{-1}][a]**	**Obs. [cm^{-1}]**
ρCH_3 (A_g)	819	810	ρCH_3 (A)	808	800
νSiC (A_g)	708	736	νSiC (A)	698	726
$\nu SiSi$ (A_g)	495	509	$\nu SiSi$ (A)	502	515
ν_sSiBr_2 (A_g)	306	318	ν_sSiBr_2 (A)	287	301

[a] Calculated wavenumbers are scaled by 0.92.

Experimental and theoretical energy differences for the complete series of methylated disilanes MeX_2SiSiX_2Me (X = H, F, Br ,Cl and I) are summarized in Table 2.

Table 2. Experimental and calculated energy differences between *anti* and *gauche* conformers of MeX_2SiSiX_2Me (X = H, F, Cl, Br and I).

Sample	**E_{gauche}–E_{anti} [kJ mol^{-1}] Raman spectroscopy**	**E_{gauche}–E_{anti} [kJ mol^{-1}] Ab initio calculations**	**Level of theory**
MeH_2SiSiH_2Me, liquid	$-0.43 \pm 0 .2$	0.40	CEP-31G**
MeF_2SiSiF_2Me, liquid	1.80 ± 0.5	3.57	6-31G**
$MeCl_2SiSiCl_2Me$[a], liquid	0.90 ± 0.2	5.03	CEP-31G**
$MeBr_2SiSiBr_2Me$, liquid	3.0 ± 0.2	5.60	CEP-31G**
MeI_2SiSiI_2Me, solution in benzene	2.7 ± 0.3	5.76	CEP-31G**

[a] taken from reference [4].

Conclusions

Calculations and Raman spectroscopy indicate that the methylated disilanes MeX_2SiSiX_2Me (X = H, F, Br and I) comprise mixtures of *anti* and *gauche* conformers in the liquid state. Apart from MeH_2SiSiH_2Me the *anti* conformation is preferred energetically. Increasing the size of the substituent from F to I slightly increases the energy difference between the rotamers; this is readily attributed to steric effects.

References:

[1] S. Mizushima, *Structure of Molecules and Internal Rotation*, Academic Press, New York, **1954**.

[2] B. Albinsson, H. Teramae, J. W. Downing, J. Michl, *Chem. Eur. J.* **1996**, *2*, 529.

[3] H. S. Plitt, J. Michl, *Chem. Phys. Lett.* **1992**, *198*, 400.

[4] M. Ernst, K. Schenzel, A. Jähn, K. Hassler, J. Mol. Struct. **1997**, *412*, 83.

[5] W. J. Stevens, H. Basch, M. Krauss, *J. Chem. Phys.* **1984**, *81*, 6026.

[6] W. J. Stevens, M. Krauss, H. Basch, P. G. Jasien, *Can. J. Chem.* **1992**, *70*, 612.

The Rotational Isomerism of $F_3CSiMe_2SiMe_2CF_3$: Do three conformers exist in the liquid state?

Robert Zink, Karl Hassler

Institut für Anorganische Chemie
Technische Universität Graz
Stremayrgasse 16, A-8010 Graz, Austria

Achim Roth, Reint Eujen

FB 9 — Anorganische Chemie
Bergische Universität — Gesamthochschule Wuppertal
Gauss-Strasse 20, D-42097 Wuppertal, Germany
E-mail: eujen@wrcd1.urz.uni-wuppertal.de

Keywords: Bis(trifluoromethyl)disilane / Rotational Isomerism / *Ortho* Conformer / Raman Spectroscopy / Ab Initio Calculations

Summary: Ab initio calculations at the MP2/6-31G* level of theory predict the existence of three enantiomeric pairs of nonequivalent conformers on the potential energy surface of the highly interesting compound $F_3CSiMe_2SiMe_2CF_3$. In agreement with theory the Raman bands at 360 cm^{-1}, 369 cm^{-1} and 380 cm^{-1} are assigned to the *gauche*, *ortho* and *anti* conformer, respectively. The relative intensities of these bands vary with temperature, and energy differences between the individual rotamers have been determined from variable temperature Raman spectra as $H_{gauche}–H_{anti}$ = 2.65 kJ mol^{-1}, $H_{ortho}–H_{anti}$ = 2.64 kJ mol^{-1} and $H_{gauche}–H_{ortho}$ = 0 kJ mol^{-1}.

Introduction

SCF/6-31G* as well as MP2/6-31G* calculations on n-Si_4Me_{10} [1] and n-C_4F_{10} [2, 3] predict the existence of three pairs of enantiomeric conformers on the potential energy surface with the silicon (n-Si_4Me_{10}) or carbon (n-C_4F_{10}) backbone forming dihedral angles of around ±60° (*gauche*), ±90° (termed *ortho* by the authors [2]) and ±165° (*anti*). Certain empirical and semiempirical calculations also predicted the existence of twisted conformations with backbone dihedral angles around 90°; however, since they made also such predictions for parent alkanes and oligosilanes they were not taken seriously [4, 5]. Recently the "three conformer idea" was vividly revitalized when J. Michl and coworkers succeeded in detecting all three rotamers in the N_2 matrix-isolation IR spectra of C_4F_{10} [2, 3]. No experimental evidence for the existence of three conformers in unstrained disilanes or oligosilane chains has been reported so far. In this work we report the conformational properties of the compound $CF_3SiMe_2SiMe_2CF_3$ studied by ab initio calculations and variable temperature Raman spectroscopy. This molecule is an interesting link between n-Si_4Me_{10} and n-C_4F_{10} and may be anticipated to show unfamiliar backbone conformations.

Synthesis of $CF_3SiMe_2SiMe_2CF_3$

A pure sample of $CF_3SiMe_2SiMe_2CF_3$ was obtained by reacting 2.8 g of $ClMe_2SiSiMe_2Cl$ with 7.45 g of CF_3Br in the presence of 7.9 g of $P(NEt_2)_3$ followed by careful fractionation at room temperature. The compound is very volatile (m.p. 42–43 °C, b.p. 132 °C) and can easily be purified by sublimation under reduced pressure.

NMR data

$\delta^1H(CH_3)$ = 0.47 ppm, $\delta^{13}C(CH_3)$ = –7.3 ppm, $\delta^{13}C(CF_3)$ = 132.2 ppm, $\delta^{19}F(CF_3)$ = –60.8 ppm, $\delta^{29}Si$ = –12.8 ppm, $^1J(SiCF_3)$ = 74.4 Hz; $\delta(^1H)$, $\delta(^{13}C)$ and $\delta(^{29}Si)$ data are referenced against TMS, $\delta(^{19}F)$ data against $CFCl_3$.

Ab Initio Calculations

An extensive search of the torsional potential of $CF_3SiMe_2SiMe_2CF_3$ was undertaken at the SCF/6-31G* level in order to identify all structurally stable conformers. Only a single rotamer, *anti*, reflecting overall C_{2h} symmetry was located. However, the potential curve is predicted to be very flat in the vicinity of conformations with $F_3CSiSiCF_3$ backbone dihedral angles of 60° and 100°. At the MP2/6-31G* level of theory three conformers with the backbone assuming dihedral angles of ±56° ($E_{rel.}$ = 6.3 kJ mol^{-1}), ±101° ($E_{rel.}$ = 4.3 kJ mol^{-1}) and ±171° ($E_{rel.}$ = 0 kJ mol^{-1}) are predicted (see Fig. 1). As we were not able to obtain harmonic frequencies at the MP2 level we performed vibrational frequency calculations at the SCF/6-31G* level for the backbone conformations with dihedral angles of 60°, 100° and 180°. These frequencies should present reasonable estimates of the relative positions of the frequencies one would obtain at the MP2 level for the true minima.

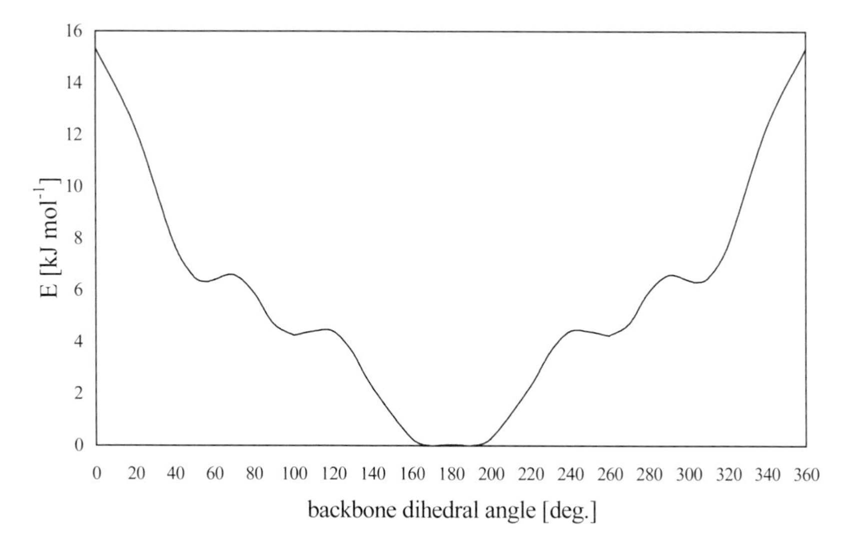

Fig. 1. Calculated torsional energy profile for $CF_3SiMe_2SiMe_2CF_3$ (MP2/6-31G*).

Raman Spectra

Table 1 summarizes scaled theoretical wavenumbers and the assignments to symmetry coordinates and experimental wavenumbers. Only modes that show Raman activity in the point group C_{2h} of the *anti* rotamer, i.e. a_g and b_g modes, are included. For the *gauche* and *ortho* conformers (point group C_2) a_u and b_u modes become a and b modes, respectively, and are also active in Raman. However, the intensities of these vibrations are very weak; therefore, the modes have been omitted in Table 1. High frequency modes (ν_sCH_3, $\nu_{as}CH_3$, δ_sCH_3 and $\delta_{as}CH_3$) and torsional vibrations have also been omitted for clarity and simplicity.

Table 1. Experimental and calculated (6-31G*/SCF) wavenumbers and potential energy distributions.

Species		Approximate description	Ab initio[a] gauche	Ab initio[a] ortho	Ab initio[a] anti	Observed Raman (solid)	PED[b]
A_g (A)[c]	ν_1	ρCH_3	897	897	897	867	61(1),11(2)
	ν_2	ρCH_3	783	783	861	782/782/837	67(2),11(1)
	ν_3	ν_sCF_3	1259	1259	1258	1206	55(3),40(5),22(8)
	ν_4	$\nu_{as}CF_3$	1192	1182	1178	1062	102(4),16(6)
	ν_5	δ_sCF_3	732	731	730	719	43(3),31(5),12(8)
	ν_6	$\delta_{as}CF_3$	525	525	525	521	72(6)
	ν_7	ρCF_3	228	228	228	229	61(7),19(13),10(10)
	ν_8	νSiC^F	354	364	377	359/369/380	24(8),17(12),15(13),11(5),11(9)
	ν_9	ν_sSiC_2	652	655	659	679	77(9),10(13)
	ν_{10}	δSiC_2	173	173	169	188	46(10),35(11),13(8)
	ν_{11}	γSiC_2	181	180	181	188	42(11),37(10),18(1),10(13)
	ν_{12}	$\delta SiSiC$	59	68	88	–/–/95	80(12),35(7)
	ν_{13}	$\nu SiSi$	474	472	470	467	39(13),23(8),20(11),14(5),11(1)
B_g (B)[c]	ν_{14}	ρCH_3	761	773	795	–/–/782	82(14)
	ν_{15}	ρCH_3	875	876	782	837/837/782	67(15),18(19)
	ν_{16}	$\nu_{as}CF_3$	1177	1167	1163	1062	103(16),17(17)
	ν_{17}	$\delta_{as}CF_3$	521	521	520	521	77(17)
	ν_{18}	ρCF_3	284	287	304	297/297/306	43(18),32(21)
	ν_{19}	$\nu_{as}SiC_2$	692	692	693	707	69(19),24(15)
	ν_{20}	ρSiC_2	117	134	180	134/134/188	55(20),39(18),28(21)
	ν_{21}	τSiC_2	120	117	119	134	44(20),39(21),24(18),11(14)

[a] scaled by 0.92 — [b] the potential energy distribution (PED > 10 %) is given for the *anti* conformer — [c] the notations A and B refer to the *ortho* and *gauche* rotamers.

As can be seen from Table 1 the strongly Raman active mode νSiC^F (this description is a rough approximation due to the strong vibrational coupling patterns, see Table 1) is predicted to be highly sensitive to the backbone conformation. Indeed three Raman bands appear at 359 cm^{-1}, 369 cm^{-1} and 380 cm^{-1} and the relative intensities of these bands clearly change with temperature. In agreement with theory the three frequencies are assigned to the *gauche*, *ortho* and *anti* conformer, respectively. Energy differences between the individual rotamers have been obtained from variable-temperature Raman spectra employing van't Hoff plots of the logarithm of intensity ratios versus the inverse temperature. The experimental values $H_{gauche}–H_{anti}$ = 2.65 kJ mol^{-1}, $H_{ortho}–H_{anti}$ = 2.64 kJ mol^{-1} and $H_{gauche}–H_{ortho}$=0 kJ mol^{-1} suffer from rather large error limits (±2 kJ mol^{-1}) since peak heights were used to determine the intensity ratios.

The mode ρCF_3 (B_g,B) is also predicted to be well separated between *anti* and *gauche* rotamers but not between *gauche* and *ortho* conformers. The two Raman bands at 297 cm^{-1} and 306 cm^{-1} are ascribed to the twisted conformation (*gauche* and *ortho*) and the *anti* rotamer, respectively. Again, the relative intensity ratio of the two bands drastically changes with temperature. This line pair could not be used for a van't Hoff plot since the wavenumbers of the *gauche* and *ortho* rotamers most likely coincide as predicted by theory.

Conclusions

A combined ab initio and variable–temperature Raman spectroscopic study provides strong evidence for the presence of three rotamers, *gauche*, *ortho* and *anti*, in the liquid state of the unstrained disilane $CF_3SiMe_2SiMe_2CF_3$, a phenomenon which has so far been only experimentally proven for the carbon chain *n*–C_4F_{10}. We are confident to obtain further proof for the existence of three conformers for $CF_3SiMe_2SiMe_2CF_3$ by matrix–isolation IR spectroscopy.

References:

[1] B. Albinsson, H. Teramae, J. W. Downing, J. Michl, *Chem. Eur. J.* **1996**, *2*, 529.
[2] B. Albinsson, J. Michl, *J. Am. Chem. Soc.* **1995**, *117*, 6378.
[3] B. Albinsson, J. Michl, *J. Phys. Chem.* **1996**, *100*, 3418.
[4] K. Morokuma, *J. Chem. Phys.* **1971**, *54*, 962.
[5] W. J. Welsh, W. D. Johnson, *Macromolecules* **1990**, *23*, 1882.

Formation, Stability and Structure of Aminosilane–Boranes

Gerald Huber, Alexander Jockisch, Hubert Schmidbaur*

Anorganisch-chemisches Institut der Technischen Universität München
Lichtenbergstr. 4, D-85747 Garching, Germany
Tel.: Int. code + (89)28913130 — Fax: Int. code + (89)28913125
E-mail: H.Schmidbaur@.lrz.tu-muenchen.de

Keywords: Silylamine / Borane / Adduct / Crystal Structure

Summary: A series of silylamines (aminosilanes) have been converted into their borane adducts using (tetrahydrofuran)– or (dimethyl sulfide)–borane as the BH_3 source. Most of the adducts are unstable in solution and undergo rapid Si–N cleavage to give silanes and aminoboranes. Together with $Me_3SiNMe_2–BH_3$ the complexes with aziridinyl- and azetidinylsilanes are the most stable. A crystal structure investigation of $Ph_3Si-N(CH_2)_2–BH_3$ has shown that the molecular geometry of the silylamine component of the adduct is virtually identical with that of the free silylaziridine with its steeply pyramidal configuration at nitrogen.

Introduction

Aminosilanes are characterized by a peculiar pattern of structure and reactivity as compared to analogous alkylamines: the ground-state configuration of Si-bound nitrogen atoms is planar or very flat, with extremely low barriers of inversion [1, 2]. The basicity and the donor properties of the amino/imino/nitrido groups are much lower than for organic amines and more often than not complex formation is followed by rapid and irreversible Si–N bond cleavage [3–6]. Knowledge of Lewis acid–base complexes of aminosilanes is therefore generally very limited.

As part of a more systematic study we have recently (re)investigated the structural chemistry of aziridinyl- and azetidinylsilanes, where extreme ring strain could be expected to modify the standard molecular geometry very considerably [7]. This work has now been extended to the corresponding borane adducts as prototypes of Lewis acid–base compounds with silylaziridines and -azetidines as the donor components. For comparison, borane adducts of other silylamines have also been studied.

Results

For the preparation of the silylamine–boranes a series of silylamines known in the literature, or prepared recently in this laboratory, were reacted with either borane–tetrahydrofuran or –dimethyl sulfide in toluene at room temperature (Table 1). A precipitate was formed in most cases, and $Ph_3SiN(CH_2)_2BH_3$ could be recrystallized using a concentration gradient in mixed solvents like dichloromethane/pentane.

Table 1. Preparation of the silylamine–boranes.

Silylamine	Borane reagent	Silylamine–borane
Me_3SiNMe_2[a]	$BH_3(Me_2S)$ or $BH_3(THF)$	$Me_3SiNMe_2(BH_3)$[b]
Ph_3SiNMe_2 [c]	$BH_3(Me_2S)$	$Ph_3SiNMe_2(BH_3)$
$Ph_3SiN(CH_2)_2$[d]	$BH_3(Me_2S)$ or $BH_3(THF)$	$Ph_3SiN(CH_2)_2(BH_3)$
$Ph_3SiN(CH_2)_3$[d]	$BH_3(Me_2S)$	$Ph_3SiN(CH_2)_3(BH_3)$
$Ph_2Si(NMe_2)_2$[e]	1 or 3 equiv. $BH_3(Me_2S)$	$Ph_2Si(NMe_2)_2(BH_3)$

[a] Commercially available — [b] Ref. [4] — [c] Ref. [8]— [d] Ref. [7] — [e] This work.

While the solid adducts can be kept at ambient temperature for quite some time, the stability in solution in most common organic solvents depends on the substituents at silicon. Thus $Me_3SiNMe_2BH_3$ and $Ph_3SiN(CH_2)_2BH_3$ are the most stable in solution. Even after 2 weeks almost no degradation is observed in C_6D_6 or $[D_8]$-toluene. $Me_3SiNMe_2BH_3$ in C_6D_6 decomposes rapidly only at temperatures above 80°C.

By contrast, in the case of the phenyl-substituted dimethylamino derivatives decomposition occurs within only a few hours. According to NMR monitoring of the decomposition pathway, the deborylated silylamine, the corresponding silane and aminoboranes are the main products, indicating Si—N cleavage by borane. (Triphenylsilyl)dimethylamine–borane, $Ph_3Si–NMe_2(BH_3)$, thus yields Ph_3SiNMe_2, Ph_3SiH, $[Me_2NBH_2]_n$ and $[H_2B–NMe_2–BH_2]H$ etc. as indicated in Scheme 1.

$Ph_3Si–N(BH_3)Me_2$ $\xrightarrow[C_6D_6,RT]{14\ d}$ $Ph_3Si–NMe_2$ + $Ph_3Si–N(BH_3)Me_2$ + $[Me_2NBH_2]_2$ + $H_2B(\mu-NMe_2)(\mu-H)BH_2$ + $Ph_3Si–H$

Scheme 1. Decomposition of (Triphenylsilyl)dimethylamine–borane.

Diaminosilanes like $Ph_2Si(NMe_2)_2$ give only the 1:1 adduct with BH_3, even if an excess of the $BH_3(L)$ reagent is applied. NMR investigations show that the BH_3 group is attached to only one of the two dimethylamino groups and that there is no rapid exchange of the borane between the two donor sites. However, the compound is also unstable in solution (Scheme 2).

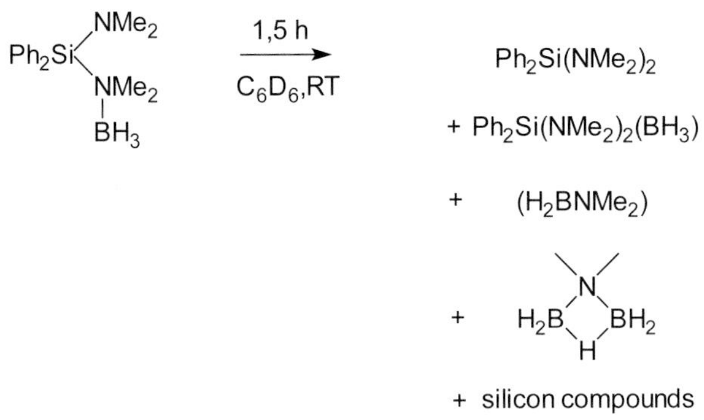

$Ph_2Si(NMe_2)(NMe_2{-}BH_3)$ $\xrightarrow[C_6D_6,RT]{1,5\ h}$ $Ph_2Si(NMe_2)_2$

$+\ Ph_2Si(NMe_2)_2(BH_3)$

$+\ (H_2BNMe_2)$

$+\ H_2B(\mu{-}NMe_2)(\mu{-}H)BH_2$

+ silicon compounds

Scheme 2. Decomposition of (Diphenylsilyl)bis(dimethylamine)–borane.

In comparison to the other phenyl–substituted compounds $Ph_3SiN(CH_2)_2{-}BH_3$, together with $Me_3SiNMe_2{-}BH_3$, is the most stable complex of the whole series. Both compounds have relatively high melting points (134°C and 112°C, respectively). This stability probably reflects the improved donor qualities of the small ring amines at silicon with their strongly pyramidal ground-state geometry at the nitrogen atoms.

A crystal structure study has shown that in fact the structure of the amine is largely retained in the BH_3 adduct as shown in Fig. 1. The crystals are orthorhombic, space group $P2_12_12_1$, with $Z = 4$ molecules in the unit cell. The molecule has no crystallographically imposed symmetry, but the geometry of the core atoms is nevertheless close to mirror symmetry.

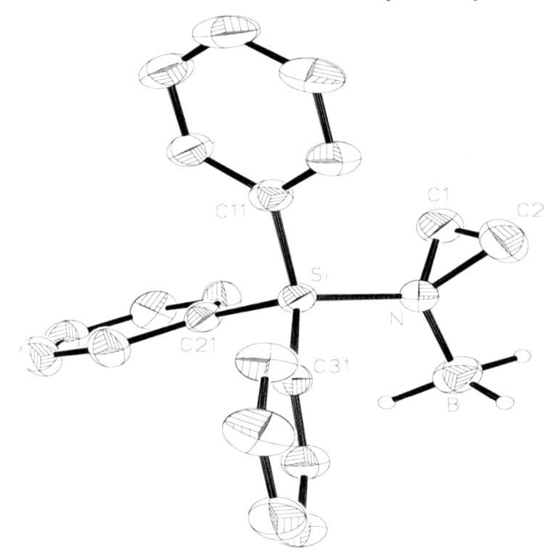

Fig. 1. Molecular structure of $Ph_3SiN(CH_2)_2{-}BH_3$ (ORTEP drawing with 50% probability ellipsoids).

The most surprising result of the structural study is the almost complete agreement of all data of the silylaziridine component of the adduct with those of the free silyl-aziridine [7]. This means that the addition of the borane acceptor to the nitrogen donor atom of the substrate occurs with only minor changes in the molecular geometry, although the B–N distance in the adduct [1.616(3) Å] suggests a strong donor–acceptor interaction (Table 2).

Table 2. Characteristic bond lengths and angles of $Ph_3SiN(CH_2)_2–BH_3$.

bond lengths [Å]		bond angles [°]			
Si–N	1.797(2)	N-Si-C11	108.44(8)	Si-N-C1	116.17(13)
Si–C11	1.872(2)	N-Si-C21	106.80(8)	Si-N-C2	123.0(2)
Si–C21	1.869(2)	N-Si-C31	111.11(8)	C1-N-B	116.2(2)
Si–C31	1.868(2)	C11-Si-C21	109.26(8)	C2-N-B	113.4(2)
N–B	1.616(3)	C11-Si-C31	109.95(9)	C1-N-C2	57.4(2)
N–C1	1.495(3)	C21-Si-C31	111.18(9)	N-C1-C2	60.6(2)
N–C2	1.476(3)	Si-N-B	117.12(14)	C1-C2-N	62.0(2)
C1–C2	1.426(4)				

It should be noted that this phenomenon arises because the free aziridine already features a strongly pyramidal configuration with the lone pair of electrons oriented in the direction required for donor–acceptor bonding. For all other aminosilanes (silylamines) the flat ground-state geometry needs to be reorganized to provide the necessary sp^3 hybridization required for bonding of the borane. This negative contribution to the energy balance of the complex formation clearly reduces the stability of the adducts of all silylamines except for the homologues with small strained nitrogen heterocycles.

Experimental details of this study are contained in the Doctorate Thesis of G. Huber, TU München 1998/9.

Acknowledgement. This work was supported by *Deutsche Forschungsgemeinschaft* and *Fonds der Chemischen Industrie*. The authors are grateful to Mr. J. Riede for collecting the X-ray data set.

References:

[1] a) K. Hedberg, *J. Am. Chem. Soc.* **1955**, *77*, 6491.
b) B. Beagley, A. R. Conrad, *J. Chem. Soc., Faraday Trans.* **1970**, *66*, 2740. The structure of $N(SiH_3)_3$ was first presented at the International Congress of Crystollography, Stockholm, July **1951**, and at the XIIth International Congress of Pure and Applied Chemistry, New York, September **1951**.

[2] M. J. Barrow, E. A. V. Ebsworth, *J. Chem. Soc., Dalton Trans.* **1984**, 563.

[3] T. D. Coyle, F. G. A. Stone in: *Boron Chemistry*, Vol. 1, (Eds. H. Steinberg, A. L. McCloskey, Pergamon, New York Press **1964**, pp. 83–166.

[4] H. Nöth, *Z. Naturforsch., Teil B* **1961**, *16*, 618.

[5] W. R. Nutt, J. S. Blanton, A. M. Boccanfuso, L. A. Silks III, A. R. Garber, J. D. Odom, *Inorg. Chem.* **1991**, *30*, 4136.

[6] A. H. Cowley, M. C. Cushner, P. E. Riley, *J. Am. Chem. Soc.* **1980**, *102*, 624.

[7] G. Huber, A. Jockisch, H. Schmidbaur, *Eur. J. Inorg. Chem.* **1998**, 107.

[8] H. Gilman, B. Hofferth, H. W. Melvin, G. E. Dunn, *J. Am. Chem. Soc.* **1950**, *72*, 5767.

Hydrosilylation and Hydroboration of Indene — a Comparison

J. Dautel, S. Abele

Institut für Anorganische Chemie der Universität Stuttgart
Pfaffenwaldring 55, D-70569 Stuttgart, Germany
Tel.: Int. code + (711)6854097 — Fax: Int. code + (711)6854241
E-mail: dautel@iac.uni-stuttgart.de

Keywords: Hydrosilylation / Hydroboration / X-Ray / ^{29}Si-NMR Spectra / ^{11}B-NMR Spectra

Summary: Hydrosilylation of indene with various hydrosilanes in the presence of Speier's catalyst leads in high yields to 1-indanyl-substituted silanes. In contrast to these results, the hydroboration of indene with $HBCl_2{\cdot}SMe_2$ results in the 2-indanyl-substituted product. Further reactions of these derivatives are investigated. The X-ray structure determination of bis(trimethylsilylamino)2-indanylborane was carried out.

Introduction

In 1993 Uozumi et al. reported on the regio- and enantioselective hydrosilylation of styrene derivatives using an optically active palladium catalyst [1]. In the context of these examinations the authors also mentioned the reaction of indene with trichlorosilane; in all cases the 1-substituted products only were got; no further characterization was given, however. In accordance with these results, our investigations on the hydrosilylation of indene with trichlorosilane, chlorodimethylsilane, and dichloromethylsilane in the presence of Speier's catalyst showed the same addition tendencies; by distillation we isolated in high yields trichloro(1-indanyl)silane (**1a**), dichloro(1-indanyl)methylsilane (**1b**), and chloro(1-indanyl)dimethylsilane (**1c**), respectively (Scheme 1).

In 1991, Hayashi et al. [2] and Burgess et al. [3] published the hydroboration of styrene derivatives with catecholborane, catalyzed by various chiral rhodium complexes. The organoboron compounds generated were not isolated, but oxidatively converted into the corresponding alcohols. In the case of indene, mixtures of 1-indanol (usually the main product) and 2-indanol could be isolated by preparative TLC. Similar investigations were carried out by Zhang et al. using catecholborane and some complexes of BH_3 as reagents [4]. In contrast to the catalyzed reactions, the uncatalyzed hydroborations with indene and related substrates yielded predominantly 2-substituted alcohols after oxidation. In good agreement with the latter work, we observed in the uncatalyzed hydroboration of indene with $HBCl_2{\cdot}SMe_2$, dimethyl sulfide–dichloro(2-indanyl)borane (**2a**) as unique product (Scheme 1).

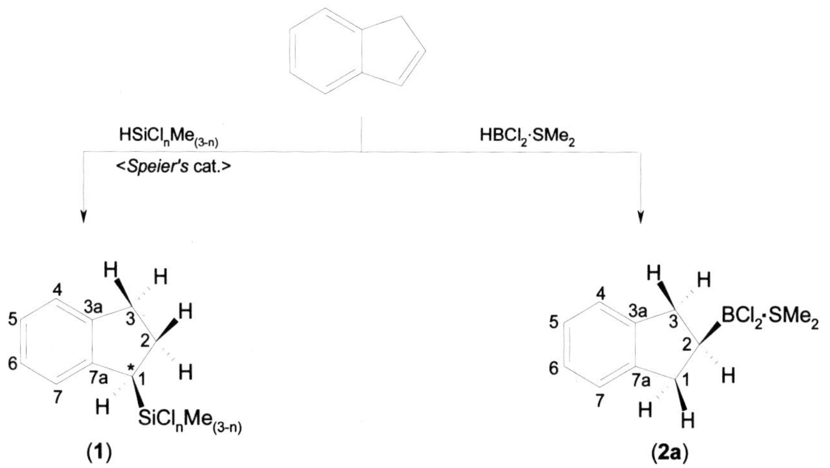

Scheme 1. Synthesis of **1a** (n = 3), **1b** (n = 2), **1c** (n = 1) and **2a**.

An accurate analysis of the ^{1}H- and ^{13}C{^{1}H}-NMR spectra enabled us to specify the presence of 1- or 2-indanyl substitution in the hydrosilylation and hydroboration products without any doubt: the 1-indanyl derivatives show no symmetry at all, resulting in 9 different ^{13}C-NMR resonances. In the aliphatic part of the ^{1}H-NMR spectra complicated patterns of ABCDE spin systems are expected. The 2-indanyl-substituted compounds, however, exhibit a mirror plane through the atom C2 and the centre of the benzene ring normal to the molecule plane. Therefore, the ring atoms show only 5 different ^{13}C-NMR signals and the ^{1}H-NMR spectra are simplified to $[AB]_2C$ spin systems in the aliphatic area by reason of symmetry. The representative examples 1a and 2a were discussed in detail (see below).

1-Indanylsilanes

Hydrosilylation of Indene

A solution of indene and Speier's catalyst was treated with trichlorosilane, chlorodimethylsilane or dichloromethylsilane, the reaction mixture was stirred at 160°C for several hours. After removal of excess hydrosilanes, trichloro(1-indanyl)silane (**1a**), dichloro(1-indanyl)methylsilane (**1b**), and chloro(1-indanyl)dimethylsilane (**1c**) were isolated by distillation as colorless liquids in yields of 60–80 %.

Characterization: **1a**: b.p. 72°C/10^{-3} torr. NMR spectra: ^{1}H: δ = 1.9 to 2.9 (m, 5H, ABCDE spin system, aliphatic H; see Fig. 1); 6.9 to 7.3 ppm (m, 4H, aromatic H). ^{13}C{^{1}H}: δ = 26.8 ($C_{2,3}$); 32.3 ($C_{2,3}$); 40.9 (C_1); 125.0 ($C_{4,5,6,7}$); 125.2 ($C_{4,5,6,7}$); 126.9 ($C_{4,5,6,7}$); 127.5 ($C_{4,5,6,7}$); 138.9 ($C_{3a,7a}$); 144.2 ppm ($C_{3a,7a}$). ^{29}Si{^{1}H}: δ = 8.6 ppm. — **1b**: b.p. 76°C/10^{-3} torr. NMR spectra: ^{1}H: δ = 0.44 (s, 3H, CH_3); 1.9 to 2.8 (m, 5H, ABCDE spin system, aliphatic H); 6.9 to 7.4 ppm (m, 4H, aromatic H). ^{13}C{^{1}H}: δ = 3.2 (CH_3); 26.8 ($C_{2,3}$); 32.5 ($C_{2,3}$); 38.6 (C_1); 124.7 ($C_{4,5,6,7}$); 125.0 ($C_{4,5,6,7}$); 126.7 ($C_{4,5,6,7}$); 126.8 ($C_{4,5,6,7}$); 141.0 ($C_{3a,7a}$); 144.0 ppm ($C_{3a,7a}$). ^{29}Si{^{1}H}: δ = 28.8 ppm. — **1c**: b.p. 70–73°C/10^{-3} torr. NMR spectra: ^{1}H: δ = 0.16 (s, 3H, CH_3); 0.21 (s, 3H, CH_3); 1.9 to 2.8 (m, 5H, ABCDE spin system, aliphatic H); 6.9 to 7.2 ppm (m, 4H, aromatic H). ^{13}C{^{1}H}: δ = 0.1 (CH_3); 0.4

(CH_3); 27.2 ($C_{2,3}$); 32.9 ($C_{2,3}$); 36.6 (C_1); 124.3 ($C_{4,5,6,7}$); 124.9 ($C_{4,5,6,7}$); 126.1 ($C_{4,5,6,7}$); 126.5 ($C_{4,5,6,7}$); 143.5 ($C_{3a,7a}$); 143.7 ppm ($C_{3a,7a}$). $^{29}Si\{^1H\}$: $\delta = 28.8$ ppm.

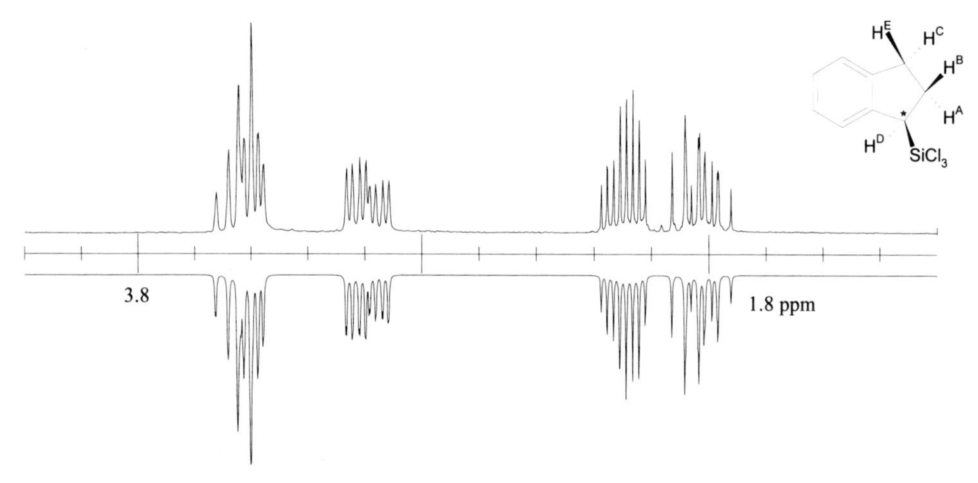

Fig. 1. Observed (upper part) and calculated [5] (lower part) ^1H-NMR spectrum (400.134 MHz) of **1a** (aliphatic region): ABCDE spin system. $\delta_A = 2.015$; $\delta_B = 2.147$; $\delta_C = 2.596$; $\delta_D = 2.805$; $\delta_E = 2.818$ ppm; $^2J_{AB\ (gem)} = -13.4$; $^3J_{AC\ (cis)} = 9.3$; $^3J_{AD\ (cis)} = 10.0$; $^3J_{AE\ (trans)} = 8.7$; $^3J_{BC\ (trans)} = 4.2$; $^3J_{BD\ (trans)} = 4.3$; $^3J_{BE\ (cis)} = 8.8$; $^4J_{CD\ (cis)} = 0.9$; $^2J_{CE\ (gem)} = -16.0$; $^4J_{DE\ (trans)} = 0.6$ Hz.

Fluoridation of 1a

Organofluorsilanes are primarily of great importance in organic chemistry. The synthesis of complicated alcohols, aldehydes or ketons by oxidative cleavage of the Si–C bond in alkyl- or alkenyltrifluorosilanes is an advantageous alternative to the classic organic route [6–8].

The reaction of **1a** with zinc(II) fluoride in diethyl ether analogously to published methods [6, 9–11], led to trifluoro(1-indanyl)silane (**1d**) as a colorless liquid in a yield of 60 %.

Characterization: b.p. 28–31°C/10^{-3} torr. NMR spectra: ^1H: $\delta = 1.7$ to 2.8 (m, 5H, $ABCDEX_3$ spin system, aliphatic H); 6.9 to 7.2 ppm (m, 4H, aromatic H). $^{13}C\{^1H\}$: $\delta = 25.6$ ($C_{2,3}$); 26.2 (q, $^2J_{CF} = 18.4$ Hz, C_1); 32.8 ($C_{2,3}$); 124.3 ($C_{4,5,6,7}$); 125.0 ($C_{4,5,6,7}$); 127.1 ($C_{4,5,6,7}$); 127.2 ($C_{4,5,6,7}$); 138.8 ($C_{3a,7a}$); 143.8 ppm ($C_{3a,7a}$). $^{19}F\{^1H\}$: $\delta = -142.2$ ppm. $^{29}Si\{^1H\}$: $\delta = -64.3$ ppm (q, $^1J_{SiF} = 292$ Hz).

Reduction of 1a

To avoid undesired Si–C bond cleavage in the reduction of **1a** to 1-indanylsilane (**1e**), the highly reactive lithium aluminum hydride has to be removed by a milder reducing agent like "Red-Al" or diisobutylalane (DIBAL-H) [10, 12, 13].

Thus, a solution of **1a** in *n*-hexane was treated with a solution of DIBAL-H in *n*-hexane at 0°C. After aqueous work-up we obtained **1e** by distillation as a colorless liquid in a yield of 90 %.

Characterization: b.p. 27–29°C/10^{-3} torr. NMR spectra: ^1H: $\delta = 1.7$ to 2.9 (m, 5H, $ABCDEX_3$ spin system, aliphatic H); 3.67 (d, 3H, $^3J_{HH} = 3.3$ Hz, SiH_3); 6.9 to 7.1 ppm (m, 4H, aromatic H). $^{13}C\{^1H\}$: $\delta = 26.2$ (C_1); 29.8 ($C_{2,3}$); 32.7 ($C_{2,3}$); 123.7 ($C_{4,5,6,7}$); 124.7 ($C_{4,5,6,7}$); 126.0 ($C_{4,5,6,7}$); 126.8 ($C_{4,5,6,7}$); 143.0 ($C_{3a,7a}$); 145.7 ppm ($C_{3a,7a}$). ^{29}Si: $\delta = -55.6$ ppm (q, $^1J_{SiH} = 196$ Hz).

Reaction of 1a with Hexamethyldisilazane

Chlorodisilazanes are built by aminolysis of chlorosilanes, by cleavage of cyclodisilazane with chlorosilazanes and by reaction of disilazanes with chlorosilanes [14–16].

The reaction of hexamethyldisilazane with trichloro(1-indanyl)silane (**1a**) resulted in dichlorotrimethylsilylamino(1-indanyl)silane (**1f**) as a colorless liquid in a yield of 75 % (Scheme 2).

$$\textbf{(1a)}\ \xrightarrow[-Me_3SiCl]{+HN(SiMe_3)_2}\ \textbf{(1f)}$$

(1a): SiCl$_3$; (1f): SiCl$_2$-N(H)SiMe$_3$

Scheme 2. Synthesis of **1f**.

Characterization: b.p. 88°C/10^{-2} torr. NMR spectra: ^{1}H: δ = 0.00 (s, 9H, SiMe$_3$); 0.98 (s, br, 1H, NH); 2.0 to 3.0 (m, 5H, ABCDE spin system, aliphatic H); 6.8 to 7.4 ppm (m, 4H, aromatic H). ^{13}C{^{1}H}: δ = 1.9 (SiMe$_3$); 27.4 ($C_{2,3}$); 32.5 ($C_{2,3}$); 39.6 (C_1); 124.9 ($C_{4,5,6,7}$); 125.0 ($C_{4,5,6,7}$); 126.6 ($C_{4,5,6,7}$); 126.8 ($C_{4,5,6,7}$); 141.3 ($C_{3a,7a}$); 144.3 ppm ($C_{3a,7a}$). ^{15}N: δ = –346.1 ppm (d, $^1J_{NH}$ = 68.5 Hz). ^{29}Si{^{1}H}: δ = –2.07 (SiCl$_2$); 7.38 ppm (SiMe$_3$).

2-Indanylboranes

Hydroboration of Indene

A solution of HBCl$_2$·SMe$_2$ in *n*-pentane was treated with indene at 0°C. After removing the solvent, dimethyl sulfide–dichloro(2-indanyl)borane (2a) could be isolated as a colorless powder, which could be purified by sublimation (35°C/10^{-3} torr). Yield: 70 %.

Characterization: NMR spectra: ^{1}H: δ = 1.45 (s, 6H, SMe$_2$); 1.9 to 3.3 (m, 5H, [AB]$_2$C spin system, aliphatic H; see Fig. 2); 7.0 to 7.2 ppm (m, 4H, aromatic H). ^{11}B{^{1}H}: δ = 11.6 ppm. ^{13}C{^{1}H,^{11}B}: δ = 19.0 (SMe$_2$); 35.0 (C_2); 36.6 ($C_{1,3}$); 124.7 ($C_{4,5,6,7}$); 126.3 ($C_{4,5,6,7}$); 144.8 ppm ($C_{3a,7a}$).

Reaction of 2a with Hexamethyldisilazane

The formation of Si–N–B bond linkages has mostly been accomplished one of following methods: so-called *cleavage* reactions, in which a silicon–nitrogen bond is cleaved by a covalent halide leading to the formation of some halosilane as a side product; and *dehydrohalogenation*, in which the elements of a hydrogen halide are eliminated [17].

By treating a solution of dimethyl sulfide–2-indanylborane-dichloride adduct (**2a**) in toluene with two equivalents of hexamethyldisilazane and usual work-up, a colorless liquid was isolated, which was converted after distillation (b.p. 94°C/10^{-2} torr) into a colorless powder in a yield of 80 %. Recrystallization in *n*-pentane (+20/–30°C) resulted colorless single crystalline cuboids.

Characterization: m.p. 53–54°C. NMR spectra: ^{1}H: δ = 0.07 (s, 18H, SiMe$_3$); 2.37 (s, br, 2H,

NH); 1.9 to 3.1 (m, 5H, $[AB]_2C$ spin system, aliphatic H); 7.1 to 7.2 ppm (m, 4H, aromatic H). $^{11}B\{^1H\}$: $\delta = 38.0$ ppm. $^{13}C\{^1H\}$: $\delta = 1.8$ ($SiMe_3$); 37.9 ($C_{1,3}$); 124.6 ($C_{4,5,6,7}$); 126.4 ($C_{4,5,6,7}$); 145.1 ($C_{3a,7a}$) ppm. The resonance of C_2 bonded to boron was not detectable. $^{29}Si\{^1H\}$: $\delta = 2.5$ ppm.

The X-ray structure determination at –100°C {orthorhombic; $P2_12_12_1$; a = 617.16 pm, b = 1222.04 pm, c = 2545.44 pm; Z = 4} showed monomeric bis(trimethylsilylamino)-2-indanylborane (**2b**) (Fig. 3).

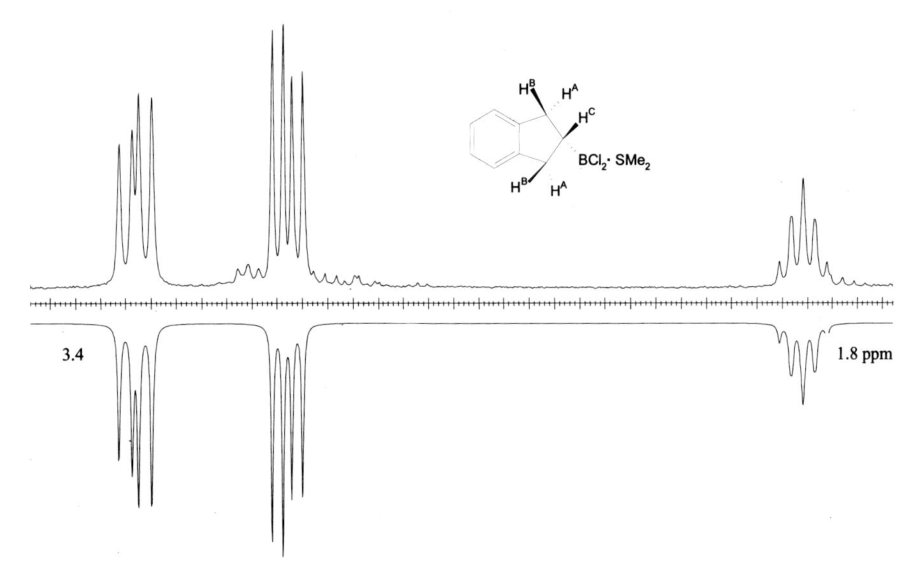

Fig. 2. Observed (upper part) and calculated [5] (lower part) ^{1}H-NMR spectrum (400.134 MHz) of **2a** (aliphatic region, SMe_2 resonance not shown): $[AB]_2C$ spin system (previously termed AA'BB'C); δ_A = 3.20 ; δ_B = 2.90; δ_C = 1.87 ppm; $^2J_{AB\,(gem)}$ = 15.2; $^3J_{AC\,(cis)}$ = 10.4; $^3J_{BC\,(trans)}$ = 8.6 Hz.

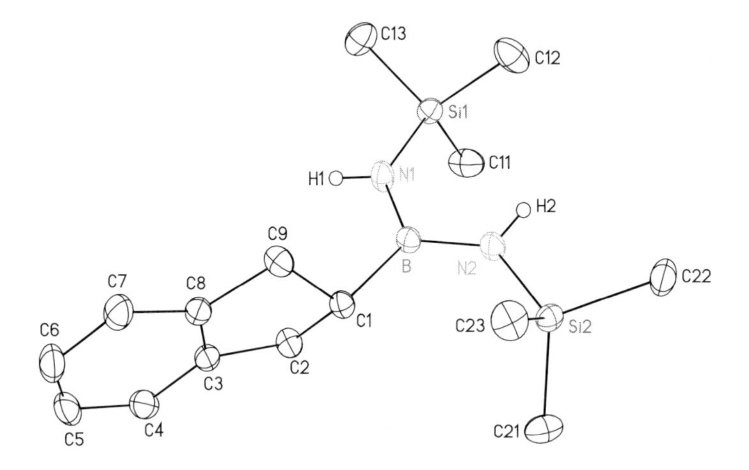

Fig. 3. Molecular structure of **2b**; characteristic structural parameters: B-C 160 pm; B-N 143 pm; N-Si 174 pm; Si-C 186 pm; C-B-N 119°; N-B-B 121°; B-N-Si 133°.

Further Silylaminoboranes

Reactions of Disilazanes with $HBCl_2{\cdot}SMe_2$

If two equivalents of hexamethyldisilazane were treated with dimethyl sulfide–boron dichloride at 0°C, a colorless solid was formed immediately – a first indication that a dehydrohalogenation took place. Bis(trimethylsilyl)aminotrimethylsilylaminoborane (**3**) could be isolated by distillation as a colorless liquid in a yield of 90 % (Scheme 3).

Characterization: b.p. 45°C/2×10^{-2} torr. NMR spectra: ^{1}H: δ= 0.10 (s, 18H, N[$SiMe_3$]$_2$); 0.31 (s, 9H, N–$SiMe_3$) 2.81 ppm (s, br, 1H, NH). The resonance of BH was not detectable. ^{11}B: δ= 33.5 ppm (d, $^1J_{BH}$ = 111 Hz). ^{13}C{^{1}H}: δ= 1.8 (N–$SiMe_3$); 4.2 ppm (N[$SiMe_3$]$_2$). ^{29}Si{^{1}H}: δ= 0.4 (N–$SiMe_3$); 4.5 ppm (N[$SiMe_3$]$_2$).

In an analogous manner, the reaction of two equivalents of divinyltetramethyldisilazane with $HBCl_2{\cdot}SMe_2$ at –78°C led to the asymmetrical bis(dimethylvinylsilyl)aminodimethylvinylsilylaminoborane (**4**) (Scheme. 3). Yield: 80 %.

Characterization: b.p. 40°C/5×10^{-2} torr. NMR spectra: ^{1}H: δ= 0.20 (s, 12H, N[$SiMe_2$]$_2$); 0.29 (s, 6H, N–$SiMe_2$); 2.42 (s, br, 1H, NH); 5.68 (m, 2H), 5.86 (m, 2H), 6.29 (m, 2H, $^2J_{gem}$ = 3.7; $^3J_{cis}$ = 14.7; $^3J_{trans}$ = 20.0 Hz, all N[Si–CH=CH_2]$_2$); 5.72 (m, 1H), 5.90 (m, 1H), 6.18 ppm (m, 1H, $^2J_{gem}$ = 4.2; $^3J_{cis}$ = 14.7; $^3J_{trans}$ = 20.0 Hz, all N–SiCH=CH_2). The resonance of BH was not detectable. ^{11}B: δ= 33.8 ppm (d, $^1J_{BH}$ = 95 Hz). ^{13}C{^{1}H}: δ= 0.1 (N–$SiMe_2$); 2.3 (N[$SiMe_2$]$_2$); 131.9 (N–SiCH=); 131.4 (N[SiCH=]$_2$); 140.0 (=CH_2); 142.2 ([=CH_2]$_2$) ppm. ^{29}Si{^{1}H}: δ= –3.5 (N–Si); 17.8 (N–Si_2) ppm.

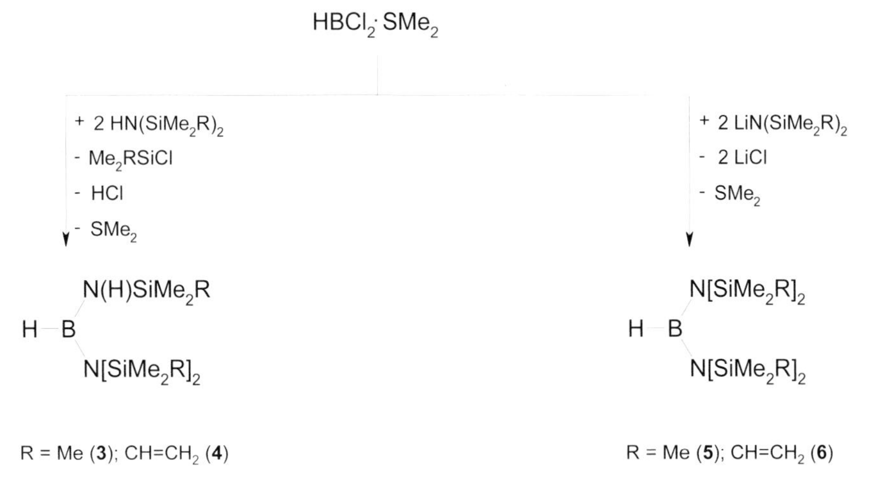

Scheme 3. Synthesis of silylaminoboranes.

Reactions of Lithiated Disilazanes with $HBCl_2{\cdot}SMe_2$

We were easily able to synthesize the corresponding symmetrical bis(silyl)aminoboranes by reaction of lithium silylamide with dimethyl sulfide–boron dichloride in convenient *one-pot* reactions (Scheme 3). The lithium amide was formed by adding *n*-butyllithium to a stirred solution of silylamine in *n*-hexane, followed by the controlled addition of the dimethyl sulfide–boron dichloride adduct. Both bis(trimethylsilyl)aminoborane (**5**) and bis(dimethylvinylsilyl)aminoborane (**6**) could be purified by distillation in yields of 60 to 75 %.

Characterization: **5**: b.p. 68°C/10^{-2} torr. NMR spectra: ^1H: δ = 0.18 ppm (s). The resonance of BH was not detectable. ^{11}B: δ = 40.2 ppm (d, $^1J_{BH}$ = 125 Hz). $^{13}C\{^1H\}$: δ = 4.0 ppm. $^{29}Si\{^1H\}$: δ = 2.4 ppm. — **6**: b.p. 70°C/10^{-3} torr. NMR spectra: ^1H: δ = 0.32 ppm (s, 12H, $SiMe_2$); 5.65 (m, 2H), 5.87 (m, 2H), 6.33 ppm (m, 2H, $^2J_{gem}$ = 3.7; $^3J_{cis}$ = 14.7; $^3J_{trans}$ = 20.6 Hz, all $CH{=}CH_2$). The resonance of BH was not detectable. ^{11}B: δ = 40.7 ppm (d, $^1J_{BH}$ = 85 Hz). $^{13}C\{^1H\}$: δ = 2.09 ($SiMe_2$); 131.4 (=CH); 141.6 ppm (=CH_2). $^{29}Si\{^1H\}$: δ = –5.8 ppm.

Acknowledgment: We thank the *Land Baden-Württemberg (Keramikverbund Karlsruhe–Stuttgart, KKS)* for generous financial support. The authors would like to thank Prof. Dr. G. Becker for supporting this work.

References:

[1] Y. Uozumi, K. Kitayama, T. Hayashi, *Tetrahedron: Asymmetry* **1993**, *4*, 2419.
[2] T. Hayashi, Y. Matsumoto, Y. Ito, *Tetrahedron: Asymmetry* **1991**, *2*, 601.
[3] K. Burgess, W. A. van der Donk, M. J. Ohlmeyer, *Tetrahedron: Asymmetry* **1991**, *2*, 613.
[4] J. Zhang, B. Lou, G. Guo, L. Dai, *J. Org. Chem.* **1991**, *56*, 1670.
[5] P. H. M. Budzelaar, gNMR V3.6.5, Cherwell Scientific, Oxford, UK, **1995**.
[6] K. Tamao, T. Kakui, M. Akita, T. Iwahara, R. Kanatani, J. Yoshida, M. Kumada, *Tetrahedron* **1983**, *39*, 983.
[7] M. Kumada, K. Tamao, J. Yoshida, *J. Organomet. Chem.* **1982**, *239*, 115.
[8] K. Tamao, M. Akita, M. Kumada, *J. Organomet. Chem.* **1983**, *254*, 13.
[9] S. E. Johnson, R. O. Day, R. R. Holmes, *Inorg. Chem.* **1989**, *28*, 3182.
[10] J. Dautel, S. Abele, W. Schwarz, *Z. Naturforsch. Teil B* **1997**, *52,* 778.
[11] J. Dautel, W. Schwarz, *"Precursors for Silicon-Alloyed Carbon Fibers"*, in: *Organosilicon Chemistry III – From Molecules to Materials* (Eds.: N. Auner, J. Weis), Wiley–VCH, Weinheim, **1998**, p. 632.
[12] G. Fritz, S. Lauble, M. Breining, A. G. Beetz, A. M. Galminas, E. Matern, H. Goesmann, *Z. Anorg. Allg. Chem.* **1994**, *620*, 127.
[13] H. Schmidbaur, J. Ebenhöch, G. Müller, *Z. Naturforsch, Teil B*, **1987**, *42*, 142.
[14] U. Wannagat, E. Bogusch, *Mh. Chem.* **1971**, *102*, 1806.
[15] U. Wannagat, J. Herzig, P. Schmidt, M. Schulze, *Mh. Chem.* **1971**, *102*, 1817.
[16] J. Silbiger, J. Fuchs, *Inorg. Chem.* **1965**, *4*, 1371.
[17] J. R. Bowser, R. L. Wells, *Appl. Organomet. Chem.* **1996**, *10*, 199.

Gallium-Silicon Heterocycles

Alexander Rodig, Wolfgang Köstler, Gerald Linti*

Institut für Anorganische Chemie der Universität Karlsruhe
Postfach 6980
Engesserstr., Geb. 30.45, 76128 D-Karlsruhe, Germany
Tel.: Int code + (721)608 2822 — Fax: Int code + (721)608 4854
e-mail: linti@achpc9.chemie.uni-karlsruhe.de

Keywords: Gallium-Silicon Heterocycles / Hypersilyl / Main Group Elements / Clusters

Summary: The reaction of $Ga_2X_4 \cdot 2$ dioxane with four equivalents of $(Me_3Si)_3SiLi(thf)_3$ (hypersilyllithium) in tetrahydrofuran affords for X = Br a 1,3,2,4-disiladigalletane together with tetrahedral gallium(I)hypersilyl. For X = Cl an anionic 1,2,3,4-silatrigalletanate, a four-membered heterocycle with a Ga_3 unit, is isolated. In addition the synthesis and structure of a silatetragallane with a *closo* cage structure are described. These primary examples of gallium/group 14 heterocycles have been investigated by X-ray single-crystal structure analysis. DFT calculations on various trigonal-bipyramidal cage compounds are compared with those on the silatetragallane.

The tris(trimethylsilyl)silyl group (= hypersilyl) has proved to be a very useful substituent in gallium chemistry. Not only trivalent gallanes like **1** to **3** [1–3] are accessible but the hypersilyl group was also excellently capable of stabilizing a number of low valent gallium compounds with gallium–gallium bonds; examples are **4** to **8** [4–6].

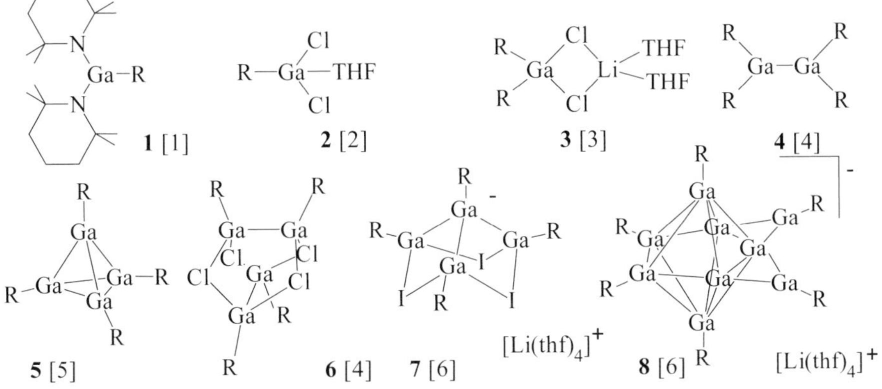

1 [1] **2** [2] **3** [3] **4** [4]

5 [5] **6** [4] **7** [6] $[Li(thf)_4]^+$ **8** [6] $[Li(thf)_4]^+$

R= $Si(SiMe_3)_3$

In all these cases the hypersilyl group remained unaffected. Consequently, the central silicon atom is bonded to only one gallium atom and no cyclic compounds are known with gallium and silicon atoms as ring building elements. In the following we will describe results, where the inherent capability of the hypersilyl ligand for Si–Si bond cleavage is used for the synthesis of gallium–silicon heterocycles, which feature a tetracoordinate silicon atom bonded to two gallium atoms.

Reactions

Recently, we have described the synthesis of **5** [5] together with $Cl_2Ga[Si(SiMe_3)_3]_2 \cdot Li(thf)_2$ (**3**) and several not further characterized byproducts by reaction of $Ga_2Cl_4 \cdot 2$ dioxane with three equivalents of $Li(thf)_3Si(SiMe_3)_3$. Another reaction course is observed, if $Ga_2Cl_4 \cdot 2$ dioxane, dissolved in tetrahydrofuran, is reacted with four equivalents of $Li(thf)_3Si(SiMe_3)_3$ (Eq. 1). In this reaction the disproportionation of the gallium(II) species leads again to several hypersilyl gallium compounds in oxidation state III, but no **5** is observed. Instead, black crystals of the unpredecented gallium–silicon heterocycle **9** are isolated. On the other hand, the analogous reaction of $Ga_2Br_4 \cdot 2$ dioxane with four equivalents of $Li(thf)_3Si(SiMe_3)_3$ affords **10** together with **5** (Eq. 2). The latter cocrystallizes with $Si(SiMe_3)_4$, which is formed during the reaction course.

Eq. 1. Formation the unpredecented gallium–silicon heterocycle **9**.

Eq. 2. Synthesis of the gallium–silicon heterocycle **10**.

It is plausible that **2** and bis(hypersilyl)gallium bromide are formed via disproportionation at first. **10** forms formally by elimination of bromotrimethylsilane from the sterically crowded $[(Me_3Si)_3Si]_2GaBr$. This takes place in the presence of excess $Li(thf)_3Si(SiMe_3)_3$, and consequently the halosilane reacts further to produce $Si(SiMe_3)_4$ (Scheme 1).

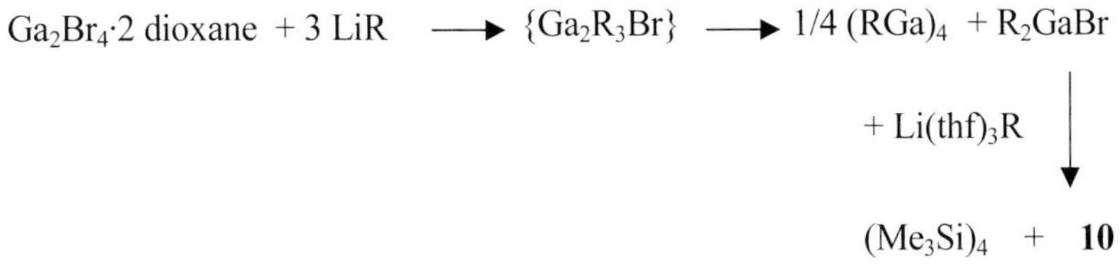

$$Ga_2Br_4\cdot 2\text{ dioxane} + 3\text{ LiR} \longrightarrow \{Ga_2R_3Br\} \longrightarrow 1/4\,(RGa)_4 + R_2GaBr \xrightarrow{+\,Li(thf)_3R} (Me_3Si)_4 + \mathbf{10}$$

Scheme 1. Proposed reaction pathways to **10**.

If a gallium subhalide of formal composition Ga_2I_3 – prepared by ultrasonic methods from the elements – is treated with $Li(thf)_3Si(SiMe_3)_3$ the silagallane **11** is isolated (Eq. 3).

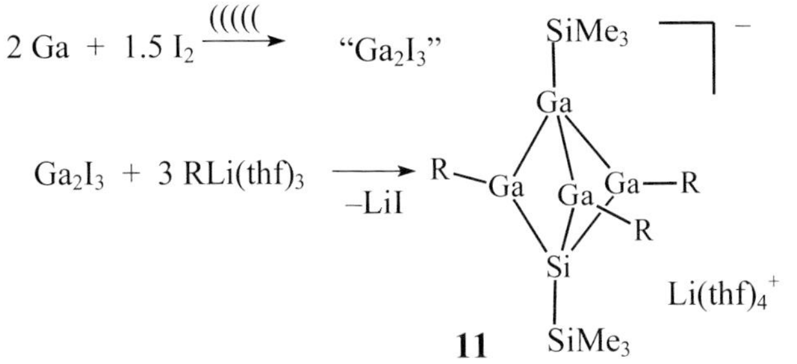

$$2\,Ga + 1.5\,I_2 \xrightarrow{(((((} \text{“}Ga_2I_3\text{”}$$

$$Ga_2I_3 + 3\,RLi(thf)_3 \xrightarrow[-LiI]{} [\mathbf{11}]^- \; Li(thf)_4^+$$

(Structure **11**: Ga–SiMe₃ apex, three Ga–R centres, Si–SiMe₃ base; labels: SiMe₃, Ga, R, Ga, Ga, Ga, R, R, Si, SiMe₃, Li(thf)₄⁺)

Eq. 3.

The pathway leading to **9** obviously also includes a step of disproportionation, but now an oligomeric Ga(I) hypersilyl reacts with $Li(thf)_3Si(SiMe_3)_3$ (scheme 2). Thus, a possible intermediate may be the three membered gallacycle $\{Ga_3[Si(SiMe_3)_3]_4\}^-$. Migration of a trimethylsilyl group from a $Si(SiMe_3)_3$ group at the four coordinate gallium center to a three coordinate one and insertion of the remaining $Si(SiMe_3)_2$ unit into a Ga–Ga bond affords **9**. Further reaction with gallium monoiodide affords **11.**

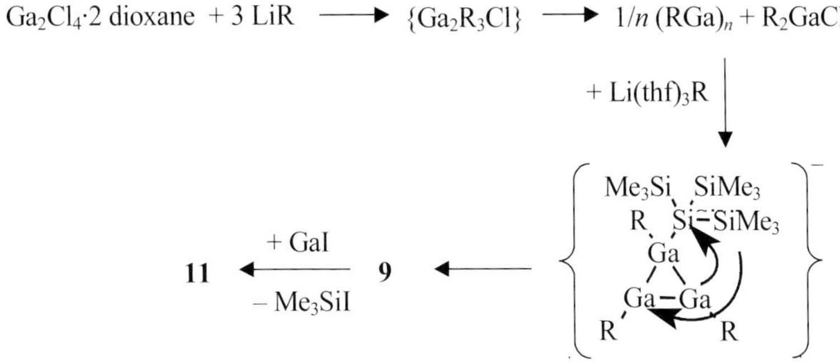

$$Ga_2Cl_4\cdot 2\text{ dioxane} + 3\text{ LiR} \longrightarrow \{Ga_2R_3Cl\} \longrightarrow 1/n\,(RGa)_n + R_2GaCl \xrightarrow{+\,Li(thf)_3R} \{\text{intermediate}\}^-$$

$$\{\text{intermediate}\}^- \longrightarrow \mathbf{9} \xrightarrow[-\,Me_3SiI]{+\,GaI} \mathbf{11}$$

(Intermediate labels: Me₃Si, SiMe₃, R, Si–SiMe₃, Ga, Ga–Ga, R, R)

Scheme 2. Proposed reaction pathways to **9** and **11**.

X-Ray Structure Analysis

9 crystallizes in the monoclinic space group $P2_1/n$ and consists of anionic Ga_3Si heterocycles (Fig 1) and isolated $Li(thf)_4$ cations. The heterocycle **9** is a diamond-shaped flat butterfly (angle between planes: 20.6°) with acute angles at the ring silicon atom and at the tetracoordinated gallium center.

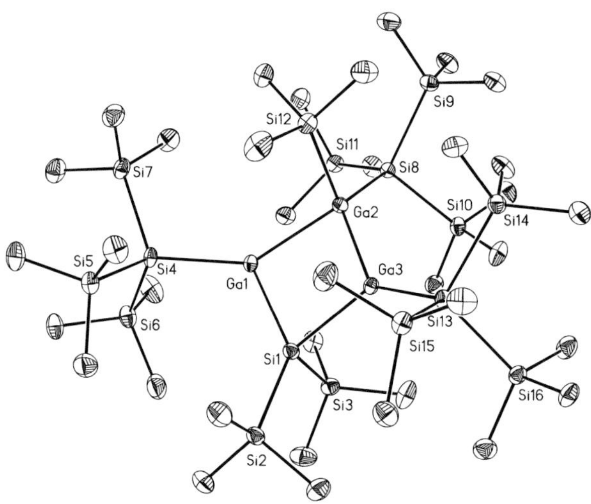

Fig. 1. View of the heterocyclic anion in **9**. Selected bond lengths [pm] and angles [°]: Ga1–Ga2 255.27(9), Ga2–Ga3 252.88(9), Ga1–Si1 247.5(2), Ga3–Si1 245.2(2), Ga1–Si4 248.4(2), Ga2–Si8 248.4(2), Ga2–Si12 243.6(2), Ga3–Si13 248.3(2), Si1–Si2 235.4(2), Si1–Si3 236.5(2), Si–$SiMe_3$ 235.5; Si1-Ga1-Ga2 94.78(4), Ga3-Ga2-Ga1 81.39(3), Si1-Ga3-Ga2 95.96(4), Ga3-Si1-Ga1 84.52(5), Si1-Ga1-Si4 119.60(6), Si4-Ga1-Ga2 144.07(5), Si12-Ga2-Si8 110.37(6), Si12-Ga2-Ga3 105.67(5), Si8-Ga2-Ga3 128.57(4), Si12-Ga2-Ga1 98.09(5), Si8-Ga2-Ga1 126.64(4), Si1-Ga3-Si13 125.63(5), Si13-Ga3-Ga2 136.59(5), Si2-Si1-Si3 106.12(8), Si2-Si1-Ga3 127.03(7), Si3-Si1-Ga3 98.00(7), Si2-Si1-Ga1 135.74(7), Si3-Si1-Ga1 97.47(7).

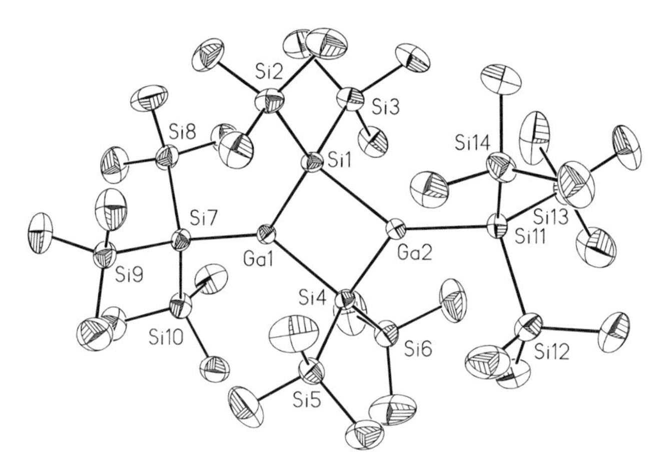

Fig. 2. View of a molecule of **10**. Selected bond lengths [pm] and angles [°]:Ga1–Si1 242.89(9), Ga1–Si4 241.2(1), Ga1–Si7 241.5(1), Ga2–Si1 242.5(1), Ga2–Si4 241.68(9), Ga2–Si11 242.5(1), Si1–Si2 232.8(1), Si1–Si3 233.9(1), Si4–Si5 233.7(2), Si4–Si6 232.4(1), Si–$SiMe_3$ 234.0; Si4-Ga1-Si1 91.59(4), Si4-Ga2-Si1 91.57(4), Ga2-Si1-Ga1 88.11(3), Ga1-Si4-Ga2 88.68(3), Si4-Ga1-Si7 132.66(3), Si7-Ga1-Si1 135.74(3), Si4-Ga2-Si11 132.77(4), Si11-Ga2-Si1 135.64(3), Si2-Si1-Si3 107.27(5), Si2-Si1-Ga2 128.62(5), Si3-Si1-Ga2 96.78(5), Si2-Si1-Ga1 127.47(5), Si3-Si1-Ga1 102.77(5), Si6-Si4-Si5 116.52(5), Si6-Si4-Ga1 123.73(5), Si5-Si4-Ga1 100.53(5), Si6-Si4-Ga2 120.08(5), Si5-Si4-Ga2 101.97(5).

10 (Fig. 2) crystallizes in the triclinic space group $P\overline{1}$ with two independent molecules in the asymmetric unit. The nearly planar four-membered gallium silicon heterocycles have an approximately quadratic alternating Ga_2Si_2 framework of distorted trigonal planar-coordinated gallium and tetrahedrally coordinated silicon centers. The molecule is less sterically crowded than **9**; therefore the Ga–Si bond lengths are shorter by approx. 5 pm.

The X-ray structure analysis of **11**, trigonal space group *P*31*c,* reveals a C_3-symmetric cluster anion (Fig. 3) with a trigonal-bipyramidal Ga_4Si core. The counterion is $[Li(THF)_4]^+$. The Ga–Ga distances between the equatorial gallium atoms are longer than those to the apical one. But in the trigonal bipyramidal gallium–iron cluster **12** (space group $P6_3/m$) the gallium–gallium distances are longer by 50 pm (Fig. 4) [7].

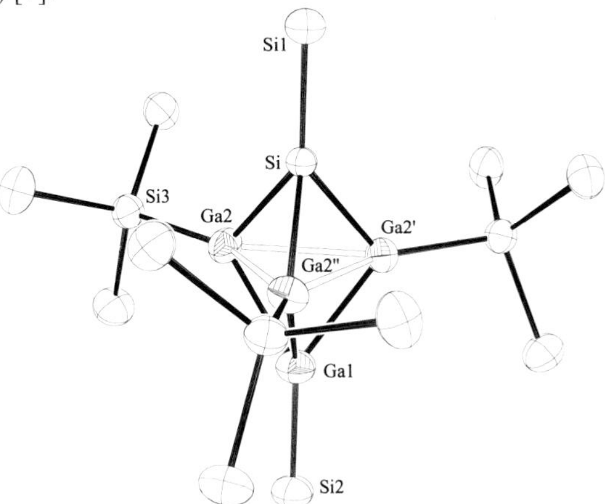

Fig. 3. View of a cluster anion of **11**. The methyl groups have been omitted for clarity. Selected bond lengths [pm] and angles [°]: Ga1–Ga2 244.0(1), Ga2–Ga2′ 279.0(1), Ga2–Si 240.2(2), Si–Si1 227.1(4), Ga1–Si2 233.8(3), Ga2-Si3 237.6(2); Ga2-Ga1-Ga2′ 69.74(4), Ga1-Ga2-Si 96.58(4), Ga2-Si-Ga2′ 71.00(6), Si2-Ga1-Ga2 138.69(2), Si3-Ga2-Si 128.73(6), Si3-Ga2-Ga1 134.64(5), Si3-Ga2-Ga2′ 151.58(6), Si3-Ga2-Ga2′′ 148.08(6)

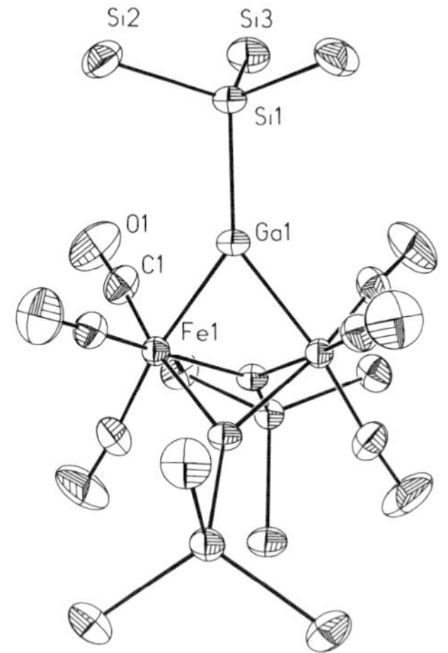

Fig. 4. View of a molecule of **12**. The methyl groups have been omitted for clarity. Selected bond lengths [pm] and angles [°]: Ga–Ga 328.9(1), Ga–Fe 238.18(7), Ga–Si1 238.7(2); Fe1-Ga1-Fe1A 74.28(4), Fe1-Ga1-Si1 142.69(2), C1-Fe1-C1 99.0(2).

Results of DFT(RI) Calculations

Is **11** to be described as a *closo* cluster with 6 cluster binding electron pairs, and **12** with only 5 cluster binding electron pairs as a *hypercloso* cluster? DFT calculations on various trigonal-bipyramidal cage compounds favor the *closo* description of **11** (Scheme 3). This is also supported by the results of a formal two electron reduction of **12**, which results in a *closo* structure, too.

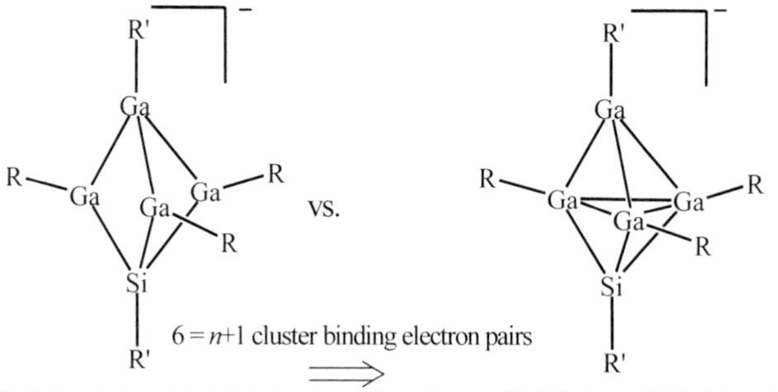

6 = n+1 cluster binding electron pairs

classical description with 6 *2e2c* bonds or *closo* cluster with 3 *2e2c*- and 3 *2e3c* bonds

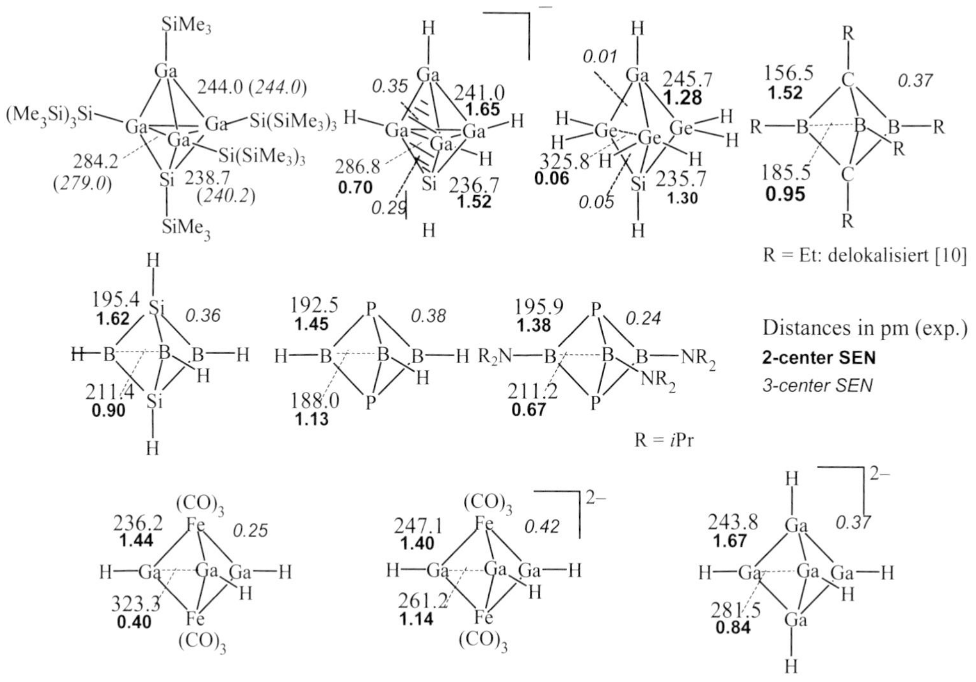

R = Et: delokalisiert [10]

R = iPr

Distances in pm (exp.)

2-center SEN

3-center SEN

Scheme 3. Results of DFT(RI) calculations on trigonal-bipyramidal group 13 clusters. BP86 functional, SV(P) base; SEN = shared electron number.

Acknowledgment: We thank Mr. H. Piotrowski, Prof. Dr. P. Klüfers and Prof. H. Nöth for collecting the crystallographic data sets. Financial support by the *Deutsche Forschungsgemeinschaft*, the *Fonds der Chemischen Industrie* and *Chemetall GmbH* is gratefully acknowledged.

References:

[1] R. Frey, G. Linti, K. Polborn, *Chem. Ber.* **1994**, *127*, 101–103.

[2] G. Linti, R. Frey, W. Köstler, Horst Urban, *Chem Ber.* **1996**, *129*, 561–569.

[3] A. M. Arif, A. H. Cowley, T. M. Elkins, R. A. Jones, *J. Chem. Soc., Chem. Commun.* **1986**, 1776–1777.

[4] G. Linti, W. Köstler, *Angew. Chem.* **1996**, *108*, 593–595; *Angew. Chem., Int. Ed. Engl.* **1996**, *35*, 550–552.

[5] G. Linti, *J. Organomet. Chem.* **1996,** *520*, 107–113.

[6] G. Linti, W. Köstler, *Angew. Chem.* **1997,** *109*, 2758–2760; *Angew. Chem., Int. Ed. Engl.* **1997**, *36*, 2644–2646 .

[7] G. Linti, W. Köstler, *Chem. Eur. J.* **1998**, *4*, 942–949.

Novel Silaheterocycles by Carbenoid and Diazo Reactions

*Volker Gettwert, Gerhard Maas**

Abteilung Organische Chemie I, Universität Ulm
Albert-Einstein-Allee 11, D-89081 Ulm, Germany
Tel.: Int. code + (731)502 2790 — Fax: Int. code + (731)502 2803
E-mail: gerhard.maas@chemie.uni-ulm.de

Keywords: Carbenes / Carbenoids / Cyclophanes / Pyrazoles / Silaheterocycles

Summary: For α-(alkynyloxy)silyl-α-diazoacetates **1**, various intramolecular reactions involving the diazo (or latent carbene) function and the C–C triple bond have been achieved. Depending on reaction conditions and substituent pattern, silaheterocycles **2**–**7** can be prepared. A thermally induced inter-/intramolecular 1,3-dipolar cycloaddition sequence of **1a**–**c** leads to [3.3](1,4)pyrazolophanes **8** and higher cyclooligomers thereof (**9**, $n \geq 3$).

Introduction

Intramolecular [3+2] cycloaddition reactions of unsaturated diazocarbonyl compounds as well as transition-metal-catalyzed intramolecular carbene-type reactions of diazo compounds constitute an important strategy in contemporary synthesis of alicyclic and heterocyclic systems [1]. In a program directed towards the synthesis of silaheterocycles according to this concept, we have used various silicon-functionalized (silyl)diazoacetates [2] as starting materials. In this communication, we report on the synthesis of silaheterocycles from α-(alkynyloxy)silyl-α-diazoacetates [3], which can be prepared easily by successive reaction of a silyl bis(triflate) with an alkyl diazoacetate and a propargyl alcohol [4].

Intramolecular Reactions of (Alkynyloxy)silyl-diazoacetates 1

Copper(I) triflate-catalyzed decomposition of diazoacetates **1** leads to silaheterocycles **2**, **3**, and **4**. Vinylcarbenes or vinylcarbene metal complexes may be assumed as reaction intermediates. With copper(I) chloride as catalyst, the dediazotization of **1** leads to the bicyclic furan **5**, which formally represents the result of an intramolecular [3+2] cycloaddition of an acylcarbene to the C–C triple bond [5].

Surprisingly, treatment of **1** with Ag_2O does not decompose the diazo ester but rather leads to the silver pyrazolate **6**. A related observation has been made recently [6]. Similary, a thermally induced intramolecular 1,3-dipolar cycloaddition reaction generates pyrazole **7**. This reaction mode is limited, however, to specific substitution patterns. In other cases, an intermolecular rather than the intramolecular [3+2] cycloaddition comes into play (see next section).

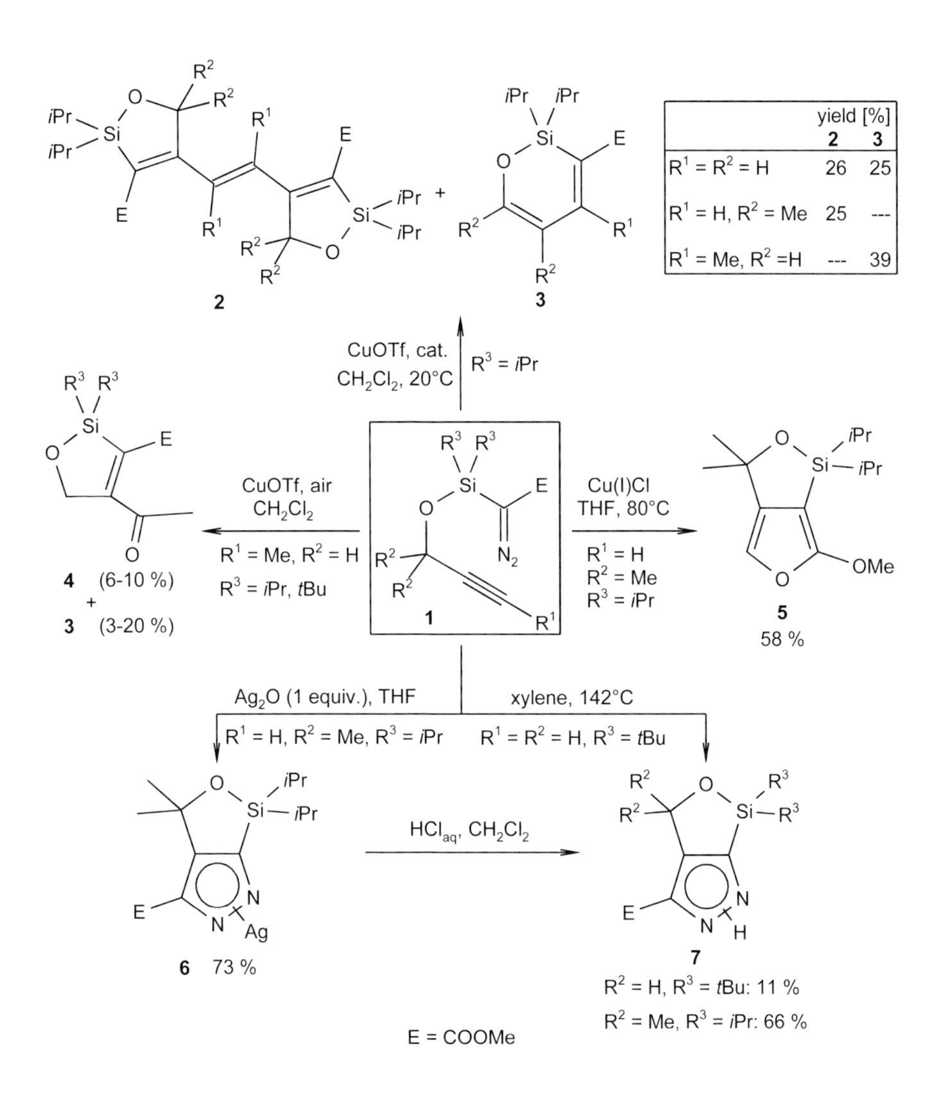

Scheme 1. Copper-catalyzed and silver-assisted intramolecular reactions of α-(alkynyloxy)silyl-α-diazoacetates **1**.

Cyclooligomerization of (Alkynyloxy)silyl-diazoacetates 1

Heating of diazoacetates **1a–c** at ca. 140–160 °C, with or without solvent, leads to [3.3](1,4)pyrazolophanes **8a–c** and the higher cyclooligomers **9a–c** (Scheme 2).

In most cases, the "cyclodimers" **8** can be isolated by crystallization, while a complete separation of the mixture of the higher "cyclooligomers" **9** was not possible.

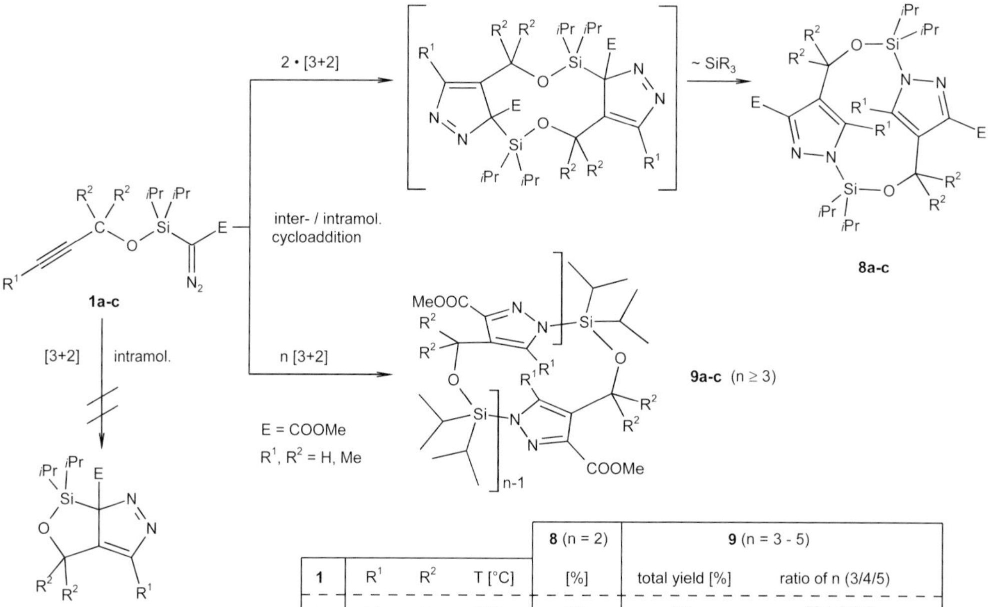

E = COOMe

R^1, R^2 = H, Me

				8 (n = 2)	9 (n = 3 - 5)	
1	R^1	R^2	T [°C]	[%]	total yield [%]	ratio of n (3/4/5)
a	H	H	142	18	76	90 / 7 / 2
b	H	Me	142	2 - 6	45-54	63 / 23 / 13
					8, **9** (n = 2 - 5)	ratio of n (2/3/4/5)
c	Me	H	162		2	26 / 5 / 67 / 1

Scheme 2. Cyclooligomerization of diazoacetates **1** by [3+2] cycloaddition reactions.

^{1}H-NMR spectra of **8b** and of the cyclooligomer mixture of **9b** are shown in Fig. 1. It is obvious that the spectrum of **9b** does not indicate the presence of a mixture of cyclooligomers, while the FD mass spectra clearly show the presence of peaks expected for cyclooligomers with 3–5 monomer units.

Cyclotrimer **9b** (n = 3) and cyclotetramer **9c** (n = 4) could be obtained in small amounts by fractionating crystallization from the respective cyclooligomer mixture. Their structures, as well as that of cyclophane **8b**, could be determinated by single-crystal X-ray diffraction and are shown in Figs. 2–4.

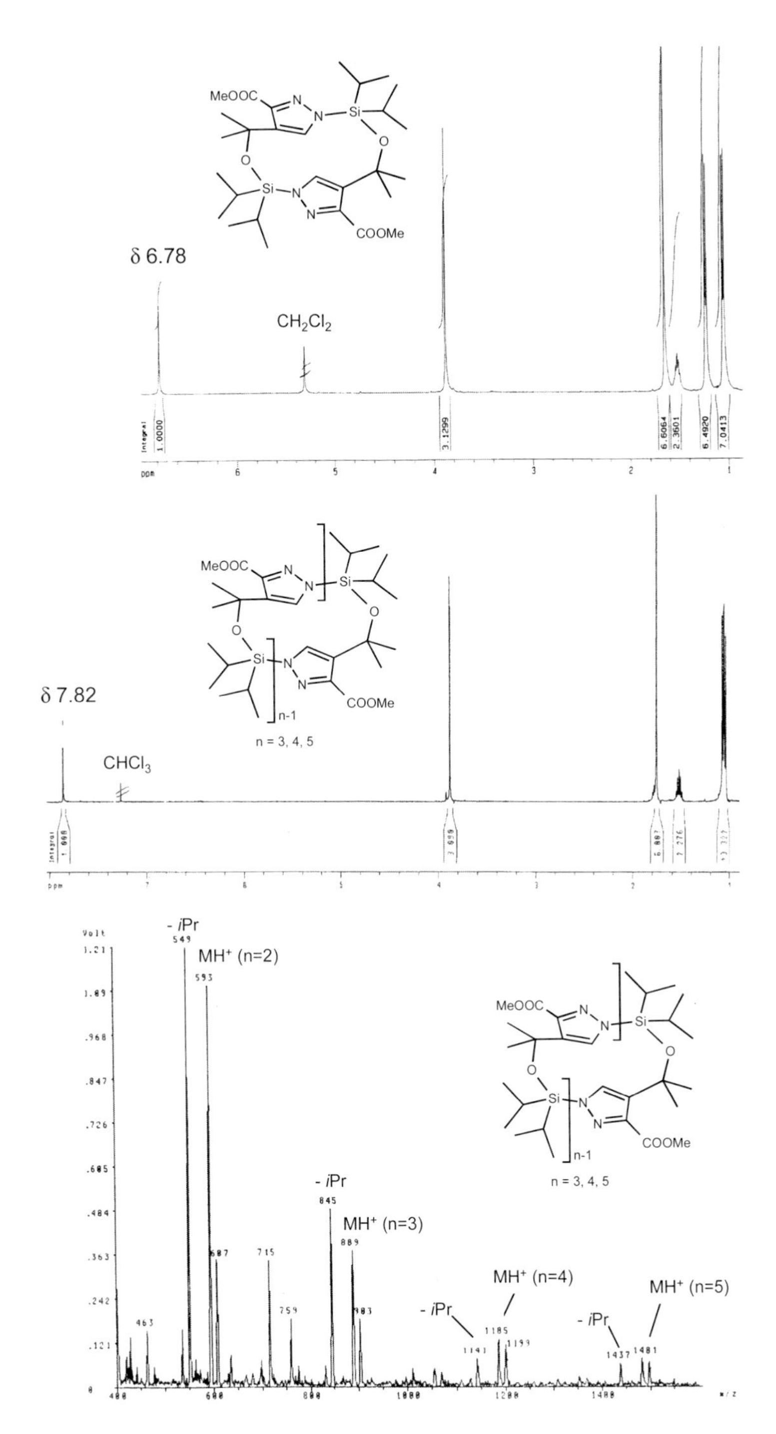

Fig. 1. Top and middle: ^{1}H-NMR spectra (500 MHz) of cyclophanes **8b** and **9b** (n = 3–5); bottom: Mass spectrum (FD, 8 kV) of cyclooligomer mixture **9b** (n = 3–5).

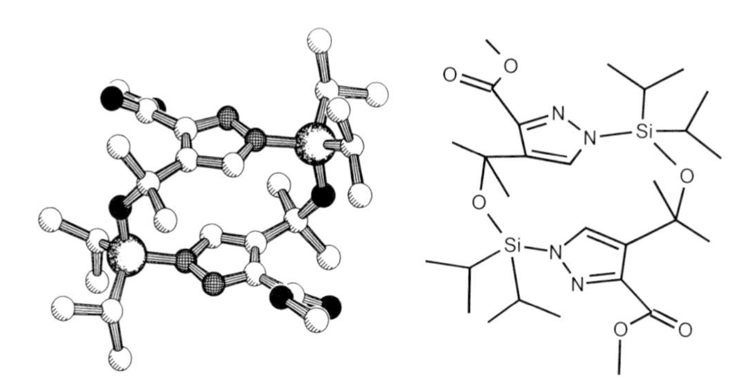

Fig. 2. Structure of **8b** (R^1 = H, R^2 = Me) in the crystal (PLUTON plot).

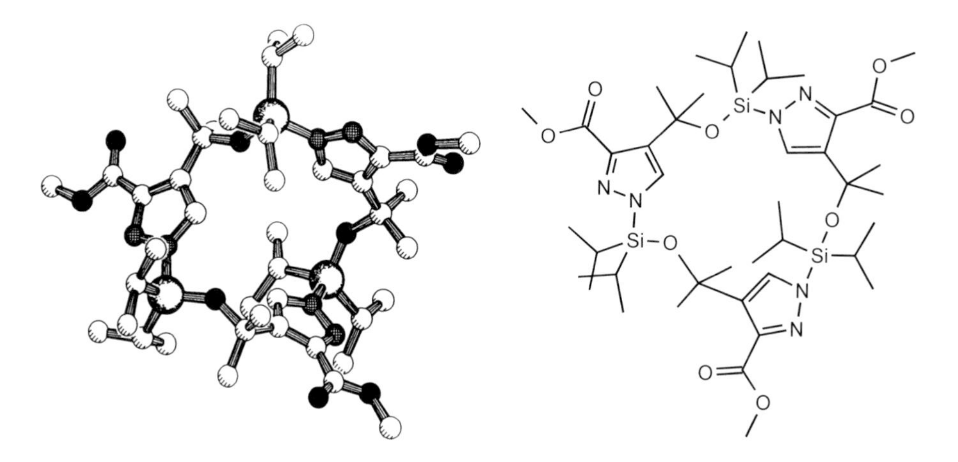

Fig. 3. Structure of **9b** (R^1 = H, R^2 = Me, n = 3) in the crystal (PLUTON plot).

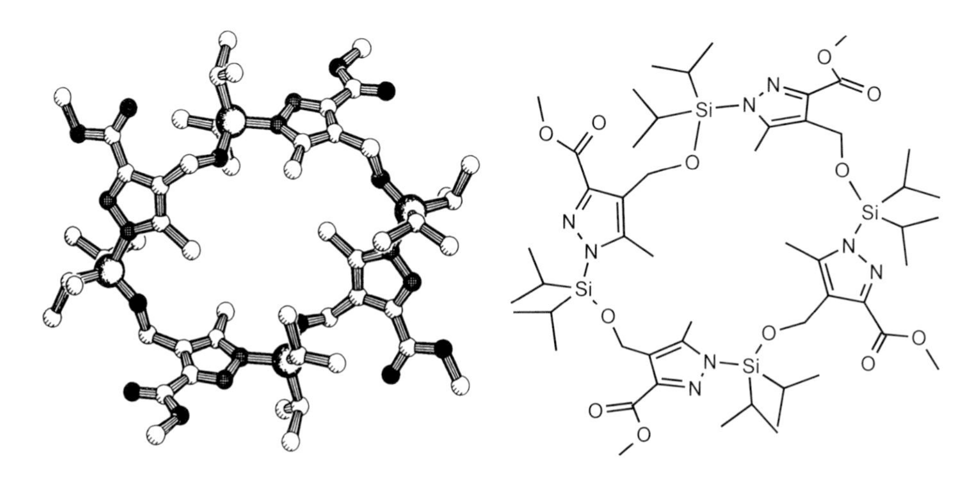

Fig. 4. Structure of **9c** (R^1 = Me, R^2 = H, n = 4) in the crystal (PLUTON plot).

References:

[1] T. Ye, M. A. McKervey, *Chem. Rev.* **1994**, *94*, 1091–1160.

[2] G. Maas, *"Carbene and Carbenoid Chemistry of Silyldiazoacetic Esters: The Silyl Group as a Substituent and a Functional Group"*, in *Organosilicon Chemistry II: From Molecules to Materials* (Eds.: N. Auner, J. Weis), VCH, Weinheim, **1996**, pp 149–159.

[3] F. Krebs, S. Bender, B. Daucher, T. Werle, G. Maas, *"Silaheterocycles from Intramolecular Reactions of Silicon-functionalized Diazoacetic Esters"*, in *Organosilicon Chemistry: From Molecules to Materials* (Eds.: N. Auner, J. Weis), VCH, Weinheim, **1994**, pp 57–59.

[4] A. Fronda, F. Krebs, B. Daucher, T. Werle, G. Maas, *J. Organomet. Chem.* **1992**, *424,* 253–282.

[5] V. Gettwert, F. Krebs, G. maas, *Eur. J. Org. Chem.* **1999**, 1213–1221

[6] A. S. Kende, M. Journet, *Tetrahedron Lett.* **1995**, *36*, 3087–3090.

Synthesis and Reactivity of Novel 1,2-Disilacyclopentanes

Carsten Strohmann*, Oliver Ulbrich

Institut für Anorganische Chemie, Universität Würzburg
Am Hubland, D-97074 Würzburg, Germany
Tel.: Int. code + (931)888 4613
E-mail: c.strohmann@mail.uni-wuerzburg.de

Keywords: Disilacyclopentanes / Trisilacyclopentanes / Bis(lithiomethyl)silanes / 1,2-Disila-4-germacyclopentanes

Summary: Disilacyclopentanes containing a heteroelement have been prepared by specific routes starting from bis(lithiomethyl)silanes or bis(lithiomethyl)disilanes. The key reaction in the sequence of synthesis is the reductive cleavage of C–S bonds of bis(phenylthiomethyl)silanes and -disilanes to obtain the corresponding difunctional lithioalkyl compounds. The crystal structures of two novel 1,2-disila-4-element-cyclopentanes and a spirocyclic disilane were obtained and show strained ring systems. Three reaction sites of 1,1,2,2-tetramethyl-4,4-diphenyl-1,2,4-trisilacyclopentane could be activated selectively.

In the last few years the design and use of various disilane compounds has gained importance because of the reactivity of the Si–Si bond and the large potential for organic synthesis involved with it. Many publications offer us numerous examples of possible reactions at the silicon–silicon bond such as addition reactions with C–C double bonds or C–C triple bonds [1, 2], addition reactions with C-element multiple bonds (e.g. aldehydes, quinones, isocyanides) [3–5] or metathesis [6, 7] and cross-metathesis [8]. In the most cases the existence of a catalyst (palladium, platinum or nickel complexes) for activation of the silicon–silicon σ bond is indispensable for a successful transformation [9–11].

In this connection cyclic disilanes show an enhanced reactivity at the silicon–silicon bond in comparison with acyclic compounds. An important aspect in view of the intensification of the reactivity is the strain of the ring system which can be observed especially in four- or five-membered ring systems. Some of the well-known ring compounds which are described by Sharma and Pannell [12] in an extensive survey of disilane chemistry are shown in Scheme 1.

Our object was the synthesis of 1,2-disilacyclopentanes of type **A** (Scheme 1) containing a heteroatom in the basic ring. One of the most striking features of these compounds is the additional influence of the heteroatom on the strain of the ring and subsequently on the reactivity of the silicon–silicon bond.

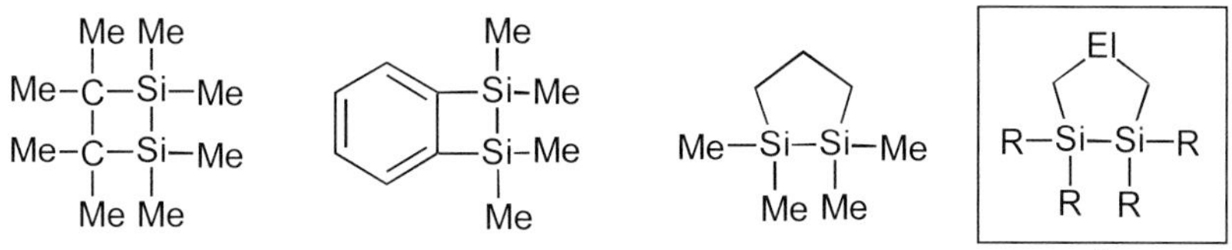

A

El: SiR_2, GeR_2, O, S, NR

R: organyl

Scheme 1. Some cyclic 1,2-Disilanes.

Despite the number of publications dealing with disilane compounds, only a limited interest has been shown in these systems containing a disilane unit and a heteroatom in the ring system, due to the lack of appropriate synthetic routes. It is known that cyclic disilanes of that type can be obtained as a product of pyrolysis [13, 14] or after trans-ylidation [15]. In this paper we will describe two synthetic strategies for the specific preparation of 1,2-disila-4-heterocyclopentanes. The deciding step in our concept is the reductive cleavage of C–S bonds and the resulting formation of dilithioalkyl compounds.

The preparation of the 1,2-disila-4-heterocyclopentanes occurs by reaction of 1,2-bis(lithiomethyl)disilanes with dichloroelement compounds or by reaction of bis(lithiomethyl)element compounds with 1,2-dichlorodisilanes (Scheme 2).

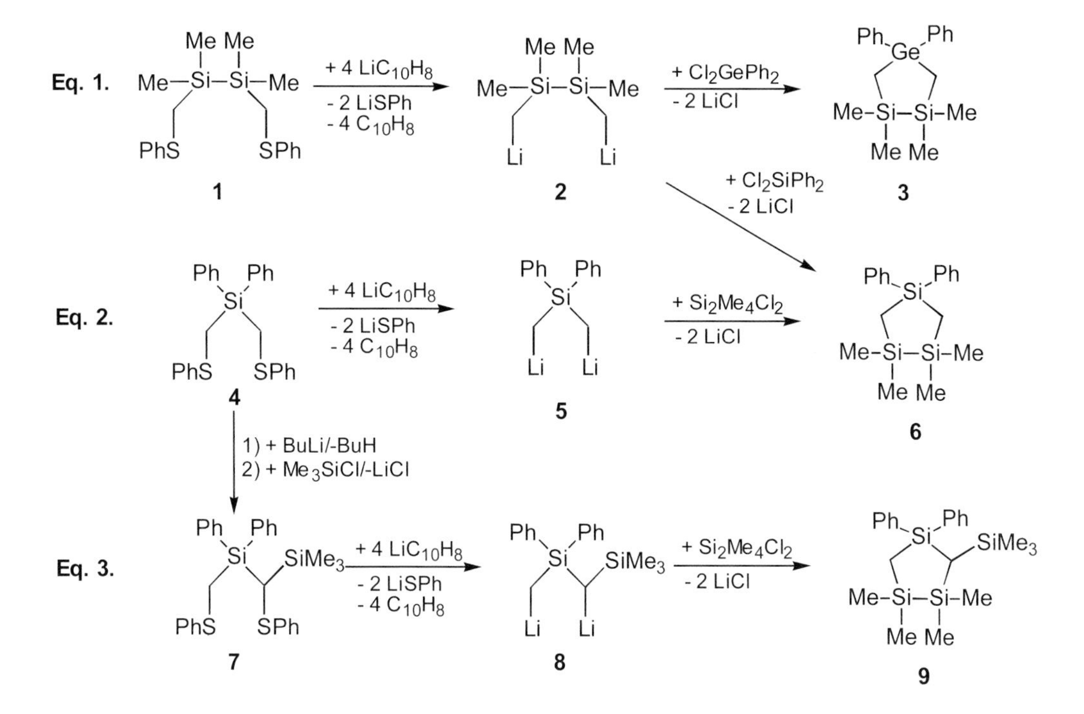

Scheme 2. Syntheses of the Trisilacyclopentanes **6** and **9** and of the 1,2-Disila-4-germacyclopentane **3**.

Bis(lithiomethyl)diphenylsilane (**5**) is readily available by a reductive cleavage of the C–S bond of bis(phenylthiomethyl)diphenylsilane (**4**) with lithium naphthalide (Eq. 2) [16–18]. Addition of 1,1,2,2-tetramethyl-1,2-dichlorodisilane to the freshly prepared bis(lithiomethyl)diphenylsilane leads to the trisilacyclopentane **6** as an oily liquid which crystallizes after distillation by Kugelrohr. The trisilacycle **9** substituted by a trimethylsilane group at the ring carbon atom is available in an analogous way (Eq. 3).

An alternative method of obtaining the desired silacycles occurs also by reductive cleavage of C–S bonds starting from 1,2-bis(phenylthiomethyl)disilane compounds. (Phenylthiomethyl)lithium reacts with 1,2-dichloro-1,1,2,2-tetramethyldisilane which results in formation of 1,2-bis(phenylthiomethyl)-1,1,2,2-tetramethyldisilane (**1**). Treatment of **1** with four equivalents of lithium naphthalide in THF at –40°C yields a red-colored solution of the dilithioalkyl compound **2**. The following reactions with dichlorodiphenylsilane and dichlorodiphenylgermane respectively result in the trisilacyclopentane **6** and the disilagermacyclopentane **3**. The germanium product is also an oily liquid which can be distilled by Kugelrohr and subsequently crystallized.

A very interesting investigation leading to a spirocyclic system with two disilane units is shown in Scheme 3. Tetrakis(lithiomethyl)silane (**11**) served well in the synthesis of **12** in a reaction with 1,2-dichloro-1,1,2,2-tetramethyldisilane [19].

$$\mathbf{10}\ \xrightarrow[-\,8\ \mathrm{DBB}]{+\,8\ \mathrm{LiDBB}}\ \mathbf{11} + 4\ \mathrm{LiSPh}\ \xrightarrow[-\,8\ \mathrm{LiCl},\ -\,4\ \mathrm{Me_3SiSPh}]{+\,2\ \mathrm{Cl_2Si_2Me_4},\ +\,4\ \mathrm{ClSiMe_3}}\ \mathbf{12}$$

Scheme 3. Synthesis of the spirocyclic system **12** (DBB = di-*tert*-butyl-biphenyl).

The structures of the derivatives **3** and **6** as well as the structure of **12** were determined (Fig. 1). The germanium compound **3** is isostructural with the silicon analogue **6**. The silicon–silicon bond lengths (2.35 Å (**3**) and 2.36 Å (**6**)) are in good agreement with other disilacyclopentanes or acyclic disilanes. The C–Si bond lengths are also in the expected scale. The endocyclic angles (Si-Si-C) are smaller than usual and indicate a distortion of the tetrahedron angle which is more distinct in **6** than in **9**. From this the conclusion may be drawn that the ring systems of these novel 1,2-disilacyclopentanes are strained.

Spirosilane **12** crystallized from ethanol in the monoclinic crystal system, space group *C2/c*. The silicon center Si(3) represents the inversion center of the molecule. The Si–C bond lengths are all in the normal range of 1.86–1.89 Å and the Si–Si bond length is 2.35 Å. The two five-membered rings are in an envelope conformation with ecliptically arranged methyl groups.

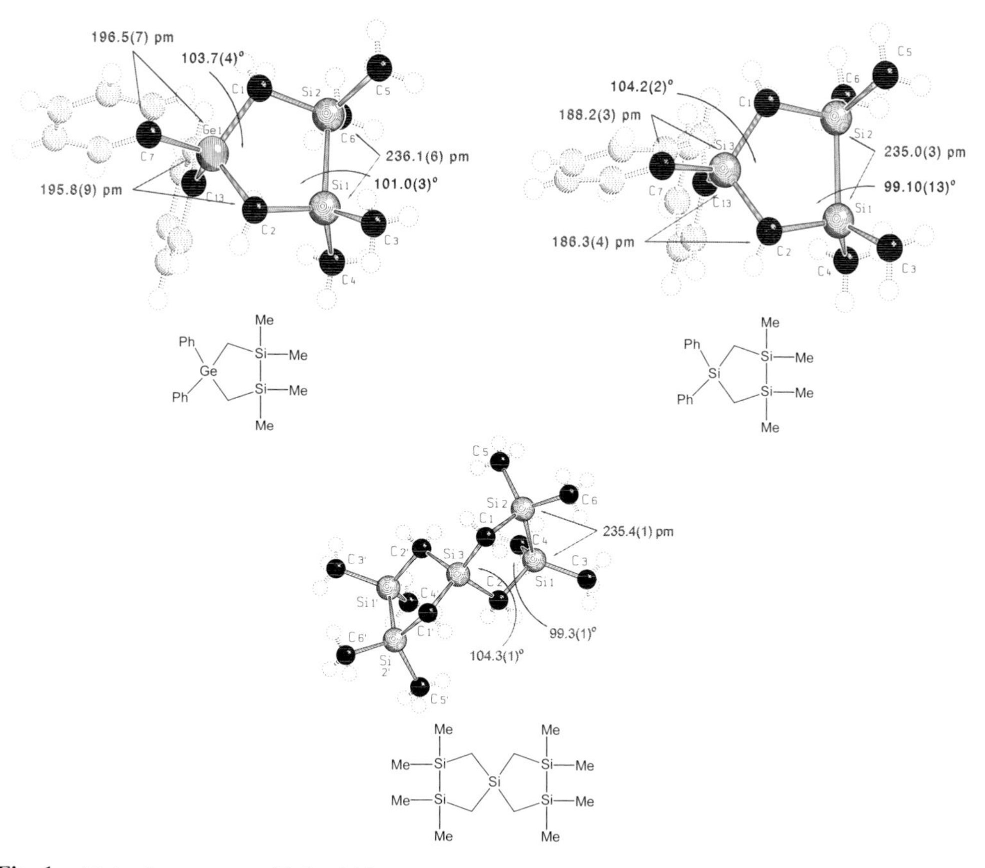

Fig. 1. Molecular structures of **3**, **6** and **12**.

To check the new cyclic compounds for reactivity the trisilacyclopentane **6** was submitted to some experiments (Scheme 4).

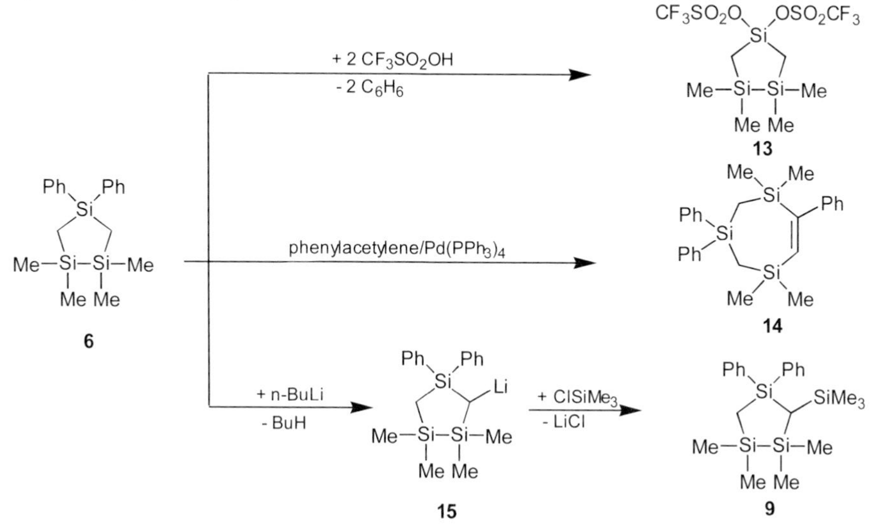

+ 2 CF_3SO_2OH
- 2 C_6H_6

phenylacetylene/$Pd(PPh_3)_4$

+ n-BuLi
- BuH

+ $ClSiMe_3$
- LiCl

6 **13** **14** **15** **9**

Scheme 4. Reactions of the trisilacyclopentane **6**.

We found that conversions can be performed selectively on three different reaction sites of the trisilacyclopentane **6**.

Both of the phenyl groups attached at the silicon atom are replaceable by triflate groups. Reaction of **6** with trifluoromethanesulfonic acid proceeds spontaneously and quantitatively at 0°C and yields **13**. Starting from **13,** functionalization of the central silicon atom is possible.

The insertion reaction of phenylacetylene into the silicon–silicon bond in the presence of tetrakis(triphenylphosphin)palladium(0) leads to the expected seven-membered ring **14**. The reaction occurs in toluene as solvent at 90°C after 24 h and provides **14** as colorless crystals after recrystallization from acetone (see structure in Fig. 2).

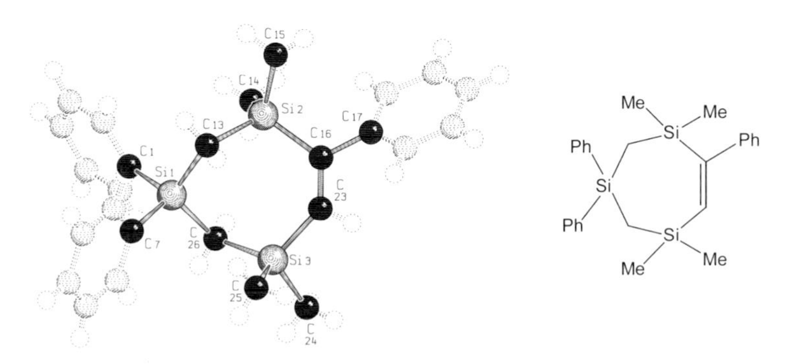

Fig. 2. Molecular structure of **14**.

We laid most importance on the question of wether it is possible to metallize a ring carbon atom, because this would confer a considerable potential in applications. Insertion of donor substituents for stabilizing catalyst intermediates by intramolecular coordination or for controlling stereoselective insertion reactions are only two aspects for further studies. The metal–proton exchange occurs already by reaction of **6** with *n*-butyllithium in THF at –78°C, proved by a subsequent trapping reaction with chlorotrimethylsilane.
Further related work on the synthesis, structure and especially the reactivity of these novel cyclic disilanes is currently under way.

Acknowledgement: We are grateful to the *Deutsche Forschungsgemeinschaft* (DFG), to *Wacker GmbH*, to *Chemetall GmbH* and to the *Fonds der Chemischen Industrie* (FCI) for financial support. We thank Prof. M. Veith (Universität Saarbrücken) for supporting this work.

Refernces:

[1] M. Ishikawa, A. Naka, J. Ohshita, *Organometallics* **1993**, *12*, 4987.
[2] K. Tamao, T. Hayashi, M. Kumada, *J. Organomet. Chem.* **1976**, *114*, C19.
[3] M. Ishikawa, H. Sakamoto, T. Tabuchi, *Organometallics* **1991**, *10*, 3173.
[4] H. Yamashita, N. P. Reddy, M. Tanaka, *Organometallics* **1997**, *16*, 5223.
[5] Y. Ito, M. Suginome, T. Matsuura, M. Murakami, *J. Am. Chem. Soc.* **1991**, *113*, 8899.
[6] Y. Uchimaru, M. Tanaka, *J. Organomet. Chem.* **1996**, *521*, 335.

[7] M. Suginome, H. Oike, P. H. Shuff, Y. Ito, *Organometallics* **1996**, *15*, 2170.
[8] N. P. Reddy, T. Hayashi, M. Tanaka, *Chem. Commun.* **1996**, 1865.
[9] M. Suginome, Y. Ito, *J. Chem. Soc. Dalton Trans.* **1998**, 1925.
[10] M. Suginome, H. Oike, S. Park, Y. Ito, *Bull. Chem. Soc. Jpn.* **1996**, *69*, 289.
[11] K. A. Horn, *Chem. Rev.* **1995**, *95*, 1317.
[12] H. K. Sharma, K. H. Pannell, *Chem. Rev.* **1995**, *95*, 1351.
[13] Y. Chen, B. H. Cohen, P. P. Gaspar, *J. Organomet. Chem.* **1980**, *195*, C1.
[14] G. Fritz, B. Grunert, *Z. Anorg. Allg. Chem.* **1981**, *473*, 59.
[15] H. Schmidbaur, W. Vornberger, *Chem. Ber.* **1972**, *105*, 3173.
[16] C. Strohmann, E. Wack, in *Organosilicon Chemistry III: From Molecules to Materials* (Eds.: N. Auner, J. Weis), Wiley–VCH, Weinheim, **1998**, p. 217.
[17] C. Strohmann, S. Lüdtke, in *Organosilicon Chemistry II: From Molecules to Materials* (Eds.: N. Auner, J. Weis), VCH, Weinheim, **1996**, p. 499.
[18] C. Strohmann, S. Lüdtke, E. Wack, *Chem. Ber.* **1996** *129*, 799.
[19] C. Strohmann, S. Lüdtke, O. Ulbrich, *Organometallics* in press.

Synthesis of Heavily Halogenated Vinylsilanes

Uwe Pätzold, Florian Lunzer, Christoph Marschner, Karl Hassler

Institut für Anorganische Chemie
Erzherzog Johann Universität
Stremayrgasse 16, A-8010 Graz, Austria
Tel.: Int. code + (316)873 8206 — FAX: Int. code + (316)873 8701
E-mail: hassler@anorg.tu-graz.ac.at

Keywords: (Trichlorovinyl)silanes / (Trifluorovinyl)silanes

Summary: By chlorination of ethyltrichlorosilane and diethyldichlorosilane with Cl_2 under UV irradiation, (pentachloroethyl)trichlorosilane and bis(pentachloroethyl)-dichlorosilane have been prepared, which react with powdered copper forming (trichlorovinyl)trichlorosilane and bis(trichlorovinyl)dichlorosilane. The derivatizations of these silanes at the silicon atom by use of phenylmagnesium bromide, trifluoromethanesulfonic acid, lithium aluminium hydride, zinc difluoride and bromoform are described. (Trifluorovinyl)-, bis(trifluorovinyl)- and tris(trifluorovinyl)silanes have been prepared by the reaction of trifluoromethanesulfonyloxysilanes and trifluorovinyl-lithium. Attempts to prepare Si polymers bearing perhalovinyl groups by dehydrogenative coupling of $RSiH_2$ with metal catalysts have not been successful.

The derivatization of silanes on the Si atom without affecting perhalogenated organic groups can be a difficult task as the latter are readily attacked by many reagents. For instance, the hydrogenation of CCl_3SiCl_3 with $LiAlH_4$ yields just 50 % of CCl_3SiH besides SiH_4 [1]. CF_3SiCl_3 is reduced more readily, but 7–9 % of SiH_2 are formed also [2]. Trihalovinyl-groups $F_2C{=}CF–$ and $Cl_2C{=}CCl–$ behave somewhat differently, as they are not so readily attacked. As our objective partly focuses on the preparation of polysilanes bearing perhalogenated organic groups, we have directed our efforts towards the preparation and functionalization of (trichlorovinyl)- and (trifluorovinyl)silanes.

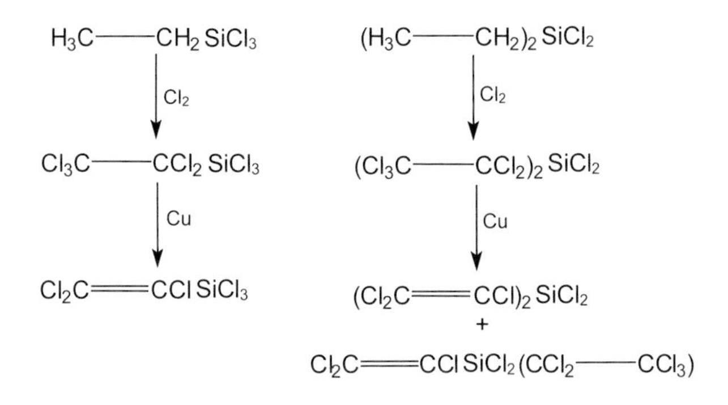

Scheme 1. Preparation of (trichlorovinyl)silanes.

By chlorination of ethyltrichlorosilane and diethyldichlorosilane with Cl_2 and subsequent reaction with powdered copper, (trichlorovinyl)trichlorosilane [3], bis(trichlorovinyl)dichlorosilane [3] and (trichlorovinyl)(pentachloroethyl)dichlorosilane can be prepared in good yields, as illustrated in Scheme 1.

(Trichlorovinyl)trichlorosilane is easily phenylated by use of phenylmagnesium bromide forming (trichlorovinyl)triphenylsilane, which can be functionalized with trifluoromethanesulfonic acid (TfOH) as illustrated in Scheme 2. The trichlorovinyl group is not attacked by CF_3SO_3H.

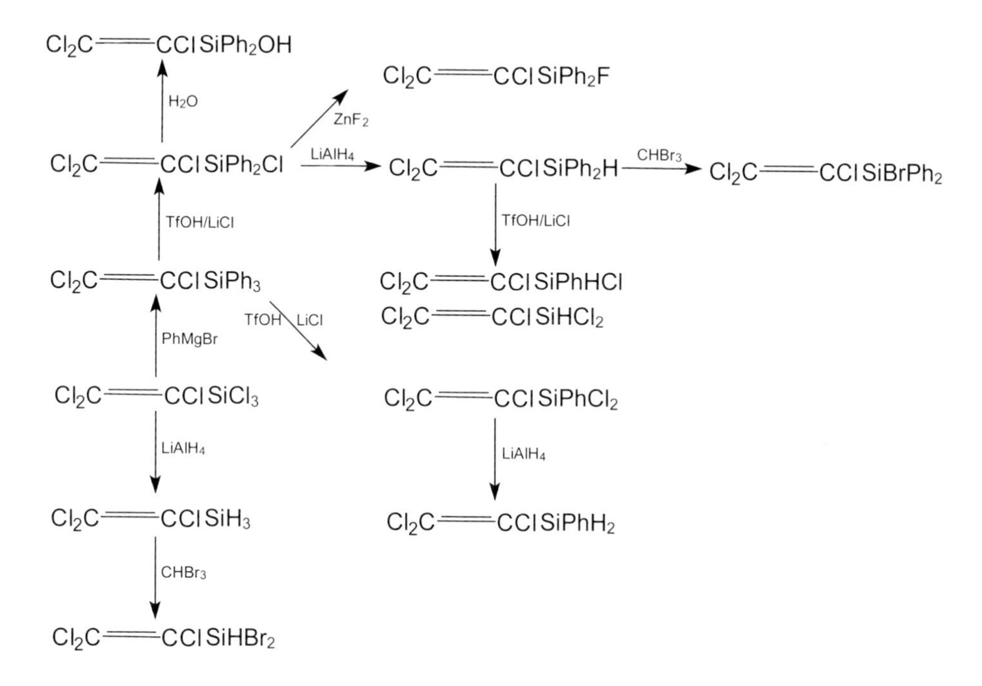

Scheme 2. Reactions of (trichlorovinyl)trichlorosilane.

(Trifluorovinyl)silanes were prepared by the reaction of trifluorovinyllithium either with $(TfO)SiH_3$ giving (trifluorovinyl)silane or with $(TfO)SiPhH_2$ giving a mixture of mono-, di- and trisubstituted vinyl derivatives that can be separated by fractional distillation (Scheme 3).

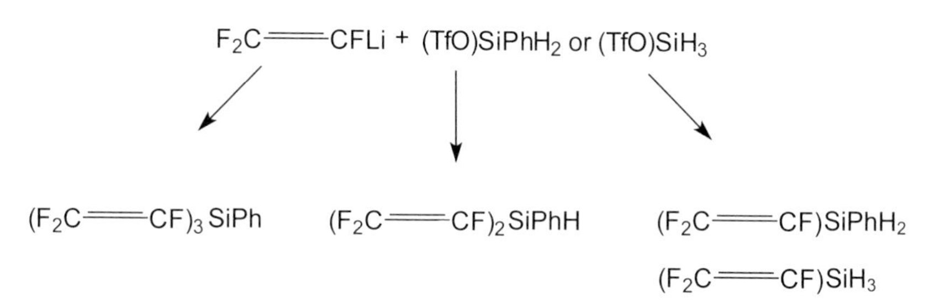

Scheme 3. Preparation of (trifluorovinyl)silanes.

Table 1 summarizes the $\delta(^{29}Si)$ chemical shifts referenced against tetramethylsilane as well as important coupling constants for all compounds.

Table 1. ^{29}Si-chemical shifts [ppm against TMS] and coupling constants [Hz] for (trichlorovinyl)- and (trifluorovinyl)silanes.

Compound	$\delta(^{29}Si)$	$^1J(SiH)$	Compound	$\delta(^{29}Si)$	$^1J(SiH)$
$Cl_2C{=}CClSiCl_3$	−12.9		$Cl_2C{=}CClSiPh_2OH$	−19.7	
$Cl_2C{=}CClSiF_3$	−85.6		$Cl_2C{=}CClSiPh_2H$	−21.3	217
$Cl_2C{=}CClSiH_3$	−58.4	220	$Cl_2C{=}CClSiPh(TfO)_2$	−41.4	
$Cl_2C{=}CClSiPh_3$	−14.1		$Cl_2C{=}CClSiPhCl_2$	−3.6	
$Cl_2C{=}CClSiPh_2OTf$	−9.3		$Cl_2C{=}CClSiPhH_2$	−34.9	218
$Cl_2C{=}CClSiPh_2F$	−13.2		$Cl_2C{=}CClSiPhH(TfO)$	−16.5	280
$Cl_2C{=}CClSiPh_2Cl$	−3.9		$Cl_2C{=}CClSiPhHCl$	−13.8	261
$Cl_2C{=}CClSiPh_2Br$	−5.2		$Cl_2C{=}CClSiH(TfO)_2$	−45.8	351
$Cl_2C{=}CClSiHCl_2$	−12.8	319	$(CCl3{-}CCl_2)SiCl_3$	−11.3	
$Cl_2C{=}CClSiBrH_2$	−32.3	262	$(CCl_3{-}CCl_2)_2SiCl_2$	−13.7	
$(Cl_2C{=}CCl)_2SiCl_2$	−13.3		$F_2C{=}CFSiH3$	−75.1	218
$Cl_2C{=}CClSiCl(CCl_2{-}CCl_3)$	−11.0		$(F_2C{=}CF)_3SiPh$	−36.9	

Attempts to polymerize $C_2Cl_3SiH_3$ by dehydrogenative coupling with $Cp_2ZrCl_2/2BuLi$, $Cp_2TiCl_2/2BuLi$ and $Cp_2ZrMe(Si(SiMe_3)_3)$ were unsuccessful due to desactivation of the catalyst by chlorination of the transition metal to give Cp_2MCl_2. The reaction of $C_2F_3SiH_3$ with $Cp_2ZrCl_2/2BuLi$ led to undefined products.

Acknowledgment: Financial support by the *Österreichische Nationalbank* is gratefully acknowledged.

References:

[1] G. Fritz, K.-H. Schmid, *Z. Anorg. Allg. Chem.* **1978**, *441*, 125.
[2] H. Beckers, H. Bürger, P. Bursch, I. Ruppert, *J. Organomet. Chem.* **1986**, *316*, 41.
[3] G. V. Motsarev, R. V. Dzhagatspanyan, A. D. Snegova, *Zh. Obshch. Khim.* **1968**, *38*, 1186.

Electrochemical Formation of Silicon–Carbon Bonds

Helmut Fallmann and Christa Grogger*

Institut für Anorganische Chemie, Technische Universität Graz
Stremayrgasse 16, A-8010 Graz, Austria
Tel.: Int. code + (316)873 8217 — Fax: Int. code + (316)873 8701
E-mail: grogger@anorg.tu-graz.ac.at

Keywords: Electrochemistry / Cyclic Voltammetry / Chlorosilanes

Summary: Using a simple undivided cell with THF/Bu_4NBF_4 as electrolyte, a stainless steel cathode and magnesium or hydrogen anodes, several organochlorosilanes R_nSiCl_{4-n} (R = Me, Ph; n = 1–3) were subjected to galvanostatic electrolysis in the presence of organic halides. With *t*-butyl chloride no Si–C bond formation was observed. Electrolyses with chlorobenzene yielded phenylated silanes when using Me_3SiCl or Me_2SiCl_2 as starting materials.

Introduction

The electrochemical coupling of chlorosilanes with organic halides appears to be an attractive pathway for the synthesis of organosilicon compounds. Regarding industrial applications, the use of our specially developed hydrogen anode [1] is of particular interest, as it produces HCl instead of metal salts. Thus, starting from a work of Bordeau et al. on the trimethylsilylation of *o*-dichlorobenzene [2], we tested the applicability of electrochemical reduction for the synthesis of phenylated and *t*-butyl-substituted silanes.

Electrolyses

Galvanostatic electrolyses were carried out at a current density of 1 mA/cm^2, using a simple, undivided cell with a stainless steel cathode and a magnesium or a hydrogen anode [1]. THF/Bu_4NBF_4 (0.2 M) was used as electrolyte.

Assuming an anionic intermediate due to a two-electron transition, there are three possible reaction products (Eq. 1). Competition between C–C, Si–C and Si–Si bond formation is dependent on the reduction potentials of the starting materials on the one hand, and their electrophilicity on the other. Thus, given that a chlorosilane is a better electrophile than an organic halide [2], cross-coupling can occur only if cathodic cleavage of the C–Cl bond is the initial reaction step. Comparing cyclic voltammetric data (Table 2, below) on the basis of this assumption, *t*-butyl chloride should be even better suited as the substrate than chlorobenzene.

$$\mathrm{>Si{-}Cl} + \mathrm{Cl{-}R} \longrightarrow \begin{cases} \mathrm{R{-}R} + n\,\mathrm{MCl}_x \\ \mathrm{>Si{-}R} + n\,\mathrm{MCl}_x \\ \mathrm{>Si{-}Si<} + n\,\mathrm{MCl}_x \end{cases} \qquad \mathrm{M = Mg, H_2,}\ \mathrm{R = Ph,}\ t\text{-Bu}$$

Eq. 1.

Coupling with Chlorobenzene

Electrolyses of several chlorosilanes with PhCl indicate that competition between Si–Si and Si–C bond formation is dependent not only on the reduction potentials of the starting silanes, but also on steric or kinetic effects on the electrode surface. Although the reduction potentials of phenylated chlorosilanes lie very close to those of methylated ones [3], Si–C bond coupling was successful with Me_3SiCl and Me_2SiCl_2 (Table 1) only.

Table 1. Electrolyses with PhCl using a Magnesium or a hydrogen anode.

Silane	Anode	Equiv. of PhCl	Main Product	Byproduct
Me_3SiCl	Mg, H	1	Me_3PhSi	C_6H_6
Me_2SiCl_2	Mg, H	1	$Me_2PhSiCl$	Me_2Ph_2Si, C_6H_6
Me_2SiCl_2	Mg, H	2	Me_2Ph_2Si	$Me_2PhSiCl$, C_6H_6
$MePhSiCl_2$	Mg	1	$(MePhSi)_n$	–
Ph_2SiCl_2	Mg	1	$[Ph_2Si]_4$	–
$PhSiCl_3$	Mg	1	Insoluble polymer	–

The formation of benzene occurs due to abstraction of H^+ or $H^\bullet$ from the solvent or the supporting electrolyte. Thus, electrolysis of PhCl in absence of any chlorosilane yielded biphenyl and benzene.

Coupling with *t*-Butyl Chloride

Electrolyses in the presence of 1 equivalent of *t*-butyl chloride did not yield the corresponding butylsilanes.

$$\mathrm{>Si{-}Cl} + t\text{-BuCl} \longrightarrow (t\text{-Bu})_2 + \mathrm{>Si{-}Cl} + \ldots$$

Eq. 2.

Due to cathodic dehalogenation as well as dehydrohalogenation of *t*-BuCl and subsequent C–C bond coupling, 2,2,3,3-tetramethylbutane, 2,2,4-trimethylpentane and homologues were detected as the only reaction products (Eq. 2).

Cyclic Voltammetry

Cyclic voltammetry studies were carried out in a THF/Bu_4NBF_4 (0.1 M) solution, using a platinum disk microelectrode (500 μm), a Pt counter electrode and a Ag/AgCl reference electrode. Ferrocene was added at the end of each experiment and used as an internal standard.

Table 2. Reduction potentials E_{pc} vs. Fc/Fc^+.

Compound	E_{pc} [V] Literature	Found[b]
HCl	– 0.6 to –1.2 [3]	– 1.2
t-BuCl	–	– 1.6
Me_2SiCl_2	– 2.6 [3][a]	– 2.5
PhCl	–	– 3.0
Me_3SiCl	> – 3.2 [1]	> – 3.5

[a] in CH_3CN/Bu_4NBPh_4 (0,1M); E_{pc} vs. SCE, v = 500 mV/s. — [b] v = 250 mV/s.

Acknowledgment: The authors would like to thank *Wacker-Chemie GmbH*, Burghausen, for the friendly donation of chlorosilanes.

References:

[1] C. Jammegg, S. Graschy, E. Hengge, *Organometallics* **1994**, *13*, 2397.

[2] D. Deffieux, M. Bordeau, C. Biran, J. Dunoguès, *Organometallics* **1994**, *13*, 2415.

[3] C. Biran et al., *J. Chi m. Phys.* **1996**, *93*, 591.

Heteroaromatic-Substituted Silanes — Synthesis, Lithium Derivatives and Anionic Rearrangements

Claudia Baum, Andrea Frenzel, Uwe Klingebiel*, Peter Neugebauer

Institut für Anorganische Chemie, Georg-August-Universität
Tammannstr. 4, D-37077 Göttingen, Germany
Tel.: Int. code + (551)39 3052 — Fax: Int. code + (551)39 3373
E-mail: uklinge@gwdg.de

Keywords: Indoylsilanes / Pyrroylsilanes / Furanylsilanes / 1,2-Silyl Shift / Macrocycle

Summary: Pure and mixed substituted indoyl-, pyrroyl-, and furanyl-silanes are formed in the reaction of heteroaromatic compounds with lithium alkyls and halosilanes (**1–3**). Chains of silyl-bridged molecules are available (**7**). The crystal structure of a lithium derivative (**10**) is presented and the 1,2-silyl group migration from nitrogen to carbon is proved. Formation of a 16-membered macrocycle (**12**) containing four indole molecules is described.

Introduction

Only a few papers on silyl derivatives of heteroaromatics, e.g. indoles and pyrroles, have been published [1–7]. After the synthesis of mono-, bis-, tris-, and tetrakis(indol-1-yl)silanes and (pyrrol-1-yl)silanes, this chemistry is extended to furan compounds [7, 8]. A further interest is to couple the heterocycles with bridging silyl groups to obtain chains and cyclic molecules.

The synthesis of 1,2-bis(silyl)indoles via 1,2-silyl-group migration of N-silylated indoles is proved by X-ray determination [9, 10].

Results and Discussion

Lithiation of furan and reaction with SiF_4 leads to tetrakis(furan-2-yl)silane (**1**) (Scheme1).

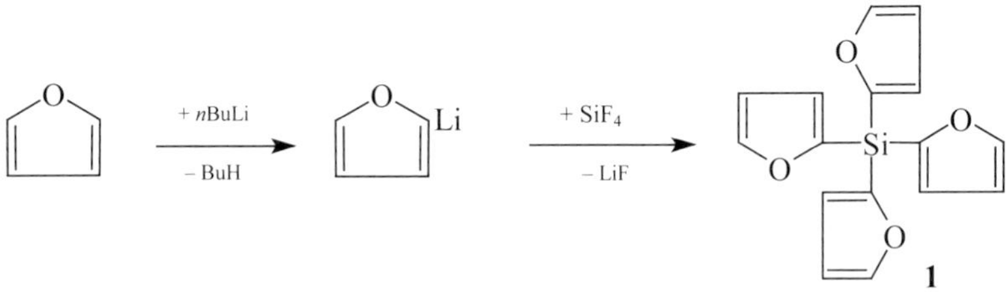

+ *n*BuLi, − BuH; + SiF_4, − LiF; Li; Si; O; **1**

Scheme 1. Synthesis of **1**.

1 (Fig. 1) crystallizes from *n*-hexane as colorless crystals.

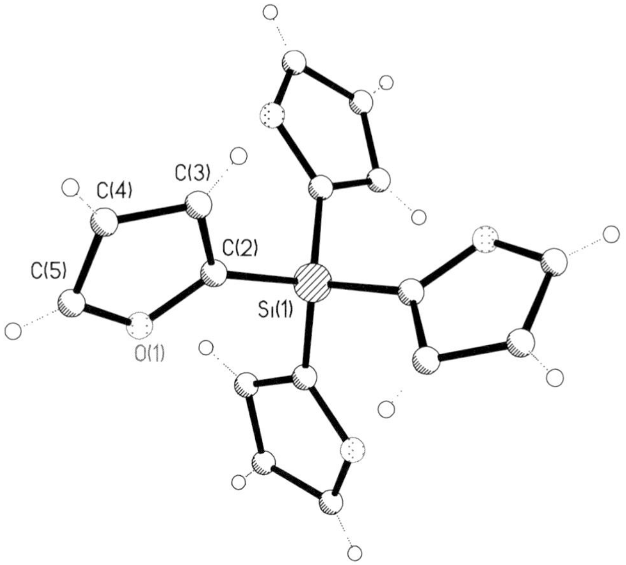

Fig. 1. Structure of **1**.

Mixed substituted indoyl-, pyrroyl-, and furanylsilanes were also achieved (Schemes 2 and 3).

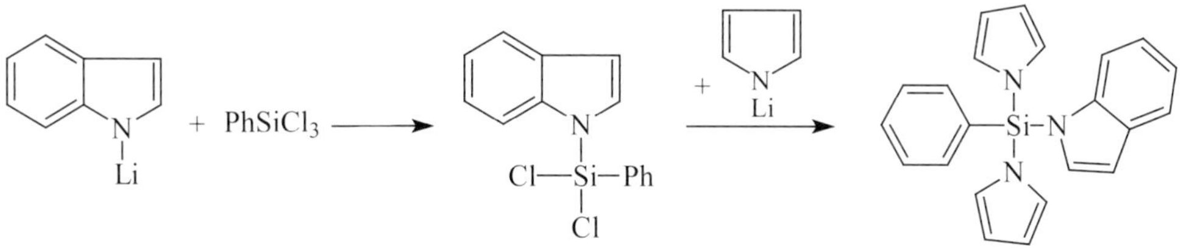

Scheme 2. Synthesis of **2.**

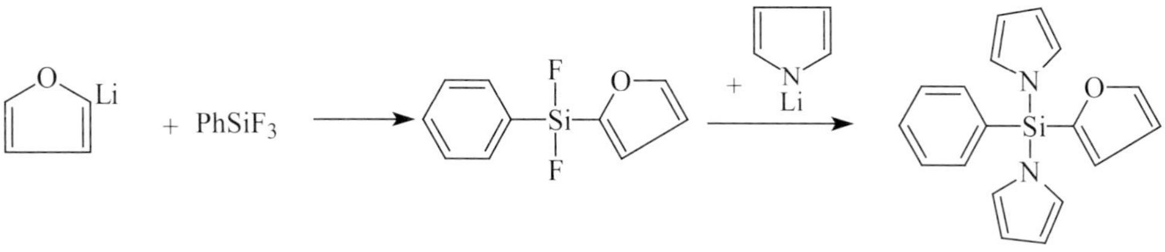

Scheme 3. Synthesis of **3.**

Starting with halosilyl-substituted compounds, chains of silyl-bridged molecules are available (Scheme 4).

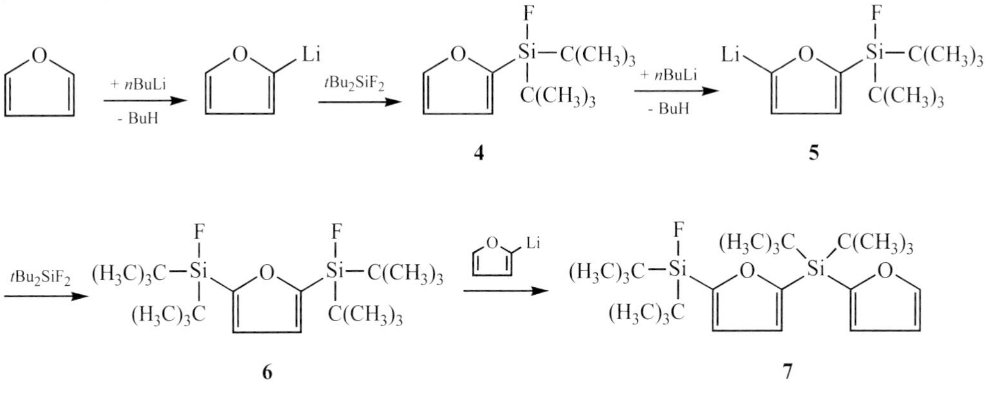

Scheme 4. Synthesis of **4**, **5**, **6**, and **7**.

To prepare cyclic molecules, **6** has been transformed (Scheme 5) into the diol **8** (Fig. 2) and diamine **9**.

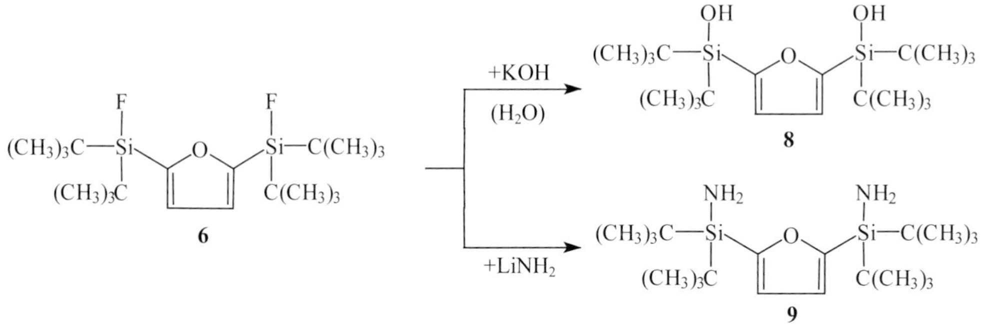

Scheme 5. Synthesis of **8** and **9**.

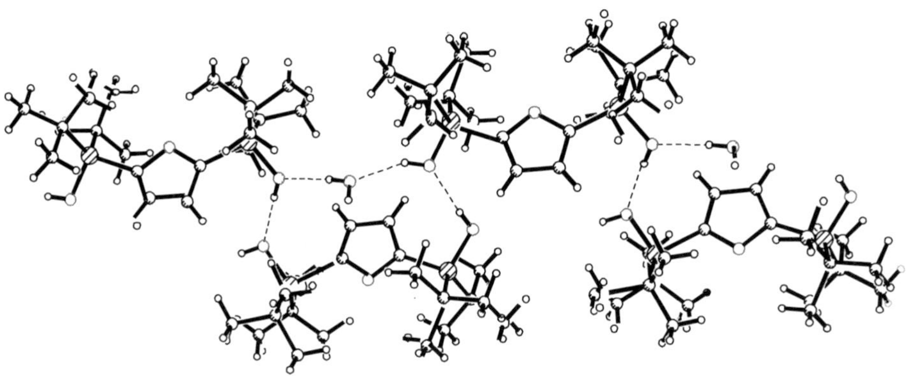

Fig. 2. Structure of **8**.

Another route to synthesize heteroaromatic silyl-bridged cyclic systems is the use of 1,2- and 1,3-bis(silyl)indoles [6, 10]. N-silyl-substituted indoles react with lithium alkyls via two pathways:

(1) In the case of bulky silyl substituents lithiation in TMEDA leads to an anionic 1,2-silyl shift. This is the first 1,2-fluorosilyl shift observed in a silyl–indole system. The second substitution occurs at the nitrogen atom (Scheme 6).

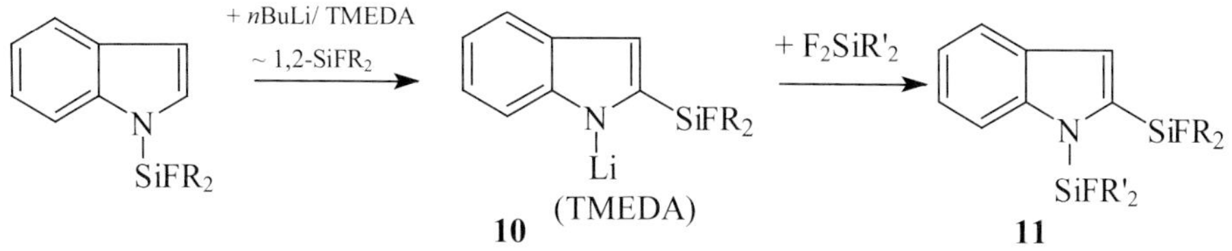

$SiFR_2$ —(+ *n*BuLi/ TMEDA, ~ 1,2-$SiFR_2$)→ **10** [N–Li (TMEDA), 2-$SiFR_2$] —(+ $F_2SiR'_2$)→ **11** [N–$SiFR'_2$, 2-$SiFR_2$]

Scheme 6. Lithiation in TMEDA: Synthesis of **10** and **11**.

In case of R= $C(CH_3)_3$ and R' = $CH(CH_3)_2$ crystal structures of **10** (Fig. 3) and **11** (Fig. 4) were determined.

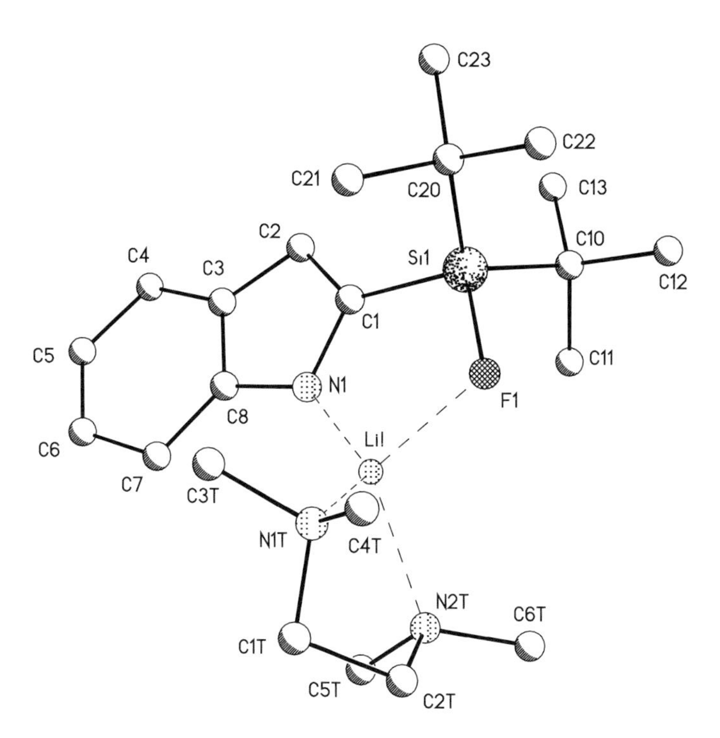

Fig. 3. Structure of **10**.

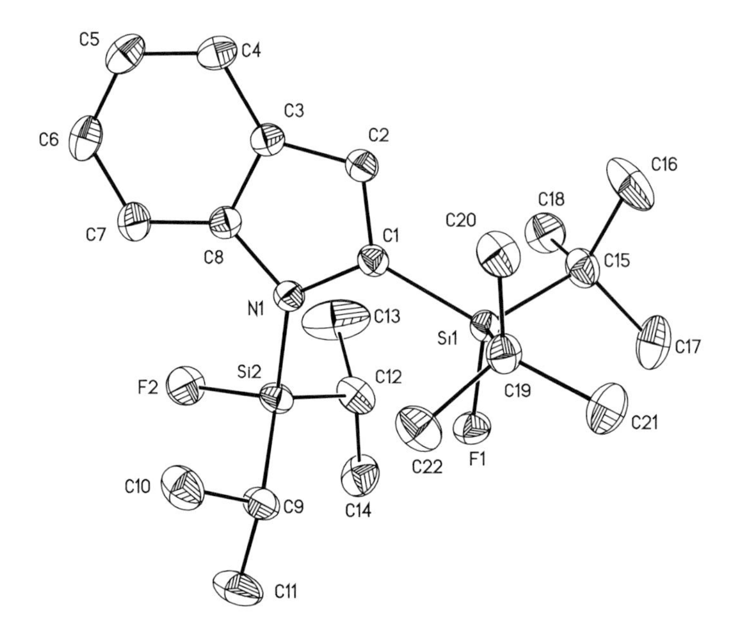

Fig. 4. Structure of **11**.

(2) Lithiation without TMEDA leads to the 1,3-bis(silyl)-substituted compound (Scheme 7).

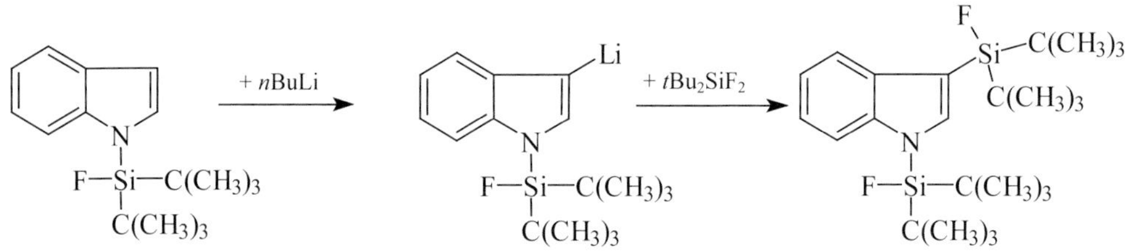

$+ n\text{BuLi}$ $+ t\text{Bu}_2\text{SiF}_2$

F–Si(C(CH$_3$)$_3$)$_2$; Li ; F–Si(C(CH$_3$)$_3$)$_2$

Scheme 7. Lithiation without TMEDA with formation of 1,3-bis(silyl)indole.

Bis(lithium di-*iso*-propylsilylindolide) reacts with two equivalents of difluorodimethylsilane with formation of a 16-membered macrocycle **12** (Scheme 8).

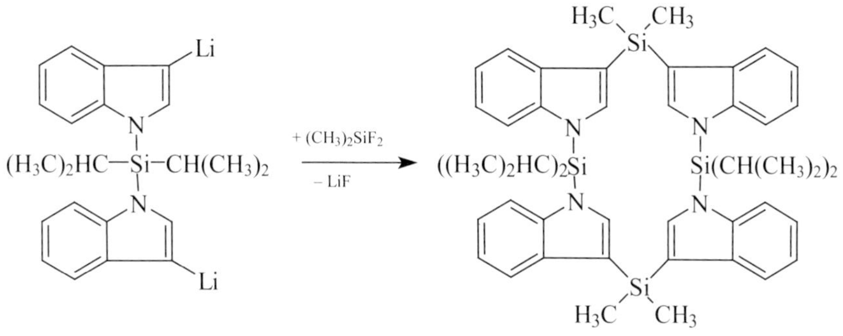

12

Scheme 8. Synthesis of **12**.

References:

[1] A. Weissberger, E. C. Taylor, *The Chemistry of Heterocyclic Compounds — Indoles*, Wiley–Interscience, New York, **1972**.

[2] D. A. Shirley, P. A. Roussel, *J. Am. Chem. Soc.* **1953**, *75*, 375.

[3] R. J. Sundberg, H. F. Russel, *J. Org. Chem.* **1973**, *38*, 3324.

[4] A. M. Barret, D. Dauzonne, I. A. O'Neil, A. Renaud, *J. Org. Chem.* **1984**, *49*, 4409.

[5] U. Frick, G. Simchen, *Synthesis Comm.* **1984**, 929.

[6] U. Klingebiel, W. Lüttke, M. Noltemeyer, H.G. Schmidt, *J. Organomet. Chem.* **1993**, *456*, 41.

[7] A. Frenzel, R. Herbst-Irmer, U. Klingebiel, M. Noltemeyer, M. Schäfer, *Z. Naturforsch., Teil B* **1995**, *50*, 1658–1664.

[8] U. Klingebiel, W. Lüttke, M. Noltemeyer, *J. Organomet. Chem.* **1993**, *455*, 51–55.

[9] A. Frenzel, U. Klingebiel, W. Lüttke, U. Pieper, *J. Organomet. Chem.* **1994,** *476*, 73–76.

[10] A. Frenzel, R. Herbst-Irmer, U. Klingebiel, M. Noltemeyer, S. Rudolph, *Main Group Chemistry, Vol. 1*, **1996**, pp 399–408.

Iodo Trimethylsilylmethylene Triphenylphosphorane — a Molecule of Theoretical and Synthetic Interest

*K. Korth, A. Schorm, J. Sundermeyer**
*H. Hermann, G. Boche**

Fachbereich Chemie, Philipps – Universität Marburg
Hans-Meerwein-Straße, D-35032 Marburg/Lahn, Germany

Keywords: Carbanion Stabilization / Phosphorus Ylides / Structure Determinations / Silyl Group Effects

Summary: In phosphorus ylides such as iodo trimethylsilylmethylene triphenylphosphorane all the carbon substituents (iodine, $SiMe_3$, PPh_3^+) are involved in carbanion stabilization. The aim of this work was to elucidate the competition of the carbon ligands in stabilizing the carbanion center. For that reason we synthesized the compounds $Ph_3P{=}C(I)SiMe_3$ **1**, $Ph_3P{=}C(I)H$ **2** and $Ph_3P{=}C(H)SiMe_3$ **3** in which the substituents (H, I, $SiMe_3$) at the ylidic carbon are systematically interchanged. As a result of our combined solid-state and quantum-chemical studies we found that hyperconjugative stabilization between the ylidic carbon and the phosphonium group is less significant due to the effect of other electron-withdrawing ligands. In particular, the stabilization effect of iodine was found to be of greater influence than expected.

Introduction

The electronic structure of phosphorus ylides has been a topic of current synthetic and theoretical interest [1–4]. Most studies describe phosphorus ylides as pyramidalized "carbanions" which are intramolecularly stabilized by a phosphonium group and additionally by suitable substituents X and Y (Fig.1, **A**). Electron-withdrawing substituents X and Y lead to a subsequent planarization (Fig.1, **B**) at the carbon, while the hyperconjugative stabilization of the phosphonium group loses significance. A further description is an almost planar carbon center (Fig.1, **C**). In that case the interaction between carbon and phosphorus is best described as a double bond. A systematic interchange of the substituents H, I and $SiMe_3$, respectively, at the methylene carbon led to compounds $Ph_3P{=}C(I)SiMe_3$ **1**, $Ph_3P{=}C(I)H$ **2** and $Ph_3P{=}C(H)SiMe_3$ **3** [5–7].

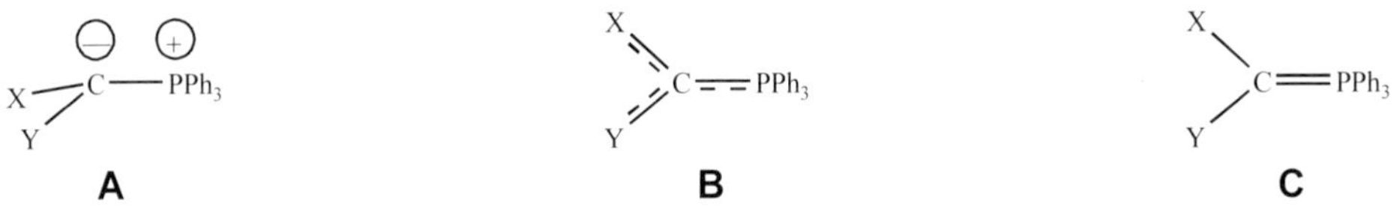

Fig. 1. Different discription of the electronic structure of phosphorus ylides.

This allows the determination of the influence of each substituent on the ylidic C_1–P bond. In particular the interaction of the carbanion center with iodine seems worth studying, because of the possible stabilization of a negative charge by iodine.

Furthermore, compound **1** is an excellent precursor for lithiation reactions and thus for further derivatization. The halogen/metal exchange reaction of **1** to give **4** is shown by the reaction of the latter with trimethylsilyl chloride to give **5** (Scheme 1).

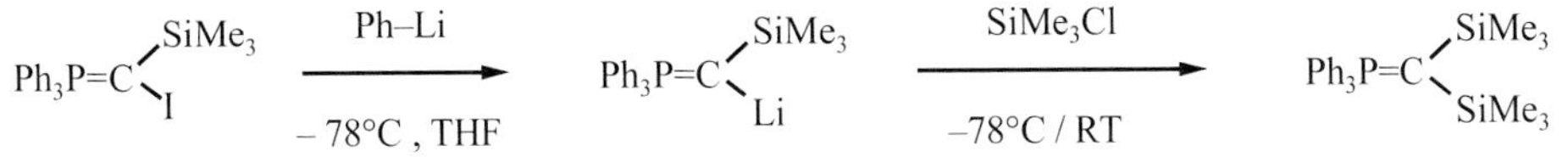

$Ph_3P{=}C(SiMe_3)(I)$ (**1**) —[Ph–Li; – 78°C , THF]→ $Ph_3P{=}C(SiMe_3)(Li)$ (**4**) —[$SiMe_3Cl$; –78°C / RT]→ $Ph_3P{=}C(SiMe_3)(SiMe_3)$ (**5**)

Scheme 1. Derivatization of the phosphorus ylid **1**.

Discussion of Results

We obtained X-ray crystal data for **1**, **2** and **3**. The structural data were compared with the results of ab initio computational studies (MP2 (full) level of theory [13]) of suitable model compounds in which the phenyl substituents at phosphorus are substituted by vinyl groups.

According to earlier theoretical results ylides have a slightly pyramidalized ylidic carbon atom (Fig. 2). The "lone pair" at carbon is oriented in an antiperiplanar fashion to one of the phosphorus ligands, the so-called "unique substituent" [4]. This conformation allows for a hyperconjugative interaction of the "lone pair" at carbon with the antibonding orbital of the $P–R_{unique}$ bond [8]. The lengthening of the $P–R_{unique}$ bond is a sensitve probe for the hyperconjugative interaction as well as the increasing C_{ylidic}-P-R_{unique} bond angle. This specific carbanion stabilization of phosphorus ylides is also well documented in the solid-state structures of **1**–**3**.

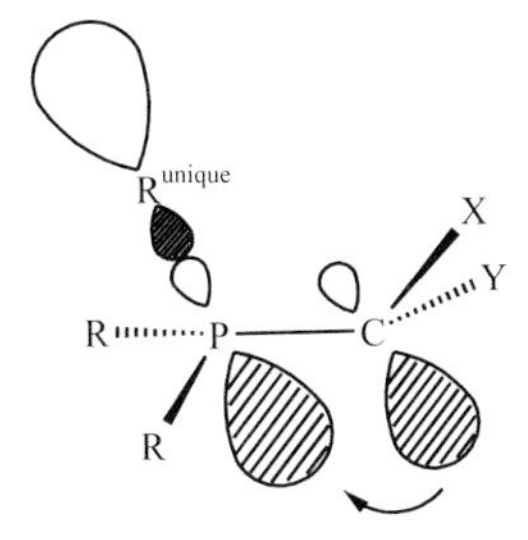

Fig. 2. The slightly pyramidalized ylidic carbon atom of phosphorus ylides.

Solid-State Structure of Iodo Trimethylsilylmethylene Triphenylphosphorane (1)

In **1** ($SiMe_3$ and I at the C_1 atom) the ylidic C_1 is substituted with a iodine and an $SiMe_3$ group besides the common PPh_3 fragment. The phenyl group C_{20}–C_{25} of **1** is ideally oriented for a hyperconjugative interaction (see Fig. 3), but there is only a negligible bond lengthening of the P–C_{20} bond. This is in agreement with observations by Schmidpeter and Nöth [9], who found that in the ylide **6** (see Scheme 2) similarly there is no bond lengthening.

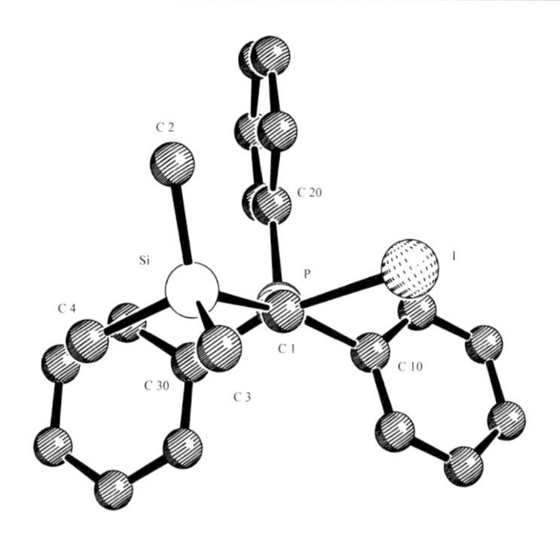

Fig. 3. Structure of **1**, viewed along C_1–P axis. For clarity hydrogene atoms are omitted.

Table 1. Selected bond lengths and angles of $Ph_3P{=}C(I)SiMe_3$ (**1**) in comparison with the calculated structure of $(vinyl)_3P{=}C(I)SiMe_3$ ($\mathbf{1}_{calc.}$).

Bond	Length [pm]	Calc. length [pm]
C_1–P	170.1(4)	168.9
C_1–I	214.1(4)	215.9
C_1–Si	185.4(4)	184.8
P–C_{10}	182.9(4)	180.5
P–C_{20}	183.0(4)	181.7
P–C_{30}	182.5(4)	180.2
Si–C_2	185.5(5)	189.1
Si–C_3	187.2(4)	189.0
Si–C_4	184.6(5)	188.6
Angle	**Value [°]**	**Calc. value [°]**
C_1-P-C_{10}	110.4(2)	114.0
C_1-P-C_{20}	120.0(2)	118.6
C_1-P-C_{30}	107.7(2)	111.5
Dihedral angle	**Value [°]**	**Calc. value [°]**
(Si-C_1-P) (P-C_1-I)	146.2	165.0

$SiMe_3$ Me_3Si P $Ph_3P{=}C$ P P $C{=}PPh_3$ P $SiMe_3$ $SiMe_3$

6

O $Ph_3P{=}C$ C Ph I

7

Scheme 2. Stabilized phosphorus ylides **6** [9] and **7** [11].

The C_1-P-C_{20} bond angle of **1** behaves quite normally and is increased as expected. The Si–C_1 bond (184.4 pm, see Table 1) is not significantly shortened and is longer than most silicon–carbon bonds in related silicon-stabilized lithium carbanions (182.3 pm) [10]. The observed bonding distance of 185.4 pm (C_1–Si) is in between the values of sp^3 and sp^2 hybridized Si–C bonds (average values: Si–C_{sp^3} bond 186.5 pm; Si–C_{sp^3} bond 184.0 pm) [14]. Probably due to steric interactions with the bulky phosphonium group the silicon center is not ideally tetrahedral. Interestingly the bond length of the Si–CH_3 bonds are found between 184.6 pm and 187.2 pm. The C–I bond is ~5 pm shorter than in **7** (see Scheme 1) [11] (C–I 219 pm), and so it is out of the range of the average value of C_{sp^3}–I bond (218–220 pm). The ylidic C_1–P bond is slightly longer than found in CH_2=PPh_3 (169.2 pm) [12]. The calculated values (see Table 1) are in good agreement with the experimental ones of the measured structure **1**. In general bond lengths differ by <2 pm, bond angles by <4°. A comparatively large difference is found for the dihedral angles (Si-C_1-P)(P-C_1-I), which could be explained by the different steric demands of the P-phenyl ligands in **1** as compared to the P–vinyl ligands in $(vinyl)_3P$=C(I)$SiMe_3$ **1**$_{calc.}$ used in the model calculations.

Solid-State Structure of Iodo Methylene Triphenylphosphorane (2)

In **2** (H and I at the C_1 atom) a stronger pyramidalization at the ylidic carbon center as compared to the silylated analogue **1** is observed (see Fig. 4 and Table 2). The dihedral angle (P-C_1-I)(P-C_1-H) (130.0°) is much smaller than the one found in **1** (151.9°). The stronger pyramidalization is also documented by NBO (natural bond orbital) analysis of the anionic carbon orbital, which shows a higher s- and a lower p-contribution in the case of **2**. There are two possibilities for an interpretation: first, there might be a steric influence because hydrogen is much smaller than $SiMe_3$ which leads to less steric hindrance; and secondly, there might be much more likely electronic reasons. From work of Schleyer et al. [2] it is known that non-metallated carbanions, stabilized by α-silicon substituents, tend to a "planarization". From our results it is not possible to distinguish between these two possibilities.

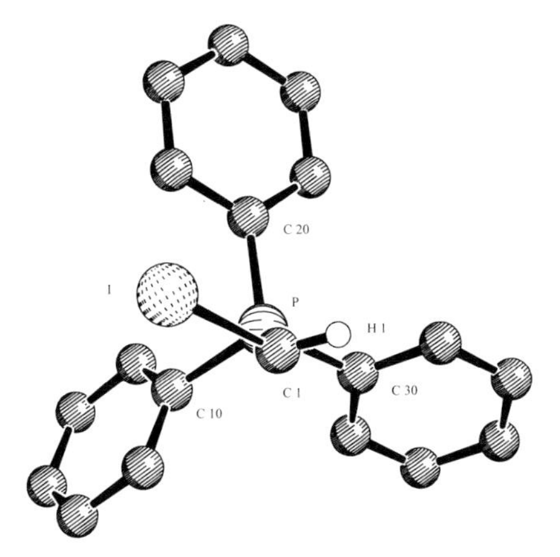

Fig. 4. Structure of **2**, viewed along C_1–P axis. For clarity hydrogene atoms are omitted.

Table 2. Selected bond lengths and angles of $Ph_3P{=}C(I)H$ (**2**) in comparison with the calculated structure of $(vinyl)_3P{=}C(I)H$ ($\mathbf{2}_{calc.}$).

Bond	Length [pm]	Calc. length [pm]
C_1–P	169.9(3)	169.2
C_1–I	212.2(3)	213.1
C_1–H_1	92(4)	108.2
P–C_{10}	180.6(2)	181.0
P–C_{20}	181.9(3)	183.4
P–C_{30}	180.4(3)	181.3
Angle	**Value [°]**	**Calc. value [°]**
C_1-P-C_{10}	110.8(1)	110.7
C_1-P-C_{20}	120.8(1)	123.6
C_1-P-C_{30}	105.3(1)	106.9
Dihedral angle	**Value [°]**	**Calc. value [°]**
(I-C_1-P) (P-C_1-H)	132.2	141.3

2 further reveals a longer P–C_{20} bond of the "unique" phenyl substituent as compared to the other phosphorus–carbon bonds (P–C_{10}, P–C_{30} and an increased bond angle C_1-P-C_{20}), both in agreement with theory [4] (see Table 2). The C–I bond (212.2 pm) in **2** is significantly shorter than in the silylated **1** (214.1 pm) and benzoylated **7** (219 pm) analogues, respectively.

Thus the stabilization effect of an iodine atom attached to an ylidic carbon atom in phosphorus ylides increases in the order **7** < **1** < **2**:

$$\underset{\mathbf{7}}{[Ph_3P{=}C(I)(CO)Ph]} < \underset{\mathbf{1}}{[Ph_3P{=}C(I)(SiMe_3)]} < \underset{\mathbf{2}}{[Ph_3P{=}C(I)H]}$$

Solid-State Structure of Trimethylsilylmethylene Triphenylphosphorane (3)

In **3** (H and $SiMe_3$ at the C_1 atom) a shortening of the C_1–Si bond of ~8 pm as compared to average values of Si–C bonds is observed. This is quite remarkable because the C_1–Si bond length (177.3 pm) is even shorter than in α-silyl organolithium compounds (~182.3 pm) [10].

There are two possible explanations: the first emphasizes that iodine stabilizes the ylidic carbon atom as discussed in the preceding section, the second concerns the much lower steric demand of hydrogen compared to iodine so the bulky trimethylsilyl group is able to move closer to the carbanionic center and stabilizes this center more efficiently. However, our findings are in good agreement with calculations of Schleyer et al. of α-silylated carbanions with and without the gegenions Li^+ and Na^+ [2].

Due to the strong carbon–silicon interactions which are documented by the extremely short C_1–Si bond, it is anticipated that the ylidic bond (C_1–P) is lengthened in the solid-state structure **3** (Table 3). In the calculation there is only a slight shortening of the C_1–Si bond in $\mathbf{3}_{calc.}$ (183.7 pm) as compared to $\mathbf{1}_{calc.}$ (184.8 pm) and so after all the lengthening effect of the C_1–P bond is not reproduced. In $\mathbf{3_{calc.}}$ the C_1–P bond is even shorter than in **1** (168.9 pm) and **2** (169.2 pm). A possible reason is the difference in electronic and steric influences of the P–phenyl ligands in **3** as compared to the P–vinyl ligands in $\mathbf{3_{calc.}}$.

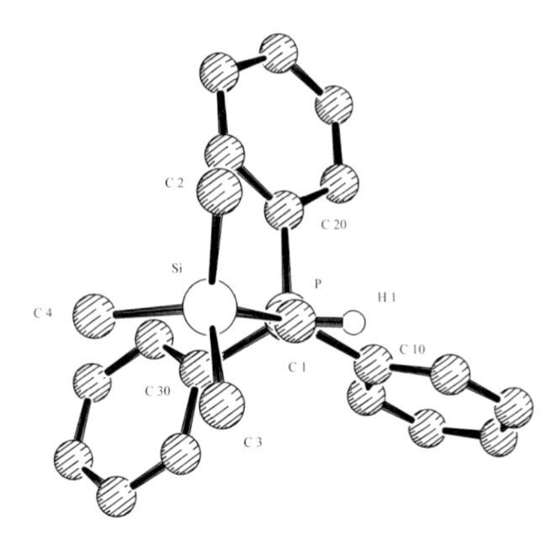

Fig. 5. Structure of **3**, viewed along C_1–P axis. For clarity hydrogene atoms are omitted.

Table 3. Selected bond lengths and angles of $Ph_3P{=}C(H)SiMe_3$ (**3**) in comparison with the calculated structure of $(vinyl)_3P{=}C(H)SiMe_3$ ($\mathbf{3_{calc.}}$).

Bond	Length [pm]	Calc. length [pm]
C_1–P	167.0(2)	167.9
C_1–Si	181.5(2)	183.7
C_1–H	–	108.5
P–C_{10}	182.1(2)	180.6
P–C_{20}	183.2(2)	182.4
P–C_{30}	181.2(2)	180.6
Si–C_2	185.5(3)	188.8
Si–C_3	187.7(3)	189.4
Si–C_4	187.1(3)	189.5
Dihedral angle	**Value [°]**	**Calc. value [°]**
(Si-C_1-P) (P-C_1-H)	161.2	176.9

References:

[1] S. M. Bachrach, C. I. Nitsche, "*The Chemistry of Organophosphorus Compounds*" (Ed.: F. R. Hartle), Wiley, New York, **1994**, p. 273.

[2] P. v. R. Schleyer, T. Clark, A. J. Kos, G. W. Spitznagel, C. Rohde, D. Arad, K. N. Houk, N. G. Rondan, *J. Am. Chem. Soc.* **1984**, *106*(22), 6467.

[3] H. Schmidbaur, *Angew. Chem., Int. Ed. Engl.* **1983**, *22*(12), 907.

[4] D. G. Gilheany, *Chem. Rev.* **1994**, *94*(5), 1339.

[5] H. J. Bestmann, A. Bomhard, R. Dostalek, R. Pichl, R. Riemer, R. Zimmermann, *Synthesis* **1992**, 787.

[6] H. J. Bestmann, H. C. Rippel, R. Dostalek, *Tetrahedron Lett.* **1989**, *30*(39), 5261.

[7] G. Stork, K. Zhao, *Tetrahedron Lett.* **1989**, *30*(17), 2173.

[8] P. v. R. Schleyer, A. J. Kos, *Tetrahedron* **1983**, *39*, 1141.

[9] H. P. Schrodel, H. Nöth, M. Schmidt-Amelunxen, W. W. Schoeller, A. Schmidpeter, *Chem. Ber./Recueil* **1997**, *130*(12), 1801.

[10] W. Zarges, M. Marsch, K. Harms, W. Koch, G. Frenking, G. Boche, *Chem. Ber.* **1991**, *124*, 543.

[11] F. S. Stephens, *J. Chem. Soc.* **1965**, 5640.

[12] H. Schmidbaur, J. Jeong, A. Schier, W. Graf, D. L. Wilkinson, G. Müller, C. Krüger, *New J. Chem.* **1989**, 13(4–5), 341.

[13] Basis set used for calculations: Cl — 6-31+G(d); H* — 6-31++G(d,p); H, C, P, Si — 6-31G(d); I — (4111/4111/1) effective core potential (M. Dolg, A. Bergner, W. Küchle, H. Stoll, H. Preuss, *J. Chem. Phys.* **1988**, *65*(6), 1321).

[14] J. Y. Corey, "*The Chemistry of Organic Silicon Compounds. The Chemistry of Functional Groups*", (Eds.: S. Patai, Z. Rappoport), Wiley, Chichester, **1989**, p. 1.

Silole-Containing π-Conjugated Systems: Synthesis and Application to Organic Light-Emitting Diodes

Kohei Tamao*, Shigehiro Yamaguchi

Institute for Chemical Research, Kyoto University,
Uji, Kyoto 611-0011, Japan
Fax: Int. code + (774)38 3186
E-mail: tamao@scl.kyoto-u.ac.jp

Manabu Uchida, Takenori Izumizawa, Kenji Furukawa

Yokohama Research Center, Chisso Corporation,
Kanazawa-ku, Yokohama 236-8605, Japan
Fax: Int. code + (45)786 5512

Keywords: Siloles / π-Conjugated Compounds / Organic Light-Emitting Diodes/ 2,5-Diarylsiloles

Summary: Various types of silole-containing π-conjugated compounds, including 3,4-diphenyl-, 3,4-dialkyl-, and 3,4-unsubstituted 2,5-diarylsiloles, are prepared based on three new synthetic methods. Some 2,5-diarylsiloles thus prepared are applied to organic light-emitting diodes as new electron-transporting materials, emissive materials, or materials for single-layer devices.

Introduction

Silole (silacyclopentadiene) has recently received much attention as a new building unit for π-conjugated compounds because of its unique electronic structure [1–7]. A notable feature of silole is its high electron-accepting properties, i.e., its low-lying LUMO, which is ascribed to the σ^*-π^* conjugation in the ring [8]. Based on this electronic structure, silole-containing π-conjugated compounds have unique photophysical properties and high potential as new materials for organic light-emitting diodes (LEDs) [9]. This new area of chemistry has been significantly dependent on the developments of new synthetic methodologies for a variety of "tailor-made" silole derivatives. In this paper, we first describe the new synthetic methods which we have developed and then focus our discussion on the application of the 2,5-diarylsilole derivatives to organic LEDs.

New Synthetic Methodologies

The first general synthetic route to 2,5-difunctionalized siloles is the intramolecular reductive cyclization of diethynylsilanes [10]. Thus, the reduction of bis(phenylethynyl)silane **1** using an excess amount (4 molar amounts) of lithium naphthalenide (LiNaph) cleanly affords 2,5-dilithiosilole **3** (Scheme 1). By trapping with various electrophiles, the 2,5-dilithiosiloles are transformed into a series of 2,5-difunctionalized siloles **4**.

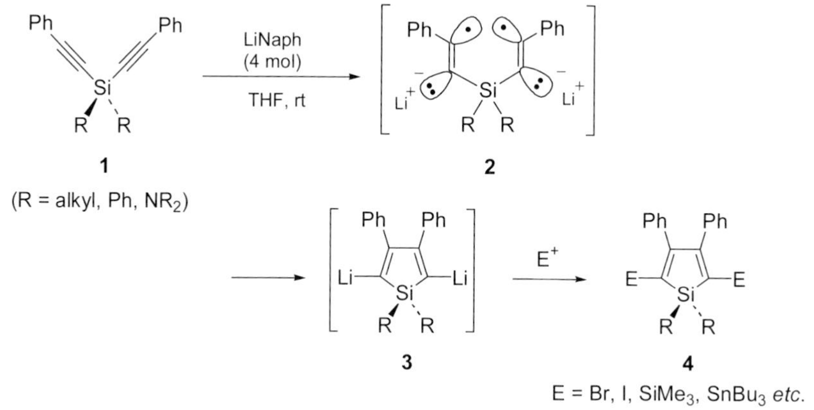

Scheme 1. Synthesis of 2,5-difunctionalized siloles **4** by reduction of bis(phenylethynyl)silane **1**.

The key point to attain high yields in the present reaction is the dropwise addition of the diethynylsilane into an "electron pool" consisting of an excess amount of reductant, and thereby both acetylene moieties are simultaneously reduced to form the bis(anion radical) intermediate **2** that undergoes radical coupling to form the 3,4-carbon–carbon bond, leaving anions at the 2,5-positions. The phenyl group at the terminal position of acetylene is also essential to obtain the dilithiosiloles. In the case of other substituents such as alkyl and silyl groups, only a complex mixture is formed, probably due to the cleavage of the Si–C_{sp} bond prior to the formation of the dilithiosilole.

This procedure can be applied to the synthesis of a number of silole-containing π-conjugated compounds [2]. Inter alia, the combination of the present cyclization with the Pd(0)-catalyzed cross-coupling reaction enables us to prepare a series of 2,5-diarylsiloles **6** in a one-pot manner from the diethynylsilanes (Scheme 2) [9, 11].

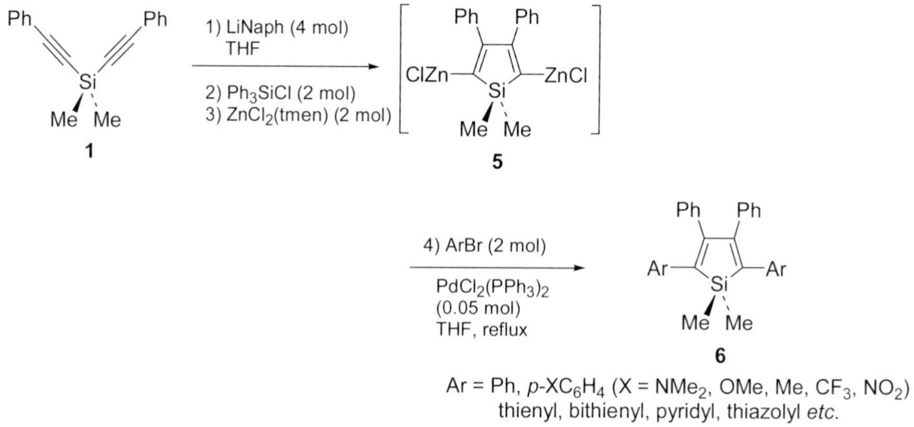

Scheme 2. One-pot synthesis of 2,5-diarylsiloles **6**.

Thus, the intramolecular reductive cyclization of diethynylsilane, followed by quenching with the remaining lithium naphthalenide with a bulky chlorosilane and transmetalation with $ZnCl_2$(tmen) (tmen = *N,N,N'N'*-tetramethylethylenediamine), affords 2,5-dizinc silole **5**, which is subsequently treated with the appropriate aryl bromides in the presence of a Pd catalyst to give the corresponding 2,5-diarylsiloles **6** in high yields.

Despite its versatility, the present method has a crucial limitation, that is, the restriction of the 3,4-substituents to only the phenyl groups as mentioned above. In order to compensate for this limitation, we have developed the following two alternative routes.

One is the synthesis of 3,4-unsubstituted siloles **8** from the corresponding tellurophenes **7** via the well-documented tellurium-lithium exchange reaction, as shown in Scheme 3 [12]. A variety of 2,5-diarylsiloles can be prepared by this method, which include unsymmetrical ones and 1,1-spirobisilole **9**.

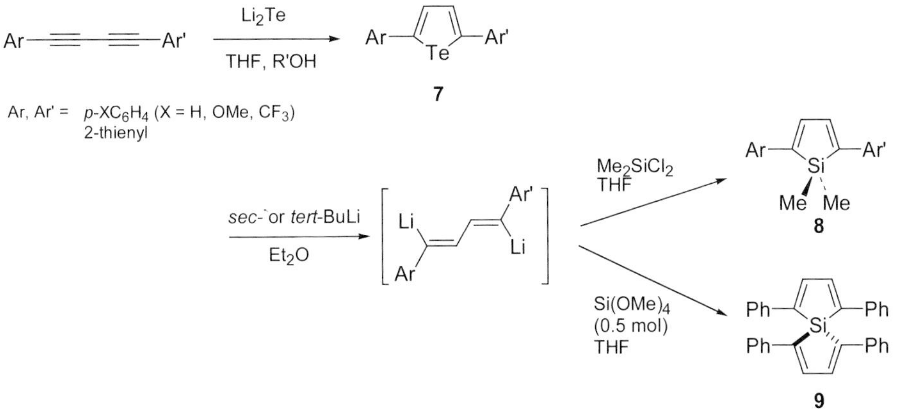

Ar—≡—≡—Ar' → (Li_2Te; THF, R'OH) → **7**

Ar, Ar' = *p*-XC_6H_4 (X = H, OMe, CF_3), 2-thienyl

7 → (*sec*- or *tert*-BuLi; Et_2O) → [dilithio diene] → (Me_2SiCl_2, THF) → **8**; or → ($Si(OMe)_4$ (0.5 mol), THF) → **9**

Scheme 3. Synthesis of 3,4-unsubstituted siloles **8** and 1,1-spirobisilole **9**.

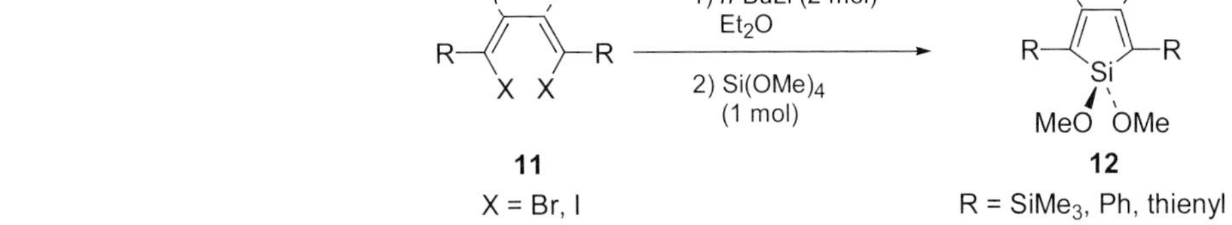

diyne or 1-alkynes → ($Ti(Oi\text{-}Pr)_4$ (1 mol), *i*-PrMgCl (2 mol); Et_2O, –50°C) → [**10**] → (I_2 (2 mol) or Br_2 (2 mol)) →

11 (X = Br, I) → (1) *n*-BuLi (2 mol), Et_2O; 2) $Si(OMe)_4$ (1 mol)) → **12** (R = $SiMe_3$, Ph, thienyl)

Scheme 4. Synthesis of 3,4-dialkyl and 3,4-unsubstituted siloles **12**.

The other route, shown in Scheme 4, involves the preparation of 1,4-diiodobutadienes **11** by the halogenolysis of the corresponding titanacyclopentadienes **10** [13], which may be a useful complement to the existing zirconacyclopentadiene route [14]. The transformation from the diiodobutadiene to siloles via the halogen–lithium exchange is a known procedure [15]. A variety of 3,4-dialkyl and 3,4-unsubstituted siloles **12** can be obtained.

In addition, 3,4-dialkyl-2,5-bis(trimethylsilyl)siloles **13**, prepared from **12** by alkylation on the silicon atom, are further transformed into 2,5-dihalosiloles **14** (Scheme 5) [16], which would be useful precursors for new silole π-conjugated compounds.

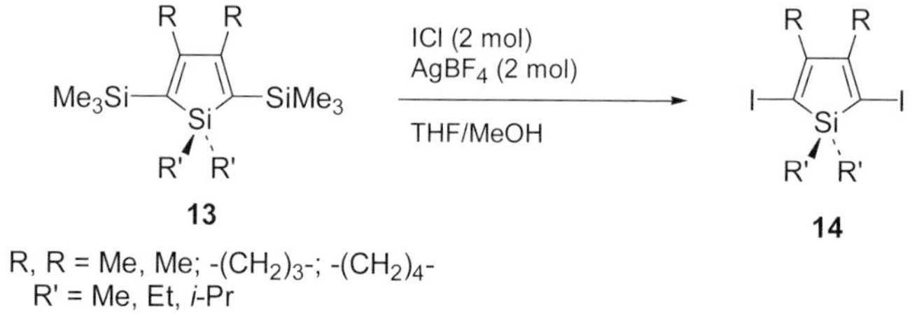

13 $\xrightarrow[\text{THF/MeOH}]{\text{ICl (2 mol), } AgBF_4 \text{ (2 mol)}}$ **14**

R, R = Me, Me; $-(CH_2)_3-$; $-(CH_2)_4-$
R' = Me, Et, *i*-Pr

Scheme 5. Synthesis of 2,5-dihalosiloles **14**.

Application to Organic LEDs

Considering the high electron-accepting properties, the silole ring would work as a core component of new efficient electron-transporting (ET) materials for organic LEDs. We have evaluated this possibility using 2,5-diarylsiloles as the silole π-electron systems [9]. Among the several 2,5-diarylsiloles examined, 2,5-di(2-pyridyl)silole (PYSPY) has been found to show very high performance as an ET material in our device having an ITO/TPD/Alq/PYSPY/Mg:Ag configuration, where the triphenylamine dimer (TPD) and tris(8-quinolinolato)aluminum (Alq) are employed as hole-transporting and emissive materials, respectively, as shown in Fig. 1.

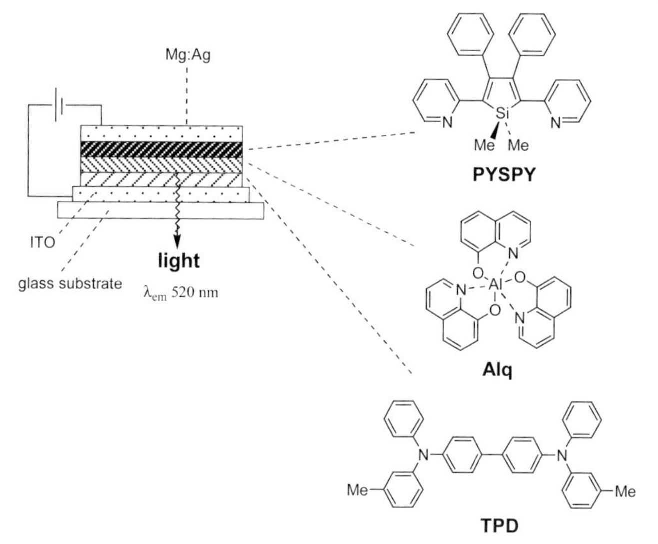

Fig. 1. Organic LEDS using PYSPY, Alq, and TPD as electron-transporting, emissive and hole-transporting materials, respectively

The device emits an yellowish-green light from the Alq layer. The threshold applied voltage is about 3 V and the maximum luminance reached 12 000 cd m^{-2} at 10 V. The luminous efficiency at 100 cd m^{-2} is 1.9 lm W^{-1}. The performance of PYSPY as an ET material exceeds that of Alq, which is one of the best ET materials reported so far.

2,5-Diarylsiloles can also be applied as efficient emissive materials and the wavelengths of their luminescence are widely tunable by changing the 2,5-aryl groups, as shown in Fig. 2.

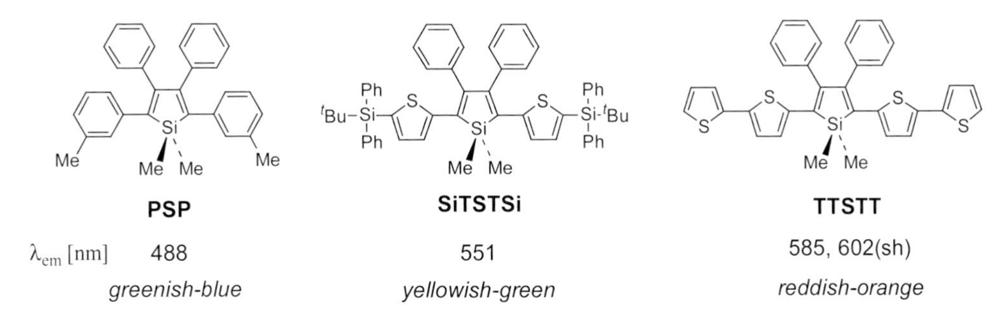

Fig. 2. Emission wavelengths of organic LEDs using 2,5-diarylsiloles as emissive electron-transporting materials. Cell configuration : ITO/TPD(500 Å)/2,5-diarylsilole(500 Å)/Mg:Ag.

Thus, in the devices having the ITO/TPD/2,5-diarylsilole/Mg:Ag configuration, three types of silole derivatives, PSP, SiTSTSi, and TTSTT, work as emissive ET materials, emitting greenish-blue, yellowish-green, and reddish-orange light, respectively. It should be noted that a bathochromic shift of the emission wavelength of about 100 nm is attained by merely changing the 2,5-aryl groups from 2-methylphenyl to bithienyl. This structural modification could be readily achieved by our one-pot synthesis shown in Scheme 2.

We have recently developed a new series of silole derivatives for the single-layer organic LEDs. Thus, diphenylamino-substituted 2,5-diarylsiloles NPSPN and NPYSPYN have been designed and synthesized as a π-electron system having both a relatively high-lying HOMO and a low-lying LUMO [17]. The single-layer devices are simply NPSPN or NPYSPYN sandwiched between the two ITO and Mg:Ag electrodes and emit a greenish-yellow light with a relatively high luminous efficiency (Fig. 3): The luminous efficiency at 100 cd m^{-2} is 0.056 and 0.26 lm W^{-1} for the case of NPSPN and NPYSPYN, respectively.

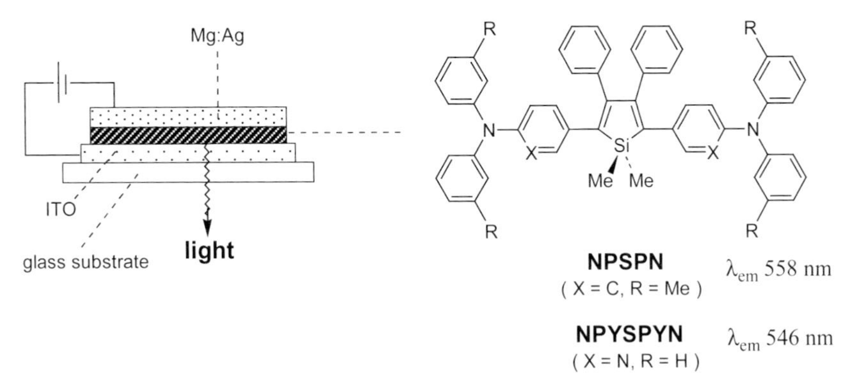

Fig. 3. Single-layer organic LEDs using NPSPN or NPYSPYN.

Conclusion

We have shown herein several new synthetic methods for the silole-based π-conjugated compounds and their possible applications to organic LEDs. The potential of the silole derivatives as new electron-transporting and emissive materials is very high. For their practical application, however, the synthesis of "tailor-made" silole compounds based on a more precise molecular design would be essential. We are now in a position to access them.

Acknowledgment: We acknowledge the *Ministry of Education, Science, Sports and Culture, Japan*, *Nagase Science and Technology Foundation*, *Ciba-Geigy Foundation* (Japan) for the Promotion of Science, and the *Japan Chemical Innovation Institute* for their financial support.

References:

[1] Recent reviews of siloles: a) J. Dubac, A. Laporterie, G. Manuel, *Chem. Rev.* **1990**, *90*, 215.
b) E. Colomer, R, J. P. Corriu, M. Lheureux, *Chem. Rev.* **1990**, *90*, 265.
c) J. Dubac, C. Guérin, P. Meunier, in: "*The Chemistry of Organic Silicon Compounds*", (Eds. Z. Rappoport and Y. Apeloig), *Vol. 2*, John Wiley, **1998**, pp. 1961–2036.

[2] Accounts: a) K. Tamao, S. Yamaguchi, *Pure Appl. Chem.* **1996**, *68*, 139.
b) S. Yamaguchi, K. Tamao, *J. Synth. Org. Chem., Jpn.* **1998**, *56*, 500.
c) K. Tamao, S. Yamaguchi, *J. Chem. Soc., Dalton Trans.* **1998**, 3693.

[3] a) J. Shinar, S. Ijadi-Maghsoodi, Q.-X. Ni, Y. Pang, T. J. Barton, *Synth. Met.* **1989**, *28*, C593.
b) T. J. Barton, S. Ijadi-Maghsoodi, Y. Pang, *Macromolecules* **1991**, *24*, 1257.
c) S. Grigoras, G. C. Lie, T. J. Barton, S. Ijadi-Maghsoodi, Y. Pang, J. Shinar, Z. V. Vardeny, K. S. Wong, S. G. Han, *Synth. Met.* **1992**, *49–50*, 293.

[4] a) G. Frapper, M. Kertesz, *Organometallics* **1992**, *11*, 3178.
b) G. Frapper, M. Kertesz, *Synth. Met.* **1993**, *55–57*, 4255.
c) J. Kürti, P. R. Surján, M. Kertesz, G. Frapper, *Synth. Met.* **1993**, *55–57*, 4338.

[5] a) Y. Yamaguchi, J. Shioya, *Mol. Eng.* **1993**, *2*, 339.
b) Y. Yamaguchi, *Mol. Eng.* **1994**, *3*, 311.
c) Y. Yamaguchi, T. Yamabe, *Int. J. Quantum Chem.* **1996**, *57*, 73.
d) Y. Matsuzaki, M. Nakano, K. Yamaguchi, K. Tanaka, T. Yamabe, *Chem. Phys. Lett.* **1996**, *263*, 119.

[6] a) S. Y. Hong, D. S. Marynick, *Macromolecules* **1995**, *28*, 4991.
b) S. Y. Hong, S. J. Kwon, S. C. Kim, *J. Chem. Phys.* **1995**, *103*, 1871.
c) S. Y. Hong, *Bull. Korean. Chem. Soc.* **1995**, *16*, 845.
d) S. Y. Hong, S. J. Kwon, S. C. Kim, D. S. Marynick, *Synth. Met.* **1995**, *69*, 701.
e) S. Y. Hong, S. J. Kwon, and S. C. Kim, *J. Chem. Phys.* **1996**, *104*, 1140.
f) S. Y. Hong, J. M. Song, *Synth. Met.* **1997**, *85*, 1113.
g) S. Y. Hong, J. M. Song, *Chem. Mater.* **1997**, *9*, 297.

[7] J. Ohshita, M. Nodono, T. Watanabe, Y. Ueno, A. Kunai, Y. Harima, K. Yamashita, M. Ishikawa, *J. Organomet. Chem.* **1998**, *553*, 487.

[8] S. Yamaguchi, K. Tamao, *Bull. Chem. Soc. Jpn.* **1996**, *69*, 2327.

[9] K. Tamao, M. Uchida, T. Izumizawa, K. Furukawa, S. Yamaguchi, *J. Am. Chem. Soc.* **1996**, *118*, 11974.

[10] K. Tamao, S. Yamaguchi, M. Shiro, *J. Am. Chem. Soc.* **1994**, *116*, 11715.

[11] S. Yamaguchi, T. Endo, K. Tamao, M. Uchida, T. Izumizawa, K. Furukawa, *74th Annual Meeting of the Chemical Society of Japan*, Kyoto, March **1997**, Abstract 4A211.

[12] M. Katkevics, S. Yamaguchi, A. Toshimitsu, K. Tamao, *Organometallics* **1998**, *17*, 5796.

[13] S. Yamaguchi, R.-Z. Jin, K. Tamao, F. Sato, *J. Org. Chem.* **1998**, *63*, 10060.

[14] a) A. J. Ashe III, J. W. Kampf, S. M. Al-Taweel, *J. Am. Chem. Soc.* **1992**, *114*, 372.
b) A. J. Ashe III, J. W. Kampf, S. M. Al-Taweel, *Organometallics* **1992**, *11*, 1491.
c) A. J. Ashe III, J. W. Kampf, S. Pilotek, R. Rousseau, *Organometallics* **1994**, *13*, 4067.
d) A. J. Ashe III, S. Al-Ahmad, S. Pilotek, D. B. Puranik, C. Elschenbroich, A. Behrendt, *Organometallics* **1995**, *14*, 2689.
e) U. Bankwitz, H. Sohn, D. R. Powell, R. West, *J. Organomet. Chem.* **1995**, *499*, C7.
f) W. P. Freeman, T. D. Tilley, L. M. Liable-Sands, A. L. Rheingold, *J. Am. Chem. Soc.* **1996**, *118*, 10457.
g) C. Xi, S. Huo, T. H. Afifi, R. Hara, T. Takahashi, *Tetrahedron Lett.* **1997**, *38*, 4099.
h) S. Yamaguchi, R.-Z. Jin, K. Tamao, *J. Organomet. Chem.* **1998**, *559*, 73.

[15] a) F. C. Leavitt, T. A. Manuel, F. Johnson, *J. Am. Chem. Soc.* **1959**, *81*, 3163.
b) F. C. Leavitt, T. A. Manuel, F. Johnson, L. U. Matternas, D. S. Lehman, *J. Am. Chem. Soc.* **1960**, *82*, 5099.
c) K. Ruhlmann, *Z. Chem.* **1965**, *5*, 354.
d) M. D. Curtis, *J. Am. Chem. Soc.* **1967**, *89*, 4241. See also [14e–14h].

[16] S. Yamaguchi, R.-Z. Jin, S. Ohno, K. Tamao, *Organometallics* **1998**, *17*, 5133.

[17] S. Yamaguchi, T. Endo, K. Tamao, M. Uchida, T. Izumizawa, K. Furukawa, manuscript in preparation.

Silylhydrazines: Lithium Derivatives, Oxidation and Condensation Reactions

Eike Gellermann, Uwe Klingebiel*, Henning Witte-Abel

Institut für Anorganische Chemie
Georg-August-Universität Göttingen
Tammannstr. 4, D-37077 Göttingen, Germany
Tel.: Int. Code + (551)39 3052 — Fax: Int. Code + (551)39 3373
E-mail: uklinge@gwdg.de

Keywords: Silylhydrazines / Lithium-Derivatives / Silylhydrazones / Crystal Structures

Summary: Condensation of stable mono(silyl)hydrazines leads to the formation of bis(silyl)hydrazines above 200°C. The degree of oligomerization of mono- and dilithiated silylhydrazines depends on the bulkiness of the substituents. Lithiated silylhydrazines react with fluorosilanes forming, for example, tris- and tetrakis(silyl)hydrazines, and six-membered rings. In oxidation reactions of bulky bis(silyl)hydrazines bis(silyl)diazenes are obtained. Lithiumderivatives of fluoro functional bis(silyl)hydrazones are precursors for 1,2-diaza-3-sila-5-cyclopentenes. Further lithiation with *t*BuLi leads to tricyclic compounds and LiF.

Silylhydrazines

In the reaction of lithiated hydrazine with bulky fluorosilanes stable mono(silyl)hydrazines are obtained. They condense to form bis(silyl)hydrazines above 200°C [1].

$$2\ R_2FSi{-}N(H){-}NH_2\ (\mathbf{1})\ \xrightarrow[-N_2H_4]{220°C}\ R_2FSi{-}N(H){-}N(H){-}SiFR_2\ (\mathbf{2})$$

Eq. 1. Synthesis of **2**.

Lithium derivatives of silylhydrazine

Depending on the bulkiness of the substituents, monolithium derivatives of silylhydrazines crystallize with the formation of hexameric, tetrameric, and dimeric units. Monolithiated bis(dimethylphenylsilyl)hydrazine **3** forms a dimeric unit in the crystal (Eq. 2) [2]. It is the only silyl hydrazide in which the lithium ions are coordinated exclusively end on (Fig. 1).

$$(PhMe_2Si)_2N{-}NH_2 \;+\; PhMe_2Si{-}NH{-}NH{-}SiMe_2Ph \xrightarrow[-\,2\,BuH]{+\,2\,BuLi} [(PhMe_2Si)_2N{-}N(H)Li{-}]_2\ \mathbf{3}$$

Eq. 2. Synthesis of **3**.

Table 1. Characteristic bond lengths and angles of **3**.

Bond lengths [pm]		Bond angles [°]	
N(1)–N(2)	151.6	N(2)-N(1)-Si(1)	107.7°
Si(1)–N(1)	170.3	Si(1)-N(1)-Si(2)	143.4°
Li(1)–N(2)	196.5	N(1)-N(2)-Li(1)	119.7°
Li(1)–C(1a)	252.1	N(2)-Li(1)-N(2a)	103.0°
Li(1)–C(14)	259.3	Li(1)-N(2)-Li(1a)	77.0°
		Σ° N(1)	359.7°

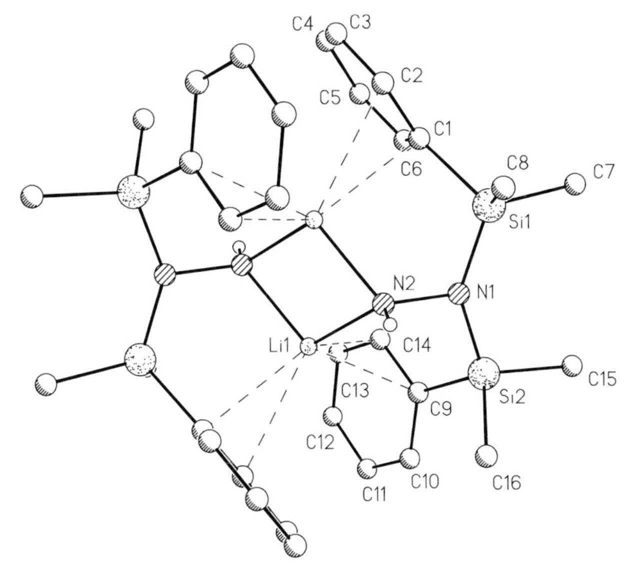

Fig. 1. Structure of **3**.

Dilithium derivatives form tetramers, trimers and monomers in the crystal [3, 4]. Bis(dimethyl-phenylsilyl)hydrazine reacts with two equivalents of *n*-butyllithium with formation of the dilithium derivative **4** (Eq. 3). The crystal structure of **4** (Table 2 and Fig. 2) shows a trimeric unit.

$$(PhMe_2Si)_2N{-}NH_2 \; + \; PhMe_2Si(H)N{-}N(H)SiMe_2Ph \xrightarrow[-\,2\,BuH]{+\,2\,BuLi} 2/3\,\left[PhMe_2Si(Li)N{-}N(Li)SiMe_2Ph\right]_3$$

4

Eq. 3. Synthesis of **4**.

Table 2. Characteristic bond lengths and angles of **4**.

Bond lengths [pm]		Bond angles [°]	
Li(1)–N(1)	200.3	N(3)-Li(1)-N(2)	45.3
Li(1)–N(2)	198.6	N(2)-Li(1)-N(1)	128.8
Li(1)–N(3)	197.6		
Li(1)–C(1)	263.6		
Si–N	169.8		
N–N	152.9		

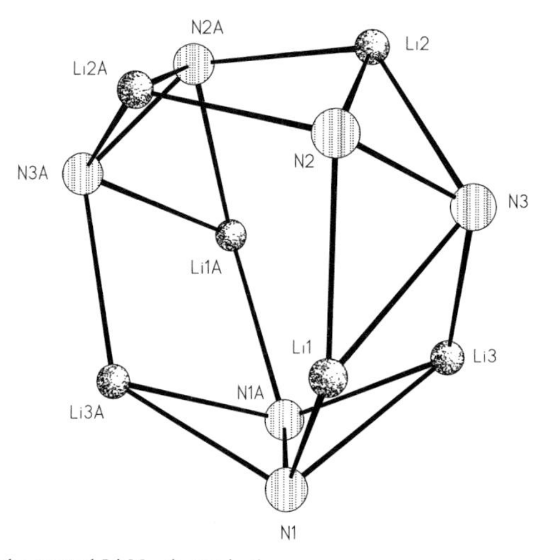

Fig. 2. Structure of the central Li_6N_6-cluster in **4**.

The central Li_6N_6-cluster of **4** shows that each of the equivalent lithium cations is bound side on as well as end on to an N–N unit [2].

Reactions

Lithium fluoride elimination from monolithiated di-*tert*-butylfluorosilylhydrazine leads to the formation of the six-membered ring **5** (Eq. 4) [5]. It is the only $(SiNHNH)_2$ six-membered ring which has been characterized by X-ray structure determination (Table 3 and Fig. 3).

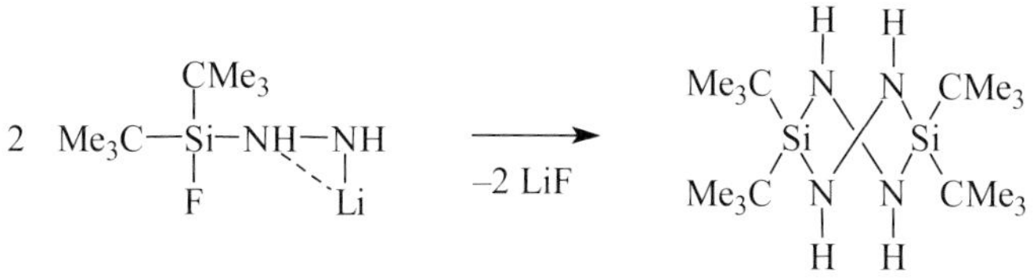

$$2\ Me_3C{-}Si(CMe_3)(F){-}NH{-}NH(Li) \xrightarrow{-2\ LiF} \mathbf{5}$$

Eq. 4. Synthesis of **5**.

Table 3. Characteristic bond lengths and angles of **5**.

Bond lengths [pm]		Bond angles [°]	
Si–N	171.5-173.3	Σ N(1)	358.5
N(1)–N(2)	144.2	Σ N(4)	357.6
N(3)–N(4)	142.9	Σ N(2)	345.0
		Σ N(3)	352.0

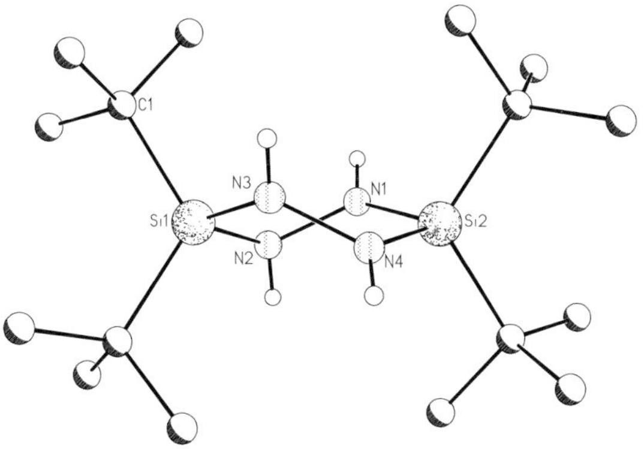

Fig. 3. Structure of **5**.

The monolithiated bis(dimethylphenylsilyl)hydrazine reacts with difluorodiisopropylsilane to give the tris(silyl)hydrazine **6** without formation of isomers (Eq. 5) [2].

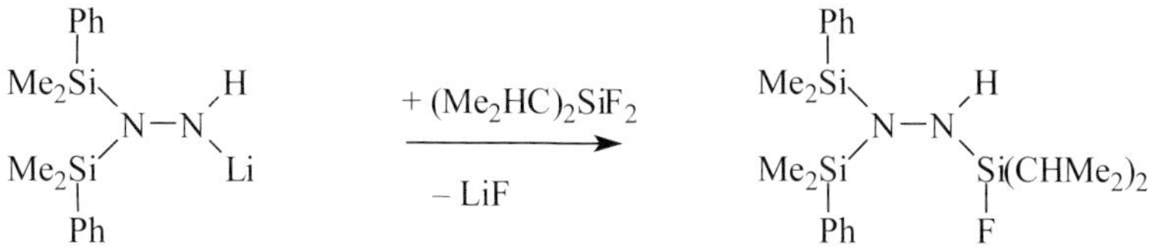

$$(PhMe_2Si)_2N{-}NHLi \xrightarrow[-\ LiF]{+\ (Me_2HC)_2SiF_2} (PhMe_2Si)_2N{-}NH{-}SiF(CHMe_2)_2\ \ \mathbf{6}$$

Eq. 5. Synthesis of **6**.

Isomerism is observed in the formation of tetrakis(silyl)hydrazines. The reaction of dilithiated bis(dimethylphenylsilyl)hydrazine with two equivalents of trifluoromethylsilane leads to the formation of the isomeric products **7** and **8** (Eq. 6) [2].

$$2\ (PhMe_2Si)(Li)N{-}N(Li)(SiMe_2Ph) \xrightarrow[-\,4\,LiBr]{+\,4\,MeSiF_3} (PhMe_2Si)(MeF_2Si)N{-}N(SiMe_2Ph)(SiF_2Me)\ (\mathbf{7}) + (PhMe_2Si)_2N{-}N(SiF_2Me)_2\ (\mathbf{8})$$

Eq. 6. Synthesis of the isomers **7** and **8**.

Diazenes

Bis(silyl)diazenes are obtained in oxidation reactions of bulky bis(silyl)hydrazines, e.g. bis(di-*tert*-butylmethylsilyl)hydrazine reacts with bromine in THF forming the first thermally stable bis(silyl)-diazene **9** (Eq. 7) [6–8].

$$(Me_3C)_2MeSi(Li)N{-}N(Li)SiMe(CMe_3)_2 + Br_2 \xrightarrow[-2\,LiBr]{THF} (Me_3C)_2MeSi{-}N{=}N{-}SiMe(CMe_3)_2\ (\mathbf{9})$$

Eq. 7. Synthesis of **9**.

Silylhydrazones

In the condensation reaction of mono(silyl)hydrazines with ketones mono(silyl)hydrazones are formed.

Lithiated mono(silyl)hydrazones can be used as precursors for the synthesis of fluoro functional bis(silyl)hydrazones. Lithium derivatives of these bis(silyl)hydrazones react spontaneously via LiF elimination to give 1,2-diaza-3-sila-5-cyclopentenenes. Further lithiation with *tert*-butyllithium leads to the formation of the tricyclic compound **10** (Eq. 8 and Fig. 4) [9].

$$\text{1-}[(Me_3C)_2MeSi]\text{-3-F-3-}CMe_3\text{-5-Me-1,2-diaza-3-sila-cyclopent-5-ene} \xrightarrow[-\,tBuH,\ -\,LiF]{+\,tBuLi} \tfrac{1}{2}\ \mathbf{10}$$

Eq. 8. Synthesis of **10**.

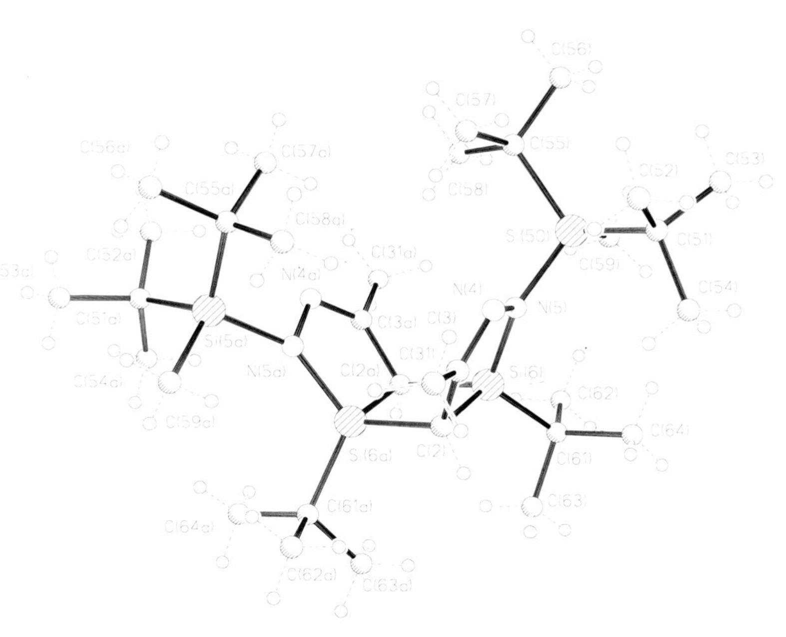

Fig. 4. Structure of **10**.

References:

[1] K. Bode, U. Klingebiel, *Adv. Organomet. Chem.* **1996**, *40*, 1–43.

[2] H. Witte-Abel, U. Klingebiel, M. Schäfer, *Z. Anorg. Allg. Chem.* **1998**, *624*, 271–276.

[3] N. Metzler, H. Nöth, H. Sachdev, *Angew. Chem.* **1994**, *106*, 1837–1839.

[4] C. Drost, U. Klingebiel, H. Witte-Abel, *Organosilicon Chemistry III: From Molecules to Materials* (Eds: N. Auner, J. Weis) VCH, Weinheim, **1998**, pp. 358–363.

[5] C. Drost, U. Klingebiel, *Chem. Ber.* **1993**, *126*, 1413–1416.

[6] H. Witte-Abel, U. Klingebiel, M. Noltemeyer, *Chem. Commun.* **1997**, 771.

[7] U. Wannagat, C. Krüger, *Z. Anorg. Allg. Chem.* **1964**, *326*, 288–295.

[8] N. Wiberg, *Adv. Organomet. Chem.* **1984**, *23*, 131–191.

[9] K. Knipping, C. Drost, U. Klingebiel, *Z. Anorg. Allg. Chem.* **1996**, *622*, 1215–1221.

Cyclosilazanes with SiH_2, $SiHal_2$, $Si(Hal)NH_2$ Groups — Experiments and Molecular Orbital Ab Initio Calculations

Bettina Jaschke, Uwe Klingebiel, Peter Neugebauer

Institut für Anorganische Chemie, Georg-August-Universität Göttingen
Tammannstr. 4, D-37077 Göttingen, Germany
Tel.: Int. code + (551)39 3052 — Fax: Int. code + (551)39 3373
E-mail:uklinge@ gwdg.de

Keywords: Amino-Chloro-Substituted Cyclodisilazanes / Halo-Substituted Cyclodisilazanes / Hydrido-Substituted Cyclodisilazanes / Ab Initio Calculations / Crystal Structures / *cis* Conformation

Summary: We present ab initio calculations and crystal structures of halo- and hydrido-substituted cyclodisilazanes as well as the synthesis and crystal structure of the first amino-chloro functional cyclodisilazane in the *cis*-conformation. Crystal structure determinations of cyclotrisilazanes are shown.

Introduction

Compared with organic-substituted cyclosilazanes only a few Si–N ring systems with inorganic substituents are known. We succeeded in the synthesis of (Cl_2SiN–) and the first (F_2SiN–) cyclodi- and cyclotrisilazanes [1–3].

Synthesis

Lithiated aminotrihalosilanes form cyclosilazanes via LiHal elimination. In the case of trimethylsilylaminotrihalosilanes, cyclosilazanes are obtained by elimination of trimethylhalosilane (Scheme 1). Using bulky substituents only four-membered rings are formed [4].

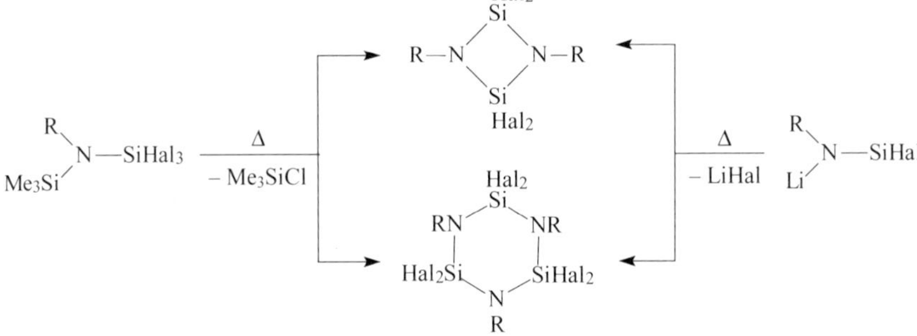

Scheme 1. Synthesis of cyclosilazanes.

Cyclodisilazanes

Molecular orbital calculations

Standard molecular orbital calculations were carried out for several cyclodisilazanes (Scheme 2) [5].

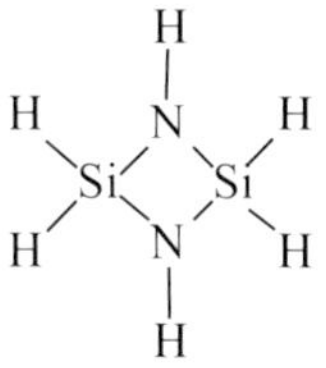

A

Si–N	173.5 pm
Si····Si	254.2 pm
Si-N-Si	94.2°
Si–H	147.8 pm
H-Si-H	106.4°

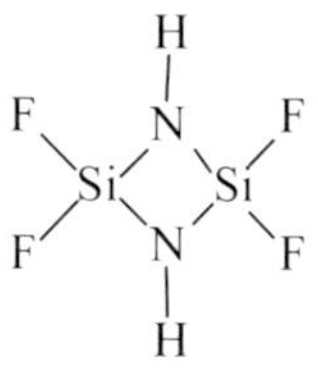

B

Si–N	171.1 pm
Si····Si	247.6 pm
Si-N-Si	92.7°

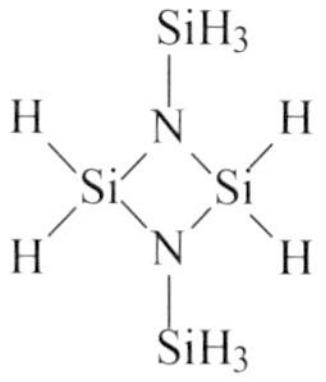

C

$Si–N_{endo}$	174.6 pm
$Si–N_{exo}$	172.4 pm
Si····Si	249.7 pm
Si-N-Si	91.3°
Si–H	147.4 pm
H-Si-H	107.6°

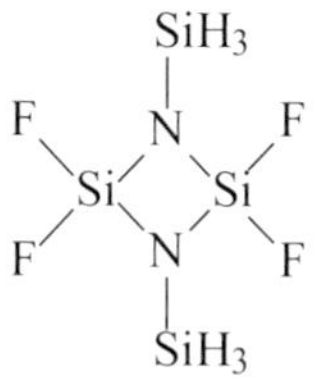

D

$Si–N_{endo}$	171.8 pm
$Si–N_{exo}$	174.5 pm
Si····Si	242.4 pm
Si-N-Si	88.4°

Scheme 2. Calculated bond lengths and angles of 1,3-cyclodisilazanes.

The calculations for 1,3-cyclodisilazanes reveal the following substituent effects on its geometry.

- Substitution of the silicon atoms in **A** or **C** with four fluorine atoms shortens the endocyclic Si–N bond, which results in a significant shortening of the Si···Si distance by 6.6 pm in **B** and by 7.2 pm in **D**.
- Substitution of the hydrogens on the nitrogen atoms in **A** or **B** by silyl groups decreases the endocyclic Si-N-Si angle, which also shortens significantly the Si···Si distance in **C** or **D**, i.e. by 4.5 pm in **C** and 5.1 pm in **D**.
- The combined effects of the four fluorines at the silicon and of the two silyl groups on the nitrogen atoms lead to a stabilisation of the ring compound and to an overall shortening of the endocyclic Si···Si distance in **D** by 11.7 pm relative to **A**.

These calculations underline the experimentally measured Si···Si distances, although the calculated ones are still considerably longer.

Crystal Structures

Many X-ray crystal structure determinations of cyclodisilazanes show that these ring systems are planar. The substituents on the Si and N atoms have substantial effects on the stabilization of the Si–N skeleton, bond length, and angles.

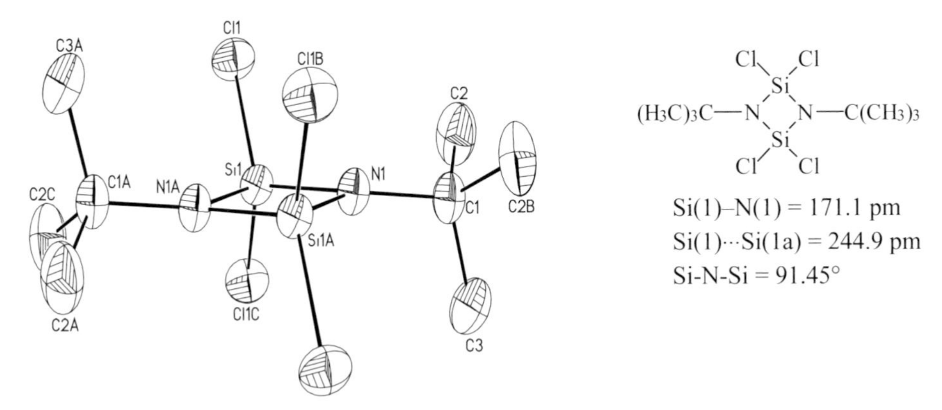

Fig. 1. X-ray crystal structure of 1,3-bis(di-*tert*-butylmethylsilyl)-2,2,4,4-tetrachlorocyclodisilazane.

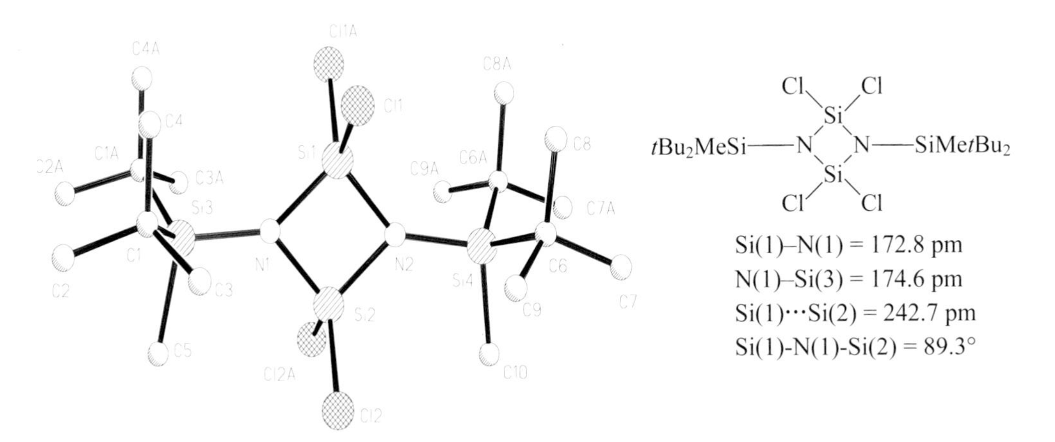

Fig. 2. X-ray crystal structure of 1,3-bis(di-*tert*-butylmethylsilyl)-2,2,4,4-tetrachlorocyclodisilazane.

Figures 1 and 2 show that the silyl substituent on the N atom effects a smaller Si-N-Si angle in the ring compared with the alkyl derivative.

Exchange of the Cl atoms by fluorine causes shorter endocyclic Si–N bonds. Both effects lead to short Si···Si distances. Figure 3 shows the X-ray crystal structure of 1,3-bis(di-*tert*-butylphenylsilyl)-2,2,4,4-tetrafluorocyclodisilazane, the compound with the shortest Si···Si distance [4]. However, there are no bonding interactions found between the silicon atoms.

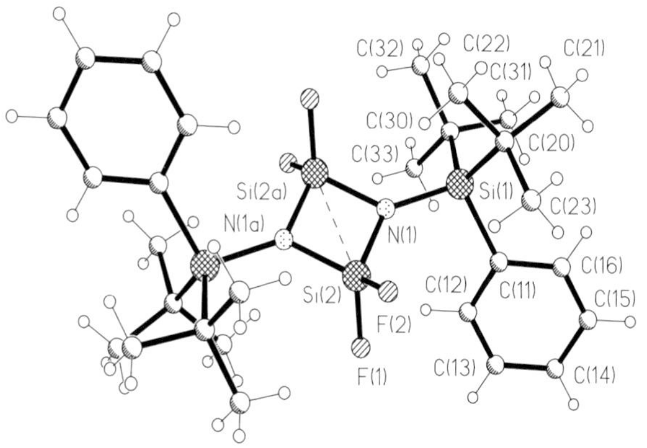

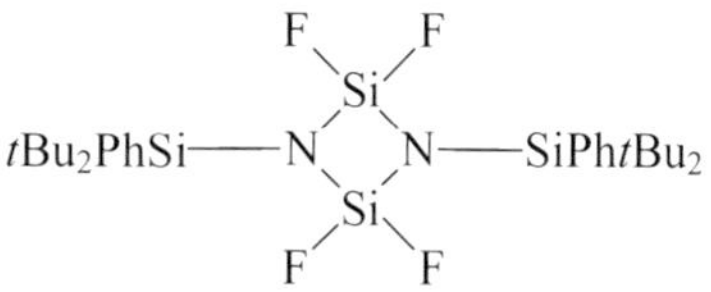

$tBu_2PhSi-N(SiF_2)_2N-SiPhtBu_2$

Si(2)–N(1) = 170.1 pm
Si(1)–N(1) = 178.3 pm
Si(2)···Si(2a) = 237.6 pm
Si(2)-N(1)-Si(2a) = 88.4°

Fig. 3. X-ray crystal structure of 1,3-bis(di-*tert*-butylphenylsilyl)-2,2,4,4-tetrafluorocyclodisilazane.

Reactions

Reaction of 1,3-bis(di-*tert*-butylmethylsilyl)-2,2,4,4-tetrachlorocyclodisilazane with $LiAlH_4$ leads to the formation of the first hydridocyclodisilazane.

$tBu_2MeSi-N(SiH_2)_2N-SiMetBu_2$

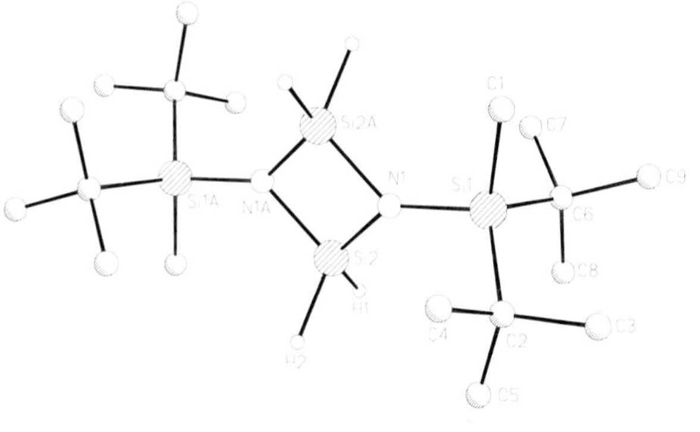

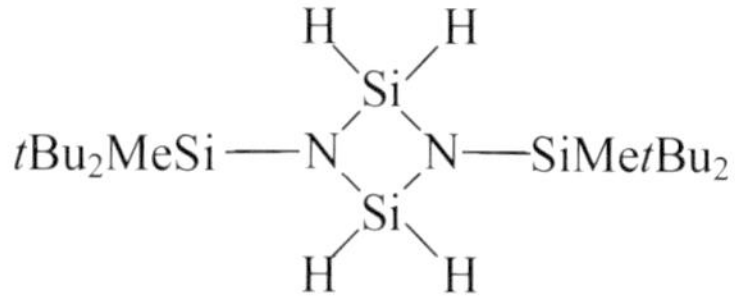

Si(1)–N(1) = 172.8 pm
Si(2)–N(1) = 173.3 pm
Si(1)···Si(1a) = 244.7 pm
Si(1)-N(1)-Si(1a) = 89.7°
Si(1)–H(1) = 156.2 pm
H(1)-Si(1)-H(2) = 93.7°

Fig. 4. X-ray crystal structure of 1,3-bis(di-*tert*-butylmethylsilyl)-2,2,4,4-tetrahydridocyclodisilazane.

The measured exocyclic Si–H bonds (Fig. 4) are longer than the calculated ones (see Scheme 2) and also than the Si–H bonds e. g. of triaminosilanes (136–138 pm) or disilanes (142–145 pm). Besides, the H-Si-H angle is considerably smaller than the calculated ones (see Scheme 2).
Ammonolysis of the 1,3-bis(silyl)-2,2,4,4-tetrachlorocyclodisilazane leads selectively to mono-substitution on each of the endocyclic Si atoms. The first amino-chloro functional cyclodisilazane is obtained.

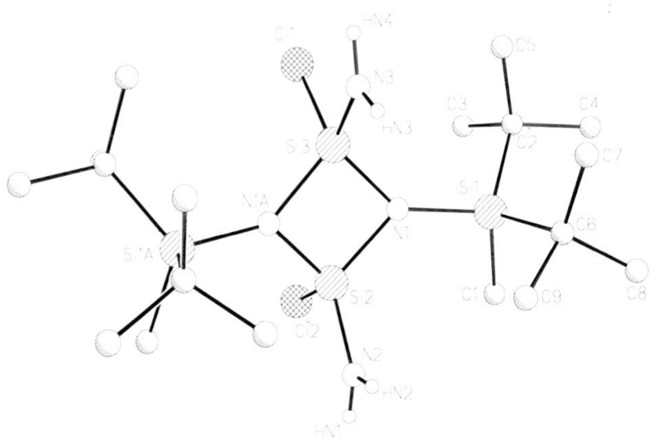

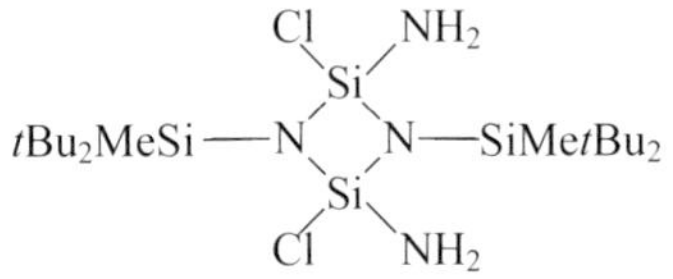

Cl NH2 / Si / tBu2MeSi—N N—SiMetBu2 / Si / Cl NH2

Si(2)–N(1) = 173.4 pm
Si(1)–N(1) = 175.0 pm
Si(2)–N(2) = 167.1 pm
Si(2)···Si(3) = 244.6 pm
Si(2)-N(1)-Si(3) = 89.87 °

Fig. 5. X-ray crystal structure of 1,3-bis(di-*tert*-butylmethylsilyl)-2,4-diamino-2,4-dichlorocyclodisilazane.

It is remarkable that this Si–N ring system is not planar and is formed exclusively in the *cis* conformation (Fig. 5). The compound contains the first planar Si–NH_2 unit. The exocyclic Si(2)–N(2) distance is the shortest which is reported in the literature.

Cyclotrisilazanes

Crystal structures

Depending on the substituents, cyclotrisilazanes are planar, or have a distorted boat, or twist conformation [6].

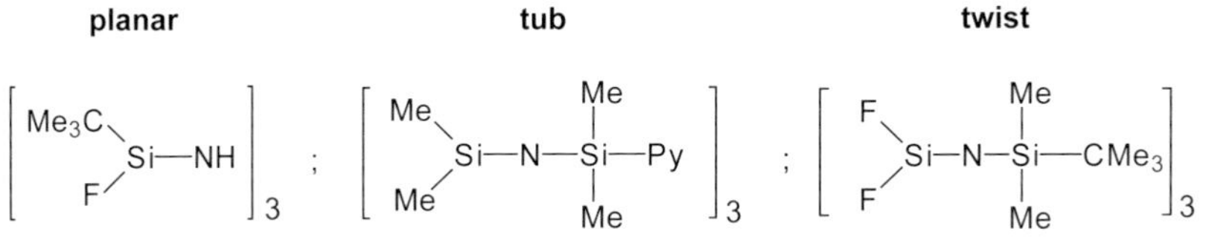

planar	tub	twist
$[(Me_3C)(F)Si–NH]_3$	$[Me_2Si–N–SiMe_2–Py]_3$	$[F_2Si–N–SiMe_2–CMe_3]_3$
Si—N : 169 pm	Me_2Si—N : 177 pm	F_2Si—N : 169 pm
Si—F : 160 pm	PySi—N : 172 pm	Me_2Si—N : 179 pm

The same effects which stabilize four-membered Si–N rings shorten endocyclic Si–N bond lengths and decrease Si-N-Si ring angles in six-membered rings.

References

[1] a) J. Haiduc and D. B. Sowerby, *Chemistry of Inorganic Homo- and Heterocycles*, Academic Press Inc. (London) LTD., **1987**, p. 221.
b) W. Fink, *Angew. Chem.* **1966**, 78, 803.

[2] S. Bartholmei, U. Klingebiel, G. M. Sheldrick, D. Stalke, *Z. Anorg. Allg. Chem.* **1988**, *556*, 129.

[3] U. Wannagat, D. Schmidt, M. Schulze, *Angew. Chem.* **1967**, *79*, 409.

[4] U. Klingebiel, J. Neemann, A. Meller, *Z. anorg. allg. Chem.* **1977**, *429*, 6.

[5] T. Müller, Y. Apeloig, I. Hemme, U. Klingebiel, M. Noltemeyer, *J. Organomet. Chem.* **1995**, *494*, 133.

[6] a) B. Tecklenburg, U. Klingebiel, M. Noltemeyer, D. Schmidt-Bäse, *Z. Naturforsch., Teil B* **1992**, *47*, 855.

b) A. Frenzel, R. Herbst-Irmer, U. Klingebiel, M. Schäfer, *Phosphoros, Sulfur, and Silicon* **1996**, 112, 155.

c) W. Clegg, G. M. Sheldrick, D. Stalke, *Acta Cryst.* **1984**, C 40, 816.

Iminosilanes: Precursors of New Rings and Unknown Ring Systems

Michael Jendras, Uwe Klingebiel*, Jörg Niesmann

Institut für Anorganische Chemie
Georg-August-Universität Göttingen
Tammannstraße 4, D-37077 Göttingen, Germany
Tel.: Int. Code + (551)39 3052 — Fax: Int. Code + (551)39 3373
E-mail: uklinge@gwdg.de

Keywords: Iminosilanes / Adducts / Stevens Migration / Cycloadditions

Summary: Our starting materials for the synthesis of iminosilanes are aminofluorosilanes that undergo fluorine-chlorine exchange with Me_3SiCl after lithiation; for this reason we are interested in the examination of such lithium derivatives. Following this preparation route we were recently able to isolate new iminosilanes. Once the unsaturated Si=N bond is synthesized there are nearly unlimited possibilities of preparing new molecules. Here Lewis base adducts with THF and pyridine are reported. The latter decompose by heating in toluene under Stevens migration. Reactions of iminosilanes with Lewis acids gave the first monomeric aminoalanes. Finally a few examples of [2+2] and [2+3] cycloadditions are described.

Synthesis of Iminosilanes via Lithiated Aminofluorosilanes

In the mid-1980s the first stable iminosilanes were prepared by two independent routes by Wiberg et al. and our group [1]. We discovered that lithium derivatives of bulky aminofluorosilanes which react with Me_3SiCl by a fluorine–chlorine exchange in the presence of the Lewis base THF are suitable precursors of iminosilanes because they allow subsequent LiCl elimination (Scheme 1). The unsaturated Si=N compounds can be purified by distillation in vacuo.

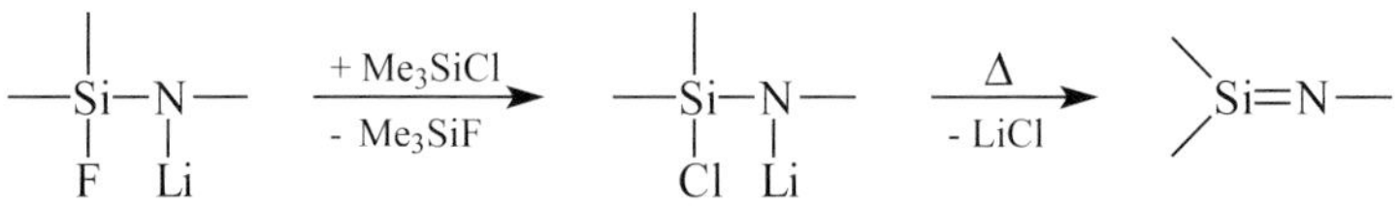

$$\text{—Si(F)—N(Li)—} \xrightarrow[-\ Me_3SiF]{+\ Me_3SiCl} \text{—Si(Cl)—N(Li)—} \xrightarrow[-\ LiCl]{\Delta} \text{>Si=N—}$$

Scheme 1. Synthesis of iminosilanes.

For a further understanding a great number of crystal structures of lithiated aminofluorosilanes was examined by X-ray structure determination and a great variety of structural types was found [2].

After addition of the Lewis base, either pyridine or triethylamine, to the lithium derivative of $(Me_3C)_2SiFNHSi(CMe_3)_2Me$ the adduct **1** or **2** is obtained.(Scheme 2).

$(Me_3C)_2Si$—N—R / F—Li(py)₂ **1** ← + pyridine — $(Me_3C)_2Si$—N—R / F Li — + NEt_3 → $(Me_3C)_2Si$—N—R / F—Li—NEt_3 **2**

R = $Si(CMe_3)_2Me$

Scheme 2. Synthesis of lithiated aminofluorosilanes.

While **1** is analogous to the already known THF adduct [3] the integrated ^{1}H-NMR spectrum of **2** shows that there is only one molecule of triethylamine coordinated to the lithium (Fig. 1).

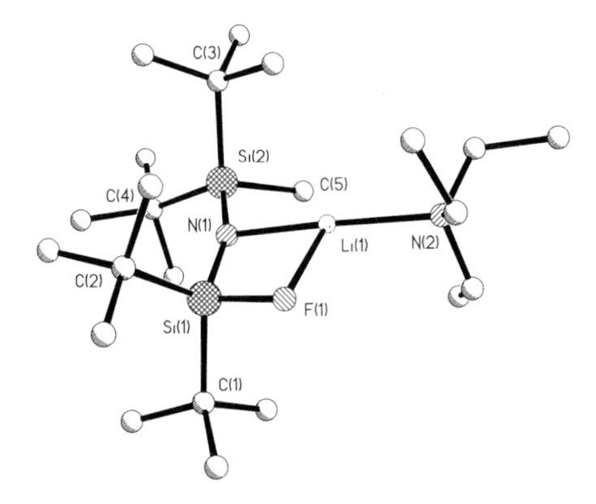

Fig. 1. Crystal structure of **2**.

Si(1)–N(1)	163.8 pm	Si(1)-N(1)-Si (2)	161.4°
Si(2)–N(1)	168.2 pm	N(1)-Li(1)-N (2)	168.8°
Si(1)–F(1)	168.3 pm	N(1)-Li(1)-F (1)	80.9°
		N(2)-Li(1)-F (1)	106.6°

The most striking features of this structure are:

- the environment of the lithium atom is almost planar (bond angle sum 356.3°)
- the Si–F bond is elongated
- there are two different Si–N bonds with a planar environment around N(1) ($\Sigma <$ N(1) = 360°)
- there is an Si-N-Si angle of 161.4° that corresponds to Wibergs THF adduct of $Me_2Si{=}N{-}Si(CMe_3)_3$ with an Si-N-Si angle of about 161° [1, 4]

Therefore **2** has to be regarded as an LiF adduct of the iminosilane.

Synthesis of $(Me_3C)_2Si{=}N{-}Si(CHMe_2)_2CMe_3$ [5]

Following the stepwise synthesis shown in Scheme 1 the free iminosilane **4** is obtained from (di-*tert*-butylfluorosilyl)(*tert*-butyldiisopropylsilyl)amine **3** as starting material.

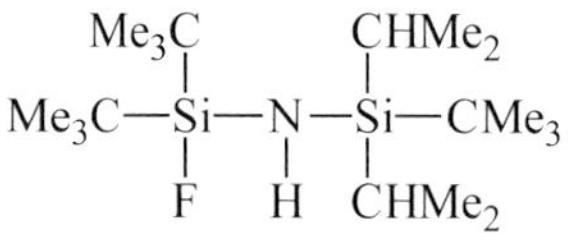

3: $\delta\ ^{29}Si$ [ppm]: 3.05 SiF; 6.85 SiC_3
bp: 97°C/0.01 mbar

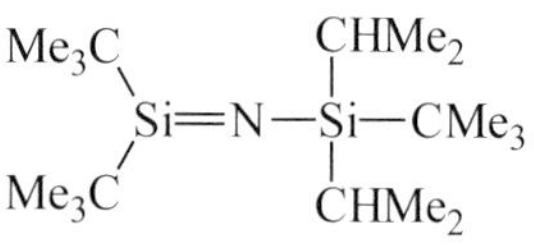

4: $\delta\ ^{29}Si$ [ppm]: –9.57 SiC_3; 76.63 Si=N
bp: 91°C/0.01 mbar

Reactions with Lewis bases and Lewis acids

According to the polarity of the Si=N bond Lewis bases (Lb) are added to the silicon atom while Lewis acids (La) react with the nitrogen atom (Scheme 3).

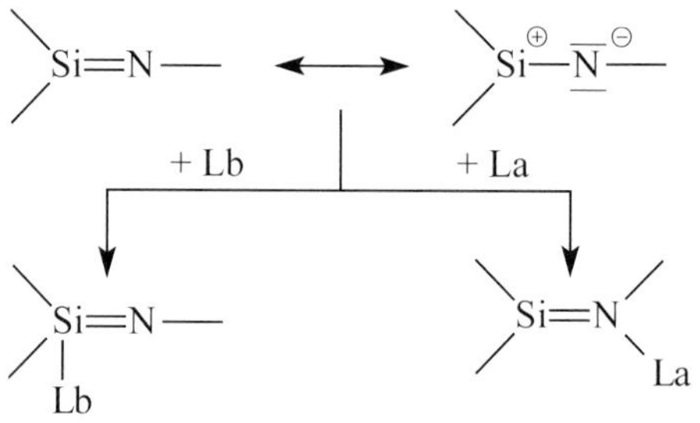

+ Lb + La

Scheme 3. Reaction of iminosilanes with Lewis bases and Lewis acids.

Lewis base adducts of the iminosilanes **4** and **5** [1] were isolated after treatment with THF (**6**, **7**) and — for the first time — with pyridine (**8**, **9**) (Scheme 4) [5].

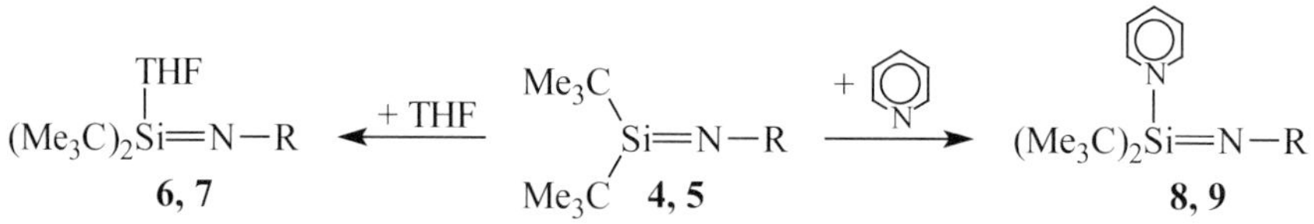

6, 7 ← + THF — **4, 5** — + pyridine → **8, 9**

4, 6, 8: R = $Si(CHMe_2)_2CMe_3$
5, 7, 9: R = $Si(CMe_3)_2Ph$

Scheme 4. Lewis base adducts of iminosilanes.

The THF adducts as well as the pyridine adducts show a strong upfield shift of the ^{29}Si-MNR signals of the unsaturated silicon atoms compared to the free iminosilanes (Table 1).

Table 1. ^{29}Si NMR chemical shifts δ [ppm] of iminosilanes (**4**, **5**) and their Lewis Base adducts (**6**–**9**).

Compound	δSi(N–Si)	δSi(N=Si)
4	–9.57	76.63
5	–13.33	80.43
6	–15.55	3.13
7	–20.26	2.68
8	–15.42	–12.20
9	–20.20	–12.59

The crystal structures of the THF adduct **6** and of the pyridine adduct **8** are shown in Fig. 2 and Fig. 3, respectively.

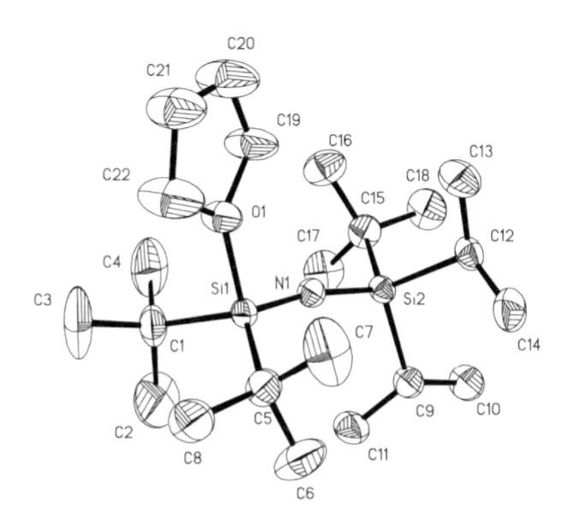

Fig. 2. Crystal structure of **6**.
Si(1)–N(1) 159.9 pm
Si(2)–N(1) 166.5 pm
Si(1)–O(1) 190.2 pm
Si(1)-N(1)-Si(2) 168.9°

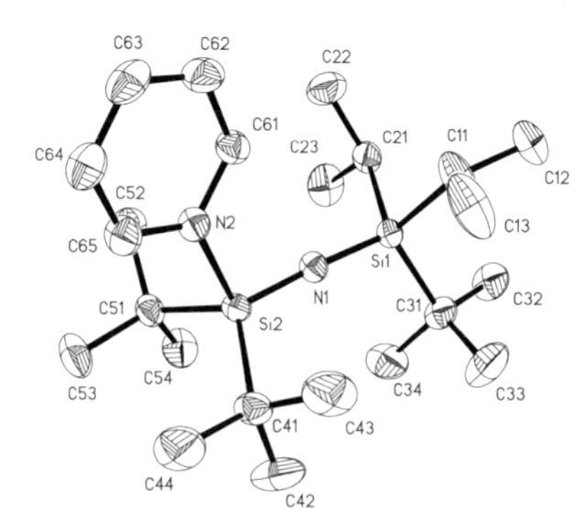

Fig. 3. Crystal structure of **8**.
Si(1)–N 1) 166.1 pm
Si(2)–N(1) 160.6 pm
Si(2)–N(2) 195.5 pm
Si(2)-N(1)-Si(1) 176.0°

If the pyridine adducts **8** and **9** are slowly heated in toluene to a temperature of 50°C they decompose under Stevens migration into **10** and **11** (Scheme 5) [5].

$$(Me_3C)_2Si(\leftarrow NC_5H_5){=}N{-}R \xrightarrow[\text{toluene}]{50\,^{\circ}C} (Me_3C)_2Si(2\text{-}C_5H_4N){-}N(H){-}R$$

8, 9 → **10, 11**

Scheme 5. Stevens migration of iminosilane–pyridine adducts.

In contrast to reactions with Lewis bases little is known about reactions of iminosilanes with Lewis acids. So addition of $AlMe_3$ and Me_2AlCl to iminosilanes at low temperature gave monomeric aminoalanes for the first time by a nucleophilic methanide-ion migration from the aluminum to the silicon atom (Scheme 6) [6].

$$\text{Si}{=}\text{N}{-}\text{Si} \quad + \quad \underset{X\,=\,Me,\ Cl}{Me_2AlX} \quad \xrightarrow{\sim 1,3\text{-}Me^{\ominus}} \quad \text{Me–Si–N(Al(Me)X)–Si}$$

Scheme 6. Formation of monomeric aminoalanes.

The products of the reaction of lithiated aminofluorosilanes with $AlCl_3$ were the first reported unsaturated silicon compounds [7–9]; they form four-membered rings with bridging chlorine atoms as exemplified in Fig. 4 (**12**). In a similar reaction with $BF_3{\cdot}OEt_2$ **13** is formed (Fig. 5).

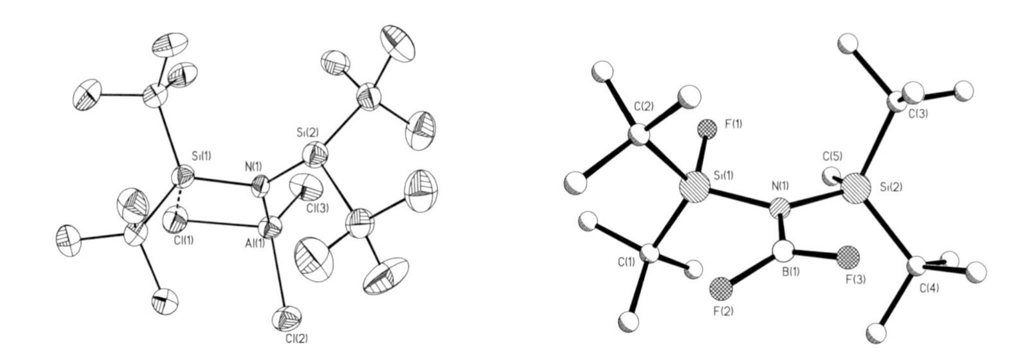

Fig. 4. Crystal structure of **12**.

Si(1)–N(1)	172.6 pm
Si(1)–Cl(1)	221.6 pm
Si(2)–N(1)	176.2 pm
Al(1)–N(1)	182.9 pm
Al(1)–Cl(1)	231.0 pm
Al(1)–Cl(2)	210.9 pm
Al(1)–Cl(3)	210.6 pm
Si(1)-N(1)-Si(2)	127.0°
Σ < C₂Si(1)N	351.9°
Σ < NAlCl(2)Cl(3)	351.7°

Fig. 5. Crystal structure of **13**.

Si(1)–N(1)	175.8 pm
Si(2)–N(1)	179.3 pm
Si(1)–F(1)	162.8 pm
N(1)–B(1)	143.0 pm
B(1)–F(2)	135.4 pm
B(1)–F(3)	134.2 pm
Si(1)-N(1)-Si(2)	129.5°
Σ < N(1)	360.0°
Σ < B(1)	359.9°

Cycloadditions

Iminosilanes offer an enormous potential for synthesis and there is already a great number of products [1]. Here, finally, a small selection of [2+2] and [2+3] cycloadditions is reported (Scheme 7).

$(Me_3C)_2Si{=}N{-}R$

$+ Me_3SiC{\equiv}CPh$ → **14**: $(Me_3C)_2Si{-}N{-}R$ / $Me_3Si{-}C{=}C{-}Ph$ (four-membered ring); **14**: R = $Si(CMe_3)_2Ph$

$+ Me_3SiC{\equiv}N$ → **15**: $(Me_3C)_2Si{-}N{-}R$ / $N{=}C{-}SiMe_3$ (four-membered ring); **15**: R = $Si(CMe_3)_2Me$

$+ 3\,C_3H_6S$, $- 3\,C_3H_6$ → **16**: $(Me_3C)_2Si{-}N{-}R$ / $S{-}S{-}S$ (five-membered ring); **16**: R = $Si(CHMe_2)_2CMe_3$

Scheme 7. [2+2] and [2+3] cycloadditions.

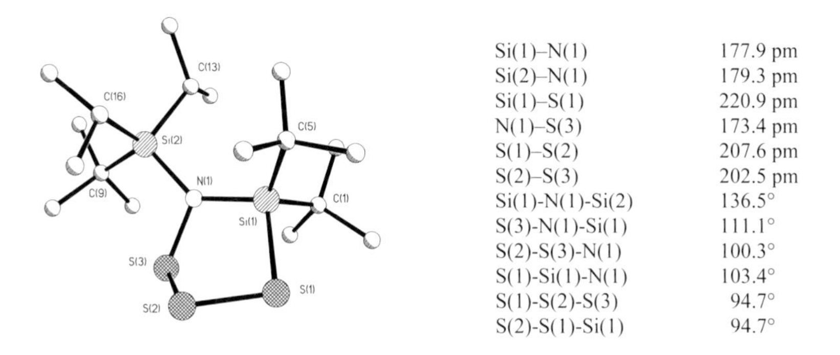

Si(1)–N(1)	177.9 pm
Si(2)–N(1)	179.3 pm
Si(1)–S(1)	220.9 pm
N(1)–S(3)	173.4 pm
S(1)–S(2)	207.6 pm
S(2)–S(3)	202.5 pm
Si(1)-N(1)-Si(2)	136.5°
S(3)-N(1)-Si(1)	111.1°
S(2)-S(3)-N(1)	100.3°
S(1)-Si(1)-N(1)	103.4°
S(1)-S(2)-S(3)	94.7°
S(2)-S(1)-Si(1)	94.7°

Fig. 6. Crystal structure of **16**.

References:

[1] I. Hemme, U. Klingebiel, *Adv. Organomet. Chem.* **1996**, *39*, 159.

[2] K. Dippel, U. Klingebiel, D. Schmidt-Bäse, *Z. Anorgan. Allg. Chem.* **1993**, *619*, 836.

[3] S. Walter, U. Klingebiel, D. Schmidt-Bäse, *J. Organomet. Chem.* **1991**, *412*, 319.

[4] N. Wiberg, K. Schurz, G. Reber, G. Müller, *J. Chem. Soc., Chem. Commun.* **1986**, 591.

[5] J. Niesmann, U. Klingebiel, M. Schäfer, R. Boese, *Organometallics* **1998**, *17*, 947.

[6] J. Niesmann, U. Klingebiel, M. Noltemeyer, R. Boese, *J. Chem. Soc., Chem. Commun.* 1997, 365.

[7] W. Clegg, U. Klingebiel, J. Neemann, G. M. Sheldrick, *J. Organomet. Chem.* **1983**, *249*, 47.

[8] W. Clegg, M. Haase, U. Klingebiel, J. Neemann, G. M. Sheldrick, *J. Organomet. Chem.*, **1983**, *252*, 281.

[9] U. Klingebiel, M. Noltemeyer, H.-G. Schmidt, D. Schmidt-Bäse, *Chem. Ber.* **1997**, *130*, 753.

Silazanes Derived from Trichlorosilane: Syntheses, Reactions and Structures

S. Abele, G. Becker *, U. Eberle, P. Oberprantacher, W. Schwarz

Institut für Anorganische Chemie der Universität Stuttgart
Pfaffenwaldring 55, D-70550 Stuttgart, Germany
Tel.: Int. code + (711)685 4172 — Fax: Int. code + (711)685 4201
E-mail: becker@iac.uni-stuttgart.de

Keywords: Lithiation / Silazanes / Aminoalanes / Crystal Structure Analyses

Summary: The reactivity of some trichlorosilane-derived silazanes has been studied. Treatment of 2-amino-1,3-bis(trimethylsilyl)-1,3-diaza-2-silacyclopentane (**2**) with two equivalents of *n*-butyllithium results in the formation of its hexameric *N*,*N*-dilithium derivative **3**, the structure skeleton of which is best described as a singly truncated but lithium-centered rhombic dodecahedron of eight lithium and five nitrogen atoms. Reduction of Si–Cl-containing silazanes with lithium alanate opens a straightforward access to aminoalanes. Bis[(*tert*-butylamino)chloroalane] (**8**), bis[(*tert*-butylamino)alane] (**9**), and 1,3-di-*tert*-butyl-1σ^4,3σ^4-diaza-2σ^4,6σ^4-dialumina[2.1.1]bicyclohexane (**10**) have been synthesized and structurally characterized.

Continuing recent investigations [1] in which the syntheses and structures of lithiated silazanes – especially of those with an Si-H group – were studied, we are now focusing our attention on compounds containing a dilithiated amino function in the neighborhood of a silicon-bound hydrogen atom and on the conversion of silazanes into aminoalanes by lithium alanate.

In the first of the two fields of research discussed here, 2-chloro-1,3-bis(trimethylsilyl)-1,3-diaza-2-silacyclopentane (**1**) turned out to be a suitable starting compound. Following a route already published by Kummer and Rochow [2] in 1963 for various chlorosilanes such as Cl_2SiMe_2, Cl_3SiMe, and $SiCl_4$, heterocycle **1** can be prepared easily in an up to 60% yield from trichlorosilane and 1,2-bis(*N*-lithium-trimethylsilylamido)ethane and is converted nearly quantitatively into the corresponding amino derivative **2** by treatment with ammonia (Eq. 1).

$$\mathbf{1}\ (\text{Si(H)Cl}) \xrightarrow[-\ NH_4Cl]{-70^\circ C,\ +\ 2\ NH_3} \mathbf{2}\ (\text{Si(H)NH}_2) \xrightarrow[-\ 2\ n\text{-BuH}]{+\ 2\ n\text{-BuLi}} \mathbf{3}\ (\text{Si(H)NLi}_2)$$

(ring N atoms bearing $SiMe_3$; Me = CH_3)

Eq. 1.

In addition to a correct elemental analysis the structure of this compound is proved by a triplet of the Si-H unit ($\delta = 4.67$; $^3J_{HH} = 4.8$ Hz; $^1J_{SiH} = 235.0$ Hz; Table 1) in the ^{1}H-NMR and a multiplet of the heterocyclic nitrogen atoms ($\delta = -352.9$) as well as a triplet of doublets of the NH_2 group ($\delta = -360.7$; $^1J_{^{15}NH} = 75.0$ Hz; $^2J_{^{15}NH} = 15.7$ Hz) in the ^{15}N-NMR spectrum (Fig. 1). Quite obviously, the steric demand of the substituted five-membered heterocycle suppresses the elimination of ammonia and the formation of the condensation product usually expected.

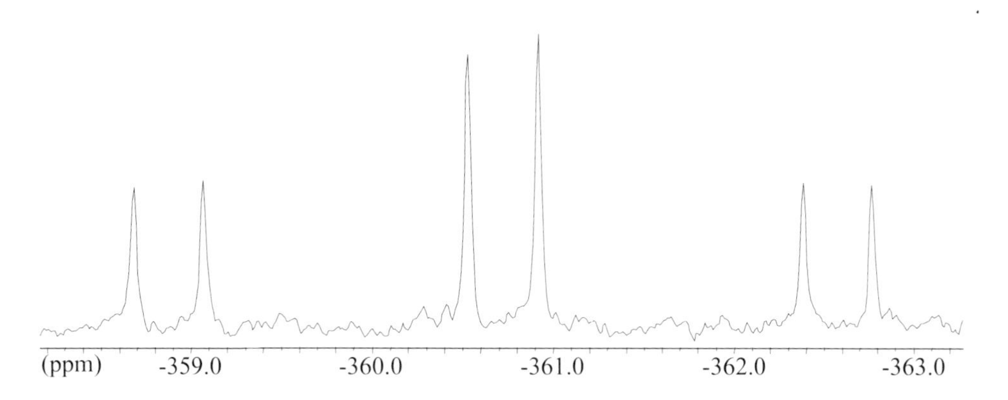

Fig. 1. Triplet of doublets of the amino group in the ^{15}N-NMR spectrum of 2-amino-1,3-bis(trimethylsilyl)-1,3-diaza-2-silacyclopentane (**2**).

Metalation of compound **2** with two equivalents of *n*-butyllithium at –60°C in *n*-pentane results in formation of the *N*,*N*-dilithium derivative **3** which has been isolated as colorless square plates after working up, concentrating and cooling the solution to –30°C (Eq. 1). As far as the heterocycle is concerned, the NMR-spectra are as to be expected (Table 1); from an $^1J_{SiH}$ parameter of 185.4 Hz the presence of strong agostic Si–H···Li interactions can be excluded.

Table 1. NMR data (δ [ppm], J [Hz])[a] and boiling points (bp, T [°C] / p [mbar]) for compounds **1**, **2**, and **3** as well as **7**.

	δCH_2[b]	δSiH	δNH_2	$\delta^{13}CH_2$	$\delta^{29}SiH$ / $^{29}SiMe_3$	$^1J_{SiH}$	bp
1	2.85	5.52	–	45.82	–16.1 / 4.95	291.5	50/0.4
2	2.87	4.67	0.63	45.96	–25.3 / 2.81	235.0	60/1.0
3	2.92 / 3.21	6.06	–	47.47	–36.9 / 3.71	185.4	–
7	2.82	5.76	–	42.83	–32.5 / –	297.6	55/0.4

[a] For ^{15}N-NMR data of compound **2** see text — [b] Center of an AA'BB' type multiplet.

A single crystal X-ray analysis ($P\bar{1}$; a = 1477.6(2), b = 1554.6(2), c = 2271.6(3) pm; $\alpha = 88.81(1)$, $\beta = 85.28(1)$, $\gamma = 85.34(1)$°; -100 ± 3°C; $Z = 12$ formula units; $R_1 = 0.063$) revealed a very complicated structure showing a hexamer to be the asymmetric part of the unit cell. The central polyhedron is best described as a strongly distorted cube of eight lithium atoms and a slightly distorted tetragonal pyramid of five nitrogen atoms penetrating each other – or in other words as a singly truncated rhombic dodecahedron (Fig. 2a). Whereas the triangular planes of the tetragonal pyramid are capped by four lithium atoms, four further atoms of this type bridge the N···N edges of its basal rectangle, which is centered additionally by a ninth lithium. The apical and one of the basal nitrogen atoms are each bound to one of the two lithium atoms forming bridges between the polyhedron and a sixth formula unit (Fig. 2b). One lithium atom of that unit is coordinated to both the nitrogen atoms of the external 1,3-diaza-2-silacyclopentane heterocycle not shown in Fig. 2b, and the second one is part of the aforementioned bridges.

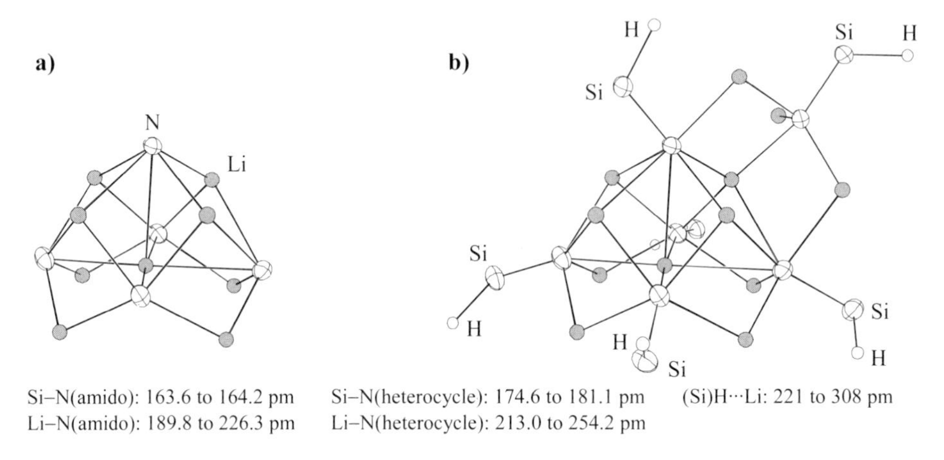

Fig. 2. (a) Singly truncated but lithium centered rhombic dodecahedron as the central unit in 2-(*N,N*-dilithium-amido)-1,3-bis(trimethylsilyl)-1,3-diaza-2-silacyclopentane (**3**); (b) central polyhedron enlarged by the sixth formula unit. For clarity the surrounding heterocycles together with their substituents have been omitted except for the Si–H groups [3]. Into the values given above two rather long Li–N distances to the centering lithium atom (244.7 and 250.7 pm) have not been included.

In the second field of research reactions of trichlorosilane with different stoichiometric amounts of *tert*-butylamine at –70 to –50°C in *n*-pentane provided a straightforward and useful access to silazanes of composition $(Me_3CNH)_{3-x}SiCl_xH$ (x = 2: **4**; x = 1: **5**; x = 0: **6**; Scheme 1).

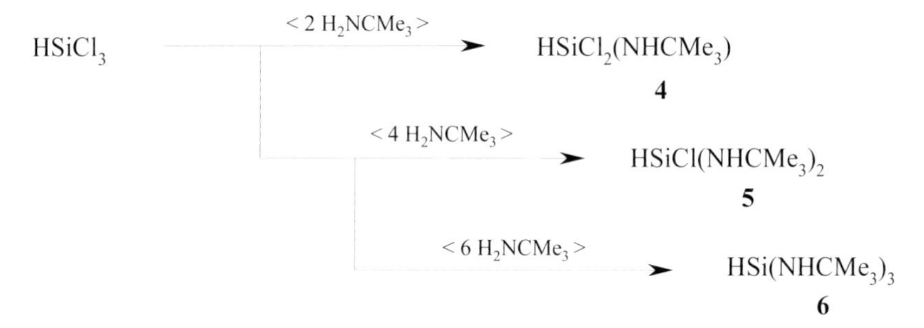

Scheme 1. Aminolysis of trichlorosilane with *tert*-butylamine in different molar ratios.

Furthermore, based on very recent work of Schmidbaur et al. [4], who reacted various 1,2-diaminoethanes with tetrachlorosilane, 1,3-di-*tert*-butyl-2-chloro-1,3-diaza-2-silacyclopentane (**7**) has been obtained from the *N,N'*-di-*tert*-butyl derivative and trichlorosilane in the presence of two equivalents of triethylamine (see Eq. 3). The compounds **4** to **7** were isolated as colorless liquids in yields of about 60 to 70% and were characterized mainly by their NMR- spectra (Tables 1 and 2).

Table 2. NMR data (δ [ppm], *J* [Hz]) and boiling points (bp, *T* [°C] / *p* [mbar]) of silazanes **4**, **5**, and **6** compared to trichlorosilane.

	δNH[a]	δSiH	$^3J_{HH}$	$\delta^{29}Si$	$^1J_{SiH}$	$\delta^{13}CMe_3$	$\delta^{13}CH_3$	bp
$HSiCl_3$	–	5.70	–	– 9.5	366.5	–	–	33/1013
4	1.34	5.50	3.9	– 26.1	321.9	51.2	32.5	55/40
5	1.18	5.34	3.1	– 37.9	274.3	49.9	33.1	50/4.0
6	0.60	4.88	2.2	– 48.4	225.2	49.0	33.9	52/0.5

[a] broad signal

As is generally known, the conversion of halosilanes into the corresponding Si–H compounds may be accomplished easily with reducing agents such as lithium alanate, dimeric diisobutylalane [5a], or sodium-bis(2-methoxyethoxo)alanate [5b]. In order to avoid the use of highly dangerous bromosilane or relatively expensive dichlorosilane as starting materials in the preparation of *tert*-butylamino- or bis(*tert*-butylamino)silane we attempted to reduce the corresponding silazanes **4** and **5** with lithium alanate. Quite remarkably, however, these reactions open an unexpectedly convenient access to aminoalanes which are prepared otherwise from alane itself and a suitable amine ([6]; for reviews in this field see *e. g.* [7]).

When suspensions of one or two equivalents of lithium alanate in diethyl ether are added to cooled solutions of (*tert*-butylamino)dichlorosilane (**4**), reactions proceed at –30°C with evolution of silane and formation of bis[(*tert*-butylamino)chloroalane] (**8**) and bis[(*tert*-butylamino)alane] (**9**), respectively. The colorless solids, which are rather sensitive to air and moisture, were first prepared by Nöth and Wolfgardt [6b] applying different reaction routes; they are isolated in up to 80% yield and recrystallized from diethyl ether (Eq. 2).

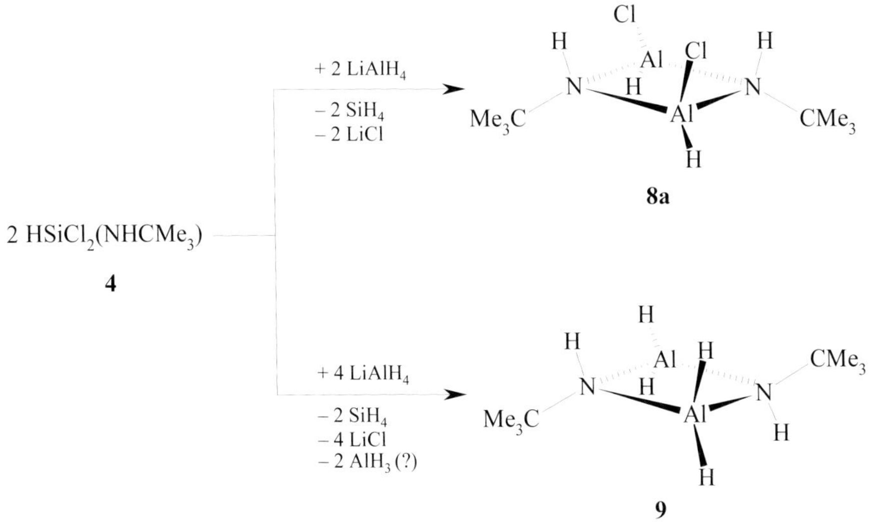

$2\ HSiCl_2(NHCMe_3)$ (**4**) $\xrightarrow[-2\ SiH_4,\ -2\ LiCl]{+2\ LiAlH_4}$ **8a**

$2\ HSiCl_2(NHCMe_3)$ (**4**) $\xrightarrow[-2\ SiH_4,\ -4\ LiCl,\ -2\ AlH_3\ (?)]{+4\ LiAlH_4}$ **9**

Eq. 2.

An analogous treatment of the heterocyclic compound **7** results in the formation of 1,3-di-*tert*-butyl-1σ^4,3σ^4-diaza-2σ^4,6σ^4-dialumina[2.1.1]bicyclohexane (**10**) (Eq. 3). The syntheses of this compound [8a] and of similar derivatives [8b], starting from the adduct $H_3Al{\leftarrow}NMe_3$ and various 1,2-bis(*tert*-butylamino)ethanes, have already been published by Raston and others some years ago.

$$HSiCl_3 \xrightarrow[-\,2\,[Et_3NH]^+\,Cl^-]{+\,(Me_3CNHCH_2)_2\,;\;+\,2\,Et_3N} \mathbf{7} \xrightarrow[-\,SiH_4;\;-\,LiCl;\;-\,LiH\,(?)]{+\,2\,LiAlH_4} \mathbf{10}$$

CMe$_3$ / N / Si(H)(Cl) / N / CMe$_3$ (**7**); Me$_3$C–N, N–CMe$_3$, H$_2$Al, AlH$_2$ (**10**)

Et = C_2H_5

Eq. 3.

X-ray structure analyses (**8**: orthorhombic, *Pbca*; a = 1239.8(2), b = 1597.6(2), c = 3224.4(5) pm; –100 ± 3°C; Z = 16 dimers; R_1 = 0.055; **9**: orthorhombic, *Pccn*; a = 1007.2(1), b = 1704.2(2), c = 797.7(1) pm; –100 ± 3°C; Z = 4 dimers; R_1 = 0.044; **10**: orthorhombic, *Pnma*; a = 1272.7(4), b = 1061.7(2), c = 1082.0(3) pm; –100 ± 3°C; Z = 4 molecules; R_1 = 0.040) reveal these aminoalanes to be built up of four-membered aluminum-nitrogen heterocycles. As for compound **8**, a special feature is observed in that the asymmetric part of the unit cell contains two crystallographically independent dimers. Whereas in the individual crystal selected from the finally isolated product one position is occupied by 1*r*,3*c*-di-*tert*-butyl-2*t*,4*t*-dichloro-1σ^4,3σ^4-diaza-2σ^4,4σ^4-dialuminacyclobutane (**8a**) only, in the second position about 19% of the 1*r*,3*c*-di-*tert*-butyl-2*t*,4*c*-dichloro isomer (**8b**), in which the two chlorine atoms show a *trans* arrangement, are present.

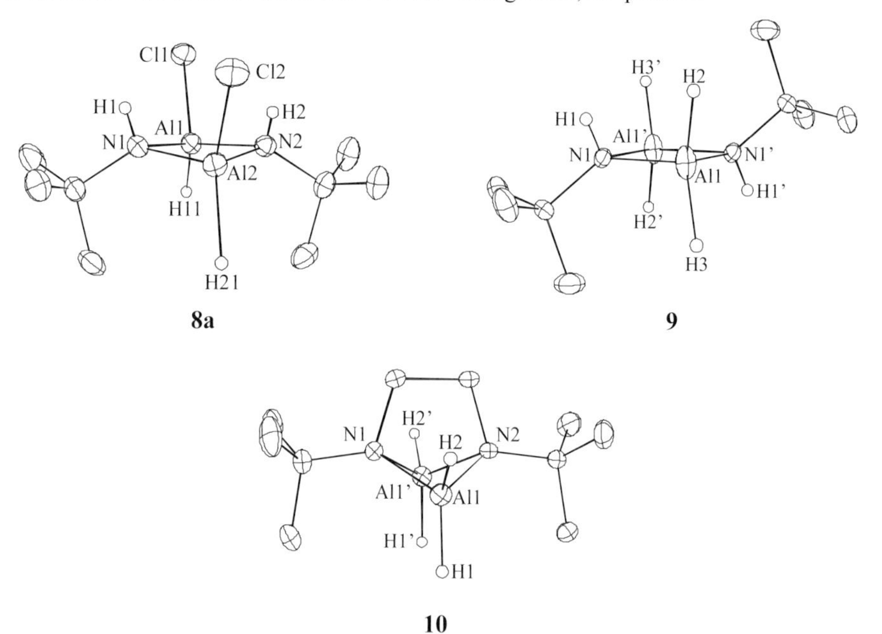

Fig. 3. Molecular structures of aminoalanes **8a** to **10** (thermal ellipsoides at 25% probability level). For a better representation of the models, hydrogen atoms of the methyl and methylene groups have been omitted [3].

In compounds **8a** and **10** the four-membered aluminum-nitrogen rings are folded at the Al···Al axes with angles of 167.6° and 126.6°, whereas the molecules of aminoalane **9** have a center of symmetry. The bond lengths between aluminum and its neighboring atoms (Table 3) do not deviate significantly from values of already published aminoalanes [9] or from theoretical calculations on, e.g., $(H_2Al–NH_2)_2$ [10]; this is also true for aminoalane **10**, the structure of which has been determined by Atwood et al. at room temperature [8a].

Table 3. Characteristic bond lengths and transannular distances (pm) in aminoalanes **8a**, **9**, and **10**.

	Al–N[a]	Al–H[b]	Al–Cl[a]	Al···Al	N···N
8a	191.6–194.0	167–176	214.1 (3×)[c]	278.1, 279.2	264.6, 265.2
9	193.9, 194.0	153, 158	–	278.8	269.7
10	195.5, 195.9	146, 148	–	273.9	243.4

Mean standard deviations: [a] (0.2) — [b] (4) — [c] Disordered part of the molecule neglected.

As published by Nöth and Wolfgardt [6b] more than twenty years ago, keeping bis[(*tert*-butylamino)chloroalane] (**8**) for eighteen hours at 150°C results in the formation of tetrakis[(*tert*-butylimino)chloroalane] and hydrogen. Due to a lack of suitable crystals we have so far been unable to perform an X-ray structure analysis and to confirm the cubane-like structure of this compound.

Acknowledgement: We thank *Land Baden-Württemberg* (Keramikverbund Karlsruhe-Stuttgart, KKS), *Fonds der Chemischen Industrie*, and *Hoechst AG* (Frankfurt/Main) for generous financial support.

References:

[1] G. Becker, S. Abele, U. Eberle, G. Motz, W. Schwarz,"*Unusual Polyhedra by Lithiation of Silazanes*", in: *Organosilicon Chemistry III: From Molecules to Materials* (Eds.: N. Auner, J. Weis), Wiley–VCH, Weinheim, **1998**, p. 342.

[2] D. Kummer, E. G. Rochow, *Z. Anorg. Allg. Chem.* **1963**, *321*, 21.

[3] *SHELXTL Plus, PC Version*; Siemens Analytical X-ray Instruments, Inc., Madison, WI, **1980**; G. M. Sheldrick, *SHELXL-93*, *Program for Crystal Structure Determinations*, University of Göttingen, Göttingen, **1993**.

[4] T. Schlosser, A. Sladek, W. Hiller, H. Schmidbaur, *Z. Naturforsch., Teil B* **1994**, *49*, 1247.

[5] a) S. Pawlenko,"*Methoden zur Herstellung und Umwandlung von Organo-Silicium-Verbindungen*", in: *Methoden der Organischen Chemie (Houben-Weyl)* (Eds.: O. Bayer, E. Müller), Bd. XIII/5 *(Organo-Silicium-Verbindungen)*, Thieme Verlag, Stuttgart, **1980**.
b) I. Kolb, J. Hetflejš, *Coll. Czech. Chem. Commun.* **1980**, *45*, 2224; J. Dautel, S. Abele, W. Schwarz, *Z. Naturforsch., Teil B* **1997**, *52*, 778.

[6] a) E. Wiberg, A. May, *Z. Naturforsch., Teil B* **1955**, *10*, 230, 232, 234.
b) H. Nöth, P. Wolfgardt, *Z. Naturforsch., Teil B* **1976**, *31*, 697.
c) G. Perego, G. del Piero, M. Corbellini, M. Bruzzone, *J. Organomet. Chem.* **1977**, *136*, 301.

d) See also for phosphino- and arsinoalanes: J. F. Janik, R. L. Wells, P. S. White, *Inorg. Chem.* **1998**, *37*, 3561.

e) See also for monomeric aminoalanes: M. A. Petrie, K. Rulandt-Senge, P. P. Power, *Inorg. Chem.* **1993**, *32,* 1135.

[7] A. Haaland,"*Normal and Dative Bonding in Neutral Aluminum Compounds*", in: *Coordination Chemistry of Aluminum* (Ed.: G. H. Robinson), VCH, Weinheim, **1993**, p. 1; G. H. Robinson,"*Organoaminoalanes: Unusual Al–N Systems*", in: *Coordination Chemistry of Aluminum* (Ed.: G. H. Robinson), VCH, Weinheim, **1993**, p. 57; M. J. Taylor, P. J. Brothers,"*Inorganic Derivatives of the Elements*", in: *Chemistry of Aluminium, Gallium, Indium and Thallium* (Ed.: A. J. Downs), Blackie–Chapman & Hall, London, **1993**, p. 111.

[8] a) J. L. Atwood, S. M. Lawrence, C. L. Raston, *J. Chem. Soc., Chem. Commun.* **1994**, 73.

b) M. G. Gardiner, S. M. Lawrence, C. L. Raston, *Inorg. Chem.* **1996**, *35*, 1349.

[9] a) H. Hess, A. Hinderer, S. Steinhauser, *Z. Anorg. Allg. Chem.* **1970**, *377*, 1.

b) A. Haaland, *Angew. Chem.* **1989**, *101*, 1017; *Angew. Chem., Int. Ed. Engl.* **1989**, *28*, 992;

c) S. J. Schauer, G. H. Robinson, *J. Coord. Chem.* **1993**, *30*, 197.

[10] R. D. Davy, H. F. Schaefer III, *Inorg. Chem.* **1998**, *37*, 2291.

Bis(trimethylsilyl)diaminosilanes: Synthesis and Reactions

Anca Oprea, Steffen Mantey, Joachim Heinicke*

Institut für Anorganische Chemie
Ernst-Moritz-Arndt-Universität Greifswald
Soldmannstr. 16, D-17487 Greifswald, Germany
Tel.: Int. code + (3834)86 4318 — Fax: Int. code + (3834)86 4319
E-mail: heinicke@mail.uni-greifswald.de

Keywords: Diaminosilanes / Dihalogenosilanes / Cross-Coupling Reactions

Summary: The synthesis of several bis(trimethylsilyl)diaminosilanes by cross-coupling reactions with lithium, their cleavage with hydrogen halides and further reactions of the dihalogenotrisilanes obtained are presented.

Introduction

All known isolable two-coordinated silylenes are cyclic monomers stabilized by bulky substituted amino groups at silicon and in most cases supplemented by a Hückel-type delocalized electron system. They were obtained by reduction of the corresponding dihalogenated derivates [1]. Attempts to synthesize analogous pyrido- or naphtho-1,3,2λ^2-diaminosiloles by the same route gave small amounts of the desired pure product or they failed [1d, 2]. This prompted us to investigate (i) the photolysis of bis(trimethylsilyl)diaminosilanes and (ii) the thermolysis of recently reported disilanes [3] and of the trisilanes as alternative methods to prepare stabilized low-coordinated silicon compounds.

The aim of this work is to present the synthesis of several bis(trimethylsilyl)diaminosilanes by cross-coupling reactions with lithium and the cleavage of the amino groups by different hydrogen halides to afford the 2,2-dihalogenohexamethyltrisilanes, which constitute suitable cyclization reagents for cyclic diaminobis(trimethyl)silanes.

Results and Discussion

Trimethylchlorosilane and dichlorodiaminosilanes **1**, obtained by reaction of $SiCl_4$ and four equivalents of the corresponding amine [4], are reacted with excess lithium in THF at room temperature to give the 2,2-diaminohexamethyltrisilanes **2** (Eq. 1). Small excess portions of Me_3SiCl are added to the reaction mixture as long as the lithium is reacting. Compound **2a** was already described by Tamao et. al. [5]. We isolated this by fractional distillation in an analytically pure form as a colorless solid (b.p. 85–90°C / 10^{-2} Torr). The compounds **2b** and **2c** are obtained in an analogous manner and form a viscous oil (b.p. 110–120°C / 10^{-2} Torr) and a colorless solid (b.p. 85–95°C / 0.1 Torr), respectively.

$$R_2SiCl_2 + >2\ Me_3SiCl \xrightarrow[THF]{>4Li} R_2Si(SiMe_3)_2$$

1 a–c **2 a–c**

a: R = Et_2N, b: R = N-piperidino, c: R = N-pyrrolidino

Eq. 1. Synthesis of 2,2-diaminohexamethyltrisilanes.

The cleavage of the amino groups is carried out in etheral solution of the diaminotrisilanes by bubbling dry hydrogen halides through the reaction mixture stirred at 0–5°C. In case of the 2,2-dichlorotrisilane this direct method sometimes gives a mixture of mono- and dichloro product. Because of this we preferred to work with a calibrated solution of hydrogen chloride in ether (ca. 1 M) (Eq. 2). The dichloro and dibromotrisilanes **3a, b** are liquids (b.p. 74°C / 5 Torr, 40°C / 10^{-1} Torr), whereas the diiodotrisilane **3c** is a colorless solid, very sensitive towards moisture and light (b.p. 70–80°C / 10^{-2} Torr). The yields of **3** were up to 65 %.

$$R_2Si(SiMe_3)_2 + >4\ HX \xrightarrow[Ether]{X = Cl, Br, I} X_2Si(SiMe_3)_2$$

2 a–c **3 a–c**

Eq. 2. Synthesis of 2,2-dihalogenohexamethyltrisilanes.

We tried the synthesis of cyclic diaminotrisilanes **5** starting from the corresponding dilithiumamides **4** and dichlorotrisilanes (Eq. 3) under various conditions (solvent: ether, THF, hexane, benzene; temperature: –78°C to reflux; reaction time: up to a week) and we could obtain only the pyrido[2,3] derivative. The reaction is carried out in boiling benzene for 8 h. The product is a pale yellow solid, which can be purified by distillation (b.p. 125–135°C/ 0.005 Torr). In other cases we observed the formation of the ring, but this could not be seperated from remaining unreacted starting material. The reason for this limitation could be the low reactivity of the bulky substituted dichlorotrisilane in combination with steric hindrance by the neopentyl group. A study of the behavior of dibromo and diiodo compounds towards **4** is in progress.

$$\mathbf{4} \xrightarrow{X_2Si(SiMe_3)_2} \mathbf{5}\ (Si(SiMe_3)_2)$$

R' = neopentyl (np)

(a, b, c)

Eq. 3. Synthesis of cyclic trisilanes.

All compounds were fully characterized by NMR, IR, MS or GC–MS and elemental analysis (C, H, N). The ^{29}Si-NMR data and selected couplings are presented in Table 1. All measurements were carried out in $CDCl_3$ with tetramethylsilane (TMS) as external reference and addition of chromiumacetylacetonate. The diiodo compound could be measured only in C_6D_6 using TMS as internal standard. We distinguished the resonances of the two silicon atoms by determining the $^1J_{Si,Si}$ and $^1J_{Si,C}$. For the aliphatic diaminotrisilanes we observed $^1J_{Si,Si}$ coupling constants around 90 Hz and $^1J_{Si,C}$ coupling constants around 42.5 Hz. After replacement of the amino groups by halides the $^1J_{Si,Si}$ decreases and the $^1J_{Si,C}$ increases with the decrease of the electronegativity of the halides. For the aromatic amino compound the coupling constants are in the middle of the values of the above-described compounds.

In the infrared spectra the frequency of the streching vibration for the Si–halide bond decreases from the dichloro- to the diiodotrisilane (Cl: 514 cm^{-1}, Br: 425 cm^{-1}, I: < 400 cm^{-1}).

Table 1. Selected NMR data of trisilanes.

Compound $R^{1,2}{}_2Si^A(Si^BMe_3)_2$	δ [ppm] Si^A	Si^B	J [Hz] $^1J_{Si,Si}$	$^1J_{Si,C}$
$R^{1,2}$ = 	–4.5	–22	90.3	42.5
$R^{1,2}$ = NEt_2	–3.2	–22.8	92.0	42.7
$R^{1,2}$ = 	–16.2	–21.6	91.5	42.8
R^1 = NEt_2 R^2 = Cl	10.8	–17.0	89	46.3
$R_2{}^{1,2}$ = 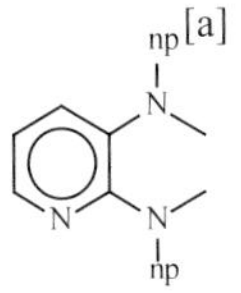	23.6	–20.8	74	43.7
$R^{1,2}$ = Cl	34.4	–11.2	75.9	47.7
$R^{1,2}$ = Br	22.6	–10.2	69.4	48.1
$R^{1,2}$ = I	–28.5	–10.7	62.4	48.4

[a] np = neopentyl

In the EI–mass spectra of the aminotrisilanes there is also an interesting aspect: the peaks for both possible silylenes can be detected, the diaminosilylene and the aminotrimethylsilylsilylene. This result is inspiring us to investigate the thermolysis or photolysis of the trisilanes, if necessary also in the presence of suitable trapping reagents. Our attempts in this direction will be reported elsewhere.

Acknowledgment: The financial support of this work by the *DFG* and the *FCI* is gratefully acknowledged. We thank Dr. M. K. Kindermann, B. Witt, I. Stoldt and Dr. A. Müller for NMR, IR and MS measurements and helpful discussions.

References:

[1] a) M. Denk, R. Lemon, R. Hayashi, R. West, A. V. Belyakov, H. P. Verne, A. Haaland, M. Wagner, *J. Am. Chem. Soc.* **1994**, *116,* 2691.
b) B. Gerhus, M. F. Lappert, J. Heinicke, R. Boese, D. Bläser, *J. Chem. Soc., Chem. Commun.* **1995**, 1931.
c) H. H. Karsch, P. A. Schlüter, F. Bienlein, M. Herker, E. Witt, A. Sladek, M. Heckel, *Z. Anorg. Allg. Chem.* **1998**, *624,* 295.
d) J. Heinicke, A. Oprea, M. K. Kindermann, T. Karparti, L. Nyulaszi, T. Vespremi, *Chem. Eur. J.* **1998**, *4,* 537.

[2] A. Oprea, J. Heinicke,"*Pyrido[b]-1,3,2λ^2-diazasilole: The First Stable Unsymmetrical Silylene*" in: *Organosilicon Chemistry III: From Molecules to Materials* (Eds.: N. Auner, J. Weis), VCH, Weinheim, **1998**, p. 50.

[3] a) S. Mantey, J. Heinicke, "*Amino-Substituted Disilanes by Reductive Coupling*" in: *Organosilicon Chemistry III: From Melocules to Materials* (Eds.: N. Auner, J. Weis), VCH, Weinheim, **1998**, p. 254.
b) J. Heinicke, S. Mantey, *Heteroatom Chem.* **1998**, in press.

[4] H. Breederveld, H. I. Waterman, *Research* **1952**, *5,* 537.

[5] K. Tamao, G. R. Sun, A. Kawachi, S. Yamaguchi, *Organometallics* **1997**, *16,* 780.

Influence of the Steric Demand of Lithium Trialkylsilylamides, -phosphanides and -arsanides on the Metathesis Reaction with Tris(trimethylsilyl)methylzinc Chloride

*Michael Wieneke, Matthias Westerhausen**

Institut für Anorganische Chemie
Ludwig-Maximilians-Universität
Butenandtstr. 5–13,haus D, D-81377 München, Germany
Tel.: Int code +(89)2180 7481 — Fax: Int code + (89)2180 7867
E-mail: maw@cup.uni-muenchen.de

Keywords: Lithium / Metathesis Reaction / Trialkylsilyl / Zinc

Summary: The reaction of lithium tri*iso*propylsilylamide with $(Me_3Si)_3CZnCl$ yields tris(trimethylsilyl)methylzinc triisopropylsilylamide. The homologous phosphanide and arsanide give $[(Me_3Si)_3CZn]_2ESi\textit{i}Pr_3$ with E = P, As. Smaller substituents at the pentele atoms lead to the formation of zincates. Tetrakis(tetrahydrofuran-*O*)lithium bis[tris(trimethylsilyl)methylzinc] tris(anilide) is isolated from the metathesis reaction of lithium anilide and $(Me_3Si)_3CZnCl$. The lithiation of Me_3SiPH_2 yields $LiP(H)SiMe_3$; however, redistribution reactions lead also to $LiPH_2$ and $LiP(SiMe_3)_2$. The reaction of this mixture with $(Me_3Si)_3CZnCl$ gives $[(thf)_4Li]^+\{[(Me_3Si)_3CZn]_4(\mu\text{-}Cl)(\mu_3\text{-}PH)_2\}^-$.

Triisopropylsilylamine (**1a**) [1], -phosphane (**1b**) [2] and -arsane (**1c**) are easily accessible by the metathesis reaction of alkali metal amide, phosphanide [3] and arsanide [4], respectively, with chlorotriisopropylsilane according to Eq. 1. Redistribution reactions were not observed. The sterically less demanding trimethylsilylphosphane (**2**) is prepared by protolysis of tris(trimethylsilyl)phosphane with methanol [5].

$$MEH_2 + ClSi\textit{i}Pr_3 \rightarrow \textit{i}Pr_3Si\text{-}EH_2 + MCl \qquad E = N{:}\ M = Na;\ E = P, As{:}\ M = (dme)Li$$

Eq. 1. Synthesis of triisopropylsilylamine (**1a**), -phosphane (**1b**) and -arsane (**1c**).

Table 1 summarizes selected spectroscopic data of triisopropylsilylamine (**1a**), -phosphane (**1b**) and -arsane (**1c**). The low-field shift of the ^{29}Si nuclei of **1b** and **1c** is remarkable in contrast to the value of $\delta(^{29}Si) = 6.14$ for amine **1a**.

Table 1. Selected spectroscopic data of triisopropylsilylamine (**1a**), -phosphane (**1b**) and -arsane (**1c**) and $(Me_3Si)_3CZnN(H)Si\textit{i}Pr_3$ (**6**), $[(Me_3Si)_3CZn]_2PSi\textit{i}Pr_3$ (**8**) and $[(Me_3Si)_3CZn]AsSi\textit{i}Pr_3$ (**9**) (chemical shifts [ppm], coupling constants [Hz], stretching vibrations [cm^{-1}]).

Compound	1a [1]	1b [2]	1c	6	8 [8]	9
NMR:						
$\delta(E^1H_2)/^1J(P,H)$	–0.06	1.00/184.0	0.35			
$\delta(^{29}Si\textit{i}Pr_3)/^1J(P,Si)$	6.14	21.33/28.8	27.00	–4.51	32.37/48.2	38.98
$\delta(^{29}SiMe_3)$				–4.85	–5.29	–5.48
$\delta(^{31}P)$		–273.9			–297.1	
IR/RE:						
$\nu_{as}(E–H)$	3473	2296	2097	3366		
$\nu_s(E–H)$	3404	2292	2086			

The lithiation of **1a** to **1c** yields quantitatively the corresponding lithium derivatives **3a** to **3c**, whereas the reaction of **2** with *n*-butyllithium [6] in tetrahydrofuran gives the redistribution reaction according to Eq. 2. Fig. 1 shows the ^{31}P-NMR spectrum of the reaction mixture. The chemical shifts of $LiP(H)SiMe_3$ (**4a**: $\delta = -288.5$, $^1J(P,H) = 158.5$ Hz), $LiPH_2$ (**4b**: $\delta = -286.6$, $^1J(P,H) = 155.2$ Hz) and $LiP(SiMe_3)_2$ (**4c**: $\delta = -302.1$) as well as the $^1J(P,H)$ coupling pattern allow an assignment. Similar dismutation reactions were reported for $LiP(H)SiH_3$ [7] and seem to be characteristic for sterically unhindered lithium silylphosphanides.

$$\underset{\mathbf{4a}}{2\ LiP(H)SiMe_3} \rightleftarrows \underset{\mathbf{4b}}{LiPH_2} + \underset{\mathbf{4c}}{LiP(SiMe_3)_2}$$

Eq. 2. Dismutation reaction of lithium trimethylsilylphosphanide **4a**.

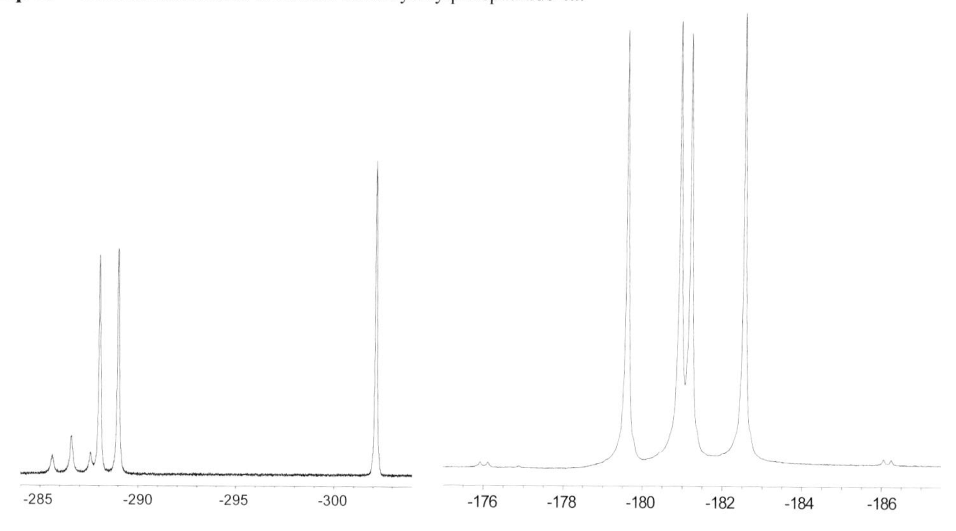

Fig. 1. Left: ^{31}P-NMR spectrum of the redistribution products of lithium trimethylsilyphosphanide **4a** (solvent: [D_8]tetrahydrofuran, 161.810 MHz). Right: ^{31}P-NMR spectrum of **10** (solvent: [D_8]tetrahydrofuran, 109.354 MHz).

$$2\ \mathrm{LiR} + 2\ \mathrm{ZnCl_2} \xrightarrow{\text{thf}} [(\mathrm{thf})_4\mathrm{Li}]^+\ [\mathrm{RZn}(\mu\text{-Cl})_3\mathrm{ZnR}]^- + \mathrm{LiCl} \xrightarrow{-4\ \text{thf}} 2\ \mathrm{RZnCl} + 2\ \mathrm{LiCl}$$

$R = C(SiMe_3)_3$ **5**

Eq. 3. Synthesis of tris(trimethylsilyl)methylzinc chloride 5.

The metathesis reaction of $LiC(SiMe_3)_3$ with anhydrous zinc(II) chloride [8] followed by a sublimation in vacuum yields tris(trimethylsilyl)methylzinc chloride **5** (Eq. 3). The metathesis reaction of lithium triisopropylsilylamide (**3a**) with **5** yields tris(trimethylsilyl)methylzinc triisopropylsilylamide (**6**) according to Eq. 4.

$$(\mathrm{Me_3Si})_3\mathrm{CZnCl} + \mathrm{LiN(H)Si^iPr_3} \rightarrow (\mathrm{Me_3Si})_3\mathrm{CZn\text{-}N(H)Si^iPr_3} + \mathrm{LiCl}$$

5 **3a** **6**

Eq. 4. Synthesis of tris(trimethylsilyl)methylzinc triisopropylsilylamide (**6**).

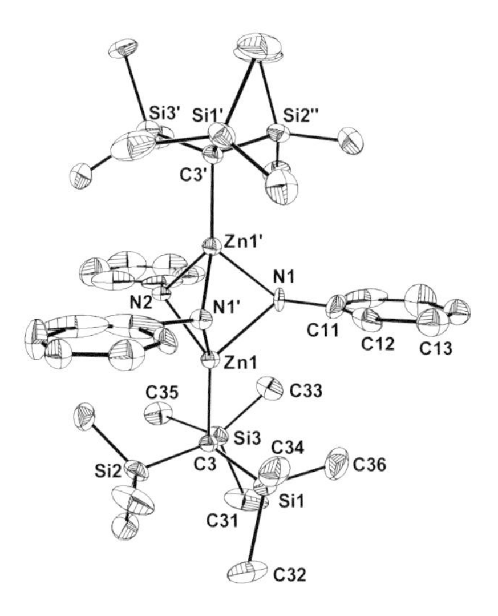

Fig. 2. Molecular structure of the anion bis[tris(trimethylsilyl)methylzinc] tris(anilide) of **7**. The hydrogen atoms are omitted for clarity. Selected bond lengths [pm]: Zn1–C3 204.6(5), Zn–N 210.6(6)–213.6(6).

If a smaller amide such as lithium phenylamide (anilide) is used in this metathesis reaction tetrakis(tetrahydrofuran-*O*)lithium bis[tris(trimethylsilyl)methylzinc] tris(μ-anilide) (**7**) is isolated according to Eq. 5. The central structural motif is a trigonal bipyramidal arrangement of the Zn_2N_3 moiety with the metal atoms in apical positions. Figure 2 shows the molecular structure of the zincate anion.

$$3\ \mathrm{LiN(H)Ph} + 2\ (\mathrm{Me_3Si})_3\mathrm{CZnCl} \xrightarrow{\text{thf}} [(\mathrm{thf})_4\mathrm{Li}]^+\ \{[(\mathrm{Me_3Si})_3\mathrm{CZn}]_2[\mu\text{-N(H)Ph}]_3\}^- + 2\ \mathrm{LiCl}$$

5 **7**

Eq. 5. Synthesis of tetrakis(tetrahydrofuran-*O*)lithium bis[tris(trimethylsilyl)methylzinc]tris[μ-anilide] (**7**)

$$2\ (Me_3Si)_3CZnCl + 2\ LiE(H)Si\mathit{i}Pr_3 \rightarrow [(Me_3Si)_3CZn]_2ESi\mathit{i}Pr_3 + 2\ LiCl + H_2ESi^{i}Pr_3$$

5 **3b/c** **8** (E = P) / **9** (E = As) **1b/c**

Eq. 6. Synthesis of **8** and **9**.

The reaction of **3b** and **3c** with $(Me_3Si)_3CZnCl$ (**5**) yields a doubly metalated phosphane **8** [9] and arsane **9**, respectively, according to Eq. 6. Selected spectroscopic data is listed in Table 1. These derivatives show a trigonal pyramidal coordination sphere at the pentel atom E with angle sums of 319.4° (E = P, **8** [9]) and 310.9° (E = As, **9**) as shown for **9** in Fig. 3.

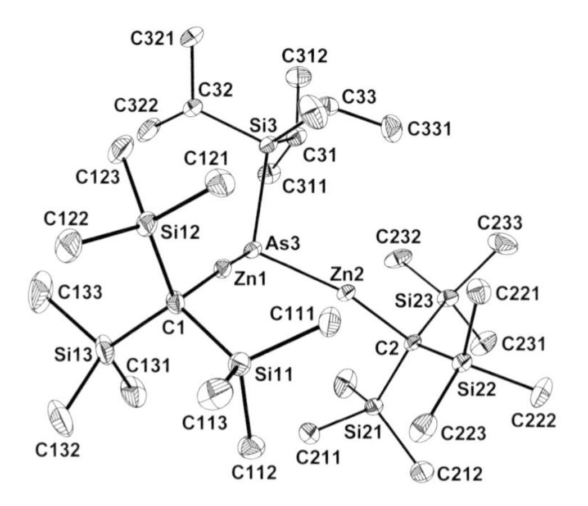

Fig. 3. Molecular structure and numbering scheme of bis[tris(trimethylsilyl)methyl-zinc] triisopropylsilylarsandiide **9**. The hydrogen atoms are omitted for clarity. Selected structural data (bond length in pm, angles in °): Zn1–As3 232.5(1), Zn2–As3 232.1(1), As3–Si3 235.4(1), Zn1–C1 197.1(4), Zn2–C2 197.1(4), Zn1-As3-Zn2 101.43(2), As3-Zn1-C1 165.1(1), As3-Zn2-C2 165.6(1).

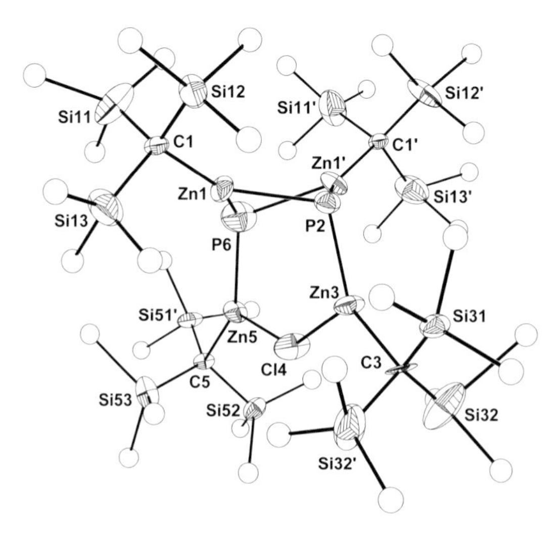

Fig. 4. Structure of the anion of **10**. The hydrogen atoms are omitted for clarity. Selected bond lengths: Zn1–P2 238.8(6), Zn1–P6 246.3(7), Zn1–C1197(1), Zn3–P2 230.3(7), Zn3–Cl4 226.7(12), Zn3–C3 199(2), Zn5–P6 232.3(10), Zn5–Cl4 221.1(12), Zn5–C5 199(2).

Neither the metathesis reaction of **5** with (dme)$LiPH_2$/(dme)$LiAsH_2$ nor that with Li_2PH [10]/Li_2AsH [11], which are known for more than fifty years, allow the isolation of $[(Me_3Si)_3CZn]_2EH$ with E = P, As. However, the reaction of **5** with lithiated trimethylsilylphosphane $LiPH_{2-x}(SiMe_3)_x$ [x = 0 (**4b**), 1 (**4a**) and 2 (**4c**)] yields lithium zincate **10**. The anion is best described as a dimeric bis[tris(trimethylsilyl)methylzinc] phosphandiide with the terminal $(Me_3Si)_3C$-Zn moieties in a *cis* configuration. Between these zinc atoms a chloride anion is bonded as shown in Fig. 4 whereas the lithium cation is coordinated tetrahedrally by four tetrahydrofuran molecules.

In Fig. 1 the ^{31}P-NMR spectrum of **10** is shown. The resonance is observed at $\delta(^{31}P) = -181.1$ with remarkably large coupling constants of $^2J(P,P) = 539.7$ Hz, $^1J(P,H) = 250.6$ Hz, $^3J(P,H) = 70,7$ Hz, and $^4J(H,H) = 10.1$ Hz. The P–H stretching modes are detected at 2354 and 2328 cm^{-1} and are shifted to rather high wave numbers. The only other zincate with a triply zinc-bonded phosphorus atom reported so far is the adamantane-like dianion $[(IZn)_6(PSiMe_3)_4]^{2-}$ with Zn–P distances between 230 and 234 pm [12].

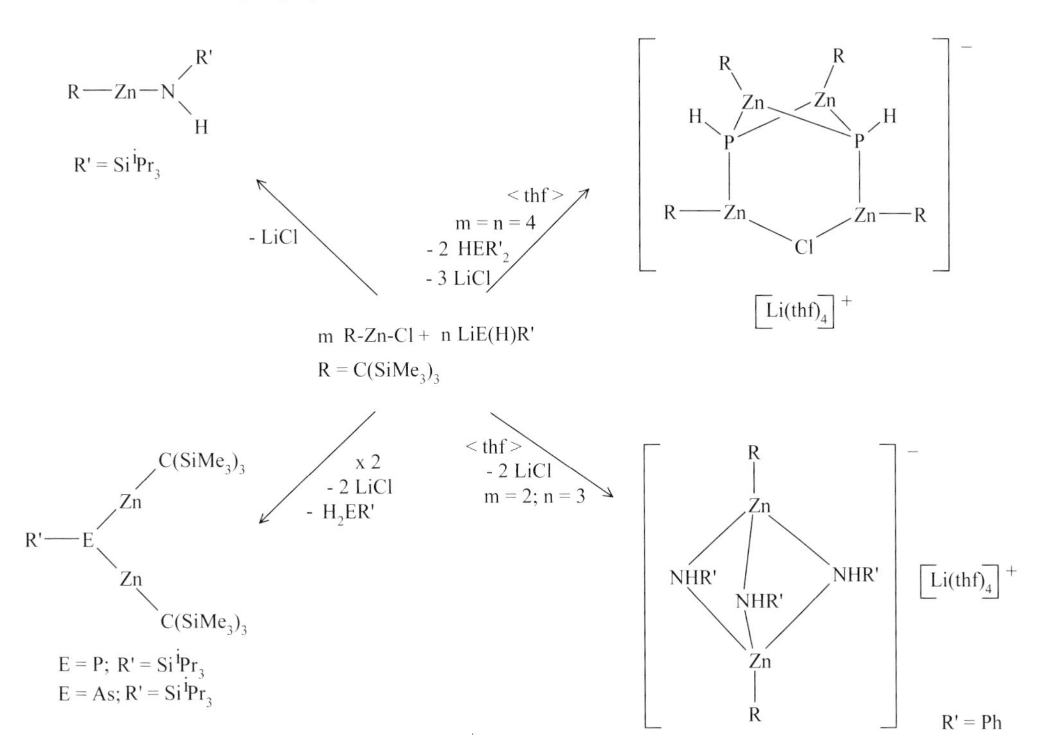

Scheme 1. Arrangement of the metathesis reactions of **5** with lithium amides, phosphanides and arsanides.

In Scheme 1 the metathesis reactions of **5** with lithium amides, phosphanides and arsanides are summarized. The sterically demanding triisopropylsilyl group leads to an alkylzinc-substituted amine and to bis(alkylzinc) substituted phosphanes and arsanes. Smaller substituents at the central atom give solvent-separated lithium zincates with a central Zn_2E_2 moiety due to oligomerization, as is also common for alkylzinc phosphanides [12,13] and arsanides [14].

Acknowledgment: We thank the *Deutsche Forschungsgemeinschaft* and the *Fonds der Chemischen Industrie* for generous financial support.

References:

[1] a) D. Gudat, H. M. Schiffner, M. Nieger, D. Stalke, A. J. Blake, H. Grondey, E. Niecke, *J. Am. Chem. Soc.* **1992**, *114*, 8857.
b) H.-J. Goetze, B. Bartylla, M. Ismeier, *Spectrochim. Acta* **1993**, *49*, 497.

[2] M. Westerhausen, R. Löw, W. Schwarz, *J. Organomet. Chem.* **1996**, *513*, 213.

[3] a) H. Schäfer, G. Fritz, W. Hölderich, *Z. Anorg. Allg. Chem.* **1977**, *428*, 222.
b) M. Baudler, K. Glinka, *Inorg. Synth.* **1990**, *27*, 228.

[4] a) G. Becker, M. Schmidt, M. Westerhausen, *Z. Anorg. Allg. Chem.* **1992**, *607*, 101.
b) G. Becker, D. Käshammer, O. Mundt, M. Westerhausen in: *Synthetic Methods of Organometallic and Inorganic Chemistry* (Ed. H. H. Karsch), (Herrmann/Brauer), Vol. 3; G. Thieme, Stuttgart, **1996**; p. 189.

[5] H. Bürger, U. Goetze, *J. Organomet. Chem.* **1968**, *12*, 451.

[6] G. Fritz, H. Schäfer, W. Hölderich, *Z. Anorg. Allg. Chem.* **1974**, *407*, 266.

[7] G. Becker, B. Eschbach, O. Mundt, M. Reti, E. Niecke, K. Issberner, M. Nieger, V. Thelen, H. Nöth, R. Waldhör, M. Schmidt, *Z. Anorg. Allg. Chem.* **1998**, *624*, 469.

[8] M. Westerhausen, B. Rademacher, W. Schwarz, J. Weidlein, S. Henkel, *J. Organomet. Chem.* **1994**, *469*, 135.

[9] M. Westerhausen, M. Wieneke, K. Doderer, W. Schwarz, *Z. Naturforsch.* **1996**, *51b*, 1439.

[10] a) C. Legoux, *Bull. Soc. Chim.* **1940**, *7*, 545.
b) C. Legoux, *Ann. Chim.* **1942**, *17*, 100.

[11] C. Legoux, *Bull. Soc. Chim.* **1940**, *7*, 549.

[12] A. Eichhöfer, D. Fenske, O. Fuhr, *Z. Anorg. Allg. Chem.* **1997**, *623*, 762.

[13] a) A. M. Arif, A. H. Cowley, R. A. Jones, A. Richard, S. U. Koschmieder, *J. Chem. Soc., Chem. Commun.* **1987**, 1319.
b) B. L. Benac, A. H. Cowley, R. A. Jones, C. M. Nunn, T. C. Wright, *J. Am. Chem. Soc.* **1989**, *111*, 4986.
c) S. C. Goel, M. Y. Chiang, W. E. Buhro, *J. Am. Chem. Soc.* **1990**, *112*, 5636.
d) S. C. Goel, M. Y. Chiang, D. J. Rauscher, W. E. Buhro, *J. Am. Chem. Soc.* **1993**, *115*, 160.
e) A. J. Edwards, M. A. Paver, P. R. Raithby, C. A. Russell, D. S. Wright, *Organometallics* **1993**, *12*, 4687.
f) A. Eichhoefer, J. Eisenmann, D. Fenske, F. Simon, *Z. Anorg. Allg. Chem.* **1993**, *619*, 1360.
g) M. A. Matchett, M. Y. Chiang, W. E. Buhro, *Inorg. Chem.* **1994**, *33*, 1109.
h) B. Rademacher, W. Schwarz, M. Westerhausen, *Z. Anorg. Allg. Chem.* **1995**, *621*, 287.
i) J. Eisenmann, D. Fenske, F. Simon, *Z. Anorg. Allg. Chem.* **1995**, *621*, 1681.
j) M. G. Davidson, A. J. Edwards, M. A. Paver, P. R. Raithby, C. A. Russell, A. Steiner, K. L. Verhorevoort, D. S. Wright, *J. Chem. Soc., Chem. Commun.* **1995**, 1989.

[14] B. Rademacher, W. Schwarz, M. Westerhausen, *Z. Anorg. Allg. Chem.* **1995**, *621*, 1439.

Silicon and Germanium Amidinates

Hans H. Karsch*[*], Peter A. Schlüter,

Anorganisch-chemisches Institut der Technischen Universität München
Lichtenbergstr. 4, D-85747 Garching, Germany
Tel.: Int. code + (89)28913132— Fax: Int. code + (89)28914421
E- mail: Hans.H.Karsch@lrz.tu-muenchen.de

Keywords: Silicon Amidinates / Germanium Amidinates / Bidentate Coordination / Pentacoordination / Hexacoordination

Summary: The synthesis of novel amidinate silicon and germanium complexes is described. It is shown that the coordination chemistry of amidinate ligands at Si(IV), Ge(II) and Ge(IV) metal centers is strongly influenced by the substituents on the nitrogen and carbon atom. The reaction of different stoichiometric amounts of lithium amidinates with $GeCl_4$ and $SiCl_4$ leads to mono-, di- and trisubstituted compounds, while the reaction of 2 equiv. NH(*t*Bu)C(Ph)NLi with $GeCl_2$·dioxane yields a tetrameric species. Therefore monoanionic, bidentate four-electron donor amidinate ligands are highly suitable for stabilizing low oxidation states and/ or high coordination numbers at group 14 element centers (Si, Ge). A newly synthesized hexacoordinated *cis*-dichloro-silicon (bis)amidinate complex possibly might be a promising candidate as a precursor for new stable Si(II) compounds.

Introduction

The coordination chemistry of amidinates is well established, not only for the transition metals [1, 2] and the lanthanoids [3], but also for the main group elements [4]. However, amidinates of silicon and germanium are almost unknown with only a few exceptions [5]. With the aim of creating an access to a new generation of Si–N compounds, we started a comprehensive study on silicon and germanium amidinate chemistry. With regard to the synthesis of new materials containing Si–N bonds, silicon amidinates seem to be promising precursors for conversion to polymeric materials of different but easily tunable structural units ressembling silicate structures but with pseudo-chalcogenide bridges (i.e. $[N–C(R)–N(R')]^{2-}$ instead of O^{2-}) and might further be converted to Si/C/N/X materials.

Results and Discussion

Reaction of Lithium Amidinates with $SiCl_4$ and $GeCl_4$

Suitable starting materials for the synthesis of silicon or germanium amidinate complexes are lithium amidinates, which are easily accessible by the reaction of lithium amides with nitriles (**A**) or by addition of organolithium compounds to carbodiimides (**B**) (Scheme 1).

$$\text{Ph—C}\equiv\text{N} + \text{R'NHLi} \longrightarrow \text{Ph—C(NHLi)=N—R'} \qquad \textbf{A}$$

R' = *i*Pr, Cy, *t*Bu, Mes

$$\text{R'—N=C=N—R'} + \text{RLi} \longrightarrow \text{R'—N}\overset{\text{R—C}}{\frown}\text{N—R'}\ (\text{Li}) \qquad \textbf{B}$$

R = Me, Mes, *t*Bu; R' = *i*Pr, *t*Bu, Cy

Scheme 1. Preparation of lithium amidinates.

Reactions of different stoichiometric amounts of the lithium amidinates (**A**) with $SiCl_4$ result in mono-, bis- and trisubstituted silicon amidinate derivates (Eq. 1).

$$x\ [\text{NH(R')C(Ph)=N]Li} + \text{XCl}_4 \xrightarrow[-\,x\ \text{LiCl}]{\text{Et}_2\text{O}} [\text{NH(R')C(Ph)N]}_x\text{XCl}_{4-x}$$

x = 1, 2, 3; X = Si, Ge

Eq. 1.

In the case of the monosubstituted amidinate complexes, a crystallographic study of **2c** (Fig. 1) confirms a monodentate coordination of the amidinate ligand to the silicon center. The C–N bonds are nearly equal. The Si–N bond (1.625(3) Å) is remarkably short and indicates some double bond character [6]. The analogous germanium structure (Fig. 2), apart from the longer Ge–N bond (1.776(2) Å), is quite similar. Though the N–H hydrogen atom has not been located in either case, it seems plausible that an N(2)–H functionality is present.

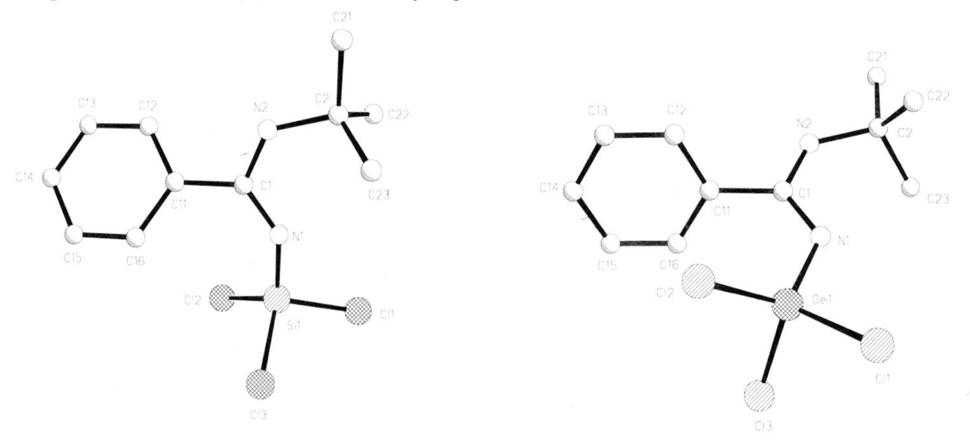

Fig. 1. Molecular structure of NH(*t*Bu)C(Ph)NSiCl_3 (**2c**). Selected bond distances [Å] and angles [°]: Si(1)–N(1): 1.625(3), C(1)–N(1): 1.300(4), C(1)–N(2): 1.333, N(1)-C(1)-N(1): 121.1.

Fig. 2. Molecular structure of NH(*t*Bu)C(Ph)NGeCl_3 (**6c**). Selected bond distances [Å] and angles [°]: Ge(1)–N(1): 1.776(2), N(1)–C(1): 1.304(3), N(2)–C(1): 1.339(2), N(1)-C(1)-N(2): 120.9(2).

Compound **2c** was used as a model system to induce an elimination of hydrochloride in order to create molecular Si/N-compounds with bridging or chelating pseudo-chalcogen structural novelties (Eq. 2). In no case was a successful formation of **5c** or oligomers thereof obtained.

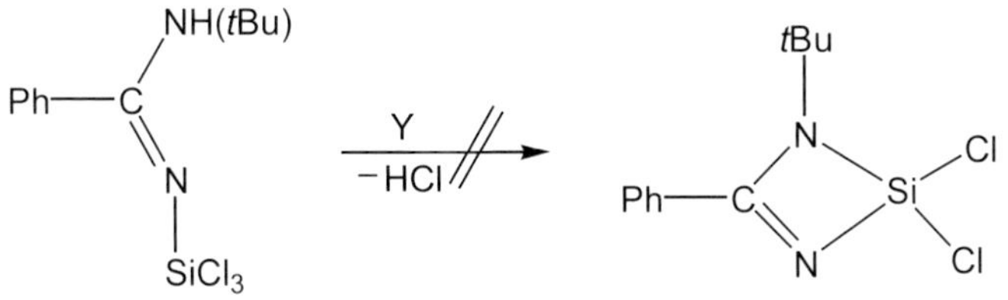

Y	Product
*n*BuLi	Substitution of the Si-Cl bond
DABCO/ toluene/ RT	No reaction
DABCO/ toluene/ 3 d/ 100°C	No reaction
*t*BuLi	Mixture of various Si-alkyl compounds
LDA	Substitution of the Si-Cl bond
Et_3N/ $SiCl_4$	No reaction
"Proton sponge"	?

Eq. 2. Experiments to induce an HCl elimination in **2c**.

The reaction of various amounts of *N,N'*-alkyl-substituted lithium amidinates **B** with $SiCl_4$ or $GeCl_4$ leads to the mono,- di- and trisubstituted amidinate complexes (Scheme 2) .

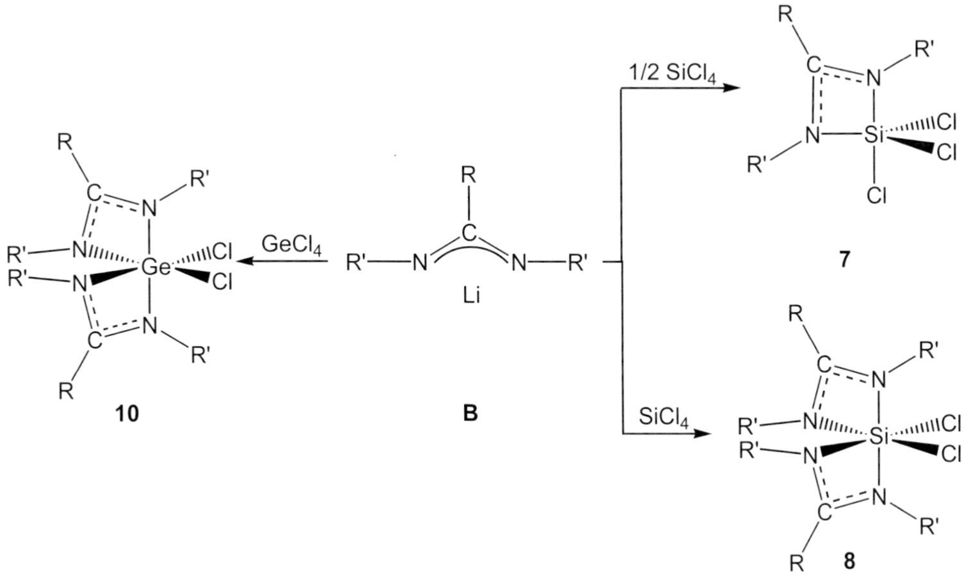

Scheme 2. Reaction of **B** with $GeCl_4$ and $SiCl_4$.

^{29}Si-NMR spectroscopic investigations confirm a bidentate coordination of the amidinate ligand in **7** and **8** indicated by a δ ^{29}Si high field shift (Table 1). These results are also confirmed by crystallographic studies on two examples.

Table 1. ^{29}Si-NMR data for the mono- and disubstituted silicon amidinates.

R'—N=C(R)—N—R'	R'	R	Number of ligands	δ ^{29}Si-NMR
7a	Cy	Me	1	–97.30
7b	*i*Pr	Me	1	–98.02
7c	Cy	Mes	1	–98.57
7d	Cy	*t*Bu	1	–89.16
7e	*t*Bu	Me	1	–99.57
8a	Cy	Me	2	–168.59
8b	*i*Pr	Me	2	–169.33
8d	Cy	*t*Bu	2	–168.76

In compound **7a** (Fig. 3) the amidinate ligand is coordinated in a bidentate fashion, with one nitrogen in an axial and one nitrogen in an equatorial position of a Ψ-trigonal bipyramid. The equatorial Si–N bond length is slightly shorter than the axial Si–N distance. The four membered Si–N–C–N ring is essentially planar.

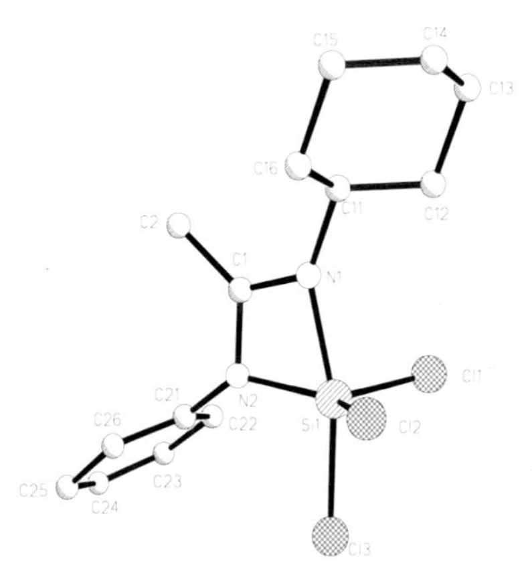

Fig. 3. Molecular structure of **7a**. Selected bond distances [Å] and angles [°]: Cl(1)–Si(1): 2.0632(9), Cl(2)–Si(1): 2.076(1), Cl(3)–Si(1): 2.1439(9), Si(1)–N(1): 1.922(2), Si(1)–N(2): 1.775(2), N(1)–C(1): 1.300(3), N(2)–C(1): 1.353(4), N(1)-C(1)-N(2): 106.3.

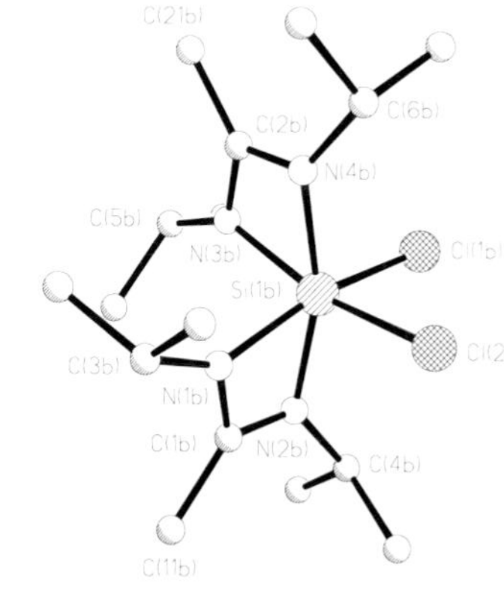

Fig. 4. Molecular structure of **8b**. Selected bond distances [Å] and angles [°]: Cl(1b)–Si(1b): 2.185(1), Cl(2b)–Si(1b): 2.197(1), Si(1b)–N(4b): 1.841(2), Si(1b)–N(2b): 1.837(2), Si(1b)–N(1b): 1.915(2), Si(1b)–N(3b): 1.914(2), N(1b)–C(1b): 1.312(4), N(2b)–C(1b): 1.333(4), N(4b)-Si(1b)-N(2b): 163.0(1).

The X-ray structure determination of the colorless crystals of **8b** (Fig. 4) shows a hexacoordinated silicon surrounded by two amidinate ligands, each placing one nitrogen in an axial and one nitrogen in an equatorial position within a distorted octahedron. The two chlorine atoms in the equatorial *cis* position complete the coordination sphere. As expected, the axial Si–N bonds are shorter than the equatorial ones. Analogously to **7a** the four-membered Si–N–C–N rings are planar. In the ^{1}H-NMR spectrum four doublets of equal intensity in the aliphatic region due to the methyl protons of the isopropyl groups, and two septets due to the CH protons of the isopropyl groups of the amidinate ligand, are observed. Temperature-dependent measurements establish a rigid molecular skeleton in the range of –80°C to +80°C. In contrast to these findings, the analogous bis(amidinate) Ge(IV) compound **10b** shows fluctional behavior: the four dublets observed at room temperature broaden on heating and coalesce to one singlet at 80°C.

Reaction of Lithium Amidinates with $GeCl_2$·dioxane

The reaction of 2 equiv. of NH(*t*Bu)C(Ph)NLi (**1c**) with $GeCl_2$·dioxane does not result in the formation of the bis(amidinate) Ge(II) complex. However, Ge_4[N(*t*Bu)C(Ph)N]$_4$ **6c,** a heterocuban type molecule with pseudochalcogenide bridges $[N\text{-}C(Ph)N(tBu)]^{2-}$ is obtained (Eq. 2). One equiv. of the lithium amidinate acts as base: the formation of the resulting amidine HN(*t*Bu)C(Ph)NH is confirmed by spectroscopic results.

$$8\ NH(tBu)C(Ph)NLi\ +\ 4\ GeCl_2\bullet dioxane \xrightarrow[-\ HN(tBu)C(Ph)NH]{THF} Ge_4[N(tBu)C(Ph)N]_4$$

6c

Eq. 3.

In contrast to Eq. 3, the reaction of two equiv. of the lithium amidinate [MeC(N*i*Pr)$_2$Li]•THF with $GeCl_2$·dioxane leads to the formation of the bis (amidinate)–germanium complex **11b** (Eq. 4).

Eq. 4. Reaction of [RC(NR')$_2$]Li with $GeCl_2$·dioxane

The X-ray structure analysis of **11b** (Fig. 5) shows, that the germanium atom is surrounded by four nitrogen atoms of the amidinate ligands with one nitrogen in an axial and one in an equatorial position of a heavily distorted Ψ-trigonal-bipyramid, with the lone pair of germanium(II) occupying the third equatorial position. In **11b** only the equatorial N atoms are planar. The equatorial Ge–N bonds are significantly shorter than the axial Ge–N bonds, which emphasizes the considerable ring strain in the spirocyclic germanium compound [7].

^{1}H-NMR-measurement in the range of –80 to 80°C shows fluctional behavior of the axial and equatorial positions: only one type of signal for the isopropyl groups can be observed.

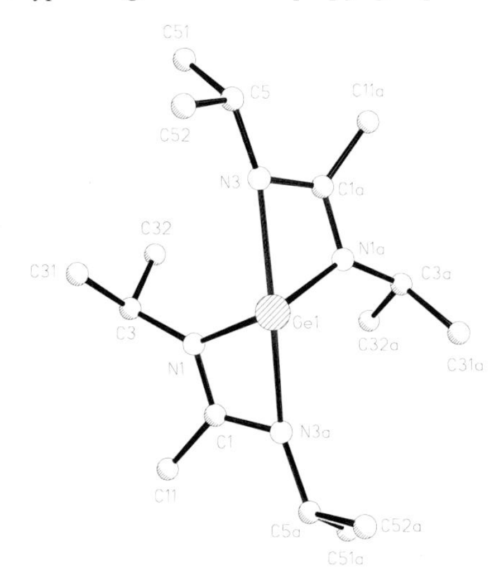

Fig. 5. Molecular structure of **11b**. Selected bond distances [Å] and angles [°]: Ge(1)–N(1): 1.982(3), Ge(1)–N(3): 2.300(2), N(1)–C(1): 1.325(4), C(1)–N(3a): 1.302(3), N(1)-Ge(1)-N(1a): 100.7(1), N(1)-Ge(1)-N(3): 96.2(9), N(3)-Ge(1)-N(3a): 145.7(1), N(1)-Ge(1)-N(3a): 61.0(9).

Conclusions

Monoanionic, mono- and bidentate four-electron donor amidinates are highly suitable ligands for stabilizing high coordination numbers and/or low oxidation states at group 14 element centers. The newly synthesized *cis*-dichloro amidinate complex **8b** may be a hopeful precursor for novel Si(II) compounds. Although the creation of compounds with dianionic pseudo-chalcogenide bridging ligands at group 14 element centers can be realized for Ge(II), there are hopeful and promising ideas for a successful realization also at Ge(IV), Si(IV) and eventually Si(II) centers.

References:

[1] D. Walther, R. Fischer, M. Friedrich, P. Gebhardt, H. Görls, *Chem. Ber.* **1996**, *129*, 1389.

[2] W. Hiller, J. Stähle, A. Zinn, K. Dehnicke, *Z. Naturforsch. Teil B.* **1989**, *44*, 999.

[3] M. Wedler, M. Noltemeyer, U. Pieper, H. G. Schmidt, D. Stahlke, F. T. Edelmann, *Angew. Chem.* **1990**, *102*, 941; *Angew. Chem., Int. Ed. Engl.* **1990**, *29*, 894.

[4] a) Y. Zhou, D. S. Richeson, *Inorg. Chem.* **1996**, *35*, 1423; b) Y. Zhou, D. S. Richeson, *Inorg. Chem.* **1996**, *35*, 2448; c) M. P. Coles, D. C. Swenson, R. F. Jordan, *Organometallics* **1997**, *16*, 5183.

[5] a) A. R. Sanger, *Inorg. Nucl. Chem. Lett.* **1973**, *9*, 351; b) K. Dehnike, *Chem. Zeit.* **1990**, *114*, 295; c) H. H. Karsch, P. A. Schlüter, F. Bienlein, M. Heckel, M. Herker, A. Sladek, E. Witt, *Z. Anorg. Allg. Chem.* **1998,** *624*, 295–309**;** d)H. H. Karsch, P. A. Schlüter, M. Reisky, *Eur. J. Inorg. Chem..* **1998**, 433–436.

[6] D. Großkopf, J. Klingebiel, J. Niesmann, "*Synthesis and Reactions of Iminosilanes*", in: *Organosilicon Chemistry II: From Molecules to Materials* (Eds.: N. Auner, J. Weis), VCH, Weinheim, **1996**, p. 127.

[7] Note added in Proof: The structures of two closely related Ge(II)amidinates exhibit a tri-coordinated Ge(II) center: S. R. Foley, C. Bensimon, D. S. Richeson, *J. Am. Chem. Soc.* **1997**, *119*, 10359.

Insertion of Aromatic Diisocyanates into the N–Si Bonds of 1,4-Bis(trimethylsilyl)-1,4-dihydropyrazine

Torsten Sixt, Karl-Wilhelm Klinkhammer, Anja Weismann, Wolfgang Kaim

Institut für Anorganische Chemie der Universität Stuttgart
Pfaffenwaldring 55, D-70550 Stuttgart, Germany
Tel.: Int. code + (711)685 4170 — Fax: Int. code + (711)685 4165
E-mail: kaim@iac.uni-stuttgart.de

Keywords: Crystal Structure / Diisocyanates / Heterocycles / Silicon–Nitrogen Bonds / Insertion

Summary: Combining bifunctional phenylenediisocyanates O=C=N–(C_6H_4)–N=C=O, C_6H_4 = *m*- or *p*-phenylene, with the electron-rich 8 π-electron heterocycle 1,4-bis(trimethylsilyl)-1,4-dihydropyrazine **1** yielded products involving the insertion of both isocyanate functions into the nitrogen–organosilicon bonds. The crystal structure of the 1:2 bis(insertion) product with *m*-phenylenediisocyanate is reported, revealing N-silylation and formation of urea functions.

Introduction

Aromatic diisocyanates are large-scale intermediate products in the industrial synthesis of polyurethanes and related polymers [1]. On the other hand, 1,4-bis(trimethylsilyl)-1,4-dihydropyrazine **1** is a thermally stable but very electron-rich molecule with 8 cyclically conjugated π-electrons in a planar six-membered heterocyclic ring (antiheteroaromaticity) [2]. The biologically relevant [3] ring system can be stabilized by 1,4-trialkylsilyl [4] or by other acceptor substituents [5]. The insertion of heterocumulenes [6] such as CO_2, COS, CS_2 or R–NCO into the N–Si bond has been studied for **1** and related molecules [7–10]. The reaction with monoisocyanates R–NCO always yielded bis-insertion products [7] so that the use of bifunctional diisocyanates was expected to produce linear oligomers or polymers [11]. After hydrolysis of the triorganosilyl groups the resulting polyurea species can be envisaged to even form aramide-like arrangements [10].

Herein we report results from the reaction between **1** and *m*- or *p*-phenylenediisocyanates OCN–(C_6H_4)–NCO.

Results and Discussion

The reaction in dilute solution between *m*-OCN–(C_6H_4)–NCO and **1** in refluxing pentane produced a precipitating, pale yellow 1:2 bis-insertion adduct **2** as the major product (m.p. 98°C).

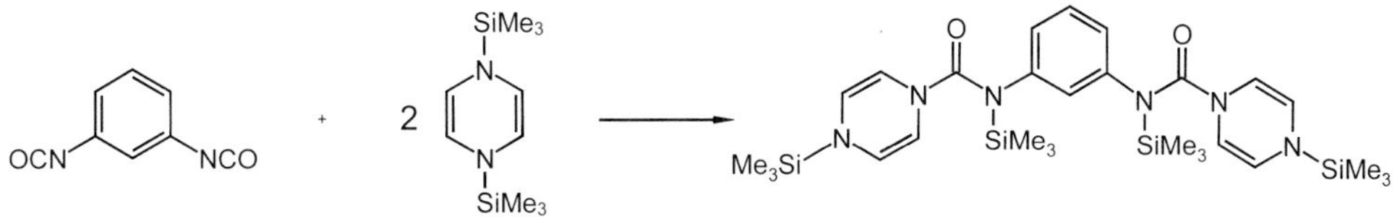

Scheme 1.

The crystallinity of **2** allowed us to determine the crystal structure. Previous ^{1}H-NMR studies [7] of 2:1 adducts between R–NCO molecules and 1,4-bis(trialkylsilyl)-1,4-dihydropyrazines have indicated the formation of both urea and isourea functionalities (Scheme 2).

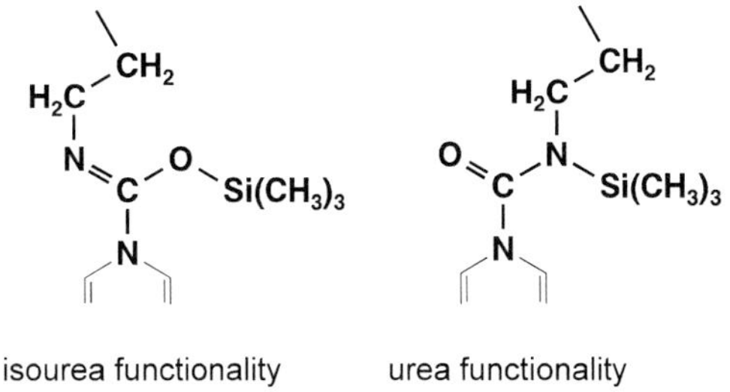

Scheme 2.

The crystal structure of **2** should answer the question of which of the various possible [7] positional and constitutional isomers is formed; a previous study of the 1:2 addition product between the alkyl diisocyanate OCN–$(CH_2)_4$–NCO and **1** has clearly shown N-silylation, i.e. the urea alternative [10].

A suitable single crystal of **2** for X-ray diffraction was obtained by cooling to –17°C of a saturated solution in 1,2-dimethoxyethane [12]. Figure 1 shows the structure of one of the crystallographically independent molecules in the crystal and the atomic numbering, the caption summarizes selected bond lengths and angles for this particular molecule. These data are very similar to those of the other molecule in the crystal, the main difference lying in the slightly different torsion angles involving the silylurea moiety.

The structure of **2** in the crystal confirms two *N*-silylurea arrangements situated in an approximate "trans" configuration relative to the central phenylene ring. As noted previously for a non-aromatic analogue [10], the bond lengths at the urea sites are quite regular, including very similar (O)C–N(dihydropyrazine) and (O)C–N($SiMe_3/C_6H_4$) bonds. The 1-trimethylsilyl-4-carbamoyl-substituted 1,4-dihydropyrazine rings are no longer planar [2] but exhibit a slight boat conformation.

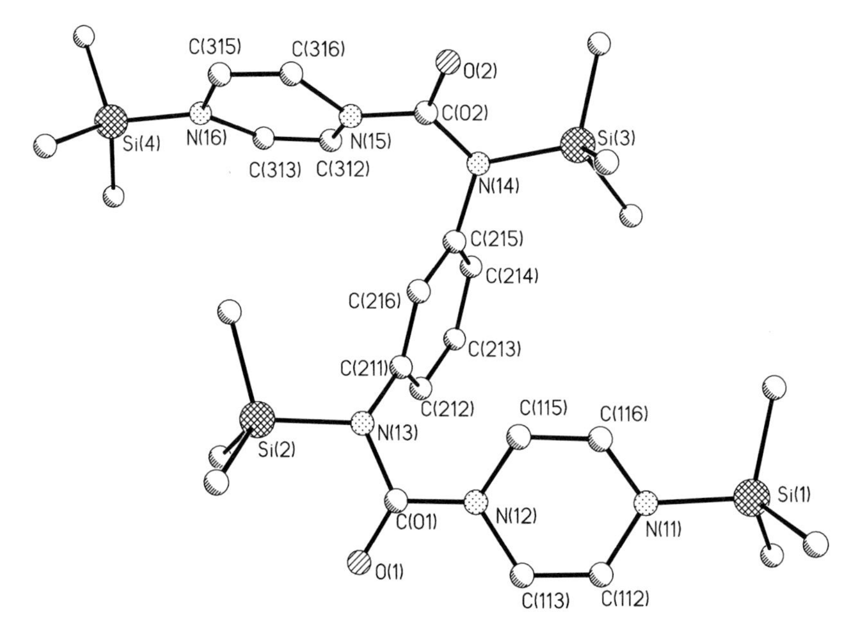

Fig. 1. Molecular structure of one of the two crystallographically independent molecules of **2**. Selected bond lengths [pm] and angles [°]: C(01)–O(1) 121.8(9), C(02)–O(2) 122.7(10), C(01)–N(12) 137.1(9), C(01)–N(13) 140.6(10), C(02)–N(14) 139.5(10), C(02)–N(15) 139.1(10), C(112)–C(113) 131.9(10), C(115)–C(116) 133.4(10), C(312)–C(313) 131.0(11), C(315)–C(316) 131.1(11), N(14)–C(215) 145.0(9), N(13)–C(211) 145.5(9), N(14)–Si(3) 179.2(6), N(13)–Si(2) 179.0(6), N(16)–Si(4) 174.4(7), N(11)–Si(1) 176.1(7); O(1)-C(01)-N(13) 122.7(7), O(1)-C(01)-N(12) 122.2(8), N(12)-C(01)-N(13) 115.0(7), O(2)-C(02)-N(15) 121.4(8), O(2)-C(02)-N(14) 121.2(8), N(14)-C(02)-N(15) 117.4(8).

For *p*-phenylenediisocyanate, *p*-OCN–(C_6H_4)–NCO, the reaction with **1** at different molar ratios and in several solvents (alkanes, chloroalkanes) always yielded an oligomeric product **3** as a yellowish, poorly soluble substance (Scheme 3).

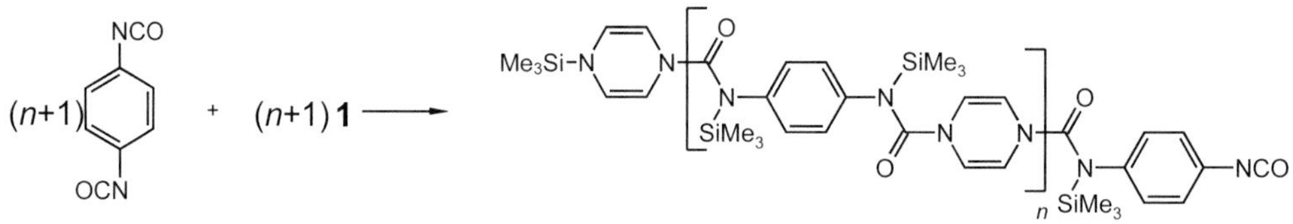

Scheme 3.

NMR data for **2** and **3** are listed in Table 1. NMR-spectroscopy in solution could be employed to detect the dynamic behavior of **2** (Fig. 2) and to analyze the average chain length of the oligomer **3**. Figure 2 illustrates how the region of 1,4-dihydropyrazine proton resonances changes for **2** between 293 K and 203 K in CD_2Cl_2.

Table 1. ^{1}H-NMR and ^{13}C-NMR data for **2** (in CD_2Cl_2) and **3** (in [D_8]THF) at 293 K.

		SiCH		Phenylene[a]		HC(Pz)		NCO	C=O
δ ^{1}H	**2**	0.03[c]	s	6.56	H^5	4.96[b]	s,br		
		0.19[e]	s	6.73	$H^{4,6}$	5.40[b]	s,br		
				7.17	H^2				
	3	0.04[c]	br	6.6 –7.7		4.77	br		
		0.12[d]	br			5.32	br		
		0.21[e]	br			5.53	br		
		0.31[e]	br			5.98	br		
δ ^{13}C	**2**	–1.58[c]		123.9	C^5	109.1			154.4[f]
		0.48[e]		125.8	$C^{4,6}$	118.2			
				129.2	C^2				
				143.3	$C^{1,3}$				
	3	–1.79[c]		125.6		108.7		121.4	148.7[f]
		0.19[e]		126.5		109.8			155.7[f]
				139.8		113.3			
				138.8		120.9			

[a] Broadened signals. — [b] For temperature dependence see Fig. 2. — [c] (Pz)N–Si<u>CH</u>.— [d] (N=C)O–Si<u>CH</u>. — [e] (O=C)N–Si<u>CH</u>.— [f] Si(Pz)<u>C</u>=O.

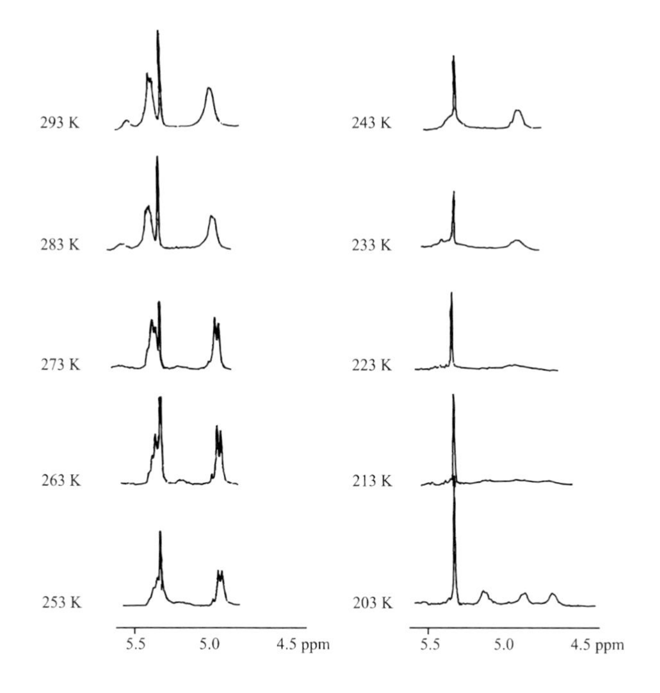

Fig. 2. Changes in the ^{1}H-NMR spectrum (1,4-dihydropyrazine proton region) of compound **2** in CD_2Cl_2 (signal at 5.32 ppm: $CHDCl_2$).

The complicated and not yet fully analyzed dynamic ^{1}H-NMR behavior of **2** is probably caused by two kinds of hindered rotations around the different carbonyl-C/nitrogen bonds (Scheme 4); both processes can occur twice in molecule **2**.

Scheme 4.

At 203 K four broad signals are visible, signifying an ABXY spin system in an almost frozen conformation; after coalescence at about 223 K the signals of an AA'BB' system emerge.

The ^{1}H-NMR spectroscopic differentiation between trimethylsilyl protons (pz)N–$SiMe_3$ (terminal) and (O=C)N–$SiMe_3$ (internal) for the oligomer **3** allowed us to obtain an estimate of about 30 for the average chain length (*n* in Scheme 3). Otherwise the resonances are of little value because of generally broadened lines.

Acknowledgment: This work was supported by the *Fonds der Chemischen Industrie.*

References:

[1] D. Dieterich, *Methoden der Organische Chemie (Houben-Weyl), Vol. E20/2*, 4th edn., Georg Thieme Verlag, Stuttgart, **1987**, p 1561; L. Born, H. Hespe, *Coll. Polym. Sci.* **1985**, *263*, 335; G. Oertel, *Kunststoffhandbuch*, *Vol. 7: Polyurethane*, Hanser Verlag, München, **1983**; M. D. Purgett, W. Deits, O. Vogl, *J. Polym. Sci., Polym. Chem. Ed.* **1982**, *20*, 2477.

[2] H.-D. Hausen, O. Mundt, W. Kaim, *J. Organomet. Chem.* **1985,** *296*, 321.

[3] W. Kaim, *Angew. Chem.* **1983**, 95, 201*; Angew. Chem., Int. Ed. Engl.* **1983**, *22*, 171.

[4] J. Baumgarten, C. Bessenbacher, W. Kaim, T. Stahl, *J. Am. Chem. Soc.* **1989**, *111*, 2126; A. Lichtblau, A. Ehlend, H.-D. Hausen, W. Kaim, *Chem. Ber.* **1995**, *128*, 745.

[5] R. Gottlieb, W. Pfleiderer, *Liebigs Ann. Chem.* **1981**, 1451.

[6] M. F. Lappert, B. Prokai, *Adv. Organomet. Chem.* **1967**, *8*, 243.

[7] W. Kaim, A. Lichtblau, T. Stahl, E. Wissing, in: *Organosilicon Chemistry II: From Molecules to Material* (Eds.: N. Auner, J. Weis), VCH, Weinheim, **1994**, p. 41.

[8] A. Ehlend, H.-D. Hausen, W. Kaim, *J. Organomet. Chem* **1995**, *501*, 283.

[9] A. Ehlend, H.-D. Hausen, W. Kaim, in *Organosilicon Chemistry II: From Molecules to Material* (Eds.: N. Auner, J. Weis), VCH, Weinheim, **1995**, p. 141.

[10] T. Sixt, F. M. Hornung, A. Ehlend, W. Kaim, in: *Organosilicon Chemistry III: From Molecules to Material,* (Eds.: N. Auner, J. Weis), Wiley–VCH, Weinheim, **1998**, p. 364.

[11] J. F. Klebe, J. B. Busch, J. E. Lyons, *J. Am. Chem. Soc.* **1964**, *86*, 4400.

[12] a) Crystal data for **2**: $C_{28}H_{48}N_6O_2Si_4$, M = 613.08 g/mol, orthorhombic ($Pca2_1$; a = 20.592(4) pm, b = 13.783(4) pm, c = 25.100(11) pm, $\alpha = \beta = \gamma$ =90°, V = 7124(4) × 10^6 pm^3, Z = 8, ρ_{calc}

= 1.143 g·cm^{-3}, μ(MoK_α)= 2.06 cm^{-1}; 7786 reflections (6762 independent, 3.24° < 2θ < 50.02° ; h = –5 to 24, k = –4 to 16, l = –28 to 29) were collected at –90°C on a 0.4 mm × 0.1 mm × 0.15 mm crystal using monochromatic Mo-$K_\alpha$radiation; $F(000)$ = 2640, $R_1[I > 2\sigma(I)]$ = 0.0638 [12b], wR_2 = 0.1386 [12c], GOF = 1.060 [12d], The structure was solved by direct methods using the SHELXTL-PLUS package [13a], the refinement was carried out with SHELXL-93 [13b] employing full-matrix least-squares methods. Anisotropic thermal parameters were refined for all non-hydrogen atoms. The hydrogen atoms were located and refined isotropically. Further information on the structure determination may be obtained from Fachinformationszentrum Karlsruhe GmbH, D-76344 Eggenstein-Leopoldshafen, Germany, on quoting the depository number CSD 408687, the names of the authors, and the book citation.

b) $R = (\Sigma||F_o| - |F_c||)/\Sigma|F_o|$.

c) $wr_2 = \{\Sigma[w(|F_o|^2 - |F_c|^2)^2]/\Sigma[w(F_o^4)]\}^{1/2}$.

d) GOF = $\{\Sigma w(|F_o|^2 - |F_c|^2)^2/(n-m)\}^{1/2}$; n = no. of reflections; m = no. of parameters.

[13] a) G. M. Sheldrick, *SHELXTL-PLUS: An Integrated System for Solving, Refining and Displaying Crystal Structures from Diffraction Data*, Siemens Analytical X-Ray Instruments Inc., Madison, WI, (USA.), **1989**.

(b) Sheldrick, G.M. *SHELXL-93*, *Program for Crystal Structure Determination*, Universität Göttingen, Germany, **1993**.

Novel Bis(1,2-*N,N*-Dimethylamino-methylferrocenyl)-Silyl Compounds

Wolfram Palitzsch, Gerhard Roewer

Institut für Anorganische Chemie, TU Bergakademie Freiberg
Leipziger Str. 29, D-09596 Freiberg, Germany
Tel.: Int. code + (3731)39 4346 — Fax: Int. code + (3731)39 4058
E-mail: palitz@orion.hrz.tu-freiberg.de

Claus Pietzsch

Institut für Angewandte Physik, TU Bergakademie Freiberg
Bernhard-Cotta-Str. 4, D-09596 Freiberg, Germany

Klaus Jacob

Institut f. Anorg. Chem., Martin-Luther-Universität Halle-Wittenberg
Geusaer Straße, D-06217 Merseburg, Germany

Kurt Merzweiler

Institut f. Anorg. Chem., Martin-Luther-Universität Halle-Wittenberg
Kurt-Mothes-Straße, D-06120 Halle, Germany

Keywords: Iron / Silicon / 1,2-*N,N*-Dimethylaminomethylferrocenyl

Summary: The treatment of silicon tetrachloride with two equivalents of 2-dimethyl-aminomethylferrocenyllithium (**1**), (FcN)Li, affords $(FcN)_2SiCl_2$ (**2**). The structure of **2** has been determined by X-ray diffraction analysis. It was used as starting chlorosilyl compound for reaction with $Na[(\eta^5\text{-}C_5Me_4Et)Mo(CO)_3]$. The reaction of **1** with 1,6-dichloro-dodecamethylhexasilane yields the novel $(FcN)(SiMe_2)_6(FcN)$ (**4**).

During the past several years various FcN derivatives have been prepared [1]. In contrast to the well-investigated compound class of 1,2-*N,N*-dimethylaminomethylferrocenyl transition metal derivatives [2] the homologous silyl complexes could attract only scarce attention so far. Few (FcN)-chlorosilanes have been prepared [3], but only one structure has been determined up to now [4]. On the other hand there has been recent interest in investigations of model compounds for the controlled hydrolysis and condensation of chlorosilanes. Auner and coworkers have reported that the intramolecular donor capabilities of the dimethylaminobenzyl ligand at silicon can stabilize 1,3-siloxanedioles [5]. From this point of view (FcN)-chlorosilane complexes are interesting as synthons for silanoles. Furthermore they should react with different nucleophilic reagents such as carbonylmetallates of transition metals to build transition metal–silicon bonds or with alkali metal–silicon compounds to build Si–Si bonds, respectively.

The reinvestigation of reactions of [2-(dimethylaminomethyl)ferrocenyl]lithium (**1**), (FcN)Li, with silicon tetrachloride has delivered new analytical data for (FcN)$SiCl_3$, especially X-ray structure analysis results showing pentacoordination of the silicon atom [4]. The treatment of silicon tetrachloride with two equivalents of (FcN)lithium (**1**) results in $(FcN)_2SiCl_2$ (**2**) as a reddish solid (Scheme 1).

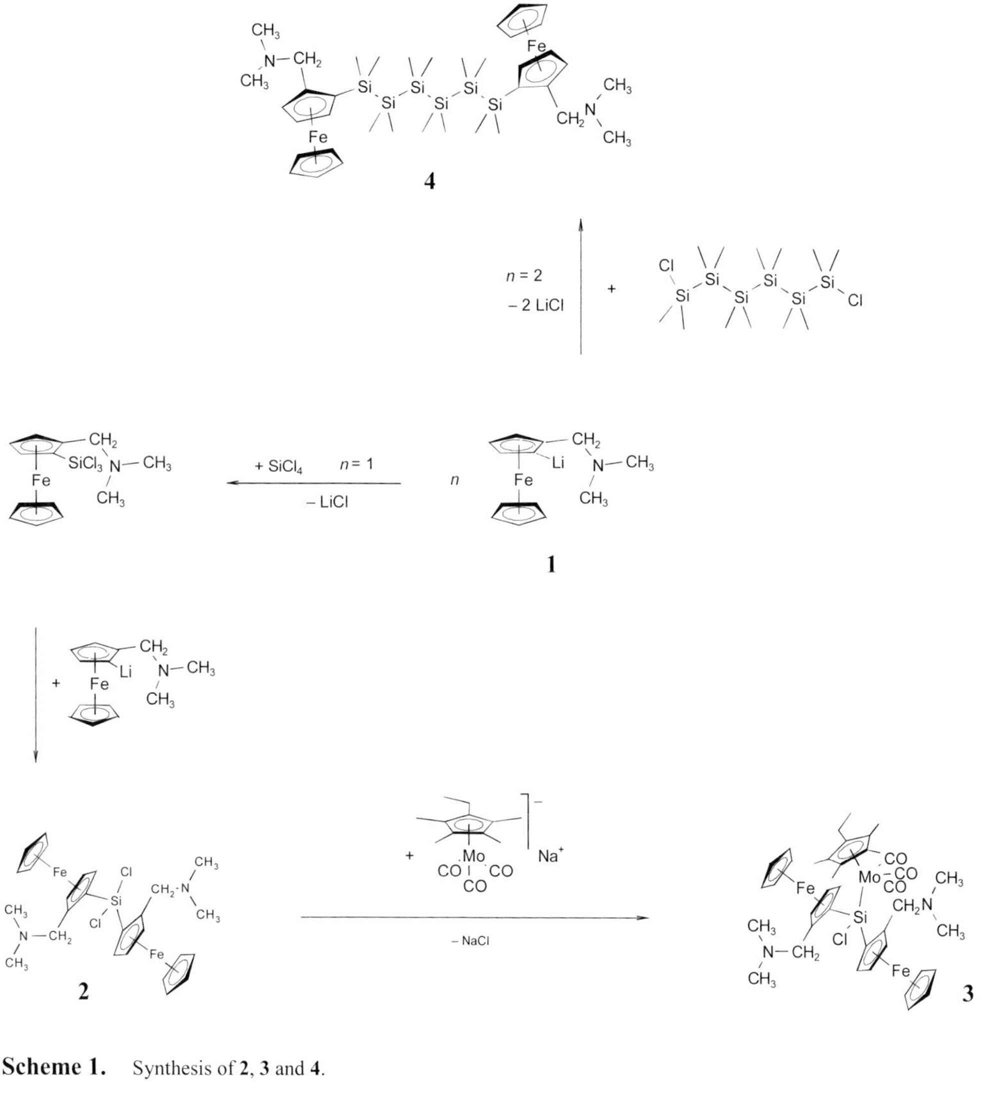

Scheme 1. Synthesis of **2**, **3** and **4**.

Recrystallization from pentane at –20°C gave analytically pure orange crystals of **2.** The ^{1}H-NMR spectrum exhibits the typical signals for the FcN ligand at 2.14 (s, $-N(CH_3)_2$), 2.96, 3.92 (d, CH_2–N), 4.14 (s, C_5H_5) and 4.31, 4.57 ppm as multiplets for the protons of the substituted cyclopentadienyl rings. The ^{29}Si-NMR signal for the $SiCl_2$ group appears at 14.02 ppm.

The solid-state structure of $(FcN)_2SiCl_2$ (**2**) is presented in Fig. 1. The coordination geometry of the iron atom in **2** is similar to that found in $(FcN)SiCl_3$ [4]. A particularly interesting feature of the structure is the distance between the silicon atom and the nitrogen atoms. But there is no evidence for a substantial Si···N bonding interaction.

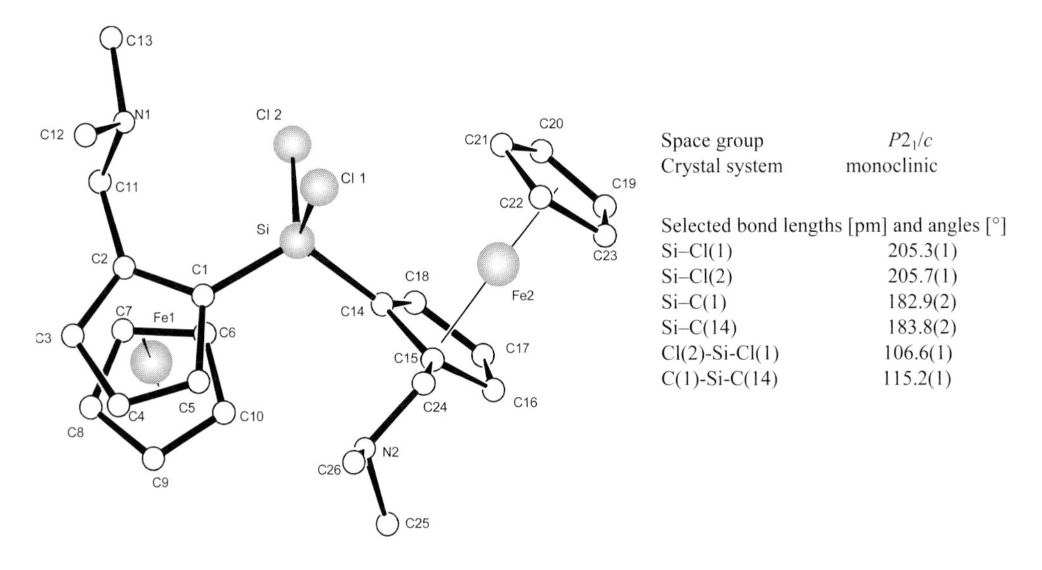

Fig. 1. Molecular structure of **2** (hydrogen atoms are omitted for clarity).

The treatment of a solution of **2** in pentane with a solution of $Na[(\eta^5\text{-}C_5Me_4Et)Mo(CO)_3]$ in THF yields $(\eta^5\text{-}C_5Me_4Et)Mo(CO)_3SiCl(FcN)_2$ (**3**, Scheme 1). The structure of **3** is clearly confirmed by spectroscopic data. The substituted cyclopentadienyl ring is indicated by the ^1H-NMR spectrum which exhibits a set of two singlets at 1.83 and 1.79 ppm (methyl groups), one triplet at 0.75 ppm and one quartet at 2.37 ppm for the ethyl group. Additionally the ^1H-NMR spectrum exhibits the typical signals for the FcN ligand at 2.14 (s, $-N(CH_3)_2$), 3.14, 3.35 (AB, CH_2–N), 4.24 (s, C_5H_5) and in the range from 4.31 to 4.4 ppm as multiplets for the protons of the substituted cyclopentadienyl ring of the ferrocenyl system. The resonance of the silicon atom is found at $\delta = -19.9$ ppm in ^{29}Si-NMR. The assigned structure was proved by the ^{13}C-NMR spectrum. The IR spectrum shows three absorptions of ν(CO) nearly 2000 cm^{-1}. This behavior is typical for metal carbonylates *cis*-$L_2M(CO)_3$.

The same synthesis route leads to compounds containing a longer silicon chain. The reaction of two equivalents of (FcN)Li (**1**) with $Cl(SiMe_2)_6Cl$ in THF/pentane yields $(FcN)(SiMe_2)_6(FcN)$ (**4**, Scheme 1). The ^1H- and ^{13}C-NMR spectra for the complex **4** are similar (Fig. 2).

Each contains the typical resonances of the 1,2-*N,N*-dimethylaminomethylferrocenyl ligand. The recorded non-equivalence of methylene group protons in 1,2-disubstituted ferrocenes agrees with results given in previous papers [3b, 6]. The ^1H-NMR resonances for the methyl groups bonded to silicon appear in a typical range at values nearly zero, but the number of six signals is suspicious, because the ^{29}Si-NMR spectrum of **4** indicates only three resonances. The silicon chain is bonded to an asymmetric carbon atom. For this reason the methyl groups should be diastereotopic.

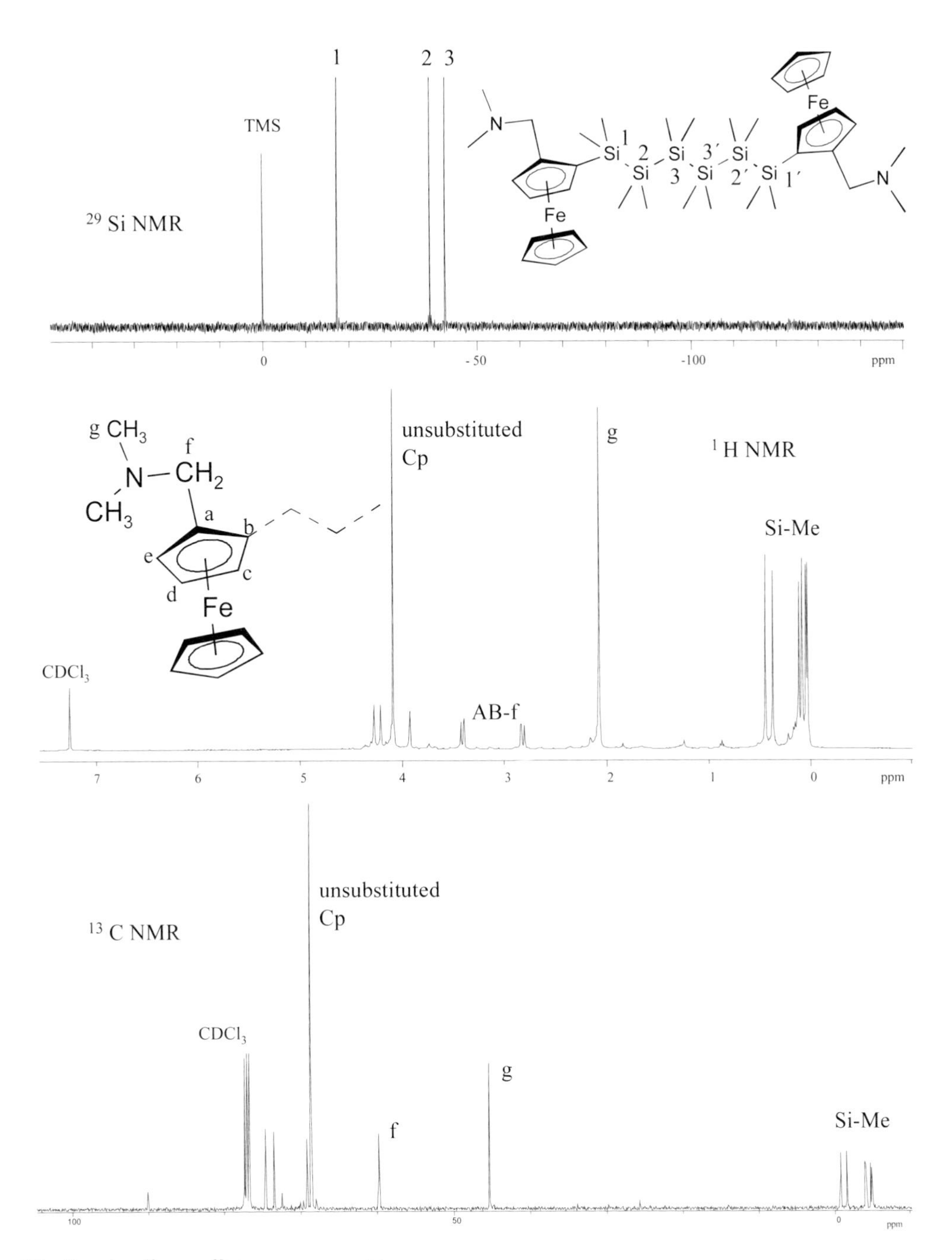

Fig. 2. ^{1}H-, ^{13}C- and ^{29}Si-NMR spectra of **4**.

The Mössbauer spectrum of compound **4** (at 100 K) exhibits clearly both Fe^{II} as Fe^{III} signals (Fig. 3).

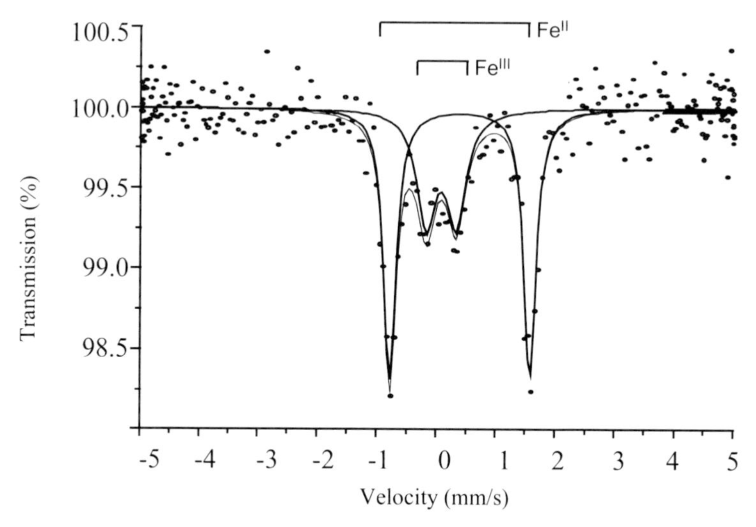

Fig. 3. Mössbauer spectrum of **4**.

The doublet with larger splitting represents Fe^{II} in the ferrocenyl groups; the doublet with the smaller splitting holds for Fe^{III}. We assume an intervalence electron transfer for the Fe^{III} part of the spectrum between the iron atoms in both ferrocenyl systems about the silicon chain of compound **4**. The relatively large line width of the Fe^{III} doublet and the lack of a singlet results from the fact that the charge transfer is slow at 100 K in relation to the time constant of the Mössbauer window ($K = 10^8\ s^{-1}$) [7]. The UV-absorption band suffers shift in different solvents (see Fig. 4). The possibility that two spin states are in thermal equilibrium could not be established by measuring the Mössbauer spectra in the temperature region 77–296 K.

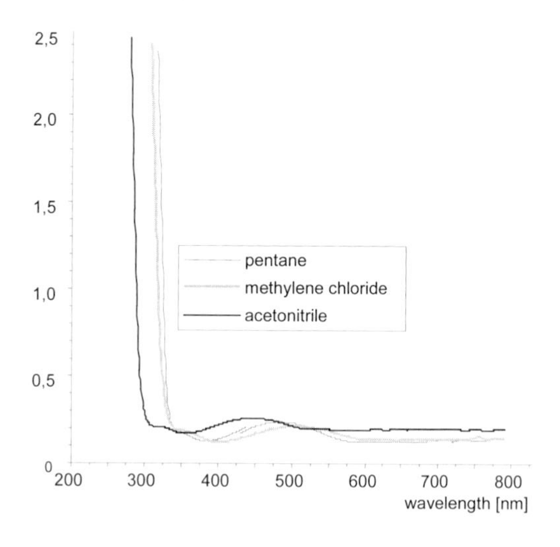

Fig. 4. UV spectra of compound **4** in different solvents.

Acknowledgment: Financial support of this work by the *Deutsche Forschungsgemeinschaft* and the *Fonds der Chemischen Industrie* is gratefully acknowledged.

References:

[1] F. T. Edelmann, K. Jacob, *J. Prak. Chem.* **1998**, *340,* 393.

[2] a) P. B. Hitchcock, D. L. Hughes, G. J. Leigh, J. R. Sanders, J. S. de Souza, *Chem. Commun.* **1996**, *16,* 1985.
b) K. Jacob, J. Scholz, C. Pietzsch, F. T. Edelmann, *J. Organomet. Chem.* **1995**, *501*, 71, and references cited therein.

[3] a) A. J. Blake, F. R. Mayers, A. G. Osborne, D. R. Rosseinsky, *J. Chem. Soc., Dalton Trans.* **1982**, *12*, 2379.
b) V. I. Sokolov, L. I. Troitskaya, L. A. Bulygina, *Metalloorg. Khim.* **1990**, *3,* 1435.
c) G. Marr, *J. Organomet. Chem.* **1967**, *9,* 147.

[4] W. Palitzsch, C. Pietzsch, K. Jacob, F. T. Edelmann, T. Gelbrich, V. Lorenz, M. Puttnat, G. Roewer, *J. Organomet. Chem.* **1998**, *554,* 139.

[5] N. Auner, R. Probst, C. R. Heikenwälder, E. Herdtweck, S. Gamper, G. Müller, *Z. Naturforsch., Teil B* **1993**, *48b,* 1625.

[6] a) G. Marr, *J. Organomet. Chem.* **1967**, *9,* 147.
b) G. Marr, R. E. Moore, B. W. Rockett, *Tetrahedron Lett.* **1968**, *2,* 2521.

[7] K. Jacob, J. Scholz, K. Merzweiler, C. Pietsch, *J. Organomet. Chem.* **1997**, *527,* 109.

Synthesis of Silicon Synthons for a Selective Oligomer Design

K. Trommer, U. Herzog, G. Roewer

Institut für Anorganische Chemie
Technische Universität Bergakademie Freiberg
Leipziger Str. 29, D-09596 Freiberg, Germany
Tel.: Int. code + (3731)39 3174 — Fax: Int. code + (3731)39 4058

Keywords: Aminosilane / Oligosilane

Summary: Amination of chlorosilanes with dialkylamine allows the synthesis of various chloro- and amino-substituted oligosilanes. The remaining reactive chloro sites enable to bond further functionalities as well as other silyl units to the silane. In this procedure the amino group is protecting potentially reactive sites at silicon atoms and can be easily exchanged by chloro substituents. By this route a desired silicon architecture can be built up.

Synthesis and Properties of Aminochlorosilanes

Tailoring of oligosilanes is of great interest for both the synthesis of predicted silane backbones and specifically functionalized reactive oligosilanes. For this purpose a protection concept is required to realize a stepwise building up of silicon skeletons. The dialkylamino group seemed to be quite useful as a protecting entity. Partially amino-substituted halogenooligosilanes can be synthesized by the reaction of dialkylamines and halogenooligosilanes (Eq. 1) [1, 2].

$$Si_2Me_nCl_{(6-n)} + 2\,x\,HNR_2 \longrightarrow Si_2Me_nCl_{(6-n-x)}(NR_2)_x + x\,HNR_2{\cdot}HCl$$

$$n = 2\text{–}5,\ 1 \leq x \leq 4$$

Eq. 1. Synthesis of partially amino-substituted chlorooligosilanes.

The initial molar ratio between dialkylamine and chlorosilane has the main influence on the product composition. Figure 1 illustrates how the product distribution of the reaction of diethylamine with pentachloro-1,2,3-trimethyltrisilane depends on the molar ratio. The various resulting product yields indicate the different reactivities of the silyl groups related to diethylamine. The following reactivity graduation of silyl groups in amino(chloro)(methyl)silanes was observed:

$$–SiMeCl_2 \quad > \quad –SiMe_2Cl \quad > \quad =SiMeCl \quad > \quad –SiMeCl(NEt_2)$$

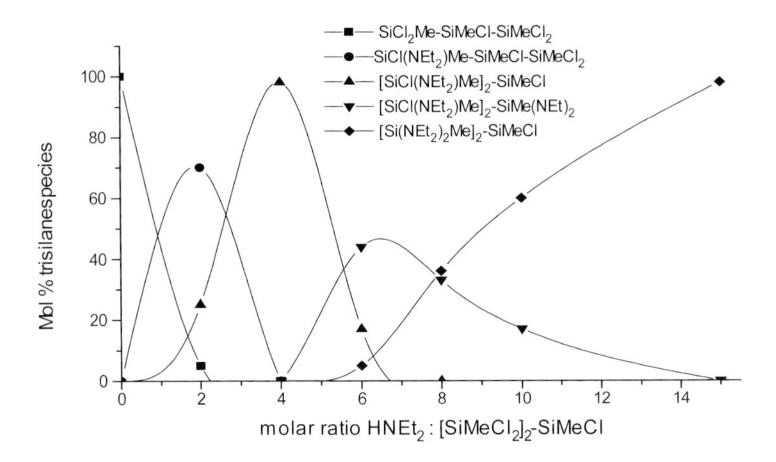

Fig. 1. Dependence of distribution of trisilane species $[SiMeX_2]_2SiMeX$ (X = Cl, NEt_2) obtained on molar ratios.

The amounts of the various products which are formed by a stepwise substitution also give hints of the possible aminochlorosilanes which can be synthesized by this method. Such conclusions are interesting especially for halogen-rich silanes. The first substitution step in all investigated methylchlorotrisilanes takes place at the chlorosilyl endgroups. Furthermore the following exchange of substituents depends on their reactivity graduation. If the steric hindrance becomes too high an intramolecular rearrangement of chloro and amino functions can occur (Eq. 2) [3].

$$[(NEt_2)MeClSi]_2SiMe(NEt_2) \xrightarrow[-HNEt_2 \cdot HCl]{HNEt_2} [(NEt_2)_2MeSi]_2SiMeCl$$

Eq. 2. Intramolecular rearrangement of chloro and amino groups.

In Figs. 2 and 3 the structural silicon units investigated and some examples of the compounds synthesized are shown.

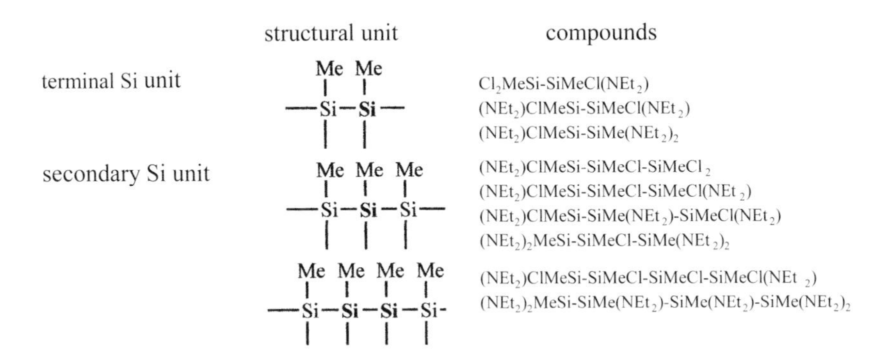

Fig. 2. Synthesized examples of oligosilanes bearing terminal and secondary silicon units.

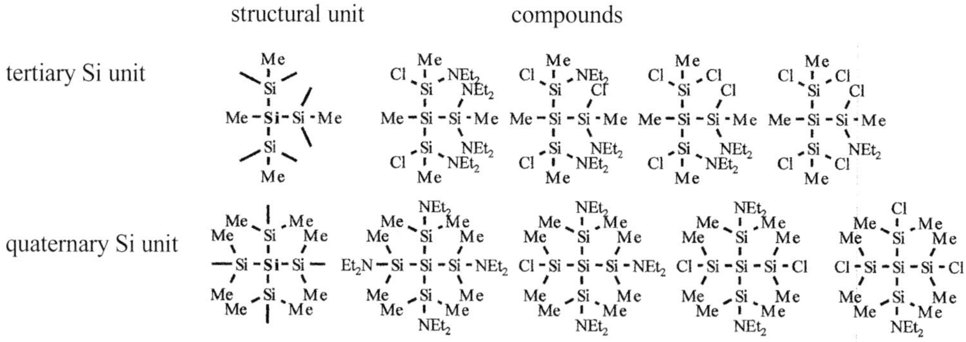

structural unit compounds

tertiary Si unit

quaternary Si unit

Fig. 3. Synthesized examples of oligosilanes bearing branched silicon units.

Reactions of Aminochlorooligosilanes

Scheme 1 illustrates the great variety of reactions which can be carried out with aminosilanes.

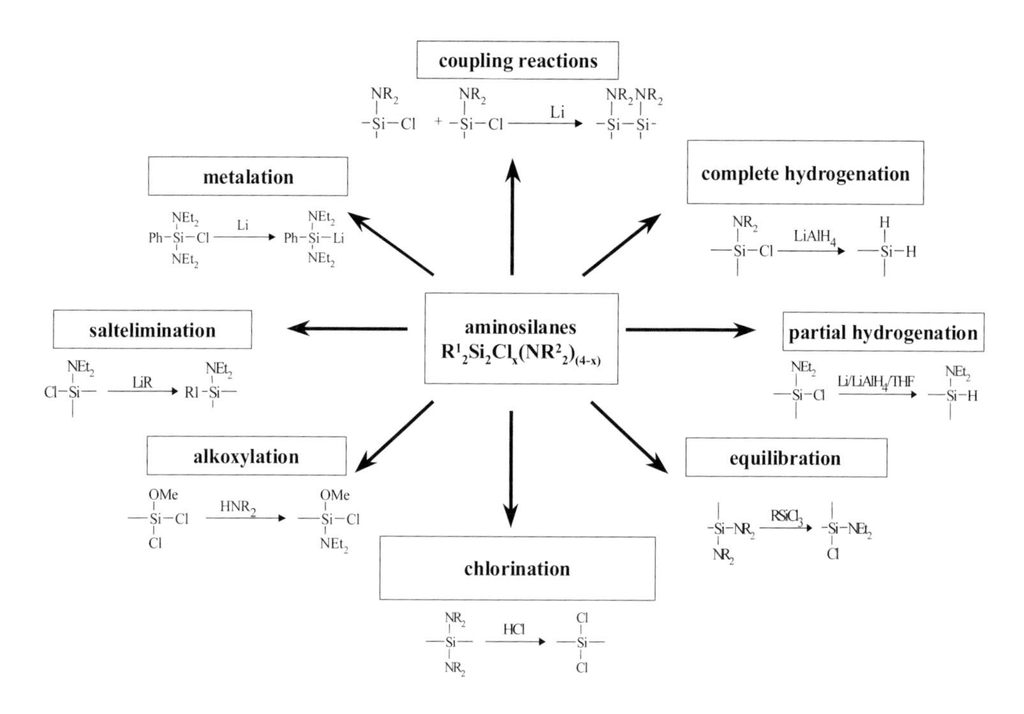

Scheme 1. Reaction facilities of aminochlorosilanes.

An enlargement of the silicon skeleton is possible by coupling reactions with lithium metal or silyllithium reagents [4–6]. The homocoupling of tris(diethylamino)-1-chloro-1,2-dimethyldisilane results in the linear hexakis(diethylamino)-1,2,3,4-tetramethyltetrasilane [7]. After treatment with HCl the hexachloro compound could be obtained. By the reaction of aminochlorophenylsilanes with lithium, mono- as well as disilyllithium reagents are formed (Scheme 2). These aminophenylsilyllithium compounds are ideal building blocks to enlarge the silicon backbone or to tailor specifically phenyl-substituted oligosilanes. Coupling experiments of the silyllithium reagents with various aminochloromethyloligosilanes resulted in both linear and branched aminomethylphenylsilanes [8]. After the resubstitution of diethylamine for chlorine the corresponding chloromethylphenylsilanes were isolated (Scheme 3).

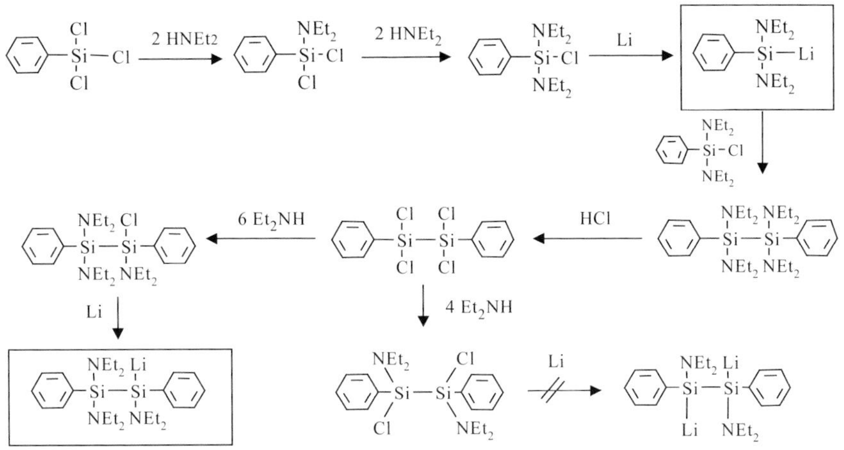

Scheme 2. Reaction pathways to generate aminophenyllithium reagents.

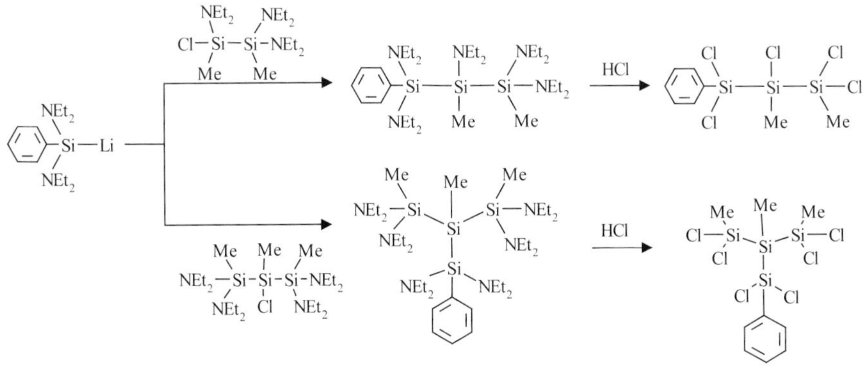

Scheme 3. Linear and branched coupling products of $Ph(NEt_2)_2SiLi$ with chlorosilanes.

Organic functionalities can also be inserted into aminochlorooligosilanes by salt elimination with lithium or Grignard reagents. This type of reaction yields various specifically organic-substituted aminosilanes which are able to react with HCl and to form the corresponding chlorosilanes (Scheme 4). Via this reaction type different olefinic functionalized disilanes were prepared [9].

Et$_2$N Me | Et$_2$N Me | Cl Me
Cl—Si—Si—NEt$_2$ $\xrightarrow{\text{LiR or BrMgR}}$ R—Si—Si—NEt$_2$ $\xrightarrow{\text{HCl}}$ R—Si—Si—Cl
Me NEt$_2$ | Me NEt$_2$ | Me Cl

Et$_2$N Me | Et$_2$N Me | Cl Me
Cl—Si—Si—Cl $\xrightarrow{\text{2 LiR or 2 BrMgR}}$ R—Si—Si—R $\xrightarrow{\text{HCl}}$ R—Si—Si—R
Me NEt$_2$ | Me NEt$_2$ | Me Cl

Scheme 4. Organic substitution of aminochlorosilanes (R = vinyl, allyl, methylvinyl).

Sometimes it is useful to have a third functionality bonded to silicon besides chloro and amino groups. Two protecting groups offer the possibility of a selective reintroduction of chlorine substituents after derivatizations with organometallic reagents. It is possible to synthesize silanes containing simultaneously alkoxy, amino and chloro functions [3]. Starting from methylchlorodisilanes the stepwise reaction with first HC(OMe)$_3$ and AlCl$_3$ and secondly HNEt$_2$ leads to the formation of (diethylamino)(methoxy)methylchlorooligosilanes. Because the Si–O bond is the most stable one in this system, attempts at applying the opposite order of synthesis failed.

Various chlorosilanes were treated in this way and their substitution pattern was investigated. The most important fact is the high regioselectivity of both reactions. The first substitution steps take place at the silyl endgroups. The next steps are controlled by the steric hindrance of the functional groups. The typical course of the substituent exchange is illustrated in Scheme 5.

Cl | Cl
—Si—Si—Si—
Cl Cl Cl
$\xrightarrow{2\times \text{I}}$
MeO | OMe
—Si—Si—Si—
Cl Cl Cl
$\xrightarrow{2\times \text{I}}$
MeO | OMe
—Si—Si—Si—
MeO Cl OMe
$\xrightarrow{\text{III}}$
MeO | OMe
—Si—Si—Si—
MeO MeO OMe

Column 1 ($\downarrow$ II at each step):

Cl | NEt$_2$
—Si—Si—Si—
Cl Cl Cl

Et$_2$N | NEt$_2$
—Si—Si—Si—
Cl Cl Cl

Et$_2$N | NEt$_2$
—Si—Si—Si—
Cl Et$_2$N Cl

Et$_2$N | NEt$_2$
—Si—Si—Si—
Et$_2$N Cl NEt$_2$

Column 2 ($\downarrow$ II at each step):

MeO | OMe
—Si—Si—Si—
Cl Cl NEt$_2$

MeO | OMe
—Si—Si—Si—
Et$_2$N Cl NEt$_2$

MeO | OMe
—Si—Si—Si—
Et$_2$N Et$_2$N NEt$_2$

Column 3 ($\downarrow$ II):

MeO | OMe
—Si—Si—Si—
MeO Et$_2$N OMe

I = + HC(OMe)$_3$ [AlCl$_3$] –HCOOMe –MeCl
II = + 2 HNEt$_2$ –HNEt$_2$·HCl
III = + MeOH / NEt$_3$ –NEt$_3$·HCl

Scheme 5. Derivatives of 1,1,2,3,3-pentachlorotrimethyltrisilane.

The results of the ^{29}Si-NMR investigations are summarized in Fig. 4. The chemical shift ranges for silyl groups in silane molecules containing methoxyfunctions are very close. Therefore often no unambiguous assignment is possible by evaluation of the NMR shifts alone.

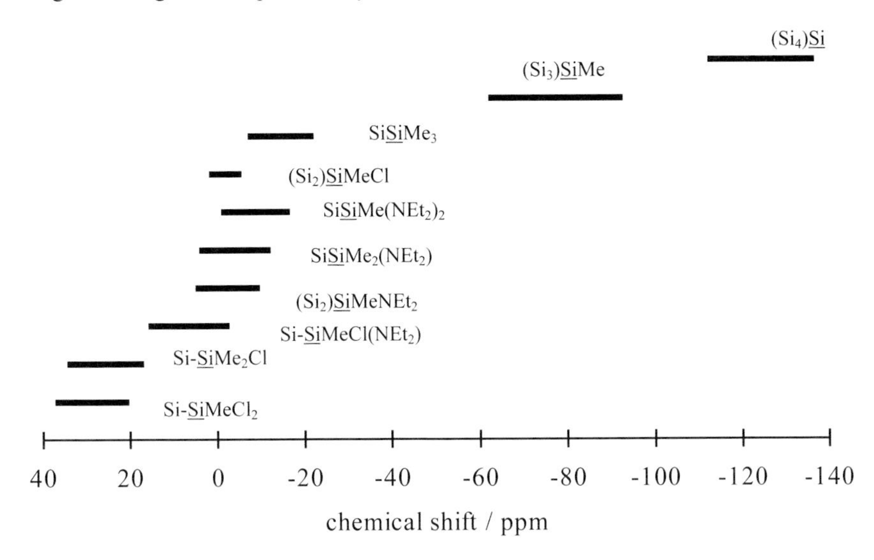

Fig. 4. ^{29}Si-NMR shift ranges of silyl groups in silane molecules containing methoxy units.

References:

[1] Y. Wan, J. G. Verkade, *Inorg. Chem.* **1993**, *463*, 73.
[2] K. Trommer, E. Brendler, G. Roewer, *J. Prakt. Chem.* **1997**, *339*, 82.
[3] U. Herzog, K. Trommer, G. Roewer, *J. Organomet. Chem.* **1998**, *552*, 637.
[4] K. Tamao, A. Kawachi, Y. Ito, *Organometallics* **1993**, *12*, 580.
[5] M. Unno, M. Saito, H. Matsumoto, *J. Organomet. Chem.* **1995**, *499*, 221.
[6] K. Tamao, G.-R. Sun, A. Kawachi, S. Yamaguchi, *Organometallics* **1997**, *16*, 780.
[7] K. Trommer, U. Herzog, G. Roewer, *J. Prakt. Chem.* **1997**, *339*, 637.
[8] K. Trommer, U. Herzog, U. Georgi, G. Roewer, *J. Prakt. Chem.* **1998**, *340*, 557.
[9] K. Trommer, U. Herzog, G. Roewer, *J. Organomet. Chem.* **1997**, *540*, 119.

Synthesis and Characterization of New Precursors Based on Organosilicon for Construction of Porous Solid Structures

S. Nitsche, E. Weber

Institut für Organische Chemie
Technische Universität Bergakademie Freiberg
Leipziger Str. 29, D-09596 Freiberg, Germany
Tel.: Int. code + (3731)39 2386 — Fax: Int. code + (3731)39 3170

K. Trommer, G. Roewer

Institut für Anorganische Chemie
Technische Universität Bergakademie Freiberg
Leipziger Str. 29, D-09596 Freiberg, Germany
Tel.: Int. code + (3731)39 3174 — Fax: Int. code + (3731)39 4058

Keywords: Aminochloroethinylsilanes / Organosilanes / Aryl Acetylides

Summary: The aim of this work consists in the gradual building up of a new class of porous precursors based on organosilicon chemistry. Therefore the rigid organic tecton 1,3,5-triethinylbenzene was linked with various chloroaminosilanes. The amino functions of the aminosilane-substituted aryl acetylides were exchanged for chlorine by treatment with HCl. These silicon–chlorine bonds allow further coupling and crosslinking reactions with bulky fragments to achieve a permanent cavity structure.

Introduction

Engineering of porous solid materials is emerging as a highly topical interest in both research and industry due to many potential applications based on their particular physical and chemical properties, e.g. as porous membranes [1], heterogeneous catalysts and adsorbing agents or to provide supportive tools as selective sensor materials for industrial pollution problems. Crystalline porous materials are suitable as coating compounds for the formation of receptor layers in the mass-sensitive sensor because they are able to accommodate selectively different guests reversibly in the empty cavities. If a crystal lattice forms these cavities this process is called "clathration". However, porous clathrate frameworks are rather labile systems because of the weak intermolecular non-covalent interactions holding the lattice structure together. Using strong covalent bonds instead of the weak interactions should lead to a permanent hollow structure.

Considering this approach we suggest the stepwise synthesis of a new class of porous solid structures that derive from assembly of individual inorganic-organic precursors. These precursors are formed by covalent linkage of geometrically well-defined rigid organic fragments and inorganic moieties containing silicon chains. First investigations and results proved that the reaction of halogenaminooligosilanes with aryl acetylides by using Li-organic intermediates is a suitable method for preparation of these precursors. The amino function acts as protecting group during the coupling step between the respective silanes and the acetylenic building block. Subsequently, it can be easily exchanged by a halogeno function via treatment with HCl. The reconversion of the reactive sites at the silicon atoms allows an enlargement of the system by coupling and crosslinking to build up a permanent cavity structure. An advantage of this synthetic approach is the great variability of both the organic and inorganic building blocks that are relatively easily available and convenient in handling.

Reaction pathways

A successful method to link inorganic and organic building units proved to be the metalation of aryl acetylide and subsequent reaction with chlorosilanes. Attempts with the relatively small organic tecton 1,3,5-triethinylbenzene have already resulted in a three-fold substitution of the acidic protons. 1,3,5-Triethinylbenzene was synthesized in two steps. At first a palladium-catalyzed coupling reaction of 1,3,5-tribromobenzene with 3 equivalents of TMS-acetylene (TMS = trimethylsilyl) in diethylamine was performed. The crude intermediate product was purified by column chromatography. In the second step the cleavage of the TMS protecting group by NaOH in THF/MeOH yielded the product as colorless needles after sublimation (Eq. 1).

1,3,5-tribromobenzene + TMS-acetylene → ($HNEt_2$ / 50°C; cat.: CuI, Pd(II)Ac_2, PPh_3) 1,3,5-tris(TMS-ethinyl)benzene → (1M NaOH / RT; MeOH / THF) 1,3,5-triethinylbenzene

Eq. 1.

To investigate changes of the substitution behavior the inorganic tectons were varied from mono- to trisilanes [2, 3]. Except of one chloro function, all silane-containing halogeno groups were protected by the diethylamino group. Therefore the corresponding chlorosilanes were treated with diethylamine (Eq. 2).

$$MeCl_2Si{-}SiMeCl{-}SiMeCl_2 \xrightarrow[n\text{-hexane}]{HNEt_2} Me(NEt_2)_2Si{-}SiMeCl{-}Si(NEt_2)_2Me$$

Eq. 2.

Coupling experiments (Eq. 3) between the lithium acetylide and different aminochloro-silanes have been shown that a complete substitution of the three acetylenic protons is successful in cases of mono- as well as di- and trisilanes.

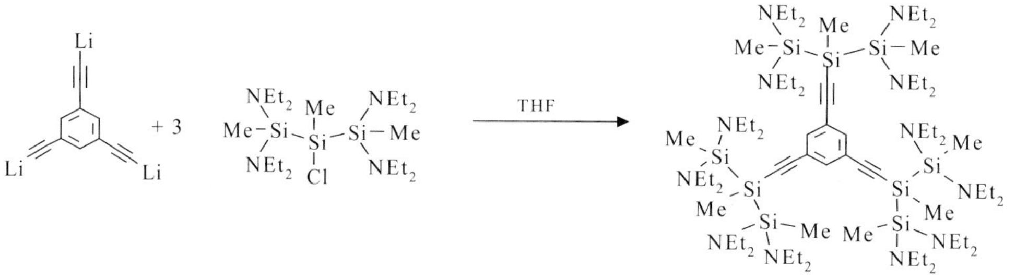

THF

Eq. 3.

The obtained aminosilane-substituted aryl acetylides were treated with HCl to convert the amino groups into reactive chloro functions (Eq. 4). Furthermore, the resulting chloro-functionalized products can be used for the next steps in order to build up porous solid structures. For this purpose some chloro positions can be protected by diethylamine again (Eq. 4).

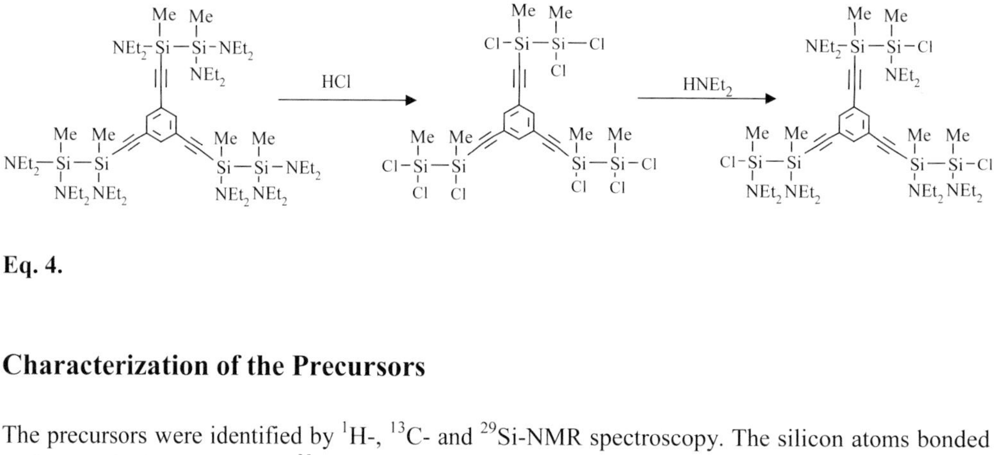

HCl

$HNEt_2$

Eq. 4.

Characterization of the Precursors

The precursors were identified by ^{1}H-, ^{13}C- and ^{29}Si-NMR spectroscopy. The silicon atoms bonded to the acetylenic groups show ^{29}Si-NMR resonances with a characteristic upfield shift of about 10 to 20 ppm compared to the chorosilanes. The shift changes of the other silicon atoms in the di- and trisilanes are insignificantly influenced by this reaction (up to 3 ppm). In the ^{13}C-NMR spectra the linkage between silicon and the C–C triple bond can be recognized from the downfield shift for the acetylenic carbons. If stereoisomers occur, two peaks for carbons bonded to silicon are visible. The isomers are also detected in the ^{29}Si-NMR spectra. The exact analytical data are given in Tables 1 and 2.

Table 1. ^{29}Si-and ^{1}H-NMR data of silane-substituted 1,3,5-triethinylbenzene [ppm].

R	δ(^{29}Si)	δ(^{1}H)				
		Si^{A}–CH_3	Si^{B}–CH_3	NCH_2CH_3	NCH_2	Ph
–Si^{A}MeCl–Si^{B}$MeCl_2$	–17.11 A 21.08 B	0.85	1.01	—	—	7.65
–Si^{A}$MeNEt_2$–Si^{B}$MeClNEt_2$[a]	–30.4; –30.9 A 3.8; 4.4 B	0.47/0.51	0.59/0.64	1.07	2.97	7.45
–Si^{A}$MeNEt_2$–Si^{B}$Me(NEt_2)_2$	–30.84 A –11.39 B	0.24	0.34	1.02	2.94	7.34
–Si^{A}Me(Si^{B}$MeCl_2)_2$	–58.09 A 27.78 B	0.53	1.05	—	—	7.58
–Si^{A}Me(Si^{B}$MeNEt_2)_2$	–74.11 A –5.46 B	0.33	0.48	1.1	3.0	7.3 / 7.4
–$SiMe(NEt_2)_2$	–28.62	0.39	—	1.17	3.05	7.52
–$SiMe_2NEt_2$	–17.59	0.20	—	0.98	2.81	7.40
–$SiPh(NEt_2)_2$	–34.52	—	—	1.03	2.98	7.32 7.58–7.75 (Si–Ph)

[a] stereoisomers

Table 2. Selected ^{13}C-NMR data of silane-substituted 1,3,5-triethinylbenzene [ppm].

R	8	7	1, 3, 5	2, 4, 6	Si^{A}–CH_3	Si^{B}–CH_3
–Si^{A}MeCl–Si^{B}$MeCl_2$	88.40	107.70	122.48	136.42	1.28	5.58
–Si^{A}$MeNEt_2$–Si^{B}$MeClNEt_2$[a]	93.80/94.0	104.84	123.82	134.83	–1.28/–0.95	2.12/2.23
–Si^{A}$MeNEt_2$–Si^{B}$Me(NEt_2)_2$	97.38	103.7	124.32	133.96	–0.59	0.12
–Si^{A}Me(Si^{B}$MeCl_2)_2$	89.44	108.19	123.12	135.95	–8.10	7.60
–Si^{A}Me(Si^{B}$MeNEt_2)_2$	96.22	106.66	124.87	133.42	–4.70	0.23
–$SiMe(NEt_2)_2$	94.94	101.42	124.40	134.34	–0.72	—
–$SiMe_2NEt_2$	98.04	104.45	126.85	137.12	2.79	—
–$SiPh(NEt_2)_2$	94.01	102.61	124.13	136.40	—	—

[a] stereoisomers

References:

[1] P. G. Harrison, R. Kannengiesser, *J. Chem. Soc., Chem. Commun.* **1996**, 415.
[2] U. Herzog, K. Trommer, G. Roewer, *J. Organomet. Chem.* **1998**, *552*, 99–108.
[3] K.Trommer, E. Brendler, G. Roewer, *J. Prakt. Chem.* **1997**, *339*, 82–84.

Synthesis and Structures of Stable Aminosilanes and their Metal Derivatives: Building Blocks for Metal-Containing Nitridosilicates

Peter Böttcher, Herbert W. Roesky

Institut für Anorganische Chemie
Universität Göttingen
Tammannstraße 4, D-37077 Göttingen, Germany
Tel.: Int. code + (551)39 3001 — Fax.: Int. code + (551)39 3373
E-mail: hroesky@gwdg.de

Keywords: Aminosilanes / Imide / Metallasilazanes / Nitrogen / Silicon

Summary: Treatment of tetrachlorodisilanes $RSiCl_2SiCl_2R$ and the trichlorosilane $(2,6\text{-}i\text{Pr}_2C_6H_3)N(SiMe_2\text{-}i\text{Pr})SiCl_3$ with sodium in liquid ammonia or an excess of liquid ammonia leads to the formation of tetraaminodisilanes $RSi(NH_2)_2Si(NH_2)_2R$ and the triaminosilane $(2,6\text{-}i\text{Pr}_2C_6H_3)N(SiMe_2\text{-}i\text{Pr})Si(NH_2)_3$ respectively, in high yields. Methyltrichlorosilane reacts with sodium in liquid ammonia to yield the cluster $(MeSi)_6(NH)_9$. The aforementioned aminosilanes are stable against condensation and are also soluble in a variety of organic solvents, thus making these compounds useful starting materials for the preparation of a variety of metallasilazanes. The metallasilazanes resulting from the aminosilanes show interesting structural features. The Si–N–M framework in these compounds is made up of either cyclic or three-dimensional polyhedral core structures. The triaminosilane and the isoelectronic silanetriol $(2,6\text{-}i\text{Pr}_2C_6H_3)N(SiMe_3)Si(OH)_3$ afford heteroadamantanes when reacted with the trimer $[MeAlN(2,6\text{-}i\text{Pr}_2C_6H_3)]_3$. The crystal structures of one of the tetraaminodisilanes, the imide and the metallasilazanes are reported.

Introduction

Our success in chemistry of silanoles [1] prompted us to develop analogous chemistry of the isoelectronic aminosilanes and thus generate possible models of mineral compounds, catalysts and ceramic materials. Until now only three stable triaminosilanes of the type $RSi(NH_2)_3$ were known in the literature [2, 3]. It was our aim to synthesize stable aminosilanes and aminodisilanes with three or more unsubstituted NH_2 groups and show with them an analogous chemistry to the silanoles.

Preparation of the Aminosilanes and the Imide

Lithiation of the anilines $RN(SiMe_3)H$ ($R = 2,6\text{-}i\text{Pr}_2C_6H_3$; C_6H_5) using *n*-butyllithium and subsequent reaction with hexachlorodisilane yielded the tetrachlorodisilanes **1a** and **2a**. Finally treatment with liquid ammonia at –78°C gave the tetraaminodisilanes **1b** and **2b** (Scheme 1) [4].

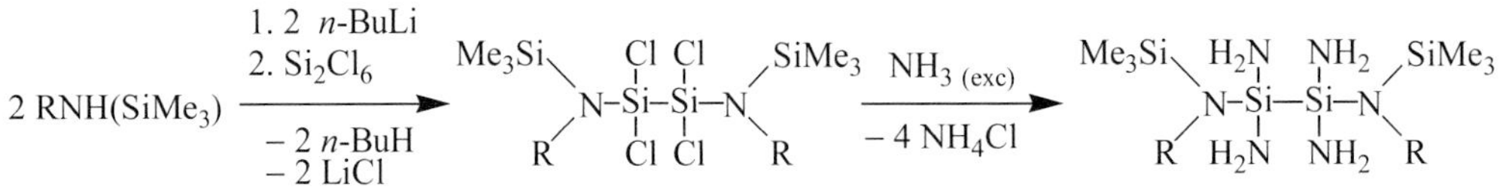

$$2\,RNH(SiMe_3) \xrightarrow[-\,2\,n\text{-}BuH,\ -\,2\,LiCl]{1.\ 2\,n\text{-}BuLi,\ 2.\ Si_2Cl_6} (Me_3Si)(R)N\text{-}SiCl_2\text{-}SiCl_2\text{-}N(R)(SiMe_3) \xrightarrow[-\,4\,NH_4Cl]{NH_{3\,(exc)}} (Me_3Si)(R)N\text{-}Si(NH_2)_2\text{-}Si(NH_2)_2\text{-}N(R)(SiMe_3)$$

1a: R = 2,6-$i$$Pr_2C_6H_3$
2a: R = C_6H_5

1b: R = 2,6-$i$$Pr_2C_6H_3$
2b: R = C_6H_5

Scheme 1. Synthesis of the tetraaminodisilanes **1b** and **2b**.

With cp* (pentamethylcyclopentadienyl) as ligand the potassium salt had to be used to generate the tetrachlorodisilane **3a**. Additional treatment with liquid ammonia or sodium in liquid ammonia at –30°C yielded the tetraaminodisilane **3b** (Scheme 2).

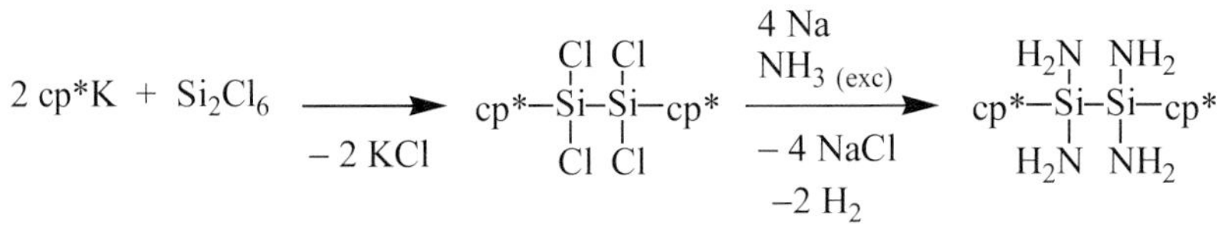

$$2\,cp^*K + Si_2Cl_6 \xrightarrow[-\,2\,KCl]{} cp^*\text{-}SiCl_2\text{-}SiCl_2\text{-}cp^*\ (\mathbf{3a}) \xrightarrow[-\,4\,NaCl,\ -2\,H_2]{4\,Na,\ NH_{3\,(exc)}} cp^*\text{-}Si(NH_2)_2\text{-}Si(NH_2)_2\text{-}cp^*\ (\mathbf{3b})$$

Scheme 2. Synthesis of the tetraaminodisilane **3b**.

Reaction of tetrachlorosilane with the lithiated aniline (2,6-$i$$Pr_2C_6H_3$)N($SiMe_2$-$i$Pr)Li, starting from the corresponding aniline and n-butyllithium, leads to the trichlorosilane **4a**, which could be converted to the triaminosilane **4b** by ammonolysis analogously to **1b** and **2b** (Scheme 3) [5].

$$(2,6\text{-}iPr_2C_6H_3)(iPrMe_2Si)NH \xrightarrow[-\,n\text{-}BuH,\ -\,LiCl]{1.\ n\text{-}BuLi,\ 2.\ SiCl_4} (2,6\text{-}iPr_2C_6H_3)(iPrMe_2Si)N\text{-}SiCl_3\ (\mathbf{4a}) \xrightarrow[-\,3\,NH_4Cl]{NH_{3\,(exc)}} (2,6\text{-}iPr_2C_6H_3)(iPrMe_2Si)N\text{-}Si(NH_2)_3\ (\mathbf{4b})$$

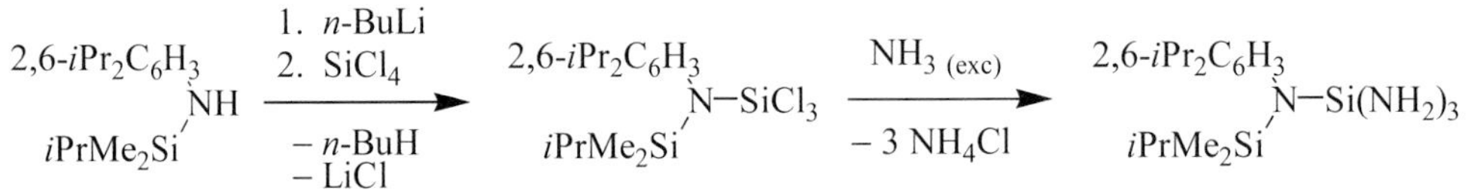

Scheme 3. Synthesis of the triaminosilane **4b**.

The aminosilanes **1b**–**4b** were formed in high yields and are stable under an inert atmosphere. The crystal structure analysis of **3b** (Fig. 1) exhibits a *trans* relationship between the η^1-bonded cp* ligands.

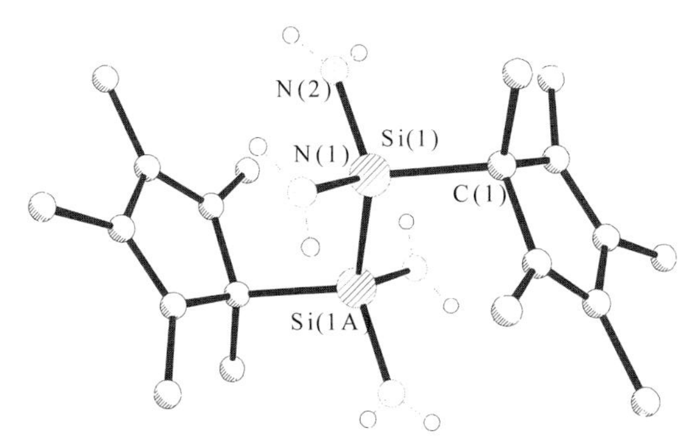

Fig. 1. Structure of **3b**.

Stirring methyltrichlorosilane with sodium in liquid ammonia at –78°C led to the formation of **5** (Scheme 4) [6].

$$6\,MeSiCl_3 + 18\,Na \xrightarrow[-\,18\,NaCl]{NH_{3(exc)}} (MeSi)_6(NH)_9$$

5

Scheme 4. Synthesis of **5**.

The X-ray crystal structure analysis of **5** (Fig. 2) shows a cage structure, in which two six-membered Si_3N_3 rings are connected by three NH-moieties.

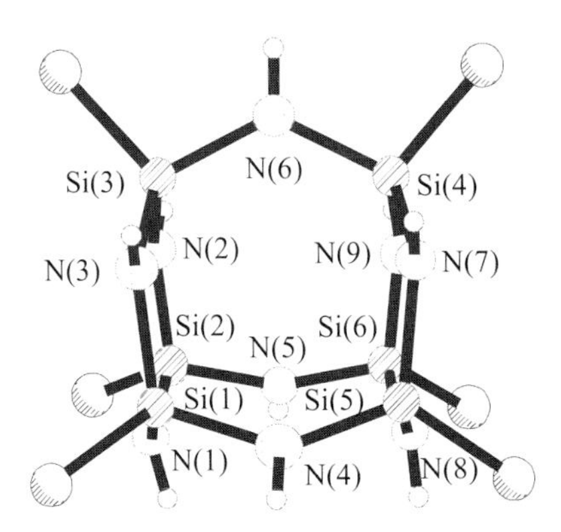

Fig. 2. Structure of **5**.

Reaction of 2b with cp*TiMe$_3$

The reaction of **2b** with an excess of cp*TiMe$_3$ in diethyl ether at room temperature yielded the titanadisilazane **6** (Scheme 5) in the form of yellow crystals [4].

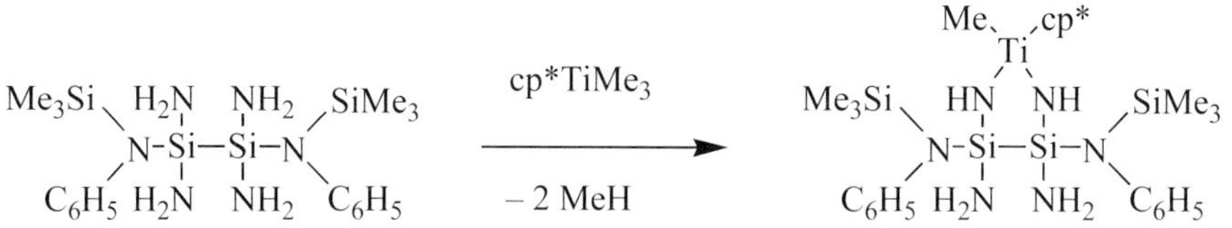

2b **6**

Scheme 5. Synthesis of the titanadisilazane **6**.

The molecular structure of **6** was fully elucidated by an X-ray crystal structure analysis and is shown in Fig. 3. The five-membered TiNSiSiN ring has an envelope conformation with the N(12) atom deviating by 39.5(6) pm from the mean plane through the other four atoms. The arrangement gives a roughly eclipsed conformation of the substituents at the two vicinal Si atoms.

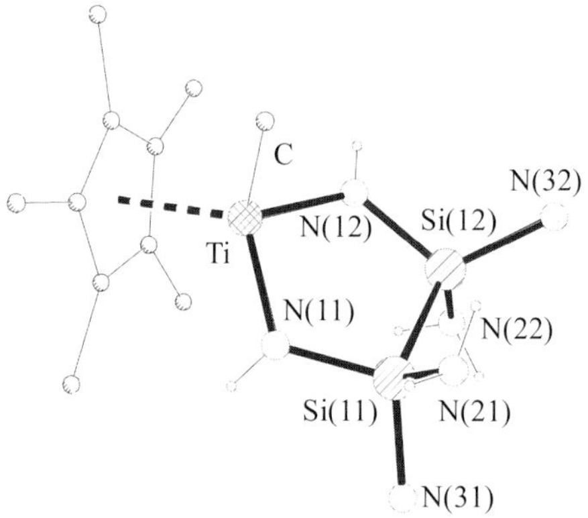

Fig. 3. Core structure of **6**.

Reaction of 4b with Group 13 Alkyls

The reaction of **4b** with $AlMe_3$ or $GaMe_3$ in 1:2 ratio at room temperature lead to the isolation of the first examples of polyhedral aluminium- and gallium-containing silazanes with $M_4N_6Si_2$ frameworks (M = Al **7**, Ga **8**) (Scheme 6) [5, 7].

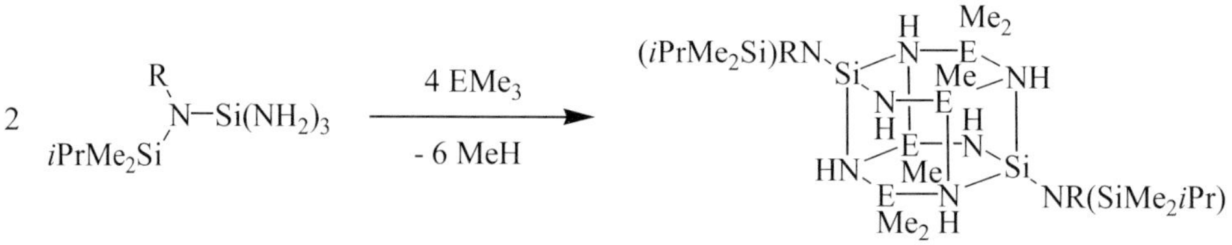

4b R = 2,6-$iPr_2C_6H_3$ **7:** E = Al **8:** E = Ga

Scheme 6. Reaction of the triaminosilane **4b** with $AlMe_3$ or $GaMe_3$.

The molecular structures of **7** and **8** were determined by single crystal X-ray diffraction studies (Fig. 4). The core structure of these compounds exhibits a drum-shaped cluster, formed by two six-membered $(AlNSi)_2$ rings which are connected by four Si–N and M–N bonds.

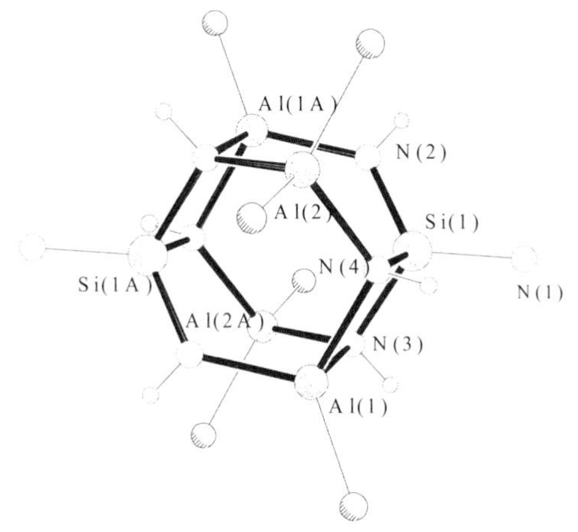

Fig. 4. Core structure of 7.

Reaction of the Trimer [MeAlN(2,6-*i*Pr$_2$C$_6$H$_3$)]$_3$ with 4b and the Silanetriol 9

Addition of **4b** or **9** to [MeAlN(2,6-*i*Pr$_2$C$_6$H$_3$)]$_3$ in toluene solution at room temperature resulted in the formation of **10** and **11** (Scheme 7) [8].

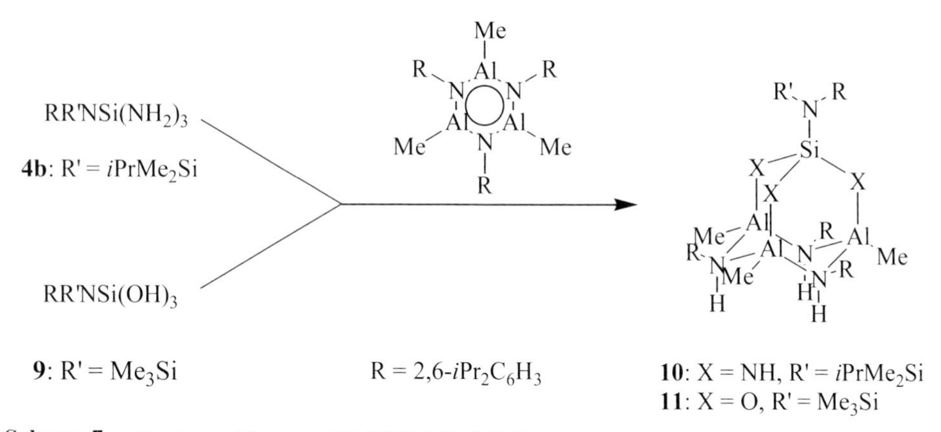

Scheme 7. Reactions of the trimer [MeAlN(2,6-*i*Pr$_2$C$_6$H$_3$)]$_3$.

The molecular structures of **10** and **11** were deduced by X-ray diffraction (Fig. 5). Compound **10** contains an $Al_3N_3SiN_3$ and **11** an $Al_3N_3SiO_3$ structural unit, both exhibiting an adamantane-like structure.

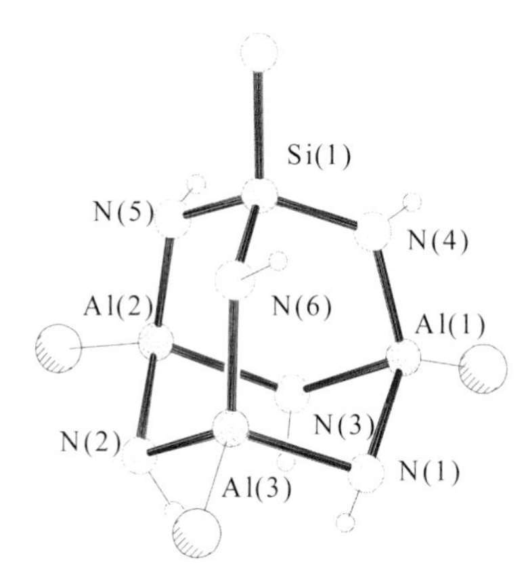

Fig. 5. Core structure of **10**.

Acknowledgment: We are thankful to the *Deutsche Forschungsgemeinschaft*, the *Bundesministerium für Bildung und Forschung*, the *Akademie der Wissenschaften* in Göttingen and *Witco GmbH* for support of this work.

References:

[1] R. Murugavel, A. Voigt, M. G. Walawalkar, H. W. Roesky, "*Silanetriols: Preparation and Their Reactions*", in: *Organosilicon Chemistry III: From Molecules to Materials* (Eds.:N. Auner, J. Weis), VCH, Weinheim, **1997**, p. 376.

[2] K. Ruhlandt-Senge, R. A. Bartlett, M. M. Olmstead, P. P. Power, *Angew. Chem.* **1993**, *105*, 459; *Angew. Chem. Int. Ed. Engl.* **1993**, *34*, 1352.

[3] K. Wraage, A. Künzel, M. Noltemeyer, H.-G. Schmidt, H. W. Roesky, *Angew. Chem.* **1995**, *107*, 2954; *Angew. Chem. Int. Ed. Engl.* **1995**, *34*, 2654.

[4] P. Böttcher, K. Wraage, H. W. Roesky, M. Lanfranchi, A. Tiripicchio, *Chem. Ber. Rec.* **1997**, *130*, 1787.

[5] C. Rennekamp, A. Gouzyr, A. Klemp, H. W. Roesky, C. Brönneke, J. Kärcher, R. Herbst-Irmer, *Angew. Chem.* 1997, *109*, 413; *Angew. Chem. Int. Ed. Engl.* 1997, *36*, 404.

[6] B. Raeke, H. W. Roesky, I. Usón, P. Müller, *Angew. Chem.* **1998**, *110*, 1508; *Angew. Chem. Int. Ed. Engl.* **1998**, *37*, 1432.

[7] C. Rennekamp, H.-S. Park, H. Wessel, H. W. Roesky, P. Müller, I. Usón, H.-G. Schmidt, M. Noltemeyer, manuscript under preparation.

[8] H. Wessel, C. Rennekamp, S.-D. Waezsada, H. W. Roesky, M. L. Montero, I. Usón, *Organometallics* **1997**, *16*, 3243.

Octachlorocyclotetrasilane, Perchloropolysilane and New Dialkoxy- and Diaminopolysilanes

Julian R. Koe, Douglas R. Powell, Jarrod J. Buffy, Robert West*

Department of Chemistry, University of Wisconsin
Madison, WI 53706, USA
Tel.: Int. code + (608)262 1873 — Fax: Int. code + (608)262 6143
E-mail: west@chem.wisc.edu

Keyword: Polysilanes / Cyclosilanes / Photolysis / Ring-Opening Polymerization

Summary: An X-ray diffraction study of crystalline octachlorocyclotetrasilane, Si_4Cl_8, shows that it has a planar, centrosymmetric structure. Photolysis of Si_4Cl_8 transforms it into the linear polymer, $(SiCl_2)_n$, when crystals of Si_4Cl_8 are photolyzed, $(SiCl_2)_n$ is obtained as single crystals. X-ray crystallography of these indicates that $(SiCl_2)_n$ has an all-*trans*, fully extended conformation. Reaction of $(SiCl_2)_n$ with alcohols, amines and water leads to replacement of the chlorine atoms and formation of novel polymers $[Si(OR)_2]_n$, $[Si(NR_2)_2]_n$, $[Si(OH)_2]_n$.

Octachlorocyclotetrasilane and the Discovery of Perchloropolysilane

Our research in this area had its origin in a reinvestigation of the compound octachlorocyclotetrasilane Si_4Cl_8 (**1**) originally prepared by Hengge and Kovar [1]. The infrared and Raman spectra of **1** were interpreted in terms of a folded, D_{2d} structure [1, 2] similar to that known for several other cyclotetrasilanes [3]. However, a semi-empirical PM3 calculation predicted a planar, D_{4h} conformation as the energy minimum for **1** [4].

We prepared the air- and moisture-sensitive yellow solid **1** by the route described in the literature [1], as illustrated in Eq. 1.

$$Ph_2SiCl \xrightarrow[THF]{Mg/MgBr_2} (Ph_2Si)_4 \xrightarrow[AlCl_3]{HCl} (SiCl_2)_4$$

Eq. 1. Synthesis of Si_4Cl_8 (**1**).

Single crystals of **1** were obtained by crystallization from toluene at –20°C. X-ray diffraction showed that the molecule has a very slightly rhomboidally compressed, square-planar structure [5], in agreement with the PM3 calculations [4] but contrasting with the conclusions from vibrational spectroscopy [1, 2]. The molecular structure of Si_4Cl_8 is shown in Fig. 1. Following a similar route we also prepared the known [1] bromine analogue Si_4Br_8; this compound likewise proved to have a planar, near-D_{4h} conformation according to an X-ray crystal structure [5].

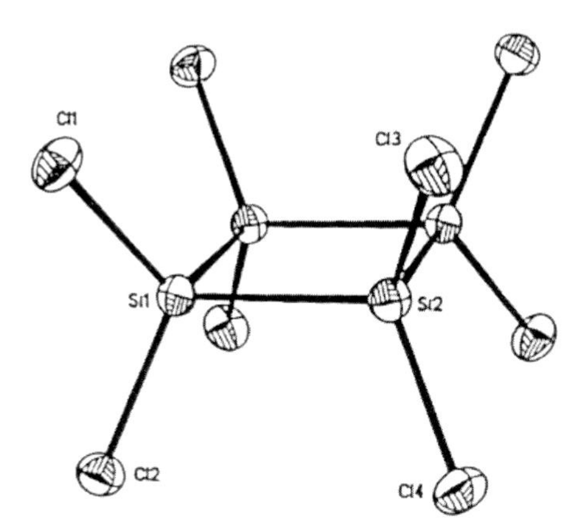

Fig. 1. Molecular structure of **1**, 50 % probability ellipsoids. Selected interatomic distances [pm] and bond angles [°]: Si–Cl, 2.039(2), 2.049(2); Si–Si 2.372(2); Si-Si-Si 90.12(5), 89.88(5); Cl-Si-Cl, 112.17(7), 111.84(7).

Although most cyclotetrasilanes for which X-ray determinations are available have folded structures (see Table 1) there are a few exceptions. $Si_4(SiMe_3)_8$ is planar, probably because of steric repulsion by the $SiMe_3$ groups [6]. And for Si_4Me_8, apparently conflicting data were obtained; X-ray crystallography showed that this compounds has a planar structure [7], although both gas-phase electron diffraction investigation [8] and vibrational data [9] indicate a folded conformation. These results are, however, not contradictory, since if the potential energy surface for bonding is rather shallow, the structure could be planar in the solid phase but bent in the vapor. The same could be true for **1**.

Table 1. Structures of some cyclotrisilanes.

Molecules	Phase	Dihedral Angle [°]	Reference
$Si_4(Me)_8$	Solid	0	7
	Gas	29.4	8
$Si(SiMe_3)_4$	Solid	0	6
Si_4PhCl_7	Solid	11.6	10
Si_4Ph_8	Solid	12.8	11
$Si(CH_2SiMe_3)_8$	Solid	36.6	3b
$Si_4Me_4(t\text{-}Bu)_4$	Solid	36.8	12
Si_4Cl_8	Solid	0	5
C_4Cl_8	Solid	19	13

In the earlier studies **1** was reported to be purified by sublimation under reduced pressure [1, 2]. When we attempted to purify **1** in this manner, we observed that the sublimed crystals were colorless, although **1** crystallized from toluene was yellow. The sublimate was also insoluble in toluene, unlike **1**. Moreover, as shown in Table 2, the IR spectrum of the sublimate was different from that of **1**, as was solid state ^{29}Si-NMR.

Table 2. Some properties of Si_4Cl_8 (**1**) and $(SiCl_2)_n$ (**2**).

	Si_4Cl_8	$(SiCl_2)_n$
IR [cm^{-1}]	83 m, 103 m, 131 s, 193 w, 205 w, 233 vvw, 299 vs, 410 w, 585 m, 607 s,631 w	87.5 w, 104 w, 137 m, 209.5 w,302 s, 580.5 w, 603 vs
^{29}Si NMR, solid	7.1	3.9
a [Å]	7.326	4.057
b [Å]	7.182	6.783
c [Å]	13.284	13.346
β [°]	104.68	(90)

A colorless, tabular crystal of the sublimate was then subjected to X-ray diffraction. To our surprise, the structure proved to be that of the linear polymer, $(SiCl_2)_n$ (Eq. 2).

$$\underset{\mathbf{1}}{Si_4Cl_8} \xrightarrow{h\nu} \underset{\mathbf{2}}{(SiCl_2)_n}$$

Eq. 2.

This observation was remarkable, since crystals of polymers that diffract as single crystals are quite rare. The structure is detailed in Fig. 2. The polymer consists of co-aligned infinite chains of $SiCl_2$ units in a fully extended, all-*trans* conformation [14].

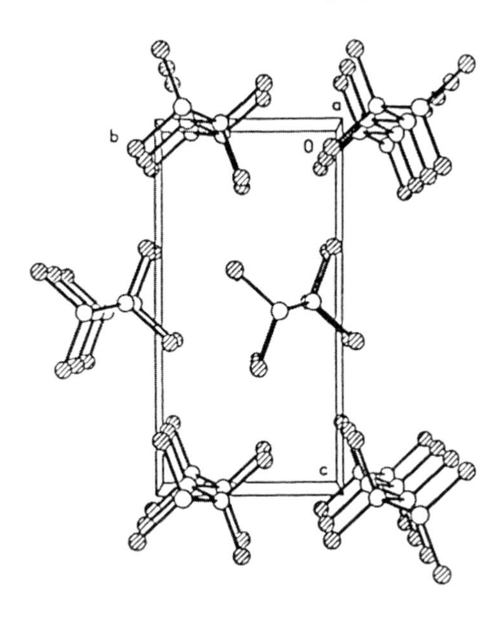

Fig. 2. Unit cell and molecular structure of **2**, 50 % probability ellipsoids. Selected interatomic distances [pm] and angles [°]: Si–Cl, 2.120(9), 2.088(9); Si–Si, 2.414(8); Si-Si-Si, 114.4(6), Cl-Si-Cl 111.0(4).

$(SiCl_2)_n$ as a true polymer has apparently not been described previously. Oligomers Si_nCl_{2n+2} have been prepared, up to $n = 10$, by the high-temperature reaction of silicon with $SiCl_4$, which presumably takes place through intermediate formation of $SiCl_2$ (Eq. 3) [15] or by an electric discharge through $SiCl_4$ [16].

$$SiCl_4 + Si \xrightarrow{1000°C} Si_nCl_{2n+2}$$

Eq. 3.

When the sublimation of **1** was repeated, but with the entire sublimation apparatur shielded from ordinary room light, the sublimate consisted of yellow crystals of **1**, as found in earlier studies. Thus the transformation of **1** to **2** is a photochemical process. How can this structural change take place? Evidence is not complete at this time, but we believe that the isomerization may be a topotactic reaction, in which the ring-opening and formation of the polymer chain take place with preservation of the crystal lattice. This requires that the lattice dimensions be rather similar. In fact they are, as shown by the data in Table 2. (The *a* constant for **2** is evidently doubled in **1**).

It is of interest to compare the structural features of **1** and **2**. The Si–Si bond length in **1** is slightly longer than normal, 237.2(2) pm, compared to the usual Si–Si bond distance of ca. 234 pm. Ring strain and consequent rehybridization may be partially responsible for the lengthening, the Si–Si distance in the related compound cyclo-Si_4Me_8 is also slightly longer than normal (mean value 236.3 pm). The Si–Si bond length in **1** is however shorter than the value predicted from the PM3 calculations, 245.5 pm [4].

In **2**, the Si–Si bond length is 241.4 pm, even greater than in **1**. The Si–Cl bond distances are also expanded in polymer **2**, to 212.0 and 208.8 pm, compared with 203.9 and 204.9 in **1**. In **1**, where the Si-Si-Si bond angles are restricted to 90°, the Cl-Si-Cl angles open somewhat to 112.7 and 111.8°. The Si-Si-Si angle for **2** is markedly spread to 114.4°, the Cl-Si-Cl angles are also greater than tetrahedral, 111.0°. The reasons underlying these bond lengths and angles are probably not steric in nature, since chlorine atoms are not especially large. We suggest that the unusual bond lengths and angles may reflect electrostatic repulsion between similar atoms, Cl···Cl and Si···Si, enhanced by the highly polar nature of the Si–Cl bond.

A possible mechanism for the ring-opening polymerization of **1** is outlined in Scheme 1. Photochemical cleavage of an Si–Si bond to form a diradical my be followed by radical attack on an Si–Si bond in a neighboring molecule to generate a radical chain process. Termination may then take place by coupling of silyl radicals, or through radical quenching by impurities. Chain-transfer by abstraction of a chlorine atom, leading to chain branching, is also possible. Formation of **2** by photolysis of polycrystalline solid **1** appears in fact to lead to a polymer containing a substantial number of crosslinks.

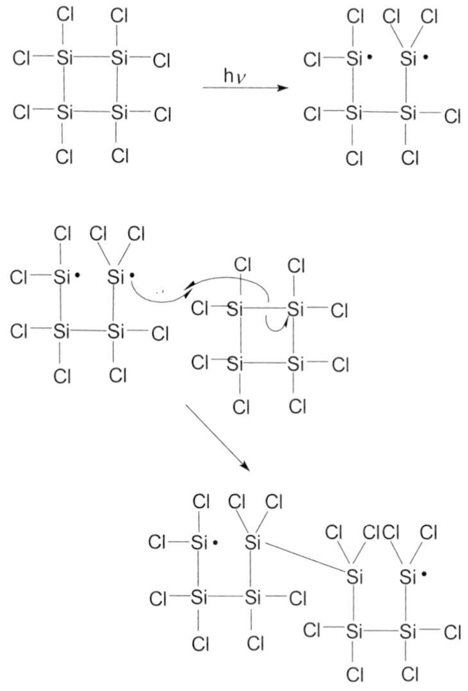

Initiation

Propagation

Scheme 1. Possible mechanism of polymerization of **1**.

Polymer **2** is insoluble in nonpolar solvents (hexane, benzene, toluene). Like **1**, it is rapidly hydrolyzed in moist air. The polymer has Lewis acid properties, for example, it catalyzes the ring-opening polymerization of THF, and the aldol condensation-polymerization of acetone. The molecular weight of **2** cannot be determined directly, but was instead estimated from the molecular weights of polymers made from it, as described below.

Polymers Derived from $(SiCl_2)_n$

A principal technical importance of **2** is that it provides a synthon for other polymers, $(SiX_2)_n$, of types which could not be synthesized heretofore. Up to now we have workend mainly with dialkoxypolysilanes, $[Si(OR)_2]_n$ and diaminopolysilanes, $[Si(NR_2)_2]_n$.

Dialkoxypolysilanes were synthesized from powdered **2** and excess alcohols, ethanol, 1-propanol, 2-propanol and 1-butanol, in toluene with triethylamine present as a base to take up HCl as it is formed (Eq. 4). Diaminopolysilanes were made by mixing **2** with excess of the dialkylamine in toluene.

$$(SiCl_2)_n \xrightarrow[-\ Et_3NH^+Cl^-]{ROH,\ Et_3N} [Si(OR)_2]_n$$

Eq. 4.

Unlike **2**, the alkoxy- and amino-substituted polysilanes are soluble in organic solvents such as toluene or hexane, so they could be more fully characterized. Their properties are summarized in Table 3. All of the poylmers show ^{29}Si-NMR resonances in the region from –6 to –23 ppm, slightly more shielded than than of **3** (–3.9 ppm). A quite surprising finding was that the bis(dialkylamino)polysilanes are bright yellow, although the dialkoxypolysilanes as well as **2** are colorless. The electronic spectra of **3** and **4** are illustrated in Fig. 3. Both polymers exhibit strong absorption bands near 350 nm, but that for **3** is relatively sharp whereas the band for diaminopolysilane **4** is extremely broad, tailing well into the visible region and hence giving rise to the yellow color.

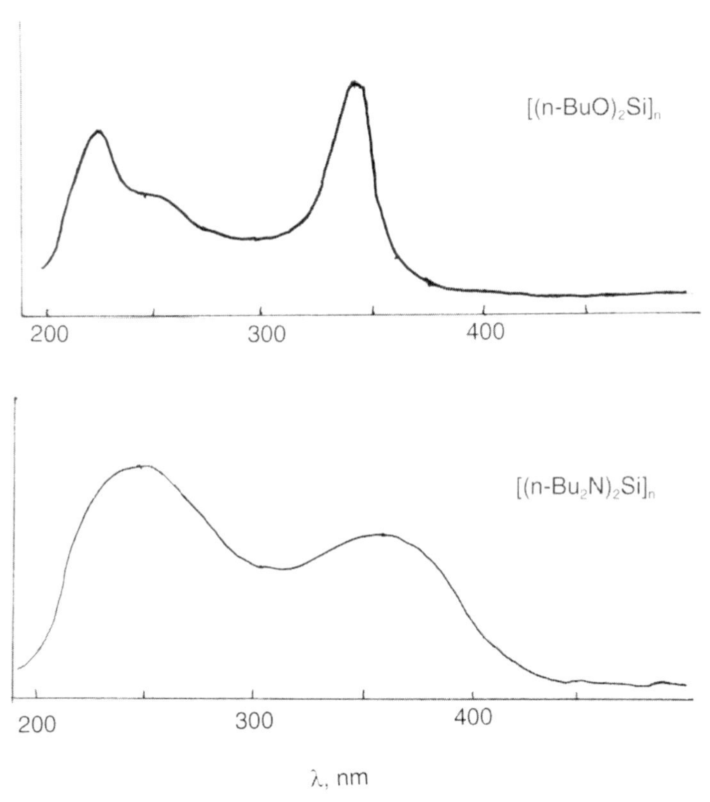

Fig. 3. Electronic spectra of polymers $[(n\mathrm{BuO})_2\mathrm{Si}_2]_n$ (above) and $(n\mathrm{Bu}_2\mathrm{N})_2\mathrm{Si}_2]_n$ (below).

Table 3. Polydialkoxysilanes and Polydialkylaminosilanes from **2**.

Polymer	^{29}Si-NMR, δ	UV, λ [nm]	M_w, GPC
$[(\mathrm{EtO})_2\mathrm{Si}]_n$	–9.8	216, 336	8 000
$[(i\mathrm{PrO})_2\mathrm{Si}]_n$	–8.2	226, 320 sh	4 900
$[(n\mathrm{BuO})_2\mathrm{Si}]_n$	–8.3	340	8 000
$[(\mathrm{Et}_2\mathrm{N})_2\mathrm{Si}]_n$	–5.6	220, 345 sh	4 000
$[(i\mathrm{Pr}_2\mathrm{N})_2]_n$	–23.6	222, 342 sh	3 000, 10 000
$[(n\mathrm{Bu}_2\mathrm{N})_2\mathrm{Si}]_n$	–7.8	230, 356	4 000

The molecular weights of the various derived polysilanes were determined by gel permeation chromatography (Table 3). The values obtained range from about to 3 000 to10 000. If we take the 8 000 value for **3** as reference, this leads to a degree of polymerization of 46. We regard this as a minimum DP value for $(SiCl_2)_n$, since some chain breaking may have taken place during the derivatization.

Polymer **2** also reacts with water, and in the process undergoes a change first to yellow, then back again to colorless. The hydrolysis of Si–Cl to Si–OH is undoubtedly accompanied by condensation between chains, as shown in Scheme 2. The yellow phase may represent a polymer with approximate composition $[Si(OH)_2]_n$, although partly crosslinked and probably still containing some Si–Cl bonds. Further crosslinking by loss of H_2O or HCl between chains should lead ultimately to a quite interesting material, a fully crosslinked, three-dimensional network with empirical composition SiO, hence a polymeric version of the well-known but poorly understood material, silicon monoxide.

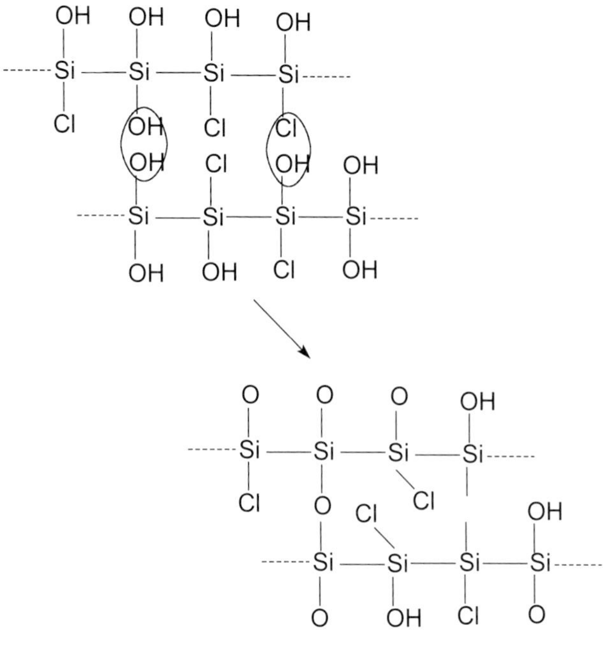

Scheme 2. Hydrolysis and condensation of **2**, with loss of water and HCl.

The synthesis of additional types of polymers from $(SiCl_2)_n$, and their possible technical applications, are being further studied in our laboratories.

Acknowledgment: This work was partially supported by a grant from the *US Office of Naval Research*. We thank Dr. Shuzi Hayase of *Toshiba Company* for crucial assistance at the beginning of this research.

References:

[1] E. Hengge, D. Kovar, Z. *Anorg. Allg. Chem.* **1979**, *458*, 163.

[2] K. Hassler, E. Hengge, D. Kovar, *J. Mol. Struct.*, **1980**, *66*, 25.

[3] a) R. West, "*Polysilanes*" in *Comprehensive Organometalic Chemistry II*, *Vol. 2*, (Eds. E.W. Abel, F. G. A. Stone, G. Wilkinson), Pergamon, Oxford **1995**, p. 79.
b) H. Watanabe, M. Kato, T. Okawa, Y. Kougo, Y. Nagai, M. Goto, *Appl. Organomet. Chem.* **1987**, *1*, 157.

[4] H. Stüger, R. Janoschek, *Phosphorus, Sulfur, Silcion* **1992**, *68*, 129.

[5] J. R. Koe, D. R. Powell, J. J. Buffy, R. West, *Polyhedron* **1998**, *17*, 1791.

[6] Y. S. Chen, P. P. Gaspar, *Organometallics* **1982**, *1*, 1410.

[7] C. Kratky, H. G. Schuster, E. Hengge, *J. Organomet,. Chem.* **1983**, *247*, 253.

[8] V. S. Mastryukov, S. A. Strelkov, L. V. Vilkov, M. Kobnits, B. Bozsondai, H. G. Schuster, E. Hengge, *J. Mol. Struct.* **1990**, *238*, 433.

[9] K. Hassler, Spectrochim, Acta **1981**, *A37*, 541.

[10] J. R. Koe, D. R. Powell, R. West, unpublished studies.

[11] L. Parkanyi, K. Sasvari, I. Barta, *Acta Crystallogr.* **1978**, *B34*, 883.

[12] C. J. Hurt, J. C. Calabrese, R. West, *J. Organomet. Chem.* **1975**, *91*, 273.

[13] T. N. Margulis, *Acta Crystallogr.* **1965**, *19*, 857.

[14] J. R. Koe, D. R. Powell, J. J. Buffy, S. Hayase, R. West, *Angew, Chem.*, *Int. Ed. Engl.* **1998**, *17*, 1441.

[15] R. Schwarz, C. Danders, *Chem. Ber.* **1947**, *80*, 444.

[16] H. Kautsky, H. Kautsky, Jr., *Chem. Ber.* **1956**, *89*, 571.

Polysilyl Dianions — Synthesis and Reactivity

Christian Mechtler, Christoph Marschner*

Institut für Anorganische Chemie, Technische Universität Graz
Stremayrgasse 16, A-8010 Graz, Austria
Tel.: Int. code + (316)873 8209 — Fax: Int. code + (316)873 8701
E-mail: marschner@anorg.tu-graz.ac.at

Keywords: Polysilanes / Silyl Anions / Cyclization / Potassium

Summary: The synthesis of α,ω-bis[tris(trimethylsilyl)silyl] alkanes was achieved by the reaction of α,ω-ditosylalkanes with tris(trimethylsilyl)silyl potassium. Reaction of these with one equivalent of potassium *tert*-butoxide resulted in clean formation of monopotassium silyl anions. Addition of another equivalent of the transmetallating agent led to the formation of the dipotassium compounds in cases where the alkyl spacer contained at least three methylene units. Partial hydrolysis of the dipotassium compounds induced an intramolecular reaction yielding a cyclic silyl potassium compound.

Introduction

Recently we developed a new method for the generation of polysilyl alkali compounds. We have demonstrated the easy conversion of tetrakis(trimethylsilyl)silane to tris(trimethylsilyl)silyl potassium upon treatment with potassium *tert*-butoxide [1] (Scheme 1). This constitutes an interesting alternative to the well known tris(trimethylsilyl)silyl lithium reagent introduced by Gilman 35 years ago [2]. We also found that this method does not exhibit a preference for the splitting of "inner" Si–Si bonds, as methyl lithium does, but usually selectively cleaves off a trimethylsilyl group. For the case of hexakis(trimethylsilyl)disilane we were able to demonstrate the exclusive formation of pentakis(trimethylsilyl)disilanyl potassium where the use of methyl lithium leads to tris(trimethylsilyl)silyl lithium and tris(trimethylsilyl)methylsilane instead. In addition we could extend the method to the use of isotetrasilanes as starting materials, leading thus to 2-potassium trisilanyls.

$$Me_3Si{-}Si(SiMe_3)_2{-}SiMe_3 \xrightarrow[-\,tBuOSiMe_3]{tBuOK,\ THF,\ RT} Me_3Si{-}Si(SiMe_3)_2{-}K$$

Scheme 1. Formation of tris(trimethylsilyl)silyl potassium.

Attempts to obtain 1,2-dipotassium tetrakis(trimethylsily)disilanyl and 1,3-dipotassium 1,1,3,3-tetrakis(trimethylsily)dimethyltrisilanyl from the respective precursors proved to be unsuccessful. In order to introduce an insulator between the two charged silicon atoms we set out to study the synthesis and behavior of α,ω-bis[tris(trimethylsilyl)silyl] alkanes.

Results and Discussion

First attempts to achieve the synthesis of α,ω-bis[tris(trimethylsilyl)silyl] alkanes by the reaction of two equivalents of tris(trimethylsilyl)silyl potassium with the respective α,ω-dibromoalkanes proved to be fruitless, leading mainly to the formation of hexakis(trimethylsilyl)disilane [3]. Since the reason for this reaction is the ease of the halogen metal exchange reaction we decided to use α,ω-ditosyl alkanes as starting materials instead. And indeed these gave rise to the formation of the desired α,ω-bis[tris(trimethylsilyl)silyl] alkanes (Scheme 2).

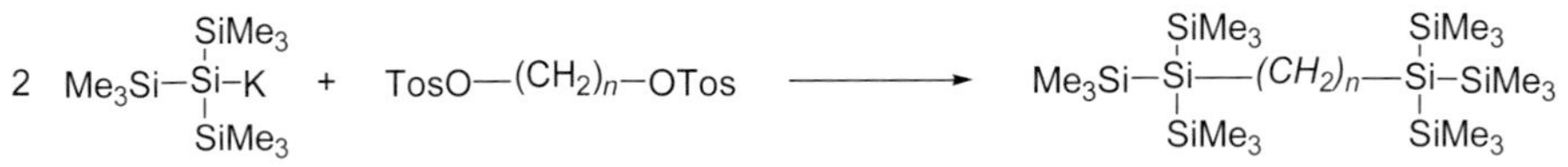

$$2\ (Me_3Si)_3Si{-}K + TosO{-}(CH_2)_n{-}OTos \longrightarrow (Me_3Si)_3Si{-}(CH_2)_n{-}Si(SiMe_3)_3$$

n = 2, 3, 4

n = 2 **1**; 3 **2**; 4 **3**

Scheme 2. Synthesis of α,ω-bis[tris(trimethylsilyl)silyl] alkanes.

Treatment of 1,2-bis[tris(trimethylsilyl)silyl] ethane (**1**) with potassium *tert*-butoxide under the conditions usually employed [1] led to the clean replacement of one trimethylsilyl group by potassium **4**. Formation of the dipotassium compound upon addition of a second equivalent of potassium *tert*-butoxide was not observed either by ^{29}Si-NMR or by GC-MS analysis of derivatized aliquots of the reaction, even after prolonged reaction times.

This was contrasted by the reactivity of the higher homologues **2** and **3**. In both cases reaction with one equivalent of potassium *tert*-butoxide also led to the formation of the mono-substituted product **5**, **6**. However, after addition of a second equivalent of potassium *tert*-butoxide in these cases the reaction proceeded a step further to yield **7**, **8**.

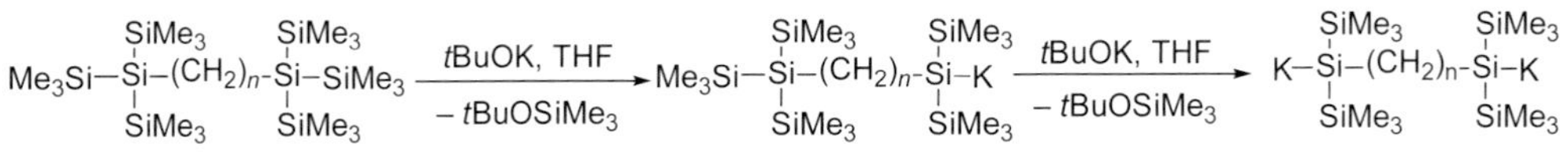

$$(Me_3Si)_3Si{-}(CH_2)_n{-}Si(SiMe_3)_3 \xrightarrow[-\ tBuOSiMe_3]{tBuOK,\ THF} (Me_3Si)_3Si{-}(CH_2)_n{-}Si(SiMe_3)_2K \xrightarrow[-\ tBuOSiMe_3]{tBuOK,\ THF} K(Me_3Si)_2Si{-}(CH_2)_n{-}Si(SiMe_3)_2K$$

n = 2 **4**; 3 **5**; 4 **6**

n = 3 **7**; 4 **8**

Scheme 3. Formation of mono and di-potassium compounds.

After a prolonged reaction time a new product was formed as indicated by ^{29}Si-NMR and GC-MS. This product also is an Si–K species and contains a newly formed Si–Si bond to give in cyclic structures **9**, **10**.

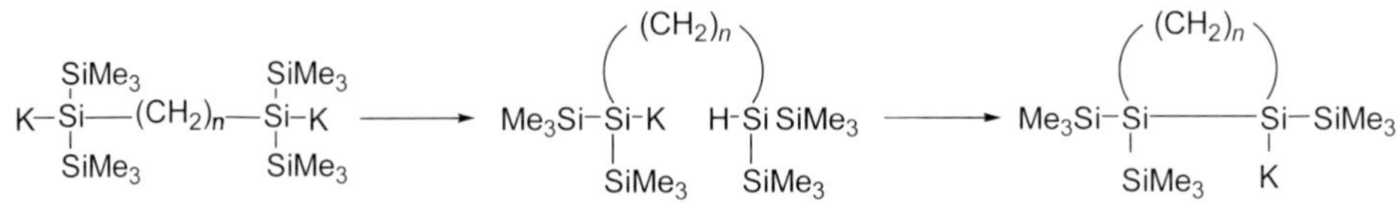

$$K(Me_3Si)_2Si{-}(CH_2)_n{-}Si(SiMe_3)_2K \longrightarrow Me_3Si{-}Si(SiMe_3)(K){-}(CH_2)_n{-}SiH(SiMe_3)SiMe_3 \longrightarrow \text{cyclo-}[Me_3Si(Me_3Si)Si{-}(CH_2)_n{-}Si(K)(SiMe_3)]$$

n = 3 **9**; 4 **10**

Scheme 4. Formation of cyclic Si–K compounds.

Our first assumption was that the formation of **9** and **10** might be due to an intramolecular attack of the Si–K moiety of the monopotassium compound onto the $Si(SiMe)_3$ group. However, this reaction would afford a cyclic structure with four trimethylsilyl groups as in **11**, **12** and trimethylsilyl potassium, none of which we were able to observe. Even if the trimethylsilyl potassium reacts immediately with the formed product this does not seem to be very likely, since it would result in the formation of hexamethyldisilane which we also did not observe. In addition we figured out that the new product is formed at the expense of the dipotassium compound and eventually found that the formation was actually caused by partial hydrolysis of the strongly basic dipotassium compound. Once the Si–H bond is formed the now more electrophilic silicon is attacked by the Si–K group and formation of the cyclic product along with trimethylsilane is accomplished. Careful addition of one equivalent of water to a reaction mixture which contained mainly the di-potassium product led to the quantitative formation of the cyclic Si–K species.

S_N2 type displacement of hydride also in this case should lead to the formation of the mentioned compound **11** or **12** with four trimethylsilyl groups, which we found to be formed on reacting **9** or **10** with trimethylsilyl chloride. We assume that this compound might be only formed as an intermediate. The hydride ion which acts as a leaving group is instantaneously attacking a trimethylsilyl group, thus, forming trimethylsilane and **9** or **10**, respectively.

This two-step process may also proceed in a concerted manner similarly to the mechanism proposed by Corriu et al. for the attack of potassium hydride on hydrosilanes (Scheme 5) [4].

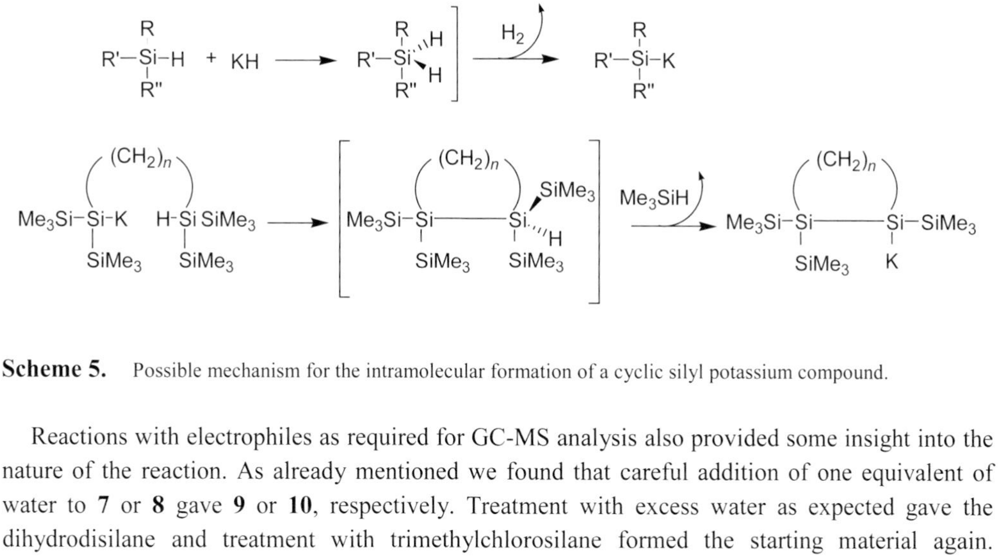

Scheme 5. Possible mechanism for the intramolecular formation of a cyclic silyl potassium compound.

Reactions with electrophiles as required for GC-MS analysis also provided some insight into the nature of the reaction. As already mentioned we found that careful addition of one equivalent of water to **7** or **8** gave **9** or **10**, respectively. Treatment with excess water as expected gave the dihydrodisilane and treatment with trimethylchlorosilane formed the starting material again. Notably, the treatment of **7** or **8** with ethyl bromide gave, besides the expected dialkylated product, also cyclic products **11** or **12**, respectively. We attribute the formation of the latter to metal–halogen exchange reaction of one Si–K moiety followed by immediate nucleophilic attack of the second Si–K group onto the newly formed silyl bromide. The same products **11** and **12** are also formed in the reaction of **9** or **10**, respectively, with trimethylchlorosilane (Scheme 6).

Scheme 6. Formation of cyclic trimethylsilane.

^{29}Si-NMR data of the central silicon atom proved to be of particular value for the assignment of the structures formed. Replacement of one trimethylsilyl group by potassium caused an expected upfield shift of about 33 ppm. Introduction of the second potassium only causes an upfield shift of another 2 ppm. Ring closure results in the in downfield shifts both at the Si–K silicon as well as on the now neighboring Si atom. This shift is caused in part by the cyclic structure and, for the $Si(SiMe_3)_2$ unit, to a greater extent by the neighboring Si–K group.

Table 1. Selected ^{29}Si-NMR data. Shifts in ppm vs external TMS standard.

Compound	δSi(α)	δSi(ω)
1	–76.5	
2	–83.0	
3	–81.4	
4	–102.4	–80.1
5	–116.0	–84.0
7	–118.1	
9	–113.5	–63.4

Acknowledgment: This work was carried out within the Sonderforschungsbereich Elektroaktive Stoffe funded by the *Fonds zur Förderung der Wissenschaftlichen Forschung in Österreich*. The authors are grateful to *Wacker–Chemie GmbH* (Burghausen) for the generous donation of chlorosilanes. Ch. Marschner thanks the *Austrian Academy of Science* for an APART scholarship.

References:

[1] C. Marschner, *Eur. J. Inorg. Chem.* **1998**, 221.

[2] H. Gilman, C. L. Smith, *Chem. Ind. (London)* **1965**, 848.

[3] H. Gilman, R. L. Harrel, *J. Organomet. Chem.* **1967**, *9*, 67.

[4] a) R. P. J. Corriu. C. Guerin, B. Kolani, *Bull. Soc. Chim. Fr.* **1985**, 973.
b) J. L Bredford, R. P. J. Corriu. C. Guerin, B. Henner, *J. Organomet. Chem.* **1989**, *370*, 9.

Synthesis and Reactions of Chlorinated Oligosilanes

Johannes Belzner*, Bernhard Rohde, Uwe Dehnert, Dirk Schär

Institut für Organische Chemie
Georg-August-Universität Göttingen
Tammannstr. 2, D-37077 Göttingen, Germany
Tel.: Int. code + (551)39 3285 — Fax: Int. code + (551)39 9475
E-mail: jbelzne@gwdg.de

Keywords: Chlorosilanes / Silylene / Disilene / Lewis Bases / Ab Initio Calculations

Summary: 1,2-Dichlorodisilanes are prepared by reaction of silyl metal compounds with dichlorosilanes and subsequent chlorination. Alternatively, 1,2-dichlorodisilanes are obtained by insertion of a silylene in the Si–Cl bond of dichlorosilanes. Dehalogenation of 1,2-dichlorodisilanes with magnesium leads to cyclic oligosilanes or vicinally dimetallated species. Ab initio calculations show that coordinated disilenes are plausible intermediates in some of these reactions.

We are interested in the synthesis of 1,2-dichlorodisilanes bearing the 2-dimethylaminomethylphenyl substituent, which may be suitable precursors for intramolecularly coordinated disilenes [1].

Preparation of Dichlorodisilanes

Two general routes were used to prepare a variety of 1,2-dichlorodisilanes.

(a) The reaction of a metalated silane such as **2** with a dichlorosilane and subsequent halogenation of the Si–H bond (Scheme 1) [1b, 2].

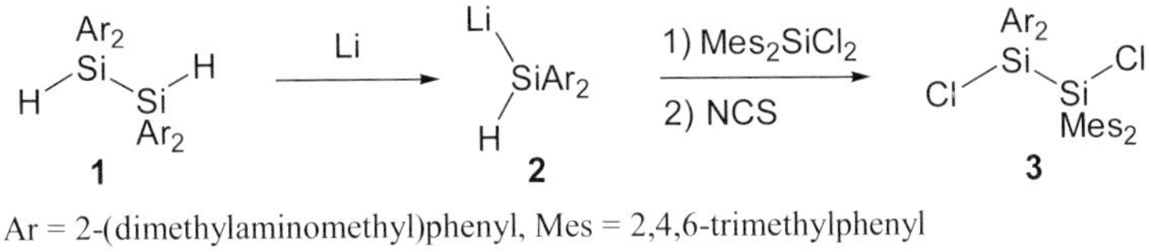

$H(Ar_2)Si{-}Si(Ar_2)H$ (**1**) $\xrightarrow{Li}$ $Li{-}SiAr_2{-}H$ (**2**) $\xrightarrow{1)\ Mes_2SiCl_2;\ 2)\ NCS}$ $Cl(Ar_2)Si{-}Si(Mes_2)Cl$ (**3**)

Ar = 2-(dimethylaminomethyl)phenyl, Mes = 2,4,6-trimethylphenyl

Scheme 1. Synthesis of **3**.

(b) Insertion of a coordinated silylene into an Si–Cl bond is a newly developed method for formation of Si–Si bonds [3], which we used to selectively obtain 1,2-dichlorodisilanes in good yield (Scheme 2). This reaction makes use of cyclotrisilanes **5** [4] or **8** as thermal silylene sources [5]. In some cases, this reaction is reversible under the conditions employed. Thus, reacting **4** and **5** in stoichiometric amounts at 60°C for 3 h, an equilibrium mixture containing 68 % **6** was obtained [6].

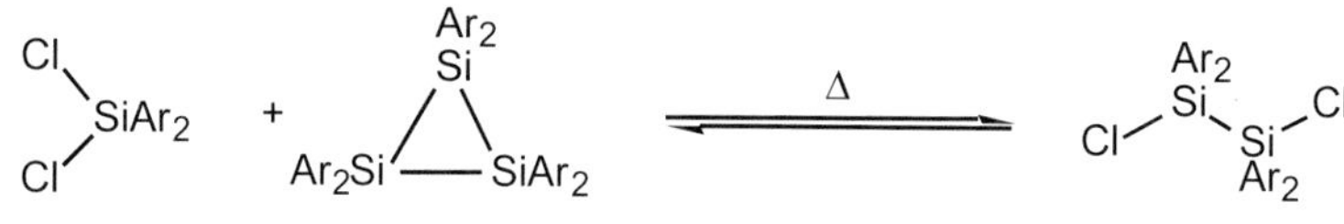

Ar = 2-(dimethylaminomethyl)phenyl

Scheme 2. Synthesis of 1,2 dichlorodisilanes via silylene insertion.

According to the ^{29}Si-NMR data only the 2-(dimethylaminomethyl)phenyl-substituted silicon center of **9** is coordinated by one or two amino groups in solution. Its ^{29}Si-NMR signal is significantly shifted to high field in comparison to $ClPh_2Si–SiPh_2Cl$ [7] as is expected for a highly coordinated silicon center [8] (Table 1).

Table 1. ^{29}Si-NMR data of selected 1,2-dichlorodisilanes in C_6D_6.

	δ(Si-1)[a]	δ(Si-2)[b]	Δδ
$(ClPh_2Si)_2$ (**14**) [7]	−6.1		—
$(ClAr_2Si)_2$[d] (**6**) [9]	−2.9		+3.2[c]
$ClAr_2SiSiPh_2Cl$[d] (**9**) [4]	−21.7	−2.3	−15.6[c]
$ClAr_2SiSiE_2Cl$[d, e] (**15**)	−4.5	+0.4	(+1.6)[c]
$ClAr_2SiSiMes_2Cl$[d, e] (**3**)	−4.8	−0.2	(−5.0)[f]
$(ClMes_2Si)_2$ (**16**) [10]	+0.2		—
$(ClArMesSi)_2$[d, g] (**7**)	−2.5 −2.6		−2.7[f] −2.8[f]

[a] δ(Si-1) = ^{29}Si-NMR shift of the 2-(dimethylaminomethyl)phenyl-substituted silicon center. — [b] δ(Si-2) = ^{29}Si-NMR shift of the silicon center substituted by Mes, Ph or E (= 2-EtC_6H_4). — [c] Δδ = δ(Si-1) – δ(**14**). — [d] Ar = 2-(dimethylaminomethyl)phenyl. — [e] The assignment is ambiguous. — [f] Δδ = δ(Si-1) – δ(**16**). — [g] Mixture of diastereomers.

Reactivity

The outcome of the reaction of the 1,2-dichlorodisilanes with magnesium depends on the substitution pattern of the disila unit. Compound **6** yielded upon treatment with magnesium cyclotrisilane **5** [5a] (Scheme 3). Analogously, cyclotrisilane **8** [5] was obtained under similar conditions from dichlorodisilane **7**, which bears, like **6**, the 2-(dimethylaminomethyl)phenyl substituent at adjacent silicon centers; the *cis,trans,trans* isomer of **8** was formed exclusively, when a mixture of diastereomers of **7** was used as starting material (Scheme 4).

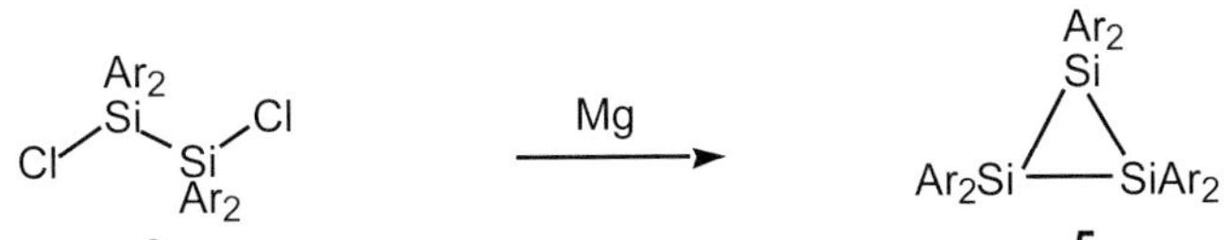

Scheme 3. Synthesis of **5**.

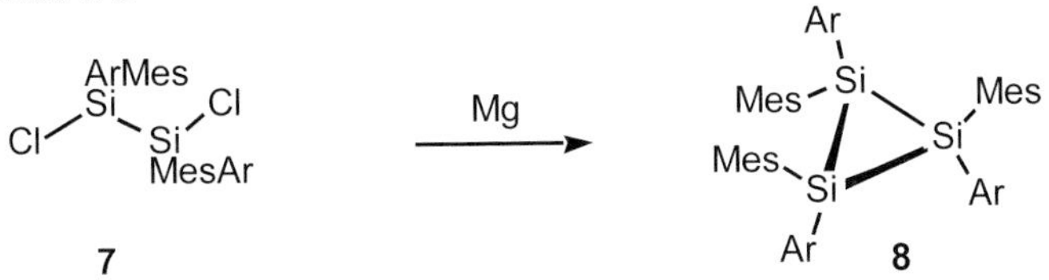

Scheme 4. Synthesis of **8**.

In contrast, 1,2-dichlorodisilane **9** yielded upon reaction with magnesium a 1:1 mixture of cyclotetrasilanes **10** and **11** (Scheme 5), which may be the products of a head-to-tail and a head-to-head dimerization of an initially formed disilene.

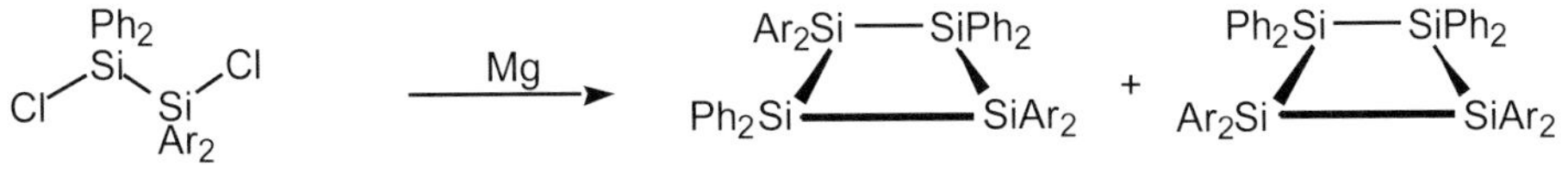

Scheme. 5. Preparation of cyclotetrasilanes **10** and **11**.

The Si_4 ring of **10** is, unlike that in octaphenylcyclotetrasilane [11], planar in the solid state; no N–Si coordination takes place (Fig. 1). In solution, one Si–Si coupling constant ($^1J_{SiSi}$ = 59.6 Hz) is observed for **10**, whereas **11** shows, due to its lower symmetry, two Si–Si coupling constants ($^1J_{SiSi}$ = 60.2 Hz, $^2J_{SiSi}$ = 32.6 Hz).

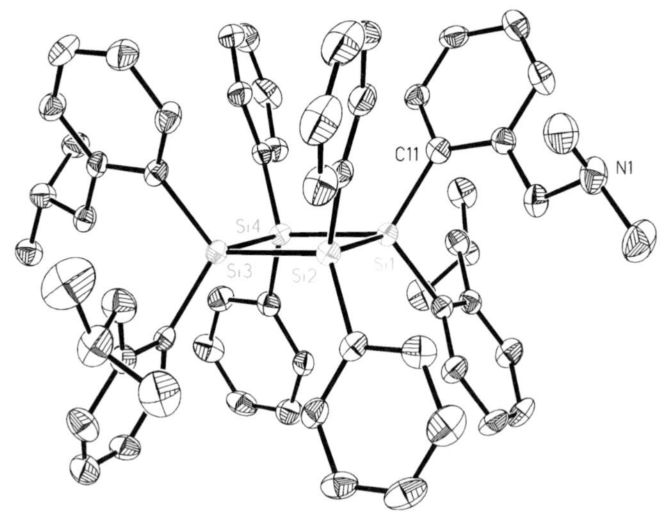

Fig. 1. X-ray crystal structure of cyclotetrasilane **10**. Selected bond lengths [pm] and angles [°]; Si(1)–Si(2) 240.2(2), Si(2)–Si(3) 240.2(2), Si(3)–Si(4) 240.1(2), Si(4)–Si(1) 239.9(2), Si(1)-Si(2)-Si(3) 90.4(6), Si(2)-Si(3)-Si(4) 89.5(6), Si(3)-Si(4)-Si(1) 90.5(6), Si(4)-Si(1)-Si(2) 89.6(6).

Dichlorodisilane **3** did not react with magnesium turnings. However, using highly active Rieke magnesium [12], an air- and moisture- sensitive compound **12** of unknown structure was obtained. When isolated **12** was treated with water or MeOH, disilane **2a** was obtained in quantitative yield (Scheme 6). The reaction of **12** with dimethyl sulfate yielded dimethyldisilane **2b** [6]. Thus, **12** appears to be a 1,2-dimetalated disilane, to which we preliminarily assign a 1,2-di-Grignard structure.

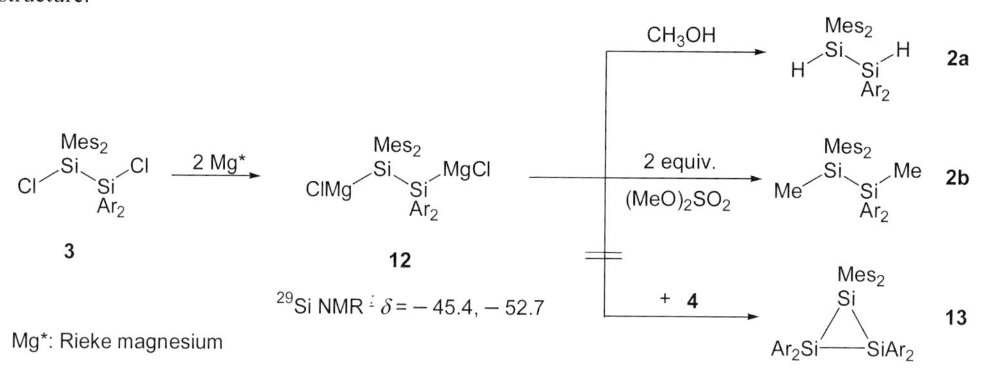

Scheme 6. Synthesis of **12**.

Coordinated Disilenes as Intermediates

Disilenes coordinated by Lewis bases have never been observed experimentally [13]. However, ab initio calculations show that complexes of disilene with one or two molecules of ammonia correspond to minima at the potential energy surface (Scheme 7).

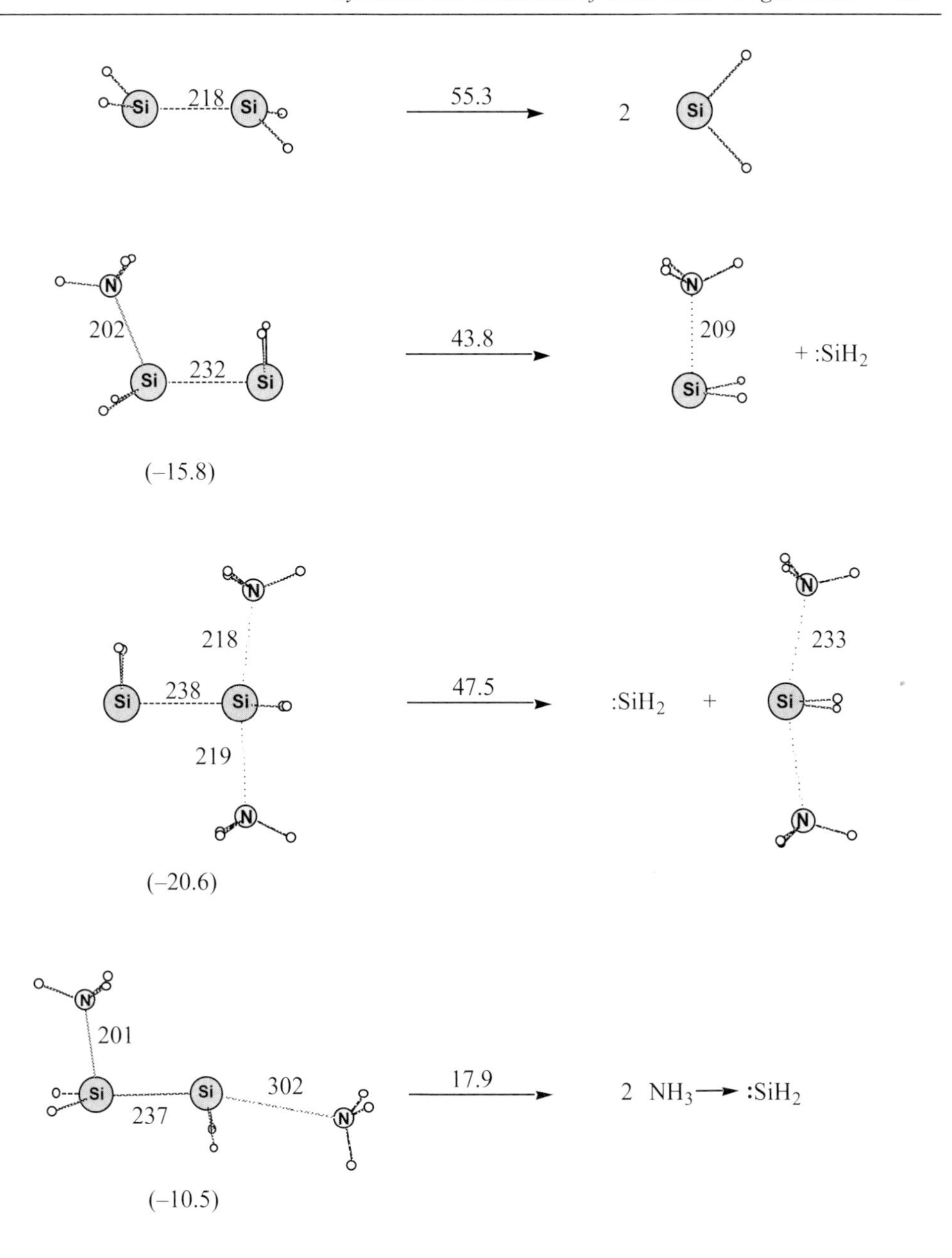

Scheme 7. Enthalpies (kcal mol^{-1}) of formation and Si–Si bond cleavage reactions of some coordinated disilenes, calculated at the B3LYP/6-311+G*//B3LYP/6-311+G* level; values in parentheses refer to the reaction enthalpy of the formation of the coordinated disilene from disilene and 1 or 2 molecules of ammonia; bond lengths in pm.

The cleavage of the Si–Si bond of a 1,2-dicoordinated disilene is appreciably less endothermic than that of disilene itself, or a mono- or geminally dicoordinated disilene. Based on these thermodynamic parameters, a plausible mechanism of the experimentally observed formation of cyclotrisilanes from dichlorodisilanes **6** and **7** could involve a vicinally dicoordinated disilene, which is in equilibrium with two silylene–Lewis base complexes. Addition of this silaylide to the coordinated disilene eventually would give the cyclotrisilane (Scheme 8.). In contrast, a geminally dicoordinated disilene is expected to be stable with regard to Si–Si bond cleavage, and accordingly no cyclotrisilane is formed.

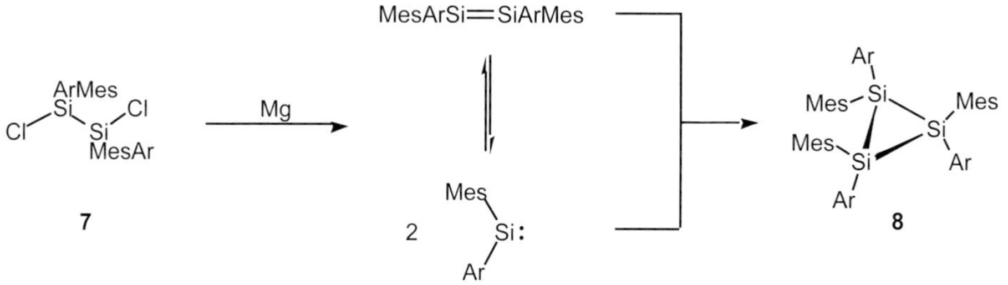

Scheme 8. Plausible mechanism of the formation of cyclotrisilane **8** from dichlorodisilane **7**.

Acknowledgment: The authors are indebted to the *Deutsche Forschungsgemeinschaft* and the *Fonds der Chemischen Industrie* for financial support of this work.

References:

[1] For the preparation of disilenes from 1,2-dichlorodisilanes, see, e. g.:
a) H. Watanabe, K. Takeuchi, K. Nakajiama, Y. Nagai, M. Goto, *Chem. Lett.* **1988**, 1343–1346.
b) M. Weidenbruch, A. Pellmann, Y. Pan, S. Pohl, W. Saak, H. Marsmann, *J. Organomet. Chem.* **1993**, *450*, 67–71.

[2] R. West, S. S. Zigler, L. M. Johnson, *J. Organomet. Chem.* **1988**, *341*, 187–198.

[3] U. Herzog, R. Richter, E. Brendler, G. Roewer, *J. Organomet. Chem.* **1996**, *507*, 221–228.

[4] a) J. Belzner, *J. Organomet. Chem.* **1992**, *430*, C51.
b) J. Belzner, H. Ihmels, B. O. Gould, B. O. Kneisel, R. Herbst-Irmer, *Organometallics* **1995**, *14*, 305–311.

[5] J. Belzner, U. Dehnert, H. Ihmels, M. Hübner, P. Müller, I. Usón, *Chem. Eur. J.* **1998**, *5*, 852–863

[6] U. Dehnert, *Dissertation*, Göttingen **1997**.

[7] H. Söllradl, E. Hengge, *J. Organomet. Chem.* **1983**, *243*, 257–269.

[8] E. A. Williams in: *The Chemistry of Organic Silicon Compounds* (Eds.: S. Patai, Z. Rappoport), Wiley, Chichester, **1989**, pp 511–554.

[9] J. Belzner, N. Detomi, H. Ihmels, M. Noltemeyer, *Angew. Chem.* **1994**, *106*, 1949–1950; *Angew. Chem. Int. Ed. Engl.* **1994**, *33*, 1854–1855.

[10] M. Weidenbruch, K. Kramer, A. Schäfer, J. K. Blum, *Chem. Ber.* **1985**, *118*, 107–115.

[11] L. Párkányi, K. Sasvári, I. Barta, *Acta Crystallogr. Sect. B* **1978**, *B34*, 883–887.

[12] R. D. Rieke, S. E. Bales, *J. Chem. Soc., Chem. Commun.* **1973**, 879–880.

[13] G. R. Gillette, G. H. Noren, R. West, *Organometallics* **1989**, *8*, 487–491.

Base-Catalyzed Disproportionation of Tetrachlorodimethyldisilane — Investigations of the Heterogeneous Catalysts

*Norbert Schulze, Gerhard Roewer**

Institut für Anorganische Chemie
Technische Universität Bergakademie Freiberg
Leipziger Str. 29, D-09596 Freiberg, Germany
Tel.: Int. code + (3731)39 3194 — Fax: Int. code + (3731)39 4058
E-mail: roewer@orion.hrz.tu-freiberg.de

Keywords: Disilane / Disproportionation / Heterogeneous Catalyst / Lewis Base

Summary: Recently we investigated the effectiveness of various Lewis base catalysts in the heterogeneously catalyzed disproportionation of 1,1,2,2-tetrachlorodimethyl-disilane. The preparation and properties of these catalysts are discussed in view of their significant catalytic parameters.

Introduction

In the synthesis field of reactive and processable polysilanes the Lewis base-catalyzed disproportionation of chloromethyldisilanes, which are byproducts of the industrial chloromethyl-silane production, proved to be a promising route to a polysilyne-type 3D architecture [1]. Concerning this reaction our investigations are focused on the effectiveness and properties of various Lewis bases in order to get a deeper insight into the disilane disproportionation mechanism.

In the temperature range of 155–180°C both trichloromethylsilane and a mixture of oligo-(chloromethyl)silanes (up to a branched heptasilane) are obtained in a first process stage (Eq. 1).

$$n\ H_3C{-}SiCl_2{-}SiCl_2{-}CH_3 \xrightarrow{\text{cat.}} (n-1)\ Cl{-}SiCl_2{-}CH_3 + H_3C{-}SiCl_2{-}\left[Si(R)(CH_3)\right]_{(n-1)}{-}SiCl_2{-}CH_3$$

$$\text{cat.} = (\equiv O)_3Si{-}(CH_2)_m{-}\text{donor} \qquad m = 1\text{–}3 \qquad R = Cl, SiCH_3R_2$$

Eq. 1. Disilane disproportionation catalyzed by Lewis bases.

Owing to a phase separation between the catalyst, the educt and reaction products the heterogeneous catalytic disproportionation offers the access to oligo- and poly(chloromethyl)silanes free of the Lewis base [2]. Some N-heterocycles grafted on the surface sites of SiO_2 give rise to a significant catalytic disproportionation process.

Preparation and Action of the Heterogeneous Catalysts

The catalyst preparation can be started generally from the heterocyclic base molecule. After the introduction of alkyltrialkoxysilyl substituents the selected N-heterocycles are grafted onto the surface of silica gel via siloxane bonds, as already published [3].

The synthesis of these (trialkoxysilyl)alkyl compounds is performed by elimination of the sodium salt of the appropriate heterocycle with chloroalkyltrialkoxysilanes or by the addition of trialkoxyvinylsilane to the heterocycle in the presence of 3 mol % of lithium metal [3].

Besides the already-known [4] grafted benzimidazole and 3,5-dimethylpyrazole system we examined Lewis bases like imidazole, indole, indazole, benzotriazole, *N,N*-dimethylaniline, 2,5-dimethylpyrrole, triazole and N-bases like *N*-methylpiperazine without a π-electron system.

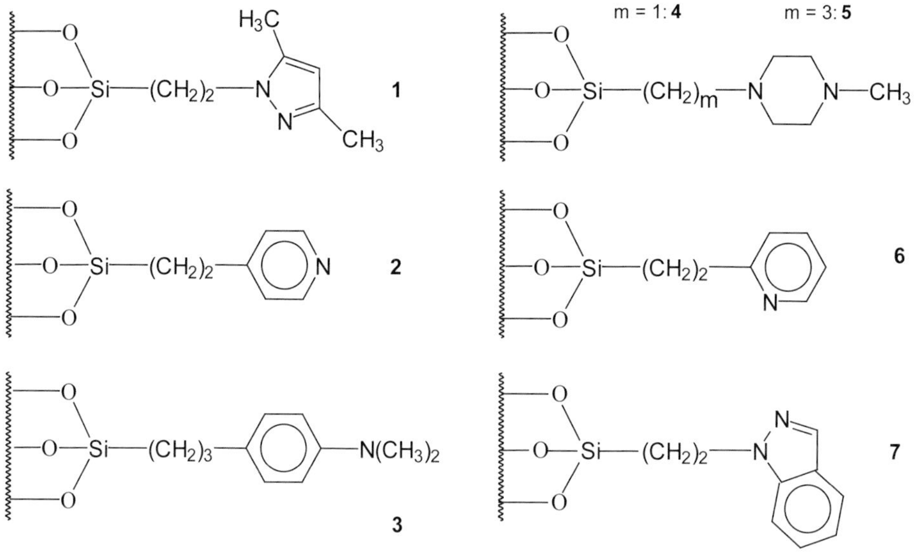

Fig. 1. Simplified structures of some grafted Lewis bases used in the disproportionation of $MeCl_2Si–SiCl_2Me$.

The proposed mechanism of this disproportionation and the formation of the oligo(chloromethyl)-silanes is described in detail in [1–4]; therefore only a short comment will be given here.

The disilane molecule is thought to be coordinated at the Lewis base by donor–acceptor interaction creating a pentacoordinated Si atom. Both the stability and the reactivity rate of this intermediate adduct depend on the steric factors and on the nucleophilicity of the donor. Dominantly the nucleophilicity determines the catalyst activity in the disproportionation reaction.

From the adduct one molecule of trichloromethylsilane is cleaved. Thus a donor-stabilized chloromethylsilylene (do→:SiClMe) is originated simultaneously. Its stepwise insertion into the Si–Cl bonds of another disilane or of an already-formed oligosilane molecule gives rise to a branched skeleton.

Our experiments have shown that even nitrogenous bases without an aromatic system, and also those which have ionization potentials significantly higher than 9 eV, catalyze the disproportionation. Heating *N,N*-dimethylaniline or *N,N,N',N'*-tetramethyl-1,4-phenylendiamine (6 mol % in each case, first ionization potentials lower than 7.5 eV [5]) with $MeCl_2Si–SiCl_2Me$ under reflux for five hours causes no or an insignificant disproportionation reaction, respectively. That means a low ionization potential as a measure of donor power is not a necessary condition for the disproportionation as supposed in [1, 4].

In Table 1 some parameters of the heterogeneous catalysts are summarized. The cleavage yield μ represents the weight ratio of the trichloromethylsilane formed to the educt $MeCl_2Si–SiCl_2Me$. In order to get comparable results the disproportionation experiment is terminated at a pot temperature of 180°C, but if μ remains lower than 15 % this temperature is not reached. By using the adsorption of suitable azo dyes [6] we were able to estimate the basicity of the grafted Lewis bases. Their loading on the silica gel support was calculated from the nitrogen content found by elemental analysis of the powdered catalysts. Values of the first ionization potential of the Lewis bases stem from other papers [7–12].

Table 1. Results of the heterogeneously catalyzed disproportionation of $MeCl_2Si$-$SiCl_2Me$ and properties of the catalyst surfaces.

No.	μ [%]	Properties of the silica gel surfaces: Basicity H_0	Specif. surface [m²/g]	Loading [mmol/g]	1st ionization potential[a,b] [eV]	^{29}Si-NMR $(RO)_3Si(CH_2)_mR'$ $(CDCl_3/TMS)$ [δ in ppm]
1	25.8	$5.0 < H_0 < 6.8$	358	0.691	8.41 [7][c]	$m = 2$; R = Et: −49.5
2	26.9	$5.0 < H_0 < 6.8$	331	0.921	9.56 [10]	$m = 2$; R = Et: −47.2
3	1.0	not determinable	381	0.396	7.38 [11]	$m = 3$; R = Me: −42.3
4	18.3	$H_0 > 6.8$	379	0.905	8.77 [12]	$m = 1$; R = Et: −52.2
5	25.5	$H_0 \approx 6.8$	380	0.514	8.77 [12]	$m = 3$; R = Et: −45.2
6	5.1	$3.3 < H_0 < 4.0$	206	1.130	9.66 [10]	$m = 2$; R = Et: −48.3
7	10.8	$5.0 < H_0 < 6.8$	340	0.660	7.74 [8]	$m = 2$; R = Et: −48.8[d]

[a] Ungrafted compound. — [b] For simplification, the appropriate alkyl group substitutes the silylalkylene spacer grafted to the silica gel. — [c] References in brackets. — [d] The isomer 2H-indazole was also indicated in minor amounts by ^{29}Si-NMR: –49.4 ppm.

Considering the same donor (e.g. imidazole; first ionization potential of the corresponding 1-ethylimidazole 8.58 eV [9]) the cleavage yield μ may also depend on the length m of the alkylene spacer between the silicon atom of the silica gel surface and the Lewis base group (Table 2).

Table 2. Results of the heterogeneously catalyzed disproportionation of $MeCl_2Si–SiCl_2Me$ and properties of the catalyst surface; ^{29}Si-NMR $(EtO)_3Si(CH_2)_m$–imidazole: m = 1: δ = –57.7 ppm; m = 2: δ = –49.1 ppm ($CDCl_3$/TMS).

m	μ	Properties of the silica gel surfaces		
		Basicity H_0	Specif. surface	Loading
	[%]		[m²/g]	[mmol/g]
1	36.3	$H_0 > 6.8$	314	0.844
2	17.7	$H_0 > 6.8$	350	0.949
3	1.5	$H_0 > 6.8$	252	1.203

m = 1–3

This dependence on the length of the alkylene spacer versus the cleavage yield μ is not a general rule as it is demonstrated by the comparison of μ values of only **4** and **5**.

Changing the Lewis base from grafted 2-pyridine **6** to the 4-pyridine derivative **2** the cleavage yield μ is increased fivefold, emphasizing the steric factors of the donor–acceptor interaction. Adding a small quantity of 4-dimethylaminopyridine (0.4 mol %, known as hypernucleophile in organic chemistry) to heated $MeCl_2Si–SiCl_2Me$ causes a steady disproportionation. Experiments with grafted 4-dimethylaminopyridine catalysts are under current investigation.

Conclusion

We performed the catalyzed disproportionation of 1,1,2,2-tetrachlorodimethyldisilane with different grafted nitrogenous donors. Apart from the steric influence their nucleophilicity plays an important role in the reaction. A π-electron system or a low first ionization potential of the donor does not seem to be a necessary condition for the disproportionation.

The prepared catalysts are characterized by their specific surface, loading and basicity H_0.

Acknowledgment: We thank the *Deutsche Forschungsgemeinschaft* and the *Fonds der Chemischen Industrie* for generous financial support.

References:

[1] R. Richter, G. Roewer, U. Böhme, K. Busch, F. Babonneau, H.-P. Martin, E. Müller, *Appl. Organomet. Chem.* **1997**, *11*, 71.

[2] U. Herzog, R. Richter, E. Brendler, G. Roewer, *J. Organomet. Chem.* **1996**, *507*, 221.

[3] R. Richter, N. Schulze, G. Roewer, J. Albrecht, *J. Prakt. Chem.* **1997**, *339*, 145.

[4] T. Lange, N. Schulze, G. Roewer, R. Richter, in: *Organosilicon Chemistry III: From Molecules to Materials* (Eds.: N. Auner, J. Weis), VCH, Weinheim, **1998**, p. 291.

[5] B. P. Bespalow, Je. W. Getmanowa, W. W. Tirow, A. A. Pankratow, *Zh. Org. Chim.* **1978**, *14*, 351; W. Kaim, H. Bock, *Chem. Ber.* **1978**, *111*, 3843.

[6] C. Walling, *J. Am. Chem. Soc.* **1950**, *72*, 1164; H. A. Benesi, *J. Am. Chem. Soc.* **1956**, *78*, 5490; W. F. Kladnig, *J. Phys. Chem.* **1979**, *83*, 765; T. Yamahaka, K. Tanabe, *J. Phys. Chem.* **1975**, *79*, 2409.

[7] W. K. Turtschanikow, K. B. Petruschenko, A. I. Wokin, A. F. Jermikow, L. A. Jeskowa, L. W. Baikalowa, J. S. Domnina, *Zh. Org. Chim.* **1989**, *25*, 1138.

[8] K. B. Petruschenko, A. I. Wokin, W. K. Turtschanikow, A. G. Gorschkow, J. L. Frolow, *Izv. Akad. Nauk SSSR Ser. Chim.* **1985**, 267.

[9] W. K. Turtschanikow, A. F. Jermikow, W. A. Schagun, L. W. Baikalowa, *Izv. Akad. Nauk SSSR, Ser. Chim.* **1987**, 2591.

[10] K. Higasi, I. Omura, H. Baba, *J. Chem. Phys.* **1956**, *24*, 623.

[11] P. G. Farrell, J. Newton, *J. Phys. Chem.* **1965**, *69*, 3506.

[12] S. F. Nelsen, J. M. Buschek, *J. Am. Chem. Soc.* **1974**, *96*, 7930.

Synthesis of Linear and Cyclic Bis(trimethylsilyl)aminooligosilanes

Waltraud Gollner, Alois Kleewein, Karin Renger, Harald Stüger*

Institut für Anorganische Chemie, Technische Universität Graz
Stremayrgasse 16, A-8010 Graz, Austria
Tel.: Int. code + (316)873 8708 — Fax.: Int. code + (316)873 8701
E-mail: Stueger@anorg.tu-graz.ac.at

Keywords: Aminochlorosilanes / Aminosilanes / Chlorosilanes

Summary: The reaction of linear (Si_nCl_{2n+2}; $n = 3, 4$) and cyclic (Si_5Cl_{10}) chlorosilanes with one or two equivalents of lithium bis(trimethylsilyl)amide ($LiN(TMS)_2$) results in the formation of the corresponding mono- or disubstituted aminochlorosilanes 1-bis(trimethylsilyl)aminoheptachlorotrisilane, 1,3-bis[bis(trimethylsilyl)amino]hexachlorotrisilane, 1-bis(trimethylsilyl)aminononachlorotetrasilane, 1,4-bis[bis(trimethylsilyl)amino]octachlorotetrasilane, bis(trimethylsilyl)aminononachlorocyclopentasilane and 1,3-bis[bis(trimethylsilyl)amino]octachlorocyclopentasilane. The hydrogenation of the resulting products with $LiAlH_4$ was examined.

Introduction

Aminosilanes are valuable compounds in modern silicon chemistry, not only as alternative precursors for chemical vapor deposition of Si_3N_4, but also because of their many interesting chemical features. The amino groups can be easily attached to silicon by the reaction of halosilanes with amines or alkali metal amides using standard literature procedures [1].

The Si–N bond in silylamines is rather stable against nucleophiles, hence allowing chemical reactions to be carried out at the Si–center without affecting the $Si–NR_2$ moiety. The stability of Si–N linkages is even enhanced when more than one Si atom is attached to nitrogen or bulky substituents such as *tert*-butyl- or aryl-groups are present. Despite the relative stability of Si–N bonds they are readily cleaved by acids and various electrophiles, both organic and inorganic, under mild conditions. The hydrogen halides or their acids, for instance, convert triorganoaminosilanes to the appropriate halosilanes R_3SiX (X = F, Cl, Br) [2, 3]. This provides an excellent route for the removal of amino substituents from silicon substrates with concurrent formation of Si-halogen bonds. The ease with which Si–N bonds can be made and broken, together with the remarkable chemical stability of aminosilanes towards various nucleophiles, provides an attractive basis for the synthesis of multifunctionalized polysilane derivatives.

The reaction of hexachlorodisilane with one or two equivalents of lithium bis(trimethylsilyl)amide ($LiN(TMS)_2$) affords the corresponding mono- or bisaminochlorodisilanes [4, 5]. Because of the stability of the Si–N bond towards $LiAlH_4$ and its facile cleavage with dry HCl, amino-substituted silanes are valuable starting materials for partially hydrogenated chlorodisilanes.

In continuation of these studies it is our aim to apply similar reaction sequences to higher linear and cyclic halosilanes. Thus partially aminated products are easily accessible by the reaction of linear chlorosilanes Si_nCl_{2n+2} or cyclic chlorosilanes Si_nCl_{2n} with one or more equivalents of $LiN(TMS)_2$. The maximum number of bis(trimethylsilyl)amino groups which can be attached to the corresponding silicon backbones is examined, as well as the hydrogenation of the resulting products.

Results and Discussion

Reaction of $LiN(TMS)_2$ with Linear Oligosilanes Si_3Cl_8 and Si_4Cl_{10}

The reaction of $LiN(TMS)_2$ (Scheme 1) with one equivalent of octachlorotrisilane or *n*-decachlorotetrasilane under mild conditions (–40°C, heptane solution) affords the monosubstituted products 1-bis(trimethylsilyl)-aminoheptachlorotrisilane (**1**) and 1-bis(trimethylsilyl)aminononachlorotetrasilane (**3**), respectively. Adding a second equivalent of $LiN(TMS)_2$ to **1** and **3** leads to 1,3-bis[bis(trimethylsilyl)amino]-hexachlorotrisilane (**2**) and 1,4-bis[bis(trimethylsilyl)amino]octachlorotetrasilane (**4**) nearly exclusively. Substitution was always observed at the terminal but never at the centrally located silicon atoms.

$$2\ LiN(TMS)_2 \xrightarrow[\text{heptane, } -40°C]{+2\ Si_3Cl_8} 2\ Cl_3SiSiCl_2SiCl_2\text{–}N(TMS)_2\ (\mathbf{1})$$

$$2\ LiN(TMS)_2 \xrightarrow[\text{heptane, } -40°C]{+1\ Si_3Cl_8} (TMS)_2N\text{–}SiCl_2SiCl_2SiCl_2\text{–}N(TMS)_2\ (\mathbf{2})$$

$$2\ LiN(TMS)_2 \xrightarrow[\text{heptane, } -40°C]{+2\ Si_4Cl_{10}} 2\ Cl_3SiSiCl_2SiCl_2SiCl_2\text{–}N(TMS)_2\ (\mathbf{3})$$

$$2\ LiN(TMS)_2 \xrightarrow[\text{heptane, } -40°C]{+1\ Si_4Cl_{10}} (TMS)_2N\text{–}SiCl_2SiCl_2SiCl_2SiCl_2\text{–}N(TMS)_2\ (\mathbf{4})$$

Scheme 1. Synthesis of linear aminochlorosilanes.

The introduction of more than two $(TMS)_2N$ groups to the trisilane failed even under harsh conditions (excess $LiN(TMS)_2$, reflux). Instead of the trisubstituted product a mixture of **1**, **2** and $LiN(TMS)_2$ was obtained.

Hydrogenation of 1, 2, 3, and 4

The hydrogenation of the remaining Si–Cl bonds in **1**, **2**, **3** and **4** (Scheme 2) using $LiAlH_4$ in diethyl ether at 0°C can be used to prepare the amino-substituted oligosilanyl hydrides

1-bis(trimethylsilyl)aminotrisilane (**5**), 1,3-bis[bis(trimethylsilyl)amino]trisilane (**6**), 1-bis(trimethylsilyl)aminotetrasilane (**7**) and 1,4-bis[bis(trimethylsilyl)amino]tetrasilane (**8**) in excellent yields. Cleavage of the Si–Si or the Si–N bond is not observed.

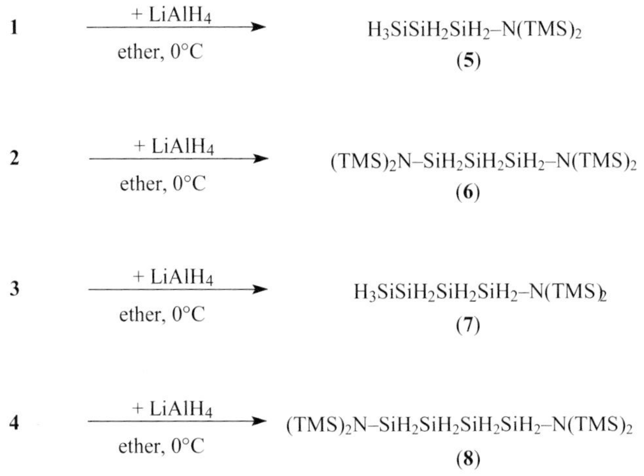

$$\mathbf{1} \xrightarrow[\text{ether, 0°C}]{+\ LiAlH_4} H_3SiSiH_2SiH_2\text{–}N(TMS)_2\ (\mathbf{5})$$

$$\mathbf{2} \xrightarrow[\text{ether, 0°C}]{+\ LiAlH_4} (TMS)_2N\text{–}SiH_2SiH_2SiH_2\text{–}N(TMS)_2\ (\mathbf{6})$$

$$\mathbf{3} \xrightarrow[\text{ether, 0°C}]{+\ LiAlH_4} H_3SiSiH_2SiH_2SiH_2\text{–}N(TMS)_2\ (\mathbf{7})$$

$$\mathbf{4} \xrightarrow[\text{ether, 0°C}]{+\ LiAlH_4} (TMS)_2N\text{–}SiH_2SiH_2SiH_2SiH_2\text{–}N(TMS)_2\ (\mathbf{8})$$

Scheme 2. Hydrogenation of the linear aminochlorosilanes 1–4.

Reaction of $LiN(TMS)_2$ with *cyclo*-Si_5Cl_{10}

The reaction of decachlorocyclopentasilane (Si_5Cl_{10}) with $LiN(TMS)_2$ (Scheme 3) under mild conditions (–70°C, heptane solution) leads to the monosubstituted product [bis(trimethylsilyl)amino]nonachlorocyclopentasilane (**9**). A significant part of Si_5Cl_{10} remains unreacted, which, however, can be removed by recrystallization.

The introduction of more than one $(TMS)_2N$ group into Si_5Cl_{10} requires more severe conditions (reflux, heptane solution), and leads to the disubstituted product 1,3-bis[bis(trimethylsilyl)amino]octachlorocyclopentasilane (**10**) nearly exclusively.

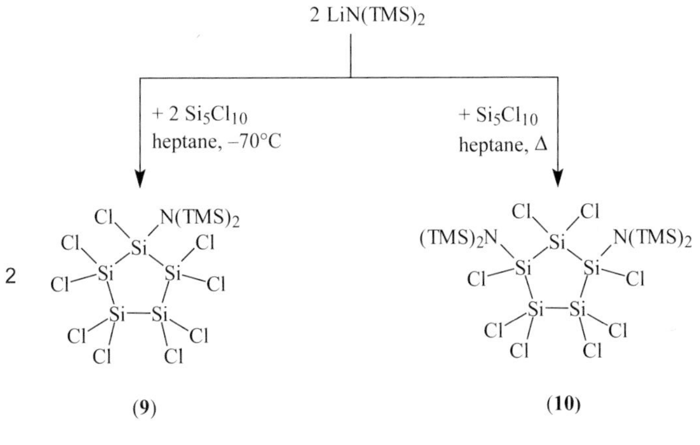

2 $LiN(TMS)_2$

+ 2 Si_5Cl_{10}, heptane, –70°C → 2 (**9**)

+ Si_5Cl_{10}, heptane, Δ → (**10**)

Scheme 3. Synthesis of aminochlorocyclopentasilanes.

Hydrogenation of 9 and 10

One could expect that the hydrogenation of the amino-substituted chlorocyclosilanes (Scheme 4) would lead to the corresponding cyclosilanyl hydrides just as observed for the linear compounds **1**–**4**. However, while the Si–N bond remains intact, the Si–Si bonds within the cyclopentasilane ring do not. The product mixtures obtained are complex and can be analyzed only by GC/MS. Treatment of **9** and **10** with $LiAlH_4$ does not only result in the formation of the expected cyclosilanes, bis(trimethylsilyl)aminocyclopentasilane (**11**) or 1,3-bis[bis(trimethylsilyl)amino]-cyclopentasilane (**12**), but also affords considerable amounts of linear silanes, arising from ring-cleavage reactions. The isolation of the pure cyclosilanes **11** and **12** from the reaction mixture could not be achieved so far.

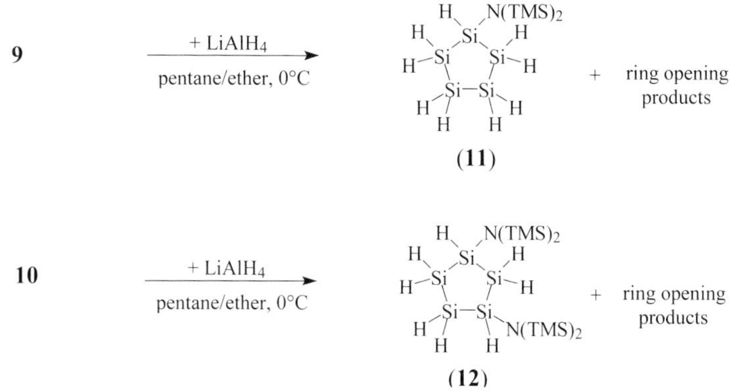

9 $\xrightarrow[\text{pentane/ether, 0°C}]{+\ LiAlH_4}$ **(11)** + ring opening products

10 $\xrightarrow[\text{pentane/ether, 0°C}]{+\ LiAlH_4}$ **(12)** + ring opening products

Scheme 4. Hydrogenation of the cyclic aminochlorosilanes **9** and **10**.

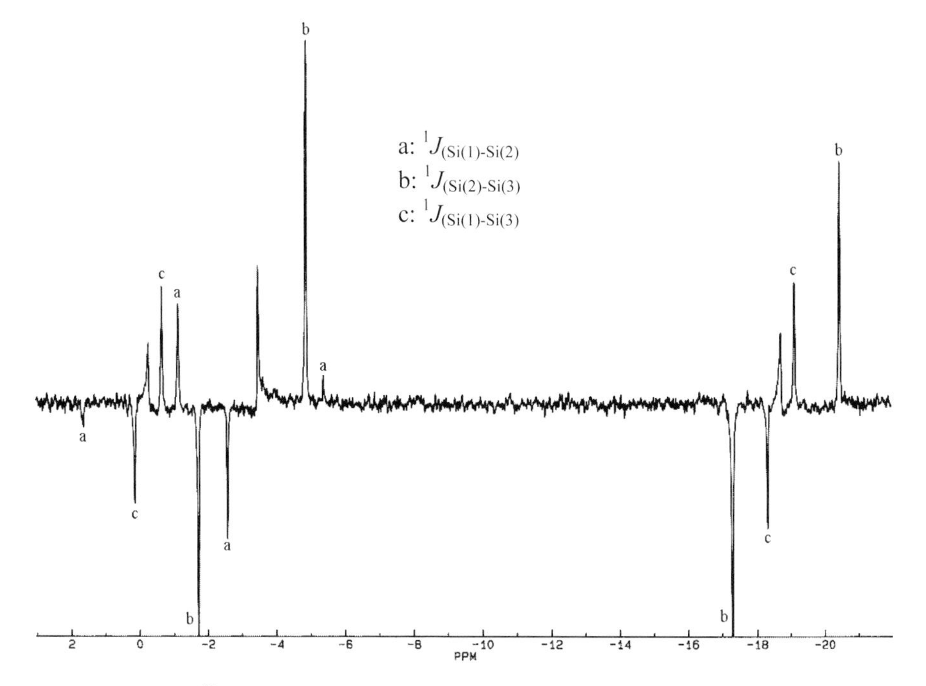

Fig. 1. INADEQUATE ^{29}Si-NMR spectrum of **1**; signal of $N[Si(CH_3)]_2$ was omitted.

NMR Spectroscopy

All aminochlorosilanes were characterized by standard spectroscopic techniques such as MS, IR, ^{1}H- and ^{29}Si-NMR as well as by elemental analysis. As can be expected, ^{29}Si-NMR spectroscopy was found to be particularly useful to prove the proposed structures. The corresponding ^{29}Si-chemical shift values depicted in Table 1 are quite unexceptional. Additionally ^{29}Si-INADEQUATE-NMR spectroscopy was employed to achieve an unequivocal assignment of the ^{29}Si-resonance lines of the aminochlorotrisilanes **2** and **3** (compare Fig. 1) and to determine the ^{29}Si–^{29}Si coupling constants.

Table 1. ^{29}Si-chemical shifts δ [ppm] vs. ext. TMS and selected ^{29}Si-^{29}Si coupling constants J [Hz] of bis(trimethylsilyl)aminochlorosilanes.

Compound		δ [ppm]	J [Hz]
$Cl_3Si^1–Si^2Cl_2–Si^3Cl_2–N(TMS^4)_2$	**1**	Si(1) –0.28 Si(2) –3.59 Si(3) –18.88 Si(4) +9.43	$^{1}J_{Si(1)-Si(2)}$ 166.17 $^{1}J_{Si(2)-Si(3)}$ 187.71 $^{2}J_{Si(1)-Si(3)}$ 47.16
$(TMS)_2N–Cl_2Si^1–Si^2Cl_2–SiCl_2–N(TMS^3)_2$	**2**	Si(1) –14.86 Si(2) –0.85 Si(3) +9.12	$^{1}J_{Si(1)-Si(2)}$ 172.52
$Cl_3Si^1–Si^2Cl_2–Si^3Cl_2–Si^4Cl_2–N(TMS^5)_2$	**3**	Si(1) –0.49 Si(2) –1.51 Si(3) –1.51 Si(4) –18.19 Si(5) +9.82	
$(TMS)_2N–Cl_2Si^1–Si^2Cl_2–SiCl_2–SiCl_2–N(TMS^3)_2$	**4**	Si(1) –16.19 Si(2) +4.24 Si(3) +9.44	
cyclo-$Si_5Cl_9N(TMS)_2$	**9**	–19.97 +3.42 +0.54 +10.46	
cyclo-$Si_5Cl_8[N(TMS)_2]_2$	**10**	–17.03 +6.04 –16.96 +9.84 +5.46 +10.02 +5.99	

Conclusions

The results of our experiments show that selective substitution of chlorine in perchlorooligosilanes by the $(TMS)_2N$ group can be extended from hexachlorodisilane to higher linear and cyclic chlorosilanes. In linear compounds substitution takes place exclusively at the terminal silicon atoms. In cyclopentasilanes formation of the 1,3-disubstitution products is preferred. Hydrogenation of the remaining Si–Cl bond in the linear aminochlorosilanes leads to the corresponding aminosilanes, while hydrogenation of the cyclopentasilane derivatives affords considerable amounts of linear silanes due to ring-cleavage reactions.

Acknowledgment: We wish to thank the *Fonds zur Förderung der wissenschaftlichen Forschung* (Wien, Austria), for financial support within the Forschungsschwerpunkt "Novel Approaches to the Formation and Reactivity of Compounds containing Silicon-Silicon Bonds" and *Wacker Chemie GmbH* (Burghausen, Germany) for the donation of silane precursors.

References:

[1] D. A. Armitage, "*Organosilicon Nitrogen Compounds*" in: *The Silicon Heteroatom Bond* (Eds.: S. Patai, Z. Rappoport), Wiley, Chichester, **1991**, p. 367.
[2] C. Eaborn, *Organosilicon Compounds*, Butterworths, London, **1960**, p. 346.
[3] E. A. V. Ebsworth, "*Volatile Silicon Compounds*" in: *International Series of Monographs on Inorganic Chemistry* (Eds.: H. Taube, A. G. Maddock), Pergamon Press, Oxford, **1963**, *Vol. 4*, p. 111.
[4] H. Stüger, P. Lassacher, E. Hengge, *J. Organomet. Chem.* **1997**, *547*, 227.
[5] H. Stüger, P. Lassacher, *Monatsh. Chem.* **1994**, *125*, 615.

Synthesis and Functionalization of Branched Oligosilanes

S. Chtchian, C. Krempner*

Fachbereich Chemie, Abteilung Anorganische Chemie
Universität Rostock
Buchbinderstr. 9, D-18051 Rostock, Germany
Tel.: Int. code + (0381)498 1841 — Fax: Int. code + (381)498 1763
E-mail: clemens.krempner@chemie.uni-rostock.de

Keywords: Oligosilanes / Silyl Anions / Bis(trimethylsilyl)methylsilyl Derivatives

Summary: Bis(trimethylsilyl)methylsilyllithium (**1**) reacts with $HSiCl_3$ and $MeHSiCl_2$ to give 1,1,1,2,3,4,5,5,5-nonamethyl-2,4-bis(trimethylsilyl)silane (**2**) and tris[bis-(trimethylsilyl)methylsilyl]silane (**3**). Bromination of **2** and **3** with $CHBr_3$ gives the bromosilanes **4** and **5**. The reaction of **4** with Na–K alloy yields 1,2-dimethyl-1,1,2,2-tetrakis[bis(trimethylsilyl)methylsilyl]disilane (**6**), as a result of reductive coupling. Contrary to **4**, bromosilane (**5**) reacts with Na–K alloy to give a solution of the silylpotassium compound (**7**). The treatment of **5** in THF with 15 % HCl leads to the formation of chlorotris[bis(trimethylsilyl)methylsilyl]silane (**9**). The molecular structures of **6** and **9** were established by single-crystal X-ray diffraction analyses.

Introduction

In recent years, much attention has been directed to highly branched oligosilanes as ligands [1] or as precursors for dendrimers [2] and other silicon-based networks [3]. We are especially interested in the stepwise building and functionalization of branched oligosilanes. Today, the formation of these materials can be obtained by salt elimination reactions of silylmetal derivatives with chlorosilanes or by heterogeneous catalytic disproportionation of methylchlorodisilanes [3].

Our strategy for preparing these materials involved the construction of suitable sterically demanding oligosilanes, containing two or three bis(trimethylsilyl)methylsilyl groups on the central silicon atom, which can be used for further functionalizations.

Results and Discussion

The required bis(trimethylsilyl)methylsilyllithium (**1**) was prepared by splitting of methyltris(trimethylsilyl)silane [4] with methyllithium in tetrahydrofuran as previously described [5]. After the solvent had been changed to *n*-pentane, **1** was allowed to react with the appropriate chlorosilanes. As shown in Scheme 1, 1,1,1,2,3,4,5,5,5-nonamethyl-2,4-bis(trimethylsilyl)-pentasilane (**2**) was prepared in 85 % yield directly by the slow addition of a two-fold molar excess of **1** to a solution of $MeSiHCl_2$ in *n*-pentane at –78°C. The 3-bromopentasilane (**4**) was obtained by bromination of **2** with tribromomethane at 120°C (yield: 92 %).

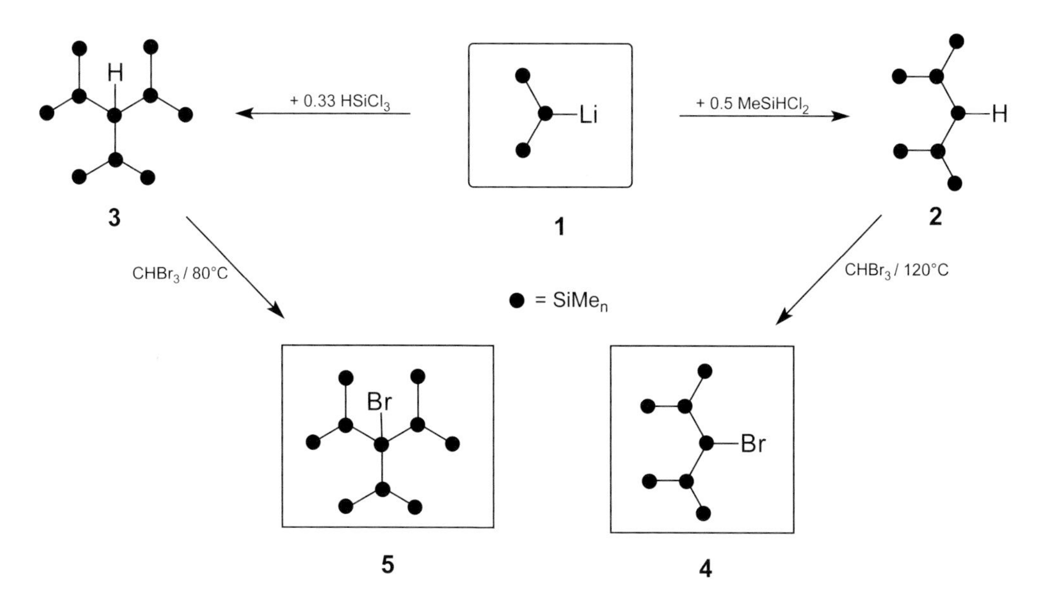

Scheme 1. Syntheses of the branched oligosilanes **2–5**.

The reaction of three equivalents of bis(trimethylsilyl)methylsilyllithium (**1**) with trichlorosilane (Scheme 1) at –78°C leads to the formation of tris[bis(trimethylsilyl)methylsilyl]silane (**3**) [6], which can be crystallized from acetone (yield: 67 %). In view of the relatively bulkiness of the three bis(trimethylsilyl)methylsilyl groups, the ease of the formation of **3** is really surprising. In analogy to the preparation of **4**, oligosilane **3** was combined with an excess of tribromomethane at 80°C for 10 days. Crystallization of the crude product from dried acetone leads to the bromo-tris[bis(trimethylsilyl)methylsilyl]silane (**5**) in a yield of 30 %.

As shown in Eq. 1, the reaction of **4** with Na–K alloy in *n*-heptane successfully yields 1,2-dimethyl-1,1,2,2-tetrakis[bis(trimethylsilyl)methylsilyl]disilane (**6**) as a result of reductive coupling. Pure **6** can be isolated by crystallization from ethyl acetate (yield: 32 %). The proposed structure is supported by NMR spectrometry and an X-ray crystal structure analysis (Fig. 1).

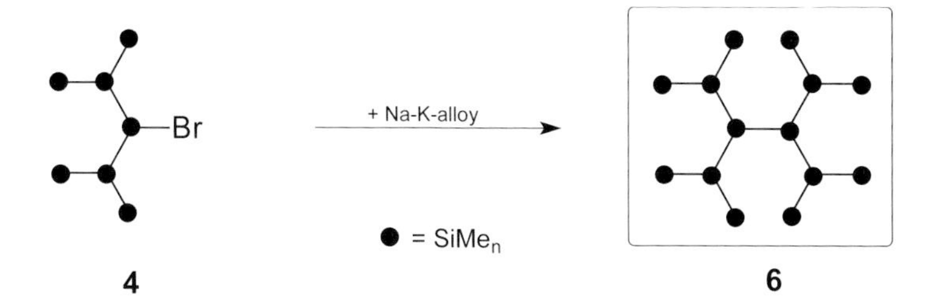

Eq. 1. Synthesis of **6**.

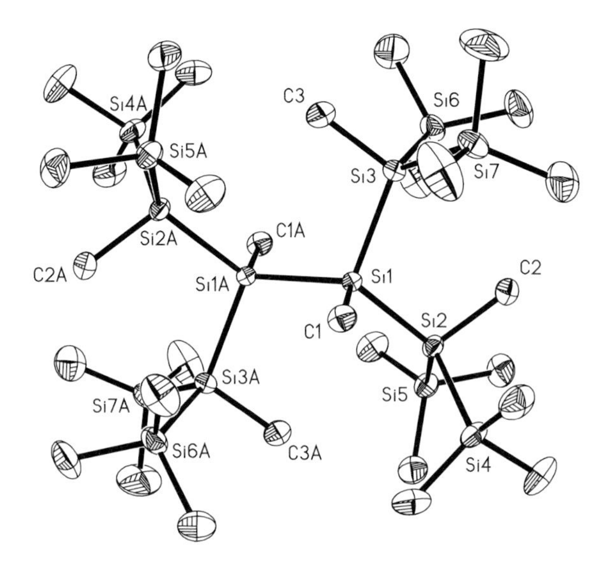

Fig. 1. Molecular structure of **6** in the crystal, hydrogen atoms omitted, selected bond lengths [Å] and angles [°]: Si1–C1 1.882(4), Si1–Si3 2.3853(15), Si1–Si1A 2.400(2), Si1–Si2 2.414(2), Si2–C2 1.920(4), Si3–C3 1.909(4), C1-Si1-Si1A 104.62(13), Si3-Si1-Si1A 108.82(7), Si2-Si1-Si1A 121.95(8), Si4-Si2-Si1 110.23(7), Si5-Si2-Si1 127.46(7), Si7-Si3-Si1 116.06(7), Si6-Si3-Si1 117.78(7).

Contrary to **4**, the more sterically hindered bromosilane **5** undergoes an Si–Br bond cleavage with Na–K alloy in N-heptane to give an orange solution of the silylpotassium compound (**7**). This highly moisture-and air-sensitive solution can be protonated, yielding the corresponding oligosilane (**4**). The formation of the silylpotassium can be proven by the reaction of **7** with an excess of D_2O leading to the deuterosilane **8** (Eq. 2).

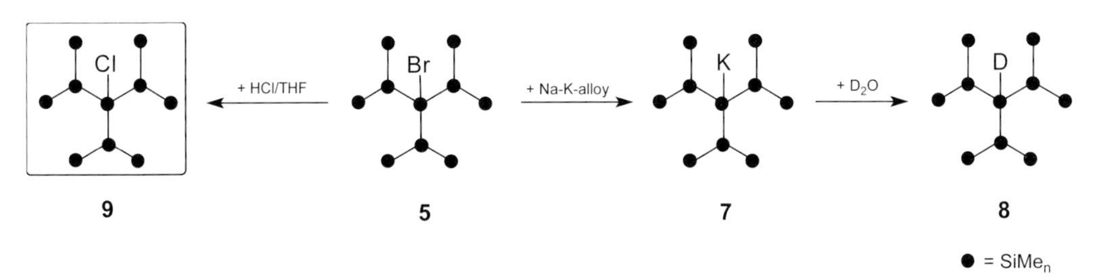

Eq. 2. Reactivity of **5**.

Surprisingly, the treatment of the bromosilane (**5**) in tetrahydrofuran with 15 % hydrochloric acid leads to the formation of chlorotris[bis(trimethylsilyl)methylsilyl]silane (**9**) in almost quantitative yield (Eq. 2). The reaction behavior of bromosilane **5** is unusual insofar as less sterically congested bromosilanes hydrolyze to silanoles or siloxanes. Nevertheless, besides **9** only traces of the corresponding silanol were observed. The molecular structure of **9** was established by a single-crystal X-ray diffraction analysis (Fig. 2).

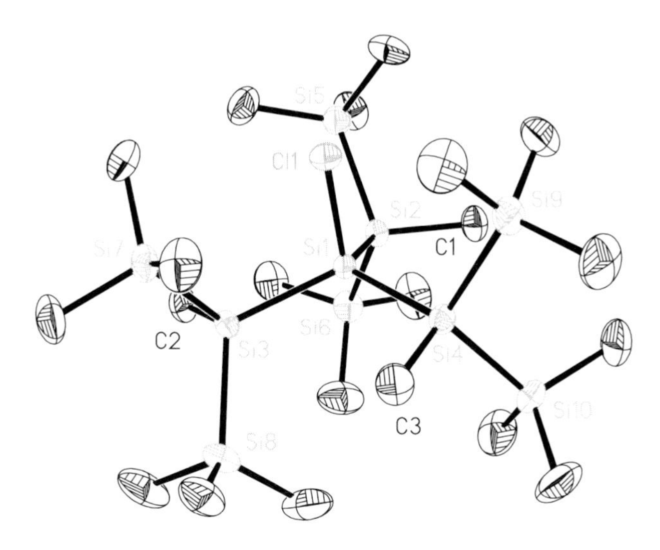

Fig. 2. Molecular structure of **9** in the crystal, hydrogen atoms omitted, selected bond lengths [Å] and angles [°]: Si1–Cl1 2.1370(11), Si1–Si3 2.3771(12), Si1–Si2 2.3799(12), Si1–Si4 2.3829(12), Si2–C1 1.900(3), Cl1-Si1-Si3 102.01(5), Cl1-Si1-Si2 101.59(5), Cl1-Si1-Si4 101.28(4), Si2-Si1-Si4 117.35(4), Si6-Si2-Si1 118.78(5), Si8-Si3-Si1 117.87(5).

Acknowledgment: We thank Prof. Dr. H. Oehme for his generous support and T. Groß, A. Spannenberg and R. Kempe for carrying out the X-ray diffraction analyses.

References:

[1] Y. Apeloig, M. Bendikov, M. Yuzefovich, M. Nakash, D. Bravo-Zhivotovskii, *J. Am. Chem. Soc.* **1996**, *118*, 48, 12228.

[2] J. B. Lambert, J. L. Pflug, J. M. Denary, *Organometallics* **1996**, *15*, 615.

[3] R. Richter, G. Roewer, U. Böhme, K. Busch, F. Babonneau, H. P. Martin, E. Müller, *Appl. Organomet. Chem.* **1997**, *11*, 71.

[4] K. Hassler, G. Kollegger, *J. Organomet. Chem.* **1995**, *485*, 233.

[5] K. M. Baines, A. G. Brook, R. R. Ford, P. D. Lickiss, A. K. Saxena, W. J. Sawyer, B. A. Behnam, *Organometallics* **1989**, *8*, 693.

[6] D. Bravo-Zhivotovskii, M. Yuzafovich, M. Bendikov, Y. Apeloig, K. Klinkhammer, R. Boese, *Poster PA54, Xith Organosilicon Symposium,* Montpellier, France, September **1996**.

Geminal Di(hypersilyl) Compounds: Synthesis, Structure and Reactivity of some Sterically Extremely Congested Molecules

T. Gross, H. Oehme*

Fachbereich Chemie der Universität Rostock
D-18051 Rostock, Germany
Tel.: Int. code + (381)498 1765 — Fax: Int. code + (381)498 1763
E-mail: hartmut.oehme@chemie.uni-rostock.de

R. Kempe

Institut für Organische Katalyseforschung an der Universität Rostock, Germany

Keywords: Hypersilyl Compounds / Tris(trimethylsilyl)silyl Derivates / X-Ray

Summary: The synthesis of derivatives of the type $[(Me_3Si)_3Si]_2CHR$ (R = OH, OCH_3, OC(O)H, OC(O)Me, NMe_2) proved to be surprisingly facile and was achieved by the reaction of tris(trimethylsilyl)silyllithium (**1**) with suitable C_1 building blocks. The results of the X-ray analyses and reactions of these di(hypersilyl) compounds are described in this paper.

Introduction

The aim of our studies was the synthesis of geminal di(hypersilyl) compounds with a central carbon atom. The compounds occupied our interest because of the expected structural peculiarities, caused by the extreme bulkiness of the two hemispherical $(Me_3Si)_3Si$ groups. As expected, the fixation of two $(Me_3Si)_3Si$ groups at one comparatively small sp^3 carbon atom is a difficult undertaking, but proved to be surprisingly facile when the second hypersilyl group is introduced by a $(Me_3Si)_3SiLi$ nucleophile approaching an α-hypersilyl carbenium ion transition state or, similarly, the carbonyl carbon atom of an acyltris(trimethylsilyl)silane. The planar surrounding of the electrophilic sp^2 carbon atom in the intermediates leads to a decrease of the steric congestion facilitating the approach of the bulky Si nucleophile.

Syntheses, Structures and Reactivities

Following the outlined concept we succeeded in synthesizing methoxy-bis[tris(trimethylsilyl)silyl]-methane (**2**) by the reaction of **1** with dichloromethyl methyl ether (molar ratio 2:1) [1]. The reaction of **1** with *tert*-butyl formate (2:1) afforded di(hypersilyl)methanol (**3**). Despite the extreme steric shielding, the alcohol **3** is reactive and was converted with acetyl chloride into the acetate **4** and underwent transesterification with methyl formate to give **5** [2]. Recently we obtained dimethylaminobis[tris(trimethylsilyl)silyl]methane (**6**) by the reaction of **1** with chloromethylene-dimethyliminium chloride (2:1) [3] (Scheme 1).

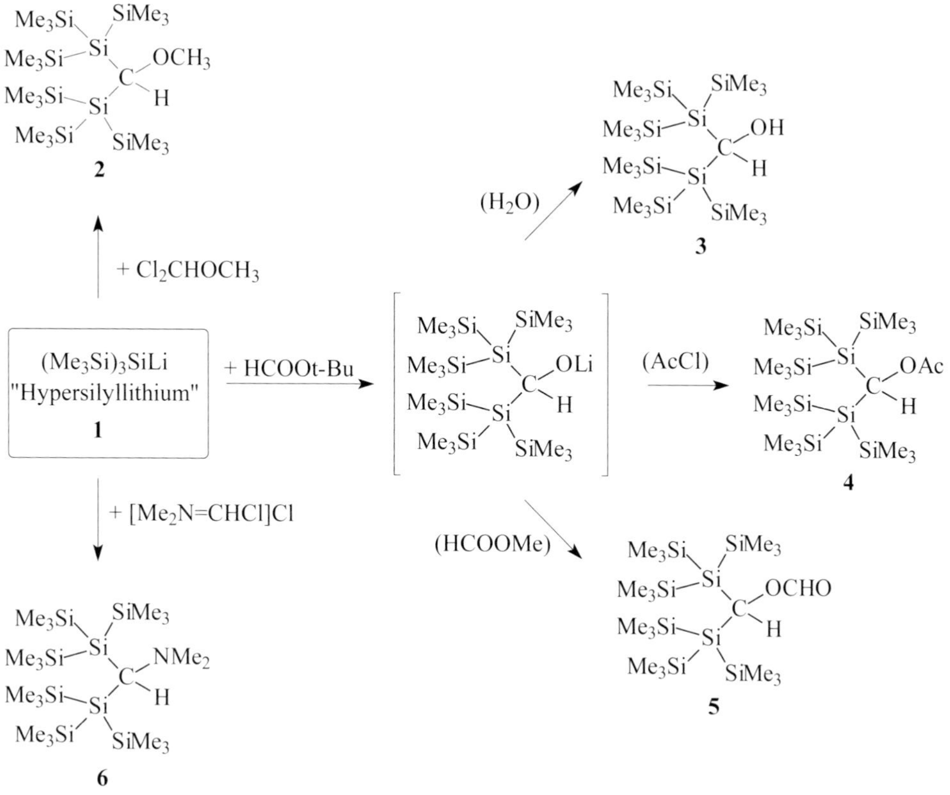

$(Me_3Si)_3SiLi$
"Hypersilyllithium"
1

$+ Cl_2CHOCH_3$ → **2**

$+ HCOOt\text{-}Bu$ → $[(Me_3Si)_3Si]_2CH(OLi)$

(H_2O) → **3**

$(AcCl)$ → **4**

$(HCOOMe)$ → **5**

$+ [Me_2N{=}CHCl]Cl$ → **6**

Scheme 1. Reactions of tris(trimethylsilyl)silyllithium (**1**) with suitable C_1 building blocks.

All the mentioned di(hypersilyl) derivatives were characterized by full spectral analyses. Contrary to our expectations the solution NMR data of **2**–**6** are very straightforward. The ^{1}H-, ^{13}C- as well as the ^{29}Si-NMR spectra show only one signal for all the trimethylsilyl groups, indicating a remarkable mobility of these substituents. On the other hand, the results of the X-ray structural analyses of **2**, **3**, **4** and **6** actually revealed tremendous distortions of the molecular skeletons.

The molecular structures of **3, 4** and **6** are demonstrated in the Figs. 1–3, respectively, and Table 1 shows the most important structural data. Two bulky substituents at one comparatively small carbon atom lead to tremendous steric distortions in all di(hypersilyl) compounds.

A comparison of the central Si-C-Si angles clearly reveals a strong dependence of these values on the spatial demand of the third C substituent. The spherical extension of the OH group in **3** is relatively small. Thus, the angle at the central sp^3 carbon atom of **3** adopts the maximum value of 135.5° (Fig. 1). The space occupied by the acetoxy group in **4** is much larger; consequently the Si-C-Si angle decreases to 131.2° (Fig. 2). The spatial demand of the dimethylamino group in **6** is slightly smaller than that of the acetoxy group in **4**, leading to an angle of 132.6° (Fig. 3).

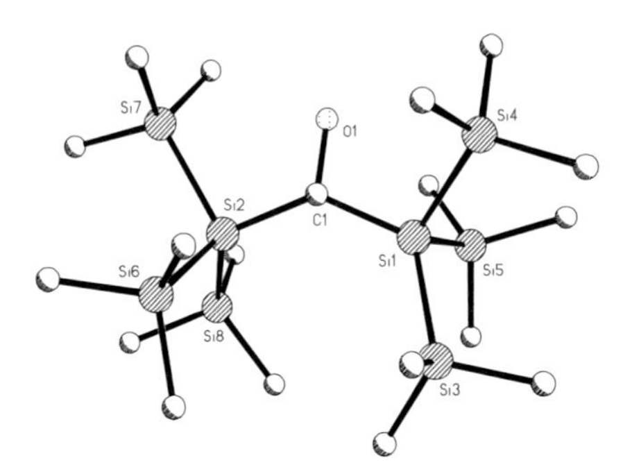

Fig. 1. Compound **3**.

Fig. 2. Compound **4**.

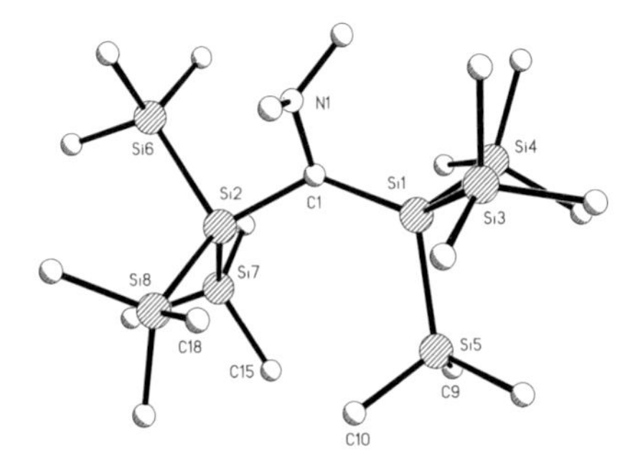

Fig. 3. Compound **6.** Selected $CSiSi_3$ tetrahedra angles [°]: C1-Si1-Si3 120.1(2); C1-Si2-Si6 98.7(2); C1-Si1-Si4 106.6(2); C1-Si2-Si7 114.8(2); C1-Si1-Si5 116.9(2); C1-Si2-Si8 122.3(2).

Table 1 Selected bond lengths and angles of compounds **3**, **4** and **6**.

Compound	Si1–C1 [Å]	Si2–C1 [Å]	Si1-C1-Si2 [°]
3	1.926(4)	1.905(4)	135.5(2)
4	1.853(2)	1.886(3)	131.3(2)
6	2.009(7)	1.947(6)	132.6(3)

Due to the pyramidal configuration of the *tert*-amino group the lengths of the two central Si–C bonds are considerably different (Si1–C1 2.009 Å, Si2–C1 1.947 Å). The distorted tetrahedral configuration of the central sp^3 carbon atom and the space demand of the dimethylamino group cause a deformation of the two $CSiSi_3$ tetrahedra in **6**. At positions, where the two $(Me_3Si)_3Si$ hemispheres contact each other, the angles are widened (C1-Si1-Si5 116.9°, C1-Si2-Si8 122.3°). Also Si3 is pushed aside by the dimethylamino group to give the maximum angle C1-Si1-Si3 of a value of 120.1°. On the other hand, the trimethylsilyl group (Si6) can evade the steric strain, giving a C1-Si2-Si6 angle of 98.7°.

As a result of the contact of the two hemispheres the trimethylsilyl groups approach to exceptionally short distances (C9–C15 3.64 Å and C10–C18 3.82 Å). Also the distances of the methyl groups to the neighboring Si-methyl groups are significantly shorter than the sum of the van der Waals radii of two methyl groups, amounting to about 4.00 Å [4]. The presented structural data of **6** are comparable with those of **3** and **4** [2].

Acid-induced sila-Peterson reaction

In presence of strong acids, 1-hydroxyalkyl-tris(trimethylsilyl)silanes undergo a rapid isomerization, a 1,2-OH/$SiMe_3$ exchange, to give 1-trimethylsilylalkylbis(trimethylsilyl)silanols (Scheme 2) [5]. The rearrangement was performed in ether and in THF; protic acids as well as Lewis acids were used as catalysts.

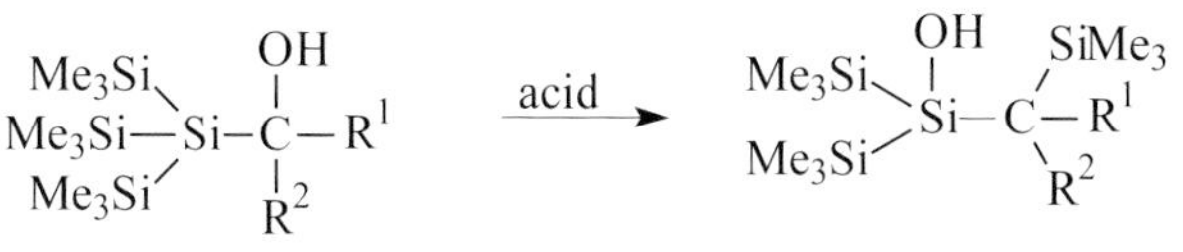

$$(Me_3Si)_3Si{-}C(OH)R^1R^2 \xrightarrow{\text{acid}} (Me_3Si)_2Si(OH){-}C(SiMe_3)R^1R^2$$

Scheme 2.

In the case of **2** this rearrangement fails. But, departing from the above-mentioned pattern, the reaction of **6** with excess of conc. sulfuric acid in THF cleanly leads to bis(trimethylsilyl)-[tris(trimethylsilyl)silylmethyl]silanol (**9**). The rearrangement proceeds slowly in THF. In ether no reaction was observed. The addition of methanol to the reaction solution gave a mixture of **9** and the methoxysilane **10** (Scheme 3).

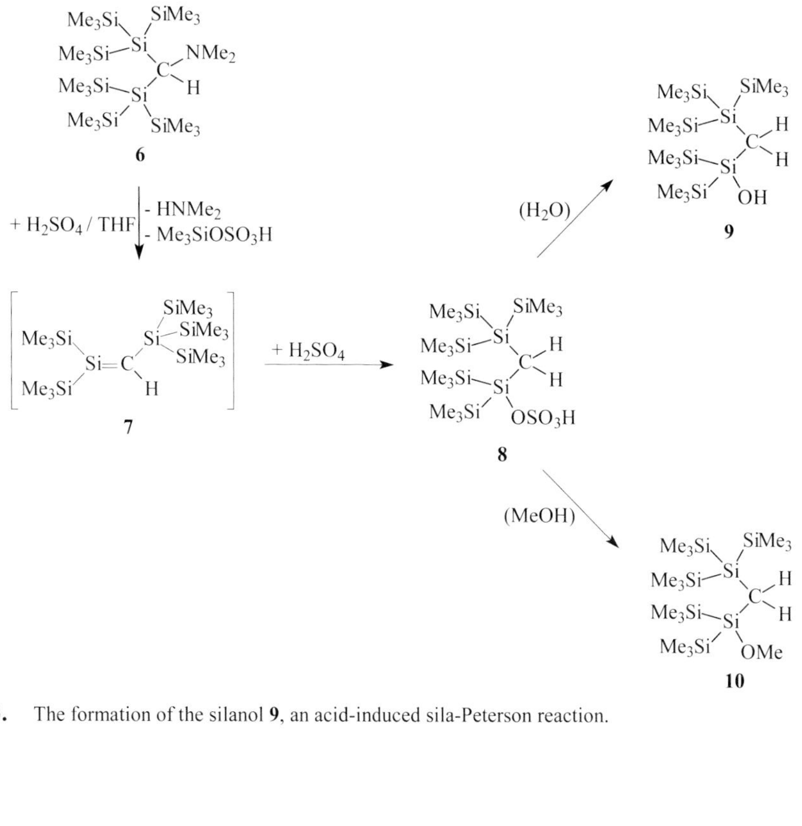

$[(Me_3Si)_3Si][(Me_3Si)_3Si]CH(NMe_2)$ (**6**)

$+ H_2SO_4$ / THF; $-$ $HNMe_2$, $-$ Me_3SiOSO_3H

$[(Me_3Si)_2Si{=}CH{-}Si(SiMe_3)_3]$ (**7**)

$+ H_2SO_4$

$(Me_3Si)_3Si{-}CH_2{-}Si(SiMe_3)_2OSO_3H$ (**8**)

(H_2O): $(Me_3Si)_3Si{-}CH_2{-}Si(SiMe_3)_2OH$ (**9**)

$(MeOH)$: $(Me_3Si)_3Si{-}CH_2{-}Si(SiMe_3)_2OMe$ (**10**)

Scheme 3. The formation of the silanol **9**, an acid-induced sila-Peterson reaction.

The mechanism of the formation of the silanol **9** (Scheme 3) is understood as the result of an acid-catalyzed formal elimination of dimethylaminotrimethylsilane generating the silene **7**, followed by the addition of H_2SO_4 at the silene double bond and subsequent hydrolysis of the silylsulfate **8**. To our best knowledge, this is the first case of an acid-induced sila-Peterson reaction.

Acknowledgment: Financial support of our work by the *Deutsche Forschungsgemeinschaft*, the *Fonds der Chemischen Industrie* and the state of *Mecklenburg-Vorpommern* is gratefully acknowledged.

References:

[1] E. Jeschke, T. Gross, H. Reinke, H. Oehme, *Chem. Ber.* **1996**, *129*, 841.

[2] T. Gross, R. Kempe, H. Oehme, *Chem. Ber./Recueil* **1997**, *130*, 1709.

[3] T. Gross, R. Kempe, H. Oehme, *Inorg. Chem. Comm.* **1998**, *1*, 128.

[4] H. Bock, K. Ruppert, C. Nähter, Z. Havlas, H.-F. Herrmann, C. Arad, I. Göbel, A. John, J. Meuret, S. Nick, A. Rauschenbach, W. Seitz, T. Vaupel, B. Solouki, *Angew. Chem.* **1992**, *104*, 564; *Angew. Chem. Int. Ed. Engl.* **1992**, *31*, 550.

[5] K. Sternberg, H. Oehme, *Eur. J. Inorg. Chem.* **1998**, 177.

The Acid-Induced Rearrangement of α-Functionalized Alkylpolysilanes

*F. Luderer, H. Reinke, H. Oehme**

Fachbereich Chemie der Universität Rostock
D-18051 Rostock, Germany
Tel.: Int. code + (381)498 1765 — Fax: Int. code + (381)498 1763
E-mail: hartmut.oehme@chemie.uni-rostock.de

Keywords: Polysilanes / Rearrangements / Silylium Ion / Carbenium Ion

Summary: In the presence of acids, α-functionalized alkylpolysilanes undergo a rapid rearrangement, a 1,2-Si,C exchange, which is interpreted as an acid-catalyzed abstraction of the respective functional group (e.g. $OSiMe_3$, OH, Cl) from the alkylpolysilane derivative affording a polysilanyl carbenium ion, which stabilizes under 1,2-migration of one trimethylsilyl group and nucleophilic attack of the conjugate base of the acid used to give the thermodynamically favored product.

1,2-Migrations of alkyl, aryl and also trimethylsilyl groups from an Si atom to a neighboring carbenium carbon atom are well known and have been described in the literature [1]. In the course of our studies on the behavior of 1-hydoxyalkyl-tris(trimethylsilyl)silanes (**1**) we found these compounds undergo a rapid isomerization when treated with protic acids or Lewis acids (Eq. 1).

$$\underset{\mathbf{3}}{R^1R^2C(SiMe_3)\text{–}Si(OH)(SiMe_3)_2} \xleftarrow{H^+} \underset{\mathbf{1}}{R^1R^2C(OH)\text{–}Si(SiMe_3)_3} \xrightarrow{BF_3} \underset{\mathbf{2}}{R^1R^2C(SiMe_3)\text{–}SiF(SiMe_3)_2}$$

Eq. 1. Rearrangement of **1** by protic acids (forming **3**) and by Lewis acids (yielding **2**).

The proposed pathway for the conversion is outlined in Eq. 2. The acid-catalyzed elimination of water from the alcohols **1** affording the carbenium ion is followed by a migration of one trimethylsilyl group from the central silicon atom to the neighboring carbon atom and attack of X^-, the conjugate base of the acid used as the catalyst, at the electrophilic silicon. The hydrolysis of the intermediates **4** gives the silanols **3** [2].

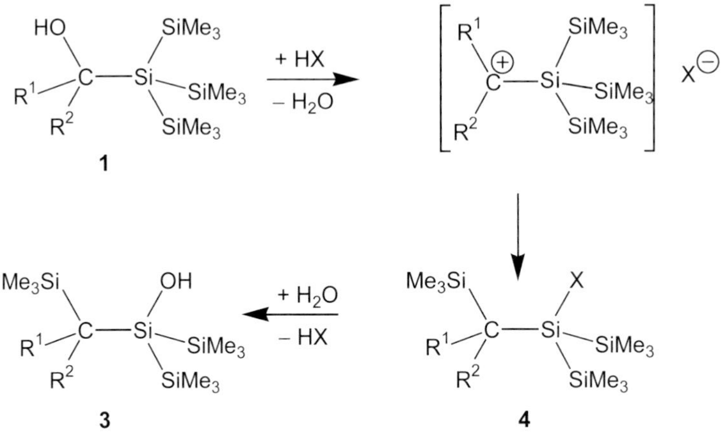

Eq. 2. Proposed pathway for the rearrangement of **1** by protic acids.

Recently we became aware of related rearrangements. α-Functionalized alkylpolysilanes such as the hydridosilanes **5** and **7** are converted by acid into the 1-trimethylsilylalkyl-siloxanes **6** or the 1-trimethylsilylalkylsilanol **8**, respectively (Eq. 3). The isomerization is performed by treatment of **5** or **7** with conc. sulfuric acid, the mechanism of the reaction being very similar to that outlined in Eq. 2.

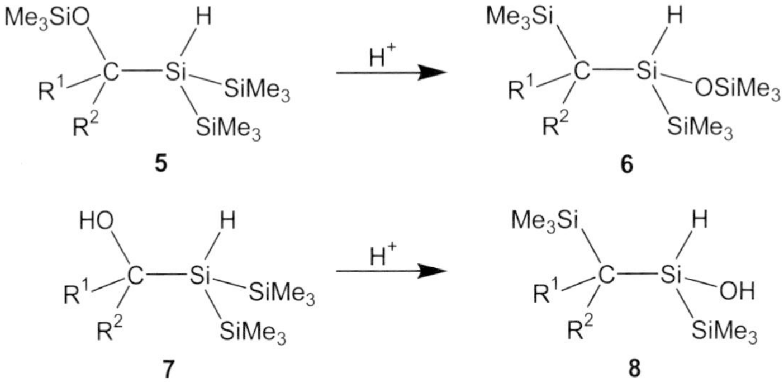

Isolated compounds: a) R^1 = Me, R^2 = Me; b) R^1 = Et, R^2 = Et

Eq. 3. Rearrangement of the hydrosilanes **5** and **7**.

The hydridosilanes **5** offer the possibility of further derivatizations and rearrangements of the resultant products. Thus, chlorination of these compounds with carbon tetrachloride is expected to lead to the chlorosilanes **9**. This, actually, was found for **5a,d,e,f,g** (Scheme 1). The chlorosilanes **9** proved to be rather labile. After prolonged standing of **9** in CCl_4 solution at room temperature a Me_3SiO/Cl interconversion occurred to give **10**. In the case of the reaction of **5b,c** with CCl_4 no chlorosilanes **9b,c** were found, but the chloroalkyldisiloxanes **10b,c** were obtained directly. Obviously, the conversion of **9b,c** , which probably act as intermediates in the formation of **10b,c**, proceeds very fast, so that their isolation failed.

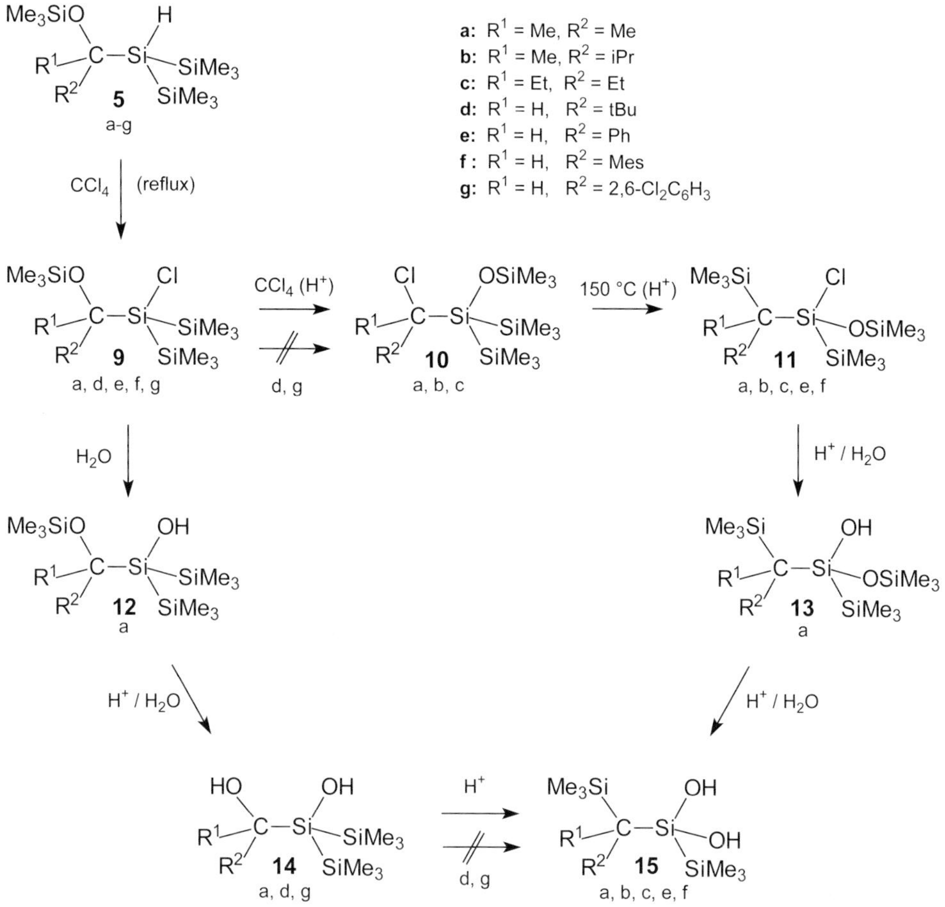

a: R^1 = Me, R^2 = Me
b: R^1 = Me, R^2 = iPr
c: R^1 = Et, R^2 = Et
d: R^1 = H, R^2 = tBu
e: R^1 = H, R^2 = Ph
f: R^1 = H, R^2 = Mes
g: R^1 = H, R^2 = 2,6-$Cl_2C_6H_3$

Scheme 1. The hydrolysis and isomerization of the chloro(1-trimethylsiloxyalkyl)bis(trimethylsilyl)silanes **9**. The compounds designated in the Scheme were isolated and fully structurally characterized.

At elevated temperature (neat, 150°C) both **9** and **10** were converted into the chloro (1-trimethylsilylalkyl)disiloxanes **11**, probably the most stable isomer of the chloro derivatives. This observed rearrangement again is understood as an acid-catalyzed isomerization of the type outlined in Eq. 2. Traces of HCl, always present in the chlorination mixture of **5** in CCl_4, are supposed to act as the catalyst.

The acid-promoted hydrolysis of **9** finally leads to the formation of the silandiols **15**. The reaction is best performed by passing a heptane/ethyl acetate solution of the chlorosilanes **9** through a silica gel column. Partial silylation of the silica gel surface by **9** releases the strong acid HCl which is obviously necessary for the observed conversion. The reaction proceeds through the respective 1-trimethylsiloxyalkylsilanols **12** and the 1-hydroxyalkylsilanols **14**, which was proved by the isolation of traces of these compounds as byproducts in the chromatographic separation and purification of **15**. Obviously, the hydrolysis of **9** is accompanied by an isomerization process, a 1,2-OH/trimethylsilyl exchange which very probably follows the same pattern discussed above for the other acid-induced conversions. As expected, the silandiols **15** were also obtained by hydrolysis of the chlorosiloxanes **11**.

It is worth mentioning that under the conditions applied the silanediols **15d,g** could not be obtained; the final products of the acid hydrolysis of **9d,g** were the 1-hydroxyalkylsilanols **14d,g**. Similarly, **9d** and **9g** were reluctant to take part in a Cl/$OSiMe_3$ exchange and the formation of **11d,g** and their resultant products. We suppose that the *tert*-butyl group in **9d** prevents the molecule from adopting a suitable conformation for the replacement of the respective substituents, and the isomerisation of **9g** fails as the result of the electronic effect of the dichlorophenyl substituent, destabilizing a necessary carbenium ion transition state. Interestingly, also in case of the related equally substituted 1-hydroxyalkyl tris(trimethylsilyl)silanes **1** (R^1 = H, R^2 = *tert*-butyl; R^1 = H, R^2 = 2,6-dichlorophenyl, respectively) no acid-induced isomerization could be performed.

The silanediols **15** proved to be stable compounds. Condensation reactions were not observed and are certainly prevented by the bulkiness of the substituents at the silanediol silicon atom. **15a** was chosen for an X-ray structure analysis, the results of which confirmed the proposed molecular structure. Bond lengths and angles are in the expected regions (Fig. 1). The crystal structure is characterized by hydrogen bonds (O–O distances of 2.187 Å) which form a dimer of **15a**.

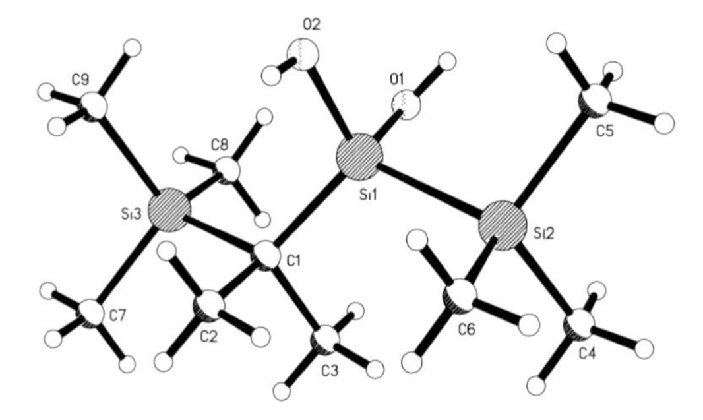

Fig. 1. Molecular structure of the silandiol **15a**. Selected bond lengths [Å] and angles [°]: Si1–Si2 2.366(2), Si1–C1 1.871(4), Si3–C1 1.898(4), Si1–O1 1.650(3), Si1–O2 1.658(3), Si3–C9 1.848(4), C1–C3 1.556(5), C1-Si1-Si2 114.70(13), Si3-C1-Si1 117.14(20), O1-Si1-O2 104.90(15).

Me_3SiO H H C Si $SiMe_3$ $SiMe_3$ *i*Pr *i*Pr — + DBPO / - PhCOOH acid → [Me_3SiO O O Ph H C Si $SiMe_3$ $SiMe_3$ *i*Pr *i*Pr]

16 **17**

→ Me_3Si O O Ph H C Si $OSiMe_3$ $SiMe_3$ *i*Pr *i*Pr

18

Eq. 4. Synthesis of the benzoyloxysilane **18**.

A third example may demonstrate the isomerization tendency of α-functionalized alkylpolysilanes. When the hydridosilane **16** was subjected to an intended radical-induced Si–Si coupling again a product of a 1,2-Si,C substituent exchange was observed. The hydridosilane **16** was heated with dibenzoyl peroxide. But instead of the expected Si,Si-coupling product we obtained the benzoyloxysilane **18** (Eq. 4). The structure of **18** was elucidated by a full spectral analysis. The results of the X-ray structural analysis are shown in Figure 2.

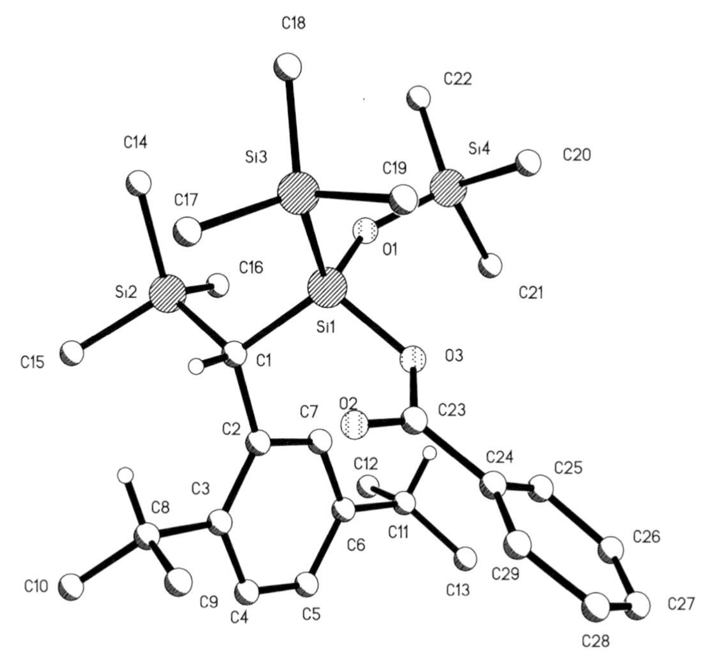

Fig. 2. Molecular structure of **18**. Selected bond lengths [Å] and angles [°]:Si1–C1 1.893(9), Si1–Si3 2.375(4), Si1–O1 1.614(7), Si1–O3 1.710(7), Si2–C1 1.880(9), Si4–O1 1.644(7), C1-Si1-O3 110.02(36), Si2-C1-Si1 114.12(48), Si1-O1-Si4 143.95(43).

Obviously, due to the steric congestion of the Si radicals formed by the interaction of **16** with dibenzoyl peroxide, the Si,Si-coupling proceeds very slowly. Thus, the reaction of the oxophilic Si radicals with the benzoyloxy radicals dominates and affords the intermediate **17**. The isomerization of **17** following the pattern described above finally leads to **18**.

As a conclusion it may be stated that the acid-induced 1,2-Si,C substituent exchange appears to be general behavior of α-substituted alkylpolysilanes. The initial step is the abstraction of a functional group, such as $OSiMe_3$ from **5** or **17**, chloride from **9** or OH from **14**, respectivly, and the formation of a carbenium ion transition state. The carbenium ion undergoes rapid transformations involving the 1,2-Si,C trimethylsilyl migration and the nucleophilic attack by the group previously eliminated or by the conjugate base of the applied acid leading to the isomerized product. But, whether the conversion proceeds through a silylium ion transition state or via a bimolecular process, involving an attack of the nucleophile at the central tetracoordinated silicon atom and thus facilitating the trimethylsilyl shift, remains an open question.

Acknowledgment: We gratefully acknowledge the support of our work by the *Deutsche Forschungsgemeinschaft*, the *Fonds der Chemischen Industrie* and the state of *Mecklenburg-Vorpommern.*

References:

[1] a) K. Tamao, M. Kumada, *J. Organomet. Chem.* **1971**, *30*, 339.
b) A. G. Brook, K. H. Pannell, E. Lebrow, J. J. Shuto, *J. Organomet. Chem.* **1964**, *2*, 491.
c) G. Märkl, M. Horn, W. Schloser, *Tetrahedron Lett.* **1986**, *27*, 4019.
d) P. D. Lickiss, *J. Chem. Soc., Dalton Trans.* **1992**, 1333; J. Chojnowski, W. Stanczyk, *Main Group Met. Chem.* **1994**, *2*, 6.
e) J. B. Lambert, L. Kania, S. Zhang, *Chem. Rev.* **1995**, *95*, 1191.
j) R. Bakhtiar, C. M. Holznagel, D. B. Jacobsen, *Am. Chem. Soc.* **1992**, *114*, 3227 and references cited therein.
g) Y. Apeloig, A. Stanger, *J.Am. Chem. Soc.* **1987**, *109*, 272.
h) Y. Apeloig, A. Stanger, *J. Am. Chem. Soc.* **1985**, *107*, 2806.

[2] a) K. Sternberg, M. Michalik, H. Oehme, *J. Organomet. Chem.* **1997**, *533*, 265.
b) K. Sternberg, H. Oehme, *Eur. J. Inorg. Chem.* **1998**, 177.

Hypersilyl Compounds of Elements of Group 15

*W. Krumlacher, H. Siegl, K. Hassler**

Institut für Anorganische Chemie
Erzherzog Johann Universität
Stremayrgasse 16, A-8010 Graz, Austria
Tel.: Int. code + (316)873 8206 — FAX: Int. code + (316)873 8701
E-mail: hassler@anorg.tu-graz.ac.at

Keywords: Hypersilyl / Group 15 Cages / Cyclotrisilane / Crystal Structure

Summary: New compounds of elements of group 15 with silicon have been prepared. By using the bulky hypersilyl substituent $-Si(SiMe_3)_3$ unusual structures could be stabilised in such a way that even the refluxing in an ether/H_2O mixture for several days resulted in only minor P–Si bond cleavage. The choice of suitable conditions led to unexpected rearrangements yielding cyclic and cage-like structures, like hexakis(trimethylsilyl)cyclotrisilane. These structures have been characterised by X-ray diffraction and ^{31}P- and ^{29}Si-NMR spectroscopy. The ^{31}P-NMR spectrum of $P_7[(SiMe_3)_3Si]_3$ shows the complex splitting pattern of an $A[MX]_3$ system.

Introduction

The supersilyl- $Si(CMe_3)_3$ and the hypersilylgroup $Si(SiMe_3)_3$ are exceptionally well suited to stabilize unusual structures that are otherwise prone to chemical attack. Here we report on the synthesis of some new cyclic and cage-like structures from elements of group 15 stabilized by the hypersilyl substituent.

Synthesis

Through the reaction of $(Na/K)_3M$ (M = P, As, Sb, Bi) with $XSi[Si(CH_3)_3]_3$ (X= I, Br, Cl, F, PhS, CF_3SO_3) we prepared various cyclic M–Si compounds such as the P_7 cage. Using suitable conditions these reactions tend to yield unexpected cyclic and cage-like structures through rearrangement, for instance hexakis(trimethylsilyl)cyclotrisilane.

A different route to prepare P–Si compounds is the reaction of trifluoromethyl-sulfonyloxytris(trimethylsilyl)silane $CF_3SO_3Si[Si(CH_3)_3]_3$ with PH_3.

Tri(hypersilyl)heptaphosphanortricyclane $P_7[(SiMe_3)_3Si]_3$

This tricyclic structure was obtained as shown in Scheme 1.

$$P_{white} + 3\ Na/K \rightarrow (Na/K)_3P$$
$$(Na/K)_3P + 3\ (Me_3Si)_3SiCl \rightarrow ((Me_3Si)_3Si)_3P_7 + 3\ Na/KCl$$

Scheme 1. Preparation route for the P_7 cage

The P_7 cage was characterized by X-ray diffraction and NMR spectroscopy (^{31}P, ^{29}Si). The $^{31}P\{H\}$-NMR spectrum shows the complex splitting pattern of an $A[MX]_3$ system, very similar to the analogous supersilyl compound [1].

Table 1. ^{31}P-NMR parameters [ppm] and coupling constants [Hz] of $P_7[(SiMe_3)_3Si]_3$ and $P_7(Si^tBu_3)_3$ (see Fig. 1).

	$P_7[(SiMe_3)_3Si]_3$	$P_7(Si\textit{t}Bu_3)_3$ [1]
δ_A (P^a)	–100,6	–113,0
δ_M (P^c)	–7,3	–31,2
δ_X (P^b)	–161,6	–175,6
$^1J_{AM}$	–349,2	–401,7
$^1J_{MX}$	–364,9	–431,1
$^1J_{XX'}$	–213,7	–205,0
$^2J_{AX}$	39,4	33,3
$^2J_{MM'}$	–13,7	–12,1
$^2J_{MX'}$	–14,6	–12,3
$^2J_{MX''}$	27,8	22,1

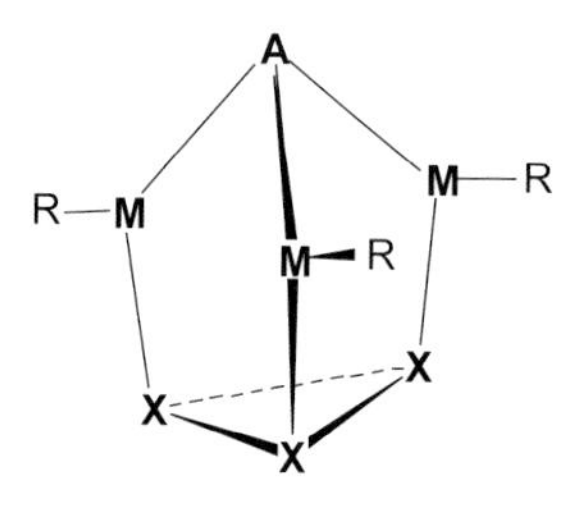

Fig. 1. Drawing of the P_7 frame with the naming scheme for Table 1.

By using the parameters listed in Table 1 an excellent agreement between the calculated and the measured spectrum (Fig. 2) could be achieved.

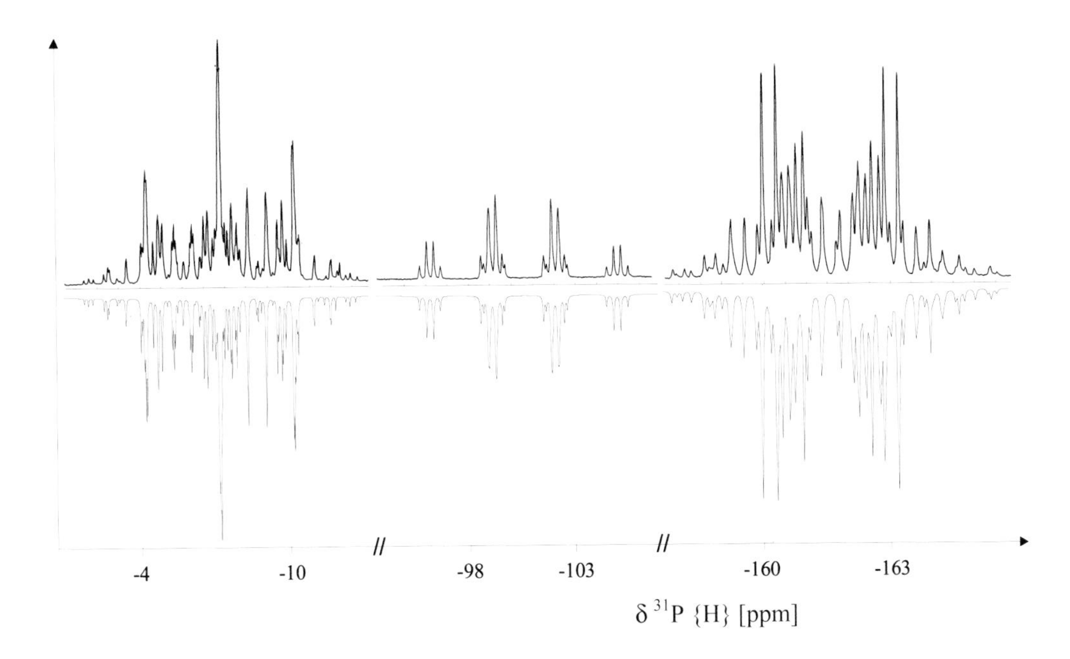

Fig. 2. A comparison of the measured $^{31}P\{H\}$-NMR spectrum (top) with the calculated one (bottom).

Colorless single crystals of $P_7[Si(SiMe_3)_3]_3$ suitable for X-ray diffraction (Fig. 3) analysis were grown in benzene. The data collection [4] was performed under a stream of N_2 at –70°C using a Siemens diffractometer with a CCD SMART detector and graphite monochromated molybdenum K_α radiation (λ = 0.71073 Å). The crystal was mounted on the tip of a glass fiber in inert oil.

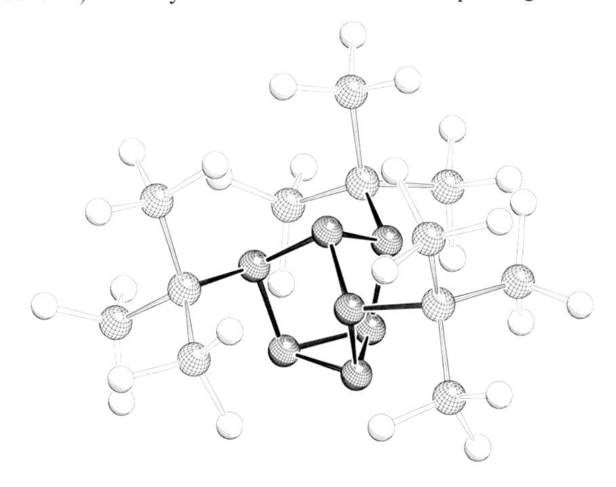

Fig. 3. Structure of tri(hypersilyl)heptaphosphanortricyclane $P_7[(SiMe_3)_3Si]_3$.

The analogous As_7 cage was synthesized in a similar way and identified by mass spectroscopy.

Hexakis(trimethylsilyl)cyclotrisilane $[Si(SiMe_3)_2]_3$

This cyclotrisilane was prepared through the reaction of Na_3As with tris(trimethylsilyl)chlorosilane at 0°C in hexane. The carefully chosen experimental conditions led to the isolation of colorless crystals which could be purified by vacuum sublimation and recrystallization in *n*-hexane.

It could successfully be characterised by X-ray diffraction (Fig. 4). Klinkhammer [2] prepared this compound using a completely different approach.

The compound was found to be stable to oxygen and moisture, indicating that the trimethylsilyl substituents can serve as effective steric blockades against external attack.

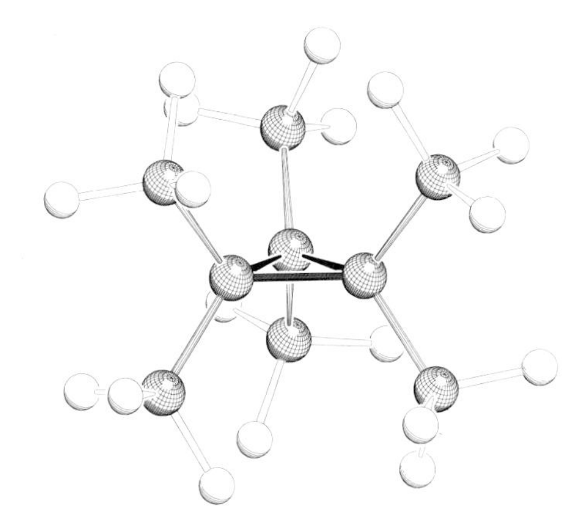

Fig. 4. Structure of hexakis(trimethylsilyl)cyclotrisilane $[Si(SiMe_3)_2]_3$.

Hypersilylphosphine $H_2PSi(SiMe_3)$

By using the sterically demanding silyl group the Si–P bond could be stabilised in such a way that even refluxing in an ether/H_2O mixture for several days resulted in only minor bond cleavage.

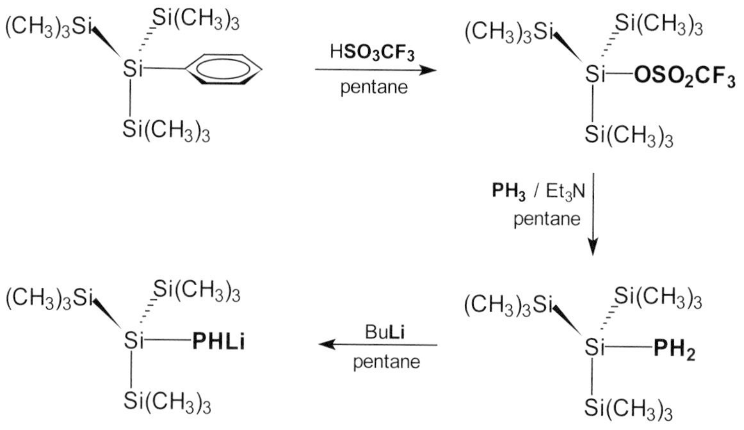

$((CH_3)_3Si)_3Si$–Ph $\xrightarrow[\text{pentane}]{HSO_3CF_3}$ $((CH_3)_3Si)_3Si$–OSO_2CF_3 $\xrightarrow[\text{pentane}]{PH_3 / Et_3N}$ $((CH_3)_3Si)_3Si$–PH_2 $\xrightarrow[\text{pentane}]{BuLi}$ $((CH_3)_3Si)_3Si$–PHLi

Scheme 2. Route for the synthesis of $((CH_3)_3Si)_3SiPH_2$ and the successive lithiation.

The synthesis of tris(trimethylsilyl)silylphosphine (Scheme 2) yielded colorless crystals that could easily be purified by vacuum sublimation. Contrary to the observations made by Uhlig [3] with the $Si(CH_3)_3$ substituent, we were not able to introduce more than one bulky hypersilyl group to PH_3.

A subsequent lithiation opens a way to prepare new compounds, for instance diphosphanes such as $(Me_3Si)_3SiPHPHSi(SiMe_3)_3$.

Acknowledgment: The authors thank the *Fonds zur Förderung der wissenschaftlichen Forschung (FWF)* Vienna for the financial support of this project (S 07905-CHE) and are grateful to *Wacker-Chemie GmbH Burghausen* for the supply of with silanes.

References:

[1] I. Kovác I. et al., *Z. Anorg. Allg. Chem.* **1993**, *619,* 453–460.

[2] K. W. Klinkhammer, *Chem. Eur. J.* **1997**, *3*,1418–1431.

[3] W. Uhlig., A. Tzschach, *Z. Anorg. Allg. Chem.* **1989**, *576*, 281–283.

[4] Crystal data for $P_7[Si(SiMe_3)_3]_3$: $C_{27}H_{81}P_7Si_{12}$ + C_6H_6, M = 1037.9 g/mol, triclinic ($P\bar{1}$; 2); a = 1400.72(1) pm, b = 1414.31(2) pm, c = 1825.98(3) pm, α = 84.216(1)°, β = 77.639(1)°, γ = 63.401(1)°, V = 3159.48 × 10^6 pm^3, Z = 2, ρ_{calc} = 1.091 g/cm^3, 27 108 reflections (18 895 independent, 1.14° < 2Θ < 30.52°; h = –19 to 20, k = –19 to 20, l = –13 to 26) were collected on a 0.4 mm × 0.4 mm × 0.3 mm crystal; $F(000)$ = 1116, $R_1[I > 2\sigma(I)]$ = 0.0385 [5a], wR_2 = 0.1149 [5b], GOF = 0.463 [5c]. The structure was solved by direct methods using XS [6] and refined by full-matrix least squares with SHELXL-93 [7], minimizing the residuals for F^2. No absorption correction was applied. Hydrogen atoms were included in the model at their calculated positions. The torsion angle of a methyl group was set to maximize the sum of the electron density at the three calculated hydrogen positions. Anisotropic displacement parameters were assigned to all non-hydrogen atoms and isotropic parameters were used for the hydrogen atoms.

[5] a) $R = (\Sigma \,||F_0| - |F_c||)/\Sigma|F_0|$.
b) $wR_2 = \{\Sigma[w(|F_0|^2 - |F_c|^2)^2]/\Sigma[w(F_0^4)]\}^{1/2}$.
c) GOF = $\{\Sigma w(|F_0|^2 - |F_c|^2)^2(n-m)\}^{1/2}$

[6] SHELXTL 5.0, Siemens Crystallographic Research System, **1994**.

[7] G. M. Sheldrick, SHELXL-93, Program for Crystal Structure Determination, Universität Göttingen, Germany, **1993**.

^{29}Si-MAS-NMR Investigations of Amino-Substituted Chloromethylpolysilanes

E. Brendler

Institut für Analytische Chemie
Technische Universität Bergakademie Freiberg
Leipziger Str. 29,D-09596 Freiberg, Germany
Tel.: Int. code + (3731)39 2266
Email: brendler@orion.hrz.tu-freiberg.de

K. Trommer, G. Roewer

Institut für Anorganische Chemie
Technische Universität Bergakademie Freiberg
Leipziger Str. 29, D-09596 Freiberg, Germany

Keywords: ^{29}Si-MAS-NMR / Polysilane / Aminosilane

Summary: ^{29}Si-CP-MAS spectra of cured and non-cured polysilane-*co*-styrene polymers are reported. Interpretation of the structural changes documented by the spectra was carried out using variable contact time and inversion recovery CP measurements for editing the spectra; the assignment of the signals is based on chemical shift ranges in oligosilanes. Using the data of model compounds, chemical shifts in hydrogen-free ceramics can be estimated and used for interpretation of spectra as demonstrated for the system $SiC_{4-n}N_n$.

Introduction

Chlorine-containing polysilanes can be modified by reaction with ammonia or amines. This process allows a hardening of the polysilane fiber surface [1] as well as influencing the properties of the resulting ceramics. Therefore the structural changes taking place during the so-called curing process are of great interest in order to design the optimum curing conditions and material properties.

Results and Discussion

Investigations were carried out on polysilane-*co*-styrene samples cured in an ammonia atmosphere at different temperatures and partial pressures of ammonia as well as samples cured with methylamine or ethylenediamine. Figure 1 shows ^{29}Si-CP-MAS spectra for an ammonia-cured and a diethylamine-cured sample in comparison to the starting polysilane-*co*-styrene. The structure of the organosilicon polymer blocks has been investigated earlier and can be described by the formula **1**, expressing that the polymer consists of an $SiMeSi_3$ network (–64 ppm) with $-SiMeCl_2$ (36 ppm) and $SiMeCl_2$–SiMeCl– substituents and –SiMeCl– spacers (24 ppm and 14 ppm for –SiMeCl–) [2].

$$H_3C-SiCl_2-[SiR(CH_3)]_x-SiCl_2-CH_3$$

1: R = Cl, $SiCH_3R_2$

The comparison of the ^{29}Si-CP-MAS spectra of the (chloromethyl)polysilane-*co*-styrene with the cured samples indicates a reaction of the curing reagent especially with the $-SiMeCl_2$ groups linked to branching sites of the silicon backbone, but also with the linear subunits $SiMeCl_2-SiMeCl-$ and $-SiMeCl-$ spacers in the silicon backbone. The intensities of these signals decreases and new signals due to nitrogen-containing chemical environments occur, which gives rise to the tasks of a *deconvolution* of the different sites and an *assignment* of the signals to certain chemical environments.

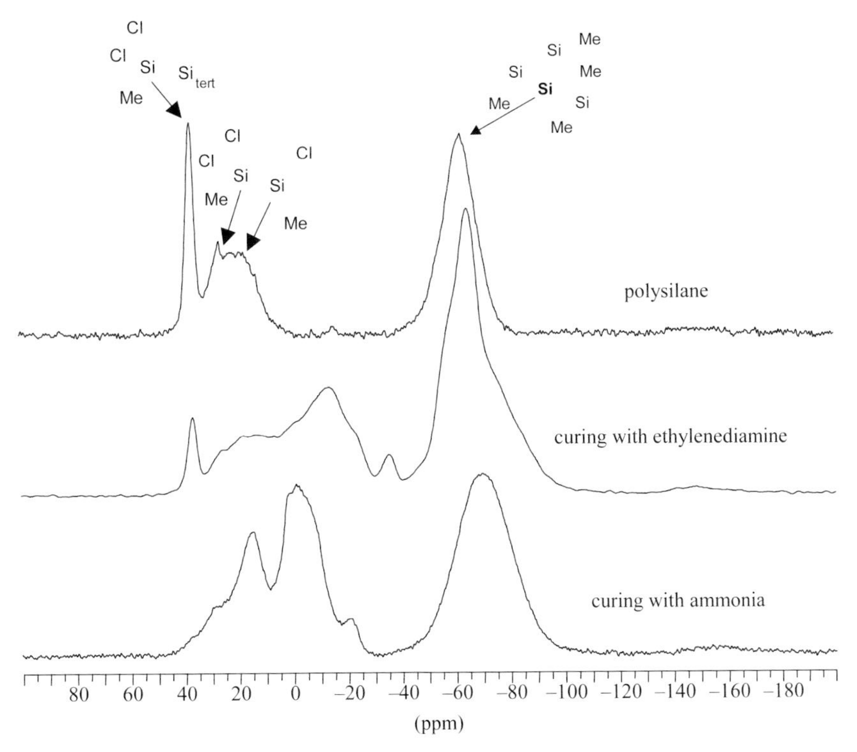

Fig. 1. ^{29}Si-CP-MAS spectra of a non-cured polysilane-*co*-styrene and of samples cured with NH_3 and ethylenediamine.

Beside dipolar dephasing experiments and variable contact time measurements the inversion recovery cross polarization technique (IRCP) proved to be a powerful method of editing spectra with strongly superimposed signals [3, 4]. Recording spectra at different inversion times allows one to distinguish between the different groups because of their different inversion behavior mainly due to differences in mobility. For the system under investigation a faster inversion behavior could be observed for the amino-substituted sites than for the chloromethyl groups. The thus achieved deconvolution of the ^{29}Si-CP-MAS spectrum of the diethylamino-substituted polysilane is shown in Fig. 2.

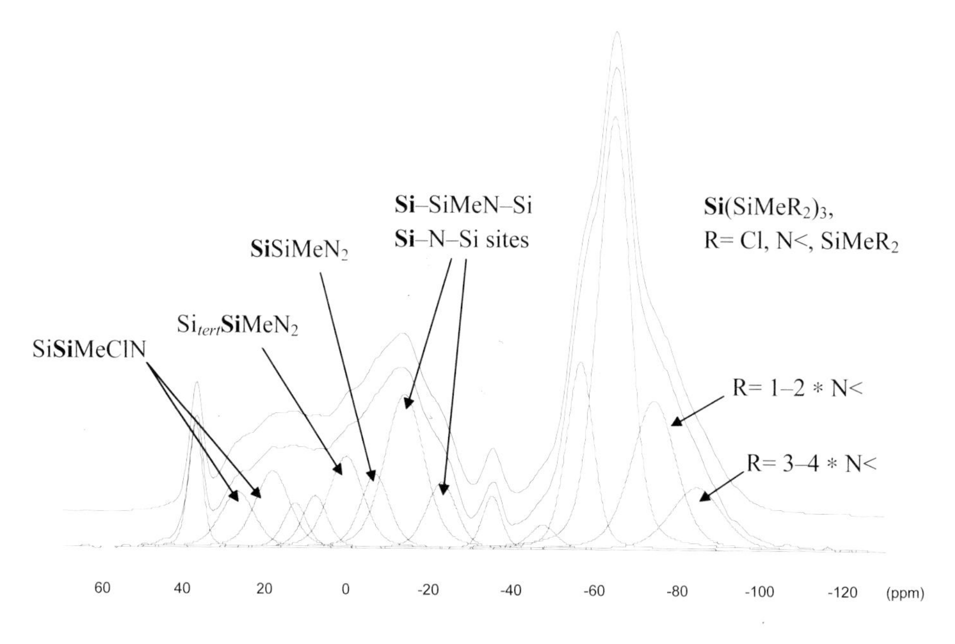

Fig. 2. Deconvolution (single peaks and upper spectrum) and interpretation of the ^{29}Si-IRCP-NMR spectrum at an inversion time τ_i= 5 ms. Sample: polysilane-*co*-styrene cured with ethylenediamine; N = amino group of ethylenediamine.

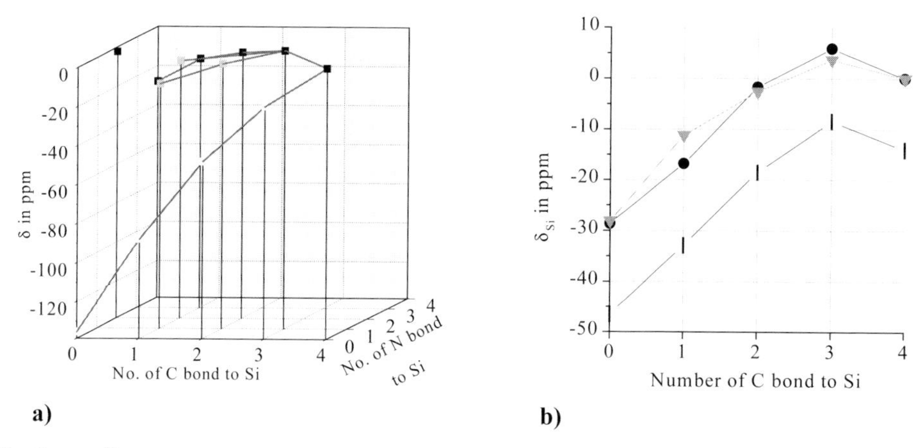

Fig. 3. (**a**) ^{29}Si chemical shifts in organosilicon compounds of type $Si(SiMe_3)_xMe_y(NEt_2)_z$ [6–8]; (**b**) Estimation for δ_{Si} of $SiC_{4-x}N_x$-sites in hydrogen-free ceramics (**I**) using the data of monosilanes $SiMe_{4-n}(NR_2)_n$ with R = CH_3(●), C_2H_5(▼).

The assignment of the signals is not straightforward. Figure 3 illustrates the difficulties of predicting ^{29}Si chemical shifts in the ternary system Si–N–C. The sagging pattern of δ_{Si} with changing substituents and the small shift differences within the series $SiC_{4-n}N_n$ lead to the fact that most of the possible chemical environments have a chemical shift between 0 and –20 ppm. Therefore the knowledge of the chemical shifts of model compounds containing the chemical environments which can be formed during the curing reaction and the changes in δ_{Si} when the chlorine substituents are replaced by amine are of essential importance for an interpretation of the spectra.

For the assignment of the resonances found in the cured sample chemical shift ranges for oligosilanes having the substituents –Cl, –Me and $-NEt_2$ were used. They are based on investigations of numerous chloromethyloligosilanes up to Si_7 and their amine and alkoxy derivatives [6, 9, 10].

A large variety of groups can be formed by a stepwise replacing of the chlorine functions by amino groups, which explains the large number of peaks obtained by the deconvolution. In general the signals are shifted upfield after replacing chlorine by an amino substituent. Replacing one chlorine in an $SiMeCl_2$ group of the tetrasilane $MeSi(SiMeCl_2)_3$ causes a downfield shift for the remaining neighboring $SiMeCl_2$ sites by 3 ppm. In the cured polymer a decrease in intensity but no significant change in δ_{Si} could be observed for the $Si_{tert.}$**Si**$MeCl_2$ sites, which allows the conclusion that the $SiMeCl_2$ sites corresponding to the signal at 36 ppm in the polymer are separated from nitrogen-bearing groups by more than one silicon atom. This corresponds also to the unchanged polymer signal at –64 ppm for the branching silicon atoms and is supported by the fact that the curing process is designed as a surface reaction and does not have the aim to convert the whole polymer.

Mixed SiMeClN< groups have their resonances between 0 and 16 ppm, ternary bond sites again at lower field than linear ones. Around 0 ppm ternary $-SiMe(N<)_2$ sites as well as –SiMeCl spacers can be observed. The signals at negative chemical shifts are due to SiMeN< spacers (–16 to –20 ppm) and $SiMe(N<)_2$ groups connected to them (about –7 ppm). Crosslinking sites –Si–NR–Si– representing silazane-like substructures can be shifted to even higher field (signal at –23 ppm).

The nature of the signal at –35 ppm is still under discussion. In comparison to the other signals it shows a faster inversion behavior and the maximum enhancement by CP lies at shorter spinlocking times, which is typical for Si–H sites.

As also reported for other preceramic systems [4] we observed, for our system, that δ_{Si} undergoes an upfield shift with the hydrogen loss during pyrolysis and formation of the ceramic system. The data of model compounds representing certain chemical environments can be used to estimate δ_{Si} of ternary hydrogen-free ceramic systems using the δ_{Si} of the binary border systems and the data of the model compounds. For the system $SiC_{4-n}N_n$ it has been done by Eq. 1.

$$\delta(SiC_{4-n}N_n) = \delta(SiMe_{4-n}(NR_2)_n) - (\delta(SiMe_4) - \delta_{SiC}) + n\cdot[(\delta(SiMe_4) - \delta_{SiC}) - (\delta(Si(NR_2)_4) - \delta Si_3N_4)]/4$$

Eq. 1.

Figure 4 shows as an example the results of the extrapolation in the system $SiC_{4-n}N_n$, discussed above. The method works as well for systems having C and O or N and O bound to silicon; in Eq. 1 δ has to be replaced by the δ_{Si} values of the corresponding model compounds. For the system Si–C–O, where the peaks for $SiC_{4-n}O_n$ sites in ceramics are well separated and therefore δ_{Si} could be determined by experiment, experimental values and estimations showed deviations of less than 3 ppm which is very small for systems with a peak width of several hundred Hertz and differences of the same dimension within the reported experimental values. Problems occur with systems containing Si–Si bonds because of the Knight shift of pure silicon. The chemical shift of **Si**$(SiMe_3)_4$ could be used instead, but until now Si–Si bonds have not been observed in the ceramics derived from the polysilane system presented so the problem is of minor importance.

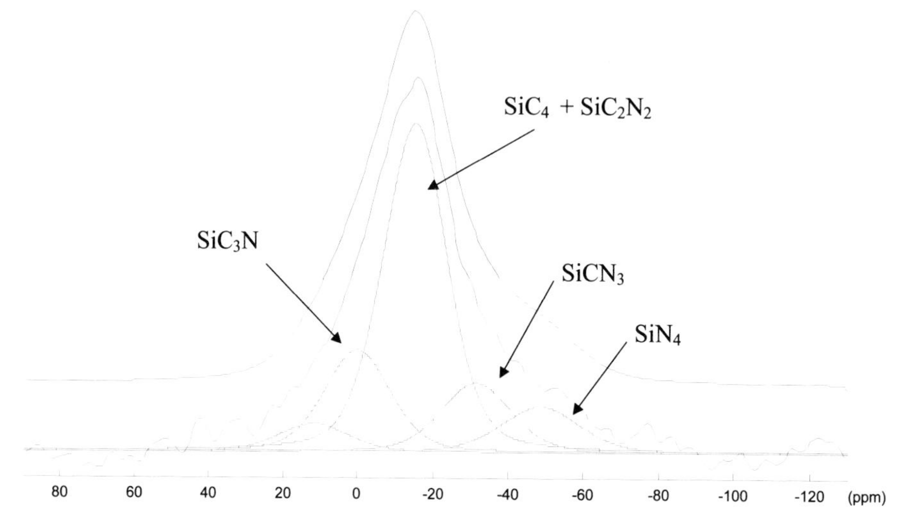

Fig. 4. ^{29}Si-MAS-NMR spectrum and deconvolution (single peaks and upper spectrum) of an Si–N–C ceramic after curing polysilane-*co*-styrene with NH_3 and pyrolysis (Ar) at 1200°C.

Using the estimation in Fig. 3(b) it was possible to carry out a deconvolution and interpretation of ^{29}Si-MAS-NMR spectra of silicon carbonitride ceramics obtained by pyrolysis of an ammonia-cured polysilane-*co*-styrene. The result is shown in Fig. 4 and represents a composition of 55 % SiC_4/SiC_2N_2 sites, 18 % SiC_3N sites and 9 % SiN_4 sites. The small low-field peak at 11.6 ppm is due to remaining chlorine (1–3 %) and small amounts of oxygen introduced by sample handling procedures.

Acknowledgment: The authors thank the *Deutsche Forschungsgemeinschaft* for financial support.

References:

[1] H.-P.Martin, E. Müller, G. Roewer, R. Richter, *Chem. Ing. Tech.* **1998**, *70*, 170.

[2] E. Brendler, K. Leo, B. Thomas, R. Richter, G. Roewer, H. Krämer, "*Disproportionation of Tetrachlorodimethyldisilane — Identification of the Primary Products*" in: *Organosilicon Chemistry II: From Molecules to Materials,* (Eds.: N. Auner, J. Weis), VCH, Weinheim, **1996**, p. 69.

[3] X. Wu, K. W. Zilm; *J. Magn. Res.* **1993**, *A102*, 205–13.

[4] F. Babonneau, R. Richter, C. Bonhomme, J. Maquet, G. Roewer, *J. Chim. Phys.* **1995**, *92*, 1745.

[5] C. Gerardin, M. Henry, F. Taulelle, *Mater. Res. Soc. Symp. Proc.* **1992**, *271*, 777.
[6] K. Trommer, U. Herzog, G. Roewer, *J. prakt. Chem.* **1997**, *339*, 637.
[7] H. Marsmann, *NMR Basic Principles and Progress 17: Oxygen 17 and Silicon 29*, Springer, Berlin, **1981**.
[8] K. Tamao, A. Kawachi, S. Yamaguchi, *Organometallics* **1997**, *16*, 780.
[9] U. Herzog, K. Trommer, G. Roewer, *J. Organomet. Chem.* **1998**, *552*, 99.
[10] U. Herzog, N. Schulze, K. Trommer, G. Roewer, *J. Organomet. Chem.* **1997**, *547*, 133.

Investigation of the Longitudinal Relaxation Time T_1 of Silanes

Christina Notheis, Erica Brendler, Berthold Thomas

Institut für Analytische Chemie
Technische Universität Bergakademie Freiberg
Leipziger Str. 29, 09596 Freiberg, Germany
Tel.: Int. code + (3731)39 2266 — Fax: *Int. code + (3731)39* 3666
E-mail: Christina.Notheis@t-online.de

Keywords: T_1 Relaxation Time / NOE Enhancement Factors / Silanes

Summary: T_1 relaxation times of different disilanes and higher silanes have been measured. A qualitative description of the different contributions to T_1 is given. For a sample containing $Ph_4Si_2Me_2$, $Ph_5Si_3Me_3$ and $Ph_6Si_4Me_4$ temperature and magnetic field dependent spectra were recorded to separate the shares of the different relaxation mechanisms. NOE enhancement factors were determined, too. Dipole-dipole interactions contribute the most important share, but spin rotation interactions are up to 25 % of the total relaxation and must not been neglected. Other mechanisms seem to be of no importance in that system.

Introduction

The low natural abundance of the ^{29}Si nucleus and its small and negative gyromagnetic ratio γ require high pulse repetition rates in order to get an acceptable signal-to-noise ratio. The negative nuclear Overhauser effect caused by the negative γ entails nulled or negative signals and has to be avoided. The generally long relaxation times T_1 lead to long experiment times. Special NMR techniques for polarization transfer like DEPT or INEPT or the addition of paramagnetic relaxation reagents could be used to shorten T_1. Both ways require certain structural properties which are not always given. Therefore, the relaxation times T_1 of several silanes have been determined in order to optimise NMR acquisition parameters for different samples.

Results and Discussion

Several contributions to the relaxational process exist. For a nucleus of a spin quantum number $I = ½$ such as ^{29}Si the total spin–lattice relaxation time is given by Eq. 1 [1], where $1/T_1$ is defined as the relaxation rate.

$$\frac{1}{T_1} = \frac{1}{T_1^{DD}} + \frac{1}{T_1^{SR}} + \frac{1}{T_1^{SC}} + \frac{1}{T_1^{CSA}}$$

Superscripts: DD = dipole–dipole interaction
SR = spin–rotation interaction
SC = scalar coupling
CSA = chemical shift anisotropy

Eq. 1.

The influence of substitution pattern and of chain length on T_1 should be detected. Therefore, symmetric disilanes of the kind 1,1,2,2-X_4-1,2-Me_2Si_2 (X = F, Cl, Br, Me, Ph) and silanes with increasing numbers of silicon and X_2MeSi substitution (X = Ph, Cl) of the sidechains were examined. Both saturation recovery (sr) and inversion recovery (ir) pulse sequences were applied and tested for their advantages and disadvantages. The big advantage of saturation recovery is a less time-consuming pulse sequence. Unfortunately, this advantage is cancelled by the less comprehensible results and the smaller dynamic range. Inversion recovery is a time-consuming method, but has the larger dynamic range and a graphic clarity, which helps to estimate the T_1 values just by a look at the spectra.

Table 1. Determined relaxation times T_1 [s].

Silane	T_1 Si^A (ir/sr)	T_1 Si^B (ir/sr)	T_1 Si^C (ir/sr)	T_1 Si^D (ir/sr)
$Ph_4Me_2Si_2$	39 / 44			
$(Ph_2MeSi^A)_2Si^BMePh$	31 / 33			
$(Ph_2MeSi^ASi^BMePh)_2$	25 / 26	16 / 16		
$(Ph_2MeSi^A)_3Si^BMe$	24 / 23	22 / 25		
Me_6Si_2	45 / 46			
$Br_4Me_2Si_2$	59 /60			
$Cl_4Me_2Si_2$	43 / 47			
$F_4Me_2Si_2$	27			
$Cl_4Me_2Si_2$[a]	51.7			
$(Cl_2MeSi^A)_2Si^BMeCl$	58	57.3		
$(Cl_2MeSi^A)_3Si^BMe$	51.7	60.3		
$Cl_2MeSi^ASi^BMeClSi^CMe(Si^DMeCl_2)$	49.8	38.1	49.9	45.2
$[(Cl_2MeSi^A)_2Si^BMe]_2$	41.8	43.4		
$[(Cl_2MeSi^A)_2Si^BMe]_2Si^CMeCl$	35.9	40.9	—[b]	
$(Me_3Si^A)_4Si^B$	63	90		

[a] Measured in a mixture of different chlorooligosilanes. — [b] Intensity too small for interpretation.

In the most cases, the differently substituted disilanes show the expected behavior (Table 1). All statements use Me_6Si_2 as the "standard" for comparison. For the brominated and chlorinated species, the efficiency of dipolar relaxation is diminished because of the smaller number of protons, but additional interactions between silicon and the quadrupolar nuclei occur. The fact that these interactions seem to be less effective in the brominated than in the chlorinated species was surprising. A comparison of the properties of an Si–Br and an Si–Cl bond such as dipolar coupling constants and bond lengths and further the quadrupole moments of both nuclei, would lead to the conclusion that bromine-substituted substances should have the more effective relaxation. It seems that dipole–dipole interactions contribute the dominant share to the total relaxation and that the quadrupolar nuclei are of minor influence. Further investigations have to be done to get a proper description of this phenomenon. The high gyromagnetic ratio of fluorine leads to a more effective dipole–dipole interaction and therefore to the small value of T_1 in $F_4Me_2Si_2$ [2]. The difference between T_1 of the phenylated and the permethylated silanes is due to the higher molecular size of the phenylated species, which is responsible for a slower molecular motion and therefore for a smaller value of the correlation time. (The correlation time τ_c is defined as the time a molecule needs to rotate through one radian.) Due to the indirect dependence of T_1 upon τ_c the relaxation time decreases with increasing size of the molecules.

Figure 1 shows the dependence of the relaxation time T_1 on the molecular size for some dichloromethyl-substituted silanes [3]. Two different trends become obvious. The comparison of silicon nuclei with the same substitution pattern but different chain length shows that T_1 increases with decreasing chain length. And T_1 decreases with increasing number of chlorine substituents, which can be explained by the growing influence of relaxation by scalar coupling. Branching of the silicon skeleton leads to an increase in T_1. This fact becomes most obvious when the relaxation time of a quaternary silicon as in $(Me_3Si)_4Si$, is determined. It shows, at 90 s, the highest value determined for a 30 % solution of a silane at room temperature in this study.

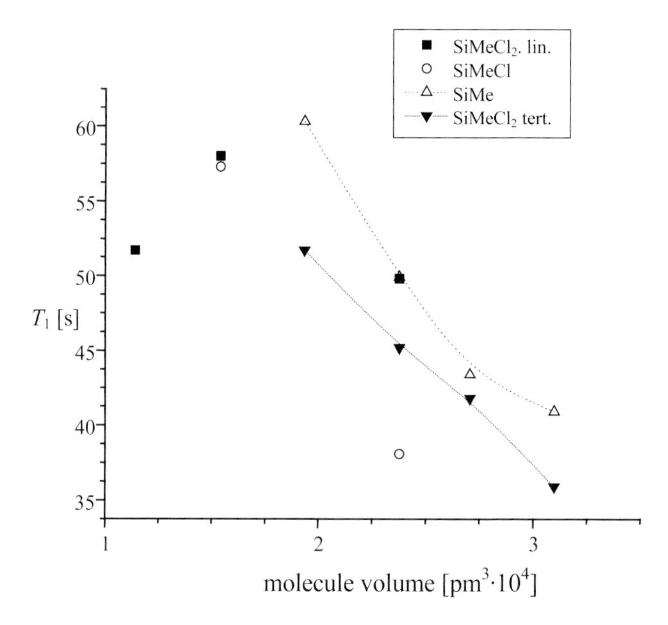

Fig. 1. Dependence of the relaxation time T_1 on the molecule size. $SiMeCl_2$ lin. stands for a dichloro-methylsilyl group next to a secondary silicon atom, $SiMeCl_2$ tert. for a dichloromethylsilyl group next to a tertiary silicon atom.

Up to this point, only a qualitative description of the different contributions was possible. To get a quantitative statement about the efficiency of the different mechanisms, temperature-dependent spectra and spectra at different magnetic fields (B_o) of a mixture of $Ph_4Si_2Me_2$, $(Ph_2MeSi)_2SiMePh$ and $(Ph_2MeSi)_3Si$ were recorded. The B_o-dependent spectra show only small differences in T_1. Since the magnitude of the changes in T_1 is the same as that of the errors in the determination of the signal intensities, we state that the dependence of T_1 on the magnetic field is small. The only share depending on B_o is T_1^{CSA}, which will be neglected in the following work. Independent ^{29}Si-CP/MAS spectra confirm a generally small anisotropy of the chemical shift. In our following investigations on diphenylmethylsilyl substituted silanes Eq. 1 can be reduced to Eq. 2.

$$\frac{1}{T_1} = \frac{1}{T_1^{\mathrm{DD}}} + \frac{1}{T_1^{\mathrm{SR}}}$$

Eq. 2.

Nuclear Overhauser enhancement factors η were determined to calculate T_1^{DD} (Eq. 3 and Eq. 4).

$$\eta = \frac{I_{\mathrm{NOE}} - I_0}{I_0}$$

I_o = signal intensity without NOE
I_{NOE} = signal intensity with the highest observed NOE

Eq. 3.

$$T_1^{\mathrm{DD}} = \frac{\gamma_{\mathrm{H}} \cdot T_1}{2\gamma_{\mathrm{Si}} \cdot \eta}$$

$\gamma_{\mathrm{H}}, \gamma_{\mathrm{Si}}$ = gyromagnetic ratios of ^{1}H or ^{29}Si
η = nuclear Overhauser enhancement factor

Eq. 4.

$1/T_1^{\mathrm{SR}}$ was calculated for different temperatures. The nuclear Overhauser enhancement factors η were set as independent of temperature, which was confirmed by our results. Table 2 lists measured and calculated relaxation times and rates for the silane mixture mentioned above.

Table 2. Relaxation rates of $Ph_4Si_2Me_2$, $(Ph_2MeSi)_2SiMePh$ and $(Ph_2MeSi)_3Si$ listed according to the shares of different relaxation mechanisms at room temperature.

Silane	T_1 [s]	$1/T_1$ [s^{-1}]	T_1^{DD} [s]	$1/T_1^{\mathrm{DD}}$ [s^{-1}]	T_1^{SR} [s]	$1/T_1^{\mathrm{SR}}$ [s^{-1}]
$Ph_4Me_2Si_2$	41.4	0.0242	55.4	0.0180	161.3	0.0062
$(Ph_2MeSi^*)_2SiMePh$	31.3	0.0319	40.2	0.0170	142.4	0.0070
$(Ph_2MeSi)_2Si^*MePh$	23.3	0.0429	33.9	0.0295	118.98	0.0084
$(Ph_2MeSi^*)_3SiMe$	22.7	0.0441	29.2	0.0343	102.6	0.0097
$(Ph_2MeSi)_3Si^*Me$	20.5	0.0488	27.3	0.0366	82.3	0.0122

Our studies show that the T_1 times determined are dependent on temperature. With increasing temperature, the relaxation times increase and therefore relaxation rates decrease. Regarding the ratio of T_1^{DD} to T_1^{SR}, it becomes obvious that this ratio is independent of temperature. Taking into consideration that both mechanisms are depending on the correlation time τ_c in the same way, this fact is not surprising.

Figure 2 shows the measured and calculated relaxation rates for the silanes under investigation.

To get an impression of the influence of the sample concentration on T_1, viscosity-dependent spectra of the trisilane Ph_2MeSi–$SiMePh$–$SiMePh_2$ were recorded and the NOE enhancement factors η determined. T_1 decreases with increasing viscosity, as expected from literature [4]. But the expected minimum and the subsequent increase in T_1 did not occur in the range we worked in. Table 3 lists the determined and calculated relaxation rates for the different silicon atoms in $(Ph_2MeSi^A)_2Si^BMePh$.

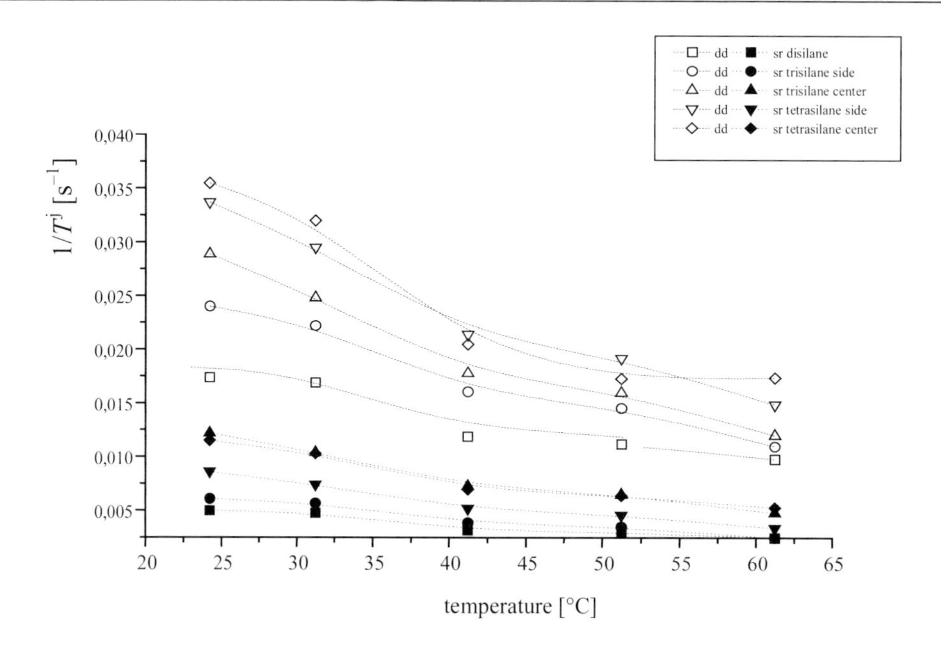

Fig. 2. Temperature dependence of the dipole–dipole (dd) and the spin–rotation (sr) relaxation rates of the different silanes.

Table 3. Determined relaxation times and calculated relaxation rates for samples of Pentaphenyltrimethyltrisilane $(Ph_2MeSi^A)_2Si^BMePh$ with different viscosities.

Viscosity [Pa s]	T_1 (Si^A) [s]	$1/T_1$ (Si^A) [s^{-1}]	$1/T_1^{DD}$ (Si^A) [s^{-1}]	$1/T_1^{SR}$ (Si^A) [s^{-1}]	T_1 (Si^B) [s]	$1/T_1$ (Si^B) [s^{-1}]	$1/T_1^{DD}$ (Si^B) [s^{-1}]	$1/T_1^{SR}$ (Si^B) [s^{-1}]
0.372	8.3	0.1199	0.0962	0.0237	6.9	0.1451	0.1037	0.0414
0.0325	17.7	0.0564	0.0452	0.0112	13.6	0.0736	0.0526	0.0210
0.0056	47.4	0.0211	0.0169	0.0042	50.8	0.0197	0.0141	0.0056

Calculating the ratio of T_1^{DD} to the total relaxation time, we determined that this ratio is independent of the samples' viscosities, but it is different for the silicon nuclei of the side groups and the center silicon nucleus. At the center silicon, relaxation via dipole–dipole interactions is up to 71 % of the total relaxation whereas it is about 80 % of the total relaxation for a side-group silicon. Since this mechanism is dependent on the number of interacting protons and the distance Si–H it is easily understood that the silicon with the smaller number of protons in a larger distance shows the less efficient dipolar relaxation.

In Fig. 3 the viscosity dependence of the individual relaxation rates is shown.

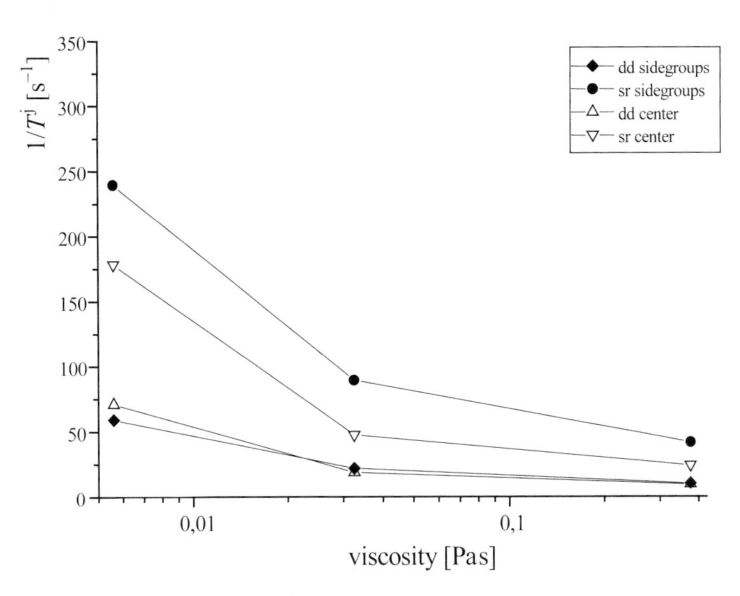

Fig. 3. Correlation of the relaxation rates $1/T_1^j$ to the viscosity of the different samples. dd stands for relaxation via dipole–dipole interactions, sr stands for spin–rotation interactions.

Acknowledgment: The authors thank the *Deutsche Forschungsgemeinschaft* (DFG) for financial support.

References:

[1] G. C. Levy, J. D. Cargioli, P. C. Juliano, T. D. Mitchel, *J. Am. Chem. Soc.* **1973**, *95*, 3445.

[2] J.-P. Kintzinger, H. Marsmann in "*NMR, Basic Principles and Progress*" *17* (Eds.: P. Diehl,E. Fluck, R. Kosfeld) Springer Verlag, Berlin **1981**.

[3] Molecular volumes have been calculated using the program version: P. Baricic, M. Mackov, J. E. Slone, *MOLGEN version 3.01*, **1993**.

[4] *Komplexes Lehrwerk Chemie LB3*, VEB Deutscher Verlag für Grundstoffindustrie, Leipzig, **1971**.

Ferrocene as a Donor in Dipolar Oligosilane Structures

*Christa Grogger, Harald Siegl, Hermann Rautz, Harald Stüger**

Institut für Anorganische Chemie, Technische Universität Graz
Stremayrgasse 16, A-8010 Graz, Austria
Tel.: Int. code + (316)873 8708 — Fax.: Int. code + (316)873 8701
E-mail: Stueger@anorg.tu-graz.ac.at

Keywords: Ferrocenylsilanes / UV Spectroscopy / Cyclic Voltammetry / Charge Transfer

Summary: The highly dipolar α,ω-disubstituted permethyloligosilanes 1-ferrocenyl-2-(2,2-dicyanoethenyl)phenyltetramethyldisilane (**1**) and 1-ferrocenyl-2-(2,2-dicyano-ethenyl)phenyldodecamethylhexasilane (**2**) were prepared in order to investigate intramolecular charge transfer phenomena. X-ray crystallography of **1** reveals the centrosymmetric point group $P2_1/c$ and a *gauche*-type arrangement of the central C–Si–Si–C bonds with a dihedral angle of 70.32°. The three waves appearing in the cyclic voltammograms of **1** and **2** are easily assigned to reduction of the $-CH{=}C(CN)_2$ acceptor and to oxidation of the ferrocenyl donor and the oligosilane spacer, respectively. UV/Vis absorption spectroscopy rules out any significant direct electron transfer from the donor to the acceptor group in **1** and **2**.

Introduction

Dipolar structures with quadratic NLO properties typically employ organic donor and acceptor groups to polarize unsaturated spacers. Recently there has been increasing interest in dipolar molecules where the donor and acceptor subunits are connected by oligosilanyl bridges [1–5] due to the outstanding electronic properties of Si–Si bonds, e.g. the facile delocalization of σ-electrons. Silicon, however, has been found to be only a weak charge transmitter in donor- and acceptor-substituted phenyl–$(Si)_n$–phenyl moieties, mainly due to the non-planar geometry of the phenyl–Si–phenyl system. The aim of the current project is to investigate whether or not direct charge transfer via Si–Si bonds can be achieved using organometallic donors containing transition elements and by optimizing the geometry of the oligosilanyl bridge. Because of its great stability and suitable redox behavior we decided to use ferrocene as the donor and to study its donor capacity when linked to properly substituted silicon substrates.

Synthesis

The reaction sequence depicted in Scheme 1 has been used to prepare the highly polar oligosilanes 1-ferrocenyl-2-(2,2-dicyanoethenyl)phenyltetramethyldisilane (**1**) and 1-ferrocenyl-2-(2,2-dicyano-ethenyl)phenyldodecamethylhexasilane (**2**) containing ferrocene as the donor and the (2,2-dicyano-ethenyl)phenyl group as the acceptor.

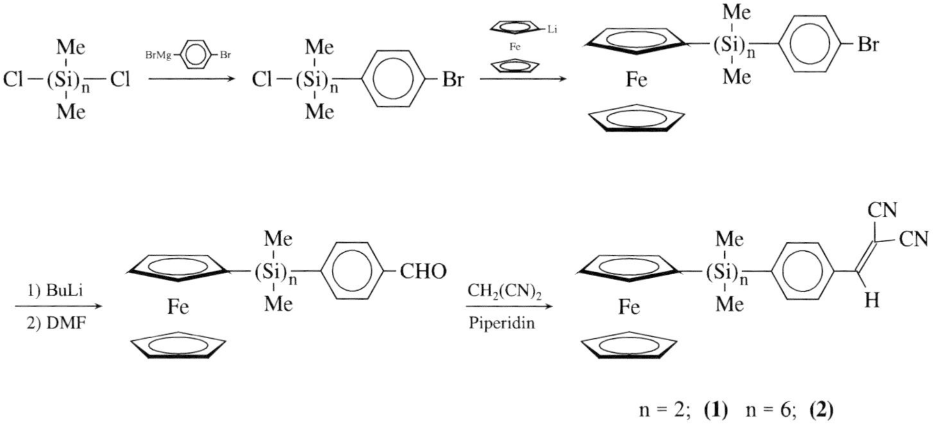

$Cl-(SiMe_2)_n-Cl$ → ($BrMgC_6H_4Br$) → $Cl-(SiMe_2)_n-C_6H_4-Br$ → (lithioferrocene) → $Fc-(SiMe_2)_n-C_6H_4-Br$

→ 1) BuLi, 2) DMF → $Fc-(SiMe_2)_n-C_6H_4-CHO$ → ($CH_2(CN)_2$, Piperidin) → $Fc-(SiMe_2)_n-C_6H_4-CH=C(CN)_2$

n = 2; **(1)** n = 6; **(2)**

Scheme 1. Synthesis of dipolar ferrocenyloligosilanes.

α,ω-Dichloropermethyloligosilanes cleanly react with 4-bromobenzene magnesium bromide to give the corresponding α-chloro-ω-(4-bromobenzene) derivatives. Subsequent reaction with 1-lithio-ferrocene allows nucleophilic displacement of the chlorine substituent by the ferrocenyl donor moiety under mild conditions without affecting the residual molecule. The (2,2-dicyano-ethenyl)phenyl acceptor, finally, can be easily introduced with BuLi/DMF followed by condensation of the resulting aldehyde with malonodinitrile using a method already published by Mignani et al. for the synthesis of similar donor/acceptor-substituted disilanes [2].

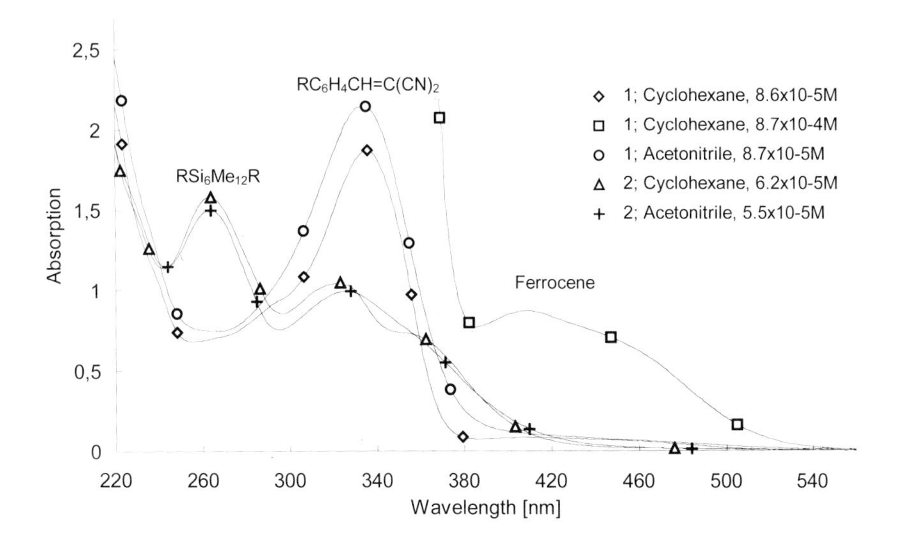

Fig. 1. UV-absorption spectra of **1** and **2** in solvents of different polarity.

UV/Vis Absorption Spectra

The UV/Vis absorption spectra of **1** and **2** depicted in Fig. 1 and Table 1 rule out any significant direct charge transfer from the donor to the acceptor group due to the following features:

- The solvatochromic shifts are small when the solvent is changed from apolar (cyclohexane) to polar (acetonitrile).
- According to literature data the position and the intensities of the absorption bands do not depend on the type of the donor. The intense absorption band at 326 nm found for 1-[4-(dimethylamino)phenyl]-2-[4-(2,2-dicyanoethenyl)phenyl]tetramethyldisilane [2], for instance, also appears in the spectrum of **1** with nearly the same λ_{max} and ε values.
- Increasing the chain length from 2 to 6 silicon atoms does not result in a marked red shift of the absorption bands, which means, that there is no increase of conjugation length on going from **1** to **2**.

The UV absorption bands can be assigned to local transitions within the Si–Si skeleton (compare entries 1 and 2 in Table 1), the (2,2-dicyanoethenyl)phenyl group (compare entry 3 in Table 1) and ferrocene (compare entry 4 in Table 1).

Table 1. UV/Vis-absorption data and assignment of **1** and **2** together with reference substances.

Entry	Compound	Solvent	λ_{max} [nm]	ε	Assignment	Ref.
1	$Me(SiMe_2)_6Me$	C_6H_{12}	220 260	14 000 21 000	σ–σ*(Si–Si)	[6]
2	$Ph(SiMe_2)_6Ph$	C_6H_{12}	265	30 500	σ–σ*(Si–Si)	[7]
3	Me_3Si–C_6H_4–$CH{=}C(CN)_2$	$CHCl_3$	322	27 950	$R–C_6H_4–CH{=}C(CN)_2$	[2]
4	Ferrocene	Ether	440 325	87 50		[8]
5	**1**	C_6H_{12}	435 sh 409 335	900 1 000 23 000	Ferrocene $R–C_6H_4–CH{=}C(CN)_2$	
	1	CH_3CN	410 sh 334	1 200 25 000	Ferrocene $R–C_6H_4–CH{=}C(CN)_2$	
6	**2**	C_6H_{12}	360 sh 325 264	10 000 16 200 21 000	$R–C_6H_4–CH{=}C(CN)_2$ σ–σ*(Si–Si)	
	2	CH_3CN	350 sh 324 263	14 400 18 000 27 000	$R–C_6H_4–CH{=}C(CN)_2$ σ–σ*(Si–Si)	

X-Ray Crystallography

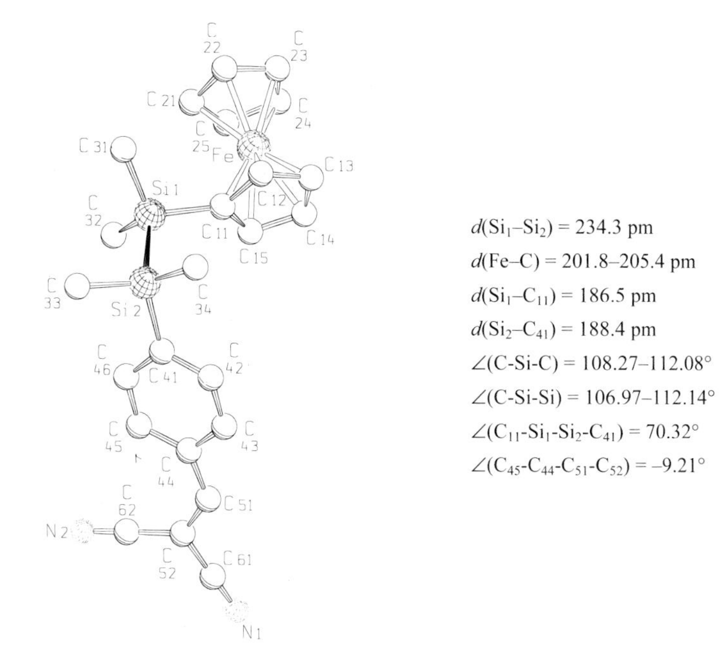

$d(Si_1–Si_2) = 234.3$ pm
$d(Fe–C) = 201.8–205.4$ pm
$d(Si_1–C_{11}) = 186.5$ pm
$d(Si_2–C_{41}) = 188.4$ pm
$\angle(C\text{-}Si\text{-}C) = 108.27–112.08°$
$\angle(C\text{-}Si\text{-}Si) = 106.97–112.14°$
$\angle(C_{11}\text{-}Si_1\text{-}Si_2\text{-}C_{41}) = 70.32°$
$\angle(C_{45}\text{-}C_{44}\text{-}C_{51}\text{-}C_{52}) = -9.21°$

Fig. 2. Molecular structure of **1**.

1 crystallizes in the centrosymmetric point group $P2_1/c$, where the unit cell contains two crystallographically independent molecules and their inverted counterparts. Most of the bond lengths and bond angles are quite unexceptional. Thus the bond length Si(1)–Si(2) = 234.3 pm is typical for disilanes bearing small substituents and agrees well with values found for other donor/acceptor-substituted disilanes [2, 3]. The geometry around the silicon atoms is approximately tetrahedral. The dihedral angle between C(52) and the phenyl ring C(41)–C(46) is only about 9°. However, it is quite noteworthy, that in contrast to 1,2-diphenyldisilanes, which usually adopt dihedral angles close to 180° [3], a *gauche*-type arrangement of the central C–Si–Si–C bonds with a dihedral angle of 70.32° is observed. Due to the resulting non-coplanarity of the donor/acceptor aromatic rings the possibilities for a direct charge transfer via the Si–Si system in **1**, therefore, will be greatly reduced, at least in the crystalline state.

Cyclic Voltammetry

Besides the reversible waves of the ferrocene donor near 0 V two irreversible waves are shown by the cyclic voltammograms of **1** and **2**. They are easily assigned to reduction of the $-CH=C(CN)_2$ acceptor and to oxidation of the oligosilane spacer by comparison with proper reference substances (see Table 2).

Table 2. Electrochemical data for **1** and **2** in CH_2Cl_2 together with references (Potentials [V] are vs. ferrocene).

	E_{pc} –CH=C(CN)$_2$	E_f[a] Ferrocene	E_{pa} –(SiMe$_2$)$_n$–
$H_2C{=}C(CN)_2$	−1.72	—	—
1	−1.67	0	1.11
Cl–Si_2Me_4–Cl	—	—	0.71
2	−1.63	0	0.69/0.95
Cl–Si_6Me_{12}–Cl	—	—	0.77/1.01

[a] $E_f = \frac{E_{pa} - E_{pc}}{2} + E_{pc}$; $\Delta E = 90$ mV.

Acknowledgement: We wish to thank the *Fonds zur Förderung der Wissenschaftlichen Forschung* (Wien, Austria), for financial support within the Spezialforschungsbereich "Elektroaktive Stoffe"and Wacker Chemie GmbH (Burghausen, Germany) for the donation of silane precursors.

References:

[1] G. Mignani, A. Krämer, G. Puccetti, I. Ledoux, G. Soula, J. Zyss, R. Meyrueix, *Organometallics* **1990**, *9*, 2640.

[2] G. Mignani, M. Barzoukas, J. Zyss, G. Soula, F. Balegroune, D. Grandjean, D. Josse, *Organometallics* **1991**, *10*, 3660.

[3] D. Hissink, P. F. van Hutten, G. Hadziioannou, *J. Organomet. Chem.* **1993**, *454*, 25.

[4] D. Hissink, J. Brouwer, R. Flipse, G. Hadziioannou, *Polymer Prepr.* **1991**, *32*, 136.

[5] D. Hissink, H. J. Bolink, J. W. Eshuis, G. G. Malliaras, G. Hadziioannou, *Polymer Prepr.* **1993**, *34*, 721.

[6] H. Gilman, W. H. Atwell, *J. Organomet. Chem.* **1964**, *2*, 369.

[7] H. Gilman, W. H. Atwell, G. L. Schwebke, *Chem. Ind.*, (London) **1964,** 1063.

[8] K. Schlögel, *Monatsh. Chem.* **1957**, *88*, 601.

Siloxene-Like Polymers: Networks

Alois Kleewein, Harald Stüger*

Institut für Anorganische Chemie, Technische Universität Graz
Stremayrgasse 16, A-8010 Graz, Austria
Tel.: Int. code + (316)873 8708 — Fax.: Int. code + (316)873 8701
E-mail: Stueger@anorg.tu-graz.ac.at

Stefan Tasch, Günther Leising

Institut für Festkörperphysik, Technische Universität Graz
Petersgasse 16/I, A-8010 Graz, Austria
Tel: Int. code + (316)873 8470 — Fax: Int. code + (316)873 8478
e-mail: f513tash@mbox.tu-graz.ac.at

Keywords: Siloxene / Polysiloxanes / Cyclosilanes / Photoluminescence / Fluorescence / Networks

Summary: In continuation of our previous contributions concerning the fluorescence behavior of polysiloxanes containing oligosilanyl substructures, the photoluminescence of hydrolysis and condensation products derived from di- and trifunctional cyclooligosilanes has been investigated. To our present knowledge there are two conditions to be met for our siloxene-like polymers in order to exhibit fluorescence: the presence of cyclosilanyl subunits and a two- or three-dimensional polymeric structure.

Introduction

Siloxene $(Si_6H_6O_3)_n$, a fascinating compound ever since its discovery by Wöhler [1], is prepared by reacting $CaSi_2$ with aqueous HCl. As the photoluminescence shown by porous silicon is attributed to the presence of siloxene layers, numerous investigations have been performed both on Wöhler (*WS*) and Kautsky siloxene (*KS*). Nevertheless, the structure of siloxene has not yet been cleared up and therefore structure models including planes, chains or cycles are discussed [2].

In order to contribute to the structure elucidation problem we started to build up siloxene-like polymers from exactly defined subunits using cyclic and linear silanes. Oligosilanes bearing methyl groups and/or functionalities like chlorine, bromine, iodine or the triflic group (trifluoro-methanesulfonyloxy = –OTf) turned out to be suitable starting materials. These functionalities are known to react readily with water in a first step to form Si–OH groups that undergo condensation reactions yielding Si–O–Si structures. As depicted in Table 1 only polymers derived from cyclic oligosilanes (entries 5–12 in Table 1) exhibit fluorescence, whereas no fluorescence was observed for linear siloxane systems (entries 1–4 in Table 1) [3].

Table 1. Fluorescence behavior of hydrolysis and condensation products of multifunctional oligosilanes (taken from reference [3]).

	Starting material	Fluorescence maximum of polymers obtained after condensation [nm]
1	$Si_2Me_4Cl_2$	None
2	$Si_3Me_5Cl_3$	None
3	$Si_4Me_6Cl_4$	None
4	$Si_4Me_4Cl_6$	None
5	*cyclo*-$Si_4Ph_4(OTf)_4$	553
6	*cyclo*-$Si_5Ph_5(OTf)_5$	522
7	*trans*-1,3-H_2-*cyclo*-Si_5Cl_8	486
8	*cyclo*-Si_5Cl_{10}	400
9	*cyclo*-Si_6Cl_{12}	432
10	*cyclo*-$Si_5Ph_5I_5$	540
11	*cyclo*-$Si_5Me_5Cl_5$	505
12	*cyclo*-$Si_6Me_6Cl_6$	436

In order to allow further conclusions concerning the origin of the photoluminescence (*PL*) in our kind of polymers, we decided to study the fluorescence behavior of hydrolysis and condensation products derived from di- and trifunctional cyclosilanes very likely consisting of silicon rings linked by oxygen to form either linear chains or two- or three-dimensional networks. As suitable starting materials for this approach the methylated chlorocyclosilanes chloroundecamethylcyclohexasilane (**1**), 1,3-dichlorodecamethylcyclohexasilane (**2**), 1,4-dichlorodecamethylcyclohexasilane (**3**) and 1,3,5-trichlorononamethylcyclohexasilane (**4**) were chosen because they are easily accessible from dodecamethylcyclohexasilane. Additionally, the *PL* of the suitably substituted cyclosilanyl monomers bis(undecamethylcyclohexasilanyl) ether (**5**), 1,3-dihydroxydecamethylcyclohexasilane (**6**) and decamethyl-7-oxahexasilanorborane (**7**) has also been investigated.

Synthesis of Cyclosilane Starting Materials

1 was prepared by demethylation of Si_6Me_{12} using $SbCl_5$ in CCl_4 following well-known literature procedures [4]. Simple hydrolysis and condensation of **1**, however, does not afford **5** in good yields. **5**, therefore, was made by reacting **1** with undecamethylcyclohexasilanol, which has to be prepared beforehand by hydrolyzing **1** in a separate step (Scheme 1) [5].

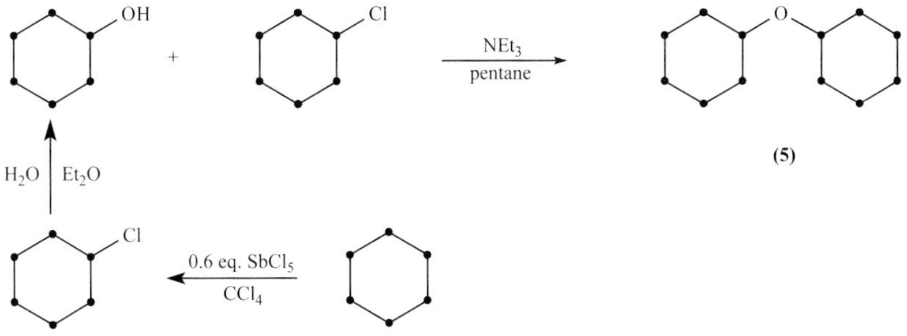

OH + Cl $\xrightarrow[\text{pentane}]{NEt_3}$ (5)

H_2O | Et_2O

Cl (1) $\xleftarrow[CCl_4]{0.6\text{ eq. }SbCl_5}$

Scheme 1. Synthesis of **5**. Throughout this article the common notation will be used in which a dot represents a silicon atom with methyl groups attached, sufficient to bring the total coordination number to four.

6 and **7** are accessible by the buffered hydrolysis of the isomeric mixture of **2** and **3** obtained from Si_6Me_{12} and 1.2 equivalents of $SbCl_5$. Subsequent vacuum distillation affords **6** and **7** as the only volatile products, because the initially formed 1,4-dihydroxydecamethylcyclohexasilane is completely converted to **7** upon heating [6] (compare Scheme 2).

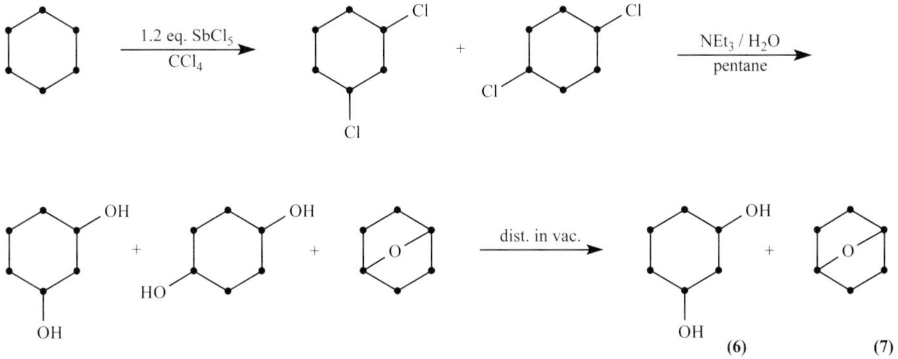

$\xrightarrow[CCl_4]{1.2\text{ eq. }SbCl_5}$ Cl Cl + Cl Cl $\xrightarrow[\text{pentane}]{NEt_3 / H_2O}$

OH OH + HO OH + O $\xrightarrow{\text{dist. in vac.}}$ OH OH (6) + O (7)

Scheme 2. Synthesis of **6** and **7**.

Both **6** and **7** can be rechlorinated using acetyl chloride or acetyl chloride/water, respectively, to give pure **2** and **3** as a mixture of the *cis*- and *trans*-isomers (Scheme 3).

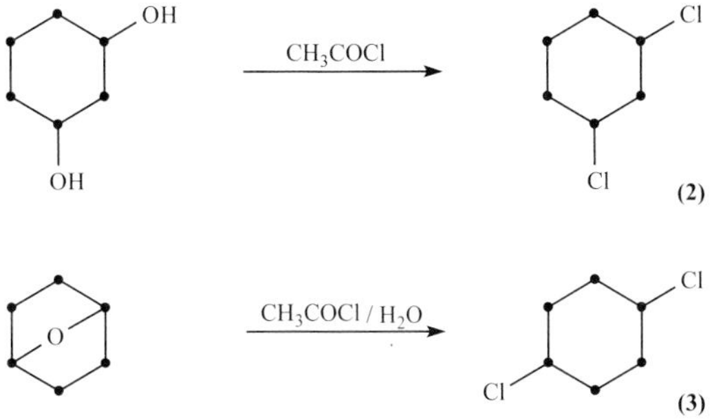

OH OH $\xrightarrow{CH_3COCl}$ Cl Cl (2)

O $\xrightarrow{CH_3COCl / H_2O}$ Cl Cl (3)

Scheme 3. Synthesis of **2** and **3**.

The trichloro compound **4** was obtained by demethylation of **2** with $SbCl_5$ as a mixture of the *trans–trans* and the *trans–cis* isomers (Scheme 4).

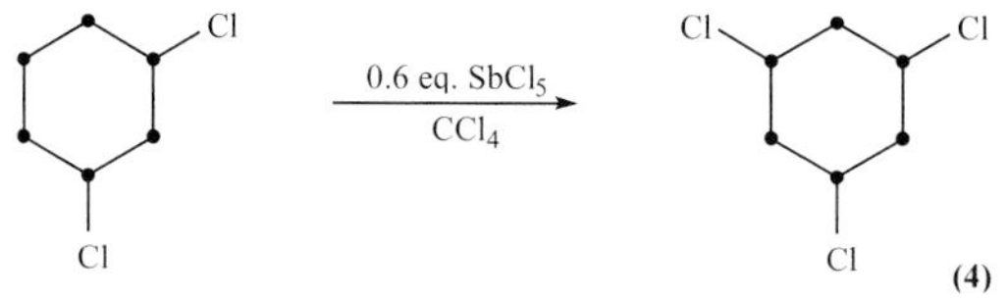

Cl Cl Cl Cl 0.6 eq. $SbCl_5$ CCl_4 Cl Cl (4)

Scheme 4. Synthesis of **4**.

Hydrolysis and Condensation

Hydrolysis of the halides **2**, **3** and **4** leads to the hydroxy compounds in a first step. As no triethylamine is used as in the synthesis of **6** and **7** (compare Scheme 2), the released hydrochloric acid acts as a catalyst for the subsequent condensation reaction. This enforces the formation of the polysiloxanes **8**, **9** and **10** (Scheme 5).

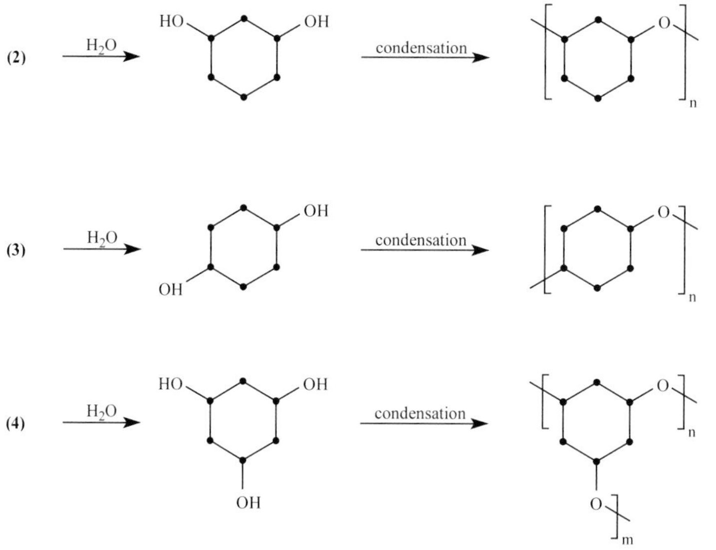

(2) H_2O HO OH condensation O n (8)

(3) H_2O OH OH condensation O n (9)

(4) H_2O HO OH OH condensation O n O m (10)

Scheme 5. Proposed structures of polymers obtained after hydrolysis and condensation of **2**, **3** and **4**.

The hydrolysis of compounds **2**, **3** and **4** was performed by the dropwise addition of water/THF (1:1) in excess to a solution of the halides in THF. After stirring for two hours the solvent was evaporated, the aqueous phase discarded and the white precipitates extracted with THF to remove low molecular fractions. Condensation was completed by heating the solids to 150°C in vacuum for 5 hours. A sample of the polymers was suspended in THF and the liquid phase was injected into a GC/MS/machine running a high-temperature program to prove the absence of monomers/oligomers. Additionally the polymers **8**, **9** and **10** were characterized by IR spectroscopy showing the absorptions listed in Table 2 in accordance with the proposed structures.

Table 2. Selected IR-absorption bands (Nujol mull) of polymers **8**, **9** and **10** [cm^{-1}].

Approx. assignment	Polymer 8	Polymer 9	Polymer 10
ν(Si–O–H)	3750 (sh)	3760 (w)	Missing
ν(Si–O–H)	3700 (sh)	3710 (w)	Missing
ν(Si–O–H)	3409 (b)	3429 (b)	3472 (b)
ν_{as}(Si–O–Si)	1034 (b)	1032 (b)	1034 (b)
ν(Si–Si)	402 (b)	409 (b)	416 (b)

Fluorescence Spectra and Conclusions

The fluorescence spectra were recorded as powder coatings on paper. Neither the monomeric structures **5**, **6** and **7**, nor the polymers **8** and **9**, show any noticeable *PL*. Just polymer **10** exhibits a broad, but very weak, *PL* centered around 420 nm. The fact that **5**, **6** and **7** do not exhibit any *PL* confirms our previous results in attributing the fluorescence of our materials to both: (1) their polymeric structure and (2) the presence of silicon rings. The absence of *PL* for the polymers **8** and **9** very likely showing linear structures (compare Scheme 5) leads us to consider both of the above conditions (1) and (2) as necessary but not sufficient. Additionally, the existence of a higher-dimensional network as present in polymer **10** seems to be necessary to bring about fluorescence in the type of compounds investigated in this study.

To determine whether the presence of the $-SiMe_2-$ units in **10** causes the dramatic decrease of the *PL* compared to our previous results, or wether it is just that the number of functionalities has to be increased, in order to achieve higher *PL* efficiencies will be the aim of further studies.

Acknowledgment: We wish to thank the *Fonds zur Förderung der wissenschaftlichen Forschung* (Wien, Austria) for financial support within the Forschungsschwerpunkt "Novel Approaches to the Formation and Reactivity of Compounds containing Silicon-Silicon Bonds" and *Wacker Chemie GmbH* (Burghausen, Germany) for the donation of silane precursors.

References:

[1] F. Wöhler, *Lieb. Ann.* **1863**, *127*, 257.

[2] P. Deak, M. Rosenbauer, M. Stutzmann, J. Weber, M. S. Brandt, *Phys. Rev. Lett.* **1992**, *69*, 2531.

[3] a.) E. Hengge, A. Kleewein, „*New Results in the Chemistry of Cyclosilanes in View of the Structure of Kautsky Siloxene*" in: *Tailor-Made Silicon Oxygen Compounds: From Molecules to Materials* (Eds.: R. Corriu, P. Jutzi), Vieweg Verlag, Braunschweig, **1996**, pp. 89–98.
b.) A. Kleewein, U. Pätzold, E. Hengge, S. Tasch, G. Leising, "*New Results in Cyclosilane Chemistry: Siloxen-Like Polymers*" in: *Organosilicon Chemistry III: From Molecules to Materials* (Eds.: N. Auner, J. Weis), Wiley–VCH, Weinheim, **1998**, pp. 327-332.

[4] F. Mitter, E. Hengge, *J. Organomet. Chem.* **1991**, *332*, 46.

[5] P. Gspaltl, *Ph.D. Thesis*, Technische Universität Graz, **1995**.

[6] A. Spielberger, P. Gspaltl, H. Siegl, E. Hengge, K. Gruber, *J. Organomet. Chem.* **1995**, *499*, 241.

Silicon-Containing Spacers for the Synthesis of Tin-Containing Multidentate Lewis Acids

Reiner Altmann, Olivier Gausset, Reinhard Hummeltenberg, Klaus Jurkschat*, Silke Kühn, Markus Schürmann, Bernhard Zobel

Fachbereich Chemie, Lehrstuhl für Anorganische Chemie II
Universität Dortmund
D-44221 Dortmund, Germany
Tel.: Int. code + (231)755 3800 — Fax: Int. code + (231)755 3797
E-mail: kjur@platon.chemie.uni-dortmund.de

Keywords: Silicon-Containing Rings / Tin-Containing Rings / Ferrocenophane / Lewis Acids

Summary: Silicon- and tin-containing rings hold potential as reagents for selective anion recognition. In this paper the syntheses and structures of the new rings **1–4**, **8–15**, and of the cyclophane **16** are presented. In these compounds the tin atoms are bridged via silicon-containing spacers. Also reported are the syntheses and structures of the novel tin- and silicon-containing ferrocenophanes **19–22**.

Introduction

The selective recognition of anions or neutral donor molecules by tailor-made host molecules has become a topic of increasing interest over the last two decades and has been reviewed recently [1]. Among these studies multidentate Lewis acids containing elements such as boron [2a], indium [2b], silicon [2c], germanium [2d], tin [2e] and mercury [2f] as well as organometallic metallocene receptor systems [3] were shown to be efficient in coordinating anions and neutral Lewis bases. One conclusion from these investigations is that the selectivity strongly depends on the preorganization of the host molecule, i.e., the more rigid the host the better is the selectivity to be expected.

In search for systems which could meet these requirements we investigate the potential of silicon-containing spacers for the syntheses of new tin-containing multidentate Lewis acidic macrocycles and report here preliminary results of our studies.

Syntheses, Results and Discussion

The reaction of bis(chloromethyl)dimethylsilane with bis[(dimethylsodiumstannyl)methyl]dimethylsilane afforded a mixture of 1,1,3,3,5,5,7,7-octamethyl-1,5-distanna-3,7-disila-cyclooctane (**1**), its dimer 1,1,3,3,5,5,7,7,9,9,11,11,13,13,15,15-hexadecamethyl-1,5,9,13-tetrastanna-3,7,11,15-tetrasilacyclohexadecane (**2**) and unidentified polymeric compounds (Eq. 1). The ring system **1** could be isolated by distillation and the macrocycle **2** by size exclusion chromatography [4].

Eq. 1. Synthesis of **1** and **2**.

The tin-bonded methyl groups of **1** can be cleaved off by reaction with $HgCl_2$, providing the corresponding multidentate Lewis acids **3** and **4**, respectively. The molecular structures of 1,5-dichloro-1,3,3,5,7,7-hexamethyl-1,3-distanna-3,7-disilacyclooctane (**3**) and 1,1,5,5-tetrachloro-3,3,7,7-tetramethyl-1,5-distanna-3,7-disilacyclooctane (**4**) show Sn(1)···Sn(1a) distances of 4.5699(7) Å for **3** and 4.413(2) Å for **4** (Fig. 1).

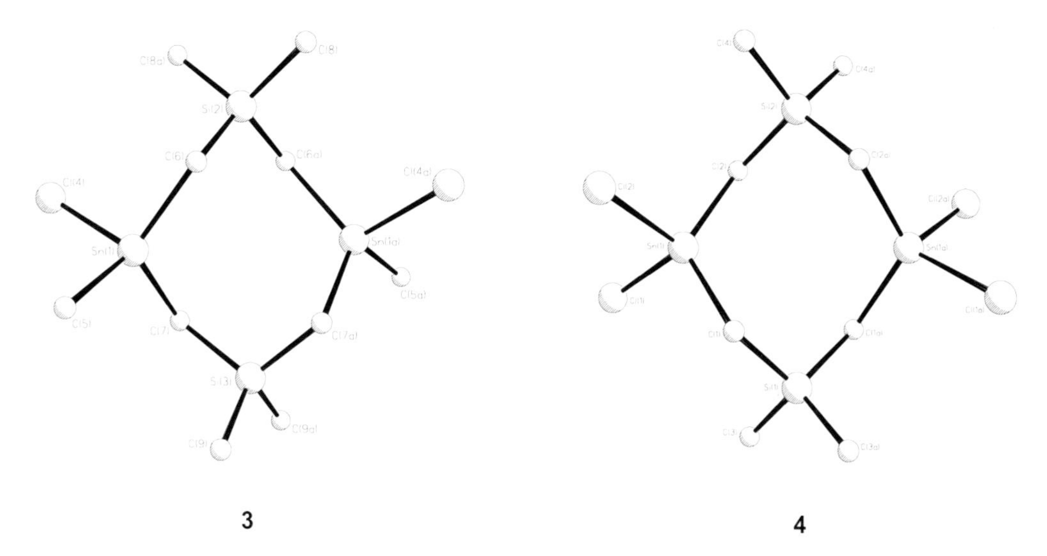

Fig. 1. Molecular structures of **3** and **4**.

Treatment of the bis(3-chloropropyl)silanes **5**, **6**, and **7** with 1,3-bis(dimethylsodiumstannyl) propane gave the 1,5-distanna-9-silacyclododecanes **8**, **9**, and **10**, respectively, as colorless oils (Scheme 1) [5]. The tin-bonded methyl groups of **8** and **10** can be cleaved off by reaction with $HgCl_2$ providing the organotin chlorides **11–14** (Scheme 1). The reaction of the 1,1,5,5,9-pentamethyl-1,5-distanna-9-silacyclododecane **9** with PCl_5 afforded the chlorosilane **15**.

Me$_2$Sn(Na)CH$_2$CH$_2$CH$_2$Sn(Na)Me$_2$ + (ClCH$_2$CH$_2$CH$_2$)$_2$Si(X)Me $\xrightarrow[-2\ NaCl]{NH_3,\ -70^\circ C}$

5, X = F
6, X = H
7, X = O^{i}Pr

8, X = F
9, X = H
10, X = O^{i}Pr

X = H, + PCl$_5$, – PCl$_3$, – HCl → **15**

+ *n* HgCl$_2$, – *n* MeHgCl; X = F, O^{i}Pr (*n* = 2, 4)

11, R = Me; X = F (*n* = 2)
12, R = Cl; X = F (*n* = 4)
13, R = Me, X = O^{i}Pr (*n* = 2)
14, R = Cl; X = O^{i}Pr (*n* = 4)

Scheme 1. Syntheses of 1,5-Distanna-9-silacyclododecanes **8**–**15**.

The reaction of (p-BrMgC$_6$H$_4$)$_2$SiMe$_2$ with Ph$_2$SnCl$_2$ afforded 1,1,15,15,29,29-hexaphenyl-8,8,22,22,36,36-hexamethyl-1,15,29-tristanna-8,22,36-trisila[1,1,1,1,1,1]paracyclophane (**16**) as a colorless solid (Eq. 2) [6].

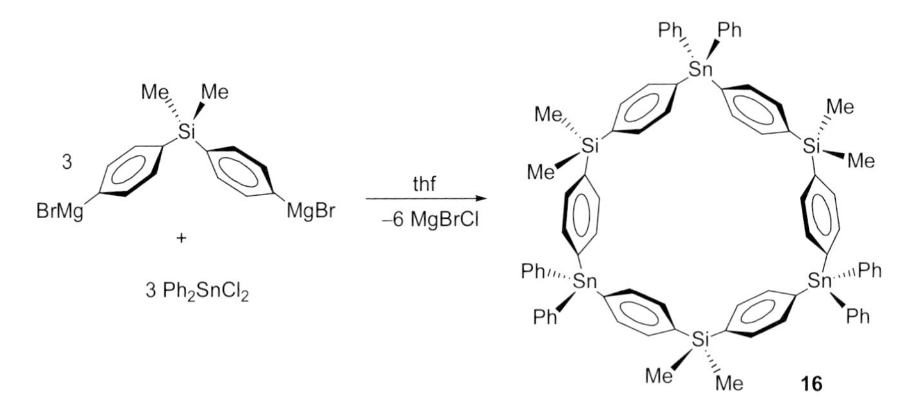

Eq. 2. Synthesis of **16**.

The cyclic structure of **16** is supported by osmometric molecular weight determination (found: 1466 g/mol) and by NMR studies. Attempts failed to cleave the exocyclic phenyl groups of **16** selectively by its reaction with Ph$_2$SnCl$_2$ or I$_2$, respectively. Further studies on this subject are envisaged.

The reaction of the Grignard reagent of 1,1'-bis[(chloromethyl)dimethylsilyl]ferrocene with the 1,1'-bis[((chlorodiorganostannyl)methyl)dimethylsilyl]ferrocenes **17** or **18** gave the ferrocenophanes **19** and **20**, respectively (Scheme 2) [7, 8]. To the best of our knowledge these compounds are the first silicon-and tin-containing ferrocenophanes. The dimers of **20** and **21** were also isolated by use of size exclusion chromatography. Their identity was confirmed unambiguously by molecular weight determination and electrospray MS.

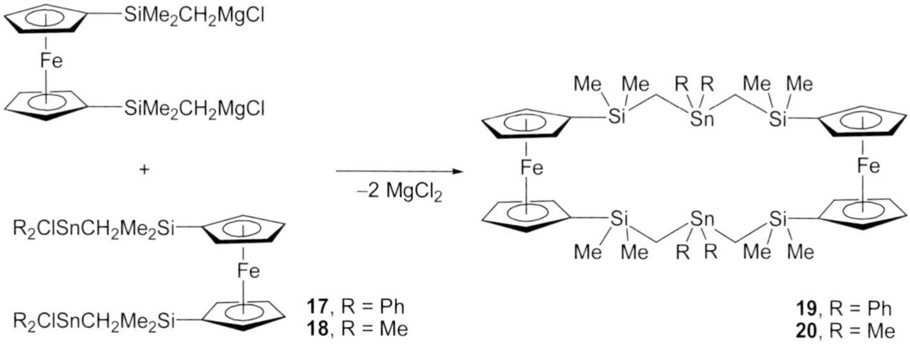

Scheme 2. Synthesis of the ferrocenophanes **19** and **20**.

Reaction of 1,1,16,16-diphenyl-3,3,14,14,18,18,29,29-octamethyl-1,16-distanna-3,14,18,29-tetrasila[5,5]ferrocenophane (**19**) with Iodine afforded the iodo-substituted derivative **21**. The molecular structures of **19** and **21** are shown in Fig. 2. Interestingly, the spacing between the tin atoms changes from Sn(1)···Sn(2) = 9.4061(5) Å in **19** to Sn(1)···Sn(2) = 6.1975(9) Å in **21**.

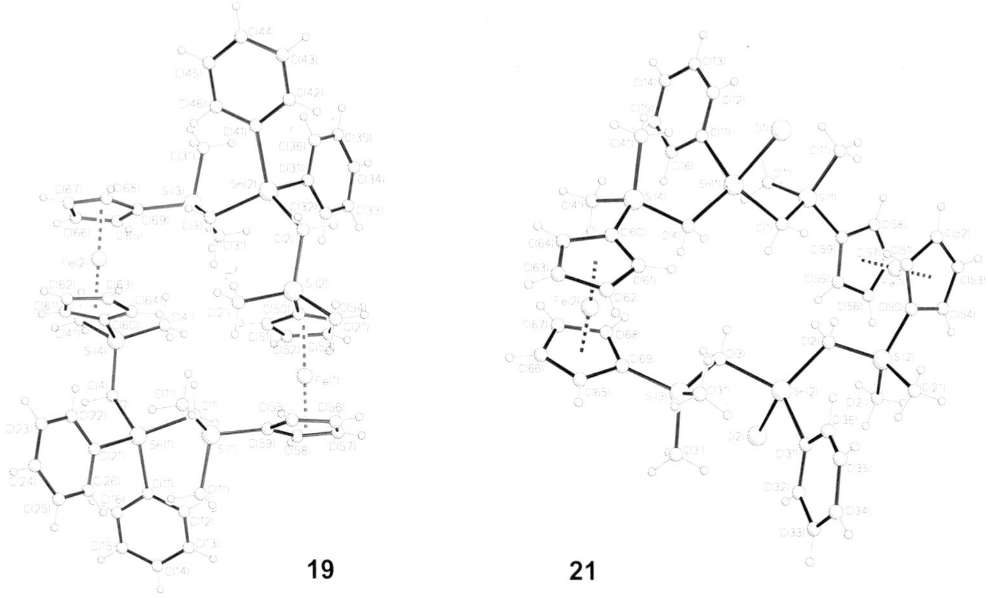

Fig. 2. Molecular structures of the ferrocenophanes **19** and **21**.

Treatment of the methyl-substituted ferrocenophane **20** with 2 mol equiv. of Me_2SnCl_2 provided 1,3,3,14,14,16,18,18,29,29-decamethyl-1,16-distanna-3,14,18,29-tetrasila[5,5]ferrocenophane (**22**) and Me_3SnCl. In contrast to the phenyl-substituted compounds **19** and **21** (Fig. 2) there is almost no change in spacing between the methyl-substituted derivatives **20** (Sn(1)···Sn(1a) = 9.102(1) Å) and **22** (Sn(1)···Sn(1a) = 8.829(3) Å) (Fig. 3).

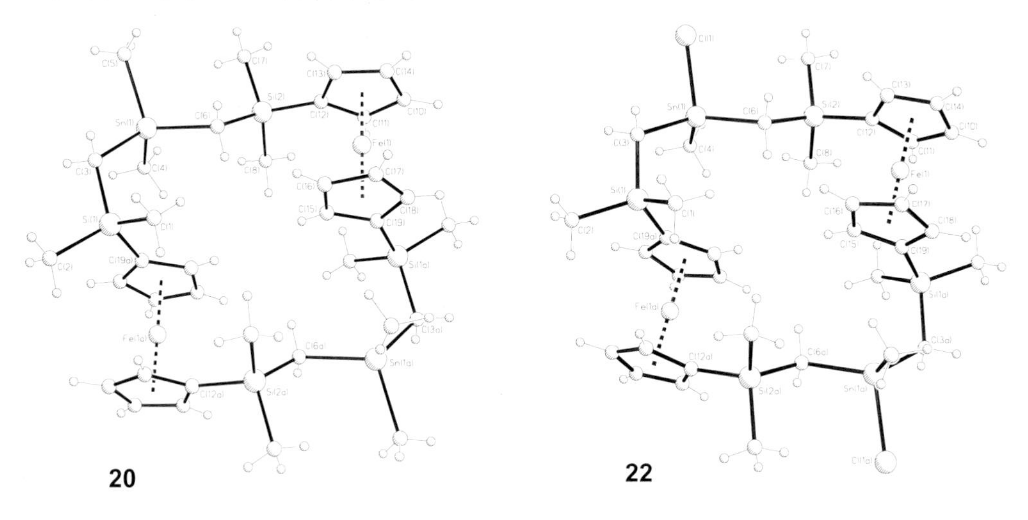

Fig. 3. Molecular structures of the ferrocenophanes **20** and **22**.

Acknowledgment: We are grateful to the *Deutsche Forschungsgemeinschaft,* the *Fonds der Chemischen Industrie*, and the *Human Capital and Mobility Programme of the European Community* for financial support.

References:

[1] a) F. P. Schmidtchen, M. Berger, *Chem. Rev.* **1997**, *97*, 1609.
b) B. Dietrich, *Pure Appl. Chem.* **1993**, *65*, 1457.

[2] a) H. E. Katz, *J. Org. Chem.* **1985**, *50*, 5027.
b) M. Tschinkl, A. Schier, J. Riede, F. P. Gabbai, *Inorg. Chem.* **1997**, *36*, 5706.
c) K. Tamao, T. Hayashi, Y. Ito, *J. Organomet. Chem.* **1996**, *506*, 85.
d) S. Aoyagi, K. Tanaka, Y. Takeuchi, *J. Chem. Soc., Perkin Trans. 2* **1994**, 1549.
e) R. Altmann, K. Jurkschat, M. Schürmann, D. Dakternieks, A. Duthie, *Organometallics* **1997**, *16*, 5716.
f) J. Vaugeois, M. Simard, J. D. Wuest, *Organometallics* **1998**, *17*, 1208.

[3] P. D. Beer, *Acc. Chem. Res.* **1998**, *31*, 71.

[4] S. Kühn, *Ph. D. Thesis*, Universität Dortmund, **1997**.

[5] R. Hummeltenberg, *Ph. D. Thesis*, Universität Dortmund, **1997**.

[6] B. Zobel, K. Jurkschat, *Main Group Met. Chem.* **1998**, *21*, 765.

[7] R. Altenmann, *Ph. D. Thesis*, Universität Dortmund, **1998**.

[8] O. Gausset, G. Delpon-Lacaze, M. Schürmann, K. Jurkschat, *Acta Crystallogr.* **1998**, *C54*, 1425.

Six-membered Stannyloligosilane Ring Systems — One Ring Size, Different Combination Pattern

*Uwe Hermann, Ingo Prass, Markus Schürmann, Frank Uhlig**

Anorganische Chemie II, Universität Dortmund
Otto-Hahn-Straße 6, D-44221 Dortmund, Germany
Tel.: Int. code + (231)755 3812
E-mail: fuhl@platon.chemie.uni-dortmund.de

Keywords: Stannylsilanes / Cyclic Stannyloligosilanes / Hydridostannylsilanes

Summary: Six-membered cyclic stannyloligosilanes **1**, **2** and **3** can be synthesized via the reaction of α,ω-difluoromethyloligosilanes with dichlorodiorganostannanes and magnesium. The availability of hydridostannylsilanes **11** and halogen derivatives **12**, **13** and **14** allows the synthesis of ring compounds with other combination patterns such as **4**, **5** and **6**. X-ray structures of two stannylsilanes and a hydridostannylsilane also shown.

Our studies are focused on synthesis and characterization of stannyloligosilanes [1–3]. Among these compounds ring systems find special interest because they are possible precursors for ring-opening polymerisation (ROP). Via this synthetic pathway it should be possible to produce polymeric stannylsilanes, a new and unexplored class of compounds.

Cyclic six-membered stannylsilanes containing tin atoms in 1- (**1**), 1,4- (**2**) or 1,2- positions (**3**) (Scheme 1) are easily available by the reaction of α,ω-difluorooligosilanes with diorgano-dichlorostannanes and magnesium.

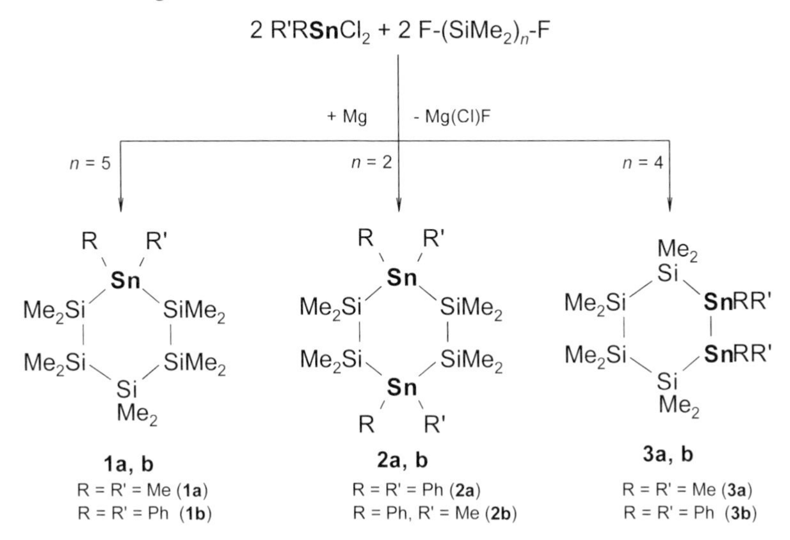

Scheme 1. Synthesis of cyclic stannyloligosilanes **1**, **2** and **3**.

The molecular structures of **1b** and **2b** (Figs. 1 and 2; Tables 1 and 2) show the expected chair conformation of the cyclohexane-type rings in the solid state.

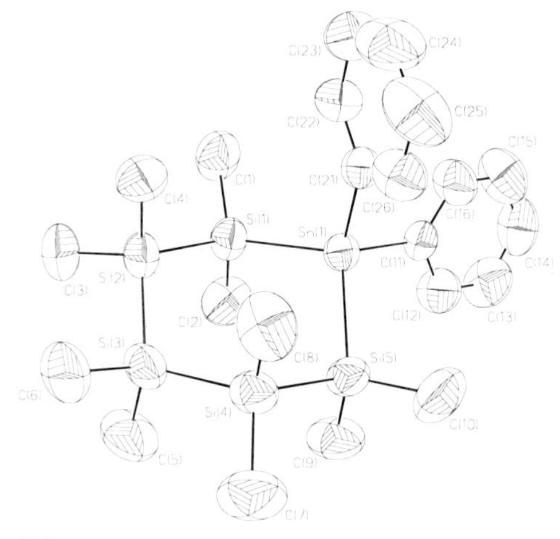

Fig. 1. Molecular structure of **1b**.

Table 1. Selected bond lengths [Å] and angles [°] of **1b**.

Bond lenghts		**Bond angles**	
Sn(1)–Si(1)	2.567(1)	C(21)-Sn(1)-C(11)	102.6(2)
Sn(1)–Si(5)	2.570(1)	Si(1)-Sn(1)-Si(5)	108.54(4)
Si(3)–Si(4)	2.334(2)		

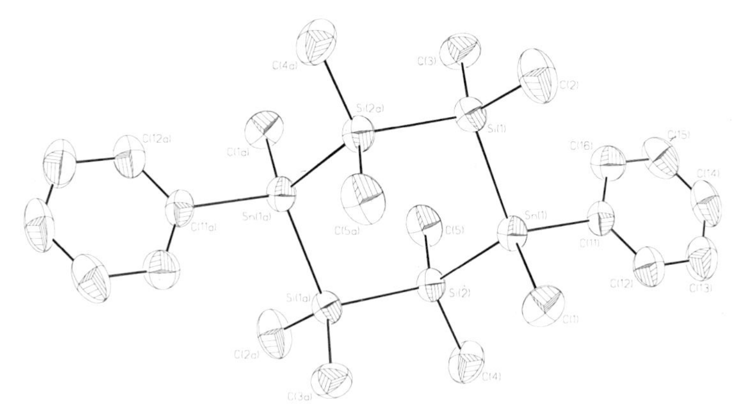

Fig. 2. Molecular structure of **2b**.

Table. 2 Selected bond lengths [Å] and angles [°] of **2b**.

Bond lenghts		**Bond angles**	
Sn(1)–Si(1)	2.567(1)	C(1)-Sn(1)-C(11)	105.3(1)
Sn(1)–Si(2)	2.570(1)	C(1)-Sn(1)-Si(1)	108.6(1)
Si(1)–Si(2)	2.330(1)	C(11)-Sn(1)-Si(1)	111.1(1)
		Si(1)-Sn(1)-Si(2)	113.93(3)

Missing compounds in the series of six-membered stannylsilanes containing tin atoms in the 1,2,3- (**4**), 1,2,4,5- (**5**) or in 1,3,5-positions (**6**) could not yet be obtained via the reaction type described. A retrosynthethic analysis (Scheme 2) reveals that the compounds should be obtainable by a Wurtz-type coupling reaction of halogen-substituted stannylsilane synthons.

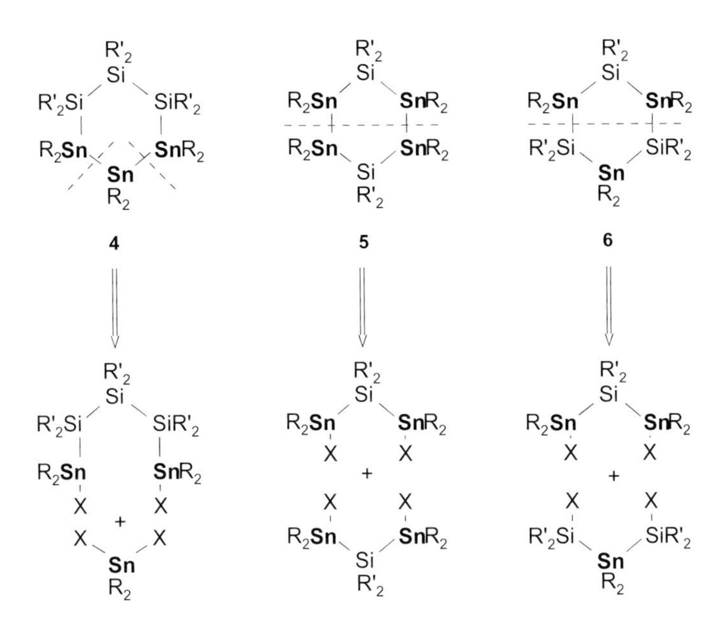

Scheme 2. Retrosynthetic analysis of cyclic stannyloligosilanes.

Attempts to substitute an organic group at the tin atom of the bis(triorganostannyl)di*methyl*silane **7** by a halogen atom always result in an Si–Sn bond cleavage. In contrast the bis(trimethylstannyl)silanes **8a,b** carrying sterically demanding substituents like isopropyl or *tert*-butyl at the silicon atom can be halogenated with $SnCl_4$ (Scheme 3).

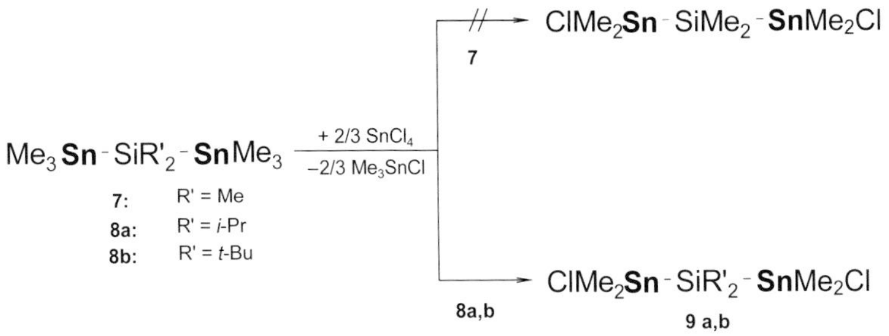

Me_3**Sn**-SiR'_2-**Sn**Me_3 —(+ 2/3 $SnCl_4$ / −2/3 Me_3SnCl)→

7: R' = Me
8a: R' = *i*-Pr
8b: R' = *t*-Bu

7 ⇸ $ClMe_2$**Sn**-$SiMe_2$-**Sn**Me_2Cl

8a,b → $ClMe_2$**Sn**-SiR'_2-**Sn**Me_2Cl **9 a,b**

Scheme 3. Functionalization of **7** and **8a,b**.

Alternatively tin and silicon atoms of stannylsilanes can be halogenated via hydride-substituted derivatives. Hydridosilylstannanes **10** are easily accessible by the reaction of diorganofluorosilanes with diorganodichlorostannanes and magnesium (Eq. 1).

$$R_2\mathbf{Sn}Cl_2 + 2\ R'_2Si(H)F \xrightarrow[-\ 2\ Mg(Cl)F]{+\ 2\ Mg} R'_2(H)Si-\mathbf{Sn}R_2-Si(H)R'_2$$

10

10a R = Me, R' = *i*-Pr
10b R = Ph, R' = *i*-Pr

Eq. 1 Synthesis of hydridosilylstannanes **10**.

Reaction of lithium hydridodi-*tert*-butylstannide with α,ω-dichlorosilanes leads to bis(hydrido-di-*tert*-butylstannyl)oligosilanes **11** (Eq. 2). Figure 3 shows the molecular structure of **11b**, the first of this new type of compounds.

$$Cl\text{-}(SiMe_2)_n\text{-}Cl + 2\ {}^tBu_2\mathbf{Sn}(H)Li \xrightarrow[-\ 2\ LiCl]{} {}^tBu_2(H)\mathbf{Sn}\text{-}(SiMe_2)_n\text{-}\mathbf{Sn}(H){}^tBu_2$$

11

11a, n = 1
11b, n = 2
11c, n = 3
11d, n = 4
11e, n = 6

Eq. 2. Synthesis of bis(hydrido-di-*tert*-butylstannyl)oligosilanes **11**.

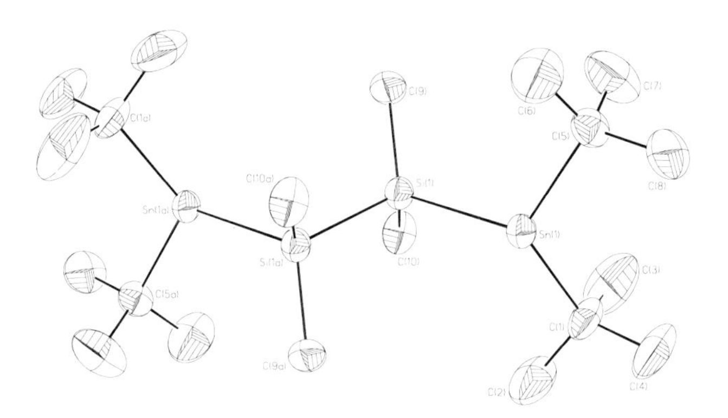

Fig. 3. Molecular structure of **11b**.

Table. 3. Selected bond lengths [Å] and angles [°] of **11b**.

Bond lenghts		Bond angles	
Sn(1)–C(5)	2.178(7)	C(5)-Sn(1)-C(1)	115.4(3)
Sn(1)–C(1)	2.188(7)	C(5)-Sn(1)-Si(1)	114.1(2)
Sn(1)–Si(1)	2.594(2)	Si(1a)-Si(1)-Sn(1)	105.2(1)
Si(1)–Si(1a)	2.338(3)		
Si(1)–C(9)	1.891(7)		

Bis(hydridodi-*tert*-butylstannyl)hexamethyltrisilane **11c** leads to a type **4** six-membered ring by reaction with bis(diethylamino)dimethylstannane (Eq. 3). Compound **4** is the first "missing link" in the series of six-membered stannylsilanes to be synthezised.

$$tBu_2(H)\mathbf{Sn}-SiMe_2-SiMe_2-SiMe_2-\mathbf{Sn}(H)tBu_2 + Et_2N-\mathbf{Sn}Me_2-NEt_2 \xrightarrow{-2\ HNEt_2} \mathbf{4}$$

4: six-membered ring $-SiMe_2-SiMe_2-SiMe_2-\mathbf{Sn}tBu_2-\mathbf{Sn}Me_2-\mathbf{Sn}tBu_2-$

Eq. 3.

Both types of hydridostannylsilane derivatives can be halogenated by the reaction with halocarbons (Eqs. 4 and 5).

$$R'_2(H)Si-\mathbf{Sn}R_2-Si(H)R'_2 \ (\mathbf{10}) \xrightarrow[-2\ CHCl_3]{+2\ CCl_4} R'_2(Cl)Si-\mathbf{Sn}R_2-Si(Cl)R'_2 \ (\mathbf{12})$$

10a, R = Me, R' = *i*-Pr
10b, R = Ph, R' = *i*-Pr

12a, R = Me, R' = *i*-Pr
12b, R = Ph, R' = *i*-Pr

Eq. 4.

$$^tBu_2(H)\mathbf{Sn}-(SiMe_2)_n-\mathbf{Sn}(H)^tBu_2 \ (\mathbf{11}) \xrightarrow[-2\ CH_2Cl_2]{+2\ CHCl_3} {}^tBu_2(X)\mathbf{Sn}-(SiMe_2)_n-\mathbf{Sn}(X)^tBu_2 \ (\mathbf{13, 14})$$

13a, n = 1, X = Cl
13b, n = 2, X = Cl
13c, n = 3, X = Cl
13d, n = 4, X = Cl
13e, n = 6, X = Cl

14a, n = 1, X = Br
14b, n = 6, X = Br

Eq. 5.

The availability of the chloro- and bromostannylsilanes **12**, **13** and **14** opens access to "missing link" compounds such as **5** and **6**. Further investigations into the preparation of derivatives **5** and **6** are in progress and will be discussed later.

Acknowledgment: The authors thank the *Deutschen Forschungsgemeinschaft* and the Federal State *Nordrhein-Westfalen* for financial support, and *ASV-Innovative Chemie GmbH* (Bitterfeld) and *Sivento Chemie GmbH* (Rheinfelden) for the generous gift of silanes. We are grateful to Prof. Dr. K. Jurkschat for support.

References:

[1] R. Hummeltenberg, K. Jurkschat, F. Uhlig, *Phosphorus, Sulphur Silicon* **1997**, *123*, 255–261.

[2] U. Hermann, I. Prass, F. Uhlig, *Phosphorus, Sulphur Silicon* **1997**, *124–125*, 425–429.

[3] C. Kayser, R. Klassen, M. Schürmann, F. Uhlig, *J. Organomet. Chem.* **1998**, *556*, 165–167.

Structure and Reactivity of Novel Stannasiloxane Complexes

Jens Beckmann, Klaus Jurkschat*, Markus Schürmann

Fachbereich Chemie, Lehrstuhl für Anorganische Chemie II
Universität Dortmund
D-44221 Dortmund, Germany
Tel.: Int. code + (231)755 3800 — Fax: Int. code + (231)755 3797
E-mail: kjur@platon.chemie.uni-dortmund.de

Keywords: Stannasiloxane / Complex / Equilibrium / X-Ray Analysis

Summary: Depending on the identity of X, stannasiloxane complexes of type [*cyclo*-$Ph_2Si(OSn$*t*$Bu_2)_2O\cdot$*t*Bu_2SnX_2] with X = OH, F exhibit different dynamic effects in solution.

Introduction

Increasing attention has been paid to organotin fragments grafted on silica surfaces, due to the significant repercussion that the cooperative effect of both tin centers and siloxane surroundings can have on their reactivity and catalytic applications [1]. Regarding the difficulties concerning the characterization of such surface species some effort has been put into the synthesis of soluble metallasiloxanes which might serve as useful model compounds [2].

Our interest is focused on the chemistry of molecular stannasiloxanes, and recently we have described a number of both cyclic and open-chain compounds with different silicon-to-tin ratios [3–6]. In this report we present the syntheses and structures of new stannasiloxane complexes and discuss their dynamic behaviour in solution.

Results and Discussion

The reaction of $Ph_2Si(OH)_2$ with (*t*$Bu_2SnO)_3$ provides [*cyclo*-$Ph_2Si(OSn$*t*$Bu_2)_2O\cdot$*t*$Bu_2Sn(OH)_2$] (**1**) in almost quantitative yield (Eq. 1) [4]. Formally, **1** consists of the six-membered ring *cyclo*-$Ph_2Si(OSn$*t*$Bu_2)_2O$ with *t*$Bu_2Sn(OH)_2$ coordinated to it.

The $Sn_3O(OH)_2$ structural motif of **1** resembles that of the “three quarter” ladder postulated for [(*t*Bu_2SnCl_2)(*t*$Bu_2SnO)_2$]. The reaction of *t*$BuSi(OH)_3$ with (*t*$Bu_2SnO)_3$ gives [*cyclo*-*t*$Bu(OH)Si(OSn$*t*$Bu_2)_2O\cdot$*t*$Bu_2Sn(OH)_2$], an analogue of **1**.

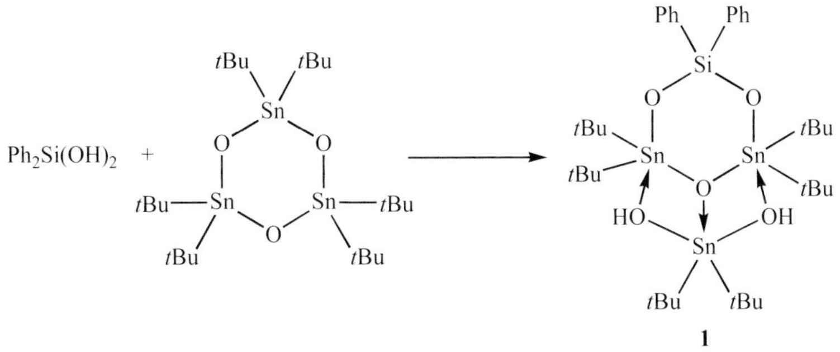

Eq. 1. Synthesis of the stannasiloxane complex [*cyclo*-$Ph_2Si(OSn\mathit{t}Bu_2)_2O \cdot \mathit{t}Bu_2Sn(OH)_2$] (**1**).

The stannasiloxane[*cyclo*-$Ph_2Si(OSn\mathit{t}Bu_2)_2O \cdot \mathit{t}Bu_2SnF_2$] (**4**) is a formal analoque of complex **1** in which the hydroxide groups are replaced by fluoride. It is obtained in good yield by the reaction of eight-membered stannasiloxane ring **2** with the tetra-*tert*-butyldifluorodistannoxane dimer **3** [5] (Scheme 1).

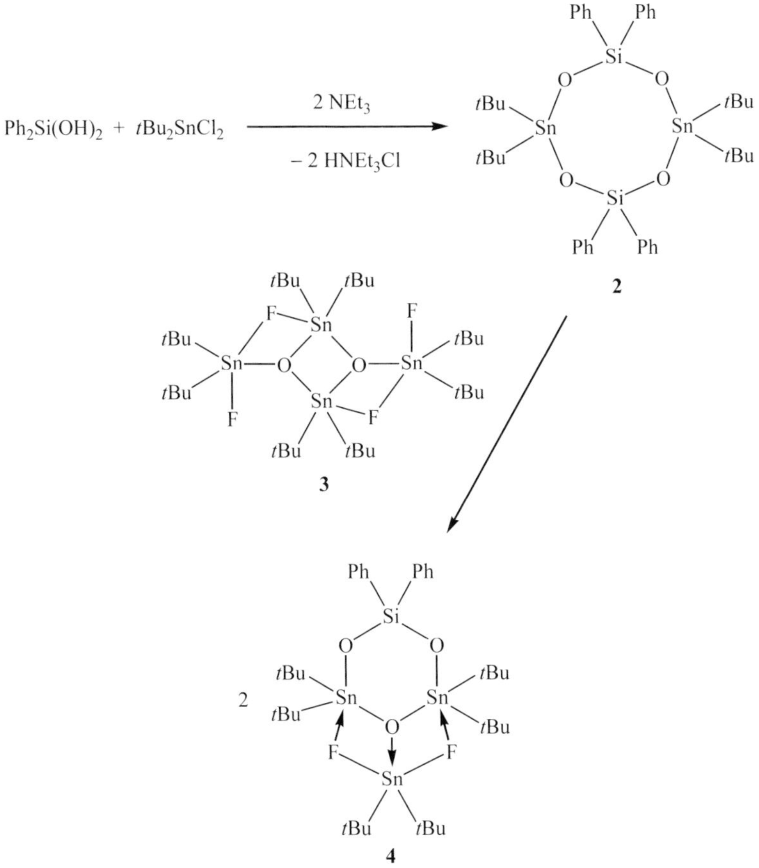

Scheme 1. Synthesis of the stannasiloxane complex[*cyclo*-$Ph_2Si(OSn\mathit{t}Bu_2)_2O \cdot \mathit{t}Bu_2SnF_2$] (**4**).

The molecular structures of **1** and **4** are shown in Fig. 1; selected bond lengths and bond angles are given in Table 1.

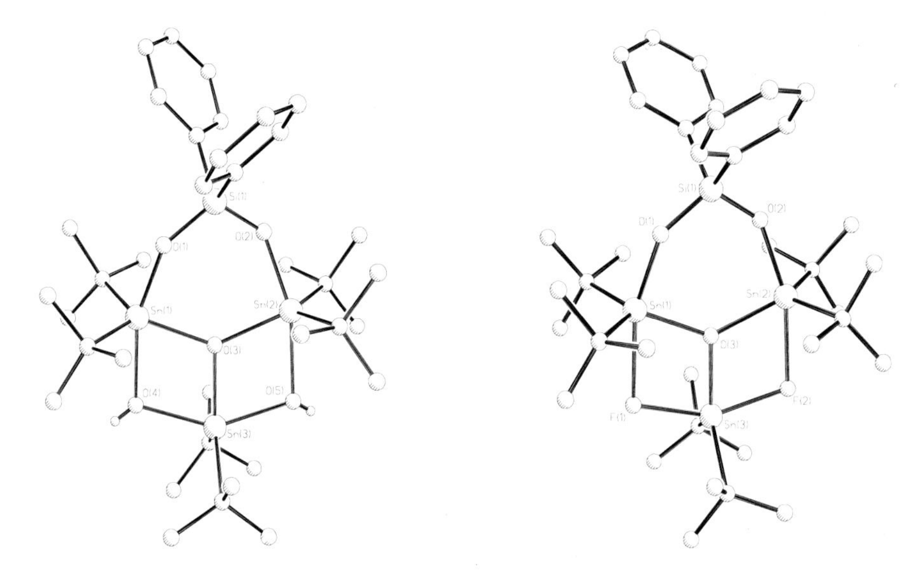

Fig. 1. Molecular structures of the stannasiloxane complexes **1** (left) and **4** (right).

A characteristic feature of both compounds is the almost planar Sn_3O_5 (**1**) or $Sn_3F_2O_3$ (**4**) skeleton. The tin atoms in **1** as well as in **4** show each a distorted trigonal-bipyramidal configuration with the *tert*-butyl groups and O(3) occupying the equatorial and O(1), O(4)/O(2), O(5) (**1**) and O(1), F(1)/O(2), F(2) (**4**) the axial positions. A major difference between the structures of **1** and **4** is the asymmetry of the Sn(1)–X(1)–Sn(3) and Sn(2)–X(2)–Sn(3) bridges. For **1** (X(1) = O(4), X(2) = O(5)) these bridges are less asymmetric (Δ = 0.155/0.162 Å) than for **4** (X(1) = F(1), X(2) = F(2); Δ = 0.325/0.280 Å).

Table 1. Selected bond lenghts [Å] and angles [°] of [*cyclo*-$Ph_2Si(OSn\textit{t}Bu_2)_2O{\cdot}\textit{t}Bu_2SnX_2$] with X = OH (**1**) and X = F (**4**).

	1 **(X(1) = O(4), X(2) = O(5)**	**4** **(X(1) = F(1), X(1) = F(2)**
Sn(1)–O(1)	2.020(2)	1.981(4)
Sn(1)–X(1)	2.261(3)	2.399(3)
Sn(1)–O(3)	2.092(2)	2.098(4)
Sn(3)–X(1)	2.106(3)	2.074(3)
Sn(3)–O(3)	2.091(2)	2.064(4)
Sn(3)–X(2)	2.103(3)	2.073(3)
O(1)-Sn(1)-X(1)	160.8(1)	162.5(1)
X(1)-Sn(3)-X(2)	146.5(1)	150.8(1)
O(2)-Sn(2)-X(2)	161.9(1)	162.6(1)
Sn(1)-O(3)-Sn(2)	136.7(1)	133.4(1)

The same holds for the solution-state structures as is evidenced by the ^{119}Sn NMR chemical shifts for **1** (Sn(3) –270.3 ppm, Sn(1)/Sn(2) –274.3 ppm) in comparison with **4** (Sn(3) –280.8 ppm, Sn(1)/Sn(2) –222.2 ppm). The asymmetry of the fluoride bridges in **4** are further supported by the $^{1}J(^{119}Sn(1)–^{19}F)$ of 1170 Hz versus $^{1}J(^{119}Sn(3)–^{19}F)$ of 2465 Hz.

In $CDCl_3$ solution the stannasiloxane complex **1** undergoes a reversible dissociation process (Eq.2) which can conveniently be monitored by ^{119}Sn NMR spectroscopy [4].

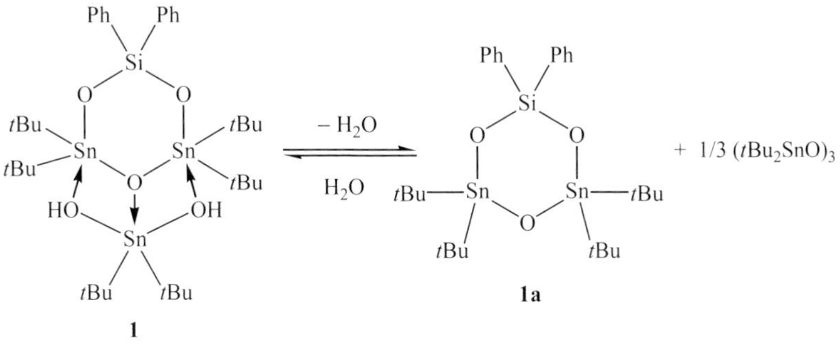

Eq. 2. Equilibrium between stannasiloxane complex **1**, six-membered cyclostannasiloxane **1a**, $(t\text{Bu}_2\text{SnO})_3$ and water in $CDCl_3$.

$t\text{Bu}_2\text{Sn(OH)}_2$, which is supposed to be the initial dissociation product along this reaction, is known for its immediate self-condensation in solution subsequently resulting in $(t\text{Bu}_2\text{SnO})_3$ and water [4].

In contrast, there is no evidence that a similar dissociation process takes place for **4**. However, a ^{119}Sn NMR spectrum of a solution of single-crystalline **4** shows two low intense signals which are assigned to eight-membered stannasiloxane ring **2** and tetra-*tert*-butyldifluorodistannoxane dimer **3,** indicating an equilibrium between stannasiloxane complex **4** and these species.

We conclude that these rearrangements in solution are driven by additional entropic effects in combination with the kinetic lability of the tin–oxygen bonds.

Interestingly, the stannasiloxane complex **4** is highly sensitive to traces of water and immediately falls apart into eight-membered stannasiloxane ring **2** and $t\text{Bu}_2\text{Sn(OH)F}$.

Acknowledgment: We are grateful to the *Deutsche Forschungsgemeinschaft* and the *Fonds der Chemischen Industrie* for financial support.

References:

[1] A. de Mallmann, N. Lot, F. Perrier, C. Lefebvre, C. Santini, J. M. Basset, *Organometallics*, **1998**, *17*, 1031.

[2] R. Murugavel, A. Voigt, M. G. Walawalkar, H. W. Roesky, *Chem. Rev.* **1996**, *96*, 2205.

[3] J. Beckmann, K. Jurkschat, D. Schollmeyer, M. Schürmann, *J. Organomet. Chem.* **1997**, *543*, 229.

[4] J. Beckmann, K. Jurkschat, B. Mahieu, M. Schürmann, *Main Group Met. Chem.* **1998**, *21*, 113.

[5] J. Beckmann, M. Biesemans, K. Hassler, K. Jurkschat, J. C. Martins, M. Schürmann, R. Willem, *Inorg. Chem.* **1998**, *37*, 4891.

[6] J. Beckmann, B. Mahieu, W. Nigge, D. Schollmeyer, K. Jurkschat, M. Schürmann, *Organometallics* **1998**, *17*, 5697.

Control of Distannoxane Structure by Silicon-Containing Spacers

Klaus Jurkschat*, Markus Schürmann, Marcus Schulte

Fachbereich Chemie, Lehrstuhl für Anorganische Chemie II
Universität Dortmund
D-44221 Dortmund, Germany
Tel.: Int. code + (231)755 3800 — Fax: Int. code + (231)755 3797
E-mail: kjur@platon.chemie.uni-dortmund.de

Keywords: Distannoxanes / Silicon Containing Spacers / Homogeneous Catalysis

Summary: In the reaction of $[Me_3SiCH_2(Cl_2)SnCH_2SiMe_2]_2Y$ (Y = C≡C, O) with $(t\text{Bu}_2SnO)_3$ the $SiMe_2$ groups control the formation of dimeric tetraorgano-distannoxanes. The syntheses and molecular structures are reported of the unsymmetric hydroxy-substituted distannoxane dimer **7** and the symmetric distannoxane dimer **8b**. Compounds of this type may serve as homogeneous catalysts in various organic reactions.

Introduction

Controlled hydrolysis of diorganotin dihalides leads to dimeric tetraorganodistannoxanes $[R_2(X)SnOSn(X)R_2]_2$ (R = alkyl, aryl; X = halogen, OH) [1, 2]. These compounds hold potential as efficient homogeneous catalysts in various organic reactions, e.g. transesterification [3], urethane formation [4], and alkyl carbonate synthesis [5].

In recent publications we reported the synthesis and structures of trimethylene-bridged double and triple ladders $\{[R(Cl)Sn(CH_2)_3Sn(Cl)R]O\}_4$ and $\{[R(Cl)Sn(CH_2)_3Sn(Cl)(CH_2)_3Sn(Cl)R]O_{1.5}\}_4$ ($R = CH_2SiMe_3$), respectively [6, 7]. Now we are interested in studying the influence of the substituents R and X as well as of the silicon-containing spacers $CH_2SiMe_2YSiMe_2CH_2$ (either rigid with Y = C≡C or non-rigid with Y = O) on the structure and properties including catalytic activity of $\{[R(X)SnCH_2SiMe_2YSiMe_2CH_2Sn(X)R]O\}_n$ ($R = CH_2SiMe_3$; X = halogen, OH, $n = 2, 4$).

Syntheses, Results and Discussion

The (triphenylstannylmethyl)dimethylsilyl-substituted acetylene derivative **2** is obtained by reaction of the di-Grignard reagent **1a** [8, 9] with Ph_3SnCl. Base-catalyzed hydrolysis of **2** provides the disiloxane **3** (Scheme 1).

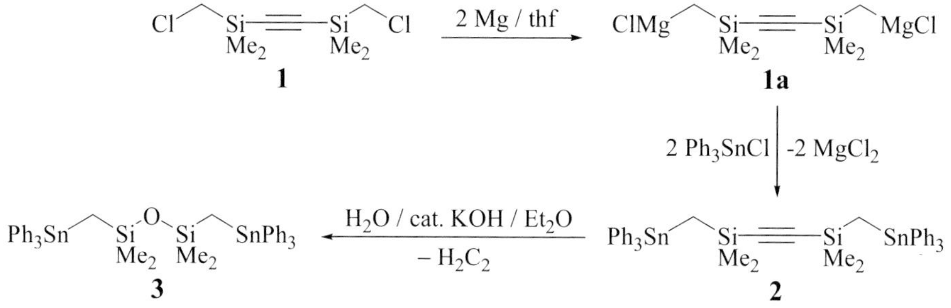

$$Cl\text{-}CH_2\text{-}SiMe_2\text{-}C{\equiv}C\text{-}SiMe_2\text{-}CH_2\text{-}Cl\ (\mathbf{1}) \xrightarrow{2\ Mg\ /\ thf} ClMg\text{-}CH_2\text{-}SiMe_2\text{-}C{\equiv}C\text{-}SiMe_2\text{-}CH_2\text{-}MgCl\ (\mathbf{1a})$$

$$\mathbf{1a} \xrightarrow[-2\ MgCl_2]{2\ Ph_3SnCl} Ph_3Sn\text{-}CH_2\text{-}SiMe_2\text{-}C{\equiv}C\text{-}SiMe_2\text{-}CH_2\text{-}SnPh_3\ (\mathbf{2})$$

$$\mathbf{2} \xrightarrow[-\ H_2C_2]{H_2O\ /\ cat.\ KOH\ /\ Et_2O} Ph_3Sn\text{-}CH_2\text{-}SiMe_2\text{-}O\text{-}SiMe_2\text{-}CH_2\text{-}SnPh_3\ (\mathbf{3})$$

Scheme 1. Synthesis of the (triphenylstannylmethyl)dimethylsilyl-substituted acetylene derivative **2** and its hydrolysis yielding the disiloxane **3**.

Compounds **2** and **3** can be converted into the tetrachloroditin compounds **4** and **5** (Eq. 1).

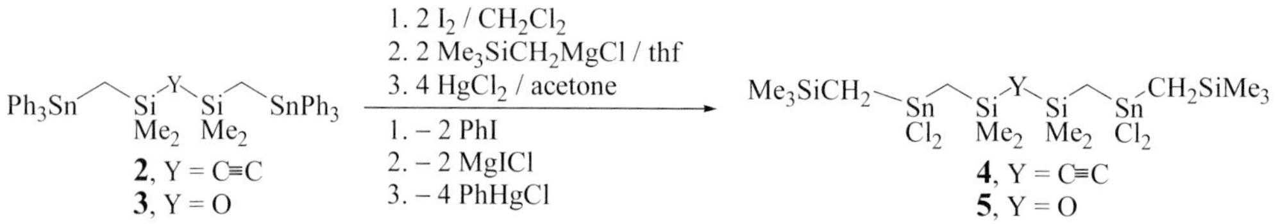

$$Ph_3Sn\text{-}CH_2\text{-}SiMe_2\text{-}Y\text{-}SiMe_2\text{-}CH_2\text{-}SnPh_3 \xrightarrow[1.\ -2\ PhI;\ 2.\ -2\ MgICl;\ 3.\ -4\ PhHgCl]{1.\ 2\ I_2\ /\ CH_2Cl_2;\ 2.\ 2\ Me_3SiCH_2MgCl\ /\ thf;\ 3.\ 4\ HgCl_2\ /\ acetone} Me_3SiCH_2\text{-}SnCl_2\text{-}CH_2\text{-}SiMe_2\text{-}Y\text{-}SiMe_2\text{-}CH_2\text{-}SnCl_2\text{-}CH_2SiMe_3$$

2, Y = C≡C; **3**, Y = O → **4**, Y = C≡C; **5**, Y = O

Eq. 1.

The reaction of **4** with $(t\text{Bu}_2\text{SnO})_3$ provides the dimeric tetraorganodistannoxane **6** rather than a tetramer with a double ladder structure [10]. Stirring of **6** in a H_2O/CH_2Cl_2 mixture yields **7** with one hydroxy-substituted distannoxane unit (Scheme 2). The alkyne units force the spacers to adopt a *cis* configuration, and consequently, **7** shows asymmetry in both the ladder core and the configuration of the spacers. Compounds of this type may serve as homogeneous catalysts in asymmetric synthesis. This strategy is currently being investigated. The molecular structure of **7** is given in Fig. 1.

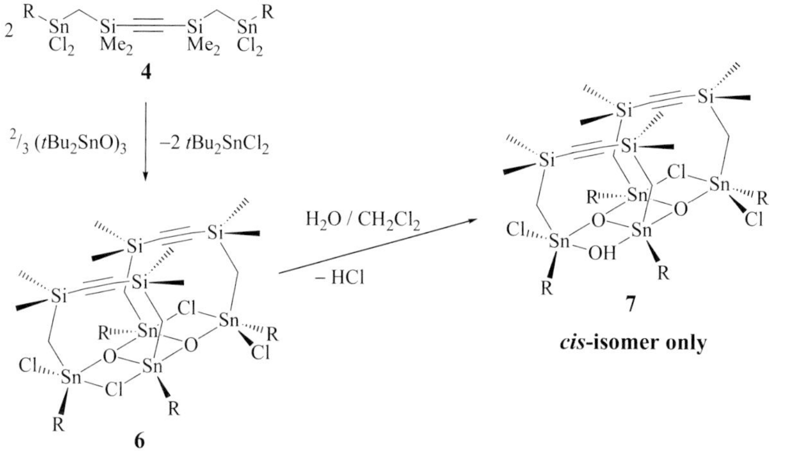

$$2\ R\text{-}SnCl_2\text{-}CH_2\text{-}SiMe_2\text{-}C{\equiv}C\text{-}SiMe_2\text{-}CH_2\text{-}SnCl_2\text{-}R\ (\mathbf{4}) \xrightarrow[-2\ t\text{Bu}_2\text{SnCl}_2]{^2/_3\ (t\text{Bu}_2\text{SnO})_3} \mathbf{6} \xrightarrow[-\ HCl]{H_2O\ /\ CH_2Cl_2} \mathbf{7}$$

7: *cis*-isomer only

Scheme 2. Results for the rigid spacer $CH_2SiMe_2C{\equiv}CSiMe_2CH_2$; R = CH_2SiMe_3.

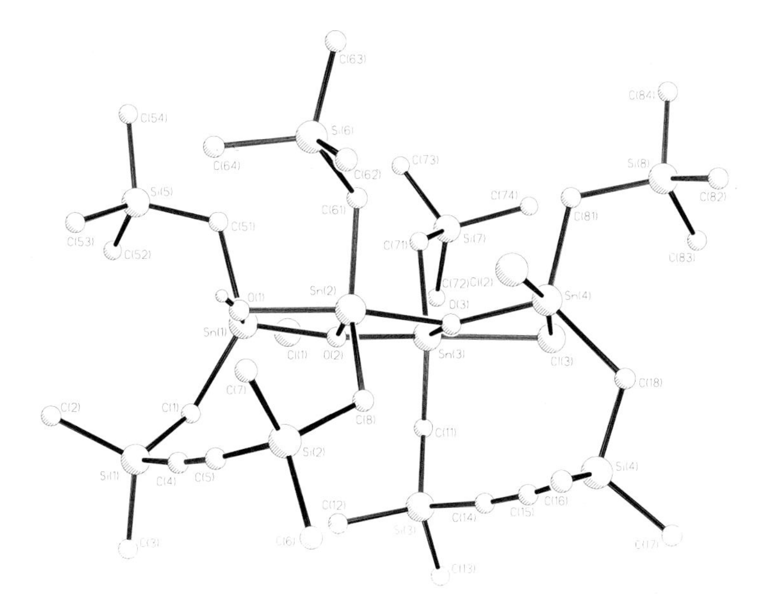

Fig. 1. Molecular structure of the unsymmetric dimeric distannoxane **7**; selected bond lengths: O(2)–Sn(3) 2.114(2), O(2)–Sn(2) 2.040(2), O(2)–Sn(1) 2.029(2), O(1)–Sn(1) 2.231(3), O(1)–Sn(2) 2.210(3), Cl(1)–Sn(1) 2.118(4), Cl(1)–Sn(3) 3.5668(11) Å.

The reaction of **5** with $(t\mathrm{Bu_2SnO})_3$ affords the dimeric tetraorganodistannoxanes **8a**/**8b**. In solution, a mixture of the *cis* isomer **8a** (45 %) and the *trans* isomer **8b** (55 %) is found (Scheme 3). In the solid state, however, only the *trans* isomer **8b** could be isolated. The molecular structure of **8b** is given in Fig. 2.

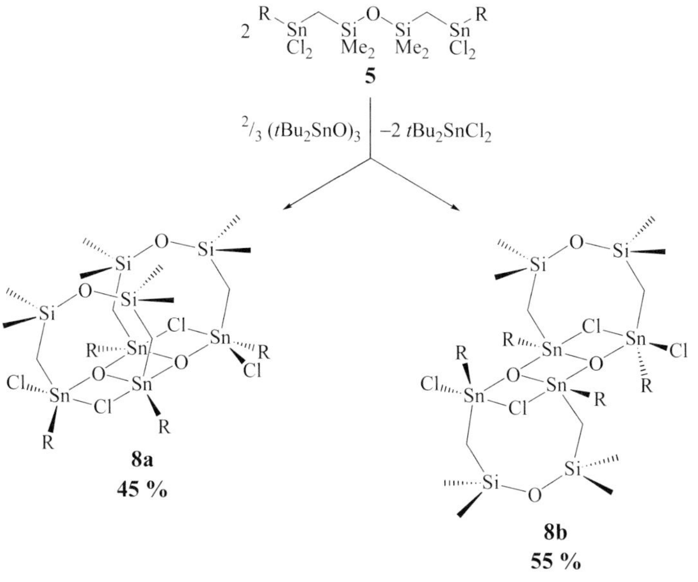

2 R–SnCl_2–CH_2–SiMe_2–O–SiMe_2–CH_2–SnCl_2–R

5

$^{2}/_{3}$ $(t\mathrm{Bu_2SnO})_3$ | -2 $t\mathrm{Bu_2SnCl_2}$

8a
45 %

8b
55 %

Scheme 3. Results for the non-rigid spacer $CH_2SiMe_2OSiMe_2CH_2$; R = CH_2SiMe_3.

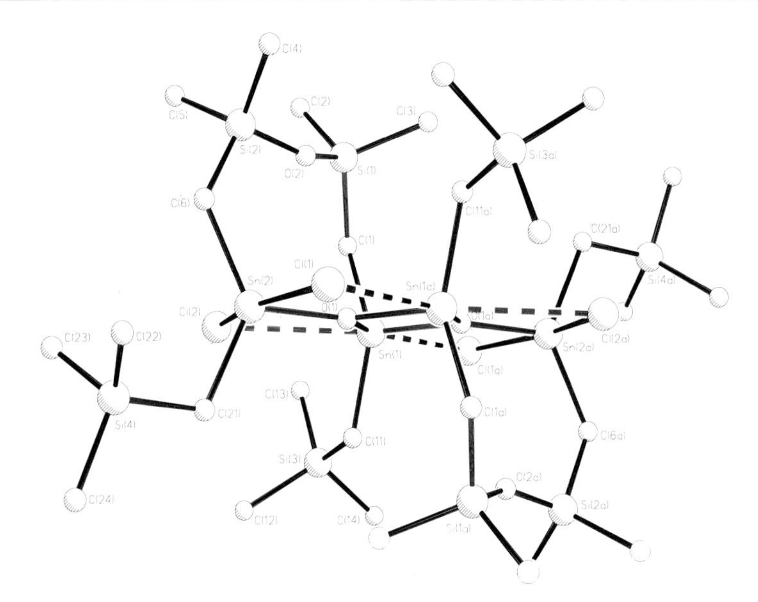

Fig. 2. Molecular structure of the dimeric distannoxane **8b**; selected bond lengths: O(1)–Sn(1) 2.099(2), O(1)–Sn(1a) 2.069(3), O(1)–Sn(2) 2.013, Cl(1)–Sn(2) 2.6328(14), Cl(1)–Sn(1a) 2.9066(14), Cl(2)–Sn(2) 2.5189(14), Cl(2)–Sn(1) 3.1985(17) Å; the molecular structure is characterized by a crystallographic center of inversion.

These preliminary results show that the silicon-containing spacers play an essential role in controlling the structure of the distannoxane dimers **7**, **8a**, and **8b**.

Acknowledgment: We are grateful to the *Deutsche Forschungsgemeinschaft* and the *Fonds der Chemischen Industrie* for financial support.

References:

[1] D. C. Gross, *Inorg. Chem.* **1989**, *28*, 2355; references cited.

[2] O. Primel, M.-F. Llauro, R. Pétiaud, A. Michel, *J. Organomet. Chem.* **1998**, *558*, 19; references cited.

[3] J. Otera, N. Dan-oh, H. Nozaki, *J. Org. Chem.* **1991**, *56*, 5307.

[4] R. P. Houghton, A. W. Mulvaney, *J. Organomet. Chem.* **1996**, *517*, 107.

[5] E. N. Suciu, B. Kuhlmann, G. A. Knudsen, R. C. Michaelson, *J. Organomet. Chem.* **1998**, *556*, 41.

[6] D. Dakternieks, K. Jurkschat, D. Schollmeyer, H. Wu, *Organometallics* **1994**, *13*, 4121.

[7] M. Mehring, M. Schürmann, H. Reuter, D. Dakternieks, K. Jurkschat, *Angew. Chem. Int. Ed. Engl.* **1997**, *36*, 1112.

[8] G. Fritz, P. Schober, *Z. Anorg. Allg. Chem.* **1970**, *372*, 21.

[9] C. Tretner, B. Zobel, R. Hummeltenberg, W. Uhlig, *J. Organomet. Chem.* **1994**, *468*, 63.

[10] M. Schulte, M. Schürmann, D Dakternieks, K. Jurkschat, *J. Chem. Soc., Chem. Commun.* **1999**, 1291.

The Influence of Intramolecular Coordination and Ring Strain on the Polymerization Potential of Cyclic Stannasiloxanes

Jens Beckmann, Klaus Jurkschat*, Nicole Pieper, Stephanie Rabe, Markus Schürmann

Fachbereich Chemie, Lehrstuhl für Anorganische Chemie II
Universität Dortmund
D-44221 Dortmund, Germany
Tel.: Int. code + (231)755 3800 — Fax: Int. code + (231)755 -3797
E-mail: kjur@platon.chemie.uni-dortmund.de

Dieter Schollmeyer

Institut für Organische Chemie, Universität Mainz
Saarstr. 21, D-55099 Mainz, Germany
Tel.: .: Int. code + (6131)39 5320 — Fax: Int. code + (6131)39 4778
E-mail: scholli@uacdr0.chemie.uni-mainz.de

Keywords: Stannasiloxane / Ring-Opening Polymerization / X-Ray Analysis / Intramolecular Coordination

Summary: Intamolecular Sn–N coordination controls the ring strain in the novel stannasiloxanes *cyclo*-$R_2Sn(OSiPh_2)_2O$ (**4**) and *cyclo*-$(R_2SnOSiPh_2O)_2$ (**5**) (R = $(CH_2)_3NMe_2$) and hence prevents their spontaneous polymerization as it is observed for related *cyclo*-*t*$Bu_2Sn(OSiPh_2)_2O$ (**1**). Intramolecular Sn–N coordination also allows isolation and characterization of the first organoelement oxides *cyclo*-$(R_2SnOSiPh_2OMO)$ (R = $(CH_2)_3NMe_2$: **8**, M = *t*Bu_2Ge; **9**, M = PhB) with three different metals in one ring.

Introduction

The synthesis of inorganic polymers is a great challenge and is motivated by the possibility of accessing new materials with tailor-made properties.

Polysiloxanes are a well-established type of polymers with widespread technical applications. One way to obtain such compounds is ring-opening polymerization (ROP) of highly strained cyclosiloxanes [1].

Comparatively scant attention has been focused on polydiphenylsiloxane $(Ph_2SiO)_n$, presumably due to difficulties in synthesizing the high molecular weight polymer which shows high crystallinity and poor solubility in most organic solvents. In general it is prepared by ring-opening polymerization of six-membered siloxane rings *cyclo*-$(Ph_2SiO)_3$ under nonequilibrium conditions using a polymerization initiator. If equilibration of these mixtures is allowed, the thermodynamically more favored eight-membered siloxane ring *cyclo*-$(Ph_2SiO)_4$ is formed as the major product [2, 3].

In an attempt to overcome these difficulties one of the Ph_2Si units in *cyclo*-$(Ph_2SiO)_3$ has formally been replaced by the substantially smaller PhB unit in order to increase the ring strain. However, the *cyclo*-$PhB(OSiPh_2)_2O$ thus obtained showed no tendency to provide polymers [4].

In addition to their use as starting compounds for polymers, modified siloxanes have attracted considerable attention as single-source precursors which can be converted thermally into ceramics by cleavage of their organic substituents. This procedure may represent a valuable alternative to common sol–gel approaches [5, 6].

Recently, we have shown that *cyclo*-$t\mathrm{Bu}_2Sn(OSiPh_2)_2O$ is a six-membered ring in solution but upon crystallization it forms the first fully characterized polystannasiloxane [7].

However, so far there is no detailed understanding of the factors which control the polymerization. Here we present the results of systematic studies on the polymerization behavior including the influence of intramolecular coordination, ring size, the identity of the organic substituents and the exchange of a ring fragment.

Organotin carboxylates are known to catalyze the polycondensation of silanols to siloxanes. Very likley, stannasiloxanes are involved in this process [8, 9].

The condensation of $Ph_2Si(OH)_2$ with $(Ph_2SnOH)_2CH_2$ under mild conditions, also presented here, is a first case study on this subject. A possible reaction mechanism is described.

Results and Discussion

The reaction of $(Ph_2SiOH)_2O$ with $t\mathrm{Bu}_2SnCl_2$ in the presence of triethylamine provides *cyclo*-$t\mathrm{Bu}_2Sn(OSiPh_2)2O$ (**1**) as crystalline compound of good solubility in common organic solvents. In solution, **1** represents a six-membered ring **1a**, but upon crystallization it turns into a polymeric chain **1b** (Scheme 1) [7].

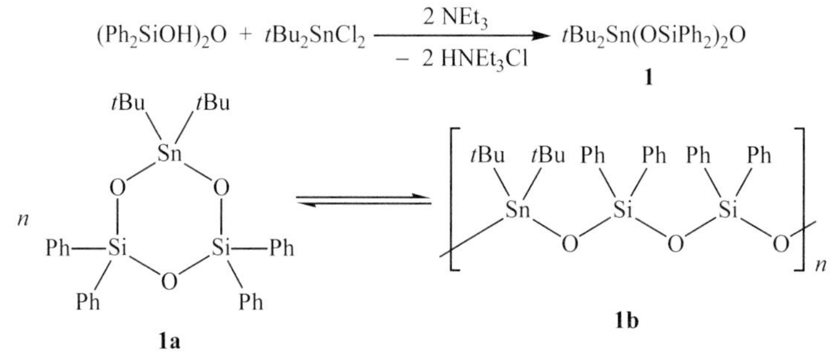

$$(Ph_2SiOH)_2O + t\mathrm{Bu}_2SnCl_2 \xrightarrow[-\ 2\ HNEt_3Cl]{2\ NEt_3} t\mathrm{Bu}_2Sn(OSiPh_2)_2O \quad \mathbf{1}$$

tBu tBu Sn O O Ph—Si Si—Ph O Ph Ph (*n* **1a**) ⇌ [tBu tBu Ph Ph Ph Ph Sn O Si O Si O]*n* (**1b**)

Scheme 1. Synthesis of the first completely characterized polystannasiloxane **1**.

Apparently, in solution the energy gain associated with the ring opening is entirely compensated by the entropy gain of a greater number of monomers.

The molecular structure shows for **1** a linear chain with an –Si–O–Si–O–Sn–O– backbone (Fig. 1). However, there is no substantial change in the conformation as compared to the parent polymer $(Ph_2SiO)_n$ [3].

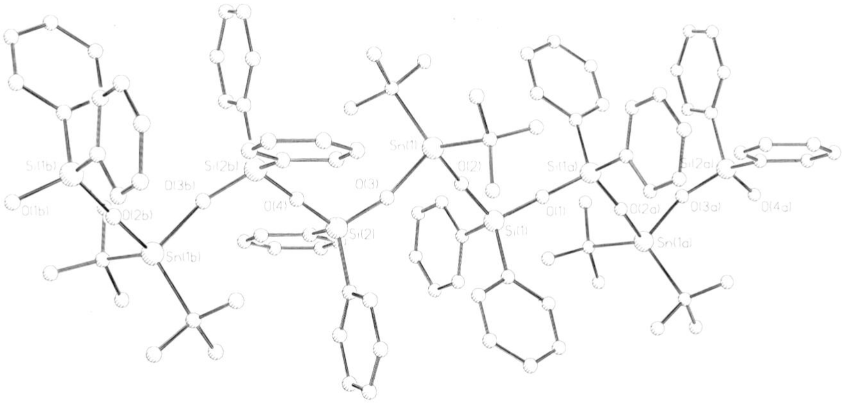

Fig. 1. Molecular structure of polystannasiloxane **1**.

The ring strain in compound **1** allows the formal insertion of a $t\text{Bu}_2\text{SnO}$ unit into the six-membered ring giving the eight-membered stannasiloxane ring **2** (Eq. 1 and Fig. 2) [7,10]. The thermodynamic driving force for this reaction might be the relief of this strain.

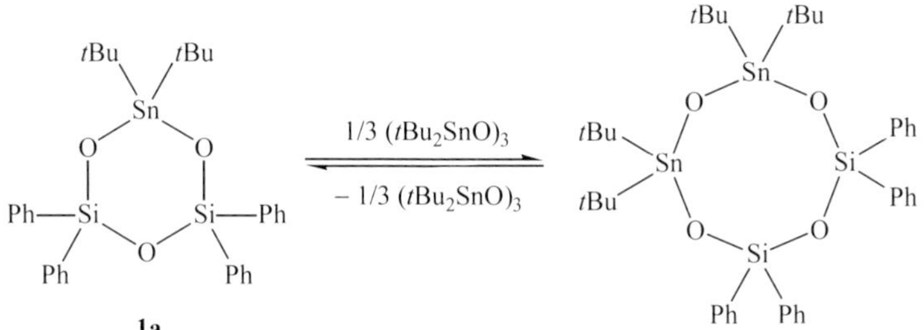

$$\mathbf{1a} \underset{-\,1/3\,(t\text{Bu}_2\text{SnO})_3}{\overset{1/3\,(t\text{Bu}_2\text{SnO})_3}{\rightleftharpoons}} \mathbf{2}$$

Eq. 1. Insertion of a $t\text{Bu}_2\text{SnO}$ unit into **1** leads to the stannasiloxane **2**.

The reaction of $\text{Ph}_2\text{Si(OH)}_2$ with $t\text{Bu}_2\text{SnCl}_2$ in the presence of triethylamine affords the stannasiloxane **3**, which is a structural isomer of **2** (Eq. 2).

$$2\ \text{Ph}_2\text{Si(OH)}_2 + 2\ t\text{Bu}_2\text{SnCl}_2 \xrightarrow[-\,4\ \text{HNEt}_3\text{Cl}]{4\ \text{NEt}_3} \underset{\mathbf{3}}{(\text{Ph}_2\text{SiOSn}t\text{Bu}_2\text{O})_2}$$

Eq. 2. Synthesis of the eight-membered stannasiloxane ring **3**.

Compound **3** represents an eight-membered ring in the solid state (Fig. 2) as well as in solution [10].

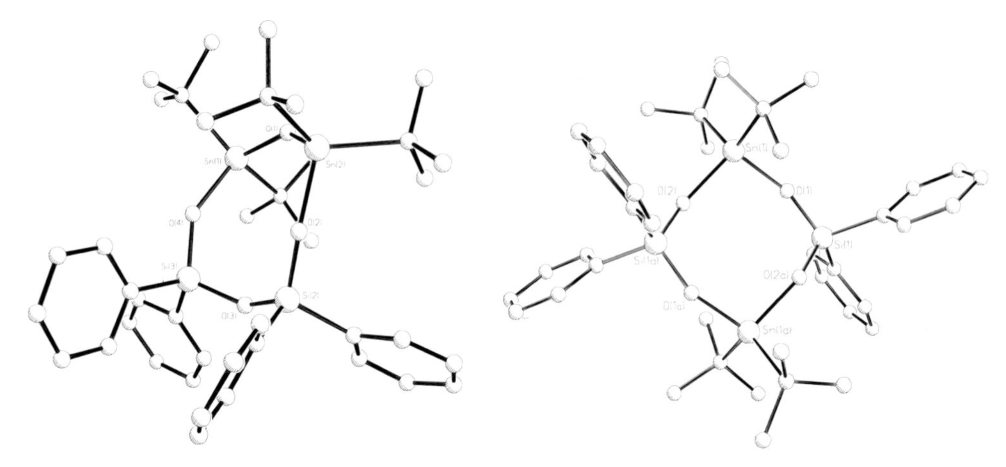

Fig. 2. Molecular structures of the isomeric stannasiloxane rings **2** (left) and **3** (right).

Depending on the stoichiometric ratio of Ph_2SiCl_2 and R_2SnCl_2 (R = $(CH_2)_3NMe_2$), their cohydrolysis provides either the six-membered stannasiloxane ring *cyclo*-$R_2Sn(OSiPh_2)_2O$ (**4**) or the eight-membered stannasiloxane ring *cyclo*-$(Ph_2SiOSnR_2O)_2$ (**5**) (Scheme 2). In both compounds the tin atoms are hexacoordinate by two intramolecular Sn–N interactions, two carbons and two oxygens, as is shown in Fig. 3 [11].

$$2\ Ph_2SiCl_2 + R_2SnCl_2 \xrightarrow[-\ 6\ NaCl\ -\ 3\ H_2O]{6\ NaOH} \underset{\mathbf{4}}{R_2Sn(OSiPh_2)_2O}$$

$$2\ Ph_2SiCl_2 + 2\ R_2SnCl_2 \xrightarrow[-\ 8\ NaCl\ -\ 4\ H_2O]{8\ NaOH} \underset{\mathbf{5}}{(Ph_2SiOSnR_2O)_2}$$

$$R = (CH_2)_3NMe_2$$

Scheme 2. Cohydrolysis of Ph_2SiCl_2 and R_2SnCl_2 affords either *cyclo*-$R_2Sn(OSiPh_2)_2O$ (**4**) or *cyclo*-$(Ph_2SiOSnR_2O)_2$ (**5**).

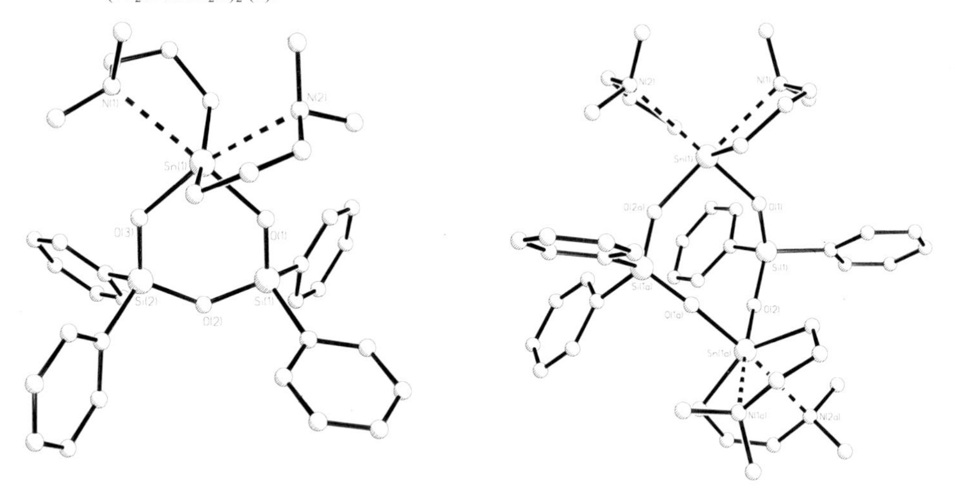

Fig. 3. Molecular structures of the stannasiloxane rings **4** (left) and **5** (right).

As result of the octrahedral configuration at tin the O-Sn-O angles in **4** and **5** are substantially smaller than the corresponding tetrahedral angles in the stannasiloxanes **1**–**3**. This effect is sufficient to prevent **4** from polymerizing.

In solution, *cyclo*-(tBu$_2$SnO)$_3$ can be inserted into the eight-membered rings **3** and **5** to give quantitatively the six-membered rings **6** and **7**, respectively (Eq. 3) [10, 12]. This reaction is entropy-driven as the number of products increases in comparison to the number of starting compounds. However, only in the case of **7** the product is also enthalpically favored in the solid state, and we succeeded in isolating the first stannasiloxane containing different types of tin atoms in one ring. In contrast, upon evaporation of a chloroform solution of **6** the latter falls apart to give the starting compounds **3** and *cyclo*-(tBu$_2$SnO)$_3$, as we could show by ^{119}Sn-MAS-NMR spectroscopy of the residue [10]. We attribute this behavior to a higher ring strain in **6** as compared to **3** and *cyclo*-(tBu$_2$SnO)$_3$. On the other hand, as a result of intramolecular Sn–N coordination, six-membered stannasiloxane **7** shows less ring strain than its eight-membered precursor **5** and hence the former can be isolated.

$$\textbf{3}, \textbf{5} \xrightarrow[-\,2/3\,(t\text{Bu}_2\text{SnO})_3]{2/3\,(t\text{Bu}_2\text{SnO})_3} 2\ \textbf{6}, \textbf{7}$$

3, R = tBu
5, R = (CH$_2$)$_3$NMe$_2$
6, R = tBu
7, R = (CH)$_3$NMe$_2$

Eq. 3. Formal insertion of tBu$_2$SnO units into eight-membered stannasiloxane rings **3** and **5**, providing the six-membered rings **6** and **7**, respectively.

The ring strain of eight-membered stannasiloxane ring **5** also allows the synthesis of the first well-defined mixed organoelement oxides **8** and **9** by reaction of **5** with tBuGe(OH)$_2$ and PhB(OH)$_2$, respectively (Eq. 4) [12].

$$\textbf{5} \xrightarrow[-\,2\,\text{H}_2\text{O}]{2\,\text{L}_n\text{M(OH)}_2} 2\ \textbf{8}, \textbf{9}$$

R = (CH$_2$)$_3$NMe$_2$
8 L$_n$M = tBu$_2$Ge
9 L$_n$M = PhB

Eq. 4. Synthesis of six-membered stannasiloxanes **8** and **9** containing heterometal units.

TGA measurements on **8** and **9** reveal loss of the organic groups upon heating, making these compounds potentially interesting for the synthesis of inorganic materials.

The intramolecular Sn-N distances in compounds **4**, **5**, **7** and **8** are between 2.620 (**4**) and 2.811 Å (**5**). The molecular structure of **8** (Fig. 4) is similar to the one of **7**.

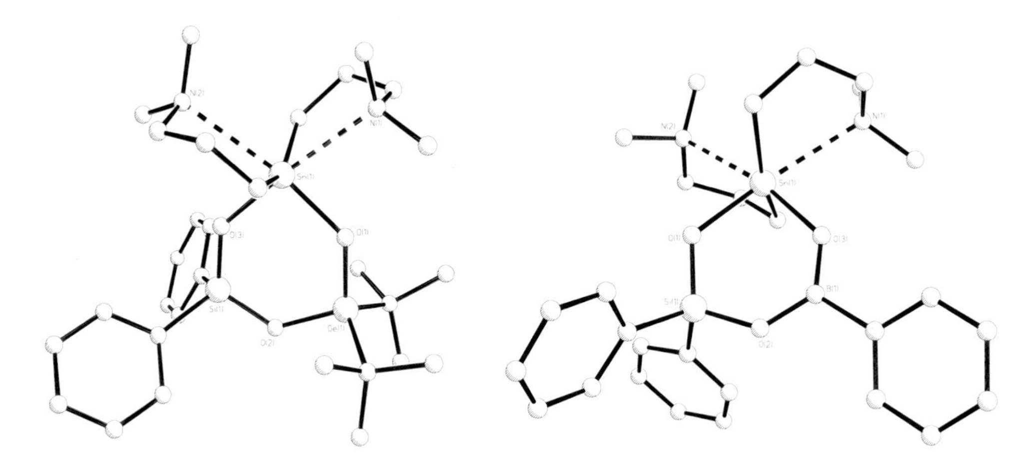

Fig. 4. Molecular structures of heterometal-stannasiloxane rings **8** (left) and **9** (right).

In order to overcome the redistribution reaction of six-membered stannasiloxane rings to give eight-membered stannasiloxane rings and *cyclo*-($t$$Bu_2SnO$)$_3$, as was observed for **6** (see above), one idea was to replace the kinetically labile oxygen bridge by a kinetically inert methylene bridge.

The reaction of $Ph_2Si(OH)_2$ with $(Ph_2SnOH)_2CH_2$ gives the desired six-membered ring **10** under mild conditions. However, on contact with water present in the reaction mixture, and with release of $(Ph_2SnOH)_2CH_2$, **10** slowly turns into the eight-membered ring **11** (Scheme 3) [13].

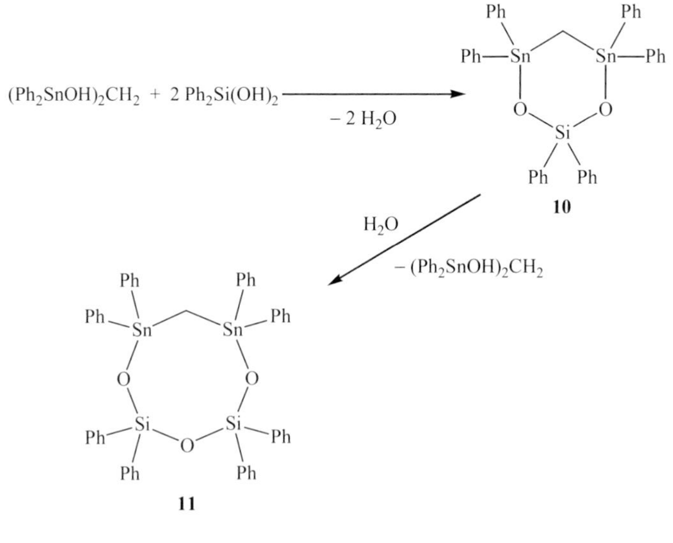

$(Ph_2SnOH)_2CH_2$ + 2 $Ph_2Si(OH)_2$ —($-$ 2 H_2O)→ **10**

Ph, Ph, Ph—Sn, Sn—Ph, O, O, Si, Ph, Ph

10

H_2O

$-$ $(Ph_2SnOH)_2CH_2$

Ph, Ph, Ph—Sn, Sn—Ph, O, O, Ph—Si, O, Si—Ph, Ph, Ph

11

Scheme 3. Reaction pathway leading to the formation the eight-membered stannasiloxane **11**.

The molecular structure in the solid state of **11** (Fig. 5) is characterized by the coexistence of two crystallographically independent and slightly different conformers.

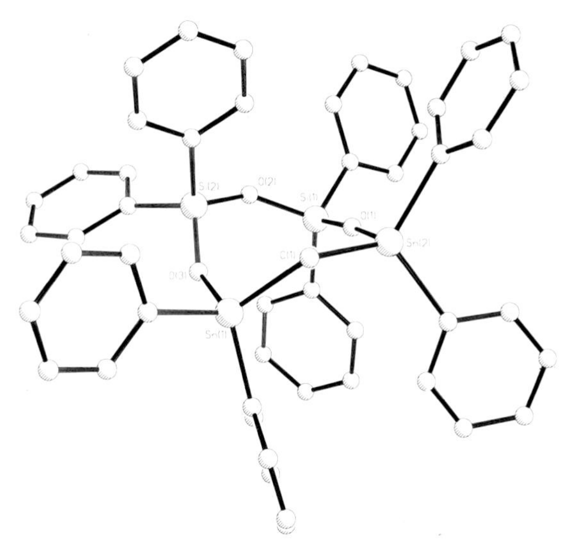

Fig. 5. Molecular structure of stannasiloxane **11**. Only one of the two cystallographically independent conformers is shown.

Ring strain in **10** might again be the reason for the ring-size enlargement to give **11**. Very likely, the ring strain combined with a small C-Sn-O angle allows **10** to be in equilibrium with its dimer **10a** (Eq. 6) prior to a further condensation to give **11**. An electrospray mass spectrum of **10** in $CH_3CN/MeOH/H_2O$ (positive mode) showed an isotopic cluster pattern for $[2\cdot\mathbf{10} + H]^+$.

2 **10** ⇌ **10a**

Eq. 5. Proposed pre-equilibium between **10** and **10a** as evidenced by electrospray mass spectrometry.

It is worth noting that an excess of $Ph_2Si(OH)_2$ can be readily condensed by a small amount of the tin compound under the same mild conditions, but condensation does not proceed without it.

Acknowledgment: We are grateful to the *Deutsche Forschungsgemeinschaft* and the *Fonds der Chemischen Industrie* for financial support.

References:

[1] I. Manners, *Angew. Chem.* **1996**, *108*, 1712.

[2] B. R. Harkness, M. Tachikawa, H. Yue, I. Mita, *Chem. Mater.* **1998**, *10*, 1700.

[3] I. L. Dubchak, T. M. Babchinister, L. G. Kazaryan, N. G. Tartakovskaya, N. G. Vasilenko, A. A. Zhdanov, V. V. Korshak, *Polym. Sci. USSR* **1989**, *31*, 70.

[4] D. A. Foucher, A. J. Lough, I. Manners, *Inorg. Chem.* **1992**, *31*, 2034.

[5] K. W. Terry, C. G. Lugmair, T. D. Tilley, *J. Am. Chem. Soc.* **1997**, *119*, 9745.

[6] A. N. Kornev, T. A. Chesnokova, V. V. Semenov, E. V. Zhezlova, L. N. Zakharov, L. G. Klapshina, G. A. Domrachev, V. S. Rusakov, *J. Organomet. Chem.* **1997**, *547*, 113.

[7] J. Beckmann, K. Jurkschat, D. Schollmeyer, M. Schürmann, *J. Organomet. Chem.* **1997**, *543*, 229.

[8] F. W. van der Weij, *Makromol. Chem.* **1980**, *181*, 2541.

[9] H. G. Emblem, K. Jones, *Trans. J. Br. Ceram. Soc.* **1984**, *79*, 56.

[10] J. Beckmann, B. Mahieu, W. Nigge, D. Schollmeyer, K. Jurkschat, M. Schürmann, *Organometallics* **1998,** *17*, 5697.

[11] J. Beckmann, K. Jurkschat, U. Kaltenbrunner, N. Pieper, M. Schürmann, *Organometallics* **1999**, *18*, 1586.

[12] J. Beckmann, K. Jurkschat, N. Pieper, M. Schürmann, *J. Chem. Soc., Chem. Commun.* **1999**, 1095.

[13] S. Rabe, *Ph. D. Thesis*, Universität Dortmund, **1999**.

Unconventional Solvents for Hydrolysis of Tetramethoxysilane

Petra E. Ritzenhoff, Heinrich C. Marsmann*

Anorganische Chemie
Universität-Gesamthochschule Paderborn
Warburger Str. 100, D-33098 Paderborn, Germany
Tel.: Int. code + (5251)60 2571
E-mail: hcm@ac16.uni-paderborn.de

Keywords: Sol–Gel Process / TMOS / Solvent / ^{29}Si-MAS-NMR Spectroscopy

Summary: Xerogels were prepared from TMOS with three different, and without, catalysts. The reaction was carried out in solvents with different polarity. The distribution of structural units was studied by ^{29}Si-MAS-NMR spectroscopy. The ratio of the Q^3/Q^4 building units increases with the dielectric constant ε_r of the solvent in all cases. Pore size distributions were determined by nitrogen adsorption. In most cases a correlation between the Q^3/Q^4 ratio and the pore size could be found.

Introduction

The advantages associated with the sol–gel process make it a suitable method for the preparation of engineering products [1, 2]. In most cases silica gels are prepared by hydrolysis of a mixture of silicic acid esters such as tetramethoxysilane (TMOS), alcohol (methanol), and water under acid or basic conditions. Different solvents are used to avoid the development of cracks during the drying step [3, 4]. Several investigations have focused on the solvent effects on the kinetics of the hydrolysis reaction and on the influence on pore size distribution and surface area [5, 6].

In this study, the effect of different solvents on the structure of the siloxane skeleton is investigated by ^{29}Si-MAS-NMR spectroscopy. Consequences of the distribution of the building units for the pore size are examined by nitrogen adsorption.

Results

The gels are produced from a mixture of TMOS, water, and solvent (Table 1) without catalyst and with dibutyl tin dilaurate (DBTDL), *N*-methylimidazole (NMIM), or HCl as catalyst.

The ^{29}Si-MAS spectra of some xerogel samples are displayed in Fig. 1. The assignment of the peaks at -92, -101, and -110 ppm to middle (Q^2), trifunctional (Q^3), and tetrafunctional (Q^4) building groups is done according to literature data [7].

Table 1. Solvents and dielectric constants [8].

	Dioxane	Acetone	Methanol	Acetonitrile	DMF	DMSO	Formamide
Dielectric constant ε_r	2.20	21.01	33.00	36.64	38.25	47.24	111.00

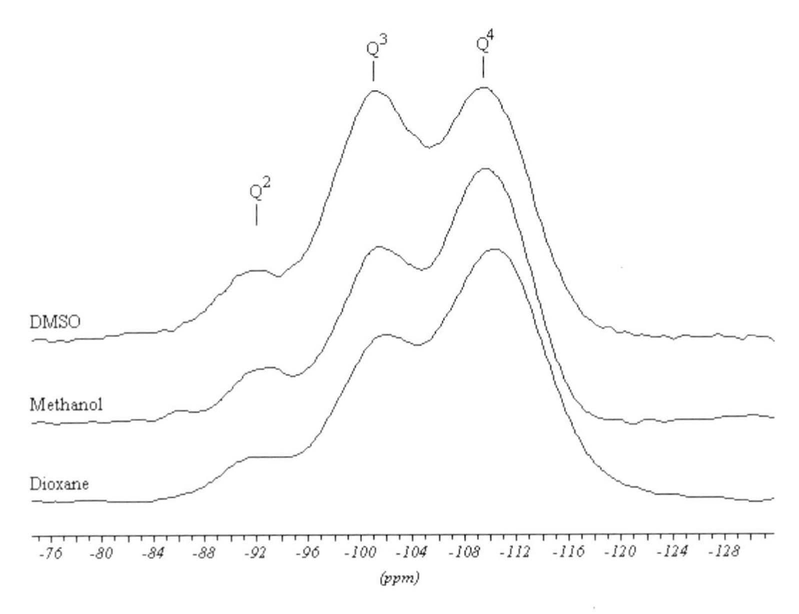

Fig. 1. ^{29}Si-MAS spectra of xerogels prepared by the hydrolysis of TMOS in different solvents with HCl as catalyst.

The catalyzed xerogels show with increasing dielectric constant ε_r of the solvent a nearly linear increase of the ratio of surface to core structure (Q^3/Q^4), with the exception of the gels prepared in formamide (Fig. 2). The highest share of trifunctional branching groups Q^3 is found in gels catalyzed by DBTDL. In the case of NMIM as catalyst the core structure units Q^4 predominate. HCl-catalyzed xerogels have ratios Q^3/Q^4 between the other catalyzed gels but some deviations from a linear increase with the dielectric constant appear. Also, for the xerogels produced without any catalyst, a certain increase is observed but no linear correlation to ε_r is found.

Table 2. Coefficients of the linear regression ($Y = A + BX$) in Fig. 2.

Catalyst	*A*	*B*	*r*
DBTDL	0.60418	0.01025	0.91167
HCl	0.40569	0.00795	0.83358
NMIM	0.31827	0.00341	0.93521

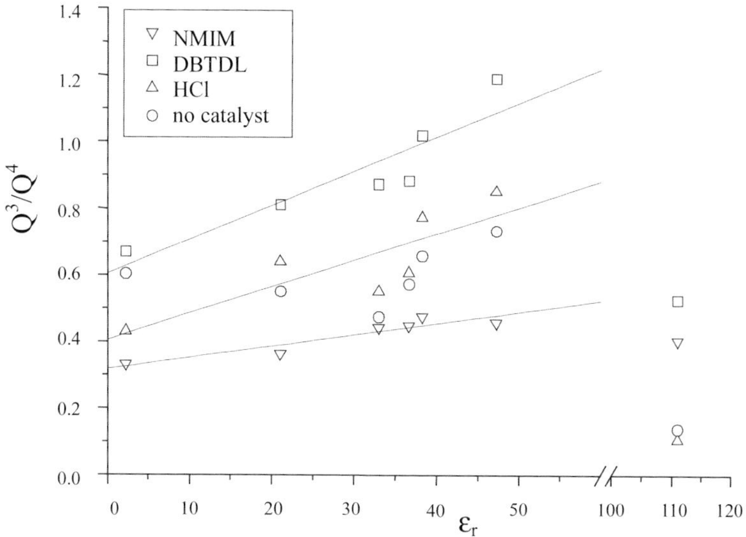

Fig. 2. Ratio of Q^3/Q^4 building units as a function of the dielectric constant ε_r of the solvent (for ε_r see Table 1, for *r* see Table 2).

From the pore size distribution examined by nitrogen adsorption it follows that xerogels with a minor content of core structure Q^4 often have no pores or micropores with width less than ~2 nm, whereas in gels with a predominating content of tetrafunctional building groups Q^4 mesopores are frequently observed.

Conclusion

This short review demonstrates how much the structure of the siloxane skeleton is affected by the polarity of the solvent employed and by the properties of the catalyst. The ratio of surface to core structure has an important influence on the pore size distribution of the silica gel.

Acknowledgement: We thank Herr V. Ehwald, *Johannes-Gutenberg Universität Mainz*, for the determination of pore size distributions.

References:

[1] B. E. Yoldas, *J. Mater. Sci.* **1979**, *14*, 1843.

[2] D. R. Ulrich, *J. Non-Cryst. Solids* **1990**, *121*, 465.

[3] A. H. Boonstra, T. N. M. Bernardts, J. J. T. Smits, *J. Non-Cryst. Solids* **1989**, *109*, 141.

[4] L. L. Hench, in: *Science of Ceramic Chemical Processing* (Eds.: L. L. Hench, D. R. Ulrich), Wiley, New York, **1986**, p. 52.

[5] I. Artaki, T. W. Zerda, J. Jonas, *Mater. Lett.* **1985**, *3*, 493.

[6] V. K. Parashar, V. Raman, O. P. Bahl, *J. Mater. Sci. Lett.* **1996**, *15*, 1403.

[7] U. Damrau, H. C. Marsmann, O. Spormann, P. Wang, *J. Non-Cryst. Solids* **1992**, *145*, 164.

[8] *CRC Handbook of Chemistry and Physics*, CRC Press, New York, 76th edition, **1995.**

Reactions of Siloxysilanes with Alkali Metal Trimethylsilanolates

*J. Harloff, E. Popowski**

Fachbereich Chemie
Universität Rostock
Buchbinderstr. 9, D-18051 Rostock, Germany

A. Tillack

Institut für Organische Katalyseforschung an der Universität Rostock
Buchbinderstr. 7-8, D-18055 Rostock, Germany
Tel.: Int. code + (381)4981822 — Fax: Int. code + (381)4981819

Keywords: Siloxysilanes / Reaction Behavior / Si–O Bond Splitting/ Alkali Metal Trimethylsilanolates

Summary: Siloxysilanes $(Me_3SiO)_3SiR$ (**1**: R = $OSiMe_3$; **2**: R = Ph; **3**: R = Me; **4**: R = H) and $(Et_3SiO)_3SiH$ (**5**) were allowed to react with alkali metal trimethylsilanolates. In the reactions of **1** and **2** with $MOSiMe_3$ (M = Li, Na, K) and of **3** with $MOSiMe_3$ (M = Li, Na) only Si–O-bond splitting takes place and corresponding $MOSiR(OSiMe_3)_2$ **6**–**13** are formed. The extent of reaction is $KOSiMe_3 >> NaOSiMe_3 > LiOSiMe_3$ and R = $OSiMe_3$ > Ph > Me. **4** reacts with $LiOSiMe_3$ to produce $LiOSiH(OSiMe_3)_2$ (**17**). In the course of the reaction of **4** with $MOSiMe_3$ (M = Na, K) the Si–O as well as the Si–H bond is broken. The reaction with $NaOSiMe_3$ produces **1**, **7**, **18**, $(Me_3Si)_2O$ and Me_3SiH, with $KOSiMe_3$ **1**, **8**, $(Me_3Si)_2O$ and Me_3SiH. **5** reacts with $KOSiMe_3$ to form $KOSiH(OSiEt_3)_2$ (**20**).

Introduction

The Si–O–Si group is split by bases because of the electrophilic character of the silicon atoms [1]. Cyclic and acyclic siloxanes react with stoichiometric amounts of organolithium compounds in a defined way with Si–O–Si splitting and formation of silanolates [2–4]. The base-catalysed polymerization of cyclosiloxanes by alkali metal trimethylsilanolates, particularly $KOSiMe_3$, starts with an Si–O bond splitting of the cyclosiloxanes forming a pentacoordinated transition state [5]. Stable pentacoordinated alkoxysilicates with monodentate alkoxy ligands are obtained in the reaction of the alkoxysilanes $Si(OR)_4$, $Ph_nSi(OR)_{4-n}$ (n = 1, 2) and $HSi(OR)_3$ with the corresponding potassium alcoholates KOR [6–8]. The reaction behavior of acyclic branched siloxanes with stoichiometric amounts of alkali metal trimethylsilanolates has not been investigated. Studies of the reaction behavior of the siloxysilanes $(Me_3SiO)_3SiR$ (**1**: R = $OSiMe_3$; **2**: R = Ph; **3**: R = Me; **4**: R = H) and $(Et_3SiO)_3SiH$ (**5**) towards the alkali metal trimethylsilanolates $MOSiMe_3$ (M = Li, Na, K) in an equimolar ratio in THF at room temperature (r.t.) in terms of Si–O bond splitting and formation of pentacoordinated siloxysilicates are the subject of the present work.

Results and Discussion

The siloxysilanes **1**–**4** have been prepared by equilibration of the chlorosilanes Cl_3SiR and $SiCl_4$ with hexamethyldisiloxane in 12.8 M sulfuric acid [9, 10] (Eqs. 1, 2).

$$4\,(Me_3Si)_2O \quad + \quad SiCl_4 \xrightarrow{12.8\ M\ H_2SO_4} \underset{\mathbf{1}\ (59\ \%)}{(Me_3SiO)_4Si} \quad + \quad 4\ Me_3SiCl$$

Eq. 1.

$$3\,(Me_3Si)_2O \quad + \quad Cl_3SiR \xrightarrow{12.8\ M\ H_2SO_4} \underset{\mathbf{2\text{–}4}\ (63\text{–}67\ \%)}{(Me_3SiO)_3SiR} \quad + \quad 3\ Me_3SiCl$$

	2	**3**	**4**
R	Ph	Me	H

Eq. 2.

Tris(triethylsiloxy)silane (**5**) was obtained by reaction of trichlorosilane with $LiOSiEt_3$ in *n*-pentane/diethyl ether (1:1) (Eq. 3).

$$Cl_3SiH \quad + \quad 3\ LiOSiEt_3 \xrightarrow[-78\ °C \rightarrow 20°C]{n\text{-pentane}/Et_2O} \underset{\mathbf{5}\ (52\ \%)}{(Et_3SiO)_3SiH} \quad + \quad 3\ LiCl$$

Eq. 3.

The siloxysilanes **1** and **2** react with $MOSiMe_3$, mainly with splitting of one Si–O bond giving the corresponding alkali metal trimethylsiloxysilanolates (Eq. 4, Table 1). The formation of trimethylsiloxysilanolates also takes place in the reaction of **3** with $LiOSiMe_3$ and $NaOSiMe_3$.

$$\underset{\mathbf{1\text{-}3}}{(Me_3SiO)_3SiR} \quad + \quad MOSiMe_3 \xrightarrow{THF} \underset{\mathbf{6\text{–}13}}{MOSiR(OSiMe_3)_2} \quad + \quad (Me_3Si)_2O$$

	6	**7**	**8**	**9**	**10**	**11**	**12**	**13**
M	Li	Na	K	Li	Na	K	Li	Na
R	$OSiMe_3$	$OSiMe_3$	$OSiMe_3$	Ph	Ph	Ph	Me	Me

Eq. 4.

The reaction of **3** with $KOSiMe_3$ produces a mixture of oligomeric and polymeric compounds and small amounts of $KOSiMe(OSiMe_3)_2$.

Table 1. ^{29}Si-NMR data of the alkali metal trimethylsiloxysilanolates $MOSiR(OSiMe_3)_2$ **6**–**13** (δ in ppm; solvents: **6**–**8** *n*-hexane, **9**–**13** THF).

	6	**7**	**8**	**9**	**10**	**11**	**12**	**13**
$\delta[Si(O_3R)]$	–95.0	–92.9	–93.3	–73.8	–72.3	–72.5	–61.6	–61.5
$\delta[Si(OC_3)]$	6.6	5.9	4.3	2.3	2.2	3.2	2.2	1.1

The extent of formation of alkali metal trimethylsiloxysilanolates has been determined by trapping of the silanolates with chlorodimethylsilane (Eq. 5, Table. 2).

$$\underset{\mathbf{6\text{–}13}}{MOSiR(OSiMe_3)_2} + ClSiMe_2H \xrightarrow{THF} \underset{\mathbf{14\text{–}16}}{HMe_2SiOSiR(OSiMe_3)_2} + MCl$$

	14	**15**	**16**
R	$OSiMe_3$	Ph	Me

Eq. 5.

Table 2. Reaction time of the reaction of siloxysilanes **1**–**3** with $MOSiMe_3$ and yields of the trapping products **14**–**16** besides the corresponding siloxysilanes.

	$(Me_3SiO)_3SiR$ R	$MOSiMe_3$ M	Reaction time [h]	Product	Yield [%]
1	$OSiMe_3$	Li	240	**14**	5.9
1	$OSiMe_3$	Na	240	**14**	15.4
1	$OSiMe_3$	K	96	**14**[a]	87.7
2	Ph	Li	240	**15**	3.0
2	Ph	Na	240	**15**	25.2
2	Ph	K	96	**15**[b]	54.6
3	Me	Li	240	**16**	3.7
3	Me	Na	240	**16**	4.5
3	Me	K	96	**16**	6.0[c]

[a] about 1 % $(HMe_2SiO)_2Si(OSiMe_3)_2$ and $(HMe_2SiO)_3SiOSiMe_3$ — [b] 1–2 % $(HMe_2SiO)_2SiPh(OSiMe_3)$ and $(HMe_2SiO)_3SiPh$ — [c] besides **16** oligomeric and polymeric compounds.

The data indicate that the extent of reaction according to Eq. 5 depends on $MOSiMe_3$ and on the substituents R of the $(Me_3SiO)_3SiR$. The following sequences are found:

$$KOSiMe_3 \gg NaOSiMe_3 > LiOSiMe_3 \text{ and } R = OSiMe_3 > Ph > Me.$$

Furthermore the trapping experiments show that in the reaction of **1** and **2** with $KOSiMe_3$ more than one Si–O bond is broken.

Treatment of the siloxysilane **4** with $LiOSiMe_3$ leads to the lithium bis(trimethylsiloxy)hydridosilanolate (**17**) (Eq. 6). The yield has been determined by trapping of **17** with chlorodimethylsilane.

$$\underset{\mathbf{4}}{(Me_3SiO)_3SiH} + LiOSiMe_3 \xrightarrow[\text{48 h r.t.}]{\text{THF}} \underset{\mathbf{17}\ (12\ \%)}{LiOSiH(OSiMe_3)_2} + (Me_3Si)_2O$$

Eq. 6.

The course of reaction of **4** with sodium and potassium trimethylsilanolate is more complicated, compared to the siloxysilanes **1**–**3**.

4 reacts with sodium trimethylsilanolate to form **1**, $NaOSiH(OSiMe_3)_2$ (**18**), **7**, Me_3SiH and $(Me_3Si)_2O$ (Scheme 1). In addition to these compounds, the starting compounds **4** (about 25 %) and $NaOSiMe_3$ are still contained in the reaction mixture.

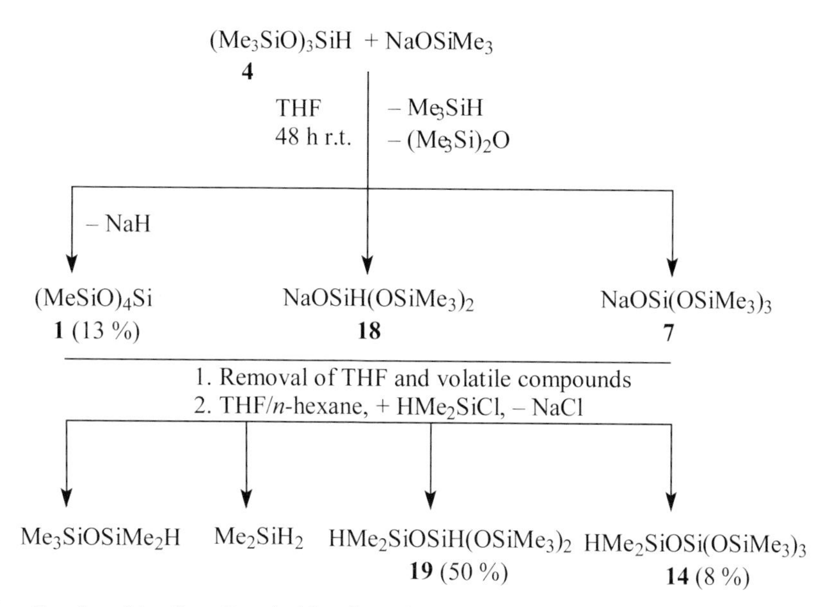

Scheme 1. Reaction of the siloxysilane **4** with sodium trimethylsilanolate.

Treatment of the mixture after removal of the solvent and the volatile compounds under reduced pressure with chlorodimethylsilane leads to pentamethyldisiloxane, dimethylsilane, and dimethylsiloxysilanes **19** and **14** (Scheme 1).

Under the same conditions as in the reaction with $NaOSiMe_3$ **4** reacts with $KOSiMe_3$ nearly completly. The formation of Me_3SiH, **1**, **8** and of an unknown product with an Si–H group {ν(Si–H) = 2080 cm^{-1}, δ[H(Si),THF] = 4.38 ppm)} is observed (Scheme 2). The H(Si)-NMR signal is very broad. The ^{29}Si–^{1}H coupling constant could not determined. Removal of the solvent and the volatile compounds from the reaction mixture yields a white solid, of which treatment with HMe_2SiCl produces Me_2SiH_2, **1** and **14**.

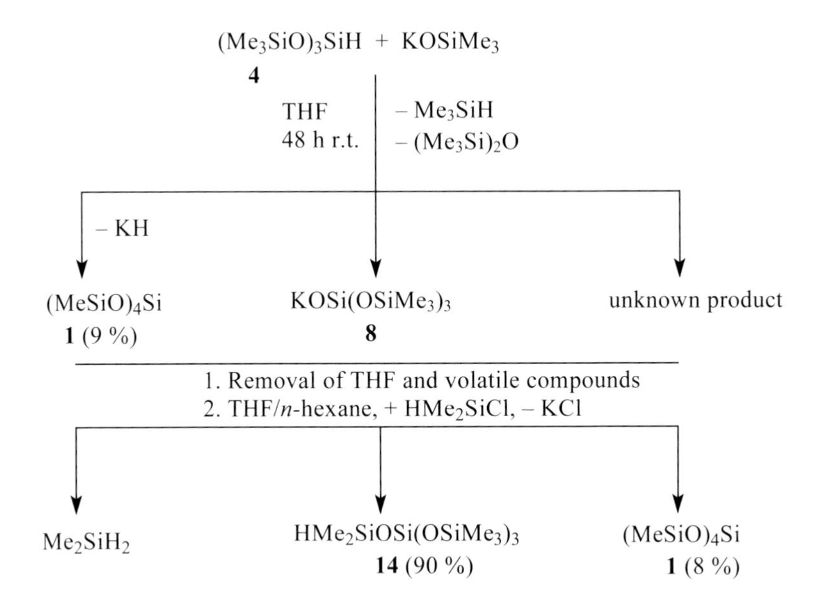

Scheme 2. Reaction of the siloxysilane **4** with potassium trimethylsilanolate.

The observed reaction products of the reactions of **4** with sodium and potassium trimethylsilanolates show the course of several competitive reactions with splitting of Si–O as well as Si–H bonds. In contrast to $(Me_3SiO)_3SiH$ (**4**) the siloxysilane $(Et_3SiO)_3SiH$ (**5**) reacts with $KOSiMe_3$ only under Si–O bond splitting (Eq. 7), presumably due to steric reasons.

$$\underset{\mathbf{5}}{(Et_3SiO)_3SiH} + KOSiMe_3 \xrightarrow[48\text{ h r.t.}]{THF} \underset{\mathbf{20}\ (68\ \%)}{KOSiH(OSiEt_3)_2} + Et_3SiOSiMe_3$$

Eq. 7.

Table 3. ^{29}Si-NMR and IR data of alkali metal hydridotrimethylsiloxysilanolates (δ in ppm (THF), J in Hz, ν in cm^{-1}).

Compound	$\delta[Si(O_3H)]$	$\delta[Si(OC_3)]$	$^1J[^{29}Si–^1H]$	ν(Si–H)
17	–79.4	2.7	253.1	2117
18	–74.2	5.0	250.1	–
20	–78.0	6.2	241.3	2097
21[a]	–73.3	2.6	238.4	2101

[a] **21**: $KOSiH(OSiMe_3)_2$

The ^{29}Si-NMR and IR data of the alkali metal hydridotrimethylsiloxysilanolates **17–21** are shown in Table 3. The compounds **6–8**, **14–17** and **21** have been synthesized by standard procedures as comparison substances for the investigations of the reaction behavior of the silyloxysilanes towards alkali metal trimethylsilanolates.

Acknowledgment: We thank the *Fonds der Chemischen Industrie* for financial support.

References:

[1] M. G. Voronkov, V. P. Mileshkevich, Y. A. Yushelevski, *The Siloxane Bond,* Plenum Press, New York, **1978.**

[2] D. Seyferth, D. L. Alleston, *Inorg. Chem.* **1963**, *2*, 418.

[3] C. L. Frye, R. M. Salinger, F. W. G. Fearon, J. M. Klosowski, T. Deyoung, *J. Org. Chem.* **1970**, *35*, 1308.

[4] Y. Pai, K. L. Serris, W. P. Weber, *Organometallics* **1986**, *5*, 683.

[5] A. R. Bassindale, P. G. Taylor, *The Chemistry of Organosilicon Compounds* (Eds.: S. Patai, Z. Rappoport), John Wiley & Sons, Chichester, **1989**, p. 839.

[6] K. C. Kumara Swamy, V. Chandrasekhar, J. J. Harland, J. M. Holmes, R. O. Day, O. Roberta, R. R. Holmes, *J. Am. Chem. Soc.* **1990**, *112*, 2341.

[7] R. J. P. Corriu, Ch. Guerin, B. J. L. Henner, Q. Wang, *Organometallics* **1991**, *10*, 2297, 3574.

[8] C. Chuit, R. J. P. Corriu, C. Reye, J. C. Young, *Chem. Rev.* **1993**, *93*, 1371.

[9] E. Popowski, H. Kelling, *Wiss. Z. Univ. Rostock, Math.-Nat. R.* **1988**, *37*, 24.

[10] U. Scheim, H. Grosse-Ruyken, K. Rühlmann, *J. Organomet. Chem.* **1986**, *312*, 27.

Cyclopentadienyl (Cp)-Substituted Silanetriols and Novel Titanasiloxanes as Condensation Products

Manuela Schneider, Beate Neumann, Hans-Georg Stammler, Peter Jutzi

Fakultät für Chemie, Universität Bielefeld
Universitätsstr. 25, D-33615 Bielefeld, Germany
Tel.: Int. code + (521)106 6181 — Fax.: Int. code + (521)106 6026
E-Mail: peter.jutzi@uni-bielefeld.de

Keywords: Silanetriols / Polyhedral Titanasiloxanes / Hydrogen-Bonded Networks / Extended Structures

Summary: A range of stable silanetriols bearing different Cp systems have been isolated. Silanetriols are useful synthons for polyhedral metallasiloxanes. Condensation of Cp silanetriols with titanium alkoxides yielded oligomeric titanasiloxanes with interesting structural features. Transformation of such polyhedra into extended structures by Cp–Si bond splitting has been pursued. Furthermore these compounds can be considered as valuable model systems for titanasilicates.

Silanetriols

Silanetriols represent versatile building blocks for the construction of polyhedral metallasiloxanes. To transform such oligomers into supramolecular structures special efforts are directed towards the preparation of stable silanetriols bearing hydrolyzable functionalities. Stabilization of silanetriols with respect to self-condensation can be achieved by steric and electronic shielding of the Si atom.

Our approach to this topic is the application of cyclopentadienyl (Cp) ligands (see Fig. 1).

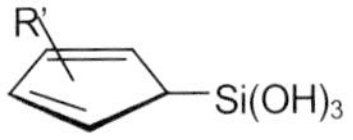

Fig. 1. Cp-substituted silanetriol.

Modification of the substitution pattern R' allows the fine tuning of steric and electronic properties of the Cp system. Thus the stable Cp-substituted silanetriols shown in Fig. 2 have been isolated and characterized.

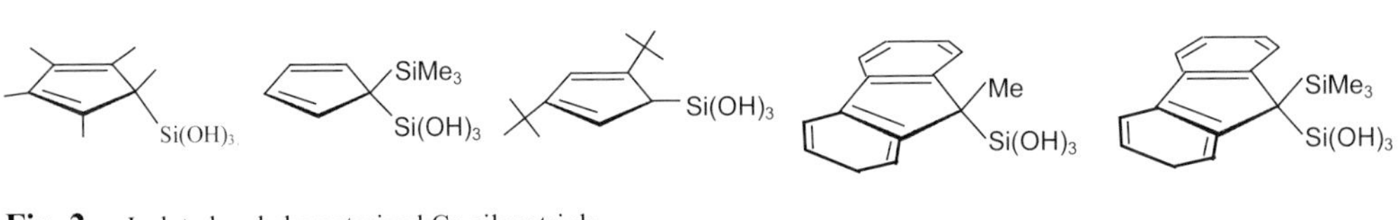

Fig. 2. Isolated and characterized Cp silanetriols.

Synthesis of these compounds was accomplished by controlled hydrolysis as depicted in Eq. 1.

R' — $SiCl_3$ → (3 Aniline, 3 H_2O, - 3 Aniline·HCl, Et_2O, 0°C) → R' — $Si(OH)_3$

Eq. 1. Synthesis of Cp silanetriols.

Depending on the Cp moiety they differ strongly in their properties and reactivity. This is illustrated by comparision of compounds **1** and **2**: whereas **1** undergoes self-condensation at room temperature, **2** cannot be forced to selfcondense even after prolonged heating. The remarkable stability of **2** in comparision with **1** can be explained by

- a reduced electrophilicity of the silicon atom (δ ^{29}Si ([D_6]DMSO) **1**: –48.7, **2**: –72.3ppm)
- a better steric shielding ($SiMe_3$ moiety instead of a Me group in the allylic position)

The different steric situation at these silanetriols is clearly seen in the X-ray structures. Based on hydrogen-bonded networks silanetriols are arranged in the solid state in a variety of supramolecular structures [1]. Whereas **1** crystallizes as a hemihydrate in a double-sheet structure [2] (Fig. 3), **2** forms a tube structure [3] (Fig. 4).

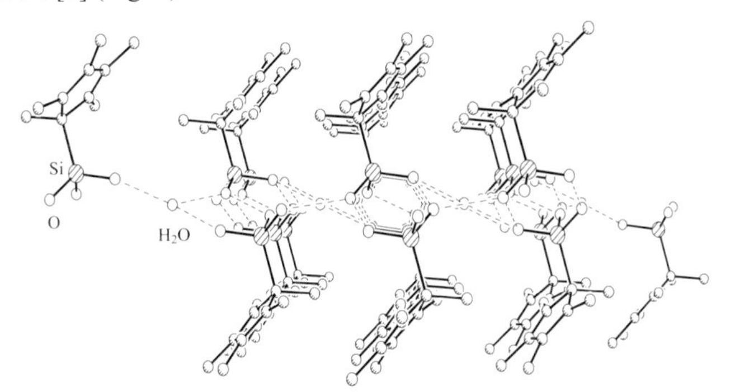

Fig. 3. Double-sheet structure of **1** · ½ H_2O.

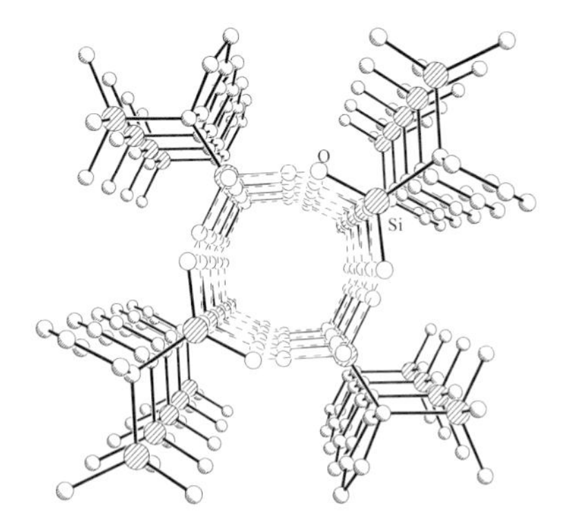

Fig. 4. Tube structure of **2**.

Polyhedral Titanasiloxanes

The synthesis of polyhedral titanasiloxanes was accomplished by condensation of silanetriols with titanium alkoxides [4] (Fig. 5).

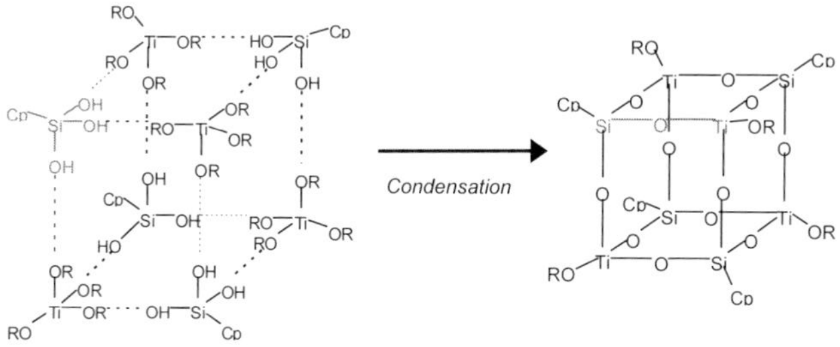

Fig. 5. Synthesis of polyhedral titanasiloxanes.

Interestingly this simple procedure does not necessarily lead to the expected symmetric frameworks. Instead even small structural changes at the reactants Ti(O**R**)$_4$ and Cp(**R**') Si(OH)$_3$ or variation of the reaction parameters result in a variety of oligomeric systems with interesting structural features. For example, Scheme 1 depicts the products obtained by the reaction of Fluorenyl(SiMe$_3$)Si(OH)$_3$ **5** with different titanium alkoxides under different reaction conditions.

Compounds **6**–**9** can be distinguished by the

- coordination number of Ti
- type of μ_n-O ($n = 2$–4) and μ_2-OR linkages of the Ti atoms
- ratio of Si to Ti

Although the cube structures **6** and **7** appear to be highly symmetric, closer inspection reveals that the cages show no symmetry at all. A strong distortion of the cube is indicated by the broad range of the Ti-O-Si angles from 137.13° to 165.09°.

The structure of **8** resembles a cube-like polyhedron with an additional "roof". The "roof" is created by insertion of the Ti(4)–(μ_3-O)$_2$–Ti(5) structural element into a cube face (Scheme 1 gives a side view). A top view of the upper part of the molecule in Fig. 6 shows the interconnection between Ti(3) and Ti(6) via μ_3-O and μ_2-O*i*Pr linkages and the differing coordination sphere at the Ti atoms. While Ti(1) and Ti(2) are tetrahedrally coordinated, Ti(3) and Ti(6) are trigonal-bipyramidally coordinated and Ti(4) and Ti(5) are square-pyramidally coordinated. The deviation of the Si:Ti ratio to 4:6 from the 1:1 stoichiometry of the reactants and the existence of Ti–O–Ti linkages indicate side reactions. We presume according to Eq. 2 the condensation of the silanetriol molecules. Hydrolysis of Ti-O*i*Pr bonds by the liberated water followed by condensation results in the formation of μ_3-O units. When the reactants are mixed according to the stoichiometry found, compound **8** is obtained in analytically pure form in quantitative yield.

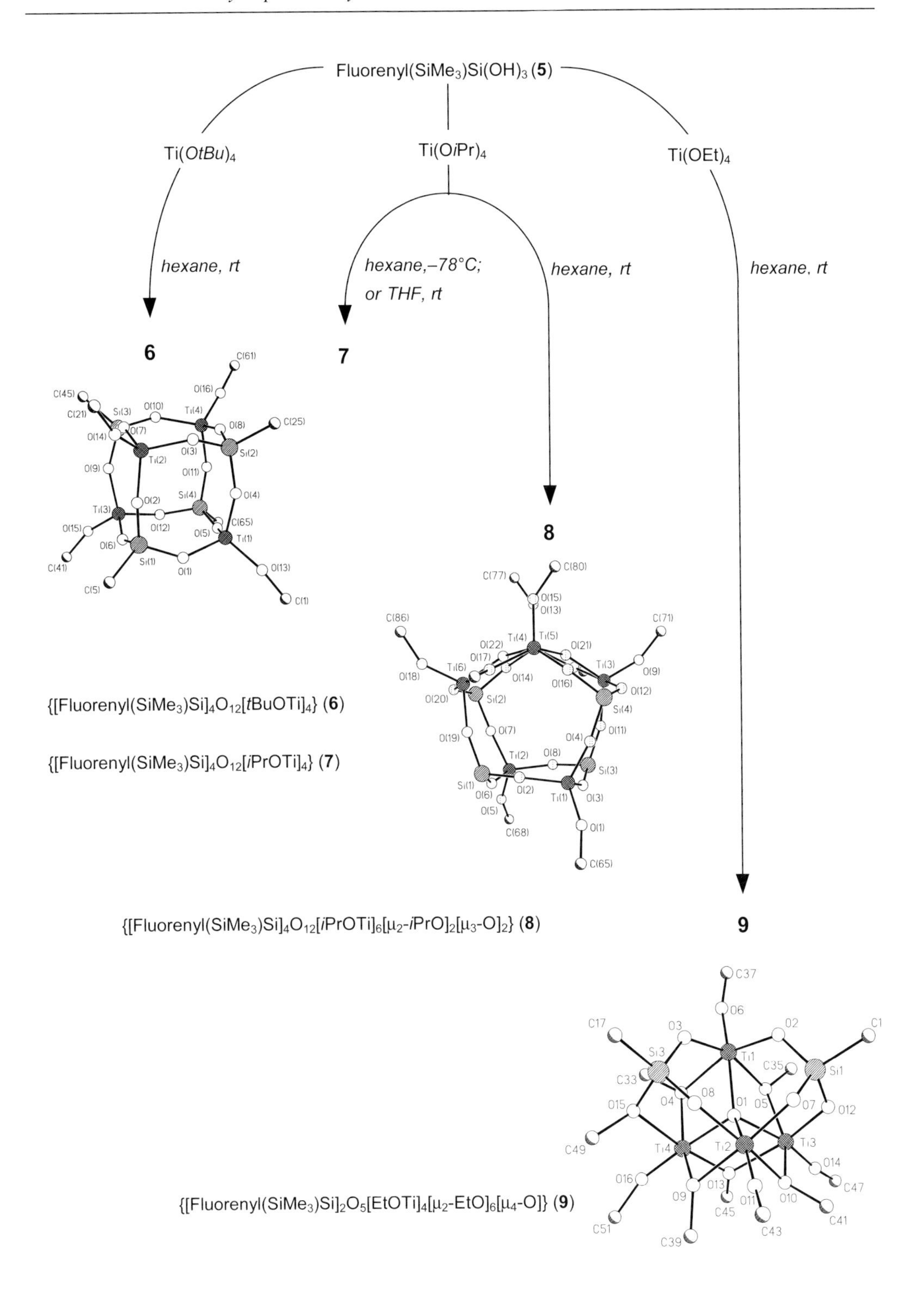

Scheme 1. Condensation reactions of Fluorenyl($SiMe_3$)Si$(OH)_3$ with titanium alkoxides.

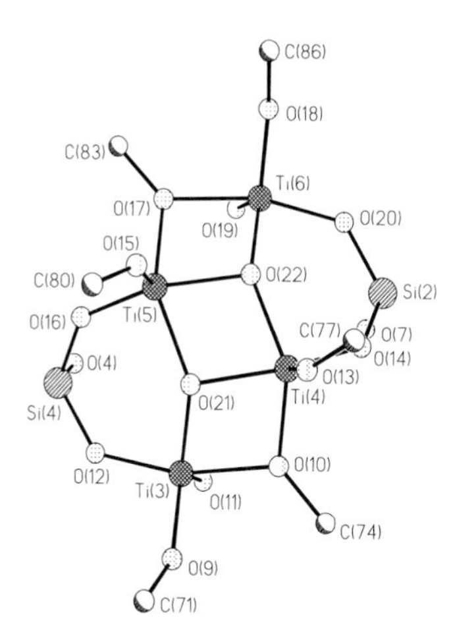

Fig. 6. Top view of the upper half of **8**.

$$4\ \text{Fluorenyl}(SiMe_3)Si(OH)_3 + 6\ Ti(O\textit{i}Pr)_4 + 2\ H_2O \longrightarrow \mathbf{8} + 14\ HO\textit{i}Pr$$

Side reaction: $2\ \text{Fuorenyl}(SiMe_3)Si(OH)_3 \longrightarrow \text{Siloxane species} + 2\ H_2O$

Eq. 2. Considerations regarding the stoichiometry of **8**.

The intriguing molecule **9** is obtained by condensation of Fluorenyl($SiMe_3$)$Si(OH)_3$ with $Ti(OEt)_4$. The structure might be described as a tetrahedron created by four Ti atoms linked by a μ_4-O unit. With one exception the tetrahedron edges are formed by μ_2-OEt bridges. Two Fluorenyl($SiMe_3$)SiO_3 groups are located on two triangular surfaces of the tetrahedron. All Ti fragments show an octahedral coordination sphere. A polyhedron representation in Fig. 7 visualizes the connection of the Ti octahedra and the Si tetrahedra. Although a 1:1 ratio of the reactants was chosen, the stoichiometry found in the cage is 4Ti:2Si. Again condensation of silanetriol molecules and hydrolysis of Ti-OEt groups followed by condensation to the μ_4-O unit is assumed (see Eq. 3). Reaction of the silanetriol, water and titanium ethoxide in the appropriate stoichiometry leads exclusively to compound **9**. A very interesting structural feature is the first observed Ti–O(Et)–Si linkage.

$$2\ \text{Fluorenyl}(SiMe_3)Si(OH)_3 + 4\ Ti(O\textit{Et})_4 + H_2O \longrightarrow \mathbf{9} + 6\ HO\textit{Et}$$

Side reaction: $2\ \text{Fuorenyl}(SiMe_3)Si(OH)_3 \longrightarrow \text{Siloxane species} + H_2O$

Eq. 3. Considerations regarding the stoichiometry of **9.**

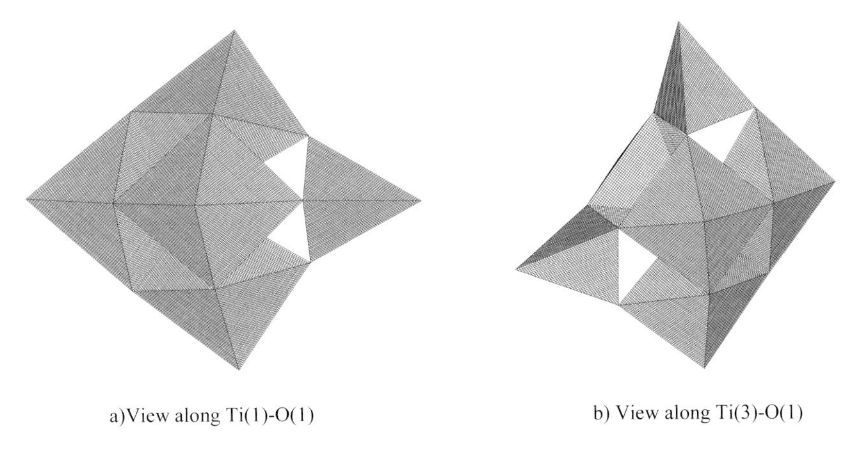

a)View along Ti(1)-O(1)
b) View along Ti(3)-O(1)

Fig. 7. Polyhedron representation of **9**.

In conclusion two aspects are important regarding the titanasiloxane structures:

- even small modifications lead to great structural effects
- structure elements such as μ_n-O (n = 2–4) and μ_2-OR linkages of Ti atoms show the analogy to Ti oxo–alkoxide chemistry

The synthesized titanasiloxane oligomers represent valuable model compounds for understanding of

- the sol-gel process to Ti–Si mixed oxides [5] (structure–function relationship)
- the catalytic mechanism and active sites at titanasilicates in epoxidation processes

Outlook

Besides allowing a fine tuning of electronic and steric properties by modification of the substitution pattern R', the Cp moiety has proven to be an excellent leaving group. Thus following the synthesis of simple polyhedral structures (e.g. **6** and **7**) a selective Cp–Si bond splitting should offer a way to new extended structures on the basis of Si–O–Ti frameworks (Fig. 8). We are currently pursuing this aspect and have obtained initial results supporting this theory.

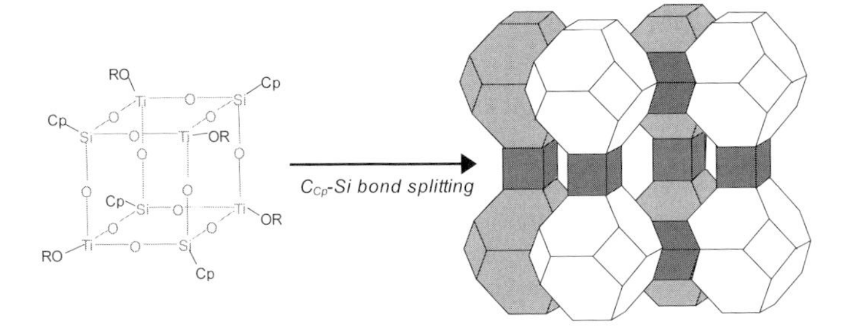

Fig. 8. Synthesis of extended structures by Cp–Si bond splitting.

Acknowledgment: This work has been supported by the *Deutsche Forschungsgemeinschaft*. Manuela Schneider thanks the Federal State *Nordrhein-Westfalen* for a research fellowship.

References:

[1] P. D. Lickiss, *Adv. Inorg. Chem.* **1995**, *42*, 147.
[2] P. Jutzi, G. Straßburger, M. Schneider, H.-G. Stammler, B. Neumann, *Organometallics* **1996**, *15*, 2842.
[3] P. Jutzi, M. Schneider, H.-G. Stammler, B. Neumann, *Organometallics* **1997**, *16*, 5377.
[4] R. Murugavel, V. Chandrasekhar, H. W. Roesky, *Acc. Chem. Res.* **1996**, *29*, 183.
[5] R. J. Davies, Z. Liu, *Chem. Mater.* **1997**, *9*, 2311.

Metal-Fragment Substituted Disilanols[†]

Wolfgang Malisch*, Heinrich Jehle, Markku Lager

Institut für Anorganische Chemie der Universität Würzburg
Am Hubland, D-97074 Würzburg, Germany
Tel.: Int. Code + (931)888 5277 — Fax: Int. Code + (931)888 4618
E-mail: Wolfgang.Malisch@mail.uni-wuerzburg.de

Martin Nieger

Institut für Anorganische Chemie der Universität Bonn
Gerhard-Domagk-Str. 1, D-53121 Bonn, Germany

Keywords: Transition Metal Effect / Metallodisilanes / Hydrolysis / Metallodisilanols / Metallosiloxanes

Summary: The metallodisilanes $C_5R_5(OC)_2Fe–Si_2Cl_5$ (**2a,b**) and $C_5R_5(OC)_2Fe–Si_2H_5$ (**3a,b**) (R = H, Me) used for the synthesis of metallodisilanols have been structurally characterized. Hydroxylation of $Cp(OC)_2Fe–Si_2Me_4Cl$ (**4**) has been realized with water in the presence of Et_3N to yield the corresponding ferriodisilanol $Cp(OC)_2Fe–Si_2Me_4OH$ (**5**). Formation of the metallodichlorotrihydroxydisilanes $C_5R_5(OC)_2M–SiCl_2–Si(OH)_3$ [R = H, M = Fe (**7a**); R = Me, M = Ru (**7b**)] is achieved by regioselective hydrolysis of the metallopentachlorodisilanes **2a** and $C_5Me_5(OC)_2Ru–Si_2Cl_5$ (**2c**). In an analogous manner $[Cp(OC)_2Fe–SiCl_2]_2$ (**11**) is converted to $[Cp(OC)_2Fe–Si(OH)_2]_2$ (**12**). Controlled base-assisted condensation of **5**, **7b** and **12** leads to the formation of the novel metallodisilanesiloxanes $Cp(OC)_2Fe–Si_2Me_4OSiMe_2H$ (**6**), $Cp^*(OC)_2Ru–SiCl_2Si(OSiMe_2H)_3$ (**8**) and $[Cp(OC)_2Fe–Si(OSiMe_2H)_2]_2$ (**13**). Specific introduction of hydroxyl groups at the α-silicon of metallodisilanes is realized by conversion of the ferriodisilanes $C_5R_5(OC)_2Fe–Si_2H_5$ (**3a,b**) into the ferriodichlorodisilanes $C_5R_5(OC)_2Fe–SiCl_2SiH_3$ (**9a,b**) with CCl_4 followed by hydrolysis to give $Cp(OC)_2Fe–Si(OH)_2–SiH_3$ (**10**).

Introduction

Due to their important role as intermediates in the technical synthesis of silicones, organosilanols have been the subject of extensive studies over several decades [2]. In context with studies on the reactivity of functionalized silicon–transition metal complexes a new type of silanols containing an Si-bonded transition metal has been established. Access to these metallosilanols is opened by the hydrolysis of metallohalosilanes [3] and by the oxygenation of metallosilanes with dimethyldioxirane [4]. Recently also a catalytic pathway for the SiH/SiOH conversion in the case of

---

[†] Part 19 of the series „*Metallo-Silanols and Metallo-Siloxanes*". In addition, part 46 of the series „*Synthesis and Reactivity of Silicon Transition Metal Complexes*". Part 18/45 see [1].

metallosilanes using the system methyltrioxorhenium/urea–hydrogen peroxide (MTO/UHP) was reported [5]. Due to the remarkably high stability of metallosilanols towards self-condensation, these compounds represent very useful precursors for controlled condensation to build up unusual arrangements of functionalized siloxanes at metal centers. We have now extended this type of chemistry to the metallodisilanes $L_nM–SiX_2–SiX_3$ [L_nM = $C_5R_5(OC)_2Fe/Ru$, $C_5R_5(OC)_2(Me_3P)Mo/W$; X = H, Cl] since they represent interesting model compounds for the study of the "transition metal effect" on the chemical properties of silicon in the α- or β-position [6, 8]. Recently first experiments resulted in the regioselective electrophilic oxygenation of metallopentahydridodisilanes $C_5R_5(OC)_2Fe–Si_2H_5$ (R = H, Me) by dimethyldioxirane to form the metallodisilanediols $C_5R_5(OC)_2Fe–Si(OH)_2–SiH_3$ [7]. We now present further interesting examples for the introduction of hydroxyl groups into metallodisilanes and bis(metallo)disilanes.

Results and Dicussion

Metallodisilanes

As starting materials the metallopentachlorodisilanes **2a–c** and the ferriodisilanes **3a,b** (Fig. 1) have been used, which are easily accessible by heterogeneous reaction of the metalates $Na[M(CO)_2C_5R_5]$ (**1a–c**) with hexachlorodisilane followed by Cl/H substitution with $LiAlH_4$ [6].

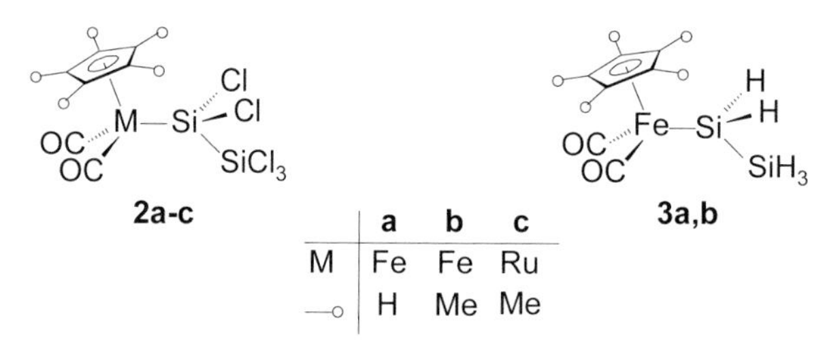

Fig. 1. The metallopentachlorodisilanes **2a–c** and the ferriodisilanes **3a,b**.

The result of the X-ray analysis of **2a** (Fig. 2) reveals different Si–Cl distances at Si1 and Si2 due to the electron-releasing capacity of the transition metal fragment. Si–Cl distances at Si2 of ca. 2.03 Å are in the range of free chlorosilanes while the Si1–Cl1 [2.0859(8) Å] and Si1–Cl2 [2.0811(9) Å] distances are significantly elongated. The Si1–Si2 distance of 2.3333(9) Å is close to that of metal-free disilanes (Me_6Si_2: 2.340 Å [9]) and to that of the disilanyl complex $C_5Me_5(OC)_2(Me_3P)W–SiCl_2–SiCl_3$ (Si1–Si2: 2.350 Å) [8].

As shown from X-ray analysis of $Cp^*(OC)_2Fe\text{-}Si_2H_5$ (**3b**), a staggered pseudo-ethane conformation concerning the Fe–Si bond is confirmed with the following pairs of *trans*-arranged substituents: Cp*/H, CO/H, CO/SiH_3 (Fig.3). Distortion with respect to the ideal conformation of about 20° for all the substituents at Si1 deduced from the dihedral angles is a consequence of the higher sterical demand of the SiH_3 group compared to the H atoms. The repulsive interaction between the silyl group and the Cp* ligand is reduced by this arrangement. The Si1–Si2 distance of 2.3355(11) Å is in the common range.

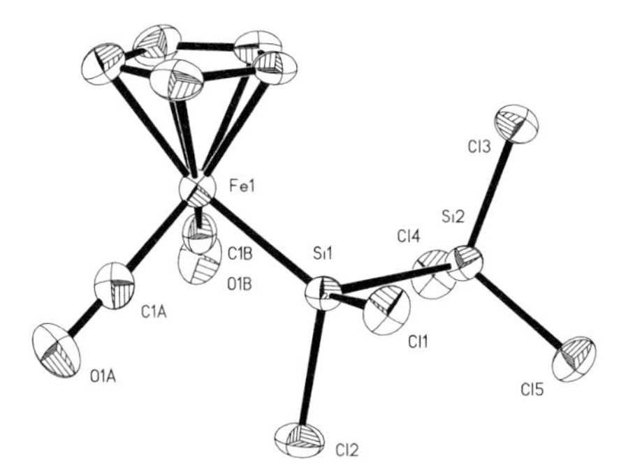

Fig 2: Molecular structure of **2a**: Selected bond lengths [Å], bond angles [°] and torsion angles [°]: Fe–Si1 2.2239(7), Si1–Si2 2.3333(9), Si1–Cl1 2.0859(8), Si2–Cl3 2.0410(9), Fe1–C1 1.770(3), C1–O1 1.137(3), C1-Fe1-Si1 88.38(8), Si2-Si1-Fe 119.20(3), Fe-C1-O1 178.9(3), C1-Fe-Si1-Si2 –80.11(8), Si1-Fe1-C1-O1 –173(10), Fe-Si1-Si2-C5 –175.05(3).

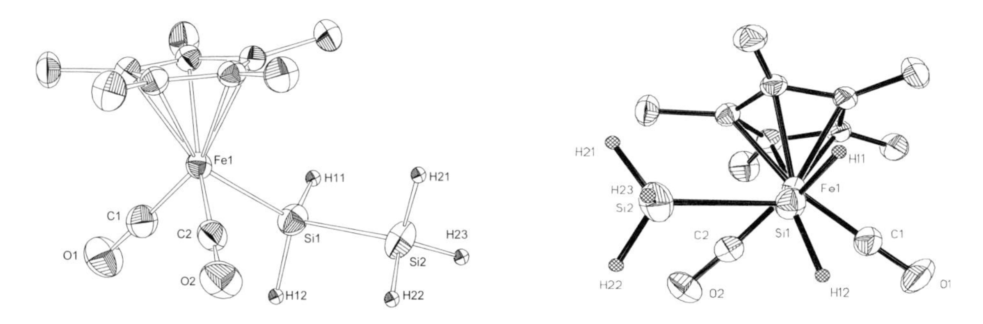

Fig. 3. Molecular structure of **3b**: Selected bond lengths [Å], bond angles [°] and torsion angles [°]: Fe–Si1 2.2980(9), Si1–Si2 2.3355(11), Si1–H11 1.43(2), Si2–H21 1.41(3), C1-Fe1-Si1 85.46(8), Si2-Si1-Fe 115.15(5), Fe-C1-O1 178.6(2), Si1-Fe1-C1-O1 –151.0(10), Si1-Fe-C2-O2 132(6), Fe-Si1-Si2-H21 –45.4(13), Cp*(Z)-Fe1-Si1-Si2 82.56, Cp*(Z)-Fe1-Si1-H11 41.72, Cp*(Z)-Fe1-Si1-H12 159.25.

Metallodisilanols

In order to get a first impression of the Cl/OH exchange activity of β-SiCl units of metallodisilanes the easily accessible $Cp(OC)_2Fe–SiMe_2–SiMe_2Cl$ (**4**) [10] was treated with 1.5 equivalents of water in the presence of Et_3N. Within a few minutes $Cp(OC)_2Fe–SiMe_2–SiMe_2(OH)$ (**5**) was formed as a microcrystalline solid. **5** is stable with respect to self-condensation and can be used for the preparation of metallodisilanesiloxanes, e.g. **6** was formed from **5** and $Me_2Si(H)Cl$ in the presence of an auxiliary base (Eq. 1).

$Cp(OC)_2Fe–Si(Me)_2–SiMe_2Cl$ (**4**) $\xrightarrow[- 2\,[Et_3NH]Cl]{+ 2\,H_2O,\ Et_3N}$ $Cp(OC)_2Fe–Si(Me)_2–SiMe_2OH$ (**5**) $\xrightarrow[- [Et_3NH]Cl]{+ Me_2Si(H)Cl,\ Et_3N}$ $Cp(OC)_2Fe–Si(Me)_2–Si(Me)_2–O–SiMe_2H$ (**6**)

Eq. 1.

This finding implies regioselective hydroxylation of the metallopentachlorodisilanes **2a,c** involving the β-silicon via base-assisted hydrolysis. Cl/OH exchange is complete within 12 h at room temperature yielding the metallodichlorotrihydroxydisilanes **7a,b** as beige, rather air-stable solids (Eq. 2). A first example of the controlled condensation is realized with **7b** to give the novel multifunctionalized (rutheniosilyl)(siloxy)trisiloxane **8** as a deep yellow solid after crystallization from *n*-pentane at –78 °C (Eq. 3).

2a,c $\xrightarrow[- 3\,[Et_3NH]Cl]{+ 3\,H_2O,\ 3Et_3N}$ **7a,b**

	a	b
M	Fe	Ru
—○	H	Me

Eq. 2.

7b $\xrightarrow[- 3\,[Et_3NH]Cl]{+ 3\,Me_2Si(H)Cl,\ 3\,Et_3N}$ **8**

Eq. 3.

Attempts to substitute in addition the chlorine atoms at the α-silicon of **7a,b** by hydroxyl groups have so far led to uncontrolled decomposition.

Hydroxylation of the α-silicon in ferriodisilanes has been tried in the case of the dichloro(ferrio)disilanes **9a,b**, obtained from the ferriodisilanes **3a,b** via regioselective chlorination with CCl_4 [6]. However, only in the case of **9a** treatment with water in the presence of Et_3N succeeded in the generation of the ferriodihydroxydisilane **10** after 14 h in 81 % yield (Eq. 4).

3a,b $\xrightarrow[- 2\,CHCl_3]{+ 2\,CCl_4}$ **9a,b** $\xrightarrow[- 2\,[Et_3NH]Cl]{+ 2\,H_2O,\ 2\,Et_3N}$ **10**

	a	b
—○	H	Me

Eq. 4.

In an analogous manner the bis(ferrio)tetrachlorodisilane (**11**) has been converted into the corresponding fully hydroxylated bis(ferrio)tetrahydroxydisilane **12**. It shows high reactivity towards $Me_2Si(H)Cl$ to give the Si–Si-based metallosiloxane **13** (Eq. 5).

Eq. 4.

The aforementioned highly functionalized metallodisilanesiloxanes are very promising compounds for further oxo-functionalizations and will be the subject of forthcoming publications.

Acknowledgment: We gratefully acknowledge financial support from the *Deutsche Forschungsgemeinschaft*; Schwerpunktprogramm "Spezifische Phänomene in der Siliciumchemie" as well as from the *Fonds der Chemischen Industrie*.

References:

[1] W. Malisch, H. Jehle, C. Mitchel, W. Adam, *J. Organomet. Chem.* **1998**, *566*, 259–263.

[2] P. D. Lickiss, *Adv. Inorg. Chem.* **1995**, *42*, 147.

[3] W. Ries, T. Albright, J. Silvestre, I. Bernal, W. Malisch, *Inorg. Chim. Acta* **1986**, *111*, 119–128.

[4] a) W. Adam, U. Azzena, F. Prechtl, K. Hindahl, W. Malisch, *Chem. Ber.* **1992**, *125*, 1409–1411.
b) S. Möller, O. Fey, W. Malisch, W. Seelbach, *J. Organomet. Chem.* **1996**, *507*, 239–244.
c) W. Malisch, R. Lankat, S. Schmitzer, J. Reising; *Inorg. Chem.* **1995**, *34*, 5701–5702.
d) W. Adam, R. Mello, R. Curci, *Angew. Chem.* **1990**, *102*, 916–917, *Angew. Chem., Int. Ed. Engl.* **1990**, *29*, 890–891.
e) W. Adam, A. K. Smerz, *Bull. Soc. Chim. Belg.* **1996**, *105*, 581–599.

[5] W. Malisch, H. Jehle, C. Mitchel, W. Adam, *J. Organomet. Chem.* 1998, *566*, 259–262.

[6] a) W. Malisch, H. Jehle, S. Möller, G. Thum, J. Reising, A. Gbureck, V. Nagel, C. Fickert, W. Kiefer, M. Nieger, *Eur. J. Inorg. Chem.* **1999**, 1597–1605.
b) S. Möller, H. Jehle, W. Malisch, W. Seelbach in *Organosilicon Chemistry III: From Molecules to Materials*, (Eds.: N. Auner, J. Weis) Wiley–VCH, Weinheim **1998**, 267–270.

[7] W. Malisch, H. Jehle, S. Möller, C. S. Möller, W. Adam, *Eur. J. Inorg. Chem.* **1998**, 1585–1587.

[8] W. Malisch, R. Lankat, W. Seelbach, J. Reising, M. Noltemeyer, R. Pikl, U. Posset, W. Kiefer, *Chem. Ber.* **1995**, *128*, 1109–1115.

[9] B. Beagly, J. J. Monaghan, *J. Mol. Struct.* **1971**, *8*, 401; W. S. Sheldrick, in: *The Chemistry of Organic Silicon Compounds* (Eds.: S. Patai, Z. Rappoport), John Wiley, New York, **1989**, 227.

[10] W. Malisch, *J. Organomet. Chem.* **1974**, *82*, 185–199.

Dicobaltoctacarbonyl-Assisted Synthesis of Ferriosilanols [†]

Wolfgang Malisch, Matthias Vögler*

Institut für Anorganische Chemie der Universität Würzburg
Am Hubland, D-97074 Würzburg, Germany
Tel.: Int. Code + (931)888 5277 — Fax: Int. Code + (931)888 4618
E-mail: Wolfgang.Malisch@mail.uni-wuerzburg.de

Keywords: Metallosilanols / Oxo Functionalization / Dicobaltoctacarbonyl

Summary: The reaction of the ferriohydridosilanes $Cp(OC)_2Fe–SiR_2H$ (**4a,c**; R = Me, Cl) and $Cp(OC)_2Fe–SiMeH_2$ (**4b**) with dicobaltoctacarbonyl yields the heterobimetallic silanes $Cp(OC)_2Fe–SiR_2–Co(CO)_4$ (**A**, R = Me; **7**, R = Cl) and $Cp(OC)_2FeSi(Me)–Co_2(CO)_7$ (**6**). Hydrolytic cleavage of the Si–Co bond in these systems forms the corresponding ferriosilanols (**5a–c**). In the case of the ferriosilane $Cp(OC)_2Fe–SiMe_2H$ (**4a**) a $Co_2(CO)_8$-catalyzed conversion to the silanol $Cp(OC)_2Fe–SiMe_2OH$ (**5a**) has been achieved.

Introduction

Among transition-metal-substituted silicon compounds, the class of metallosilanols is of particular interest, since electron-rich transition metal substituents have a strong stabilizing effect on the Si–OH group. Unlike most organosilanols, the metallosilanols show no tendency towards self-condensation and even compounds with two or three Si–OH groups have been isolated. In the past two pathways for the synthesis of ferriosilanols have been established (Scheme 1).

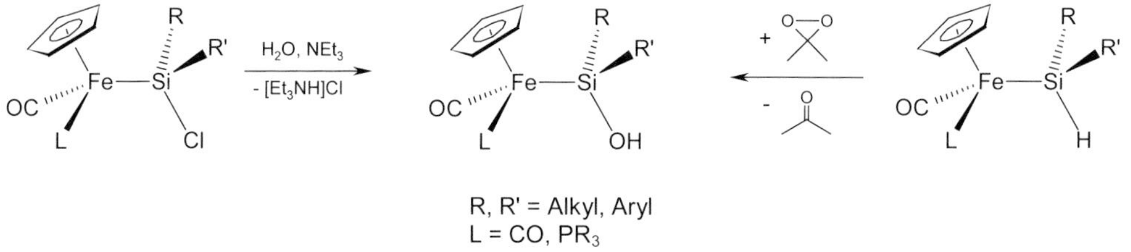

R, R' = Alkyl, Aryl
L = CO, PR_3

Scheme 1. Two pathways for the synthesis of ferriosilanols.

[†] Part 20 of the series "*Metallo-Silanols and Metallo-Siloxanes*". In addition, part 47 of the series "*Synthesis and Reactivity of Silicon Transition Metal Complexes*". Part 19/46 see [1]

The hydrolysis of ferriochlorosilanes in the presence of an auxiliary base leads to the corresponding ferriosilanols in high yield [2]. However, this reaction is limited to systems with L = CO, because the reactivity of the electron-rich $Cp(OC)(R_3P)Fe$-substituted chlorosilanes is insufficient for a Cl/OH exchange.

Another useful route for the synthesis of metallosilanols is based on the reaction of metallohydridosilanes with 1,1-dimethyldioxirane, resulting in the insertion of oxygen into the Si–H-bond [3]. This procedure is supplementary to the hydrolysis route, since it allows the formation of silanols with a phosphine–iron substituent.

Results

A new synthesis of ferriosilanols under extremely mild conditions has been worked out involving the hydrolysis of $(OC)_4Co$–silanes which are formed by the well-known reaction between silanes and dicobaltoctacarbonyl (Eq. 1) [4].

$$2\ R_3Si{-}H \ + \ Co_2(CO)_8 \longrightarrow 2\ R_3Si{-}Co(CO)_4 \ + \ H_2$$

Eq. 1.

A prominent characteristic of the Si-Co-bond is its high reactivity towards protic reagents [5].

$$R_3Si{-}Co(CO)_4 \ + \ H{-}X \longrightarrow R_3Si{-}X \ + \ HCo(CO)_4 \quad (X = Cl,\ F,\ OH)$$

Eq. 2.

In a first experiment the $Co_2(CO)_8$-assisted route of Si–H metallation followed by hydrolysis has been successfully employed for the conversion of triphenylsilane to triphenylsilanol. The addition of water to a solution of **2** in diethyl ether led to the quantitative formation of the silanol **3** within 30 min. (Eq. 3).

$$\underset{\mathbf{1}}{Ph_3Si{-}H} + Co_2(CO)_8 \xrightarrow[-\ HCo(CO)_4]{} \underset{\mathbf{2}}{Ph_3Si{-}Co(CO)_4} \xrightarrow[-\ HCo(CO)_4]{+\ H_2O} \underset{\mathbf{3}}{Ph_3Si{-}OH}$$

Eq. 3.

The result of Eq. 3 encouraged us to extend the dicobaltoctacarbonyl-assisted hydrolysis to the preparation of ferriosilanols, taking advantage of the easy cleavage of the Si–Co bond in the intermediates, whereas the more stable Fe–Si bond remains intact. In this context the ferriosilane **4a** and one equivalent of dicobaltoctacarbonyl were combined in diethyl ether. Due to the instability of the intermediate **A** the resulting solution was treated in situ with an excess of H_2O, which resulted in the formation of the ferriosilanol **5a** in moderate yields (Eq. 4).

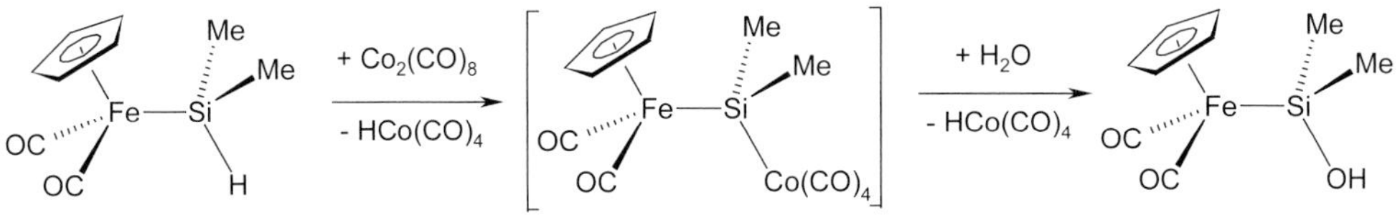

Eq. 4. Synthesis of the ferriosilanol **5a**.

In a similar manner the ferriodihydridosilane **4b** was converted to the corresponding silandiol **5b** (Eq. 5). Unlike species **A** in Eq. 4 the trimetallosilane **6**, which features a structure with a bridged $Co_2(CO)_7$ unit and a direct Co–Co bond, could be isolated and characterized [6]. The hydrolysis of **6** in diethyl ether forms the silanediol **5b** in 72 % yield within 30 min.

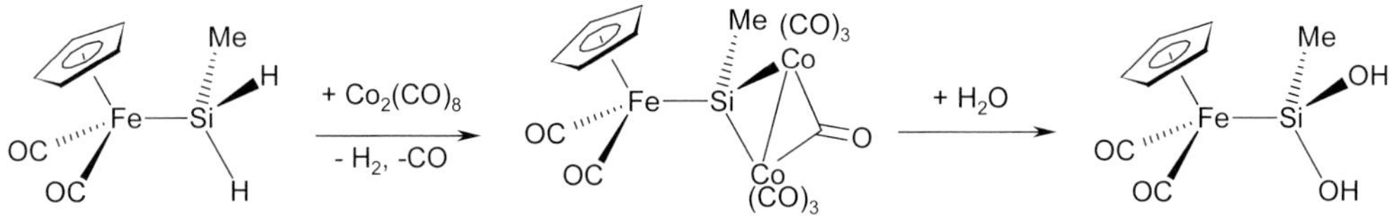

Eq. 5. Synthesis of the ferriosilanediol **5b**.

The hydrolytic cleavage of the Si–Co bond under extremely mild conditions offers a chance of selective SiH/SiOH transformation in chlorosilanes, yielding unknown chlorosilanols. This type of conversion has been examined for the dichlorosilane $Cp(OC)_2Fe–SiCl_2H$ (**4c**). Treatment of **4c** with $Co_2(CO)_8$ produces the hetero-bismetallic silane **7**, which was also isolated and fully characterized. However, all attempts to selectively hydrolyze the Si–Co bond in **7** have failed so far and only the silanetriol **5c** could be obtained (Eq. 6).

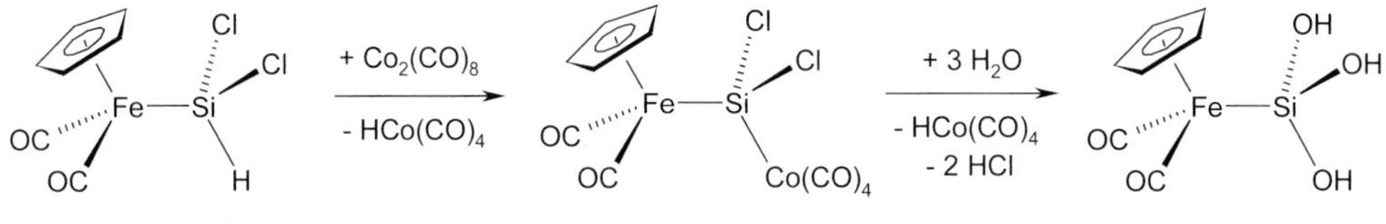

Eq. 6. Synthesis of the ferriosilanetriol **5c**.

Regeneration of the metallation agent dicobaltoctacarbonyl from $HCo(CO)_4$ which is formed in the hydrolysis steps of Eqs. 4 and 6 suggests the possibility of a catalytic pathway for silane/silanol transformations (Scheme 2).

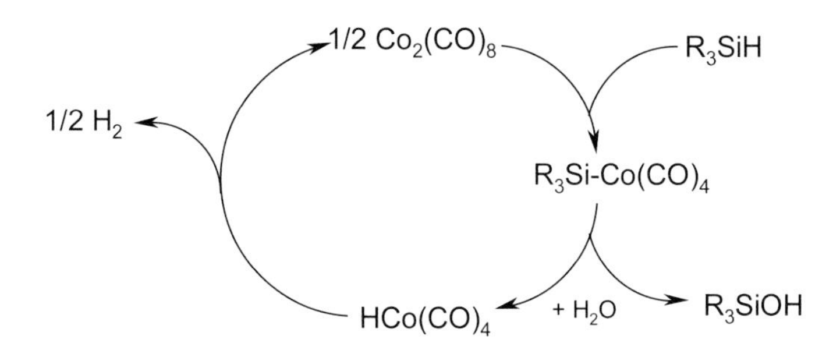

Scheme 2. Catalytic pathway for silane/silanol transformations.

The ferriosilane $Cp(OC)_2Fe–SiMe_2H$ (**4a**) is the first example in which this concept has been realized. When 5 mol % $Co_2(CO)_8$ are added to a solution of **4a** and H_2O in diethyl ether, the ferriosilane **4a** is converted to the corresponding silanol **5a** within 3 h. The conversion of other silanes via this potentially useful procedure is currently under investigation.

Acknowledgment: This work has been generously supported by the *Deutsche Forschungsgemeinschaft* (Schwerpunktprogramm: "Spezifische Phänomene in der Siliciumchemie").

References:

[1] W. Malisch, H. Jehle, M. Lager, M. Nieger "*Metal-Fragment Substituted Disilanols*" in *Organosilicon Chemistry IV: From Molecules to Materials* (Eds.: N. Auner, J. Weis), Wiley–VCH, Weinheim, **1999**, pp. 437.

[2] W. Ries, T. Albright, J. Silvestre, I. Bernal, W. Malisch, C. Burschka, *Inorg. Chim. Acta* **1986**, *111*, 119.

[3] W. Malisch, K. Hindahl, H. Käb, J. Reising, W. Adam, F. Prechtl, *Chem. Ber.* **1995**, *128*, 963.

[4] J. F. Harrod, A. J. Chalk, *J. Am. Chem. Soc.* **1965**, *87*, 1133.

[5] B. J. Aylett, J. M. Campbell, *J. Chem. Soc.* A **1969**, 1910.

[6] W. Malisch, H. U. Wekel, I. Grob, *Z. Naturforsch., Teil B* **1982**, *37*, 601.

Novel Silanols and Siloxanes Substituted with the Ferriomethyl Fragment $Cp(OC)_2FeCH_2$[†]

Wolfgang Malisch*, Marco Hofmann
Institut für Anorganische Chemie der Universität Würzburg
Am Hubland, D-97074 Würzburg, Germany
Tel.: Int. Code + (931)888 5277 — Fax: Int. Code + (931)888 4618
E-mail: Wolfgang.Malisch@mail.uni-wuerzburg.de

Martin Nieger
Institut für Anorganische Chemie der Universität Bonn
Gerhard-Domagk-Straße 1, D-53121 Bonn, Germany

Keywords: Ferriosilanols / Oxo Functionalization / Ferriosiloxanes

Summary: Metalation of the chloromethylsilanes $ClCH_2SiMe_{3-n}H_n$ (n = 1–3) (**2a–c**) with the sodium ferrates $Na[Fe(CO)_2C_5R_5]$ (R = H, Me) (**1a**,**b**) leads to the ferriomethyl-substituted silanes $C_5R_5(OC)_2Fe–CH_2SiMe_{3-n}H_n$ (R = H, Me; n = 1–3) (**3a–e**). Oxo functionalization of **3a**,**c**,**d** with dimethyldioxirane yields the ferriomethyl-silanols $Cp(OC)_2Fe–CH_2SiMe_{3-n}(OH)_n$ (n = 1–3) (**4a–c**). **4a**,**b** are in addition obtained via hydrolysis of the ferriomethylchlorosilanes $Cp(OC)_2Fe–CH_2SiMe_{3-n}Cl_n$ (n = 1, 2) (**5a**,**b**) in the presence of Et_3N. **4a**,**b** are stable towards self-condensation below 0°C and can be isolated and structurally characterized. The existence of the silanetriol $Cp(OC)_2Fe–CH_2Si(OH)_3$ (**4c**) could only be proved by subsequent condensation with the organochlorosilanes RMe_2SiCl (R = H, Me) to give the tetrasiloxanes $Cp(OC)_2Fe–CH_2Si(OSiMe_2R)_3$ (**6f**,**g**). The analogous reaction of **4a**,**b** leads to the di- and trisiloxanes $Cp(OC)_2Fe–CH_2SiMe_{3-n}(OSiMe_2R)_n$ (n = 1, 2) (**6b–e**).

Metal-fragment substituted silanols $L_nM–SiR_{3-n}(OH)_n$ (n = 1–3) have attracted interest in previous years [2, 3], especially with regard to the stabilizing effect of the electron-donating metal fragment. It guarantees isolation even of metallo-silanetriols, offering easy access to diverse metal-substituted oligo- and polysiloxanes via controlled co-condensation with functionalized organosilanes. This class of compounds represents attractive models for transition metal complexes anchored on silica surfaces [4].

Up to now, only metallo-silanols with a direct metal–silicon bond have been realized, containing mainly metal fragments of the type $C_5R_5(OC)_2(Me_3P)M$ (M = Mo, W; R = Me, H) [2, 3] or $C_5R_5(OC)_2M$ (M = Fe, Ru; R = Me, H) [3].

---

† Part 21 of the series "*Metallo-Silanols and Metallo-Siloxanes*" In addition, Part 48 of the series "*Synthesis and Reactivity of Silicon Transition Metal Complexes*". Part 20/47, see [1].

Our interest has now focused on silanols having the silicon separated from the metal fragment by a methylene group, which promise a higher tendency for self-condensation, an essential prerequisite to obtain access to metal-fragment substituted polysiloxanes.

In this context we have synthesized ferriomethyl-substituted silanols and siloxanes starting with the ferriomethylhydridosilanes **3a–e** [5], generated by nucleophilic metalation of the chloromethylsilanes **2a–c** with the sodium ferrates **1a,b** (Eq. 1).

1	a	b
—o	H	Me

2	a	b	c
n	1	2	3

3	a	b	c	d	e
n	1	2	3	1	3
—o	H	H	H	Me	Me

Eq. 1. Synthesis of **3a–e**.

The compounds **3a–e** are isolated in good yields as dark brown liquids (**3a–d**) or as an orange-brown solid (**3e**), which are soluble in nonpolar solvents, thermally moderately stable, but light-sensitive.

As previously demonstrated, dimethyldioxirane is a mild and selective oxidizing reagent for the conversion of Si–H-functionalized organo- or metallo-silanes to the corresponding silanols [3b, 6]. This method was applied to generate the ferriomethyl-substituted silanols, silanediols and silanetriols **4a–c** from **3a–c** in acetone at –78°C (Scheme 1, route a). Additionally, **4a,b** were prepared by hydrolysis of the ferriomethylchlorosilanes **5a,b** in diethyl ether at 0°C in the presence of the auxiliary base Et_3N (Scheme 1, route b).

	a	b	c
n	1	2	3

Scheme 1. Synthesis of **4a–c**.

The ferriomethylsilanol **4a** is isolated in good yield as yellow-brown waxy solid, which is rather soluble in aliphatic solvents and can be stored for several weeks at –20°C without any self-condensation. At room temperature, however, this process leads to the binuclear disiloxane-bridged iron complex **6a** (Scheme 2, route a). Additionally, **6a** is available by controlled co-condensation of **4a** with the ferriomethylchlorosilane **5a** in the presence of Et_3N (Scheme 2, route b).

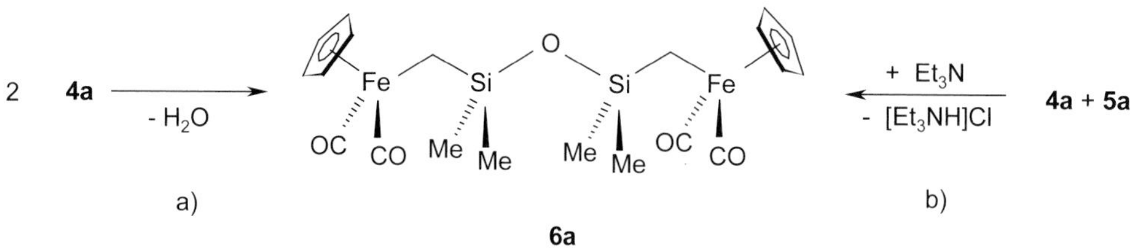

2 **4a** $\xrightarrow{-H_2O}$ (a) **6a** (b) $\xleftarrow[-\ [Et_3NH]Cl]{+\ Et_3N}$ **4a + 5a**

Scheme 2. Synthesis of **6a**.

The disiloxane **6a** shows a ^{29}Si-NMR line shifted approximately 10 ppm to higher field in comparison to the silanol **4a** [12.84 (**6a**) versus 21.50 ppm (**4a**)].

A single-crystal X-ray diffraction analysis of **4a** (Fig. 1a) reveals a pseudo-octahedral conformation at the iron atom. The Fe-C-Si bond angle of 120.36(8)° is in good accordance with that of $Cp(OC)(Ph_3P)Fe–CH_2SiMe_3$ [122.8(2)°] [7]. The Si–O bond distance of 1.6599(13) Å lies in the normal range of other organo- and metallosubstituted silanols [2c, 6c, 8]. **4a** shows intermolecular hydrogen bonding in the solid state resulting in the formation of tetrameric units indicated by the oxygen–oxygen distance of 2.7320(16) Å (Fig. 1b).

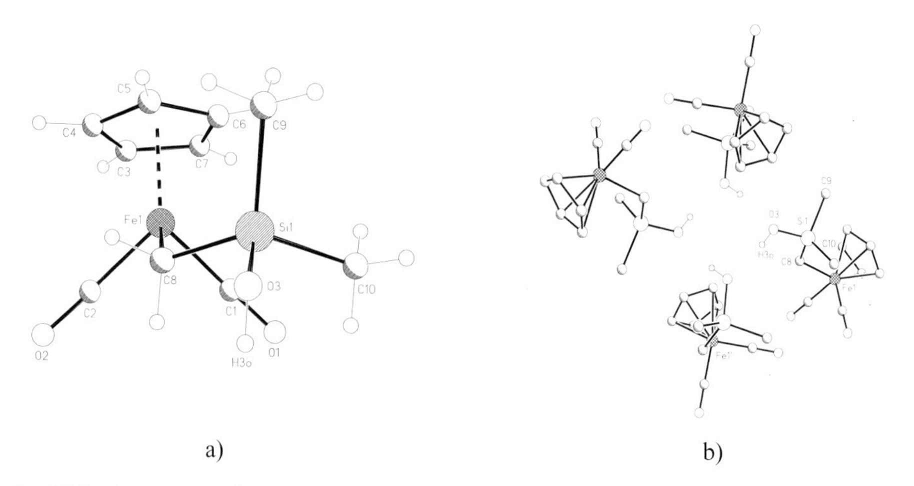

Fig. 1. Molecular structure of $Cp(OC)_2Fe–CH_2SiMe_2OH$ (**4a**); in b) hydrogen atoms (except HO) are omitted for clarity.

The ferriomethylsilanediol **4b**, isolated as a microcrystalline, light yellow powder, which shows good solubility in polar solvents like acetone and acetonitrile is stable with respect to self-condensation only at temperatures below –20°C. This behavior is even more pronounced in the case of the silanetriol **4c**, whose isolation under analogous conditions failed until now. However, the existence of **4c** can be proved by subsequent base-assisted condensation with the chlorosilanes RMe_2SiCl (R = H, Me) leading to the corresponding ferriomethyl-substituted tetrasiloxanes **6f,g**. The silanol **4a** and the silanediol **4b** can be transformed in the same manner to give the di- and trisiloxanes **6b–e** (Eq. 2).

$$Cp(OC)_2Fe{-}CH_2{-}Si(Me)_{3-n}(OH)_n + n\,ClSiMe_2R \xrightarrow[-\,n\,[Et_3NH]Cl]{+\,n\,Et_3N} Cp(OC)_2Fe{-}CH_2{-}Si(Me)_{3-n}(OSiMe_2R)_n$$

4	a	b	c
n	1	2	3

R = H, Me

6	b	c	d	e	f	g
n	1	1	2	2	3	3
R	H	Me	H	Me	H	Me

Eq. 2. Synthesis of **6b–g**.

The ferriomethylsiloxanes **6a–g** are isolated in good yields as brown to orange-brown oils which are soluble in all common organic solvents.

Investigations currently in progress are directed towards the synthesis of iron-substituted oligo- and polysiloxanes produced via self-condensation of the silanedi- and -triols **4b**,**c**.

Acknowledgment: This work has been generously supported by the *Deutsche Forschungsgemeinschaft* (Schwerpunktprogramm: "Spezifische Phänomene in der Siliciumchemie").

References:

[1] W. Malisch, M. Vögler, "*Dicobaltoctacarbonyl-Assisted Synthesis of Ferriosilanols*", in: *Organosilicon Chemistry IV: From Molecules to Materials* (Eds.: N. Auner, J. Weis), Wiley–VCH, Weinheim, **1999**, p. 442.

[2] a) W. Malisch, R. Lankat, S. Schmitzer, J. Reising, *Inorg. Chem.* **1995**, *34*, 5701.
b) W. Malisch, R. Lankat, O. Fey, J. Reising, S. Schmitzer, *J. Chem. Soc., Chem. Commun.* **1995**, 1917.
c) W. Malisch, S. Schmitzer, R. Lankat, M. Neumayer, F. Prechtl, W. Adam, *Chem. Ber.* **1995**, *128*, 1251.

[3] a) W. Malisch, S. Möller, R. Lankat, J. Reising, S. Schmitzer, O. Fey, in: *Organosilicon Chemistry II: From Molecules to Materials* (Eds.: N. Auner, J. Weis), VCH, Weinheim, **1996**, p. 575.
b) S. Möller, O. Fey, W. Malisch, W. Seelbach, *J. Organomet. Chem.* **1996**, *507*, 239.

[4] a) W. A. Herrmann, A. W. Stumpf, T. Priermeier, S. Bogdanovic, V. Dufaud, J.-M. Basset, *Angew. Chem.* **1996**, *108*, 2978.
b) Y. I. Yermakov, B. N. Kuznetsov, V. A. Zakharov, *Catalysis by Supported Complexes*, Elsevier, New York, **1981**.
c) S. N. Borisov, M. G. Voronkov, E. Y. Lukevits, *Organosilicon Heteropolymers and Hetero Compounds*, Plenum, New York, **1970**.

[5] For the synthesis of **3a**,**b** see: a) K. H. Pannell, *J. Organomet. Chem.* **1970**, *21*, P17.
b) J. E. Bulkowski, N. D. Miro, D. Sepelak, C. H. Van Dyke, *J. Organomet. Chem.* **1975**, *101*, 267.
c) C. L. Randolph, M. S. Wrighton, *Organometallics* **1987**, *6*, 365.

[6] a) W. Adam, U. Azzena, F. Prechtl, K. Hindahl, W. Malisch, *Chem. Ber.* **1992**, *125*, 1409.
b) W. Malisch, K. Hindahl, H. Käb, J. Reising, W. Adam, F. Prechtl, *Chem. Ber.* **1995**, *128*, 963.
c) W. Adam, R. Mello, R. Curci, *Angew. Chem.* **1990**, *102*, 916; *Angew. Chem. Int. Ed. Engl.* **1990**, *29*, 890.
d) W. Adam , A.K. Smerz, *Bull. Soc. Chim. Belg.* **1996**, *105*, 581.

[7] S. G. Davies, I. M. Dordor-Hedgecock, K. H. Sutton, M. Whittaker, *J. Am. Chem. Soc.* **1987**, *109*, 5711.

[8] a) Z. H. Aiube, N. H. Buttrus, C. Eaborn, P. B. Hitchcock, J. A. Zora, *J. Organomet. Chem.* **1985**, *292*, 177.
b) P. D. Lickiss, N. L. Clipston, D. W. H. Rankin, H. E. Robertson, *J. Mol. Struct.* **1995**, *344*, 111.
c) H. Puff, K. Braun, H. Reuter, *J. Organomet. Chem.* **1991**, *109*, 119.

Cyclosiloxanes with Si–Si–O and Si–O Groups

Christian Wendler, Helmut Reinke, Hans Kelling

Fachbereich Chemie der Universität Rostock
Buchbinderstrße 9, D-18051 Rostock, Germany
Tel.: Int code + (381)498 1755 — Fax: Int code + (381)498 1763
E-mail: hans.kelling@chemie.univ-rostock.de

Keywords: Cyclosiloxanes / Disilanyloxy Compounds / Synthesis / Structure

Summary: Mono-, bi- and tricyclic siloxanes containing at least one Si–Si–O group have been synthesized by cyclocondensation of chlorodisilanes with α,ω-siloxanediols or tetramethyldisilane-1,2-diol respectively. The reactions of 1,1,2,2-tetrachlorodimethyldisilane and of hexachlorodisilane give cyclization in 1,2- as well as in 1,1-position at the Si–Si group, depending on the ring width of the resulting cyclosiloxane.

Introduction

Investigations on synthesis, structure and reactivity of siloxanes [1, 2] have been continued by attempts to synthesize cyclosiloxanes with at least one Si–Si–O group instead of Si–O. Only a few such compounds are known until now [3–6]. Such siloxanes could be interesting due to the additional reactivity of the Si–Si bond.

Results and Discussion

As initial products containing the Si–Si unit we used 1,2-dichlorotetramethyldisilane, 1,1,2,2-tetrachlorodimethyldisilane, hexachlorodisilane and the previously unknown tetramethyldisilane-1,2-diol, which we just synthesized by cautious hydrolysis of 1,2-dichlorotetramethyldisilane. Both chloromethyldisilanes are available starting from the so-called “disilane fraction” of the direct chloromethylsilane synthesis (Hüls-Silicone) by methods already described in the literature [7, 8] and partly improved by us.

As basic products containing the Si–O unit we used the siloxanediols HO–Me_2Si–(O–$Me_2Si)_n$–OH ($n = 1, 2, 3$).

We found as the most convenient method for the cyclosiloxane synthesis the cyclocondensation of the chlorodisilanes with the siloxanediols or the disilanediol in conditions following the dilution principle using pyridine as HCl acceptor.

Reactions of 1,2-dichlorotetramethyldisilane with the siloxanediols according to Eq. 1 give the expected seven- to eleven-membered rings **I**, **II** and **III**, cf. [5, 6]. Reactions of 1,1,2,2-tetrachlorodimethyldisilane with one equivalent of siloxanediol give the monocyclic siloxanes **IV**, **V**, **VI** , **VII** (Scheme 1), for disiloxanediol ($n = 1$) as well with endocyclic as with exocyclic Si–Si, and for tri- and tetrasiloxanediol ($n = 2, 3$) with exocyclic Si–Si only.

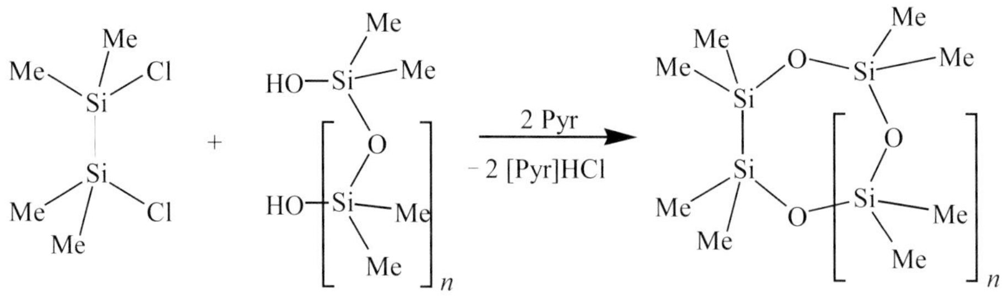

I II III

n 1 2 3

Eq. 1. Reactions of 1,2-dichlorotetramethyldisilane with the siloxanediols.

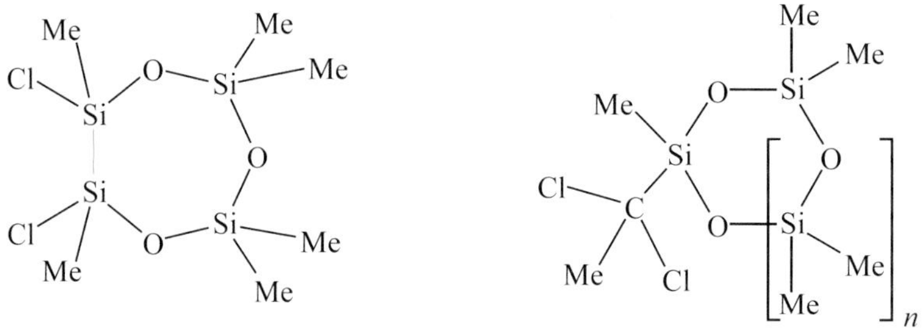

IV

V VI VII

n 1 2 3

Scheme 1. Products of the reactions of 1,1,2,2-tetrachlorodimethyldisilane with one equivalent of siloxanediol.

The reaction of 1,1,2,2-tetrachlorodimethyldisilane with two equivalents of siloxanediol gives in a similar way the bicyclic cyclosiloxanes **VIII**, **IX**, **X**, for disiloxanediol ($n = 1$) with endocyclic Si–Si only, and for tri- and tetrasiloxanediol ($n = 2, 3$) exclusively with exocyclic Si–Si (Scheme 2). The cyclocondensation in the 1,1-position at Si–Si seems to be favored, especially for at least eight-membered rings.

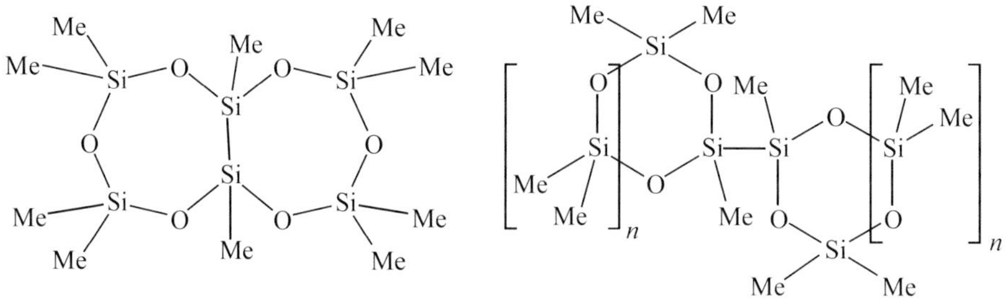

VIII

IX X

n 2 3

Scheme 2. Products of the reactions of 1,1,2,2-tetrachlorodimethyldisilane with two equivalents of siloxanediol.

The reaction of hexachlorodisilane with three equivalents of siloxanediol gives for all siloxanediols the tricyclic siloxanes with endocyclic Si–Si in all three rings (**XI**, **XII**, **XIII**) as well as the tricyclic siloxanes with Si–Si endocyclic in only one ring and exocyclic in the others (**XIV**, **XV**, **XVI**), according to Scheme 3.

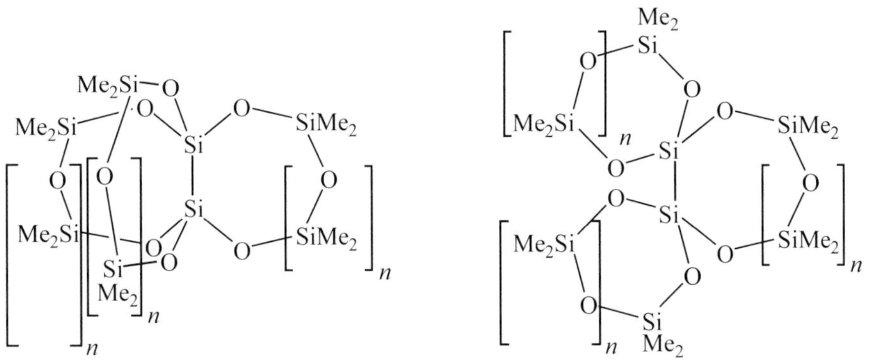

	XI	XII	XIII
n	1	2	3

	XIV	XV	XVI
n	1	2	3

Scheme 3. Products of the reaction of hexachlordisilane with three equivalents of siloxanediol.

Some cyclocondensation results for the reaction of tetramethyldisilane-1,2-diol with chlorosilanes are given in Scheme 4. With tetrachlorodimethyldisilane (molar ratio 2:1) the bicyclic siloxane **XVII** is formed, and with hexamethyldisilane (molar ratio 3:1) the tricyclic siloxane **XVIII**, both containing Si–Si–O units only. **XVII** has already been synthesized by another way [4]. The 1:1 reaction with $SiCl_4$ gives the ten-membered ring **XIX** only.

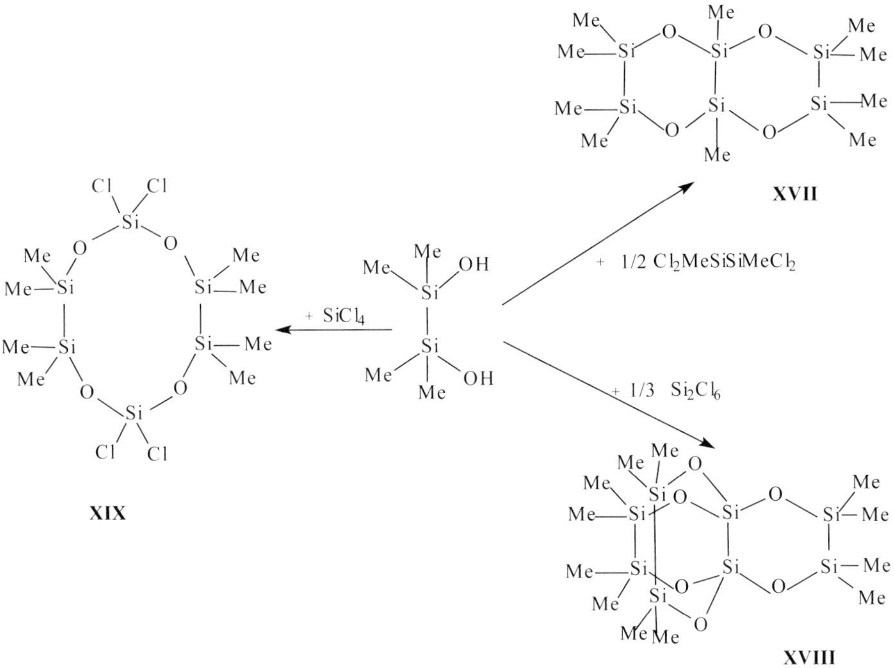

Scheme 4. Cyclocondensation products.

All cyclosiloxanes have been characterized by ^{1}H- and ^{29}Si-NMR, mass spectra and elemental analysis. The ^{29}Si-NMR signals for the Me_2SiO groups are within the expected area for common cyclosiloxanes (–7 ppm for six-membered rings, about –20 ppm for the higher-membered rings). The Si–Si– signals are high-field shifted with increasing number of O-substituents (for $OMe_2SiSiMe_2O$ 5 to 0 ppm, for $O_2MeSiSiMeO_2$ –21 to –28 ppm, for O_3SiSiO_3 –68 to –100 ppm).

For the two crystalline tricyclosiloxanes **XI** and **XVIII** the structure was confirmed by X-ray analysis. The three seven-membered rings in **XI** show only small deviations from planarity (Fig.1).

The three six-membered rings in **XVIII** have a twist like conformation with only small dihedral angles for the two central Si–O bonds (e.g. Si(1)–O(1) and Si(1A)–O(1B)) compared with the higher dihedral angle for the two peripheral Si–O bonds (e.g. Si(2)–O(1) and Si(2B)–O(1B)) (Fig. 2).

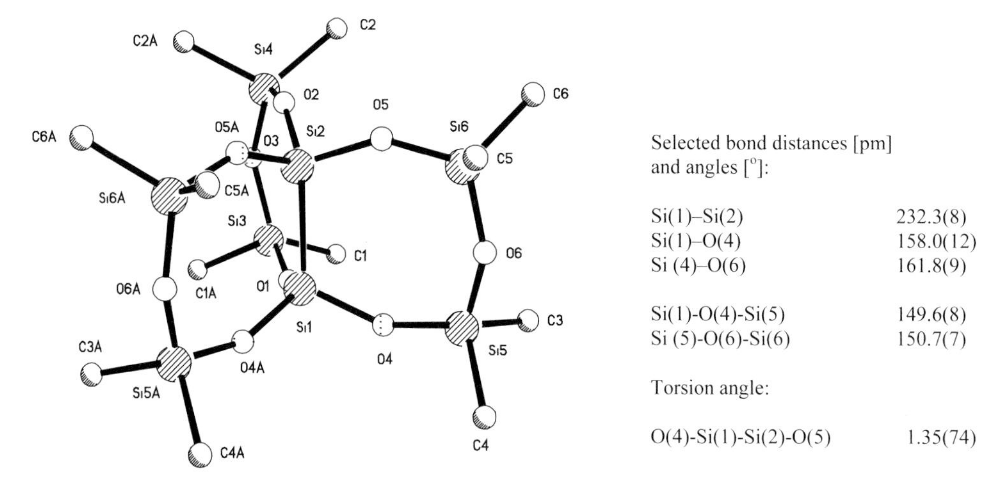

Selected bond distances [pm] and angles [°]:

Si(1)–Si(2)	232.3(8)
Si(1)–O(4)	158.0(12)
Si (4)–O(6)	161.8(9)
Si(1)-O(4)-Si(5)	149.6(8)
Si (5)-O(6)-Si(6)	150.7(7)

Torsion angle:

O(4)-Si(1)-Si(2)-O(5)	1.35(74)

Fig. 1. Crystal structure of **XI**.

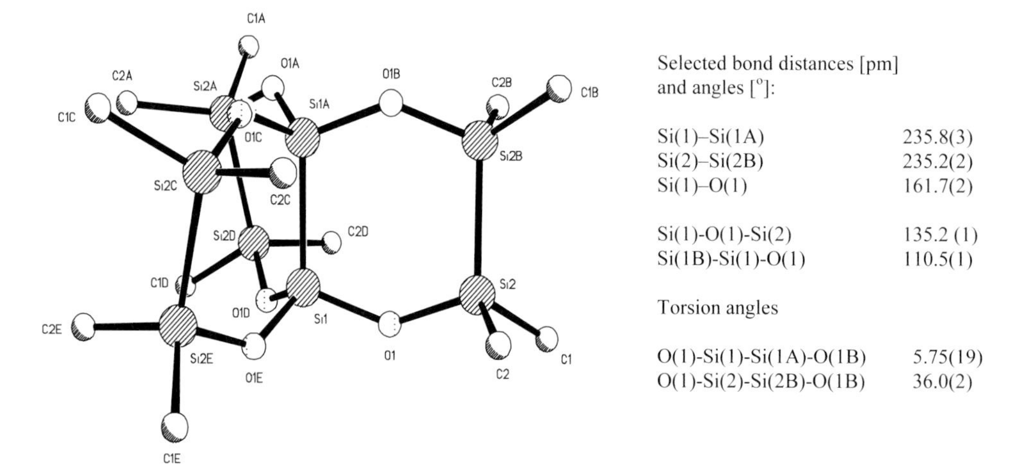

Selected bond distances [pm] and angles [°]:

Si(1)–Si(1A)	235.8(3)
Si(2)–Si(2B)	235.2(2)
Si(1)–O(1)	161.7(2)
Si(1)-O(1)-Si(2)	135.2 (1)
Si(1B)-Si(1)-O(1)	110.5(1)

Torsion angles

O(1)-Si(1)-Si(1A)-O(1B)	5.75(19)
O(1)-Si(2)-Si(2B)-O(1B)	36.0(2)

Fig. 2. Crystal structure of **XVIII**.

Acknowledgement: We thank the *Fonds der Chemischen Industrie* for financial support.

References:

[1] H. Kelling, D. Lange, A. Surkus, "*On the Acid Catalyzed Reaction of Siloxanes with Alcohols*", in: *Organosilicon Chemistry I: From Molecules to Materials* (Eds.: N. Auner, J. Weis), VCH, Weinheim, **1994**, p. 67.

[2] S. Wandschneider, *Ph. D. Thesis*, Universität Rostock, **1998**.

[3] E. Hengge, *Top. Curr. Chem.* **1974**, *51*, 1.

[4] M. Kumada, K. Tamao, *Adv. Organomet. Chem.* **1968**, *6*, 19.

[5] G. A. Razuvaev, V. V. Semenov, T. N. Bretnova, A. N. Kornev, *Gen. Chem. USSR* **1986**, *56*. 884.

[6] J. Y. Becker, M. Shen, R. West, *J. Electroanal. Chem.* **1996**, *417*, 77.

[7] M. Kumada, H. Sakurai, T. Watanabe, *J. Organomet. Chem.* **1967**, *7*, 15.

[8] H. Matsumoto, T. Motegi, M. Hasagawa, Y. Nagai, *J. Organomet. Chem.* **1977**, *142*, 149.

Syntheses, Structures, and Properties of Zwitterionic Monocyclic λ^5Si-Silicates

Reiner Willeke, Ruth E. Neugebauer, Melanie Pülm, Olaf Dannappel, Reinhold Tacke*

Institut für Anorganische Chemie, Universität Würzburg
Am Hubland, D-97074 Würzburg, Germany
Tel.: Int. code + (931)888 5250 — Fax: Int. code + (931)888 4609
E-mail: r.tacke@mail.uni-wuerzburg.de

Keywords: Pentacoordinate Silicon / Zwitterionic λ^5Si-Silicates / Monocyclic λ^5Si-Silicates / Crystal Structures

Summary: The syntheses and properties of derivatives of a new class of zwitterionic monocyclic λ^5Si-silicates with an SiO_2C_2F framework are reported. These molecular pentacoordinate silicon compounds contain bidentate diolato(2–) ligands that derive from benzohydroximic acid, oxalic acid, or glycolic acid. In addition, the crystal structures of these compounds are described.

Introduction

Over the past few years, we have synthesized and structurally characterized a series of zwitterionic acyclic and spirocyclic λ^5Si-silicates, such as compounds **1** [1] and **2** [2] (Scheme 1). These zwitterions contain a pentacoordinate (formally negatively charged) silicon atom and a tetracoordinate (formally positively charged) nitrogen atom. We have now succeeded in preparing a series of related zwitterionic monocyclic λ^5Si-silicates. We report here on the syntheses, properties, and crystal structures of compounds **3a/3b**–**5a/5b** (Scheme 1) [3] (for reviews dealing with pentacoordinate silicon compounds, see [4–10]).

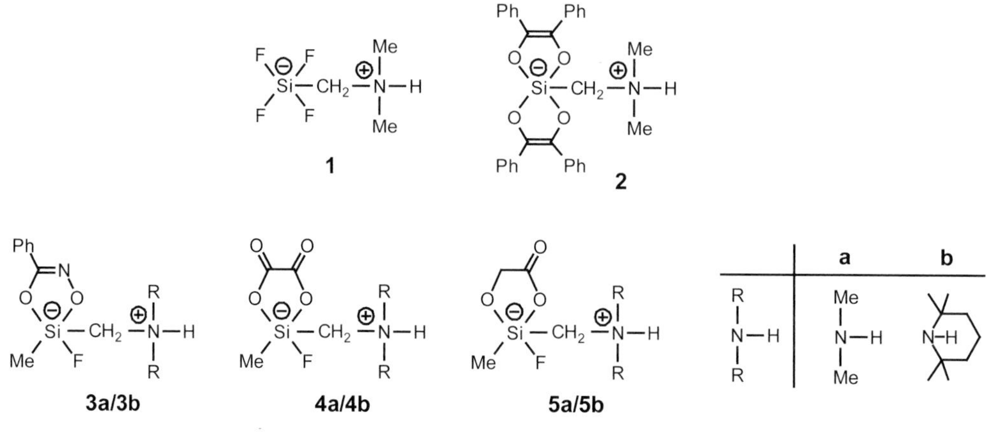

Scheme 1. Zwitterionic λ^5Si-silicates **1**, **2**, and **3a/3b**–**5a/5b**.

Results and Discussion

The zwitterionic monocyclic λ^5*Si*-silicates **3a**–**5a** were synthesized according to Scheme 2 by reaction of [(dimethylammonio)methyl]trifluoro(methyl)silicate (**6a**) with one molar equivalent of the *O,O′*-bis(trimethylsilyl) derivatives of benzohydroximic acid, oxalic acid or glycolic acid (formation of two molar equivalents of Me_3SiF). Compounds **3b**–**5b** were prepared analogously starting from trifluoro(methyl)[(2,2,6,6-tetramethylpiperidinio)methyl]silicate (**6b**) (Scheme 2).

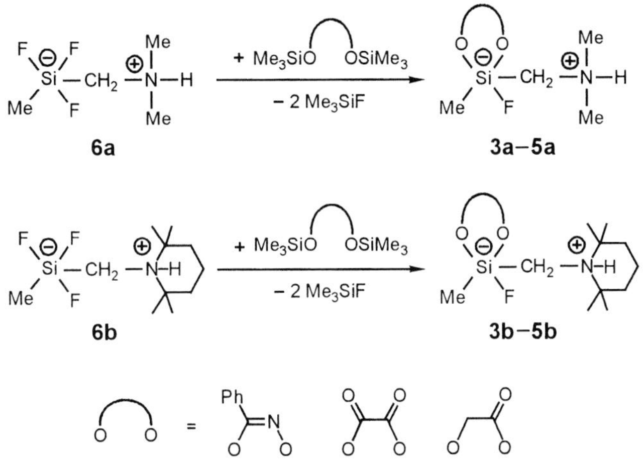

Scheme 2. Syntheses of the zwitterionic λ^5*Si*-silicates **3a/3b**–**5a/5b**.

All reactions were carried out in acetonitrile at room temperature and compounds **3a/3b–5a/5b** were isolated in good yields (75–98 %) as colorless crystalline solids. Their identities were established by elemental analyses (C, H, N), solution-state NMR experiments (^1H, ^{13}C, ^{19}F, ^{29}Si), and mass-spectrometric studies. In addition, all compounds were structurally characterized by single-crystal X-ray diffraction.

The zwitterionic λ^5*Si*-silicates **3a/3b**–**5a/5b** crystallized in the space group *Pbca* (**3a**), $P2_1/c$ (**3b, 4a, 5a**), or $P2_1/n$ (**4b, 5b**). The structures of these zwitterions in the crystal are depicted in Figs. 1–3. Selected geometric parameters of the *Si*-coordination polyhedra of **3a/3b**–**5a/5b** are given in Table 1.

Table 1. Selected geometric parameters for **3a/3b–5a/5b** (distances in pm, angles in deg.).

Compd.	Si–O1	Si–O2	Si–F	Si–C1	Si–C2	O1-Si-F
3a	184.78(14)	170.3(2)	168.44(12)	189.8(2)	186.2(2)	170.29(7)
3b	187.83(14)	170.6(2)	167.12(12)	191.3(2)	186.5(2)	170.94(7)
4a	191.8(2)	173.92(14)	166.61(14)	190.6(2)	185.4(2)	172.43(5)
4b	187.10(9)	172.8(1)	168.24(8)	189.15(11)	186.11(13)	172.62(4)
5a	186.79(13)	168.27(13)	170.53(11)	190.4(2)	186.4(2)	173.68(6)
5b	187.34(13)	167.30(14)	168.85(11)	190.4(2)	187.3(2)	175.81(6)

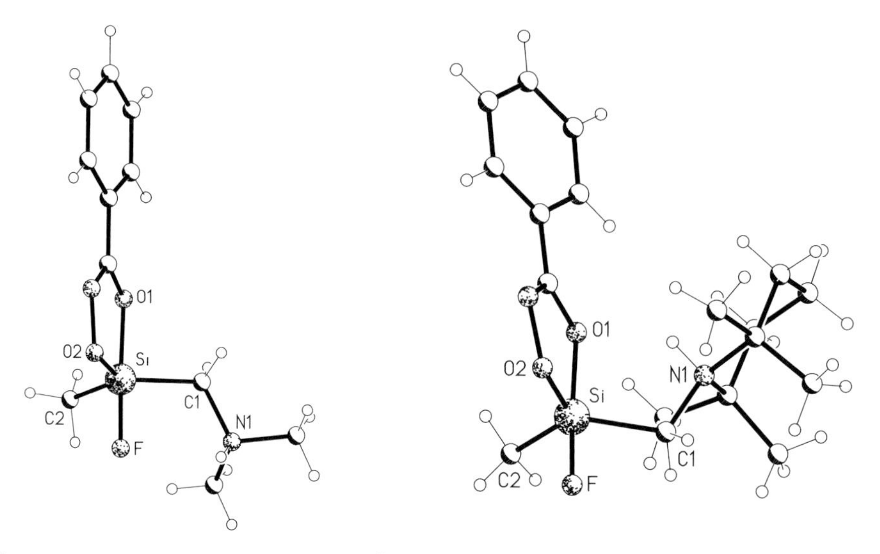

Fig. 1. Molecular structures of the zwitterionic λ^5Si-silicates **3a** (left) and **3b** (right) in the crystal.

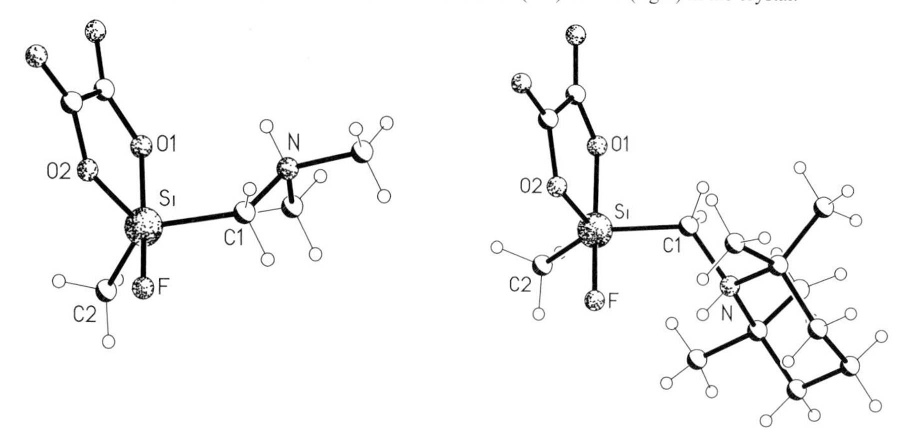

Fig. 2. Molecular structures of the zwitterionic λ^5Si-silicates **4a** (left) and **4b** (right) in the crystal.

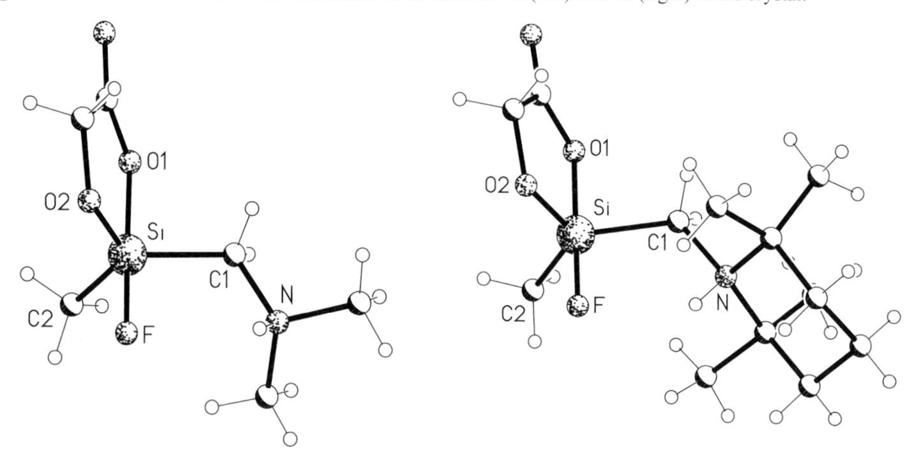

Fig. 3. Molecular structures of the zwitterionic λ^5Si-silicates **5a** (left) and **5b** (right) in the crystal.

As can be seen from Figures 1–3 and Table 1, the coordination polyhedra around the silicon atoms of **3a/3b–5a/5b** are distorted trigonal bipyramids, the axial positions being occupied by the fluorine atom and an oxygen atom (O1). In all cases the axial Si–O distances are significantly longer than the equatorial ones (Si–O1 > Si–O2). In addition, distinct differences between the two equatorial Si–C distances were observed (Si–C1 > Si–C2). As would be expected from the presence of the potential NH donor functions and potential oxygen and fluorine acceptor atoms, N–H···O and/or N–H···F hydrogen bonds are found in the crystals of **3a/3b–5a/5b**.

As shown by multinuclear NMR studies, compounds **3a/3b–5a/5b** also exist in solution (**3a–5a**, **4b**: [D_6]DMSO; **3b**, **5b**: $CDCl_3$). Due to the chiral nature of these zwitterions, their $SiCH_2N$ protons are diastereotopic (**3a–5a**: ABX spin systems in the ^{1}H-NMR spectra, with F as the X nucleus; **3b–5b**: ABMX spin systems in the ^{1}H-NMR spectra, with N*H* as the M nucleus and F as the X nucleus).

Acknowledgment: We thank the *Deutsche Forschungsgemeinschaft* and the *Fonds der Chemischen Industrie* for financial support and *Bayer AG* (Leverkusen and Wuppertal-Elberfeld, Germany) and *Merck KGaA* (Darmstadt, Germany) for support with chemicals.

References:

[1] R. Tacke, J. Becht, O. Dannappel, R. Ahlrichs, U. Schneider, W. S. Sheldrick, J. Hahn, F. Kiesgen, *Organometallics* **1996**, *15*, 2060, and references cited therein.

[2] M. Pülm, R. Tacke, *Organometallics* **1997**, *16*, 5664, and references cited therein.

[3] Compound **4a** has already been described: R. Tacke, O. Dannappel, M. Mühleisen, "*Syntheses, Structures, and Properties of Molecular λ⁵Si-Silicates Containing Bidentate 1,2-Diolato(2–) Ligands Derived from α-Hydroxycarbocyclic Acids, Acetohydroximic Acid, and Oxalic Acid: New Results in the Chemistry of Pentacoordinate Silicon*", in: *Organosilicon Chemistry II: From Molecules to Materials* (Eds.: N. Auner, J. Weis), VCH, Weinheim, **1996**, pp. 427–446.

[4] S. N. Tandura, M. G. Voronkov, N. V. Alekseev, *Top. Curr. Chem.* **1986**, *131*, 99.

[5] W. S. Sheldrick, "*Structural Chemistry of Organic Silicon Compounds*", in: *The Chemistry of Organic Silicon Compounds, Part 1* (Eds.: S. Patai, Z. Rappoport), Wiley, Chichester, **1989**, pp. 227–303.

[6] R. R. Holmes, *Chem. Rev.* **1990**, *90*, 17.

[7] C. Chuit, R. J. P. Corriu, C. Reye, J. C. Young, *Chem. Rev.* **1993**, *93*, 1371.

[8] C. Y. Wong, J. D. Woollins, *Coord. Chem. Rev.* **1994**, *130*, 175.

[9] E. Lukevics, O. A. Pudova, *Chem. Heterocycl. Compd.* **1996**, *353*, 1605.

[10] R. R. Holmes, *Chem. Rev.* **1996**, *96*, 927.

Isomerization of Chiral Zwitterionic Monocyclic λ^5Si-Silicates

Ruth E. Neugebauer, Rüdiger Bertermann, Reinhold Tacke*

Institut für Anorganische Chemie, Universität Würzburg
Am Hubland, D-97074 Würzburg, Germany
Tel.: Int. code + (931)888 5250 — Fax: Int. code + (931)888 4609
E-mail: r.tacke@mail.uni-wuerzburg.de

Keywords: Pentacoordinate Silicon / Zwitterionic λ^5Si-Silicates / Monocyclic λ^5Si-Silicates / Chirality / Isomerization / Epimerization / NMR Spectroscopy

Summary: A series of chiral zwitterionic monocyclic λ^5Si-silicates with an SiO_2C_2F framework have been synthesized and characterized. Various time- and temperature-dependent solution-state NMR experiments have been performed to distinguish between different potential mechanisms (intramolecular isomerization and intermolecular ligand exchange) that could change the absolute configuration of the chiral trigonal-bipyramidal λ^5Si-silicate skeleton of the title compounds.

Introduction

The monocyclic pentacoordinate silicon compounds **1a**, **1b**, **2a**, **2b**, and **3** are derivatives of a new class of zwitterionic (molecular) λ^5Si-silicates with an SiO_2C_2F framework (in this context, see also [1, 2]). We report here on the syntheses of these compounds and NMR-spectroscopic studies concerning their behavior in solution. The aim of these investigations was to distinguish between different potential mechanisms (intramolecular isomerization and intermolecular ligand exchange) that could change the absolute configuration of the chiral trigonal-bipyramidal λ^5Si-silicate skeletons of the title compounds (for reviews dealing with pentacoordinate silicon compounds, see [3–9]).

	R
a	H
b	Me

1 **2**

3

Results and Discussion

The zwitterionic monocyclic λ^5*Si*-silicates **1a** and **1b** were synthesized by reaction of the zwitterionic λ^5*Si*-trifluorosilicate **4** with one molar equivalent of the *O,O'*-bis(trimethylsilyl) derivatives of glycolic acid and 2-methyllactic acid, respectively (Scheme 1). Compounds **2a** and **2b** were prepared analogously starting from the zwitterionic λ^5*Si*-trifluorosilicate **5** (Scheme 1). As shown for compound **1a** in Fig. 1, the zwitterionic λ^5*Si*-silicates **1a**, **1b**, **2a**, and **2b** are chiral and exist as pairs of enantiomers [(*A*)- and (*C*)-enantiomers]. All compounds were isolated as racemic mixtures.

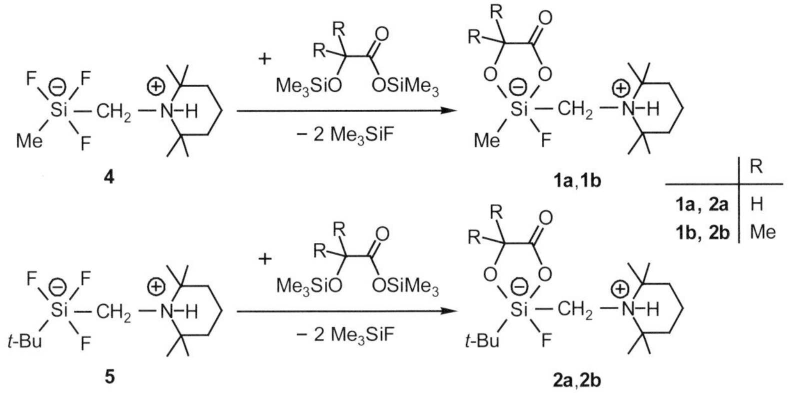

	R
1a, 2a	H
1b, 2b	Me

Scheme 1. Syntheses of the zwitterionic monocyclic λ^5*Si*-silicates **1a**, **1b**, **2a**, and **2b** (products isolated as crystalline racemic mixtures).

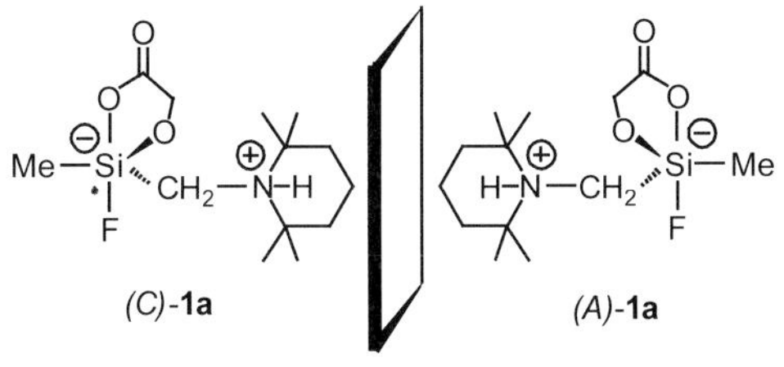

Fig. 1. Structures of the enantiomers of the zwitterionic monocyclic λ^5*Si*-silicate **1a**.

Reaction of the optically active zwitterionic λ^5*Si*-trifluorosilicate (*S*)-**6** with one molar equivalent of the *O,O'*-bis(trimethylsilyl) derivative of (*S*)-mandelic acid gave a mixture of two diastereomeric products, the zwitterionic monocyclic λ^5*Si*-silicates (*S,S,A*)-**3** and (*S,S,C*)-**3** (Scheme 2). The isomer (*S,S,A*)-**3** was obtained as a diastereomerically pure crystalline solid.

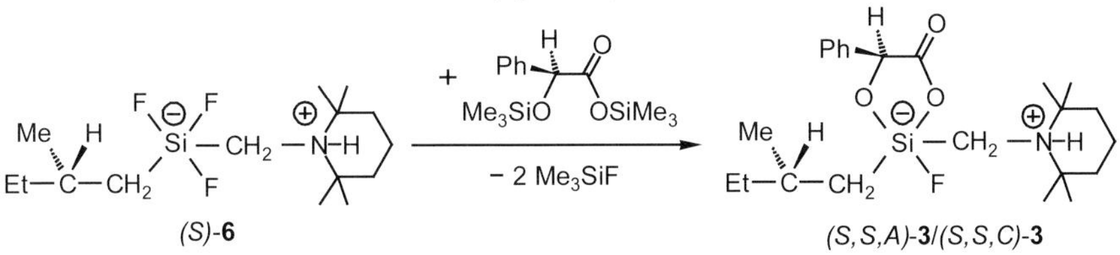

Scheme 2. Syntheses of the zwitterionic monocyclic λ^5*Si*-silicates (*S,S,A*)-**3**/(*S,S,C*)-**3** (mixture of diastereomers; (*S,S,A*)-**3** isolated as a diastereomerically pure crystalline product).

All syntheses were carried out in acetonitrile at room temperature and compounds **1a**, **1b**, **2a**, **2b**, and (*S,S,A*)-**3** were isolated in good yields (78–86 %) as colorless crystalline solids. Their identities were established by elemental analyses (C, H, N), solution-state NMR experiments (^{1}H, ^{13}C, ^{19}F, ^{29}Si), and mass-spectrometric studies (APCI MS). In addition, all compounds were structurally characterized by single-crystal X-ray diffraction (data not given; for the crystal structure of **1a**, see [2]). The *Si*-coordination polyhedra of compounds **1a**, **1b**, **2a**, **2b**, and (*S,S,A*)-**3** were found to be somewhat distorted trigonal bipyramids, with the fluorine atom and carboxylate oxygen atom in the axial positions.

As shown by $^{19}F\{^{1}H\}$-NMR studies of the racemic compound **1b** (282 MHz, 22°C, $CDCl_3$), the respective enantiomers are configurationally stable on the NMR time scale. While the $^{19}F\{^{1}H\}$-NMR spectrum of racemic **1b** under achiral conditions shows only one resonance signal, a set of two signals was observed upon addition of 0.75 molar equivalents of the chiral solvating agent (*R*)-**7** (Fig. 2). This phenomenon results from the existence of diastereomeric solvates [(*A*)-**1b** · (*R*)-**7**, (*C*)-**1b** · (*R*)-**7**] in solution.

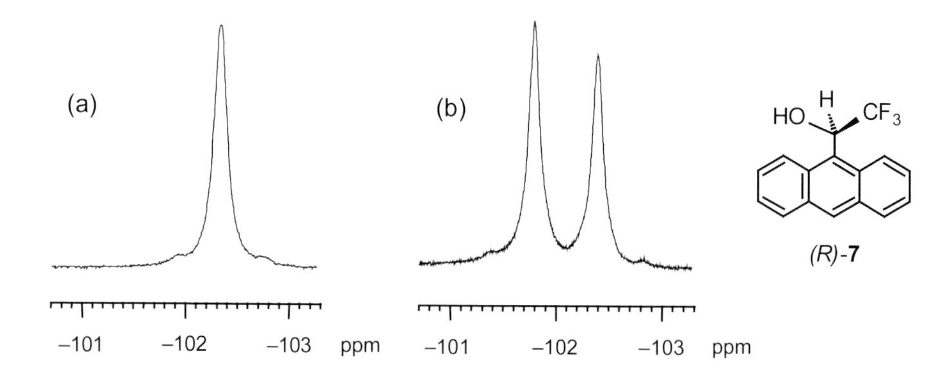

Fig. 2. $^{19}F\{^{1}H\}$-NMR spectra of the racemic zwitterionic monocyclic λ^5Si-silicate **1b** in solution (a) without addition of the solvating agent (*R*)-**7** and (b) after addition of 0.75 molar equivalents of (*R*)-**7** (282 MHz, 22 °C, $CDCl_3$).

As shown by ^{1}H-NMR studies of the zwitterionic λ^5Si-silicate (*S,S,A*)-**3** (300 MHz, –20°C, CD_2Cl_2), this optically active zwitterion undergoes an isomerization in solution yielding the corresponding diastereomer (*S,S,C*)-**3** [detection of a thermodynamic equilibrium, (*S,S,A*)-**3** $\rightleftharpoons$ (*S,S,C*)-**3**; ratio of diastereomers ca. 1:1] (Fig. 3). This isomerization (epimerization) process involves an inversion of absolute configuration of the chiral trigonal-bipyramidal λ^5Si-silicate framework of the two diastereomers. As the kinetics of this epimerization do not depend significantly on the concentration of **3** (concentration range studied 2.9–61 mmol L^{-1}), the existence of an intramolecular process can be assumed. In the case of compounds **1a**, **1b**, **2a**, and **2b**, an analogous process would lead to a conversion of the respective (*A*)- and (*C*)-enantiomers into each other.

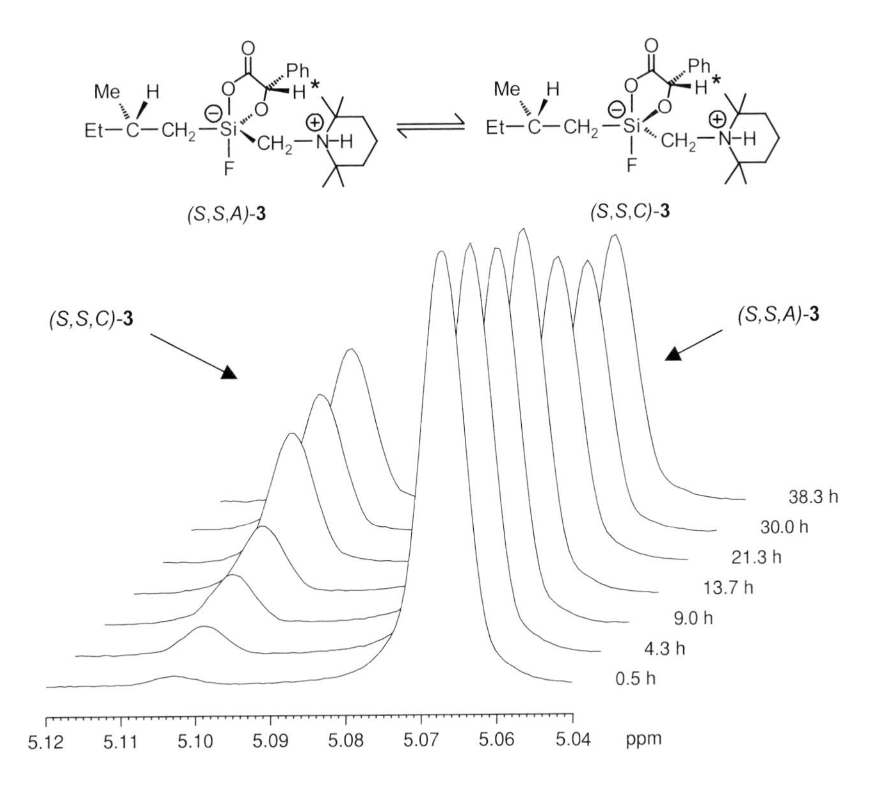

Fig. 3. Partial ^{1}H-NMR spectra of the zwitterionic λ^5*Si*-silicate (*S,S,A*)-**3**/(*S,S,C*)-**3** in solution demonstrating the kinetics of the epimerization process (*S,S,A*)-**3** $\rightleftharpoons$ (*S,S,C*)-**3** (300 MHz, –20°C, CD_2Cl_2). The kinetics were followed after dissolving a sample of diastereomerically pure (*S,S,A*)-**3** at –20°C. The protons observed are marked with an asterisk.

^{19}F{^{1}H}-NMR studies with equimolar mixtures of **1a** and **2b** (282 MHz, 0–21°C, CD_2Cl_2) have demonstrated that these zwitterions undergo an intermolecular exchange of their diolato(2–) ligands in solution leading to the formation of compounds **1b** and **2a** (Scheme 3 and Fig. 4).

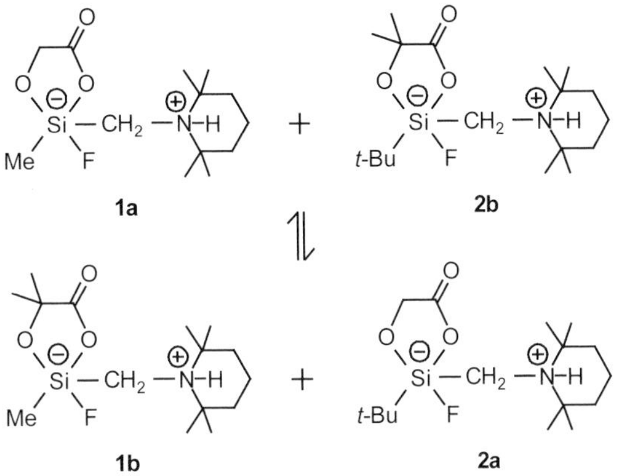

1a + **2b** ⇌ **1b** + **2a**

Scheme 3. Intermolecular exchange of the diolato(2–) ligands of **1a** and **2b**.

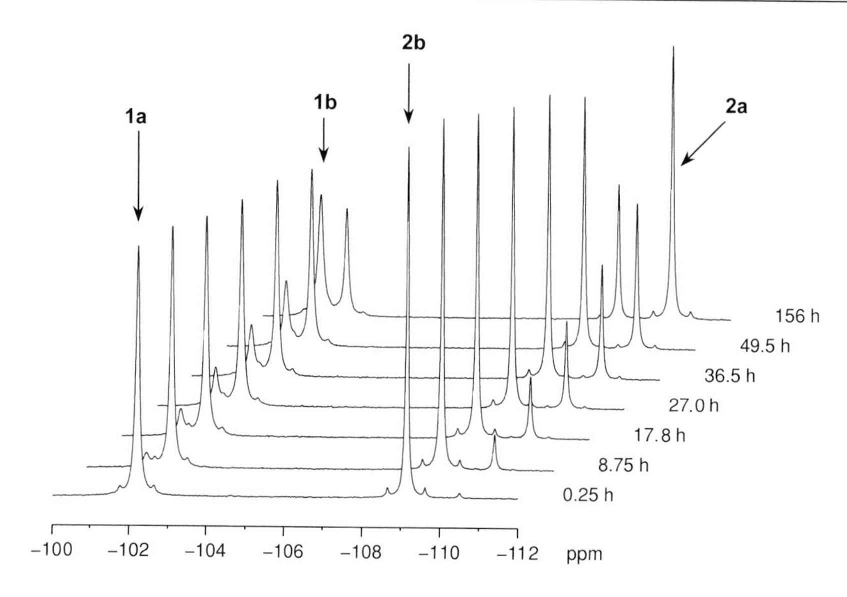

Fig. 4. Kinetics of the intermolecular ligand exchange observed after dissolving an equimolar mixture of the zwitterionic λ^5Si-silicates **1a** and **2b** in CD_2Cl_2 (c = 157 mmol L^{-1}). The kinetics were followed by $^{19}F\{^1H\}$-NMR experiments (282 MHz, 0°C).

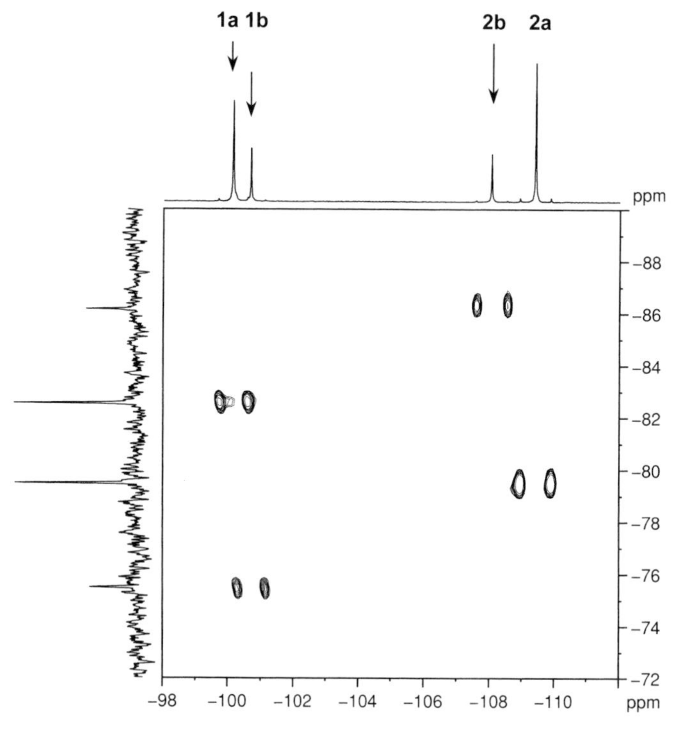

Fig. 5. 2D $^{19}F,^{29}Si$ inverse correlation of an equilibrium mixture of the zwitterionic λ^5Si-silicates **1a**, **1b**, **2a**, and **2b** in solution (^{19}F, 282 MHz; ^{29}Si, 59.6 MHz; CD_3CN, 22°C). The projection on the F1 axis shows the $^{29}Si\{^1H,^{19}F\}$-NMR spectrum, whereas the projection on the F2 axis shows the $^{19}F\{^1H\}$-NMR spectrum. Note that the 2D spectrum shows doublets in F2 with the spin coupling constant $^1J(^{29}Si,^{19}F)$.

The resulting mixtures of **1a**, **1b**, **2a**, and **2b** (molar ratio ca. 0.46:1:1:0.46; thermodynamic equilibrium) could also be detected by a 2D $^{19}F,^{29}Si$ inverse correlation acquired with *z*-gradient selection (Fig. 5).

In contrast to the above-mentioned intramolecular epimerization studied at –20°C, no intermolecular ligand exchange could be detected at this particular temperature; i.e., the intermolecular process (which might also lead to an inversion of absolute configuration of the chiral trigonal-bipyramidal λ^5*Si*-silicate skeleton) is significantly slower than the intramolecular epimerization. Further $^{19}F\{^1H\}$-NMR experiments using different solvents ($CDCl_3$, CD_2Cl_2, CD_3CN, CD_3CN/water) demonstrated that the kinetics of the intermolecular ligand exchange and the final ratio (thermodynamic equilibrium) of compounds **1a**, **1b**, **2a**, and **2b** do not depend significantly on the solvent used (Fig. 6).

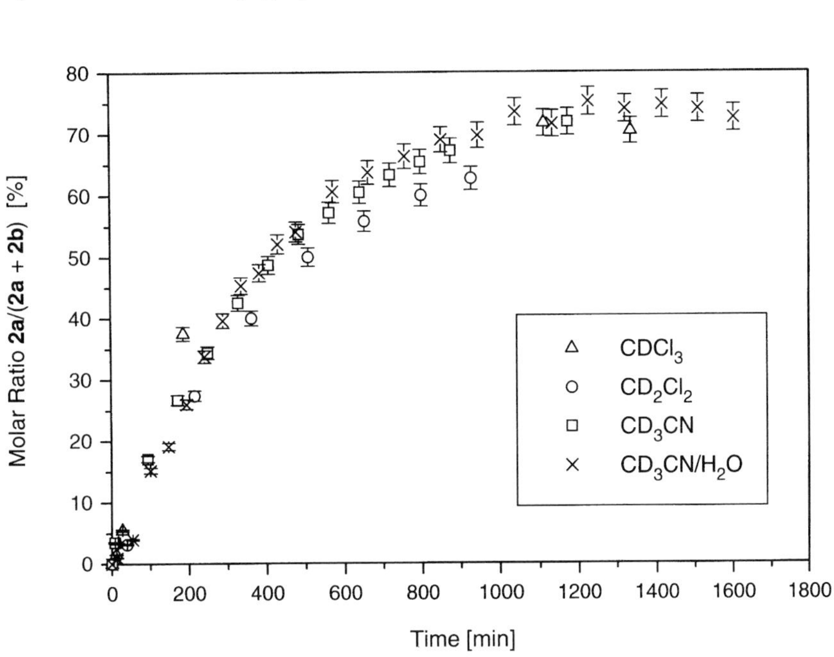

Fig. 6. Kinetics of the intermolecular ligand exchange observed after dissolving an equimolar mixture of the zwitterionic λ^5*Si*-silicates **1a** and **2b** in various solvents ($CDCl_3$, CD_2Cl_2, CD_3CN, CD_3CN/H_2O; c = 100 mmol L^{-1}). The data given were obtained by $^{19}F\{^1H\}$-NMR experiments (282 MHz, 22°C).

In conclusion, the intramolecular epimerization observed for the diastereomers (*S,S,A*)-**3** and (*S,S,C*)-**3** might be considered as a general process that allows an inversion of absolute configuration of the chiral trigonal-bipyramidal λ^5*Si*-silicate framework of the title compounds. As shown in Fig. 7, this process is significantly faster than the above-mentioned intermolecular ligand exchange that might also involve inversion of absolute configuration (although not proven experimentally). The molecular mechanisms of both processes are still unknown.

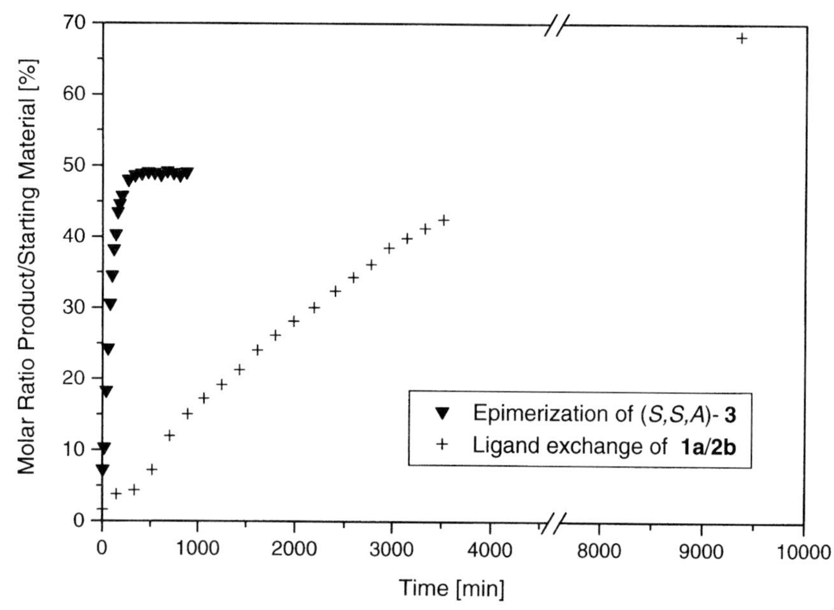

Fig. 7. Comparison of the kinetics of the intramolecular epimerization of (*S,S,A*)-**3** (c = 37 mmol L^{-1}) and the intermolecular ligand exchange of an equimolar mixture of racemic **1a** and **2b** (c = 157 mmol L^{-1}) in CD_2Cl_2 at 0°C. The data given were obtained by ^{1}H-NMR (300 MHz) (epimerization) or ^{19}F{^{1}H}-NMR (282 MHz) experiments (ligand exchange).

Acknowledgment: We thank T. Paschold (*Institut für Anorganische Chemie, Universität Würzburg*) for providing samples of compounds **2a** and **2b**. Financial support of this work by the *Deutsche Forschungsgemeinschaft* and the *Fonds der Chemischen Industrie* is gratefully acknowledged.

References:

[1] R. Tacke, O. Dannappel, M. Mühleisen, "*Syntheses, Structures, and Properties of Molecular λ^5Si-Silicates Containing Bidentate 1,2-Diolato(2–) Ligands Derived from α-Hydroxycarbocyclic Acids, Acetohydroximic Acid, and Oxalic Acid: New Results in the Chemistry of Pentacoordinate Silicon*", in: *Organosilicon Chemistry II: From Molecules to Materials* (Eds.: N. Auner, J. Weis), VCH, Weinheim, **1996**, pp. 427–446.

[2] R. Willeke, R. E. Neugebauer, M. Pülm, O. Dannappel, R. Tacke, "*Syntheses, Structures, and Properties of Zwitterionic Monocyclic λ^5Si-Silicates*", in: *Organosilicon Chemistry IV: From Molecules to Materials* (Eds.: N. Auner, J. Weis), Wiley–VCH, Weinheim, **1999**, pp. 456–459.

[3] S. N. Tandura, M. G. Voronkov, N. V. Alekseev, *Top. Curr. Chem.* **1986**, *131*, 99.

[4] W. S. Sheldrick, "*Structural Chemistry of Organic Silicon Compounds*", in: *The Chemistry of Organic Silicon Compounds, Part 1* (Eds.: S. Patai, Z. Rappoport), Wiley, Chichester, **1989**, pp. 227–303.

[5] R. R. Holmes, *Chem. Rev.* **1990**, *90*, 17.

[6] C. Chuit, R. J. P. Corriu, C. Reye, J. C. Young, *Chem. Rev.* **1993**, *93*, 1371.
[7] C. Y. Wong, J. D. Woollins, *Coord. Chem. Rev.* **1994**, *130*, 175.
[8] E. Lukevics, O. A. Pudova, *Chem. Heterocycl. Compd.* **1996**, *353*, 1605.
[9] R. R. Holmes, *Chem. Rev.* **1996**, *96*, 927.

Zwitterionic Spirocyclic λ^5Si-Silicates Containing Diolato(2–) Ligands Derived from Acetohydroximic Acid and Benzohydroximic Acid

*Andreas Biller, Brigitte Pfrommer, Melanie Pülm, Reinhold Tacke**

Institut für Anorganische Chemie, Universität Würzburg
Am Hubland, D-97074 Würzburg, Germany
Tel.: Int. code + (931)8885250 — Fax: Int. code + (931)8884609
E-mail: r.tacke@mail.uni-wuerzburg.de

Keywords: Pentacoordinate Silicon / Zwitterionic λ^5Si-Silicates / Crystal Structures / Acetohydroximato(2–) Ligand / Benzohydroximato(2–) Ligand

Summary: Derivatives of a new class of zwitterionic spirocyclic λ^5Si-silicates with an SiO_4C framework have been prepared. The syntheses and the properties of these molecular pentacoordinate silicon compounds are reported. The crystal structures of two of the zwitterions are described.

Introduction

Over the past few years, a series of zwitterionic (molecular) spirocyclic λ^5Si-silicates with an SiO_4C framework have been synthesized and structurally characterized [1–12]. The λ^5Si-silicates **1** [1], **2** [3], **3** [9], and **4** [11] are typical examples of this particular type of compound (Scheme 1).

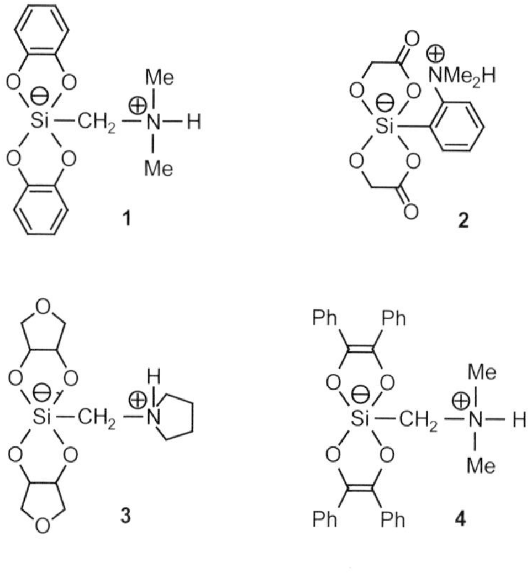

Scheme 1. Typical examples of zwitterionic spirocyclic λ^5Si-silicates.

In continuation of our systematic studies on zwitterionic pentacoordinate silicon compounds, we have now succeeded in synthesizing the molecular λ^5*Si*-silicates **5**–**10** (for reviews dealing with pentacoordinate silicon compounds, see [13–19]). These zwitterions belong to a new class of spirocyclic pentacoordinate silicon compounds with an SiO_4C skeleton. They contain diolato(2–) ligands derived from acetohydroximic acid and benzohydroximic acid, respectively (tautomers of acetohydroxamic acid and benzohydroxamic acid). Preliminary data concerning **5** have already been published elsewhere [8]. We report here on the syntheses and properties of compounds **5**–**10**. In addition, the crystal structures of **6** and **8** are described.

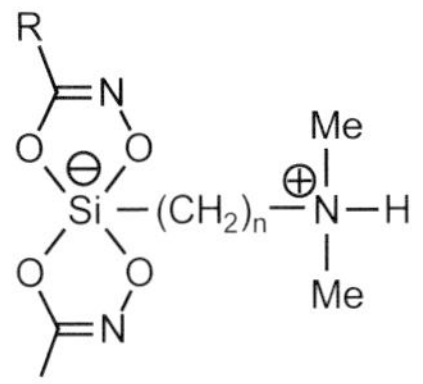

5–10

	R	n
5	Me	1
6	Ph	1
7	Me	2
8	Ph	2
9	Me	3
10	Ph	3

Results and Discussion

Following the strategy used for the synthesis of compound **5** [8], the zwitterionic λ^5*Si*-silicates **6**–**10** were prepared according to Scheme 2 (method a).

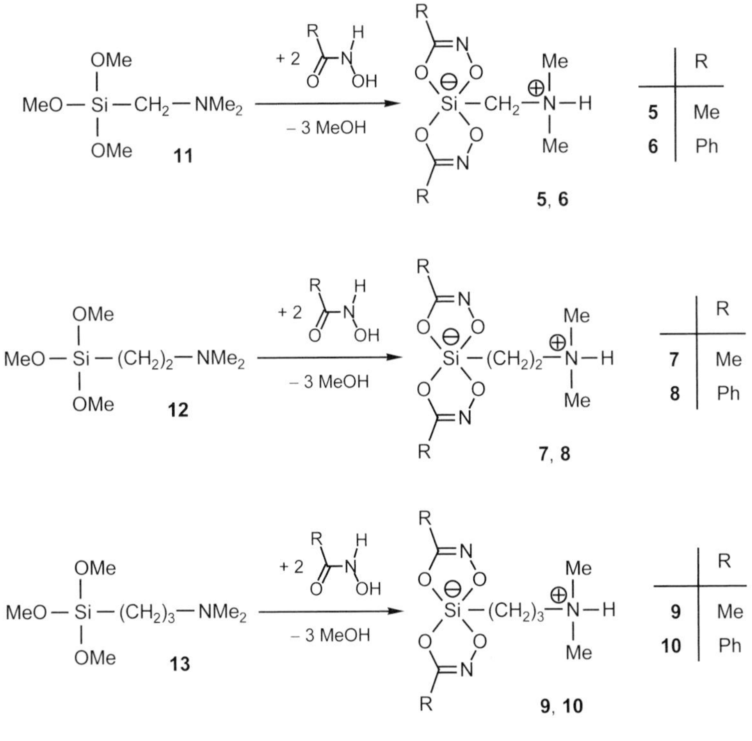

$MeO-Si(OMe)_2-CH_2-NMe_2$ (**11**) + 2 RC(O)N(H)OH → (– 3 MeOH) **5, 6**

	R
5	Me
6	Ph

$MeO-Si(OMe)_2-(CH_2)_2-NMe_2$ (**12**) + 2 RC(O)N(H)OH → (– 3 MeOH) **7, 8**

	R
7	Me
8	Ph

$MeO-Si(OMe)_2-(CH_2)_3-NMe_2$ (**13**) + 2 RC(O)N(H)OH → (– 3 MeOH) **9, 10**

	R
9	Me
10	Ph

Scheme 2. Syntheses of the zwitterionic λ^5*Si*-silicates **5**–**10** according to method a.

Compounds **5** and **6** were obtained by reaction of the silane $(MeO)_3SiCH_2NMe_2$ (**11**) with two molar equivalents of acetohydroxamic acid and benzohydroxamic acid, respectively (formation of three molar equivalents of MeOH). The derivatives **7**–**10** were synthesized analogously, starting from the silanes $(MeO)_3Si(CH_2)_2NMe_2$ (**12**) and $(MeO)_3Si(CH_2)_3NMe_2$ (**13**).

The λ^5Si-silicates **5** and **6** were prepared alternatively (method b) by reaction of the silane $Ph(MeO)_2SiCH_2NMe_2$ (**14**) with two molar equivalents of acetohydroxamic acid and benzohydroxamic acid, respectively (Scheme 3). This method involves cleavage of two Si–O bonds and one Si–C bond (formation of two molar equivalents of MeOH and one molar equivalent of benzene).

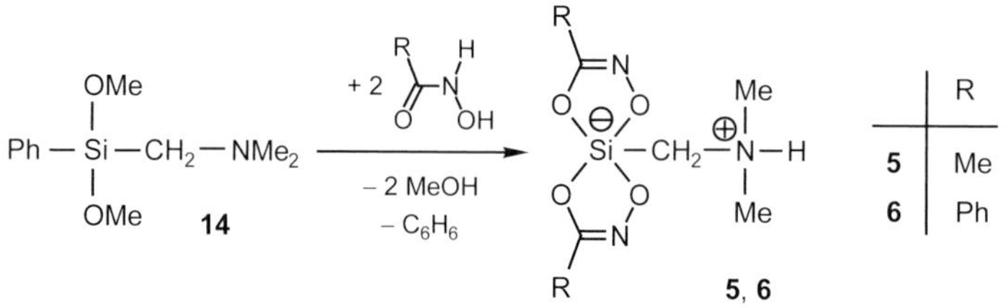

	R
5	Me
6	Ph

Scheme 3. Syntheses of the zwitterionic λ^5Si-silicates **5** and **6** according to method b.

All syntheses were carried out in acetonitrile at room temperature and compounds **5**–**10** were isolated in good yields (67–87 %) as high-melting crystalline solids. Their identities were established by elemental analyses (C, H, N), solution-state NMR (^1H, ^{13}C, ^{29}Si; [D_6]DMSO) and solid-state CP/MAS NMR experiments (^{13}C, ^{29}Si), and mass-spectrometric studies (FAB MS).

In addition, the crystal structures of **6** and **8** were determined by single-crystal X-ray diffraction. Compounds **6** and **8** crystallized in the space group *C*2/*c* and *P*2$_1$/*c*, respectively. The structures of the zwitterions in the crystals of **6** and **8** (two crystallographically independent molecules; molecule A and molecule B) are depicted in Figs. 1 and 2, respectively.

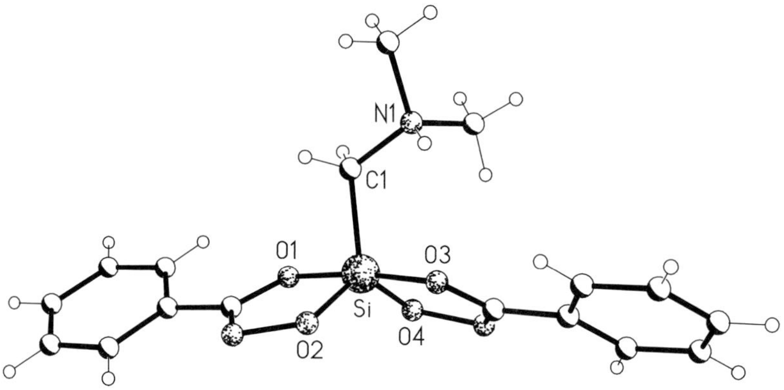

Fig. 1. Molecular structure of the zwitterion in the crystal of **6**.

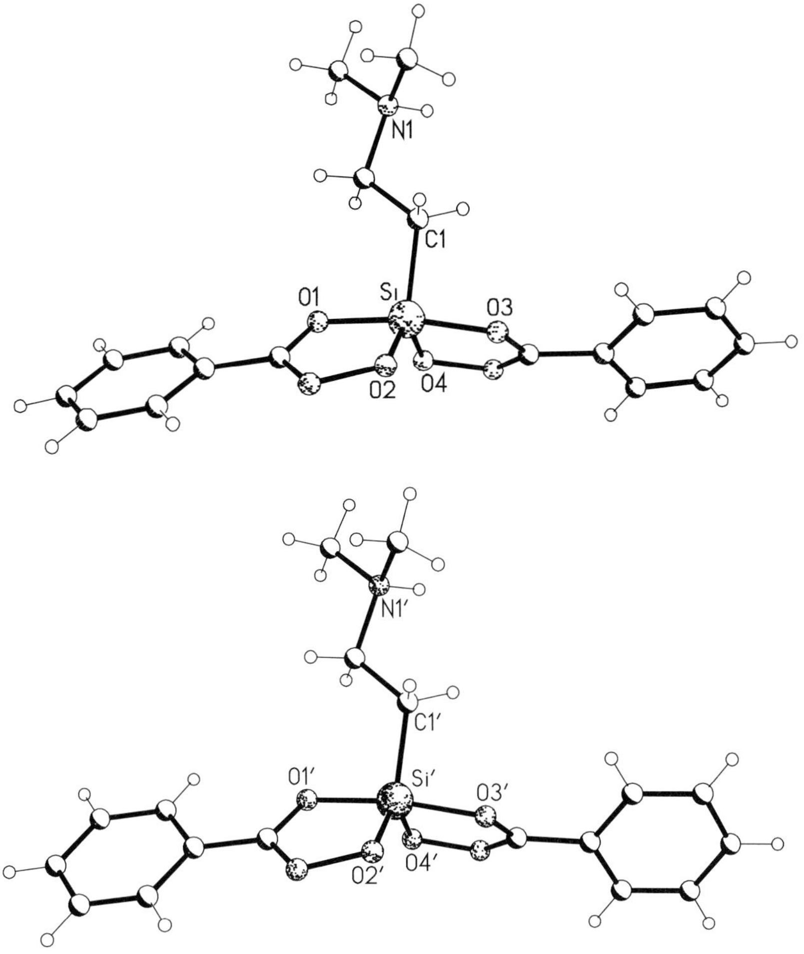

Fig. 2. Molecular structures of the zwitterions in the crystal of **8** (molecule A, above; molecule B, below).

The coordination polyhedra surrounding the silicon atoms in these zwitterions are distorted trigonal bipyramids with the C*O* oxygen atoms in the axial sites and the N*O* oxygen atoms in the equatorial positions. Selected geometric parameters for the *Si*-coordination polyhedra of **6** and **8** are given in Table 1. In general, these data are very similar to those observed for the related zwitterionic λ^5*Si*-silicate **5** [8].

Table 1. Selected geometric parameters for **6** and **8** (distances in pm, angles in deg.).

Compd.	Si–O1	Si–O2	Si–O3	Si–O4	Si–C1	O1-Si-O3
6	176.61(14)	170.46(14)	176.85(14)	168.67(14)	189.7(2)	173.16(7)
8[a]	176.4(4)	169.8(4)	179.0(4)	169.4(4)	188.0(5)	171.0(2)
8[b]	176.0(4)	169.3(4)	180.1(4)	170.4(4)	186.6(6)	171.4(2)

[a] Molecule A. — [b] Molecule B.

Acknowledgment: Financial support of this work by the *Deutsche Forschungsgemeinschaft* and the *Fonds der Chemischen Industrie* is gratefully acknowledged.

References:

[1] R. Tacke, A. Lopez-Mras, J. Sperlich, C. Strohmann, W. F. Kuhs, G. Mattern, A. Sebald, *Chem. Ber.* **1993**, *126*, 851.

[2] R. Tacke, A. Lopez-Mras, W. S. Sheldrick, A. Sebald, *Z. Anorg. Allg. Chem.* **1993**, *619*, 347.

[3] R. Tacke, A. Lopez-Mras, P. G. Jones, *Organometallics* **1994**, *13*, 1617.

[4] R. Tacke, M. Mühleisen, P. G. Jones, *Angew. Chem.* **1994**, *106*, 1250; *Angew. Chem., Int. Ed. Engl.* **1994**, *33*, 1186.

[5] M. Mühleisen, R. Tacke, *Chem. Ber.* **1994**, *127*, 1615.

[6] M. Mühleisen, R. Tacke, *Organometallics* **1994**, *13*, 3740.

[7] R. Tacke, M. Mühleisen, A. Lopez-Mras, W. S. Sheldrick, *Z. Anorg. Allg. Chem.* **1995**, *621*, 779.

[8] R. Tacke, O. Dannappel, M. Mühleisen, *"Syntheses, Structures, and Properties of Molecular λ^5Si-Silicates Containing Bidentate 1,2-Diolato(2–) Ligands Derived from α-Hydroxycarboxylic Acids, Acetohydroximic Acid, and Oxalic Acid: New Results in the Chemistry of Pentacoordinate Silicon"*, in: *Organosilicon Chemistry II: From Molecules to Materials* (Eds.: N. Auner, J. Weis), VCH, Weinheim, **1996**, pp. 427–446.

[9] R. Tacke, O. Dannappel, *"New Zwitterionic Silicon–Oxygen Compounds Containing Pentacoordinate Silicon Atoms: Experimental and Theoretical Studies"*, in: *Tailor-made Silicon–Oxygen Compounds – From Molecules to Materials* (Eds.: R. Corriu, P. Jutzi), Vieweg, Braunschweig/Wiesbaden, **1996**, pp. 75–86.

[10] R. Tacke, J. Heermann, M. Pülm, *Organometallics* **1997**, *16*, 5648.

[11] M. Pülm, R. Tacke, *Organometallics* **1997**, *16*, 5664.

[12] R. Tacke, J. Heermann, M. Pülm, I. Richter, *Organometallics* **1998**, *17*, 1663.

[13] S. N. Tandura, M. G. Voronkov, N. V. Alekseev, *Top. Curr. Chem.* **1986**, *131*, 99.

[14] W. S. Sheldrick, *"Structural Chemistry of Organic Silicon Compounds"*, in: *The Chemistry of Organic Silicon Compounds, Part 1* (Eds.: S. Patai, Z. Rappoport), Wiley, Chichester, **1989**, pp. 227–303.

[15] R. R. Holmes, *Chem. Rev.* **1990**, *90*, 17.

[16] C. Chuit, R. J. P. Corriu, C. Reye, J. C. Young, *Chem. Rev.* **1993**, *93*, 1371.

[17] C. Y. Wong, J. D. Woollins, *Coord. Chem. Rev.* **1994**, *130*, 175.

[18] E. Lukevics, O. A. Pudova, *Chem. Heterocycl. Compd.* **1996**, *353*, 1605.

[19] R. R. Holmes, *Chem. Rev.* **1996**, *96*, 927.

Isoelectronic Zwitterionic Pentacoordinate Silicon Compounds with $\textit{Si}O_4C$ and $\textit{Si}O_5$ Frameworks

Birgitte Pfrommer, Reinhold Tacke*

Institut für Anorganische Chemie, Universität Würzburg
Am Hubland, D-97074 Würzburg, Germany
Tel.: Int. code + (931)888 5250 — Fax: Int. code + (931)888 4609
E-mail: r.tacke@mail.uni-wuerzburg.de

Keywords: Pentacoordinate Silicon / Zwitterionic $\lambda^5\textit{Si}$-Silicates / $\textit{Si}O_4C$ Skeleton / $\textit{Si}O_5$ Skeleton / Crystal Structures / Dynamic Behavior / Chirality / Isomerization

Summary: The zwitterionic spirocyclic $\lambda^5\textit{Si}$-silicates **5** and **6** ($\textit{Si}O_4C$ skeletons) and their respective isoelectronic analogs **7** and **8** ($\textit{Si}O_5$ skeletons) were synthesized and structurally characterized by single-crystal X-ray diffraction. In addition, the dynamic behavior of the isoelectronic CH_2/O analogs **6** and **8** in solution ($[D_6]$DMSO) was studied by VT ^{1}H-NMR experiments. The molecular structures of the CH_2/O pairs **5**/**7** and **6**/**8** in the crystal are very similar, and the dynamic behavior of the CH_2/O analogs **6** and **8** in solution (reversible coalescence phenomena) is also characterized by distinct similarities. The results of the VT ^{1}H-NMR studies can be interpreted in terms of an intramolecular process that leads to an inversion of absolute configuration of the chiral zwitterions (Λ-**6** $\rightleftharpoons$ Δ-**6**, $\Delta G^{\ddagger} = 89\pm2$ kJ mol^{-1}; Λ-**8** $\rightleftharpoons$ Δ-**8**, $\Delta G^{\ddagger} = 84.6\pm0.6$ kJ mol^{-1}).

Introduction

In contrast to the well-established chemistry of zwitterionic spirocyclic $\lambda^5\textit{Si}$-silicates with an $\textit{Si}O_4C$ framework [1–11] (such as compounds **1** [1], **2** [9], and **3** [7]), only little is known about related zwitterions with an $\textit{Si}O_5$ skeleton (such as compound **4** [11]). We report here on the syntheses, structures, and properties of the zwitterionic $\lambda^5\textit{Si}$-silicates **5** and **6** ($\textit{Si}O_4C$ skeletons) and their respective isoelectronic analogs **7** and **8** ($\textit{Si}O_5$ skeletons). The studies presented here were carried out with a special emphasis on the comparison of the isoelectronic CH_2/O pairs **5**/**7** and **6**/**8** (for reviews dealing with pentacoordinate silicon compounds, see [12–18]).

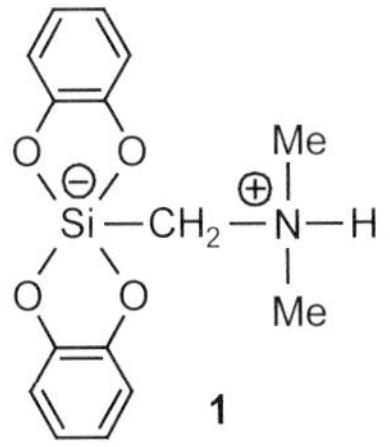

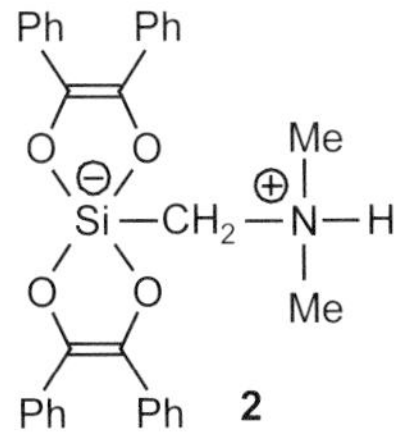

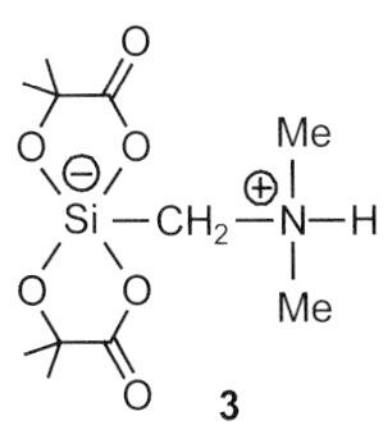

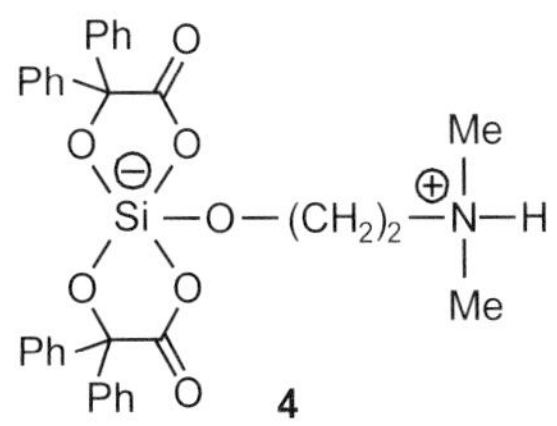

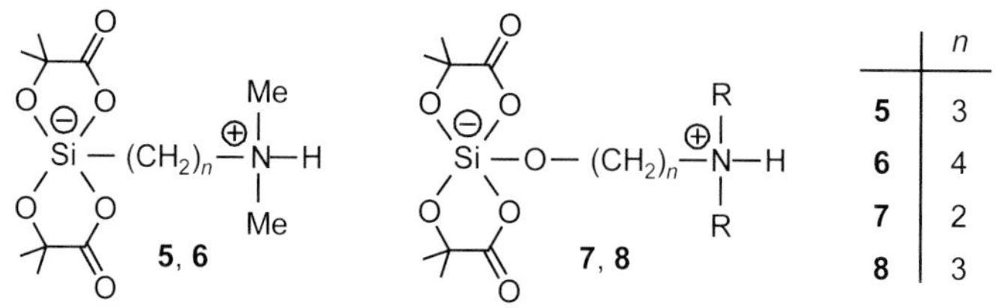

	n
5	3
6	4
7	2
8	3

Results and Discussion

The zwitterionic λ^5Si-silicates **5** and **6** were synthesized by reaction of [3-(dimethylamino)propyl]-trimethoxysilane and [4-(dimethylamino)butyl]trimethoxysilane, respectively, with two molar equivalents of 2-methyllactic acid in acetonitrile at room temperature (Scheme 1).

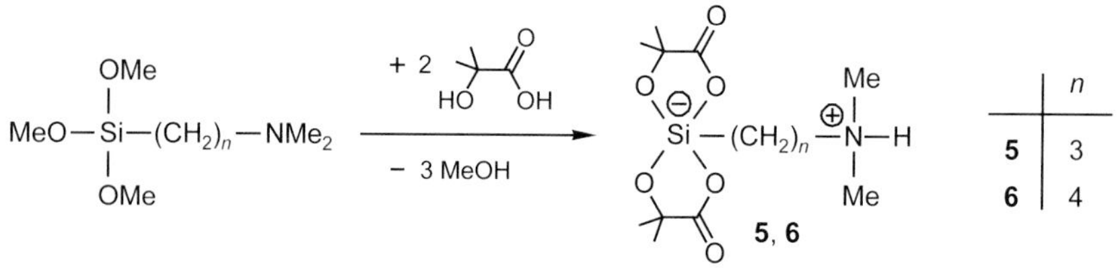

	n
5	3
6	4

Scheme 1. Syntheses of compounds **5** and **6**.

The isoelectronic analogs **7** and **8** were prepared by treatment of tetramethoxysilane with 2-methyllactic acid and the corresponding aminoalcohol (molar ratio 1:2:1) in boiling acetonitrile (removal of the methanol formed by distillation) (Scheme 2). Alternatively, compounds **7** and **8** were obtained by reaction of tetrakis(dimethylamino)silane with two molar equivalents of 2-methyllactic acid and one molar equivalent of the respective aminoalcohol in boiling acetonitrile (Scheme 2).

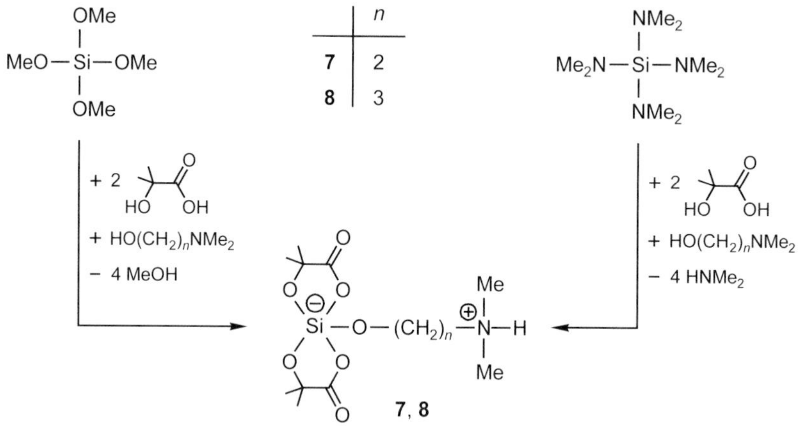

	n
7	2
8	3

Scheme 2. Syntheses of compounds **7** and **8**.

The λ^5Si-silicates **5**–**8** were isolated in good yields (71–90 %) as colorless crystalline solids. Their identities were established by elemental analyses (C, H, N), solution-state (^1H, ^{13}C, ^{29}Si; $[D_6]DMSO$) and solid-state (^{29}Si-CP/MAS) NMR experiments, and mass-spectrometric investigations (FAB MS). In addition, compounds **5**–**8** were structurally characterized by single-crystal X-ray diffraction.

The title compounds crystallize in the space group $P2_1/n$ (**5**) or $P2_1/c$ (**6**–**8**). The structures of the zwitterions in the crystal are shown in Fig. 1. As the two crystallographically independent molecules (molecules A and B) observed in the crystal of **6** and **8** are very similar, only one of them (molecule A) is depicted.

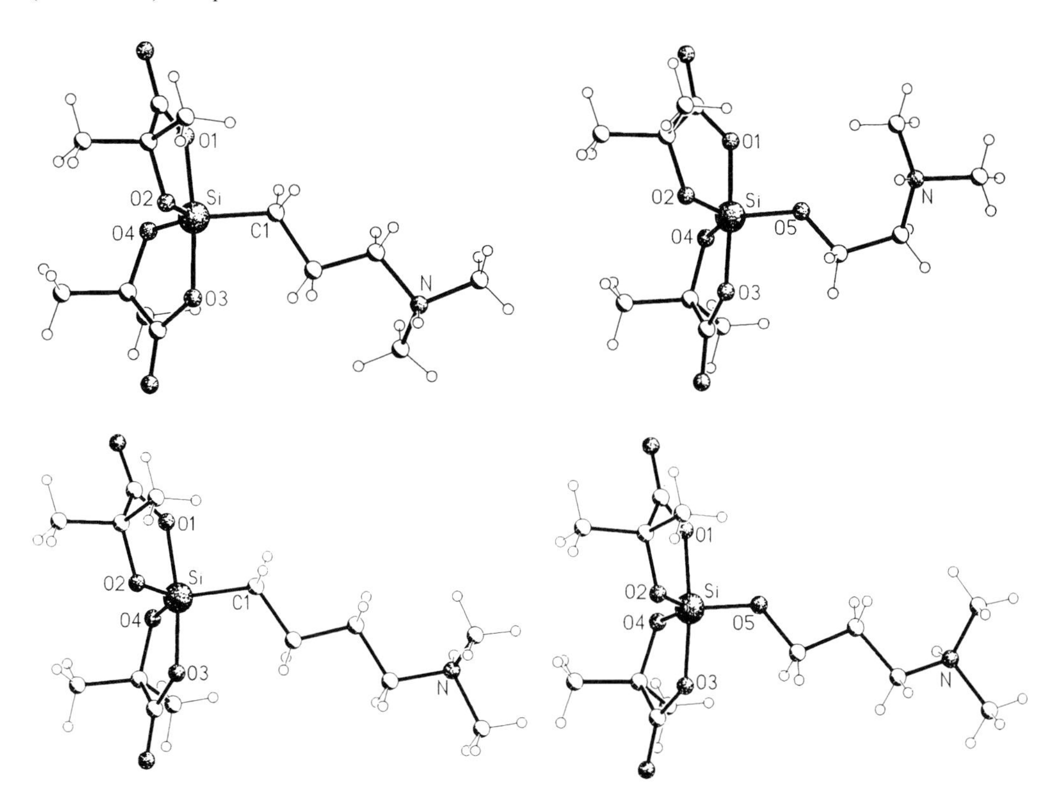

Fig. 1. Molecular structures of **5** (left above), **6** (molecule A, left below), **7** (right above), and **8** (molecule A, right below) in the crystal.

The coordination polyhedra around the central silicon atoms of **5**–**8** are distorted trigonal bipyramids, with the carboxylate oxygen atoms in the axial positions. In all cases, the axial Si–O distances are significantly longer than the equatorial ones. Selected geometric parameters for the coordination polyhedra of **5**–**8** are given in Table 1. As can be seen from Fig. 1 and Table 1, the structures of the SiO_4C and SiO_5 skeletons of the isoelectronic CH_2/O pairs **5**/**7** and **6**/**8** are very similar.

The isoelectronic CH_2/O analogs **6** and **8** were studied by solution-state ^1H-NMR experiments in the temperature range of 27–128°C (**6**) and 27–124°C (**8**) (solvent $[D_6]DMSO$). In accordance with the chiral nature of the λ^5Si-silicate framework of these compounds, diastereotopism for the two methyl groups of the 2-methyllactato(2–) ligands was observed. For both zwitterions two separated singlets for these methyl groups were detected at room temperature, indicating that the Λ- and Δ-enantiomers of compounds **6** and **8** are configurationally stable on the NMR time scale under these conditions. Upon heating, line broadening and coalescence phenomena were observed, which were found to be completely reversible on subsequent cooling. These results can be interpreted in terms of an intramolecular process that leads to a conversion of the respective Λ- and Δ-enantiomers into each other (in this context, see [7]). The free energies of activation for this process amount to 89±2 kJ mol^{-1} (**6**) and 84.6±0.6 kJ mol^{-1} (**8**) [19].

Table 1. Selected geometric parameters for compounds **5**–**8** (distances in Å, angles in deg.).

	5	**6**[a]		**7**	**8**[a]	
Si–O1	1.832(2)	1.841(3)	1.836(4)	1.798(2)	1.804(2)	1.785(2)
Si–O2	1.672(2)	1.663(3)	1.665(4)	1.659(2)	1.665(2)	1.662(2)
Si–O3	1.804(2)	1.786(3)	1.810(4)	1.773(2)	1.785(2)	1.810(2)
Si–O4	1.670(2)	1.685(3)	1.671(3)	1.653(2)	1.660(2)	1.668(2)
Si–C1	1.871(4)	1.874(4)	1.868(4)			
Si–O5				1.643(2)	1.633(2)	1.634(2)
O1-Si-O3	173.40(11)	170.1(2)	169.82(14)	176.83(8)	172.68(12)	172.08(12)

[a] Left column: molecule A; right column: molecule B.

In conclusion, the molecular structures of the isoelectronic CH_2/O pairs **5**/**7** and **6**/**8** in the crystal are very similar, and the dynamic behavior of the isoelectronic CH_2/O analogues **6** and **8** in solution is also characterized by distinct similarities.

Acknowledgment: Financial support of this work by the *Deutsche Forschungsgemeinschaft* and the *Fonds der Chemischen Industrie* is gratefully acknowledged.

References:

[1] R. Tacke, A. Lopez-Mras, J. Sperlich, C. Strohmann, W. F. Kuhs, G. Mattern, A. Sebald, *Chem. Ber.* **1993**, *126*, 851.

[2] R. Tacke, A. Lopez-Mras, P. G. Jones, *Organometallics* **1994**, *13*, 1617.

[3] R. Tacke, M. Mühleisen, P. G. Jones, *Angew. Chem.* **1994**, *106*, 1250; *Angew. Chem., Int. Ed. Engl.* **1994**, *33*, 1186.

[4] M. Mühleisen, R. Tacke, *Organometallics* **1994**, *13*, 3740.

[5] R. Tacke, M. Mühleisen, A. Lopez-Mras, W. S. Sheldrick, *Z. Anorg. Allg. Chem.* **1995**, *621*, 779.

[6] R. Tacke, O. Dannappel, M. Mühleisen, "*Syntheses, Structures, and Properties of Molecular λ^5Si-Silicates Containing Bidentate 1,2-Diolato(2–) Ligands Derived from α-Hydroxycarboxylic Acids, Acetohydroximic Acid, and Oxalic Acid: New Results in the Chemistry of Pentacoordinate Silicon*", in: *Organosilicon Chemistry II: From Molecules to Materials* (Eds.: N. Auner, J. Weis), VCH, Weinheim, **1996**, pp. 427–446.

[7] R. Tacke, O. Dannappel, "*New Zwitterionic Silicon-Oxygen Compounds Containing Pentacoordinate Silicon Atoms: Experimental and Theoretical Studies*", in: *Tailor-made Silicon-Oxygen Compounds – From Molecules to Materials* (Eds.: R. Corriu, P. Jutzi), Vieweg, Braunschweig/Wiesbaden, **1996**, pp. 75–86.

[8] R. Tacke, J. Heermann, M. Pülm, *Organometallics* **1997**, *16*, 5648.

[9] M. Pülm, R. Tacke, *Organometallics* **1997**, *16*, 5664.

[10] R. Tacke, J. Heermann, M. Pülm, I. Richter, *Organometallics* **1998**, *17*, 1663.

[11] R. Tacke, M. Mühleisen, *Inorg. Chem.* **1994**, *33*, 4191.

[12] S. N. Tandura, M. G. Voronkov, N. V. Alekseev, *Top. Curr. Chem.* **1986**, *131*, 99.

[13] W. S. Sheldrick, "*Structural Chemistry of Organic Silicon Compounds*", in: *The Chemistry of Organic Silicon Compounds, Part 1* (Eds.: S. Patai, Z. Rappoport), Wiley, Chichester, **1989**, pp. 227–303.

[14] R. R. Holmes, *Chem. Rev.* **1990**, *90*, 17.

[15] C. Chuit, R. J. P. Corriu, C. Reye, J. C. Young, *Chem. Rev.* **1993**, *93*, 1371.

[16] C. Y. Wong, J. D. Woollins, *Coord. Chem. Rev.* **1994**, *130*, 175.

[17] E. Lukevics, O. A. Pudova, *Chem. Heterocycl. Compd.* **1996**, *353*, 1605.

[18] R. R. Holmes, *Chem. Rev.* **1996**, *96*, 927.

[19] Data obtained by line-shape analysis.

The Zwitterionic Spirocyclic λ^5Si-Silicate [(Dimethylammonio)methyl]bis-[salicylato(2–)-O^1,O^3]silicate: Experimental and Computational Studies

*Melanie Pülm, Reiner Willeke, Reinhold Tacke**

Institut für Anorganische Chemie, Universität Würzburg
Am Hubland, D-97074 Würzburg, Germany
Tel.: Int. code + (931)888 5250 — Fax: Int. code + (931)888 4609
E-mail: r.tacke@mail.uni-wuerzburg.de

Keywords: Pentacoordinate Silicon / Zwitterionic λ^5Si-Silicates / Salicylato(2–) Ligand / Crystal Structures / Ab Initio Studies / Stereochemistry

Summary: The zwitterionic spirocyclic λ^5Si-silicate [(dimethylammonio)methyl]-bis[salicylato(2–)-O^1,O^3]silicate (**6**) was synthesized by two different methods, and the crystal structures of the acetonitrile solvate **6a**·$0.7CH_3CN$ and the isomeric compound **6b** were determined by single-crystal X-ray diffraction. In addition, ab initio studies of **6a**, **6b**, and the related anionic model species hydridobis[salicylato(2–)-O^1,O^3]silicate(1–) (**7**) were carried out (SCF/SVP geometry optimizations). The experimentally established structures of the two crystallographically independent zwitterions (molecules A and B) in the crystal of **6a**·$0.7CH_3CN$ are characterized by trigonal-bipyramidal *Si*-coordination polyhedra, with the carboxylate oxygen atoms in the axial sites. As opposed to this, one carboxylate oxygen atom and one alcoholate oxygen atom occupy the axial sites of the trigonal-bipyramidal *Si*-coordination polyhedron of the isomeric zwitterion **6b**. The experimentally established structures of the zwitterions in the crystals of **6a**·$0.7CH_3CN$ and **6b** are similar to those calculated for **6a** and **6b**, respectively (SCF/SVP geometry optimizations). The energetically preferred isomer of the anionic model species **7**, the anion **7a**, is characterized by a trigonal-bipyramidal *Si*-coordination polyhedron, with both carboxylate oxygen atoms in the axial sites, thus corresponding to the structure of **6a**.

Introduction

In the course of our studies on zwitterionic (molecular) λ^5Si-silicates, we have synthesized a series of spirocyclic compounds containing two diolato(2–) ligands that derive from α-hydroxycarboxylic acids [1–10]. The λ^5Si-silicates **1** [1], **2**·DMF [5], and **3**·H_2O [8] are typical examples of this particular type of compound. The spirocyclic frameworks of these zwitterions contain two five-membered SiO_2C_2 ring systems. Recently, we have also reported on the zwitterionic spirocyclic λ^5Si-silicates **4** and **5** containing two six-membered SiO_2C_3 ring systems [11]. We have now succeeded in preparing the related compound [(dimethylammonio)methyl]bis[salicylato(2–)-

O^1,O^3]silicate (**6**). We report here on two different syntheses of the zwitterionic λ^5Si-silicate **6** and the crystal structure analysis of its acetonitrile solvate **6a**·0.7CH_3CN. In addition, the crystal structure of the isomeric zwitterionic λ^5Si-silicate **6b** is described. Furthermore, the results of ab initio studies of **6a**, **6b**, and the related anionic model species hydridobis[salicylato(2–)-O^1,O^3]silicate(1–) (**7**) are reported. The studies presented here were carried out with a special emphasis on the structural chemistry of spirocyclic pentacoordinate silicon compounds containing two bidentate salicylato(2–)-O^1,O^3 ligands (for reviews dealing with pentacoordinate silicon compounds, see [12–19]).

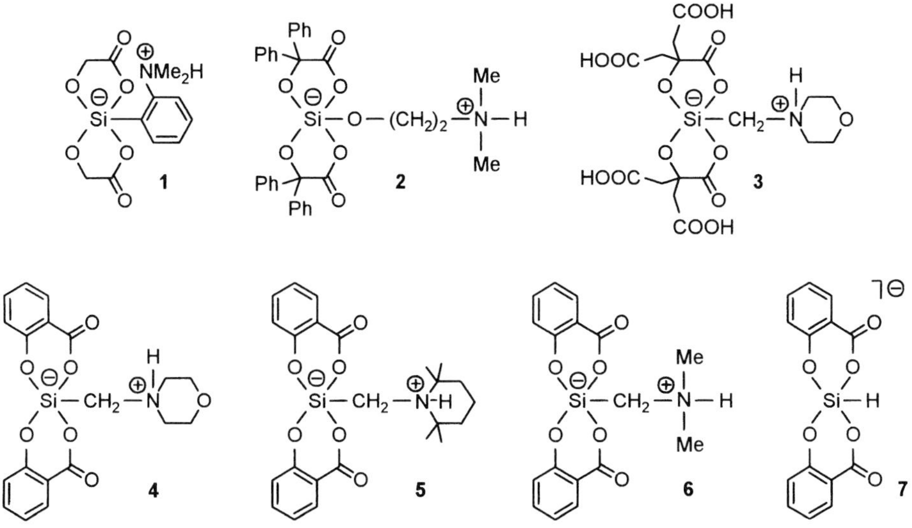

Results and Discussion

Following the strategy described for the preparation of compound **4** [11], the zwitterionic λ^5Si-silicate **6** was synthesized by reaction of [(dimethylamino)methyl]dimethoxy(phenyl)silane (**8**) [9] with salicylic acid (molar ratio 1:2) in acetonitrile at room temperature (method a) (Scheme 1). This synthesis involves cleavage of two Si–O bonds and one Si–C bond (formation of two molar equivalents of methanol and one molar equivalent of benzene).

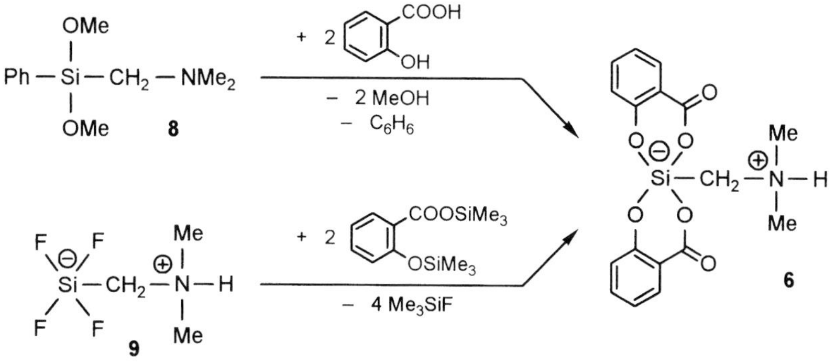

Scheme 1. Syntheses of the zwitterionic λ^5Si-silicate **6**.

Alternatively, compound **6** was synthesized by reaction of [(dimethylammonio)methyl]-tetrafluorosilicate (**9**) [20] with 2-(trimethylsilyloxy)benzoic acid trimethylsilyl ester (molar ratio 1:2) in acetonitrile at room temperature (method b) (Scheme 1). This reaction involves cleavage of four Si–F bonds (formation of four molar equivalents of fluorotrimethylsilane). Compound **6** was isolated in good yields (method a, 74 %; method b, 82 %) as the crystalline acetonitrile solvate **6a**·$0.7CH_3CN$ (bulk) which was characterized by elemental analyses (C, H, N) and ^{13}C and ^{29}Si solid-state VACP/MAS NMR experiments [21]. In addition, a few crystals of a second solid phase (isomer **6b**) were isolated.

Compounds **6a**·$0.7CH_3CN$ and **6b** were structurally characterized by single-crystal X-ray diffraction [22]. There are two crystallographically independent zwitterions (molecules A and B) in the crystal of **6a**·$0.7CH_3CN$. The structures of these molecules are very similar (Fig. 1, Table 1).

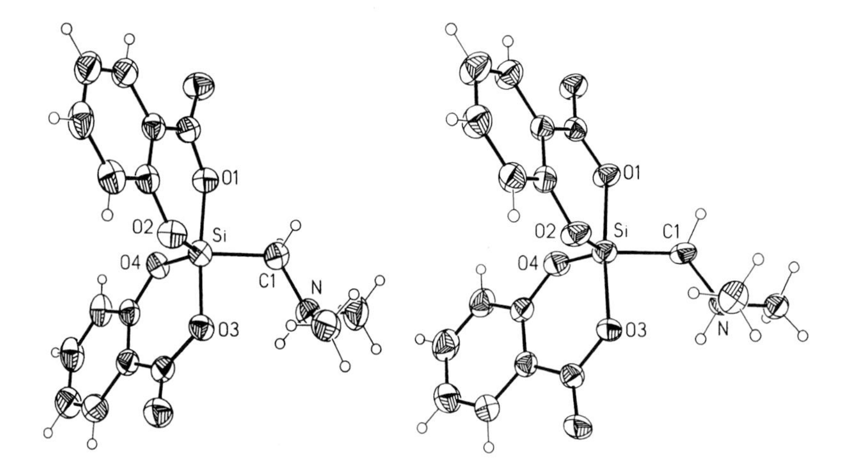

Fig. 1. Structures of molecule A (left) and molecule B (right) in the crystal of **6a**·$0.7CH_3CN$ (probability level of displacement ellipsoids 50 %), showing the atomic numbering scheme. For selected bond distances and angles, see Table 1.

The coordination polyhedra surrounding the silicon atoms of **6a** are distorted trigonal bipyramids, with the carboxylate oxygen atoms in the axial positions. The axial Si–O distances differ only slightly from one another (molecule A: 1.760(2) Å, 1.781(2) Å; molecule B: 1.769(2) Å, 1.797(2) Å) and are significantly longer than the equatorial ones (molecules A and B: 1.662(2) Å–1.670(2) Å). The *Si*-coordination polyhedron of the isomeric compound **6b** can also be described as a distorted trigonal bipyramid (Fig. 2, Table 2); however, in this case the two carboxylate oxygen atoms occupy one axial and one equatorial site leading to a significant differentiation between the two axial (1.738(2) Å, 1.782(2) Å) and between the two equatorial Si–O distances (1.697(2) Å, 1.662(2) Å). In contrast to the almost planar five-membered SiO_2C_2 ring systems of **1**, **2**·DMF, and **3**·H_2O, both six-membered SiO_2C_3 rings of **6a**·$0.7CH_3CN$ and one of the two SiO_2C_3 rings of **6b** (ring with O1 and O2) are puckered, the silicon atoms deviating from the main plane given by the respective carbon and oxygen atoms. Similar structural features were also observed for the related compounds **4**, **4**·CH_3CN, and **5** [11]. The other SiO_2C_3 ring of **6b** (ring with O3 and O4) is almost planar.

Table 1. Selected experimentally established and calculated distances [Å] and angles [°] for **6a** as obtained by single-crystal X-ray diffraction (**6a**·0.7CH$_3$CN, molecules A and B) and SCF/SVP geometry optimizations.

	6a·0.7CH$_3$CN (A)	6a·0.7CH$_3$CN (B)	6a (calcd.)
Si–O1	1.781(2)	1.797(2)	1.725
Si–O2	1.668(2)	1.662(2)	1.659
Si–O3	1.760(2)	1.769(2)	1.780
Si–O4	1.670(2)	1.666(2)	1.662
Si–C1	1.903(3)	1.887(3)	1.936
O1-Si-O2	94.57(10)	94.96(9)	95.2
O1-Si-O3	175.21(11)	174.63(9)	171.2
O1-Si-O4	87.94(9)	86.58(10)	92.3
O1-Si-C1	84.25(11)	85.47(10)	84.6
O2-Si-O3	87.35(10)	88.96(9)	89.8
O2-Si-O4	116.83(11)	115.03(11)	115.5
O2-Si-C1	123.78(11)	124.08(12)	119.8
O3-Si-O4	95.12(9)	95.09(10)	92.1
O3-Si-C1	91.03(11)	89.30(10)	86.6
O4-Si-C1	119.28(11)	120.79(12)	124.6

As expected from the presence of the potential NH donor groups and the potential oxygen acceptor atoms, the zwitterions in the crystals of **6a**·0.7CH$_3$CN and **6b** form N–H···O hydrogen bonds. In the crystal of **6a**·0.7CH$_3$CN, tetramers consisting of two molecules A and two molecules B were observed, whereas the crystal lattice of **6b** is built up by centrosymmetric dimers.

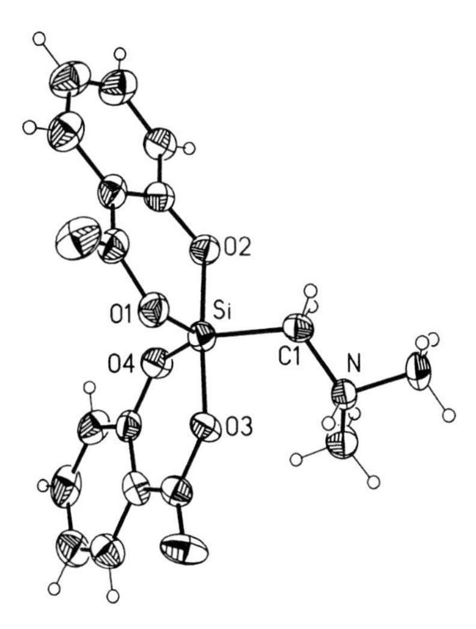

Fig. 2. Structure of **6b** in the crystal (probability level of displacement ellipsoids 50 %), showing the atomic numbering scheme. For selected bond distances and angles, see Table 2.

Table 2. Selected experimentally established and calculated distances [Å] and angles [°] for **6b** as obtained by single-crystal X-ray diffraction and SCF/SVP geometry optimizations.

	6b (exp.)	**6b (calcd.)**
Si–O1	1.697(2)	1.667
Si–O2	1.738(2)	1.719
Si–O3	1.782(2)	1.776
Si–O4	1.662(2)	1.666
Si–C1	1.880(2)	1.939
O1-Si-O2	94.96(8)	95.5
O1-Si-O3	85.89(7)	91.7
O1-Si-O4	115.31(8)	115.3
O1-Si-C1	116.88(9)	115.6
O2-Si-O3	175.41(9)	170.8
O2-Si-O4	89.23(8)	90.7
O2-Si-C1	85.13(8)	86.4
O3-Si-O4	94.49(7)	91.3
O3-Si-C1	90.46(8)	85.4
O4-Si-C1	127.79(9)	129.1

The structures of the isomeric zwitterions **6a** and **6b** and five different structures of the anion **7** were studied by quantum-chemical methods. For this purpose, geometry optimizations at the SCF level with an optimized SVP basis set [23] were performed using the TURBOMOLE program system [24]. Generally, the results of the computational studies were in good agreement with the experimentally established data.

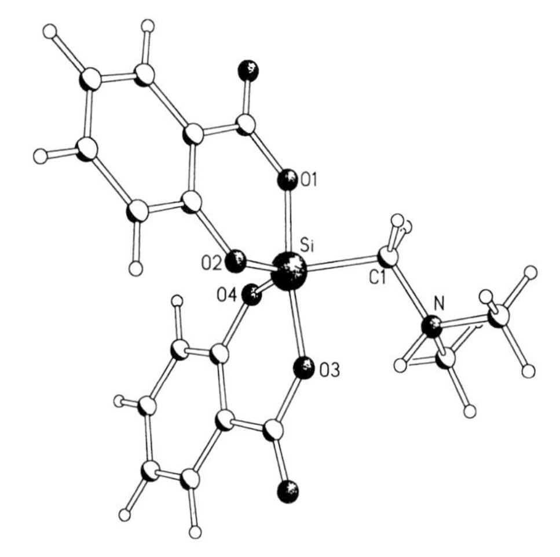

Fig. 3. Calculated structure of **6a** as obtained by SCF/SVP geometry optimizations. For selected bond distances and angles, see Table 1.

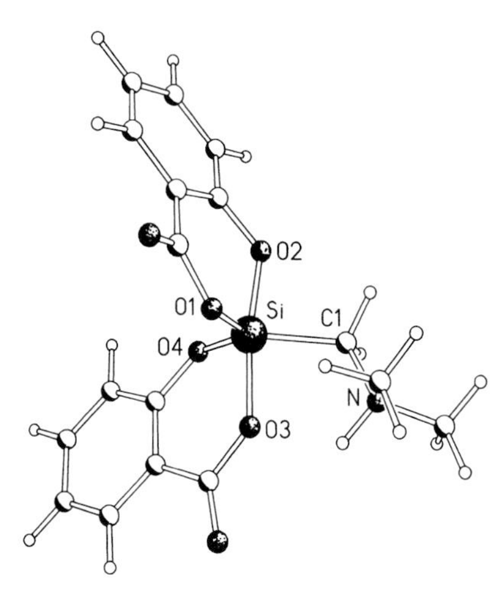

Fig. 4. Calculated structure of **6b** as obtained by SCF/SVP geometry optimizations. For selected bond distances and angles, see Table 2.

The calculated structures of the isomeric zwitterions **6a** and **6b** are depicted in Figs. 3 and 4. The calculated energies of these isomers are very similar (energy difference 1.1 kJ mol^{-1}) [25]. As can be seen from the relevant geometric data (Table 1), the calculated *Si*-coordination polyhedron of **6a** and the experimentally established coordination polyhedra of **6a**·0.7CH_3CN (molecules A and B) are rather similar. Except for the missing differentiation between the two calculated equatorial Si–O distances and for the missing puckering of one of the two six-membered SiO_2C_3 rings in the crystal structure of **6b**, the calculated and experimentally established geometries of the *Si*-coordination polyhedron of **6b** are also quite similar (Table 2).

SCF/SVP geometry optimizations of the anion **7** demonstrated the distorted trigonal bipyramid, with both carboxylate oxygen atoms in the axial positions (**7a**, local minimum), to be the energetically preferred structure [26]. The alternative trigonal-bipyramidal geometries **7b** (local minimum) and **7c** (local minimum) and the two square-pyramidal geometries **7d** (transition state) and **7e** (transition state) were found to be energetically less favorable. The calculated structures of **7a**–**7e** are depicted in Fig. 5 and the relevant geometric data are summarized in Table 3.

Interestingly, the energy differences between **7a**, **7b**, and **7c** are rather small (**7a**/**7b**, 4.0 kJ mol^{-1}; **7a**/**7c**, 6.2 kJ mol^{-1}). This is in excellent agreement with the experimentally proved existence of the two isomers of **6** (**6a** and **6b** correspond to the model species **7a** and **7b**, respectively) [27]. In contrast, the three trigonal-bipyramidal structures of bis[glycolato(2–)-O^1,O^2]hydridosilicate(1–) (**10**) differ significantly in their energy (**10a**/**10b**, 22.3 kJ mol^{-1}; **10a**/**10c**, 42.6 kJ mol^{-1}) [28].

10

This is in agreement with the experimental finding that the crystal structures of related zwitterionic spirocyclic λ^5Si-silicates, such as **1**–**3**, only fit with the structure of the model species **10a** (both carboxylate oxygen atoms in the axial sites of a distorted trigonal-bipyramidal *Si*-coordination polyhedron) [1–10]. On the other hand, the energy differences between the local minimum **7a** and the two transition states **7d** and **7e** (**7a**/**7d**, 30.8 kJ mol^{-1}; **7a**/**7e**, 33.8 kJ mol^{-1}) are very similar to those calculated for the local minimum **10a** and the related square-pyramidal transition states **10d** and **10e** (**10a**/**10d**, 29.6 kJ mol^{-1}; **10a**/**10e**, 31.4 kJ mol^{-1}) [9, 28].

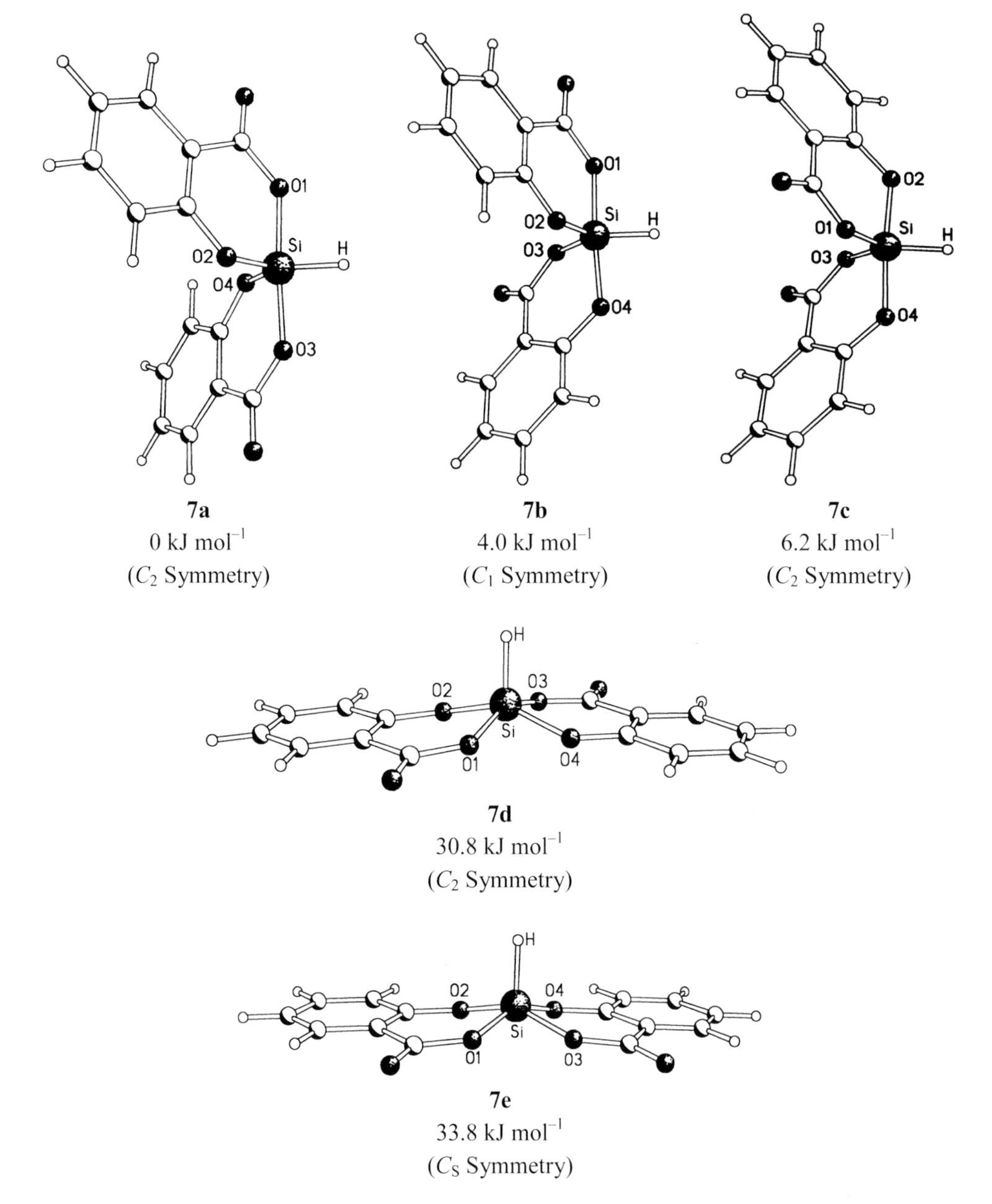

Fig. 5. Calculated structures and relative energies of **7a**–**7e** as obtained by SCF/SVP geometry optimizations.

Table 3. Selected calculated interatomic distances [Å] and angles [°] for **7a**–**7e** as obtained by SCF/SVP geometry optimizations.

	7a	**7b**	**7c**	**7d**	**7e**
Si–H	1.454	1.455	1.457	1.460	1.461
Si–O1	1.769	1.763	1.691	1.739	1.729
Si–O2	1.681	1.684	1.758	1.718	1.728
Si–O3	1.769	1.688	1.691	1.739	1.729
Si–O4	1.681	1.764	1.758	1.718	1.728
O1-Si-O2	93.0	93.0	92.6	90.3	90.3
O1-Si-O3	176.0	91.1	112.4	154.0	83.5
O1-Si-O4	89.3	175.5	89.9	83.0	152.1
O1-Si-H	88.0	88.0	123.8	103.0	103.9
O2-Si-O3	89.3	112.1	89.9	83.0	152.1
O2-Si-O4	112.6	89.2	175.6	150.0	82.5
O2-Si-H	123.7	124.3	87.8	105.0	104.0
O3-Si-O4	93.0	92.6	92.6	90.3	90.3
O3-Si-H	88.0	123.6	123.8	103.0	103.9
O4-Si-H	123.7	87.5	87.8	105.0	104.0

Another interesting aspect concerns the Si–O distances of the local minima of the anions **7** and **10**. The calculated axial Si–O distances of **7a**, **7b**, and **7c** are significantly longer than the equatorial ones, regardless of whether the axial sites are occupied by carboxylate or alcoholate oxygen atoms (Table 3). In contrast, the differences between the axial and equatorial Si–O distances of **10a**, **10b**, and **10c** depend significantly on the nature of the oxygen atoms in the respective axial and equatorial positions [9].

Acknowledgment: We thank the *Deutsche Forschungsgemeinschaft* and the *Fonds der Chemischen Industrie* for financial support and *Bayer AG* (Leverkusen and Wuppertal-Elberfeld, Germany) and *Merck KGaA* (Darmstadt, Germany) for support with chemicals. In addition, we thank Dr. R. Herbst-Irmer (University of Göttingen) for helpful discussions concerning the crystal structure analysis of compound **6b** (twinning).

References:

[1] R. Tacke, A. Lopez-Mras, P. G. Jones, *Organometallics* **1994**, *13*, 1617.

[2] R. Tacke, M. Mühleisen, P. G. Jones, *Angew. Chem.* **1994**, *106*, 1250; *Angew. Chem., Int. Ed. Engl.* **1994**, *33*, 1186.

[3] M. Mühleisen, R. Tacke, *Chem. Ber.* **1994**, *127*, 1615.

[4] M. Mühleisen, R. Tacke, *Organometallics* **1994**, *13*, 3740.

[5] R. Tacke, M. Mühleisen, *Inorg. Chem.* **1994**, *33*, 4191.

[6] R. Tacke, O. Dannappel, M. Mühleisen, *"Syntheses, Structures, and Properties of Molecular λ^5Si-Silicates Containing Bidentate 1,2-Diolato(2–) Ligands Derived from α-Hydroxy-carboxylic Acids, Acetohydroximic Acid, and Oxalic Acid: New Results in the Chemistry of*

Pentacoordinate Silicon", in: *Organosilicon Chemistry II: From Molecules to Materials* (Eds.: N. Auner, J. Weis), VCH, Weinheim, **1996**, pp. 427–446.

[7] R. Tacke, O. Dannappel, *"New Zwitterionic Silicon-Oxygen Compounds Containing Pentacoordinate Silicon Atoms: Experimental and Theoretical Studies"*, in: *Tailor-made Silicon-Oxygen Compounds — From Molecules to Materials* (Eds.: R. Corriu, P. Jutzi), Vieweg, Braunschweig/Wiesbaden, **1996**, pp. 75–86.

[8] R. Tacke, J. Heermann, M. Pülm, *Organometallics* **1997**, *16*, 5648.

[9] R. Tacke, R. Bertermann, A. Biller, O. Dannappel, M. Pülm, R. Willeke, *Eur. J. Inorg. Chem.* **1999**, 795.

[10] R. Tacke, B. Pfrommer, M. Pülm, R. Bertermann, *Eur. J. Inorg. Chem.* **1999**, 807.

[11] R. Tacke, J. Heermann, M. Pülm, I. Richter, *Organometallics* **1998**, *17*, 1663.

[12] S. N. Tandura, M. G. Voronkov, N. V. Alekseev, *Top. Curr. Chem.* **1986**, *131*, 99.

[13] W. S. Sheldrick, *"Structural Chemistry of Organic Silicon Compounds"*, in: *The Chemistry of Organic Silicon Compounds, Part 1* (Eds.: S. Patai, Z. Rappoport), Wiley, Chichester, **1989**, pp. 227–303.

[14] R. R. Holmes, *Chem. Rev.* **1990**, *90*, 17.

[15] C. Chuit, R. J. P. Corriu, C. Reye, J. C. Young, *Chem. Rev.* **1993**, *93*, 1371.

[16] C. Y. Wong, J. D. Woollins, *Coord. Chem. Rev.* **1994**, *130*, 175.

[17] E. Lukevics, O. A. Pudova, *Chem. Heterocycl. Compd.* (Engl. Transl.) **1996**, *32*, 1381.

[18] R. R. Holmes, *Chem. Rev.* **1996**, *96*, 927.

[19] D. Kost, I. Kalikhman, *"Hypervalent Silicon Compounds"*, in: *The Chemistry of Organic Silicon Compounds, Part 2* (Eds.: Z. Rappoport, Y. Apeloig), Vol. 2, Wiley, Chichester, **1998**, pp. 1339–1445.

[20] R. Tacke, J. Becht, O. Dannappel, R. Ahlrichs, U. Schneider, W. S. Sheldrick, J. Hahn, F. Kiesgen, *Organometallics* **1996**, *15*, 2060.

[21] After removal of the acetonitrile of **6**·0.7CH_3CN at room temperature in vacuo, a solvent-free sample of **6** was characterized. ^{13}C and ^{29}Si VACP/MAS NMR spectra were recorded at room temperature on a Bruker DSX-400 NMR spectrometer with bottom layer rotors of ZrO_2 (diameter 7 mm) containing ca. 200 mg of sample [100.6 MHz (^{13}C); 79.5 MHz (^{29}Si); external standard TMS (δ 0); spinning rate 6000 Hz; contact time 5 ms; 90° ^1H transmitter pulse length 3.6 μs; repetition time 4 s]. ^{13}C VACP/MAS NMR (1358 transients): δ 44.8 (NCH_3), 48.4 (NCH_3), 51.3 ($SiCH_2N$), 114.2 [C-1, diolato(2–) ligand (DL)], 117.0 (C-1, DL), 119.6 (2 C, C-3, DL), 121.0 (C-5, DL), 121.5 (C-5, DL), 130.0 (C-6, DL), 133.8 (C-6, DL), 135.3 (C-4, DL), 136.0 (C-4, DL), 157.6 (C-2, DL), 160.5 (C-2, DL), 164.4 (C=O), 166.0 (C=O); ^{29}Si VACP/MAS NMR (104 transients): δ –120.0. Analysis Calcd. for $C_{17}H_{17}NO_6Si$: C, 56.81; H, 4.77; N, 3.90. Found: C, 56.4; H, 5.0; N, 4.0.

[22] (a) Crystal data for **6a**·0.7CH_3CN: $C_{17}H_{17}NO_6Si \cdot 0.7CH_3CN$, M = 388.14 g mol^{-1}, triclinic, space group $P\bar{1}$, a = 10.3274(11) Å, b = 14.136(2) Å, c = 14.260(2) Å, α = 68.269(13)°, β = 80.919(14)°, γ = 86.906(13)°, V = 1909.6(4) $Å^3$, Z = 4, D_{calc} = 1.350 g cm^{-3}. A colorless single-crystal (0.2 mm × 0.2 mm × 0.05 mm; obtained by crystallization from acetonitrile) was mounted in inert oil (RS 3000, Riedel-de Häen), and 30545 reflections were collected on a Stoe IPDS diffractometer at T = 173(2) K [graphite-monochromated Mo-K_α radiation (λ = 0.71073 Å)]; $F(000)$ = 814. 6417 independent reflections and 33 restraints were used for the refinement of 520 parameters. The structure was solved by direct methods (program SHELXS-97) [29] and refined on F^2 (program SHELXL-97) [30]. The non-hydrogen atoms

were refined anisotropically; for the refinement of the hydrogen atoms a riding model was employed. $S = 0.872$ ($S = \{\Sigma[w(F_o^2 - F_c^2)^2] / (n - p)\}^{0.5}$; n = no. of reflections; p = no. of parameters); $w^{-1} = \sigma^2(F_o^2) + (aP)^2 + bP$, with $P = (\text{Max } F_o^2, 0 + 2F_c^2) / 3$, a = 0.0625, and b = 0); $R1$ [$I > 2\sigma(I)$] = 0.0442 ($R1 = \Sigma||F_o| - |F_c|| / \Sigma|F_o|$); $wR2$ (all data) = 0.1107 ($wR2 = \{\Sigma[w(F_o^2 - F_c^2)^2] / \Sigma[w(F_o^2)^2]\}^{0.5}$); max./min. residual electron density +0.557/–0.251 e Å^{-3}. Crystallographic data (excluding structure factors) for the structure reported in this paper have been deposited with the Cambridge Crystallographic Data Centre as supplementary publication no. CCDC-108614. Copies of the data can be obtained free of charge on application to The Director, CCDC, 12 Union Road, Cambridge CB2 1EZ, UK [Fax: (Int. code) + (1223) 336 033; e-mail: deposit@ccdc.cam.ac.uk].
(b) Crystal data for **6b**: $C_{17}H_{17}NO_6Si$, M = 359.41 g mol^{-1}, monoclinic, space group $P2_1/c$, a = 14.476(3) Å, b = 8.793(2) Å, c = 14.490(3) Å, β = 116.76(3)°, V = 1646.9(6) Å^3, Z = 4, D_{calc} = 1.450 g cm^{-3}. A colorless crystal (0.3 mm × 0.2 mm × 0.1 mm; obtained by crystallization from acetonitrile) was mounted in inert oil (RS 3000, Riedel–de Häen), and 20375 reflections were collected on a Stoe IPDS diffractometer at T = 173(2) K [graphite-monochromated Mo-K$_\alpha$ radiation (λ = 0.71073 Å)]; $F(000)$ = 752. 2999 reflections were used for the refinement of 232 parameters. The structure was solved by direct methods (program SHELXS-97) [29] and refined on F^2 (program SHELXL-97) [30]. The crystal was non-merohedrally twinned with an angle of rotation around [1 0 0] of 180°. The occupation factor for the two domains refined to 0.50. The non-hydrogen atoms were refined anisotropically; for the refinement of the hydrogen atoms a riding model was employed. $S = 0.921$ ($S = \{\Sigma[w(F_o^2 - F_c^2)^2] / (n - p)\}^{0.5}$; n = no. of reflections; p = no. of parameters); $w^{-1} = \sigma^2(F_o^2) + (aP)^2 + bP$, with $P = (\text{Max } F_o^2, 0 + 2F_c^2) / 3$, a = 0.0275, and b = 0); $R1$ [$I > 2\sigma(I)$] = 0.0348 ($R1 = \Sigma||F_o| - |F_c|| / \Sigma|F_o|$); $wR2$ (all data) = 0.0708 ($wR2 = \{\Sigma[w(F_o^2 - F_c^2)^2] / \Sigma[w(F_o^2)^2]\}^{0.5}$); max./min. residual electron density +0.195/–0.169 e Å^{-3}. Crystallographic data (excluding structure factors) for the structure reported in this paper have been deposited with the Cambridge Crystallographic Data Centre as supplementary publication no. CCDC-108615. Copies of the data can be obtained free of charge on application to The Director, CCDC, 12 Union Road, Cambridge CB2 1EZ, UK [Fax: (Int. code) + (1223)336-033; e-mail: deposit@ccdc.cam.ac.uk].

[23] SVP basis set used: Si, (10s 7p 1d)/[4s 3p 1d]; C, N, and O, (7s 4p 1d)/[3s 2p 1d]; H, (4s 1p)/[2s 1p]. For details of the SVP basis set optimization, see [20].

[24] R. Ahlrichs, M. Bär, M. Häser, H. Horn, C. Kömel, *Chem. Phys. Lett.* **1989**, *162*, 165. Computational results were obtained by using software programs from Biosym/MSI (San Diego).

[25] Calculated SCF energies [Hartree]: **6a**, –1445.46947 (C_1 symmetry); **6b**, –1445.46906 (C_1 symmetry). For these calculations, conformations were chosen which allow the existence of N–H···O3 hydrogen bonds.

[26] Calculated energies (SCF + single point MP2 + E_{vib0} energies [Hartree]): **7a**, –1275.68438 (C_2 symmetry); **7b**, –1275.68287 (C_1 symmetry); **7c**, –1275.68203 (C_2 symmetry); **7d**, –1275.67265 (C_2 symmetry); **7e**, –1275.67149 (C_S symmetry).

[27] As the energy difference between the two isomers **7a** and **7c** is also rather small, related zwitterionic λ^5*Si*-silicates containing *Si*-coordination polyhedra, with both carboxylate oxygen atoms in the equatorial sites, might also exist in crystalline phases.

[28] **10a**, C_2 symmetry (both carboxylate oxygen atoms in the axial sites); **10b**, C_1 symmetry (one carboxylate oxygen atom and one alcoholate oxygen atom in the axial sites); **10c**, C_2 symmetry (both alcoholate oxygen atoms in the axial sites); **10d**, C_2 symmetry (hydrogen atom in the apical site); **10e**, C_S symmetry (hydrogen atom in the apical site). Data were taken from [9].

[29] SHELXS-97, University of Göttingen, Germany, **1997**; G. M. Sheldrick, *Acta Crystallogr., Sect. A* **1990**, *46*, 467.

[30] SHELXL-97, University of Göttingen, Germany, **1997**.

Oxygen, Phosphorus or Sulfur Donor Ligands in Higher-Coordinated Organosilyl Chlorides and Triflates

U. H. Berlekamp, A. Mix, P. Jutzi, H. G. Stammler, B. Neumann*

Fakultät für Chemie, Universität Bielefeld
Universitätsstr. 25, D-33615 Bielefeld, Germany
Tel.: Int. code + (521)106 6181 — Fax.: Int. code + (521)106 6026
E-mail : uwe_heinrich.berlekamp@uni-bielefeld.de

Keywords: Higher-Coordinated Silyl Chlorides / Higher-Coordinated Silyl Triflates / Silyl Cations / Reactivity of Silyl Cations

Summary: Aryl ligands carrying oxygen, phosphorus or sulfur donor atoms in the side chain are introduced at silicon. Here we report on the synthesis of silyl chlorides and silyl triflates. The interaction between the silicon centre and the donor atom are discussed on the basis of NMR and X-ray crystallographic data. In the reaction of bis[2-(methoxymethyl)phenyl]methylsilyl triflate **8** with nucleophiles, reactivity of oxonium as well as of siliconium ions is observed.

The ability of silicon to extend its coordination sphere in organic compounds has been intensively studied in the last decade by Corriu et al. [1–4] and some other groups [5, 6]. In these investigations mainly the 2-(dimethylaminomethyl)phenyl ligand **A** (Fig. 1) was used [1–6]. The results obtained promoted the understanding of nucleophilic substitution processes at silicon; furthermore, reactive silicon compounds with a silicon–element double bond [2] or species containing siliconium ion have been stabilized with this kind of ligand [4, 6].

In contrast to the great number of pentacoordinated organosilanes with the amino-functionalized aryl ligand, only a few pentacoordinated species with ligands bearing other donor elements in the side chain have been described [7–9]. Therefore, we focused our interests on the coordination behavior of organosilanes containing phenyl ligands with either hard or with soft donor elements like oxygen, phosphorus or sulfur (**B**) (Fig. 1). Furthermore, we were interested in synthesizing organosilicon compounds containing siliconium ions stabilized by these donor functionalized phenyl ligands.

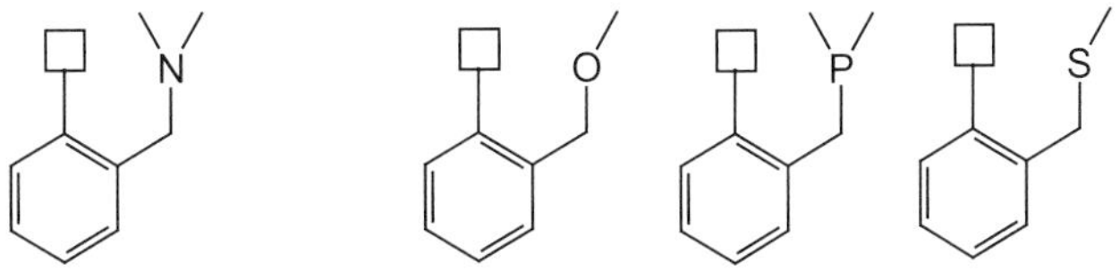

Fig. 1. Ligand systems.

The organochlorosilanes **1–4** (Scheme 1) were synthesized by reaction of the lithiated aryl ligand with $RSiCl_3$ (R = H or Me).

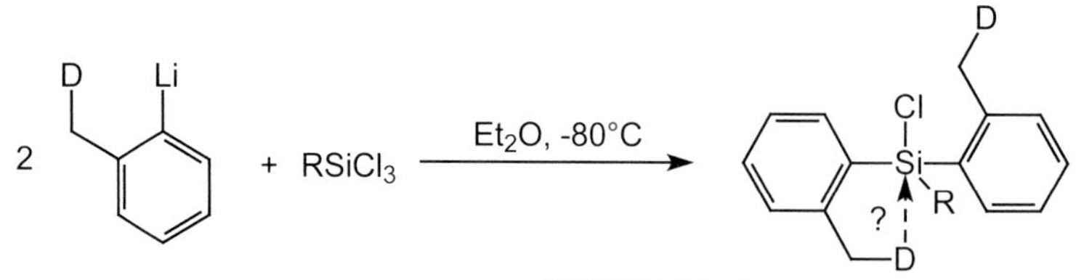

	1	2	3	4
R	H	H	Me	Me
D	PMe_2	SMe	OMe	PMe_2

Scheme 1. Synthesis of the silyl chlorides **1–4**.

The extend of the coordination at silicon is displayed by the ^{29}Si-NMR resonances (Table 1) in comparison to diphenylchlorosilane or diphenylmethylchlorosilane.

Table 1. ^{29}Si-NMR data of the chlorosilanes **1–4**.

	1	2	3	4	Ph_2SiHCl	$Ph_2SiMeCl$
^{29}Si NMR [ppm]	–13.4	–18.2	3.4	10.6	–5.4	10.6

The solid-state structure of bis(2-(methylthiomethyl)phenyl)chlorosilane **2** (Fig. 2) shows an interaction between one sulfur atom and the silicon in a trigonal bipyramidal environment. The Si–S bond length of 313.30(16) pm is shorter than the sum of the van der Waals radii (390 pm).

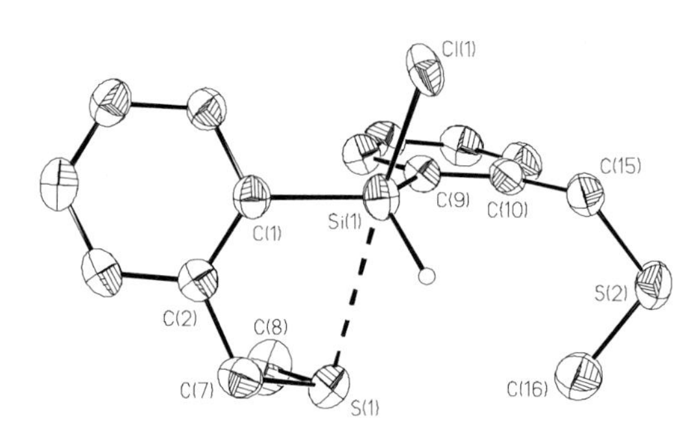

Fig. 2. Structure of **2**: selected bond lengths [pm]: Si(1)–S(1): 313,30(16), Si(1)–Cl(1):217,64(15); selected bond angles [°]: Cl(1)-Si(1)-S(1): 166,76 (6).

The silyl triflates **5-9** (Scheme 2) were synthesized by reacting the organosilanes or organochlorosilanes **3** and **4** with trimethylsilyl triflate with formation of trimethylsilane or trimethylsilyl chloride, respectively [10].

	5	**6**	**7**	**8**	**9**
R	H	H	H	Me	Me
D	OMe	PMe_2	SMe	OMe	PMe_2

Scheme 2. Synthesis of the silyl triflates **5-9** by chloride (hydride) abstraction.

The ^{29}Si-NMR data (Table 2) for **5-9** indicate a strong donor interaction with the silicon centre. The silyl triflates containing phosphorus or sulfur donor atoms show a dynamic coordination behavior at room temperature.

Table 2. ^{29}Si-NMR data of the silyl triflates **5-9**.

		5	**6**	**7**	**8**	**9**	**Ph_2SiH·Trfl.** [11]
	R	H	H	H	Me	Me	
	D	OMe	PMe_2	SMe	OMe	PMe_2	
^{29}Si [ppm]		–47.2	–67.8	–47.9	–25.0	–2,7	–2.1
$^1J_{SiH}$ [Hz]		272	257	279			257
$^1J_{SiP}$ [Hz]			19.8			27.7	

Conductivity measurements of the silyl triflates **5-9** show their dissociation into the higher-coordinated silyl cation and the triflate anion.

The solid-state structure of the organosilyl triflate **8** (Fig. 3) shows an interaction of both oxygen atoms with the silicon atom in a trigonal bipyramidal environment. The Si–O bonds lengths (195.95 (18) and 196.99(18) pm) are comparatively long.

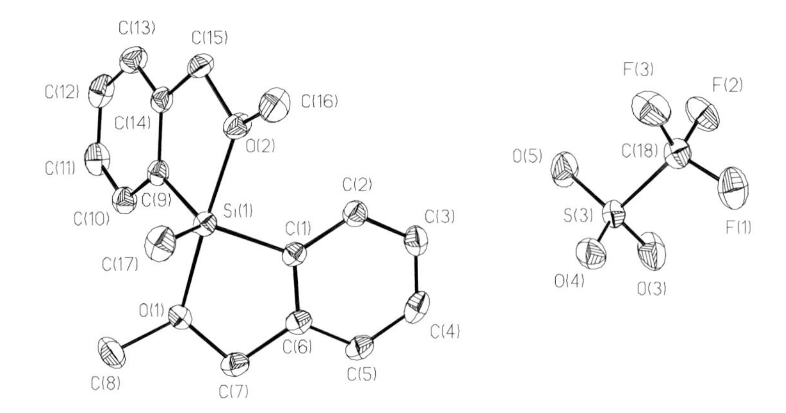

Fig. 3. X-ray structure of **8**: selected bond lengths [pm]: Si(1)–O(1): 195.95(18), Si(1)–O(2): 196.99(18), Si(1)–C(17):186.3(3); selected bond angles [°]: O(1)-Si(1)-O(2): 174.30(8), Σ_{Si}: 360°.

The silyl triflate **8** (Scheme 3) shows an "ambireactive" behavior of the cation. Reaction with a methylmagnesium chloride or water leads to the methyl-substituted product and to the siloxane, respectively, as expected for an electrophilic silicon center. An oxonium ion reactivity is observed in the reaction with neutral Lewis bases such as triethylamine and trimethylphosphine.

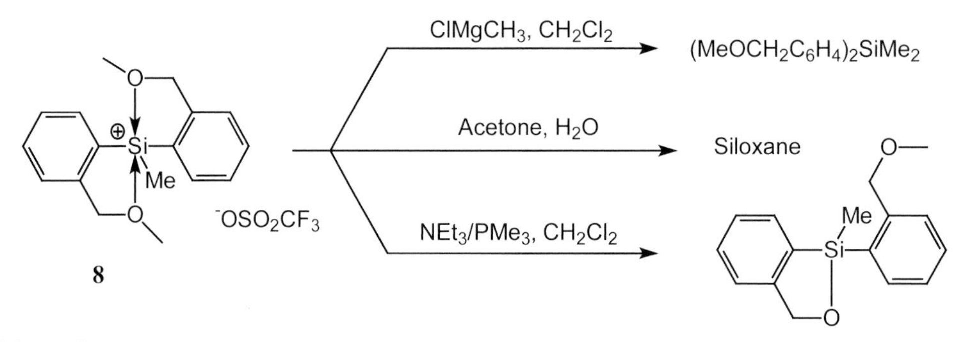

Scheme 3. Reaction of **8** with nucleophiles.

Acknowledgment: We thank the *Fonds des Verbandes der Chemischen Industrie* for supporting this research project.

References:

[1] R. J. P. Corriu, A. Kpoton, M. Poirier, G. Royo, J. Y. Corey, *J. Organomet. Chem.* **1984**, *277*, C25.

[2] P. Arya, J. Boyer, F. Carre, R. J. P. Corriu, *Angew. Chem.* **1989**, *101*, 1069.

[3] R. J. P. Corriu, A. Kpoton, M. Poirier, G. Royo, A. de Saxce, J. C. Young, *J. Organomet. Chem.* **1990**, *395*,1; C. Chiut, R. J. P. Corriu, C. Reye, J. C. Young, *Chem. Rev.* **1993**, *93*, 1371.

[4] C. Chuit, R. J. P. Corriu, A. Mehdi, C. Reye, *Angew. Chem.* **1993**, *105*, 1372.

[5] N.Auner, R. Probst, F. Hahn, E. Herdtweck, *J. Organomet. Chem.* **1993**, *459*, 25.

[6] J. Belzner, D. Schär, B. Kneisel, R. Herbst-Irmer, *Organometallics* **1995**, *14*, 1840.

[7] A. Mix, U. H. Berlekamp, H. G. Stammler, B. Neumann, P. Jutzi, *J. Organomet. Chem.* **1996**, *521*, 177.

[8] N. V. Timosheva, T. K. Prakasha, A. Chandrasekaran, R. O. Day, R. R. Holmes, *Inorg. Chem.* **1996**, *35*, 3614.

[9] G. Müller, M. Waldkirch, A. Pape, "*Phosphine Coordination to Silicon Revisited*", in: *Organosilicon Chemistry III: From Molecules to Materials* (Eds.: N. Auner, J. Weis), Wiley–VCH, Weinheim, **1998**, p. 452.

[10] U. H. Berlekamp, P. Jutzi, A. Mix, B. Neumann, H.-G. Stammler, W. W. Schoeller, *Angew. Chem.* **1999**, *111*, 2071; *Angew. Chem. Int. Ed.* **1999**, *38*, 2048.

[11] W. Uhlig, *Chem. Ber.* **1992**, *125*, 47.

Coupling Constants through a Rapidly Dissociating-Recombining N→Si Dative Bond in Pentacoordinate Silicon Chelates

Inna Kalikhman*, Sonia Krivonos, Daniel Kost*

Department of Chemistry, Ben Gurion University of the Negev
Beer-Sheva 84105, Israel,
Tel.: Int. code + (7)646 1192 Fax: Int. code + (7)647 2943
E-mail: kostd@bgumail.bgu.ac.il

Thomas Kottke, Dietmar Stalke

Institut für Anorganische Chemie der Universität Würzburg
D-97074 Würzburg, Germany,
Tel.: Int. code + (931)888 5246 Fax: Int. code + (931)888 4605
E-mail: dstalke@wacx51.chemie.uni-wuerzburg.de

Keywords: Pentacoordinate Silicon / Multinuclear NMR / Stereodynamics

Summary: The preparation of new pentacoordinate silicon complexes **4**–**6** from polychloro- or fluoro-silanes and O-silylated-*N*,*N*-dimethylhydrazides is described, along with a typical trigonal-bipyramidal (TBP) crystal structure. Evidence is presented that **4**–**6** undergo rapid N→Si dissociation–recombination. Coupling constants are reported extending across the dative bond, even at temperatures in which N–Si dissociation is rapid. A Karplus-type correlation between dihedral angles and vicinal coupling constants is shown.

Introduction

We have recently reported the first observation and measurement of spin–spin interactions in neutral hexacoordinate silicon chelates [1] which extend across the N→Si coordinative bond and over two, three, and even four bonds. These coupling constants were highly sensitive to small geometrical modifications in the complex. The major geometrical requirement found for spin–spin coupling over two bonds (N→Si–F or N→Si–H) in hexacoordinate complexes **1a–1c** was that the corresponding bond angle be 90° or very close to it.

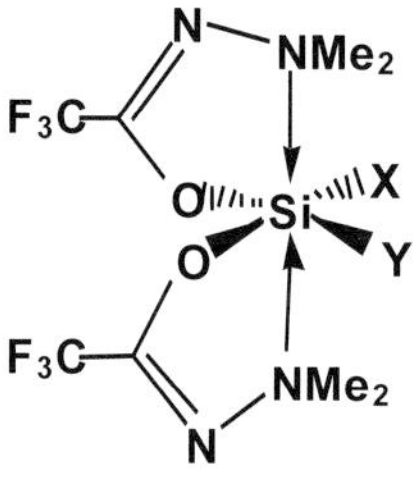

1a	X = Cl,	Y = H
1b	X = F,	Y = Ph
1c	X = Y = F	

It was of interest to study also similar coupling constants in analogous pentacoordinate complexes, in order to (a) learn whether the two complex types behave similarly in terms of spin–spin interactions across the dative bond, and (b) further study the sensitivity of coupling constants to small geometrical changes. The one-bond coupling constants via the dative N–Si bond $^{1}J(^{15}N \rightarrow ^{29}Si)$, have so far been reported only for ^{15}N-enriched silatranes and silocanes [2].

We now report the first measurements of coupling constants extending over one, two and three bonds across the dative bond in pentacoordinate silicon chelates. Coupling via the N→Si bond was found to persist even at temperatures at which rapid dissociation and recombination of this bond take place [3, 4].

Synthesis

The general synthetic method for the preparation of penta- [4] and hexacoordinate [5] neutral silicon complexes from O-silylated hydrazides **2** has been described previously. The syntheses used in the present study are shown in Scheme 1. In both reactions the pentacoordinate complex is an intermediate on the way to the corresponding hexacoordinate species. However, the equilibrium condition is sufficiently favorable to allow the isolation of **4** as the major product simply by using an excess of **3**. By contrast, the intermediate **6** is unstable, and is transformed spontaneously in solution at 300 K within 2–3 hours to **7**. At lower temperatures the solution NMR spectra of **6** could be studied for several hours before it completely transformed to **7**.

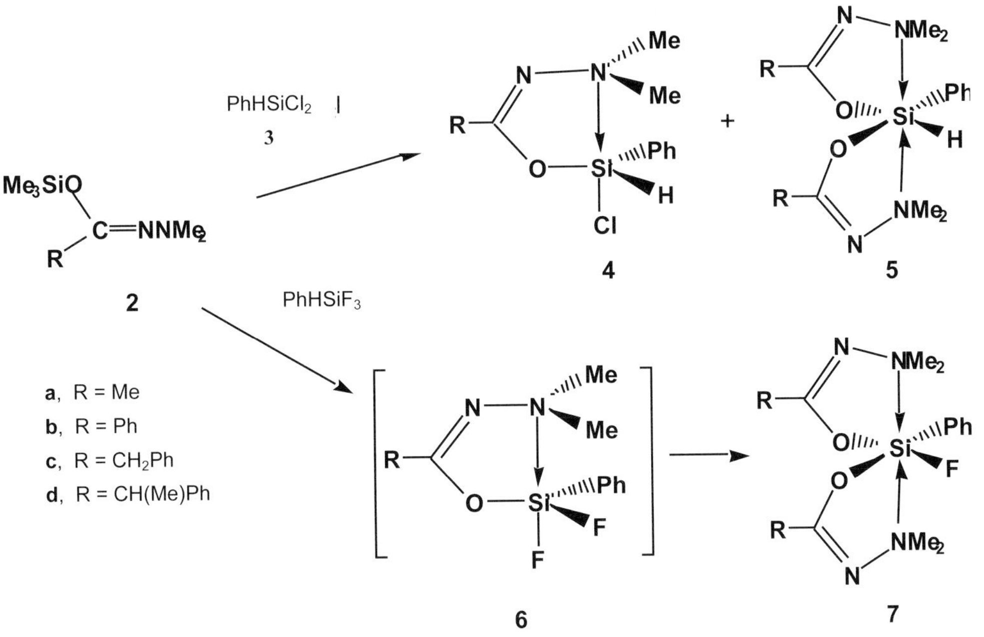

Scheme 1. Synthesis of penta- and hexacoordinate neutral silicon complexes from O-silylated hydrazides.

Crystal and Solution Structure

A single-crystal structure was determined for compound **4b** (Fig. 1).

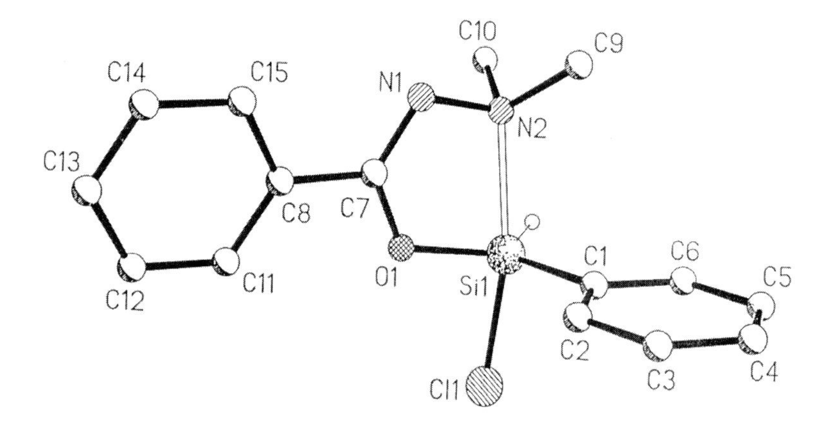

Fig. 1. Crystal structure of **4b**.

The structure provides evidence for the pentacoordination at silicon, and the associated near trigonal-bipyramidal (TBP) geometry in the solid state.

Table 1. NMR data: chemical shifts (δ) in ppm, coupling constants in Hz.

Compound	T [K]	Chemical shift (multiplicity, *J*) ^{1}H		^{13}C	^{29}Si
		SiH	NMe$_2$	NMe$_2$	
4a	273	6.16	1.87 2.47	48.03 49.60	–62.4 (d, 318.9)
	315	6.14	2.23	48.81	–61.0 (d, 318.9)
4b	263	6.24	2.02 2.62	48.22 49.81	–62.6 (d, 321.9)
	330	6.21	2.32	48.71	–59.5 (d, 321.9)
4c	273	6.11	1.90 2.51	47.91 49.54	–61.8 (d, 318.9)
	320	6.09	2.40	48.78	–60.7 (d, 321.0)
4d	253	6.05	1.83 2.42	48.16 49.82	–62.7 (d, 320.0)
		6.06	1.81 2.38	47.80 49.66	–63.0 (d, 320.0)
	330	6.08	2.19	48.61	–61.0 (d, 319.5)
		6.07	2.19	48.42	–61.3 (d, 319.5)
6a	230		2.53 (t, 0.8)	48.06 (t, 1.5)	–96.5 (t, 239)
	300		2.51	48.06 (t, 1.5)	–95.1 (t, 241)
6b	263		2.50 (t, 1.4)	48.70 (t, 2.0)	–94.7 (t, 242)
	300		2.52 (t, 1.4)	48.70 (t, 2.0)	–94.1 (t, 245)
6d	200		2.45 2.38	47.91 47.81	–96.2 (t, 238)
	300		2.37 (t, 1.1)	47.90 (t, 1.9)	–94.4 (t, 244)

Evidence for a similar molecular structure in solution is obtained from an examination of the multinuclear NMR spectra of compounds **4**:

(a) The ^{29}Si chemical shifts of these complexes are all within 2–3 ppm from an average –60 ppm (Table 1), typical of pentacoordinate silicon complexes [6].

(b) The one-bond coupling constant between silicon and the hydrido nucleus [^{1}J(Si–H)], which is highly sensitive to the hybridization of the silicon atom, measures ca. 320 Hz in **4a**–**4d**, again in full agreement with a pentacoordinate TBP-type silicon complex.

(c) In solutions of each of the **4** complexes at low temperature the two *N*-methyl groups are diastereotopic. This is evidence for restricted rotation about the N–N bond, most likely resulting from coordination of the dimethylamino nitrogen to silicon.

6 is likewise characterized by a typical $\delta(^{29}\text{Si}) \cong -95$ ppm.

Stereodynamics; evidence for N→Si dissociation-recombination

4 and **6** undergo ligand-site exchange reactions, monitored by NMR (Table 1). Introduction of chiral carbon centers in **4d** and **6d** enables the assignment of the exchange process to N→Si dissociation–recombination and not pseudorotation (Scheme 2): in **4d**, at low temperatures, two diastereomers due to the presence of two chiral centers on C and Si are evident by the doubling of resonances: two C-Me singlets, two C-H quartets, and four N-Me singlets. Upon heating, the N-Me signal-pairs coalesce, without concomitant coalescence of the C-H and C-Me signals, i.e., without loss of the diastereomers. This means that no inversion at silicon takes place in this process, and hence coalescence must be due to N→Si dissociation, followed by rotation about the N–N bond and recombination.

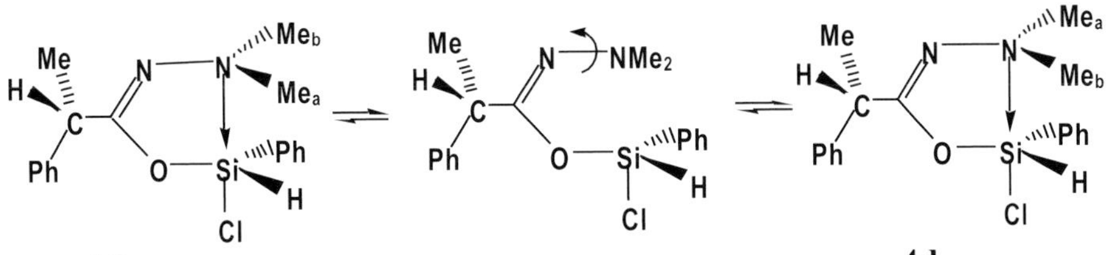

Scheme 2. Exchange of the *N*-methyl groups via N→Si dissociation-recombination.

Conversely, in **6a**–**6d** pseudorotation at silicon is rapid even at low temperatures, resulting in equivalent F ligands and hence in absence of diastereomers in **6d**. The rapid pseudorotation at silicon is evident from the appearance of triplets due to equivalent fluorines in the ^{1}H-, ^{13}C-, ^{15}N-, and ^{29}Si-NMR spectra of **6a** ($^{4}J(^{19}\text{FSiNC}^{1}\text{H}) = 0.8$, $^{3}J(^{19}\text{FSiN}^{13}\text{C}) = 1.5$, $^{2}J(^{19}\text{FSi}^{15}\text{N}) = 11.2$, and $^{1}J(^{19}\text{F}^{29}\text{Si}) = 239$ Hz, respectively). The N-Me groups are diastereotopic at low temperature only in **6d**, due to the presence of the chiral carbon and N→Si coordination. Coalescence in **6d** must therefore also result from N→Si dissociation-recombination.

Coupling Constants via Dative Bond

Despite the evidence for rapid N→Si dissociation in **4** and **6**, coupling constants through this bond are observed (Table 2). Thus, for example, in the undecoupled ^{13}C NMR spectrum of **4b** at 243K in $CDCl_3$ solution one of the N-Me signals is coupled through three bonds by the Si-H nucleus (^3J = 1.7 Hz). Likewise, the natural abundance ^{15}N NMR spectrum of **4c** clearly shows two-bond coupling to Si-H, and the elusive ^{15}N–^{29}Si one-bond coupling can be seen in the proton decoupled version of that spectrum (Fig. 2).

Table 2. Coupling constants (Hz) across the N→Si coordinative bond in **4**.

Compound	R	^{13}C (243K) $^3J(^{13}CNSi^1H)$	^{15}N{H} (300K) $^1J(^{15}N^{29}Si)$	^{15}N (300K) $^2J(^{15}NSi^1H)$
4a	Me	1.6	3.2	10.8
4b	Ph	1.7	2.4	10.8
4c	$PhCH_2$	1.7	3.3	10.6
4d	PhMeCH	1.8		

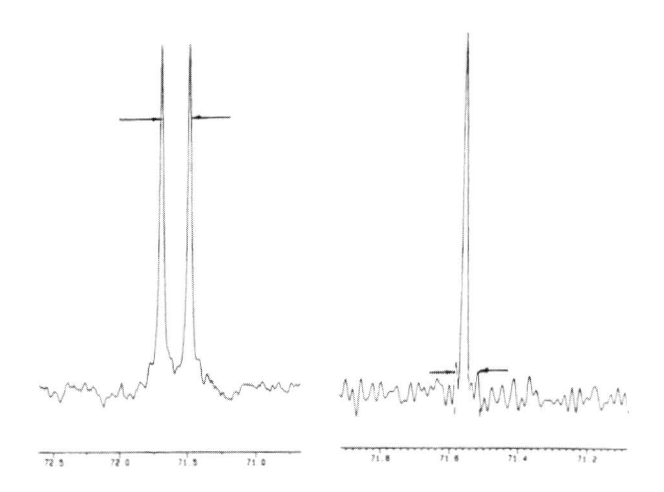

Fig. 2. Natural abundance ^{15}N-NMR spectra of **4c** (N-Me region, CDCl3, 300 K). Left, undecoupled; right, ^{1}H-decoupled, featuring N–Si coupling. ^2J(N–Si–H) = 10.6 Hz; ^1J(N–Si) = 3.3 Hz.

The Karplus correlation, connecting vicinal coupling constants through the N→Si bond in hexacoordinate complexes [1] with the corresponding dihedral angles, can now be extended also to pentacoordinate complexes on the basis of the present results (Table 3). It appears that the same Karplus-type correlation accommodates both hexa- and penta-coordinate complexes, as is depicted in Fig. 3.

Table 3. Three-bond coupling constants in **1a** and **4b** and the corresponding H-Si-N-C dihedral angles.

Compound	H-Si-N-C [°]	$^3J(^1H–^{13}C)$ [Hz]
1a	157.4	2.4
	31.9	1.1
	49.3	>0
	–76.1	0
4b	97.5	0
	28.3	1.7

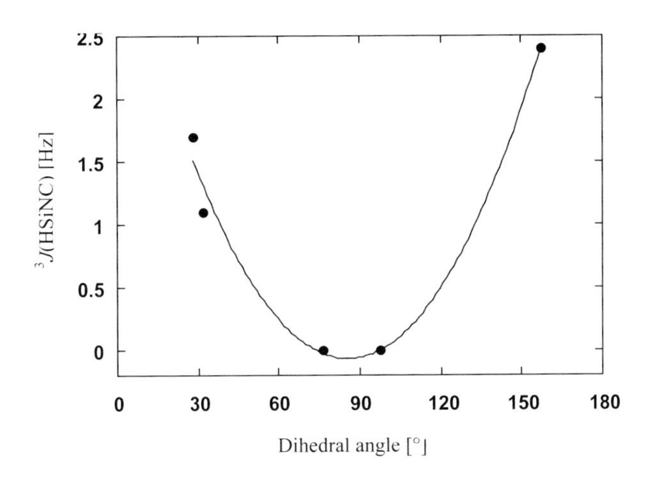

Fig. 3. Vicinal coupling constants as a function of crystallographic dihedral angles.

Acknowledgment: Financial support from the *Israel National Science Foundation*, administered by the *Israel Academy of Sciences and Humanities*, is gratefully acknowledged.

References:

[1] I. Kalikhman, S. Krivonos, D. Stalke, T. Kottke, D. Kost, *Organometallics* **1997**, *16*, 3255.

[2] E. Kupce, E. Liepins, A.Lapsina, I. Urtane, G. Zelcans, E. Lukevics, *J. Organomet. Chem.* **1985,** *279*, 343.

[3] R. J. P. Corriu, A. Kpoton, M. Poirier, G. Royo, J. Y. Corey, *J. Organometal. Chem.* **1984,** *277*, C25.

[4] I. Kalikhman, S. Krivonos, A. Ellern, D. Kost, *Organometallics* **1996**, *15* , 5073.

[5] D. Kost, I. Kalikhman, and M. Raban, *J. Am. Chem. Soc.* **1995**, *117*, 11512.

[6] D. Kost and I. Kalikhman, "*Hypervalent Silicon Compounds*", in: *The Chemistry of Organic Silicon Compounds* (Eds.: Z. Rappoport, Y. Apeloig), *Vol. 2*, Wiley,Chichester, **1998**, pp. 1339–1445.

Salen–Silicon Complexes — a New Type of Hexacoordinate Silicon

Jörg Haberecht, Frank Mucha, Uwe Böhme, Gerhard Roewer

Institut für Anorganische Chemie
Technische Universität Bergakademie Freiberg
Leipziger Str. 29, D-09596 Freiberg, Germany
Tel.: Int. code + (3731)39 3174 — Fax: Int. code + (3731)39 4058
E-mail: roewer@orion.hrz.tu-freiberg.de

Keywords: Azomethine Compounds / Silicon / Silicon Complexes

Summary: The reaction of tetradentate azomethine ligands H_2salen* [*N,N'*-ethylene-bis(2-hydroxyacetophenoneimine)] or H_2salen‡ [*N,N'*-ethylene-bis(3,5-di-*tert*-butyl-salicylideneimine)] with chlorosilanes produces silicon compounds with hexacoordinated silicon (salen)SiXY (X = Cl; Y = H, Cl, CH_3, C_6H_5). Both ligands X and Y can be exchanged for fluorine. The crystal structure analysis of (salen*)SiF_2 reveals the truly hexacoordination of the silicon atom. The chloro compounds (salen*)$SiCl_2$ undergo Wurtz-type coupling reactions leading to oligosilanes with hypervalent silicon. Coupling of (salen*)$SiCl_2$ with acetylides creates compounds with an Si–C≡C–Si backbone.

Hypervalent silicon compounds attract interest from both the structural and reactivity point of view [1]. The azomethine *N,N'*-ethylene-bis(2-hydroxyacetophenoneimine) (salen*H_2; **1**), was formed by condensation of ethylenediamine with 2-hydroxyacetophenone. We set out to synthesize hexacoordinate silicon complexes containing the salen* ligand. The anion salen*$^{2-}$ is able to chelate the silicon atom through four donor atoms. There are some rare examples of salen–silicon compounds known from the literature [2], but characterization of these compounds seems to be doubtful [3]. Structural aspects are uncertain due to the lack of crystal structure data.

We were able to prepare salen* silicon complexes by reaction of the salen* ligand as free acid (Scheme 2) or sodium salt (Scheme 1) with the corresponding chloro silicon compound. It is also possible to use Si_2Cl_6 as starting compound to get (salen*)$SiCl_2$ (Scheme 2).

The extremely high-field shift of the ^{29}Si-NMR signals indicates the presence of hexacoordinate silicon atoms (Table 1). The formation of two isomers (*cis*- **6** and *trans*- **7**, Scheme 2 and Table 1) is indicated by the NMR spectra. The reaction of hexachlorodisilane with H_2salen* yielded only one isomer, as proven by the NMR data.

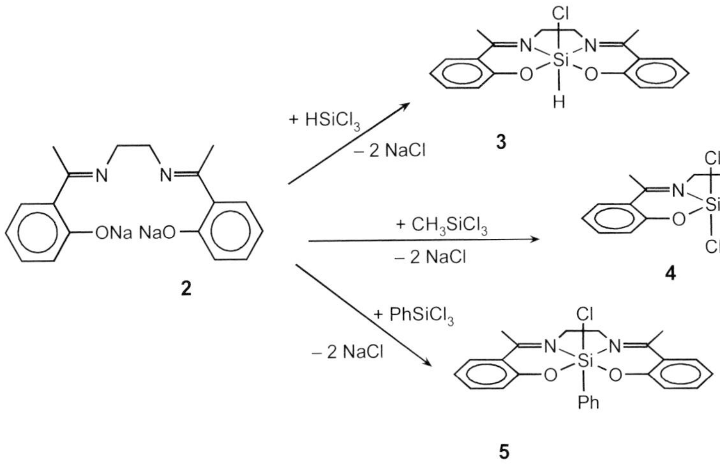

Scheme 1. Reactions of Na_2salen* with chlorosilanes.

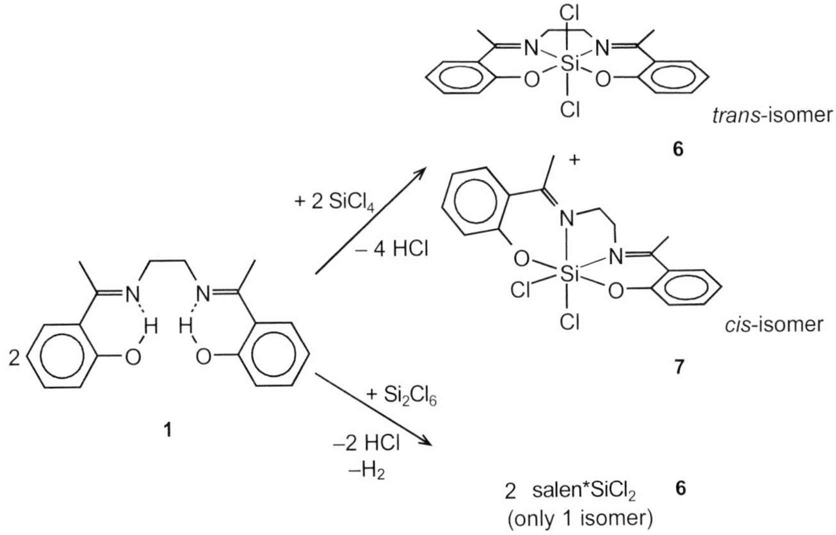

Scheme 2. Reactions of H_2salen* with silicon tetrachloride and hexachlorodisilane.

The fluoro derivative (salen*)SiF_2 (**8**) was prepared by reaction of (salen*)$SiCl_2$ (**6**) with ZnF_2 (Eq.1).

$$\underset{\mathbf{6}}{(\text{salen*})SiCl_2} + ZnF_2 \xrightarrow[\text{3 h; RT}]{\text{THF}} \underset{\mathbf{8}}{(\text{salen*})SiF_2} + ZnCl_2$$

Eq. 1. Synthesis of (salen*)SiF_2 (**8**).

The X-ray structure analysis of (salen*)SiF_2 (**8**) clearly demonstrates the octahedral coordination of the silicon atom (Fig. 1) [4]. There are a number of crystal structures of bis-chelate compounds with hexacoordinate silicon. Most of these had essentially a tetrahedral arrangement around silicon with the coordinated nitrogen donor atoms "capping" the tetrahedra at relatively large distances (N–Si between 2.5 and 3.0 Å) [5].

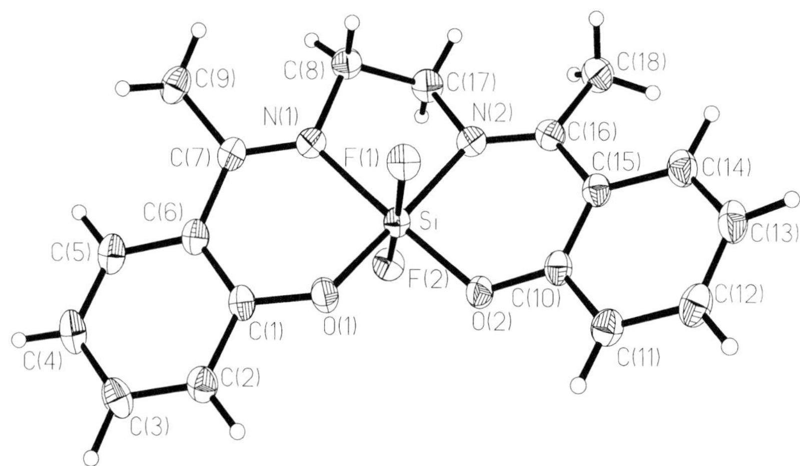

Fig. 1: Crystal structure of **8**. Monoclinic, space group *P*21/*a*, *R* = 0.0431 for 3358 reflections ($F_o > 2\sigma F_o$) and 0.0482 for all 3784 reflections. Selected bond distances [Å] and angles [°]: Si–F(1) 1.677(1), Si–F (2) 1.670(1), Si–O(1) 1.721(1), Si–O(2) 1.724(1), Si–N(1) 1.931(2), Si–N(2) 1.937(2), F(1)-Si-F(2) 172.40(5), 0(1)-Si-N(2) 178.00(6), O(2)-Si-N(1) 176.75(6), F(1)-Si-O(1) 91.68(6), F(1)-Si-O(2) 93.85(6), F(2)-Si-O(1) 93.17(6), F(2)-Si-O(2) 92.00(6), F(1)-Si-N(1) 86.93(5), F(1)-Si-N(2) 87.36(6), F(2)-Si-N(1) 86.96(5), F(2)-Si-N(2) 87.60(6), N(1)-C(8)-C(17)-N(2) 46(1).

The Si-F distance in **8** corresponds well with bond lengths found in other hypervalent silicon compounds (1.60–1.73 Å) [1d, 6]. The distances Si–O and Si–N are remarkably short [6]. The distortion of the octahedral coordination environment around silicon probably originates from the conformation of the chelating salen* ligand.

To obtain more soluble compounds we used the ligand system *N,N'*-ethylene-bis(3,5-di-*tert*-butylsalicylideneimine) (salen$^{‡}H_2$, **9**). Analytical data are shown In Table 1. Salen‡–silicon complexes are available by the same procedures as the salen* complexes.

R' = R" = Cl **10**
R' = R" = F **11**
R' = Me; R" = Cl **12**
R' = Ph; R" = Cl **13**

Wurtz condensation of (salen*)$SiCl_2$ by alkaline metals gives polysilanes containing main chain hexacoordinate silicon. Coupling with acetylendiyls affords polycarbosilanes with a Si–C≡C–Si backbone (Scheme 3).

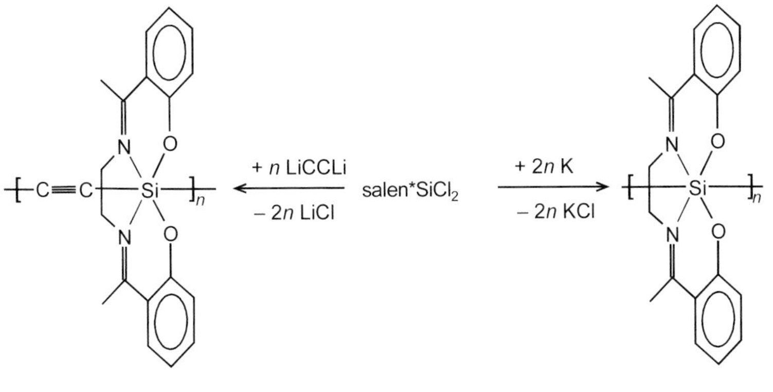

+ *n* LiCCLi
– 2*n* LiCl
salen*$SiCl_2$
+ 2*n* K
– 2*n* KCl

14 **15**

Scheme 3. Metal condensation reactions of salen*$SiCl_2$.

Till now the reaction of salen‡$SiCl_2$ with nucleophilic reagents led always to the decomposition of the salen‡ ligand. This effect probably results from the attack of the nucleophilic reagent at the azomethine proton.

The reaction of the salen‡Si(Cl)Me-derivative **12** with ZnF_2 also gives the corresponding salen–SiF_2 compound.

Table 1. Selected NMR [ppm] and IR [cm^{-1}] data.

Compound	^{13}C-NMR[a]				^{29}Si-NMR[a]	ν(C=N)
	C_1	C_7	C_8	R'		
1	163.2	172.6	50.3			1607
3	157.6	171.0	47.1		–131.6	1618
4	158.4	175.9	45.3	5.9	–150.7	1629
5	159.7	179.7	46.9	128.2–136.2	–173.1	1612
6	157.3	176.4	46.1		–186.1	1630
7	158.1	171.6	46.7		–188.0	1630
8	160.3	170.7	46.4		–187.9[b]	1613
9	158.0	167.6	59.6			1633
10	157.8	165.4	52.0		–187.6	1619
11	159.3	164.7	54.0		–187.1[c]	1648
12	156.0	173.2	53.9	2.4	–104.1	1619
13	157.7	174.1	52.8	128.2–137.1	–118.5	1617
14	168.4	169.1	51.4		–88[d]	1607
15	156.9	158.0	47.0	80.0	–130[d]	1613

[a] Compounds **1**–**8**, **14**, **15** recorded in [D_6]DMSO, compounds **9**–**13** recorded in $CDCl_3$ (for numbering see Fig. 2). — [b] $^1J_{Si-F}$ = 179.2 Hz. — [c] $^1J_{Si-F}$ = 172.6 Hz. — [d] Broad signal in solid state.

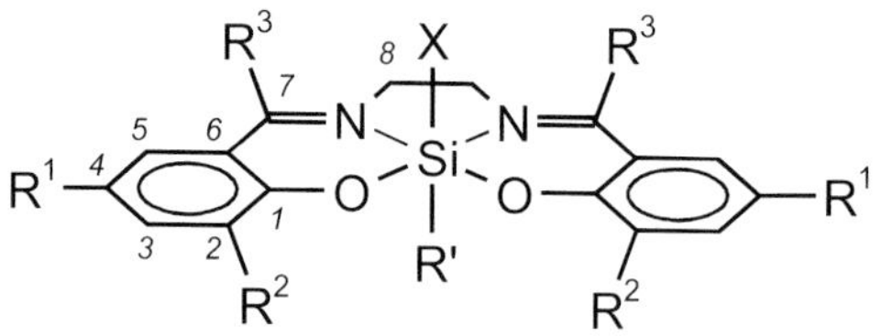

Fig. 2. Numbering of the Compounds **1–15**.

References:

[1] a) R. J. P. Corriu, J. C. Young, in: *The Chemistry of Organic Silicon Compounds*, (Eds.: S. Patai, Z. Rappoport), Wiley, Chichester, **1989**, p. 1241.
b) R. R. Holmes, *Chem. Rev.* **1990**, *90*, 17 and **1996**, *96*, 927.
c) R. J. P. Corriu, *J. Organomet. Chem.* **1990**, *400*, 81.
d) C. Chuit, R. J. P. Corriu, C. Reye and J. C. Young, *Chem. Rev.* **1993**, *93*, 1371.

[2] a) B. N. Ghose, *Acta Chimica Hungarica* **1985**, *118*, 191.
b) K. S. Siddiqi, F. M. A. M. Aqra, S. A. Shah, S. A. A. Zaidi, *Polyhedron* **1993**, *12*, 1967.

[3] Compare for instance the wrong data of elemental analysis in [2b]. The calculated sum formula was given with $C_{14}H_{14}N_2Cl_2Si$ and the found data fitted this formula well. The correct formula should be $C_{16}H_{14}N_2O_2Cl_2Si$.

[4] F. Mucha, U. Böhme, G. Roewer, *Chem. Commun.* **1998**, 1289.

[5] a) C. Breliere, F. Carre, R. J. P. Corriu, M. Poirier, G. Royo, J. Zwecker, *Organometallics* **1989**, 8, 1831.
b) F. Carre, G. Cerveau, C. Chuit, R. J. P. Corriu, C. Reye, *New J. Chem.* **1992**, *16*, 63.
c) F. Carre, C. Chuit, R. J. P. Corriu, A. Mehdi, C. Reye, *Organometallics* **1995**, *14*, 2754.

[6] I. Kalikhman, S. Krivonos, D. Stalke, T. Kottke, D. Kost, *Organometallics* **1997**, *16*, 3255.

Photoluminescence of Organically Modified Cyclosiloxanes

Udo Pernisz

Dow Corning Corporation
Central Research & Development,
Midland,MI 48686, USA
Tel: Int. code + (517)496 6087 — Fax: (Int. code) + (517)496 5121
e-mail: udo.pernisz@dowcorning.com

Norbert Auner

Institut für Anorganische Chemie
Johann Wolfgang Goethe-Universität Frankfurt
Marie-Curie-Str.11, D-60439 Frankfurt am Main, Germany
Tel: Int. code + (69)798 29591 — Fax: Int. code + (69)798 29188
E-mail: auner@chemie.uni-frankfurt.de

Keywords: Photoluminescence / Cyclosiloxane / Silacyclobutenes / Phosphorescence

Summary: The phenomenon of intense blue photoluminescence observed in organo-silicon molecules containing aromatic moieties (π-electron systems) was investigated by measuring the steady-state emission and excitation spectra as well as the decay time constants of the phosphorescence emission of several compounds that are considered representatives of two classes of materials that exhibit the effect. These are the 2,3-diphenyl-substituted silacyclobutenes, and several phenylated cyclosiloxanes. The fundamental aspect regarding the question of the origin of this photoluminescence and its appearance in the visible (blue) part of the spectrum is discussed also in the light of the fact that a similar emission occurs with silsesquioxane structures heat-treated in oxygen such that their internal molecular (nano-)voids, or cages, are preserved.

Introduction

The observation of intense blue photoluminescence in phenyl-containing silicon compounds [1] such as the silaspirocycles **1** and also in the 2,3-diphenyl-1-silacyclobut-2-enes **2** by themselves (in the following referred to as *t*olane *c*yclo*a*dducts, or TCA) upon excitation with UV light (337 nm) led to the investigation of this phenomenon in a range of related materials containing siloxane and aromatic moieties. The origin of the photoluminescence is assumed to be essentially associated with the π-electron system of the substituents at the carbon or silicon atoms since the compounds themselves usually exhibit photoluminescence although for small molecules such as benzene this emission occurs generally in the UV.

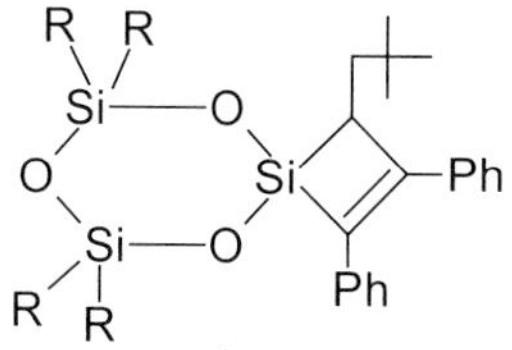

1 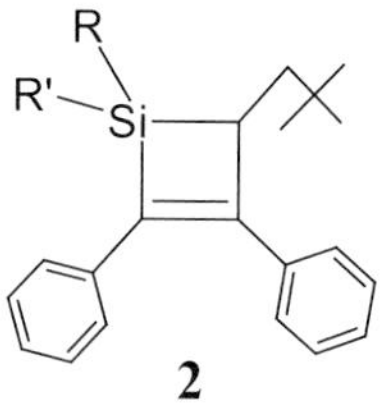 **2**

The fact that the photoluminescence was observed in the visible, especially in the blue part of the spectrum, prompted an investigation aimed at understanding the role of the silicon atom in this effect when it is directly bound to such an aromatic moiety. This was done with the assumption that the silicon causes a shift of the emission from the UV sufficiently large to be observed in the blue.

The study of the effects the silicon bond has on the electronic structure of aromatic molecules, that is, on the molecular orbitals of the organosilicon system, can be expanded to other substituents on the Silicon, including different π-conjugated groups. Another example is the presence of the siloxyl group or, generally, linear or cyclic siloxanes with which the aromatic part interacts via the silicon atom to which it is bound. The cyclosiloxanes are of interest also in the wider context of the photoluminescent properties of silica materials containing various defects, including mechanical ones (strained bond-angels) and chemical ones (non-bridging oxygen defects, and silicon radicals due to dangling bonds).

Suitable molecules covering the range of interest were investigated by measuring the emission and excitation spectra of the steady-state photoluminescence and of its phosphorescence component (at a fixed time delay after excitation with a pulsed light source) as well as the time dependence of the phosphorescence in the time regime above *ca.* 5 µs.

Experimental Details

The compounds investigated fall into two classes. The first one is based on a silacyclobutene subunit which is obtained through cycloaddition reaction of one mol-equivalent dichloro-neopentylsilene (obtained from the equimolar mixture of trichlorovinylsilane and *tert*-butyllithium) to tolane [2]. Starting from 1,1-dichloro-2,3-diphenyl-1-silacyclobut-2-ene (Cl_2TCA), a whole series of differently substituted derivatives is available by usual organometallic routes [3] where the methyl (Me), ethyl (Et), phenyl (Ph), hydroxyl (OH), and other groups replace the chlorines. Utilizing the silicon dichloro functionality and the outstanding chemical and thermal stability of the silacyclobutene, the TCA building block is also easily incorporated into cyclosiloxanes yielding silaspirocyclic compounds.

The second class of compounds comprises the substituted cyclosiloxanes. A series of partially phenylated methylcyclotrisiloxanes and -cyclotetrasiloxanes with the sum formula $Ph_{2n}Me_{2(k-n)}(SiO)_k$ was prepared and analyzed where n = 0, 1, 2, 3 with k = 3, and n = 4 with k = 4 (*i.e.*, perphenylcyclotetrasiloxane). Another series dealt with the stereoregularly-built phenylated trimethylsiloxy-cyclosiloxanes of different ring sizes of which **3** shows the cyclotetrasiloxane as an example (the phenyl group on the front silicon is given as Ph). This group of compounds, $[PhSiO(OSiMe_3)]_n$, with n = 4, 6, 8, 12, was synthesized following the literature [4].

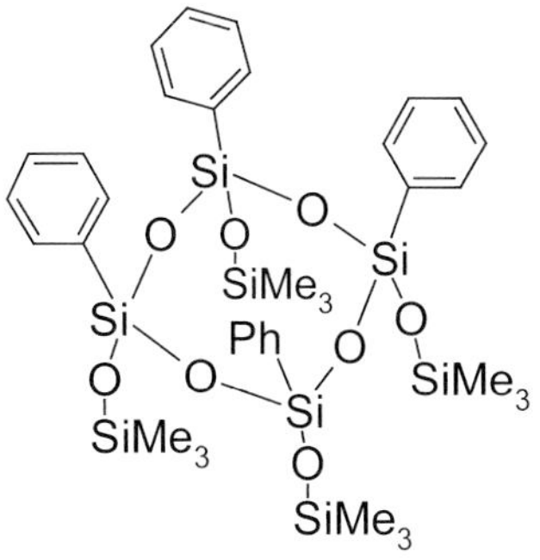

3

Photoluminescence spectra were obtained with a SPEX Fluorolog 2 instrument (Jobin/Yvon) in which the monochromators were equipped with single gratings of 1200 lines/mm, blazed at wavelengths of λ = 330 nm and λ = 500 nm for excitation and emission, respectively. The instrument's focal length was f = 0.22 m; for the commonly selected slit widths of 0.25 mm, the spectral bandwidth of the instrument was typically 1 nm. The powdered solid samples were inserted into the center tube (*ca.* 3 mm inner diameter) of a quartz dewar with which measurements could be performed at room temperature and at the temperature of liquid nitrogen (77 K).

Phosphorescence excitation and emission spectra were obtained by using a flash lamp of 3 μs duration (at half maximum) for excitation of the sample at selected wavelengths, recording the emission intensity at longer wavelengths after an adjustable delay time during an appropriately set time window. Similarly, phosphorescence intensity decay time measurements were performed by varying the delay time at fixed excitation and emission wavelengths.

Results

Photoluminescence data are presented as emission and excitation spectra obtained under steady-state irradiation (thus containing both fluorescence and phosphorescence components), and as phosphorescence spectra for the two compound classes investigated.

Silacyclobutene Derivatives (R_2TCA)

The visual observation of blue photoluminescence is confirmed by the spectral distribution of the emitted light measured under continuous excitation at a wavelength λ_{xc} = 320 nm. Figure 1 shows the luminescence emission spectra of five TCA derivatives investigated for which R = methyl (Me), ethyl (Et), phenyl (Ph), hydroxyl (OH), and phenylethynyl (Ph–C≡C–); both Cl are replaced with the same substituent for each compound. The measurements were made at room temperature. The Me_2-, Et_2-, and $(OH)_2$TCA samples have very similar emission characteristics with the maximum located at around 400 nm; there is a negligible effect on its position, and only a small effect on intensity. For the phenyl- and phenylethynyl-TCA, however, the effect of the silicon substituend is much more pronounced. The latter one appears to quench the photoluminescence emission, perhaps causing a small shift of the emission maximum to longer wavelengths; for the Ph_2TCA a strong shift of the maximum to *ca.* 420 nm is observed as well as an increase in the intensity over the alkyl-substituted compounds.

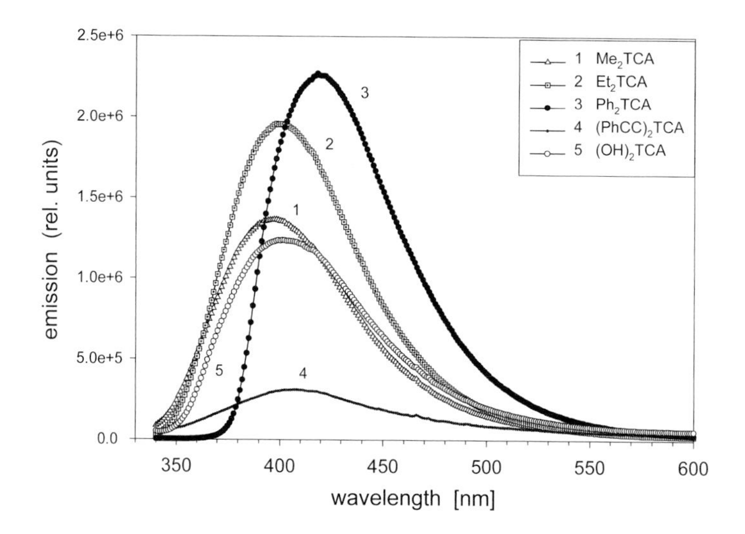

Fig. 1. Photoluminescence emission spectra of 2,3-diphenyl-4-neopentyl-1-silacyclobute-2-enes (**2**, TCA), measured as solids at room temperature. Excitation wavelength λ_{xc} = 320 nm. 1: R = Me, Me_2TCA; 2: R = Et, Et_2TCA; 3: R = Ph, Ph_2TCA; 4: R = –C≡C–Ph, $(PhCC)_2$TCA; 5: R = OH, $(OH)_2$TCA.

These observations indicate that the photoluminescence is associated mainly with the *cis*-stilbene part of the silacyclobutene molecule but that the group on the silicon atom affects the characteristics of the electronic transitions (oscillator strengths, non-radiative paths) by modifying the electron density of the σ-bonds at the Silicon.

The primary effect of the silicon atom on the molecular orbital structure of the *cis*-stilbene moiety is seen in the shift of the absorption process from the far UV of the *cis*-stilbene compound itself to the near UV of the dialkyl–TCA compounds where the maximum of the excitation spectrum typically occurs at 340–350 nm while the maximal absorption of the *cis*-stilbene is observed at ≈280 nm (in solution) [5] a shift in energy of 0.83 eV (19 kcal/mol). This phenomenon is generally observed when a silicon atom replaces a carbon atom in a π-electron system [6] and is explained by an inductive effect resulting in considerable electron donation from the silicon substitutent. It can also be assessed by comparing the dimethyl- and diphenylsilacyclobutenes **2** with the corresponding 1,2-diphenylcycloalkene and with *cis*-stilbene itself [7] at the temperature of liquid N_2. While the absorption peaks of the latter compounds are at 306 nm and 283 nm, respectively, the maximum of the photoluminescence excitation spectrum for the former (Me_2TCA) was observed at 345 nm.

In the case of the diphenylsilacyclobutene, an additional long-wavelength shift was seen that is shown in Fig. 2 where both the excitation spectrum and the emission spectrum of Ph_2TCA are plotted (*cf.* trace 3 in Fig. 1 for the latter). The excitation maximum occurs at 385 nm (3.22 eV); the Stokes shift of the Ph_2TCA (35 nm) thus becomes 0.27 eV.

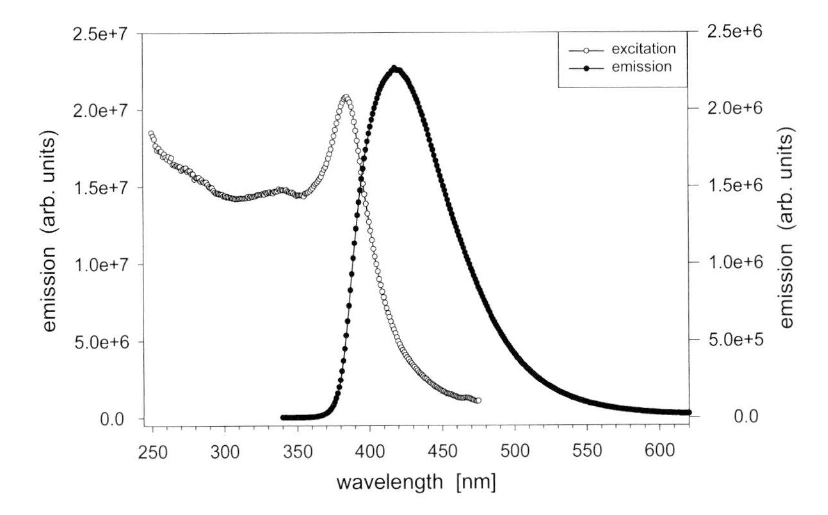

Fig. 2. Excitation and emission spectra of the Ph_2TCA, measured as a solid at room temperature. Left trace: Excitation spectrum; its emission was monitored at a wavelength of λ_{ms} = 505 nm. Right trace: Emission spectrum excited at a wavelength of λ_{xc} = 320 nm.

Another factor in the emission process appears to be the ring structure itself through which the silicon atom is attached to the phenyl groups. This is observed when the substituted TCA compounds **2** are compared to their linear analogs of comparable electronic configuration such as trimethylsilyl-*cis*-stilbene (**4**, R = Me) for which no photoluminescence could be observed upon excitation with light of 337 nm. In contrast, one observes strong blue emission from the triphenylsilyl-*cis*-stilbene (**4**, R = Ph) under these conditions, confirming the previously stated role of the silicon atom in lowering the energy levels of aromatic compounds bound to it. For **4**, R = Me, Ph this is shown in Fig. 3 for an excitation wavelength of λ_{xc} = 320 nm with the emission spectra recorded at room temperature. The data from the silacyclobutenes **2** are taken from Fig. 1. Most notably, the emission from **4**, R = Me is orders of magnitude lower than what is obtained from the Me_2TCA. With the phenyl analoga, the emission intensity for the silacyclobutene Ph_2TCA is significantly higher than it is for the silapropene **4**, R = Me despite the fact that one more phenyl group is attached to the silicon atom. This suggests that the four-membered ring indeed plays an essential role in the mechanism by which the silicon affects the luminescent behavior of the stilbene, that is, the electronic structure of the molecular orbitals.

R R R Si Ph Ph

4

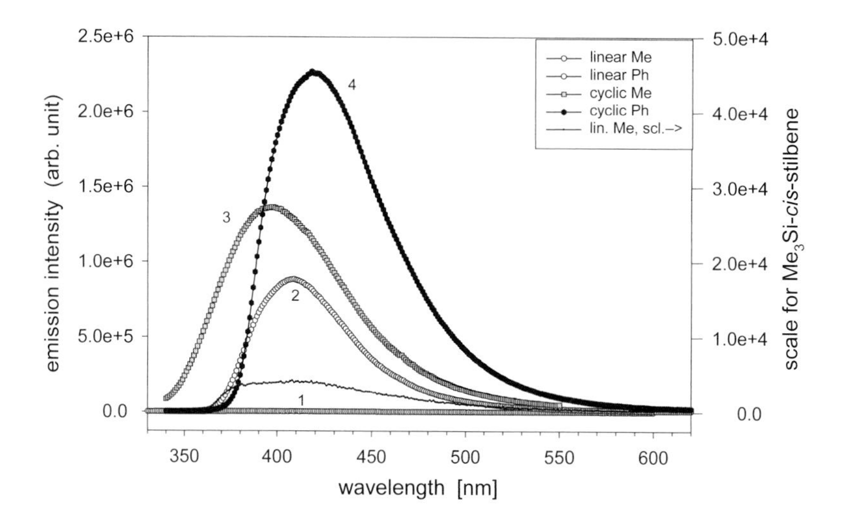

Fig. 3. Photoluminescence emission spectra of (*E*)-1-trimethylsilyl-1,2-diphenylethene (trace 1; the right-hand scale applies to the fine line plotted 50× magnified) and (*E*)-1-triphenylsilyl-1,2-diphenylethene (trace 2), and of Me_2TCA (trace 3) and Ph_2TCA (trace 4). Measurements made at room temperature, spectrum excited at λ_{xc} = 320 nm.

This observation, however, depends on the wavelength at which the molecule is excited. For a wavelength of λ_{xc} = 365 nm (instead of λ_{xc} = 320 nm), the silapropenes **4** with R = Me, Et, Ph show all three intense blue photoluminescence. Both the excitation spectra and the emission spectra of the three compounds are plotted in Fig. 4. Again, the phenyl-substituted molecule shows the strongest emission intensity in either mode.

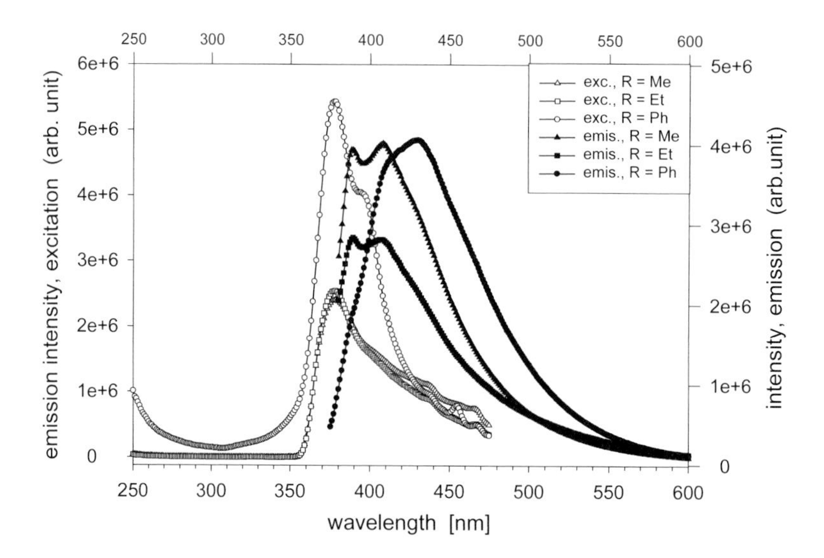

Fig. 4. Excitation and emission spectra of triorganosilyl-substituted *cis*-stilbenes **4** excitation wavelength λ_{ms} = 505 nm; emission measured at λ_{xc} = 365 nm, at room temperature with samples in quartz cuvette.

The data on the series of TCA compounds described so far were obtained in steady-state mode, allowing a collection time of *ca.* 1 s for each wavelength. If a long-lived phosphorescent state were present, it would have contributed to the spectral features recorded. A measurement at room temperature of the delayed emission after excitation by a monochromatic light pulse showed no significant signal (except for scattered light around the lamp's maximum at 467 nm). However, at the temperature of liquid N_2, strong phosphorescence is observed with a maximum at 560 nm, see Fig. 5. The delay time was 50 μs (for a light pulse width, at half maximum, of ≈3 μs), the emission was collected during a 10 ms window. This emission is ascribed to a triplet state of the Me_2TCA molecule that is located 2.21 eV above the ground state.

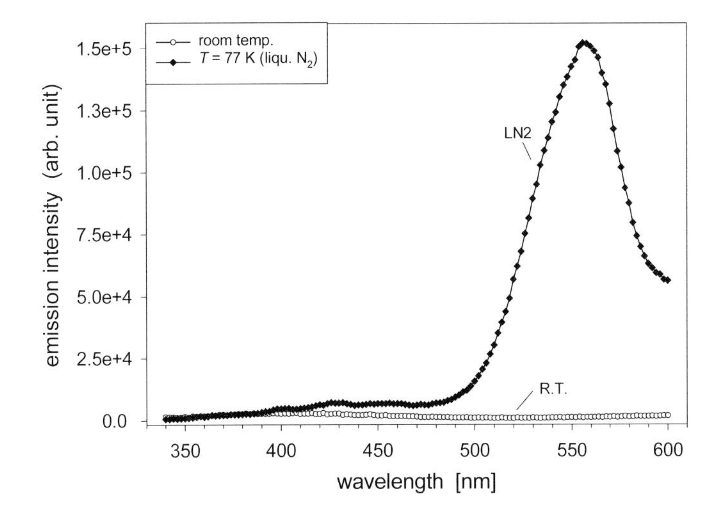

Fig. 5. Phosphorescence emission spectra of solid Me_2TCA, measured with a delay time of t_d = 50 μs after the excitation light flash (3 μs duration at half max.) using a sample window of t_{smp} = 10 ms; lower trace (open circle): at room temperature; upper trace (black diamond): at liquid N_2 temperature, T = 77 K.

Perphenylated Cyclosiloxanes

Since the spirocycles **1** with the siloxane structure as well as the bis-TCA-disiloxanediol **5**, which is obtained from $(HO)_2TCA$ by heating between 80°C and 110°C in toluene solution as the condensation product, exhibited intense photoluminescence similar to the cyclobutenes described above, the perphenylated cyclotri- and -tetrasiloxane compounds were also investigated.

OH HO
Ph Si O Si Ph
Ph Ph
5

Of interest was the question how ring size and number of phenyl groups on the ring would affect the emission characteristics. Fig. 6 shows the steady-state photoluminescence spectra of the two perphenylated cyclotri- and -tetrasiloxanes at room temperature, identified by n = 3 and n = 4, respectively, according to the sum formula $Ph_{2n}Me_{2(k-n)}$cyclo<k>siloxane where <k> = 3 and 4 stand for D_3 and D_4. The plot also shows data for the partially phenylsubstituted cyclotrisiloxanes with n = 0, 1, and 2; the case n = 0 describes the permethylated molecule. Three observations can be made: (i) only the phenyl-substituted cyclosiloxanes (n > 0) show photoluminescence (the increase of the signal below 350 nm is due to stray light from the excitation monochromator only incompletely rejected by the emission monochromator); (ii) the photoluminescence intensity increases with phenyl content with the vibrational lines emphasizing different parts of the emission band; (iii) the larger ring size shifts the maximum of the emission towards longer wavelengths, from 352 nm to 407 nm. The second point demonstrates that the effect is essentially due to the aromatic substitutents on the siloxane ring which pulls the benzene emission from the UV into the visible, but suggests also that the strain in the siloxane ring affects the molecular orbitals of the molecule.

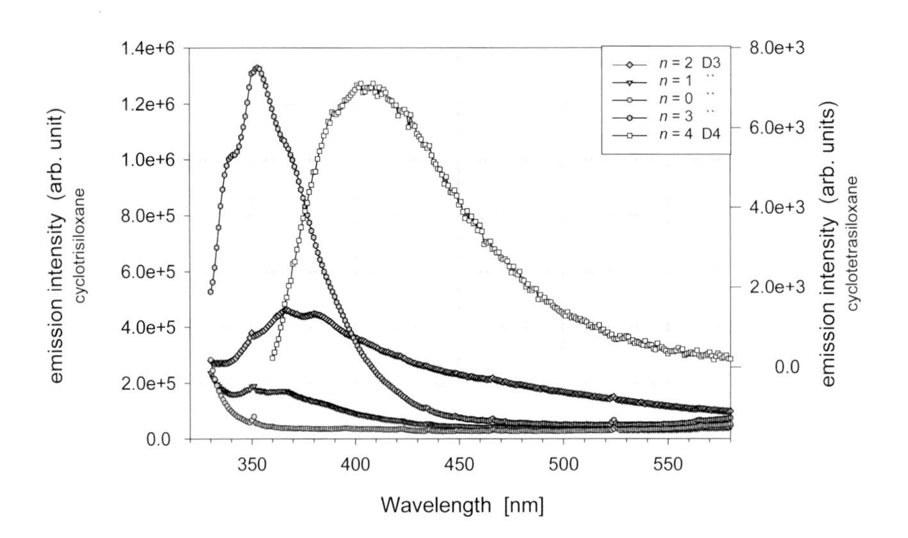

Fig. 6. Photoluminescence emission of $Ph_{2n}Me_{2(3-n)}$cyclotrisiloxane excited at a wavelength of λ_{xc} = 320 nm, spectral bandwidth 0.8 nm. Solid sample, measured at room temperature (no second-order filter).

Effect of Ring Size on Phenyl-substituted Cyclosiloxanes.

The stereoregularly-built phenylated cyclosiloxanes, **3**, with a trimethylsiloxy group attached opposite each phenyl group has been described in the literature and was accordingly synthesized. The four cyclics prepared were the eight-, and twelve- sixteen- and twentyfour-membered rings; in the context of this work, they will be referred to as D4, D6, D8, and D12, using the letter D as an abbreviation for the functionalized siloxy group in the siloxane chain. For comparison, a commercially available small linear compound, tetraphenyldisiloxanediol, was also included in the photoluminescence study. Although not a cyclic, it is referred to as D2 in the following.

Since the phenyl groups in the stereoregular cyclosiloxanes chosen for this investigation lie essentially on the same side of the siloxane ring (except for the D12 ring in which sequences of three phenyls alternate in partial up and down orientation [4]), it appears possible that an effect of energy transfer between the phenyl groups is observed in the photoluminescence emission, caused either by direct interaction between the π–electrons of the aromatic rings, or mediated by the siloxane groups separating adjacent phenyl groups attached to silicon atoms. It was also expected that the emission maximum could shift more to longer wavelength but that this effect should eventually saturate and spectral characteristics become independent of ring size.

Figure 7 shows the photoluminescence emission spectra of the four stereoregular phenyl-cyclosiloxanes, D4–D12, together with the tetraphenyldisiloxanediol, D2, measured at liquid N_2 temperature. The graph demonstrates that the ring structure has a pronounced effect on the spectral characteristics of the photoluminescence; however, it is not immediately obvious in what way the cyclosiloxane size controls each spectrum. One notes the very strong emission above 370 nm of the D2; this feature is also obtained in the other molecules with decreasing intensity although there is a D8–D6 reversal in this sequence. The strong wide peak around 460 nm of the D8 ring is surprising.

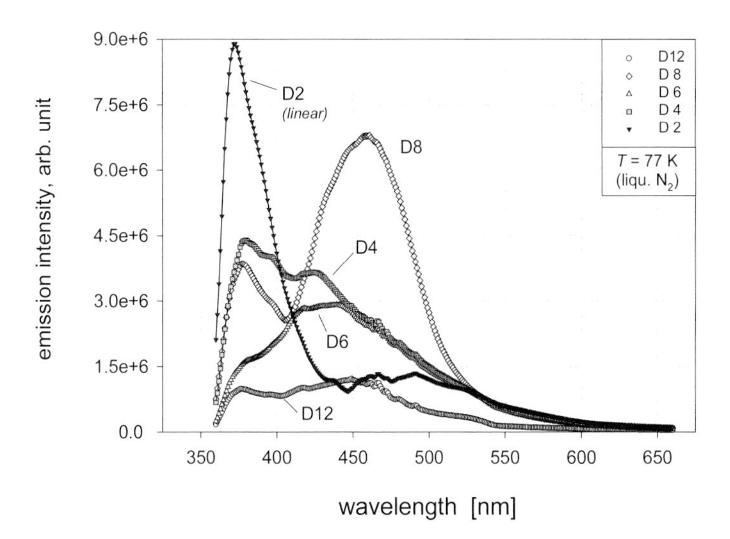

Fig.7. Photoluminescence emission spectra of the stereoregular phenylcyclosiloxanes $[PhSiO(OSiMe_3)]_n$ where $n = 4, 6, 8$, and 12, together with the tetraphenyl-1,3-disiloxanediol, measured as solids (powder) at liquid nitrogen temperature, $T = 77$ K. Excitation wavelength $\lambda = 280$ nm, spectral bandwidth $\Delta\lambda = 2$ nm, edge filter (50% transmission at 370 nm). Legend: D2 – tetraphenyldisiloxanediol; D4–D12 – phenylcyclo-siloxane, ring sizes of $n = 4$–12.

The luminescence excitation spectra shown in Fig. 8 do not clarify the picture although one can observe the same rank ordering in the emission intensity (measured at $\lambda_{ms} = 505$ nm) as with the emission spectra except for the D2 molecule having the lowest value. This appears plausible considering that if interaction effects were essentially localized between two adjacent aromatic groups the behavior of the large ring should be very similar to that of the short linear molecule.

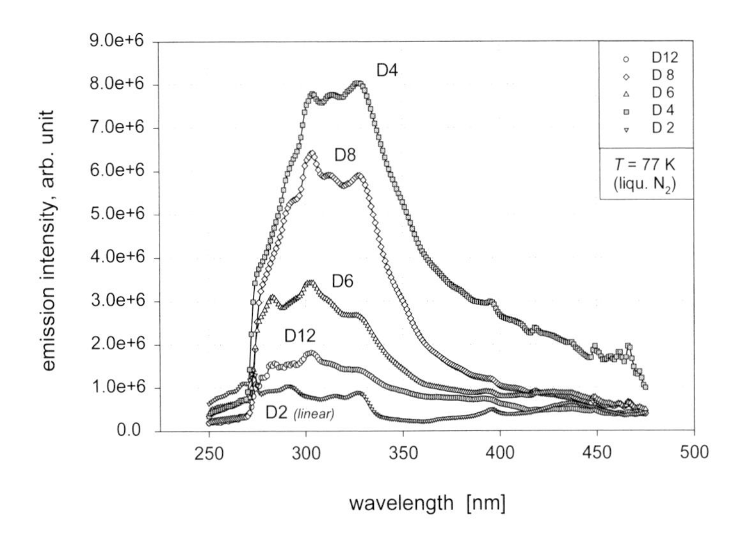

Fig. 8. Photoluminescence excitation spectra of the stereoregular phenylcyclosiloxanes $[PhSiO(OSiMe_3)]_n$ where $n = 4$, 6, 8, and 12, together with the tetraphenyl-1,3-disiloxanediol, measured as solids (powder) at liquid nitrogen temperature, $T = 77$ K. Emission wavelength $\lambda = 505$ nm, spectral bandwidth $\Delta\lambda = 1$ nm, edge filter (50% transmission at 495 nm). Legend: D2 – tetraphenyldisiloxanediol; D4–D12 – phenylcyclosiloxane, ring sizes of $n = 4$–12.

The other feature common to all spectra in Fig. 8 is the sequence of small peaks and shoulders which can be analyzed into at least two series of vibrational bands, one with an energy separation of 0.36 eV (2930 cm^{-1}), and the other consisting of lines separated by 0.13 eV (1050 cm^{-1}). Two more line series are discernible but were not further identified and assigned. The two major series are ascribed to the stretch vibration of the C–H group on the phenyl ring, and to the Si–O–Si vibration from the siloxane ring, according to IR data found in the literature [8]. The appearance of the Si-O vibronic structure indicates that the siloxane ring indeed participates in the energy relaxation processes between the absorption and emission of photons in these compounds.

While the measurements of the photoluminescence intensity under steady-state excitation as discussed above did not reveal a convincing systematic effect from the cyclosiloxane ring size, it was possible to achieve a classification of the emission characteristics from the phosphorescence excitation measurements. In this experiment, the emission intensity is recorded after a fixed delay time (Δt = 50 μs was selected) at a fixed emission wavelength (505 nm) as a function of the excitation wavelength. Measurements of the time dependence of this emission were also made at a fixed excitation wavelength $\lambda_{xc} = 320$ nm, and for two select emission wavelengths of $\lambda_{ms} = 380$ nm and $\lambda_{ms} = 458$ nm.

Figure 9 shows the phosphorescence excitation spectra of the five compounds. Intensity is no longer an obvious distinction, except for the D12 molecule for which the data are shown scaled up by a factor of 100, and for D2 where the maximum occurs at ca. 260 nm, dropping off in emission intensity towards longer wavelengths. However, the characteristic features in each spectrum can be sorted according to the symmetry of the siloxane ring: most obviously, D4 and D8 fall into one group, and D6 with D12 fall into the other; D2 shares most of the peaks common to both groups with more uniform intensity distribution. It thus appears that 3-fold and 4-fold symmetry of the siloxane ring emphasizes different features in the (vibronic) structure of the long-lived phosphorescence of these phenylated siloxane compounds, and that the disiloxanediol constitutes the building block that exhibits the basic interactions by which the absorbed photon energy relaxes into the emitting state from which the energy is released at long times.

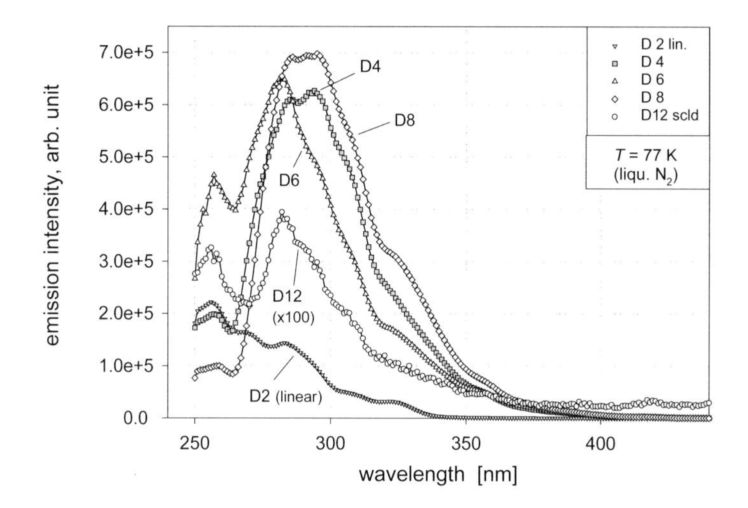

Fig. 9. Photoluminescence excitation spectra of the stereoregular phenylcyclosiloxanes $[PhSiO(OSiMe_3)]_n$ where $n = 4$, 6, 8, and 12, together with the tetraphenyl-1,3-disiloxanediol, measured as solids (powder) at liquid nitrogen temperature, $T = 77$ K. Emission wavelength $\lambda = 505$ nm, spectral bandwidth $\Delta\lambda = 8$ nm, edge filter (50% transmission at 495 nm). Excitation flash duration 3 μs (at half maximum), delay time $t_{dl} = 10$ μs, sampling window $t_{smp} = 10$ ms, cumulative emission collected from 10 flashes per data point. Legend: D2 – tetraphenyldisiloxanediol; D4–D12 – phenylcyclosiloxane, ring sizes of $n = 4$–12. Note groupings of spectra according to symmetry of ring: D4 and D8 show the same spectral features, while D6 and D12 have their own, different set of features.

This symmetry model was confirmed with a measurement of the phosphorescence time dependence for the five compounds; the data are shown in Fig. 10. Two features can be observed. The first one is a build-up of the delayed emission intensity for $t < \approx 50$ μs during which time the energy from a molecular state that absorbs 3.87 eV photons ($\lambda_{xc} = 320$ nm) is pumped into the emitting state at 3.26 eV ($\lambda_{ms} = 380$ nm, *cf.* Fig. 7) at a higher rate than it emits energy. The second one is the ordering of the spectra by the appearance of a second such peak for molecules with higher symmetry, *i.e.*, D2, D4, and D8, a feature that is missing from the traces of the molecules with 3-fold symmetry, *i.e.*, D6, and D12. It is remarkable that D2 shares the double-peak feature with the molecules of 4-fold symmetry; this suggests that the 3-fold symmetry of the siloxane ring suppresses a second electronic state to which the phenyl-silicon system has otherwise access, and from which long-lived delayed emission can occur.

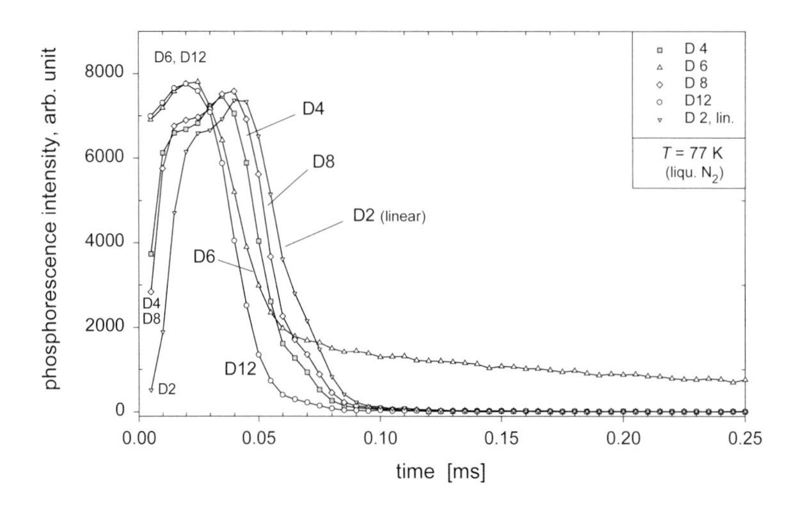

Fig. 10. Phosphorescence time dependence of the stereoregular phenylcyclosiloxanes $[PhSiO(OSiMe_3)]_n$ where $n = 4$, 6, 8, and 12, measured together with the tetraphenyl-1,3-disiloxanediol, as solids (powder) at liquid nitrogen temperature, $T = 77$ K. Excitation wavelength $\lambda_{xc} = 320$ nm, emission wavelength $\lambda_{ms} = 380$ nm, spectral bandwidth $\Delta\lambda = 8$ nm, edge filter (50% transmission at 370 nm). Excitation flash duration 3 μs (at half maximum), delay time increments $t_{dl} = 5$ μs, sampling time window $t_{smp} = 10$ μs, cumulative emission collected from 10 flashes per data point. Legend: D2 – tetraphenyldisiloxanediol; D4–D12 – phenylcyclosiloxane, ring sizes of $n = 4$–12 as indicated. Note groupings of curves according to symmetry of rings: D4 and D8 show the same spectral features as the linear disiloxane unit, namely a second maximum after the first peak, while D6 and D12 have only one maximum. Note also that there is a long-lived emission for $T > 0.1$ ms from the D6 ring ($n = 6$).

The obvious grouping of the D12 ring with the D6 ring is somewhat puzzling since its behavior could be expected to be equally a multiple of the D4 ring (or should perhaps show features of both symmetries). However, the x-ray structure of the D12 molecule [4] shows clearly that the orientation of the phenyl groups varies widely around the ring, the phenyls no longer pointing clearly to one side of the siloxane ring as it is the case with the smaller rings. It rather appears that series of three phenyls appear to form subgroups that resemble the grouping of the D6 ring more than the D4 ring; such an effect is not visible for the D8 ring. Another possible consequence of this disorientation — which is due to the high flexibility of the large siloxane backbone — is the decreased phenyl-phenyl interaction around the ring which may explain the surprisingly low phosphorescence emission observed for the D12 ring, *cf.* Figure 9.

It is also noted that the D6 molecule shows a substantial emission with much longer time constant than the comparatively fast decay seen between 30 μs and 80 μs for the other molecules. This comparatively long-lasting phosphorescence is responsible for the large emission intensity at 460 nm observed from D6 in the phosphorescence emission spectrum, see Fig. 11. The origin of the electronic state from which this emission occurs is currently not understood, that is, whether it is due to another symmetry effect specific to the D6 structure, or perhaps to an unknown impurity, is not clear. However, the evidence presented below suggests that all the stereoregular phenylcyclosiloxanes show an emission in that region.

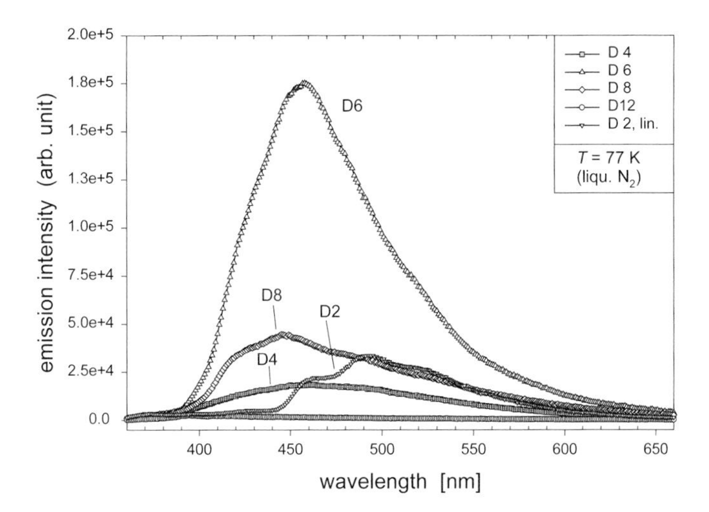

Fig. 11. Phosphorescence emission spectra of the stereoregular phenylcyclosiloxanes [PhSiO(OSiMe$_3$)] where $n = 4$, 6, 8, and 12, measured together with the tetraphenyl-1,3-disiloxanediol, as solids (powder) at liquid nitrogen temperature, $T = 77$ K. Excitation wavelength $\lambda_{xc} = 320$ nm, spectral bandwidth $\Delta\lambda = 8$ nm, edge filter (50% transmission at 370 nm). Excitation flash duration 3 μs (at half maximum), delay time $t_{dl} = 50$ μs, sampling time window $t_{smp} = 10$ ms, cumulative emission from 10 flashes per point. Legend: D2 – tetraphenyl-disiloxanediol; D4–D12 – phenylcyclosiloxane, ring sizes of $n = 4$–12 as indicated. Note the large maximum obtained from D6 ($n = 6$) at 460 nm.

The rapid decay after the peak emissions at 20 μs or 35 μs which applies to all five species, is characterized by time constants between approximately 10 μs to 20 μs, somewhat dependent on the molecule. A semilogarithmic plot of the data shows the additional emission from the D6 molecule to be an exponential phosphorescence decay characterized by a time constant of 0.25 ms as seen in Fig. 12. The decay into the noise of the emission intensity from the other rings indicates a similar time constant. The behavior becomes more consistent if the phosphorescence time dependence is measured at an emission wavelength of $\lambda_{ms} = 458$ nm (instead of at 380 nm). This is the wavelength at which the strong phosphorescence emission was observed from the D6 ring, *cf.* Fig. 11. The data are plotted semilogarithmically in Fig. 13. The surprising feature is the emergence of an additional emission peak or at least a shoulder between 0.1 ms and 0.2 ms, most pronounced for the D4 ring, but discernible for all ring systems; the exception is the linear molecule D2 which nevertheless shows a second, long emission decay time constant similar to that of the other compounds. This feature means that excitation energy is slowly transferred into an energy level from which it is emitted with a time constant of 250 μs, a value which applies approximately to all five molecules. This energy level was not accessible at a wavelength of 380 nm which was used for the experiment shown in Fig.s 10 and 12. Thus, the phosphorescence is emitted from two energy levels at $E_{ph,1} = 2.71$ eV with a lifetime of $\tau_1 = 250$ μs, and at $E_{ph,2} = 3.26$ eV with $\tau_1 \approx 15$ μs.

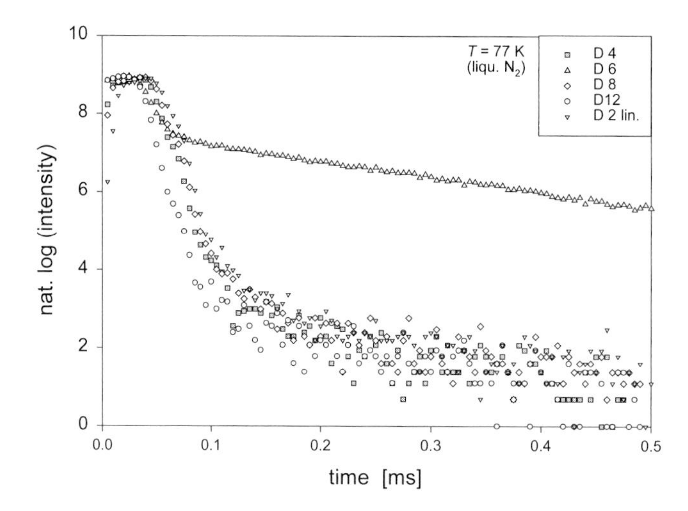

Fig. 12. Semilogarithmic plot of the phosphorescence time dependence of the stereoregular phenylcyclosiloxanes $[PhSiO(OSiMe_3)]_n$ where $n = 4$, 6, 8, and 12, measured together with the tetraphenyl-1,3-disiloxanediol, as solids (powder) at liquid nitrogen temperature, T = 77 K. Excitation wavelength λ_{xc} = 320 nm, emission wavelength λ_{ms} = 380 nm, spectral bandwidth $\Delta\lambda$ = 8 nm, edge filter (50% transmission at 370 nm). Excitation flash duration 3 μs (at half maximum), delay time increments t_{dl} = 5 μs, sampling time window t_{smp} = 10 μs, cumulative emission collected from 10 flashes per data point. Legend: D2 – tetraphenyl-disiloxanediol; D4–D12 – phenylcyclosiloxane, ring sizes of n = 4–12 as indicated.

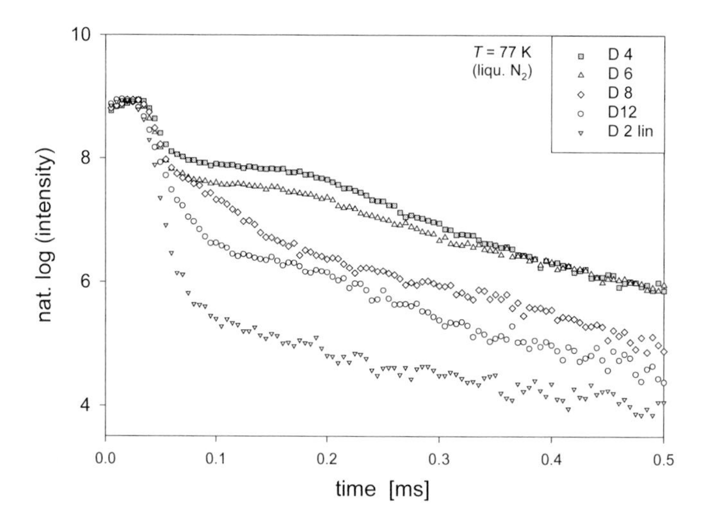

Fig. 13. Semilogarithmic plot of the phosphorescence time dependence of the stereoregular phenylcyclosiloxanes $[PhSiO(OSiMe_3)]_n$ where $n = 4$, 6, 8, and 12, measured together with the tetraphenyl-1,3-disiloxanediol, all measurement conditions as described in the legend to Fig. 12 above, except for the emission wavelength λ_{ms} = 458 nm.

Summary and Conclusions

The photoluminescence emission obtained from π-electron systems such as stilbene or benzene is red-shifted into the blue part of the visible spectrum if a silicon atom is attached to the conjugated moiety. Additional substitutions on the silicon atom affect the spectral features of the molecule as demonstrated with the derivatives of TCA **2** such as the diphenylsilacyclobutene compound. Diol derivatives **5** of this molecule that were also investigated include the spirocyclics **1** and the cyclosiloxanes.

With a related class of compounds, the phenylated cyclosiloxanes **3**, the effect of siloxane ring size on the spectral characteristics of the blue photoluminescence was investigated in more detail. It was shown that the cyclosiloxane symmetry is the controlling parameter for the spectral distribution of the phosphorescence intensity as well as for the temporal behavior of the emission. The specific features in the phosphorescence excitation spectrum and the phosphorescence emission time dependence that are modified by the size and symmetry of the siloxane ring are also present in the linear phenyldisiloxanediol which indicates that the molecular structure responsible for the observed effects consists of two phenyl groups each attached to a silicon atom in a siloxane configuration. From this observation, as well as from the fact that the Si–O vibronic signature is observed in the excitation spectrum of each molecule with phenylsiloxane structure, it is concluded that the siloxane unit participates in the energy absorption, relaxation, and emission processes in such compounds. This is remarkable since the siloxane moiety itself does not exhibit a comparable photoluminescence phenomenon as demonstrated by the hexamethylcyclotrisiloxane in comparison with the hexaphenylcyclotrisiloxane and the octaphenylcyclotetrasiloxane. And the linear disiloxanes $R_2Si(X)OSi(X)R_2$ (X = Cl, OH; R = Me, Ph) and the trisiloxanediols $R_2Si(OH)OSiR_2OSi(OH)R_2$ also exhibit blue photoluminescence in the case that R = Ph but not for R = Me, and even the simple di- and triphenyl-substituted molecules Ph_2SiOH_2 and Ph_3SiX (X = Cl, OH, N_3, Me, H, ...) show photoluminescence. Thus the effect of the silicon atom bonded to the phenyl group on the shift of the photoluminescence from the UV into the blue part of the visible spectrum is further modified by the interaction with the siloxane bridges between the phenyl groups.

Acknowledgment: We thank M. Backer for the syntheses of the silacyclobutenes and the silacyclobutene-substituted D3 and D4 compounds during his Ph.D. thesis work, and B. Herrschaft for the X-ray crystallographic analysis. Discussions with H. Roskos and M. Thomson as well as synthetic support by O. I. Shchegolikhina, B. Goetze, and M. Grasmann are gratefully acknowledged. The experiments have benefitted from the skills of A. A. Hart, F. N. Noble, and L. R. Jodarski who carried out most of the photoluminescence measurements reported here.

References.

[1] a) U. Pernisz, N. Auner, M. Backer, *Polym. Preprints* **1998**, *39(1)*, 450.
b) U. Pernisz, N. Auner, M. Backer, "*Photoluminescence of Phenyl- and Methylsubstituted Cyclosiloxanes*" in: *Silicones and Silicone-Modified Materials*, (Eds.: S. J. Clarson, J. J. Fitzgerald, M. J. Owen, S. D. Smith) ACS Symp. Ser. No. 729, Dallas, TX, **1998**, Ch. 7.

[2] N. Auner, C. Seidenschwarz, N. Seewald, E. Herdtweck, *Angew. Chem.* **1991**, *103*, 1172; *Angew. Chem. Int. Ed. Engl.* **1991**, *30*, 1151.

[3] M. Backer, M. Grasmann, W. Ziche, N. Auner, C. Wagner, E. Herdtweck, W. Hiller, M. Heckel, "*Silacyclobutenes – Synthesis and Reactivity*", in: *Organosilicon Chemistry II: From Molecules to Materials* (Eds.: N. Auner, J. Weis), VCH, Weinheim **1996**, pp. 41–47.

[4] O. I. Shchegolikhina, V. A. Igonin, Y. A. Molodtsova, Y. A. Pozdniakova, A. A. Shdanov, T. V. Strelkova, S. V. Lindeman, *J. Organomet. Chem.* **1998**, *562(1-2)*, 141–151.

[5] G. H. Brown, "*Photochromism*", in: *Techniques of Chemistry* (Ed.: A. Weissberger), *Vol. III* Wiley Interscience, New York, **1971**, p.476.

[6] H. Bock, K. Wittel, M. Veith, N. Wiberg, *J. Am. Chem. Soc.* **1976**, *98*, 109.

[7] G. Hohlneichner, M. Müller, M. Demmer, J. Lex, J. H. Penn, L.-X. Gan, P. D. Loesel, *J. Am. Chem. Soc.* **1988**, *110*, 4483.

[8] E. D. Lipp, A. L. Smith, in: *The Analytical Chemistry of Silicones* (Ed.: A. L. Smith), Chemical Analysis 112; John Wiley & Sons, New York, **1991**, ch. 11, p. 305.

Octasilsesquioxanes as Traps for Atomic Hydrogen at Room Temperature

Michael Päch, Reinhard Stößer*

Fachbereich Chemie, Humboldt-Universität zu Berlin
Hessische Straße 1-2, D-10115 Berlin, Germany
Tel.: Int. code + (30)2093 7376 — Fax: Int. code + (30)2093 7375
e-mail: reinhard=stoesser@chemie.hu-berlin.de

Keywords: Atomic Hydrogen / Octasilsesquioxanes / Radiation Chemistry / EPR / ^{29}Si-Superhyperfine Structure

Summary: The radiation chemical behavior of octasilsesquioxanes and related compounds containing an Si_8O_{12} unit has been investigated. After irradiation using ^{60}Co γ-radiation they show an ESR spectrum which is inequivocally attributed to atomic hydrogen. Experiments using deuterated compounds revealed both the source of trapped $H^\bullet/D^\bullet$ and mechanistic details of the trapping process. Unexpectedly, the presence of radical scavengers was found a) to increase markedly the radiation chemical yield of trapped $H^\bullet$ for the irradiation of solids and b) to be a prerequisite for successful trapping of $H^\bullet$ in irradiated solutions of Si_8O_{12} compounds. Evidence for the entrapment of $H^\bullet$ inside the Si_8O_{12} unit was derived from the observation of the ^{29}Si superhyperfine interaction.

Introduction

Compounds containing the cube-shaped Si_8O_{12} unit (Fig. 1) have continuously attracted attention since their discovery some 50 years ago. The attractiveness of polyhedral polysilsesquioxanes and related compounds has its roots in their physical and chemical properties as well as in their aesthetic geometry. In 1994 Sasamori et al. [1] discovered another interesting feature of one of these compounds: they found the trimethylsiloxylated derivative of the Si_8O_{12} cage able to stabilize radiolytically generated hydrogen atoms at room temperature. This is one of a very few examples of hydrogen atoms observable at ambient temperatures. It should be noted that, so far, all attempts have failed to trap $H^\bullet$ in fullerenes, which in turn are known to be molecular traps e.g. for He, N and lanthanides. The most surprising fact they reported on [1] was that the "jacketed hydrogens" are even stable in etheral solution. These first results raised a lot of questions; some of the answers found by us [2, 3] are presented in the following contribution.

What is the source of the trapped hydrogen atoms? Is there any spectroscopic proof for the incorporation of hydrogen atoms in the Si_8O_{12} cages? Does the trapping of hydrogen atoms proceed intra- or intermolecularly? What influence do radical scavengers have during radiolysis? Is it possible to trap atomic hydrogen not only in solid silsesquioxanes but also in their solutions?

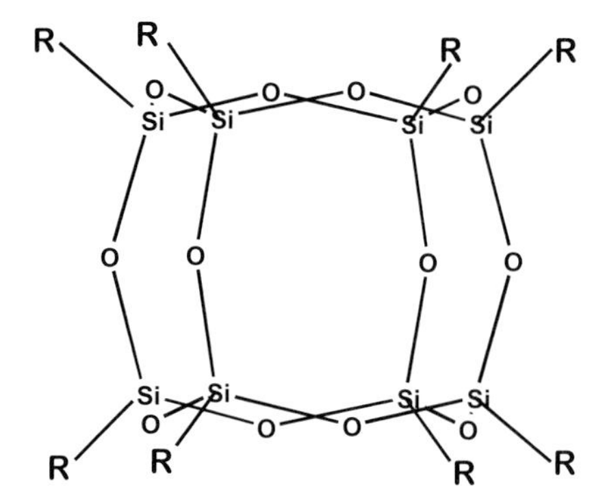

Fig 1. General structure of octasilsesquioxane compounds. R stands for H, D, Me, Et, Pr, c-C_6H_{12}, C_2H_3, C_6H_5, $OSiMe_3$, $OSi(CD_3)_3$, 3-chloropropyl. Cages with mixed substitution patterns as well as heterosilsesquioxanes have been studied successfully [3].

The Source of Trapped Hydrogen Atoms

To make sure that the trapped H atoms derive from the substituents of the silsesquioxane cages, d_{72}-Q_8M_8 was prepared in a multistep procedure starting from CD_3MgI [3] and γ-irradiated. After being irradiated in the solid state at ambient temperature, it shows the typical EPR spectrum of atomic deuterium, as depicted in Fig. 2b. Up to now, EPR spectroscopy is the only means of detecting trapped H atoms with sufficient sensitivity.

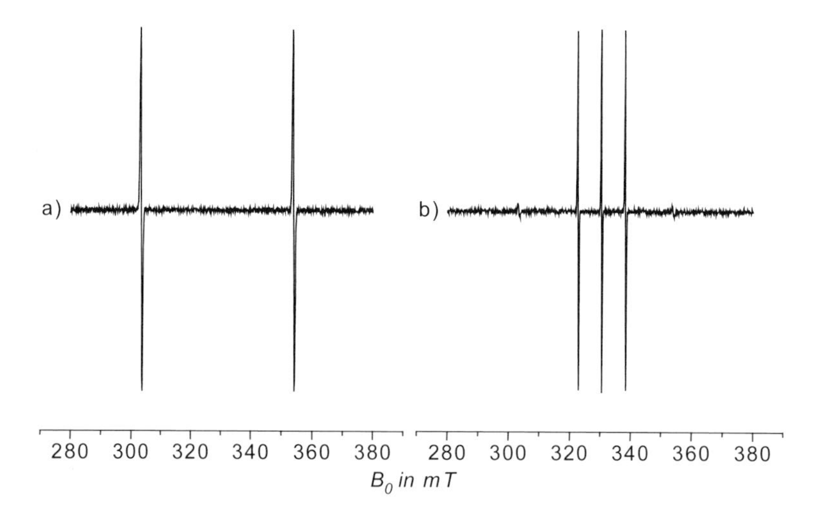

Fig. 2. EPR spectra of a) h_{72}-Q_8M_8 and b) d_{72}-Q_8M_8 irradiated as solids. The a) two and b) three lines are due to hyperfine interaction with nuclear spins of $I = ½$ (H) and $I = 1$ (D) respectively.

This result provides convincing proof for the trapped atoms originating from the substituents in samples irradiated as solids.

^{29}Si-Shfs Proves Trapping Inside the Cages

The deoxygenation of solutions of irradiated silsesquioxanes, e.g. in C_2Cl_4, led to very well resolved EPR spectra (Fig. 3). Linewidths (ΔB_{pp}) in solution were in the range of 0.002–0.009 mT, whereas in solids linewidths of about 0.10–0.15 mT were observed. The spectral patterns can be accounted for quantitatively by weighted addition of the theoretically expected superhyperfine spectra of $H^{\bullet}$ trapped in Si_8O_{12} isotopomers possessing no ^{29}Si (68.2 %), one ^{29}Si (26.7 %) or two ^{29}Si nuclei (4.6 %) respectively (see Table1).

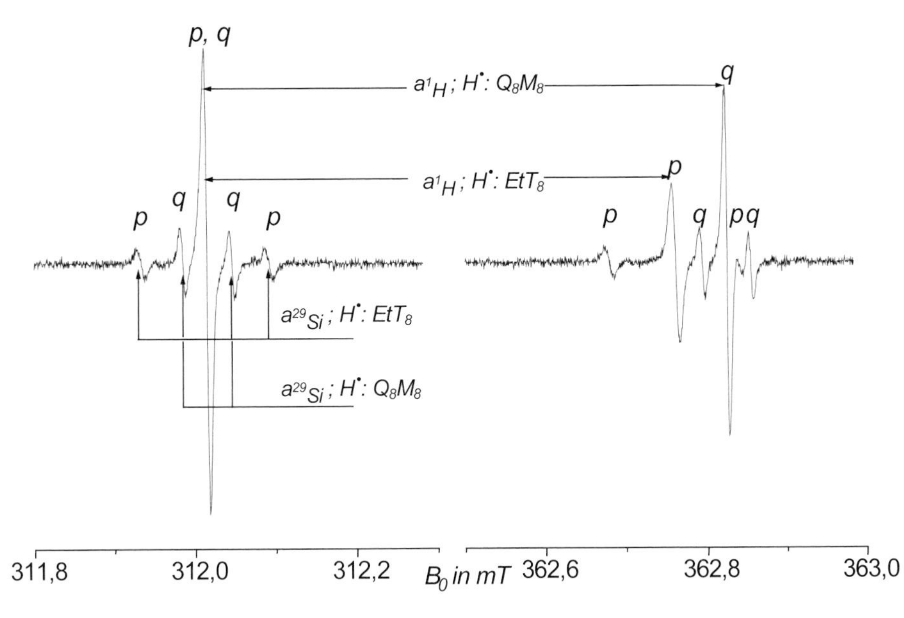

Fig. 3. EPR-spectra of a deoxygenated solution. of γ-irradiated Q_8M_8 and EtT_8 in C_2Cl_4 at 293 K. ΔB_{pp} is in the order of 2–9 μT. *p* and *q* denote signals due to $H:PrT_8$ and $H:Q_8M_8$ respectively.

This result provides a convincing spectroscopic proof for the hydrogen atoms being trapped in the Si_8O_{12} cages tumbling in the fluid solution. Therefore the shfs becomes isotropic and their coupling constants were found to depend both on the temperature and the group electronegativity of R. Furthermore, it may be concluded that the rhombohedric distortion of the cages in solids is removed upon their dissolution. The Pauli exclusion principle [4] is assumed to be responsible for the superhyperfine interaction between the electron and the ^{29}Si nuclear spins.

Table 1. Intensities of individual superhyperfine transitions.

No. of shfs component	Calculated intensity [%]	Measured intensity [%]
1	1.23	1.145
2	13.6	13.63
3	70.7	70.49
4	13.2	13.63
5	1.25	1.145

Radiation Chemical Aspects

Most surprisingly, the relative yield of $H_{tr}^{\bullet}$ is considerably increased by radical scavenging additives (e.g. NO, O_2, I_2, 2,6-di-*tert*-butyl-4-methylphenol) present during γ-irradiation. In their presence, the dose dependence of [$H_{tr}^{\bullet}$], e.g. in PrT_8, becomes almost linear whereas in the absence of any scavengers it reaches a much lower and quasi-stationary level. Only for HT_8 and MeT_8 are the yields of $H_{tr}^{\bullet}$ not noticeably affected by scavengers. This is probably because the scavengers cannot enter the crystal lattices of HT_8 and MeT_8.

Concerning the mechanism, it could be shown that intermolecular processes are involved in trapping. This was done by means of a crossover experiment in which d_{72}-Q_8M_8 and h_{72}-Q_8M_8 were used. In the presence of radical scavengers, trapping of $H^{\bullet}$ was even possible in solution, e.g. in *c*-C_6H_{12} and *t*BuOMe. By using C_6D_{12}, it could be shown that most but not all H atoms trapped in solution originate from the solvent. This can be seen from Fig. 4.

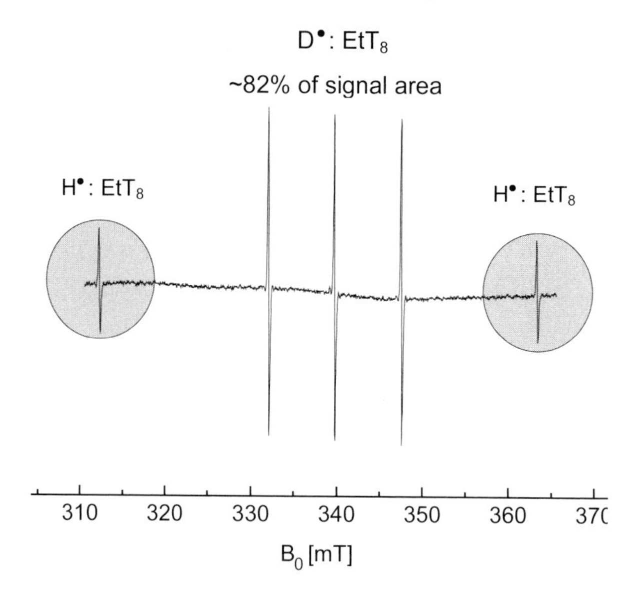

Fig. 4. ESR spectrum of solid EtT_8, previously irradiated in C_6D_{12} solution. I_2 was added before irradiation.

Trapping in solutions and presumably in solids too was shown to be a net effect, i.e. there are also radiation-induced processes which remove trapped hydrogen atoms from the cages.

Resumé

It was shown that organosilicon chemistry can contribute in a unique manner to the physical chemistry of atomic hydrogen: due to their organophilicity and good solubility, octasilsesquioxanes allowed the observation of the superhyperfine interaction of ^{29}Si and render themselves objects worthy of further studies in radiation chemistry as well as in physical chemistry and synthesis. Most recently, promising results were obtained in attempts to circumvent the tedious γ-irradiations.

Acknowledgment: Sincere thanks are extended to Dr. E. Janata and D. Gassen, *Hahn–Meitner-Institut Berlin GmbH*, for their technical assistance with the γ-irradiation.

References:

[1] R. Sasamori, Y. Okaue, T. Isobe, Y. Matsuda, *Science* **1994**, *265*, 1603.

[2] M. Päch, R. Stößer, *J. Phys. Chem.* **1997**, *101*, 8360.

[3] M. Päch, *Ph.D. Thesis*, Humboldt-Universität zu Berlin, **1997.**

[4] F. J. Adrian, *J. Chem. Phys.* **1960**, *32*, 972; S. N. Foner, E. L. Cochran, V. A. Bowers, C. K. Jen, *J. Chem. Phys.* **1960**, *32*, 963.

Highly Functionalized Octasilsesquioxanes Synthesis, Structure and Reactivity

Michael Rattay, Peter Jutzi

Fakultät für Chemie, Universität Bielefeld
Universitätsstr. 25, D-33615 Bielefeld, Germany
Tel.: Int. code + (521)106 6181 — Fax.: Int. code + (521)106 6026
E-Mail: peter.jutzi@uni-bielefeld.de

Dieter Fenske

Institut für Anorganische Chemie, Universität Karlsruhe
Engesserstraße, D-76128 Karlsruhe, Germany

Keywords: Octasilsesquioxanes / Synthesis / Structure / Hydroformylation / Catalysis

Summary: The synthesis and structure of octakis[4-(trimethylsilylethynyl)phenyl]-octasilsesquioxane (**1**), the first representative of a new class of octasilsesquioxanes bearing ridid π-conjugated substituents, is described. Furthermore we present the synthesis and structure of octakis(tetracarbonylcobaltio)octasilsesquioxane (**2**). In combination with triphenylphosphine, **2** shows catalytic activity in the hydroformylation of 1-hexene.

Introduction

The three-dimensional cube-like silicon–oxygen framework and the capability of bearing eight substituents give rise to the broad interest in octasilsesquioxanes and their chemistry.

Eightfold organo-functionalized octasilsesquioxanes are promising precursors for the preparation of highly siliceous materials or organolithic macromolecular compounds [1]. They are useful core molecules in dendrimer chemistry[2] and alternative precursors for various applications in micro-electronics[3, 4]. Moreover, they are useful building blocks for the design of tailor-made organic–inorganic hybrid materials.

Octakis(4-(trimethylsilylethynyl)phenyl)octasilsesquioxane (1)

The short reaction time, the mild reaction conditions as well as the high yields and the selectivity proved the hydrolytic oligocondensation of zwitterionic λ^5-spirosilicates to be an appropriate method for the synthesis of organo-functionalized octasilsesquioxanes [5]. The synthesis of **1** required a modification of the reaction conditions and an extension to ionic λ^5-spirosilicates.

Pyridiniumbis(2,3-naphthalenediolato)-4-(trimethylsilylethynyl)phenylsilicate (**3**), a new ionic λ^5-spirosilicate, was synthesized in seven steps starting from 4-bromoacetophenone in an overall yield of 34 %.The hydrolytic oligocondensation of **3** (Scheme 1) led to the formation of a colorless precipitate, which showed a good solubility in most of the common solvents. Analytically pure **1** was obtained by crystallization from toluene in a yield of 88 %. **1** was characterized by two sharp signals at –17.2 and –78.2 ppm in the ^{29}Si NMR and a characteristic vibration at 1128 cm^{-1} in the IR spectrum, which represents the asymmetric Si–O–Si stretching mode of the octasilsesquioxane framework.

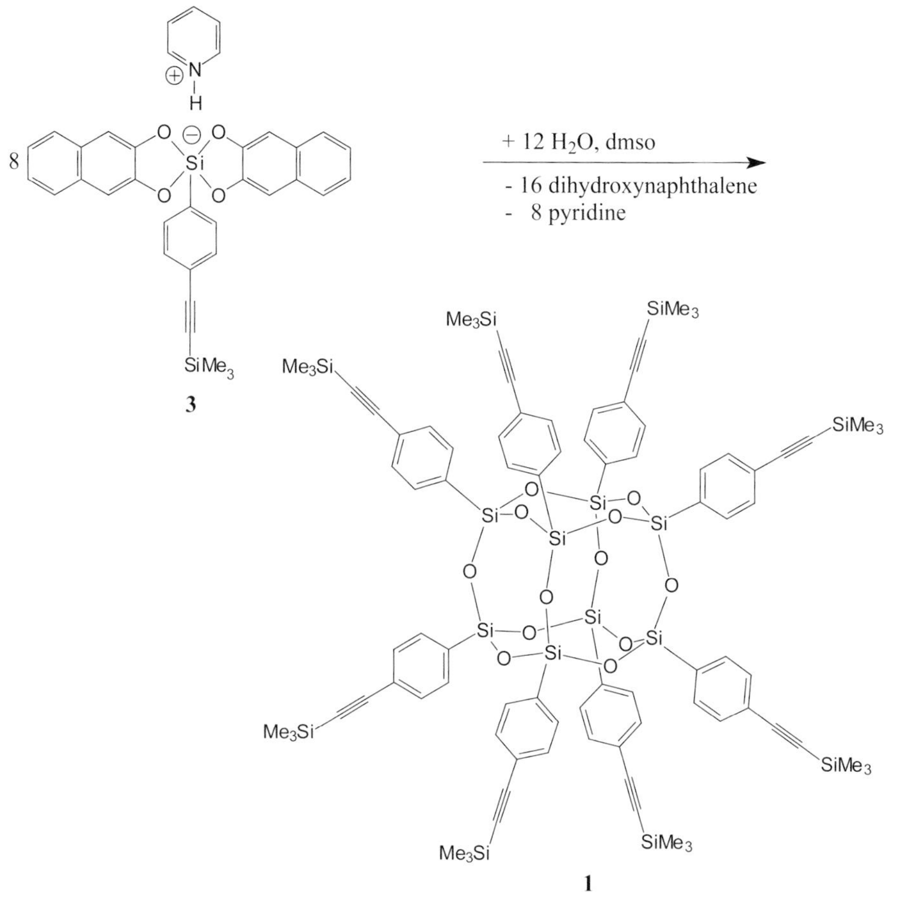

8 [3] + 12 H_2O, dmso → - 16 dihydroxynaphthalene - 8 pyridine → [1]

Scheme 1. Synthesis of **1** via hydrolytic oligocondensation of the ionic λ^5-spirosilicate 3 (dmso = dimethylsulfoxide).

The crystal structure of **1** is shown in Fig. 1; the atom labeling is omitted for clarity. The molecule contains a cubic Si_8O_{12} framework with a 4-(trimethylsilylethynyl)phenyl substituent bound to each of the silicon atoms. The silicon atoms of the siloxane cage are approximately tetrahedrally coordinated. The whole molecule shows C_i symmetry. In contrast to the solid state, **1** shows perfect O_h symmetry in solution, which was verified by NMR spectroscopy.

According to UV/Vis spectra the inductive effect of the siloxane framework is comparable to those of halogen substituents. Neither inter- nor intramolecular interaction of the conjugated π-systems was observed in solution.

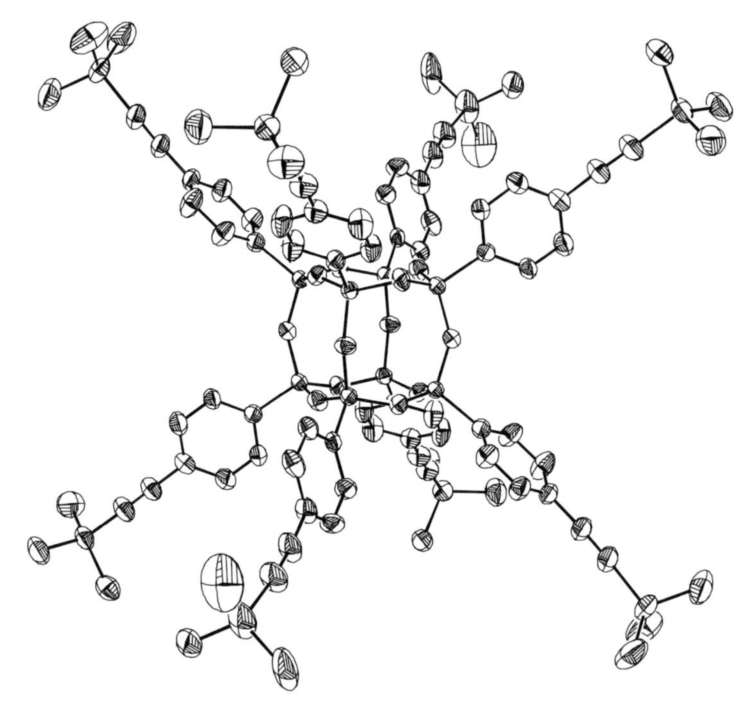

Fig. 1. Molecular structure of **1**.

The mass spectrometric analysis of the raw material showed that at least two more polyhedral oligosilsesquioxanes had been formed. By means of MALDI TOF spectrometry they were identified as hexakis[4-(trimethylsilylethynyl)phenyl]hexasilsesquioxane (**4**) and tetrakis[4-(trimethylsilylethynyl]tetrasilsesquioxane (**5**). However, neither a separation by crystallization nor a selective accumulation of one of these species has succeeded so far.

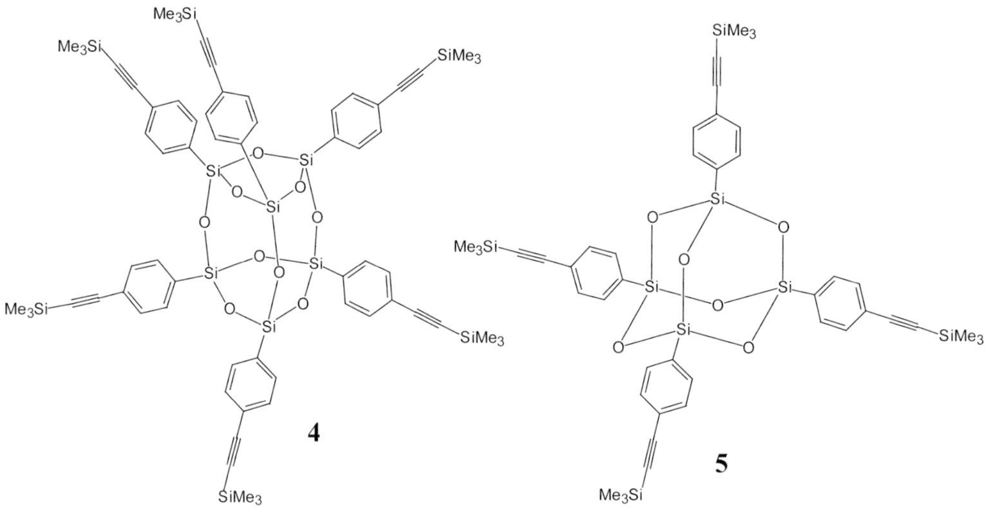

Fig. 2. Byproducts of the hydrolytic oligocondensation of **3**.

Octakis(tetracarbonylcobaltio)octasilsesquioxane (2)

The reaction of octahydridosilsesquioxane with octacarbonyldicobalt in a 1:4 stoichiometry leads to the formation of **2** in quantitative yield [6].

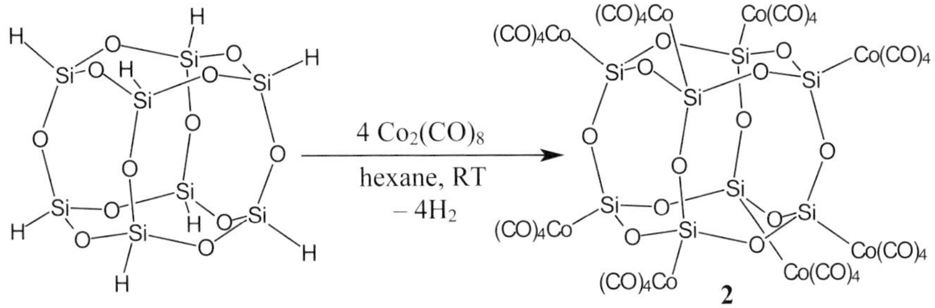

$$\mathrm{H_8Si_8O_{12}} \xrightarrow[\text{hexane, RT}\;\; -\,4\mathrm{H_2}]{4\,\mathrm{Co_2(CO)_8}} [(\mathrm{CO})_4\mathrm{Co}]_8\mathrm{Si_8O_{12}}\ (\mathbf{2})$$

Scheme 2. Synthesis of octakis(tetracarbonylcobaltio)octasilsesquioxane (**2**).

2 crystallizes in the cubic space group $P2_13$. The molecular structure of **2** in the solid state with the atom labeling scheme is shown in Fig. 3.

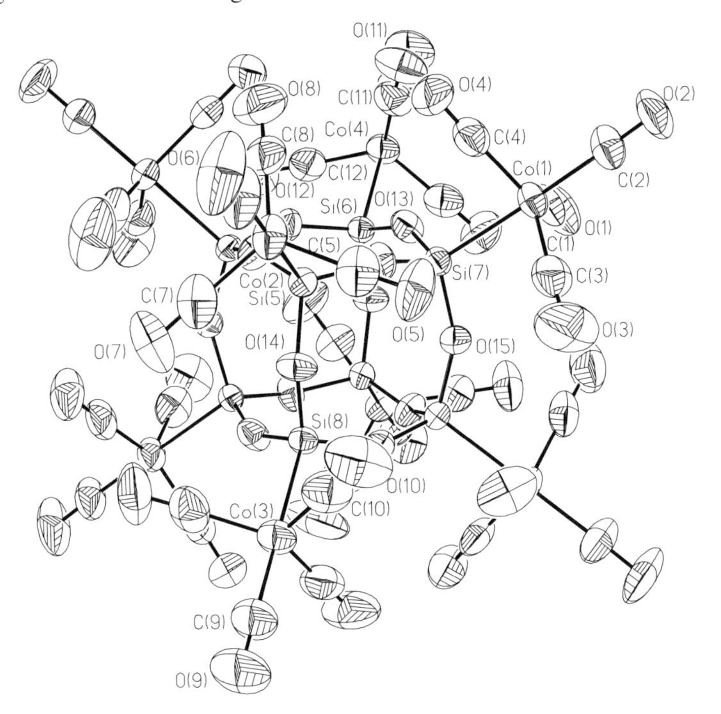

Fig. 3. Molecular structure of **2**.

The molecule contains a cubic Si_8O_{12} framework with a $Co(CO)_4$ substituent bound to each of the silicon atoms. The silicon atoms are approximately tetrahedrally surrounded by three oxygen atoms of the Si-O framework and the cobalt atom of the $Co(CO)_4$ substituent. **2** shows a remarkable stretching of the Si–O framework along its C_3 symmetry axis which results in an arrangement of overall C_3 symmetry.

The coordination geometry at the cobalt centers corresponds to a distorted trigonal bipyramidal arrangement with significant deviation from local C_{3v} symmetry. Subsequently the equatorial carbonyl groups bend out of the trigonal plane towards the silicon atoms. The bond lengths as well as the bond angles within the $Co(CO)_4$ units do not differ significantly from those in $[Co(CO)_4]H_7Si_8O_{12}$ [7]. As a consequence intramolecular interactions of the eight $Co(CO)_4$ groups of **2** are negligible. This is reflected in the high flexibility of the Si_8O_{12} framework, which seems to be able to compensate for the sterical demand of eight $Co(CO)_4$ substituents.

The insolubility of **2** in common solvents does not allow a routine NMR analysis (^{13}C, ^{29}Si) in solution. In the ^{29}Si-CP-MAS NMR spectrum of **2** a signal at –55 ppm is observed. The signal does not show any cross-polarization but exhibits a remarkable anisotropy which is in good agreement with the determined solid-state structure.

Due to its highly metal functionalized Si–O framework **2** can be seen as a model compound for Si-O-supported transition metal catalysts. In first experiments we have studied the catalytic activity of **2** in the hydroformylation of 1-hexene. The experiments were performed in toluene at a temperature of 120°C and a reaction time of 18 h. The initial CO/H_2 pressure at room temperature was 70-80 bar. The use of a catalyst formulation of **2** and triphenylphosphane in a 1:8 stoichiometry led to complete conversion of 1-hexene to the corresponding aldehydes. NMR and GC analyses of the hydroformylation products showed a 3:1 mixture of 1-heptanal and 2-methylhexanal had been formed. Filtration of the reaction mixture led to the isolation of a brownish solid, which still showed catalytic activity. According to IR spectroscopic results it is supposed that the catalytically active species formed in situ is a substitution product of **2** and triphenylphosphine. However, the mechanistic pathway of this catalysis is not yet understood. Experiments leading to a further understanding are under investigation.

Acknowledgment: This work was supported by the *Deutsche Forschungsgemeinschaft*. We thank Prof. Dr. H.C. Marsmann for the recording of the ^{29}Si CP-MAS spectrum and Dr. Waidelich from *Perseptive Biosystems Wiesbaden* for the recording of the MALDI TOF spectra.

References:

[1] P. G. Harrison, *J. Organomet. Chem.* **1997**, *542*, 141.
[2] A. R. Bassindale, T. E. Gentle, *J. Mater. Chem.* **1993**, *3*, 1319.
[3] N. P. Hacker, *MRS Bull.* **1997**, *22*, 33.
[4] A. Schmidt, S. Babin, K. Böhmer, H. W. P. Koops, *Microelectron. Eng.* **1997**, *35*, 129.
[5] R. Tacke, A. Lopez-Mras, W. S. Sheldrick, A. Seebald, *Z. Anorg. Allg. Chem.* **1993**, *619*, 347.
[6] M. Rattay, D. Fenske, P. Jutzi, *Organometallics* **1998**, *17*, 2930.
[7] G. Calzaferri, R. Imhof, K. W. Törnroos, *J. Chem. Soc., Dalton Trans.* **1993**, 3741.

Synthesis and Characterization of Chloro-, Allyl- and Ferrocenyl-Substituted Silsesquioxanes

Aslihan Mutluay, Peter Jutzi

Fakultät für Chemie, Universität Bielefeld
Universitätsstr. 25, D-33615 Bielefeld, Germany
Tel.: Int. code + (521)106 6181 — Fax.: Int. code + (521)106 6026
E-Mail: peter.jutzi@uni-bielefeld.de

Keywords: Dendrimers / Ferrocenes / Hydrosilylation / Silsesquioxanes

Summary: Starting with octa(vinyldimethylsiloxy)octasilsesquioxane (**$Q_8M_8^V$**) as the central core molecule, allyl- and chloro-substituted dendritic macromolecules have been prepared by the divergent route. The sequence of hydrosilylation with dichloromethylsilane followed by allylation of all Si–Cl groups with CH_2=$CHCH_2MgBr$, was used for building up each generation (**G1–G4**). We also functionalized the dendrimer periphery via hydrosilylation reaction of allyl-functionalized dendrimers with dimethylsilylferrocene. These novel molecules have been characterized by $^1H/^{13}C/^{29}Si$-NMR, IR spectroscopy and MALDI-TOF spectrometry. The ferrocenyl-substituted molecules have also been analyzed by cyclovoltammetry.

Introduction

Treelike molecules are attracting increasing attention because of their unique structure and properties. Since the first report on dendrimers has been given by F. Vögtle and co-workers [1] in 1978, several synthetic pathways to dendrimers have been developed and a number of core molecules and monomers have been used to prepare different dendrimers [2]. The surface of dendrimers can be modified with many organotransition-metal complex fragments. Ferrocenyl-based dendrimers can be used in the chemical modification of electrodes, in the construction of amperometric biosensors or as multi-electron reservoirs [3].

Chloro and Allyl Substituted Silsesqiuoxane Dendrimers

In our strategy for the synthesis of carbosilane dendrimers, hydrosilylation and allylation reaction sequences were chosen starting with octa(vinyldimethylsiloxy)octasilsesquioxane (**$Q_8M_8^V$**) as core molecule (Scheme 1).

As a first step in realizing dendrimers with terminal allyl and chloro groups we prepared the first-generation octa[2-(dichloromethylsilylethyl)dimethylsiloxy]octasilsesquioxane (**G1-Cl**) via Pt-catalyzed hydrosilylation of the core molecule **$Q_8M_8^V$** with dichloromethylsilane. Then each Si–Cl function is converted to an allyl group in order to form the first generation of an allyl-substituted dendrimer (**G1-Allyl**). The resulting compound **G1-Allyl** is used as a new core and the reactions of the first generation are repeated to yield the second generation with 32 Si–Cl or Si–Allyl functional groups. The third and fourth generations are made in the same manner (Scheme 2).

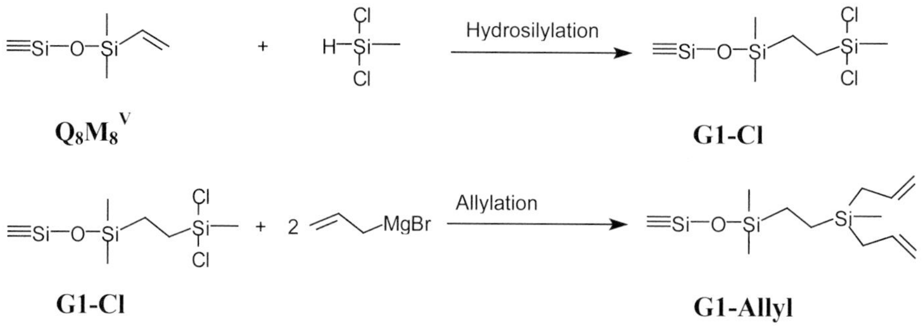

≡Si–O–Si(vinyl) + H–SiCl₂– —Hydrosilylation→ ≡Si–O–Si–CH₂CH₂–SiCl₂–

$Q_8M_8^V$ **G1-Cl**

≡Si–O–Si–CH₂CH₂–SiCl₂– + 2 CH₂=CHCH₂MgBr —Allylation→ ≡Si–O–Si–CH₂CH₂–Si(allyl)₂–

G1-Cl **G1-Allyl**

Scheme 1. Sequence for the synthesis of the first silsesquioxane dendrimer generation.

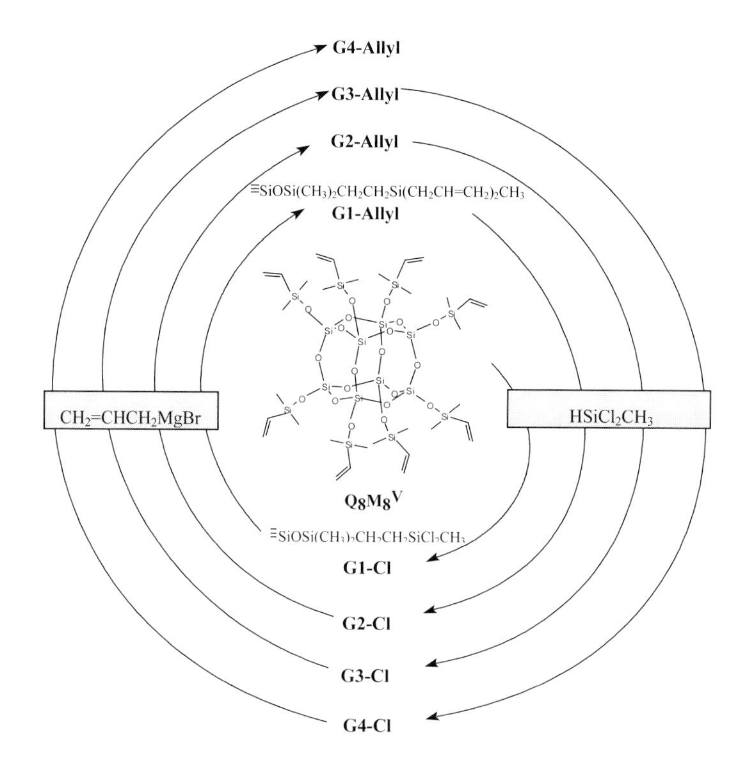

Scheme 2. Repeated sequence for the dendrimer synthesis (G*n*-Cl = *n*th generation of chloro-substituted dendrimer; G*n*-Allyl = *n*th generation allyl-substituted dendrimer; n = 1–4).

An alternative reaction of **G1-Cl** with allylic alcohol leads to a compound that also exhibits allyl groups on the periphery (**G1-OAllyl**, Scheme 3).

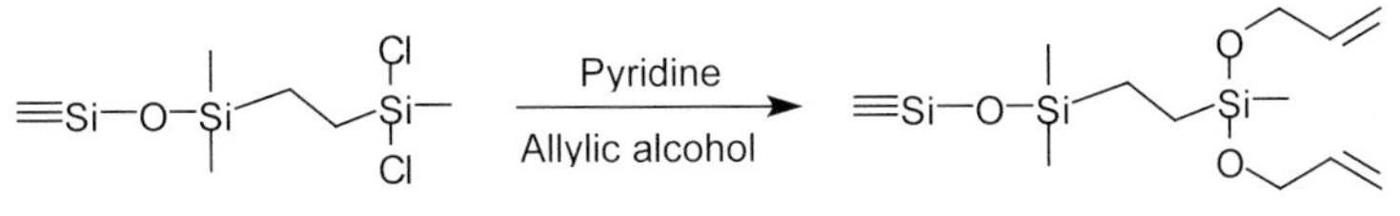

Pyridine
Allylic alcohol

G1-Cl **G1-OAllyl**

Scheme 3. Alcoholysis of **G1-Cl** with allylic alcohol.

The products were obtained in nearly quantitative yields as colorless viscous oils and dissolve readily in common organic solvents. All dendrimers have been fully characterized using ^{1}H-NMR, ^{13}C-NMR, ^{29}Si-NMR and IR spectroscopy. We found that the platinum-catalyzed hydrosilylation reaction gives the β-addition product with high regiospecificity. No other isomers were detected by ^{1}H-NMR and ^{13}C-NMR analysis.

Spectroscopic determinations of G*n* generations (*n* = 1–4) clearly indicate the presence of chloro and allyl groups, respectively, on the periphery of the dendrimers. The NMR and IR spectra reflect the transition from **Gn-Cl** to **Gn-Allyl**. For example the IR spectrum of **G2-Allyl** shows essentially complete disappearance of the Si–Cl peak (540 cm^{-1}) and the presence of a characteristic peak for the allylic double bond (1630 cm^{-1}) (Fig. 1).

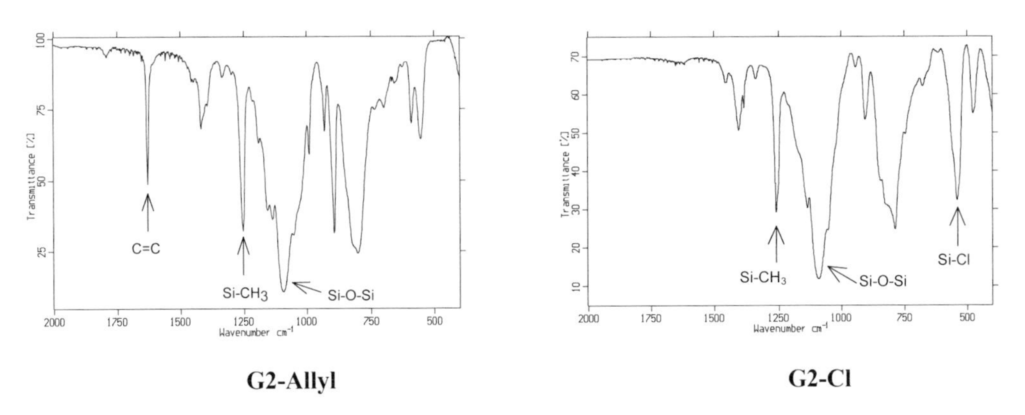

Fig.1. IR spectra of silsesquioxane dendrimers **G2-Allyl** and **G2-Cl**.

All substances possess NMR spectral patterns similar to one of the previously obtained analogues [4]. Upon conversion from chloro-substituted to allyl-substituted silsesquioxane dendrimers (**G1–G3**) the ^{1}H-NMR signal of the methyl groups ($Si(CH_3)Cl_2$) changes from δ = 0.75 ppm to δ = –0.04 ppm in the latter ($Si(CH_3)CH_2CH{=}CH_2$). The ^{1}H-NMR spectrum of the fourth generation of the chloro-substituted dendrimer **G4-Cl** still shows allylic signals that can be assigned to incomplete conversion.

Matrix-assisted laser desorption ionization time-of-flight (MALDI-TOF) mass spectrometry is the best method to find evidence of minor impurities. So whereas the MALDI-TOF spectra of **G1-Allyl** and **G2-Allyl** show the high purity of the samples, for the next-bigger branched compound **G3-Allyl** additional peaks were detected, probably corresponding to a defect structure. All MALDI-TOF analyses of the allyl-substituted dendrimers **G1-G3-Allyl** afforded the $[M+Na]^+$ or $[M+H]^+$ ion, or both (Table 1).

Table 1. Selected MALDI-TOF mass spectrum data [*m*/*z*] of G*n*-Allyl dendrimers.

Compound	M_W	Found	Calcd.	
G1-Allyl	2232.8	2257.3	2255.8	$[M+Na]^+$
		2233.7	2233.8	$[M+H]^+$
G2-Allyl	4256.9	4279.6	4279.9	$[M+Na]^+$
G3-Allyl	8297.7	8320.1	8320.7	$[M+Na]^+$

Ferrocenyl-Substituted Silsesquioxane Dendrimers

The pure ferrocenyl-substituted silsesquioxane dendrimers are synthesized by hydrosilylation reaction of allyl-substituted dendrimers, **G1-Allyl** and **G1-OAllyl**, with dimethylsilylferrocene in toluene (Scheme 4).

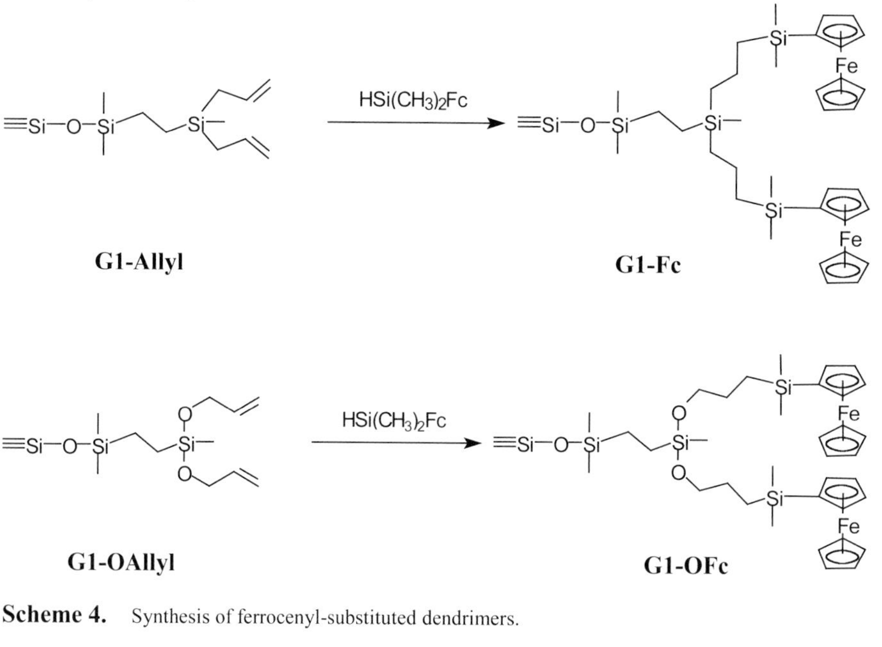

$HSi(CH_3)_2Fc$

G1-Allyl **G1-Fc**

$HSi(CH_3)_2Fc$

G1-OAllyl **G1-OFc**

Scheme 4. Synthesis of ferrocenyl-substituted dendrimers.

The products are viscous, orange-brown oils that are characterized by ^{1}H-, ^{13}C- and ^{29}Si-NMR and IR spectroscopy. The electrochemistry of the new organometallic dendrimers has also been examined. Figure 2 shows typical cyclic voltammograms of the ferrocene-terminated dendrimers **G1-Fc** and **G1-OFc**. They exhibit a single peak with a symmetrical wave shape that is characteristic of a reversible redox process. The $E_{1/2}$ values are 558 mV for **G1-Fc** and 559 mV for **G1-OFc** vs. decamethylferrocene. The fact that only a single redox process is observed implies simultaneous multi-electron transfer of all ferrocene centers at the same potential. This means that the sixteen ferrocene moieties are independent of each other. As proved by controlled potential coulometry the **G1-Fc** and **G1-OFc** complexes undergo a single-stepped $16e^-$ oxidation.

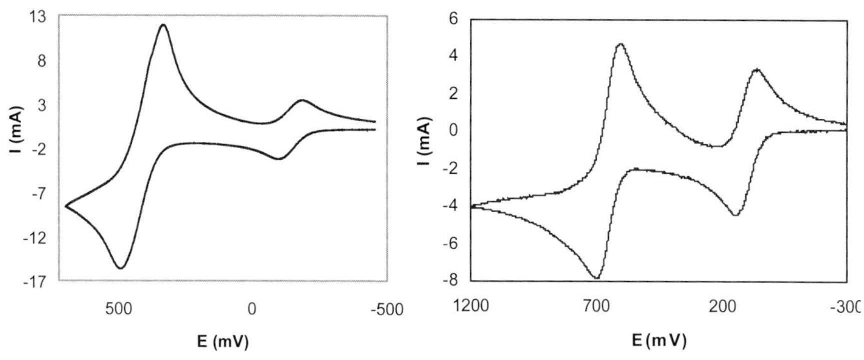

Fig.2. Left: Cyclic voltammogramm of **G1-Fc** (in 0.1 mol/L $[Bu_4N][PF_6]$ CH_2Cl_2 solution, half-wave potential $E_{½}$ = 558 mV vs. DMFc) Right: Cyclic voltammogramm of **G1-OFc** (0.1 mol/L $[Bu_4N][PF_6]$ in $CH_2Cl_2CH_3CN$ solution (4:1), half-wave potential $E_{½}$ = 559 mV vs. DMFc). The first oxidation wave belongs to decamethylferrocene that has been used as a reference in the electrochemical studies.

Acknowledgment: The MALDI-TOF spectra were recorded in co-operation with Dr. R. Krüger, Dr. G. Schulz and Dr. H. Much (*BAM, Berlin*). The cyclic voltammograms were recorded in co-operation with Prof. Dr. P. Zanello (*University of Siena, Italy*).

References:

[1] E. Buhlein, W. Wehner, F. Vögtle, *Synthesis* **1978**, 155.

[2] G. R. Newkome, C. N. Moorefield, F. Vögtle, *Dendritic Molecules*, VCH, Weinheim, **1996**.

[3] a) P. D. Hale, J. Inagaki, H. Karan, Y. Okamoto, T. A. Stotheim, *J. Am. Chem. Soc.* **1989**, *111*, 3482.
b) M. E. Wright, M. S. Sigman, *Macromolecules* **1992**, *25*, 6055.
c) A. Togni, G. Rihs, *Organometallics* **1993**, *12*, 3368.
d) F. Moulines, L. Djakovitch. R. Boese, B. Gloaguen, W. Thiel, J. L. Fillaut, M. H. Delville, D. Astruc, *Angew. Chem.* **1993**, *155*, 1132.
e) D. Astruc, *New J. Chem.* **1991**, *16*, 305.

[4] P. Jutzi, C. Batz, A. Mutluay, *Z. Naturforsch. Teil B* **1994**, *49*, 1689.

Octakis(Dimethylphosphinoethyl)octasila-sesquioxane: Synthesis, Characterization and Reactivity

Sabine Lücke, Kai Lütke-Brochtrup, Karl Stoppek-Langner

Anorganisch-Chemisches Institut, Westfälische Wilhelms-Universität
Wilhelm Klemm-Str. 8, D-48149 Münster, Germany
Tel.: Int. code + (251)833 6098 — Fax: Int. code + (251)833 6098
E-mail: slucke@uni-muenster.de

Keywords: Silasesquioxanes / Complex Ligands / σ-Donor

Summary: The addition of $HPMe_2$ to the vinyl-T_8 silasesqiuoxane quantitatively yields the novel octakis(dimethylphosphinoethyl)octasilasesquioxane **1** which can be converted quantitatively into the thio-derivative **2** via mild oxidation with sulfur in CS_2. Reaction with $W(CO)_5THF$ gives the octatungsten derivative **3** as an interesting example for supermolecular transition metal complexes.

Introduction

Silasesquioxanes T_n (n even and ≥ 8) with polyhedral Si–O arrangements represent a class of molecular compounds showing a variety of useful chemical and physical properties. Of particular interest are the octasilasesquioxanes T_8 which, e.g., exhibit structural analogies to the α-cage of zeolites or can serve as model systems to mimic the surface reactivity of derivatized silicas [1]. On the other hand, T_8 silasesquioxanes are a rather new class of multidentate complex ligands, provided that terminal functional groups with σ-donor properties are introduced into the molecule. This contribution reports on the preparation and characterization of the octakis(dimethyl-phosphinoethyl)octasilasesquioxane **1** and the reaction of this new polyhedral siloxane with sulfur and $W(CO)_5THF$, respectively. The IR spectra of the products show the characteristic strong band of an Si–O–Si asymmetric stretching vibration ($\nu_{Si-O-Si} = 1120\ cm^{-1}$) in all cases.

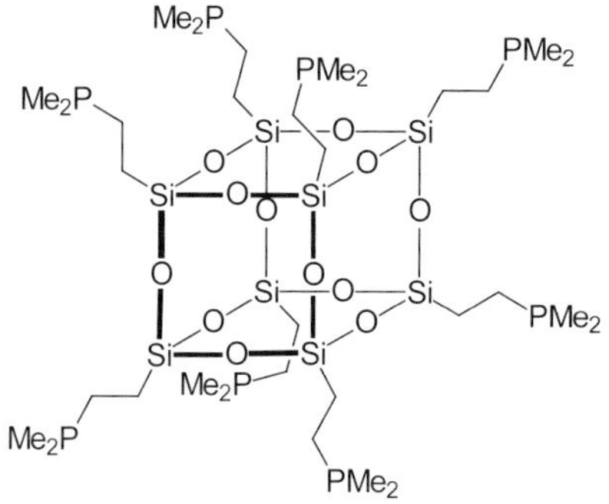

1

Experimental

The first experimental step is the formation of the known vinyl-T_8 cage using the $FeCl_3$-catalyzed hydrolytic polycondensation of vinyltrichlorosilane in a biphasic solvent system (Scheme 1) [2]. The cubic silasesquioxane product is obtained in the form of a white powder, easily soluble in pentane. The X-ray structural data have been reported elsewhere [3].

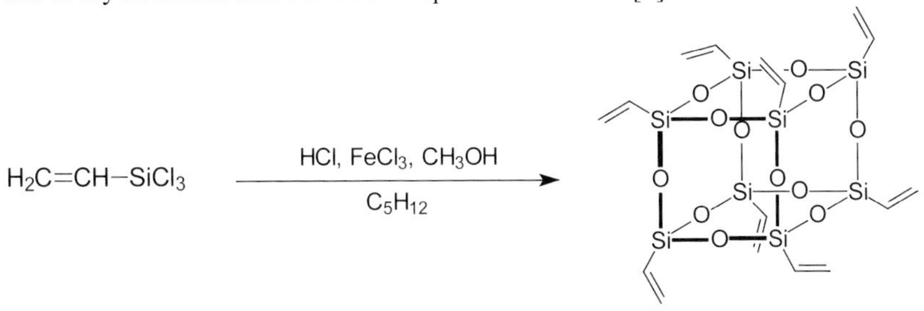

$H_2C{=}CH{-}SiCl_3 \xrightarrow[C_5H_{12}]{HCl,\ FeCl_3,\ CH_3OH}$

Scheme 1. Synthesis of the vinyloctasilasesquioxane.

The introduction of the Me_2P fragment was carried out via addition of dimethylphosphane, prepared by hydrolytic cleavage of the $Me_4P_2S_2$ in the presence of $P(n\text{Bu})_3$, to the vinyl-T_8, leading quantitatively to the silasesquioxane **1** (Scheme 1).

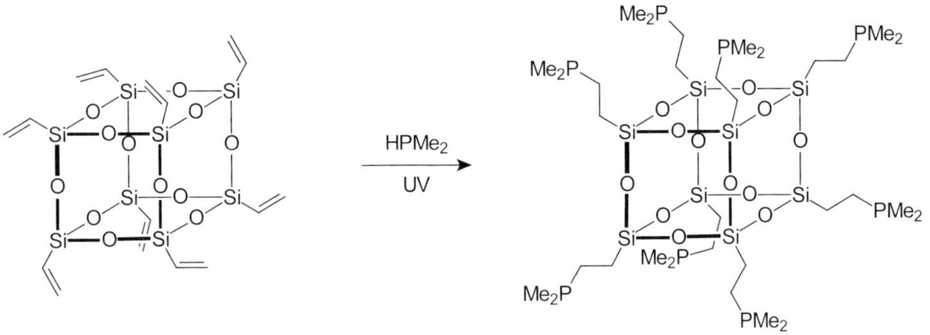

$\xrightarrow[UV]{HPMe_2}$ **1**

Scheme 2. Synthesis of the octakis(dimethylphosphinoethyl)-octasilasesquioxane **1**.

Due to the microcrystalline character of **1** and the high sensitivity of this molecular siloxane to oxygen, no crystals suitable for X-ray structure analysis could be obtained so far; the exclusive formation of **1** , however, has been ascertained by spectroscopic investigation. NMR : δ_H = 0.6–0.9 ppm (m, spinsystem AA′BB′, br, 2H; $Me_2PCH_2C\underline{H}_2$), 1.0 ppm (d, $^2J_{PH}$ = 1.2 Hz, 6H, $P(C\underline{H}_3)_2$), 1.2 - 1.4 ppm (m, br., 2 H, $Me_2PC\underline{H}_2CH_2$), δ_C = 7,1 ppm ($^2J_{PC}$ = 9.7 Hz; $Me_2PCH_2\underline{C}H_2$), 13.5 ppm ($^1J_{PC}$ = 13.5 Hz; $P(\underline{C}H_3)_2$), 24.1 ppm ($^1J_{PC}$ = 10.7 Hz; $Me_2P\underline{C}H_2CH_2$); δ_P = –45.5 ppm.

In order to obtain a more stable derivative of **1**, the reaction of the octakis(dimethylphophinoethyl)octasilasesquioxane with sulfur in CS_2 (Scheme 3) was carried out. Using this reaction, the new silasesquioxane **2** can be obtained quantitatively in form of a red microcrystalline precipitate. Spectroscopic data of **2**: NMR: δ_H = 1.7 ppm (d, $^2J_{PH}$ = 12.7 Hz; $P(C\underline{H}_3)_2$), δ_C =3.7 ppm ($Me_2PCH_2\underline{C}H_2$), 20.0 ppm ($^1J_{PC}$ = 54 Hz; $P(\underline{C}H_3)_2$), 28.3 ppm ($^1J_{PC}$ = 55 Hz; $Me_2P\underline{C}H_2CH_2$); δ_P = 40.9 ppm.

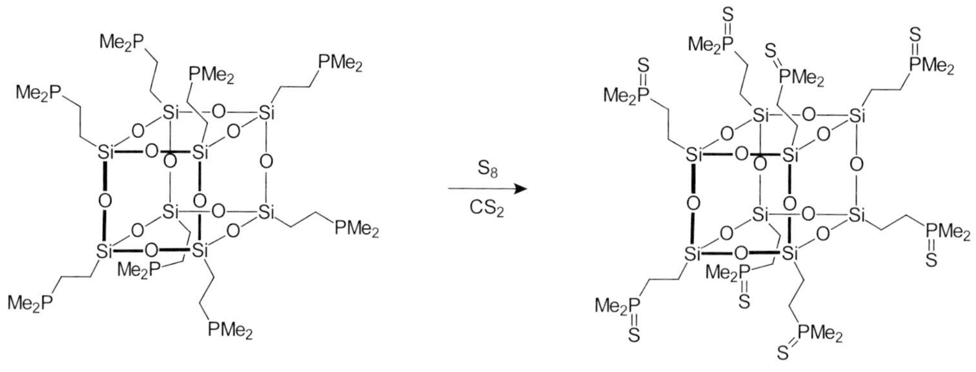

2

Scheme 3. Synthesis of the octakis(dimethyl(thio)phosphinoethyl)octasilasesquioxane **2**.

The ligand properties of **1** were investigated utilizing complex formation with $W(CO)_5THF$ generated from the tungsten hexacarbonyl $W(CO)_6$ (Scheme 4) in a falling-film photoreactor. The crude product is a pale yellow and microcrystalline material, soluble in $CDCl_3$. Crystals suitable for X-ray structural studies have not been obtained so far.

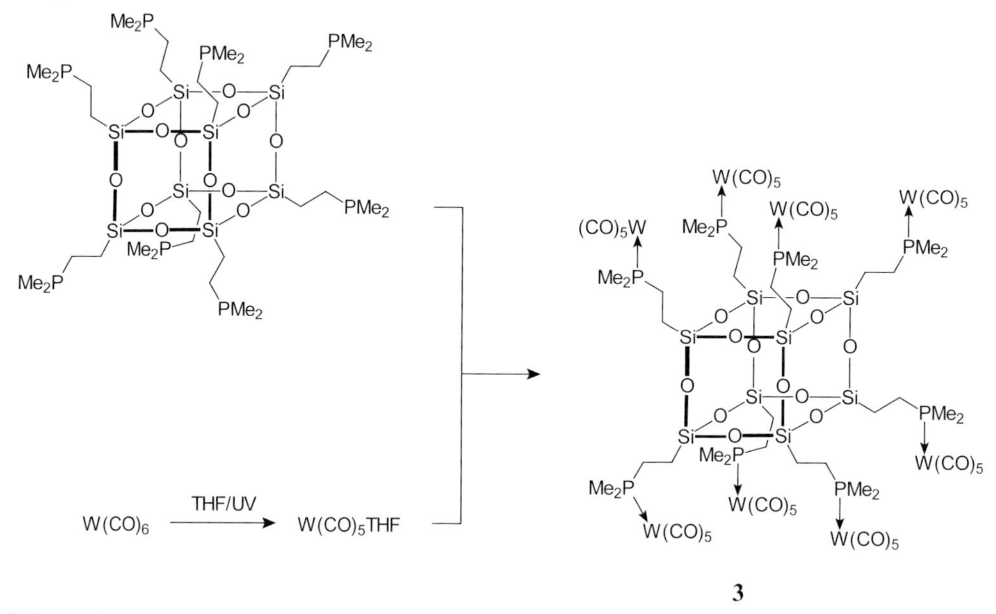

Scheme 4. Synthesis of the octakis[tungsten(pentacarbonyle)dimethylphosphinoethyl]octasilasesquioxane **3**.

According to the NMR spectroscopic data, no uncoordinated PMe_2 groups (δ = –45.5 ppm) remain in the reaction product. NMR: δ_H = 1.1 ppm ($^2J_{PH}$ = 5.5 Hz; $P(C\underline{H}_3)_2$); δ_P = –20.0 ppm ($^1J_{WP}$ = 250 Hz); δ_C = 17.3 ppm ($Me_2PCH_2\underline{C}H_2$), 25.6 ppm ($P(\underline{C}H_3)_2$), 29.7 ppm ($Me_2P\underline{C}H_2CH_2$).

Furthermore, the vibrational spectra indicate that eight equivalent $W(CO)_5$ fragments are attached to the cubic Si–O framework. This conclusion is evident from group theoretical considerations leading to three distinct vibrational modes in the corresponding spectra (local symmetry C_{4v}, races 2 A_1 + E).

Conclusion

The UV-catalyzed reaction of Me_2PH with the vinyl-T_8 silasesquioxane yields quantitatively the novel octakis(dimethylphosphinoethyl)octasilasesquioxane as the first reactive siloxane-based molecule with polydental σ-donor properties emphasized by the easily performed conversion into the thio derivative or the formation of an octatungsten complex with $W(CO)_5{\cdot}THF$. Future work, therefore, has to focus on both the introduction of other phosphane ligands PR_2 into the Si–O cage and the reaction with suitable transition metal fragments.

References

[1] a) S. Lücke, K. Stoppek-Langner, M. Läge, B. Krebs, *Z. Anorg. Allg. Chem.* **1997**, *623*, 1243–1246.
b) U. Dittmar, B. J. Henden, U. Flörke, C. Marsmann, *J. Organomet. Chem.* **1989**, *373*, 153–163.

[2] P. A. Agaskar, *Inorg. Chem.* **1991**, *30*, 2707.

[3] M. G. Voronkov, T. N. Martynova, R. G. Mirskov, V. I. Belyl, *Zh. Obshch. Khim.* **1979**, *49*, 1522.

Synthesis of Homo- and Mixed-Functionalized Octa-, Deca- and Dodeca-Silsesquioxanes by Cage Rearrangement and their Characterization

Eckhard Rikowski

Old address:
Universität-GH Paderborn
Anorganische und Analytische Chemie
Warburger Straße 100
D-33098 Paderborn, Germany

New address (since March 1998):
GKSS-Forschungszentrum
Institut für Chemie
Max-Planck-Straße
D-21502 Geesthacht, Germany
Tel.: Int. code + (4152)872467
E-mail: eckhard.rikowski@gkss.de

Dedicated to Professor Marsmann on the occasion of his 60th birthday

Keywords: Silsesquioxanes / Cage Rearrangement / NP-HPLC / ^{29}Si-NMR

Summary: The homo- and mixed-functionalized polyhedral silsesquioxanes (n-C_3H_7)$_8$($SiO_{1.5}$)$_8$, (n-C_3H_7)$_{10}$($SiO_{1.5}$)$_{10}$, (Cl–C_3H_6)(n-C_3H_7)$_7$(($SiO_{1.5}$)$_8$, (Cl–C_3H_6)-(n-C_3H_7)$_9$($SiO_{1.5}$)$_{10}$, 1,2-(Cl–C_3H_6)$_2$(n-C_3H_7)$_6$($SiO_{1.5}$)$_8$, 1,3-(Cl–C_3H_6)$_2$(n-C_3H_7)$_6$-($SiO_{1.5}$)$_8$, 1,7-(Cl–C_3H_6)$_2$(n-C_3H_7)$_6$($SiO_{1.5}$)$_8$, 1,2,7-(Cl–C_3H_6)$_3$(n-C_3H_7)$_5$($SiO_{1.5}$)$_8$, 1,3,6-(Cl–C_3H_6)$_3$(n-C_3H_7)$_5$($SiO_{1.5}$)$_8$ and 1,2,3-(Cl–C_3H_6)$_3$(n-C_3H_7)$_5$($SiO_{1.5}$)$_8$ can be synthesized by simultaneous cage rearrangement of (n-C_3H_7)$_8$($SiO_{1.5}$)$_8$ and (Cl–C_3H_6)$_8$($SiO_{1.5}$)$_8$ in a ratio of 7:1. The cage rearrangement mixture can be separated by NP-HPLC. Cage rearrangement of a mixture of (n-C_3H_7)$_8$($SiO_{1.5}$)$_8$ and (Cl–C_3H_6)$_8$($SiO_{1.5}$)$_8$ in a ratio of 1:7 leads to a mixture of the homo-functionalized silsesquioxanes (Cl–C_3H_6)$_n$($SiO_{1.5}$)$_n$ with n = 8, 10, 12 and the mixed-functionalized silsesquioxane (Cl–C_3H_6)$_9$(n-C_3H_7)($SiO_{1.5}$)$_{10}$. This mixture can also be separated by NP-HPLC. The new mixed-functionalized silsesquioxanes (Cl–C_3H_6)(n-C_3H_7)$_9$-($SiO_{1.5}$)$_{10}$ and (Cl–C_3H_6)$_9$(n-C_3H_7)($SiO_{1.5}$)$_{10}$ show inverse ^{29}Si-NMR spectra. These decasilsesquioxanes are also characterized by their ^{29}Si-2D-INEPT-INADEQUATE-NMR spectra.

Introduction

Polyhedral silsesquioxanes $R_n(SiO_{1.5})_n$ (with R = H, halogen, organic residue and n = 6, 8, 10, 12, 14, 16, 18, 20) present themselves as models for surface-modified silica gels. Most of the known polyhedral silsesquioxanes are octasilsesquioxanes $R_8(SiO_{1.5})_8$. Only a few silsesquioxanes of other cage sizes have been synthesized so far, compared to the large number of octasilsesquioxanes. A new method to prepare deca- and dodecasilsesquioxanes in a systematic strategy was found by the cage rearrangement of octasilsesquioxanes [1,2].

However, the greatest number of silsesquioxanes are homo-functionalized, which means that all residues R at the silsesquioxane cage $R_n(SiO_{1.5})_n$ are of the same type. In mixed-functionalized silsesquioxanes $R_xR'_y(SiO_{1.5})_n$ (with $x + y = n$), R and R' are different. Compared to the number of homo-functionalized silsesquioxanes mixed-functionalized silsesquioxanes are rather rare. With the exception of $(C_6H_5)H_9(SiO_{1.5})_{10}$ [3] all other mixed-functionalized silsesquioxanes show the octa-cage size. Two new mixed-functionalized silsesquioxanes of the deca-cage type described in this article can also be synthesized by cage rearrangement.

Results

Cage rearrangement of silsesquioxanes is an interesting method of synthesizing homo-functionalized silsesquioxanes (see Eq. 1.).

$$a\,(RSiO_{1.5})_8 \longrightarrow b\,(RSiO_{1.5})_8 + c\,(RSiO_{1.5})_{10} + d\,(RSiO_{1.5})$$

Eq. 1. Cage rearrangement of silsesquioxanes.

Simultaneous rearrangement of the two silsesquioxanes $(Cl–C_3H_6)_8(SiO_{1.5})_8$ and $(NCS–C_3H_6)_8(SiO_{1.5})_8$ in a ratio of 1:1 leads to a very complicated mixture of octa-, deca- and dodecasilsesquioxanes with a statistical distribution of the residues $Cl–C_3H_6$ and $NCS–C_3H_6$ [2]. The separation of this mixture by NP-HPLC has not been successful so far.

However, simultaneous rearrangement of the two silsesquioxanes $(Cl–C_3H_6)_8(SiO_{1.5})_8$ and $(n\text{-}C_3H_7)_8(SiO_{1.5})_8$ in the ratio of 1:7 or 7:1 leads to complicated mixtures of octa-, deca- and dodecasilsesquioxanes. But here it is possible to separate these mixtures using NP-HPLC (see Table 1 and Fig. 1).

Table 1. Retention times of the mixture of the rearrangement of octa[(3-chloropropyl)-silsesquioxane] and octa(*n*-propyl)silsesquioxane in a ratio of 7:1.

Silsesquioxane	Retention time [min]
$(Cl–C_3H_6)_8(SiO_{1.5})_8$	7.5
$(Cl–C_3H_6)_9(n\text{-}C_3H_7)(SiO_{1.5})_{10}$	11.9
$(Cl–C_3H_6)_{10}(SiO_{1.5})_{10}$	12.0
$(Cl–C_3H_6)_{12}(SiO_{1.5})_{12}$	18.5

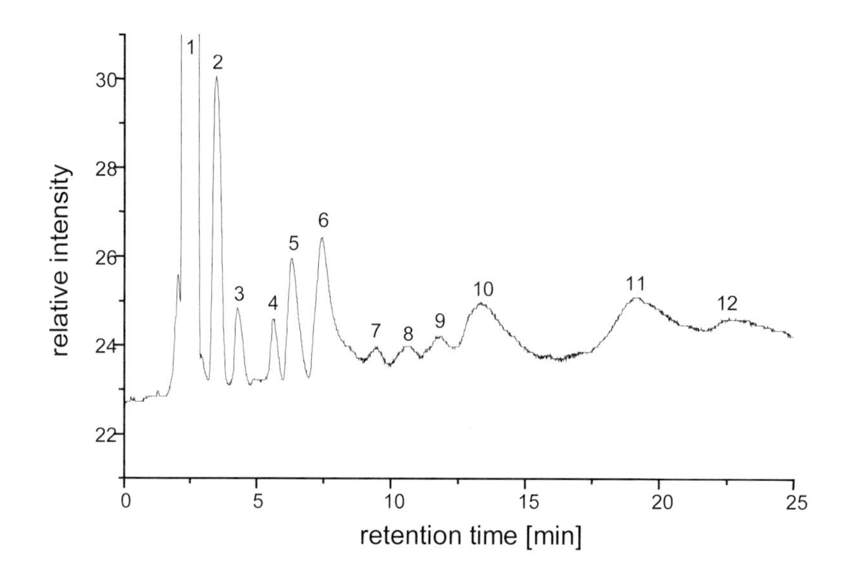

Fig. 1. Separation of the mixture of the rearrangement of octa[(3-chloropropyl)silsesquioxane] and octa(n-propyl)-silsesquioxane in a ratio of 1:7. Column : Merck Hibar; 250-25 Si60 (5μm), eluent : n-pentane, flow rate: 25 mL/min. Peaks: 1: $(n\text{-}C_3H_7)_8(SiO_{1.5})_8$, $(n\text{-}C_3H_7)_{10}(SiO_{1.5})_{10}$, 2: $(Cl{-}C_3H_6)(n\text{-}C_3H_7)_7(SiO_{1.5})_8$, 3: $(Cl{-}C_3H_6)(n\text{-}C_3H_7)_9(SiO_{1.5})_{10}$, 4: p-$(Cl{-}C_3H_6)_2(n\text{-}C_3H_7)_6(SiO_{1.5})_8$, 5: m-$(Cl{-}C_3H_6)_2(n\text{-}C_3H_7)_6(SiO_{1.5})_8$, 6: o-$(Cl{-}C_3H_6)_2(n\text{-}C_3H_7)_6(SiO_{1.5})_8$, 7–12: higher silsesquioxanes.

The homo-functionalized silsesquioxanes $(n\text{-}C_3H_7)_8(SiO_{1.5})_8$, $(n\text{-}C_3H_7)_{10}(SiO_{1.5})_{10}$, $(Cl{-}C_3H_6)_8(SiO_{1.5})_8$, $(Cl{-}C_3H_6)_{10}(SiO_{1.5})_{10}$ and D_{2d}-$(Cl{-}C_3H_6)_{12}(SiO_{1.5})_{12}$ presented in Fig. 1 and Table 1 can also be obtained by the classical cage rearrangement of silsesquioxanes described in [1, 2].

The mixed-functionalized octasilsesquioxanes 1,2-$(Cl{-}C_3H_6)_2(n\text{-}C_3H_7)_6(SiO_{1.5})_8$, 1,3-$(Cl{-}C_3H_6)_2(n\text{-}C_3H_7)_6(SiO_{1.5})_8$, 1,7-$(Cl{-}C_3H_6)_2(n\text{-}C_3H_7)_6(SiO_{1.5})_8$, 1,2,7-$(Cl{-}C_3H_6)_3(n\text{-}C_3H_7)_5$ $(SiO_{1.5})_8$, 1,3,6-$(Cl{-}C_3H_6)_3(n\text{-}C_3H_7)_5(SiO_{1.5})_8$ and 1,2,3-$(Cl{-}C_3H_6)_3(n\text{-}C_3H_7)_5(SiO_{1.5})_8$ (see Fig. 1) have already been prepared by cohydrolysis of $(Cl{-}C_3H_6)SiCl_3$ and $(n\text{-}C_3H_7)SiCl_3$ in a ratio of 1:7 [4, 5]. In this regard, cage rearrangement of silsesquioxanes is an additional method of preparing mixed-functionalized octasilsesquioxanes.

In Fig. 2 the structures and ^{29}Si-NMR spectra of the new mixed-functionalized decasilsesquioxanes $[Cl{-}(CH_2)_3)]_9[CH_3(CH_2)_2)](SiO_{1.5})_{10}$ and $[Cl{-}(CH_2)_3)][CH_3(CH_2)_2)]_9(SiO_{1.5})_{10}$ are presented. Only one other mixed-functionalized decasilsesquioxane has been described so far [3].

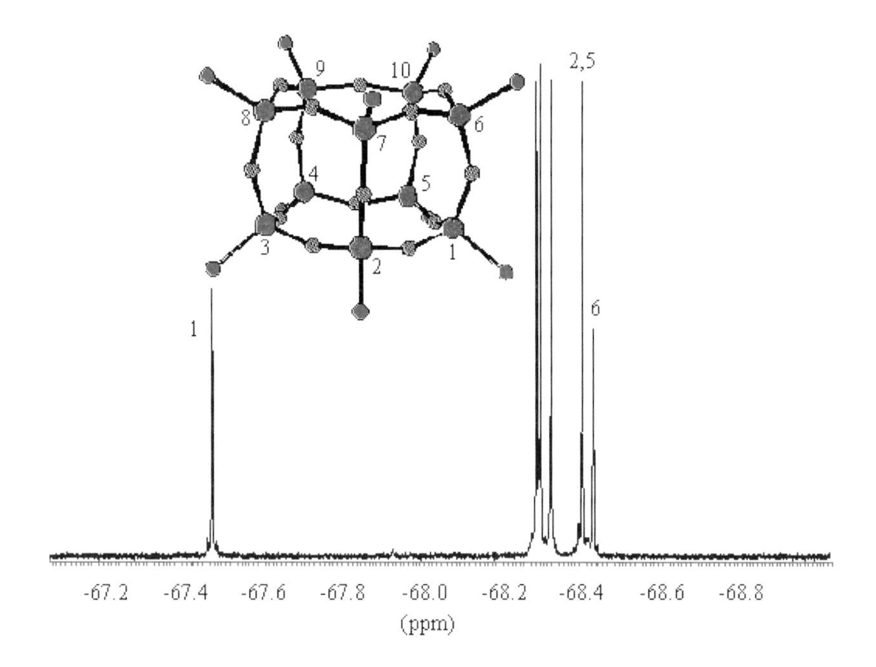

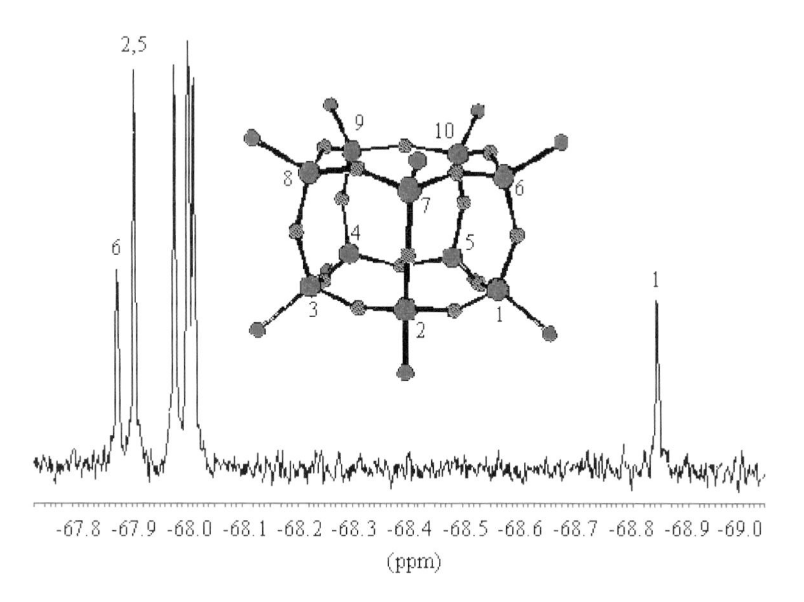

Fig. 2. ^{29}Si-NMR spectra of $[Cl{-}(CH_2)_3)]_9[CH_3(CH_2)_2)](SiO_{1.5})_{10}$ and $[Cl{-}(CH_2)_3)][CH_3(CH_2)_2)]_9(SiO_{1.5})_{10}$.

The substitution pattern is recognizable by the ^{29}Si-2D-INEPT-INADEQUATE-NMR spectra (Fig. 3).

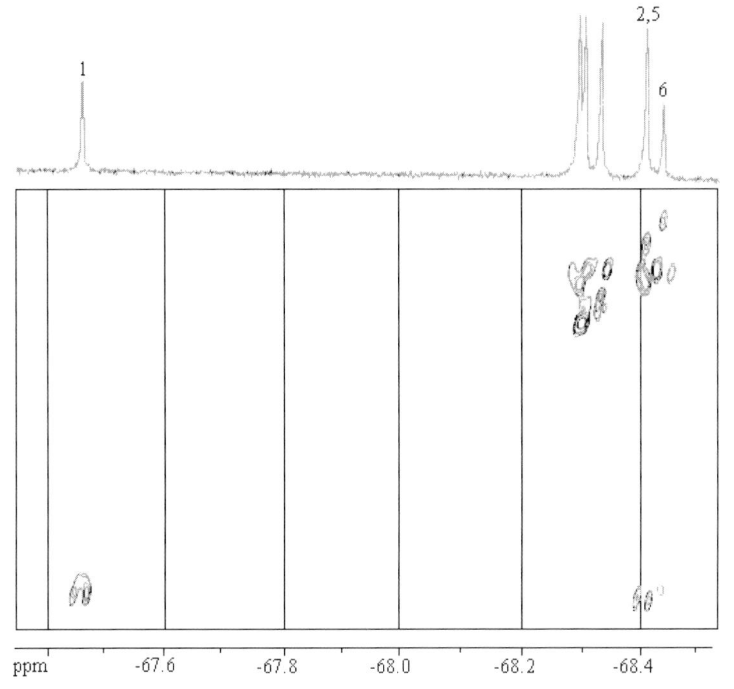

Fig. 3. ^{29}Si-2D-INEPT-INADEQUATE-NMR spectrum of $[Cl–(CH_2)_3)]_9[CH_3(CH_2)_2)](SiO_{1.5})_{10}$.

Acknowledgment: I thank Prof. Dr. Marsmann for the opportunities of free research during the time I was working on my PhD–thesis at the *Universität-GH Paderborn*. Some of the results are presented here. I thank him for many NMR spectra, the *DFG* for financial support and *Hüls AG* for a gift of chemicals.

References:

[1] H. C. Marsmann, U. Dittmar, E. Rikowski, "*New Functionalized Silsesquioxanes by Substitutions and Cage Rearrangement of Octa[(3-chloropropyl)silsesquioxane]*", in: *Organosilicon Chemistry II: From Molecules to Materials* (Eds.: N. Auner, J. Weis), VCH, Weinheim, **1996**, p. 691.

[2] E. Rikowski, H. C. Marsmann, *Polyhedron* **1997**, *16*, 3357.

[3] G. Calzaferri, C. Marcolli, R. Imhof, K. W. Törnroos, *J. Chem. Soc., Dalton Trans.* **1996**, 3313.

[4] B. J. Hendan, H. C. Marsmann, *J. Organomet. Chem.* **1994**, *483*, 33.

[5] B. J. Hendan, H. C. Marsmann, "*Silsesquioxanes of Mixed Functionality — Octa[(3-chloropropyl)-n-propylsilsesquioxanes] and Octa[(3-mercaptopropyl)-n-propyl-silsesquioxanes] as Models of Organomodified Silica Surfaces*", in: *Organosilicon Chemistry II: From Molecules to Materials* (Eds.: N. Auner, J. Weis), VCH, Weinheim, **1996**, p 685.

Synthesis of Polyhedral Hexa-, Octa-, Deca- and Dodeca-Silsesquioxanes Separation by NP-HPLC, SEC and LAC Characterization by MALDI-TOF-MS

Ralph-Peter Krüger, Helmut Much, Günter Schulz

Bundesanstalt für Materialforschung und –prüfung
Unter den Eichen 87, D-12205 Berlin, Germany
Tel.: Int. code + (30)8104 3373 — Fax: int. code + (30)8104 1637
E-mail: ralph.krueger@bam.de

Eckhard Rikowski

GKSS-Forschungszentrum
Institut für Chemie
Max-Planck-Straße, D-21502 Geesthacht, Germany
Tel.: Int. code + (4152)/872467
E-mail: eckhard.rikowski@gkss.de

Dedicated to Professor Heinrich C. Marsmann on the occasion of his 60th birthday

Keywords: Silsesquioxanes / NP-HPLC / LAC / SEC / MALDI-TOF-MS

Summary: Polyhedral silsesquioxanes $R_n(SiO_{1.5})_n$ with n = 6, 8, 10, 12 and R = CH_3, C_2H_5, n-C_3H_7, n-C_4H_9, i-C_4H_9, n-C_5H_{11}, n-C_6H_{13}, n-C_7H_{15}, n-C_8H_{17}, n-C_9H_{19}, n-$C_{10}H_{21}$, C_6H_5, Cl–C_3H_6, Br–C_3H_6, I–C_3H_6, NCS–C_3H_6, C_6F_5–C_3H_6 and Me_3SiO can be synthesized by hydrolysis of $RSiCl_3$ or $RSi(OMe)_3$, modification of existing silsesquioxanes or cage rearrangement of octasilsesquioxanes to the greater deca- and dodecasilsesquioxanes. Some of the mixtures of silsesquioxanes produced by cage rearrangements can be separated by NP-HPLC, SEC (size exclusion chromatography) and LAC (liquid affinity chromatography). These mixtures can also be analyzed by MALDI-TOF-MS using a special matrix.

Introduction

Polyhedral silsesquioxanes present themselves as models for surface-modified silica gels [1]. With the exception of some special cases octasilsesquioxanes normally are obtained by hydrolysis of trichlorosilanes or trialkoxysilanes. Cage rearrangement of octasilsesquioxanes is a new method to prepare mixtures of octa-, deca- and D_{2d}-dodecasilsesquioxanes [2, 3]. These mixtures have to be separated to analyze the single compounds. A new possibility for analyzing silsesquioxanes or silsesquioxane mixtures is given by MALDI-TOF-MS.

Results

The silsesquioxanes or silsesquioxane mixtures analyzed by NP-HPLC, SEC, LAC and MALDI-TOF-MS can be obtained by the following methods: hydrolysis of trichloro- or trialkoxysilanes (Eq. 1), hydrosilylation of octa(hydridosilsesquioxanes) (Eq. 2), modification of existing silsesquioxanes (Eq. 3), trimethylsilylation of tetraethylammonium silicates [4] (Eq. 4) and trimethylsiloxylation of hydridosilsesquioxanes [5] (eq. 5).

$$8\ RSiX_3 + 12\ H_2O \xrightarrow{H_2PtCl_6} (RSiO_{1.5})_8 + 24\ HX$$

R = H-, ethyl-, propyl-, *n*-butyl-, *iso*-butyl-, 3-chloropropyl-; X = Cl, OMe, OEt

Eq. 1. Synthesis of octasilsesquioxanes by hydrolysis of trichloro- or trialkoxysilanes.

$$H_8(SiO_{1.5})_8 + 8\ H_2C{=}CHR \xrightarrow{H_2PtCl_6} (RCH_2CH_2)_8(SiO_{1.5})_8$$

alkenes: 1-pentene, 1-heptene, 1-octene, 1-nonene, 1-decene, allylpentafluorobenzene

Eq. 2. Synthesis of octa-silsesquioxanes by hydrosilylation.

$$(Cl\text{-}C_3H_6)_8(SiO_{1.5})_8 + 8\ X^- \longrightarrow (X\text{-}C_3H_6)_8(SiO_{1.5})_8 + 8\ Cl^- \qquad \text{with } X = Br, I, NCS$$

Eq. 3. Synthesis of octasilsesquioxanes by modification of existing silsesquioxanes.

$$6\ Si(OC_2H_5)_4 + 6\ N(C_2H_5)_4OH + 9\ H_2O \longrightarrow [N(C_2H_5)_4]_6Si_6O_{15} + 24\ C_2H_5OH$$

$$[N(C_2H_5)_4]_6Si_6O_{15} + 6\ (CH_3)_3SiCl \xrightarrow{H_2PtCl_6} [(CH_3)_3SiO]_6(SiO_{1.5})_6 + 6\ N(C_2H_5)_4Cl$$

Eq. 4. Preparation of $(Me_3SiO)_6(SiO_{1.5})_6$.

$$(CH_3)_3SiCl + (CH_3)_3NO \xrightarrow{H_2PtCl_6} Cl(CH_3)_3Si \cdot ON(CH_3)_3$$

$$H_n(SiO_{1.5})_n + 2n\ Cl(CH_3)_3Si \cdot ON(CH_3)_3 \xrightarrow{H_2PtCl_6} [(CH_3)_3SiO]_n(SiO_{1.5})_n + n\ (CH_3)_3NO \cdot HCl + n\ (CH_3)_3SiCl + n\ (CH_3)_3N$$

Eq. 5. Synthesis of $(Me_3SiO)_n(SiO_{1.5})_n$ with $n = 8, 10, 12$.

Mixtures of octa-, deca-, and D_{2d}-dodecasilsesquioxanes are obtained by cage rearrangements of octasilsesquioxanes (Eq. 6) [3].

$$a\ (RSiO_{1.5})_8 \longrightarrow b\ (RSiO_{1.5})_8 + c\ (RSiO_{1.5})_{10} + d\ (RSiO_{1.5})_{12}$$

Eq. 6. Synthesis of octa-, deca- and D_{2d}-dodecasilsesquioxanes by cage rearrangement.

These mixtures have to be separated to characterize the single compounds. NP-HPLC [3], SEC, LAC and MALDI-TOF-MS have been tested as methods of analyzing silsesquioxane mixtures [6, 7].

NP-HPLC has shown good results in separating functionalized silsesquioxanes (see Table 1). Separation occurs by interaction of functional groups of the silsesquioxane with the phase material. An increasing number of functional groups leads to increased retention times.

Table 1. Functionalized silsesquioxanes $[X\text{-}(CH_2)_3]_n(SiO_{1.5})_n$ separated by NP-HPLC[a].

	Retention time [min]		
X	***n* = 8**	***n* = 10**	***n* = 12**
Cl	7.5	12.0	18.5
Br	7.2	11.5	17.2
I	6.0	10.5	–
NCS	6.8	10.9	18.0
C_6F_5	7.4	11.8	–

[a] column: Merck Hibar 250-25 Si60 (5 µm), eluent: *n*-hexane/$CHCl_3$ = 1:1 (v/v), flow rate: 25 mL/min.

Samples were analyzed using an SEC system consisting of a Hewlett Packard liquid chromatograph 1090 and a Wyatt interferometric refractometer detector. The mobile phase was toluene pumped at a flow rate of 1 mL/min. 100 µL of sample solution (5 mg/mL) were injected. The stationary phase consisted of 3 columns (300 mm × 8 mm ID) filled with PSS SDV gel (10^5, 10^3, 10^2 Å; 5 µm). The software of PSS Mainz (WIN GPC) was used for data acquisition.

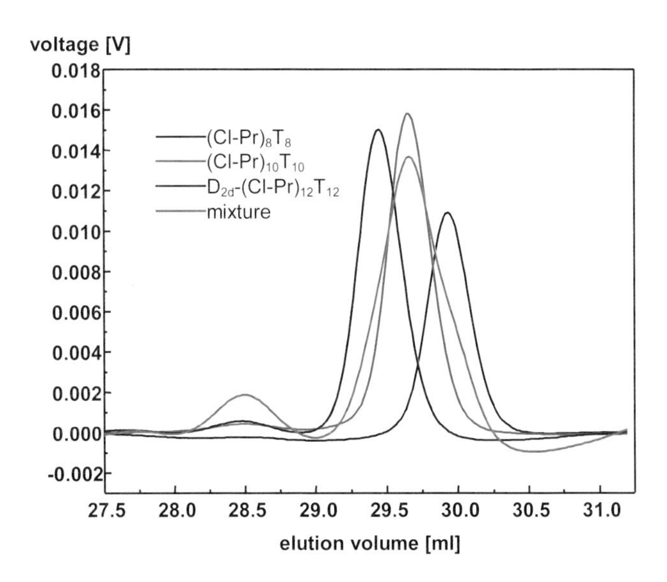

Fig. 1. Silsesquioxanes T_8-, T_{10}-, T_{12}-polyhedra (pure compounds) and mixture of these silsesquioxanes in the SEC-mode: no separation of the mixture.

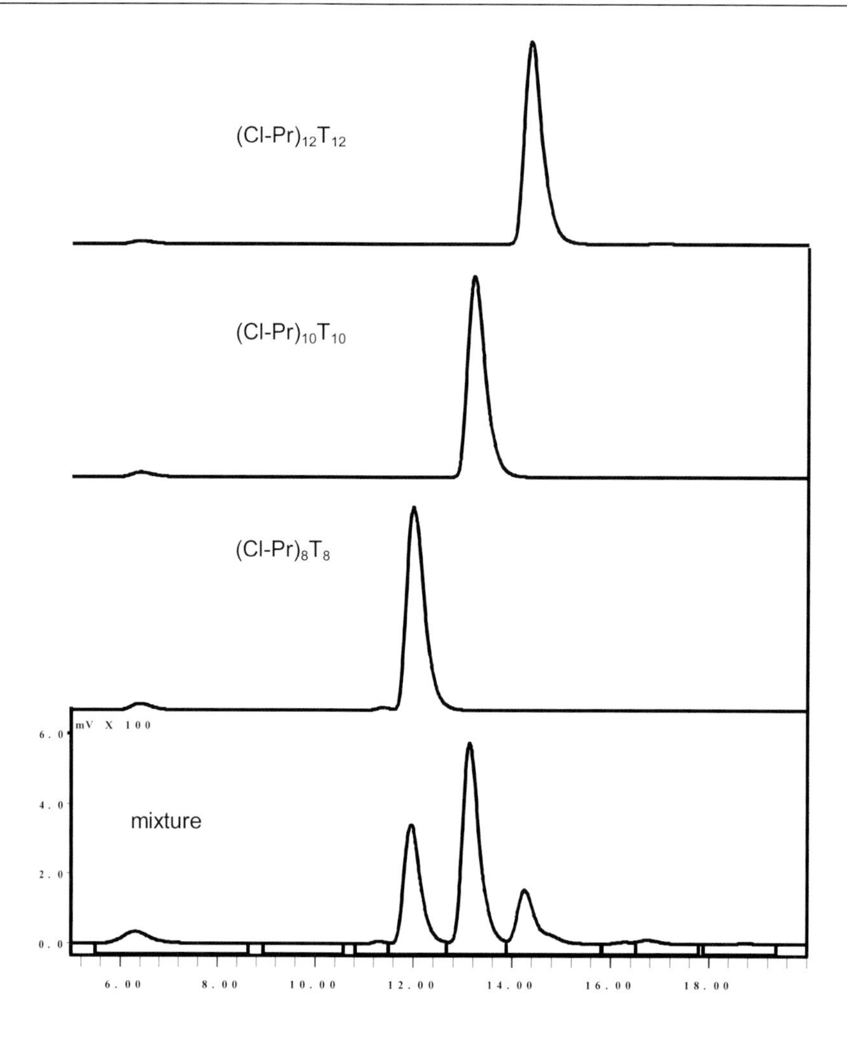

Fig. 2. Separation of silsesquioxanes by LAC. Column: Machery-Nagel; 1. Nucleosil 100 Å (5μm), 2. Nucleosil 300 Å (5μm); flow rate: 0,5 mL/min; eluent: 70 vol % toluene, 30 vol % *iso*-octane, $T = 30°C$.

The silsesquioxane mixtures or the separated silsesquioxanes can be analyzed by MALDI-TOF-MS (Table 2). MALDI-TOF mass spectrometry was carried out on a Kratos Kompact MALDI III. 0.5 μL of a solution (25 mg/mL) of 2,4,6-trihydroxyacetophenone or 20 mg/mL of 2-nitrophenyl octyl ether and 10 mg/mL silver trifluoroacetate were mixed on a stainless steel sample slide. The solvent was evaporated in a stream of air at ambient temperature. Bovine insulin was used for calibration. Conditions for the measurements: polarity positive, flight path reflection, 20 kV acceleration voltage, nitrogen laser ($\lambda = 337$ nm).

Table 2. MALDI-TOF-MS data of silsesquioxanes.

Silsesquioxane	Mass peak of the Ag-adduct in MALDI-TOF-MS
$(CH_3)_8(SiO_{1.5})_8$	644.5
$(C_2H_5)_8(SiO_{1.5})_8$	755.6
$(n\text{-}C_3H_7)_8(SiO_{1.5})_8$	867.8
$(n\text{-}C_3H_7)_{10}(SiO_{1.5})_{10}$	1058.1
$(n\text{-}C_4H_9)_8(SiO_{1.5})_8$	979.7
$(i\text{-}C_4H_9)_8(SiO_{1.5})_8$	980.4
$(n\text{-}C_5H_{11})_8(SiO_{1.5})_8$	1091.8
$(n\text{-}C_6H_{13})_8(SiO_{1.5})_8$	1204.3
$(n\text{-}C_7H_{15})_8(SiO_{1.5})_8$	1316.6
$(n\text{-}C_8H_{17})_8(SiO_{1.5})_8$	1429.4
$(n\text{-}C_8H_{17})_{10}(SiO_{1.5})_{10}$	1758.9
$(n\text{-}C_8H_{17})_{12}(SiO_{1.5})_{12}$	2089.4
$(n\text{-}C_9H_{19})_8(SiO_{1.5})_8$	1539.4
$(n\text{-}C_9H_{19})_{10}(SiO_{1.5})_{10}$	1897.0
$(n\text{-}C_{10}H_{21})_8(SiO_{1.5})_8$	1651.0
$(n\text{-}C_{10}H_{21})_{10}(SiO_{1.5})_{10}$	2036.8
$(C_6H_5)_8(SiO_{1.5})_8$	1140.4
$(Me_3SiO)_6(SiO_{1.5})_6$	954.9
$(Me_3SiO)_8(SiO_{1.5})_8$	1238.6
$(Me_3SiO)_{10}(SiO_{1.5})_{10}$	1520.8
$(Me_3SiO)_{12}(SiO_{1.5})_{12}$	1802.6
$(Cl\text{–}C_3H_6)_8(SiO_{1.5})_8$[a]	1061.4
$(Cl\text{–}C_3H_6)_{10}(SiO_{1.5})_{10}$[a]	1319.3
$(Cl\text{–}C_3H_6)_{12}(SiO_{1.5})_{12}$	1407.5

[a] Detected as Na adduct.

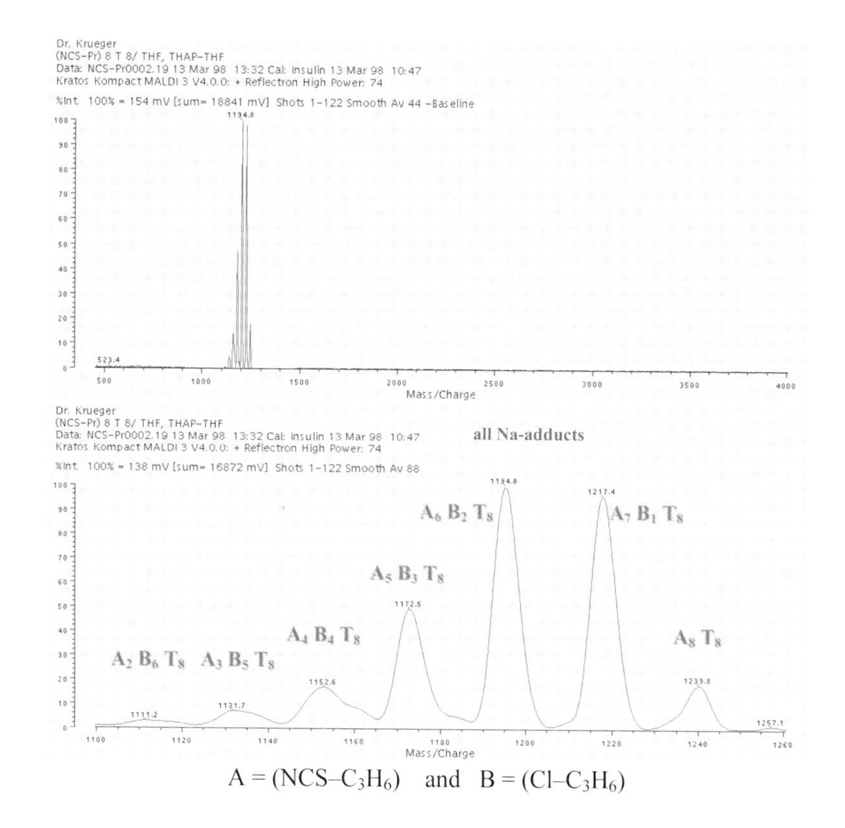

A = $(NCS\text{–}C_3H_6)$ and B = $(Cl\text{–}C_3H_6)$

Fig. 3. MALDI-TOF-MS spectra of a mixture of mixed-functionalized silsesquioxanes.

A mixture of mixed-functionalized silsesquioxanes prepared by substitution reaction (Eq. 7) can be analyzed with MALDI-TOF-MS (Fig. 3).

$$(Cl\text{–}C_3H_6)_8(SiO_{1.5})_8 \; + \; y \; NaNCS] \longrightarrow (Cl\text{-}C_3H_6)_x(NCS\text{–}C_3H_6)_y(SiO_{1.5})_8 \; + \; y \; NaCl$$

$$\text{with } x + y = 8$$

Eq. 7. Preparation of mixed-functionalized silsesquioxanes by substitution reaction.

Acknowledgment: We thank the *Deutsche Forschungsgemeinschaft* for financial support and *Hüls AG* for a gift of chemicals. All the syntheses of the silsesquioxanes have been done at the University of Paderborn in Prof. Marsmann's working group.

References:

[1] U. Dittmar, B. J. Hendan, U. Flörke, H. C. Marsmann, *J. Organomet. Chem.* **1995**, *489*, 185.

[2] H. C. Marsmann, U. Dittmar, E. Rikowski, "*New Functionalized Silsesquioxanes by Substitutions and Cage Rearrangement of Octa[(3-chloropropyl)silsesquioxane]*", in: *Organosilicon Chemistry II: From Molecules to Materials* (Eds.: N. Auner, J. Weis), VCH, Weinheim, **1996**, p. 691.

[3] E. Rikowski, H. C. Marsmann, Polyhedron **1997**, *16*, 3357.

[4] D. Hoebbel, G. Engelhardt, A. Samoson, K. Újszászy, Yu. Smolin, *Z. Anorg. Allg. Chem.* **1987**, 552, 236.

[5] P. A. Agaskar, V. W. Day, W. G. Klemperer, *J. Am. Chem. Soc.* **1987**, 109, 5554.

[6] R.-P. Krüger, H. Much, G. Schulz, *Int. J. Polymer Analysis and Characterization* **1996**, *2*, 221.

[7] O. Wachsen, K. H. Reichert, R.-P. Krüger, H. Much, G. Schulz, *Polym. Degradation and Stability* **1997**, *55*, 225.

Mixed Functionalized Octa(organylsilsesquioxanes) with Different Side Chains as Ligands for Metal Complexes

Stefan M. J. Brodda, Heinrich C. Marsmann*

Anorganische und Analytische Chemie
Universität-Gesamthochschule Paderborn
Warburger Straße 100, D–33098 Paderborn, Germany
Tel.: Int. code + (5251)60 2571
E-mail: hcm@ac16.uni-paderborn.de

Keywords: Silsesquioxanes / Cohydrolysis / Normal-Phase HPLC

Summary: Mixed functionalized octa(*n*-alkyl/halogenalkylsilsesquioxanes) with different side chains has been prepared by acid-catalyzed cohydrolysis of *n*-alkyl- and halogenalkyltrichlorosilanes. The silsesquioxane mixtures were separated in preparative amounts with normal-phase HPLC. Mixed functionalized octa(*n*-alkyl/halogenalkylsil-sesquioxanes) with different side chains are suitable examples of an organically modified silica gel. They can be modified by nucleophilic substitution of the halogen atom with $-PPh_2$ and rhodium–phosphino complexes are be formed.

Introduction

Organically modified silica gel finds a number of applications as stationary phases in chromatography or as anchors for transition metal complexes used as immobilized catalysts. The metal complexes are attached to the silica gel by coordinating them with phosphine (PR_2–), mercapto (–SH) or other groups, which are linked to the support over an alkyl chain ("spacer") [1–3]. To understand the mechanism of catalysis by supported metal complexes it is necessary to know more about the coordination of the silica-supported catalyst on the surface and the influence of the spacer length on the catalytic activity [4].

The synthesis of mixed functionalized octa(organylsilsesqioxanes) by cohydrolysis has already known for a long time [5]. But since the application of normal-phase HPLC it has been possible to separate the silsesquioxane mixture. Thus these substances are accessible to preparative chemistry and to spectroscopic examination [6, 8]. In the following the preparation of some new mixed functionalized octa(*n*-alkyl/halogenalkylsilsesquioxanes) by cohydrolysis and their modification are reported.

Results

Mixed functionalized octa(*n*-alkyl/halogenalkylsilsesquioxanes) with different side chains were synthesized by hydrochloric acid catalyzed cohydrolysis of *n*-alkyl- and halogenalkyltrichlorosilanes in the molar ratio 7:1 [6, 8].

Table 1. Educts of the cohydrolysis and the resulting mixed functionalized octa(*n*-alkyl/halogenalkylsilsesquioxane)-systems **1**–**4**..

n-Alkyltrichlorosilane	Halogenalkyltrichlorosilane	Silsesquioxane system	
CH_3SiCl_3	$ClCH_2SiCl_3$	$[ClCH_2]_n[CH_3]_{8-n}[Si_8O_{12}]$	(**1**)
$C_2H_5SiCl_3$	$Cl(CH_2)_2SiCl_3$	$[Cl(CH_2)_2]_n[H(CH_2)_2]_{8-n}[Si_8O_{12}]$	(**2**)
$C_3H_7SiCl_3$	$Cl(CH_2)_3SiCl_3$	$[Cl(CH_2)_3]_n[H(CH_2)_3]_{8-n}[Si_8O_{12}]$	(**3**)
$C_4H_9SiCl_3$	$Br(CH_2)_4SiCl_3$	$[Br(CH_2)_4]_n[H(CH_2)_4]_{8-n}[Si_8O_{12}]$	(**4**)

The silsesquioxane mixtures of **2**, **3** and **4** were separated in preparative amounts with normal-phase HPLC on a Merck Hibar® RT 250-25 Si 60 Lichrospher® 5 µm column and with an eluent, which is composed of 90 vol % *n*-pentane and of 10 vol % dichloromethane.

The HPLC chromatograms of **2**, **3** and **4** have similar structures (see Fig. 1). The retention times increase with the number of halogen atoms and their local concentration in the molecules.

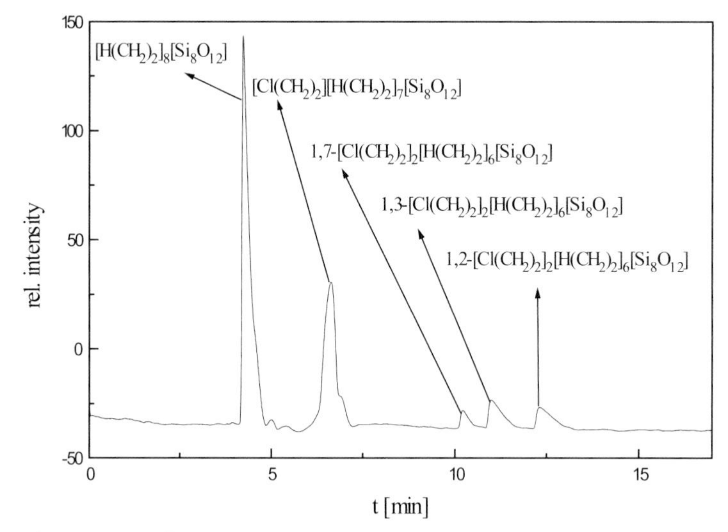

Fig. 1. HPLC chromatogram of 2.

On the other hand the retention times decrease with the length of the side chain (see Table 2).

Table 2. Retention times (t_r) of the compounds of the mixed-functionalized octa(*n*-alkyl/halogenalkylsilsesquioxane) systems **2**–**4**.

No.	Silsesquioxane	t_r [min]	No.	Silsesquioxane	t_r [min]
2	$[H(CH_2)_2]_8[Si_8O_{12}]$	4:14	**3**	$[H(CH_2)_3]_8[Si_8O_{12}]$	3:45
	$[Cl(CH_2)_2][H(CH_2)_2]_7[Si_8O_{12}]$	6:39		$[Cl(CH_2)_3][H(CH_2)_3]_7[Si_8O_{12}]$	4:36
	1,7-$[Cl(CH_2)_2]_2[H(CH_2)_2]_6[Si_8O_{12}]$	10:13		1,7-$[Cl(CH_2)_3]_2[H(CH_2)_3]_6[Si_8O_{12}]$	7:51
	1,3-$[Cl(CH_2)_2]_2[H(CH_2)_2]_6[Si_8O_{12}]$	11:01		1,3-$[Cl(CH_2)_3]_2[H(CH_2)_3]_6[Si_8O_{12}]$	8:28
	1,2-$[Cl(CH_2)_2]_2[H(CH_2)_2]_6[Si_8O_{12}]$	12:19		1,2-$[Cl(CH_2)_3]_2[H(CH_2)_3]_6[Si_8O_{12}]$	9:40
4	$[H(CH_2)_4]_8[Si_8O_{12}]$	3:33			
	$[Br(CH_2)_4][H(CH_2)_4]_7[Si_8O_{12}]$	4:25			
	1,7-$[Br(CH_2)_4]_2[H(CH_2)_4]_6[Si_8O_{12}]$	6:59			
	1,3-$[Br(CH_2)_4]_2[H(CH_2)_4]_6[Si_8O_{12}]$	7:30			
	1,2-$[Br(CH_2)_4]_2[H(CH_2)_4]_6[Si_8O_{12}]$	8:21			

A normal-phase HPLC separation of **1** was not possible, because it was insoluble in every usual solvent.

The compounds of **2**, **3** and **4** were identified by ^{29}Si-NMR spectroscopy, because octa(organylsilsesquioxanes) with two different substitutents on the silicon atoms ($R_nR'_{8-n}T_8$) have a characteristic signal pattern, which depends on the degree of substitution [6, 8].

Through nucleophilic substitution of the halogen atoms of **2**, **3** and **4** with potassium diphenylphosphide, phosphino ligands can be formed [7, 8].

Ligand displacement reactions with these phosphino ligands and transition metal complexes showed that only from the mono- and the two *ortho*-functionalized octa(alkyl/halogenalkylsilsesquioxanes) it is possible to build defined complexes (Eq. 1) [8].

$$[Rh(CO)_2Cl]_2 + 2\ 1,2\text{-}[PPh_2(CH_2)_3]_2[H(CH_2)_3]_6[Si_8O_{12}] \rightarrow$$

$$2\ CO + 2\ [Rh(CO)(1,2\text{-}[PPh_2(CH_2)_3]_2[H(CH_2)_3]_6[Si_8O_{12}])Cl]$$

Eq. 1.

References:

[1] Z. M. Michalska, *J. Mol. Catal.* **1977/78**, *3*, 125–134.

[2] L. Z. Wang, Y. Y. Jiang, *J. Organomet. Chem.* **1983**, *251*, 39–44.

[3] A. L. Prignano, W. C. Trogler, *J. Am. Chem.. Soc.* **1987**, *109*, 3586–3595.

[4] M. Capka, Z. M. Michalska, J. Stoch, *J. Mol. Catal.* **1981**, 323.

[5] T. N. Martynova, T. I. Chupakhina, *J. Organomet. Chem.* **1988**, *345*, 11.

[6] B. J. Hendan, H. C. Marsmann, *J. Organomet. Chem.* **1994**,*483*, 33–38.

[7] U. Dittmar, *Dissertation*, Universität-Gesamthochschule-Paderborn, **1993**.

[8] B. J. Hendan , *Dissertation*, Universität-Gesamthochschule-Paderborn, **1995**.

Polymerization of Silanes with Platinum Metal Catalysts

Christian Mechtler, Christoph Marschner*

Institut für Anorganische Chemie, Technische Universität Graz
Stremayrgasse 16, A-8010 Graz, Austria
Tel.: Int. code + (316)873 8209 — Fax: Int. code + (316)873 8701
E-mail: marschner@anorg.tu-graz.ac.at

Keywords: Polysilanes / Polymerization / Transition Metal Catalysis

Summary: It was found that some late transition metal complexes such as $(Ph_3P)_3RhCl$, $Pd(allyl)_2Cl_2$ and $Pd(dba)_2$ effectively catalyze the polymerization of 1,2-disubstituted disilanes leading to polymers with molecular weight up to $M_W = 1.9 \times 10^4$.

Introduction

Due to their interesting physical properties polysilanes have received some attention in recent years [1]. While Wurtz-type coupling of dichlorosilanes is still the main route to this type of materials some other methods have emerged during the last few years [2]. Among these the transition metal-catalyzed polymerization of hydrosilanes was established mainly by the groups of Harrod [3] and Tilley [4]. This approach, however, focused on the use of early transition metal catalysts, in particular group 4 metallocenes. Late transition metal compounds, although well known for their abilities to promote Si–C and Si–O bond formation, were found to be only of limited use for the synthesis of polysilanes [5].

Results and Discussion

During the last years several studies from our laboratory have concentrated on group 4 metallocene catalyzed polymerization of polysilanes starting from 1,2-disubstituted disilanes [6]. In an attempt to investigate the use of these disilanes employing late metal catalysis we screened a number of metal complexes. We found that the reaction of neat 1,2-dimethyldisilane with Wilkinson's catalyst as well as with $Pd_2(allyl)_2Cl_2$ or $Pd(dba)_2$ led to the formation of poly(methylsilane) and methylsilane (Eq. 1). Molecular weights (M_w) of the polysilanes ranged from 1.7×10^3 to 1.9×10^4 with a polydispersity of around 1.5, as indicated by GPC.

$$n\ MeH_2SiSiH_2Me \xrightarrow{\text{cat.}} (MeHSi)_n + n\ MeSiH_3$$

Eq. 1: Late transition metal-catalyzed polymerization reaction of MeH_2SiSiH_2Me.

Analysis of the early stages of polymerization by means of GC/MS revealed an interesting difference between the two reactions catalyzed either by $Pd_2(allyl)_2Cl_2$ (Fig. 1) or by Cp_2ZrMe_2 (Fig. 2) [7]. While in the latter case a high amount of branched material (products with shorter retention time on the GC) was produced even at the oligomer stage of the reaction, the palladium-catalyzed reaction was found to give mainly linear products. This behavior is also reflected by the solubility of the polymers formed. The highly branched material from the metallocene-catalyzed reaction was soluble only to a molecular weight of around 2000 while all the polymers formed by late metal catalysts were easily soluble even at higher molecular weight.

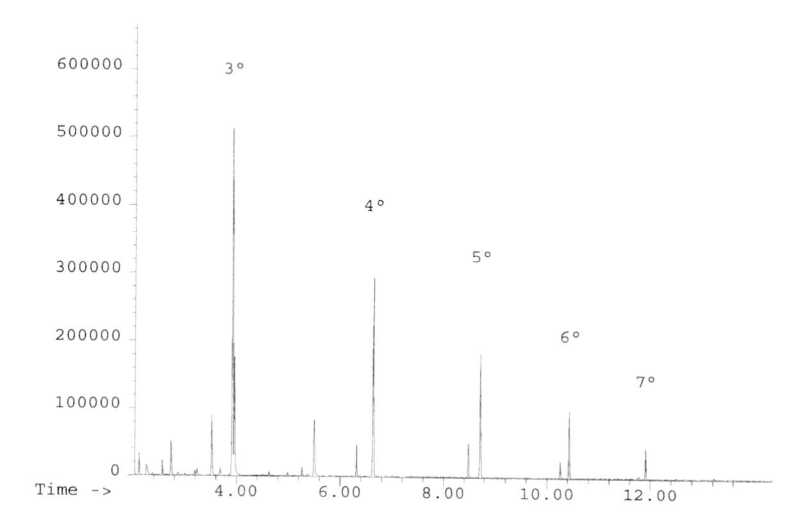

Fig. 1. Oligomeric distribution; $MeH_2SiSiH_2Me/Pd_2(allyl)_2Cl_2$; $t_R = 25$ min.; $n° =$ chain length.

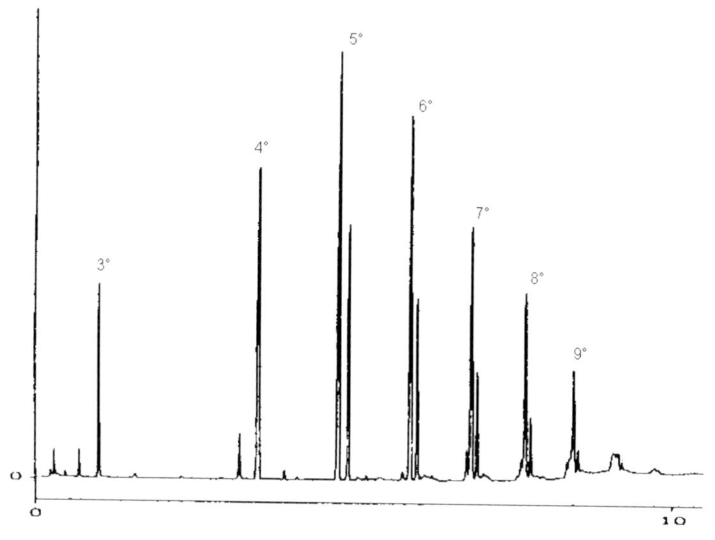

Fig. 2. Oligomeric distribution; $MeH_2SiSiH_2Me/Cp_2ZrMe_2$; $t_R = 3$ min.; $n° =$ chain length [7].

Analysis of the ^{29}Si-NMR spectrum (Fig. 3) indicates the formation of a completely atactic polymer.

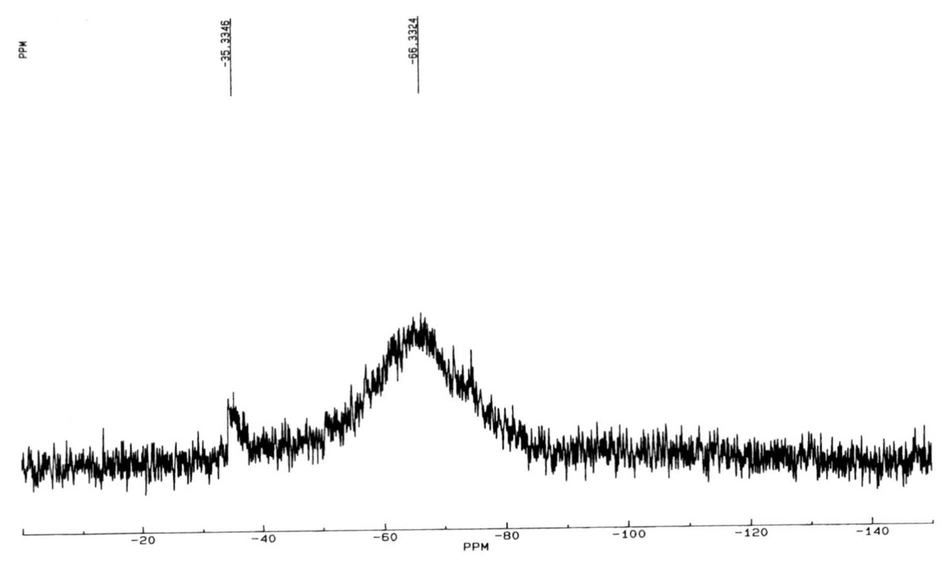

Fig. 3. ^{29}Si-NMR spectrum of $(MeSiH)_n$ formed from polymerization of MeH_2SiSiH_2Me catalyzed by $(PPh_3)_3RhCl$ after 3 days of reaction.

Reaction of phenylsilane employing the same conditions led, as already known [8], to the formation of oligomers accompanied by the redistribution products diphenylsilane and silane (Table 1). However, the degree of Si–Si bond formation versus redistribution varied considerably, with Wilkinson's catalyst mostly favoring the formation of diphenylsilane. 1-Methyl-2-phenyldisilane, which we successfully utilized in the metallocene-catalyzed reaction, also led to the formation of a polysilane. However, although GC/MS shows some oligomers with incorporated phenylsilyl units, diphenylsilane is detected as well, indicating the redistribution reaction to be active also in this case. The same was found for the copolymerization of 1,2-dimethyldisilane with phenylsilane.

Table 1. Polymerization reactions.

Entry	Catalyst	Monomer	Polymerization
1	$(PPh_3)_3RhCl$	$PhSiH_3$	–
2	$(PPh_3)_3RhCl$	MeH_2SiSiH_2Me	+
3	$(PPh_3)_3RhCl$	MeH_2SiSiH_2Ph	+
4	$(PPh_3)_3RhCl$	$PhSiH_3 + MeH_2SiSiH_2Me$	+
5	$Pd_2(allyl)_2Cl_2$	$PhSiH_3$	–
6	$Pd_2(allyl)_2Cl_2$	MeH_2SiSiH_2Me	+
7	$Pd_2(allyl)_2Cl_2$	$PhSiH_3 + MeH_2SiSiH_2Me$	+
8	$Pd(dba)_2$	MeH_2SiSiH_2Me	+

Conclusion

While polymerization reactions employing disilanes for the bis-silylation of unsaturated compounds have been known for some while [9], our system seems to be the first one to yield a high molecular weight polymer catalyzed by late transition metal complexes.

Based on the available data this appears to be a redistribution reaction. Tanaka and coworkers reported a similar reaction a few years ago [10]. During their investigation of the reaction of $Pt(PEt_3)_3$ with $Me_2HSiSiHMe_2$ they observed redistribution of the disilane to dimethylsilane and higher oligosilanes up to the heptasilane. Since the initial stage of our reaction looks very similar to Tanaka's we assume a similar mechanism to be working. The fact that our reaction does not stop at the oligomer stage may be attributed to a number of factors. Firstly, and probably most importantly, we carried out our reactions in neat silane, which provides the highest possible concentration and was found to be favorable for the formation of higher molecular weight polymers also in other cases. Secondly, the MeHSi unit seems to have some favorable properties compared to Me_2Si in terms of both sterics and electronics. Finally, the metal complex may play an important role as well. For example, we found Wilkinson's catalyst to be very effective while $(Ph_3P)_3RhH$ was not active at all. Whether metal silylenes as suggested by Tanaka play an important role in this reaction or not remains to be investigated.

In summary we have found a promising route for the synthesis of polymethylsilane catalyzed by palladium or rhodium catalysts. In contrast to the metallocene-derived polymers of the same type which are highly branched, our polymers feature a high degree of linearity and are thus more soluble. While the reaction worked well for the synthesis of polymethylsilane, attempts to incorporate phenylsilyl units either by means of copolymerization or using MeH_2SiSiH_2Ph as the starting material were not successful.

Acknowledgment: Financial support by the *Fonds zur Förderung der wissenschaftlichen Forschung in Österreich* (SFB *Electroactive Materials* at the TU-Graz, Projekt 904) is gratefully acknowledged. The *Wacker Chemie GmbH*, Burghausen kindly provided various organosilanes as starting materials. C. Marschner thanks the *Austrian Academy of Sciences* for an APART (Austrian Program for Advanced Research and Technology) scholarship.

References

[1] R. D. Miller, J. Michl, *Chem. Rev.* **1989**, *89*, 1359.
[2] R. West, in: *The Chemistry of Organic Silicon Compounds* (Eds. S. Patai, Z. Rappoport) Wiley, New York, **1989**, *Vol. II*, Ch. 19.
[3] C. Aitken, J. F. Harrod, E. Samuel, *J. Organomet. Chem.* **1985**, *279*, C11.
[4] T. Imori, T. D. Tilley, *Polyhedron* **1994**, *13*, 2231, and references cited therein.
[5] K. Brown-Wensley, *Organometallics* **1987**, *6*, 1590.
[6] E. Hengge, M. Weinberger, *J. Organomet. Chem.* **1992**, *433*, 21.
[7] E. Hengge, M. Weinberger, *J. Organomet. Chem.* **1993**, *443*, 167.
[8] I. Ojima, S.-I. Inaba, T. Kigore, Y. Nagai, *J. Organomet. Chem.* **1973**, *55*, C3.
[9] K. A. Horn, *Chem. Rev.* **1995**, *95*, 1317.
[10] H. Yamashita, M. Tanaka, *Bull. Chem. Soc. Jpn.* **1995**, *68*, 403, and references cited therein.

Polymorphism and Molecular Mobility of Poly(silane)s and Poly(silylenemethylene)s

Christian Mueller, Claudia Schmidt

Institut für Makromolekulare Chemie, Universität Freiburg
Stefan-Meier-Str. 31, D-79104 Freiburg, Germany
schmidtc@ruf.uni-freiburg.de

Florian Koopmann, Holger Frey

Freiburger Materialforschungszentrum, Universität Freiburg
Stefan-Meier-Str. 21, D-79104 Freiburg, Germany

Keywords: Poly(silane)s / Poly(carbosilane)s / Phase Behavior / Pressure–Volume–Temperature Measurement (*pVT*) / Deuterium NMR

Summary: Symmetrically substituted poly(di-*n*-alkylsilane)s with side chains of from 4 to 6 methylene units and poly(di-*n*-butylsilylenemethylene) were investigated using ^{2}H solid-state NMR to obtain information on the molecular mobility of main and side chains in the different crystalline and mesomorphic phases. In addition, pressure–volume–temperature (*pVT*) measurements were carried out to investigate the dependence of the phase behavior on both temperature and pressure.

Introduction

Poly(silane)s [poly(silylene)s] with organic side chains are inorganic–organic hybrid polymers that exhibit unusual properties due to their one-dimensional Si chains, such as strong UV absorptions and photoconductivity [1]. Furthermore, poly(di-*n*-alkylsilane)s with linear side chains in the range of 4 to 14 methylene units exhibit an order–disorder transition from the crystalline phase to a hexagonal, columnar mesophase (μ) with conformational disorder, resulting in thermo- and piezochromism [2]. Recently, for the homologous series of poly(di-*n*-alkylsilylenemethylene)s with *n*-alkyl side chains ranging from ethyl to *n*-hexyl similar conformational disorder has been reported. For these latter polymers generally disordering occurs only 5 to 10°C before isotropization [3].

In the present work we have studied the C1-deuterated poly(silane)s PD4S, PD5S, PD6S and poly(di-*n*-butylsilylenemethylene) P4SC (Scheme 1). All samples possess molecular weights above 100 000 g/mol. The CD_2 group in the vicinity of the silicon atoms of the polymer backbone is used as a probe for local mobility in solid-state ^{2}H-NMR spectroscopy.

$$-\!\!\left[\mathrm{Si}(\mathrm{CD_2R})_2\right]_n\!\!- \qquad -\!\!\left[\mathrm{Si}(\mathrm{CD_2R})_2\text{-}\mathrm{CH_2}\right]_n\!\!-$$

Scheme 1. C1-deuterated poly(silane)s (R = *n*-propyl to *n*-pentyl) and poly(di-*n*-butylsilylenemethylene) with R = *n*-propyl.

Results and Discussion

Figure 1 shows the ^{2}H-NMR spectra of slowly crystallized samples of the C1-deuterated poly(silane)s, [D$_4$]PD6S and [D$_4$]PD5S, and of poly(di-*n*-butylsilylenemethylene), [D$_4$]P4SC. For PD4S (not shown) spectra similar to those of PD5S were obtained. Deuteron-NMR line shapes are sensitive to rotational motions and to phase transitions whenever a change of the molecular dynamics is involved. For the samples studied, at temperatures well below the disordering temperature, T_d, powder spectra with quadrupolar splittings ($\Delta\nu$) of about 125 kHz (top row of Fig. 1) are obtained, which are attributed to the crystalline, immobile phase. At temperatures above T_d, motionally narrowed Pake spectra are observed with splittings of approximately 23 kHz and 12 kHz for the poly(silane)s and the poly(carbosilane), respectively, which are characteristic of the μ phase. The transition to the mobile mesophase occurs over a broad range of temperatures, in which both types of spectra are superimposed (rows 2 and 3 of Fig. 1). The pure mesophase is observed only for the poly(silane)s, while P4SC becomes isotropic before the transition to the μ phase is complete (bottom row of Fig. 1).

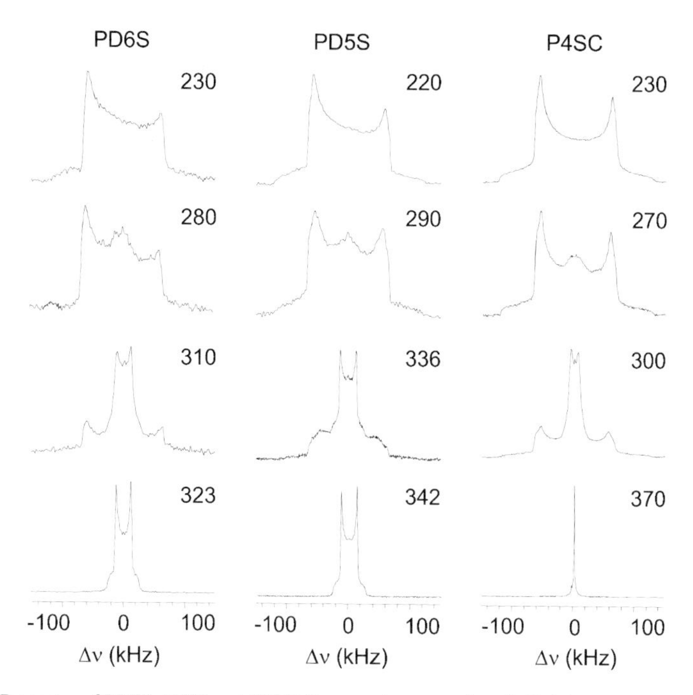

Fig. 1. ^{2}H-NMR spectra of PD6S, PD5S and P4SC. Temperatures are given in Kelvin.

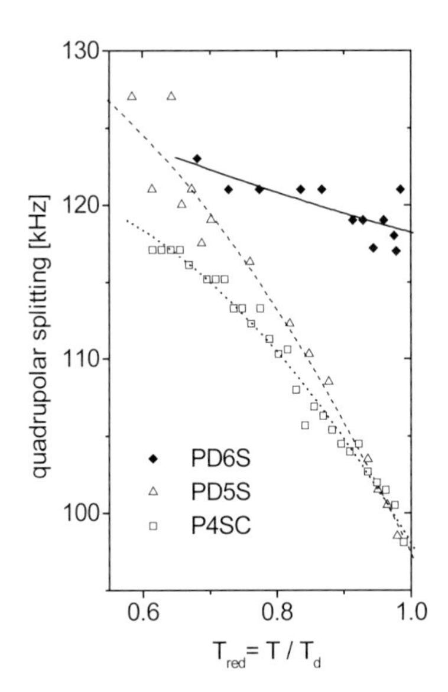

Fig. 2. Δν of PD6S, PD5S and P4SC.

Clearly, PD6S and PD5S, which are known to exhibit different conformations of the backbone in the crystalline phase (all-*trans* and 7_3-helix, respectively [1]) differ significantly in their spectra. In Fig. 2, this difference is illustrated by plotting the decrease of the quadrupolar splitting Δν of the Pake spectra shown in Fig. 1 as a function of the reduced temperature $T_{red} = T/T_d$. In contrast to PD6S, PD5S (and PD4S) exhibit a significant decrease of Δν with temperature. P4SC, the polymer with the different type of backbone, behaves similarly to PD5S. A more rapid decrease of Δν can be explained by higher mobility already in the crystalline phase. The difference between PD6S and the other poly(silane)s studied is explained by the different crystalline structures. Interestingly, the spectra of P4SC are comparable to those of the poly(silane)s. In contrast to the poly(silane)s, the mobile phase, characterized by a quadrupolar splitting of about 12 kHz, is observed only in a very small temperature interval and coexists with the rigid crystalline phase until isotropization occurs. However, P4SC shows a reduction of Δν in the same range as PD5S.

The spectra in the μ phase of all investigated poly(silane)s are similar, which is indicative of a similar type of motion in the mesophase. These spectra are motionally narrowed by a factor of about 1/5. If the polymer backbone were completely rigid, only free rotation of the side chains with a tetrahedral geometry would be observable, commonly characterized by a factor of 1/3. The considerably lower factor of 1/5 is evidence that the silicon-based polymer main chains are also involved in molecular motions in the mesophase. Thus, the mesophase is characterized by conformational exchange in side chains and backbone combined with reptational and rotational motions of the main chain. The factor of motional narrowing in the case of P4SC is about 1/10, significantly lower than for the poly(silane)s, indicating a higher flexibility of the Si–C backbone in the μ phase.

In addition to the ^{2}H-NMR experiments, *pVT* measurements have been carried out. Both isobaric and isothermal experiments were performed for all polymers. Figures 3 and 4 (left) show the isobaric measurements on the poly(di-*n*-alkylsilane)s, measured with a heating rate of 1 K/min. In the case of PD6S only one first-order phase transition from the all-*trans* crystalline to the mesomorphic phase is observable. No additional pressure-induced transitions are detected. With increasing pressure the phase transition temperature, T_d, is shifted to higher values and a decrease in the change of specific volume at the disordering transition, ΔV_{sp}, is observed.

PD4S and PD5S are known to exhibit piezochromism due to pressure-induced changes of their backbone conformation [4]. This is also reflected by the *pVT* data obtained. When raising the pressure from 10 to 95 MPa, PD5S shows the well-known phase transition from 7_3-crystalline to mesophase. From ca. 100 MPa up to 160 MPa there is no significant change of V_{sp} that would indicate a phase transition. At pressures above 160 MPa a new transition develops.

For PD4S the 7_3–μ transition vanishes too, but another phase transition develops simultaneously. With increasing pressure the variation of V_{sp} at the pressure-induced transition increases. In an isobaric experiment, a second high-temperature, high-pressure phase transition is observed (barely visible in Fig. 4, left). This new transition is characterized by a small change of V_{sp} and the most distinct increase of the phase transition temperature with increasing pressure. In Fig. 4 (right; pressure-up runs are plotted with open squares) isothermal measurements for PD4S are shown. Full pressure cycles were run from 10 MPa up to 200 MPa and down to 10 MPa. The isotherm at 110°C shows no difference between the pressure-up and pressure-down runs. The polymer remains in the μ phase for all pressures. The 30°C run reveals a solid–solid transition beginning at about 110 MPa which is completely irreversible upon decreasing the pressure. Analogous phase behavior is observed at 30°C for PD5S (not shown). Based on the isobaric and isothermal experiments, we attribute the detected transitions of PD4S and PD5S at higher pressure to a partial conversion of the 7_3-helical to the all-*trans* conformation. The high-temperature, high-pressure phase transition of PD4S is not yet understood.

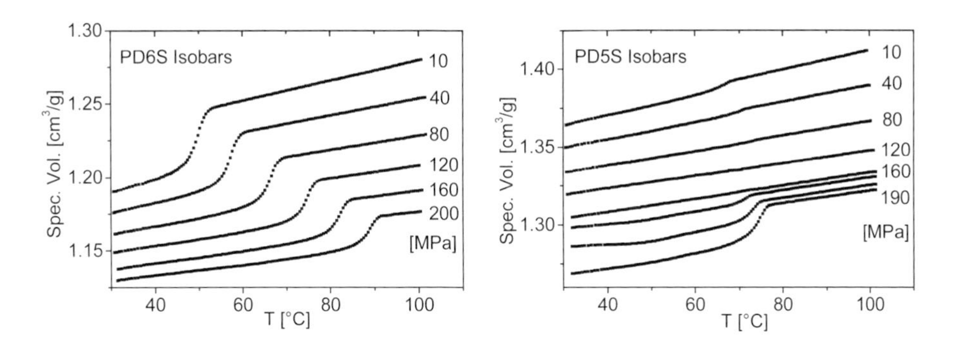

Fig. 3. Isobaric *pVT* curves of poly(di-*n*-hexylsilane) and poly(di-*n*-pentylsilane).

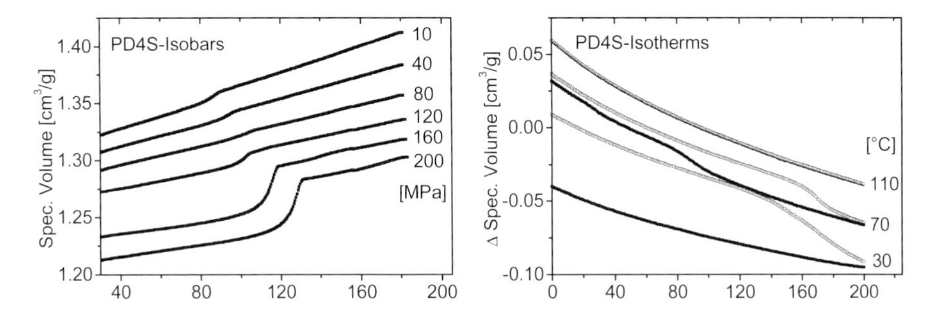

Fig. 4. Isobaric (left) and isothermal (right) *pVT* curves of poly(di-*n*-butylsilane).

In Fig. 5 the isobaric and isothermal *pVT* curves of P4SC are shown. Because of the very small changes in specific volume the first phase transition can hardly be seen in the isobaric experiment. The two transitions are resolved only in the heating runs. The phase transition temperature, T_d, is shifted to higher values and a decrease in ΔV_{sp} is observed for both phase transitions. The two transitions are separated by only about 7°C for all pressures investigated. In the isothermal experiment (Fig. 5, right) pressure-induced crystallization can be seen for temperatures above 120°C, concurrently with a decrease in V_{sp}. Crystallization experiments performed with different pressure cycles show no distinct changes in the phase behavior; in particular, the pure μ phase could not be observed.

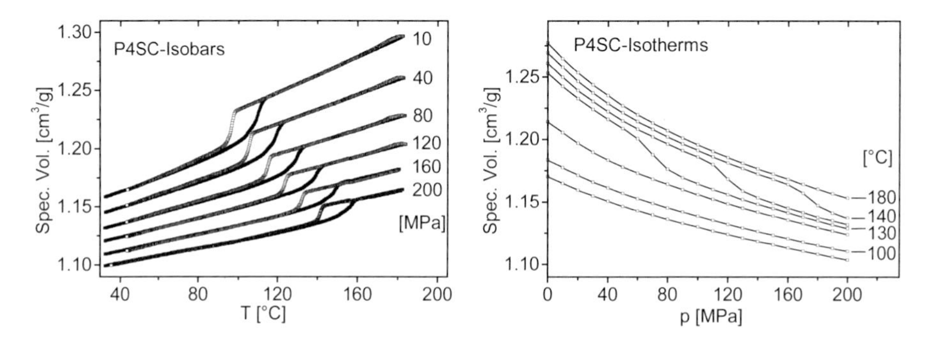

Fig. 5. Isobaric and isothermal *pVT* measurements of poly(di-*n*-butylsilylenemethylene): heating rate 1K/min.

In conclusion, the phase behavior of symmetrically di-*n*-alkyl-substituted poly(silane)s and poly(silylenemethylene)s is similar: both classes of polymers form the same type of mesophase, although it cannot be obtained in pure form for the poly(silylenemethylene) studied. The mobility in the mesophase, characterized by the quadrupole splitting, appears to depend strongly on the chemical structure of the backbone. For different poly(silane)s, the mobility in the crystalline phase is obviously influenced by the conformation of the backbone (all-*trans* vs. 7_3-helical) and therefore depends on the length of the side-chains. Application of pressure to poly(silane)s with 7_3-helical backbones leads to the formation of high-pressure crystalline modifications with all-*trans* backbone conformation. The *pVT* studies made it possible to define the precise conditions for the pressure-induced phase transitions.

Acknowledgment: We gratefully acknowledge financial support by the *Deutsche Forschungsgemeinschaft* (Schwerpunkt “Spezifische Phänomene in der Silicium-Chemie”).

References

[1] R. Miller, J. Michl, *Chem. Rev.* **1989**, *89*, 1359.

[2] F. C. Schilling, F. A. Bovey, A. J. Lovinger, J. M. Zeigler, *Macromolecules* **1986**, *19*, 2660.

[3] F. Koopmann, H. Frey, *Macromolecules* **1996**, *29*, 3701.

[4] F. C. Schilling, F. A. Bovey, D. D. Davis, A. J. Lovinger, R. B. Macgregor, C. A. Walsh, J. M. Zeigler, *Macromolecules* **1989**, *22*, 4648.

Synthesis, Functionalization and Cross-Linking Reactions of Poly(silylenemethylene)s

Wolfram Uhlig

Laboratorium für Anorganische Chemie
Eidgenössische Technische Hochschule Zürich
ETH-Zentrum, CH-8092 Zürich, Switzerland
Tel.: Int. code + 16322858 — Fax: Int. code + 16321149

Keywords: Organosilicon Polymer / Poly(silylenemethylene) / Silyl Triflate

Summary: Novel poly(silylenemethylene)s have been prepared by ring-opening polymerization of 1,3-disilacyclobutanes followed by a protodesilylation reaction with triflic acid. Reactions of the triflate derivatives with organomagnesium compounds, $LiAlH_4$, amines or alcohols gave functional substituted and branched poly(silylene-methylene)s, which may serve as suitable precursors for silicon carbide and Si/C/N-based materials.

Introduction

In recent years, poly(silylenemethylene)s and related polymers received increasing attention as novel materials [1] and as organic precursors for silicon based ceramics [2, 3]. However, the early work in the area of poly(silylenemethylene)s was limited to polymers with methyl or phenyl groups on silicon and was hindered by the lack of suitable synthetic routes to functional substituted poly(silylenemethylene)s. Recently, Interrante reported the preparation and characterization of chloro- and alkoxy-substituted high molecular weight poly(silylenemethylene)s [4]. Our investigations of the cleavage of silicon–element bonds by triflic acid have shown that the protodesilylations of amino-, phenyl-, *para*-tolyl-, and *para*-anisylsilanes lead to pure silyltriflates in high yields, which react with nucleophiles without any exchange processes [5]. The work of Interrante stimulated us to investigate the synthesis of novel functional substituted and branched poly(silylenemethylene)s with high molecular weights and with a regular alternating arrangement of silicon and carbon atoms in the polymer backbone using silyl triflate intermediates.

Results and Discussion

The ring-opening polymerization of 1,3-disilacyclobutanes was described to be the most convenient method for the preparation of high molecular weight poly(silylenemethylene)s having a regular alternating arrangement of silicon and carbon atoms in the polymer main chain [4]. As indicated in Scheme 1, we obtained poly(methylphenylsilylenemethylene) **1** by the catalytic ROP of 1,3-dimethyl-1,3-diphenyl-1,3-disilacyclobutane [6]. As expected from the non-stereospecific nature of the polymerization process, as well as from the fact that the starting monomer was obtained as a *cis–trans* mixture, the asymmetrically substituted polymer **1** was found by NMR spectroscopy to adopt an atactic configuration. **1** is a suitable starting polymer for the preparation of numerous functional substituted poly(silylenemethylene)s. The protodesilylation reaction with triflic acid leads to the triflate derivative **2**, which can be converted with nucleophiles into the polymers **3**–**8** as shown in Scheme 2.

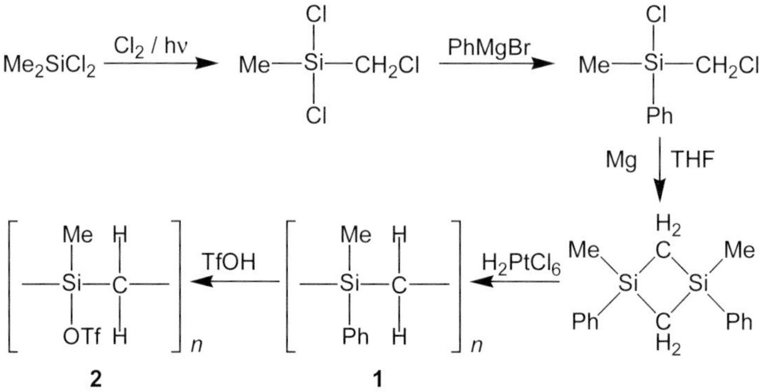

$Me_2SiCl_2 \xrightarrow{Cl_2 / h\nu} MeSiCl_2(CH_2Cl) \xrightarrow{PhMgBr} MeSiCl(Ph)(CH_2Cl) \xrightarrow{Mg / THF}$ 1,3-dimethyl-1,3-diphenyl-1,3-disilacyclobutane $\xrightarrow{H_2PtCl_6} [-SiMe(Ph)-CH_2-]_n$ (**1**) $\xrightarrow{TfOH} [-SiMe(OTf)-CH_2-]_n$ (**2**)

Scheme 1. Ring-opening polymerization (ROP) of 1,3- dimethyl-1,3-diphenyl-1,3-disilacyclobutane.

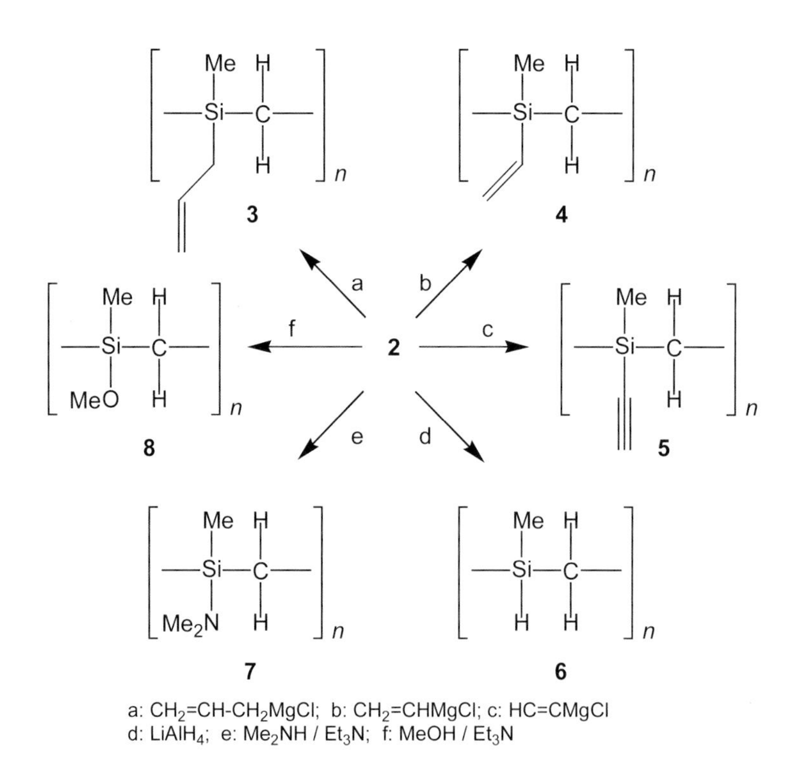

Scheme 2. Synthesis of functionalized poly(silylenemethylene)s **3**–**8**.

Table 1. Molecular weights, polydispersities and NMR data of the polymers **1–16**.

No.	Formula	$\delta^{29}Si$ [ppm]	M_w	M_w/M_n	Features of Polymers
1	$[MePhSi–CH_2]_n$	–4.7	82 500	2.4	Slightly yellow solid
2	$[Me(OTf)Si–CH_2]_n$	+39.8	–	–	Yellow, highly viscous oil; hydrolytically sensitive
3	$[MeAllSi–CH_2]_n$	+2.1	65 000	2.9	Slightly yellow, viscous oil
4	$[MeViSi–CH_2]_n$	+1.9	38 900	3.2	Yellow, highly viscous oil
5	$[Me(HC{\equiv}C)Si–CH_2]_n$	–17.9	35 500	3.0	Slightly brown, viscous oil
6	$[MeHSi–CH_2]_n$	–13.5	34 600	3.4	Yellow, highly viscous oil
7	$[Me(NMe_2)Si–CH_2]_n$	+7.4	–	–	Yellow, highly viscous oil; hydrolytically sensitive
8	$[Me(MeO)Si–CH_2]_n$	+14.0	–	–	Yellow, highly viscous oil
10	$[MePhSi–CH_2–Me(p\text{-}Anis)Si–CH_2]_n$	–6.2 br	88 000	2.7	Slightly yellow solid
11	$[MePhSi–CH_2–Me(OTf)Si–CH_2]_n$	–3.8 +38.6	–	–	Yellow, highly viscous oil; hydrolytically sensitive
12	$[Me(CH_2)Si–CH_2]_n$	+1.5	65 000	3.4	Yellow solid
13	$[Me(HC{\equiv}C)Si–CH_2–MePhSi–CH_2]_n$	–16.5 –5.0	53 000	2.9	Slightly brown, highly viscous oil
14	$[Me({\equiv}C)Si–CH_2–MePhSi–CH_2]_n$	–18.0 –6.9	109 000	3.1	Slightly brown solid
15	$[Me(Si–)Si–CH_2–MePhSi–CH_2]_n$	–21.0 br –4.3 br	48 700	4.9	Yellow solid
16	$[MePhSi–CH_2–Me(–NH)Si–CH_2]_n$	–6.3 br +3.3 br	–	–	Yellow-brown solid

The polymers **3–8** were characterized by ^{29}Si-, ^{13}C-, and ^{1}H-NMR spectroscopy and elemental analysis. The molecular weights in the range of $M_w \approx 5\times10^4$ g/mol were similar to those of the starting polymer **1** (see Table 1), but they were only determined in the case of polymers which were hydrolytically stable. Otherwise perturbations of the measurements by siloxane formation could not be excluded. From the similar molecular weights of **1** and of the products we conclude that no backbone cleavages occur during the protodesilylation and substitution processes. We next attempted to synthesize poly(silylenemethylene)s with two different functional groups. Therefore, we tried to prepare a poly(silylenemethylene) with two different leaving groups at the silicon atoms. Schmidbaur found that *para*-anisyl–silicon bonds were cleaved much faster by triflic acid than phenyl–silicon bonds [7]. We synthesized in a two-step reaction 1,3-dimethyl-1-phenyl-3-*para*-anisyl-1,3-disilacylobutane **9** from an amino–substituted parent compound (Scheme 3). The ring-opening polymerisation gives the corresponding poly(silylenemethylene) **10** and the reaction with triflic acid leads to a polymer **11**, in which only the *para*-anisyl groups are replaced by triflate substituents. The signals of the ^{29}Si-NMR spectrum are narrow and indicate a high regularity of the structure.

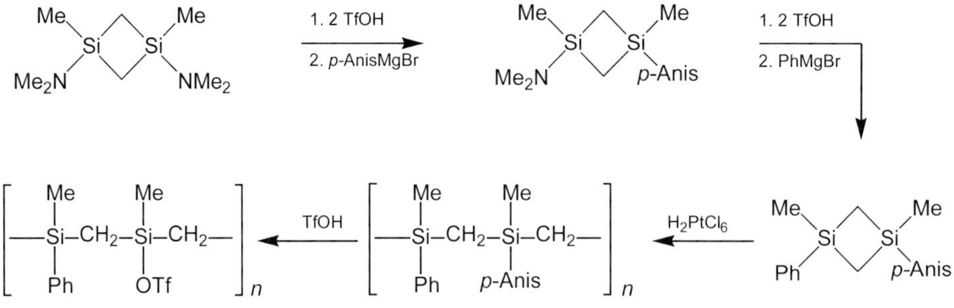

Scheme 3. Synthesis and reactivity of poly(silylenemethylene) **10** containing different leaving groups.

Purely linear poly(silylenemethylene)s tend to depolymerize on heating and afford little or no ceramic yields. They are not useful as SiC precursors. This observation is understandable, if we take into consideration the requirements for an idealized preceramic polymer. The polymer should contain cages and rings (network polymer) to decrease the elimination of volatile fragments resulting from backbone cleavage [1]. Therefore, we investigated several cross-linking reactions of linear, functional substituted poly(silylenemethylene)s. The platinum-catalyzed cross-linking was carried out between the vinyl-substituted poly(silylenemethylene) **4** and the poly(silylenemethylene) **6** containing silicon–hydrogen bonds. The hydrosilylation reaction can proceed only intermolecularly and results in polymer **12.** The NMR spectra are consistent with the purely β-hydrosilylation product (Scheme 4). Thermal degradation of **12** was investigated by thermogravimetric analysis (TGA) in an argon atmosphere. Decomposition started at ca. 300°C and the maximum rate of decomposition was found to be between 450 and 500°C. A char yield of 19 % was observed.

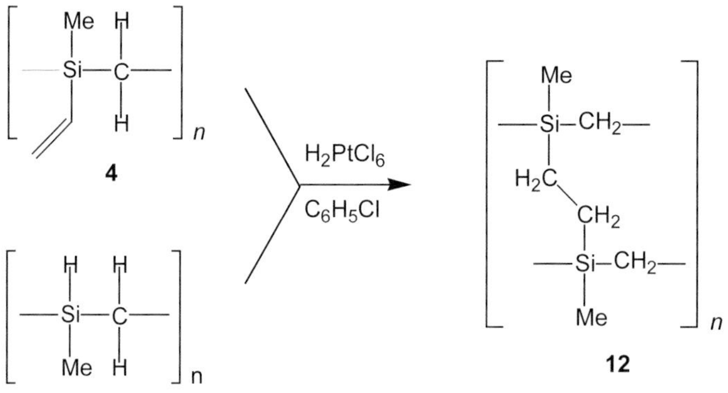

Scheme 4. Hydrosilylation reaction between polymers **4** and **6**.

Recently, our interest was directed to another type of intermolecular cross-linking reaction. We used the triflate-substituted poly(silylenemethylene) **11** as starting compound. The conversion with ethynylmagnesium chloride gave a polymer **13** with ethynyl side groups. **13** could be metalated using methylmagnesium chloride. In a last step this magnesium compound was coupled with the triflate derivative **11** to give a poly(silylenemethylene) **14** containing ethynyl bridges between the silicon–carbon chains (Scheme 5). Some time ago, we found that triflate substituted polysilanes and polysilynes can be reduced with potassium–graphite to give highly branched derivatives [8]. Potassium–graphite C_8K was found to be a very effective reducing agent for the formation of silicon–silicon bonds [9]. Consequently, we tested the applicability of this cross-linking principle for the synthesis of branched poly-(silylenemethylene)s. The reductive coupling of **11** with potassium–graphite proceeds at room temperature in high yields and short times leading to the polymeric network **15** shown in Scheme 5. The reaction must be carried out using the "inverse reducing method" (addition of potassium–graphite to a solution of the triflate substituted polymer). Otherwise (addition of the polymer solution to potassium–graphite), the triflate anion can be reduced by an excess of C_8K. It must be emphasized that the reduction can proceed either intramolecularly or intermolecularly. Therefore, relatively broad ^{29}Si-NMR signals and a broad molecular weight distribution were found for compound **15.** The thermal degradation of the polymers was investigated by TGA and ceramic residues of 61 % (**14**) and 52 % (**15**) were found. Poly(silylenemethylene)s containing amino side groups ("polycarbosilazanes") are interesting precursors for Si_3N_4 and Si/C/N ceramics [3, 10]. Therefore, we tried to use ammonia as a branching agent. The polycarbosilazane **16** is formed from the triflate-substituted polymer **11** and a solution of ammonia and triethylamine (1:2) in ether (Scheme 5).

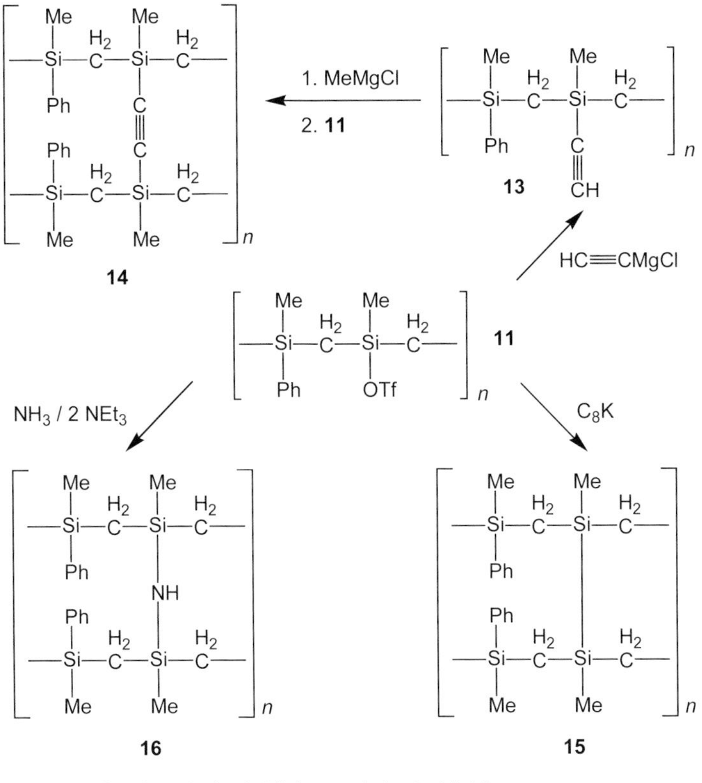

Scheme 5. Syntheses of the branched poly(silylenemethylene)s **14–16**.

Acknowledgement: This work was supported by *Wacker-Chemie GmbH* (Burghausen). Furthermore, the author acknowledges Prof. R. Nesper for support of this investigation.

References:

[1] H. R. Allcock, *Adv. Mater.* **1994**, *6*, 106.
[2] M. Birot, J.-P. Pillot, J. Dunoguès, *Chem. Rev.* **1995**, *95*, 1443.
[3] H.-P. Baldus, M. Jansen, *Angew. Chem.* **1997**, *109*, 339.
[4] L. V. Interrante, Q. Liu, I. Rushkin, Q. Shen, *J. Organomet. Chem.* **1996**, *521*, 1.
[5] W. Uhlig, *Chem. Ber.* **1996**, *129*, 733.
[6] W. Uhlig, *Z. Naturforsch., Teil B* **1997**, *52*, 577.
[7] C. Rüdinger, H. Beruda, H. Schmidbaur, *Chem. Ber.* **1992**, *125*, 1401.
[8] W. Uhlig, *Z. Naturforsch.Teil B* **1996**, *51*, 703.
[9] A. Fürstner, H. Weidmann, *J. Organomet. Chem.* **1988**, *354*, 15.
[10] R. Riedel, A. Kienzle, W. Dressler, L. Ruwisch, J. Bill, F. Aldinger, *Nature (London)* **1996** *382*, 796.

Highly Branched Polycarbosilanes: Preparation, Structure, and Functionality

Christian Drohmann, Olga B. Gorbatsevich, Aziz M. Muzafarov, Martin Möller*

Abteilung Organische Chemie III / Makromolekulare Chemie
Universität Ulm
D-89069 Ulm, Germany
Tel.: Int. code + (731)50 22870 — Fax: Int. code + (731)50 22883
E-mail: martin.moeller@chemie.uni-ulm.de

Keywords: Hyperbranched Polymers / Carbosilanes / Platinum

Summary: Hyperbranched polycarbosilanes were synthesized by hydrosilylation of AB_n functional monomers with A = H–Si and B = $-CH{=}CH_2$. Variation of the reaction conditions yielded polymers with different molecular weight distributions. The prepared polymers are transparent colorless viscous liquids with low glass transition temperatures. Polymers with short spacers between their functional groups possess globular shape as proved by size exclusion chromatography (SEC) coupled with a multi-angle laser light-scattering (MALLS) detector and viscodetector.

Highly branched polymers have attracted much a large scientific interest during the last few years. These new materials can be subdivided into two catagories. *Dendrimers* are perfectly defined concerning molecular weight and branching. Their synthesis is performed either by a divergent or a convergent approach and requires elaborated protection–activation concepts. *Hyperbranched polymers* are obtained by polycondensation of AB_n functional monomers in a one-pot synthesis providing a much simpler route to dendritically branched high-molecular-weight macromolecules. While the synthetic concept excludes crosslinking in principle, it does not allow one to prepare macromolecules of defined primary structure [1, 2]. Increasing conversion of an AB_n polyaddition reaction where all functional groups are equal in reactivity leads to an increasingly broader dispersion of the molecular weight. In the limiting case, i.e. at full conversion, the polydispersity reaches infinity. In addition hyperbranched polymers exhibit a distribution in their branching structure extending from totally linear to perfect dendrimers [3–5]. In principle this is also the case for the very elegant selfbranching polymerization reactions [6, 7] which have been developed recently by J. Fréchet and K. Matyjazewski with cationic and radical (SCVP, ATRP) reactive chain ends [8, 9].

However, because of the one-pot synthesis and the mostly simple synthetic procedures, hyperbranched polymers can be prepared in large volumes and are of general interest in coating technology and as viscosity modifiers [10]. Recent efforts are directed towards achieving control of the branching structure. A number of approaches have been developed in order to prepare hyperbranched polymers with (i) a minimized polydispersity and (ii) improved branching structure, i.e. approaching the perfection of well-defined dendrimers where end groups are only located in the outer shell.

Among these approaches the most successful strategy is based on a seeding technique: a B_m seed is reacted with an AB_n monomer under conditions where the B groups of the seed are always in large excess over the B groups of the added monomer. In this way reaction of monomers with each other is minimized and the number of growing macromolecules is in the ideal case equal to the number of seed molecules [11, 12]. Modifications of this concept are the semicontinuous addition [13] and the intermediate immobilization of the core molecules which allows easy separation of all macromolecules which were not linked to the original seed [14]. Another idea was presented by Suzuki et al. as "multibranching polymerization" [15].

The present work is directed towards the preparation of hyperbranched polycarbosilanes with an optimized structure. The synthesis of hyperbranched polycarbosilanes has been reported for a variety of different monomers, e.g. for allyltris(dimethylsiloxy)silane [16], methyldiundecenoxysilane [17], methyldivinylsilane, methyldiallylsilane [18], and for triallylsilane [18, 19]. Perfect silicon-containing dendrimers were prepared by alternating hydrosilylation with siloxane formation [20–22] or also by reaction sequences including Grignard reaction [23, 24]. Compared to other syntheses of hyperbranched polymers, hydrosilylation differs by the fact that the reaction involves a catalyst which interacts strongly with the vinyl groups [25]. In the case of platinum catalysts, the actual catalytic species are assumed to be small colloidal particles which are formed in the initial stage of the reaction even when the catalyst is added as a molecular complex [26–28].

The present work is focused on the formation of hyperbranched polycarbosilanes by hydrosilylation of olefinic substituted silanes methyldivinylsilane (MDVS, **1**), methyldiallylsilane (MDAS, **2**), triallylsilane (TAS, **3**), and methyldiundecenylsilane (MDUS, **4**) and the physical properties of the resulting polymers [29–31]. The monomers were synthesized by reaction of chlorosilanes with the corresponding amount of alkenylmagnesium halide.

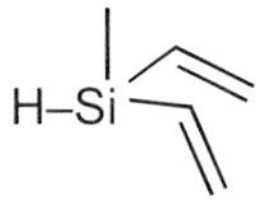

H–Si H–Si H–Si H–Si

1 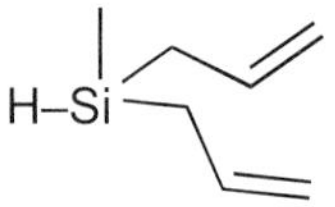 **2** 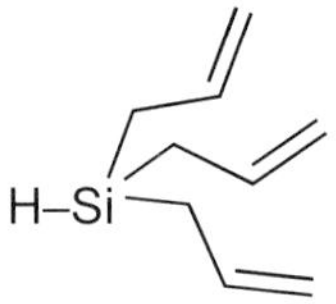 **3** 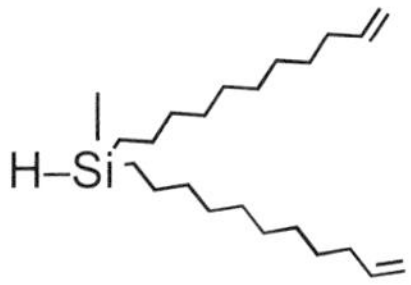 **4**

Variation of the monomer structure is expected to affect the structure of the final polymer. Comparison with regularly branched dendritic polycarbosilanes will make it possible to evaluate the characteristic structure–property relationships.

The polyaddition by means of a platinum catalyst yielded hyperbranched macromolecules with one terminal SiH group and n+1 vinyl groups or $2n$+1 vinyl groups in the case of the triallyl compound, n being the degree of polymerization (Scheme 1).

The dependence of the resulting molecular weight distribution on monomer concentration, catalyst, catalyst concentration, and temperature was investigated systematically; relative variations in the molecular weight distribution were monitored by size exclusion chromatography.

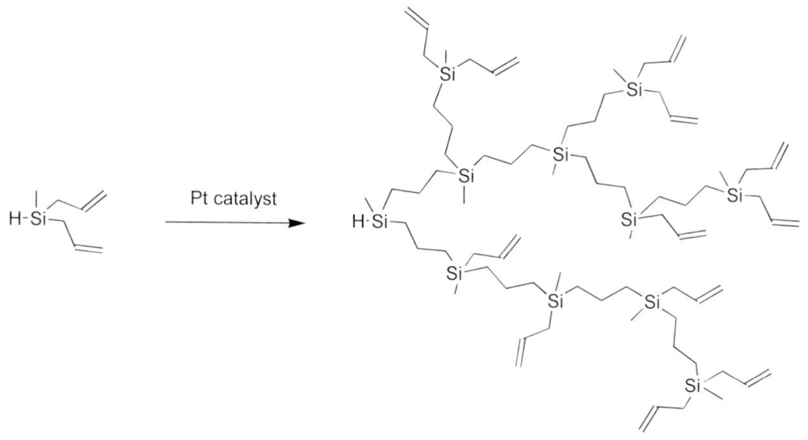

Scheme 1. Synthesis of hyperbranched macromolecules: one terminal SiH group, n+1 vinyl groups (n = 11).

For the short-chain monomers catalysis by H_2PtCl_6 (Speier's catalyst) was found to yield crosslinked products even at very low catalyst concentrations due to side reactions. In contrast, the Pt(0) and Pt(II) complexes yielded soluble products for a large range of reaction temperatures and catalyst concentrations. A different behavior was observed for the polyaddition of methyldiundecenylsilane. Bulk polymerization without detectable crosslinking could be achieved only with H_2PtCl_6 as the catalyst.

In principle it is not expected that the polyaddition of AB_2 monomers gives a crosslinked product [2]. The observed crosslinking and gelation indicate the occurrence of rearrangement or exchange reactions, e.g. formation of R_2SiH_2 groups with the very reactive catalysts. At least partly this can be favored by local jumps in temperature due to the large heat of reaction of the hydrosilylation [26]. The cleavage of silicon–carbon bonds by platinum has already been described in [32].

Generally, the choice of the catalyst had little influence on the molecular weight distribution as long as all other reaction conditions were maintained (see Fig. 1).

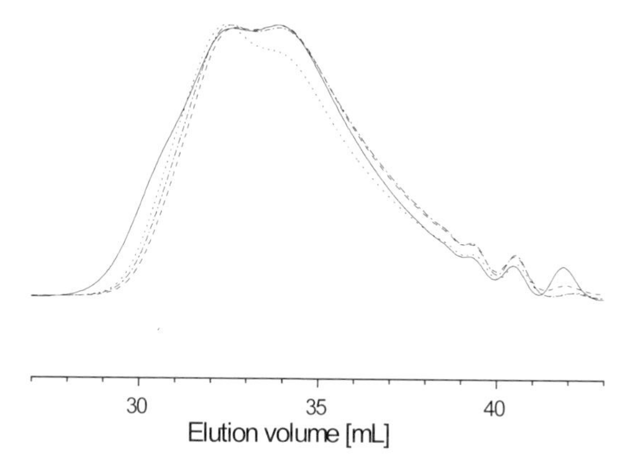

Fig. 1. SEC elugrams of hyperbranched polymethyldiundecenylsilane synthesized using different catalysts (platinum concentration = 0.0006 wt. % (PC072, PC085, Speier's catalyst), 0.0002 wt. % (Pt^{2+}); solvent: n-hexane; T = 20°C; reaction time 7 d). PC072 (———), PC085 (-----), Pt^{2+} (········); and Speier's catalyst (—·—·—).

In all cases a molecular weight distribution with nearly the same maximum and a shoulder corresponding to a higher-molecular-weight fraction was obtained. The resolution of additional high-molecular-weight peaks and/or shoulders demonstrates that the molecular weight distribution is not monomodal as one might expect for a random polyaddition without side reactions.

Higher catalyst concentrations enhanced the formation of a distinct high-molecular-weight fraction for polymethyldiallylsilane (Fig. 2) as well as for polytriallylsilane. In contrast, for polymethyldiundecenylsilane the molecular weight decreased with increasing amount of catalyst. This fact can be explained by a catalyst concentration-dependent nucleation of polymer molecules [30], i.e. higher molecular weights can be achieved when fewer seeds are formed at the beginning of the reaction.

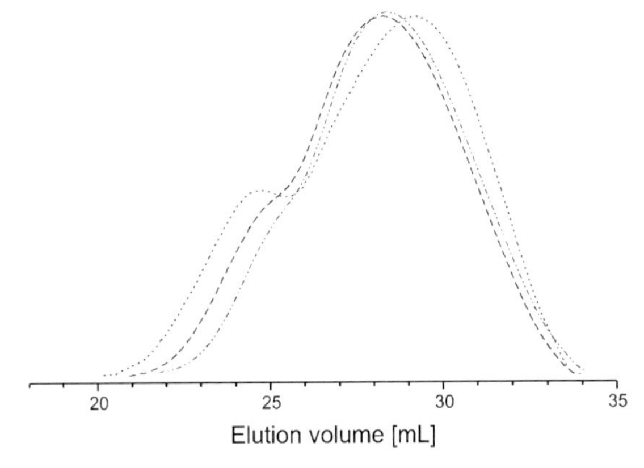

Fig. 2. SEC elugrams of hyperbranched polymethyldiallylsilane synthesized using different amounts of catalyst (PC072, 20°C, bulk, platinum concentration = 0.004 % (———), 0.02 % (– – – –), 0.04 % (-······-), 0.4 % (·······)).

Raising the reaction temperature accelerated the polymerization but at the same time also the polymodal character of the molecular weight distribution became much more pronounced. Dilution of the monomer slowed down the rate of reaction and the molecular weight built up.

The hyperbranched polymethyldivinylsilane exhibited a molecular weight distribution with a strongly developed peak in the low-molecular-weight region. This species was isolated by preparative SEC. It could be proven by GC–MS that this fraction consisted of trimers, tetramers, and pentamers with many isomers. This raised the question whether and to which extent cyclization occurred [33, 34]. Cyclization occurs if an A group and a B group of the same molecule react with each other intramolecularly. Because every molecule possesses only one A group the resulting molecule will only have B groups further on. Based on the molecular weight and the quantity of each species, the amount of SiH was calculated for the case of only open chain species to be ca. 30 % of the initial number of Si–H groups. In contrast NMR and IR spectroscopy demonstrated that hydrosilane species were almost absent. Comparison of the ^{1}H-NMR signals for vinyl and SiH showed that the amount of SiH was at the detection limit (<4 %). Qualitatively this was confirmed by IR. Cyclization can explain the conversion of SiH groups. In the case of the vinylic monomer this appears particularly likely of the possible six-ring formation.

In the case of the diallyl and the divinyl monomers, subsequent addition of further monomer did not affect the molecular weight. The molecular weight distribution remained rather invariant. This indicates formation of new polymer molecules instead of further growth of the preformed polycarbosilanes. A different result was observed for the polyaddition of methyldiundecenylsilane. The successive addition of new monomer yielded polymers with increasingly higher molecular weights, up to ten millions (determined by SEC online light scattering). Figure 3 depicts a series of SEC elugrams for polymethyldiundecenylsilane which were recorded after six subsequent addition steps of monomer without adding new catalyst. It might be concluded that the monomers with the short alkenyl groups yielded sterically more crowded polymers where the addition reaction was increasingly hindered at higher molecular weights.

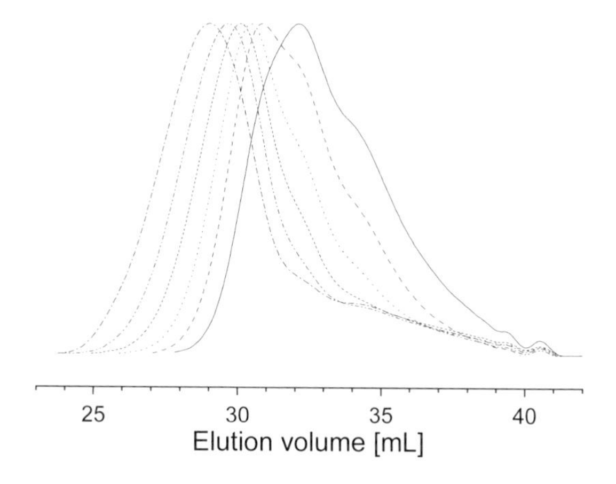

Fig. 3. SEC elugrams showing the molecular weight increase by addition of further monomer after full conversion after 1 (——), 2 (–––), 3 (····), 4 (---), 5 (–··–··–), and 6 (–·–·–) addition steps.

This is not the case for methyldiundecenylsilane. Furthermore, it may also be considered that the catalytic species can be entrapped or complexed by the growing polymer because of the strong interaction of platinum with olefinic groups. In this case the exchange of catalyst between different molecules will be slower than the addition of new monomer to the growing molecules.

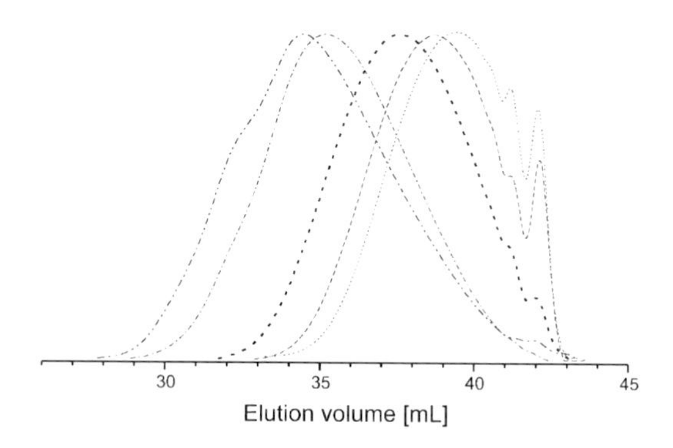

Fig. 4. SEC elugrams of hyperbranched polymethyldiallylsilane after different reaction times (PC072: 0.02% Pt, 20°C, bulk). 2 h (········); 3 h (––––); 5 h (- - - -); 7 h (–·–·–); 9 h (–··–··–); 24 h (——).

The question was raised of how the polyaddition process proceeds. Therefore, the evolution of the molecular weight distribution and the conversion of the Si–H groups after different reaction times was determined by SEC and by IR spectroscopy, respectively. The data in Fig. 4 allow one to distinguish different stages of the process. At the first stage the monomer concentration gradually decreased and low-molecular oligomers were formed in the reaction mixtures. The corresponding molecular weight distributions were monomodal and shifted to higher molecular weights with increasing time. Parallel IR experiments allowed one to follow the monomer conversion directly. After full monomer consumption a shoulder at higher molecular weights developed. As discussed above, formation of a distinct high-molecular-weight fraction indicates occurrence of a side reaction. This side reaction, however, is slow compared to the hydrosilylation reaction, and does not affect the molecular weight distribution initially. It can be assumed that formation of a high-molecular shoulder and crosslinking are caused by the same side reaction.

The polyaddition of methyldiundecenylsilane was followed directly by ^{1}H-NMR. As the reaction proceeded the signals at 5.82 (–C$\underline{H}$=), 4.98 (=C$\underline{H}_2$), and 3.92 (SiH) ppm decreased. At the same time a signal appeared at 5.44 ppm. This signal can be assigned to 1,2-substituted double bonds (–C$\underline{H}$=C$\underline{H}$–) which result from a slow isomerization reaction. The Si–H signal vanished at the end of the reaction, and the signals of the vinyl group decreased to half of the intensity. The conversion p of the H–Si groups can be described by the Eq. 1 [31].

$$p = \frac{6 \cdot [HSi] - [CH_2] - [CH_{outer}] - \frac{3}{2} \cdot [CH_{inner}]}{3 \cdot [HSi] - [CH_2] - [CH_{outer}] - \frac{3}{2} \cdot [CH_{inner}]}$$

Eq. 1.

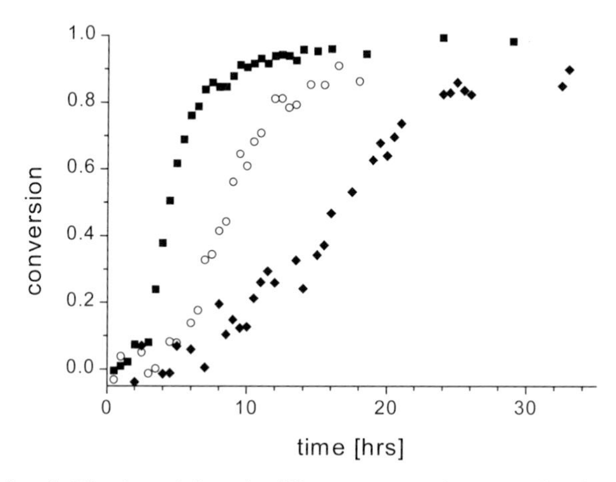

Fig. 5. Conversion of methyldiundecenylsilane in different concentrations as a function of time determined by ^{1}H-NMR (solvent: *n*-hexane; PC085 catalyst (platinum concentration = 1×10^{-3} wt. % (5×10^{-5} mol L^{-1}); T = 20°C; monomer concentrations: 0.665 mol L^{-1} (32 wt. %, ■), 0.401 mol L^{-1} (19 wt. %, O), 0.222 mol L^{-1} (12 wt. %, ◆)).

Figure 5 depicts the dependence of the conversion p on time for three monomer concentrations as evaluated from the ^{1}H-NMR spectra. In all cases the absolute amount of monomer was 3 mL. The amount of solvent was varied in order to change the monomer concentration. The catalyst concentration was kept constant at 1×10^{-3} wt.% (5×10^{-5} mol L^{-1}) platinum. All curves have the same S-shaped form. The initial retardation time was longer the lower the monomer concentration. The delay in the polymerization can be explained by the time needed to form the active catalytic species [28]. Figure 6 depicts a plot of $[-\ln[(2-2p)/(2-p)]]$ versus reaction time. It was shown that Eq. 2

$$-\ln\frac{2-2p}{2-p}=k_n[\mathrm{M}]_0 t$$

Eq. 2.

describes the conversion for an ideal AB_2 polyaddition with a rate of reaction $-d[A]/dt = k_n[A][B]$. $[M]_0$ designates the initial monomer concentration.

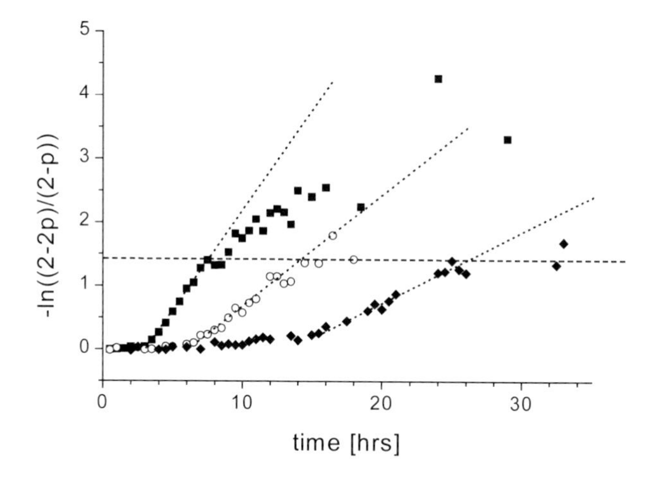

Fig. 6. Plot of $[-\ln[(2-2p)/(2-p)]]$ vs. time for the three reactions described in Fig. 5; the dotted lines designate the linear dependence at the beginning of the hydrosilylation reaction; the dashed line indicates the value at which the experimental values start to deviate from the theoretical ones.

For all three concentrations the slope in the first stage is zero, i.e. no reaction takes place. During a second stage, a constant slope is observed. If the slope is divided by the initial monomer concentration, the rate constant of the reaction is obtained to be k_n = 1.39 L mol^{-1} s^{-1} consistently for all three reactions.

At $[-\ln[(2-2p)/(2-p)]] \approx 1.3$ corresponding to a conversion $p = 0.84$ the reaction slowed down for all three cases. At this conversion the degree of polymerization is ca. 6. Thus, after six addition steps the reaction already slows down. Reasons for this decrease in the reaction rate could be increasing steric hindrance, isomerization or cyclization.

The time dependence of the molecular weight is consistent with the data on the rate of hydrosilylation. Figure 7 shows SEC elugrams for the polymer that was prepared in the most concentrated solution. The other polymers yielded similar elugrams. Only after an initial induction stage of several hours did the molecular weight start to increase significantly. The diagram in Fig. 8

represents the time dependence of the weight-average of the molecular weight for the three different concentrations. It must be noted that in this case also multimodal molecular weight distributions were observed at high conversion, indicating the occurrence of an additional process by which polymer molecules were linked to each other. This effect was most distinct for the most concentrated solution.

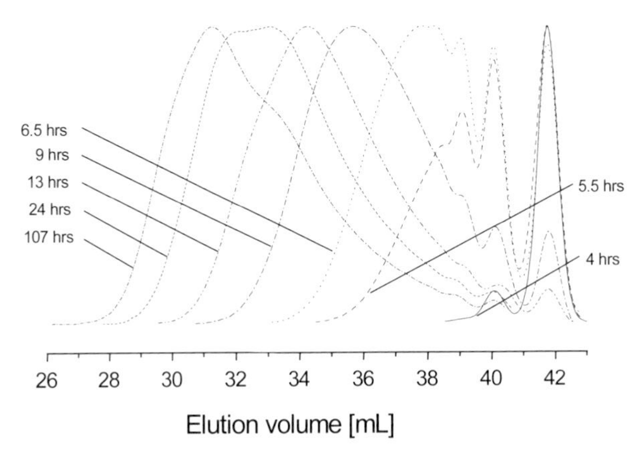

Fig. 7. SEC elugrams of polymethyldiundecenylsilane at different times following the course of polyaddition of MDUS (c = 0.665 mol L^{-1}(32 wt. %); solvent: *n*-hexane; PC085 catalyst (platinum concentration = 1×10^{-3} wt. % (5×10^{-5} mol L^{-1})); $T = 20°C$).

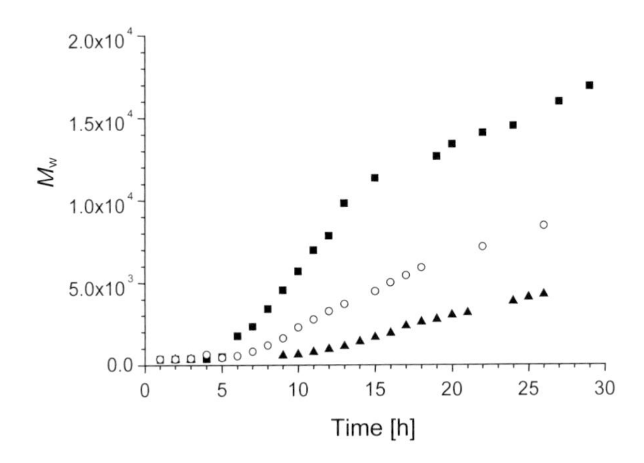

Fig. 8. Weight-average M_w of polymethyldiundecenylsilane as a function of time (solvent: *n*-hexane; PC085 catalyst (platinum concentration = 1×10^{-3} wt. % (5×10^{-5} mol L^{-1})); $T = 20°C$; monomer concentrations: 0.665 mol L^{-1} (32 wt. %, ■), 0.401 mol L^{-1} (19 wt. %, O), 0.222 mol L^{-1} (12 wt. %, ◆)).

The polymers were viscous liquids that were eventually colored yellowish by residual catalyst. They were very soluble in toluene, chloroform, THF, *n*-hexane, petrolether or acetone, but did not dissolve in water and lower alcohols like methanol or ethanol. The effect of improved solubilities of dendritic molecules compared to linear polymers has been mentioned before and attributed to their architecture and their functional groups [35].

Because of the hyperbranched dense molecular structure the absolute molecular weights cannot be determined by SEC calibrated against linear polystyrene standards. Therefore, we fractionated hyperbranched carbosilane polymers by preparative SEC and used the thus-obtained narrow molecular weight fractions to calibrate the analytical SEC [30, 31, 36]. Absolute values of molecular weight in the high-molecular-weight region were detected by means of an online multi-angle laser light scattering detector (MALLS) combined with a detector for the intrinsic viscosity. Calibration in the low-molecular-weight region was performed by assigning the elution volumes to the separate oligomeric peaks and therefore to the molecular weights of these species.

Due to the dense structure of the hyperbranched polymers their size, i.e. their radius of gyration, is small compared to linear polymers. Therefore, the relation between radius of gyration and elution volume could be determined only for high-molecular-weight polymethyldiundecenylsilane (PMDUS) (Fig. 9).

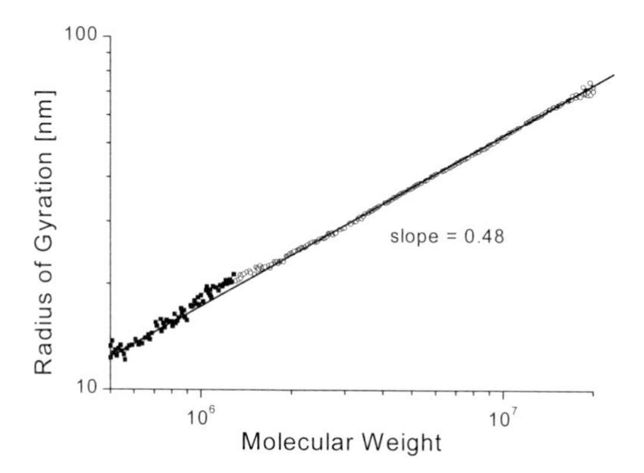

Fig. 9. Radius of gyration of polymethyldiundecenylsilane as a function of molecular weight (double logarithmic plot).

The exponent ν in Eq. 3 for the molecular weight dependence of the radius of gyration,

$$<R_g^2>^{1/2} = A \cdot M^{\nu}$$

Eq. 3.

is known to be characteristic for the molecular shape. It gives structural information about the macromolecules, i.e. for a hard sphere ν equals 1/3 and for a coil it has values between 0.5 (θ, coil) and 0.6. In the case of PMDUS in a good solvent the slope of the double-logarithmic plot of radius of gyration vs. molecular weight yielded a value of 0.48, demonstrating that the hyperbranched PMDUS molecules are considerably compacted compared to a linear coil molecule.

Complementary structural information can be obtained by the Mark–Houwink relation between viscosity and molecular weight (Eq. 4).

$$[\eta] = K \cdot M^{a}$$

Eq. 4.

The exponent a is equal to zero for hard spheres and for coil molecules it is 0.5 at θ conditions.

Online viscosity detection made it possible to observe the $[\eta] - M^a$ dependence experimentally.

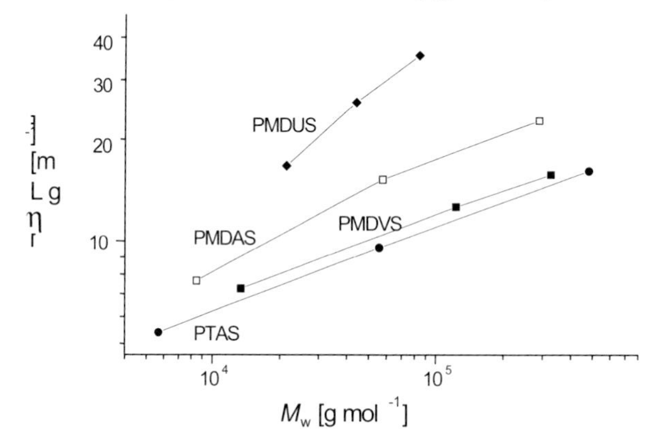

Fig. 10. Mark–Houwink plots of polymethyldivinylsilane(■), polymethyldiallylsilane (□), polytriallylsilane (●), and polymethyldiundecenylsilane (◆).

For polymers from MDVS, MDAS or TAS a value of 0.23 was obtained, demonstrating a rather dense structure of the high-molecular-weight fractions. In contrast, under the same conditions the Mark–Houwink exponent of PMDUS was 0.55, demonstrating a less dense structure compared to the polymers made from short-chain monomers. This is consistent with the relatively low intrinsic viscosities. The AB_3-functional PTAS yielded the lowest intrinsic viscosities. With increasing chain length the intrinsic viscosities increased for the AB_2-functional PMDVS, PMDAS, and PMDUS.

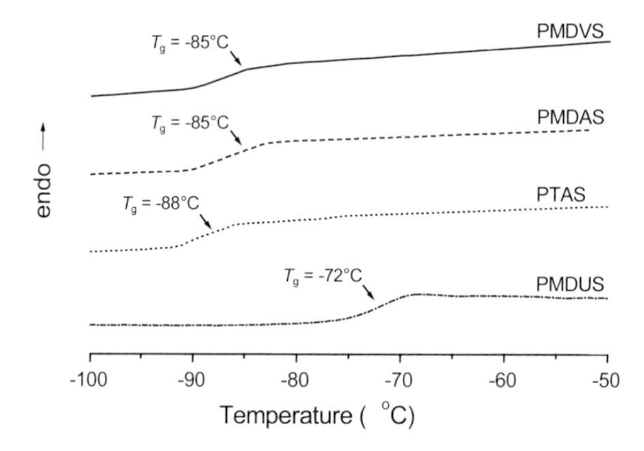

Fig. 11. DSC heating scans for PMDVS (——), PMDAS (–––––), PTAS (- - - -), and PMDUS(- · · · -).

The glass transition temperatures (T_g) for linear polymers decrease below a minimum mass [37] because of the increasing contribution of end groups. For hyperbranched polymers T_g is strongly dependent on the end group functionalities [38]. It might be expected that it will also depend on the branching density, which corresponds to the number of end groups. The glass transition temperatures of hyperbranched polycarbosilanes were shown to have low values [18]. The highly branched PTAS had indeed the lowest T_g (–88°C). The glass transition temperature of PMDVS and PMDAS was –85°C. For the less dense PMDUS the value was –72°C.

References:

[1] P. J. Flory, *J. Am. Chem. Soc.* **1952**, *74*, 2718.

[2] P. J. Flory, *Principles of Polymer Chemistry*, Cornell University Press, Ithaca, NY, USA, **1953**.

[3] D. Hölter, A. Burgath, H. Frey, *Acta Polym.* **1997**, *48*, 30.

[4] D. Hölter, H. Frey, *Acta Polym.* **1997**, *48*, 298.

[5] U. Beginn, C. Drohmann, M. Möller, *Macromolecules* **1997**, *30*, 4112.

[6] A. H. E. Müller, D. Yan, M. Wulkow, *Macromolecules* **1997**, *30*, 7016.

[7] D. Yan, A. H. E. Müller, K. Matyjaszewski, *Macromolecules* **1997**, *30*, 7024.

[8] J. M. J. Fréchet, M. Henmi, I. Gitsov, S. Aoshima, M. R. Leduc, R. B. Grubbs, *Science* **1995**, *269*, 1080.

[9] S. G. Gaynor, S. Edelman, K. Matyjaszewski, *Macromolecules* **1996**, *29*, 1079.

[10] E. Malmström, A. Hult, *J. Macromol. Sci. — Rev. Macromol. Chem. Phys.* **1997**, *C37*, 555.

[11] R. Hanselmann, D. Hölter, H. Frey, *PMSE Prepr. (Am. Chem. Soc., PMSE Div.)* **1997**, *77*, 165.

[12] W. Radke, G. Litvinenko, A. H. E. Müller, *Macromolecules* **1998**, *31*, 239.

[13] E. Malmström, M. Johansson, A. Hult, *Macromolecules* **1995**, *28*, 1698.

[14] P. Bharati, J. S. Moore, *J. Am. Chem. Soc.* **1997**, *119*, 3391.

[15] M. Suzuki, A. Ii, T. Saegusa, *Macromolecules* **1992**, *25*, 7071.

[16] L. J. Mathias, T. W. Carothers, *J. Am. Chem. Soc.* **1991**, *113*, 4043.

[17] A. M. Muzafarov, M. Golly, M. Möller, *Macromolecules* **1995**, *28*, 8444.

[18] A. M. Muzafarov, O. B. Gorbatsevich, E. A. Rebrov, G. M. Ignat'eva, T. B. Chenskaya, V. D. Myakushev, A. F. Bulkin, V. S. Papkov, *Polym. Sci.* **1993**, *35*, 1575.

[19] C. Lach, P. Müller, H. Frey, R. Mülhaupt, *Macromol. Rapid Commun.* **1997**, *18*, 253

[20] E. A. Rebrov, A. M. Muzafarov, V. S. Papkov, A. A. Zhdanov, *Dokl. Akad. Nauk SSSR* **1989**, *309*, 376.

[21] H. Uchida, Y. Kabe, K. Yoshino, A. Kawama, T. Tsumuraya, S. Masamune, *J. Am. Chem. Soc.* **1990**, *112*, 7077.

[22] A. Morikawa, M. Kakimoto, Y. Imai, *Macromolecules* **1991**, *24*, 3469.

[23] A. W. van der Made, P. W. N. M. van Leeuwen, J. C. de Wilde, R. A. C. Brandes, *Adv. Mat.* **1993**, *5*, 446.

[24] D. Seyferth, D. Y. Son, A. L. Rheingold, R. L. Ostrander, *Organometallics* **1994**, *13*, 2682.

[25] I. Ojima, "*The Hydrosilylation Reaction*" in: *The Chemistry of Organic Silicon Compounds*, (Eds.: S. Patai, Z. Rappoport), John Wiley, Chichester, **1989**, p. 1479.

[26] J. L. Speier *Adv. Organomet. Chem.* **1979**, *17*, 407.

[27] L. N. Lewis, N. Lewis, *J. Am. Chem. Soc.* **1986**, *108*, 7228.

[28] L. N. Lewis, *J. Am. Chem. Soc.* **1990**, *112*, 5998.

[29] C. Drohmann, O. B. Gorbatsevich, A. M. Muzafarov, M. Möller, *Polym. Prepr.* **1998**, *39(1)*, 471.

[30] O. B. Gorbatsevich, A. M. Muzafarov, C. Drohmann, M. Möller, in preparation.

[31] C. Drohmann, M. Möller, O. B. Gorbatsevich, A. M. Muzafarov, in preparation.

[32] I. S. Akhrem, N. M. Chistovalova, M. E. Vol'pin, *Russ. Chem. Rev.* **1983**, *52*, 542.

[33] A. M. Muzafarov, E. A. Rebrov, V. S. Papkov, *Russ. Chem. Rev.* **1991**, *60*, 807.

[34] F. Chu, C. J. Hawker, P. J. Pomery, D. J. T. Hill, *J. Polym. Sci. A: Polym. Chem.* **1997**, *35*, 1627.

[35] K. L. Wooley, J. M. J. Fréchet, C. J. Hawker, *Polymer* **1994**, *35*, 4489.

[36] G. Schulz, J. Falkenhagen, R.-P. Krüger, H. Much, E. Rikowski, W. Schnabel, A. Alonso Gutierrez, C. Drohmann, M. Möller, paper presented at 11th International Symposium on Polymer Analysis and Characterization (ISPAC-11), 25 May–27 May **1998** at Santa Margherita Ligure, Genova, Italy.

[37] K. L. Wooley, C. J. Hawker, J. M. Pochan, J. M. J. Fréchet, *Macromolecules* **1993**, *26*, 1514.

[38] M. N. Bochkarev, M. A. Katkova, *Russ. Chem. Rev.* **1995**, *11*, 1035.

Photoconductivity of C_{60}-doped Poly(disilanyleneoligothienylene)s

Masaya Kakimoto*, Hideki Kashihara, Tohru Kashiwagi, Toshihiko Takiguchi

Osaka R&D Laboratories, Sumitomo Electric Industries, Ltd.
1-1-3 Shimaya, Konohana-ku, Osaka 554, Japan
Tel.: Int. code + (6)64665768 — Fax: Int. code + (6)64661274
E-mail: kakimoto@okk.sei.co.jp

Keywords: Oligothiophene / Fullerene / Photoconductivity / Polysilane / Electrophotography

Summary: The photoconductive properties of poly[tetraethyldisilanylene-oligo-(2,5-thienylene)]s [PDS(Th)$_m$; $m = 2 - 4$], which have a disilanylene group and an oligothienylene unit in a polymer backbone, and fullerene (C_{60}) doping effects were investigated. The polymer PDS(Th)$_4$ was photoconducting when irradiated in visible light and the quantum efficiency for photocarrier generation was 2 % at 480 nm. C_{60} doping enhanced the efficiency effectively to 85 % at 470 nm ($E = 3\times10^5$ V cm^{-1}) by a photoinduced charge transfer mechanism. PDS(Th)$_4$ also has a high-hole drift mobility in the order of 10^{-4} cm^2 V^{-1} s^{-1} at room temperature, which is almost as high as that of poly(methylphenylsilane). We have also investigated their electrostatic properties as single-layered, positively charged photoreceptors for electrophotography.

Introduction

Recently, there has been growing interest in the synthesis and properties of silicon-containing polymers such as polysilanes and polymers with the regular alternating arrangement of an Si–Si bond and π-electron system in a main chain. These polymers can be used as photoresists, conducting and photoconducting materials, and ceramic precursors [1]. Although polysilanes show high hole-drift mobilities in the order of 10^{-4} cm V^{-1} s^{-1}, the photocarrier generation efficiency of polysilanes is low (ca. 1 % at high electric fields for poly(methylphenylsilane)) upon UV irradiation [2]. On the other hand, Malliaras et al. [3] and Ishikawa and co-workers [4, 5] have reported the synthesis of polymers with a regular alternating arrangement of a silylene or disilanylene unit and an oligothienylene unit in the polymer backbone, whose absorption and luminescence are in the visible region.

In this paper, we report the photoconductivity in the visible light region of poly[tetraethyldisilanyleneoligo(2,5-thienylene)]s [PDS(Th)$_m$; $m = 2 - 4$], their quantum efficiency of photocarrier generation and hole-drift mobility including effects of C_{60}-doping [6]. We have also investigated their electrostatic properties as single-layered, positively charged photoreceptors for electrophotography.

Results and Discussion

Photocarrier Generation of Poly(disilanyleneoligothienylene)s

Polymers were synthesized in good yields according to the Scheme 1. The molecular weights (M_w) are 10 000 – 50 000, for example PDS(Th)$_2$: M_w = 51 000, M_w/M_n = 3.4; PDS(Th)$_3$: M_w = 47 000, M_w/M_n = 3.4; PDS(Th)$_4$: M_w = 34 000, M_w/M_n = 2.4, respectively. Preparation of these monomers and polymers was already reported in the previous literature [5].

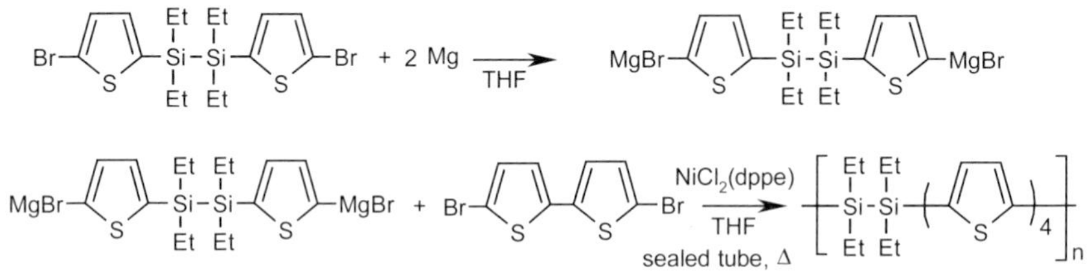

Scheme 1. Synthesis of PDS(Th)$_4$.

Absorption spectra of the PDS(Th)$_m$ (m = 2 – 4) are shown in Fig. 1. The absorption maxima of the polymers shifts to a longer wavelength as the number of thienylene rings increases.

The wavelength dependence of the photocurrent quantum efficiency of PDS(Th)$_m$ is shown in Fig. 2. Comparing the photocurrent quantum efficiency curves with their absorption spectra, the position of each maximum peak for the photocurrent quantum efficiencies appears at 30 – 40 nm longer wavelength than that of the respective absorption maximum. These peaks correspond to the absorption edge of the polymers. This means that almost all incident light would be absorbed in the vicinity of the illuminated surface of the films in the region of the absorption bands of the polymers.

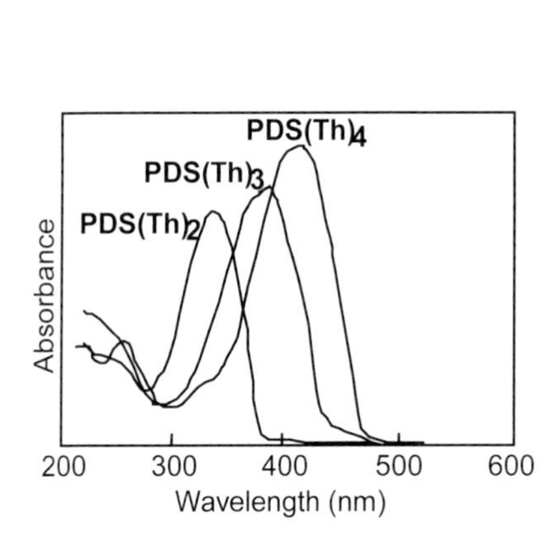

Fig. 1. Absorption spectra of PDS(Th)$_m$, m = 2 – 4.

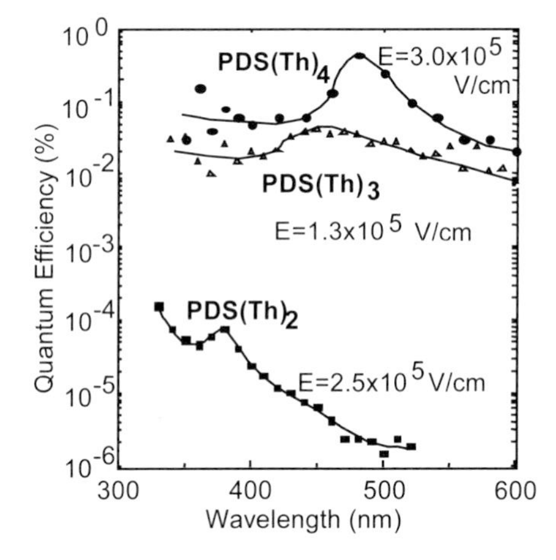

Fig. 2. Quantum efficiency of photocarrier generation of PDS(Th)$_m$, m = 2 – 4.

The generated excitons must be quenched because the density of the excitons in this area is probably very high. Consequently, the peaks of the photocurrent quantum efficiency appear at the edge of the absorptions, which establish bulk excitation in the whole film.

Interestingly, the photocurrent quantum efficiency of both $PDS(Th)_3$ and $PDS(Th)_4$ is determined to be in the order of $10^{-1} - 10^{-2}$ % at an electric field of about 10^5 V cm^{-1}, while that of $PDS(Th)_2$ is measured to be in the order of 10^{-5} %, which is smaller by two or three orders of magnitude. These results may be ascribed to their different electronic structures. Tanaka et al. have reported the electronic structure of simplified poly(disilanyleneoligothienylene)s $[SiH_2SiH_2(Th)_m]_n$ with $m = 1 - 5$. Their results show that the value of the π–π* band gap decreases as the number of thiophene rings (m) increases, whereas that of the σ–σ* band gap does not change significantly [7]. As a result, the band gap of the π–π* becomes smaller than that of the σ–σ*, when m is larger than 3. Therefore, photocarrier generation from the polymers having increased π-conjugation would be expected to be high. We suppose that in the case of $PDS(Th)_3$ and $PDS(Th)_4$, π-conjugation length is enough to generate photocarriers effectively, but in the case of $PDS(Th)_2$, both π-conjugation and σ-conjugation length are not sufficient. $PDS(Th)_3$ and $PDS(Th)_4$ are indeed photoconducting in the visible light region, but their quantum efficiency is only 2.0 % at high electric field ($E = 6\times10^5$ V m^{-1}).

Hole-Drift Mobility of $PDS(Th)_4$

Figure 3 shows the field dependence of the hole-drift mobilities of $PDS(Th)_4$ at room temperature (293 K), which were measured by the time-of-flight (TOF) method. The logarithm of the mobility depends on the square root of the field strength, and the mobility varies from 1.1×10^{-4} cm^2 V^{-1} s^{-1} at 2×10^5 V cm^{-1} to 2.1×10^{-4} cm^2 V^{-1} s^{-1} at 6×10^5 V cm^{-1}. These results are of considerable interest because these values are on the same order as the hole-drift mobility of polysilanes previously reported [2]. To our knowledge, this is the highest mobility in σ–π conjugated polymers reported so far. The carrier mobility of polythiophene and alkylpolythiophenes are reported to be in the order of 10^{-5}–10^{-4} cm^2 V^{-1} s^{-1} [8, 9]. On the other hand, Garnier et al. [10] reported that the hole mobilities of some oligothiophenes as thin-film field-effect transistors, and sexithiophene has the highest mobility of 2×10^{-3} cm^2 V^{-1} s^{-1} among them. Recently, Dodabalapur et al. [11] reported a higher mobility, $(1-3)\times10^{-2}$ cm^2 V^{-1} s^{-1} in the same construction using sexithiophene.

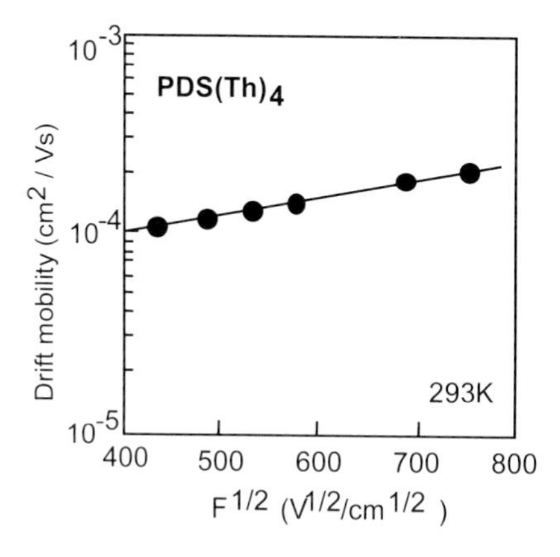

Fig. 3. Field dependence of the hole-drift mobilities in $PDS(Th)_4$ at room temperature.

However, quaterthiophene and alkyl-substituted quaterthiophene showed a lower hole mobility (2×10^{-7} and 5×10^{-5} $cm^2 V^{-1} s^{-1}$, respectively) than those of the polythiophene and alkylpolythiophene. Comparing these results with the high-drift mobility of PDS(Th)$_4$, it seems likely that some effect of σ–π conjugation between the oligothiophenes and the Si-Si unit, or some π-conjugation interaction between oligothiophene units, is present in our system.

Effect of C_{60} Doping

An increase in the photocarrier generation of poly(methylphenylsilane) (PMPS) by fullerene (C_{60}) doping has been previously reported by Wang et al. [12]. In this case, no change is observed in its absorption spectra, but the photoluminescence is effectively quenched [13]. In order to increase the photocarrier generation efficiency of the present polymer in the visible light region, we examined the effect of the C_{60} doping.

First, we prepared a film of PDS(Th)$_4$ doped with 1.5 wt % of C_{60} on a quartz plate, and measured the changes in the absorption and photoluminescence spectra. No change was observed in the absorption spectra, indicating that no charge transfer occurs in the dark, but the photoluminescence of PDS(Th)$_4$ at about 530 nm was strongly quenched by addition of a small amount of C_{60}, similar to the case of PMPS.

The ionization potential of PDS(Th)$_4$ measured by a low-energy photoelectron emission analyzer was found to be 5.5 eV, and the bandgap of PDS(Th)$_4$ was estimated to be 2.6 eV from its absorption spectrum. On the basis of these results, the energy diagram of PDS(Th)$_4$ and C_{60} is shown in Fig. 4. As for PMPS, the valence band (V.B.) of PDS(Th)$_4$ is located 0.2 eV lower than the LUMO of C_{60}, and consequently direct charge transfer probably does not occur. However, on exposure to light whose energy exceeds the bandgap, an electron may be transferred from the excited state of PDS(Th)$_4$ to C_{60}. Therefore, the photocarrier generation efficiency would be expected to be increased with fullerene doping.

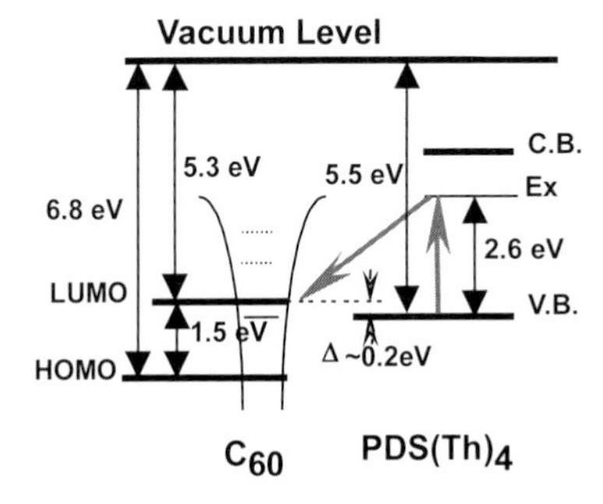

Fig. 4. Energy diagram of PDS(Th)$_4$ and C_{60}.

The photocurrent quantum efficiencies of undoped PDS(Th)$_4$ and PDS(Th)$_4$ doped with 1.5 wt % C_{60} versus wavelength are shown in Fig. 5. The photocurrent quantum efficiency of the doped sample is greatly enhanced, compared with that of the undoped one. In fact, the photocurrent quantum efficiency of undoped PDS(Th)$_4$ was only 0.5 % ($E = 3\times10^5$ V cm^{-1}) upon irradiation at 480 nm. The photocurrent quantum efficiency of the C_{60}-doped PDS(Th)$_4$, however, increased to 11.5 % ($E = 1.5\times10^5$ V cm^{-1}) upon irradiation at 470 nm. Furthermore, Fig. 6 shows the field dependence of the photocurrent quantum efficiency of $PDS(Th)_4$ and $PDS(Th)_4$ doped with C_{60}. The quantum efficiency of the doped sample increases with the field strength, and reaches a maximum value of 85 % at a field of 3×10^5 V cm^{-1}.

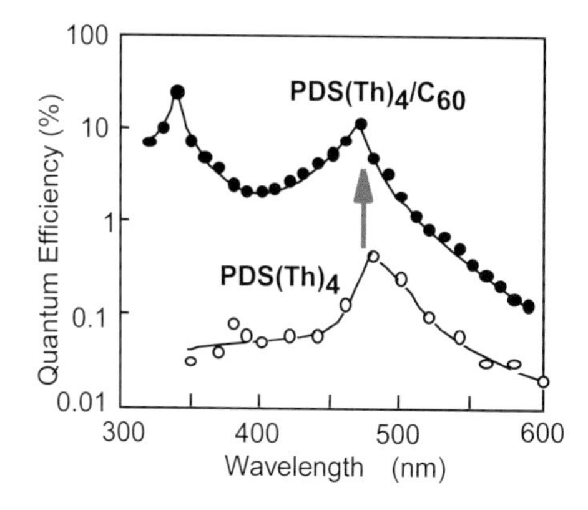

Fig. 5. Wavelength dependence of photocurrent quantum efficiency of $PDS(Th)_4$ and $PDS(Th)_4/C_{60}$.

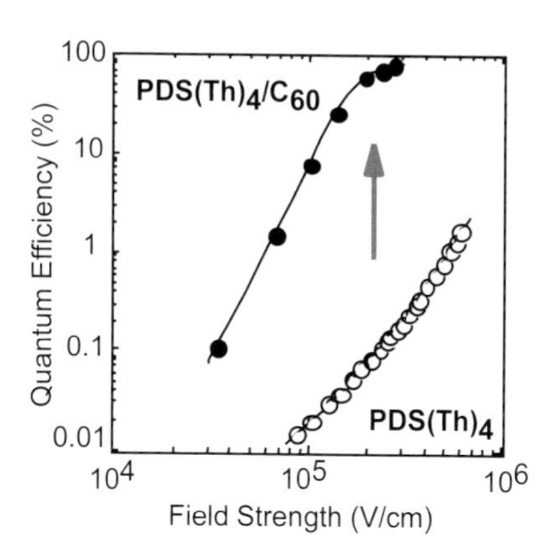

Fig. 6. Field dependence of photocurrent quantum efficiency of $PDS(Th)_4/C_{60}$ compared with that of $PDS(Th)_4$.

This enhancement of the photoconductivity caused by C_{60} doping may be ascribed the photoinduced electron transfer. We have reported that while the photoinduced electron transfer from polysilanes with alkyl substituents to C_{60} does not occur, the electron transfer from polysilanes with aromatic side groups (e.g. PMPS) to C_{60} is effective [13]. Yoshino et al. have also reported that the photoconduction of poly(3-alkylthiophene) is remarkably enhanced upon C_{60} doping [14]. Sariciftci et al. reported the similar photoinduced electron transfer in oligothiophene/C_{60} composite films [15]. Therefore, we believe that the photoinduced electron transfer occurs between π-orbitals of the thienylene rings and those of C_{60}.

Electrostatic Properties as Photoreceptors for Electrophotography

The above results suggest that poly(disilanyleneoligothienylene)s are expected to be useful photoconductive materials that can be applied to electrophotography. We have prepared the single-layered photoreceptors of $PDS(Th)_4$ and C_{60}-doped $PDS(Th)_4$ ($PDS(Th)_4/C_{60}$), which were coated about 10 μm deep on an aluminum substrate, which is illustrated in Fig. 7, and investigated their electrostatic properties. The photoinduced discharge was measured using a systematic electrophotographic analyzer (EPA-8200, Kawaguchi Electric Works). The light source was a halogen lamp (100 W). The surface potential changes are shown in Fig. 8. The surfaces of both samples were easily charged to +750 V and photo-discharge was observed when they were exposed in a white light. The photo-induced discharge curve of $PDS(Th)_4/C_{60}$ shows an excellent photoresponse for a blue light as a positive charge mode photoreceptor. This result consistent with the high hole-drift mobility and high photocarrier-generation efficiency of $PDS(Th)_4/C_{60}$.

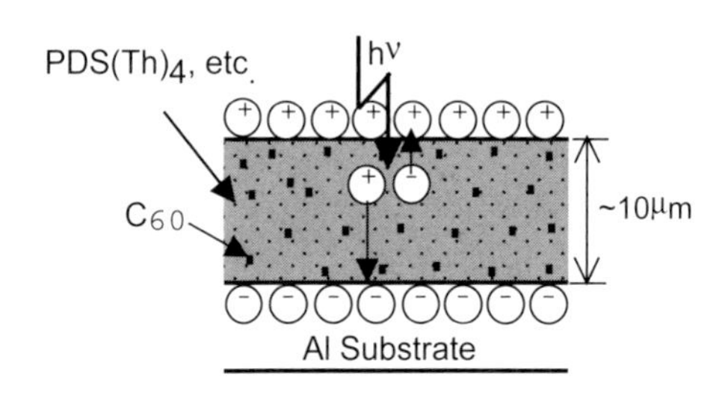

Fig. 7. Proposed structure of the single-layered, positively charged photoreceptor using silicon-based polymers.

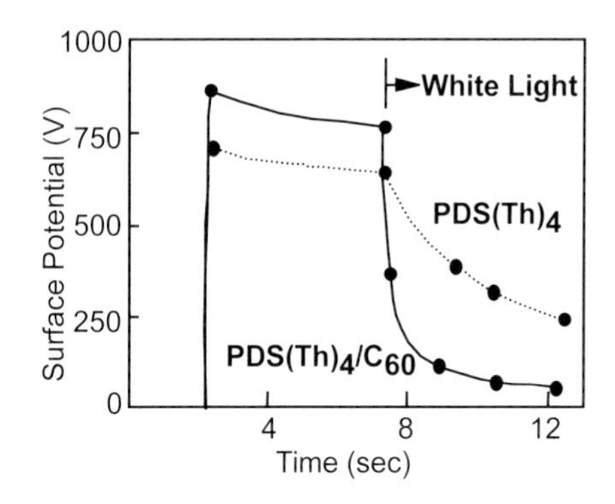

Fig. 8. Photoinduced discharge curves for $PDS(Th)_4$ and $PDS(Th)_4/C_{60}$ photoreceptors using 100 W halogen lamp (200 lux).

Conclusion

The photoconductive properties of poly(disilanyleneoligothienylene)s [$PDS(Th)_m$, $m = 2–4$] with or without C_{60} doping have been investigated. $PDS(Th)_4$ showed the photoconduction at 480 nm, providing the first example of visible photoconduction in silicon-based polymers, and its hole-drift mobility was in the order of 10^{-4} cm^2 V^{-1} s^{-1}, at room temperature. The photocurrent quantum efficiency for $PDS(Th)_4$ was increased to 85 % by C_{60} doping ($E = 3\times10^5$ V cm^{-1}, 470 nm), and this doping effect may be explained by the photoinduced charge transfer mechanism. Furthermore, we have prepared the single-layered photoreceptors of C_{60}-doped $PDS(Th)_4$ and investigated their electrostatic properties. The photoinduced discharge curve of surface potential shows an excellent photoresponse as a positive charge mode photoreceptor. This result consistent with the high hole drift mobility and high photocarrier generation efficiency of $PDS(Th)_4/C_{60}$.

Acknowledgment: The authors wish to thank Professor Mitsuo Ishikawa of Kurashiki University of Science and the Arts for his helpful advice concerning polymer synthesis. We also wish to thank Professor Masaaki Yokoyama of Osaka University for the carrier drift mobility measurements and helpful advice. This work was performed by *Sumitomo Electric Industries, Ltd.*, under the management of the *Japan Chemical Innovation Institute* as a part of the Industrial Science and Technology Frontier Program supported by *the New Energy and Industrial Technology Development Organization.*

References:

[1] R. D. Miller, J. Michl, *Chem. Rev.* **1989**, *89*, 1359; and references therein.

[2] R. G. Kepler, J. M. Zeigler, L. A. Harrah, S. R. Kurtz, *Phys. Rev. B* **1987**, *35*, 2818.

[3] G. G. Malliaras, J. K. Herrema, J. Wildeman, R. H. Wieringa, R. E. Gill, S. S. Lampoura, G. Hadziioannou, *Adv. Mater.* **1993**, *5*, 721.

[4] J. Ohshita, D. Kanaya, M. Ishikawa, *Appl. Organomet. Chem.* **1993**, *7*, 269.

[5] A. Kunai, T. Ueda, K. Horata, E. Toyoda, J. Ohshita, M. Ishikawa, K. Tanaka, *Organometallics* **1996**, *15*, 2000.
[6] M. Kakimoto, H. Kashihara, T. Kashiwagi, T. Takiguchi, J. Ohshita, M. Ishikawa, *Macromolecules* **1997**, *30*, 7816.
[7] K. Tanaka, H. Ago, T. Yamabe, M. Ishikawa, T. Ueda, *Organometallics* **1994**, *13*, 3496.
[8] A. Tsumura, H. Koezuka, T. Ando, *Synth. Met.* **1988**, *25*, 11.
[9] A. Assadi, C. Svensson, M. Willander, O. Inganas, *Appl. Phys. Lett.* **1988**, *53*, 195.
[10] F. Garnier, *Pure Appl. Chem.* **1996**, *68*, 1455; and references therein.
[11] A. Dodabalapur, L. Torsi, H. E. Katz, *Science* **1995**, *268*, 270.
[12] Y. Wang, R. West, C. -H. Yuan, *J. Am. Chem. Soc.* **1993**, *115*, 3844.
[13] K. Yoshino, K. Yoshimoto, M. Hamaguchi, T. Kawai, A. A. Zakhidov, H. Ueno, M. Kakimoto, H. Kojima, *Jpn. J. Appl. Phys.* **1995**, *34*, L141.
[14] K. Yoshino, X. H. Yin, S. Morita, T. Kawai, A. A. Zakhidov, *Solid State Commun.* **1993**, *85*, 85.
[15] N. S. Sariciftci, L. Smilowitz, A. J. Heeger, F. Wudl, *Science* **1992**, *258*, 1474.

Application of Functionalized Polysilanes in Organic Light-Emitting Diodes

Florian Lunzer, Christoph Marschner

Institut für Anorganische Chemie, Technische Universität Graz
Stremayrgasse 16IV, A-8010 Graz, Austria
Tel.: Int. code + (316)873 8209 — Fax: Int. code + (316)873 8701
E-mail: marschner@anorg.tu-graz.ac.at

Markus Wuchse, Stefan Tasch, Günther Leising

Institut für Festkörperphysik, Technische Universität Graz
Petersgasse 16/II, A-8010 Graz, Austria
Tel.: Int. code + (316)873-8460 Fax: Int. code + (316)873-8478
E-mail: leising@stg.tu-graz.ac.at

Keywords: Light-Emitting Diodes / Hole Transport Layer / Dehydropolymerization / Polysilanes

Summary: The use of differently substituted polysilanes as hole transport layers (HTL) in organic light-emitting diodes (OLEDs) was investigated. While hydrogen-substituted polysilanes deteriorate the diode behavior amino-substituted polysilanes lead to a significant improvement of the diode performance compared to a single-layer OLED.

Introduction

Organic light-emitting diodes (OLEDs) based on thin conjugated polymer films have attracted much interest recently because of their possible application in large-area flat-panel displays [1]. The use of double layer structures employing a hole transport layer (HTL) leads to improved device performance (Fig. 1).

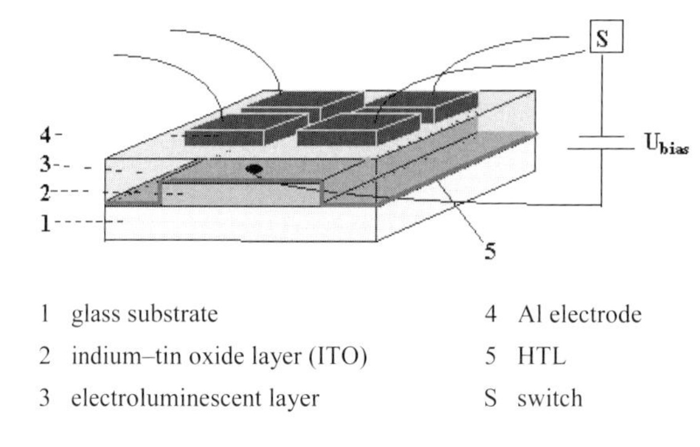

Fig. 1. Scheme of a two-layer OLED device.

The HTL enhances exciton formation and recombination in the emissive layer by blocking the electrons away from the ITO anode and efficiently injecting holes into the electroluminescent layer. Polysilanes which are insulators for electrons have a good hole mobility of around 10^{-4} cm^2/V s due to σ-conjugation of the electrons along the polymer chain. Moreover they are usually transparent in the whole visible region. These properties make them interesting candidates for application as hole transport layers in LEDs.

Up to now only permethylated or arylated polysilanes have been used as HTLs [2, 3]. The effects of other substituents on polysilanes synthesized by catalytic dehydrocoupling reaction on the device performance are investigated.

Synthesis of Polysilanes

Polyphenylsilane (**1**) and poly(1-methyl-2-phenyldisilane) (**2**) were synthesized by dehydrocoupling reaction of phenylsilane and 1-methyl-2-phenyldisilane, respectively, catalyzed by Cp_2ZrCl_2/2 BuLi (Eqs. 1, 2) [4].

Polyaminophenylsilane (**3**) was synthesized by copolymerization of phenylsilane with ammonia catalyzed by Cp_2TiMe_2 (Eq. 3) [5].

Eq. 1.

$$PhSiH_3 \xrightarrow{\text{cat}} \underset{\mathbf{1}}{H[SiPhH]_nH}$$

Eq. 2.

$$H_2MeSiSiPhH_2 \xrightarrow{\text{cat}} \underset{\mathbf{2}}{H[SiMeH]_n[SiPhH]_mH}$$

Eq. 3.

$$PhSiH_3 + NH_3 \xrightarrow{\text{cat}} \underset{\mathbf{3}}{H_2N[SiPhNH_2]_nNH_2}$$

OLED Fabrication

Polysilanes are spin-coated onto a transparent ITO substrate, then the luminescent layer — tris(8-hydroxyquinoline)aluminum (Alq_3) — and the Al electrode are evaporated onto the substrate (Fig. 2).

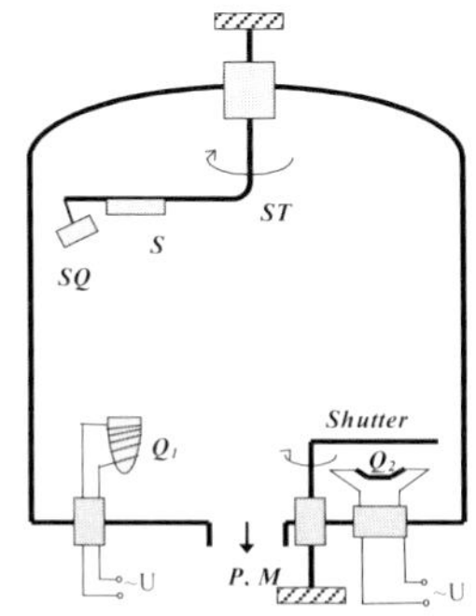

P, M	to turbomolecular rotarypump and manometer
Q1	source for organic materials
Q2	source for metals
S	substrate
ST	turnable substrate holder
SQ	microbalance for evaporation control

Fig. 2. Scheme of the evaporation setup.

Results

While polyphenylsilane (**1**) and poly(1-methyl-2-phenyldisilane) (**2**) rather deteriorate the diode behavior polyaminophenylsilane (**3**) strongly reduces the onset voltage of the devices and more than doubles the power efficiency and brightness compared to the single-layer LEDs. (Figs. 2, 3) These results compare favorably to the data Kido et al. reported for double-layer devices using Alq_3 as emissive layer and polymethylphenylsilane (PMPS) as HTL [2]. In this case a luminescence of 115 cd/m^2 was measured at a current of 10 mA. For the device Al/Alq_3/polyaminophenylsilane/ITO, 130 cd/m^2 were obtained at the same current density.

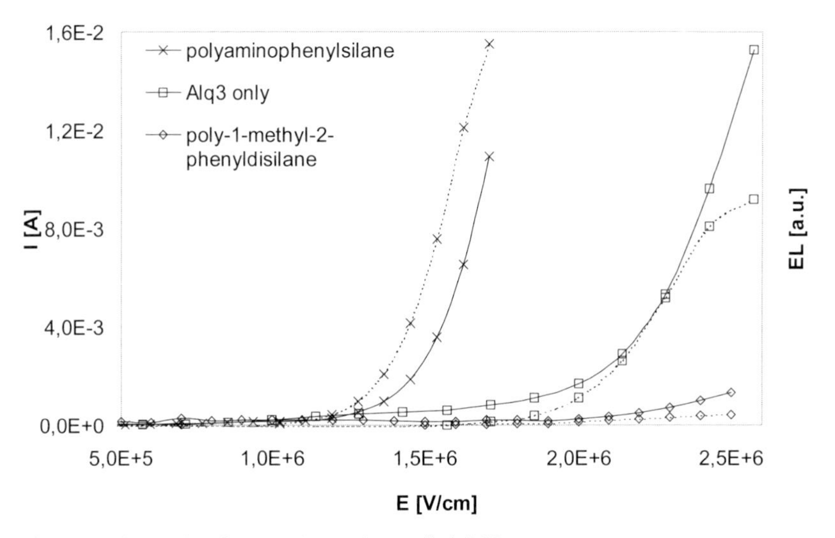

Fig. 3. Luminescence (·········) and current (———)vs applied field.

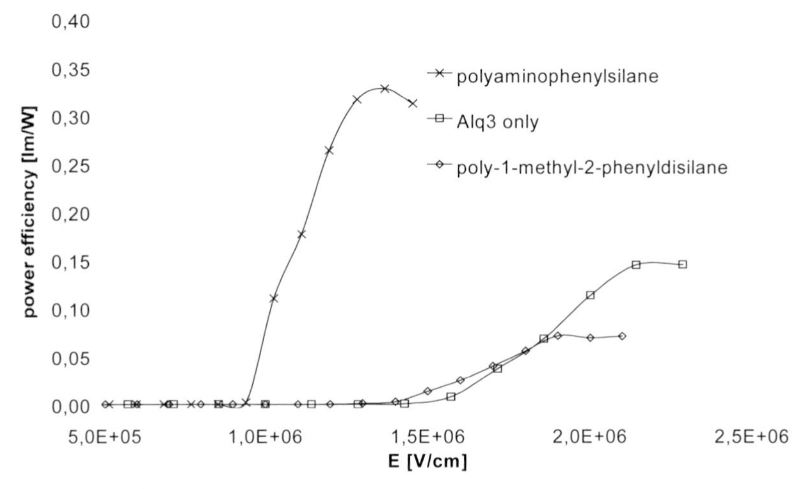

Fig. 4. Power efficiency of the prepared diodes.

As the diodes suffer considerable heating under work conditions the better performance of the latter polysilane may partly be due to the better thermal stability of aminated polysilanes compared to hydrogen-substituted polymers. Thermogravimetric measurements showed a stability up to 250°C for polyaminophenylsilane while hydrogen-substituted polymers undergo a significant weight loss at temperatures above 100°C.

The electronic properties of the polysilanes in discussion which might also be a reason for the differences in their behavior as HTLs; the workfunctions and the hole conductivity are currently under investigation.

Acknowledgment: This work was supported by the *Sonderforschungsbereich Elektroaktive Stoffe*, Graz. We want to thank the *Wacker-Chemie GmbH*, Burghausen for the friendly donation of chlorosilanes. C. Marschner thanks the *Austrian Academy of Sciences* for an APART scholarship (Austrian Program for Advanced Research and Technology).

References:

[1] A. Kraft, A. C. Grimsdale, A. B. Holmes, *Angew. Chem.* **1998**, *110*, 416.
[2] J. Kido, K. Nagai, Y. Okamoto, T. Skotheim, *Appl. Phys. Lett.* **1991**, *59*, 2760.
[3] H. Suzuki, H. Meyer, S. Hoshino, D. Haarer, *J. Appl. Phys.* **1995**, *78*, 2684.
[4] T. Imori, T. D. Tilley, *Polyhedron* **1994**, *13*, 2231 and references cited therein.
[5] H. Q. Liu, J. F. Harrod, *Organometallics* **1992**, *11*, 822.

Commercial Hybrid Organic–Inorganic Polymers

Barry Arkles

Gelest, Inc.
612 William Leigh Drive
Tullytown, PA 19007

Keywords: Polymers / Hybrid Organic–Inorganic Polymers / Polymer Network / IPN / Hybrid Polymer Technologies / Contact Lenses / Coatings / Surfactants / Wetting Agents / Silicones / Silicone Polyimides / Siloxane Resins

Summary: The ability to understand and control synthesis of both inorganic and organic polymers has led to the design of a wide range of novel structures. Block polymer, interpenetrating polymer network (IPN), biomimetic, ormosil and nanocomposite approaches have been applied to generate materials with unique structure–property relationships. These properties allow applications in optical and automotive coatings, injection molding and extrusion grades of thermoplastics, and preceramic polymers. Examples of current industrial technology are provided in the context of an overview of hybrid organic–inorganic polymers.

Introduction

Recent attention directed to hybrid organic–inorganic polymers suggests a new field of development. Surprisingly, successful commercial hybrid organic–inorganic polymers have been part of manufacturing technology since the 1950's. During any particular time frame, the current level of understanding of polymer chemistry and structure–property relationships led to the generation of hybrid materials with unique properties. Today's focus on hybrid organic–inorganic polymers promises to yield many new materials. Those materials that demonstrate viability in the crucible of the marketplace must either possess unique properties that enable new end-use applications or are well outside the cost-performance envelope of existing commercial materials. Looking at current hybrid polymers that have won niches in the marketplace suggests directions for future winners.

There is no accepted definition for the bewildering variety of materials that are described as hybrid organic–inorganic polymers. A tentative definition is that hybrid organic–inorganic polymers consist of clear regions or morphologies in which organic structures (C, H, N, O) dominate and separate regions in which distinct structures imposed by heteroatoms dominate and that physical properties of the polymers which are not a linear or geometric average of the regions can be observed. For example, poly(vinyltrimethylsilane) and poly(trimethylsilylpropyne), two polymers of interest in permselective membrane technology, are not considered hybrid polymers since there is no distinct region of the polymer which is associated with the silicon heteroatom. On the other hand, dimethylsiloxane-bisphenol A carbonate block polymers are considered hybrid polymers since they display independent glass transition temperatures associated with the inorganic and organic regions.

Two structural paradigms help visualize hybrid systems. Proceeding from an organic polymer perspective, the opportunity for hybrid systems clearly exists in copolymer, graft, block and interpenetrating polymer network morphologies (Fig. 1). Although there are no good examples, homopolymers derived from macromers also have the potential for generating hybrids.

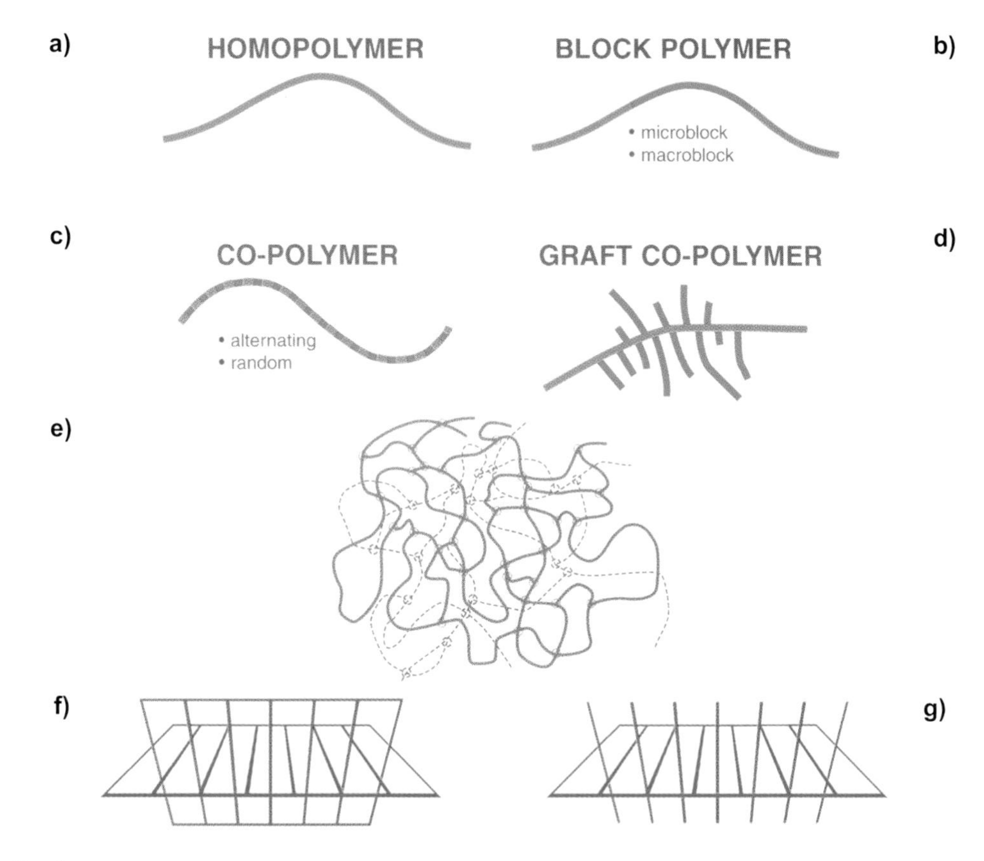

Fig. 1. Hybrid organic–inorganic polymer systems can be devised for all structural paradigms of polymer chemistry including (a) homopolymer, (b) block copolymer, (c) copolymer, (d) graft copolymer, and (e) interpenetrating polymer networks, including, shown as geometrical abstractions, (f) true and (g) semi-interpenetrating versions, where crosslinks are depicted as junctions of horizontal and vertical lines.

The only inorganic polymers that have either achieved or been seriously considered for commercial applications have been derived from the group IVA and IVB elements, of which silicon is preeminent. Consequently, the inorganic polymer perspective proceeds from the introduction of organic substituents into an amorphous polyoxymetalate structure that is associated with silicates and siloxanes. A scheme for naming these structures derives from the number of oxygens bound to each metal atom (Fig. 2). A metal with 4 oxygen substitutions is termed a Q resin. A metal with three oxygens and one organic substitution is termed a T resin. A metal with two oxygens and two organic substitutions is termed a D resin. For silicon a pure Q resin is SiO_2 or quartz, a T-resin is a silsesquioxane, and a D resin is a linear siloxane.

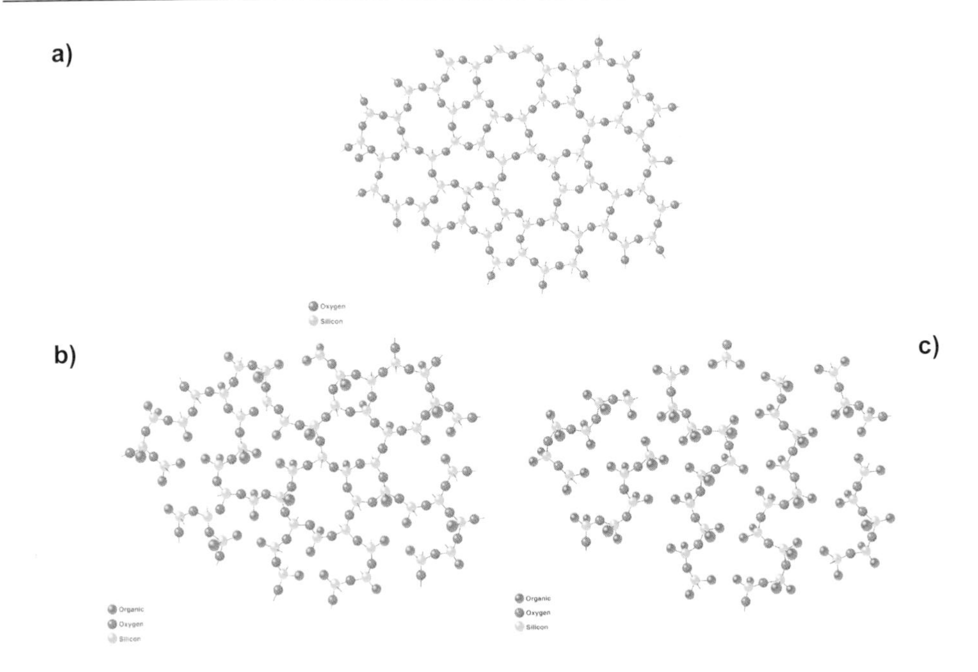

Fig. 2. Hybrid organic–inorganic polymers can be visualized as successive organic substitutions of polyoxymetallates. For example, a Q-structure (a), where 4 oxygens are bound to a metal atom (silicon dioxide or quartz in the example of silicon), gives rise to a less rigid T-resin (b) when there is one organic substitution on each metal atom and linear D-resins (c) when there are two organic substitutions on each metal atom (exemplified in the case of silicon by silicone oils); (red: organic substituent, blue: oxygen, yellow: silicon).

Copolymer Hybrids

Polyacrylates provide the backbone for both tin and silicon based polymers that find uses as disparate as materials for oxygen permeable contact lenses and marine anti-foulant coatings. Other low volume applications include photolithography for conventional graphics as well as microelectronics, oxygen permeable films for membrane-enrichment technology and food packaging, and as charge carriers and dispersants for pigments utilized in reprographics. A review of hybrid organic–inorganic copolymers derived from vinylic metal compounds, written outside the context of commercial application, demonstrates how few of the systems have demonstrated utility [1].

Acrylate Functional Silane Copolymers — Contact Lenses

Contact lenses represent a successful application of hybrid technology that evolves in response to continued challenges in product design and function. Maximum contact lens comfort is achieved with materials that allow the eye to "breathe". The primary design parameters that lead to selection of silicone materials are permeability (Dk) and equivalent oxygen percentage (EOP), a finished lens measurement that considers the oxygen demand of ocular tissue. A minimum EOP of 5–7 % has been proposed for finished lenses, equivalent to the oxygen available to the eye during sleep (when the eyelid covers the eye.) An EOP below 2 % causes corneal edema. Figure 3 shows how the EOP relates to permeabilities for various lens materials at different thicknesses [2].

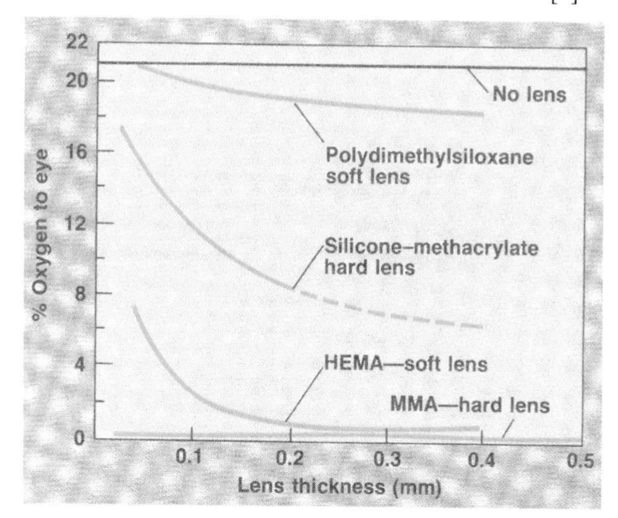

Fig. 3. Equivalent oxygen permeability comparison of contact lens materials. For reference air = 21 % O_2. (from Hill)

Other parameters requiring consideration in polymer design are wettability, dimensional stability and refractive index. The enhancement of oxygen permeability in siloxane lens systems is associated with high relative proportions of silicon–oxygen and silicon–carbon bonds. These long bonds lead to a free volume element which in the case of polydimethylsiloxane is 5–6 times greater than that for polymethacrylates [3].

The first methacrylate-silicone hybrid to achieve widespread commercialization was introduced by Syntex (now Wesley–Jessen) for rigid gas permeable (RGP) lenses based on the technology disclosed by Gaylord [4]. In the simplest example, methacryloxypropyltris(trimethylsiloxy)silane (**1**) or similar highly siloxy substituted methacrylates are copolymerized with other methacrylate monomers, by which means the oxygen permeability of siloxanes is combined with the mechanical and optical properties of the methacrylates [5].

$$H_2C{=}\underset{CH_3}{\underset{|}{C}}-\overset{O}{\overset{\|}{C}}-O(CH_2)_3\overset{OSi(CH_3)_3}{\overset{|}{\underset{OSi(CH_3)_3}{\underset{|}{Si}}}}-OSi(CH_3)_3$$

1

While RGP lenses offer the greatest oxygen permeability, visual acuity, and durability of any lens system, the extended break-in period for wearer comfort compared to hydrogel systems based on hydrated crosslinked hydroxyethylmethacrylate (HEMA), has led to an erosion of market share for these materials. Despite greater initial wearer comfort with HEMA lenses, the reduced oxygen permeability disallows extended periods of wear due to problems associated with corneal edema. Compared to MMA, the incorporation of siloxanes into HEMA systems is more challenging due to large differences in the solubility parameters, which frequently leads to polymer domain separation and opacity. Group transfer polymerization (GTP), by virtue of the ability to form hybrid block polymer systems, is a promising technology for the next generation of soft oxygen-permeable contact lenses. This is a method in which repeated addition of methacrylate monomers are made to a "living" polymer chain bearing a silylated ketene acetal (Eqs. 1 and 2) [6, 7].

$$(H_3C)_2C{=}C(OSi(CH_3)_3)(OCH_3) + H_2C{=}C(CH_3){-}C(=O){-}OCH_3 \xrightarrow{\text{catalyst}} H_3C{-}C(CO_2CH_3)(CH_3){-}CH_2{-}C(CH_3){=}C(OSi(CH_3)_3)(OCH_3)$$

Eq. 1.

$$n\ H_2C{=}C(CH_3){-}C(=O){-}OCH_3 + H_3C{-}C(CO_2CH_3)(CH_3)\ CH_2{-}C(CH_3){=}C(OSi(CH_3)_3)(OCH_3) \longrightarrow H_3C{-}C(CO_2CH_3)(CH_3){-}\left(CH_2{-}C(CO_2CH_3)(CH_3)\right)_n{-}CH_2{-}C(CH_3){=}C(OSi(CH_3)_3)(OCH_3)$$

Eq. 2.

The technology disclosed by Seidner incorporates methacryloxyethoxytrimethylsilane (**2**) (blocked HEMA) and methacryloxytris(trimethylsiloxy)silane as comonomers [8].

$$H_2C{=}C(CH_3){-}C(=O){-}O{-}CH_2CH_2O{-}Si(CH_3)_3$$

2

Contact lenses based on GTP technology are being introduced under the tradename Lifestyle® by Permeable Technologies (Fig.4).

Fig. 4. (Photo)-Rigid oxygen permeable contact lenses and a new generation extended wear hydrogel lens incorporate a variety of organosilicon substituted monomers (courtesy L. Seidner, Permeable Technologies).

Acrylate Functional Tin Copolymers — Marine Coatings

Until recently, the largest volume hybrid was tributyltinmethacrylate copolymers utilized as marine anti-foulant coatings. The effectiveness of the tin hybrid polymers is demonstrated in Fig. 5, which shows the lack of marine organism growth on a treated metal coupon centered over an untreated control. The economic importance of the technology can be demonstrated in the case of a very large cargo container operating at 15 knots for 300 days/year consumes $7 million in fuel. Increased drag due to marine fouling can easily increase fuel consumed to maintain speed by 30 % or over $2 million. First patented in 1965 [9], the potential of the tin hybrid polymers was not appreciated until the mid 70's [10] and did not achieve EPA approval until 1978. The enabling patent technology is credited to A. Milne [11]. The success of organotin copolymers is attributed to the dry polymer film integrity coupled with the slow electrolyte-catalyzed hydrolysis of the tin-oxygen bond which releases trialkyltin compounds (Scheme 1).

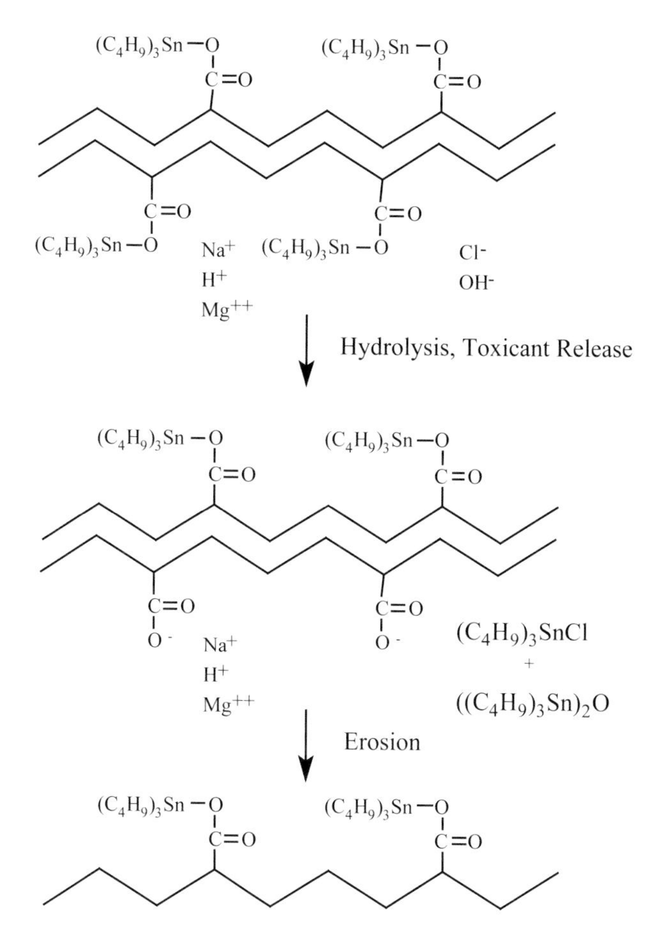

Scheme 1. Organotin copolymers releases trialkyltin compounds wich behave as toxicants.

The trialkyltins behave as toxicants while simultaneously forming a hydrophilic erodeable layer. The erodeable film is key since, after being removed by the physical action of seawater, a fresh active surface is exposed to the marine environment. The release of the tributyltin compounds is relatively linear. At the end of the release, the polymer film is totally removed and repaint of vessels can proceed without extensive surface preparation [12]. This class of polymer is currently offered by Elf Atochem and is marketed under the trade name Biomet® 300. The use of this technology has been curtailed by environmental regulations over recent years. Reevaluation, however, indicates that the tributyltinmethacrylates may actually be the anti-biofouling alternative with least environmental impact [13].

Fig. 5. Tributyltinmethacrylate copolymers provide protection against marine biofouling as shown by this coated test coupon mounted on an uncoated metal control and submerged in Biscayne Bay, Florida for twenty-four months (courtesy M. Gitlitz, Elf Atochem).

Hybrid Grafts

Commercial hybrid grafts include structures in which organic polymers are grafted onto a polysiloxane chain, such as polyethyleneoxide graft polydimethyl-siloxane surfactants, or structures in which alkoxysilanes are grafted onto organic polymers such as crosslinkable polyethylenes.

Surfactants and Wetting Agents

Polydimethylsiloxanes have many unique surface applications associated with the lowest surface tension of any non-fluorinated polymer and the high flexibility of the siloxane backbone [14].

Polydimethylsiloxanes achieve high concentrations at boundaries between different phases which accounts for applications such as release agents and defoamers. The combination of organic polyalkylene oxide pendants on the inorganic backbone of polydimethylsiloxane adds the ability to control wetting and surface tension. A typical structure is shown below and is alternately referred to as a block, graft, or pendant copolymer system. In earlier technology versions, the propyl link between the alkylene oxide and polysiloxane was replaced by a direct Si–O–C bond which is not hydrolytically stable.

$$H_3C-Si(CH_3)_2-O\left(Si(CH_3)_2-O\right)_m\left(Si(CH_3)(CH_2CH_2CH_2O(CH_2CH_2O)_xH)-O\right)_n Si(CH_3)_3$$

3

They are widely used as surfactants particularly in emulsification and polyurethane foam stabilization. They orient at air/liquid and solid/liquid boundaries in a manner suitable for lubrication of polymer interfaces.

There are several short reviews of the synthetic chemistry [15, 16] and physical properties for siloxane-alkylene oxide copolymers **3**. The synthesis generally proceeds through two steps, the production of polymethylhydrosiloxane-dimethylsiloxane copolymers and the subsequent platinum catalyzed hydrosilylation of an allyl terminated polyether (Eq. 3).

$$H_3C-Si(CH_3)_2-O\left(Si(CH_3)_2-O\right)_m\left(Si(CH_3)(H)-O\right)_n Si(CH_3)_3 \xrightarrow[Pt^0]{A} H_3C-Si(CH_3)_2-O\left(Si(CH_3)_2-O\right)_m\left(Si(CH_3)(CH_2CH_2CH_2O(CH_2CH_2O)_xR)-O\right)_n Si(CH_3)_3 \;\; \mathbf{3}$$

Eq. 3. Synthesis of **3** ($A = H_2C{=}CHCH_2O(CH_2CH_2O)_xR$).

Trisiloxanes are a special class of surfactants when $m = 0$. They orient to present a "cloud" of seven methyl groups to give a projected surface energy of about 20 dynes/cm, about 10 dynes/cm less than of the hydrophobic surface energy component of hydrocarbon surfactants. The small size of the trisiloxane unit compared to hydrocarbons also contributes to more rapid wetting and film spread. A significant body of enabling technology for the hydrolytically stable siloxane-alkylene oxide copolymers is credited to W. Reid [17] at Union Carbide (now Witco–OSi). The company markets the products under the name Silwet®. The technology of these products is the subject of an excellent review [18].

Crosslinkable Polyethylene (XLPE) Through Siloxane Bond Formation

Polyethylene has a desirable balance of electrical, mechanical and processing properties that has led to acceptance as a wire and cable insulation material. The continuous operating temperature of polyethylene has been increased from 70°C to more than 90°C by peroxide and radiation crosslinking. These methods, which have shortcomings in production efficiency, product homogeneity and process safety, have been largely supplanted by methods which incorporate pendant alkoxysilanes. The alkoxysilanes crosslink by a hydrolytic mechanism to form silsesquioxane networks. The technology is widely accepted in wire and cable insulation, including telephone and medium voltage power cables. It has also extended its application areas to heat-shrinkable tubing and compression resistant foam.

There are two major embodiments of the technology. One is a graft technology in which vinyltrimethoxysilane is peroxide grafted to a polyethylene backbone prior to or concomitant with crosshead extrusion of the cable. If the grafting is done prior to the crosshead extrusion of the cable, the process is referred to as the Sioplas® Process and represents a two-step post-polymerization process technology[19, 20]. If the grafting is done concomitant with crosshead extrusion of the cable, the process is referred to as the Monosil® Process and represents a one-step post-polymerization process technology [21, 22]. In both processes, a tin catalyzed moisture crosslinking after the final extrusion completes the process (Eq.4).

$$\text{—}CH_2CH_2Si(OCH_3)_3 + (CH_3O)_3SiCH_2CH_2\text{—} \xrightarrow{H_2O\ /\ R_2Sn(OR)_2} \text{—}CH_2CH_2Si(O\text{—})_2\text{—O—}Si(O\text{—})_2CH_2CH_2\text{—}$$

Eq. 4.

Companies that practice the graft technology include Quantum Chemicals. Cable processors that practice both graft and crosslink technology include BICC, Alcan, and Okonite.

The success of silane graft technology for crosslinkable HDPE was based in part upon the ability to produce product in conventional thermoplastic process equipment with relatively little new capital investment. As product benefits became recognized and markets grew, it became possible to consider copolymerization technology which has a much greater capital barrier for commercialization. The enabling technology was patented by Mistubishi [23]. A peroxide initiated polymerization at 2500 atm of 0.5 to 3 % vinyltrimethoxysilane with ethylene generates copolymers with properties similar to graft copolymers (Eq. 5). In some cases, high temperature properties appear further enhanced, presumably due to the elimination of reactive tertiary carbon sites.

$$99\ H_2C{=}CH_2 + 1\ H_2C{=}CH\text{—}Si(OCH_3)_3 \xrightarrow[220°\ ROOR]{2500\ atm} \text{—}CH_2CH_2CH_2CH(Si(OCH_3)_3)CH_2CH_2\text{—}$$

Eq. 5.

Companies which have commercialized copolymer technology include Neste (VISICO®), Union Carbide (SI–LINK®) and AT Polymers (Aqua–Link®).

Moisture Cure RTV Sealants

A technology which is parallel in concept to crosslinkable HDPE is based on an end-group functionalization of polypropylene oxide [24]. The result is a liquid polymer which cures in the presence of moisture to form conformable low-cost sealants and caulks (Eq. 6). The technology has been commercialized by Kaneka under the name MS polymer.

$$CH_3Si(OCH_3)_2(CH_2)_3O(CH_2CH(CH_3)O)_n(CH_2)_3Si(OCH_3)_2CH_3 \xrightarrow[R_2Sn(OOOCR)_2]{H_2O} \left[-O-\underset{O-}{\overset{|}{Si}}(CH_3)(CH_2)_3O(CH_2CH(CH_3)O)_n(CH_2)_3\underset{O-}{\overset{|}{Si}}(CH_3)-O-\right]_{network}$$

Eq. 6.

Hybrid Block and Macromer Copolymers

Silicone-Polycarbonate Block Copolymers

The earliest well-characterized block copolymers to achieve commercial acceptance were the bisphenol A polycarbonate-polydimethylsiloxane polymers (**4**) developed by H. Vaughn at General Electric [25]. The resins exhibited a structure in which there was a Si–O–C transition between inorganic and organic blocks.

$$\left(-C_6H_4-C(CH_3)_2-C_6H_4-O\overset{O}{\overset{\|}{C}}O- \right)_m \left(-Si(CH_3)_2-O- \right)_n$$

4

Under the tradename Copel® LR Resin, silicone polycarbonates found applications ranging from oxygen enrichment membranes (including heart-lung machines), aerospace canopies and interlayers for "bullet-proof" glazing. They are prepared by the phosgenation of bisphenol A and chlorine terminated polydimethylsiloxane oligomer in methylene chloride with pyridine as a base acceptor (Eq. 7).

$$HO-C_6H_4-C(CH_3)_2-C_6H_4-OH \text{ (excess)} + Cl-Si(CH_3)_2-O\left(Si(CH_3)_2-O\right)_n-Si(CH_3)_2-Cl \xrightarrow{Cl-C(=O)-Cl,\ \text{pyridine}} \left(\left(C_6H_4-C(CH_3)_2-C_6H_4-OC(=O)\right)_m\left(Si(CH_3)_2-O\right)_n\right)$$

4

Eq. 7.

These materials have been described as random block copolymers in which the blocks are polydisperse with a fairly low degree of polymerization. They demonstrate outstanding physical properties. For example, a 50–55 wt.% bisphenol A copolymer has a tensile strength of 3000 psi (20.7 MPa), an elongation of 300 %, and an oxygen permeability of 50 $\times 10^{-9}$ $cm^3/(s, cm^2, cm\ Hg\Delta P)$ and can be cast to films <5000 Å thick. The Achille's heel of these polymers is their lack of long-term hydrolytic stability. More current versions of the technology eliminate the Si–O–C transition, replacing it with a stable urethane transition [26]. The technology is currently practiced in microelectrodes and "on-chip" oxygen electrodes such as those introduced by i–Stat. An example of the polymer structure, with the additional introduction of polar cyano groups which facilitate ion transport, is shown below [27].

$$\left[-NH-(CH_2)_3-\left[\left(Si(CH_3)_2-O\right)_x\left(Si(CH_3)(CH_2CH_2CH_2CN)-O\right)\right]_m-Si(CH_3)_2-(CH_2)_3-NH-C(=O)-\left(O-C_6H_4-C(CH_3)_2-C_6H_4-OC(=O)\right)_z\right]_n$$

5

Silicone-Polyimides

General Electric also developed a series of silicone polyimides. While the chemistry was first reported in 1966 [28], these early versions did not find commercial application until the mid 1980's when National Starch acquired the technology indirectly and marketed a series of resins through its Ablestik division for applications in microelectronics. These include die-bonding adhesive and dielectric packaging applications and have been sold under such tradenames as Rely–Imide®, Tabcoat® and Conductimer®. In early versions of the technology, the polymer structures were relatively simple and in their fully imidized forms, e.g. **6**, they were essentially thermoset in nature.

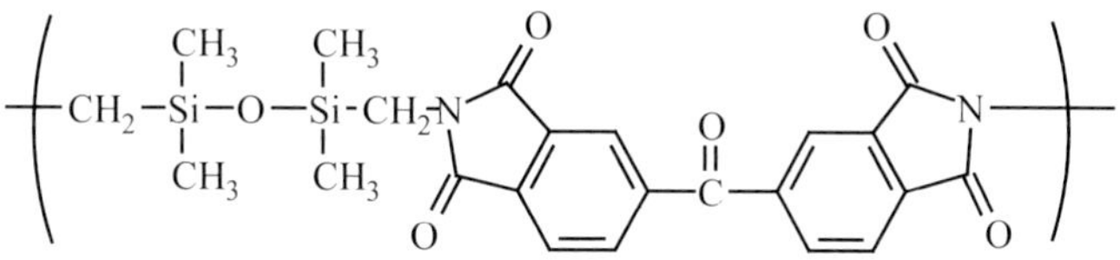

6

By carefully adjusting block size and comonomer composition, J. Cella at General Electric developed a series of thermoplastic silicone-polyimides (**7**) with excellent electrical and thermal properties as well as low smoke generation [29]. These materials are marketed under the name Siltem® as wire coatings and enamels, and can be found in a variety of aerospace applications including passenger aircraft.

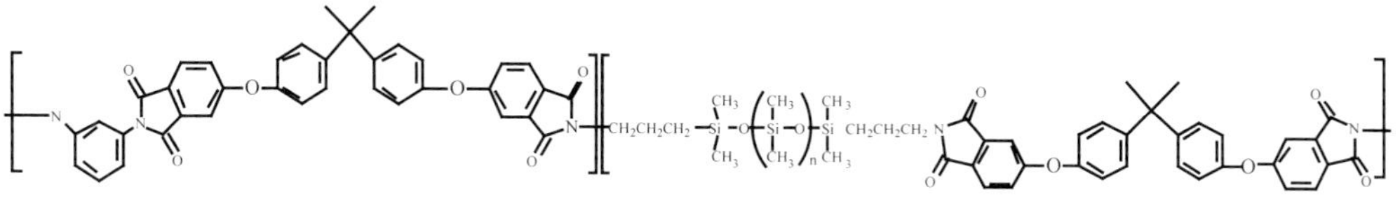

7

Diene-Siloxane Resins

Diene-Hydrosiloxane resins offer a radically different approach to hybrid organic–inorganic resins for electrical applications. The enabling technology was developed by R. Liebfried [30] at Hercules and was introduced to the market under the tradename Sycar®. The technology utilizes multifunctional hydrosiloxanes and diene monomers in a two-stage reaction technology dependent on hydrosilylation (Eq. 8). In the first stage, one of two olefin positions of the diene is reacted to give an "A-stage" resin which contains an approximate molar equivalent of unsaturated groups to reactive silicon hydrides. The surprising result is that this "A-stage" resin is a stable liquid. At elevated temperatures (>150°C), the reaction completes. Because both the diene and siloxane monomers have low dipole moments, the result is an impregnating resin or encapsulant with a dielectric constant of 2.6 and water absorption below 0.05 % after immersion in boiling water.

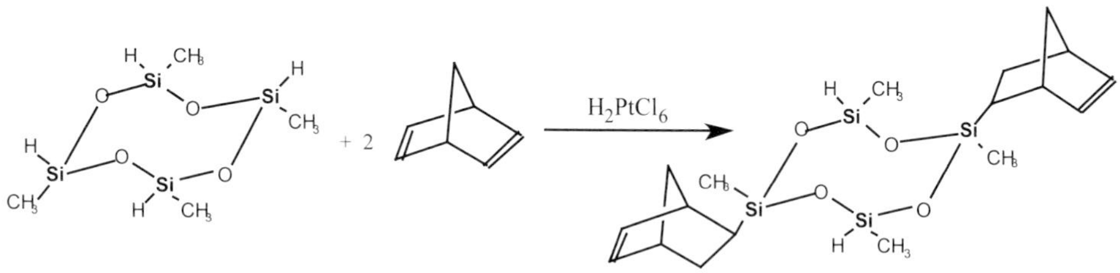

$$+\ 2 \xrightarrow{H_2PtCl_6}$$

Eq. 8.

This technology was recently acquired by National Starch/Ablestik and is finding applications as a laminating resin for printed circuit boards and for chip on board (COB) direct encapsulation of chips, without conventional packaging [31].

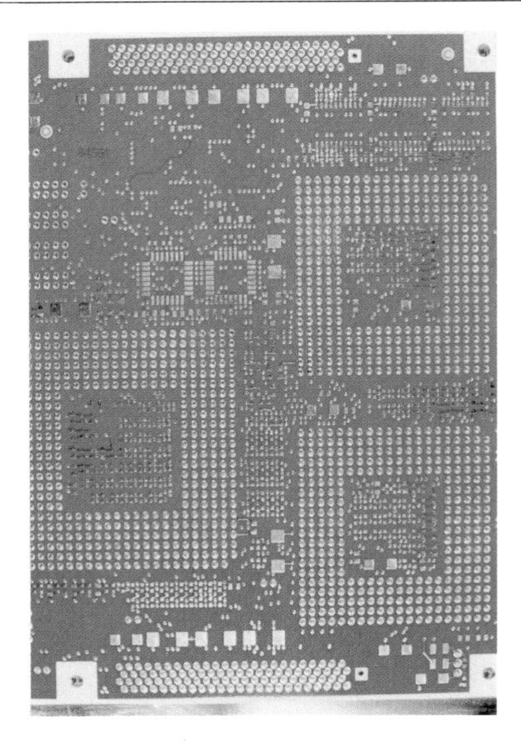

Fig. 6. The Diene-Siloxane resins have low dielectric constants allowing fabrication of multilayer printed circuit boards for use in computer applications (courtesy R. Liebfried, Hercules).

Interpenetrating Polymer Network Technology

True IPNs: Silicone-Urethanes

Fatigue strength, toughness, flexibility, and low interaction with plasma proteins are bioengineering criteria that are satisfied by a variety of silicone-urethane hybrids. The earliest example of hybrid organic–inorganic interpenetrating polymer technology commercialization is a bioengineering material introduced by Kontron Cardiovascular as Cardiothane® 51. The material was originally described as an aromatic polyether-polydimethylsiloxane copolymer [32]. More properly, the material can be described as an IPN containing domains of pure silicone and urethane [33]. Network formation is driven by an acetoxy cure silicone reaction. The material found primary use for the casting of medical device components such as blood pumps and intra-aortic balloons (Fig. 7).

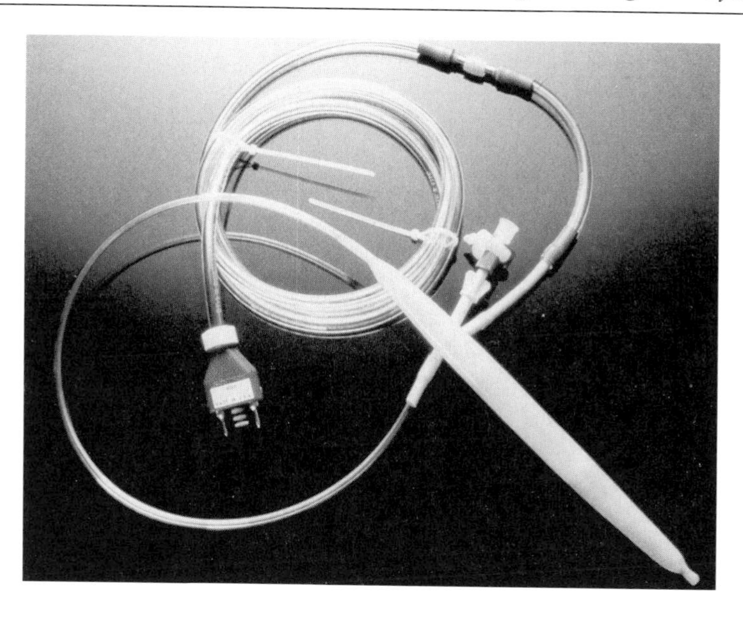

Fig. 7. Intra-aortic balloon pumps, requiring mechanical and fatigue strength along with physiological inertness, are fabricated from silicone-urethane IPNs (courtesy Kontron Cardiovascular).

Semi-IPNs: Silicone-Urethanes, Silicone-Polyamides

Classical or true IPNs are based on two thermosetting polymers that form crosslinks without combining with each other except by entanglement. Semi-IPNs are based on combinations of crosslinkable and nonreactive linear polymers in which mutual entanglement is maintained by the crosslinked resin. Silicone-thermoplastic semi-IPNs are formed by a reactive processing method in which a silicone is cross-linked in a linear high molecular weight thermoplastic. The enabling technology was disclosed by B. Arkles [34] and is offered under the tradename Rimplast® by LNP Corp. In the simplest embodiment of this method, separate extrusions of a vinyl functional siloxane and a hydride functional siloxane with a base resin are prepared. The materials are pelletized and mixed together with a platinum catalyst to form a homogeneous blend. During subsequent extrusion or injection molding of fabricated parts, a vinyl-addition crosslinking reaction is activated. The result is a silicone semi-IPN with hybrid properties: the mechanical properties of a thermoplastic and release, oxygen-permeability and blood–polymer interactions associated with silicones [35]. Network formation is frequently manifested in crystalline thermoplastics by changes in crystalline behavior including both Tm and anisotropy. Applications of silicone-urethane semi-IPNs include catheter tubing (Fig. 8). Applications of silicone-polyamide semi-IPNs include capstans for high speed paper and tape transport as well as optical component devices in photographic equipment.

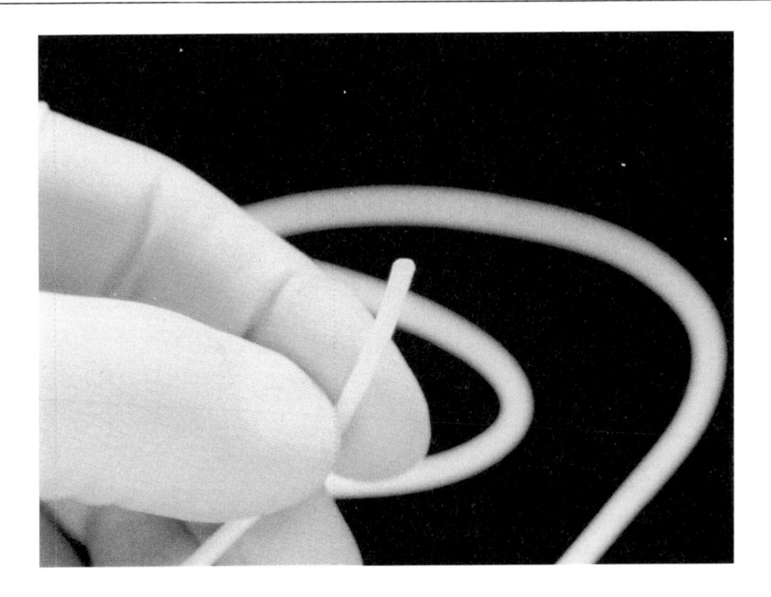

Fig. 8. Catheter tubing extruded from silicone-urethane semi-IPNs maintains physiological properties of silicones plus high speed processing and mechanical advantages of thermoplastic elastomers.

Hybrid Organic–Inorganic Resins

Hybrid organic–inorganic resins may be regarded as copolymer, graft or interpenetrating polymer network technologies. They typically utilize all three elements. Interestingly, the most prominent examples of commercialization represent the earliest and latest embodiments of hybrid organic–inorganic polymer technology.

The simple hydrolysis reaction products of organotrichlorosilanes were studied by K. Andrianov in the late 1930's [36]. These were the first examples of silsesquioxane or silicone T-resins. (J. F. Hyde was the first to directly polymerize and study linear diorganosiloxane polymers.) In the early 1950's, S. Brady and coworkers at Dow Corning produced a series of low molecular weight phenylsilsesquioxane-alkylsilsesquioxane copolymers of low molecular weight and high hydroxyl functionality [37]. A typical structure is shown below.

$$\left[-O-\underset{\underset{|}{O}}{\overset{CH_3CH_2CH_2}{Si}}-O-\underset{\underset{|}{O}}{\overset{C_6H_5}{Si}}-O-\underset{OH}{\overset{C_6H_5}{Si}}-O- \right]_{3\text{-}8}$$

8

The phenylsilsesquioxane copolymers can be reacted into a variety of standard alkyd resin formulations [38]. The result is a range of high performance weather resistant coatings used, for example, on superstructures for naval aircraft. Similarly, alkoxy functional silsesquioxane resins are incorporated into a variety of polyester formulations used in coil coating applications [39].

Commercial abrasion resistant coatings (ARCs) show a broad elaboration of silsesquioxane or T-resin technology. Simple unmodified methylsilsesquioxane resins of controlled molecular weight and hydroxyl content were the earliest examples of abrasion resistant coatings. Current ARC technology rarely uses these materials, although they retain important roles in electrical component coatings such as resistors and molding compounds as well as spin-on dielectrics in microelectronic interlayer dielectric and planarization applications. Tradenames include Techneglass® (NEC) and Accu–Spin® (Allied–Signal). The materials are prepared by controlled hydrolysis of alkyltrialkoxysilanes (Eq. 8).

$$H_3C-Si(OC_2H_5)_3 \xrightarrow{H_2O} \left[-O-Si(OH)(OCH_3)-O-Si(OCH_3)(OH)-O- \right]$$

Eq. 8.

While these materials cannot be defined as hybrid organic–inorganic polymers, many of their modifications are hybrids. For example, cohydrolysis with tetraethoxysilane hydrolysis incorporates Q-units into silsesquioxanes. In other technologies silica sols of low dimensionality are generated and alkyltrialkoxysilanes are reacted with them under hydrolytic conditions [40, 41]. Alternately, cohydrolysates with functional trialkoxysilanes can introduce specific reactivity or other physical properties. The ceramic community conceptualizes these resins as organic modified silicas and have dubbed them "Ormosils [42]". A large number of ormosils have been offered as scratch resistant coatings and antireflection layers for eyewear [43 – 45]. Many of the antireflection coatings substitute metals such as titanium for silicon in the Q portion of the structure. Companies which offer lens coatings include Essilor (Silor®), SDC, American Optical, and Gentec. In other variations ormosils modified by organic resins in a graft interpenetrating polymer network technology [46] are offered as protective coatings by Ameron under the tradename PSX®.

A new variant on the technology is a thermal or UV driven cure system which allows the formation of hybrid systems from modified T-resins. The technology exploits the rearrangement reaction of siloxanes substituted at the β-position with electron withdrawing groups (Eq. 9). The technology disclosed by B. Arkles and D. Berry [47, 48] initially described thermally driven chloroethylsilsesquioxane rearrangements, but more recent work has focused on acetoxyethylsilsesquioxanes. Copolymer versions of the technology lead readily to hybrid structures. Pure or modified silicon dioxide patterns can be written by thermal-cure micro-contact printing or UV laser [49].

$$-O-Si(CH_2CH_2Cl)(O-)-O-Si(CH_2CH_2Cl)(OH)-O- \xrightarrow[-\ CH_2=CH_2,\ HCl]{h\nu} -O-SiCl(O-)-O-Si(O-)(O-)-O- \xrightarrow[-HCl]{H_2O} \text{SiO}_2 \text{ network}$$

Eq. 9.

The technology has been introduced to the market under the tradename Seramic® SI. It is finding applications in the formation of dielectric structures in technologies such as flat-panel displays (Fig. 9).

Fig. 9. Micro-contact printing generated silicon dioxide patterns on silicon from β-chloroethylsilsesquioxanes give the ability to "build-up" rather than "etch-back" to create dielectric structures (courtesy R. Composto, Laboratory for Research on the Structure of Matter, U. of Pennsylvania).

Clear Scratch-Resistant Automotive Coatings

A dramatic new high volume application for hybrid organic–inorganic polymers is in large volume automotive coatings. In current automotive finish technology, coloration is provided in a pigmented basecoat. A clear scratch-resistant overcoat is applied which must not only satisfy optical and mechanical requirements, but must increasingly provide protection from environmental factors such as UV and chemical attack. In a technology introduced by Dupont as Generation 4® (Fig. 10), this is accomplished by utilizing two hybrid polymers systems which crosslink simultaneously during cure to form a polymer network which is partially grafted and partially interpenetrating. I. Hazan disclosed the enabling technology [50]. A high crosslink density acrylate tetrapolymer core which includes an alkoxysilane substitution, typically methacryloxypropyltrimethoxysilane, and residual unsaturation is generated which provides a high modulus scratch-resistant function. The high crosslink density core polymer is dispersed in a second low crosslink density terpolymer also containing an alkoxysilane which primarily provides film forming properties. Melamine-formaldehyde resins and catalysts are blended into the polymer dispersion allowing the following cure reactions (Eqs. 10–12).

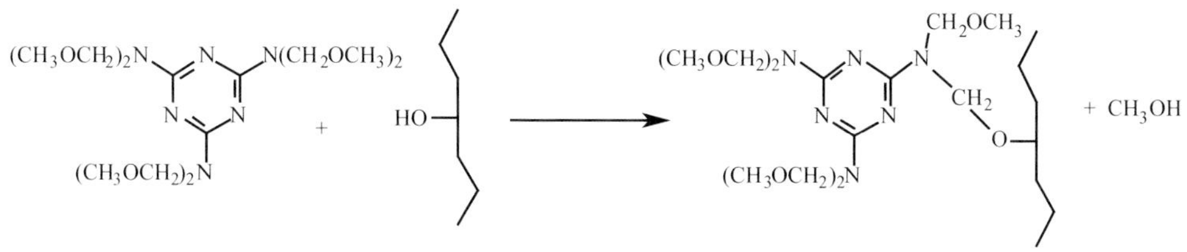

Eq. 10.

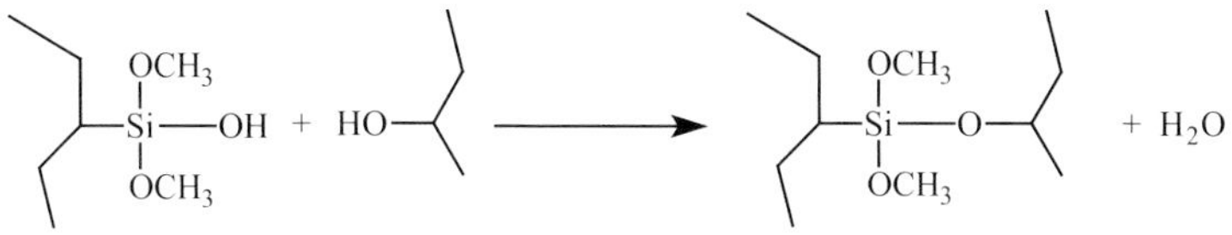

Eq. 11.

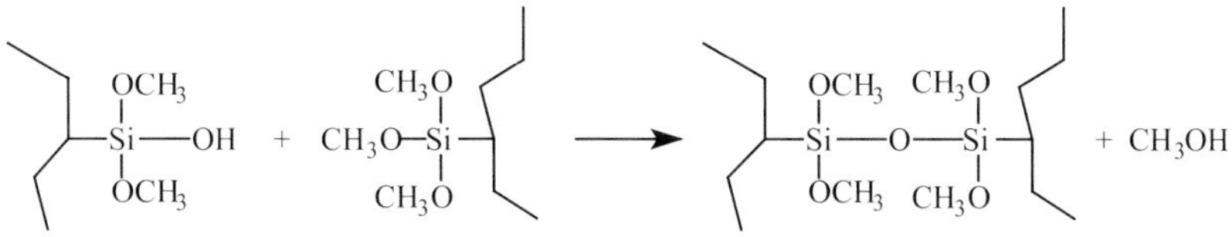

Eq. 12.

The superior scratch and environmental etch resistance of these coatings led to their acceptance as topcoats for 8 of the top 10 sellers for 1997, including Ford Taurus, Toyota Camry and Honda Civic/Del Sol (Fig. 11).

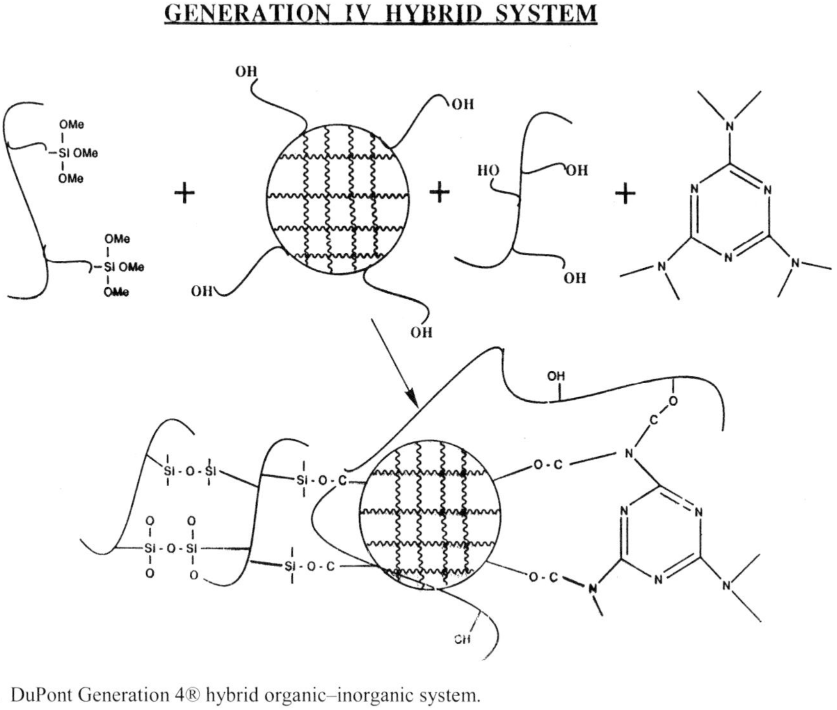

GENERATION IV HYBRID SYSTEM

Fig. 10. DuPont Generation 4® hybrid organic–inorganic system.

Fig. 11. Automotive finishes such as that on the Honda Civic have depth, luster and environmental resistance that is the result of a complex mixture of hybrid organic–inorganic materials cured into mixed graft interpenetrating polymer network systems.

Conclusion

At any point in the development of polymer chemistry the technical appreciation and understanding of the then current organic polymer chemistry and inorganic polymer chemistry led to the development of hybrid polymer technologies. The development of new hybrids depends on the ability to establish and understand structure–property and reactivity relationships. The viability of hybrids in the market place depends on achieving a set of properties, including economics outside the envelope created by more conventional materials. With increasing research focus on hybrids and a marketplace which has developed exceedingly complex demands, hybrid polymer systems are expected to have ever increasing impact.

References:

[1] C. Allen et al., *Heteroatom Chem.* **1993**, *4*, 159.
[2] R. Hill, *Contact Lens Manufactures Association*, **1981**, *Paper 10-81*, 240.
[3] K. Ziegel, F. Erich, *J. Poly. Sci., Part A2* **1970**, *8*, 2015.
[4] N. Gaylord, US Patent 3 808 178, **1974**; US Patent 4 120 570, **1978**.
[5] B. Arkles, *CHEMTECH* **1983**, *13*, 542.
[6] O. Webster, D. Sogah, in: *Silicon Chemistry* (Eds.: J. Corey, E. Corey), Gaspar Horwood–Wiley, **1988**, p. 41.
[7] W. Hertler, in: *Silicon in Polymer Synthesis* (Ed.: H. Kricheldorf), Springer, Berlin, **1996**, p. 69.
[8] L. Seidel et al., US Patent 5 244 981, **1993**.
[9] J. Leebrick, US Patent 3 167 473, **1965**.
[10] D. Atherton, J. Verbogt, M. A. M. Winkeler, *J. Coat. Tech.* **1978**, *51(657)*, 88.
[11] A. Milne, US Patent 4 021 392, **1977**.
[12] M. H. Gitlitz, *J. Coat. Tech.* **1981**, *53(678)*, 46.
[13] A. M. Rouhi, *Chem. Eng. News* **1998**, *41*, *April*, 27.
[14] M. J. Owen in: *Siloxane Polymers* (Eds.: S. Clarson, J. Semlyen), Horwood–Prentice Hall, **1993**, p. 309.
[15] C. Burger et al., in: *Silicon in Polymer Synthesis* (Ed.: H. Kricheldorf), Springer, Berlin, **1996**, p. 138.
[16] A. Noshay, J. McGrath in: *Block Copolymers*, Academic Press, **1977**, p 400.
[17] B. Kanner, W. G. Reid, I. Peterson, *Ind. Eng. Chem. Prod. Res. Dev.* **1967**, *6*, 88.
[18] R. M. Hill in: *Speciality Surfactants* (Ed.: I. Robb), Blackie Publishers, London, **1997**, pp. 143.
[19] H. Scott, US Patent 3 646 155, **1972**.
[20] B. Thomas, M. Bowrey, *Wire J.* **1977**, *10(5)*, 88.
[21] P. Swarbrick et al., US Patent 4 117 195, **1978**.
[22] M. Kertscher, *Rev. Gen. Caoutch. Plast.* **1978**, *55*, 57.
[23] T. Isaka et al., US Patent 4 413 066, **1983**.
[24] T. Watanabe et al., *Jpn. Kokai Tokkyo Koho*, JP 09 124 922; *Chem. Abstr.* **1998**, *127*, 35652.
[25] H. Vaughn, US Patent 3 189 662, **1965**; US Patent 3 419 634, **1968**; US Patent 3 419 635, **1968**.
[26] J. Riffle, R. Freelin, A. Banthia, J. McGrath, *J. Macromol. Sci. Chem.* **1981**, *A15*, 967.
[27] P. Kinson, C. Orland, J. Klebe, "*Chemical Modification of Siloxane Copolymers for Use as Membranes in pH or K^+ Electrodes.*" at 11th Organosilicon Symposium, **1977**.
[28] A. Berger, US Patent 3 274 155, **1966**.
[29] J. Cella, US Patent 4 808 686, **1989**; *ACS Polymer Reprints* **1985**, *26*.
[30] R. Liebfried, US Patent 4 900 779, **1990**.
[31] J. Bard, R. Brady, "*A New Moisture-Resistant Liquid Encapsulant*", in: *42nd ETC Conference Proc.* **1992**, p. 1018.
[32] E. Nyilas, US Patent 3 562 352, **1971**.
[33] R. Ward, E. Nyilas, in: *Organometallic Polymers* (Eds.: C. Carraher, J. Sheats, C. Pittman), Academic Press, **1978**, p .
[34] B. Arkles, US Patent 4 714 739, **1987**; US Patent 4 970 263, **1990**.

[35] B. Arkles, J. Crosby, in: *Silicon-Based Polymer Science* (Eds.: J. Zeigler, F. W. G. Fearon), American Chemical Society, **1990**, p. 181.

[36] K. Andrianov, *Metalorganic Polymers*, Wiley, **1965**; USSR Patent 55 899, **1937**.

[37] E. Warrick, in: *Forty Years of Firsts*, McGraw-Hill, **1990**, p. 212.

[38] L. Brown, in: *Treatise on Coatings, Vol. 1, Part III, Film-Forming Compositions* (Eds.: R. Myers, M. Dekker) **1973**, p. 513.

[39] W. Finzel, H. Vincent, *Silicones in Coatings*, Federation of Societies for Coatings Technology, Blue bell, PA, **1996**.

[40] Clark, US Patent 4 027 073, **1977**.

[41] D. Olsen et al., US Patent 4 491 508, **1985**.

[42] H. Schmidt in: *Better Ceramics Trough Chemistry* (Eds.: C. J. Brinker et al.), North. Holland, New York, **1984**, p. 327.

[43] G. Phillip, H. Schmidt, US Patent 4 746 366, **1988**; G. Phillip, H. Schmidt, in: *Ultrastructure Processing of Advanced Ceramics* (Eds.: J. MacKenzie, D. Ulrich), Wiley, New York, **1988**, p. 651.

[44] K. Mori et al., US Patent 4 895 767, **1990**.

[45] E. Yajima et al., US Patent 5 165 992, **1992**.

[46] R. Foscante et al., US Patent 4 250 074, **1981**; US Patent 5 275 645, **1994**; US Patent 5 618 860, **1997**; US Patent 5 804 616, **1998**.

[47] B. Arkles, D. Berry, L. Figge, US Patent 5 853 808, **1998**.

[48] B. Arkles, D. Berry, L. Figge, R. Composto, T. Chiou, H. Colazzo, *J. Sol-Gel Sci Tech.* **1997**, *8*, 465.

[49] J. Sharma, D. Berry, R. Composto, H. Dai, *J. Mater. Res.* **1999**, *14*, 990.

[50] I. Hazan, M. Rummel, US Patent 5 162 426, **1992**.

Inorganic–Organic Copolymers — Materials with a High Potential for Chemical Modification

Klaus Rose, Sabine Amberg-Schwab, Matthias Heinrich*

Fraunhofer-Institut für Silicatforschung
Neunerplatz 2, D-97082 Würzburg, Germany
Tel.: Int. code + (931)4100 626 — Fax: Int. code + (931)4100 559
E-mail: rose@isc.fhg.de

Keywords: Inorganic–Organic Copolymer / Hybrid Material / ORMOCER® / Sol–Gel / Functional Coatings

Summary: The surface properties of coatings derived from inorganic–organic copolymers were adjusted by the proper choice of monomeric organoalkoxysilanes of the general type $R'_nSi(OR)_{4-n}$ (n = 1 or 2). Special compounds with functional groups in R' were incorporated into an inorganic backbone via hydrolysis and condensation reactions during sol–gel processing forming an inorganic–organic hybrid material. Perfluorinated alkyl chains in R' reduce the surface energy, thus facilitating anti-adhesive behaviour of the resulting coating against polar and nonpolar substances. Due to the presence of ionic compounds, e.g. ammonium moieties, the specific surface resistance is decreased from 10^{15} to 10^8 Ω. Thus electrical charging of the surface is inhibited and the attraction of dust particles is avoided. For a special application in sensor technology a polyacryloxysiloxane based coating modified with secondary amines is used as a CO_2-sensitive layer on silica optical fibers. The reaction of amino groups with CO_2 can be detected by optical means.

Introduction

In many applications the mechanical, chemical or physical properties of material surfaces play an important role. If, for any reason, the requirements cannot be met by the bulk of the material, the application of coatings and surface modification is a convenient method to improve the properties. Increasing demands on new materials caused growing interest in inorganic–organic hybrid materials. They exhibit the potential to combine the inherent properties of inorganic materials like glass or ceramics with those of organic polymers [1]. During the last decade the sol–gel process has been established as a versatile method for the preparation of new inorganic–organic hybrid materials, also called ORMOCER®s (ORMOCER®: trademark of Fraunhofer-Gesellschaft zur Förderung der angewandten Forschung e.V. in Germany) [2]. Since the synthesis is carried out in solution, the resulting sols can be used as a lacquer in order to apply the hybrid material onto substrates for the improvement of surface properties. In this paper we describe how incorporated functional groups contribute to achieve the desired properties of the resulting coating material.

Material Synthesis

The sol–gel route, originally directed towards the synthesis of purely inorganic materials, is increasingly being extended to the preparation of inorganic–organic copolymers [3], especially in coating technology. Starting from hydrolyzable molecular compounds such as alkoxy compounds of silicon, for instance $R_2Si(OEt)_2$ or $RSi(OEt)_3$, hydrolysis and condensation is induced by the addition of water. Thus an inorganic silicone-like prepolymer or silica-like network is formed at low temperatures bearing functional organic groups R (Scheme 1).

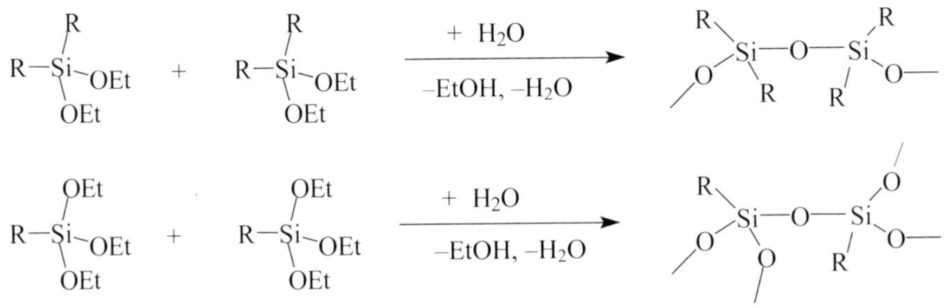

$$R_2Si(OEt)_2 + R_2Si(OEt)_2 \xrightarrow[-EtOH,\ -H_2O]{+\ H_2O} -O-SiR_2-O-SiR_2-O-$$

$$RSi(OEt)_3 + RSi(OEt)_3 \xrightarrow[-EtOH,\ -H_2O]{+\ H_2O} -O-SiR(O-)-O-SiR(O-)-O-$$

Scheme 1. Sol–gel synthesis of inorganic–organic hybrid materials.

The role of the organic groups R can be twofold. They can act as modifiers of the inorganic backbone, introducing properties which are well known from organic compounds or polymers as described later in this paper. Moreover, in the case of reactive methacryl or epoxy groups, formation of an additional organic network is possible by means of standard polymerisation techniques, generating an interpenetrating network on a molecular scale. In the case of ORMOCER® coating materials these polymerization reactions are used as curing reactions after the coating procedure, induced by UV radiation or thermal treatment (Fig. 1).

Fig. 1. Polymerizable ORMOCER® components.

Due to the great variety of available organic substituents, the resulting material can be tailored for many applications. Moreover, the direct linkage of the functional groups to the silicon, which is a part of the inorganic matrix, protects these groups from diffusion or bleeding and subsequent loss of activity, thus maintaining the properties of the material for the lifetime of the coated substrate. As a result of the synthesis which is carried out in a homogeneous solution, the formed material also exhibits a great homogeneity and can be described as a molecular composite material.

Generation of Low Surface Energy

Water-repellent, hydrophobic coatings exhibit a high surface interfacial tension and a low surface energy respectively, thus preventing wetting and spreading of water, but forcing drop formation, dripping and draining away. Oleophobic surfaces have even more decreased surface energy and therefore they show additional oil or fat repellency. Whereas substances bearing long alkyl chains are sufficient for a hydrophobic effect, fluorinated alkyl chains are necessary in order to achieve oleophobic behavior.

Hydrophobic as well as oleophobic properties of ORMOCER®s were achieved by using various silanes bearing different lengths of fluoralkyl chains which were incorporated by adding them to the mixture of the precursor compounds prior to the hydrolysis. In Table 1 the fluoralkyl-substituted silanes are shown.

Table 1. Silanes with fluoralkyl chains.

Fluoralkyl silane		F atoms	CF_2 groups	CF_3 groups
$CF_2H(CF_2)_3{-}CH_2{-}O{-}C(=O){-}NH{-}(CH_2)_3{-}Si(OEt)_3$	**1**	8	4	0
$CF_3(CF_2)_5{-}(CH_2)_2{-}Si(OEt)_3$	**2**	13	5	1
$CF_2H(CF_2)_7{-}CH_2{-}O{-}C(=O){-}NH{-}(CH_2)_3{-}Si(OEt)_3$	**3**	16	8	0

Compound **2** is available commercially, whereas compounds **1** and **3** were synthesized by the reaction of the fluorinated alcohol with isocyanatopropyltriethoxysilane according to Scheme 2.

$$R{-}CH_2{-}OH + OCN{-}(CH_2)_3{-}Si(OEt)_3 \longrightarrow R{-}CH_2{-}O{-}\overset{\overset{\displaystyle O}{\|}}{C}{-}NH{-}(CH_2)_3{-}Si(OEt)_3$$

Scheme 2. Synthesis of the fluorinated silane; R = fluorinated alkyl chain.

The measurement of the contact angle of liquids on surfaces is an excellent method to investigate the wetting behavior of polar and nonpolar substances. Generally, an increase of the contact angle is caused by reduced wettability due to reduced surface energy. Typically H_2O and CH_2I_2 are chosen as test substances to evaluate the surface properties regarding wetting behavior and adhesion and to determine the hydrophobic as well as oleophobic properties of the surfaces. In Table 2 the contact angles of the test substances are given on surfaces of inorganic–organic hybrid materials which are modified with different fluoralkyl-substituted silanes.

The results indicate that the silane additive with a short fluoralkyl chain containing four CF_2 units shows only a limited effect towards water and hydrocarbons which can only be improved if high amounts are added.

In order to achieve a significant effect exhibiting not only hydrophobic but also oleophobic behavior of the coating, the presence of long-chain fluoralkyl groups is necessary in the coating composition. Due to the incorporation of silanes bearing six- or eight-membered fluoralkyl chains, high contact angles in the range of 90–95° towards both H_2O and CH_2I_2 are realized, indicating an anti-adhesive and repellent effect of the surface. This effect, which is well known from polytetrafluorethylene (PTFE), can be achieved with a maximum amount of only 1 mol % of the fluorinated compound.

Table 2. Contact angles of test substances on fluoralkyl-modified ORMOCER®s in comparison to uncoated substrates.

Fluoralkylsilane	Amount [mol %]	Contact angle of H_2O [°]	Contact angle of CH_2I_2 [°]
1	1	57	not measured
1	20	82	74
2	1	93	90
2	20	95	91
3	1	92	82
PTFE		98	75
Uncoated glass		20	45

The fact that modification with **2** is more effective than with **3,** especially in the case of nonpolar substances, is remarkable, because the fluorine content of **2** is lower, compared to **3**. From the literature it is well known that CF_3 groups decrease the surface energy more efficiently than CF_2 groups [4]. In the present case this effect seems to be more powerful than that of a high fluorine content.

From Electrical Insulation to Conductivity

Common polymers are usually good electrical insulators with a surface resistance of $>10^{14}\,\Omega$. Therefore, electrical charges cannot be removed and the surface will be electrostatically charged. As a result dust particles are attracted and due to abrasive properties of these particles the surface is scratched during cleaning. But also safety hazards can arise in industrial production processes if the environment is electrostatically charged. Electrical charges are removed easily if the surface conductivity is increased, i.e. the surface resistance has to be decreased to $\leq 10^{11}\,\Omega$ [5].

By using special monomeric alkoxysilane compounds, bearing ionic ammonium moieties, antistatic inorganic–organic coatings are available with a significantly decreased surface resistance. In Fig. 2 some of the ammonium silane compounds which were used for the modification of the inorganic–organic hybrid material are shown.

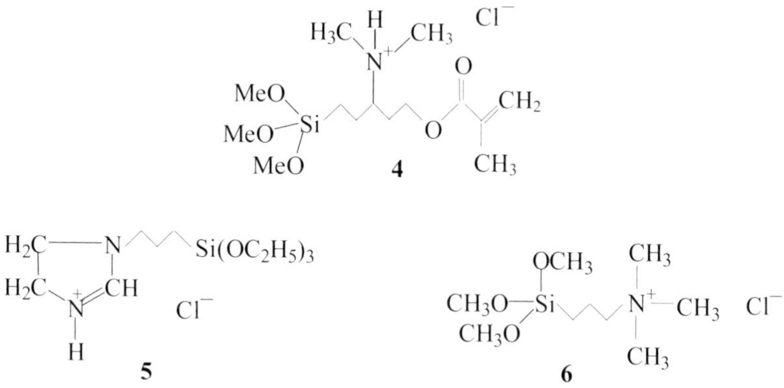

Fig. 2. Examples of quarternary ammonium compounds.

The best results were obtained, when compound **6** was incorporated into the ORMOCER® by covalent Si–O–Si bonds generated via sol–gel synthesis. A surface resistance of $10^8\,\Omega$ was achieved with 30 mol% of **6**. Consequently, even after intensive rubbing, the ORMOCER® surface is not electrically charged and adhesion of dust particles is avoided.

Chemically Sensitive Materials

Fiber-optic sensors are developed for environmental monitoring, process control and medical diagnostics. The functional operation of this sensor type is based on the interaction of light propagating through the fiber with a reagent that interacts with environmental substances to be sensed. This special reagent is immobilized in a coating on the optical fiber and, due to its reaction with environmental substances, the optical properties of the coating will change and can be detected. The special mechanism of this detection method is based on the penetration of a small portion of the guided light (evanescent wave) into the layer surrounding the fiber which contains the sensitive component [6].

To answer the demands of the fiber technology for a highly flexible coating, curable by UV radiation within a few seconds, acrylate-modified linear polysiloxanes were synthesized (Scheme 3). Amino or thioether moieties, which are necessary for the desired chemical sensitivity, form a part of the molecular structure and they are incorporated during the first step of the synthesis procedure. The synthesis of the coating materials in the first step comprises the Michael-type addition of $H_2N(CH_2)_3SiMe(OEt)_2$ or $HS(CH_2)_3SiMe(OMe)_2$ to one acrylic C=C bond of a monomeric multifunctional acrylate compound. In the second step hydrolysis is conducted and the linear polysiloxane is formed with free acrylic moieties, available for UV curing.

In the described material different properties have been integrated, such as capability of UV curing, availability of chemically sensitive moieties and a flexible structure suitable for coating silica optical fibers.

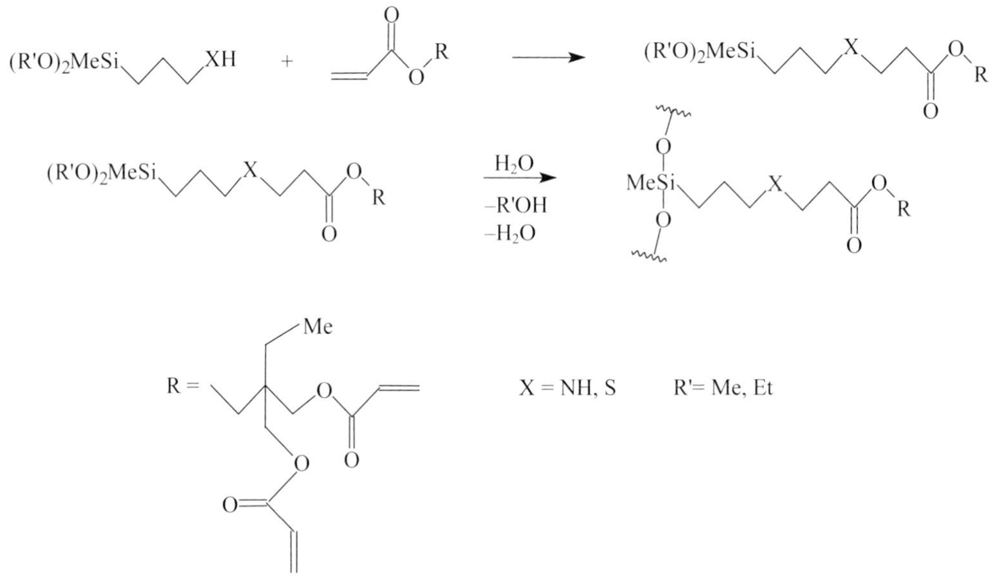

$$(R'O)_2MeSi\text{–}(CH_2)_3\text{–}XH + CH_2{=}CH\text{–}C(O)OR \longrightarrow (R'O)_2MeSi\text{–}(CH_2)_3\text{–}X\text{–}CH_2CH_2\text{–}C(O)OR$$

$$(R'O)_2MeSi\text{–}(CH_2)_3\text{–}X\text{–}CH_2CH_2\text{–}C(O)OR \xrightarrow[-R'OH,\ -H_2O]{H_2O} \text{–O–}MeSi(\text{–}(CH_2)_3\text{–}X\text{–}CH_2CH_2\text{–}C(O)OR)\text{–O–}$$

R = ; X = NH, S ; R'= Me, Et

Scheme 3. Synthesis of the chemically sensitive polyacryloxysiloxane.

By treatment with gaseous CO_2, a layer containing the amino derivative (X = NH) exhibited new signals at 1560 and 1540 cm^{-1} in the infrared spectra due to formation of carbamic acid derivatives [7] (Scheme 4). By treating a cured layer containing the thioether moiety (X = S) with gaseous SO_2 a new signal at 320 nm was observed by UV–Vis spectroscopy; this is attributed to the formation of a thioether/SO_2 adduct [8] (Scheme 4).

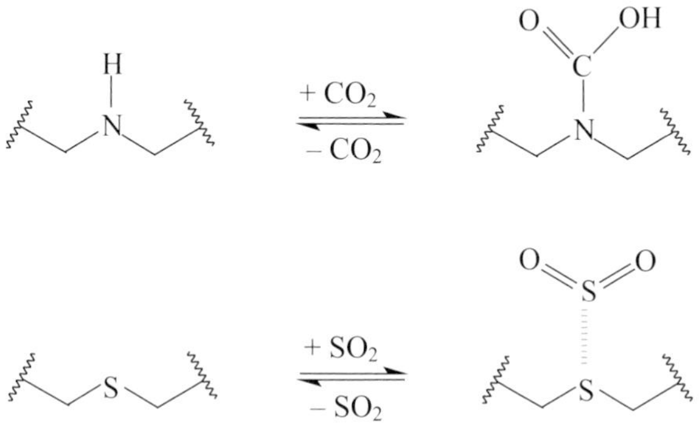

$$\text{–CH}_2\text{–NH–CH}_2\text{–} \underset{-CO_2}{\overset{+CO_2}{\rightleftharpoons}} \text{–CH}_2\text{–N(COOH)–CH}_2\text{–}$$

$$\text{–CH}_2\text{–S–CH}_2\text{–} \underset{-SO_2}{\overset{+SO_2}{\rightleftharpoons}} \text{–CH}_2\text{–S}\cdots SO_2\text{–CH}_2\text{–}$$

Scheme 4. Reversible reactions of functional groups with gases.

Effects of gaseous CO_2 on the output intensity of a silica fiber coated with the amine-modified polysiloxane have been determined and measured at a wavelength of 1200 nm.

The results demonstrate that there is a significant increase of the attenuation of the guided light due to the influence of gaseous CO_2 on the output intensity of a silica fibre [9].

Conclusion

Based on a versatile synthesis method, the sol–gel process, and the proper choice of various monomeric starting compounds, functionalized inorganic–organic hybrid materials are available. By using functionalized alkoxysilanes which form the inorganic oxidic matrix, the functional moieties are fixed in the siloxane network. With the few examples described in this paper we have demonstrated the high potential of chemical modification of inorganic–organic copolymers, opening a wide field of applications.

Acknowledgment: A part of the work was funded by the *European Community*. Moreover, we thank the *Institute of Radio Engineering and Electronics* for fruitful collaboration in the field of fiber optic chemical sensors.

References:

[1] B. M. Novak, *Adv. Mater.* **1993**, *5,* 6.
[2] U. Schubert, N. Hüsing, A. Lorenz, *Chem. Mater.* **1995**, *7,* 2010.
[3] C. J. Brinker, G. Scherer, *Sol–Gel Science, The Physics and Chemistry of Sol–Gel Processing*, Academic Press, New York, **1990**.
[4] M. J. Owen, *Ind. Eng. Chem. Prod. Res. Dev.* **1980**, *19,* 97.
[5] K. H. Kochem, H.-U. ter Meer, H. Mellbauer, *Kunststoffe* **1992**, *82,* 575.
[6] G. Gauglitz, *Nachr. Chem. Tech. Lab.* **1995**, *43,* 316.
[7] Y. Yoshida, S. Ishii, T. Yamashita, *Chem. Lett.* **1984**, 1571.
[8] V. Matejec, K. Rose, M. Hayer, M. Pospisilova, M. Chomat, *Sens. Act. B* **1997**, *38–39*, 438.
[9] K. Rose, V. Matejec, M. Hayer, M. Pospisilova, *J. Sol–Gel. Sci. Technol.* **1998**, *13,* 729.

High Performance Silicon Polymer with Organoboron Structure

Toshiya Sugimoto, Motokuni Ichitani, Koji Yonezawa, Kazuhiro Okada

Minase Research Institute, Sekisui Chemical Co., Ltd.
2-1 Hyakuyama, Shimamoto-Cho, Mishima-gun, Osaka 618-8589, Japan
Tel.: Int. code + (75)962 8813 — Fax: Int. code + (75)961 5353
E-mail: sugimo08@smile.sekisui.co.jp

Keywords: Diethynylbenzene-silylene / Carborane / Hybrid / Crosslinking / Thermal Stability

Summary: Carborane-hybridized silicon polymers were synthesized from the hydrosilylation reaction between diethynylbenzene-silylene polymers with reactive vinyl side groups and 1,7-bis(dimethylsilyl)carborane and 1,7-bis(diphenylsilyl)carborane. Precise NMR spectroscopic studies show that the structures of these polymeres were quite complicated: unreacted vinyl groups and pendant carborane substitution exist along with crosslinked carborane. These polymers show plastic moldability. Thermal treatment of the hybrid polymers gives excellent stability in both thermal and mechanical properties of the polymers. Their unreacted moieties seem to act as crosslinkable thermosetting features to make these polymers more stable.

We have been joining in the Japanese national project to develop a high performance silicon polymer [1]. Our goal is to develop a thermally durable silicon polymer by using a stable organic-inorganic hybrid structure. Processiable materials with lightness, thermal durability, and toughness have been wanted. Such materials should have plastic moldability and ceramic durability.

To improve the thermal durability and inflammability of the materials, we have been investigating organoboron structures to hybridize with silicon polymers. In this field, the preceramic character of the borazine structure [2] and its catalytic function for anti-oxidation [3] are already known. Lately Chujo reported hydroboration polymerization to obtain boron-based polymers [4].

$R{-}BH_2$ + X → X B R

However, attempts to obtain silicon–boron hybrid polymer systems have merely been experimentally investigated. Recently, we found that the cage carborane is the best candidate for this purpose.

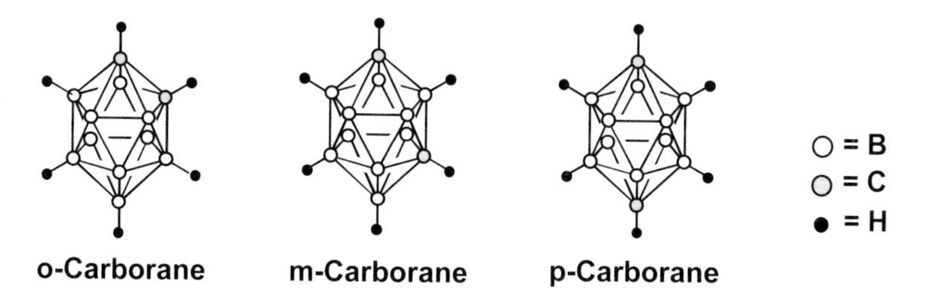

In the US and the Russian munitions field, carborane: dicarba-*closo*-dodecaborane(12) [5] has been used to improve the properties of silicones. Some American companies have synthesized carborane–siloxane polymers as high temperature elastomers [6]. Lately, Keller's group has reported a diacetylene–carborane–siloxane system showing one of the highest stabilities in silicon polymers [7].

$$\left[CB_{10}H_{10}C - Si(R^1)(R^2) - \left(O - Si(R^3)(R^4) \right)_m \right]_n$$

$$\left[C{\equiv}C - C{\equiv}C - Si(Me)_2 - O - Si(Me)_2 - CB_{10}H_{10}C - Si(Me)_2 - O - Si(Me)_2 \right]_n$$

We have been studying non-siloxane silicon polymers to obtain thermally stable and mechanically strong material. Ethynylene–silylene polymers have been found with excellent properties [8]. In addition, we have tried to introduce boron structures into silicon polymers to get further durability.

$$HO-SiR_2-CB_{10}H_{10}C-SiR_2-OH + Cl-SiR_2-C_6H_4-SiR_2-Cl \longrightarrow \left[SiR_2-CB_{10}H_{10}C-SiR_2-O-SiR_2-C_6H_4-SiR_2-O \right]$$

Eq. 1. Condensation of a bifunctional carborane.

We have designed several strategies in order to introduce carborane structure into silicon polymers. In the main chain of the polymer, carborane was substituted with two reactive groups and condensed with a bifunctional compound utilizing conventional methods. Various trials revealed that the carborane structures reduced the reactivities of these functional groups to result in lower yields and molecular weights. Kalachev reported that decaborane could be introduced into the ethynyl group to form a carborane structure [7]. We tried to use this method using diethynylbenzene–disiloxane polymers, but found that cleavage of the main chain was concerted with the insertion. As though these chains were short, the polymers showed good thermal stabilities.

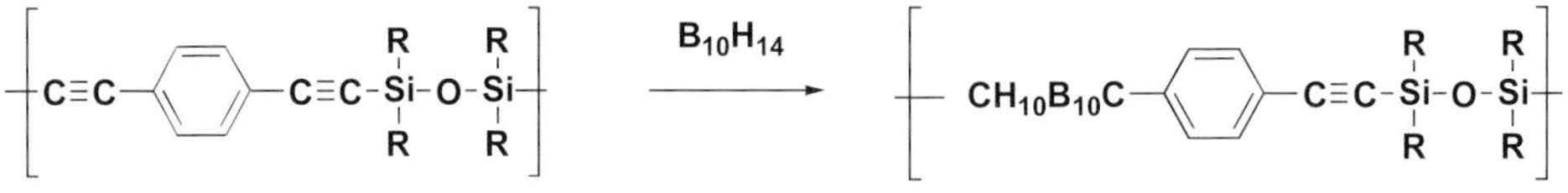

Eq. 2. Insertion of borane into acetylenic bonds.

While main chain strategies gave inefficient results, introduction of carborane to side chains is a relatively easy way to make hybrid systems. As Kunai reported, the versatile hydrosilylation reaction is the most reliable way to build hybrid polymers [9]. Reaction between diethynylbenzene–silylene polymers with reactive vinyl side groups **1** and 1,7-bis(dimethylsilyl)carborane $[HSi(CH_3)_2\text{-}(CB_{10}H_{10}C)SiH(CH_3)_2]$ **2a** (R = CH_3) in the presence of hydroplatinic acid catalyst gave carborane-hybridized silicon polymers **3a** with 67 % yield (Scheme 1). GPC measurement of **3a** provides a 21 600 M_W vs. polystyrene standard, while the original polymer **1** had 6 400. 1,7-Bis(diphenylsilyl)carborane **2b** (R = Ph) was also reacted with polymer **1** to get a similar hybrid **3b** with a lower yield and M_W, only 43 % and 5 800 respectively. Here hybrid silicon polymers with carborane side groups have been synthesized successfully.

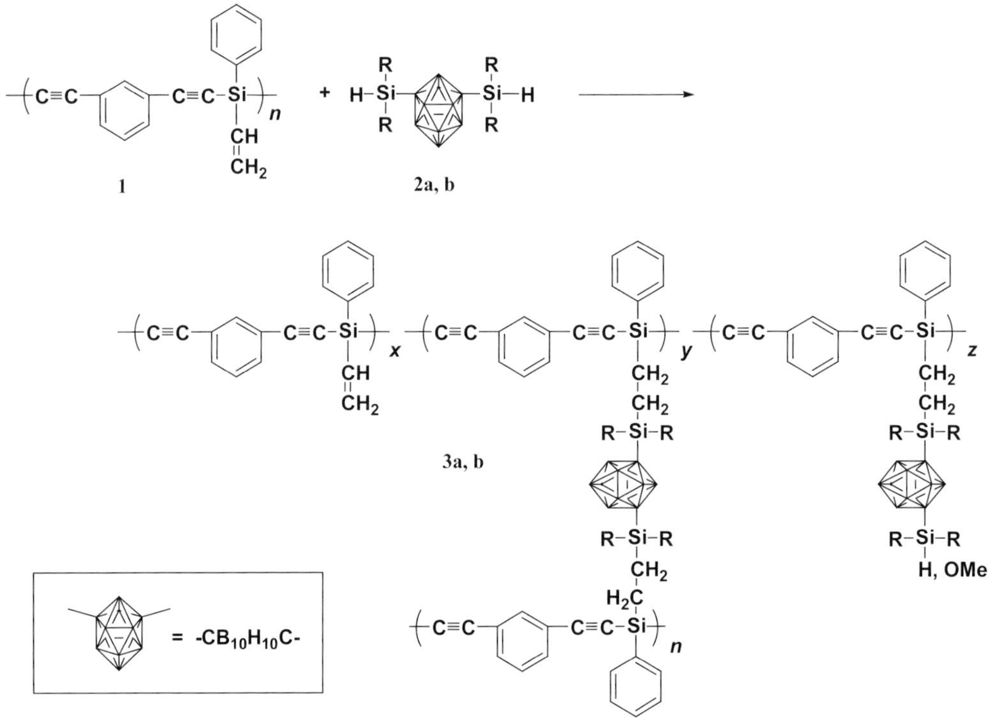

Scheme 1. Synthesis of carborane-hybridized silicon polymers **3**.

For comparison, 1,1'-bis(dimethylsilyl)ferrocene **4a**, 1,1'-bis(diphenylsilyl)ferrocene **4b** and 1,4-bis(dimethylsilyl)benzene **5** were also prepared and reacted with vinyl polymer **1** to form hybrid polymers **6a**, **6b** and **7** respectively. In the case of **6a** and **7**, higher reactivities of the silanes lead to gelation. While electron deficient carborane seems to reduce Si–H reactivity, the conjugation effect of the aromatic system on the silicon atoms may gain the Si–H reactivity. Phenyl-substituted **6b** resulted in a lower yield and M_W than methyl **6a**. Bulkiness of the substituent on the silyl group of the reactants seems to mainly contribute the product yields, suggesting that the dimethyl silyl group of the carborane is more suitable to approach the vinyl group. These features are summarized in Table 1.

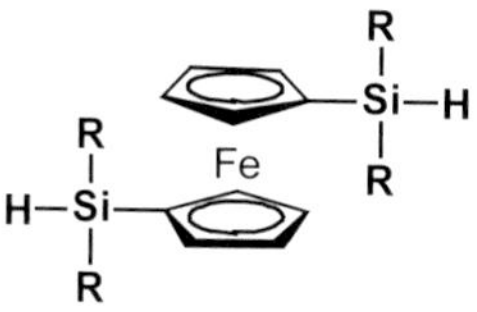

4a, b

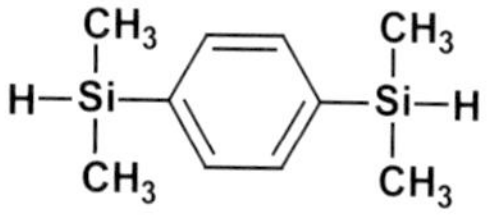

5

Table 1. Properties of hybrid polymers.

		Yield	GPC[a]		DSC	TGA in air	
Polymer	**Solvent**	**[%]**	M_w	M_w/M_n	T_g[°C]	T_{d5}[b] [°C]	W_{800}[c][d] [%]
1	THF		6 400	2.2	53	560	22 +
3a	THF	73	21 000	3.5	108	780	95 –
3a	toluene	81	24 400	1.7	104	782	95 –
3b	THF	43	5 800	1.7	77	473	35 +
3b	toluene	57	9 900	2.4	not distinctiv	482	82 –
6a	THF	50[e]	(50 800	3.3)[e]	none	360	40 +
6b	THF	67	8 200	1.8	123	412	31 +
7	THF	67[e]	(14 500	2.7)[e]	89	463	34 +

[a] Relative to polystyrene standard — [b] Temperature at 5 % weight loss — [c] Residue at 800°C — [d] (+): Burned out, (–) Ignited then extingueshed — [e] Insoluble gel was formed. Data from soluble fraction.

IR and NMR spectroscopies show a certain amount of carborane in the silicon polymers. The B–H stretching band (2594cm^{-1}) of the carborane cage is observed in the infrared spectrum of the hybrid polymer **3a**. ^{1}H-NMR analysis is also carried out to provide a B–H resonance (0.6–4.0 ppm) and a hydrosilylated ethylene unit (0.1–0.4 ppm). Unreacted Si–H (4.0 ppm) is hardly observed but unreacted vinyl (5.5–6.4 ppm) and solvent-capped Si–OCH_3 (3.5 ppm) peaks are found. These results show that carborane introduction is certainly occurring, but some units are partially bound to the main chain like a pendant group. By a careful NMR integration study of these peaks in **3a**, the ratio of the unreacted vinyl group (x), the carborane unit substituted on one side (z) and moiety crosslinked on both sides (y) is calculated. In the ^{1}H-NMR, comparison of vinyl and B–H protons could give (x):($y + z$), and the ratio of B–H to Si–H with Si–OCH_3 corresponds to (y):(z). Carborane content $(y + z)/(x + y + z)$ is calculated around 33 % in the THF system and 53 % in the toluene system. It is hard to figure out (y):(z) because Si–H and Si–OCH_3 peaks are very small and overlap in the B–H region. Rough estimation of (y):(z) gave 8:2 in the THF system and 7:3 in the toluene system, respectively [10].

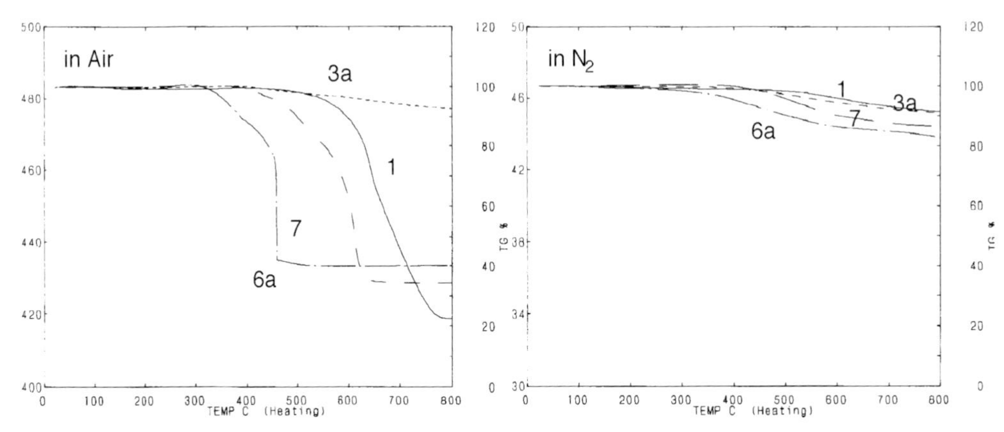

Fig. 1. Thermogravimetric curves of the hybrid polymers.

As for the thermal analysis shown in Fig.1, the original vinyl polymer **1** itself has relatively high thermal stability; the 5 % weight loss temperature is 560ºC in air. Furthermore, the carborane hybrid **3a** shows almost no weight loss up to around 600ºC under an air atmosphere, and the ceramic yield at 800ºC is as high as 95 %. While the ferrocene or benzene analogues **6a, 6b** and **7** show poorer performance than the original, the carborane hybrid system is exceptional. While the toluene systems brought good results corresponding to their high carborane contents in both the methyl- and phenyl-substituted products (**3a**,**3b**), the phenyl polymer **3b** (THF) does not survive at the higher temperature, reflecting its lower carborane content. Thus the carborane structure does have an important role in their thermal durability, and introduction of around 30 % is necessary to achieve these thermal stabilities. Under nitrogen, TGA shows less difference; all of them had high performance with more than 90 % ceramic yields at 800ºC.

Although the polymers **3a** and **3b** have complicated structures, this partial hybridization has a positive effect in processing. According to DSC analyses, they showed melting behavior; pseudo-T_g values are around 100ºC. We can mold this material powder into small bar samples. Fully crosslinked gels **6a** and **7** are insoluble in organic solvents and **6a** is unprocessable. The other hybrids are also processable. These molded samples were submitted to thermal and mechanical tests. A bending test reveals that the carborane polymer **3a** has relatively high strength with a flexural modulus of 1.86 GPa at room temperature.

A most remarkable point about these carborane hybrid polymers is that they have crosslinkable groups for further thermal treatment. Unreacted vinyl, Si–H, solvent-capped Si–OCH_3 and C–C triple bonds are condensed with the others during heating above 300ºC. This is a kind of thermosetting procedure giving mechanical strength and thermal stability. An IR spectral study recorded the consumption of acetylenic and/or Si–H bonds around 2155 cm^{-1} during the treatment. The flexural modulus of thermally treated **3a** was remained at 0.95 Gpa even at 250ºC, which is comparable with carbon engineering plastics.

A flammability test under a 50 % oxygen atmosphere was also carried out. The vinyl polymer **1**, ferrocene **6b** and benzene polymers **7** are burned out when attacked with flame. The carborane hybrids **3a, 3b** are both ignited but extinguished soon in the case of the higher ($y + z$) contents. The introduction of carborane brings non-flammability to the hybrid polymers along with thermal and mechanical strength. Transformation of the boron structure to boron oxide was observed around

400–500°C in XPS studies of the thermally treated materials. This ceramic feature may bring durability to the polymer.

Our attempt to obtain a silicon–boron hybrid polymer system was successful. Their complicated crosslinkable structure acts as a thermosetting resin to give thermal and mechanical stability with good processability. Carborane hybridization gives exceptional non-flammability. This is a kind of organic-inorganic hybrid system that has plastic versatility and ceramic stability.

Acknowledgment: This work was performed by *Sekisui Chemical Co., Ltd.* under the management of the *Japan Chemical Innovation Institute* (former known as the *Japan High Polymer Center*) as a part of the Industrial Science and Technology Frontier Program supported by the *New Energy and Industrial Technology Development Organization.*

References:

[1] T. Ogawa, M. Murakami, *Chem. Mater.* **1996**, *8*, 1260.
T. Iwahara, J. Kotani, K. Ando , K. Yonezawa, *Chem. Lett.*, **1995**, 425.
M. Itoh, K. Inoue, K. Iwata, M. Mitsuzuka, T. Kakigano, *Macromolecules*, **1994**, *27*, 7917.

[2] D. Seyferth, H. Plenio, *J. A. Ceram. Soc.* **1990**, *73*(7), 2131.
D. Seyferth, K. Buchner, W. S. Rees, Jr., W. M. Davis, *Angew. Chem. Int. Ed. Engl.* **1990**, *73*(3), 2131.

[3] A. Morikawa, T. Amano, *Kobunshi Kagaku* **1973**, *30*(340), 479.
N. Pourahmady, P. I. Bak, *J. M. S.-Pure Appl. Chem.* **1994**, *A31*(2), 185.

[4] Y. Chujo, I. Tomita, Y. Hashiguchi, H. Tanigawa, E. Ihara, T. Saegusa, *Macromolecules* **1991**, *24*, 345.

[5] M. F. Hawthorne, *Current Topics in the Chemistry of Boron* (Ed.: G.W., Kabalka), The Royal Society of Chemistry **1994**, p. 207.

[6] S. Papetti, B. B. Schaeffer, A. P. Gray, T. L. Heying, *J. Polym. Sci.: Part A-1* **1966**, *4*, 1623.
E. N. Peters, *J. Macromol. Sci.-Rev. Macromol. Chem.* **1979**, *C17*(2), 173.
D. D. Stewart, E.N. Peters, C. D. Beard, G. B. Dunks, E. Hedaya, G. T. Kwiatkowski, R. B. Moffitt, J. J. Bohan, *Macromolecules* **1979**, *12*(3), 373.
T. K. Dougherty, *US Pat. 5, 264, 285*, **1993**.

[7] L. J. Henderson, T. M. Keller, *Macromolecules* **1994**, *27*, 1660.

[8] R. J. P. Corriu, N. Devylder, C. Guerin, B. Henner, A. Jean, *Organometallics* **1994**, *13*, 3194.
T. Fujisaka, B. Yamaguchi, K. Okada, *27th Organosilicon Symposium* **1994**, p. 17.

[9] A. Kunai, E. Toyoda, I. Nagamoto, T. Horio, M. Ishikawa, *Organometallics*, **1996**, *15*, 75.

[10] M. Ichitani, K. Yonezawa, K. Okada, T. Sugimoto, *Polymer Journal* **1999**, *31*, 11.

Polydimethylsiloxane (PDMS): Environmental Fate and Effects

Nicholas J. Fendinger

The Procter and Gamble Company
Sharon Woods Technical Center
11511 Reed Hartman Highway
Cincinnati, OH 45241 U.S.A.
Tel.: Int. code + (513)626 2257 — Fax: Int. code + (513)626 1375
E-mail: fendinger.nj@pg.com

Keywords: Polydimethylsiloxane / PDMS / Environment / Environmental Effects / Degradation

Summary: Polydimethylsiloxanes (PDMS) are used in many industrial products and processes and in a variety of consumer applications, such as coatings, polishes, detergents, personal care products, foods, and medicines. Described in this paper are the results of a five-year industry-sponsored research program to increase the understanding of PDMS environmental fate and effects. The research investigated PDMS behavior during wastewater treatment and demonstrated how PDMS breaks down in soil and sediment, advancing the understanding of the environmental fate of this material. New environmental effects tests demonstrated that no adverse effects to aquatic and terrestrial organisms are anticipated from PDMS or its breakdown products, at concentrations many times higher than could possibly occur in the environment from typical applications. Laboratory and field measurements demonstrate that PDMS does not bioaccumulate.

Introduction

Polydimethylsiloxanes (PDMS; $Me_3SiO(SiMe_2O)_nSiMe_3$) are an important class of polymers that have many applications in industrial processes, coatings, polishes, detergents, personal care products, foods, and medicines. Given the many uses and anticipated growth of PDMS volumes, the PDMS Fluids Environmental Task Force was formed in 1992 under the auspices of the Silicones Environmental Health and Safety Council (SEHSC) to develop key PDMS environmental fate and effects data. The silicone industry undertook this program voluntarily as an environmental stewardship initiative. This program was unique because it involved all the major US silicone manufacturers (Bayer, Dow Corning, General Electric, Goldschmidt, Huls America, Rhone-Poulenc, Shin-Etsu, Wacker Silicones, and Whitco) and a consumer product manufacturer (Procter and Gamble). The objectives of this extensive environmental safety program were to: (1) better define environmental exposure in relevant compartments, (2) investigate PDMS degradation in soil and sediment, and (3) determine if environmental effects occur and if they are ecologically relevant. More recently the SEHSC research activities and those by the Centre Européen des Silcones (CES) and Silicone Industry Association of Japan (SIAJ) have been coordinated globally by the Global Silicones Council (GSC). The results of this globally coordinated environmental research program are presented in this paper. More detailed information on the results presented here can be found in Fendinger et al. [1].

PDMS Use and Environmental Loading

Polydimethylsiloxanes are the most commonly used organosilicone polymer with approximately 140 metric tons consumed annually in the US. The way PDMS is used determines how it enters the environment and its ultimate fate [2]. For example, most PDMS consumed in the US (60 %) is used to manufacture elastomers, pressure-sensitive adhesives, and modified polysiloxane polymers. The remaining fraction of PDMS is used in industrial applications, coatings, household, personal products, foods, and drugs. PDMS used as textile or paper coatings, dielectric fluids, and heat-transfer fluids constitute about 18 % of the total PDMS used and are largely recycled, landfilled or incinerated after use. PDMS used as industrial process aids or in household and personal care products are disposed of primarily down the drain after use as a component of wastewater. These uses represent only about 10 % of the total volume of PDMS used but have the greatest potential to enter the aquatic and terrestrial environments. The focus of this review is on the fraction of PDMS that enters wastewater treatment systems from down-the-drain disposal and is released into the environmental as a result of wastewater treatment practices. Additional PDMS will enter the environment from its use in polishes and lubricants that make up the remaining 10 % of the total volume. This use distribution is probably similar to that found in other industrialized countries.

Eco-Relevant Physico-Chemical Properties

For the purposes of this review, PDMS will be defined as fully methylated polysiloxanes with a molecular weight of 800 to 300 000. Eco-relevant properties of these materials are summarized in Table 1.

Table 1. Eco-relevant physico-chemical properties of PDMS [1].

Viscosity	0.65 to >500 000 cs
Chemical/thermal stability	High
Solubility in aliphatic hydrocarbons	High
Vapor pressure	$<<<1.0\times10^{-4}$ mm Hg at 25°C
Water solubility	Insoluble
Log K_{oc}	5.8
Log K_{ow}	Approximately 5

PDMS Fate During Wastewater Treatment

The eco-relevant properties of PDMS suggested that it would be highly removed during wastewater treatment as a result of sorption onto sludge solids. The high molecular weight combined with the chemical and thermal stability also suggests that biodegradation will not be a significant removal mechanism. These conclusions regarding the fate of PDMS during wastewater treatment are supported by laboratory and field monitoring studies. For example, Palmer [3] investigated the potential for the aerobic and anaerobic biodegradation of PDMS in waste-activated sludge. At the conclusion of a 2-month incubation period >90 % of the ^{14}C-labeled PDMS remained associated with sludge solids with no evidence of PDMS biodegradation as indicated by $^{14}CO_2$ evolution. Additional work by Hobbs et al. [4] also showed that no carbon dioxide or volatile organic products were formed from PDMS dosed into activated sludge over a 70-day incubation period.

A pilot-scale activated sludge plant was used by Watts et al. [5] to investigate both the fate of PDMS during wastewater treatment and characterize any potential impacts that PDMS may have on the wastewater treatment process. Mass balance experiments conducted with ^{14}C- labeled PDMS across the treatment process demonstrated that essentially all of the PDMS partitioned onto sludge solids during the treatment process. Sludge-bound PDMS was shown by chromatographic analysis to remain unchanged during both aerobic treatment and anaerobic sludge digestion processes [6]. It was also determined that PDMS had no effect on the operating parameters of the treatment plant (pH, suspended solids, sludge volume index, and specific oxygen update) or physiological activity of the sludge microflora. This work demonstrated that PDMS partitions onto sludge, remains unchanged during the treatment process and does not impact the operating parameters of the treatment plant.

The findings of the laboratory and pilot-scale treatment studies were also verified by actual field monitoring conducted at eight municipally owned wastewater treatment plants across North America [7]. Newly developed gel permeation chromatography/inductively coupled plasma (GPC/ICP) analytical techniques were used to provide structural information and measure PDMS concentrations in influent wastewater, sludge, and treated effluent at each of the plants studied. In general the PDMS measured in influent wastewater to the treatment plants had a greater polydispersity than the 350 Cst PDMS standard and reflected the many sources of material that would contribute to the molecular distribution measured in wastewater (Fig. 1). In addition, the GPC profile for PDMS measured in influent matched the profile observed in digested sludge indicating that PDMS remained unchanged during the treatment/sludge stabilization processes. PDMS influent wastewater concentrations ranged from 87.4 to 373.5 μg/L and averaged 244±112.7 μg/L for plants that received wastewater from predominantly domestic sources. The level of PDMS measured in influent wastewater corresponds closely with the predicted influent concentration calculated from the amount of PDMS disposed down the drain (14 000 metric tons/year), the population of the United States, and per capita water consumption. The agreement between measured and predicted influent wastewater concentrations is expected, given that PDMS is not anticipated to degrade during conveyance in wastewater collection systems. This agreement also provides assurance that PDMS used in applications that are disposed of down the drain is well accounted for. Treated effluent levels of PDMS from the plants monitored were generally less than the quantitation level of the GPC/ICP analytical technique of 5 μg/L and demonstrate that PDMS is highly removed (>94 % at all the plants monitored) during wastewater treatment. The trace amount of PDMS present in the effluent is expected to be associated with the suspended solids. PDMS sludge levels ranged from 290 to 5155 mg/kg (dry weight) and varied as a function of influent wastewater PDMS concentration and sludge processing method.

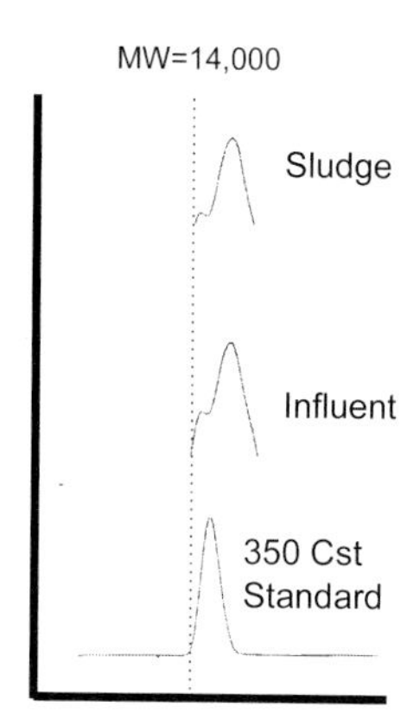

Fig. 1. Representative gel permeation chromatograms of a 350 Cst PDMS standard and PDMS extracted from influent and sludge collected from the Kenton (OH) municipally owned wastewater treatment plant.

PDMS Environmental Mass Flow

The wastewater treatment monitoring study, combined with results from laboratory- and pilot-scale wastewater treatment studies, can be used to determine the relative importance of various environmental compartments on the mass flow of PMDS that is disposed of down the drain. The first assumption in this assessment is that all PDMS used in a down-the-drain product will reach either a wastewater treatment plant or a home septic system. This is a good assumption in the US where >98 % of all wastewater receives some form of treatment prior to discharge [8]. The relative proportion of wastewater that undergoes treatment at a wastewater treatment plant to that portion that is discharged to a septic tank is approximately 3 to 1. PDMS that reaches a wastewater treatment plant will either be sorbed to sludge solids or discharged with liquid effluent to a receiving body of water. PDMS discharged to a home septic system will either be sent to a wastewater treatment plant following removal of the septage from the tank, or incorporated into the soil environment through direct disposal of septage or effluent. Based on results from the wastewater treatment plant monitoring and pilot-scale treatment plant studies, between 94 and 97 % of PDMS will be sorbed to sludge during wastewater treatment with less than 3–6 % discharged to surface waters as a component of effluent suspended solids. The disposal of sludge will therefore control the anticipated environmental mass-flow and fate of PDMS disposed of down the drain. The relative importance of various sludge disposal methods in the US is illustrated in Fig. 2. Incineration and landfilling of sludge are not expected to result in significant environmental dispersion of PDMS. For example, PDMS incineration products have been shown to be carbon dioxide and silica, and PDMS that is landfilled with sludge is expected to remain a non-leachable component of the sludge. Direct land application of sludge is therefore the predominant means by which PDMS is dispersed in the environment.

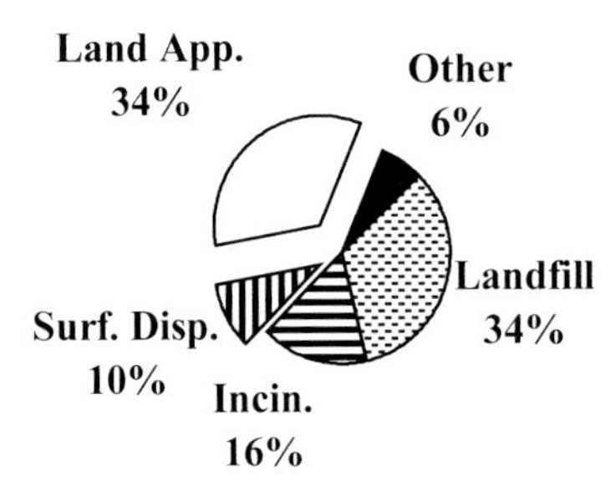

Fig. 2. Relative importance of various sludge disposal methods in the US [8]. "Surface Disposal" includes application of sludge to non-agricultural soil and where piles of sludge are left on the land; "Land Application" is where sludge is applied to agricultural and non-agricultural soils for a beneficial purpose; "Other" includes ocean disposal of sludge, which is no longer practiced in the US.

PDMS Fate in Soil

Studies conducted by Lehmann et al. [9] and Carpenter et al. [10] have confirmed previous work that PDMS will degrade on soil under drying conditions. Lehmann et al. conducted their studies using a sandy clay loam soil collected in Bay County, MI (Londo Soil). ^{14}C-labeled PDMS applied to this soil was found to be converted to water-extractable components when the soil gradually dried from 12 % to 3 % moisture. Simultaneously with the formation of water-desorbable products was an observable decrease in molecular weight of material extracted by tetrahydrofuran (THF). Identification of dimethylsilanediol as the major degradation product was accomplished using gas chromatography/mass spectrometry. PDMS added to soil maintained under moist conditions degraded at a slower rate, as indicated by a 3 % conversion to water-extractable products after a six-month incubation period. In separate experiments, Carpenter et al. [10] added non-labeled PDMS to an EPA standard soil and then monitored changes in PDMS concentration and molecular weight using GPC/ICP instrumental methods of analysis. Recovery of parent PDMS from the standard soil matrix decreased from near 100 % at day 0 to less than 25 % after 20 days, incubation at room temperature. Accompanying this trend was an increase in the amount of water-extractable silicon detected by ICP and a shift to lower molecular weight materials was observed in GPC/ICP chromatograms.

In order to demonstrate the potential of PDMS degradation on a variety of soils, Lehmann et al. [11] conducted additional experiments on seven soils having widely differing properties. In these experiments moist soil was spiked with ^{14}C-labeled PDMS, then allowed to gradually dry at room temperature over a two-week period. On all soils, PDMS was found to degrade to lower molecular weight materials, as indicated by longer GPC retention times, and to water-extractable products (Fig. 3).

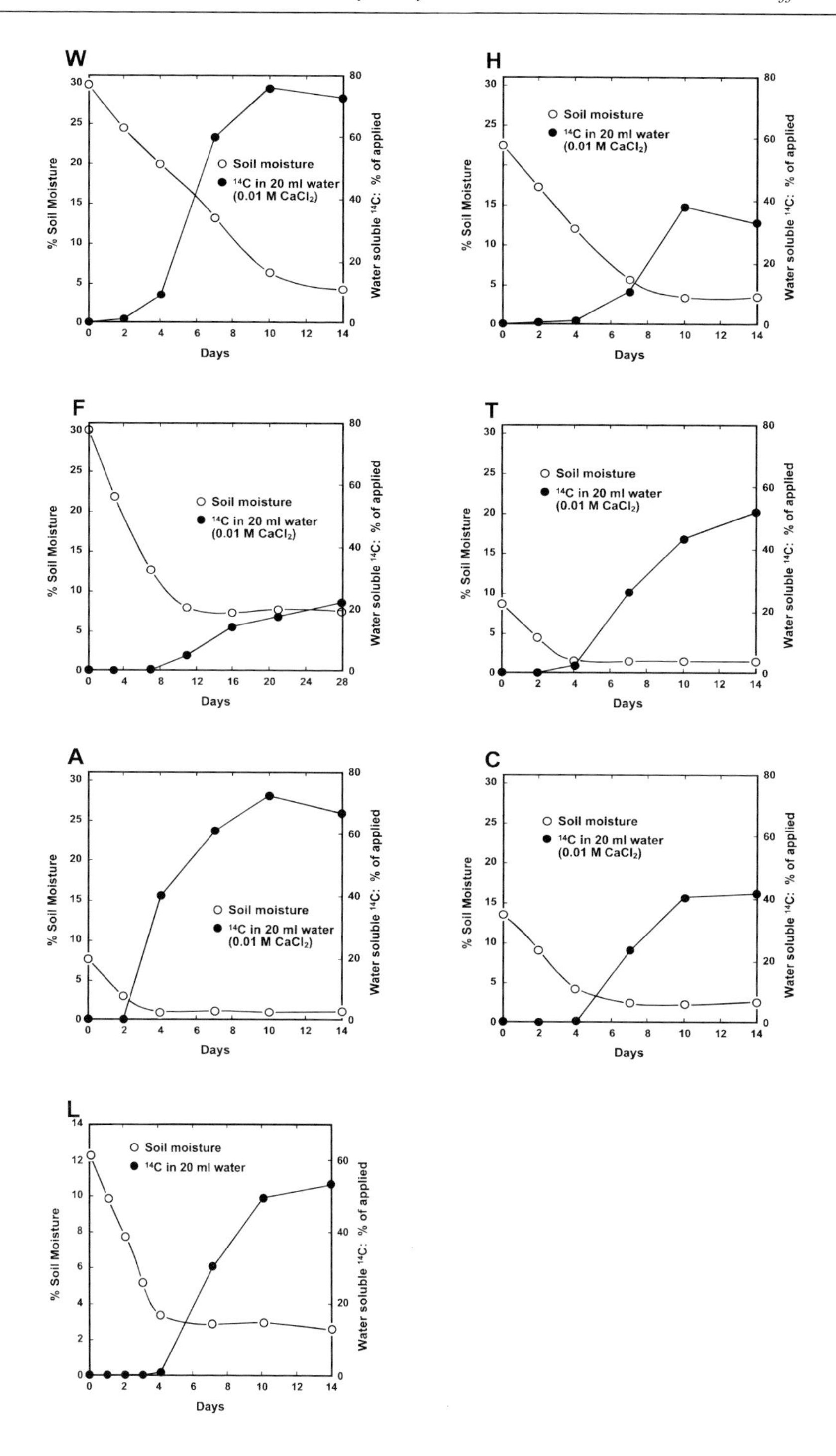

Fig. 3. PDMS degradation on a variety of soils (A: Appling; C, Cashmont; F, Fargo; H, Hastings; L, Londo; T, Tuscola; W, Wahiawa) to materials extracted with 0.01 M $CaCl_2$ as a function of decreased water content.

Further PDMS degradation experiments conducted by Xu et al. [12] on specific clays demonstrated that all the clay minerals examined catalyze PDMS degradation to an extent. The clay minerals selected for this study represent the major clay types found in most soils. The rates and products of PDMS degradation on clays were determined utilizing ^{14}C-labeled PDMS and incubation at 22°C and 32 % relative humidity. All the clay minerals studied promoted PDMS degradation to lower molecular weight materials and water-extractable materials over a period of hours for the most active clays to days for the less active clays. Kaolinite, beidellite, and nontronite promoted PDMS degradation at the fastest rates while geothite and smectites promoted PDMS degradation at slower rates. The involvement of both Brønsted and Lewis acid sites in PDMS degradation kinetics was demonstrated. The rate of PDMS clay degradation increased as a function of increased Brønsted acidity at lower humidity and more acidic exchangeable cations. Surface normalized degradation rate constants on kaolinite, prophyllite, and talc demonstrated the involvement of Lewis acid sites. The environmental significance of this work is that the widespread occurrence of the minerals studied ensures that PDMS will degrade in all soils as long as critical soil moisture contents are favorable. Optimum moisture levels for PDMS degradation on soil are generally less than 5 % (w/w).

In order to further demonstrate the practical nature of PDMS degradation under various climatic and soil regimes, Gupta et al. [13] modified an existing soil moisture model and used PDMS degradation rate data on soil to investigate the probability of PDMS degradation under field conditions. Soil moisture predictions were done using the Simultaneous Heat and Water (SHAW) model originally developed by Flerchinger and Sexton [14] and long-term climate data for Columbus, OH; Athens, GA; San Juan, PR; and Los Angeles, CA. Predicted soil moisture values from the SHAW model were then linked to laboratory PDMS degradation rate data as a function of moisture determined on Londo Soil. The predicted amount of PDMS remaining in soil for any day after an initial addition follows the trend: Los Angeles < San Juan < Athens < Columbus < St. Paul. This trend is expected because the order predicted correlates with the extent of soil drying and duration of dry soil conditions at each location. At all five locations, less than 5 % of the applied PDMS is predicted to remain in the soil surface after one year. However, at depths deeper than 2.5 cm the majority of PDMS remains after one year.

The predicted loss of PDMS from soils is supported by field observations made by Fendinger et al. [7] in agricultural fields that were amended with sludge from wastewater treatment plants. By examining the history of sludge applications to each field a calculated PDMS loading was determined based on the amount of sludge applied, number of applications, and current level of PDMS found in the sludge. The predicted PDMS concentrations were then compared with the measured concentration in each field. Measured PDMS concentrations were in some cases as much as 50 % lower than the predicted concentration based on historical PDMS loadings to the fields. This observation provides indirect evidence of PDMS degradation. Moreover, DMSD was detected in soils that either had high PDMS loadings or were collected from more arid climates and provided additional evidence of PDMS soil degradation in the field.

Additional PDMS soil monitoring work was conducted by McAvoy et al. [15] on experimental field plots established by USDA to study the impact of long-term extremely high loadings of sludge solids to a variety soils. Results from this work also showed that loss of PDMS in sludge amended soils is dependent on soil moisture conditions. For example, under extremely arid conditions (Moreno, CA) soil concentrations of PDMS were near or below detection after a total sludge amendment of 2880 mt/ha over a 20-year period. In addition, a lower molecular weight silicone compound was present in the soil at concentrations of up to 20 μg/g in the arid soil, providing additional evidence of PDMS degradation. At the other extreme, a one time amendment of 500 mt/ha in a rainforest environment (Washington State) showed some but not significant loss of PDMS from soil over a 20-year period. A lower molecular weight peak was not detected in the rainforest soil.

Fate of PDMS Degradation Products in Soil

PDMS degradation products were found to bind to soil surfaces, volatilize, and biodegrade. Lehmann et al. [16] established the potential for DMSD biodegradation by generating PDMS degradation products in situ by drying soil previously spiked with ^{14}C-labeled PDMS. The soil was then remoistened and up to 3 % of the initial ^{14}C spike for alfalfa treated soil appeared as $^{14}CO_2$ following a four month incubation period indicating slow biological oxidation of the degradation products formed.

The biodegradation of DMSD was confirmed and a pathway determined by Sabourin et al. [17] using ^{14}C-labeled DMSD spiked directly on soil. This work established DMSD soil biodegradation rates that ranged from 0.05 to 0.5 %/wk on four soils spiked with 100 ppm DMSD. Other work, conducted by Lehmann et al. [18] determined the rate of DMSD biodegradation on four additional soils of 0.4 to 1.6 %/wk with total conversion of ^{14}C-DMSD to $^{14}CO_2$ of 9–35 % over a 30-week period. A comparison of biodegradation and volatilization rates for specific soils suggested that volatilization would dominate in all of the soils studied [19]. Volatilized DMSD is expected to undergo demethylation reactions with OH radicals in the atmosphere, based on the atmospheric fate observed for other volatile organosilicone materials [20].

PDMS Fate in Sediment

Effluents from wastewater treatment plants will contain low levels of PDMS (<5 μg/L) sorbed onto effluent solids. These solids will eventually deposit into bottom sediments. The potential for PDMS biodegradation in sediments was examined by Christianson [21] in sediment/water columns amended with 500 ppm of 350-Cst ^{14}C-labeled PDMS. The cores were aerated and the overlying water and off-gas were monitored for the formation of soluble or volatile materials. Results showed no evidence of biodegradation or biotransformation, indicated by the lack of formation of volatile or solubilized ^{14}C materials, over a 56-day period.

Carpenter et al. [22] examined the biodegradation of ^{14}C-labeled PDMS in freshwater sediments over longer incubation times and found 5–10 % of the original PDMS hydrolyzed to DMSD and about a 0.25 % conversion to $^{14}CO_2$ following one year. The rate and extent of PDMS degradation were related to microbial activity in sediment because conversion rates in sterile controls were significantly lower. The degradation rates measured in sediment correspond closely to the PDMS degradation rates measured in moist soil by Lehmann et al. [9] with 3 % DMSD and 0.13 % CO_2 formation over a six-month incubation time.

PDMS Occurrence in the Environment

Water, soil, sediment and biota have been extensively monitored by a number of investigators over the last twenty years. A complete compilation of PDMS concentrations measured in these environmental compartments is found in Fendinger et al. [1]. These data were used by Fendinger et al. [1] to establish worst-case concentrations in each relevant environmental compartment. The results of this analysis along with PDMS concentrations measured in remote soils and sediments are summarized in Table 2.

Table 2. Worst-case water, soil, sediment, and biota PDMS concentrations from remote locations and areas influenced by wastewater treatment discharges, sludge loadings or industrial activity.

Environmental Compartment	PDMS Concentration[a]	Comment/Reference
Surface water	<<2.5 µg/L	PDMS is generally not detected in surface water[b].
Sediment with significant anthropogenic input	26 mg/kg	Based on reported analyses conducted at 412 locations[b].
Sediment from remote areas	<0.1 mg/kg	SEHSC PDMS Environmental Fact Sheet [c].
Soil previously amended with sludge	17 mg/kg	Based on reported analyses conducted at 27 locations[b].
Soil from remote areas	<0.2 mg/kg	SEHSC PDMS Environmental Fact Sheet [c].
Biota	<0.7 mg/kg	At or near the detection limit[b].

[a] 90th percentile concentrations for each compartment. — [b] Ref. [1]. — [c] Ref. [19].

Environmental Fate Overview

In summary, the environmental fate of PDMS is determined by a number of factors that include use patterns, material properties, and the environmental compartment in which the material is introduced. These considerations have been described in detail and are summarized in Fig. 4.

Environmental Effects

Environmental effects testing of PDMS on aquatic and terrestrial organisms using both standard laboratory methods and more realistic exposure scenarios has been performed for a variety of PDMS formulations and breakdown products. A complete summary of all testing conducted to date is found in Fendinger et al. [1]. Early studies focused on standard aquatic laboratory test methods, employing extremely high test concentrations dosed in the water column. Testing at high concentrations in the water column would represent exposure scenarios similar to what would be expected from a spill or accidental release of large quantities of PDMS. However, PDMS is not water-soluble, and the results of these early aquatic studies were difficult to interpret in the context of more typical exposure scenarios. Since the majority of PDMS in the environment is located in soil and sediment, more recent studies have been conducted in which soil- and sediment-dwelling organisms were exposed to soil and sediment spiked with PDMS, frequently at levels well in excess of actual environmental concentrations.

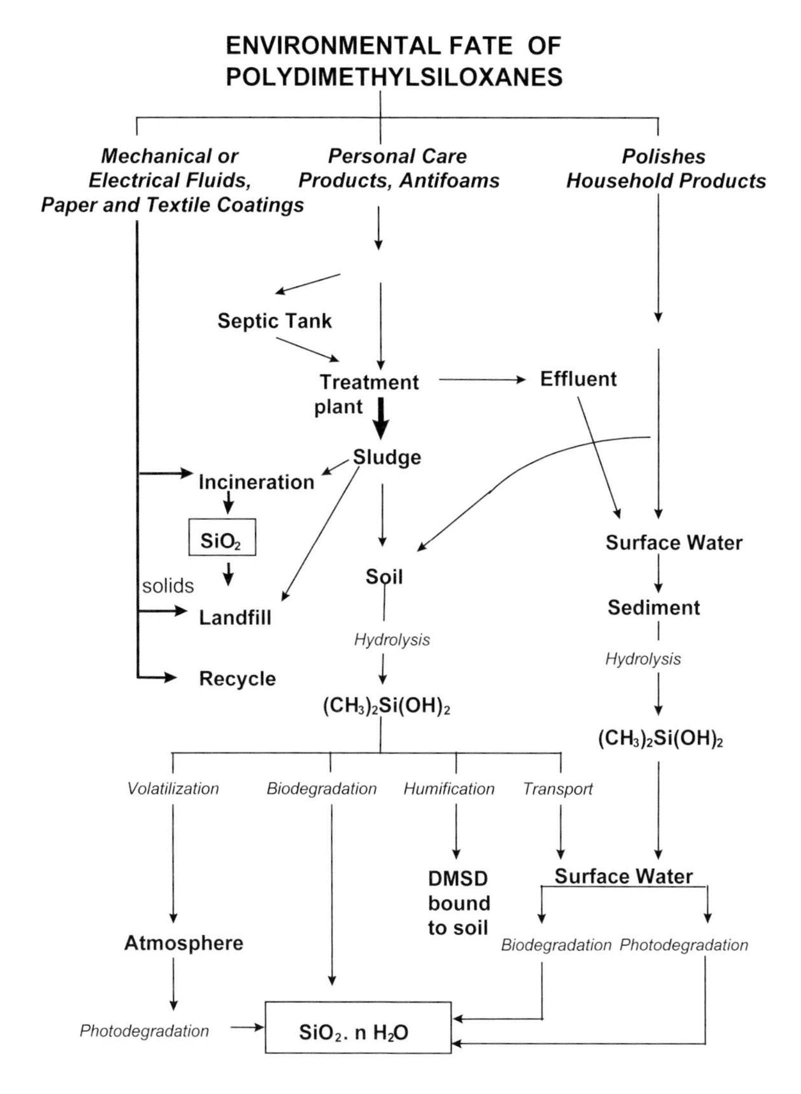

Fig. 4. PDMS environmental fate overview [23].

Results from this extensive effects testing demonstrates that PDMS has a relatively low toxicity to freshwater, marine and terrestrial organisms. As a result, in most cases no dose/response relationships or toxicity differences between species exist. The toxicity test methods that most realistically simulate PDMS exposure in the environment produce the most reliable measure of potential PDMS environmental effects. For example, sediment-bound PDMS will be the primary route of exposure in the aquatic environment, because PDMS has not been measured in overlying water (due to its negligible water solubility and potential for sorption onto sediments). Therefore, studies in which PDMS was dosed as a component of sediment are the most realistic exposure scenario. PDMS loadings to the terrestrial environment occur as a result of sludge loading to soil, either through agricultural or landscaping practices. Therefore, PDMS terrestrial effects studies that simulate an agricultural soil amended with sludge represent the most realistic test matrix. Aquatic and terrestrial effects studies that meet these exposure criteria are summarized in Table 3.

Table 3. Results from PDMS terrestrial and aquatic effects studies that utilized realistic exposure systems [1].

Organism	Test conditions	NOEC (mg/kg)	End points
Terrestrial			
Eisenia foetida	PDMS dosed in high-organic soil	1100[a]; with no evidence of bioaccumulation	Number and viability of cocoons, survival and growth of adults and offspring
Folsomia candida	PDMS dosed in OECD standard soil matrix	250[b]	Survival and reproduction.
Aquatic/Sediments			
Chironomus tentans	PDMS dosed into high-, medium- and low-organic content sediments	350-560[a]; with no evidence of bioaccumulation	Survival and growth
Daphnia magna	*Daphnia magna* cultured over sediment amended with PDMS.	572[a]	Growth, survival, number of offspring, mortality
Hyalella azteca	PDMS dosed in pond sediment	2200[a]	Survival
Ampelisca abdita	PDMS dosed in marine sediment	2300[a]	Survival

[a] Indicates highest dose tested with no effects observed in test.— [b] Indicates nominal concentration.

Potential for Bioaccumulation

PDMS fluids have a log octanol/water partition coefficient of approximately 5 that suggests PMDS may partition into organisms either directly or through food sources. However, work conducted by Annelin and Frye [24] demonstrated that linear PDMS oligomers with a molecular weight greater than 600 were not taken up by rainbow trout over exposure periods that ranged from 35 to 70 days. Additional work by Bruggeman et al. [25] showed that uptake of PDMS with a molecular weight greater than approximately 600 was not detectable in guppies following six weeks of feeding with PDMS-amended food. Hobbs et al. [26] also showed that ^{14}C-labeled PDMS (MW approx. 10 000) tissue concentrations in bluegill sunfish were not related to duration or level of exposure. These experiments demonstrate that PDMS with more than eight units do not get bioabsorbed. Subsequent work with earthworms [27], guppies [28], bullhead catfish [29], and benthic invertebrates [30, 31] also demonstrates that PDMS does not bioaccumulate. Additionally, field monitoring of PDMS concentrations in fish collected from impacted areas does not show any significant accumulation [1].

Conclusions

The SEHSC research efforts combined with those by the GSC resulted in the development of a comprehensive environmental database for PDMS, a material that has both industrial and consumer product applications. The results of this five-year research effort demonstrated that most PDMS used in down-the-drain applications enters the terrestrial environment as a component of wastewater treatment plant sludge, with aquatic sediments receiving between 3 to 6 % of this mass. Laboratory and field studies demonstrated the potential for PDMS degradation in soils and sediments. The rate of degradation is slow in sediment and wet soil and increases as a function of decreased moisture level to half-life values of days in dry soil. No adverse ecological effects are indicated from the effects testing.

References:

[1] N. J. Fendinger, R. G. Lehmann, E. M. Mihaich, "*Polydimethylsiloxane*", in: *The Handbook of Environmental Chemistry, Vol. 3*: Anthropogenic Compounds Part H, Organosilicon Materials (Ed.: G. Chandra), Springer-Verlag, Berlin, **1997**, p. 181.

[2] R. B. Allen, P. Kochs, G. Chandra, "*Industrial Organosilicon Materials, Their Environmental Entry and Predicted Fate*", in: *The Handbook of Environmental Chemistry*, Vol. 3: Anthropogenic Compounds Part H, Organosilicon Materials (Ed.: G. Chandra), Springer-Verlag, Berlin, **1997**, p. 1.

[3] D. T. Palmer (Dow Corning), Study no. SC 900091, **1992**.

[4] E. J. Hobbs, M. L. Keplinger, J. C. Calandra, *Envrion. Res.* **1975**, 397, 10.

[5] R. J. Watts, S. Kong, C. S. Haling, L. Gearhart, C. Frye, B.W. Vigon, *Water Res.* **1995**, *2405*, 79.

[6] R. G. Lehman, C. L. Frye, D. A. Tolle, T. C. Zwick, *Water, Air, Soil, Polut.* **1996**, *237*, 87.

[7] N. J. Fendinger, D. C. McAvoy, W. S. Eckhoff, B. B. Price, *Environ. Sci. Technol.* **1997**, *1555*, 31.

[8] U.S. Environmental Protection Agency, 40 CFR Parts 257, 403, and 503, **1993**.

[9] R. G. Lehmann, S. Varaprath, C. Frye, *Environ. Tox. Chem.* **1994**, *1061*, 13.

[10] J. C. Carpenter, J. A. Cella, S. B. Dorn, *Environ. Sci. Technol.* **1995**, *864*, 29.

[11] R. G. Lehmann, S. Varaprath, R. B. Annelin, J. Arndt, *Environ. Tox. Chem.* **1995**, *1299*, 14.

[12] S. Xu, R. G. Lehmann, J. R. Miller, G. Chandra, *Environ. Sci. Technol.* **1998** (in press).

[13] S. C. Gupta, U. B. Singh, J. F. Moncrif, G. N. Flerchinger, R. G. Lehmann, N. J. Fendinger, *Silicone Environmental Health and Safety Council*, **1998**.

[14] G. N. Flerchinger, K. E. Sexton, *Trans. Am. Soc. Agric. Eng.* **1989**, *565*, 32.

[15] D. C. McAvoy, K. M. Kerr, N. J. Fendinger, paper presented at *Soc. Env. Tox. Chem. 17th Annu. Mtg.* **1996**.

[16] R. G. Lehmann, S. Varaprath, C. Frye, *Environ. Tox. Chem.* **1994**, *1753*, 13.

[17] C. L. Sabourin, J. C. Carpenter, T. K. Leib, J. L. Spivack, *Appl. Env. Microbiol.* **1996**, *4352*, 62.

[18] R. G. Lehmann, J. R. Miller, H. P. Collins, *Water, Air, Soil Polut.* **1998** (in press).

[19] R. G. Lehmann, J. R. Miller, *Environ. Tox. Chem.* **1996**, *1455*, 15.

[20] R. Atkinson, *Environ. Sci. Technol.* **1991**, *863*, 25.

[21] K. Christianson, *Silicone Environmental Health and Safety Council*, **1994**.
[22] J. C. Carpenter, T. K. Leib, C. L. Sabourin, J. L. Spivack, paper presented at *Soc. Env. Tox. Chem. 17th Annu. Mtg.* **1996**.
[23] *Silicone Environmental Health and Safety Council*, Reston, VA, **1998**.
[24] R. B. Annelin, C. L. Frye, *Sci. Tot. Env.* **1989**, *1*, 83.
[25] W. A. Bruggeman, D. Weber-Fung, A. Opperhuizen, J. van der Steen, A. Wijbenga, O. Hutzinger, *Toxicol. Env. Chem.* **1984**, *287*, 7.
[26] E. J. Hobbs, M. L. Keplinger, J. C. Calandra, *Env. Res.* **1975**, *397*, 10.
[27] N. Garvey, M. K. Collins, E. M. Mihaich paper presented at *Soc. Env. Tox. Chem. 17th Annu. Mtg.* **1996**.
[28] A. Opperhuizen, E. W. van der Velde, F. Gobas, D. Liem, J. van der Steen, O. Hutzinger, *Chemosphere* **1985**, *1871*, 14.
[29] R. B. Annelin (Dow Corning), Study no. I-0005-0669, **1979**.
[30] A. E. Putt, paper presented at *Soc. Env. Tox. Chem. 17th Annu. Mtg.* **1996**.
[31] J. Kukkonen, P. F. Landrum, *Environ. Toxicol. Chem.* **1995**, *523*, 14.

Silicones for the Textile Industry

Michael Meßner

Wacker Chemie GmbH
Friedrich-von-Heyden-Platz 1, D-01612 Nünchritz, Germany
Tel.: Int. code + (35265)73 350 — Fax.: Int. code + (35265)73 003
E-mail: michael.messner@wacker.de

Keywords: Silicones / Aminosilicones / Textiles / Auxiliaries / Softeners

Summary: The development of silicone application in the textile industry and the chemistry of silicone softeners is reviewed.

Today silicones find applications in most of the major industries. The textile industry was one of the first branches that discovered the unique features of silicones. Even in the very beginning of technical silicone chemistry first attempts were made to use the potential hydrophobicity of the chlorosilanes obtained by the Müller–Rochow process for textile finishing [1]. For this purpose, gaseous chlorosilanes were brought into contact with the fiber and then neutralized with gaseous ammonia. This treatment never found widespread application because of the loss of tensile strength of the fabric caused by the generated hydrogen chloride. So the career of silicones in textile application actually started when they became available in the form of emulsions.

About 10 % of the silicone production is used for apparel and technical textiles. In this field silicones are mainly used to functionalize textiles. The characteristic hydrophobicity of silicones was firstly utilised in the mid-1950s. The products used in this period were based on combinations of hydrogen siloxanes and OH-terminated dimethylsiloxane fluids. Catalyzed with metal compounds like tin laurate, these products lead to a crosslinked silicone film on the fiber surface. Compared to the previously used impregnating agents like paraffin waxes or stearylchromium compounds, the silicones provide a much better hydrophobicity and even rubbing fastness and wash resistance are improved dramatically.

The peak of the use of silicones as hydrophobizing agents for textiles was reached in the 1970s. Then a new generation of hydrophobizing agents based on fluorocarbons appeared on the market. In the following two decades the silicones were replaced step-by-step by fluorocarbon-based products, because the latter do provide additional advantages like oleophobicity, excellent soiling properties and enhanced wash resistance to the treated fabric.

The use of silicones in textile coating can be subdivided into two areas. One field refers to sophisticated applications like airbag or bakery release coatings. Here the silicones are applied by special equipment as liquid silicone rubbers or room-temperature vulcanizing systems [2] . But the applications are not limited to such pure silicone coatings. Often they are used in combinations with organic polymers like acrylic binders where the silicones work as additives for enhancing hydrophobicity and flexibility of the coating.

The second, fastest-growing silicone application in the textile business uses the softening properties of silicones, especially aminofunctional fluids. For more than 50 years this application was totally occupied by a wide range of products which are derived from fatty acids and amines [3] Even today the majority of textile softeners are based on this chemistry. But the drop in prices for amino-functional fluids entails a continuous substitution of the classical softeners by aminosilicones.

The term "Softening textiles" may appear a little strange at first sight and it may be asked if this is a real necessity. To answer this question we have to get an insight into the processes of textile manufacturing and especially the chemistry connected with apparel production.

As shown in Fig. 1 apparel production can be roughly subdivided into the following steps: thread fabrication in the spinning mill; fabric production in the weaving mill; and finishing. Textile finishing is subdivided into the stages pretreatment; dyeing and printing; and the actual finishing step. In the last step the fabric acquires all the features which are desired by consumers and which make a textile wearable. Finally the ready-made fabric is tailored in the clothing factory.

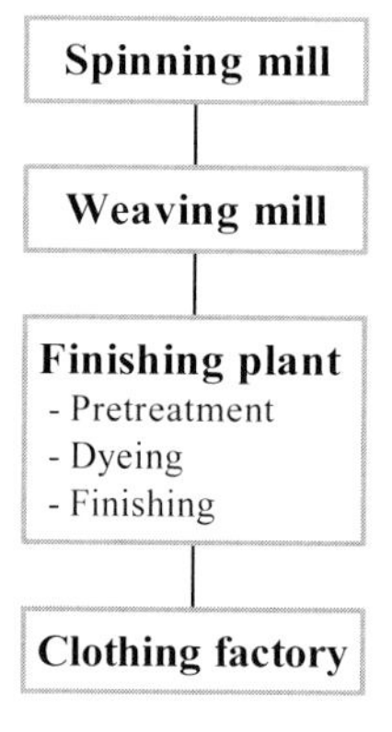

Fig. 1. Apparel production.

For a closer look at the chemistry connected with textile finishing the whole schematic processing of a cotton fabric is displayed in Fig. 2.

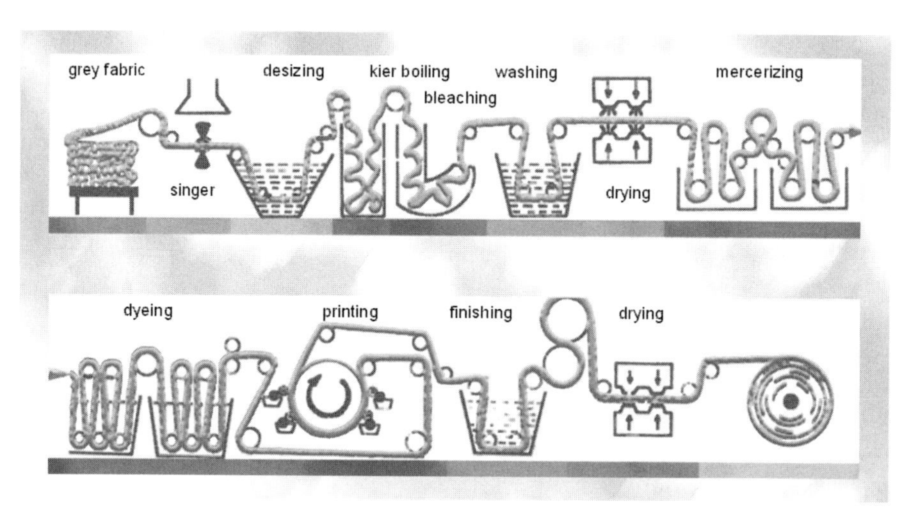

Fig. 2. Processing of woven cotton.

In the first step woven cotton must be freed of residuals from the weaving process like fiber fragments which stand out from the surface. This down is removed by passing the gray fabric through a so-called "singer" where the down is simply burnt off with a flame. Otherwise the down would tend to give an uneven appearance of the dyed fabric.

The next step also serves to undo the consequences of the weaving process. Without using a sizing agent [4] the high mechanical stress to which threads are exposed during the weaving process would lead to flaws. The sizing agents work on the one hand like a glue for the threads and on the other hand as gliding agent. The chemicals used as sizing agents are based on starch, modified cellulose and polyacrylics. The desizing of polyacrylics and modified cellulose is usually done by a washing process. But we have to keep in mind that removal is often incomplete and sizing residuals remain on the fabric. This often leads to problems in the following steps, especially in the finishing step, where the highly anionic charged sizing residuals come in contact with cationic charged compounds. In consequence this often results in coagulation of pad-bath ingredients and spots on the fabric.

So-called "kier-boiling", to obtain a chemical breakdown of natural waxes and other residues like dark-colored pieces of the seed capsules of the cotton plant, is usually carried out in a sodium hydroxide solution at temperatures above 80°C. The resulting fabric loses its hydrophobicity and becomes receptive to the bleach.

To achieve a white fabric the cotton must be bleached. In the past chlorine-generating salts like sodium hypochlorit were used, but for environmental reasons today's most common bleaching agent is hydrogen peroxide. Unfortunately hydrogen peroxide shows insufficient stability. Even small amounts of heavy metal ions like iron work as catalysts for an undesired decomposition of the bleaching agent. The OH radicals generated cause serious damage of the fabric. To overcome this problems various types of complexing agents are used.

Prior to the mercerizing treatment cotton is highly crystallized and in this form unable to absorb higher amounts of dyestuff. But for achieving deep and dark colors this absorption is absolutely necessary. For this need the cotton fabric is contacted under tension with concentrated cold sodium hydroxide solution. The fiber absorbs the caustic, swells and becomes receptive to dyestuff. Higher amounts of caustic remain in the fiber and in this way the mercerized fabric becomes a source for caustic in further steps carried out at lower pH levels. This is the main reason for the addition of higher amounts of acetic acid in the pad bath of the finishing step.

For dyeing and printing of cotton a multitude of dyes with totally different chemical structures are available. Substantially the dyestuffs for cotton are classified in direct, reactive, sulfur and ingrain dyes and pigments for printing. All these dyes need for their application special auxiliaries and conditions.

Without further treatment the fabric would only fulfill the optical demands of consumers. In this form the cotton fabric would crease strongly and show an unacceptable soiling tendency. Both are consequences of the decrystallization during mercerizing. In the finishing step the fabric is treated with a mixture of chemicals to eliminate the mentioned drawbacks and to enhance the wearability. For improvement of the wrinkle resistance, resins mainly based on ethyleneurea and formaldehyde are used. They work as crosslinkers for the cellulose and close the pores. In addition to this chemicals which provide water and oil repellency, flame resistance and many other features to the fabric can be applied. Unfortunately most of these chemicals create a harsh, rigid and unpleasant feeling in the treated fabric. Without applying softeners the feel of the finished fabric would remind one of an emery paper.

For this purpose aminosilicones are the best textile softeners. The adherence of the cationic sidegroups to the anionic fiber surface on the one hand and the resulting orientation of the silicone backbone on the other leads to a unique feel of the treated textile. Performance features such as softness, elasticity or emulsifiability can easily be varied by choosing the viscosity, functionality and nitrogen content of the amino fluid.

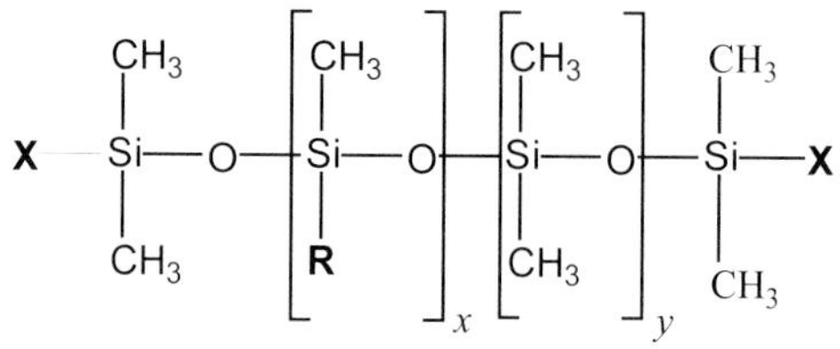

$$X-Si(CH_3)_2-O-[Si(CH_3)(R)-O]_x-[Si(CH_3)_2-O]_y-Si(CH_3)_2-X$$

R = $H_2N\text{-}CH_2CH_2CH_2\text{-}$ or
$H_2N\text{-}CH_2CH_2\text{-}NH\text{-}CH_2CH_2CH_2\text{-}$

X = reactive: $\text{-}O\text{-}CH_3$, $\text{-}O\text{-}CH_2CH_3$
= non reactive: $\text{-}CH_3$

Fig. 3. Basic structure of common aminosilicones.

The main products on the market can be described by the structure displayed in Fig. 3 or very similar ones. Diamino- and aminopropyl groups are commonly used as sidechain functionalities. The backbone functionalities are represented by trimethylsilyl groups for nonreactive fluids or hydrolyzable alkoxy groups for reactive ones. Typical viscosities are 300 to 10 000 mPa s, nitrogen contents below 1%.

The features of the different silicones depend mainly on their functionality. If we take a look at the most widespread type, the nonreactive aminoethylaminopropyl-functional fluids, which were introduced onto the market very early, we see a product covering most of the main requirements such as good softness, emulsifiability, wash resistance and medium elasticity. Alkoxy-terminated diamino fluids show enhanced elasticity. For both types, only the yellowing properties are insufficient. This means that after a drying step at temperatures of above 150°C a visible yellowish discoloration of the cotton appears. These criteria gained importance in the mid 1980s when the textile mills started to elevate the temperatures during the final drying step. This was done to reduce the content of free formaldehyde of the finished cotton when formaldehyde became under toxicological pressure.

The yellowing properties change dramatically when the aminoethylaminopropyl group is replaced by the aminopropyl group. Figure 4 shows the relative whiteness of differently treated fabrics versus the drying temperatures applied. Relative whiteness in this respect means the percentage ratio of the amino fluid treated sample compared to a simply water-treated sample.

From this graph two conclusions can easily be derived:

- the whiteness of a treated fabric is directly connected with the amine content of the applied silicone;
- the structure of the amino functional side chain has a dramatic impact on the whiteness of treated cotton.

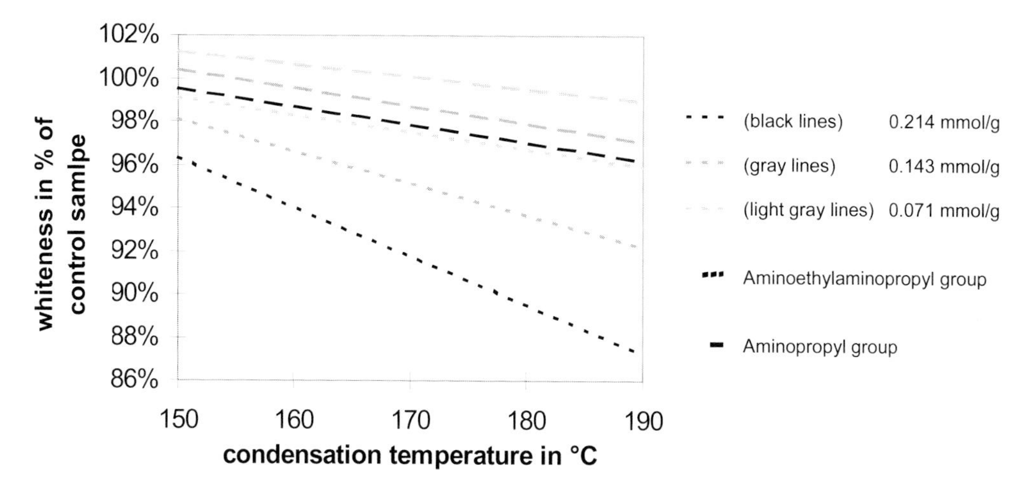

Fig. 4. Yellowing properties of aminosilicones.

Many efforts have been made to improve the yellowing properties and the hydrophilicity of silicone softeners. In Fig. 5 a small selection of suggested structures are therefore collected.

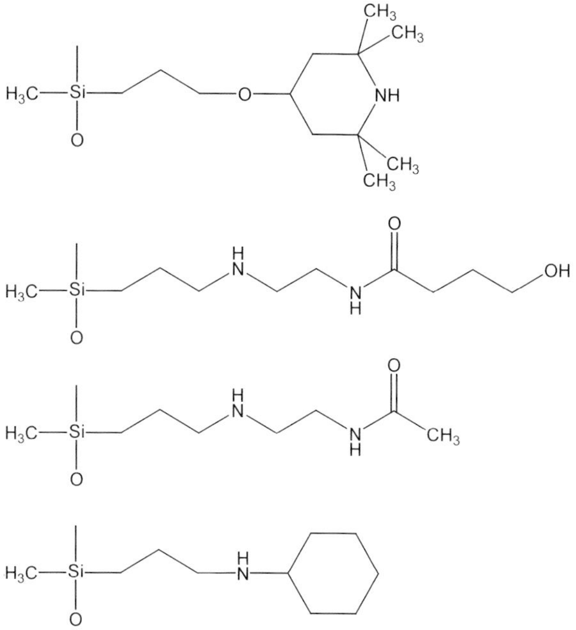

[5]

[6]

[7]

[8]

Fig. 5. Suggested sidegroups for non-yellowing silicones.

As mentioned before, the consumption of aminosiloxanes for textile application has increased continuously over the last 20 years. This growth was almost exclusively caused by the substitution of classical softeners in continuous applications as described above. A further expansion will depend on the ability of silicon chemists to enable their products be applicable in exhaustion processes like jet dyeing where today's aminosilicones are not suitable.

References:

[1] W. Noll, *Chemie und Technologie der Silicone, 2nd ed.*, VCH, Weinheim, **1968**, p. 506.
[2] B. Böddeker, paper presented at *Silicone-Coated Fabrics: Developments, Applications, Outlook*, 6th International Conference on Textile Coating and Laminating, Düsseldorf, **1996**
[3] P. Hardt, *Umweltfreundliche Textilweichmacher*, Melliand Textilberichte, **1990**, p. 699.
[4] D. Behr, *Was versteht man unter Schlichten?,* Maschen-Industrie, **1996**, p. 593.
[5] Rhone-Poulenc, US Patent 5 540 952.
[6] Dow Corning, EP-A 0 342 830.
[7] Wacker, US Patent 5 302 659.
[8] Wacker, US Patent 4 874 662.

Poly-Functionalization of Polysiloxanes — New Industrial Opportunities

Stefan Breunig, Josette Chardon, Nathalie Guennouni, Gerard Mignani*, Philippe Olier, Andre van der Spuy

Rhodia, Usines Silicones
55 rue des Frères Perret, BP 22-69191
Saint Fons Cedex, France
Tel: Int code + (4)72737618 — Fax: Int. code + (4)72936818
E-mail: gerard.mignani@rhodia.eu.com

Carol Vergelati

Rhone-Poulenc, Centre de Recherches des Carrières
85 avenue Frères Perret, BP 62-69192
Saint Fons Cedex, France.

Keywords: Silicones / Functionalizations / Platinum Catalysts / Textile Applications

Summary: Polysiloxanes containing grafted and/or end-capped functional groups, due to their specific properties, receive attention in various industrial applications. This paper presents the synthesis and development of new polyfunctional silicones. For example, the grafting of selected organofunctional groups onto silicone backbones allows us to increase the hydrophobic–oleophobic permanence or hydrophilic softness properties for textile treatments. To achieve this, new synthetic routes using novel organometallic catalysts and specific monomers have been developed.

Introduction

Polysiloxanes are versatile molecules that are used in a wide range of applications. Indeed nearly all surface treatment applications have a use for these unique polymers. This arises from their outstanding properties such as chemical inertness, high-temperature resistance, UV protection and IR stability, as well as a high flexibility of the polymer chain, especially at low temperatures, and excellent surface activity [1]. Our most fruitful approach to modified polysiloxane polymers is the use of reactive polysiloxanes with specific organic moieties grafted onto the silicone backbone. We have developed this strategy to prepare new functional polysiloxanes with specific additional properties. With an appropriate choice of the grafted chains, these polymers can be used to increase the softness of textile materials or to obtain hydrophobic–oleophobic effects. If we wish to obtain a permanent effect, we graft functions onto the silicone chain which allow for strong interactions with the textile substrate. Implementing this strategy has allowed us to produce promising results in a large number of industrial applications.

Organo-Polysiloxane Polymer Synthesis

The most widely used methods of preparing organo-polysiloxane polymers are based either on a reequilibration, on a hydrosilylation, or on a dehydrogenation process. Polydimethylsiloxanes with functional groups **F** may be classified in four categories [2] (Fig. 1).

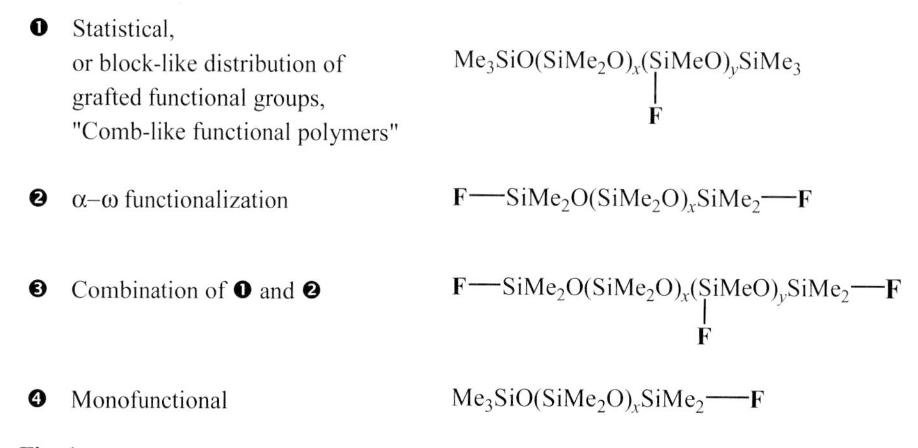

Fig. 1. Polydimethylsiloxanes with functional groups **F**.

In the case of the hydrosilylation pathway, an unsaturated organic compound, containing the appropriate functional group, is reacted with a polydimethylsiloxane that carries Si–H bonds in the chain or at the chain ends. The hydrosilylation route has proved itself to be the most convenient pathway if we wish to control the tacticity of the functionalization in the final polymer.

New Organometallic Complexes for the Hydrosilylation Reaction

The use of organometallic complexes, containing various kinds of ligands, is advantageous in hydrosilylation reactions [3]. Their catalytic properties allow for lower temperatures and cleaner reactions. Useful catalytic complexes of Pt, Rh, Co, Ru, Pd, Ni, Cu and lanthanides have been described. Platinum-based complexes, like the “Karstedt” catalyst, are the most frequently used compounds. The major part of the current research, publications and patents are concerned with the thermal activation of these catalysts. Our recent work focused on the development of a series of new homogeneous platinum catalysts for the hydrosilylation reaction containing quinone ligands (e.g. **1**, Scheme 1) [4]. The synthesis of this kind of catalyst is performed using a ligand exchange reaction starting from an industrial Karstedt complex and the quinone ligand. The isolated yields are >70 %, and the complexes obtained are highly soluble in a silicone medium. Moreover, they are not water-or air-sensitive.

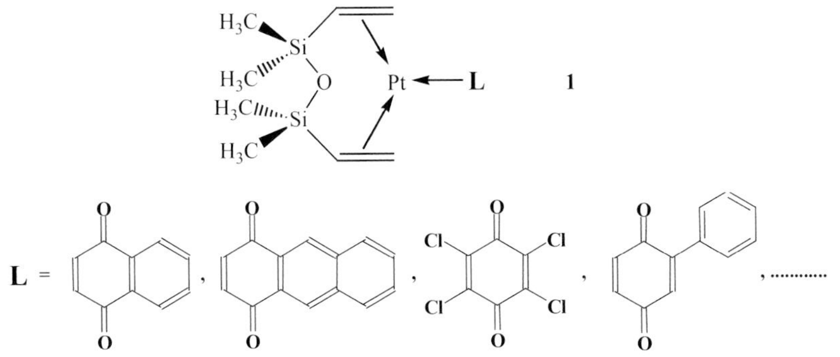

Scheme 1. New homogeneous platinum catalysts for the hydrosilylation reaction.

The X-ray data of the complex **1** with L = naphthoquinone shows a η^2 coordination of the quinone ligand (PtC distances in the Pt(0)–(–CH=CH–) unit : 2.11(1) and 2.12(1) Å) [5]. The ^{195}Pt NMR analysis of this complex is compatible with a Pt(0) oxidation state (Pt(0)–naphthoquinone: δ: –5575 ppm, Kartstedt catalyst: δ: –6131ppm /$CDCl_3$ / Ref: H_2PtCl_6).

For specific applications, such as paper release resins, we have developed two types of photochemically activated platinum hydrosilylation catalysts. The complexes **1** are highly useful in silicone resin formation using Si–H and Si–Vi or Si–H and Si–epoxy functional polysiloxanes under thermal or UV irradiation conditions.

New Fluorinated Polysiloxanes with High Performance in Textile Applications

Fluorinated polysiloxanes based on a linear polysiloxane backbone with grafted aliphatic fluoro-carbon groups have attracted increasing attention in recent years due to their heat and solvent resistance. Another interesting property is their low surface tension [6]. Many fluorinated polysiloxanes, containing perfluoroalkyl [7], perfluoroalkyl ether [8] and perfluoroalkyl ester moieties [9], have been studied in detail. Some of these fluorinated silicone polymers exhibit mesomorphic surface textures. These structures are attributed to the organization of the fluorocarbon–hydrocarbon grafts along the silicone backbone forming well-defined supramolecular architectures on the surface [10]. The organization of these perfluorinated chains is crucial to obtain a very low surface energy.

In the case of a large number of industrial textile applications the need to obtain the following properties has been identified :

- Strong oleophobic and hydrophobic character
- Good durability / permanence of these properties after machine washing
- Good chemical stability in the washing medium
- Good softness and non-yellowing properties
- Applicability in the form of a stable emulsion or microemulsion
- Safe and reproducible processing with low cost
- Good patent position
- Attractive price

Our strategy is to utilize linear silicone polymers grafted with multiple specific moieties, one type exhibiting oleophobic–hydrophobic properties, such as –R[Rf]$_2$ groups (with Rf = perfluoroalkyl group), and the second type exhibiting "reactive" properties versus the textile surface (= "A"-group) using hydrogen-bridge bonding and/or chemical bonding:

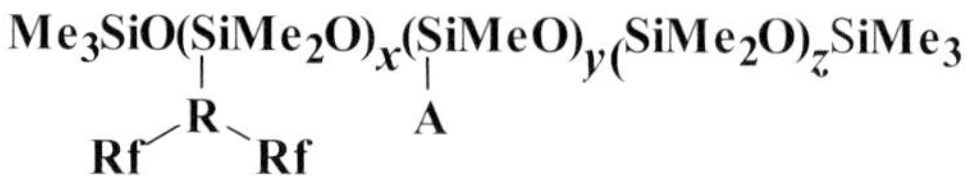
Me3SiO(SiMe2O)x(SiMeO)y(SiMe2O)zSiMe3
R
Rf
Rf
A

In order to maximize the oleophobic–hydrophobic properties we have developed new bisperfluoroalkyl compounds and introduced them as side-chains attached to malonate or malonamide spacers as depicted in Scheme 2 [11]. Using the Bis-Rf group we induced a high crystallinity.

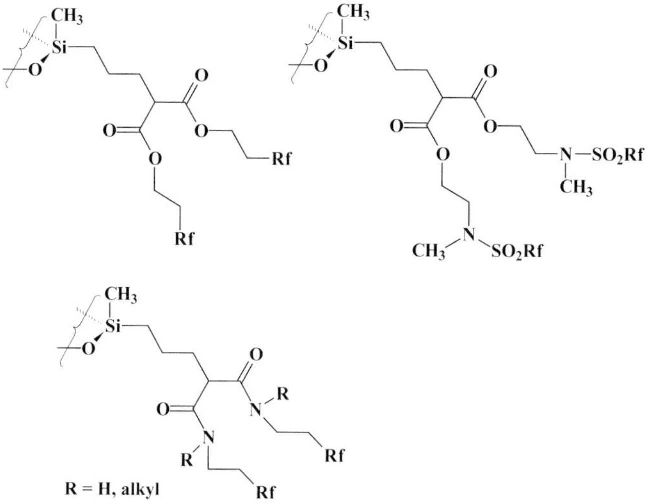
CH3
Si
O
Rf
N—SO2Rf
CH3
SO2Rf
R
N
R = H, alkyl

Scheme 2. Bisperfluoroalkyl compounds.

Concerning the reactive "A"-group, several different moieties have been introduced (Scheme 3).

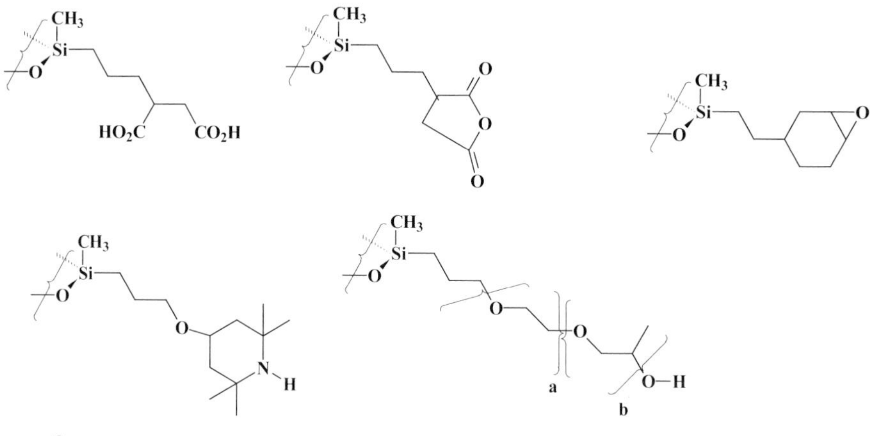
CH3
Si
O
HO2C
CO2H
N
H
O—H
a
b

Scheme 3. Examples of reactive "A"-groups.

Preparation of New Fluorinated Polysiloxanes

The synthesis of bifunctional silicone polymers was performed by a step by step hydrosilylation process using a platinum catalyst and the corresponding monomers (Scheme 4). We synthesized new hydrosilylable bisperfluoroalkyl monomers (e.g. monomer **a**) by trans-esterification of bis-alkyl malonate with perfluoroalcohols using specific Ti(IV)-based catalyst complexes.

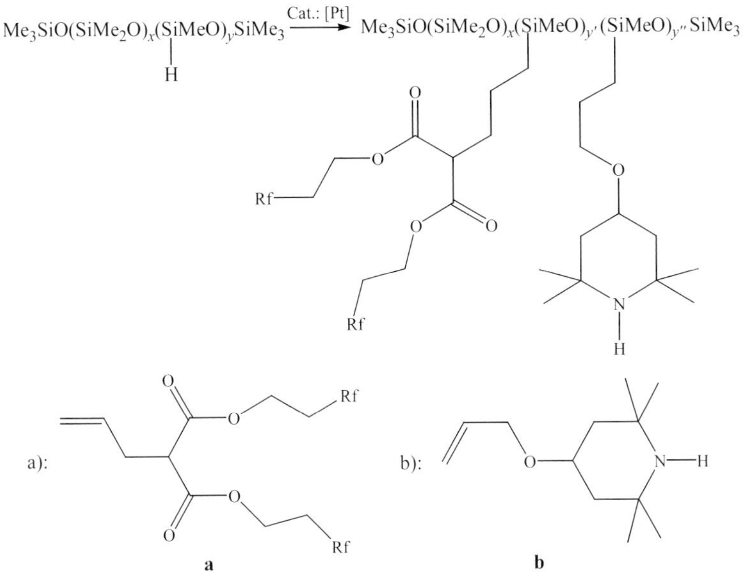

$$Me_3SiO(SiMe_2O)_x(SiMeO)_ySiMe_3 \xrightarrow{\text{Cat.: [Pt]}} Me_3SiO(SiMe_2O)_x(SiMeO)_{y'}(SiMeO)_{y''}SiMe_3$$

a): **a**

b): **b**

Scheme 4. Synthesis of bifunctional silicone polymers (Rf = C_8F_{17}, C_nF_{2n+1} and $N(CH_3)SO_2C_nF_{2n+1}$, **a** and **b** are monomers (hydrosilylation agents).

We adopted the classification shown in Table 1 for the functional groups grafted on the silicone backbone or other silicone moieties.

Table 1. Classification of functional groups.

Functional group	Class	Functional group	Class
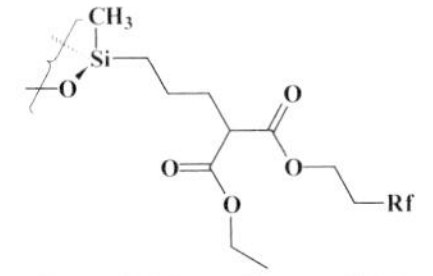	MRf	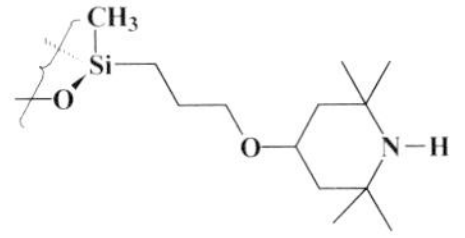	HALS
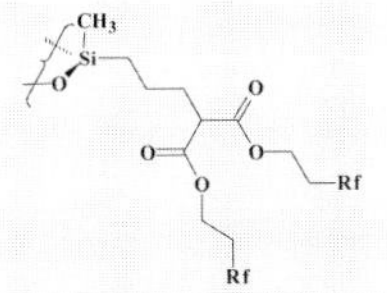	BRf	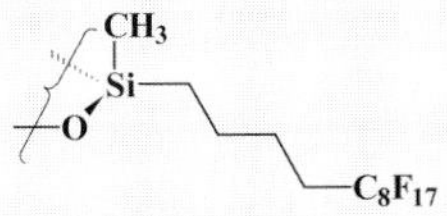	Rf

Continuation Table 1

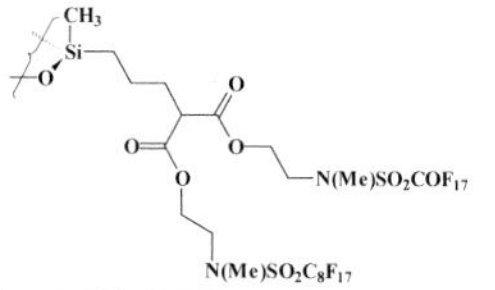 CH_3 Si O O O O O $N(Me)SO_2COF_{17}$ $N(Me)SO_2C_8F_{17}$	BSRf	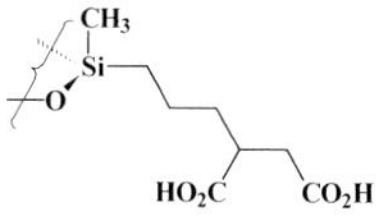 CH_3 Si O HO_2C CO_2H	BAC
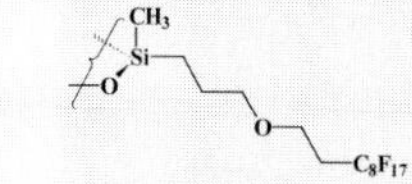 CH_3 Si O O C_8F_{17}	ARf	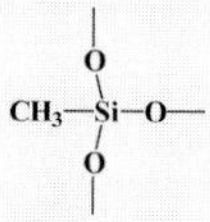 CH_3–Si–O– (O above and below)	**T**
$–SiMe_2O–$	D	$Me_3Si–O–$	M

Physicochemical Characterizations

The new functional polysiloxanes were characterized by NMR (^1H, ^{13}C, ^{29}Si) and IR techniques as well as by elemental analysis. The monomers (Table 2) and the functionalized polymers (Table 3) were also examined by thermal analysis.

Table 2. Thermal analysis of some monomers.

Product	T_m [°C]
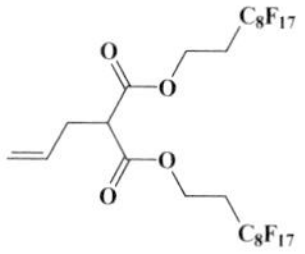 O O O O C_8F_{17} C_8F_{17}	38
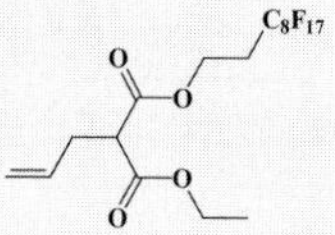 O O O O C_8F_{17}	–20
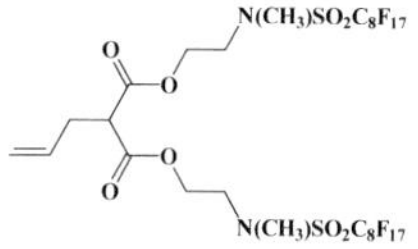 O O O O $N(CH_3)SO_2C_8F_{17}$ $N(CH_3)SO_2C_8F_{17}$	95
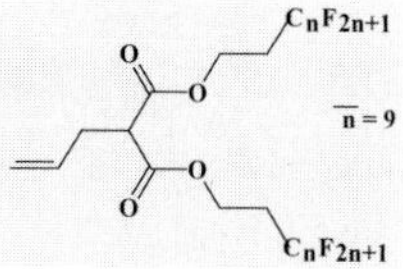 O O O O C_nF_{2n+1} $\bar{n} = 9$ C_nF_{2n+1}	40–80, maximum 68
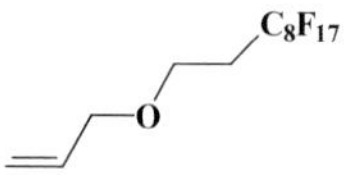 O C_8F_{17}	Liquid
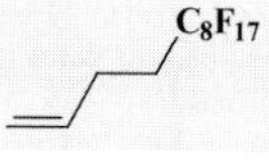 C_8F_{17}	Liquid

Table 2 indicates that the value of the melting point depends on both:

- the kind of the perfluorinated alcohol (a solid form is only obtained with Bis-Rf moieties).
- the choice of the spacer (e.g. malonic alkyl ester groups; sulfamido esters are very interesting spacers if one wishes to obtain a solid material).

Table 3. Thermal analysis of some functionalized polymers.

Product	Characteristics
$MD_xD(MRf)_yM$	T_1: –16.7°C / T_2: 20.2°C
$MD_xD(BAC)_{y'}D(MRf)_{y''}M$ with $y = y' + y''$	T_1: –16.6°C / T_2: 18.5°C
$MD_{8x}D(BAC)_{y'}D(BRf)_{y''}M$	T_1: 17.8°C / T_2: 42.7°C
$MD_{8x}D(BSRf)_{y'}D(HALS)_{y''}M$	93°C
$MD_{8x}D(MSRf)_{y'}D(HALS)_{y''}M$	–35° to 10°C, Maximum: –10°C
$MD_{8x}D(HALS)_{y'}D(BRf)_{y''}M$	T_1: –18.6°C / T_2: 40.2°C

The X-ray results in Table 4 confirm the possibility of a self-organization of the bisperfluoroalkyl ester moieties. The mono-perfluoroalkyl ester moieties, on the contrary, are unable to self-organize in the same way and do not lead to solid products at room temperature. The organization observed in the case of the bisperfluoroalkyl ester leads to better performances and has been found to be thermodynamically stable.

Table 4. X-ray analyses.

Product	Crystallinity
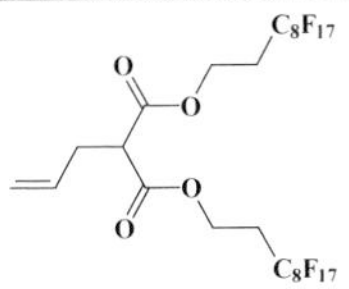 C_8F_{17}, O, O, O, O, C_8F_{17} (structure)	69 % (by crystallization) 71 % (40°C) 72 % (40°C, 15 days)
$MD_xD(BAC)_{y'}(D(MRf)_{y''}M$	Amorphous
$MD_xD(BAC)_{y'}(D(BRf)_{y''}M$	53 % 53 % (40°C, 15 day)

A minimum fluorine content is required to obtain a low surface energy level. Values of γ_s for entry 1 are higher than for other molecules (Table 5). Dewetting values for γ_s are minimized when Bis-Rf moieties are used. Stable coatings exhibiting a low surface energy can be obtained with a lower fluorine content than with conventional "Fluorosilicones" using Bis-Rf moieties.

Table 5. Surface tension properties.

	% Fluorine (w/w)	Structure	γ_s Wetting / γ_s Dewetting	Wetting		Dewetting	
				γ_{sd}	γ_{sp}	γ_{sd}	γ_{sp}
1	13.8	$MD_{8x}D(BAC)_{y'/2}D(BRf)_{y''/2}M$	29.9/42.8	29.6	0.3	35.8	7
2	42	$MD_{8x}D(BAC)_{y'}D(BRf)_{y''}M$	12.3/33.5	11.1	1.2	24.7	8.8
3	31.2	$MD_{8x}D(BAC)_{y'}D(BRf)_{y''}M$	11.6/16.8	11.2	0.4	13.5	3.3
4	34	$MD_{8x}D(HALS)_{y'}D(BRf)_{y''}M$	11.2/43	10.4	0.8	32.3	10.7
5	38.2	$MD_{8x}D(HALS)_{y'}D(ARf)_{y''}M$	13.2/38.7	12.1	1.1	28.2	10.5
6	59	$MD_{x}D(BAC)_{y'}D(BRf)_{y''}M$	12.1/21.8	11.3	0.8	19.3	2.5
7	52.2	$MD_{x}D(BAC)_{y'}D(Rf)_{y''}M$	11.7/29.1	11.1	0.6	22.9	6.2
8	41	$MD_{x}D(BAC)_{y'}D(MRf)_{y''}M$	8.6/49.3	7.3	1.3	30.4	18.9
9	54.1	$MD_{8x}D(BRf)_{y'+y''}M$	12.9/11.6	11.6	1.3	14.7	1.8

Application Results with Textile Support

Textile Application Results Using Solvent Perfluorinated Silicones Solutions (Toluene)

Oil repellency and water repellency are only satisfactory with Bis-Rf moieties (Table 6). A minimum is required. Poor results were observed with O-Rf and Mono-Rf moieties. A slight influence of the nature of the anchoring function was observed.

Table 6. Textile application results with solvent perfluorinated silicone solutions.

	Structures	Oil repellency[a]	Water repellency[a]
1	$MD_{x}D(BAC)_{y}D(BRf)_{y/3}M$	2	1
2	$MD_{x}D(HALS)_{z}D(Bis\text{-}Suf\text{-}Rf)_{z}M$	6	6
3	$MD_{x}D(HALS)_{z}D(Bis\text{-}Rf)_{z}M$	5	7
4	$MD_{x}D(BAC)_{z}D(Bis\text{-}Rf)_{z}M$	6	5
5	$MD_{x}D(HALS)_{z}D(O\text{-}Rf)_{z}M$	0	0
6	$MD_{x/4}D(BAC)_{0.8x}D(Mono\text{-}Rf)_{3x}M$	4	1
7	$MD_{x/4}D(MonoRf)_{2x}M$	4	1
8	$MD_{x}D(Bis\text{-}Rf)_{2x}M$	6	7
9	$MD_{3x}D(Bis\text{-}Ac)_{3y}D(Rf)_{2x}M$	1	0
10	$MD_{3x}D(Bis\text{-}Ac)_{3y}D(Mono\text{-}Rf)_{2x}M$	6	3

[a] Oil and Water repellency tests [12]: 700 ppm of fluorine onto PA66 fabric. Heat treatment: 4 min at 140°C.

Textile Application Results Using Emulsion Medium

The following parameters have been examined:

(a) Influence of the nature of the textile support

Perfluorinated silicone polymer tested:

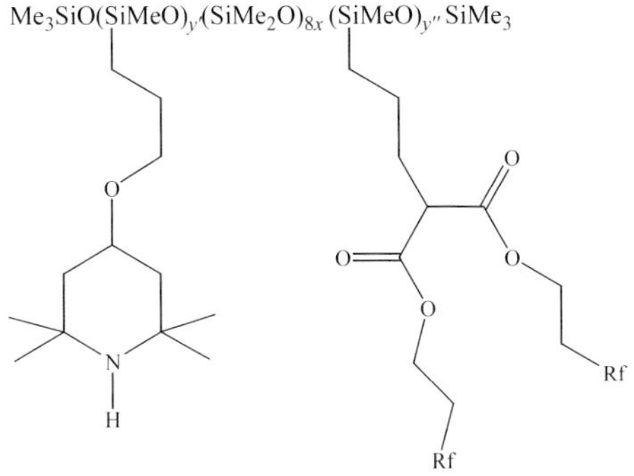

$Me_3SiO(SiMeO)_{y'}(SiMe_2O)_{8x}(SiMeO)_{y''}SiMe_3$

We do not observe any loss of performance after five washing cycles (Table 7). These results are comparable with those obtained with some monoperfluorocarbon compounds (e.g. polyacrylates). However, no heat treatment is required to regenerate the initial organophobicity and hydrophobicity properties, contrary to some monofluorocarbon polymers (e.g. polyacrylates). The same efficiency is observed in the case of PA microfibers, PET–cotton and cotton supports.

Table 7. Influence of the nature of the textile support.

Supports[a]	**Initial**			**5 Laundry washes**			**5 Laundry washes + ironing**		
	OR[b]	**Spray test**[c]		**OR**[b]	**Spray test**[c]		**OR**[b]	**Spray test**[c]	
		Rate	**Water [%]**		**Rate**	**Water [%]**		**Rate**	**Water [%]**
PA microfibers	6	80	3	6	80	5	6	70	7
PET–Cotton	6	80	3	6	80	5	5	70	7
Cotton	6	70	4	5	80	5	5	70	6

[a] Tested with emulsion of the perfluorinated silicone polymer, 2500 ppm fluorine. — [b] Oil repellency [12]. — [c] Spray test [12].

(b) Influence of amount of fluorine applied

Perfluorinated silicone polymer tested:

$Me_3SiO(SiMeO)_{y'}(SiMe_2O)_{8x}(SiMeO)_{y''}SiMe_3$

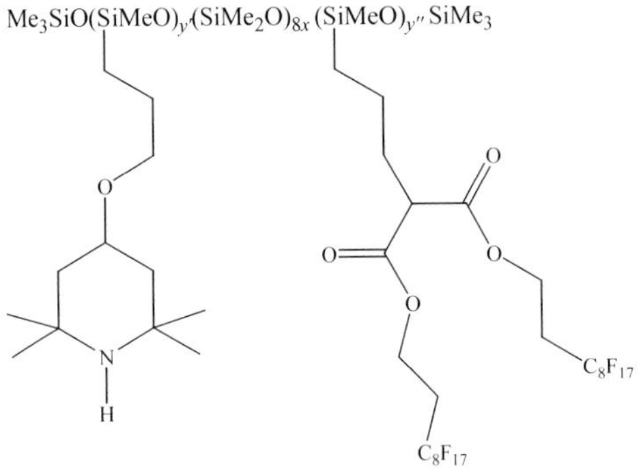

Table 8. Influence of amount of fluorine applied.

Fluorine[a]	Initial			5 Laundry washes			5 Laundry washes + ironing		
[ppm]	OR[b]	Spray test[c]		OR[b]	Spray test[c]		OR[b]	Spray test[c]	
		Rate	Water [%]		Rate	Water [%]		Rate	Water [%]
500	2	70	6	1	70	14	2	70	5
1000	3	80	4.5	3	70	10	4	70	5
2000	5	80	4	5	80	7	5	80	5
3000	6	80	3	6	80	5	6	70	7

[a] Tested on PET–cotton support. — [b] Oil repellency [12]. — [c] Spray test [12].

Table 8 indicates that a fluorine content of between 2000 and 3000 ppm deposited on the surface is most efficient for oil and water repellency.

(c) Influence of the type of the anchoring function

Perfluorinated silicone polymer tested:

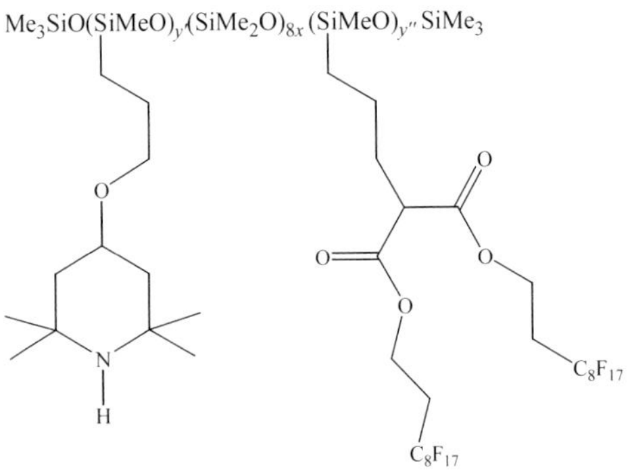

$Me_3SiO(SiMeO)_{y'}(SiMe_2O)_{8x}(SiMeO)_{y''}SiMe_3$ **1**

$Me_3SiO(SiMe_2O)_{8x}(SiMeO)_ySiMe_3$ **2**

$y = y' + y''$

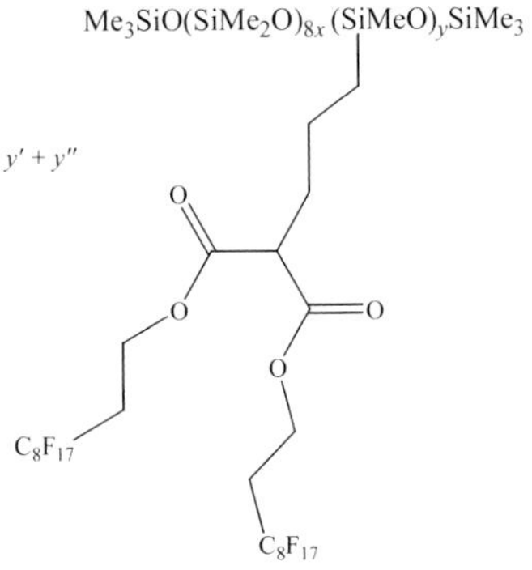

Table 9. Influence of the type of the anchoring-function.

Polymers[a]	Initial			5 Laundry washes			5 Laundry washes + ironing		
	OR[b]	Spray test[c]		OR[b]	Spray test[c]		OR[b]	Spray test[c]	
		Rate	Water [%]		Rate	Water [%]		Rate	Water [%]
1	6	80	3	6	80	5	6	70	7
2	3	70	7	0	50	25	0	50	17

[a] Tested on PET–cotton support with 3000 ppm fluorine. — [b] Oil repellency [12]. — [c] Spray test [12].

Permanency is only achieved with HALS anchoring functions, which is dramatically demonstrated in Table 9. We also note that a small amount of HALS function is sufficient to obtain good permanency.

Conclusions and Perspectives

Polysiloxanes with bis-perfluoroalkyl groups attached to the silicone backbone via spacers such as ester or sulfone-ester groups provide good hydrophobic and oleophobic properties. The correlation between the self-organization of fluoroalkyl groups grafted onto the silicone chain and hydrophobic–oleophobic properties has been demonstrated. Studies are in progress concerning the characterization of the supramolecular architecture responsible for these excellent hydrophobic–oleohobic properties.

New Polysiloxanes with Hydrophilic-Softening and Non-Yellowing Properties for Textile Treatments

At Rhodia our synthesis efforts have also been concentrated on developing new functional silicone oils that provide good softening and non-yellowing properties. A typical structure is denoted by [13]:

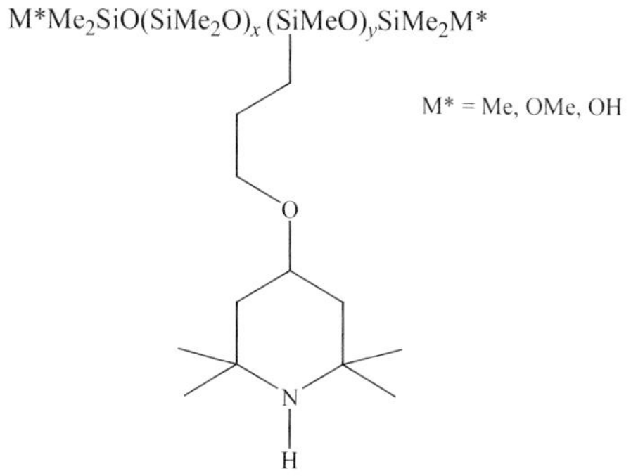

$M^*Me_2SiO(SiMe_2O)_x(SiMeO)_ySiMe_2M^*$

$M^* = Me, OMe, OH$

For a large number of textile applications, we need to obtain additional properties such as a high hydrophilicity. Hydrophilic properties are most conveniently introduced by grafting of polyether chains. Therefore, we developed new functional polysiloxanes containing polyether and HALS moieties on the same backbone [14]:

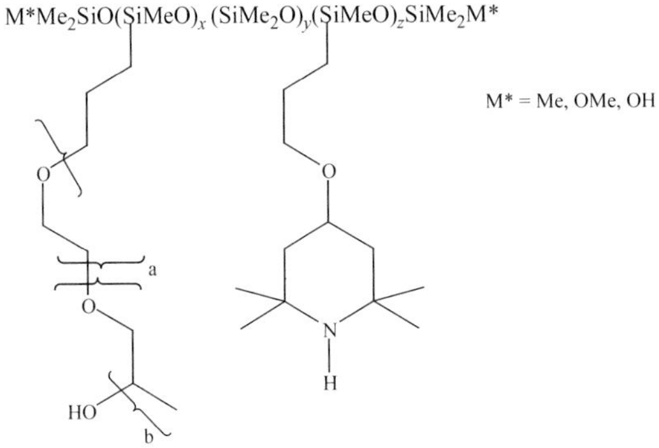

$M^*Me_2SiO(SiMeO)_x(SiMe_2O)_y(SiMeO)_zSiMe_2M^*$

$M^* = Me, OMe, OH$

Preparation of polysiloxanes with polyether and HALS grafts is by two routes:

(a) Equilibration route:

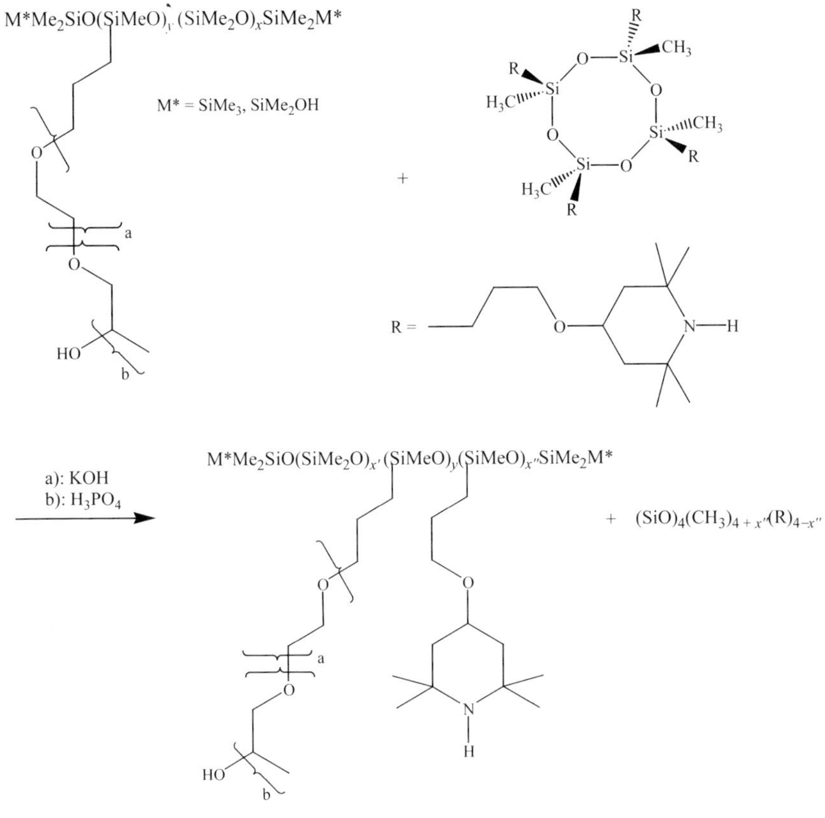

$M^*Me_2SiO(SiMeO)_{y'}(SiMe_2O)_xSiMe_2M^*$

$M^* = SiMe_3, SiMe_2OH$

+

R =

a): KOH
b): H_3PO_4

$M^*Me_2SiO(SiMe_2O)_{x'}(SiMeO)_y(SiMeO)_{x''}SiMe_2M^*$ + $(SiO)_4(CH_3)_{4+x''}(R)_{4-x''}$

(b) Hydrosilylation route:

Table 10. Textile application results.

Products	Textile test result		Non-yellowing test[c]
	Hydrophilic properties[a]	Softness[b]	
$M^*D_xD(HALS)_yM^*$ $M^*= SiMe_3, SiMe_2OMe$	~ 20s	+ + +	+ + +
$M^*D_xD(HALS)_{y'}$ D(Poly-ether)$_{y''}M^*$ $M^*= SiMe_3, SiMe_2OMe$	~9–10s	+ + –	+ + +
MD_xM	~20s	– – –	

[a] Starting fabric PET-Cotton: 6 s, TEGEWA test [15]. — [b] Handle evaluation. — [c] Yellow index evaluation measured with spectrocolorimeter.

Conclusions and Perspectives

Table 10 demonstrates that these polyfunctional silicone polymers containing polyether and HALS moieties have good hydrophilic, softness and non-yellowing properties in the case of textile treatments.

General Conclusions and Perspectives

Functional silicone oils represent an excellent means to solve a number of modern industry's problems. The functionalization of polydimethylsiloxane backbones with multiple grafts, compatible or not with each other, opens a wide field of advanced materials synthesis. Tailor-made functional polymers with designed properties according to given specifications can be synthesized. To achieve this goal we have developed a number of different functionalization strategies, relying basically on redistribution, hydrosilylation and dehydrogenation reactions. Advanced knowledge of the reaction mechanisms of the SiH-group chemistry allows us to rapidly evaluate the feasibility of the synthesis of new structures. Furthermore, knowledge of the limits of the system helps to achieve results in a more economic way. Hydrosilylation and equilibration processes are the most frequently applied methods in the synthesis of functional silicone oils and silicone networks for coating applications.

There are many applications of polyfunctional silicones imaginable. In the present paper we have shown our ability to modulate the properties of these molecules, as well as their softening, non-yellowing, oleophobic and hydrophobic characteristics. These structures have found application in fiber treatments strongly enhancing the quality of the final products. Polyfunctional silicones are an important and promising subject of investigation, from the industrial as well as the academic point of view.

References:

[1] W. Noll, *Chemistry and Technology of Silicones*, Academic Press, New York; **1968**; *Les Silicones*: *Productions et Applications*, Rhône-Poulenc Silicones Department / Techno Nathan **1988**.

[2] C. Burger, F.-H. Kreuzer, "*Polysiloxanes and Polymers Containing Siloxane Groups*", in: *Silicon in Polymer Synthesis* (Eds.: H. R. Kricheldorf), Springer Verlag, Berlin, **1996**.

[3] a) L. H. Sommer, E. W. Pietrusza, F. C. Whitmore, *J. Am. Chem. Soc.* **1947**, *69*, 188.
b) V. B. Pukhnarevich, E. Lukevics, L. T. Kopylova, M. G. Voronkov, *Perspectives of Hydrosilylation* (Ed.: E. Lukevics), Riga, Latvia, **1992**, 383.
c) B. Marciniec. *Appl. Homogen. Catal. Organomet. Compd.*, **1996**, *1*, 487.

[4] Rhône-Poulenc Patent.

[5] Unpublished results.

[6] M. Owen. *"Silicone Surface Activity"* in *Silicon-Based Polymer Science*, Advances in Chemistry Series No. 224 (Eds.: J. M. Ziegler F. W. Fearon), American Chemical Society, Washington DC, **1990**.

[7] M. M. Doeff, E. Lindner, *Macromolecules* **1989**, *22*, 951; H.Kobayashi, M. J. Owen *Macromolecules* **1990**, *23*, 4929.

[8] R. Dorigo, D. Teyssié, J. Yu, S. Boileau, *Polym. Prep.* **1990**, *31-32*, 420.

[9] S. Boileau, E. Beyon, P. Babin, B. Benneteau, J. Dunogues, D. Teyssié, *Eur. Coat. J.* **1996**, *7–8*, 109.

[10] E. Beyou, P. Babib, B. Bennetau, J. Dunogues, D. Teyssiè, S. Boileau, *J. Polym. Sci., Part A, Polym. Chem.* **1994**, *32*, 1673; E. Beyou, B. Bennetau, J. Dunogues, P. Babin, D. Teyssié, S. Boileau, J. M. Corpart, *Polym. Internat.* **1995**, *38*, 237.

[11] Rhône-Poulenc Patent.

[12] *Technical Manual of the American Association of Textile Chemists and Colorists, Vol. 64,* **1989** (Library of Congress Catalogue Number 54-34349). Oil repellency: AATCC 118; Water repellency: isopropanol test; Spray test: AATCC 22.

[13] Rhône-Poulenc Patent.

[14] Rhône-Poulenc Patent.

[15] TEGEWA, *Textilberichte* **1987**, *68*, 581.

The Use of a Silicone-Bonded Azo Dye as Chemical Proton Detector

U. Müller, A. Utterodt[*][a]

Institut für Organische Chemie
Martin-Luther-Universität Halle-Wittenberg
Geusaer Straße, D-06217 Merseburg, Germany
Tel.: Int. code + (3461)462009 — Fax.: Int. code + (3461)462081
E-mail: utterodt@gmx.de

Keywords: Proton / Acid / Detection / Diazonium Salt / Azo Dye

Summary: Quantitative determination of protons in nonaqueous solvents is of special interest in many fields of investigation like cation-induced photopolymerization and photocatalysis. In this way we checked an azo dye as a proton detector for the determination of protons in several organic solvents (methanol, acetonitrile, dimethoxyethane). To prevent characteristic problems of the photochemical investigations (quenching, sensitization) it was advantageous to separate the probe dye from the sample by fixing into a polymer matrix of a silicone network. For this purpose a 4-(*N*-allyl-*N*-methyl)aminoazobenzene was added to a H-functionalized siloxane by Pt catalysis. The remaining Si–H groups were crosslinked to a siloxane monomer containing vinyl ether groups. This polymer allows the application of the probe system in organic solvents without any restriction. The detection can be done by common UV–Vis spectroscopy. In spite of a low microviscosity of siloxanes, the equilibrium of the protonation is reached in a few seconds. Experiments were carried out with HCl gas, HCl in water and solutions of HCl, *p*-toluenesulfonic acid, and tetrafluoroboric acid diethyl ether complex in 1,2-dimethoxyethane. In contrast to a homogeneous solution of the dye and an onium salt misinterpretations by photogenerated carbocation species can be excluded.

Quantitative determination of protons in non-aqueous solvents is of special interest in many fields of investigation, such as cation-induced photopolymerization and photocatalysis. In this way we checked an azo dye as proton probe for the determination of protons in several organic solvents (methanol, acetonitrile, dimethoxyethane) [1]. The reaction of the dye with a proton is shown in Scheme 1. A linear dependence of the acid concentration was found for the absorption of the protonated species. The calibration is necessary for each solvent.

---

[a] Present adress: Institut für Physikalische Chemie, Universität Jena, D-07743 Jena, Germany.

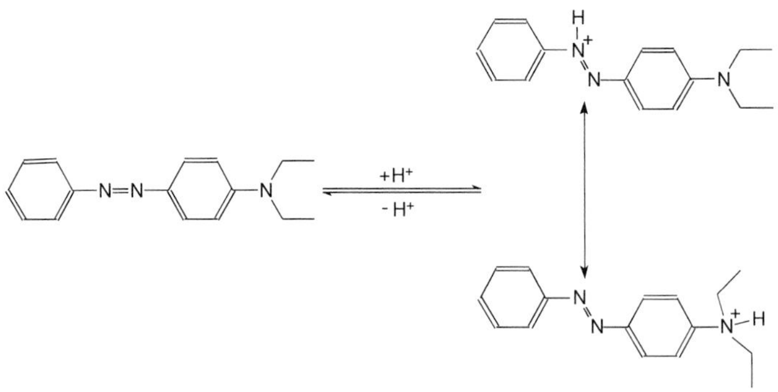

Scheme 1. Proton-dependent reactions of the applied dye.

Using the azo dye we studied the quantum yields of proton formation in the direct photolysis of diazonium salts in homogeneous solution. In this way we found that the amounts of diazonium compound (-DS) photolysed under air are similar to the amounts of protons generated. But the expected ratio $\phi(\text{-DS}) \approx \phi(H^+)$ was observed only with donor- or chlorine-substituted diazonium salts. Mostly a ratio $\phi(\text{-DS}) / \phi(H^+)$ larger than 1 was found. Substituents with a strong electron withdrawl (4-F-, 3-CF_3- and 3,5-$(CF_3)_2$) caused the largest effects. Presumably, charge transfer of the azo dye with stronger electrophilic diazonium salts is the reason for this discrepancy. The observed bathochromic shift of the absorption maxima of the azo compound by increasing concentrations of diazonium salts supports this thesis; see Fig. 1. Even electrophilic substitution reactions of azo compounds with diazonium salts are known [2].

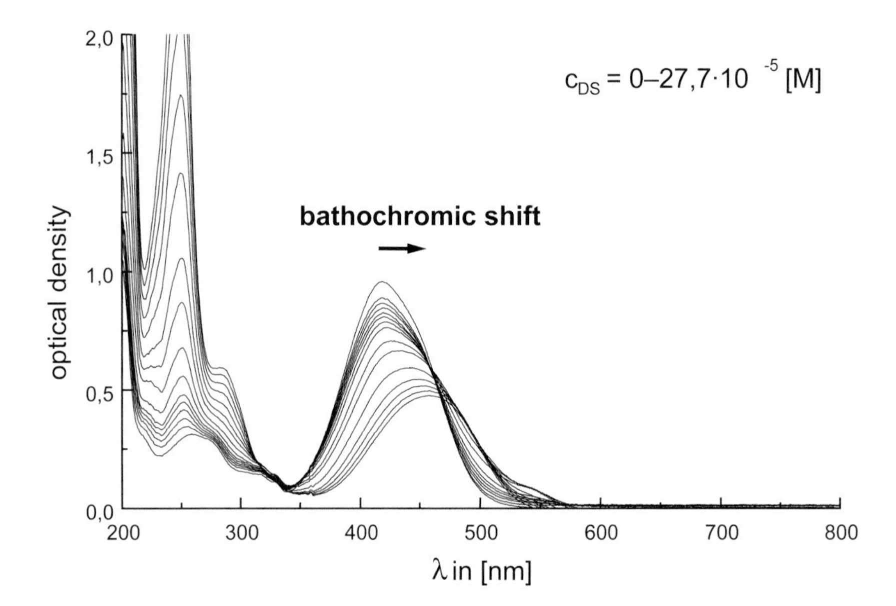

Fig. 1. Changes of the absorption behavior of 4-(*N,N*-diethylamino)azobenzene in acetonitrile ($c_{Azo} = 2.64 \times 10^{-5}$ mol/L) with increasing concentrations of 3,5-di(trifluoromethyl)benzenediazonium salt ($\lambda_1^{max} = 286.4$ nm; $\lambda_2^{max} = 248.8$ nm; $\varepsilon(365\text{nm}) = 80$ L·mol^{-1}·cm^{-1}).

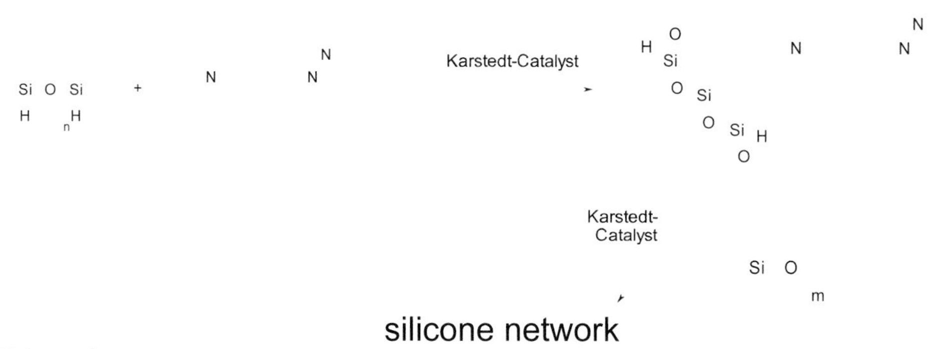

Scheme 2. Principle of synthesis.

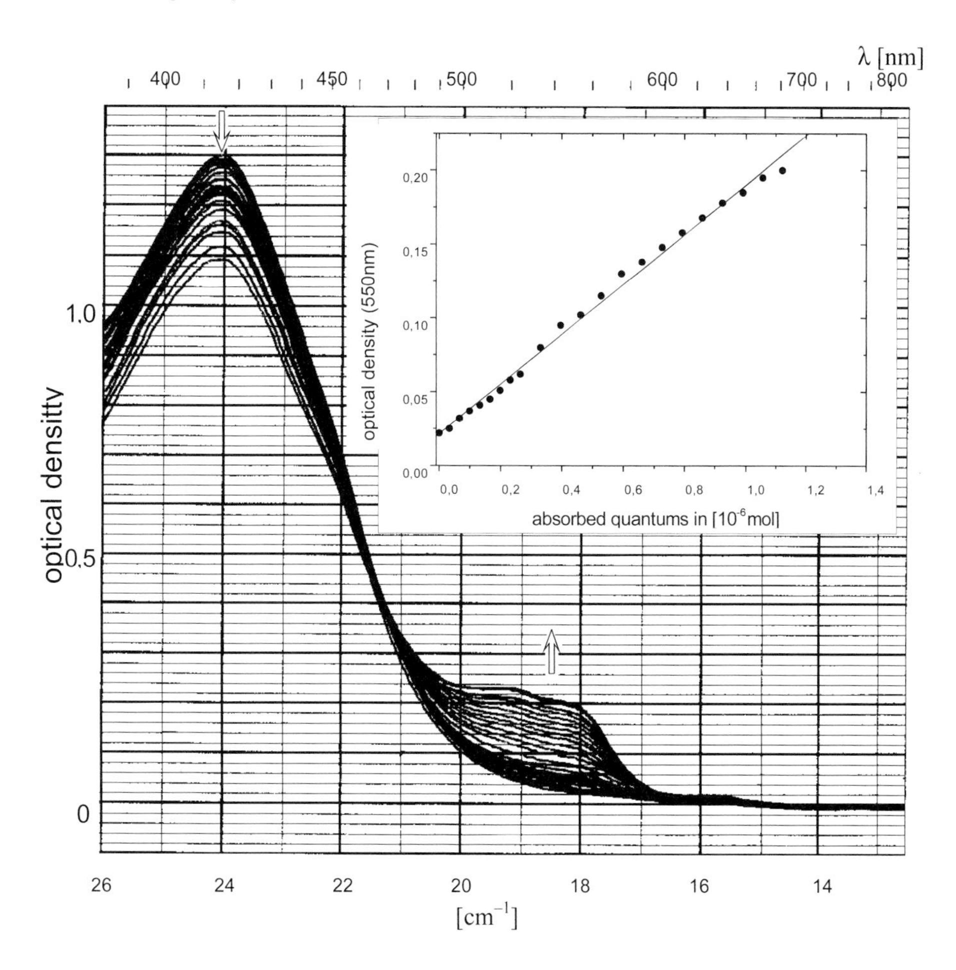

Fig. 2. Changed absorption of the acidochromic silicone layer in 1,2-dimethoxyethane during photolysis of a 4-octadecyloxybenzenediazonium hexafluorophosphate ($c_0 = 6.8 \times 10^{-4}$ mol/L; $\lambda^{exc} = 313$nm).

To prevent thermal interactions of the proton source (onium salts) and specific problems of the photochemical investigations (quenching, sensitization) it was advantageous to separate the azo dye from the sample by fixing it into a polymer matrix of a silicone network. For this purpose a H-functionalized siloxane (NM 203, Chemiewerk Nünchritz) was added to the 4-(*N*-allyl-*N*-methyl)-aminoazobenzene with an Si–H-conversion of 0.2 %. The remaining Si–H groups were partly crosslinked with 0.2 % silicone monomer containing double bonds (Dehesive® 920, Wacker-Chemie) corresponding to Scheme 2. The transparent polymer synthesized allows the application of the probe system in each solvent and can observed easily by common UV–Vis spectroscopy. Because of the low microviscosity of siloxanes the equilibrium of the protonation is reached very fast. Depending on the density of the silicone network the equilibration time was observed within 1–5 s. For an absorbance of 1 a 50-nm layer of acidochromic silicone was fixed on a 200-nm glass carrier. Using an irradiation and detection setup in a 90° arrangement, photochemical reaction and detection can be automated by a sequential process.

The investigation of the photolysis of a diazonium salt in a solution of 1,2-dimethoxyethane using the azo dye silicone layer is shown in Fig. 2. Clearly, there is no shift of the azo dye absorption depending on the composition of the solution during the photolysis. The measurement gives a linear slope of the absorbance of the protonated azo dye versus the light absorbed by the diazonium salt. The real H^+ concentration had not been determined up to now, because a non aqueous solution of the relevant acid (HPF_6) was not available for calibration. Investigation is still in progress.

Acknowledgment: The authors are very grateful for the financial and material support of the *Deutsche Forschungsgemeinschaft* and *Wacker-Chemie GmbH.*

References:

[1] U. Müller, A. Utterodt, *J. Inf. Recording* **1998**, *24*,159.
[2] N. J. Bunce, *J. Chem. Soc., Perkin Trans. I* **1974**, *8*, 942.

New Insights in the Influence of Oxygen on the Photocrosslinking of Silicone Acrylates

U. Müller

Institut für Organische Chemie
Martin-Luther-Universität Halle-Wittenberg
Geusaer Straße, D-06217 Merseburg, Germany
Tel.: Int. code + (3461)462009 — Fax.: Int. code + (3461)462081
E-mail: u.mueller@chemie.uni-halle.de

Keywords: Photocrosslinking / Initiator Photolysis / Silicone Acrylate

Summary: Many initiators used in radical photopolymerization are carbonyl compounds. It is well known that the IR frequency of the carbonyl group depends strongly on the substituent. During photolysis of the initiator radicals are formed, which initiate the crosslinking reaction. All in all, this results in a disappearance of the carbonyl group or in a shift of the frequency of the C=O stretching vibration. Therefore, such types of reactions are suitable to monitor the disappearance of the initiator during its photolysis. Moreover, this type of reaction is suitable to study the influence of oxygen on the initiation process. This work presents results of the initiator photolysis (benzoin type) in a silicone acrylate layer measured by means of real time infrared (RTIR), IR, and GC/MS techniques. Moreover, correlations between the initiator photolysis and crosslinking rate of the silicone acrylates are shown.

Silicone acrylates represent not only a special type of polymer, they are also useful as model systems for the investigation of the photocrosslinking process in bulk [1]. The crosslinking process can be described by means of a radical chain process. The radicals are formed in a photochemical step from the photoinitiatior, which is dissolved in the silicone acrylate. From calorimetric results it is inferred that oxygen inhibits and terminates the chain process. The rate of the crosslinking reaction is proportional to the intensity of the incident light, and to the quantum yield of the initiator photolysis. The inhibition time of the crosslinking is indirectly proportional to the intensity of the incident light, and to the quantum yield of the initiator photolysis. Moreover, a decrease in the air pressure results in an increase in the reaction rate as well as a decrease in the inhibition time.

Many initiators used in radical photopolymerization are carbonyl compounds. It is well known that the frequency of the carbonyl group changes strongly depending on the substituents. Such types of reactions are suitable to monitor the disappearance of the initiator, owing to initiator photolysis [2].

Scheme 1 shows some sequential reactions of the radicals formed by α-cleavage of benzoin isopropyl ether (BIPE), a typical photoinitiator in the radical-induced crosslinking.

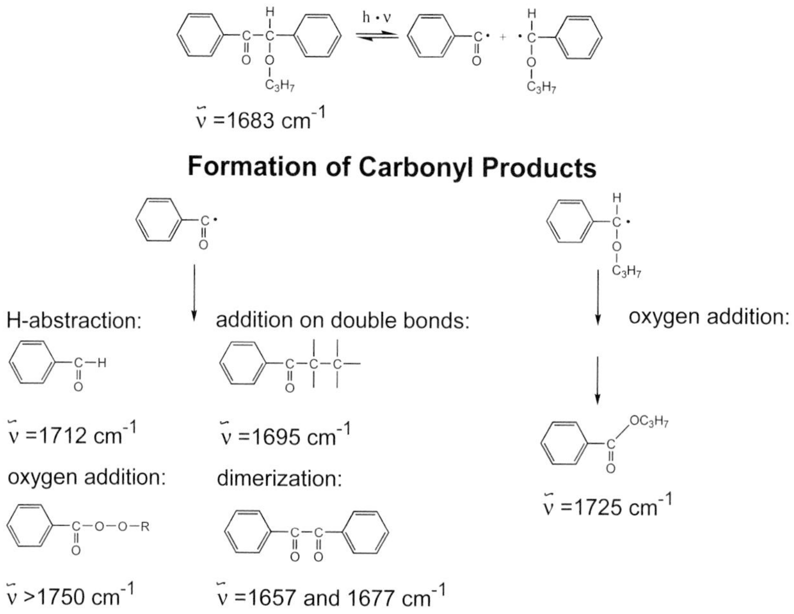

h · ν

$\tilde{\nu}$ = 1683 cm^{-1}

Formation of Carbonyl Products

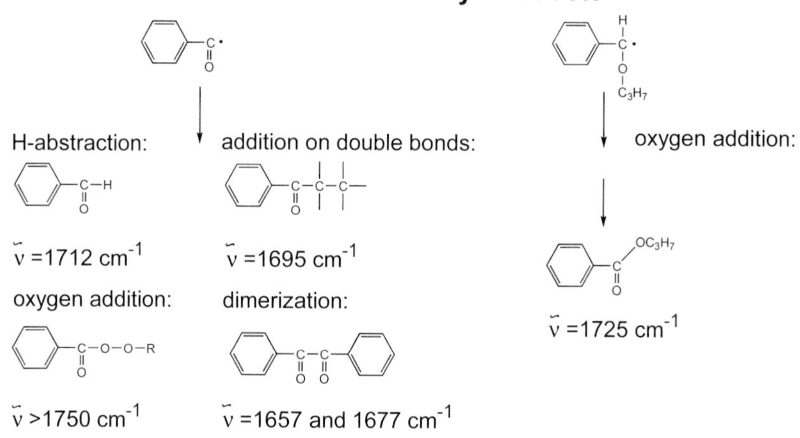

Scheme 1. Carbonyl products as a result of initiator photolysis in a silicone acrylate.

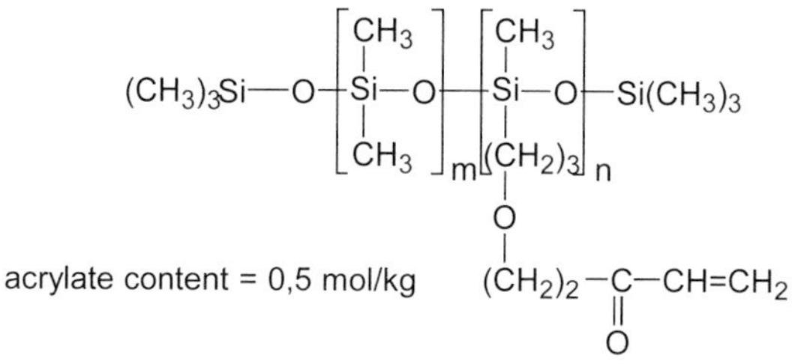

$(CH_3)_3Si-O-[Si(CH_3)_2-O]_m-[Si(CH_3)((CH_2)_3-O-(CH_2)_2-C(=O)-CH=CH_2)-O]_n-Si(CH_3)_3$

acrylate content = 0,5 mol/kg

Scheme 2. The silicone acrylate used.

In silicone acrylates under inert conditions benzoyl addition on double bonds has been observed [2]. Using the real-time infrared technique (RTIR) one can show that the rate of initiator photolysis (R_α) is proportional to the rate of the formation of addition product. Moreover, the polymerization rate (R_P) is directly proportional to the rate of initiator photolysis [2] (Eq. 1), where [M] is the molar concentration of double bonds, [M$_o$] is the initial concentration of double bonds, t is the time, $k(x)$, and $k'(x)$ are conversion (x) dependent quantities, I_o is the intensity of the incident light, and R_α is the reaction rate of the α-cleavage; α and β are exponents.

$$R_P = (-dx/dt)\cdot[M_o] = k(x)\cdot[M]^\alpha\cdot I_o^\beta = k'(x)\cdot[M]^\alpha\cdot(R_\alpha)^\beta$$

Eq. 1.

$R_P \sim R_\alpha$ agrees well with Eq. 1, which describes the rate of the photocrosslinking process (R_P) under stationary irradiation conditions [1, 2]. Only this general expression reflects the real situation in the polymeric system investigated in all detail, because the conditions (viscosity, concentration of oxygen, monomer concentration etc.) for the reaction partners are changing all the time. Moreover, the linear relation between R_P and R_α shows that the light intensity exponent α is in the order of 1. Such a value is typical for a polymerization with a first-order termination step. Such termination behavior was observed for the polymerization of silicone acrylates where traces of oxygen influence the radical termination [1–4].

Moreover, using the RTIR technique one can show that oxygen reacts as a strong radical scavenger [2]. The oxygen addition sequence product of the alkoxy benzyl radical — the benzoic ester (see Scheme 1) — is demonstrable by IR spectroscopy and by GC/MS measurements [2]. Interestingly, no carbonyl oxygen addition sequence product of the benzoyl radical was found. Nevertheless, the benzoyl radicals produced should be completely consumed by oxygen, because addition products on double bonds are not detectable. Presumably, the benzoyl peroxy product decomposed under decarboxylation [5].

The results show that the rate of initiator photolysis is higher under air than in the absence of oxygen. The rate of initiator decay is a function of the initiator used and the oxygen content in the layer. The calculated quantum yield of initiator decay in the silicone acrylate layer correlates with the quantum yield determined in a hexamethyldisiloxane solution; see Fig. 1. Nevertheless, the absolute values differ in both systems.

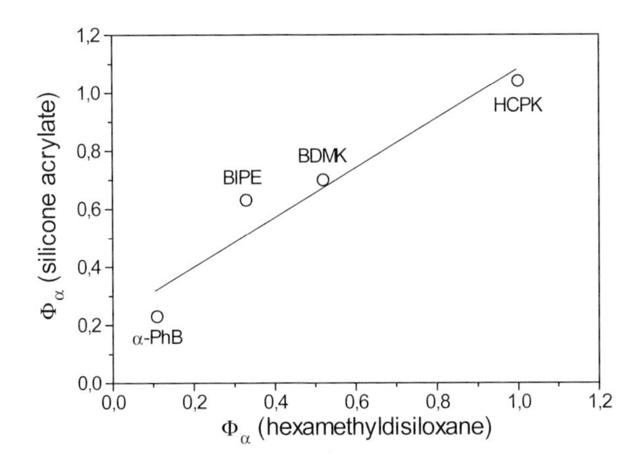

Fig. 1. Plot of quantum yield of the initiator decay Φ_α determined in oxygen-free hexamethyldisiloxane vs. Φ_α determined in silicone acrylate (α-PhB = α-phenylbenzoin, BDMK = benzildimethylketal, HCPK = 1-hydroxycyclohexyl phenyl ketone; all values from [2, 3]).

In a thin layer ($\leq$100 μm) under air the polymerization is inhibited all the time. Moreover, the RTIR study proves that the acrylate double bonds were not consumed during the irradiation. This fact shows that the inhibition process starts before the initiator radical formed can be added to a double bond. The inhibition of the polymerization is caused by the oxygen scavenging of initiator radicals.

$$\Phi \cdot \eta_{abs} \cdot I_0 = k_{ox} \cdot [R\cdot] \cdot [O_2]$$

Eq. 2.

$$d[O_2]/dt = -D \cdot q/V \cdot d[O_2]/dl$$

Eq. 3.

The polymerization starts after consumption of the oxygen in the layer. During the inhibition period the formation of initiating radicals is identical to the consumption of these radicals by oxygen (Eq. 2), where $[O_2]$ is the oxygen concentration in the layer. But the oxygen concentration is a function of diffusion and a function of the diffusion pathway length l (Eq. 3), where Φ is the quantum yield of primary radical formation, $[R\cdot]$ is the primary radical concentration, η_{abs} is the absorbed part of of the incident light I_o, V is the volume of the sample, q is the surface area of the sample and D is the oxygen diffusion coefficient.

In the bulk of this pathway, the polymerization is inhibited all over the time. On account of this oxygen diffussion, one observe an unhardened surface or in thin layers (≤ 100 μm) the complete inhibition of the polymerization. Nevertheless, the acrylate double bonds are not consumed during the irradiation. A crosslinking is possible after oxygen removal (laminate, or inert atmosphere) [1] provided that the initiator is not completly consumed by oxygen.

Acknowledgments: The author is very grateful to *DAAD (Procope-Project)* for financial support. Special thanks also to Dr. C. Decker (Mulhouse, France) for help and discussion of the presented results.

References:

[1] U. Müller, *J. Macromol. Sci. — Pure Appl. Chem.* **1996**, *A33*, 33.
[2] U. Müller, *J. Photochem. Photobiol. A: Chem.* **1997**, *102*, 237.
[3] U. Müller, C. Vallejos, *Angew. Makromol. Chem.* **1993**, *206*, 171.
[4] U. Müller, S. Jockusch, H.-J. Timpe, *J. Polym. Sci., Part A: Polym. Chem.* **1992**, *30*, 2755.
[5] I. Lukác, C. Kósa, Macromol. *Rapid Commun.* **1994**, *15*, 929.

Synthesis of New α,ω-Olefinic Silicon-Containing Compounds via Cross-Metathesis

O. Nuyken*, K. Karlou-Eyrisch

Lst. f. Makromolekulare Stoffe, TU München
Lichtenbergstr.4, D-85747 Garching, Germany
Tel.: Int. code + (089)289 13571 — Fax.: Int. code + (089)289 13562
E-mail: nuyken@ch.tum.de

J. Weis, B. Deubzer, C. Herzig

Wacker-Chemie GmbH, Werk Burghausen, Germany
D-84480 Burghausen

Keywords: Reactive Silicone / Metathesis / Carbene Complexes / Tolman Angle

Summary: We studied the synthesis of a new generation of silicon compounds containing two or four terminal double bonds. The synthesis was realized via transition-metal-catalyzed cross-metathesis of silicon-containing cycloolefins with ethylene (ethenolysis). Silicon-containing α,ω-dienes may be useful, e.g. as crosslinking agents in polymers and as precursors for α,ω-difunctional compounds. In general, the problem in the cross-metathesis of cycloolefins with acyclic olefins (here: ethylene) is the occurence of side reactions which lead to homo-metathesis products besides the desired cross-metathesis products. Therefore, the choice of a suitable catalyst and a cycloolefin for which ring-opening is energetically favorable as well as the optimization of the reaction conditions are crucial for the selectivity of the cross-metathesis.

Introduction

The concept of "Reactive Silicones" comprises a great number of different silicon compounds containing various functional groups. Examples of such organofunctional silicones include hydroxy-, amino-, epoxy-, carboxy-, acryloxyalkyl- or vinyl-functionalized polydimethylsiloxanes.

α,ω-Organodifunctional silicones are of growing scientific and economic interest as reactive modification reagents in organic polymer synthesis. Sil(ox)anes containing olefinic substituents are of remarkable synthetic value in organic chemistry, mainly due to their ability to react with numerous different electrophiles to form substituted products [1]. At present, the technically used vinyl- and 5-hexenylsilicon building blocks are synthesized via hydrosilylation of acetylene or 1,5-hexadiene, respectively. Grignard reactions offer an alternative approach to prepare these reactive compounds. However, due to their low efficiency Grignard reactions are of no commercial interest. Another problem is the formation of large amounts of salts.

Olefin metathesis is a catalytic reaction in which alkenes are converted to new products via the cleavage and re-formation of carbon–carbon double bonds (Scheme 1).

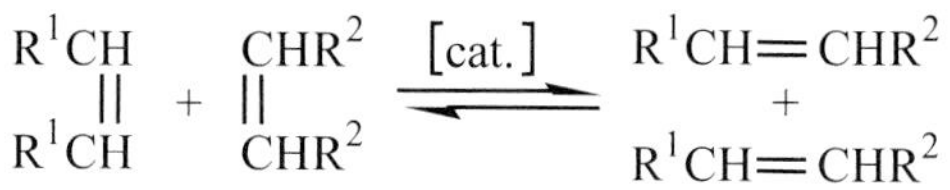

$$\begin{array}{c} R^1CH \\ || \\ R^1CH \end{array} + \begin{array}{c} CHR^2 \\ || \\ CHR^2 \end{array} \underset{}{\overset{[cat.]}{\rightleftharpoons}} \begin{array}{c} R^1CH{=}CHR^2 \\ + \\ R^1CH{=}CHR^2 \end{array}$$

Scheme 1. Olefin metathesis.

The first conventional catalysts based on tungsten and molybdenum complexes exhibited very low activity in the homo-metathesis of allylic-substituted silanes [2].

Marciniec et al. showed that acyclic olefins containing silicon undergo ruthenium-catalyzed metathesis reactions. For example, the homo-metathesis of vinyltriethoxysilane leads to *z*-1,2-bis(triethoxysilyl)ethylene in yields of more than 80 % [3]. These authors published the first cross-metathesis reactions of vinylsilanes in 1988 and subsequently also of alkoxy- and alkyl-substituted vinylsilanes and terminal olefins. In these reactions the conventional ruthenium and rhodium complexes $RuCl_2(PPh)_3$ and $RhCl(PPh_3)_3$ were used. A major drawback was the necessity to employ temperatures higher than 100°C [4]. Besides the desired cross-metathesis products, the dimeric silane was obtained in significant amounts as the result of homo-metathesis, when equimolar amounts of educts were used. In the cross-metathesis of alkyl-substituted vinylsilanes homo-metathesis could only be suppressed by employing a large excess of the corresponding coupling partners [5]. Finkel'shtein et al. reported the ring closure and acyclic diene-metathesis of bisallyl-substituted silanes and disilanes [6]. Triethoxysilylnorbornene was also converted by Finkel'shtein et al. to ethoxysilyl-containing polymers via ROMP [7]. Recently, Blechert et al. reported the synthesis of a series of asymmetric silaolefins via metathesis [8].

In this paper we report the cross-metathesis of cycloolefins with ethylene (ethenolysis). We present the selective synthesis of linear silicon-containing α,ω-olefins starting with norbornene derivatives containing alkyl-, amine- or alkoxy-substituted silicon moieties and ethylene (Scheme 2).

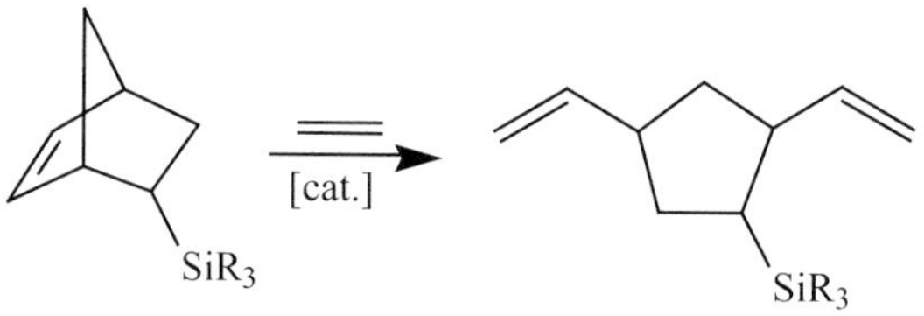

[cat.]

SiR_3 SiR_3

$SiR_3 = Si(OCH_2CH_3)_3$ (**I**)
$Si(CH_3)_2OSi(CH_3)_3$ (**II**)

Scheme 2. Synthesis of silicon-containing α,ω-olefins.

Ruthenium Catalysts

Following the use of "well-defined" catalysts by Grubbs and Schrock, the metathesis reaction became a focus of catalyst and synthesis research. This interest is due to the enormous potential of this reaction in the synthesis of polymers (ROMP, ADMET, alkyne-metathesis). Moreover, this reaction offers an elegant way to synthesize new olefins via cross-metathesis with two different olefinic compounds.

Ruthenium–carbene complexes of the type [$RuCl_2(=CHR)(PR^*_3)_2$] are extraordinarily useful catalysts for olefin metathesis. Furthermore, they have unique properties, e.g. high tolerance towards atmospheric oxygen, functional groups and moisture.

For the cross-metathesis reactions we used two different homogeneous ruthenium catalysts: one multicomponent catalyst prepared in situ, and the other catalyst consisting of one single component. Both are based on ruthenium(II) and the latter can be isolated in pure form. Figure 1 shows the ruthenium catalysts that were used in cross-metathesis reactions to synthesize silicon-containing α,ω-dienes.

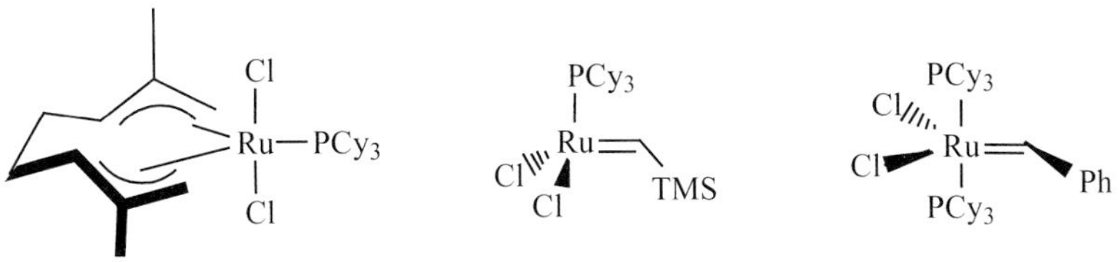

1d **1dTMS** **2**

PCy_3 : tricyclohexylphosphane; TMS : trimethylsilylene

Fig. 1. Ruthenium(II)-based metathesis catalysts.

For the catalytic properties of the complexes both electronic and steric effects of the ligands are of great importance [9]. The cross-metathesis reaction was performed with ruthenium catalysts carrying the strongly basic ligand tricyclohexylphosphane which has a relatively large Tolman angle (pK_a = 9.7, Tolman angle 170°). Electron-rich ligands stabilize the formation of the metal carbene. The Tolman angle influences the equilibrium constant for the binding of the ligand to the metal. For example, complexes containing trimethylphosphane ligands (Tolman angle 118°) are only weakly dissociated in solution whereas complexes with ligands of high steric demand, e.g. tricyclohexylphophane ligands (Tolman angle 170°), are strongly dissociated.

Starting with a dimeric bis(allyl)ruthenium complex, the precatalyst **1d** was synthesized by complexation with tricyclohexylphosphane. After addition of trimethylsilyldiazomethane, the basic bisallylic ligands were displaced from the coordination sphere of the metal to form the catalytically active ruthenium(II)–alkylidene species **1dTMS** [10]. Although the ruthenium–alkylidene complex generated in situ is easy to obtain, the structure of catalytically active species is poorly defined. Therefore, a control over the activity is difficult, and reproducible experimental results are almost impossible to achieve.

Alternatively, the cross-metathesis reactions were performed with the single-component Grubbs ruthenium–phenylidene complex **2** [11]. Similarly to **1dTMS**, the catalyst is a ruthenium(II)-alkylidene complex with strongly basic tricyclohexylphosphane ligands of high steric demand. The possibility of isolating the ruthenium-alkylidene complex **2** and its stability are based on the bulky and electron-rich phosphane ligands and on the phenylidene group coordinated to the ruthenium moiety. In contrast to the multicomponent system, this highly active and well-defined catalyst offers great advantages with respect to reproducibility and reaction control, i.e. activity and selectivity.

Ethenolysis of Norbornenes with Silicon Moieties

For the selectivity of the cross-metathesis reaction with ethylene (ethenolysis) the choice of a suitable catalyst and a cycloolefin for which ring-opening is energetically favorable as well as the optimization of the reaction conditions are crucial. Cycloolefins with a norbornene skeleton were chosen for their high ring strain. The silicon-containing educts can be obtained by hydrosilylation of norbornadiene with the corresponding silanes.

The results of the investigated cross-metathesis reactions of silicon containing norbornene derivatives with ethylene showed a poor reactivity for some substituents. Reactive chlorosilanes act as catalyst poisons and silazanes diminish the catalyst's activity by coordination of nitrogen to the metal center. Starting from norbornene derivatives with alkoxy-substituted silicon moieties, a selective synthesis of new silicon-containing α,ω-dienes could be accomplished (Scheme 2).

A quantitative yield could be achieved by employing catalysts **1dTMS** and **2**. In comparison to **1dTMS** smaller amounts of catalyst **2** were required to obtain the same results. Addition of trimethylsilyldiazomethane enhances the carbene formation from **1d** and hence the activity of the catalyst synthesized in situ. However, at the same time the stability of the catalyst is diminished. Repeated reaction of the educt–product mixture with ethylene by addition of new catalyst leads to completion.

The structurally defined Grubbs ruthenium catalyst can be isolated in pure form, allows more efficient control over the reaction and leads to quantitative yields in cross-metathesis reactions.

Ethenolysis of Silicone-Bridged Norbornene

We were able to synthesize silicon compounds with two terminal double bonds selectively by ring-opening metathesis reaction of norbornene derivatives in the presence of ethylene.
Furthermore, it was possible to use those highly selective cross-metathesis reactions to synthesize reactive silicones with four terminal double bonds (Scheme 3).

CH3 CH3 Si O Si n CH3 CH3 — 2 ═ [Ru] → CH3 CH3 Si O Si n CH3 CH3

$n = 0\text{-}13$

Scheme 3. Ethenolysis of silicone-bridged norbornene.

These dimethylsiloxane-bridged α,ω-dienes can be converted to the previously described reactive silicones by reaction with cyclic dimethylsiloxanes (e.g. D_4) as shown in Scheme 4. Since this reaction is possible without cleavage of auxiliary groups these α,ω-dienes are of great interest. Reactive terminal C=C double bonds are generally required to get fast reactions with hydrosiloxanes. For example, it is possible to build up a polymer network from Si–H-functionalized siloxanes via noble-metal-catalyzed polyaddition reactions. Another specific feature of these compounds containing vinylcyclopentane moieties is the low tendency of the terminal double bonds to isomerize to inner double bonds, in contrast to conventional allylic groups.

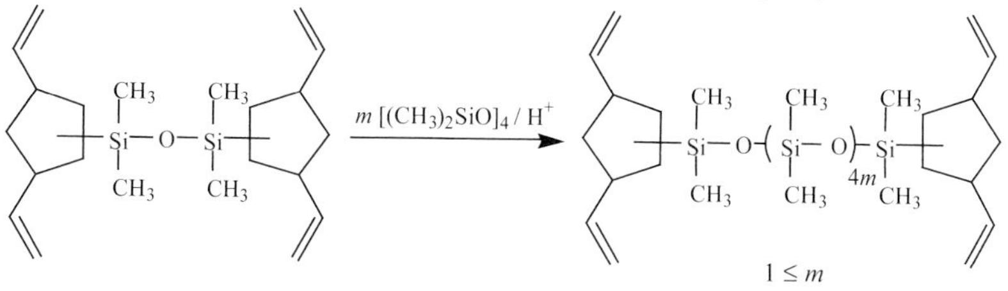

$$m\,[(CH_3)_2SiO]_4 / H^+ \qquad 1 \le m$$

Scheme 4. Equilibrium of different dimethylsiloxane-bridged divinylcyclopentanes with four terminal C=C double bonds and D_4.

Since ethenolysis leads to selective formation of terminal double bonds, the determination of the catalyst activity is possible. For this purpose we developed a special apparatus which allows time-dependent measurements of ethylene consumption by means of a mass-flow controller [12]. Based on these measurements we calculated the catalyst turn-over frequency (TOF).

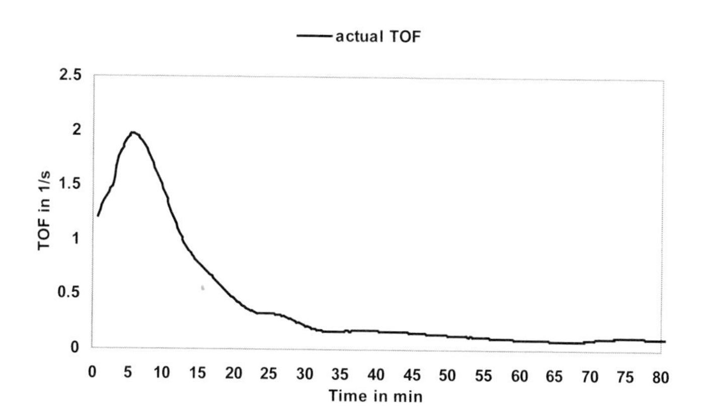

Fig. 2. Activity–Time diagram: cycloolefin **II**, catalyst **2**, [Ru] = 22.4×10^{-6} mol/L, [**II**] = 47 mmol/L, T = 40°C, p_{ethene} = 2.0 bar (0.29 mol/L), 400 mL toluene.

References:

[1] W. A. Colvin, *Silicon Reagents in Organic Synthesis*, Academic Press, London, **1988**.
[2] B. Marciniec, *Principles and Advances in Molecular Catalysis*, Wroclaw Univ. Press, Poznan–Wroclaw, **1993**.
[3] B. Marciniec, J. Gulinski, *J. Organomet. Chem.* **1984**, *C19*, 266.
[4] Z. Foltyniwicz, B. Marciniec, *J. Organomet. Chem.* **1989**, *376*, 15.
[5] B. Marciniec, C. Pietraszuk, *J. Organomet. Chem.* **1993**, *447*, 163.
[6] S. Finkel'shtein, E. B. Portnykh, N. V. Ushakov, U. M. Vdovin, *Izv. Akad. Nauk SSSR Ser. Khim.* **1981**, 641.
[7] N. V. Ushakov, S. Finkel'shtein, E. B. Portnykh, U. M. Vdovin, *Izv. Akad. Nauk SSSR Ser. Khim.* **1981**, 2835.

[8] M. F. Schneider, N. Lucas, J. Velder, S. Blechert, *Angew. Chem., Int. Ed. Engl.* **1997**, *3*, 36.
[9] A. Demonceau, A. W. Stumpf, E. Saive, A. F. Noels, *Macromolecules* **1997**, *30*, 3127.
[10] W. A. Herrmann, W. C. Schattenmann, O. Nuyken, S. C. Glander, *Angew.Chem.* **1996**, *108*, 1169.
[11] P. Schwab, R. H. Grubbs, J. W. Ziller, *J. Am. Chem. Soc.* **1996**, *118*, 100.
[12] K. Karlou-Eyrisch, PhD Thesis, TU München, **1998**.

The Science of Building Materials and the Repair of Buildings: Masonry Protection with Silicones

Helmut Weber

Wacker-Chemie GmbH
Geschäftsbereich S-C-B
Hanns-Seidel-Platz 4, D-81737 München, Germany
Tel.: Int. code + (89)627 92011 — Fax: Int. code + (89)620 92018

Keywords: Damages to Building Materials / Water Absorption / Water Vapor / Diffusion / Masonry Protection / Silicon Resin Network

Summary: The repair of buildings is a major factor in a country's economy. In Germany alone, each year roughly DM 250 billion is set aside for it. Preservation and repairs account for roughly 60 % of all building work there. The conditions obtaining when new buildings are erected differ from those when preservation and repair work are carried out. Whereas with new buildings it is possible to make provision for all technical, physical and chemical needs at the planning stage, there is no such scope for standing buildings. Consequently, there are no made-to-measure solutions for building repairs. Each building is different and requires specific measures. These can vary considerably and depend essentially on the type of building, the nature of the building materials used, the use to which the building has been put and its location. This paper is an attempt to shed more light on this complex situation and to show possible outlets for applying organosilicon chemistry to the repair and protection of buildings.

Composition of Building Materials

Building materials – primarily the mineral kind – have many different components. These are essentially the binder, the aggregate and the pores or pore systems formed by compaction.

There are all kinds of binders in use, but carbonates are among the most common. These are particularly vulnerable to attack by acid as they are in fact salts of the weak carbonic acid. They are thus soluble in all acids, even carbonic acid itself. Perhaps the most familiar carbonate binder is lime, which hardens to calcium carbonate. Sulfates, such as calcium sulfate (gypsum), also act as binders but these are mostly present in small amounts or by way of impurity. They have the advantage of being resistant to acids but unfortunately are also fairly soluble in water. Sulfate binders may therefore only be used, or have only proved their worth, wherever moisture levels are extremely low. Ranking high in the list of established binders are the cements, which are employed in all kinds of different compositions. Chief among them is Portland cement, which is a mixture of silicate, aluminate and ferrite phases. It was the development and use of Portland cement that made it possible to erect water-resistant buildings, since the cement, in contrast to the carbonates, has the ability to harden hydraulically, i.e., by incorporation of water. It can therefore harden under water; this property led to a whole host of new applications in the building trade. In standing buildings, the various binders are generally present in completely different mixing ratios. Not surprisingly, their properties and durability are also completely different.

Aggregates, like binders, vary a great deal in composition and also differ for instance in their resistance to acids. There are dense, heavy types with closed surfaces, such as quartz sand and lime sand, and there are lightweight types with a porous structure and open surface; these have come increasingly to the fore in recent years. Examples are expanded clays, perlite and vermiculite. Lightweight aggregate may be used to make building materials with superior thermal insulation, enhanced flexibility and less susceptibility to cracking.

A third property of building materials is achieved by compaction or using binders and aggregates. Here we are concerned with the pore system and individual pores of the building material. They govern the extent of water absorption, pollutant absorption and, for example, water vapor permeability and compressive strength, tensile strength in bending — to name but a few physical properties. Pores are named after their shape as through-pores, closed pores, blind pores, bottleneck pores, and connecting pores. But this classification is pretty superficial and fails to describe the physical significance of pores and pore systems. A better insight is gained by classifying them according to size. For example, pores with a radius of $<10^{-7}$ m are called micropores, those with a radius of 10^{-4} to 10^{-7} m are macropores or capillary pores and those with a radius $>10^{-4}$ m are air pores. The radius or size of the respective pores determines the extent of water absorption by the building materials as ultimately every pore size is associated with a special water-absorption mechanism that is characteristic of the building material. A further important aspect of porosity is the pore volume of the building material. This encompasses all the pores and describes the maximum possible amount of water that can be absorbed by a building material.

Absorption of Water and Pollutants

Water absorbency is one of the most critical properties of building materials and it often governs durability and resistance to corrosion. Water may be absorbed as either liquid or vapor. Liquid water is absorbed by the pores either spontaneously, in which case the process is generally referred to as capillary absorption, or under pressure, e.g., of vadose water. Vapor absorption mainly entails the mechanisms of condensation, condensation in capillaries and hygroscopicity. It is always a function of the prevailing relative air humidity. Normally, for condensation to occur, the relative air humidity must be 100 %. But in capillaries it can occur at lower values. The smaller the pores are, the more able they are to fill up with water when the relative air humidity is far below 100 %. Hygroscopicity (the tendency to absorb moisture) is linked to salt content. It is triggered at a relative humidity of around 50 % and increases dramatically as the relative humidity rises. Salts differ enormously in their hygroscopicity. As a rule of thumb, their hygroscopicity is proportional to their solubility in water. Extremely soluble salts, such as calcium nitrate, and ammonium nitrate, are highly hygroscopic. Figure 1 shows the relationship between the various water-absorption mechanisms and pore size.

Micropores Gel pores	Macropores Capillary pores	Air pores
Capillary condensation Condensation	Capillarity Condensation Hygroscopicity	Vadose water Water absorption under pressure Condensation Hygroscopicity

10^{-9} 10^{-8} 10^{-7} 10^{-6} 10^{-5} 10^{-4} 10^{-3} 10^{-2}

Pore radius (m)

Fig. 1. Relationship between water absorption mechanism and pore size.

The Gravest Types of Building Damage

Most people know about the close link between water absorption and building damage. Increased absorption of water by building materials and a higher incidence of and susceptibility to damage are usually observed together; ultimately, most damage to buildings is put down to absorption of water. In other words, were it not for the water absorption, such damage would not occur in the first place. Frequently, the progress and extent of the damage enhanced by uptake of pollutants. Pollutants dissolved in water, i.e., acidic gases, and salts, are major culprits. The gravest types of building damage are:

- Water Absorption:
  - Frost damage
  - Hygroscopic swelling and shrinking
  - Reduction in thermal insulation
  - Colonization by microorganisms

- Water absorption and absorption of waterborne pollutants (pollutants are soluble salts and gases):
  - Crystallization damage
  - Hydration damage
  - Increase in equilibrium moisture content through hygroscopicity
  - Corrosion of steel in reinforced concrete, e.g., by chloride ions
  - Freeze–thaw salt damage
  - Chemical conversion of binders, soluble salts (e.g., lime in plaster)

- Absorption of gaseous pollutants:
  - Carbonation of reinforced concrete by CO_2.

Environmental Protection and Masonry Protection

Masonry protection is a subcategory of environmental protection. Environmental protection itself may be divided into active and passive types. In the active type, appropriate steps are taken to lower the level of pollutant emissions, e.g., CO_2, SO_2, NO_x and organic solvents. This also applies to the levels of all kinds of waste that should be minimized. In passive environmental protection, on the other hand, measures are taken to reduce the extent to which buildings are exposed to water and pollutants. This is masonry protection in its truest form. Masonry protection prevents pollutants from damaging both the building materials and the buildings. By this definition, therefore, to reduce the water absorption of a building material is to engage in passive environmental protection. In the light of this definition, masonry protection is crucially important.

Demands on Masonry Protection Agents from Physico-chemical And Ecological Aspects

The prime duty of masonry protection agents is to eliminate those factors which are conducive to damage. This means, for example, that water absorption and ultimately the uptake of pollutants must be reduced, at least as far as the water-soluble pollutants are concerned. There is the ecological viewpoint to be considered as well of course, i.e., the masonry protection agents must be so designed that they do not release any products that could have an adverse effect on the environment. The most important demands may be summarized as follows.

- Protection against harmful influences, primarily against water absorption
- No noticeable reduction in the drying properties of building materials
- High affinity for the building material and corresponding durability
- No formation of by-products that could damage the building
- Ease of application and processing
- Good price/performance ratio
- Favorable reversibility behavior
- Solvent-free or low-solvent, preferably dilutable with water
- Easy to package, to cut down on containers

Of all these properties, let us now consider two in particular, namely water absorption and water vapor permeability. These are two physical properties that have an incisive effect on water balance in buildings.

In buildings, water absorption in the facade area is described by the water absorption coefficient (w value). This returns the capillary water absorption of building materials as a function of time and is defined in DIN 52 617 as w [$kg/m^2 h^{0.5}$].

In other words, the water absorption coefficient is a measure of the capillary absorption behavior of a building material and returns the value for absorption expressed in terms of square meters after an hour's exposure. A w value of 0.5 therefore means that the building material can absorb roughly half a liter of water per square meter in one hour.

The water vapor permeability is described by the so-called diffusion-equivalent air layer thickness (s_d value). This s_d value is determined in accordance with DIN 52 615 and is defined as that thickness of a layer of air at rest which has the same diffusion properties as an equivalent thickness of building material. Moreover, the s_d value is the product of the diffusion resistance coefficient, μ, and the thickness, s, of the building material (Eq. 1).

$$s_d = \mu\, s \;\; [\text{m}]$$

Eq. 1.

Clearly, building materials and facades will have the best possible protection and thus have the best possible water balance if both the water absorption coefficient, w, and the diffusion resistance, s_d, are as small as possible, i.e., in mathematical terms, tend towards zero. These relationships were first expressed by Künzel in his theory of facade protection, which is illustrated in Fig.2.

Water release	>>	Water absorption
Drying		Saturation
Diffusion		Capillary water uptake
s_d value		w value

Ideal masonry protection means:
S_d value → 0 $\quad$ w → 0

Demands on facade protection according to Künzel

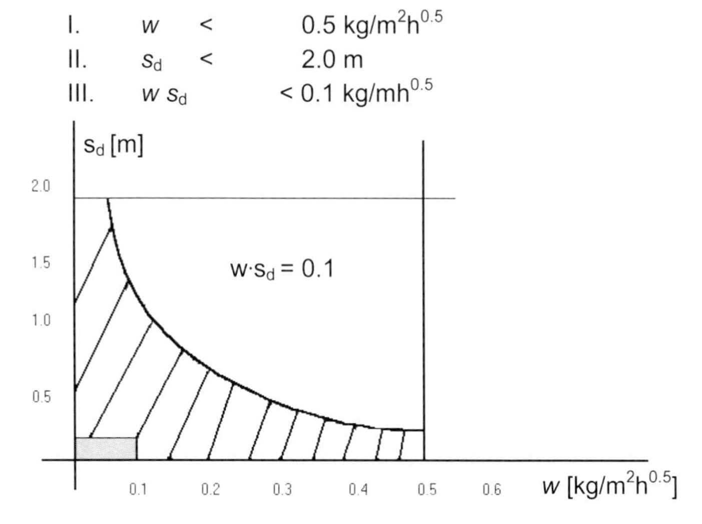

Fig. 2. Künzel's theory of facade protection.

Masonry Protection through Surface Protection

There are various methods of applying masonry protection agents. But it must be remembered that, for example, the possibility of carrying out masonry protection measures on standing buildings is generally restricted to surface protection as it is not usually possible to impregnate across the entire thickness of the walls. Surface protection may be subdivided as follows: masonry protection measures that work on the principle of water-repellent impregnations; masonry protection measures that entail coating the surface with paints for protective and decorative reasons; and masonry protection provided by injecting and impregnating the cross-sectional areas of the wall, as measures for combating rising damp or, in individual cases, for consolidating the outer layer of building material.

Organosilicon Compounds as Masonry Protection Agents

Organosilicon compounds, or silicones for short, have been used in masonry protection for roughly 50 years. The base materials are predominantly trifunctional silanes with whose help it is possible to build up three-dimensional crosslinked structures that are generally known as silicone resins or silicone resin derivatives. The chemistry of silicone masonry protection is therefore essentially the chemistry of silicone resins. Silicone masonry protection agents include all products built up from silicon T units (silicone resin crosslinking agents). But they are all used, irrespective of the number of units. The following compounds, which differ mainly in molecular structure, generally serve as masonry water repellents: silanes, siloxanes, siliconates and silicone resins (Fig. 3). Figure 4 shows the various molecular units that ultimately form the silicone resin network.

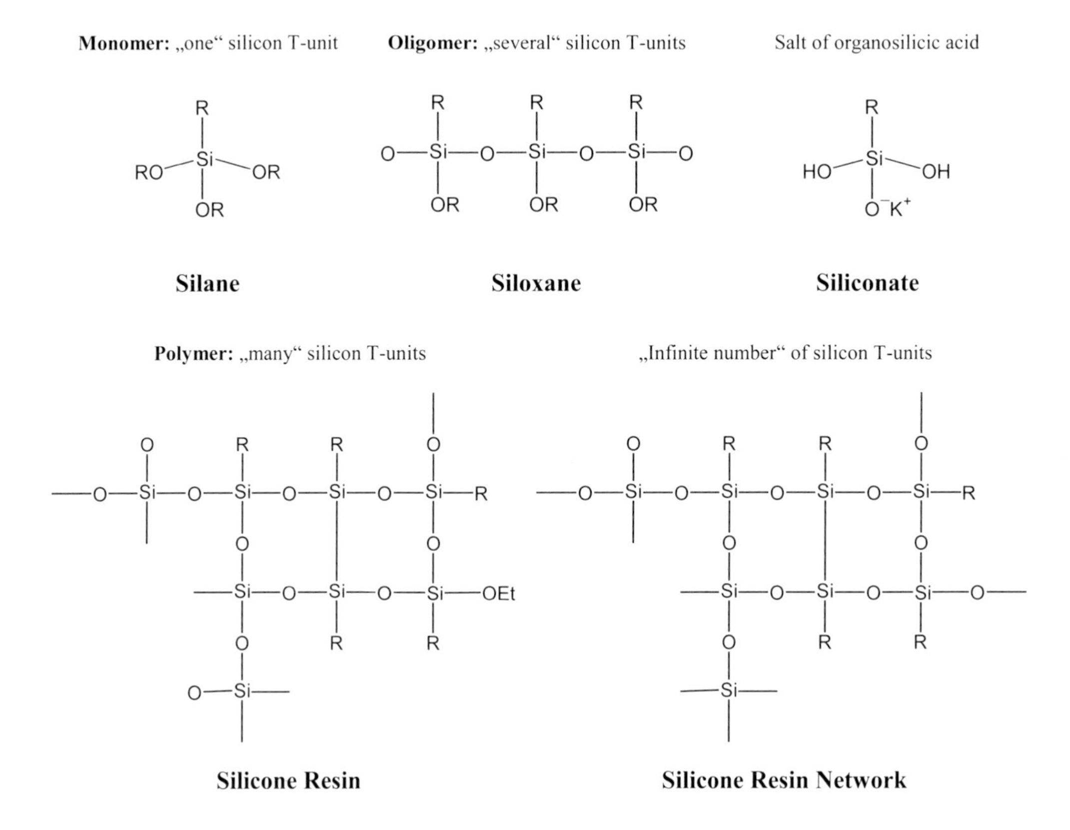

Fig. 3. Silanes, siloxanes, silicone resins, and siliconates.

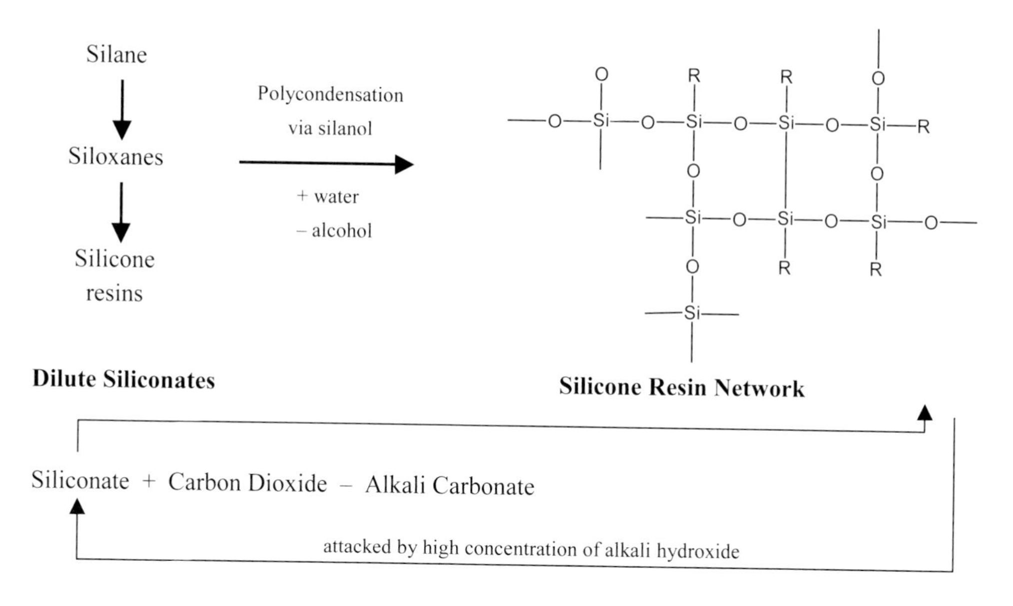

Fig. 4. Silanes, siloxanes, silicone resins and siliconates react to form a silicone resin network.

The various products, properties and applications are shown in the following summaries.

Summary 1: Silanes

- In building, always: alkylalkoxysilanes.
- Simplest silicon T-units (monomers): *one* organic group and *three* alcohol groups, e.g. Fig. 5.
- Contain 50–80 wt % bound alcohol.

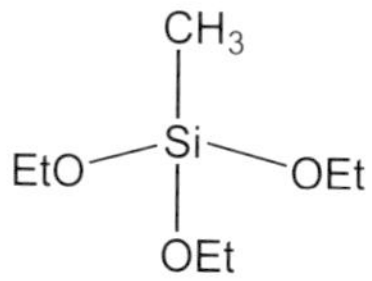

CH_3–Si(OEt)–(OEt)–(OEt)

Fig. 5. Methyltriethoxysilane.

Properties

- Clear, low-viscosity, more or less volatile liquids.
- Readily soluble in most organic solvents.
- High content of bound alcohol.
- Long-chain silanes are stable against alkaline attack.
- Particularly long reaction time until the silicone resin network forms.

Applications

- Impregnation of reinforced concrete with long-chain silanes.
- Silanes are generally applied undiluted.

Summary 2: Siloxanes

- In building, always: alkylalkoxysiloxanes.
- Consist of three to six silicon T-units (oligomers), e.g., Fig.6.
- Contain 30–40 wt. % bound alcohol.

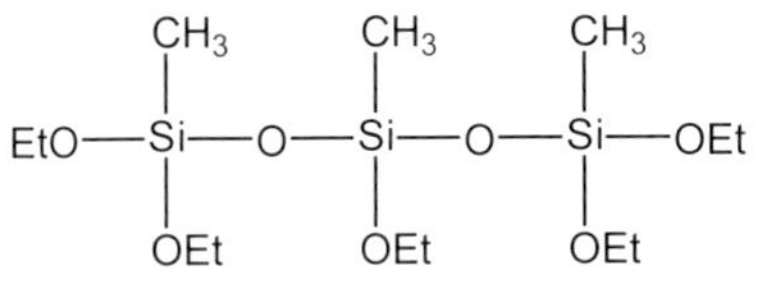

Fig. 6. Methylethoxysiloxane.

Properties

- Clear, mobile, non-volatile fluids.
- Readily soluble in most organic solvents.
- Moderate content of bound alcohol.
- Long-chain siloxanes are stable to alkaline attack.
- Rapid reaction to form silicone resin network.

Applications

- In combination with silanes as impregnating agents and primers for concrete, natural stone, sand–lime brick or brick, as oil-proofing agents, anti-graffiti agents and stone strengtheners, as chemical damp-proofing agents.
- Siloxane/silane mixtures are generally applied as aqueous or solvent-based dilute solutions.

Summary 3: Siliconates

- Salts of organosilicic acids.
- In building: potassium methylsiliconate (Fig. 7).
- Contain over 50 wt. % bound alkali hydroxide (e.g., KOH).

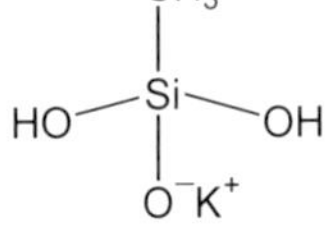

Fig. 7. Potassium methylsiliconate.

Properties

- 40–50 wt. % aqueous solution, highly alkaline.
- Low resistance to alkaline attack.
- Reacts with atmospheric carbon dioxide (carbonation) to form a silicone resin network (waterglass chemistry).

Applications

- Impregnation of building elements produced in-plant from clay, aerated concrete and gypsum; chemical damp-proofing agents to promote protection against capillary moisture.
- Siliconates are generally used as highly diluted solutions.

Summary 4: Silicone Resins

- In building, always: methylsilicone resins (Fig. 8).
- Consist of 30 to 80 silicon T-units (polymers).
- Contain 2–10 wt. % bound alcohol.

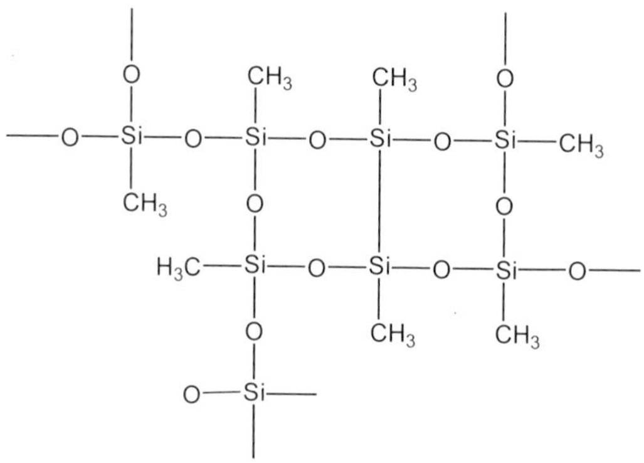

Fig. 8. Solid methylsilicone resin.

Properties

- Viscous to solid (varies with degree of condensation), colorless.
- Soluble in organic solvents (alcohols → aliphatics → aromatics).
- Very low content of bound alcohol.
- Rapid reaction to form silicone resin network.

Applications

- Binders for the surface-coatings and electrical and electronics industries.
- Binders for silicone resin plasters and facade paints in the building industry.
- Generally used as silicone resin emulsions.

Summary 5: Application Areas for Silicone Masonry Water Repellents

Silicone masonry water repellents are ideal for reducing capillary water absorption and retaining full water vapor permeability. The products are prepared in different ways. There are low-viscosity impregnating agents for conferring water repellency or oil repellency, agents for chemical damp-proof courses and consolidation, cream-like products for protecting concrete, and emulsions and powders that are used either as additives or binders for plasters, paints and building materials. Such systematic development has now made it possible to rectify most, if not all, problems with damp masonry by recourse to diverse silicone masonry water repellents. An overview and summary of the various application areas, listed by building chemistry, building materials industry and masonry coatings, are provided below.

In the Building Chemicals Industry:

- As impregnating agents for concrete, natural stone, lime–sand brick or terracotta bricks.
- As oil-proofing agents.
- As anti-graffiti agents.
- As stone strengtheners.
- As chemical damp-proofing agents.

In the Building Materials Industry:

- As integral or surface water repellents for in-plant waterproofing of building materials made of:
    - Gypsum,
    - Clay,
    - Aerated concrete,
    - Mineral fibers,
    - Lightweight aggregate.
- As integral water repellents for building mixtures based on lime, cement and gypsum.

In the Building Coatings Industry:

- As water repellent additives for paints and plasters.
- As primers.
- As binders for silicone resin paints and silicone resin plasters.

What "Super" Spreads? — An Analysis of the Spreading Performance of the Single Components of the Trisiloxane Superspreader Type M_2D^*-$(CH_2)_3$-$(OCH_2CH_2)_n$-OCH_3

R. Wagner

GE Bayer Silicones
Building R20, D-51368 Leverkusen, Germany
Tel.: Int. code + (214)3067715 — Fax: Int. code + (214)3056411
E-Mail: roland.wagner@gepex.ge.com

Y. Wu, G. Czichocki, H. v. Berlepsch, B. Weiland, F. Rexin

Max-Planck-Institut für Kolloid- und Grenzflächenforschung
Am Mühlenberg 2, D-14476 Golm, Germany

L. Perepelittchenko

Institute for Petrochemical Synthesis
Leninskii Prospect 29, 117912 Moscow, Russia

Keywords: Trisiloxane Surfactants / Superspreading / Phase Behavior

Summary: Surfactants of the formula $[(CH_3)_3SiO]_2CH_3Si(CH_2)_3(OCH_2CH_2)_nOCH_3$, n = 3 to 9, have been synthesized. The temperature and concentration dependent spreading performance on hydrophobized Si-wafer surfaces has been investigated. Pronounced spreading velocity maxima were found at 6°C (n = 4), 26°C (n = 6) and 40°C (n = 8). A single superspreading component does not exist. The ability to spread is closely related to the phase transition temperature for L_α (lamellar phase) → 2Φ (two-phase system).

Introduction

Aqueous solutions of certain trisiloxane surfactants wet rapidly low-energy surfaces (water contact angle > 90°) [1]. The spreading rate of a "superspreader" solution significantly exceeds that expected for a purely liquid diffusion controlled process [2, 3].

It has been shown recently that this unique property is part of the much broader concept of a surfactant-enhanced spreading. Aqueous solutions of ethoxylated alcohols (C_iE_j) rapidly wet moderately hydrophobic solid surfaces [4, 5]. The same effect was observed for certain double-chained anionic and cationic hydrocarbon surfactants [6]. A detailed analysis of spreading data obtained at a Si-surfactant solution/liquid hydrocarbon interface led to the conclusion that this rapid spreading can be understood as a Marangoni flow driven process [7]. Although the role of dispersed surfactant phases [4, 5] or microstructures has been demonstrated, a correlation between spreading performance and extensively studied phase behavior [8–10] could not be established.

Despite the efforts made, apparently simple questions concerning the rapid spreading of siloxane surfactants on low-energy surfaces remained unsolved. Are single components of the complex surfactant mixtures responsible for the superspreading effect? Which physicochemical property qualifies certain molecules? Does a synergistic effect exist?

Therefore, we synthesized the single components of the intensively investigated superspreader Silwet L77® (Union Carbide). Further, the concentration-dependent spreading performance of these well defined molecules was investigated.

Materials and Methods

In order to synthesize defined 1,1,1,3,5,5,5-heptamethyltrisiloxanyl (MD*M) derivatives of the general formula $[(CH_3)_3SiO]_2CH_3Si(CH_2)_3(OCH_2CH_2)_{3\text{-}9}OCH_3$ a complex reaction sequence was applied (Scheme 1).

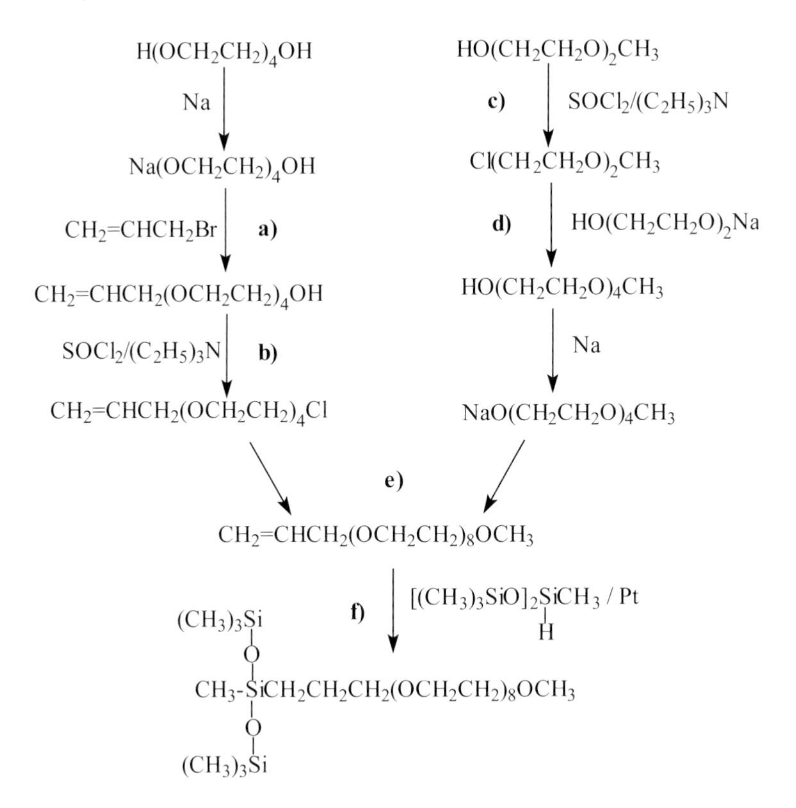

Scheme 1. Reaction sequence yielding the octaethylene glycol derivative of 1,1,1,3,5,5,5-heptamethyltrisiloxane.

Due to the high boiling points of the longer-chained derivatives (n = 6 to 9) elimination of ethylene or diethylene glycol moieties occurs during the distillations and minor amounts of silicon-containing species bearing shorter oligoethylene glycol chains are found (Table 1) [11].

Table 1. Hydrosilylation yields, color, purity and contamination compositions of the siloxanyl-modified oligoethylene glycols.

	Yield [%]	Color	Purity [% GC]	Contaminations Si [%]	Contaminations Without Si [%]	B.p. [°C/mmHg)
EO3	62	Colorless	>>99			136–140/0.35
EO4	44	Colorless	>>99			167–173/0.3
EO5	32	Colorless	99		1	187–195/0.3
EO6	28	Pale yellow	97.5	1 (EO5 type)	1.5	216–226/0.7
EO7	55	Pale yellow	96	1.5 (EO6 type)	2.5	237–242/0.7
EO8	33	Yellow	95	1.5 (EO6 type)	3.5	249–256/0.75
EO9	24	Pale brown(Pt)	90	10 (EO7+EO8 type)		272–281/0.75

Temperature-dependent spreading experiments were carried out at 6°C, 26°C and 40°C on trimethylsilyl-modified silicon wafers (10 μl surfactant solution drops). Prior to these experiments the wafer pieces were energetically characterized by contact angle measurements versus water, hexadecane, pentadecane and tetradecane (Table 2). The data for the strictly non-polar alkanes ($\gamma_{lv} = \gamma_{lv}^{LW}$) were used to calculate γ_{sv} (Neumann [12], Eq. 1) and γ_{sv}^{LW} (Good [13], Eq. 2). The practical identity between γ_{sv} and γ_{sv}^{LW} indicates that this surface is of a non polar character.

$$\cos\Theta = \frac{(0.015\gamma_{sv} - 2)\sqrt{\gamma_{sv} * \gamma_{lv}} + \gamma_{lv}}{\gamma_{lv}(0.015\sqrt{\gamma_{sv} * \gamma_{lv}} - 1)}$$

Eq. 1

$$1 + \cos\Theta = 2\sqrt{\frac{\gamma_{sv}^{LW}}{\gamma_{lv}^{LW}}}$$

Eq. 2

Table 2. Surface energy of the modified silicon wafer calculated from contact angles

	Liquid H_2O	Liquid $C_{14}H_{30}$	Liquid $C_{15}H_{32}$	Liquid $C_{16}H_{34}$
γ_{lv} [mN/m] (20°C)	72.6	26.6	27.3	27.6
Θ [°]	91	29	34	38
γ_{sv} [mN/m] (Neumann)		23.2	23.7	24.0
γ_{sv}^{LW} [mN/m] (Good)		23.0	23.6	23.9

Figures 1a to 1d summarize the time dependence of the spreading areas and the corresponding initial spreading velocities.

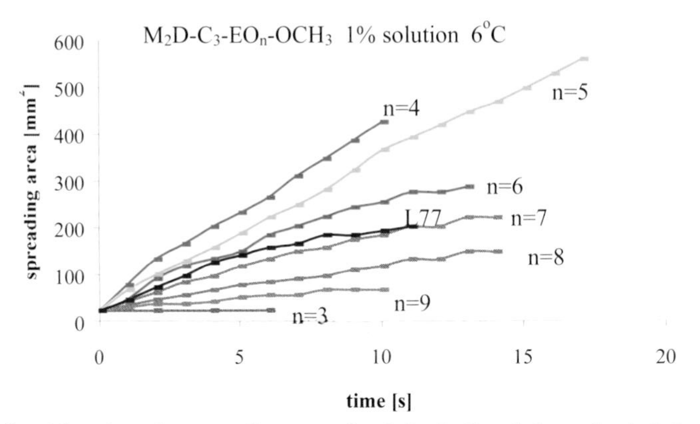

Fig. 1a. Time-dependent spreading areas for defined oligoethylene glycol derivatives and Silwet L77, T = 6°C, c = 1 wt. %.

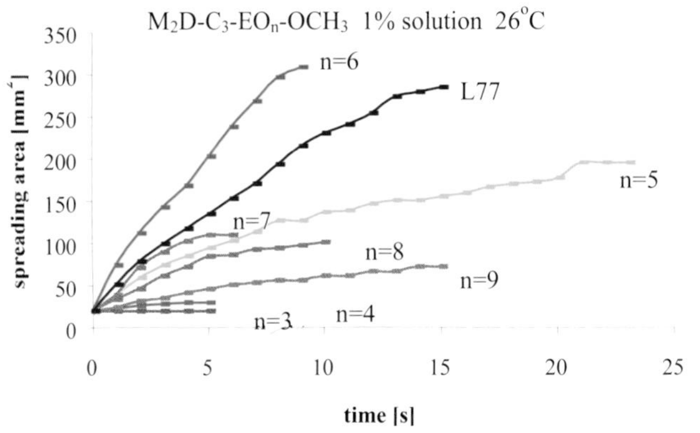

Fig. 1b. Time-dependent spreading areas for defined oligoethylene glycol derivatives and Silwet L77, T = 26°C, c = 1 wt. %.

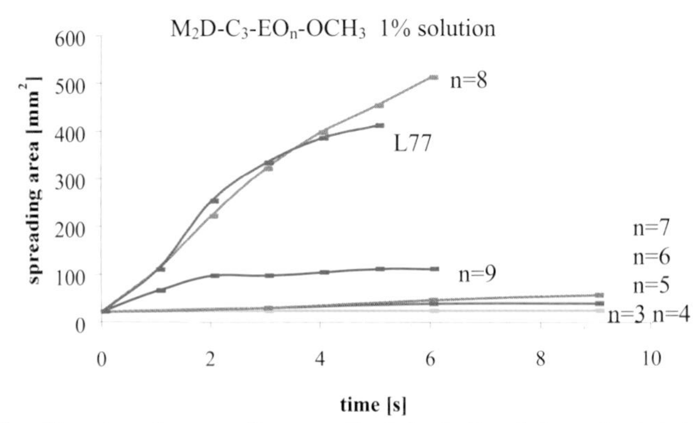

Fig. 1c. Time-dependent spreading areas for defined oligoethylene glycol derivatives and Silwet L77, T = 40°C, c = 1 wt. %.

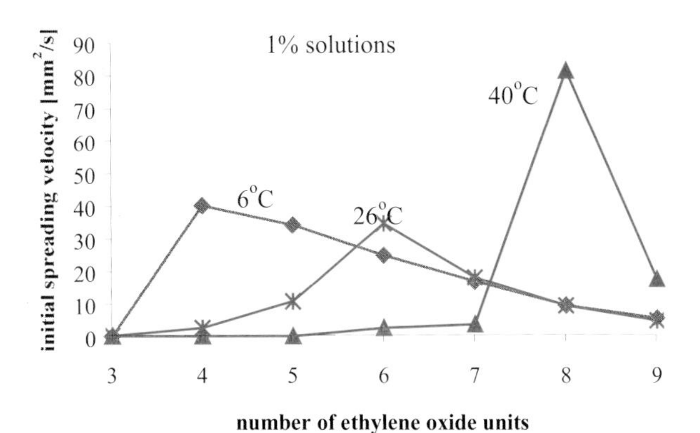

Fig. 1d. Initial spreading velocities of defined oligoethylene glycol derivatives, $c = 1$ wt. %.

Figure 2 depicts the temperature dependent phase behavior of the binary systems surfactant–water at a constant concentration of 1 wt. %.

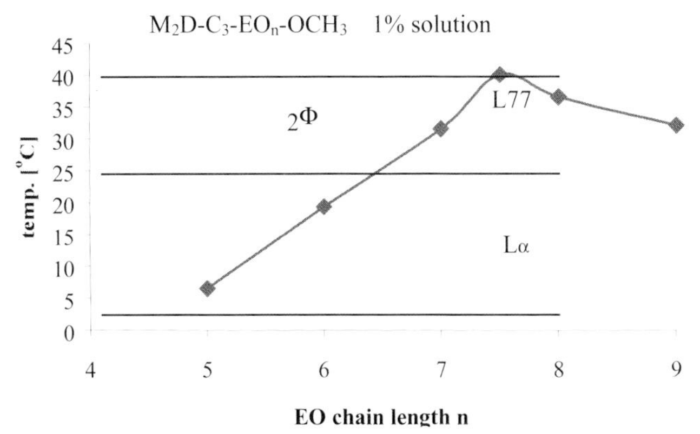

Fig. 2. Phase behaviour of 1wt.% surfactant solutions.

Results

The data in Tables 1a to 1d prove the existence of a pronounced chain length dependence of the spreading performance.

At 6°C (Figs. 1a and 1d) derivative EO4 is the best spreader. The oligoethylene glycol chain extension up to EO9 causes a dramatic reduction of the ability to spread. On the other hand the derivative with the shortest chain (EO3) did not spread at all.

The spreading data for 26°C (Figs. 1b and 1d) immediately disclose that a single superspreading compound does not exist. The spreading velocity maximum has shifted with increasing temperature towards the longer-chained derivative EO6. Successive shortenings or extensions of the oligo-ethylene glycol chain have the same velocity-retarding effect.

A further increase of the spreading temperature (Figs. 1c and 1d) yields the now expected phenomenological result. The spreading velocity maximum shifts towards derivative EO8.

The chain length dependence of the initial spreading velocity was found to be less pronounced for low temperatures (Fig. 1d). Although the maximum at EO4 for 6°C exists the difference from longer-chained derivatives is relatively small. The opposite is true for the highest temperature under investigation. At 40°C the shortening or extension of the oligoethylene glycol chain for a single moiety (maximum at EO8) causes a dramatic reduction of the ability to spread. At high temperatures an extremely narrow "chemical structure window" exists for the occurrence of the superspreading effect.

The phenomenological description of the temperature influence on the spreading performance does not disclose the driving force of the process. Since this physicochemical property has to be temperature-dependent the phase behavior of the surfactant–water binary systems was investigated (Fig. 2). Generally, at low temperatures an extended L_α region (lamellar phase) was found. At higher temperatures a chain length dependent phase transformation towards a 2Φ system (separated water and oil phases) takes place.

A comparison of phase transition and spreading experiment temperatures helps decisively to disclose the principle. The phase transition temperature for the solution of derivative EO8 is 36.5°C. At 40°C (superspreading behavior) it is located in the 2Φ region close to the phase transition temperature. The same pattern holds for the EO6 solution. Here the phase transition temperature is 19.3°C. At 26°C spreading temperature this system has just undergone the phase transition towards the 2Φ state. Derivative EO4 does not deviate from this rule although the phase transition temperature could not be measured directly. This substance showed superspreading behavior at 6°C. At this temperature EO4 solutions are located in the 2Φ region close to the phase transition temperature of –5 to 0°C.

The data emphasize that the ability to spread on low-energy, non-polar solid substrates is a property of the trisiloxane moiety. The length of the oligoethylene glycol chain adjusts the phase state at a given temperature. From the fact that 2Φ systems close to the phase transition temperature were found to be the best superspreader mixtures a serious question arises. Is superspreading a time-dependent phenomenon? Over longer periods of time the initially microdisperse 2Φ systems should form less spreading active separated layers. Since the well known hydrolysis of the trisiloxane moiety cannot easily be suppressed these materials are unsuitable materials for corresponding time-dependent experiments [14, 15].

Acknowledgement: The project "Polyhydroxylated silicon compounds" is supported by the *Deutsche Forschungsgemeinschaft* (Reg. No. WA 1043/1-2).

References:

[1] S. Zhu, W. G. Miller, L. E. Scriven, H. T. Davis, *Colloids Surfaces* **1994**, *90*, 63.
[2] F. Tiberg, A. M. Cazabat, *Europhys. Lett.* **1994**, *25*, 205.
[3] F. Tiberg, A. M. Cazabat, *Langmuir* **1994**, *10*, 2301.
[4] T. Stoebe, Z. Lin, R. M. Hill, M. D. Ward, H. T. Davis, *Langmuir* **1996**, *12*, 337.
[5] T. Stoebe, Z. Lin, R. M. Hill, M. D. Ward, H. T. Davis, *Langmuir* **1997**, *13*, 7270.
[6] T. Stoebe, Z. Lin, R. M. Hill, M. D. Ward, H. T. Davis, *Langmuir* **1997**, *13*, 7276.
[7] T. Stoebe, Z. Lin, R. M. Hill, M. D. Ward, H. T. Davis, *Langmuir* **1997**, *13*, 7282.
[8] M. He, R. M. Hill, Z. Lin, L. E. Scriven, H. T. Davis, *J. Phys. Chem.* **1993**, *97*, 8820.

[9] Z. Lin, R. M. Hill, H. T. Davis, L. E. Scriven, Y. Talmon, *Langmuir* **1994**, *10*, 1008.
[10] M. He, Z. Lin, L. E. Scriven, H. T. Davis, *J. Phys. Chem.* **1994**, *98*, 6148.
[11] R. Wagner, Y. Wu, G. Czichocki, H. v. Berlepsch, B. Weiland, F. Rexin, *Appl. Organomet. Chem.*, **1999**, in press.
[12] O. Driediger, A. W. Neumann, P.-J. Sell, *Kolloid-Zeitschr. Z. Polym.* **1965**, *201*, 101.
[13] C. J. van Oss, M. J. Chaudhury, R. J. Good, *Chem. Rev.* **1988**, *88*, 927.
[14] M. Knoche, H. Tamura, M.J. Bukovac, J. Agric; *Food Chem.* **1991**, *39*, 202.
[15] K. D. Klein, W. Knott, G. Koerner, "*Silicone Surfactants — Development of Hydrolytically Stable Wetting Agents*" in: *Organosilicon Chemistry II: From Molecules to Materials* (Eds. N. Auner, J. Weis), VCH, Weinheim, **1996**.

Trisiloxane Surfactants — Mechanisms of Spreading and Wetting

Joachim Venzmer

Central Research Department, Th. Goldschmidt AG
Goldschmidtstraße 100, D-45127 Essen/Germany
Tel.: Int. Code + (201)1732302 — Fax: Int. Code + (201)1731993
E-mail:joachim.venzmer@de.goldschmidt.com

Stephen P. Wilkowski

Goldschmidt Chemical Corp., 914 E. Randolph Rd., Hopewell, VA , USA 23860
Tel.: Int. Code + 804-541-8658 — Fax: Int. Code + 804-541-2783
E-mail: swilkows@goldschmidtusa.com

Keywords: Superspreading / Trisiloxane / Surface Tension / Silicone / Vesicles

Summary: Aqueous solutions of some nonionic trisiloxane surfactants exhibit superior wetting properties on hydrophobic substrates ("superspreading"). Although this phenomenon has received extensive coverage in the scientific literature in recent years, its mechanism and the reason why only some silicone surfactants show this behavior remained unclear. Apart from the characteristic of a low surface tension, which is typical of most silicone surfactants, it will be shown why it is crucial that the surfactants form bilayer aggregates (vesicles, lamellar phases) in aqueous solution. Based on molecular packing considerations, there is a fundamental difference between the behavior at the solid/liquid interface of a micelle-forming surfactant and of a vesicle-forming surfactant. The model presented here is supported by measurements of the spreading kinetics.

Introduction

A broad range of industrial applications take advantage of the unique surface-active properties of organosilicones. In aqueous systems, hydrophilically substituted trisiloxane derivatives function as excellent wetting agents. Therefore, trisiloxane surfactants have been used as adjuvants in agricultural applications for a number of years [1]. They have been shown to increase foliar uptake ... ductions in surface tension and large surface area spreading of spray solutions ... surfactants contain a hydrophilic headgroup such as a polyether group, which ... trisiloxane hydrophobe via an alkyl spacer, as illustrated in Fig. 1.

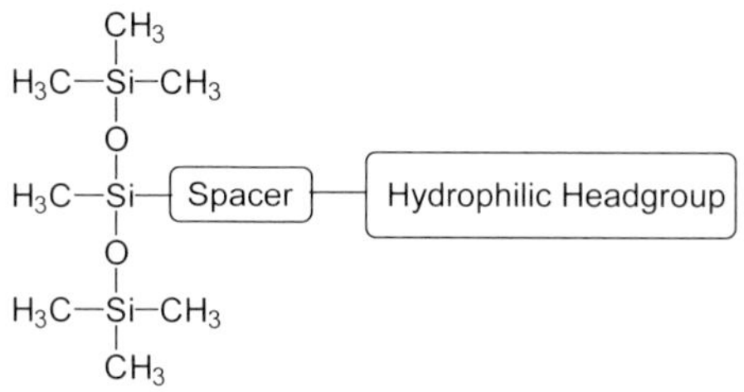

Fig. 1. Schematic structure of trisiloxane surfactants.

The wetting exhibited by some trisiloxane surfactants is so extensive and rapid that it has been referred to as "superspreading" [3]. Typically, a small drop (50 μL) of a diluted aqueous solution (0.1 % w/w) of such a trisiloxane spreads out on a hydrophobic surface such as a polypropylene sheet into a thin, wetting film approx. 80 mm in diameter within tens of seconds. This is about 20 times the area wetted by a 1 % w/w solution of a conventional organic surfactant such as a nonylphenol ethoxylate.

While the phenomenon of superspreading is easily observed, explanations for this behavior have been more elusive. Over the years, a number of mechanistic theories on superspreading have been put forth. Early concepts attributed the molecular shape of the trisiloxanes to be critical to superspreading [4]. The T-shaped configuration of the trisiloxane surfactant was described to roll or "zipper" at the spreading front of the droplet. Later, it was observed that most superspreading solutions are slightly turbid [5]. The turbidity was said to indicate a "dispersed surfactant-rich phase" responsible for spreading. Recently, it has been proposed that superspreading requires extremely rapid diffusion and specific interfacial tension dynamics [6].

These theories, however, do not explain why only some trisiloxane surfactants exhibit superspreading behavior, and why minor variations in the hydrophilic head group can lead to dramatically reduced spreading. This effect is demonstrated in Table 1 for the four specific trisiloxane surfactants used in our studies.

Table 1. Spreading behavior of trisiloxane surfactants.

	Surfactant[a]	Spreading on polypropylene (0.1 % w/w; 50 μL), [mm]
1	M(D'EO$_{7.5}$OMe)M	80
2	M(D'EO$_{6}$PO$_{3}$OH)M	76
3	M(D'EO$_{7.5}$PO$_{3.5}$OH)M	56
4	M(D'EO$_{10}$PO$_{2}$OH)M	16

[a] M–D'–M = trisiloxane hydrophobe; EO = ethylene oxide; PO = propylene oxide; Me = methyl endgroup; OH = hydroxyl endgroup.

In this paper it will be shown how these differences in spreading performance can be explained by a simple model based on molecular packing considerations. In addition, the spreading kinetics and the influence of concentration, humidity and salt addition on the spreading process will be discussed, supporting our model.

Theoretical Considerations

The classical method of describing the interaction of a liquid drop on a solid substrate utilizes Young's equation [7]:

$$\gamma_s = \gamma_{sl} + \gamma_l \cos\theta$$

where:

γ_l	=	the surface tension of the liquid
γ_s	=	the surface tension of the solid
γ_{sl}	=	the interfacial tension between the liquid and the solid
θ	=	contact angle of the drop of liquid on the solid

This relation can also be expressed in terms of a spreading coefficient $S = \gamma_s - (\gamma_l + \gamma_{sl})$ [8], where the requirement for complete wetting is a positive spreading coefficient.

Applying these equations to the examples in Table 1 leads to the following problem. The surface tensions of the four trisiloxane surfactant solutions are all very similar (23–24 mN/m, static value, drop volume tensiometer), and the surface tension of the substrate is also constant (approx. 30 mN/m). Therefore, the critical parameter which determines the sign and magnitude of the spreading coefficient must be the interfacial tension between the surfactant solution and the substrate. Unfortunately, this value cannot be measured directly and must be calculated from surface tension and contact angle. The contact angles in these systems are all very low and do not provide a means to differentiate between trisiloxanes. Therefore, this approach cannot explain the differences in spreading behavior directly. However, differences in interfacial tensions between these systems can be explained based on molecular packing considerations as described later in this text.

A simple but effective concept described by Israelachvilli et al. predicts the aggregation of surfactants in solution. This model is based on the molecular shape of the surfactant molecules. In this model, the ratio of the size of the hydrophobic and the hydrophilic portion of the molecules is expressed in terms of a critical packing parameter P [9]:

$$P = \frac{V}{a_o\, l_c}$$

where:			
	V	=	lipophilic chain volume
	a_o	=	interfacial area hydrocarbon/water
	l_c	=	length of lipophilic chain

According to this concept, cone-shaped surfactant molecules with the hydrophilic headgroup larger than the hydrophobic portion ($P < 1$) form curved aggregates such as spherical or cylindrical micelles in solution. When the hydrophobic and hydrophilic portions of the surfactant are of equal size ($P \approx 1$), the spontaneous mean curvature is zero, resulting in the formation of surfactant bilayer aggregates. These could be either flat lamellae (lamellar or L_α-phase), spherically closed bilayers called vesicles or liposomes, or the so-called sponge or L_3-phase, which is characterized by two continuous water phases separated by a single bilayer of surfactant molecules [10, 11].

Originally, this model was proposed for classic surfactants with alkyl chains as the hydrophobe. However, this model also appears to be applicable to silicone surfactants. The four trisiloxane surfactants used in this study exhibit distinct differences in their phase behavior (Table 2) as a function of length and hydrophilicity of the polyether headgroup.

Table 2. Phase behavior of silicone surfactants.

	Surfactant	Phase Behavior (1 to 5 % w/w)
1	$M(D'EO_{7.5}OMe)M$	complex (L_α, L_3, vesicles) < 44°C [5]
2	$M(D'EO_6PO_3OH)M$	complex (L_α, L_3, vesicles) 17–42°C
3	$M(D'EO_{7.5}PO_{3.5}OH)M$	only slight shear birefringence at 52–60°C
4	$M(D'EO_{10}PO_2OH)M$	Only cloud point at 55°C

Applying this model, the hydrophilic parts in surfactants **1** and **2** must have the same cross-sectional area as the trisiloxane hydrophobes in order to form surfactant bilayers and exhibit this complex phase behavior. In the surfactants with the longer and more hydrophilic polyether headgroups (e.g. **4**), the critical packing parameter must be less than 1 and accordingly no bilayer aggregates are formed; the trisiloxane derivative **4** has only a classical cloud point. Correlating this phase behavior with the spreading properties shown in Table 1 suggests that the superspreading phenomenon is connected with the presence of surfactant bilayer aggregates [12].

The way in which surfactants interact with solid substrates has recently been clarified with the help of atomic force microscopy [13]; these results provide a basis for our explanation of the different spreading properties of the trisiloxanes. When micelle-forming surfactants adsorb on a hydrophobic substrate, hemi-micelles are formed corresponding to the shape of the molecules (Fig. 2A). This arrangement forces hydrophilic headgroups into contact with the hydrophobic substrate, an orientation less than ideal for lowering interfacial tension. In contrast, when bilayer-forming surfactants with a critical packing parameter of $P \approx 1$ interact with hydrophobic substrates, such curved aggregates cannot develop; instead a smooth coverage of the substrate has been found with only the hydrophobic portion of the surfactant being in contact with the surface (Fig. 2B). Therefore, the interfacial tension at this interface can be expected to be lower than that with micelle-forming surfactants. Considering Young's equation, this should be a thermodynamic reason for a more positive spreading coefficient in the case of bilayer-forming surfactants.

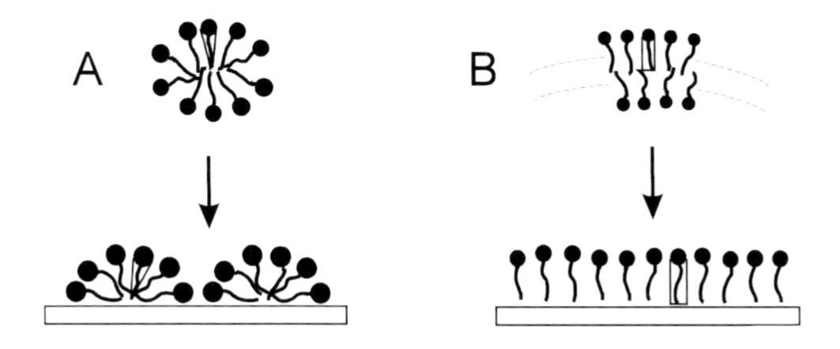

Fig. 2. Interaction of surfactants with solid substrates (according to [13]): A) micelle-forming surfactants; B) bilayer-forming surfactants.

At the air/water interface there is also a fundamental difference between the behavior of micelle- and of bilayer-forming surfactants. In a system containing micelles, these thermodynamically stable aggregates are in equilibrium with single surfactant molecules in solution and a surfactant monolayer at the air/water interface (Fig. 3A). In the case of bilayer assemblies such as vesicles, these aggregates are thermodynamically unstable. Since the surfactants in this system are more hydrophobic, they are less soluble and at the air/water interface there is a so-called insoluble surfactant monolayer. When a vesicle comes in contact with the surface, it spontaneously bursts, depositing large amounts of surfactant at the surface. These surfactant molecules cannot go into solution again; instead they are available at the surface, e.g. for processes such as spreading. Therefore, superspreading with a high demand of surfactant at the surface could be obtained by using bilayer-forming surfactants.

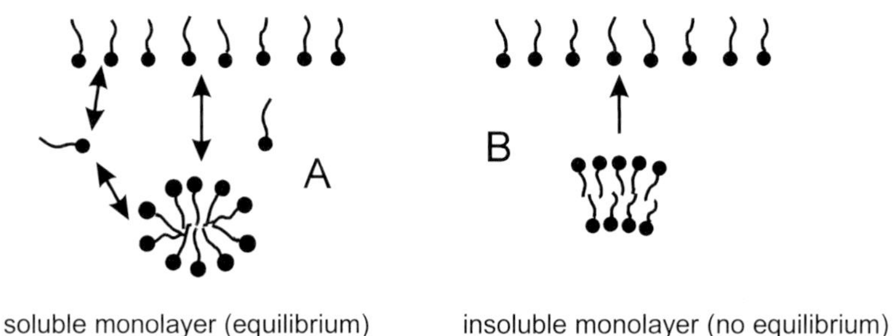

Fig. 3. Interaction of surfactants with air/water interface: A) micelle-forming surfactants; B) bilayer-forming surfactants.

It can be concluded that based on these molecular packing considerations there are the following two basic requirements for superspreading:

1. A siloxane hydrophobe is necessary for a low surface tension.

2. The hydrophobic and hydrophilic portions of the surfactant should be of equal size ($P \approx 1$) enabling the formation of bilayer aggregates. This should lead to a lower interfacial tension substrate/water and a faster spreading kinetics.

The following experiments describe some phenomena which can be observed with superspreading trisiloxane surfactants. They can all be explained with the help of these theoretical considerations and therefore further support these statements.

Results and Discussion

Spreading Kinetics

The spreading mechanism can be expected to be reflected in the spreading kinetics. Therefore, the spreading velocity during the first 30 s of the spreading process has been followed as a function of concentration. With all superspreading surfactants, the same interesting kinetic dependence has been observed, as shown in Fig. 4 for the trisiloxane derivative **2**. First, at low concentrations, the spreading velocity is proportional to surfactant concentration, but above a certain limit (0.1 % w/w) the velocity becomes constant. This kinetic behavior can only be explained by the existance of two rate-determining processes. First, at low concentrations, the spreading rate is limited by the transport of the surfactant aggregates to the surface, which is proportional to the concentration of the aggregates. Above a certain limit, however, the amount of the surfactant at the surface is not rate-determining any more: rather it is the reorientation of the surfactant molecules at the surface, i.e. the transformation of bilayer aggregates such as vesicles to a monolayer. This interpretation of the kinetic data could be confirmed by the comparison of static and dynamic surface tension as a function of concentration: whereas the static value is constantly low at all concentrations below 0.05 % w/w, the dynamic surface tension still decreases significantly between 0.05 and 0.1 % w/w (data not shown).

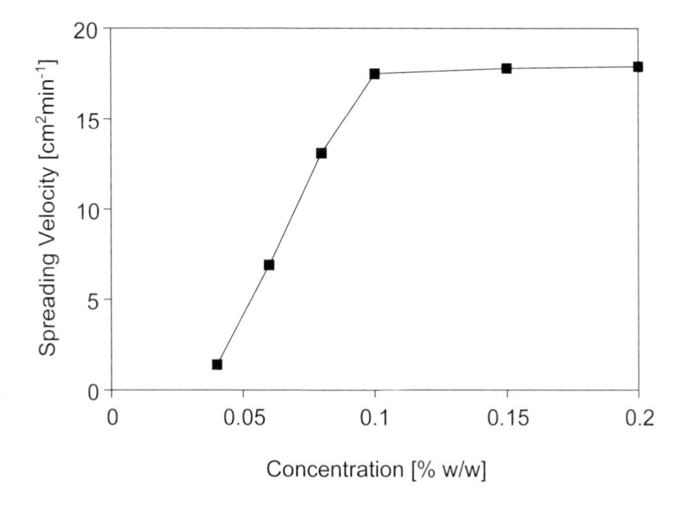

Fig. 4. Spreading velocity on polypropylene as function of concentration for M(D'EO_6PO_3OH)M (**2**).

Concentration and Humidity Effects

Choosing a suitable trisiloxane surfactant concentration for aqueous formulations, one comes across a surprising effect: the spreading area at 0.5 or 1 % w/w trisiloxane is much lower than at 0.1 % w/w; therefore, 0.1 % w/w is often used in the application of trisiloxane surfactants. Explaining this strange behavior is again possible by considering the molecular packing and provides additional insight into the spreading mechanism. Looking at the dependence of spreading area on surfactant concentration under laboratory conditions (50 % relative humidity), one finds the above-mentioned peculiar dependence (Fig. 5, •). First the spreading area increases with concentration, but at less than 0.1 % w/w trisiloxane surfactant, the spreading area decreases again. In contrast, when the spreading experiment is performed at 100 % relative humidity, the spreading area is proportional to surfactant concentration over the entire concentration range (Fig. 5, ■). This is the maximum spreading area one can theoretically reach, as shown already by Zhu [14], considering that via the spreading process the droplet has been transformed into a pancake of water with one monolayer of surfactant at each of the air/water and substrate/water interfaces.

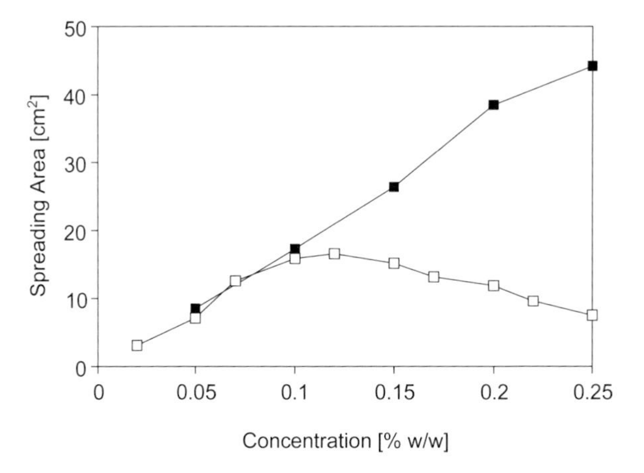

Fig. 5. Spreading area as function of concentration for M(D'EO$_{7.5}$OMe)M (**1**): • = 50 % relative humidity; ■ = 100 % relative humidity.

The decrease in spreading area under laboratory conditions must somehow be a result of evaporation; this connection could be clarified by studying the spreading kinetics. When looking at the spreading area as a function of time for different concentrations (Fig. 6), the spreading velocity at the beginning of the spreading process is independent of concentration above 0.1 % w/w, as already discussed above. However, after 10–20 s, the spreading process suddenly stops, the earlier the higher the surfactant concentration. At this point only 10 % of the water is evaporated, as proven by performing the spreading experiment on an analytical balance. This means there should be a sufficient amount of water left for further spreading; so the reason for the sudden stop in spreading is not evaporation as such, but a secondary effect caused by evaporation. This can be the formation of a lamellar phase at the surface or the spreading front of the drop, which then stops spreading because of its high viscosity. This interpretation is supported by the observation that this influence of humidity is more pronounced with surfactants having an even more extended lamellar phase such as the trisiloxane derivative analogous to **1** but with an acetyl endgroup (result not shown).

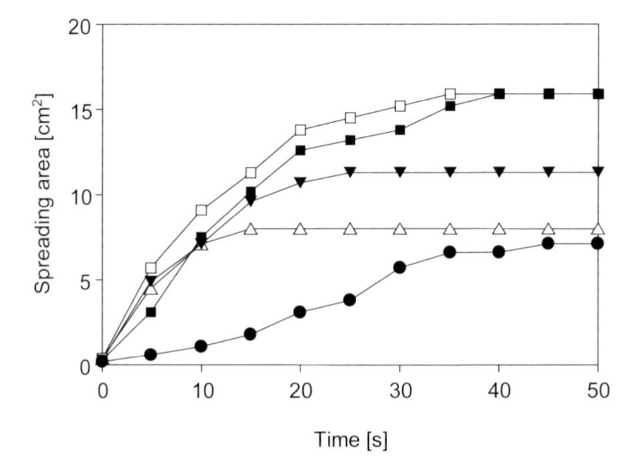

Fig. 6. Spreading area as function of time for M(D'EO$_{7.5}$OMe)M (**1**): ● = 0.05 %; ■ = 0.10 %; • = 0.15 %; ▼ = 0,20 %; Δ = 0.25 %.

Salt Effects

It is known that by addition of electrolytes one can reduce the cloud point of nonionic surfactants. This effect is caused by a reduction of the hydration sphere with increasing ionic strength and disruption of H-bonding which holds the polyether group in solution. In other words, by addition of salt the hydrodynamic volume of the polyether headgroup is reduced [15] and the critical packing parameter increases. Applying this principle to the four trisiloxane surfactants used in this study shows that the trisiloxanes **1** and **2** with an ideal packing parameter for the formation of bilayer aggregates become too hydrophobic upon addition of salt; they turn insoluble and superspreading is prevented. In the case of the other, more hydrophilic, trisiloxane derivatives having critical packing parameters less than 1, the addition of salt leads to dramatic improvement in spreading (Fig. 7). The size of the polyether headgroup is reduced by the addition of salt to such an extent that the requirements for superspreading are met.

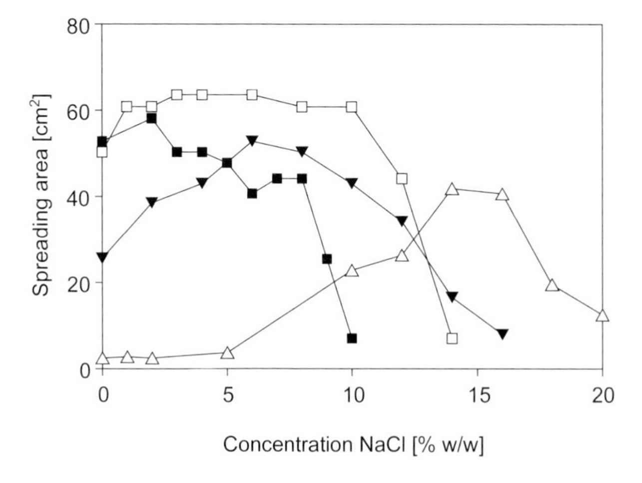

Fig. 7. Spreading area as function of salt concentration: • = M(D'$EO_{7.5}$OMe)M (**1**); ■ = M(D'EO_6PO_3OH)M (**2**), ▼ = M(D'$EO_{7.5}PO_{3.5}$OH)M (**3**); Δ = M(D'$EO_{10}PO_2$OH)M (**4**).

Conclusion

Superspreading requires low surface tension, but not all trisiloxane surfactants that possess very low surface tensions superspread. For superspreading to occur, it is necessary for surfactant molecules to form bilayer aggregates. Based on molecular packing considerations, this leads to low substrate/water interfacial tensions and a large amount of surfactant at the surface enabling high spreading velocities. The results of experiments investigating the effect of surfactant concentration, electrolyte addition and the spreading kinetics further support this model for superspreading.

Acknowledgment: We would like to thank Dr. Kai Köppen for performing the kinetic measurements and most of the spreading tests, Wayne Kennedy for dynamic surface tension measurements, Prof. Heinz Rehage (University Essen) for many fruitful discussions and the Video Enhanced-Microscopy investigations in his laboratory, and Dr. Krystina Kratzat for help with the polarization microscopy.

Previously published: J. Venzmer, S. P. Wilkowski, "*Trisiloxane Surfactants — Mechanisms of Spreading and Wetting*", *Pesticide Formulations and Application Systems, 18, ASTM STP 1347,* (Eds: J. D. Nalewaja, G. R. Goss, R. S. Tann), American Society for Testing and Materials, **1998**.

References

[1] J. A. Zabkiewicz, R. E. Gaskin, J. M. Balneaves, "*Effect of Additives on Foliar Wetting and Uptake of Glyphosate into Gorse (Ulex europaeus)*" in: *"Symposium on Application and Biology"* (Ed: E. S. E. Southcombe), British Crop Protection Council, Monograph No. 28, **1985**, pp. 127–134.

[2] R. J. Field, N. G. Bishop, "*Promotion of Stomatal Infiltration of Glyphosate by the Organosilicone Surfactant Silwet L-77 Reduces the Critical Rainfall Period*", *Proceedings of the European Weed Research Society,* **1988**, pp. 205–206.

[3] S. Zhu, W. G. Miller, L. E. Scriven, H. T. Davis, "*Superspreading of Water-Silicone Surfactant on Hydrophobic Surfaces*" , *Colloids and Surfaces A* **1994**, *90*, 63–78.

[4] K. P. Anathapadmanabhan, E. D. Goddard, P. Chandar "*A Study of the Solution, Interfacial and Wetting Properties of Silicone Surfactants*", *Colloids and Surfaces* **1990**, *44*, 281–297.

[5] M. He, R. M. Hill, Z. Lin, L. E. Scriven, H. T. Davis, "*Phase Behavior and Microstructure of Polyoxyethylene Trisiloxane Surfactants in Aqueous Solution*", *J. Phys. Chem.* **1993**, *97*, 8820-8834.

[6] T. Svitova, H. Hoffman, R. M. Hill, "*Trisiloxane Surfactants: Surface/Interfacial Tension Dynamics and Spreading on Hydrophobic Surfaces*", *Langmuir* **1996**, *12*, 1712–1721.

[7] T. Young, "*An Essay on the Cohesion of Fluids*", *Phil. Trans. Roy. Soc. London* **1805**, *95*, 65–87.

[8] S. Ross, P. Becher, "*The History of the Spreading Coefficient*", *J. Colloid Interface Sci.,* **1992**, *149*, 575–579.

[9] J. N. Israelachvilli, D. J. Mitchell, B. W. Ninham, "*Theory of Self-Assembly of Hydrocarbon Amphiphiles into Micelles*", *J. Chem. Soc., Faraday Trans. II* **1976**, *72*, 1525–1568.

[10] D. J. Mitchell, G. J. T. Tiddy, L. Waring, T. Bostock, M. P. McDonald, "*Phase Behavior of Polyoxyethylene Surfactants with Water*", *J. Chem. Soc., Faraday Trans. I* **1983**, *79*, 975–1000.

[11] R. Strey, R. Schomäcker, D. Roux, F. Nallet, U. Olsson, "*Dilute Lamellar and L_3 Phases in the Binary Water–$C_{12}E_5$ System*", *J. Chem. Soc., Faraday Trans.* **1990**, *86*, 2253–2261.

[12] R. M. Hill, M. He, H. T. Davis, L. E. Scriven, "*Comparison of the Liquid Crystal Phase Behavior of Four trisiloxane Superwetter Surfactants*", *Langmuir* **1994**, *10,* 1724–1734.

[13] S. Manne, H. E. Gaub, "*Molecular Organization of Surfactants at Solid-Liquid Interfaces*", *Science* **1995**, *270*, 1480–1482.

[14] X. Zhu, "*Surfactant Fluid Microstructure and Surfactant Aided Spreading*", *Ph. D. Thesis*, University of Minnesota, Minneapolis, **1992**.

[15] F. E. Bailey, J. V. Koleske, "*Poly(ethylene Oxide)*", Academic Press, New York, **1976**, p. 97.

Silicone Rubbers
Innovative — High Performance — Efficient

Klaus Pohmer, Helmut Steinberger

Bayer AG, Inorganics Business Group, Silicones Business Unit
Olof-Palme-Str. 15, D-51368 Leverkusen, Germany
Tel.: Int code + (214)303 1916 — Fax: Int code + (214)303 1934
E-mail: klaus.k.p.pohmer@bayer-ag.dbp.de

Keywords: Silicone Rubber (VMQ) / High Consistency Rubber (HCR) / Liquid Silicone Rubber (LSR) / Room-Temperature Vulcanizing (RTV) / Liquid Injection Molding (LIM)

Summary: Silicone rubbers are speciality rubbers which because of their outstanding properties, such as low-temperature flexibility, chemical and thermal resistance, and virtually constant mechanical properties over a wide temperature range, are regarded as high–performance products whose market is growing on average by 5 to 10 % per year. Strictly speaking there now are four different systems, although the boundaries are rather fluid: room-temperature-vulcanizing one-component systems (RTV-1K), which are mainly used as sealants; room-temperature-vulcanizing two-component systems (RTV-2K), which are used predominantly as embedding compounds; high-temperature-vulcanizing liquid silicones (LSR), which are normally processed by the injection molding process to molded articles; and high-temperature-vulcanizing solid silicones (HTV or HCR), which are largely extruded.

Silicone rubbers have a market volume world-wide of approximately 150 000 t/a. Liquid silicones currently account for around 10 % of this, and their consumption compared to solid silicones is rising at an above-average rate. If we look at the rubber market as a whole (see Fig. 1), silicone rubbers with a share of around 1 % are still specialities which because of their outstanding properties are used in many applications.

Compared with conventional organic rubbers, the synthesis route from raw material to processable rubber is much more complex and costly, despite the fact that the natural material, silica sand, is available in virtually unlimited quantities. One of the reasons for this is the high energy requirement in the first production stage, the production of silicon by reduction of silicon dioxide with carbon. This is followed by direct synthesis of chlorosilanes (Rochow–Müller) which is carried out on an industrial scale in place of a three-stage Grignard synthesis. In the third stage, a siloxane synthesis takes place through hydrolysis or methanolysis of the chlorosilanes. The actual polymerization occurs by equilibration of the siloxanes with alkali metal hydroxides. The resulting polymers are processed in a first compounding stage with reinforcing fillers (normally silicas) to silicone rubber base compounds. The processable rubber mixture is produced in the sixth stage by introducing crosslinking agents, pigments or similar additives.

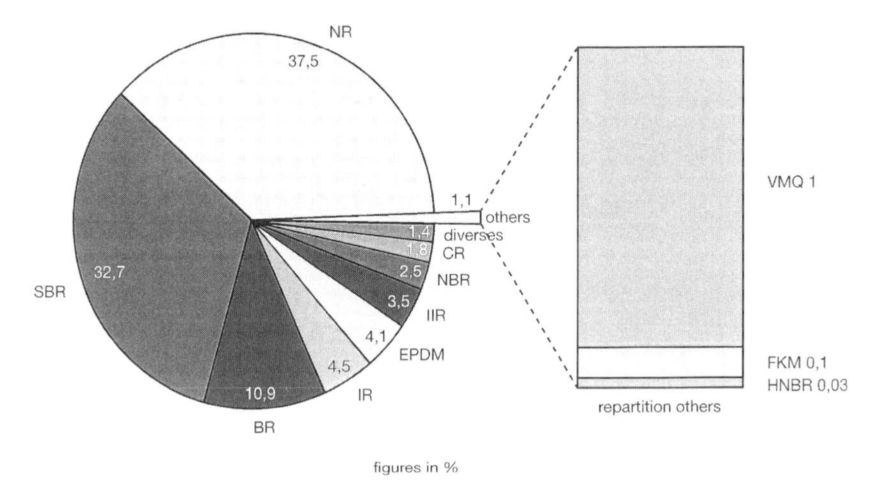

Fig. 1. World-wide rubber consumption according to grade (total market in 1997: 16.6 million t/a) [1].

Difunctional dichlorosilanes are used as raw materials for the synthesis of silicone rubbers (see Fig. 2). The outstanding raw material in terms of quantities is dichlorodimethylsilane [2]. Dichloromethylvinylsilane and, especially for crosslinking agents, dichloromethylsilane are used in smaller quantities. High-transparency speciality polymers with glass transition temperatures below –100°C are produced from dichloromethylphenyl- or dichloro-diphenylsilane as copolymers. Dichlorotrifluoropropylmethylsilane is the educt for particularly swell-resistant polymers.

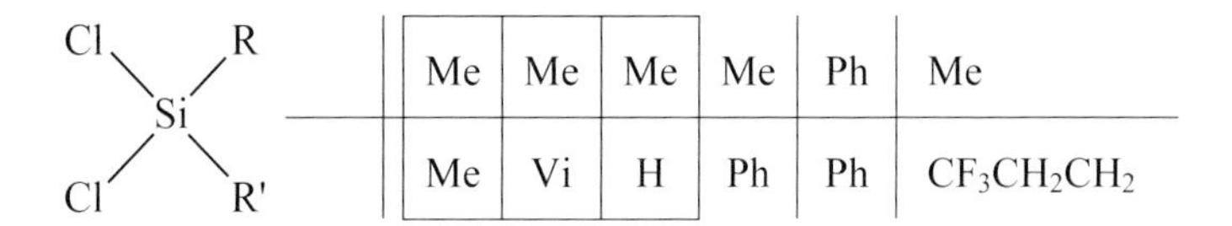

Me	Me	Me	Me	Ph	Me
Me	Vi	H	Ph	Ph	$CF_3CH_2CH_2$

Fig. 2. Organofunctional siloxanes for the production of silicone rubbers (Vi = vinyl).

Salient features of the properties of elastomer articles made from silicone rubber include:

- high thermal stability, even under long-term exposure, up to approx. +180°C
- resistance and flexibility even at low temperatures, up to approx. –50°C (even lower with speciality grades)
- constant rubber-mechanical properties over a wide temperature range (approx. –50°C to +180°C)
- outstanding ageing properties
- excellent weatherability based on high ozone and UV stability
- excellent dielectric properties over a wide temperature range (approx. –50°C to +180°C)
- good insulation properties (electrical and thermal)
- favorable fire behavior (low flammability, no toxic decomposition products)
- generally good physiological compatibility

These properties mean that silicone rubbers are in direct competition with organic speciality rubbers and fluororubbers. In quantity terms, the various properties come under the definition of a performance index. A performance index can be produced by evaluating various properties qualitatively using a grading system of marks and adding up the marks awarded. In one example, heat resistance, low-temperature flexibility, oil resistance, mechanical properties and ozone resistance were evaluated on a scale with marks ranging from 1 (unsatisfactory) to 10 (excellent) which were added together to give a performance index (see Fig. 3). If the performance index obtained is set against the price for one liter of rubber, the price/performance ratio of the various rubbers can be compared with one another.

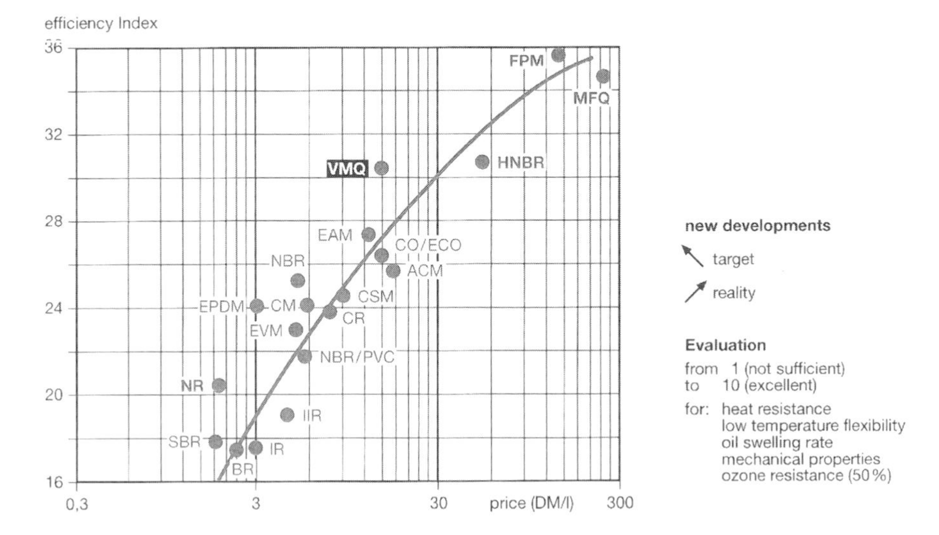

Fig. 3. Price/performance ratio of various rubbers [3].

Figure 3 shows that silicone rubbers using this performance evaluation come into the upper bracket. The prices for silicone rubbers, however, are in the middle bracket and thus are much lower compared with speciality rubbers such as HNBR, FPM and MFQ.

Strengths and weakness of a rubber can be especially good in a "spider" or "grid" or even "pole coordinate diagram". Relevant properties are evaluated in terms of quality as "good", "average" or "poor" or, if they can be measured, evaluated quantitatively by concrete data and plotted in a two-dimensional multi-coordinate system.

If we compare the properties of various speciality rubbers against one another, silicones (VMQ) exhibit particular strengths in thermal stability, low-temperature flexibility, processability and price, but are weak in terms of abrasion, acid resistance and oil swelling. The particular strengths of the hydrogenated nitrile–butadiene rubbers (HNBR) lie in their good results for tensile strength and abrasion; a typical weakness is — as with all organic rubbers — poor low-temperature flexibility. Fluororubbers (FKM) are especially heat and acid-stable and also have a very low swell rate in oils; the downside is low-temperature flexibility, processability and price, which compared to other rubbers is extremely high (see Fig. 4).

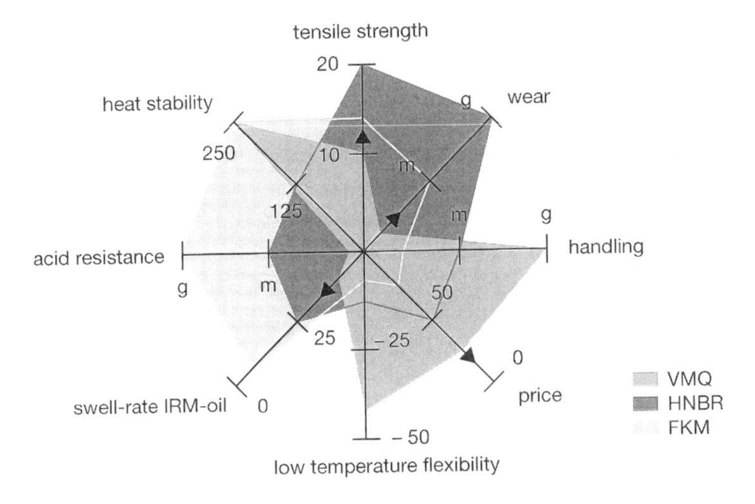

Fig. 4. Strengths and weaknesses of VMQ, HNBR and FKM.

Leaving aside room-temperature-vulcanizing one-component silicones (RTV-1K), which are used largely as jointing compounds, there are three different types of silicone rubbers:

- high-temperature-vulcanizing solid rubbers (HTV) or hot vulcanisates (HV), frequently termed high-consistency rubber (HCR) in Anglo-American literature
- high-temperature-vulcanizing liquid silicone rubbers (LSR) and
- room-temperature-vulcanizing two-component systems (RTV-2K)

This sub-division has developed historically on the basis of viscosity and crosslinking temperatures. The advent of addition-crosslinking solid rubbers and low-temperature-vulcanizing LSR grades (LTV) has meant that the boundaries are now fluid. The choice of name is largely decided in the industry on the basis of certain application criteria and is therefore scientifically not really comprehensible. The scope of the viscosity level for the individual silicone rubbers is thus quite broad as Table 1 clearly shows.

Table 1. The silicone rubber family [4].

Type	HTV	LSR	RTV-2K
Viscosity of compound [Pa·s] ($1 s^{-1}$)	200 000 ± 50 000	500	50 ± 40
Viscosity of polymer [Pa·s]	20 000	30	5 ± 4
content of reinforceing filler [%] (50 Shore A)	30	20	10
Typ of crosslinking[a]	R, A	A	A, C

[a] R = radical-induced, A = addition-cured, C = condensation-cured.

The following systems are normally used today for crosslinking silicone rubbers:

- polycondensation,
- radical polymerization, initiated
  — by peroxide decomposition or
  — by radicals from energy-rich radiation
- hydrosilylation reaction (polyaddition)

In the condensation reaction (see Scheme 1), the hydroxyl groups of the polymer react with the alcoholate groups of the crosslinking agent, a silica ester, by linking via siloxane bonds, releasing the corresponding alcohol. Polycondensation is used in RTV-1K and RTV-2K systems.

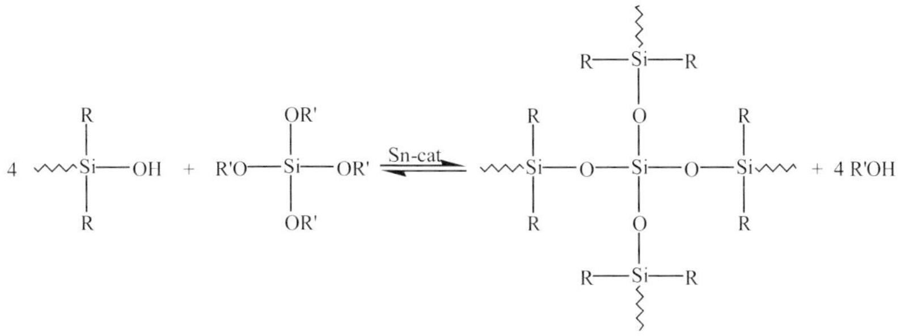

$$4\ \sim Si(R)_2{-}OH + Si(OR')_4 \underset{}{\overset{\text{Sn-cat}}{\rightleftharpoons}} Si[O{-}Si(R)_2\sim]_4 + 4\ R'OH$$

Scheme 1. Crosslinking by polycondensation [5].

In radical polymerization (see Scheme 2), radicals are formed in a first stage determining the rate. Examples used industrially for radical sources, apart from electron radiation, are aroyl peroxides such as bis(2,4-dichlorobenzoyl) peroxide or bis(2,4-methylbenzoyl) peroxide and alkyl peroxides such as dicumyl peroxide or 2,5-dimethyl-2,5-di-*tert*-butyl peroxyhexane. In the second stage, the actual crosslinking reaction, a radical addition based on the usual pattern for a radical chain reaction, takes place via the double bonds of the vinyl groups in the polymer. Radical polymerization is today used virtually exclusively for crosslinking solid silicones.

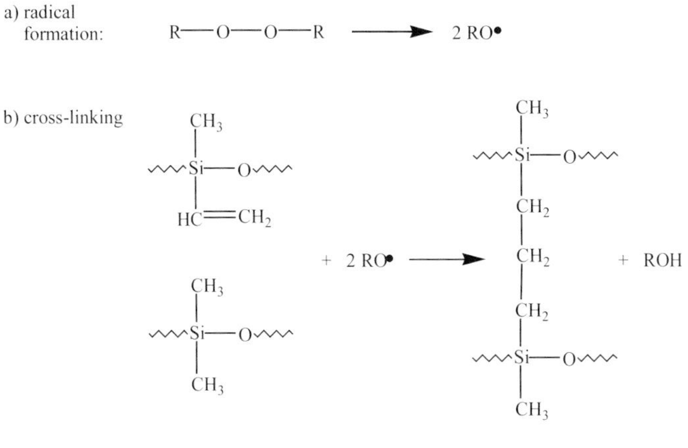

a) radical formation:

$$R{-}O{-}O{-}R \longrightarrow 2\ RO\bullet$$

b) cross-linking

$$\sim Si(CH_3)(HC{=}CH_2){-}O\sim + \sim Si(CH_3)_2{-}O\sim + 2\ RO\bullet \longrightarrow \sim Si(CH_3){-}CH_2{-}CH_2{-}CH_2{-}Si(CH_3){-}O\sim + ROH$$

Scheme 2. Crosslinking by radical polymerization [6].

The hydrosilylation reaction (see Scheme 3) is strictly speaking a platinum-catalysed addition of a Si-H group of silicon-hydrogen-functionalised crosslinking agents to the double bond of vinyl-group-functionalised silicone polymers.

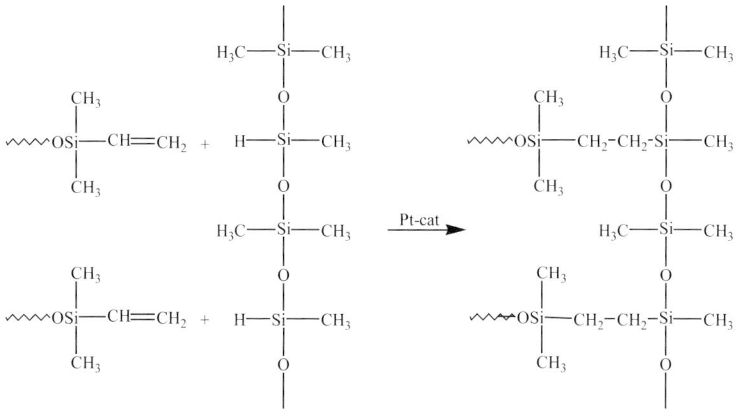

Scheme 3. Crosslinking by hydrosilylation [7].

The catalyst cycle [8] consists of four stages (see Scheme 4):

(1) The platinum–vinyl complex is formed by reaction of the polymer containing vinyl groups and a platinum compound , e.g. $PtCl_4$ or H_2PtCl_6.

(2) An Si–H bond of the crosslinking molecule reacts with the platinum–vinyl complex under oxidative addition; the oxidation stage on the platinum changes from 0 to +II.

(3) The resulting π complex is rearranged to an σ complex.

(4) With the reductive elimination of an elastomer molecule, the oxidation state of the platinum changes again from +II to 0.

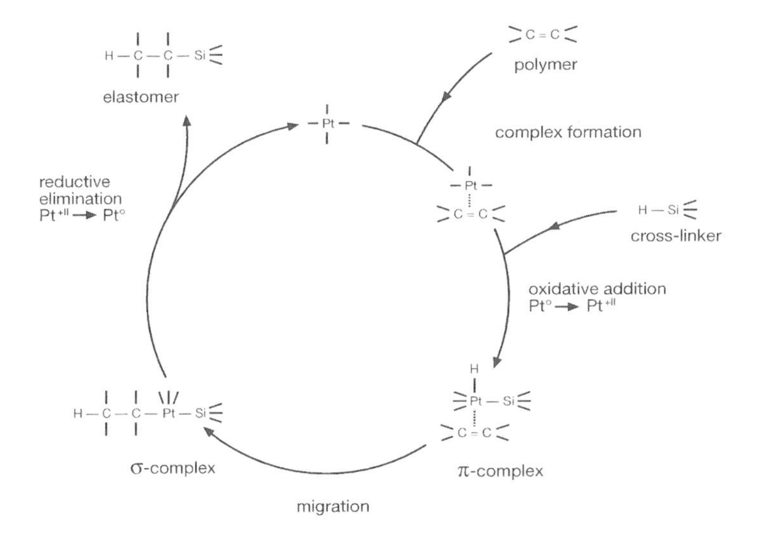

Scheme 4. Catalysis cycle for the hydrosilylation reaction [8].

As a comparison of torque as a function of time (see Fig. 5) shows, addition crosslinking is around 2.5 to 5 times faster than radical polymerization. This has a significant effect, especially with thin molded articles, on the production cycle time because with thin parts the crosslinking rate is directly proportional to the quantity of catalyst. In articles with a thick wall, the effect of heat transmission, which because of the thermal insulation effect of polymers is known to be poor, is predominant.

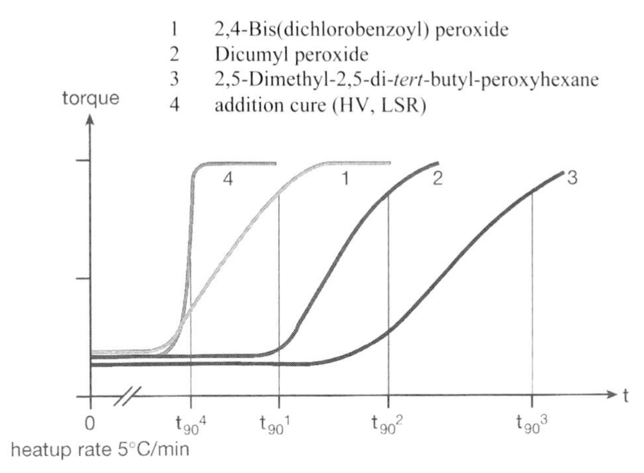

Fig. 5. Comparison of crosslinking rates of variously catalyzed silicone rubbers.

The viscosity and crosslinking rates affect the processing method (see Table 2). RTV-1K systems are used in open processing because of their slow reaction characteristics and are therefore for the most part injection-molded or spread-applied. RTV-2K systems can be slowly or rapidly crosslinked with low viscosity and for this reason are primarily used as embedding compounds. Pumpable LSR is extremely suitable for fully automated large-scale production of molded articles because of its high crosslinking rate. An average to high crosslinking rate with high viscosity makes HTV systems particularly suitable for extrusion.

Table 2. Comparision of processing methods for silicone rubbers.

	Cure Temperature			
	High		**Low**	
	HTV	**LSR**	**RTV-2K**	**RTV-1k**
Viscosity	High	Medium to low (pumpable)	Low	Medium
Processing	Press molding Injection molding Extrusion	Injection molding Coating (Extrusion)	Injection molding Casting Coating Dipping	Machine-free processing Injection molding Coating
Reaction type	Peroxide Addition	Addition	Condensation Addition	Condensation
Reaction speed	Medium to fast	Fast to very fast	Low to fasr	Low

Viscosity and reactivity therefore also directly affect the molding temperature (see Fig. 6) and the logarithmus of the viscosity itself is linear with the average chain length of the silicone polymers used.

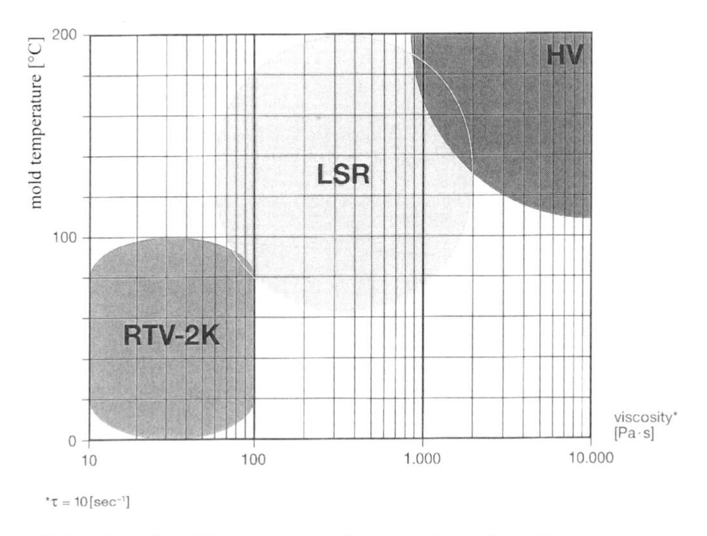

Fig. 6. Comparison of viscosity and molding temperature for processing various silicones.

Because of their high viscosity, solid rubbers are normally compounded in trough kneaders, on internal mixers or on the roll. Cables, hoses and profiles are produced using an extruder. Molded articles can be produced with HTV silicones in compression, transfer or injection molding processes (see Fig. 7).

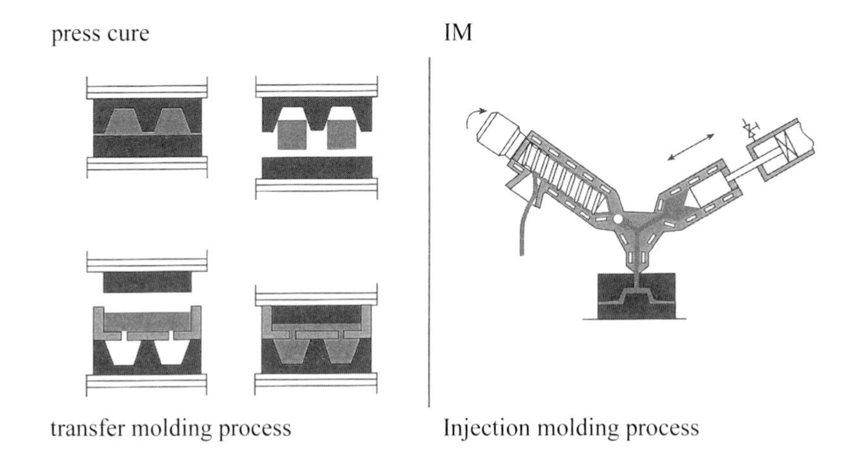

Fig. 7. Production of molded articles from solid silicone rubbers.

Traditional processing [9] (see Fig. 8) of liquid silicones uses the liquid injection molding process where the two components are conveyed and metered hydraulically with a two-component metering and mixing unit via metering pumps. The two components are combined in the mixing block and mixed in the static or dynamic mixer. The finished mixture is injection-molded via a screw or ram directly into a mold fitted into the injection-molding machine and vulcanized. Additives and pigment pastes can be metered as third or fourth components by volumetric metering before the mixing block or screw safely and accurately for the process in quantities between 0.3 and 5 %.

Bulky molded articles [10] can be delivered by special metering and mixing units (see Fig. 8.) where the large volume of rubber can be supplied surge-free over the whole metering period. Injection in this case is directly from the metering and mixing unit into the mold; a separate injection unit as in the production of small molded articles is not required here.

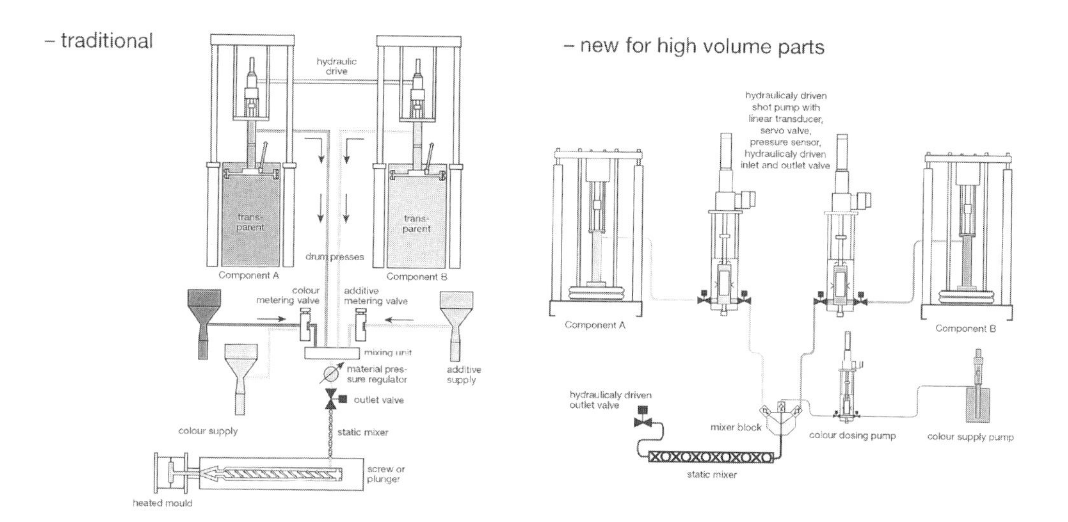

Fig. 8. Features of LSR processing [9,10].

The properties and processing methods discussed mean that silicone rubbers have a great many potential applications. The following examples will highlight this variety of applications:

- Baby care: pacifiers
  teething rings
  nipple shields

- Household applications: gaskets for sanitaryware
  seals for domestic appliances
  fittings for showerheads
  window profiles
  tubes and valves for coffee machines
  door profiles

- Medical technology: respiratory bellows and masks
  sealing and O rings
  membranes

tubes and tube couplings for catheters
injection plungers
stoppers for injection bottles
compounds for dental impressions

- Automotive:
  shock absorbers and rubber springs
  cable grommets
  radiator hoses
  vent flap seals
  O rings
  windscreen wipers
  headlight seals
  safety caps
  valve cover gaskets
  weather packs and strips
  central locking membranes
  spark plug sockets
  turbo charger hose

- Electronics:
  anode safety caps
  seals
  switch covers and membranes
  keypads for computers, telephones etc.
  electrical insulation tube

- Electrical:
  cable fittings (plugs, cuffs, sealing ends)
  insulators (hollow, rod-type suspending, rail insulators)
  surge absorbers
  safety cables

- Design:
  molding for prototyping
  centrifugal molding for the production of fashion applications

To summarize the points discussed, silicone rubbers are innovative. An example of this are inserts in LSR showerheads which do not become blocked and can be produced cost-effectively in the injection-molding process as composite elements with thermoplastic nylon 6.6 30 % glass fiber. Silicone rubbers, however, are also high performance, and this is evident particularly in the many applications where they have been used for years at high temperatures, for exsample seals in irons or other domestic appliances. And not least, silicone rubbers are also efficient. An example of this molds for prototyping without which cost-effective production of individual items would not in fact be possible.

There are other potential developments in the future for all the positive silicone properties. Certainly there is still a need for development in tensile strength, swelling resistance in oil and abrasion resistance. Price is rapidly becoming a major factor in all applications, and there is no doubt that the price of silicone rubbers will ease further in the future as we have seen in the past, provided that the raw materials used do not rise in price. And finally, the prices which for years have been falling have meant that the market for silicone rubbers (depending on the type of rubber) has grown by around 5 to 10 % a year.

References:

[1] Bayer AG, Rubber Business Group, in-house material.

[2] J. Burkhardt, "*Chemistry and Technology of Polysiloxanes*", in: *Silicones — Chemistry and Technology* (Eds.: G. Koerner, M. Schulze, J. Weis), Vulkan Verlag, Essen, **1991**, p. 21.

[3] W. Obrecht, J. P. Lambert, M. Happ, C. Oppenheimer-Stix, R. Dunn, R. Krüger, W. Nentwig, N. Rooney, R. T. LaFlair, U. U. Wolf, J. Duffy, G. J. Wilson, R. Steiger, A. Marbach, K. M. Diedrich, J. Ackermann, H. D. Thomas, S. D. Pask, H. Buding, A. Ostrowicki "*Rubber, 3. Synthetic*", in: *Ullmann's Ecyclopedia of Industrial Chemistry, A23*, VCH, Weinheim, **1993**, p. 239.

[4] H. Steinberger, "*Flüssiger Siliconkautschuk, immer noch ein Zukunftswerkstoff*" in: *Fachtagung Siliconkautschuke 22.–23. Oktober 1997, Würzburg*", Süddeutsches Kunststoff-zentrum, Würzburg, **1997**.

[5] W. Hechtl, "*Chemistry and Technology of Room-Temperature-Vulcanizable Silicone Rubber*", in: *Silicones — Chemistry and Technology*, (Eds.: G. Koerner, M. Schulze, J. Weis), Vulkan Verlag, Essen, **1991**, p. 45.

[6] H. G. O. Becker, W. Berger, G. Domschke, E. Fanghänel, J. Faust, M. Fischer, F. Gentz, K. Gewald, R. Gluch, R. Mayer, K. Müller, D. Pavel, H. Schmidt, K. Schollberg, K. Schwetlick, E. Seiler, G. Zeppenfeld, *Organikum*, VEB Deutscher Verlag der Wissenschaften, Berlin, **1976**, p. 196.

[7] D. Wrobel, "*Structure and Properties of Hot-Vulcanized Silicone Rubbers*", in: *Silicones — Chemistry and Technology*, (Eds.: G. Koerner, M. Schulze, J. Weis), Vulkan Verlag, Essen, **1991**, p. 61.

[8] A. W. Parkins and R. C. Poller, "*An Introduction to Organometallic Chemistry*", Macmillan, London, **1986**, 241.

[9] K. Pohmer, G. Schmidt, H. Steinberger, T. Bründl, T. Schmidt, *Kunststoffe* **1997**, *87*, 1396; *Kunststoffe Plast Europe*, **1997**, October/46.

[10] K. Pohmer, N. Spirig, *Kunstoffe-Synthetics*, **1998**, 01/12.

The Fine Art of Molding: Flexible Molds of RTV-2 Silicone Rubber

Georg Kollmann

Business Unit Elastomers / Business Team Technical Specialities
Wacker-Chemie GmbH, D-84480 Burghausen, Germany
Tel.: Int. code + (8677)83 3436 — Fax: Int. code + (8677)83 5735
E-mail: georg.kollmann@wacker.de

Keywords: RTV-2 / Silicone Rubber / Molding / Mold Making / Negative Molds / Copies / Reproduction Materials

Summary: Invented in the 1950s, room-temperature-vulcanizing two-component (RTV-2) silicone rubbers have found many different applications since. One of the most important is their use as a material for making flexible negative molds, which are perfectly suitable to obtain faithful copies of even intricate models out of the common reproduction materials. Due to their easy processing, high elasticity, short-term resistance to temperatures up to 350 °C, excellent release properties and, last but not least, only minor physiological and environmental effects in comparison with other flexible mold materials, RTV-2 silicone rubbers have established themselves world-wide as the top material for making flexible negative molds. This paper will first focus on the chemistry and special features of RTV-2 silicone rubber, and then offer a closer look at molding techniques and applications.

Introduction

Two-component silicone rubber that vulcanizes at room temperature (RTV-2 silicone rubber) was invented in the 1950s. Since that time it has been developed into high-tech materials that have found many different applications. The most important are:

- Potting or coating of electric and electronic components
- Molded parts such as cable joints and terminations, high-voltage insulators or pads for the transfer-printing process
- Flexible negative molds.

The term "molding" used within the scope of this paper means "making faithful copies of models (patterns) out of reproduction materials with the intermediate step of making a negative mold".

The model for the mold can be an original, a preliminary version of the final work in a different material such as modeling clay, wax, plasticine, a photosensitive resin (stereolithographic master), a laser-sintered thermoplastic material etc, or itself a copy.

Constituents of RTV-2 Silicone Rubber

The most importent constituents are:

- silicone polymers
- silicone crosslinkers
- catalysts
- cocatalysts
- inhibitors
- fillers
- softeners
- pigments
- additives

The polymers used in RTV-2 silicone rubbers are mostly linear, i.e. non-branched long-chain polydiorganosiloxanes formed by alternating silicon and oxygen atoms. Methyl groups saturate both remaining bonds at the silicon. Reactive groups terminate the chains (Fig. 1).

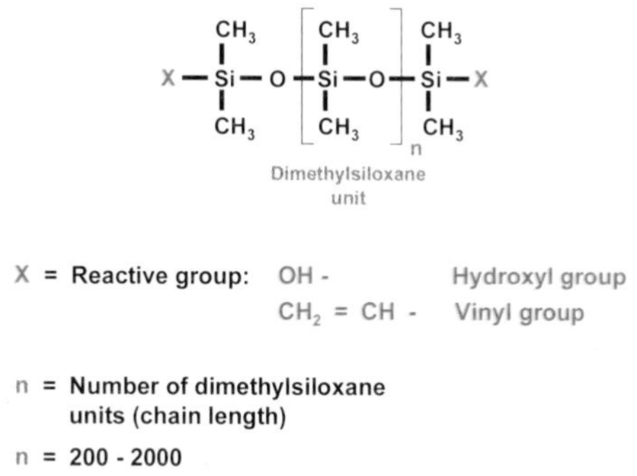

$$X-\underset{CH_3}{\overset{CH_3}{Si}}-O-\left[\underset{CH_3}{\overset{CH_3}{Si}}-O\right]_n-\underset{CH_3}{\overset{CH_3}{Si}}-X$$

Dimethylsiloxane unit

X = Reactive group: OH - Hydroxyl group
$CH_2 = CH$ - Vinyl group

n = Number of dimethylsiloxane units (chain length)
n = 200 - 2000

Fig. 1. Chemical structure of α,ω-functional linear polydimethyl siloxanes.

From the reaction of such polymers with silicone crosslinkers, which need to have at least three reactive groups, three-dimensional networks with rubber-like elasticity are obtained. The chemical nature of the silicone crosslinker to be used depends on the crosslinking system, further details of which are given in the next Chapter.

To achieve a reaction rate which is sufficiently high at room temperature, the use of catalysts and, if required, cocatalysts is imperative, the chemical nature of which also depends on the crosslinking system.

The reactivity of the catalyst is often adjusted for practical needs by adding special substances that reduce the system activity, so-called inhibitors.

Since the mechanical strength of pure polydimethylsiloxane/crosslinker networks is far too low for most technical applications, fillers are added, which are usually classified as either "inert", i.e. not reinforcing, or "active", i.e. reinforcing.

The surface of inert fillers such as powdered quartz, diatomaceous earth, calcium silicates, calcium carbonate or iron oxides, just to mention the most frequently used, does not show major chemical or physical interaction with the polydimethylsiloxane network. Thus, this kind of filler just contributes hardness, swelling resistance to solvents and a certain, albeit rather low, level of mechanical strength to the system.

By contrast, active fillers, exclusively selected from different kinds of pyrogenic silica with BET surfaces of 90–350 m^2/g, are able to strongly interact with both themselves and the dimethylsiloxane polymers, which is due to their high surface energy and ability to form hydrogen bridges. This interaction results in a marked increase in mechanical strength, mainly abrasion and tear resistance, of the cured rubber.

Further possible constituents of RTV-2 systems are:

- silicone fluids, i.e. polydiorganosiloxanes that have no reactive groups and are used to reduce the viscosity of the uncured rubber and/or the hardness of the vulcanizate
- pigments used to either color the rubber or, if the minor component is pigmented, to allow a visual control of the mixing process of both components
- additives such as heat stabilisers, antistatic agents etc.

Crosslinking Systems of RTV-2 Silicone Rubber

There are two different crosslinking systems available to vulcanize RTV-2 silicone rubbers:

- Condensation cure
- Addition cure

Condensation Cure

The dimethylsiloxane polymers in condensation-curing systems have hydroxyl functions as terminal reactive groups.

The crosslinkers are mostly alkoxysilanes (silicic-acid alkyl esters) (Fig. 2).

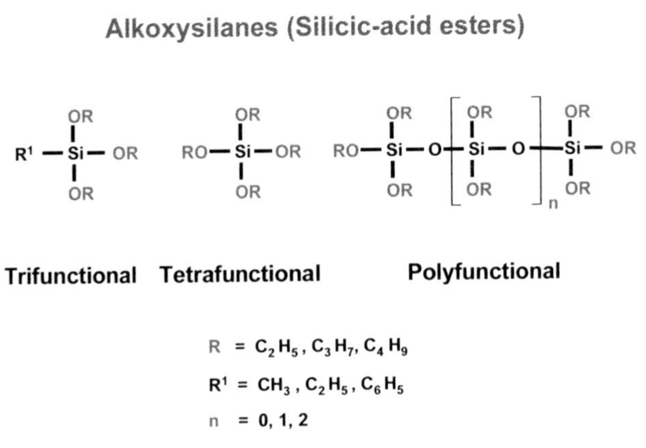

Alkoxysilanes (Silicic-acid esters)

$R^1-Si(OR)_3$ — **Trifunctional**

$Si(OR)_4$ — **Tetrafunctional**

$RO-Si(OR)_2-O-[Si(OR)_2-O]_n-Si(OR)_2-OR$ — **Polyfunctional**

$R = C_2H_5, C_3H_7, C_4H_9$

$R^1 = CH_3, C_2H_5, C_6H_5$

$n = 0, 1, 2$

Fig. 2. Chemical structure of crosslinkers used in condensation-curing RTV-2 silicone rubber systems.

As catalysts, almost exclusively dialkyltin dicarboxylates are used. Since the catalytically active species is the dialkyltin dihydroxide, the original catalyst compound must be first hydrolysed to become effective. It is therefore imperative that a certain level of moisture, i.e. water in vapor form as a cocatalyst, is present in both the rubber system and the ambient atmosphere.

The characteristics of a condensation-curing system are:

- The reaction is isothermal.
- A volatile alcohol is formed as condensation product.
- A mass defect occurs due to the volatilization of the alcohol, which results in a "chemical" shrinkage of the cured rubber in the range of 0.2–2 % (linear), depending on the concentration of reactive groups.
- The curing reaction reverts at temperatures exceeding 90°C, which makes it impossible to cure such a system by applying heat.
- Inhibition of cure is only possible in the case of a lack of moisture in the rubber or the ambient atmosphere.

The mechanism of condensation cure is given in Scheme 1.

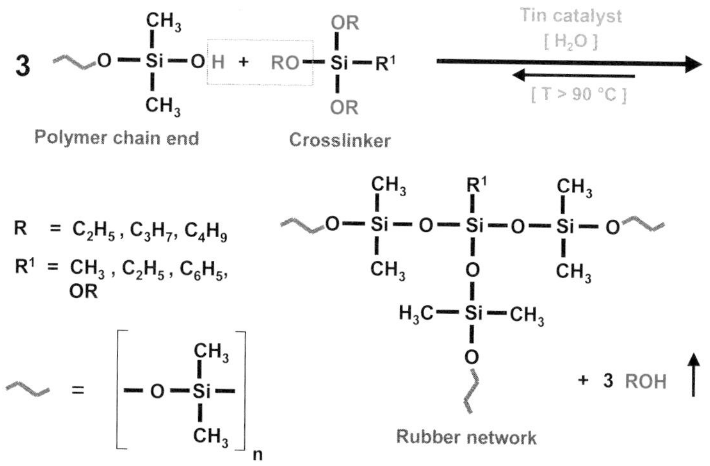

Tin catalyst [H_2O]
[T > 90 °C]
Polymer chain end
Crosslinker
R = C_2H_5, C_3H_7, C_4H_9
R^1 = CH_3, C_2H_5, C_6H_5, OR
+ 3 ROH
Rubber network

Scheme 1. Reaction mechanism of condensation-curing RTV-2 silicone rubber systems.

Addition Cure

The dimethylsiloxane polymers in addition-curing systems have vinyl functions as terminal reactive groups.

The crosslinkers are mostly linear polysiloxanes, with hydrogen functions positioned in the chain (Fig. 3).

As catalysts, almost exclusively organoplatinum complexes are used, which require no further activation.

The characteristics of an addition-curing system are:

- The reaction is isothermal.
- No volatile reaction product is formed.
- Hence no mass defect occurs. Only very small "chemical" shrinkage of the vulcanizate in the range of < 0.1 % (linear) result, due to minimal volume contraction during crosslinking.
- The curing reaction does not revert at temperatures exceeding 90°C. It is therefore possible to cure such a system at temperatures up to 200°C.
- Inhibition of cure is caused by various widley used substances such as sulfur compounds, amines, organometallic compounds, unsaturated hydrocarbons etc.

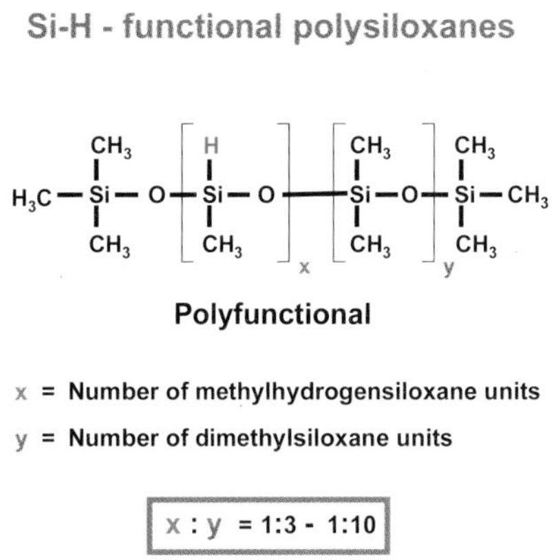

Fig. 3. Chemical structure of crosslinkers used in addition-curing RTV-2 silicone rubber systems.

The mechanism of addition cure is given in Scheme 2.

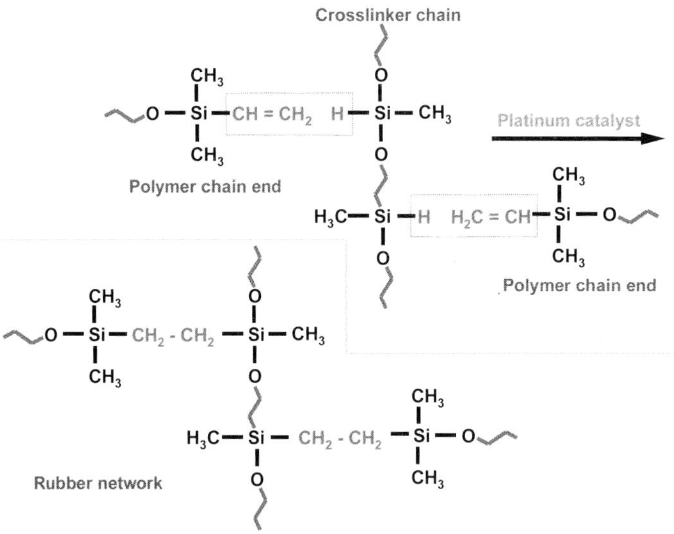

Scheme 2. Reaction mechanism of addition-curing RTV-2 silicone rubber systems.

Properties of RTV-2 Silicone Rubber

To obtain stable systems, it is imperative to split up the rubber constituents into two components in such a way that unwanted reactions will not occur during their storage.

Condensation-curing rubbers consist of a "rubber base" and a "curing agent" ("hardener", "catalyst").

Polymers, fillers, softeners and some additives are normally contained in the rubber base, whilst crosslinkers and tin catalysts, as well as extenders and dyes or pigments (if added for visual mixing control), are included in the curing agent.

By contrast, the two components of addition-curing systems, usually named "A" and "B", never contain crosslinker and catalyst in the same component, as this would result in the evolution of hydrogen gas.

As a consequence of the different compositions of the components, different application properties for the two curing systems are also obtained.

Since crosslinker and catalyst in condensation-curing systems are both contained in the curing agent, every alteration of its proportion will automatically result in a simultaneous variation of the quantities of both constituents. Thus, increasing or reducing the proportion of the curing agent with the objective of just enhancing or decreasing the reactivity of the system means that the concentration of crosslinker also and, consequently, the crosslinking density will inevitably be altered correspondingly, which might affect the vulcanizate properties. Any variation of the proportion of the curing agent should therefore only be made within the limits that are recommended by the manufacturer. In many cases, however, curing agents with different activities are available.

Due to the presence of crosslinker and catalyst in different components, addition-curing RTV-2 silicone rubbers must only be mixed in the specified weight ratios to obtain safe and reproducible cure, i.e. the prescribed mixing ratio of the components must never be varied. It is, however, possible to alter the reactivity of the system by varying the processing temperature or adding additional catalyst or inhibitor, by which the properties of the cured rubber will not be markedly affected.

The effect of temperature on the curing time is relatively small for condensation-curing rubbers, i.e. a system with low reactivity at room temperature, and thus with long pot life, cannot be accelerated considerably by applying heat, whereas addition-curing rubbers, even if they have long pot lives at room temperature, will cure within a few minutes at temperatures near to 200°C.

In Table 1, the fundamental differences between condensation and addition-curing RTV-2 silicone rubber are summarised again.

Table 1. Fundamental differences between condensation and addition-curing RTV-2 silicone rubber.

	Condensation	**Addition**
Components	Rubber base + curing agent	Components A + B
Crosslinker / catalyst	Both part of curing agent	In different components
Metering of components	Variable within limits; curing agents with different reactivity available	Mixing ratio must not be varied
Effect of temperature	Relatively small: long pot life = long curing time	Large: long pot life at RT/short curing time at high temperatures
	Reversion, hence no curing possible at temperatures > 70°C	No reversion, hence curing possible at temperatures up to 200°C
Linear chemical shrinkage	0.2–2 %, depending on chain length of polymers and proportion of crosslinker	< 0.1 %
Inhibition of cure	Lack of moisture	Various widely used substances

For RTV-2 silicone rubbers used for mold making, there are further distinguishing features, besides the crosslinking system, that are very important:

- consistency of the ready-to-use material
- pot life and curing time
- hardness of the cured rubber
- tear strength of the cured rubber
- resistance of the cured rubber to the reproduction materials processed in the rubber molds

The possible consistencies of RTV-2 silicone rubber are listed in Table 2 together with their corresponding ranges of viscosity. The higher the value of viscosity, the less flowable the system.

Table 2. Consistencies of RTV-2 silicone rubber.

Consistency	**Range of viscosity [mPa s]**
Pourable	1 000 – 150 000
Spreadable	200 000 – 800 000
Spreadable, non-sag	> 1 000 000
Kneadable	≫1 000 000

Pourable compounds with standard mechanical strength, i.e. without high tear strength, usually have viscosities in the range of 5 000 to 40 000 mPa s and are characterized by both very good flow and self-deaerating performance.

By contrast, pourable compounds that produce vulcanizates with high tear strength, normally have viscosities in the range of 20 000 to 150 000 mPa s and reduced flow. The air entrapped during the mixing process can only be completely removed by evacuating the mixture.

Whilst spreadable systems will, although slowly, sag and flow off non-horizontal surfaces, spreadable non-sag rubbers will not, at least in a thickness of 10 mm or less, which allows rubber layers to be applied to vertical surfaces if, e.g., skin molds are to be made.

Pot lives and curing time may vary, depending on the system reactivity, within wide limits. Typical figures for the commonly used classes of reactivity are given in Table 3.

Table 3. Pot life and curing time of RTV-2 silicone rubber.

Reaction rate	Pot life [min]	Curing time (demoldable after) [h]
"Fast"	2 – 10	0.5 – 2
"Normal"	20 – 60	4 – 12
"Slow"	90 – 180	15 – 24

The vulcanizate hardness of the rubber grades used for mold making is exclusively specified according to the indentation standard method Shore A (measuring range: 0–100 points). The higher the Shore A value, the harder the rubber. Table 4 shows typical figures for the commonly used classes of hardness. Most mold-making rubbers have a Shore A hardness between 8 and 70 points.

Table 4. Hardness of RTV-2 silicone rubber.

Vulcanizate hardness	Range of indentation hardness Shore A [Points]
"Soft"	0 – 30
"Medium"	31 – 50
"Hard"	51 – 80

One of the most important properties of RTV-2 rubbers used for mold making is the tear strength of the cured rubber, which describes its performance in the case of mechanical damage and simultaneous tensile load. Whilst the "standard" qualities will easily be torn in two in such case, even if thick layers are involved, the so-called "high-strength" grades will resist by producing a "knotty" tear, i.e. tear propagation stops after a short distance because the tearing force is dissipated within the rubber network.

"High-strength" RTV-2 rubber systems always contain active, i.e. reinforcing fillers that inevitably cause, compared to the grades with "standard" mechanical properties, increased viscosity and therefore reduced flow and self-deaeration tendency. Therefore, pourable "high-strength" systems with very low viscosity and good self-deaeration performance are not possible.

Processing of RTV-2 Silicone Rubber for Mold Making

In mold making, the following processing steps are to be taken:

- preparing the model
- preparing the components
- metering the components
- mixing the components
- removing the air entrapped during mixing
- making the mold

Preparing the Model

To prevent the mold from adhering to the model, it is necessary to seal porous or, in the case of condensation-curing rubber, moisture-absorbing surfaces as well as surfaces consisting of silicate materials such as glass or porcelain prior to applying the silicone rubber. Suitable coating materials are polyvinyl alcohol, methyl cellulose (wallpaper adhesive), soft soap or Vaseline.

If addition-curing grades are to be used, it must be ensured by preliminary tests that contact inhibition will not occur. If inhibition is found, appropriate remedy has to be taken by replacing the contact material or covering it with a non-inhibiting protective coating.

Preparing the components

It is indispensable to thoroughly stir up pourable RTV-2 silicone rubber bases and components, with the exception of the transparent or translucent grades, prior to use in order to redisperse homogeneously any inactive filler that might have settled during storage. Otherwise severe curing problems may result.

"High-strength" grades, which tend to thicken somewhat on long storage due to their content of active filler, should be stirred vigorously to restore their original consistency.

Metering the components

It is absolutely essential to meter the components accurately, since only by precisely following the mixing ratio is it possible to obtain reproducible pot lives and curing times and, even more important, vulcanizates whose properties come up to specification.

Metering can be done either by weight using a balance, or by volume using a calibrated vessel, pipette, disposable syringe or an automatic mixing and metering device.

Mixing the components

It is imperative to mix the two components, rubber and curing agent ("catalyst") in the case of condensation-curing grades and A and B in the case of addition-curing grades, thoroughly to obtain a homogeneous mixture.

Pourable and spreadable products can be mixed by hand with a spatula or, for larger amounts, with a mechanical stirrer or automatic mixing and metering equipment.

The kneadable compounds are mixed on a triple roll mill or in a kneader.

Removing the air entrapped during mixing

The mixing process, If not carried out under vacuum, unavoidably introduces a certain amount of air into the rubber mix.

To obtain vulcanizates without any air bubbles, pourable grades have to be deaerated in a desiccator or vacuum cabinet at reduced pressure.

By contrast, spreadable, non-sag or kneadable compounds cannot be deaerated by evacuation. In this case, first a thin, bubble-free layer of a pourable grade is applied to the surface of the model followed by the air-bubble-containing spreadable or kneadable compound.

Making the mold

Free-flowing RTV-2 silicone rubbers are, after deaeration, just poured over the model, whereas spreadable and non-sag compounds are applied using a spatula. Kneadable grades in the form of a sheet are pressed on the model surface by hand or using a roller.

Molding Techniques using RTV-2 Silicone Rubber

The earliest molds that have been found are rigid and made of mineral materials such as clay or plaster. Even today, rigid molds made of these materials or organic resins are frequently used if models without undercuts are to be reproduced.

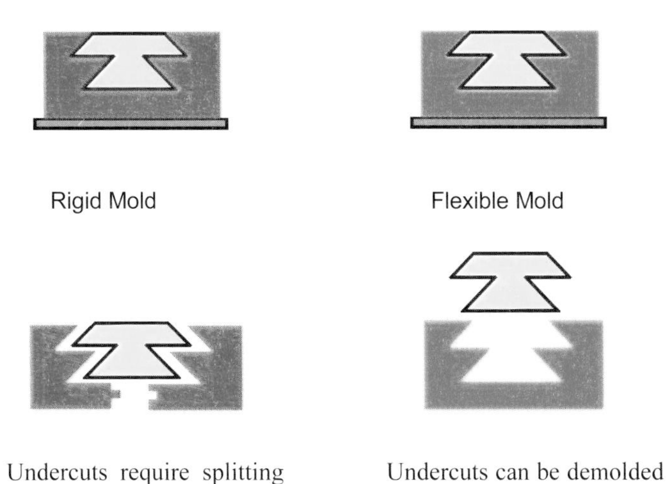
Rigid Mold
Flexible Mold
Undercuts require splitting or destruction of the mold for demolding
Undercuts can be demolded by just stretching the mold

Fig. 4. Advantages of flexible molds.

The development of molds exhibiting flexible properties was a great step forward in molding techniques. It was in medieval times, when mold makers discovered that gelatin extracted from animal bones could be used for flexible molds. Now, for the first time, it was possible to remove a mold from a model with even deeper undercuts without destroying the model or being forced to split up the rigid mold into several parts. The advantage of flexible molds is illustrated by Fig. 4.

While gelatin has only limited elasticity and very poor mechanical strength, natural rubber is highly resilient, and its appearance in Europe as a mold-making material in the early 19th century prompted a surge of molding activity. Stabilised with ammonia, it is used as a watery latex emulsion, which is brushed onto the surface of the model. By applying several layers, extremely strong, elastic molds can be made, which, however, have drawbacks. They deteriorate quickly in storage, turning brittle within a few months as the latex is oxidized by the air.

Modern chemistry has created a number of new mold-making materials with improved performance. These include organic resins for rigid molds as well as thermoplastic vinyl rubbers, polyurethane elastomers and, finally, RTV-2 silicone rubbers. The latter have become more and more used, in spite of their relatively high price, because they are the only materials that offer the ideal combination of properties essential for high-performance molding applications.

The outstanding properties of RTV-2 silicone rubbers as mold-making materials are:

- very easy to process, as neither heat nor pressure is required for cure, i.e. no costly equipment has to be used
- extremely flexible when cured and, in addition, long-term elastic.
- high-strength grades available, which are resistant to tearing and tear propagation
- extreme reproduction fidelity of surface details
- excellent release properties in combination with almost all materials commonly used for reproductions
- outstanding heat resistance (200°C for long-term, 350°C for short-term exposure)
- quite safe regarding handling precautions and environmental compatibility, e.g. compared with polyurethanes

It would, of course, go far beyond the scope of this paper to deal with all the established molding methods. The following exposition will therefore be limited to the most important and frequently used mold-making techniques.

One-part block mold

Objects such as stucco work, seals or medals where only one side is to be copied, and which are relatively small, are usually reproduced using a one-part block mold. This involves placing the model in a box and covering it with free-flowing silicone rubber (pouring technique). Alternatively it is pressed into a pre-formed block of kneadable silicone rubber (impression technique).

An important alternative to making two-part block molds is to make a one-part mold and cut it along a parting line after curing to obtain a two-part mold. This procedure is nearly exclusively used to make the molds used in the vacuum casting process within the scope of "rapid prototyping". As there is a risk of damaging the model or of cutting the parting line incorrectly, transparent silicone rubber grades are preferably used.

Block molds have the advantage that they are not very labor-intensive and the mold has inherent stability, so no support is needed. Drawbacks of block molds are the relatively large quantities of silicone rubber required and the mold's weight.

The detailed steps of making a two-part block mold by cutting a one-part block mold open are given in Fig. 5.

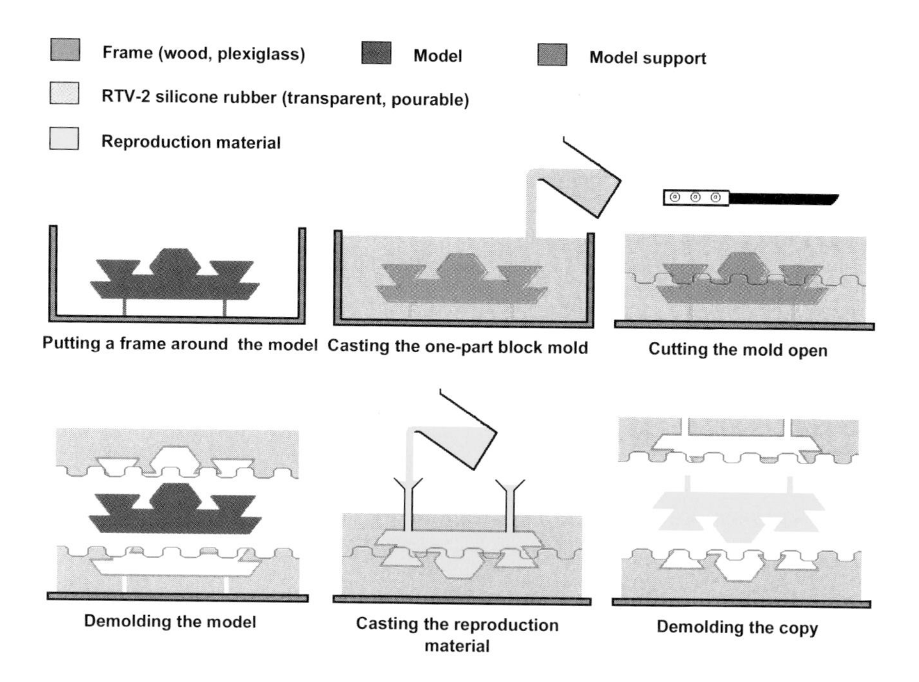

Fig. 5. Two-part block mold made by pouring a one-part block mold and cutting it open along a parting line.

One-part skin mold

If only one side of large models needs to be molded, or if the model has deep undercuts, block molding is impractical or impossible. The alternative is to make a one-part skin mold between 5 and 10 mm thick.

The advantages of a skin mold lie in its lower weight and material consumption; it is, however, much more labor-intensive.

Skin molds can be made either by the pouring or by the spreading technique. If the model is not too large and is suitable for horizontal molding, the method of choice is usually pouring the skin mold.

The surface of the model is covered with a uniform layer of clay or plasticine. The thickness of the spacing layer determines the thickness of the final mold. The next step is to build up a rigid layer on top of the spacer. This support is either poured (plaster or casting resin) or built up in layers by spreading the material (polyester, polyurethane or epoxy resin reinforced with fiberglass matting). Once the support has set, it is separated from the spacer, which is then discarded. The support, which must contain air-escape and feeding holes, is now fitted over the model. The silicone rubber is poured into the hollow space between the support and the model and allowed to cure. The support is then lifted off and laid on its back. The skin mold is peeled off the model and placed in the support, with the sprues locating it in exactly the right position. Fig. 6 shows the details of this technique.

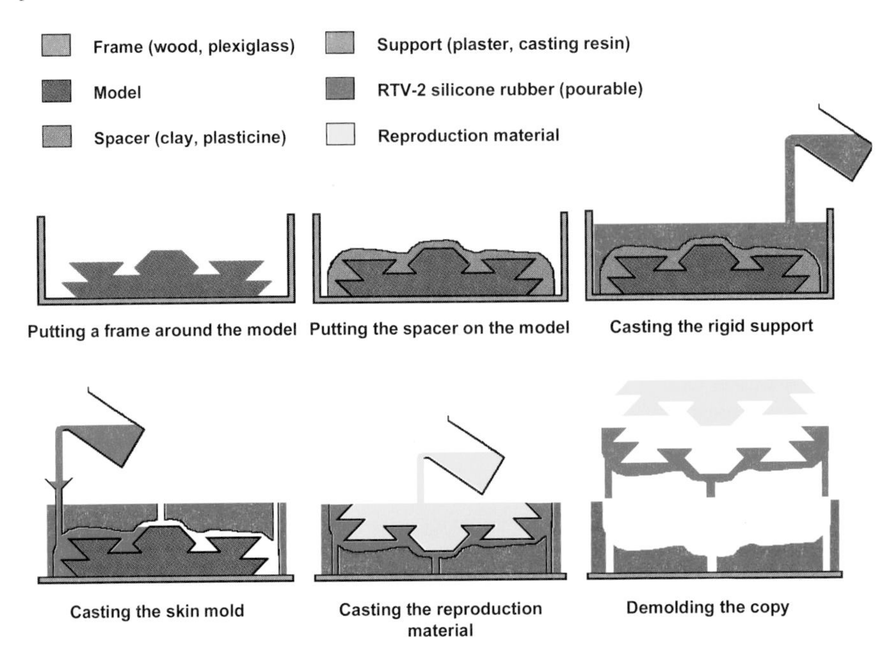

Fig. 6. One-part skin mold made by the pouring technique.

If the model is large, or if it must remain vertical, the spreading technique has to be used. This is a layering method, where the first step is to apply a thin layer of pourable silicone rubber to the model using a brush. This "fine layer", which must be free from air bubbles, ensures that all the details of the original are reproduced with accuracy. When the first layer has partially cured, a top layer of non-sag silicone rubber is applied using a spatula. The surface of the top layer must be smoothed and locks constructed. The last stage is to make the support using jute strips soaked in plaster or a fiberglass-reinforced organic resin. The demolding stages are identical to those described above for the pouring technique.

Other molding techniques

Besides the techniques mentioned above, others are known to make two-part and multi-part block or skin molds by pouring, spreading or impression. Since these methods are much more complicated and labor-intensive, they are replaced by the simpler ones wherever possible.

Reproduction Material

The maximum number of copies that can be made from a silicone rubber mold depends primarily on the reproduction material used.

The most common reproduction materials suitable to be processed in molds made of RTV-2 silicone rubber are:

- Mineral materials such as plaster, white cement, concrete, synthetic stone or ceramics.
- Waxes
- Casting resins such as unsaturated polyester resins, polyurethane resins and foams, epoxy resins and, within limits, acrylic resins
- Low-melting metal alloys with melting points below 350°C.

The resistance of the molds to the reproduction material may vary within wide limits. The maximum number of copies that can be obtained depends on the

- chemical nature of the reproduction material used
- setting temperature of the reproduction material
- volume and shape of the casting
- physical properties of the silicone rubber
- use of additional release agents
- frequency of replication
- time the casting is left in the mold before it is demolded
- kind of mold, i.e. the mold-making technique used
- maintenance given to the mold

Characteristic data on how many copies can be obtained with the individual reproduction materials are given in Fig. 7.

As can be taken from the chart, casting resins cause much more damage to silicone rubber molds than do mineral materials and waxes, which is due to their chemical aggressiveness enhanced by the exothermal setting reaction.

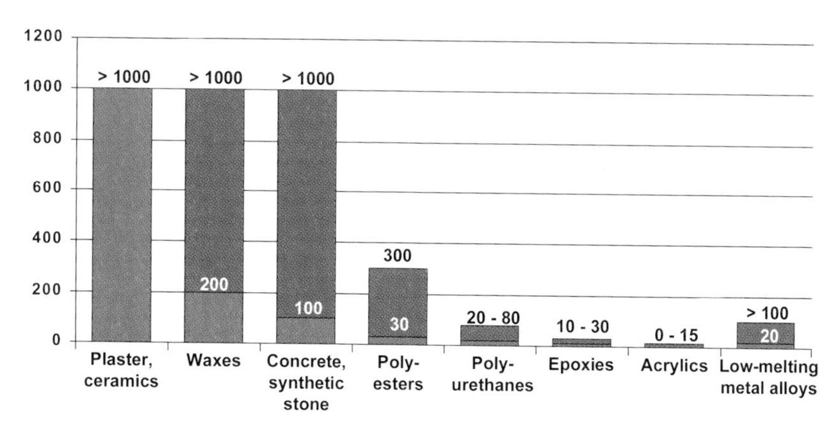

Fig. 7. Number of copies that can be obtained using the different reproduction materials.

Though thermal decomposition of silicone rubber already starts at 200°C, yet it is possible to obtain a considerable number of copies using low-melting metal alloys. This is due to the formation of a layer of silica at the mold surface produced by pyrolysis of the silicone rubber. Thanks to its very low thermal conductivity, this layer acts as a protective coating that prevents the mold from further short-term thermal decomposition.

Main Molding Applications

The most important fields of application for molds made of RTV-2 silicone rubber are:

- Duplication of prototype master models (made by special techniques such as laser stereolithography, laser powder sintering, LOM etc.) in polyurethane or epoxy resins by the vacuum casting process within the scope of "rapid prototyping".
- Serial or small-scale production of
  - stucco parts and facade ornaments out of plaster
  - decorative parts out of cast concrete or synthetic stone
  - ornamental door panels and window frames out of polyester or polyurethane resins
  - fronts of kitchen and bath furniture out of polyester or polyurethane resins
  - reproduction furniture and ornamental mirror and picture frames out of polyurethane resins or rigid foams
  - hand-painted figurines and sculptures out of polyester resins
  - ornamental candles out of wax
  - costume jewellery and decorative buttons out of polyester or epoxy resins
  - imitation leather (the surface structure of original leather is copied by a layer of flexible polyurethane which is then transferred to the surface of cheap, low-grade leather)
- Reproductions for museums and artists
  - duplicates of priceless originals kept in the strong room
  - copies to be exhibited at other museum sites
  - limited numbers of copies of figurines or sculptures created by artists
- Production molds for chinaware and sanitary ware out of plaster.
- Wax copies of models required for the lost wax (investment) casting process for high-melting metals or alloys with melting points higher than 350 °C such as bronze, silver or gold that cannot be cast into molds of silicone rubber. (The wax copy made from the silicone rubber mold is encased in a high-temperature resistant mineral material such as fireclay. When this is fired, the wax melts and flows away, leaving a rigid negative mold. The molten metal is then poured into the mold cavity. For demolding the copy, the rigid mold has to be destroyed.)
- Galvanoplastics. (The surface of the silicone rubber mold is first made conductive by applying a coat of colloidal silver or of graphite. Then the mold is electroplated.)

Summary

For decades now, RTV-2 silicone rubber has been the top material available for mold making, which is proved by its continuously increasing consumption for this application (current global use is more than 20 000 tons per year). This is due to the unique combination of properties offered by this material:

- easy processing
- high elasticity
- resistance to high temperatures
- faithful reproduction of surface details
- excellent release properties and resistance to commonly used reproduction materials
- only minor physiological and environmental effects

Since molding applications are expected to become more and more important, the future of RTV-2 silicone rubber as a mold-making material can be considered very promising indeed.

Molecular Reactors Based on Organosilicon μ-Networks

Manfred Schmidt*, Olaf Emmerich, Christopher Roos, Karl Fischer,

Institut für Physikalische Chemie, Johannes Gutenberg-Universität
Jakob-Welder-Weg 11, D-55128 Mainz, Germany
Tel.: Int. code + (6131)393770 — Fax: Int. code + (6131)392970
E-mail: mschmidt@mail.uni-mainz.de

Frank Baumann, Bernd Deubzer, Johann Weis

Wacker-Chemie GmbH
Postfach 1260, D-84480 Burghausen, Germany

Keywords: Organosilicon μ-Networks / Molecular Boxes / Noble Metal Cluster / Molecular Reactors

Summary: Organosilicon μ-networks are utilized as nanoreactors consisting of single crosslinked molecules. Functionalized core–shell particles with a core comprising reactive Si–H moieties are utilized to reduce noble metal salts like $HAuCl_4$ to elemental gold clusters, which are topologically immobilized in the core of the networks. The preparation of hollow micro-networks leads to molecular containers which can be loaded with reactants.

Introduction

Nearly monodisperse organosilicon micro-networks are easily prepared by condensation of organotrimethoxysilanes $R–Si(OCH_3)_3$ (with R representing functional groups like hydrogen, methyl, vinyl, allyl, methacryl or benzyl chloride groups) in water in the presence of surfactants [1]. The resulting hydrosol is converted into the corresponding organosol by quantitative reaction of all "surface Si–OH" groups with a monofunctional methoxytrialkylsilane which is soluble in all common organic solvents like toluene, cyclohexane, THF, chloroform, etc. Subsequent addition of two monomers with different functional groups allows for the preparation of core–shell structures, because under appropriate reaction conditions no new particles are formed when the second monomer is added [2]. Likewise, trimethoxyalkylsilanes and dimethoxydialkylsilanes may be co-condensed in order to produce elastomeric micro-networks which do swell in good solvents depending on the crosslinking density [3]. Based on the organosilicon micronetworks described above molecular reactors are synthesized by two different strategies.

Active Micro-Reactors

In the present context active micro-reactors are defined as functionalized particles the functional groups of which constitute one of the reaction agents. This is realized by the synthesis of a core–shell micro-network with Si–H moieties in the core surrounded by an "inert" $Si\text{-}CH_3$ network shell. Addition of noble metal ions, such as Au^{3+}, Ag^{+}, Pt^{4+}, Pd^{2+}, leads to the formation of topologically trapped noble metal clusters in the core of the micro-networks because the SiH groups act as the reducing agent. Depending on the diffusion rate of the metal ions into the core of the micro-networks, which is controlled by the crosslinking density and thickness of the shell, the formation of a single cluster per particle (slow diffusion limit, Fig. 1) or the nucleation of several small clusters (fast diffusion limit, Fig. 2) is observed in analogy to the formation of metal clusters in block copolymer micelles [4, 5]. Also, apparently empty, i.e. metal-cluster-free particles are found. It is not yet clear why no clusters are found in a minor, sometimes also major, fraction of the micronetworks. The potential application of noble-metal-filled organosilicon micronetworks in catalysis is the subject of current investigations.

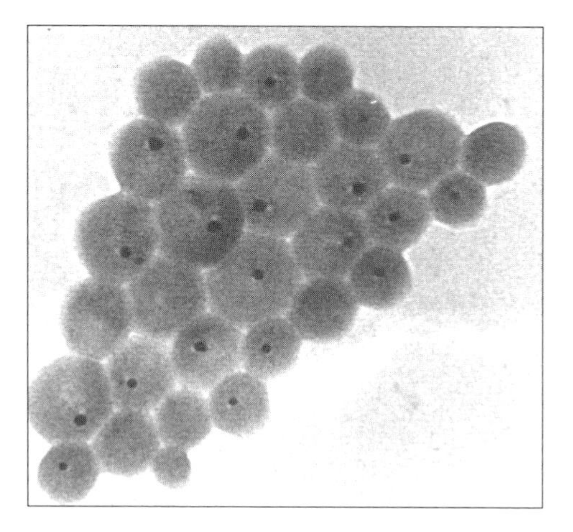

Fig. 1. TEM picture of one gold cluster topologically trapped within a micro-network.

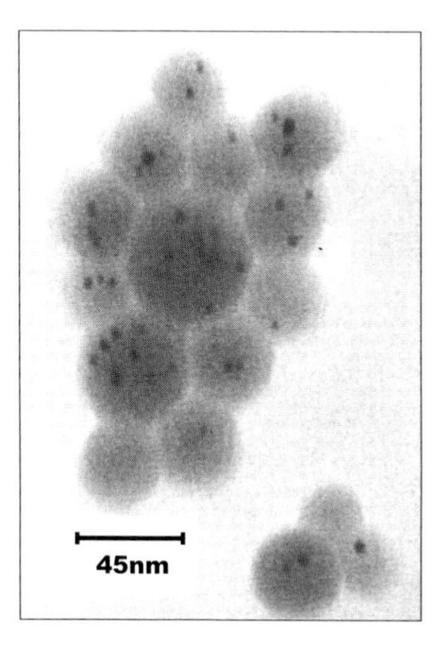

Fig. 2. TEM picture showing several gold clusters within a micro-network.

Passive Micro-Reactors

The active micro-reactors described above cannot be recycled because the SiH moieties cannot be renewed. Recyclable micro-networks may be realized in the form of passive micro-reactors which do not actively take part in the reaction but merely provide the confined reaction space. For this purpose hollow micro-networks are synthesized: first, a micro-emulsion of linear poly(dimethylsiloxane) (PDMS) of low molar mass (M_w = 2000–3000 g/mol) is prepared and the endgroups are deactivated by reaction with methoxytrimethylsilane. Subsequent addition of trimethoxymethylsilane leads to core–shell particles with the core formed by linear PDMS surrounded by a crosslinked network shell. Due to the extremely small mesh size of the outer network shell the PDMS chains become topologically trapped and do not diffuse out of the micro-network over periods of several months (Fig. 3). However, if the mesh size of the outer shell is increased by condensation of trimethoxymethylsilane and dimethoxydimethylsilane the linear PDMS chains readily diffuse out of the network core and are removed by ultrafiltration. The remaining "empty" or "hollow" micro-network collapses upon drying (Fig. 4). So far, shape-persistent, hollow particles are prepared of approximately 20 nm radius, which may be viewed as structures similar to crosslinked vesicles. At this stage the reactants cannot be concentrated within the micro-network in respect to the continuous phase.

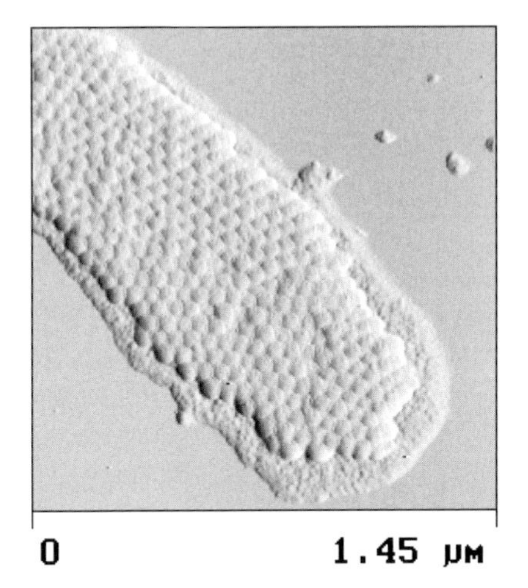

Fig. 3. AFM picture of organosilicon microspheres with a core of linear PDMS chains.

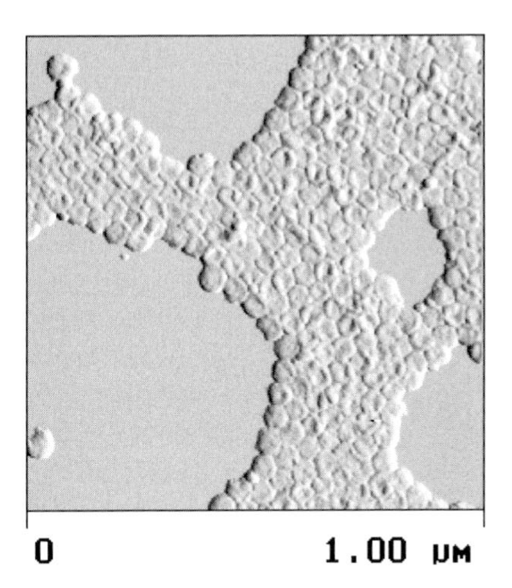

Fig. 4. AFM picture of a hollow, collapsed organosilicon micro-network.

Therefore, the current strategy is based on an even more complex architecture of the micro-network. After formation of the PDMS-containing micro-emulsion a thin shell of a functionalized monomer (e.g. *p*-chloromethylphenyltrimethoxysilane) is formed, followed by a second elastomeric shell of trimethoxymethylsilane and dimethoxydimethylsilane. Again, the linear PDMS chains diffuse out of the micro-network. So far, the experiments are successfully completed. Future work comprises the reaction of *p*-chloromethylphenyltrimethoxysilane, which is exclusively located at the inner rim of the hollow micro-networks, with trimethylamine, which will hydrophilically modify the inner surface by formation of ionic ammonium groups. This last step has not yet been realized experimentally due to an unknown hindrance to the trimethylamine reaching the inner surface of the micro-network. Future plans comprise dissolution of the hydrophilized micro-network in an unpolar solvent. Upon addition of water it is believed that, via a small but finite equilibrium concentration of water in the organic solvent, a small water pool is formed inside the micro-network. Likewise hydrophilic chemicals may enter the water pool and form the desired product within a confined reaction space.

Acknowledgment: We are indebted to Dr. S. Sheiko and Prof. M. Möller, *University of Ulm, Germany*, for performing the AFM measurements. Financial support of the *Fonds der Chemischen Industrie*, from the *BMBF* (project no.: 03D0040 A9) and from *Wacker Chemie GmbH* is gratefully acknowledged.

References:

[1] F. Baumann, M. Schmidt, B. Deubzer, M. Geck, J. Dauth, *Macromolecules.* **1994**, *27*, 6102.

[2] F. Baumann, B. Deubzer, M. Geck, J. Dauth. S. Sheiko, M. Schmidt, *Adv. Mater.* **1997**, *9*, 955.

[3] F. Baumann, B. Deubzer, M. Geck, J. Dauth, M. Schmidt, *Macromolecules* **1997**, *30*, 7568.

[4] J. P. Spatz, A. Roescher, M. Möller, *Adv. Mater.* **1996**, *8*, 337.

[5] M. Antonietti, E. Wenz, L. Bronstein, M. Seregina, *Adv. Mater.* **1995**, *7*, 1000.

Fine Silica: From Molecule to Particle: Quantum Chemical Modeling and Vibration Spectra Verification

V. Khavryutchenko

Institute of Surface Chemistry, National Academy of Sciences of Ukraine
Kiev, 252028 Ukraine
E-mail: vkhavr@compchem.kiev.u

Keywords: Silica Clusters / Meta-Silicon Acid / Quantum-Chemical Modeling / Semiempirical AM1 and PM3 Approximations / Vibrational Spectroscopy

Summary: This article presents the method of particle formation by condensing small molecules to particles. The quantum-chemical level simulation was performed to study step-by-step first steps of fumed silica synthesis from molecules to silica particles. Vibrational spectroscopy (infrared and inelastic neutron scattering) was used to verify simulated structures.

The first steps of synthesis play an important role in the formation of fumed silica particles in the flame. To study these processes we used a supercluster approach in quantum chemistry. To calculate large silica particles we used a sophisticated semiempirical method, PM3. Each system was fully optimized; additionally force field and dipole moment derivatives have been evaluated. Both experimental IR and neutron inelastic scattering (INS) spectra were used to verify computational models.

The quantum-chemical level simulation was performed to study first steps of fumed silica synthesis from molecules to small silica protoparticles. Vibrational spectroscopy (infrared and inelastic neutron scattering) was used to verify simulated structures. There is an application of the thesis: "compute verifiable, verify computable" because a lot of modeling methods may produce a lot of results, but how is one to select the more realistic structure? In our opinion the best method of verification is vibrational spectroscopy, especially for small particles and amorphous substances. To perform space structure calculations and evaluations of force field and dipole moments the sophisticated semiempirical quantum-chemical method AM1, with a PM3 parameter set, was used as a software CLUSTER-Z1 and COSPECO applied to personal computers [1].

It was postulated that the silicon dioxide molecule is a first intermediate product in the route from silicon tetrachloride to fumed silica. Due to high temperatures in the flame the ortho-silicon acid or meta-silicon acid will not be the main product of silicon tetrachloride hydrolysis or/and oxidation. The silicon dioxide molecule is well known to be stable at high temperatures or in a matrix-isolated state. Therefore, we are able to use silicon dioxide molecules as initial silicon-containing substances to simulate silica particle formation. Meta-silicon acids are able to react with other meta-silicon acid molecules and produce bigger clusters. Three meta-silicon acid molecules form cyclic clusters with four-coordinated silicon atoms and two-coordinated oxygens (Fig. 1). At high temperatures this cluster may loose some water molecules from hydroxyl groups and regenerate active sites with three-coordinated silicon atoms and one-coordinated oxygens, leading to the polymerization process.

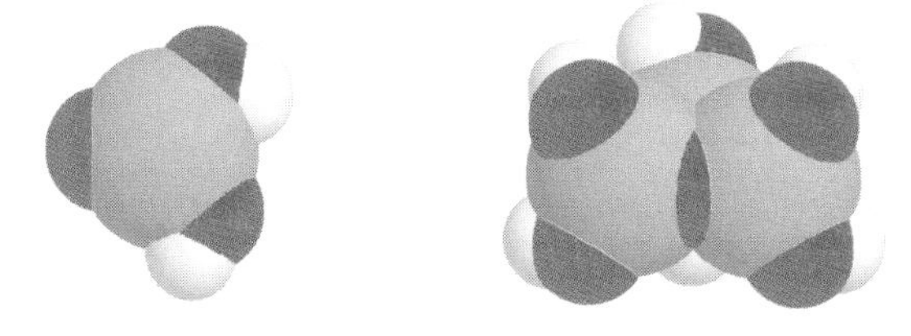

Fig. 1. Metasilicon acid and its cyclic trimer.

To simulate the high-temperature polymerization process from silicon dioxide molecules to small silica particles a somewhat simplified approach was used. A set of silicon dioxide molecules is placed in a cubic lattice with a large molecule–molecule distance, and is submitted to the optimization process. As the chemical reaction is following a potential energy valley, the optimiszation process pathway is able to represent some of the possible reaction pathways. The system, which consist of some tens or more atoms, may exist in a large set of space structures, as usual. But as a result of the optimization process, only one possible space structure will be realized, and this structure will be firstly reached on the way to the global minima of the potential energy for the system under study. That is important to mention, when using this method.

The first system under study consists of 27 silicon dioxide molecules as a cubic lattice with distances between silicon atoms of 7 Å. This atomic space structure was taken as a starting point (see Fig. 2, left). In Fig. 2, right, the result of the space structure optimization process is presented.

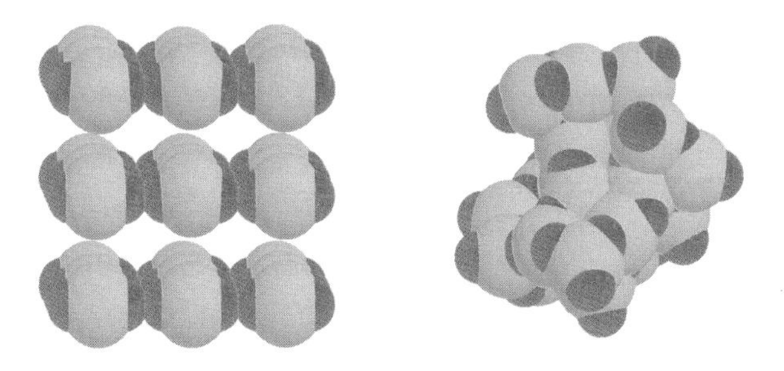

Fig. 2. Initial and final structure for $27SiO_2$ molecules system.

The vibration spectrum of the small cluster $27SiO_2$ is not able to reproduce the vibrational INS spectrum of bulk quartz glass, but the spectrum from inner cluster atoms with standard coordination numbers is similar to experimental data (Fig. 3). The bands near 1000–1200 cm^{-1} are connected with antisymmetrical silicon–oxygen band vibrations; the bands near 700–800 cm^{-1} are connected with symmetrical silicon–oxygen band vibrations, and the bands near 300–500 cm^{-1} are connected with oxygen–silicon–oxygen angle vibrations. The most intense band in cluster vibration spectrum at 180 cm^{-1} may be assigned to surface vibration of one-coordinated oxygen. The band at 100 cm^{-1} in the calculated spectrum and at 50 cm^{-1} in the experimental quartz glass INS spectrum, may be assigned to rotations of the silicon–oxygen tetrahedron.

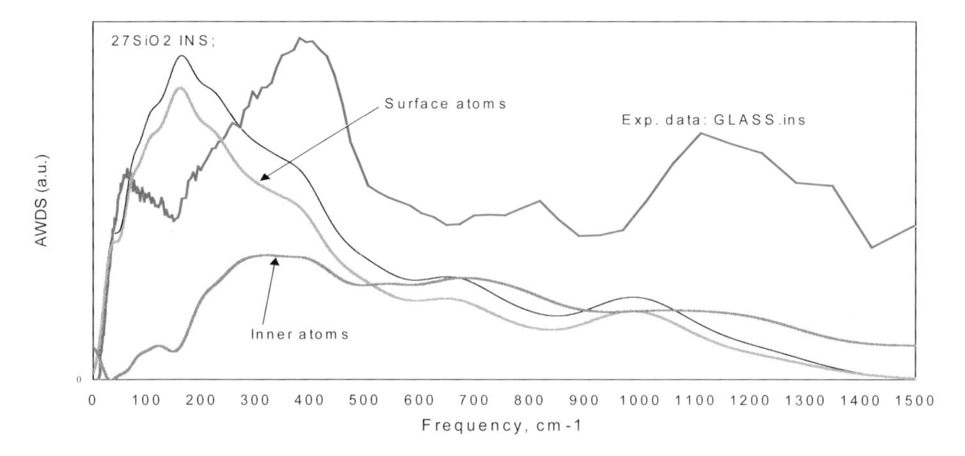

Fig. 3. Calculated vibrational spectra for cluster $27SiO_2$ and experimental inelastic neutron scattering spectrum for quartz glass.

The main active centers are the group SiO^- with a one-coordinated oxygen, and Si^+ with three-coordinated silicon, but the more chemical active center Si^+ may interact with neighboring oxygen. As a result the number of three-coordinated silicon atoms is less than that of one-coordinated oxygen atoms and the groups SiO^- with one-coordinated oxygen mainly cover the pure silica surface. More active silicon atoms, bound to three-coordinated oxygen atoms, are inside the particle body. Therefore, fast reactions may be with surface SiO^- one-coordinated oxygen groups. Other reactions may need more time. Due to the high chemical reactivity, these protoparticles interact without reaction barriers, with formation of bigger particles.

This protoparticle is able to react with water molecules. Firstly, the water molecules are adsorbed on the active centers SiO^- with one-coordinated oxygen, and form H-bonds (see Fig. 4, right). If a hydrogen atom approaches the silica oxygen, overcoming the reaction barrier, then two surface hydroxyl groups will be formed (see Fig. 4, left). Water molecules may saturate all active centers, and as a result particles with a hydroxylated surface will be formed.

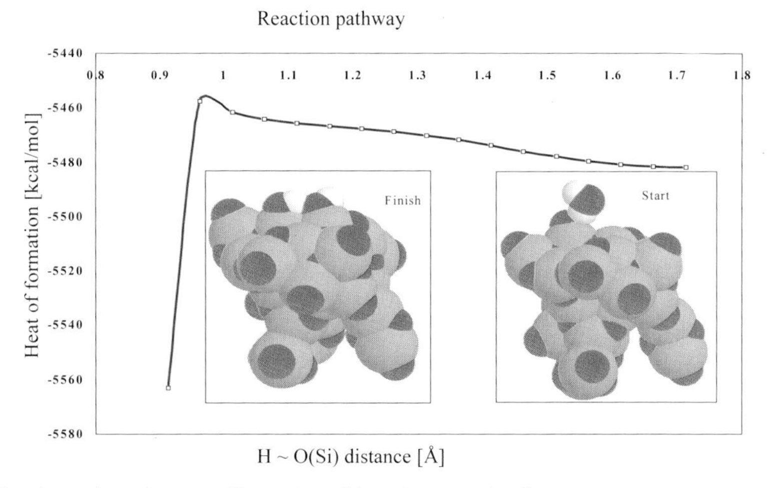

Fig. 4. Reaction pathway between silica protoparticle and water molecule.

In the flame synthesis, which may include silicon dioxide molecule formation and condensation to protoparticles, surface termination by hydroxyls may protect the particles against coagulation and the formation of large bulky solids. Next steps will be particle–particle interactions and the formation of a space superstructure of particles. These processes, and space structures formed as a result, are presented in Fig. 5.

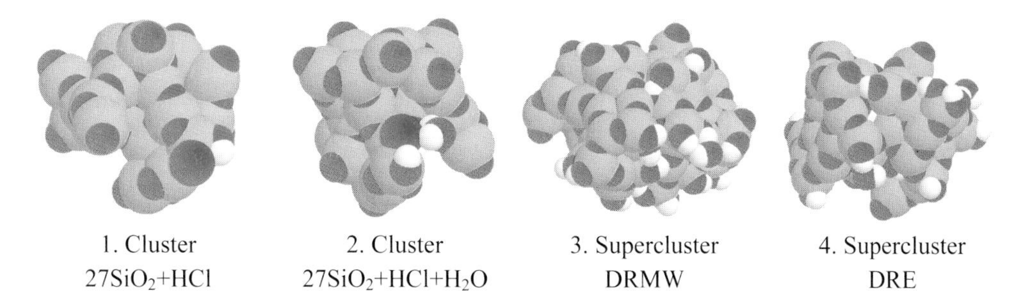

Fig. 5. (1) Product of reaction simulation between silica cluster $27SiO_2$ and HCl; (2) Product of reaction simulation between silica cluster $27SiO_2$ and HCl with H_2O; (3) Product of reaction simulation between two silica clusters $27SiO_2+3{\cdot}H_2O$ and $27SiO_2+6{\cdot}H_2O$; (4) Product of reaction simulation between two silica clusters $27SiO_2+3{\cdot}H_2O$ and $27SiO_2+5{\cdot}H_2O$.

Hydrochloric acid is able to react with a hydroxyl-free silica surface without a barrier, with formation of hydroxyl groups and weakly bonded chlorine (Fig. 5(1)). The distance between chlorine and silicon for the system under study is 2.417 Å, compared to the experimental value of 2.017 Å and calculated value 2.054 Å for $SiCl_4$. The formal charge on chlorine for the system under study is $-0.754\ e^-$, compared with $-0.354\ e^-$ for $SiCl_4$. The Wiberg index (bond order) for the chlorine–silicon bond for the system under study is 0.3231, compared to a value of 0.8468 for $SiCl_4$. A similar phenomenon is presented for systems like water and hydrochloric acid/water (Fig. 5(2)). These data show that the chlorine atom is bonded to the silica cluster mainly as an ion, and it is possible to understand the system under study as an ion-pair, which may dissociate into ions. As a result the charged particles will be repelled electrostatically.

Partially hydroxylated silica clusters are able to interact at long distances by dipole–dipole interactions, followed by attraction. At short distances, reaction may occure without any chemical reaction barrier of ordinary siloxane bonds, if silica particles contain one-coordinated oxygen and/or three-coordinated silicon atoms at their surfaces. In any other cases, without a coverage of chemically active groups, silica particles are able to interact by surface hydroxyl groups with formation of H- onds (Figs. 5(3) and 5(4)).

Reference:

[1] V. Khavryutchenko, *CLUSTER-Z1 & COSPECO, Personal Computer Software for Quantum Chemistry and Computation Vibration Spectroscopy*, Computation Chemistry Group, Kiev, Ukraine, **1990–1998**.

Fumed Silica Structure and Particle Aggregation: Computational Modelling and Verification by Vibrational Spectroscopy

V. Khavryutchenko*

Institute of Surface Chemistry, National Academy of Sciences of Ukraine
Kiev, 252028 Ukraine
E-mail: vkhavr@compchem.kiev.u

E. Nikitina

Institute of Applied Mechanics, Russian Academy of Sciences
Leninsky Prospekt, 31A, Moscow, 117334 Russia

H. Barthel, J. Weis

Wacker-Chemie GmbH, Werk Burghausen
D-84480 Burghausen, Germany

E. Sheka

Russian People's Friendship University, General Physics Department
ul.Ordjonikidze, 3, Moscow, 117302 Russia

Keywords: Silica Clusters / Fumed Silica Particles / Quantum-Chemical Modeling / Semiempirical AM1 and PM3 Approximations / Vibrational Spectroscopy

Summary: This article presents the method of particle formation by dividing a bulk to fine particles. The quantum-chemical level simulation was performed to study step-by-step the conversion process from small parts of a crystal body to small sized silica particles. Vibrational spectroscopy (infrared and inelastic neutron scattering) was used to verify simulated structures.

There are two possible ways to produce fine particles: by condensing small molecules to particles or by dividing large solids into fine particles. This article presents the second of these two methods of particle formation (dividing a bulk to fine particles) and describes what will happen with finely divided parts of a large crystal structure.

The quantum-chemical level simulation was performed to study step-by-step the conversion process from small parts of a crystal body to small sized silica particles. Vibrational spectroscopy (infrared and inelastic neutron scattering) was used to verify simulated structures. There is an application of the thesis: "compute verifiable, verify computable" because a lot of modeling methods may produce a lot of results but how is one to select the more realistic structure? In our opinion the best method of verification is vibrational spectroscopy, especially for small sized particles and amorphous substances. To perform space structure calculations and evaluation of force field and dipole moments the sophisticated semiempirical quantum-chemical method AM1, with a PM3 parameter set, was used as personal computer applied software CLUSTER-Z1 and COSPECO [1].

As a starting structure we used a crystobalite-like structure as this crystalline form is more stable at the high temperature of the flame synthesis. As is well known, the particle surface is covered by hydroxyl groups which terminate all broken bonds on a particle body. To present the particle structure as a whole the interatomic distance distribution function as a sum of sets of Gauss functions with a small broadening parameter of 0.05 Å was used. To present the particle chemical behavior as a whole the coordination number distribution for all atom types has been used.

In Fig. 1 the space structure, the interatomic distance and coordination number distribution function for the most regular diamond-like silicon dioxide structure is presented. This cluster was constructed as 3×3×3 extended crystobalite cells and it containings 708 atoms. This cluster has been used as a reference for highly regular structures.

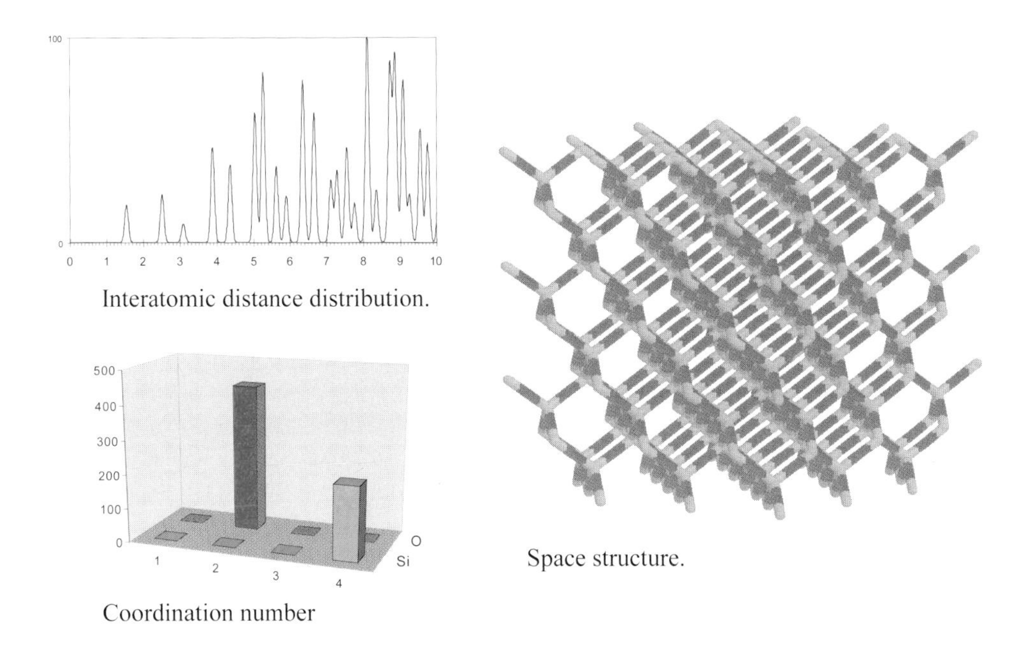

Fig. 1. The space structure, the interatomic distance and coordination number distribution function for the most regular diamond-like silicon dioxide structure.

There are three typical silica clusters: Si28, Si28-17 and $28SiO_2$, and their characteristics are presented in Fig. 2. The first cluster (Si28) simulates a regular crystobalite silica structure, the second cluster (Si28-17) simulates the result of a step-by-step dehydroxylation process of the fully hydroxylated cluster Si28 and the third cluster ($28SiO_2$) simulates the result of polymerization of 28 free silicon dioxide molecules. In our opinion the structures of all these clusters are amorphous (see the second line of pictures in Fig. 2, compared to Fig. 1).

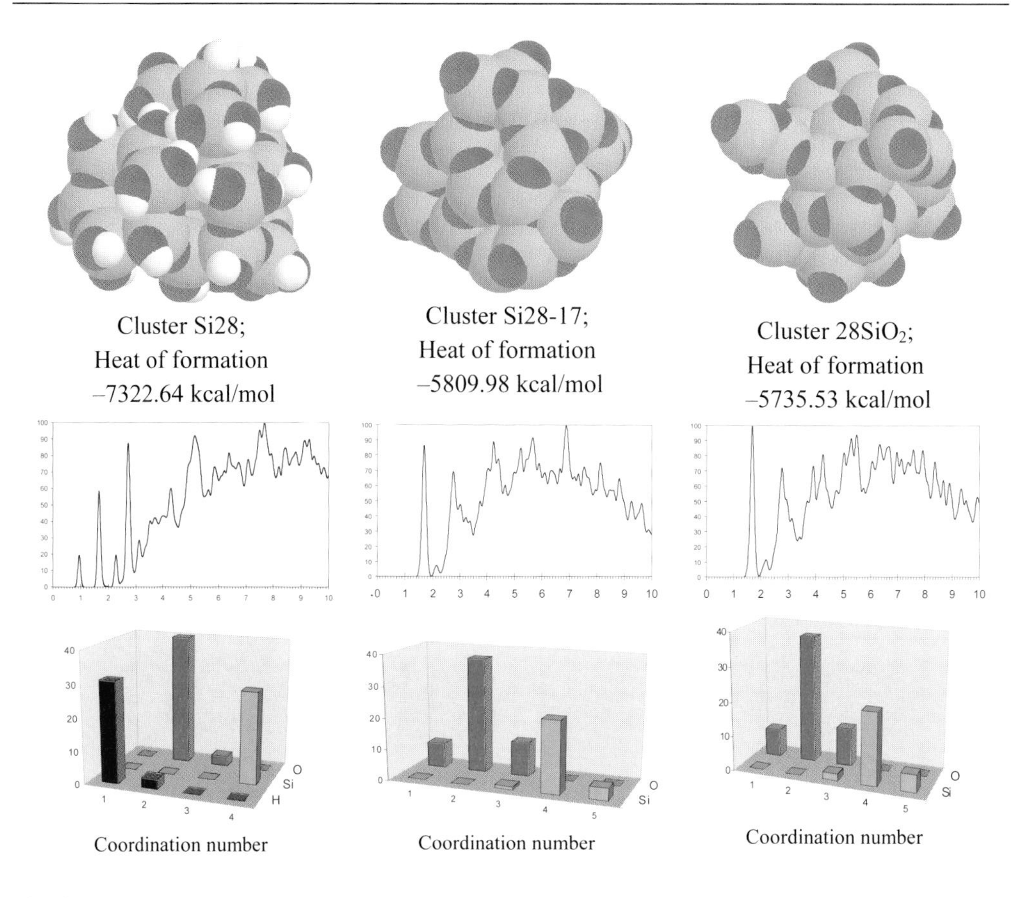

Fig. 2. Typical silica clusters and their characteristics.

The first cluster (Si28) has a regular crystobalite-like silica structure, but due to surface effects its space structure is amorphous as a result of Si-O-Si angle deformation (Fig. 3(1)). For this small cluster the standard value for the angle Si-O-Si of 147° [2], in the crystal state, is broadened to values from 130 up to 175°. Calculated vibrational spectra (both IR and inelastic neutron scattering) for cluster Si28 are in good agreement with experimental spectra (Figs. 3(2) and 3(3)).

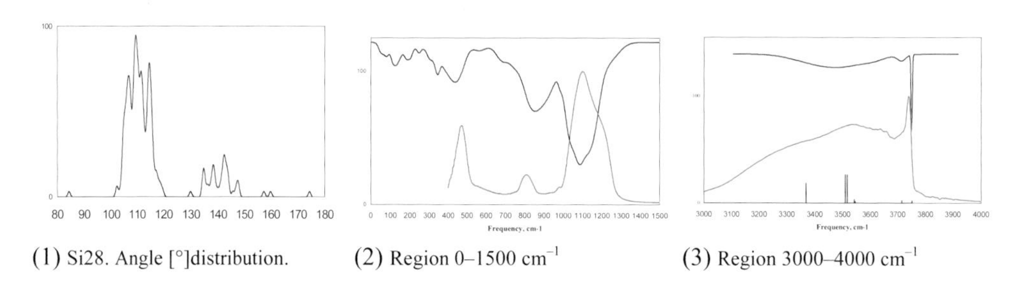

(1) Si28. Angle [°]distribution.
(2) Region 0–1500 cm^{-1}
(3) Region 3000–4000 cm^{-1}

Fig. 3. (1) Angle distribution for cluster Si28, broadening parameter 0.5; (2) IR spectra: calculated IR spectrum for cluster Si28; (3) IR spectra: experimental DRIFT spectrum for Wacker fumed silica HDK T-30.

Bonded by H-bonds, some hydroxyl groups produce a compound structure of O–H bond vibrations in the region 3000–3800 cm^{-1}. Though of equal chemical composition, the heat of formation for cluster Si28-17 is –5809.98 kcal/mol compared to the heat of formation of –5735.53 kcal/mol for the cluster $28SiO_2$. These data show that the cluster Si28-17 structure is more stable than that formed by polymerisation of 28 free silicon dioxide molecules. At high temperatures less stable structures, like the friable packed cluster $28SiO_2$ will be converted to more stable structures like a densely packed cluster Si28-17. Therefore, independently of the production method (from free molecules or from a large crystal body) the protoparticles of fumed silica, obtained by the flame method, will be amorphous and covered by a hydroxyl shell.

Silica particles are able to undergo interactions. All clusters studied have significant dipole moments and at long distances they may interact with mutual attraction and orientation. At short distances the result of cluster–cluster interactions depends on the chemical behavior of the surface. Silica surfaces without terminal hydroxyl groups react without any chemical reaction barrier and produce bigger particles, without borders between clusters, and the system is connected by siloxane bonds (Fig. 4(1)), whereas this pathway may lead to bulky glass.

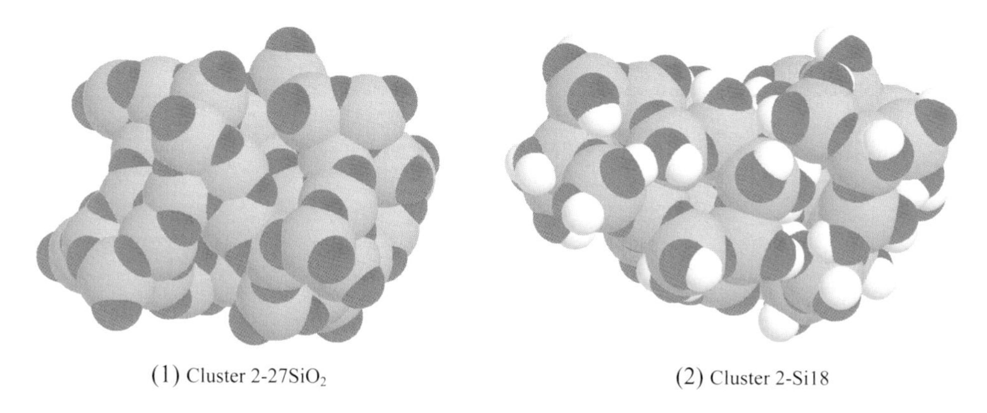

(1) Cluster 2-$27SiO_2$ (2) Cluster 2-Si18

Fig. 4. Results of cluster–cluster interaction: (1) Two fully dehydroxylated clusters $27SiO_2$ are fused to form one particle without activation barrier. The system is connected by siloxane bonds. (2) Two fully hydroxylated clusters Si18 react to form a particle agglomerate. This system is connected by H-bonds and dipole interactions.

Another result may occur if the particles are covered by hydroxyl shells. These are able to interact at long distances with mutual attractions and orientations but at short distances hydroxyls interact by weak H-bonds and dipole–dipole interactions. Due to long distance dipole–dipole interactions and weak H-bonding, hydroxylated particles form friable aggregates with a mass fractal structure (Fig. 5).

Fig. 5. Fumed silica particle. HDK T30 Wacker Chemie, specific area 300 m^2/g.

References:

[1] V. Khavryutchenko, *CLUSTER-Z1 & COSPECO, Personal Computer Software for Quantum Chemistry and Computation Vibration Spectroscopy*, Computation Chemistry Group, Kiev, Ukraine, **1990–1998**.

[2] A. F.Wells. *Structural Inorganic Chemistry, Vol. 3*; Mir, Moscow (Russian edition), **1988**, p. 117.

Mechanical Behavior of Polydimethylsiloxanes under Deformation. Quantum-Chemical View

E. Nikitina*

Institute of Applied Mechanics, Russian Academy of Sciences
Leninsky Prospekt, 31A, Moscow, 117334 Russia
Tel.: Int. code + (095)9385518, Fax: Int. code + (095)9380711
E-mail: nxreatex@cityline.ru

V. Khavryutchenko

Institute of Surface Chemistry, National Academy of Sciences of Ukraine
Kiev, 252028 Ukraine

E. Sheka

Russian People's Friendship University, General Physics Department
ul.Ordjonikidze, 3, Moscow, 117302 Russia

H. Barthel, J. Weis

Wacker-Chemie GmbH, Werk Burghausen
D-84480 Burghausen, Germany

Keywords: Quantum-Chemical / Modeling / Polydimethylsiloxane / Mechanical Properties / Uniaxial Tension

Summary: A response of a silicone polymer fragment to external stresses is considered in terms of a mechanochemical reaction. The quantum-chemical realization of the approach is based on a coordinate-of-reaction concept for which purpose a mechanochemical internal coordinate (MIC) specifying a deformational mode is introduced. The related force of response is calculated as the total-energy gradient along the MIC, while the atomic configuration is optimised over all other coordinates under the MIC constant-pitch elongation. The approach is applied to a set of linear silicone oligomers Si_n with $n = 4$, 5, and 10, subjected to uniaxial tension followed by the rupture of the molecule and a post-fracture relaxation. Peculiarities of the mechanical behavior of the oligomer are analyzed as well as the oligomer strength and the related Young's moduli. A cooperative radical-driven mechanism of silicone polymer fracture is suggested.

Silicone elastomers span a large field of experimental studies and technical applications [1] while microscopical attempts to study their mechanical properties are not known at all. Thus, the reasons for the high elasticity of the polymers are discussed only qualitatively. The quantitative origin of the chemical bond scission as well as the bulk body fracture are obscure at atomic level. At the same time, a theoretical study of the polymer behavior under external stress is quite developed. Two approaches are known to enable quantitative consideration of the mechanical properties of polymers. The first, which might be called as a *dynamic approach*, is based on the calculation of the energy of elastic deformation of a polymer by evaluating the related force constants of the body in either a classical physical [2] or a quantum-chemical (QCh) [3, 4] manner. The approach is mainly aimed at the determination of elastic constants such as Young's moduli. In the classical physical form it has received a wide recognition as applied to organic polymers (see [2] and references therein). The other one, an *atomically microscopic approach* was proposed in the early 1970s [5] and was aimed at a QCh consideration of a chemical bond breaking. Two seven-membered oligomers of polyethylene with different peripheral chemical groups subjected to uniaxial tension have been considered. It was the first time that a stress–strain interrelation was obtained quantum-chemically. Since then two other attempts at the QCh consideration of such interrelations have been made with respect to the ethane molecule [6] and a solitary Si–O bond [7]. In spite of a clearly seen restriction of the first results, those obtained have evidenced quite clearly a doubtless advantage of the QCh approach to the mechanical phenomena treatment. Active development of quantum-chemical tools as well as steady building-up of computer facilities have made it possible to come back to the problem and to raise the question about a transformation of the modern computational chemistry into a *computational mechanochemistry*. General concepts of such a transformation have been considered in [8], and applied to the deformation of some aliphatic silica model systems.

In the current paper the suggested quantum-mechanochemical (QMCh) approach is applied to three linear polydimethylsiloxane (PDMS) oligomers Si_n described by a formula

$$(CH_3)_3SiO\text{–}[(CH_3)_2SiO]_{(n-2)}\text{–}Si(CH_3)_3$$

with $n = 4$, 5, and 10, subjected to uniaxial tension. An extended computational experiment has been carried out [9] by using purposely designed QMCh software DYQUAMECH [10] exploiting high-level semiempirical techniques (PM3 in the current study). The study is based on the introduction of a particularly specified *mechanical internal coordinate* (MIC) [8], which describes the deformational mode. The calculations were done following the coordinate-of-reaction concept performing the total-energy minimization at every step of the constant-pitch elongation of the MIC over a full set of internal coordinates excluding the MIC. The residual total-energy gradient along the MIC is attributed to the force of response, related to the deformation of atomic configuration. The fact that the calculations were performed in internal coordinates has given a superior possibility of connecting a physical deformation with intimate properties of the molecules studied.

Atomic reconstruction of the structures of the oligomers completing by an Si–O bond rupture has been traced at all stages of deformation. A common behavior of all three species has been observed. Three stages of deformation differing in by structural transformation have been detected (see Fig.1). In fact, the conformational transformations in the molecular structure caused by changing either torsional (stage 1) or bond (stage 2) angles are responsible for the deformation resulting in approximate doubling of the initial MIC length. Both stages practically require no energy. The stages are followed by chemical bond stretching (stage 3) resulting in a Si–O bond rupture due to mechanically stimulated reaction. The main losses of energy are connected with this stage.

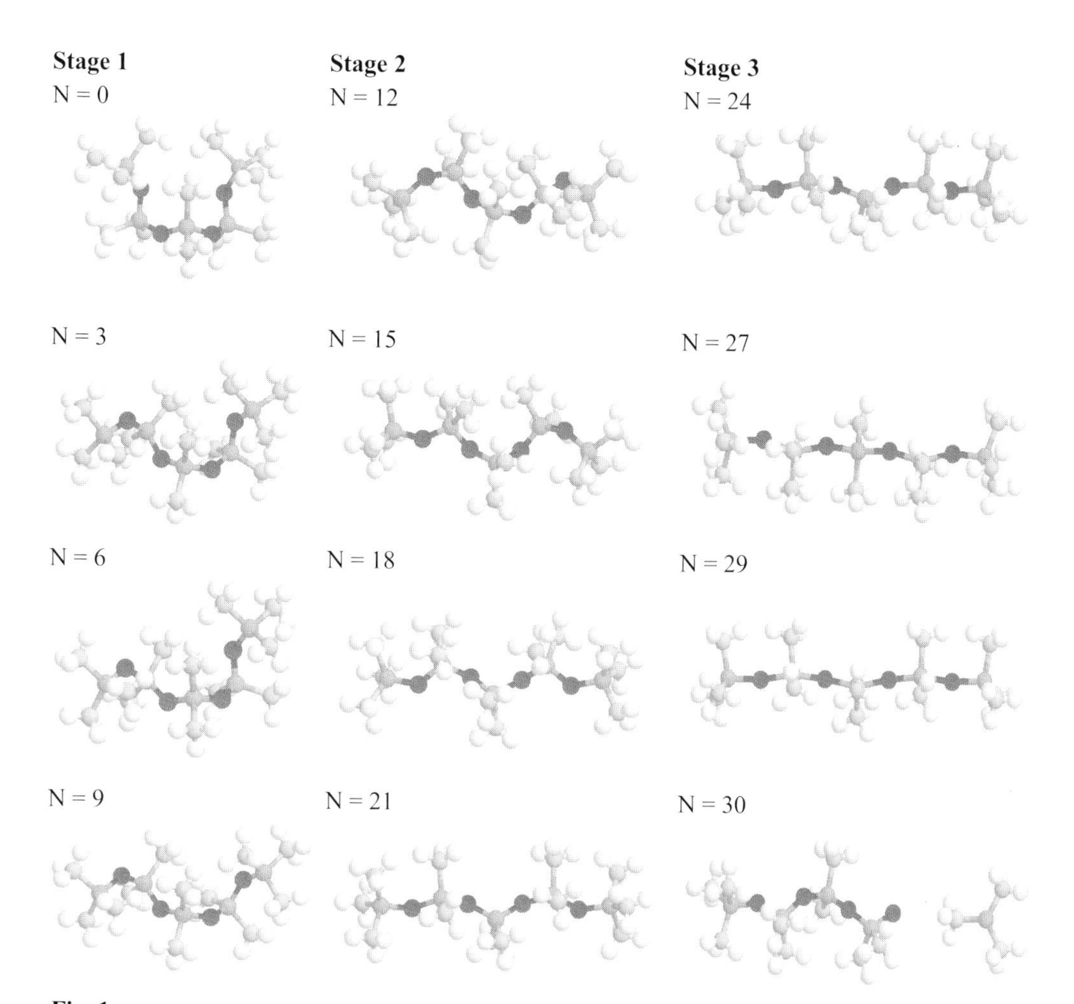

Fig. 1. Successive steps of deformation of Si5. N = number of steps with a 0.3 Å pitch.

The reaction products just formed have been analyzed from the viewpoint of electronic properties. As occurred, two non-charged molecular fragments (Eq. 1) with non-saturated oxygen and silicon atoms have been produced in any case, when a Si-O bond is broken. These atoms possess a spin density equal to unity that convincingly indicates their radical nature (see Table 1).

$$\sim\!O\!-\!Si(CH_3)_2\!-\!O\!-\!Si(CH_3)_2\!-\!CH_3 \longrightarrow \sim\!O\!-\!Si(CH_3)_2\!-\!O^{*} + {}^{*}Si(CH_3)_2\!-\!CH_3$$

Eq. 1. Formation of two non-charged molecular fragments.

Table 1. Atomic characteristics at Si–O bond breaking (molecule Si4).

No.	Atom	Charge	Spin density	Sum over Wiberg's coeff.	Free valency index
Non-stretched molecule					
1	Si	0.64572	0	3.82118	0.17882
2	O	–0.54664	0	2.05719	–0.05719
3	Si	0.83811	0	3.72894	0.27106
4	O	–0.54791	0	2.06706	–0.06706
5	Si	0.83729	0	3.73185	0.26815
6	O	–.54690	0	2.05832	–0.05832
7	Si	0.65205	0	3.81552	0.18448
Molecule after rupture					
1	Si	0.64739	0.00012	3.81862	0.18138
2	O	–0.53777	–0.00004	2.06212	–0.06212
3	Si	0.83361	0.00262	3.73258	0.26742
4	O	–0.54458	0.00736	2.07398	–0.07398
5	Si	0.83317	0.13955	3.72999	0.27001
6	O*	–0.34427	–0.97822	1.04708	0.95292
7	Si*	0.29654	1.07241	2.98015	1.01985

The observed structura–deformational stages have been analyzed from the energy viewpoint. Energy–elongation, force–elongation, and stress–strain interrelations have been studied. As seen in Fig. 2, the conformation-driven stages 1 and 2 are characterized by energy parameters that are very close to zero parameters indeed. Stage 3, caused by chemical bonds elongation, involves the regions of elastic and plastic deformation as well as molecule breaking.

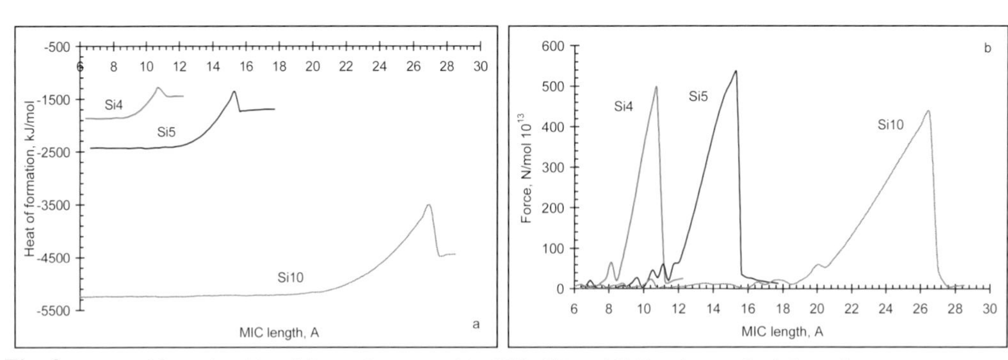

Fig. 2. Heat of formation (a) and force of response (b) of Si4, Si5, and Si10 under tensile deformation.

Analyzing the deformational energy–stress and stress–strain interrelations shown in Fig. 3, the limit values of the oligomer strength as well as their Young's moduli have been determined (see Table 2). It is obvious that the conformation–driven stages of deformation are responsible for the high-elasticity or rubbery-like state of the oligomers as well for the inner friction causing a peculiar stress-strain hysteresis (see Fig. 3).

Table 2. Calculated mechanical characteristics for PDMS oligomers.

Oligomer	F_{max} [kJ/mol A]	A_{def} [kJ/mol]	σ_{max} [Gpa]	E [Gpa] [a] "w"	E [Gpa] [a] "σ"
Si4	495	894	42	296	230
Si5	533	1098	45	297	228
Si10	435	1473	37	213	157

[a] "w" and "σ" denote values obtained by treating the elastic deformation energy curves and stress-strain curves, respectively.

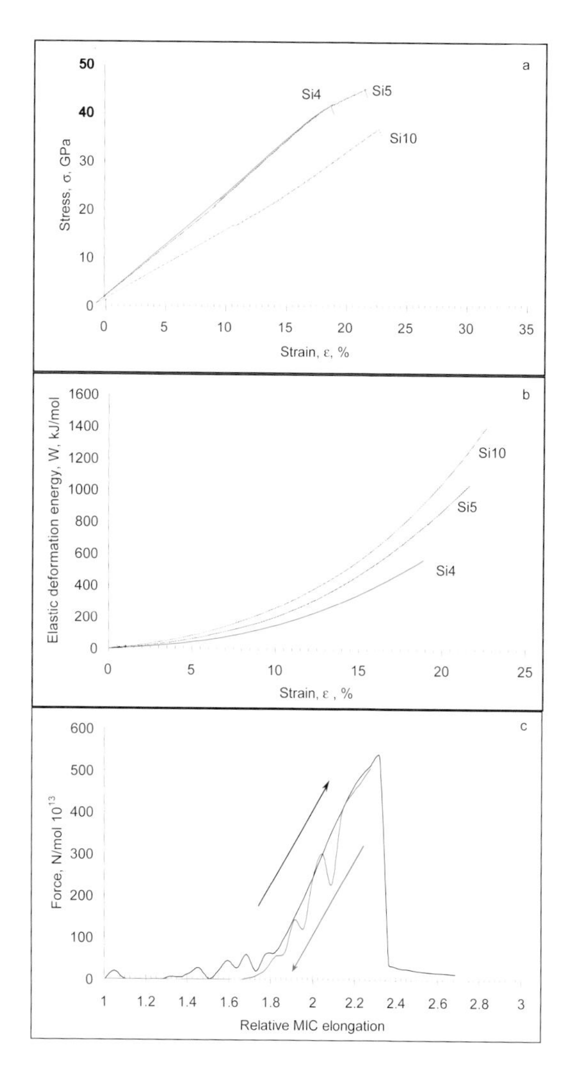

Fig. 3. Stress (a) and energy of elastic deformation (b) versus strain for Si4, Si5, and Si10 (c) Stress-strain hysteresis.

Comparing the calculated and experimental data for the oligomer strength and Youngs' moduli, a many-order inconsistency has been exhibited similarly to that known for organic polymers [2]. The finding has stimulated us to suggest that a Zhurkov's et al. [11] radical–driven mechanism of polymer fracture, which is accepted for organic species, be extended for silicone species as well. The mechanism throws light upon a reinforcement of the latter by filling them with highly dispersed silica. The polymer immobilization caused by adsorption might stop the fracture reaction, thus reinforcing the polymer.

References:

[1] J. E.Mark, B. Erman, *Rubberlike elasticity*, Wiley, New York, **1988**.

[2] K. Tashiro, *Progr. Polym. Sci.* **1993,** *18*, 377.

[3] H. E. Klei, J. P. Stewart, *Int. J. Quant. Chem. Symp.* **1986**, *230*, 529.

[4] S. G. Wierschke, *Mater. Res. Soc. Symp. Proc.* **1989**, *134*, 313.

[5] D. S. Bordeaux, *J. Polym. Sci., Polym. Phys. Ed.* **1973**, *11*, 1285.

[6] V. S. Yushchenko, T. P. Ponomareva, E. D. Shchukin, *J .Mat. Sci.* **1992**, *27*, 1659.

[7] V. S. Yushchenko, E. D. Shchukin, M. Hotokka, *J. Mat. Sci.* **1993**, *28*, 920.

[8] V. Khavryutchenko, E. Nikitina, A. Malkin, E. Sheka, *Phys. Low-Dim. Struct.* **1995**, *6* 65.

[9] E. A. Nikitina, V. D. Khavryutchenko, E. F. Sheka, H. Barthel, J. Weis (in preparation).

[10] V. D. Khavryutchenko and A. V. Khavryutchenko Jr., DYQUAMECH *Dynamical-Quantum Modelling in Mechanochemistry Software for Personal Computers* (Institute of Surface Chemistry, National Academy of Science of the Ukraine, Kiev, **1993**.

[11] S. N. Zhurkov, V. I. Vettegren, V. E. Korsukov, I. I. Novak, *Fracture* **1965**, 545.

Intermolecular Interactions of Polydimethylsiloxane Oligomers with Hydroxylated and Silylated Fumed Silica: Quantum Chemical Modeling

*E. Nikitina**

Institute of Applied Mechanics, Russian Academy of Sciences
Leninsky Prospekt, 31A, Moscow, 117334 Russia
Tel.: Int. code + (095)9385518, Fax: Int. code + (095)9380711
E-mail: nxreatex@cityline.ru

V. Khavryutchenko

Institute of Surface Chemistry, National Academy of Sciences of Ukraine
Kiev, 252028 Ukraine

E. Sheka

Russian People's Friendship University, General Physics Department
ul.Ordjonikidze, 3, Moscow, 117302 Russia

H. Barthel, J. Weis

Wacker-Chemie GmbH, Werk Burghausen
D-84480 Burghausen, Germany

Keywords: Quantum-Chemical Modeling / Semiempirical AM1 and PM3 Approximations / Polydimethylsiloxane Oligomer Adsorption / Hydroxylated Fumed Silica / Silylated Fumed Silica

Summary: To gain a better understanding of the microscopic intermolecular interactions between fumed silica particles and polydimethylsiloxane (PDMS) polymers, a quantum-chemical modelling in the framework of modified semiempirical AM1 and PM3 methods has been performed for a series of superclusters simulating fragments of a real particle surface interacting with five-membered PDMS oligomers. Impacts of chemical composition and structural configuration of both substrates and oligomers on the adsorption of the latter has been studied.

Introduction

Fumed silica is widely used for the reinforcement of polydimethylsiloxane (PDMS) elastomers. The intermolecular interaction of the filler surface with the PDMS matrix controls this process [1, 2] so that understanding which factors influence the interaction at the filler/PDMS interface has become a crucial point for further development of the technology. Among the factors of interest there are the chemical composition of the filler surface, the configurations of the siloxane chains of both the filler and the PDMS component, and the presence of other chemical additives at the filler/PDMS interface such as water, etc. To answer these questions, an extended quantum-chemical (QCh) study has recently been performed. This paper briefly reviews the results obtained and is organized in the following way. The next Section gives a general description of the basic concepts of the approach used. The following section is devoted to the description and substantiation of the model used. The results related to the impact of the chemical composition of the filler surface on the interaction at the interface are then given. Analogous results relevant to the impact of the siloxane chain configurations are discussed next. The role of H-bonding at the interface is considered in the perultimate section alongside discussion of the impact of the water molecule on the interaction. The conclusion summarizes the essentials obtained.

Basic Concept

The study performed has been designed so as to be optimal for the succeess of the following scheme:

- elaborating the basic model of the silica filler/PDMS polymer interface, making it possible to highlight the mechanism of the interaction occurring there as well as to take into account varying chemical compositions and structural configurations of the constituents;
- carrying out a large series of QCh calculations of the models selected;
- verification of the results obtained.

Let us go briefly through all these stages.

Models for the Study

As known (see [3] and references therein), such a complicated system as a mixture of the silica filler with PDMS polymer can be divided into two microphases with different local mobilities of the PDMS chains, namely:

- immobilized chain units adsorbed at the filler surface, and
- mobile fragments of chain outside the adsorption layer (see Scheme in Fig. 1(a)).

When the models for this study were being suggested, the importance of the real atomic structure in and outside the place of contact was obvious, although understood only schematically. Therefore, the study we began with an atomization of the interface between the filler and the PDMS polymer. Figure 1(a) presents a sketch of the space housing a large fragment of hydroxylated fumed silica and a piece of a scrambled polymer. According to profound studies by different techniques [4-6], the area of the closest (adsorption) contact between the filler and PDMS polymer was determined to involve a few (4–6) monomer units, which is visualized in Fig. 1(a). Obviously, to describe processes occurring in the contact area means to disclose the mechanisms governing the intermolecular interaction. The picture shown in Fig. 1(a) gave an idea of how to construct a model structure to be appropriated for simulating the place of contact. That should consist of a piece of the filler (substrate) large enough to accommodate a PDMS oligomer a few monomer units in length. A supercluster containing 48 $SiO_{4/2}$ units (referred to below as the Si48OH supercluster) has been chosen as a basic model for the substrate. A set of linear PDMS oligomers described by a general formula $(CH_3)_3SiO[(CH_3)_2SiO]_{x-2}Si(CH_3)_3$ with $x = 2, 3, 4, 5$ (SiX below) presented the PDMS component of the interface in the study. The largest of them (Si5) is shown in Fig. 1(b).

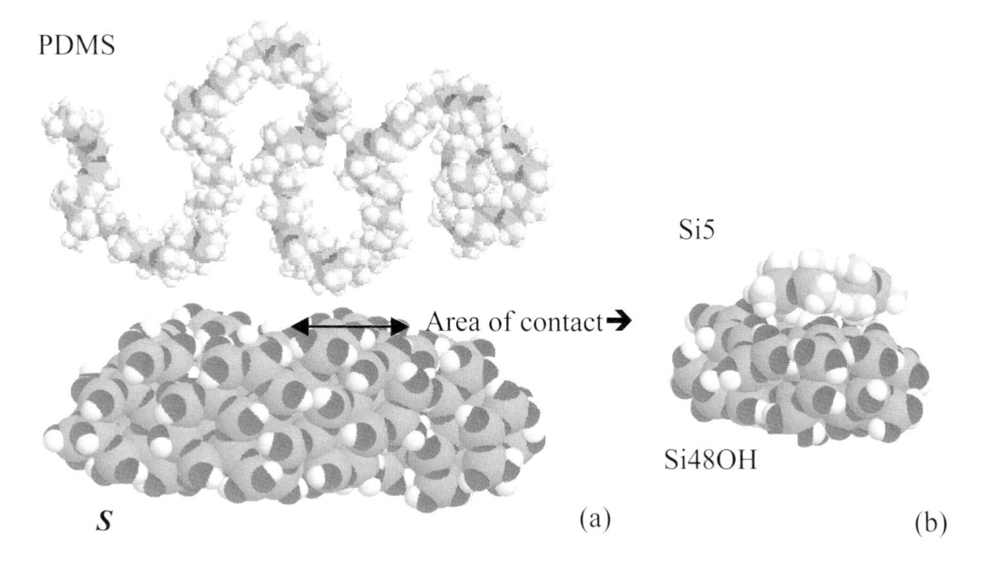

Fig. 1. Schematic representation of PDMS/fumed silica interface. A general view (a) and suggested model of contact area (b).

Substrate Models

A series of the substrate models basing on the Si48OH supercluster consisting of 222 atoms has been studied. Figure 2 presents a general view of the equilibrated structures of the cubic (c) and tetragonal (t) configurations of the basic supercluster Si48OH obtained when applying the AM1 and PM3 approximations incorporated in the DYQUAMOD software [7] used for simulation.

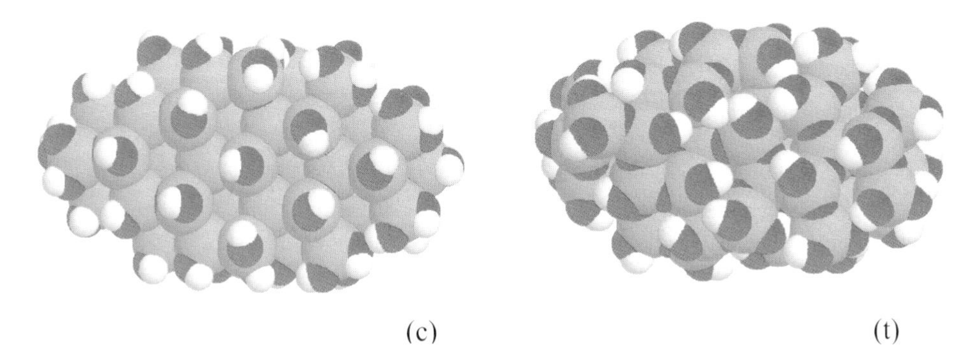

Fig. 2. Equilibrated structure of the basic supercluster Si48OH [10, 11] in cubic (c) and tetragonal (t) configurations.

With respect to silica and silica-like species, these identical techniques distinguished by parameters only provide a unique opportunity to simulate two equilibrated configurations of siloxane chains differing by the value of the Si-O-Si angles. Thus, the AM1 method maintains well the average value of the angle ranged between 170° and 180° so that the related siloxane chain can be called straightened, while the values ranging between 135° and 160° are typical for the PM3 techniques that lead to bent siloxane chains. As known, straightened and bent configurations of the siloxane chains are actually observed, being characteristic for the high-temperature cubic and low-temperature tetragonal polymorphs of cristabolite-like silica crystalline solids [8, 9].

Whether the filler surface is hydroxylated or silylated affects the interaction significantly. That is why the substitution of the surface hydroxyl groups of as-prepared hydroxylated fumed silica by trimethylsiloxy (TMS) units (surface silylation) is widely used in the silicone industry.

Two clusters modeled the silylated substrates: Si2Me, containing 2 (isolated), and Si6Me, containing 6 (associated) TMS groups, respectively. The equilibrated structures of these two silylated clusters, which represent a silica surface area with different degrees of silylation constructed from the same basic Si48OH cluster, are shown in Fig. 3.

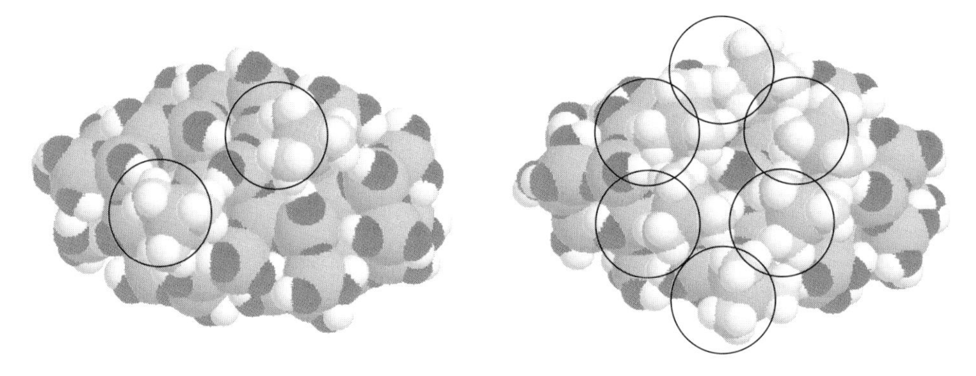

Fig. 3. Equilibrated structure of the silylated supercluster Si2Me (a) and Si6Me (b) [10, 11].

For cluster Si6Me two modifications differed in the configuration that the siloxane chains had taken, as well. The proposed basic supercluster and its silylated forms are large enough to position Si*X* PDMS oligomer models with $x \leq 5$ thus allowing study of intermolecular interaction in the ad-systems under varying conditions. The detailed characteristics of the substrates are given elsewhere [10–13].

PDMS Oligomer Models

Figure 4 presents equilibrated structures of a set of four linear PDMS oligomers Si*X* with $x = 2–5$. As previously, straightened (s) and bent (b) configurations of the molecules were simulated by the application of either the AM1 or the PM3 technique, respectively. Figure 5 presents a general view of the heat of formation for both sets. A clearly seen homological character of the quantity is observed for both calculated sets. The difference between the values obtained by techniques AM1 and is about 15 %.

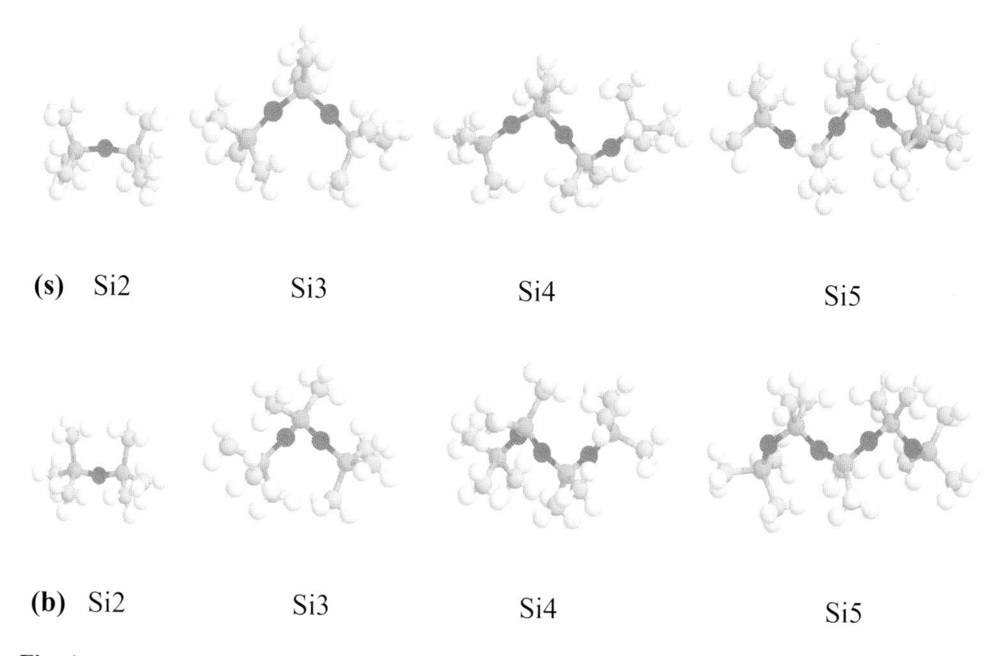

Fig. 4. Equilibrated structure of the linear PDMS oligomers SiX with straightened (s) and bent (b) configurations of the siloxane chains.

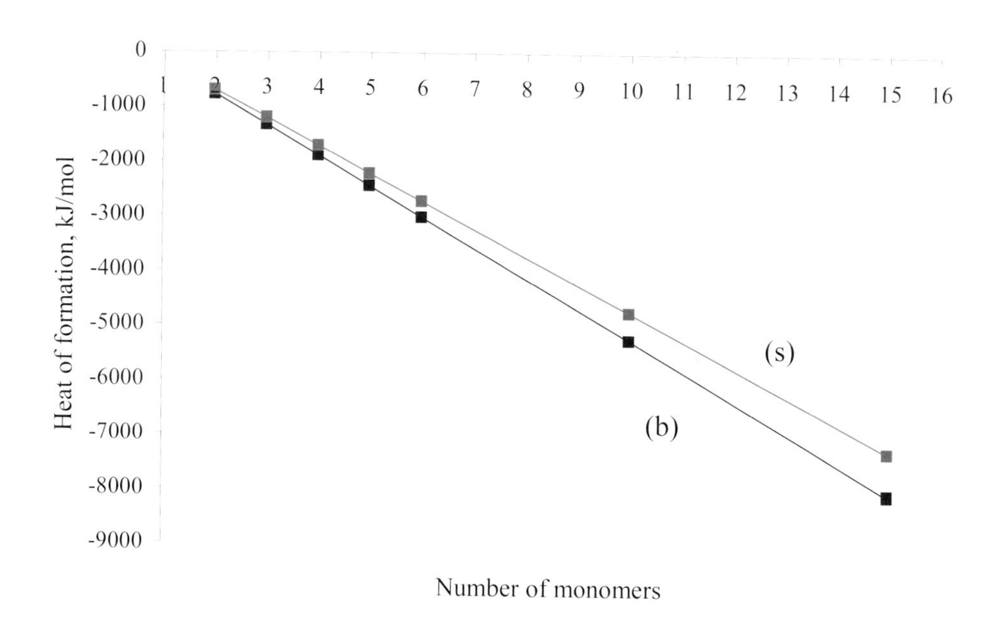

Fig. 5. Heat of formation of the linear PDMS oligomers SiX with straightened (s) and bent (b) configurations of the siloxane chains.

Ad-System Models

Eight superclusters comprising eight chemically composed ad-systems based on two substrates (Si48OH and Si6Me) and four oligomers (Si*X* with *X* = 2, 3, 4, and 5) present the starting basic models of the study. When being subjected to the optimization procedure by application of the above two QCh techniques, each basic supercluster can be split into a number of final configurations forming the following families:

AM1 family:

Conformationally pure subset: I (cubic) + II (straightened)
Conformationally mixed subset: I (cubic) + II (bent)
I (tetragonal) + II (straightened)
I (tetragonal) + II (bent)

PM3 family:

Conformationally pure subset: I (tetragonal) + II (bent)
Conformationally mixed subset: I (tetragonal) + II (straightened)
I (cubic) + II (bent)
I (cubic) + II (straightened).

Here notations I and II are related to the substrate and oligomer components of the studied ad-system, respectively.
The above family branching seems to be quite reasonable from the experimental viewpoint due to the high conformational lability of both silica species and PDMS compounds. A fairly complete vision of the problem as a whole can be obtained when considering properly selected sets of the family members, which will be presented below.

Calculation technique and data presentation

The computational part of the study was carried out by using DYQUAMOD software [7] consisting of *sp*-basis QCh software CLUSTER-Z1 [14] further developed for supercluster calculations, and COSPECO computational spectroscopy software [15] aimed at the calculation of vibrational spectra. CLUSTER-Z1 is based on a set of semiempirical methods, among which AM1 and PM3 were used in the study. Computations were run on SUN SPARKStation workstations, types 2, 10, and 20 types as well as on personal computers with 2×Pentium PRO 200 processors.

In the current case, parallel employment of the above two QCh methods is of particular interest for the studied system. The problem is that the techniques are absolutely identical conceptually, but differ in application to large atomic configurations of the species consisting of siloxane chains (such as silica and PDMS). That is to say, these two methods provide two different average values of the Si-O-Si valence angle of 180° and 146°, respectively, due to a slight difference in the method parameters. A very high lability of the angle lays the foundation for the change so that the finding is not a simple drawback of the computations but reflects the intrinsic nature of the species. Actually, as known, both PDMS oligomers and silica are quite conformationally labile, which leads to a large range for the angle values of 110–180° determined experimentally [8]. The well-known polymorphism of solid silica originats from that fact. Therefore, a parallel application of both QCh methods gives a unique possibility of simulating conformationally different structures for both substrate and adsorbate and of highlighting the changes in the adsorption process caused by structure transformations. Thus, a computational "artefact" becomes a powerful tool for a microscopic study.

The results obtained are presented as conventional atomic images provided by RasMol molecular visualization program [16] supplemented by tabulated values of the heats of formation, as well as by a set of distribution functions describing "densities of states" for atomic charges, valence bond lengths, and bond angles [17]. When plotting, the above quantity values, l, are broadened by the Gaussian function $\exp(-l^2/2\lambda^2)$ where the broadening parameter λ was specified for every quantity separately. Thus obtained distribution functions made allowance for a detailed description of structural distortions of both substrates and adsorbates caused by intermolecular interactions.

Results and Discussion

General Remarks on the Adsorption Energy

Altogether 18 ad-systems were studied which are listed in Table 1 alongside the main energy characteristics. A set of the systems studied has made allowance for examining the intermolecular interaction occurring on the fumed silica/silicone oligomer interface in detail. The related values of the adsorption energy are determined from Eq. 1, where $\Delta H_{ad\text{-}syst}$, ΔH_{sub} and ΔH_{olig} determine the heat of formation of the system considered, and these of the substrate and oligomer involved, respectively.

$$E_{ads} = \Delta H_{ad\text{-}syst} - (\Delta H_{sub} + \Delta H_{olig})$$

Eq. 1.

As seen from Table 1, the E_{ads} data form two large groups related to the AM1 and PM3 families with serveral fold difference in the energy for the same ad-systems in favor of the PM3 one. This finding should be emphasized particularly when comparing with the difference in the heat of formation of free substrates [10, 11] and oligomers (see Fig. 5) estimated by both techniques.
The difference does not exceed 15–20 % in favor of the PM3 family. This is a real measure of the effect caused by the different parameterization involved in each technique. Therefore, the many-fold effect in the adsorption energy is obviously related to the intermolecular interaction. As has become known recently [18], the PM3 technique noticeably overestimates the H···H interaction. Actually, in fact, the E_{ads} values determined by the AM1 technique are in good agreement with the experimental data [2]. Both facts can explain the observed difference if one suggests that this very interaction plays the most important role in the adsorption process. What follows will support the suggestion well.

Table 1. Energy characteristics of silica/PDMS oligomer ad-complexes.

No.	Ad-system[a]	$E_{ads.}$ [kcal/mol]	$E_{ads.}$ [kJ/mol]
AM1-family			
1	Si48OH(c) + Si5(s)	–10.4	*–43.6*
2	Si48OH(c) + Si5(s), on diols	–8.9	*–37.1*
3	Si6Me(c) + Si5(s)	–6.9	*–28.7*
4	Si48OH(c) + Si5(b)	–8.7	*–36.4*
5	Si48OH(t) + Si5(s)	–9.1	*–38.3*
6	Si48OH(t) + Si5(b)	–12.7	*–53.1*
5	Si6Me(t) + Si5(b)	–17.5	*–73.0*
PM3-family			
6	Si48OH(t) + Si5(b)	–60.7	–253.8
7	Si48OH(t) + Si5(b), H-bonded	–63.1	*–264.2*
8	Si48OH(t) + Si5(b), on diols	–45.9	*–192.3*
9	Si2Me(t) + Si5(b)	–48.9	*–204.8*
10	Si6Me(t) + Si5(b)	–44.9	*–187.7*
11	Si48OH(t) + Si5(s)	–69.1	*–289.3*
12	Si48OH(c) + Si5(b)	–50.0	*–209.2*
13	Si48OH(t) + Si2(b)	–23.9	*–100.1*
14	Si48OH(t) + Si3(b)	–45.1	*–188.9*
15	Si48OH(t) + Si4(b)	–52.0	*–217.6*
16	Si6Me(t) + Si2(b)	–18.3	*–76.4*
17	Si6Me(t) + Si3(b)	–32.2	*–134.7*
18	Si6Me(t) + Si4(b)	–36.8	*–153.9*

[a] (c) and (t) denote cubic and tetragonal silica structure; (s) and (b) denote straightened and bent siloxane chain configurations of oligomer molecules.

As a consequence what has been said above, below we shall restrict ourselves to discussion predominantly of the AM1-family data, which are well complete to highlight the main peculiarities of the interaction mechanism at the interface. Particular attention should be paid to the ad-systems from the conformationally mixed subset. Those included in the AM1 family are related to the ad-systems where the equilibrated straightened configuration of the siloxane chain of either substrate or oligomer molecule or even of both components was replaced by the bent one. As was said previously, in the case of the "bent" substrate, the latter has been taken in the equilibrated configuration obtained by the PM3 calculation. The configuration was kept fixed during the further optimization of the adsorbed molecule position by applying the AM1 technique. Thus, the ΔH_{sub} value in Eq. 1 corresponds to the calculation performed at the fixed geometry. Similarly, when a"bent" oligomer molecule was considered, the value of all bond angles Si-O-Si was kept at 146 degree while all other internal coordinates were included in the optimization. Correspondingly, the ΔH_{olig} value in Eq. 1 was calculated under the same conditions.

To study the behavior of a "homological series" of adsorbate on a set of substrates, six ad-systems, namely, Si48OH + SiX as well as Si6Me + Si*X*, where $X = 2, 3, 4, 5$ forming two sets with respect to the substrate, were investigated. The relevant energies of adsorption are given in Table 1. Figure 6 presents the energy dependence on the number of monomer units. As seen from the Figure, the observed dependence can be approximated well to linear in both cases, which is very consistent with a homological approach to the studied systems.

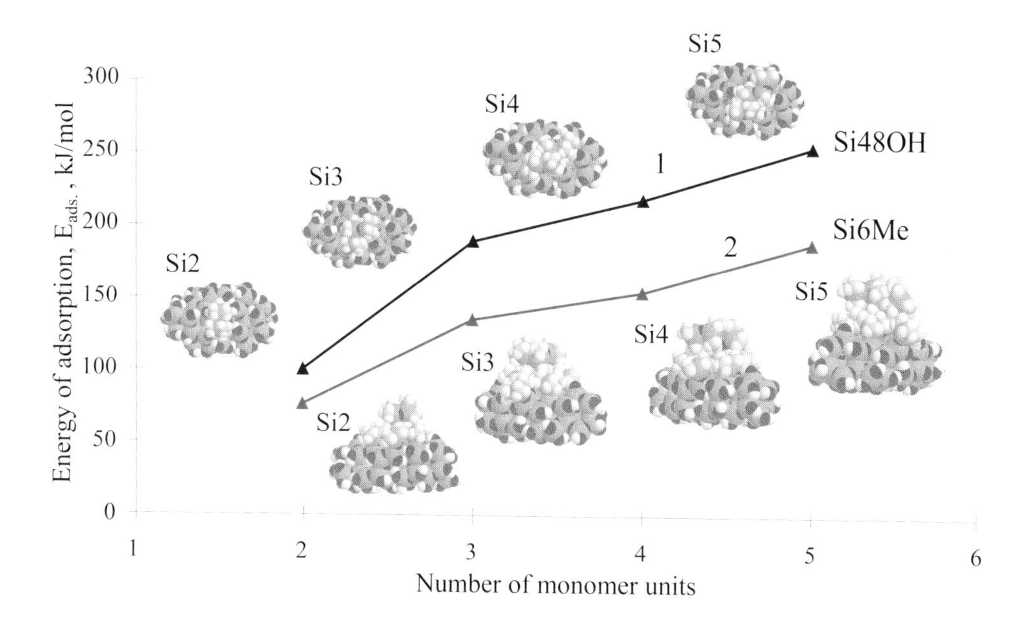

Fig. 6. Energy of adsorption of a series of linear PDMS oligomers SiX adsorbed on the hydroxylated Si48OH (Curve 1) and silylated Si6Me (Curve 2) substrates. PM3 family.

Impact of the Chemical Composition of the Substrate Surface Terminating Groups

Hydration/Silylation

A common picture of the adsorption of a PDMS oligomer molecule on hydroxylated and silylated surface is presented in Fig. 7 for the Si5 oligomer. The ad-system is taken from the conformationally pure set so that both the substrates and adsorbed molecule are taken in a straightened siloxane chain configuration. To simplify the following discussion, the Si48OH + Si5 system in Fig. 7(a) is considered as a reference one with a relative adsorption energy $E_{ads}^{(r)}$ equal to 1.

When the surface is silylated, the $E_{ads}^{(r)}$ reduces to 0.66. The value is kept at the level of 0.65–0.75 for other SiX oligomers as well as follows from Table 1 and Fig. 6. It is important that the analogous $E_{ads}^{(r)}$ value of ~0.75 for the conformationally pure subset of the PM3 family falls in this interval as well. When the silylation coverage is not enough for housing the Si5 molecule wholly (see ad-system 9, Si2Me+Si5, in Table 1 with only two TMS groups on the surface), the $E_{ads}^{(r)}$ of ~0.80 is in between the values characteristic for the entirely hydroxylated and silylated silica surface. Corresponding ad-system structures are represented in Fig. 8.

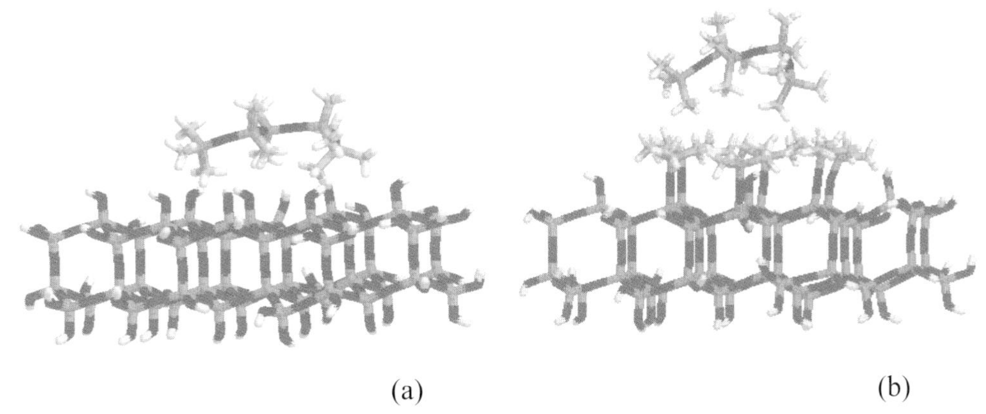

Fig. 7. Equilibrated structure of the ad-systems of the Si5 molecule adsorbed on the hydroxylated Si48OH (a) and silylated Si6Me (b) substrate. AM1 family.

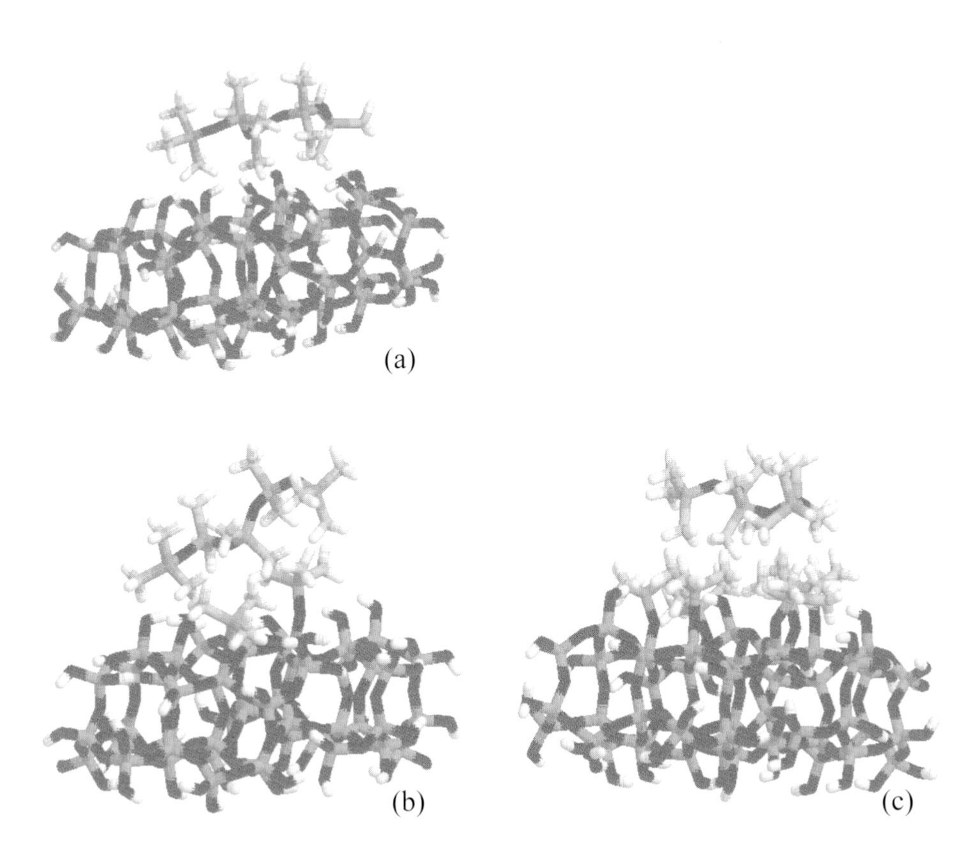

Fig. 8. Equilibrated structure of the ad-systems of the Si5 molecule adsorbed on the hydroxylated Si48OH (a); silylated Si2Me (b) and Si6Me (c) substrate. PM3 family.

Figure 9 presents a distribution of the closest intermolecular H···H distances for the ad-systems in Fig. 7.

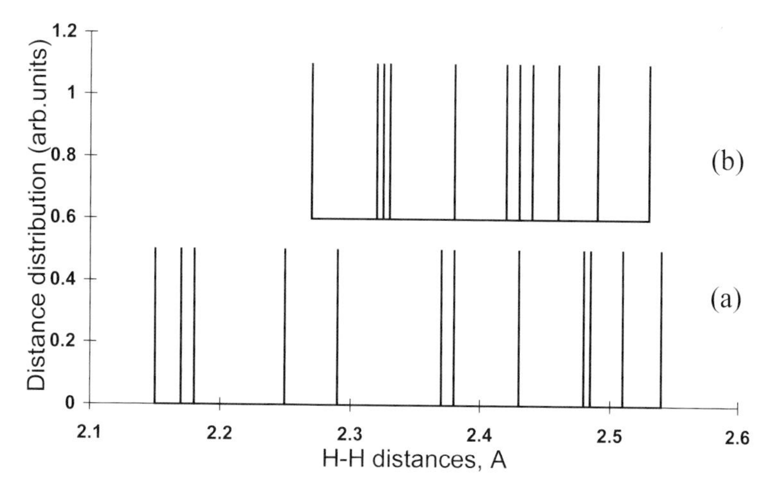

Fig. 9. Distribution of a set of shortest intermolecular H···H distances [12] for the ad-systems of Si5 adsorbed on the hydroxylated (a) and silylated (b) substrate. AM1 family.

As seen from the Figure, the Si5 molecule is moved out from the silylated surface, which seems to be the main reason for the adsorption energy reducing. Actually, Eq. 2, which is fulfilled when the shortest H···H distances (R_{sh}) for the systems under consideration are taken, satisfies the van der Waals law. Therefore, the interaction is mainly non-specific and governed by the weak dispersive attraction.

$$(R_{sh(Hydr)}/R_{sh(Sil)})^6 \approx 0.7$$

Eq. 2.

Silanols/Silanediols

Hydroxyl units terminating the fumed silica surface are mainly spread over the particle surface as silanols (up to 85 %) [19]. However, they can form silanediols groups as well, which are mainly situated at the particle edges, thus providing a particular "chemical heterogeneity" in the surface composition. The basic supercluster Si48OH reproduces such heterogeneity and makes it possible to study its impact on the interaction at the interface. Figure 10 demonstrates the situation when the Si5 molecule interacts predominantly with either a silanol- or a silanediol-covered surface. In the studied case the $E_{ads}^{(r)}$ reduces to 0.85 showing that the "chemical effect" consists in the molecule moving away from the surface due to less suitable conditions for molecule accommodation. The finding confirms again that the van der Waals interaction takes place in this case.

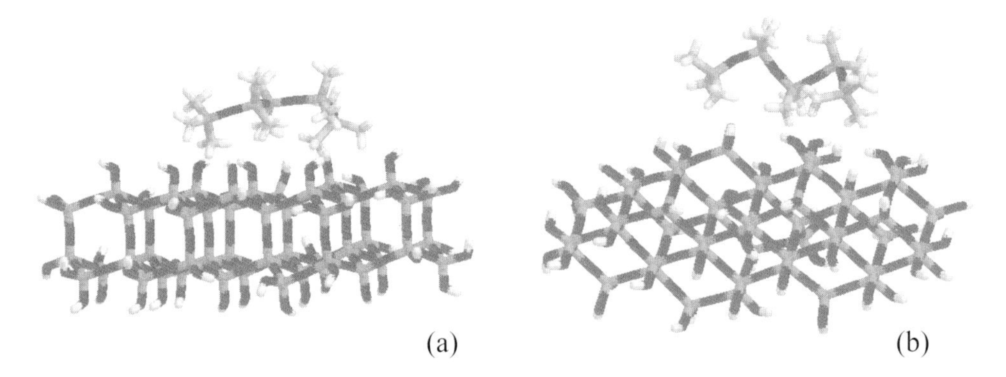

Fig. 10. Equilibrated structure of the ad-systems of the Si5 molecule adsorbed on the hydroxylated substrate over the surface covered by silanols (a) and silanediols (b) predominantly. AM1-family.

Impact of the Siloxane Chain Configuration

The analysis has been performed for the data related to the configurationally mixed subset of the AM1 family. The Si48OH + Si5 ad-system was chosen for the consideration; the main results are shown in Fig. 11. Figure 11(a) relates to the case when the substrate is bent and the Si5 molecule remains straightened. The relevant $E_{ads}^{(r)}$ value is 0.88 and is reduced by 12 % from the reference value for the system in Fig. 7(a). When the molecule configuration is also bent (see Fig. 11(b)) the contact between the adsorbate, and the surface becomes more favorable and the $E_{ads}^{(r)}$ becomes 1.67, thus growing by almost 70 %. This finding shows that the adsorption favors the bent configuration for both substrate and oligomer. As was shown recently by a profound analysis of the Si48OH substrate structure based on a comparative study of calculated and experimental vibrational spectra [11], the bent configuration of the substrate is actually the real case. The bent configuration of the free oligomer molecule is also energetically favorable.

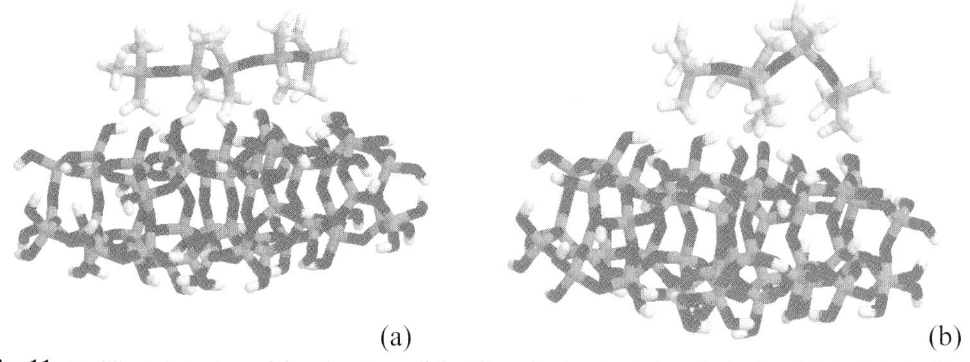

Fig. 11. Equilibrated structure of the ad-systems of the Si5 molecule adsorbed on the hydroxylated substrate. Bent configuration of the siloxane chains of the substrate while the molecule is in a straightened (a) and bent (b) configuration. AM1 family.

H-Bonding Contribution into the Intermolecular Interaction

In the case of the AM1 family ad-systems, no H-bonds have been detected in the reference systems at any starting positions of the oligomer molecule with respect to the substrate. However, analysing the ad-system by the PM3 technique (see Fig. 12(a)), a shortening of the least intermolecular O···H distance from 3.54 to 2.60 Å has been found. The finding has stimulated us to make a search for an intermolecular H-bond creation in the latter case. In fact, when sliding the oligomer molecule over the substrate, one can find a position when a hydroxyl of the substrate and an oxygen atom from a siloxane chain come close enough to form a standard O-H···O bond with a standard O···H distance of 1.7 Å. Such a configuration is shown in Fig. 12(b).

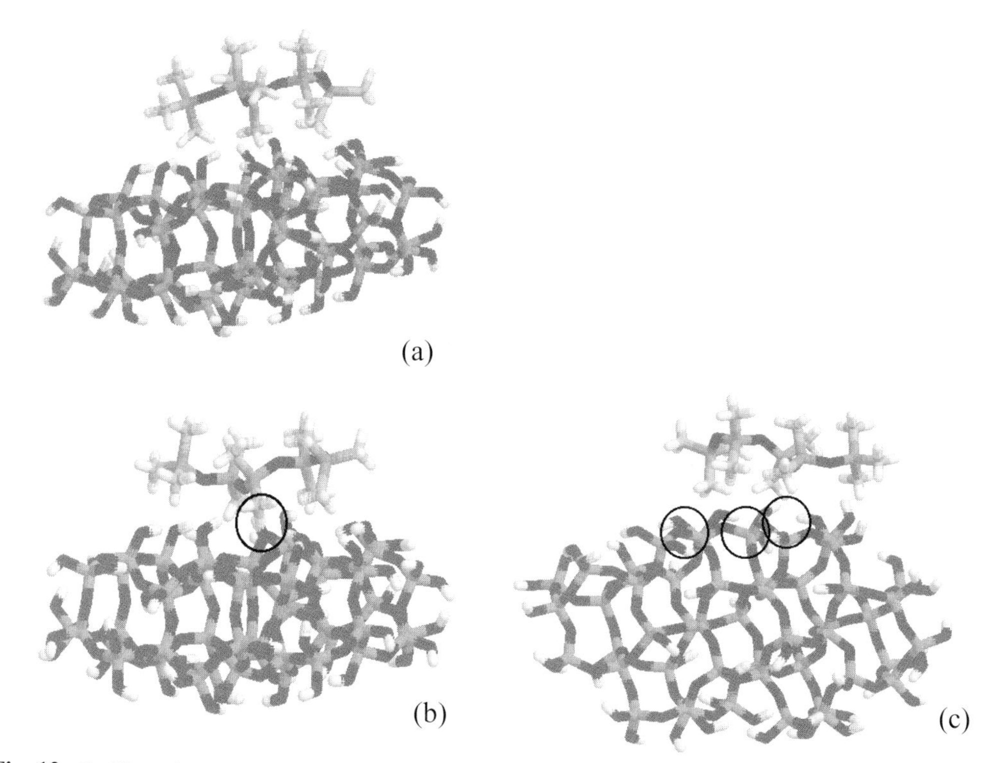

Fig. 12. Equilibrated structure of the ad-systems of the Si5 molecule adsorbed on the hydroxylated substrate. Bent configuration of the siloxane chains of both the substrate and the molecule. (a) No H-bonds are available. (b) One intermolecular H-bond is formed. (c) Three (intramolecular) H-bonds are formed on the substrate surface. PM3 family.

As previously, we shall refer to the related adsorption energy $E_{ads}^{(r)}$ while the reference energy $E_{ads}^{(r)} = 1$ is related to the basic configuration in Fig. 12(a). The ad-system in Fig. 12(b) is characterized by the $E_{ads}^{(r)} = 1.04$. It is important to note that, as seen from Table 1, the H-bond formation lifts the adsorption energy by ~10 kJ/mol, which is characteristic for a standard H-bond heat of formation. However, the relative energy increase constitutes only 4 %, just showing convincingly that the main part of the intermolecular interaction is provided by other mechanisms. As for the mixed ad-system studied, no intermolecular H-bonds were detected. The findings brought us to the conclusion that the H-bonding occurs at the fumed silica/PDMS oligomer interface only occasionally and plays a minor role even if available.

Another case of the H-bonding availability has been found by investigating the case when the Si5 molecule is placed over an area covered mainly by silanediol groups. When both substrate and oligomer molecule are in the bent configuration, a formation of H-bonds between the surface hydroxyl units is observed due to a compression of the ad-molecule (see Fig. 12(c)). To distinguish this case from the previous one it will be referred to below as attributed to the intramolecular H-bonding.

To investigate the question of H-bonding in silica–PDMS species precisely, a set of computations has been performed. Those are related to small systems where H-bonding could be exhibited explicitly. The data obtained are summarized in Table 2, to be discussed below.

Table 2. H-bonding contribution into the intermolecular interaction[a].

No.	System	$\Delta H_{syst.}$	$E_{ads.}$	$E_{adsHB} - E_{ads.}$	$E_{adsHB}/ E_{ads.}$
1	M3QOH	–519.6 / *–2174.4*			
2	Si2	–184.4 / *–771.3*			
3	Si5	–584.4 / *–2445.7*			
4	2×M3QOH (HB)	–1047.7 / *–4384.0*	–8.4 / *–35.3*	–2.2/*–9.3*	1.35
5	2×M3QOH (Inv.)	–1045.4 / *–4374.7*	–6.2 / *–25.9*		
6	Si2 + M3QOH (HB)	–711.4 / *–2976.8*	–7.4 / *–31.1*	–0.9/*–4.0*	1.15
7	Si2 + M3QOH (Inv.)	–710.4 / *–2972.8*	–6.5 / *–27.1*		
8	Si5 + M3QOH (HB)	–1111.9 / *–4653.2*	–7.9 / *–33.2*	–0.9/*–4.1*	1.14
9	Si5 + M3QOH (Inv.)	–1111.0 / *–4649.1*	–6.9 / *–29.1*		
10	Si5 + Si48OH (HB)	–13242.3 / *–55413.9*	–63.1 / *–264.2*	–2.5/*–10.4*	1.04
11	Si5 + Si48OH (Inv.)	–13239.9 / *–55403.5*	–60.7 / *–253.8*		

[a] The energy is given in kcal/mol / *kJ/mol.*

Rows 4 and 5 relate to an M3QOH species suggested to simulate an individual hydroxyl group. Depending on the mutual disposition of the molecules in a dimer, two limiting possibilities can be formed: the former involves an intermolecular H-bond (HB), while there is no H-bond in the latter configuration (Inv.). As seen from Table 2 the intermolecular interaction energy is greater only by 35 % in the presence of the H-bond. The main part is due to dispersive interaction.

The interaction of the individual center with an oligomer molecule can be organized in two ways as well. Rows 6 and 7 present a case for the Si2 oligomer. As previously with the dimer, the M3QOH molecule can interact with Si2, both forming an intermolecular H-bond (HB) and without it (Inv.). The contribution of the H-bond formed to the energy of the intermolecular interaction constitutes only 15 %. A similar situation with the Si5 oligomer is presented in rows 8 and 9 of Table 2. As previously, the H-bond contribution to the intermolecular interaction energy is 14 %. Finally, in rows 10 and 11 the data for large silica cluster (Si48OH) are presented. Obviously, the contribution of H-bonding to the interaction energy is even smaller (4 %).

The study performed has shown convincingly that H-bonding does not play any significant role in the adsorption of a PDMS oligomer on a fumed silica surface. However, the question arises how to explain a vividly seen intensification of a low-frequency IR sideband located in the region of

3600–3200 cm^{-1} (see Fig. 13). The sharp band at 3750 cm^{-1} is related to the O–H stretching vibrations of the isolated hydroxyl units on the surface.

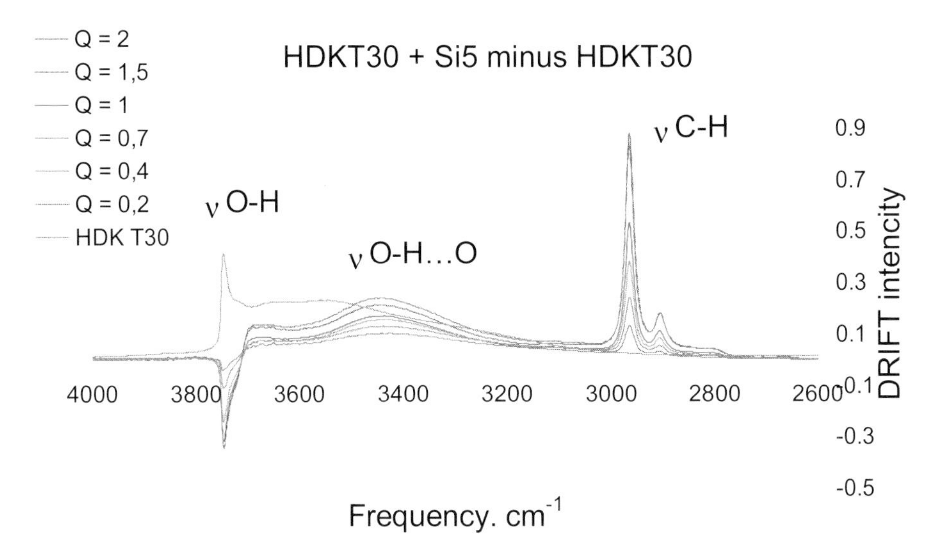

Fig. 13. Adsorption-induced change in DRIFT-spectra (difference spectra) of the fumed silica–Si5 oligomer mixture with increasing adsorbate coverage Q of Si5.

As known, the band at 3750 cm^{-1} losses its intensity until it vanishes completly when the Si5 coverage of the hydroxylated surface grows. The explanation is found by looking at what is going on at the fumed silica interface when the Si5 molecule is adsorbed. Figure 12b reproduces an equilibrated structure of the Si48OH + Si5 ad-complex when an intermolecular H-bond is formed. Evidently, H-bond formation becomes more frequent when the adsorbate coverage increases. Figure 12c shows a case when the adsorbed molecule is located over the surface fragment covered predominantly by silanediol units. As shown earlier, the presence of the adsorbed molecule stimulates the formation of H-bonds between hydroxyl units of the surface. Both processes will decrease the number of free hydroxyls on the surface and, simultaneously, will increase the number of H-bonds formed, which will cause intensification of the band at 3600–3200 cm^{-1} discussed above.

Concluding Remarks

Concluding the discussion of the results obtained during the study performed, let us have a general look at the energetic and structural behaviour of the Si48OH + Si5 and Si6Me + Si5 ad-systems, which accumulate all the main features of the processes, highlighted by calculations. Table 3 contains the adsorption energy values alongside experimental data available for filled PDMS polymers. Only the AM1-family data are presented in the table ranging over all systems studied, both conformationally pure and conformationally mixed.

Table 3. H-bonding contribution into the intermolecular interaction[a].

Ad-system	E_{ads} total [kJ/mol]		E_{ads} per monomer unit [kJ/mol]		
	Calculated	Experimental	Calculated	Experimental	Experimental
Si5 on Si48OH (PDMS/hydroxylated silica)	–43.6	–32.0[a]	–8.7	–8.0 ··· –12.0[c]	–10.5[d]
Si5 on Si6Me (PDMS/silylated silica)	–28.7	~–20.0[b]	–5.7		<-8.5[d]

[a] Activation energy for adsorption–desorption processes at PDMS interface by broad-band dielectric spectroscopy [4]. — [b] Estimated according to data on ^{2}H-NMR solid-state spectroscopy [6]. — [c] ^{1}H-NMR study [6]. — [d] Inverse gas chromatography study of hexamethyldisiloxane [20].

As seen from Table 3, the range of the data spreading constitutes about 50 and 60 % of the average value for the hydroxylated- and silylated-substrate systems, respectively, which is a measure of the range of the quantity confidence interval. The range is not large and correlates well with the dispersion of the experimental data obtained for the relevant systems [2]. As for the absolute values for the energy, they are in good agreement with the data known for the filled PDMS polymers. The findings allow the conclusion that the suggested model for the contact area at the filler/PDMS polymer interface is quite appropriate.

The Si5 ad-molecule behaviour is well illustrated by the distribution functions shown in Fig. 14. The data are grouped for the AM1 and PM3 families separately. A comparison of the data for a free molecule with those related to the molecule in the adsorbed state has revealed a quite noticeable change in all the fundamental characteristics of the molecule in spite of the fact that intermolecular interaction is rather weak.

Summarizing the results obtained, the following conclusions can be made.

The intermolecular interaction at the fumed silica/PDMS polymer interface is mainly non-specific. Only in particular conditions can H-bonding at the interface take place. Two possibilities have been highlighted when either *intermolecular* (Fig. 12(b)) or *intramolecular* (Fig. 12(c)) H-bonds are formed although the contribution of thus-formed H-bonds the interaction energy does not exceed 4 %. Therefore, the H-bonding does not play an important role in the interaction occurring at the filler/PDMS polymer interface. However, such formation of H-bonds whose number increases when the ad-molecule coverage grows, fully explains the peculiar behavior of the vibrational spectra of the relevant ad-systems in the region of the O–H stretching vibrations subjected to changes when the coverage increases.

Changing the chemical composition of the substrate surface-terminating groups causes a reduction of the adsorption energy in the region of 25–35 % of the reference value when hydroxyls are substituted by TMS groups and of 15 % when silanediols are the main terminating units.

The siloxane chain configuration has an impact on the adsorption energy favouring the case when the configuration of both substrate and adsorbed molecule are bent. The finding correlates well with the accepted view on the favorable configuration of the fumed silica particle and PDMS polymer.

The QCh study performed has provided a large bulk of results enabling one to highlight the main peculiarities of the intermolecular interaction at the fumed silica/ PDMS polymer interface from getting reliable values of the adsorption energy to explaining a particular feature related to the optical spectra behavior.

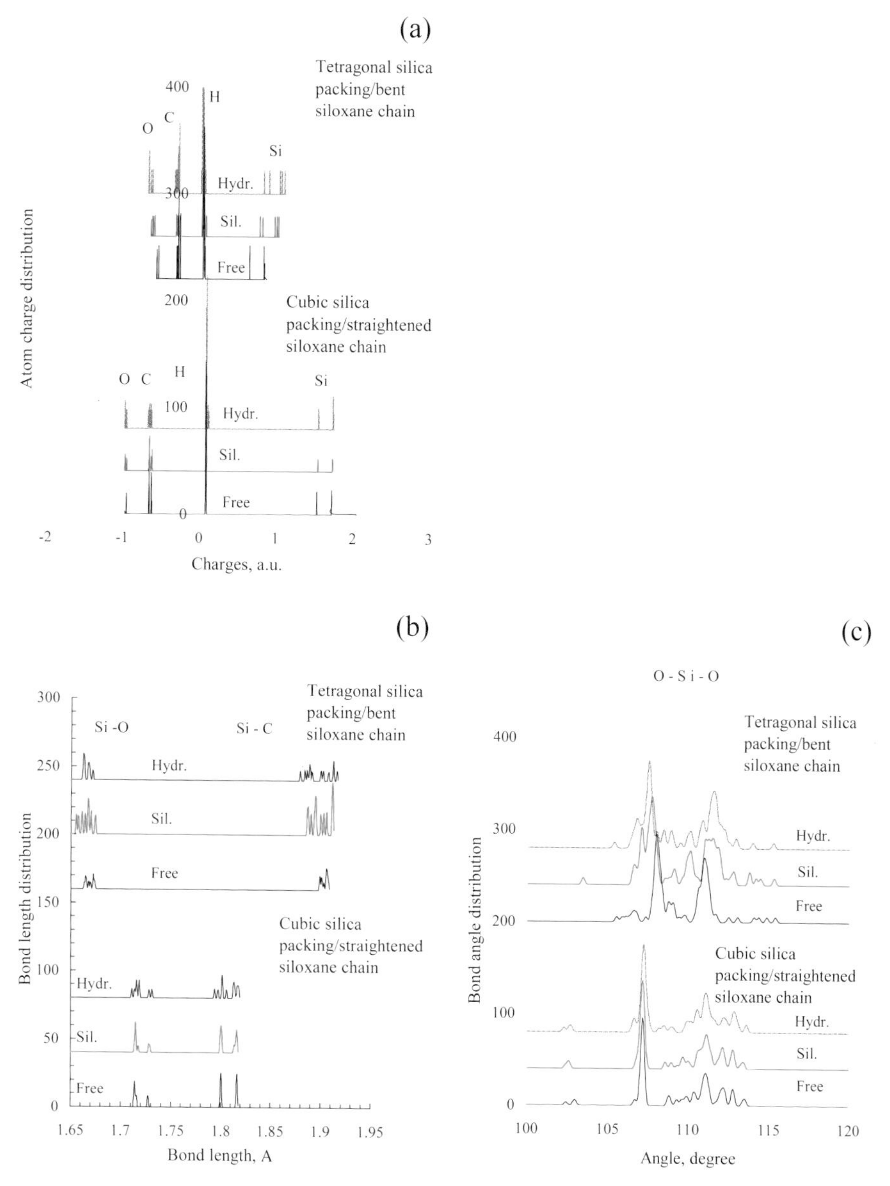

Fig. 14. Distribution functions for the ad-systems of the Si5 molecule adsorbed on the Si48OH and Si6Me superclusters. (a) Atom charge. (b) Si-O and Si-C bond lengths. (c) Bond angles Si-O-Si. The broadening parameters λ constitute 0.001 a.u., 0.0005 Å, and 0.1°, respectively.

References:

[1] B. Barthel, L.Rösch, J. Weis, A. Khalfi, H. Balard, E. Papirer, *Composite Interfaces* **1999**, *Vol 6, No. 1*, 27–34.

[2] A. Khalfi, E. Papirer, H. Balard, H. Barthel, M. Heinemann, *J. Coll. Interf. Sci.*, **1996**, *184*, 586.

[3] V. M. Litvinov, in: *Organisilicon Chemistry II: From Molecules to Materials.* (Eds. N. Auner, J. Weis), VCH, Weinheim, **1996**, p. 779.

[4] K. U. Kirst, F. Kremer, V. M. Litvinov, *Macromolecules* **1993**, *26*, 975.

[5] V. M. Litvinov, *Polymer Sci.USSR* **1998,** *30*, 2250.

[6] V. M. Litvinov, H. W. Spiess, *Macromol. Chem.* **1992**, *193*, 1181.

[7] V. D. Khavryutchenko, A. V. Khavryutchenko, Jr., *DYQUAMOD Dynamical-Quantum Modelling Software for Personal Computers,* Joint Institute for Nuclear Researches, Dubna, and Institute of Surface Chemistry, National.Academy of Sciens of Ukraine, Kiev, **1993**.

[8] F. Liebau, in: *Structural Chemistry of Silicates*, Springer-Verlag Heidelberg, **1985**.

[9] V. Gavrilenko, A. Grekhov, D. Korbutjak, V. Litovchenko, in: *A Handbook of Semiconductor Optical Properties*, Naukova Dumka, Kiev, **1987** (in Russian).

[10] V. D. Khavryutchenko, E. A. Nikitina, E. F. Sheka, H. Barthel, J.Weis, *Phys. Low-Dim. Struct.* **1998**, *Vol 5, No. 6*, 1–30.

[11] V. D. Khavryutchenko, E. A. Nikitina, E. F. Sheka, H. Barthel, J. Weis, *Phys. Low-Dim. truct.*, to be published

[12] V. D. Khavryutchenko, E. A. Nikitina, E. F. Sheka, H. Barthel, J. Weis, *Surf. Rev. Lett.* **1997**, *Vol 4, No. 5*, 879–883.

[13] E. Nikitina, V. Khavryutchenko, E. Sheka, H. Barthel, J.Weis, in: *Book of Abstracts ICSCS-9, 9th Int. Conference on Surface and Colloid Science,* Sofia, Bulgaria, 6–12 July, **1997**, 578.
E. Nikitina, V. Khavryutchenko, E. F. Sheka, H. Barthel, J.Weis, *Colloid. Surf.* **1999**, accepted for publication.

[14] V. A. Zayetz, *CLUSTER-Z1 Quantum Chemical Software*, Institute of Surface Chemistry, National Academy of Sciens of Ukraine, Kiev, **1990**.

[15] V. D. Khavryutchenko, *COSPECO Computational Vibrational Spectroscopy Software*, Institute of Surface Chemistry, National Academy of Sciens of Ukraine, Kiev, **1990**.

[16] R. Sayle, *RasWin Molecular Graphics*, Windows version 2.6 Shareware software, **1995**, **1996**.

[17] V. D. Khavryutchenko, E. F. Sheka. *Phys. Low-Dim. Struct.*, **1995,** *Vol 4, No. 5*, 349.

[18] The Origin of the Problems with PM3 Core Repulsion Function:
I. Gabor (a), B. Csonka (a,b), J. G. Janos (b)
(a) Department of Inorganic Chemistry, Technical University of Budapest, H-1521 Budapest, Hungary.
(b) Laboratoire de Chimie Thaorique, Unicersite Henri Poincare, Nancy 1, B.P. 239, Vandoeuvre-Nancy Cedex, France; http://web.inc.bmc.hu/~csonka/crf.htm.

[19] A. A.Chuyko, Yu. I. Gorlov, *Khimia Poverkhnosti Kremnezema: Stroenie Poverkhnosti, Aktivnyr Tsentry, Mekhanizmy Sorbtsii* (The Chemistry of the Silica Surface: Structure of the Surface, Active Centers, Mechanisms of Sorption), Naukova Dumka, Kiev, **1992**.

[20] H. Balard, E. Papirer, A. Khalfi, H. Barthel, J. Weis, in: *Organosilicon Chemistry IV: From Molecules to Materials* (Eds.: N. Auner, J. Weis), Wiley–VCH, Weinheim, **1999**, p. 773.

Adsorption of Polydimethylsiloxane on Hydrophilic and Silylated Fumed Silica. Investigations by Differential Scanning Calorimetry

A. Altenbuchner, H. Barthel, T. Perisser, L. Rösch, J. Weis*

Wacker-Chemie GmbH,
Werk Burghausen
D-84480 Burghausen, Germany

Keywords: Fumed Silica / Polydimethylsiloxane / PDMS / Differential Scanning Calorimetry (DSC) / Glass Transition Temperature

Summary: Fumed silica is a synthetic amorphous silicon dioxide produced by burning silicon tetrachloride in an oxygen–hydrogen flame. Surface areas range from 50 m^2/g up to 400 m^2/g. Fumed silica is used as an active filler providing a high performance in reinforcing silicone rubbers. Adsorption interactions of the polydimethylsiloxane (PDMS) chains within the elastomeric network on the filler surface play an important role in the mechanisms of reinforcement. The aim of this paper is to study the impact of surface silylation on the adsorption behavior of PDMS polymers on hydrophilic and silylated surfaces. Differential scanning calorimetry (DSC) was used to monitor the impact of adsorption on the mobility of the PDMS chains. Typical DSC signals, like the glass transition, are observed for only the free polymer, but not for polymer chains adsorbed at the silica surface. The PDMS adsorption capacity of silicas at different degrees of silylation has been determined. On the free silica surface, adsorption results in about four to five layers of immobilised PDMS chain segments.

Introduction

The performance of fumed silica as an active reinforcing filler in silicone rubbers is strongly related to its adsorption interactions with the polydimethylsiloxane (PDMS) chains of the elastomeric network. However, fumed silica is also an excellent and efficient thickening agent. Therefore, to control its rheological action in industrial applications, fumed silica is submitted to surface silylation [1]. Inverse gas chromatography (IGC) at infinite dilution and at finite concentations has been demonstrated to provide important information on the adsorption interactions of PDMS chain segments, or oligomers, on hydrophilic and silylated silica surfaces at a molecular level [2]. Semiempirical quantum-chemical modeling revealed that the main mechanism of interaction is non-specific and related to dispersion energies [3], mainly. The aim of this paper is to study the impact of surface silylation on the adsorption behavior of PDMS polymers on hydrophilic and silylated surfaces. Solid-state ^{1}H-NMR [4] showed that adsorption on a silica surface results in a dramatic loss of mobility of PDMS polymers.

Differential scanning calorimetry (DSC) is a suitable technique to monitor temperature-dependent enthalpic changes in a system related to phase transition. It is well known that PDMS polymer shows a glass transition temperature T_g at about –120°C, and also phase transitions of crystallization and melting. Solid state ^{1}H-NMR [4] revealed a marked shift in its T_g, to about 0°C, when a PDMS polymer is adsorbed strongly on a silica surface. In this study we investigated the use of DSC to monitor the mobility of the PDMS chains on hydrophilic and silylated silica surfaces.

Experimental

Hydrophilic fumed silicas with BET surface areas of 126 m²/g (Wacker HDK S13) and 309 m²/g (Wacker HDK T30) have been used. Controlled silylation was performed by adding water, methanol and trimethylchlorosilane aerosols (molar ratio 1:1:1) to the mechanically stirred silica, followed by heating to 300°C for 2 h. The degree of silylation was determined by the carbon content (%C) of the silica, using a CS/244 Leco, and by the amount of residual silanol groups on the silica surface (%SiOH), before and after silylation, using an acid–base titration with aqueous NaOH in a water/methanol mixture (50:50), following a procedure according to Sears [5]. Data of hydrophilic and silylated silicas are presented in Table 1.

Table 1. Hydrophilic and silylated fumed silica samples.

Silica	Samples [a]	SiOH content [b] [%]	Carbon content [%]	BET surface area (N_2 at 78 K) [c] [m²/g]
HDK S13	S13-100	100	0.00	126
HDK S13	S13TMS66	66	0.57	113
HDK S13	S13TMS59	59	0.69	111
HDK S13	S13TMS50	50	0.74	112
HDK S13	S13TMS40	40	0.95	111
HDK S13	S13TMS15	15	2.00	96
HDK T30	T30-100	100	0.00	309
HDK T30	T30TMS84	84	0.57	300
HDK T30	T30TMS74	74	0.87	280
HDK T30	T30TMS61	61	1.44	279
HDK T30	T30TMS48	48	1.83	266
HDK T30	T30TMS41	61	2.05	263
HDK T30	T30TMS30	84	2.74	233
HDK T30	T30TMS15	15	4.67	183

[a] S13TMS*xx* and T30TMS*xx* series: S13 and T30 silicas, trimethylsiloxy (TMS) silylated, characterized by %SiOH = *xx*. — [b] Hydrophilic silicas S13-100 and T30-100: 1.8 SiOH/nm². — [c] Silylation of the silica surface decreases its surface energy and, therefore, the BET surface area also decreases, as it is related at the same time to the geometrical surface and the strength of the adsorption interaction itself. It has been shown elsewhere that silylation does not alter the geometrical surface area of a fumed silica markedly [6].

As PDMS, a trimethylsiloxy-endcapped silicone fluid with a viscosity of 1000 mPa s at 25°C was used (Wacker AK1000). For covering silica by PDMS, a mixture of PDMS/tetrahydrofuran

(THF) (25:75) was sprayed onto the stirred silica, followed by removal of THF at 120°C. Surface coverage of PDMS on hydrophilic or silylated silica was performed at 5 different levels at least, varying between 5 and 40 %wt. The PDMS coverage on silica was determined by %C. The analysis of %SiOH of the PDMS-covered silica confirmed very pure physisorption of PDMS: PDMS-covered silicas provide the same %SiOH as the parent silica, irrespective of the degree of silylation. Differential scanning calorimetry (DSC) was performed at –160 to +260°C with a Mettler 30, using silica amounts of about 10–20 mg silica. Measurements were performed at a scan rate of +10°C/min; rates of +2 and +20°C/min did not show markedly different results. Each scan was repeated at least 3 times; the experimental error of the resulting data in Δc_p (heat capicity), $-\Delta H_c$ (heat of crystallization) and ΔH_m (heat of melting) were <0.01 J/K g, <0.2 J/g and <0.1 J/g, respectively.

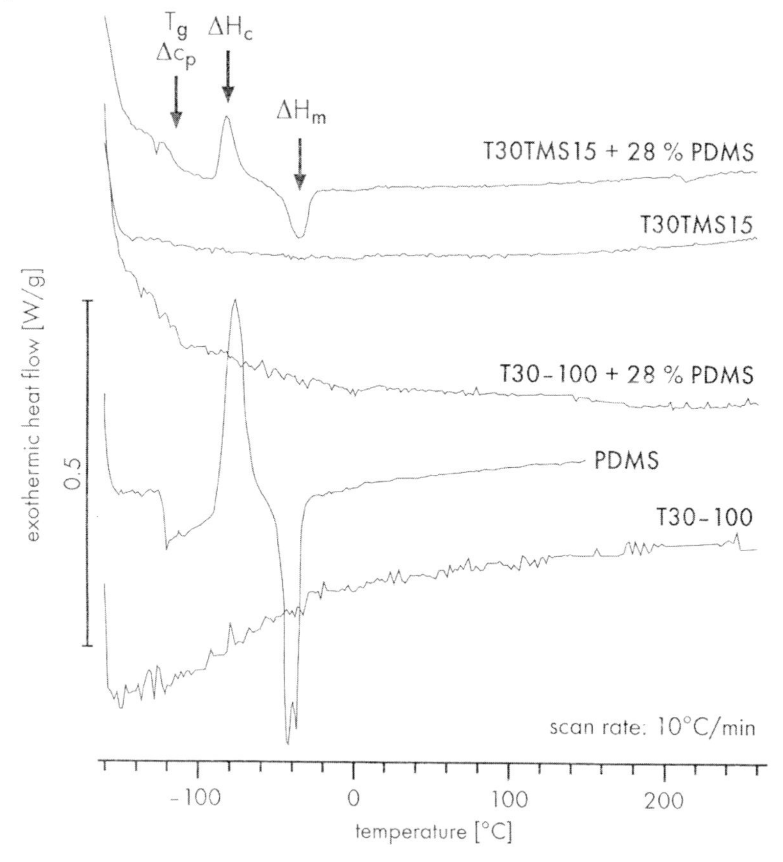

Fig. 1. DSC diagrams of pure HDK T30 (T30-100), of free PDMS polymer (PDMS), of T30-100 with 28 % PDMS, of T30TMS15 and of T30TMS15 with 28 % PDMS.

PDMS as a free silicone fluid exhibits three clearly distinct signals (Fig. 1):

(1) –119 ± 1°C: step in heat capacity Δc_p at T_g, the glass transition temperature;
(2) –80 ± 1°C: peak related to $-\Delta H_c$, the exothermic heat of crystallization:
(3) –42/-31 ± 1°C: twin peaks related to ΔH_m, the endothermic heat of melting at the pour point.

All three signals reflect the mobility of the PDMS chain: At –160°C, the starting temperature of our DSC scans, PDMS consists of amorphous and crystalline domains. At temperatures above T_g, enhanced chain mobility in the amorphous phases leads to a change in Δc_p, provided there are more than 15 dimethylsiloxy (DMS) units in the PDMS chain, which is true for a free PDMS polymer with a viscosity of 1000 mPa s (related to a chain length of about 200 DMS units). As samples are shock-frozen, there is always an excess of amorphous phase, which, with increasing mobility, crystallizes at –80°C: $-\Delta H_c$. At –35°C the system melts: ΔH_m.

Results and Discussion

Hydrophilic silicas

Silicas covered by low loadings of PDMS do not show any of these three distinct signals, revealing the immobilization of the polymer chains in the adsorbed state on the surface. It has been reported from solid-state ^{1}H-NMR that PDMS chains, strongly adsorbed on a silica surface, show a T_g shifted from –120°C (free PDMS) to about 0°C [4]. In our case, DSC monitors the disappearence of the Δc_p at –120°C. There is some qualitative indication of an ill-defined step at around 0°C in most of our DSC diagrams, which could be related to a T_g of adsorbed PDMS, shifted by more than +100°C, due to adsorption [7]. However, a quantitative exploitation of that region is not possible. Adsorption also suppresses the crystallization and the melting of PDMS, and again DSC is monitoring the loss of those both phase transitions of the polymer. However, overloading of the silica surface by PDMS results in the reappearance of those three typical DSC signals, indicating the occurrence of free and non-adsorbed polymer. We propose that the maximum amount of PDMS on the silica surface that does not show any DSC signals of the free polymer may be taken as a measure of the adsorption capacity of the silica surface towards PDMS (Fig. 2).

a)

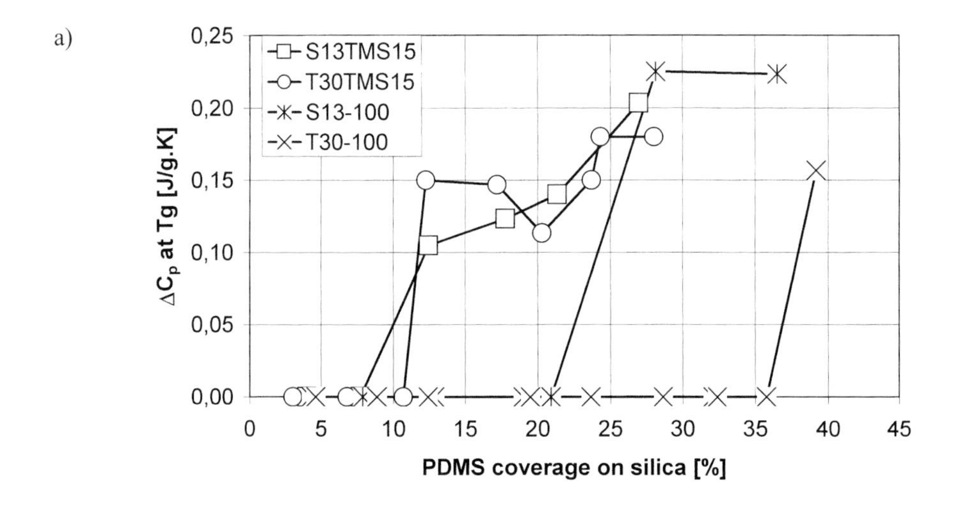

b)

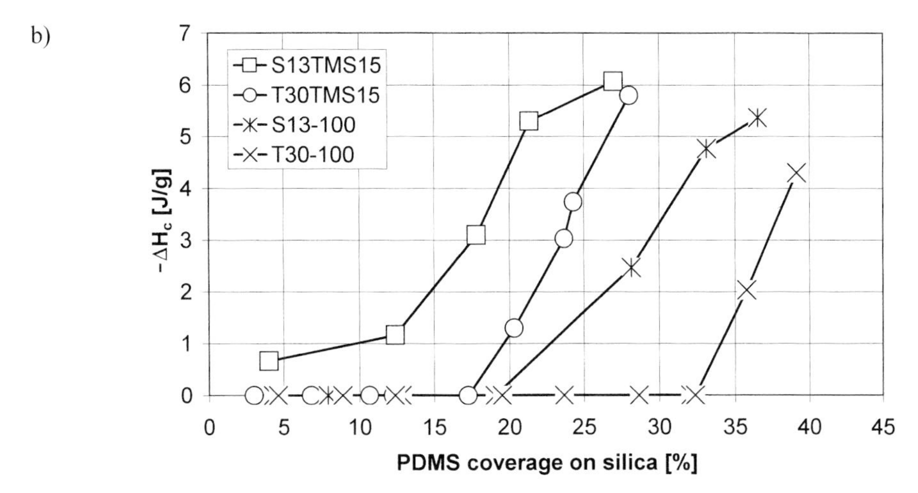

Fig. 2. (a) Δc_p and (b) $-\Delta H_c$ versus PDMS coverage on hydrophilic and silylated silicas S13, T30, S13TMS15 and T30TMS15.

Owing to their BET surface areas, the adsorption capacity of T30 should be $f = 309/126 = 2.5$ times larger than that of S13. Our data show markedly lower factors, revealing a PDMS-accessible surface area for T30 of about 160 m^2/g with regard to Δc_p or 210 m^2/g with regard to $-\Delta H_c$ and ΔH_m. It has been reported elsewhere that microporosity weakens the reinforcing ability of fumed silica in a silicone rubber [8], by lowering the adsorption capacity of the silica surface towards PDMS chains. The surfactant hexadecylammonium bromide, CTAB, has been proposed to model the adsorption of polymers on microporous surfaces. Comparing the factor f between the CTAB surface area of T30 and S13 (molecular area $a_{CTAB} = 0.4$ nm^2), $f = 245/126 = 1.9$, with that resulting from PDMS adsorption, reveals a larger space requirement of an adsorbed DMS unit within the PDMS chain as one would expect from a monolayer coverage of individual DMS units only ($a_{DMS} = 0.3$ nm^2). The same relationship generally holds also for silylated T30TMS15 and S13TMS15. Obviously, silylation has no marked influence on the microporosity of T30. It has been proposed that microporosity on a silica T30 occurs as surface patterns that are poor in SiOH groups and therefore are not involved in the silylation reactions [2].

Silylated Silica

Silylating the silica by trimethylchlorosilane decreases the surface silanol content of the silica, and also the free space of oxide surfaces. As seen from Figs. 2, 3 and 4, silylation shifts the onset of DSC signals from about 20 % of PDMS coverage to <10 % PDMS coverage in the case of S13, and from >35 % PDMS coverage to about 10 % PDMS coverage for T30. This shift generally proceeds with increasing silylation, and it is observed for all three kinds of DSC signals, namely Δc_p, $-\Delta H_c$ and ΔH_m.

Δc_p versus PDMS coverage provides a step-like behavior, which is readily understandable remembering that at least 15 DMS units should be in a chain in the non-adsorbed amorphous state to yield a T_g. With regard to $-\Delta H_c$ and ΔH_m, both transitions show no signals below a characteristic coverage of PDMS. But above that characteristic coverage, the signals increase linearly with the PDMS content. That fits again with an interpretation involving an adsorption-related immobilization, suppressing any DSC signals. If PDMS is present in amount larger than the maximum adsorption capacity of the silica surface, then any mobile chain units contribute to crystallization or melting enthalpies.

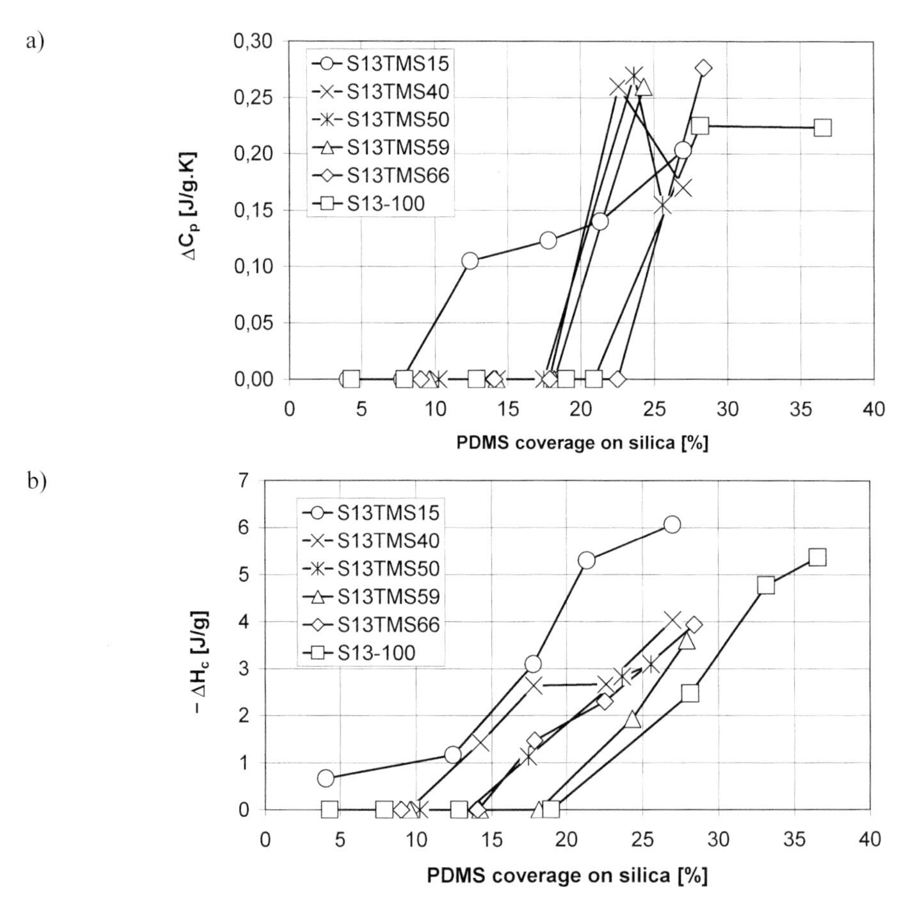

Fig. 3. (a) Δc_p and (b) $-\Delta H_c$ versus PDMS coverage on hydrophilic and silylated silicas S13: S13-100 and S13TMS*xx*.

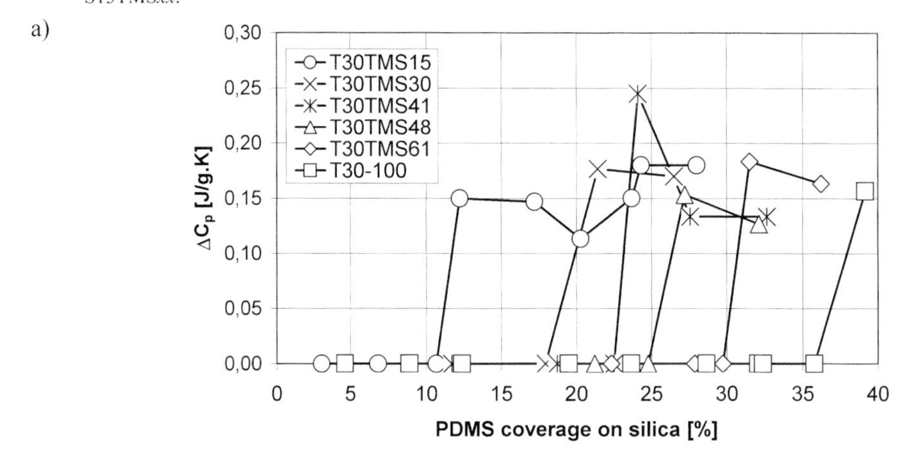

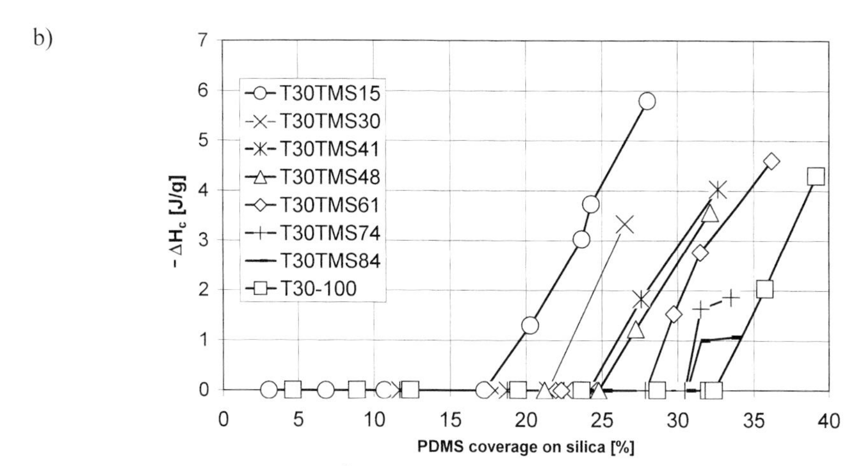

Fig. 4. (a) Δc_p and (b) $-\Delta H_c$ versus PDMS coverage on hydrophilic and silylated silicas T30: T30-100 and T30TMS*xx*.

The temperatures of all DSC signals of PDMS-overloaded silicas fit well with that temperatures of free PDMS (see subscript to figure 1): Δc_p / T_g at -118 ± 2°C, $-\Delta H_c$ at -81 ± 3°C and ΔH_m at -39 ± 2°C. This shows clearly that the signals are related to non-adsorbed and free polymer segments.

Expressing the degree of silylation as %SiOH, the "degree of silylation–PDMS adsorption capacity" relationship is reversed for S13, with regard to S13TMS59 and S13TMS66 (Fig. 3b). However, it is true if using %C as the leading parameter of the degree of silylation (see Table 1).

Plotting the PDMS adsorption capacity of the silica surface — using the lowest coverage by PDMS without any detectable DSC signals — versus the degree of silylation, expressed by %SiOH, reveals a marked change of the adsorption behavior at 40–50 % of silylation (see Fig. 5). A similar transition has been observed for direct interactions of silica particles themselves [8]. By IGC using PDMS oligomers — i.e. on a molecular level of adsorption — it was proposed that at >50 % of silylation only isolated islands of TMS groups exist and, therefore, the free SiO_2 surface dominates properties. But at <50 % of silylation the free SiO_2 surface forms isolated islands only, and the TMS groups dominate surface properties [2]. In our case of polymer adsorption, below 40–50 % of silylation the adsorption capacity of silica versus PDMS increases strongly with %SiOH, i.e. the amount of free SiO_2 surface. Above 50 % of silylation, the adsorption capacity of the silica versus PDMS remains mainly constant. Obviously, isolated islands of TMS groups do not markedly interfere with the adsorption of long-chain polymers, whereas a dense coverage of TMS groups hinders the adsorption of PDMS, and the amount of free SiO_2 patches controls the adsorption capacity.

The change of adsorption capacity becomes more and more distinct in the series $\Delta H_m < -\Delta H_c < \Delta c_p$ and it is more pronounced with S13 than with T30. The phase transition "melting" seems to be less sensitive to immobilization by adsorption than "crystallization" or "glass transition", the latter being the most sensitive one. On a structurally homogeneous surface, as on S13, steric hindrance by trimethylsiloxy units controls the accessible free SiO_2 surface and therefore dominates the interactions with the PDMS chains. However, on a surface exhibiting additional patterns of microporosity, as on T30, surface structure and steric hindrance control adsorption [2]. Hence, in the latter case of T30, the immobilization of PDMS by adsorption is less clearly described by the degree of silylation of the surface, alone.

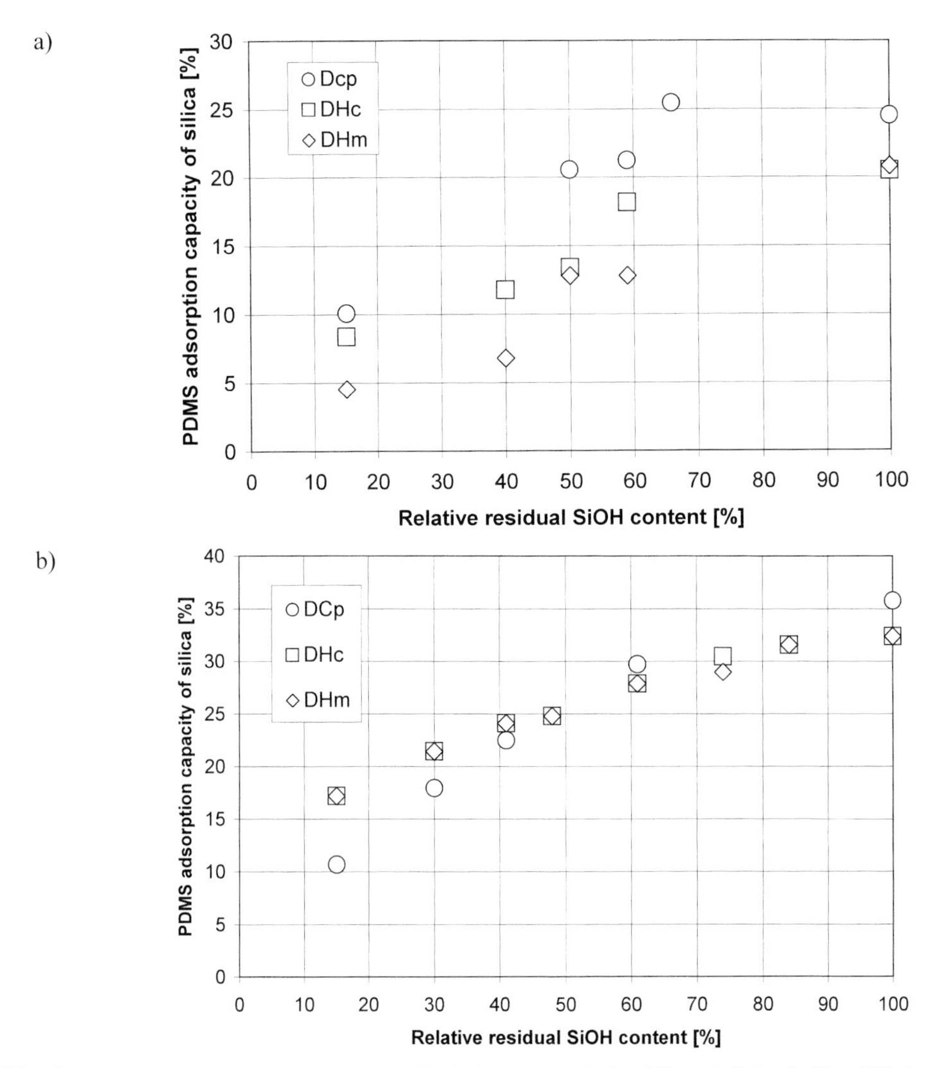

Fig. 5. (a) S13 and (b) T30 versus relative residual SiOH content hydrophilic and silylated silicas [%]: Δc_p (Dcp), $-\Delta H_c$ (–DHc) and ΔH_m (DHm).

Using the adsorption capacity of the silica surface with regard to PDMS, as calculated from the largest PDMS coverage without DSC signals, it is possible to calculate the adsorption capacity of DMS units, too. Together with a molecular area $a_{DMS} = 0.3\ nm^2$, a DMS-related surface area may be estimated. It turns out that this DSC-measured DMS surface area is a multiple of the BET surface area. Obviously, adsorption of the PDMS polymer leads to a multilayer coverage of subsequently immobilized layers of DMS units on the silica surface (Table 2). However, attempts to relate that multilayer coverage to the area of the free silanol surface, evaluated from %SiOH times the total BET surface area (N_2 at 78 K) remain ambiguous: the number of surface layers is spread over a range of 2–9. Recently, hexamethyldisiloxane (Si2) has been used to determine the accessible free SiO_2 surface of hydrophilic and TMS-silylated silicas [2]. Relating the DSC-measured DMS surface area to this Si2 surface area yields a number of immobilized DMS layers of 4–5, in excellent agreement with data from solid-state ^{1}H-NMR [4].

Table 2. DSC-measured DMS surface area (a_{DMS} = 0.3 nm^2), DMS layers related to the SiOH surface area (= %SiOH × BET), Si2 surface area measured by IGC [2] and immobilized layers of DMS units: all DSC data based on the signals of PDMS crystallization, $-\Delta H_c$.

Silica	DSC: DMS surface area [m^2/g]	SiOHsurface (%SiOH×BET) [m^2/g]	Layers of DMS on SiOH surface	Surface area using Si2 [m^2/g]	Layers of DMS on an Si2 surface area
T30-100	791	309	2.6	194	4.1
T30TMS84	771	260	3.0	195	4.0
T30TMS74	744	229	3.3	145	5.1
T30TMS61	681	188	3.6	142	4.8
T30TMS48	605	148	4.1	139	4.4
T30TMS41	589	127	4.7	131	4.5
T30TMS30	523	93	5.6	112	4.7
T30TMS15	420	46	9.1	100	4.2

Conclusion

Application of differential scanning calorimetry to PDMS provides suitable signals to monitor the mobility of polymer chains: at the glass transition temperature $T_g = -119 \pm 1$°C a step in the heat capacity Δc_p, at $T = -80 \pm 1$°C a peak related to the exothermic heat of crystallization $-\Delta H_c$ and at the pour point with $T = -42 \pm 1$°C a peak related to the endothermic heat of melting ΔH_m. All DSC signals turned out to be highly sensitive to the immobilization of the PDMS chain due to adsorption: only the free, non-adsorbed polymer shows the signals as mentioned above, and adsorption on the silica surface leads to their disappearance. By loading silicas, of different degrees of silylation by trimethylchlorosilane, with increasing amounts of PDMS, the adsorption capacity of the silica with regard to PDMS has been determined. Our investigations show that DSC provides data suitable to describe the adsorption behavior of PDMS polymers on hydrophilic and silylated fumed silica surfaces. Adsorption of PDMS on the free SiO_2 surface of silica leads to 4–5 layers of immobilized dimethylsiloxy units. With regard to the degree of silylation there is a marked change in the adsorption behavior at about 40–50 % of silylation.

References:

[1] H. Barthel, L. Rösch, J. Weis, *Organosilicon Chemistry II: From Molecules to Materials* (Eds.: N. Auner, J. Weis), VCH, Weinheim, **1996**, p.761.

[2] a) H. Balard, E. Papirer, A. Khalfi, H. Barthel, J. Weis, *Organosilicon Chemistry IV: From Molecules to Materials* (Eds.: N. Auner, J. Weis), Wiley–VCH, Weinheim, **1999**, p.773.
b) A. Khalfi, "Etude des Propriétés de Surface de Silices de Combustion Silanisées, par Chromatographie Gazeuse Inverse, en Vue de leur Utilisation pour le Reforcement du PDMS", Thesis, L'Université de Haute Alsace, Mulhouse, France, **1998.**

[3] E. Nikitina, V. Khavryutchenko, E. Sheka, H. Barthel and J. Weis, *Organosilicon Chemistry IV: From Molecules to Materials* (Eds.: N. Auner, J. Weis), Wiley–VCH, Weinheim, **1999**, 745.

[4] V. M. Litvinov, *Organosilicon Chemistry II: From Molecules to Materials* (Eds.: N. Auner, J. Weis), VCH, Weinheim, **1996**, p. 789.

[5] G. W. Sears, *Anal. Chem.* **1956**, *28*(12), 1981.

[6] H. Barthel, *Proc. Fourth Symp. on Chemically Modified Surfaces* (Eds. H. A. Mottola and J. R. Steinmetz), Elsevier, Amsterdam, **1992**, p. 43.

[7] V. M. Litvinov, private communication, **1998**.

[8] H. Barthel, *Colloids Surf. A: Physicochem. and Eng. Aspects* **1995**, *101*, 217–226

[9] H. Barthel, F. Achenbach, H. Maginot, *Proc. Int. Symp. on Mineral and Organic Functional Fillers in Polymers (MOFFIS 93)*, Université de Namur, Belgium, **1993**, p. 301.

Initial and Silylated Silica Surfaces: Assessing Polydimethylsiloxane – Silica Interactions Using Adsorption Techniques

H. Balard*, E. Papirer, A. Khalfi

Institut de Chimie des Surfaces et Interfaces (CNRS)
15, rue Starky BP 2488, 68096 Mulhouse, France
Tel: Int. code + 3 89 60 87 00 – Fax: Int. code + 3 89 60 87 99
E-mail: h.balard@mulhouse-univ.fr

H. Barthel, J.Weis

Wacker-Chemie GmbH
Johannes-Hess-Str. 24, D-84489 Burghausen, Germany
Tel: Int. code + (8677)83 4833 – Fax Int. code + (8677)83 3093
E-mail: herbert.barthel@wacker.de

Keywords: Silica / Silylation / Inverse Gas Chromatography / Surface heterogeneity / PDMS oligomers

Summary : Inverse gas chromatography (IGC), at infinite dilution conditions, is a very sensitive method for the monitoring of surface property changes of divided solids submitted to thermal or chemical treatments. For instance, the evaluation of the surface activity of fumed silica samples, prepared by controlled silylation processes, is a key for the tailoring of silicas used as fillers in the PDMS elastomer industry. The aim of the present work is to demonstrate how IGC measurements, using PDMS oligomers as specific molecular probes, are suitable for evidencing the strong influence of the degree of silylation, but also of the surface roughness, upon the surface properties of fumed silica samples. Simple models describing both the unmodified and silylated pyrogenic silica surfaces are proposed.

Introduction

Fumed silicas are convenient reinforcing fillers for polydimethylsiloxane (PDMS) elastomers. Unfortunately, because of the high reactivity of the bare silica surface, uncured mixes of fumed silica and PDMS harden slowly during storage, hindering any further use. A controlled and partial silanization of the silica surface prevents such hardening by lowering its PDMS adsorption capacity at room temperature. But one has to find the right balance between the surface deactivation and its residual adsorption activity.

Testing this residual activity is therefore of major importance. We demonstrated previously [1–3] that inverse gas chromatography (IGC), at infinite dilution and at finite concentration conditions, is a very sensitive and powerful method for the characterization of filler surfaces. More recently, we showed that PDMS oligomers, used as molecular probes, are much more appropriate than are classical IGC probes like *n*-alkanes, chloroalkanes or ether, for evidencing the surface property changes of a fumed silica surface modified by a controlled silylation process [4]. The aim of the present work is to demonstrate how IGC, using again PDMS oligomer probes, allows one to monitor precisely the influence of the presence of trimethyl groups on the surface properties of silylated silica samples, in a large range of silylation ratios, and permits one finally to propose a model of the surface topology of a fumed silica surface.

Silicas and Silylated Silicas

Hydrophilic Silicas

Two hydrophilic silicas from Wacker-Chemie, HDK S13 denoted as S13 and HDK-T30 denoted as T30, were used for the preparation of silylated silica samples. Their characteristics are reported in Table 1.

Table 1. Characteristics of the S13 and T30 silicas.

Silica	S_{spe} N_2 [m^2/g]	S_{spe} CTAB [m^2/g]	D_S CTAB	D_S SAXS	[OH/nm^2]
S13	131	133	2.0	2.0	1.83
T30	300	245	2.5	2.2	1.78

The chosen silica samples differ not only by their respective specific surface areas, but also by their surface roughness. If the HDK-S13 silica surface, presenting a fractality parameter of 2, can be considered to be flat at the molecular scale, the microporous nature [5] of the HDK-T30 silica surface generates a rough surface having a fractality parameter approaching 2.5.

Silylated Silicas

The two hydrophilic silicas were silylated by reaction with trimethylchlorosilane (TMCS). The silylation was performed on a silica sample, impregnated with 50 % (w/w) of a water methanol mix (1/1), by spraying variable amounts of silane in order to control the TMS coverage ratio. Then, the treated silica samples were dried at 300°C for 2 h. The characteristics of the modified silicas are reported in Table 2.

Table 2. Main molecular characteristics of S13TMS*x* and T30 TMS*x* silica samples.

Silica sample[a]	n_{OH} SiOH/nm^2	n_g (N_2) TMS/nm^2	n_g(CTAB) TMS/nm^2	$n_g + n_{OH}$	τ[b] [%]
S13	1.83	0.0	0.0	1.83	0
S13TMS90	1.65	0.27	0.27	1.97	12
S13TMS81	1.48	0.49	0.49	1.97	21
S13TMS73	1.34	0.56	0.56	1.90	24
S13TMS61[c]	1.12	0.96	0.96	2.08	41
S13TMS44	0.81	1.09	1.09	1.90	47
S13TMS41	0.75	1.23	1.23	1.98	52
S13TMS40	0.73	1.32	1.32	2.05	53
S13TMS32	0.59	1.58	1.58	2.17	68
T30	1.70	0.00	0.00	1.70	0
T30TMS85	1.45	0.32	0.38	1.83	17
T30TMS76	1.29	0.57	0.69	1.98	30
T30TMS63	1.07	0.80	0.97	2.04	42
T30TMS60	1.02	0.80	0.96	1.98	41
T30TMS57	0.97	0.88	1.06	2.03	46
T30TMS52	0.88	0.99	1.19	2.07	51
T30TMS50	0.85	1.07	1.30	2.15	56
T30TMS39	0.66	1.12	1.35	2.01	58
T30TMS31	0.53	1.32	1.61	2.14	69

[a] The silica samples are quoted in the following systematic way: symbol of the initial silica followed by TMS (for trimethylsilyl graft) and *x* (percentage of residual silanol groups, determined by acid–base analysis with regard to the initial surface density of the unmodified silica) For example, S13TMS32 means a silica S13 silylated using TMCS, having a residual density of silanol groups equal to 32 %. — [b] The relative coverage ratios of the surface by TMS groups are calculated taking a TMS molecular area equal to 43 Å^2. — [c] This silica sample was prepared using trimethylethoxysilane instead of TMCS as silylation reagent.

It is worth to point out that :

(1) The total number of functional groups (SiOH + TMS) per nm^2 remains quite constant taking into account the measurement errors on both elemental carbon and SiOH determinations: a proof of the absence of any secondary chemical reaction.

(2) For a given surface density of residual silanol groups, the surface density of TMS groups (taking into account the nitrogen BET surface area) is systematically lower for the T30TMS*x* silica samples than for the S13TMS*x* ones. But, if the TMS surface density is calculated using the CTAB specific surface area, then they become quite comparable for a given density in residual silanol groups. The optimal surface density, of about 1.6 TMS group per nm^2, is reached for the samples (S13TMS32 and T30TMS39) having the highest degree of silylation.

This latter observation is clearly highlighted when plotting the TMS surface densities, calculated using both the specific surface areas calculated from nitrogen and CTAB adsorption measurements, versus the surface density of residual silanol groups as depicted on Fig. 1.

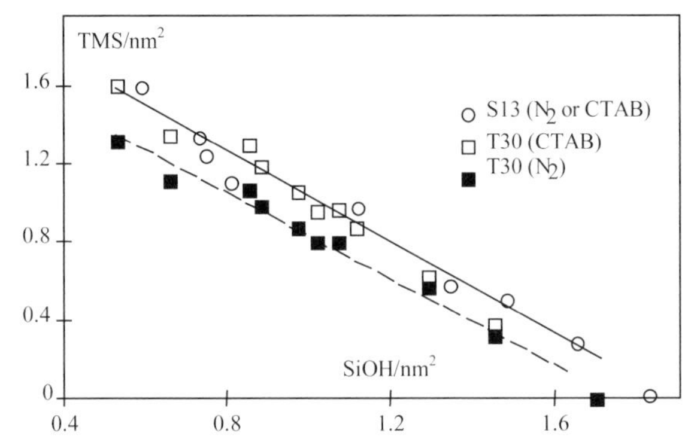

Fig. 1. Variation of the surface density of TMS groups, calculated using either N_2 or CTAB specific surface area determination methods, with the surface density of residual silanol groups on S13TMS*x* and T30TMS*x* silica samples.

The difference in surface roughness between the S13 and T30 silicas permits one to explain easily their different behaviors towards the reaction of silylation. In the case of the S13 silica, which is flat at the molecular level, all silanol groups are accessible to the TMCS silylation agent and the maximum of the grafting ratio is only limited by the bulkiness of the TMS groups. On the contrary, on the rough surface of T30, part of the silanol groups are located in the microporous structure and, therefore, cannot be silylated for steric reasons: only the outer surface (estimated from CTAB adsorption measurements) of the particles may in fact be silylated. Hence, the CTAB molecule and its trimethylammonium radical head, having a bulkiness close to that of the TMCS reagent, gives a good evaluation of the accessible surface susceptible to silylation. It becomes obvious that both studied silicas behave identically when the surface density of TMS groups is calculated using the CTAB specific surface area.

Study of the Surface Properties using Inverse Gas Chromatographic Techniques

The interaction capacity of these silylated silicas, with PDMS monomer units, was evaluated using inverse gas chromatography both at infinite dilution and in finite concentration conditions, using as probes PDMS oligomers having the general formula $(CH_3)_3SiO[Si(CH_3)_2]_n$-$CH_3$, denoted thereafter as Si(n+1), e.g. Si2 for the hexamethyldisiloxane.

IGC at Infinite Dilution Conditions (IGD-ID)

Free Energies of Adsorption of PDMS Oligomers

IGC-ID gives easy access to the standard variations of free energies of adsorption of PDMS oligomers (Si2, Si3, Si4, Si5 and Si6) from their net retention volumes, calculated knowing their net retention times and the flow rate of the carrier gas. The free energies of adsorption are related to the net retention volumes by Eq.1.

$$\Delta G_a^0 = -RT \ln V + C$$

Eq. 1.

where R is the ideal gas constant, T the absolute temperature, V the net retention volume of the probe, i.e., the gas volume necessary to push the probe through the chromatographic column, and C is a constant depending on the choice of the reference state of the adsorbed molecule.

Figure 2 displays the evolution of the standard free energies of adsorption of the oligomer probes on the silylated S13TMSx and T30TMSx silica samples.

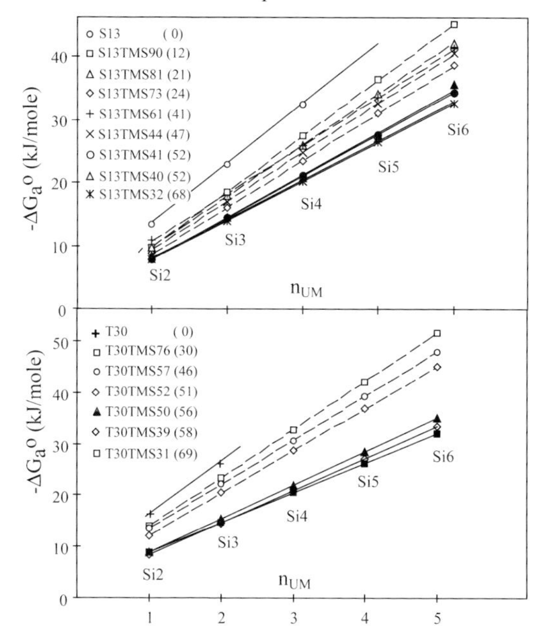

Fig. 2. Variation of the standard free energies of adsorption ΔG_a^0 of linear PDMS oligomers with the number of monomer units of the probes for the silylated S13TMSx and T30TMSx silica samples (measured at 170°C). The coverage ratios of the silica surface by the TMS groups [%] are indicated in parentheses.

As expected, on the most interactive surfaces, those of the unmodified silicas, only the 2 or 3 first oligomers are eluted and the corresponding "oligomer straight line", obtained when plotting ΔG_0^a versus the number of monomer units of the injected oligomers, lies apart, above all the other straight lines corresponding to the silylated silica samples that present a lower capacity of interaction. But, a completely unexpected result clearly appears since silylated silica samples may now be classified in 2 well-separated groups. The gap between these groups corresponds to a very slight change of the surface coverage by the TMS groups, occurring at a coverage ratio of around 50 %.

This jump is evidenced when plotting the variation of the free energy of adsorption of Si3 versus the coverage ratio as depicted on Fig. 3.

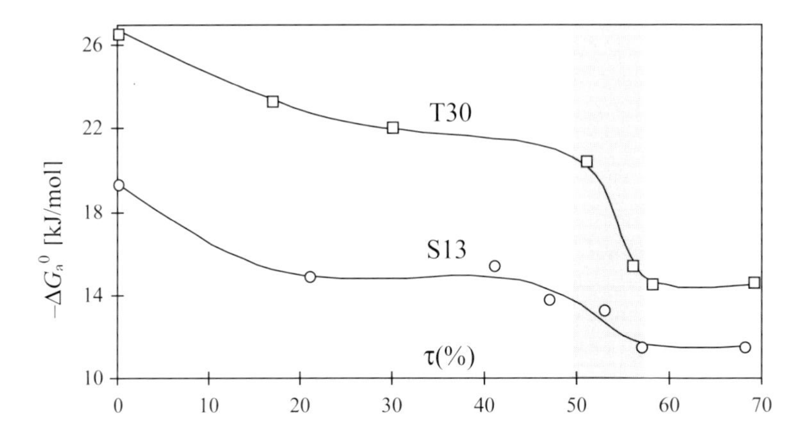

Fig. 3. Variation of the standard free energies of adsorption (ΔG_a^0) of Si3 oligomer with the TMS coverage ratios of the silica surface of silylated S13TMS*x* and T30TMS*x* silica samples (measured at 170°C).

Increment of Free Energies of Adsorption per DMS Monomer Unit.

Extending the approach of Dorris and Gray [6] to the PDMS oligomers, we may calculate an increment of free enthalpy of adsorption corresponding to a monomer unit, ΔG_a^{UM}, according to:

$$\Delta G_a^{UM} = -RT \ln (V_{n+1}/V_n)$$

Eq. 2.

where V_n and V_{n+1} are the retention volumes of oligomers having respectively n and n+1 monomer units.

Figure 4 displays the variation of ΔG_a^{UM} versus the coverage ratio of the surface by the TMS groups for both S13TMS*x* and T30TMS*x* silica samples.

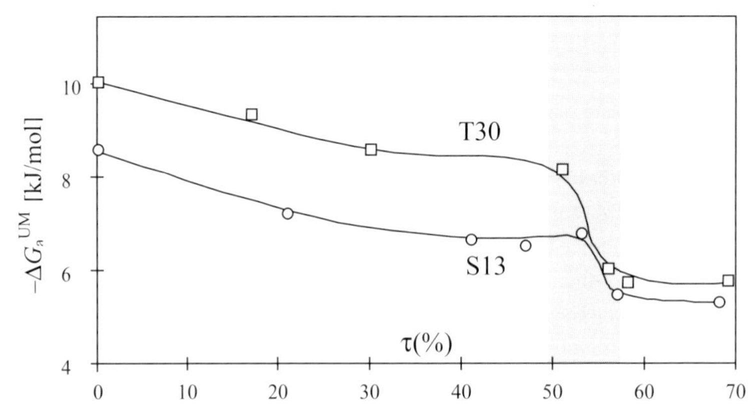

Fig. 4. Variation of the increment of the free energy of adsorption per monomer unit (ΔG_a^{UM}) of linear PDMS oligomers with the TMS coverage ratios of the silica surface of silylated S13TMS*x* and T30TMS*x* silica samples (measured at 170°C).

After a slight decrease of ΔG_a^{UM} when going from the initial unmodified silica to a silica having a coverage ratio of about 25 %, that is due probably to the disappearance of the adsorption sites having the highest capacity of interaction, a jump in the surface properties is clearly evidenced around a surface coverage ratio of about 50 %, as previously observed for the variations of the free energies of adsorption.

In order to explain such an evolution, we propose a very simple model describing the evolution of the surface properties of a silica surface submitted to a controlled silylation process. To describe a silylated surface, we first distribute randomly the TMS grafts on a model surface made of a hexagonal bi-dimensional network of sites as previously proposed by Sindorf and Maciel [7]. Such a silylated surface corresponding to a 30 % surface coverage by the TMS groups is shown on Fig. 5.

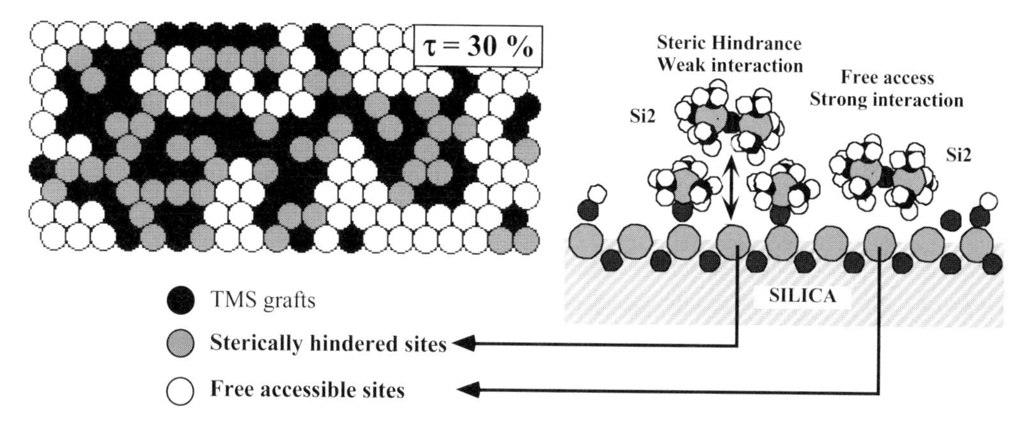

Fig. 5. Schematic representation of a TMS silylated silica surface having a surface coverage ratio of 30 % by the TMS groups and of the interaction of the bulky Si2 probe with the two types of adsorption sites on a silylated silica surface.

On this figure, one can distinguish three main types of species: the grafted TMS groups (black circles), the free accessible sites of adsorption (white circles) and sterically hindered sites of adsorption (gray circles) with which the oligomer probe, here the Si2, can only interact weakly because of the steric hindrance induced by two neighboring TMS groups as depicted on the right-side of Fig. 5.

In a chromatographic process, a given molecule is submitted to a high number of sorption and desorption steps. If the motion of a molecule along the chromatographic column is mainly ensured by the gas flow, the molecule can also migrate on the solid surface itself. If the surface is energetically perfectly homogeneous, this phenomenon will have no influence on the measured retention time. On the contrary, on a heterogeneous surface, the sites having the highest energy of interaction will act as potential wells quenching the migrating adsorbed molecule. The residence time τ_r of the molecule on any site is proportional to the exponential of its energy of interaction according to Eq. 3.

$$\tau_r = t_o \exp(\varepsilon/RT)$$

Eq. 3.

Hence, the sites having the highest energies of interaction will mainly contribute to the global retention time of the probe, which is a complex function of the energy of interaction of the sites present on the surface, their number and finally their capture radius, because it is obvious that a strong interactive site can only capture a molecule that is adsorbed in its neighboring environment. The globally measured retention time t_r is then given by Eq. 4.

$$t_r = \Sigma\ (n_S, \varepsilon_S, r_S)$$

Eq. 4.

where n_S is number of sites having the energy of interaction ε_S and a capture radius r_S. All those parameters are at present unfortunately unknown.

Finally, the molecule after a given time will desorb and migrate in the gas phase. Therefore, the molecule only visits a very reduced part of the solid surface. In other words, the probe will be most sensitive to the surface heterogeneity over a very short distance that may be estimated very roughly to several molecular lengths, depending strongly, of course, on the temperature of measurement.

On a non-silylated initial silica, the adsorbed molecule can freely migrate between its site of adsorption and that of desorption. Introducing TMS grafts on the surface, acting as purely geometrical barriers, will necessarily restrict its mobility. With increasing coverage ratios of the silica surface by the TMS, this limiting effect will show up as illustrated on Fig. 6, which displays the evolution of the surface characteristics with increasing silylation ratios.

It is seen that, for coverage ratios equal or higher than 50 %, TMS groups cover the majority of the surface and that the residual sites of adsorption form only some "holes" in this continuous TMS coverage. Consequently, the molecule can no longer migrate on the surface towards the more interactive sites and therefore a sudden decrease of the retention time is observed. In fact, this transition between a discontinuous coverage by TMS groups to a continuous one corresponds to the disappearance of the connectivity of the surface between the adsorption sites, over a short distance.

When looking at Fig. 6, it appears that another transition should be observed for a coverage ratio of 25 %, when the free accessible sites of adsorption begin to form islands in a continuous coverage made of both sterically hindered surface sites and TMS groups. In fact, this particular coverage ratio corresponds to the disappearance of the long-range connectivity between free accessible adsorption sites.

In other words, a molecule can no longer migrate all around the silica surface without being influenced by the presence of TMS groups. As pointed out above, the residence time of an isolated molecule is too short for it to be sensitive to a change of the long-range connectivity. Therefore, IGC-ID experiments are unable to detect this transition. To perform this detection, one may use an object much greater in size than that of the isolated probe: a population of molecules fulfills this conditions and, in fact, this transition may thus be evidenced. The method is based on the study of the interaction of a population of molecules with the solid surface, as is the case when performing IGC at finite concentration conditions (IGC-FC), or determinations of wettability by 2-propanol/water mixtures or viscosity measurements, as we shall see it below.

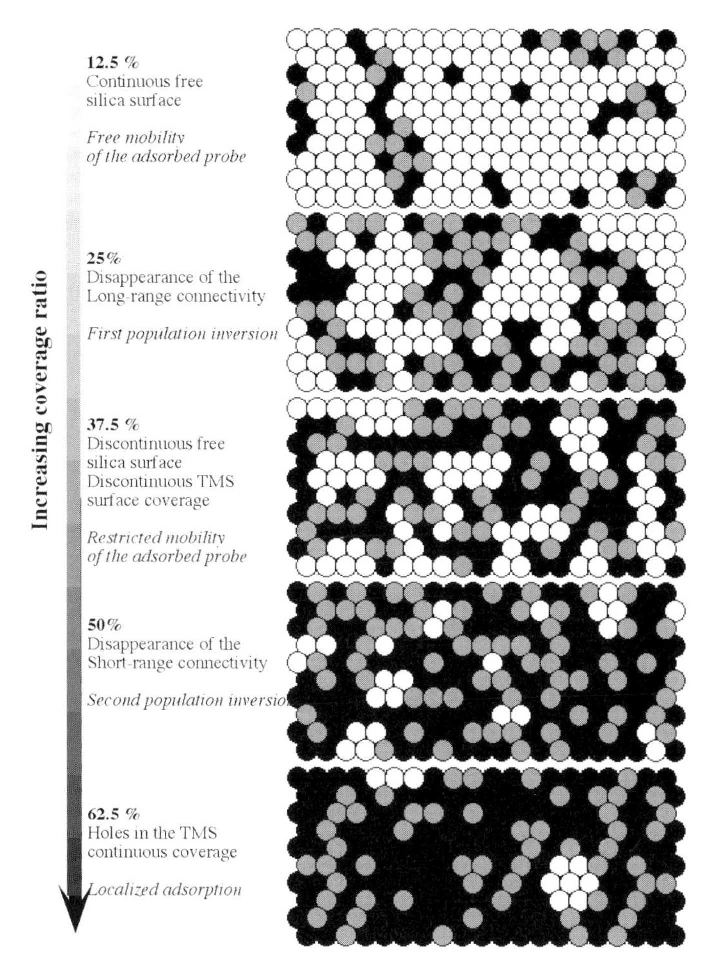

Fig. 6. Schematic representation of the silylated silica surface for increasing coverage ratios by TMS groups: 12.5, 25, 37.5, 50, and 62.5 %.

Inverse Gas Chromatography at Finite Concentration Conditions (IGC-FC)

Characteristic Constants of the Isotherms

In contrast to IGC-ID, a measurable amount of probe is injected in the column resulting in a large range of coverage ratios of the solid surface by the probe. Injections of liquid probe of a few microliters are required for ICG in finite concentration conditions. The injected volume is mainly dependent on the surface area developed by the solid in the chromatographic column, and on the measurement temperature.

For this study of the silylated silica surfaces, we have chosen the PDMS oligomer having the lowest degree of polymerization — Si2. This is a bulky molecule that will be mainly sensitive to the surface heterogeneity due to the progressively increasing density of TMS groups because of its lack of polarity [8].

Different IGC techniques for the determination of the adsorption isotherms have been reviewed by Conder [9]. The simplest one, "the elution characteristic point method" (ECP), allows the acquisition of the desorption isotherm from a single chromatographic experiment. Using this method, the first derivative of the adsorption isotherm can be readily calculated starting from the retention times and the signal heights of characteristic points taken on the diffuse descending front of the chromatogram, according to Eq. 5.

$$\left(\frac{\partial N}{\partial P}\right) = \frac{J D t'_{\mathrm{r}}}{mRT}$$

Eq. 5.

where N is the number of absorbed molecules, P the pressure of the probe at the output of the column, t'_{r} the net retention time of a characteristic point on the rear diffuse profile of the chromatogram, J the James and Martin coefficient, D the output flow rate and m the mass of adsorbent. In practice, the experimental conditions have to be carefully controlled in order to obtain a physically meaningful first derivative of the isotherm.

From the adsorption isotherm, it is possible to assess the specific surface area of the solid and the BET constant corresponding to the given organic probe.

The evolution of the surface specific area with the coverage ratio of the silica surface by the TMS groups is displayed Fig. 7.

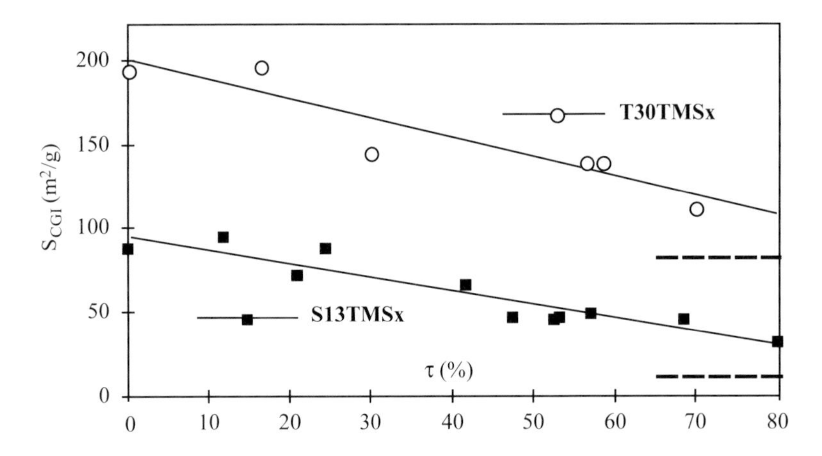

Fig. 7. Evolution of the specific surface areas of the S13TMS*x* and T30TMS*x* silica samples (measured at 63°C, using IGC-CF and Si2) with the silica surface coverage ratios by the TMS groups.

In the case of the S13TMS*x* silicas, it is observed that the specific surface area measured using Si2 decreases linearly with increasing coverage ratios and would approach zero for a theoretical coverage ratio of 100 %: a proof that, in these experimental conditions, the Si2 probe interacts only with the free residual silica surface and provides a convenient method to estimate the latter. The T30TMS*x* silica samples follow the same trend, but the value extrapolated at a coverage ratio of 100 % is not equal to zero and corresponds approximately to one-third of the initial specific surface area measured with the same molecule. This indicates that, due to the influence of the surface roughness, the TMS groups are unable to form a regular and continuous coverage over all the silica T30 surface. This point will be discussed again below.

Figure 8 displays the evolution of the BET constant with the coverage ratio of the silica surface by the TMS groups.

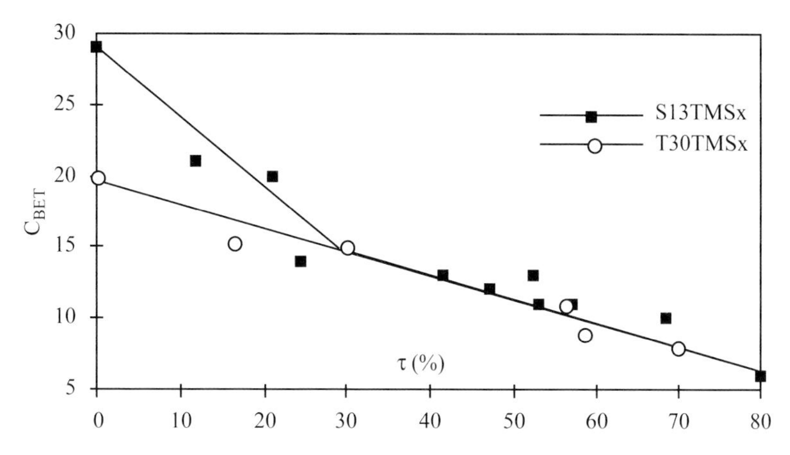

Fig. 8. Evolution of the BET constants of Si2, measured using IGC-CF at 63°C, on the S13TMS*x* and T30TMS*x* silica samples with the silica surface coverage ratios by the TMS groups.

Again the two series of silylated silica samples behave differently. The BET constants of the T30TMS*x* series decrease linearly with increasing coverage ratios, whereas a break point is observed around a coverage ratio equal to 25 % for the S13TMS*x* series. The initial S13 silica exhibits a slightly higher BET constant value than the more interactive T30 silica. A reasonable explanation is that the surface roughness of the T30 silica prevents the formation of a monolayer as dense as the one formed on the more regular S13 silica. At this point, it is important to underline that, as previously predicted, a population of molecules is able to detect the first transition, but only on a flat and homogeneous surface, a condition contained implicitly in our model and that is not fulfilled by the T30 rough silica surface.

We will now consider how both silylation and surface roughness influence the surface heterogeneity of unmodified and silylated silica samples, heterogeneity that can be estimated from the shape analysis of the IGC desorption isotherm.

Surface Heterogeneity: Distribution Functions of the Adsorption Energies

All approaches described in the literature [10] are based on a physical model that supposes that an energetically heterogeneous surface, with a continuous distribution of adsorption energies, may be described, in the simplest way, as a superposition of a series of homogeneous adsorption patches. Hence, the amount of adsorbed molecules (probes) is given by an integral equation (Eq. 6).

$$N(P_m, T_m) = N_0 \int_{\varepsilon\,min}^{\varepsilon\,max} \theta\,(\varepsilon, P_m, T_m)\,\chi(\varepsilon)\,d\varepsilon$$

Eq. 6.

where: $N(P_m, T_m)$ is the number of molecules adsorbed at the pressure P_m and temperature T_m of measurement, N_0 is the number of molecules needed for the formation of a monolayer, $\theta(\varepsilon, P_m, T_m)$ is the local isotherm, ε the adsorption energy of a site, and $\chi(\varepsilon)$ is the distribution function (DF) of the sites seen by the probe. The range of adsorption energies is included between minimal (ε_{min}) and maximal (ε_{max}) values.

From a mathematical point of view, solving the above integral equation is not a trivial task because it has no general solution. But the real distribution function can be approached according to the Rudzinski–Jagiello approximation [11] using a signal treatment method based on Fourier transforms allowing a simple method for the filtration of the isotherm data (elimination of the experimental noise contribution). The remarkable robustness of this new approach, versus noise and irregular sampling, was carefully tested and the energetic surface heterogeneities of a series of solids [2, 12, 13] was evaluated. This method will now be applied for the determination of the surface heterogeneity of the two series of silylated silica samples.

Figure 9 shows the distribution functions of Si2 of the two initial unmodified S13 and T30 silicas.

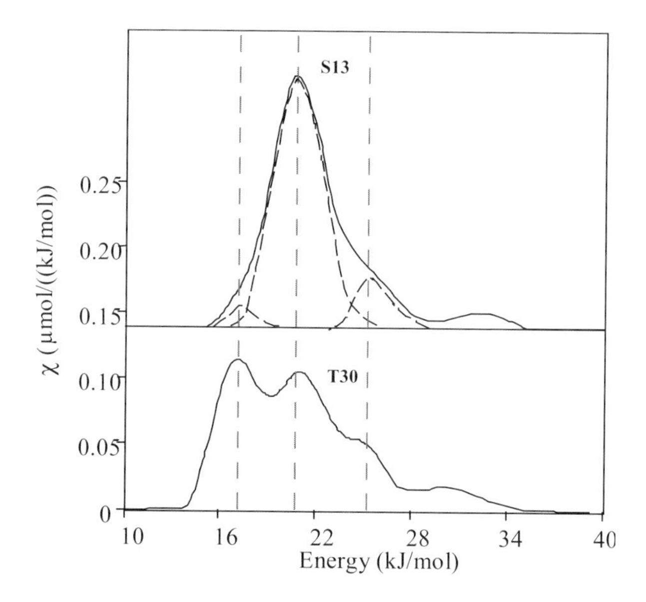

Fig. 9. Distribution functions of the adsorption energies of Si2 on S13 and T30 silicas, measured at 63°C.

It appears immediately that the S13 silica can be considered as almost homogeneous from the point of view of Si2, if we excluded a weak peak laying around 32 kJ/mol. This is absolutely not the case for the T30 silica, which exhibits mainly a trimodal distribution function with a major component appearing at the same energy value as that on the S13 silica.

Taking into account that Si2 exchanges mainly non-specific interactions [8], the variation of the surface functionality cannot explain such polymodality. But, on the other hand, it is a bulky molecule which is certainly sensitive to the surface geometry. Hence, the peak at high adsorption energy (32 kJ/mol) and the shoulder at moderate energy (30 kJ/mol) may be attributed to sites in which the probe may be more or less inserted: in functions respectively between aggregates and between primary particles junctions as depicted schematically on Fig. 10.

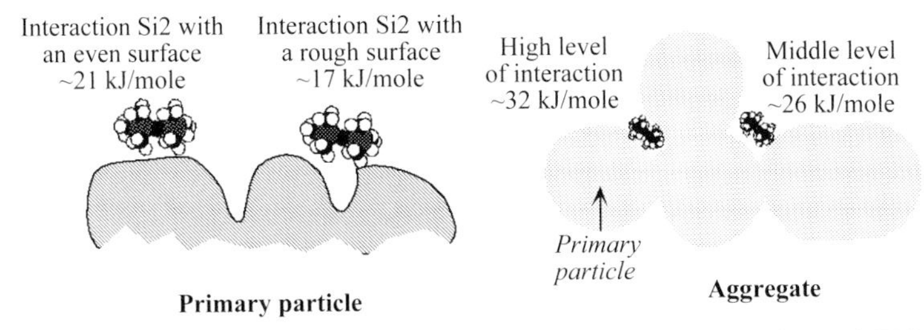

Fig. 10. Schematic representation of the different possible types of adsorption sites on the rough T30 silica surface.

The main peak at 21 kJ/mol is related to the flat silica surface, the same one that forms the majority of the S13 silica surface. Finally, according to this assumption, the low-energy peak may be attributed to the interaction of the probe with the entries of the micropores: a location that does not permit a good fit of the molecule with the solid surface and therefore lowers its energy of interaction. It is worth pointing out that components having much lower intensities can be present in the S13 silica distribution function (dashed curves) and proves that this silica is not perfectly homogeneous.

Let us examine, now, the influence of the silylation on the distribution functions of the adsorption energies for the two series of silylated silicas.

Figure 11 displays the distribution functions of the adsorption energies of the Si2 probe normalized to the total specific surface area, for the S13TMS*x* silica samples.

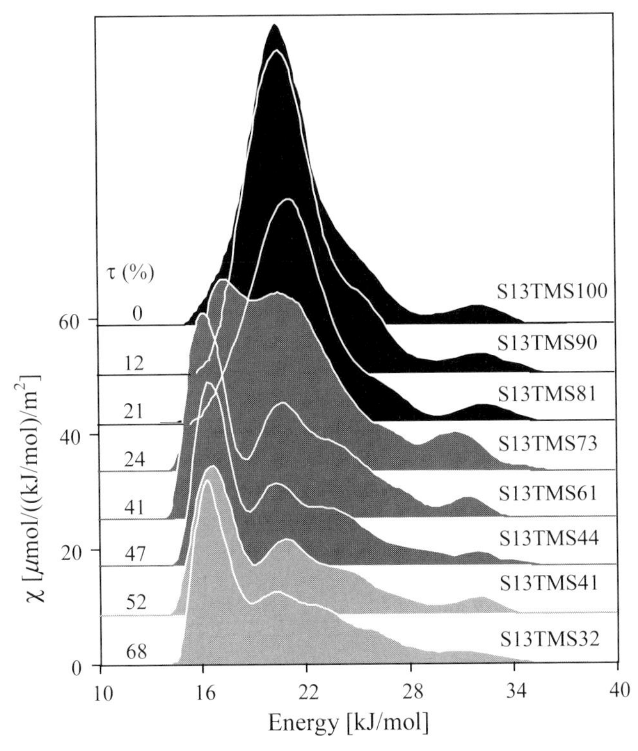

Fig. 11. Surface area (130 m^2/g) on S13TMS*x* silylated silica samples, measured at 63°C.

As expected, the total surface area under the curves diminishes with increasing coverage ratios. But, if silylated silica having a coverage ratio lower than 25 % exhibits a monomodal distribution curve like the initial unmodified silica, a new peak appears at lower energy (around 17 kJ/mol) when we go through this critical coverage ratio where the maximum shifts towards 16 kJ/mol when increasing the silylation ratio, whereas its relative intensity definitely increases. Referring to our previous simple model of silylated silicas, this peak could be attributed to the less energetic sterically hindered sites whose number increases rapidly above the first critical coverage ratio at 25 %.

It is also noteworthy that the relative contribution of the uncovered silica surface remains relatively important, giving evidence of a perfect random distribution of the TMS groups on the silica surface and correlatively of the initial silanol groups.

Again, this observation proves the ability of a population of molecules to detect the disappearance of the long range connectivity. What about the influence of the surface roughness on the evolution of the distribution function when the silylation ratio increases? The distribution functions of the adsorption energies of Si2 are reported on Fig. 12.

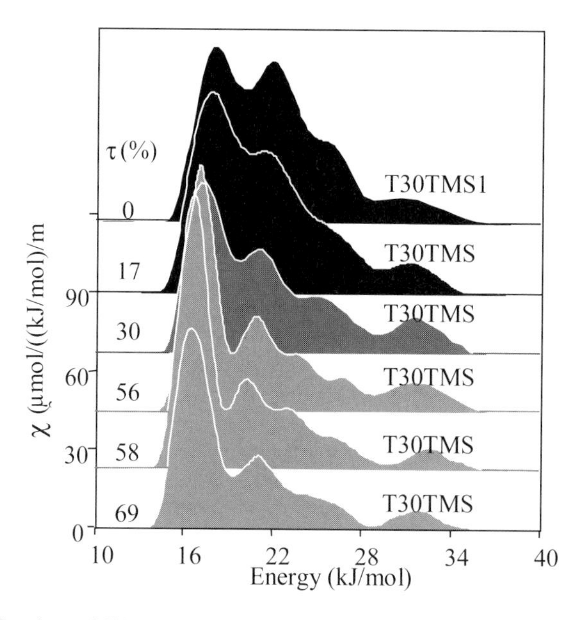

Fig. 12. Distribution functions of Si2 adsorption energies, normalized to the total accessible specific surface area (245 m^2/g) on T30TMS*x* silylated silica samples, measured at 63°C.

As previously, for the S13TMS*x* series, we observe a regular decrease of the surface area of the normalized distribution functions with increasing coverage ratios, but, on the contrary, no modification of their polymodalities is noted. Only the peak appearing at the lowest energy sees its intensity increase, whereas its maximum shifts progressively from 18 kJ/mol to 16.5 kJ/mol. That means in fact that the peak related to the sterically hindered sites overlies progressively the peak corresponding to the lowest energy of the initial T30 silica. The contribution of the high-energy component remains important even at high coverage ratios indicating that a part of the surface is not covered by the TMS groups.

As seen previously for the BET constant evolution, the roughness of the surface prevents the observation of the first transition at a 25 % coverage ratio probably because the presence of the micropores favors a persistence of the long-range connectivity that can only disappear on regular surfaces. This interpretation is also supported by the fact, seen previously from the analysis of the T30TMS*x* analytical data, that only the outer part of the silica particles responds to the silylation process.

We shall see later how this observation can be related to the formation process of the fumed silicas in the flame.

Influence of the Silylation on the Macroscopic Properties of the Modified Silicas

Study of the Wettability of the Silylated Silicas

The wettability of the S13TMS*x* and T30TMS*x* silica samples was studied using Steven's test [14]. This test is based, on the one hand, on the variation of the superficial tension of 2-propanol/water mixes with composition and, on the other hand, on the ability of a mix having a given composition to wet, or not, the powder of interest. If the powder is not wetted, this indicates that its surface energy is lower than the superficial tension of the solvent mix and it begins to be wetted when the superficial tension of the liquid mix and of the silica sample are equal. Hence, starting from pure water, one searches for the composition having the highest water content that is able to wet the powder. Due to the complexity of the interfacial phenomena between a mix of two solvents and the solid surface, this method leads to a rough estimation of the actual surface energy of the solid. But, it is very sensitive and therefore interesting for the sake of comparison of the surface properties of a series of solids submitted to a given treatment: here the controlled silylation process. The evolution of the surface energies with coverage ratios is shown in Fig. 13 for the S13TMS*x* and T30TMS*x* silica series.

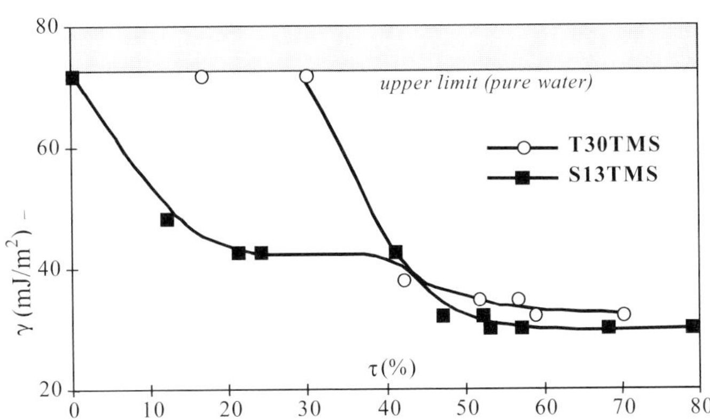

Fig. 13. Evolution of the surface energy of the S13TMS*x* and T30TMS*x* silica samples with the TMS surface coverage ratio according to Steven's test [14].

Again, it is observed that the behaviors of the two silica series differ completely besides their wettability by the 2-propanol/water mixes. In the case of the S13TMS*x* series, the surface energy decreases rapidly until a coverage ratio of 25 % is reached. This happens when the long range surface connectivity disappears. Thereafter, it decreases again rapidly around 50 % with the disappearance of the short-range connectivity.

In contrast, in the same domain of coverage ratios, the T30TMS*x* silicas remain completely hydrophilic and wetted by the pure water and then their surface energies decrease quickly and quite join those of the S13TMS*x* silicas for coverage ratios exceeding 40 %. This observation confirms that the surface roughness allows the persistence of the long-range connectivity up to high grafting ratios. We shall now examine the influence of the degree of silylation on the viscosity of silylated silicas/silicone oil mixes.

Study of the Viscosity of Silylated Silicas/Silicone Oil Mixes

Mixes of S13TMS*x* and T30TMS*x* silylated silicas and a silicone oil (Wacker-Chemie AK1000) were submitted to dynamic rheological measurements using an oscillating cone/plane rheometer. In Fig. 14, the evolution of the viscosity, measured at 25°C and 0.278 Hz, with the silica surface coverage ratio is shown.

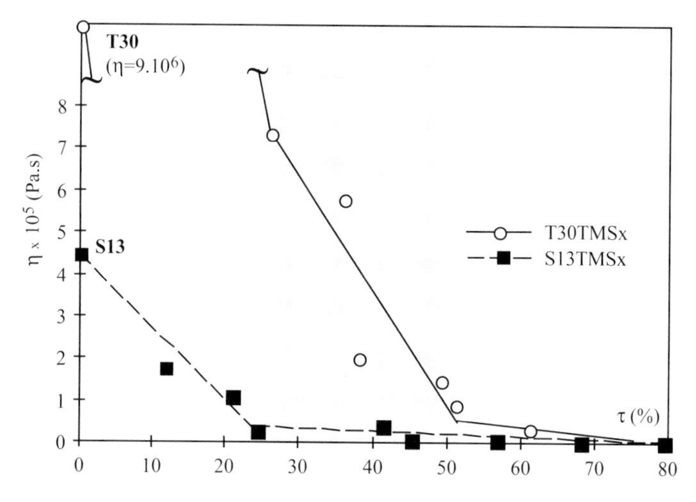

Fig. 14. Variation of the dynamic viscosity of an AK1000 silicone oil containing 8 % of S13TMS*x* or T30TMS*x* silica samples, measured at 0.278 Hz, with the TMS coverage ratios.

Again, the two series of silylated silicas behave differently. For the S13TMS*x* series, it is observed that the dynamic viscosity decreases linearly until a coverage ratio of approximately 25 % is reached, and varies very slightly when the grafting ratio increases. That means that, according to our model, the main contribution to the viscosity is linked with the presence of the free accessible sites or, in other words, to the sterically unhindered silanol groups that permit strong particle — particle interactions through hydrogen bonds. This contribution diminishes, as expected, with the dramatic decrease in relative superficial density of these free accessible sites at the first transition corresponding to the disappearance of the long-range connectivity . Over this critical coverage ratio, the residual silanol groups that are mainly sterically hindered by the neighboring TMS groups permit only weak particle–particle interactions.

In contrast, for the T30TMS*x* series, two steps are observed:

(1) In a first step, there is a very strong decrease in the viscosity from the value obtained with the initial unmodified silica to the one having a coverage ratio of only 12 %. This jump in rheological properties is attributed to the fact that the external, and therefore accessible, silanol groups are at first silylated on the rough surface of the T30 silica. This will affect strongly and quickly the hydrogen bond interaction capacity between two particles.

(2) Then, in a second step, the decrease in the viscosity goes on more slowly up to a coverage ratio of 50 %, at which both S13TMS*x* and T30TMS*x* exhibit quite the same thickening effect. This particular behavior is understandable if we assume a very heterogeneous distribution of the TMS groups and correlatively the presence on the silica surface of important unsilylated patches that allow the formation of relatively strong hydrogen bonds between silica particles. This heterogeneous distribution has to be related to the microporous structure of the T30 particles that leads to a persistence of the long-range connectivity above a coverage ratio of 25 %.

We shall discuss now how the results obtained using IGC techniques, wettability and viscosity measurements are related to the process of formation of the fumed silica particles in the burner flame.

Discussion

Fumed silicas are produced in a burner fed with a mix of tetrachlorosilane, hydrogen and oxygen. Ulrich [15] described the process of formation of the silica particles, based on the instantaneous formation of "protoparticles", which collide and coalesce leading to the formation of the primary particles. The rate of coalescence depends on the viscosity of the molten silica, which is very high at the flame temperature, about 1500 K. These primary particles further stick together to form aggregates. Modifying the composition of the feeding mix will change the flame temperature. The lower is the latter, the higher is the specific surface area of the silica produced. Barthel et al. [5] proved that, correlatively, the surface roughness measured in terms of surface fractality increases.

The increase of the specific surface area and of the surface roughness with the flame temperature can be easily explained if we assume that decreasing the temperature will lower the degree of coalescence of the protoparticles in the primary particle. Therefore, a primary particle formed at high temperature will exhibit a flat surface and possess a lower specific area, whereas a primary particle formed at lower temperature will look like a raspberry having a high surface roughness and, correlatively, a higher specific area. The evolution of the surface structure of the primary particle with the flame temperature is displayed in Fig. 15.

The surface structure of the primary particle will certainly have some consequences for its surface functionality. While leaving the flame and cooling down, the surface of the silica particles will be "hydrated" by water present in the combustion gases. For a non-porous regular particle, the surface temperature will be homogeneous and the hydration process will proceed in a homogeneous manner. It was seen previously that the distribution function of Si2 adsorption energy (Fig. 9) contains the same components as the T30 silica distribution function, but with a much lower intensity. These components may be related to "scars" originating from the coalescence of the protoparticles.

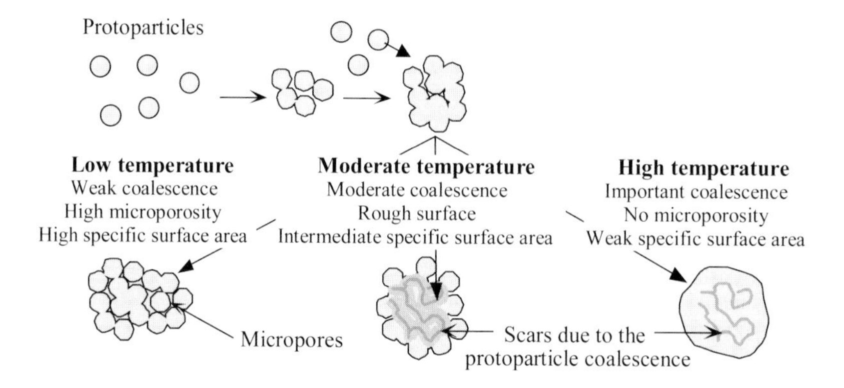

Fig. 15. Schematic representation of the surface topology of the fumed silica primary particles, depending on the flame temperature.

In other words, when protoparticles coalesce, "hot silica", compared with the surface temperature, will be pushed towards the primary particle surface and will lead to poorly functionalized domains when comparing with the external surface. This phenomena is schematized in Fig. 16.

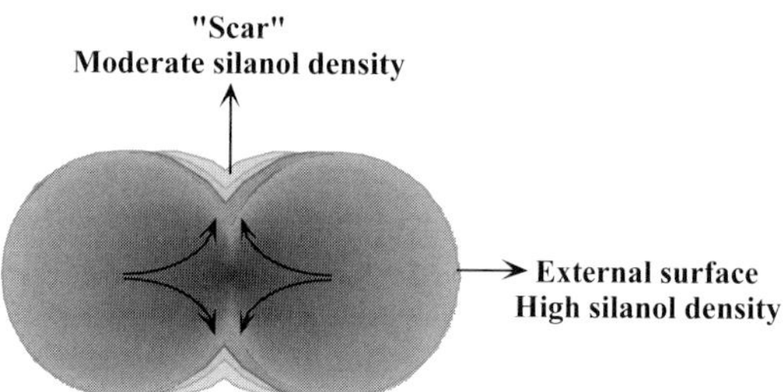

Fig. 16. Schematic representation of the formation of a "scar" by coalescence of two protoparticles.

Hence, a silica that exhibits no surface roughness could be, nevertheless, imperfectly homogeneous from a chemical point of view.

In contrast, for a microporous particle having a high surface roughness, an important temperature gradient will exist between the bottom of the micropores and the outer part of the particle surface, since the high viscosity of the molten silica will prevent temperature homogenization near the surface through local convection currents. The latter will cool down more rapidly than the former and will be more exposed to the combustion gases. This will lead to a higher local surface density of silanol groups whereas the border of the micropore will be lacking in silanol groups. This local variation will obviously have a major influence on the TMS group formation and distribution, which will be regular on a flat surface, whereas they will form groves separated by channels that follow the micropore entrances as depicted in Fig. 17.

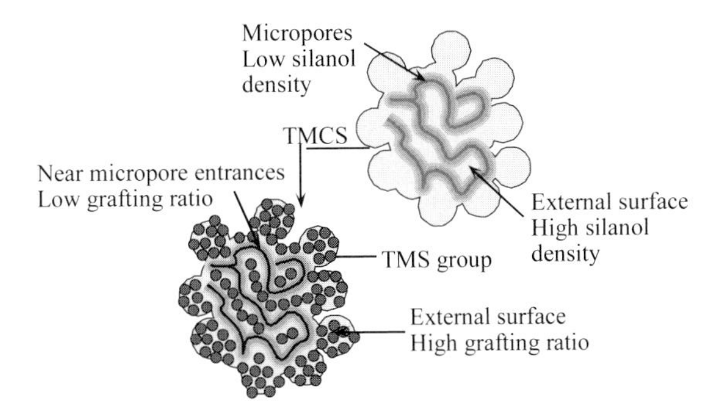

Fig. 17. Schematic representation of the surface topology of a silylated microporous fumed silica.

Hence, it is now easy to understand why the T30 silica surface roughness will induce the persistence of the long-range connectivity as has been evidenced both by IGC-FC experiments and by wettability or viscosity measurements.

Conclusion

This study demonstrates the astonishing potential of IGC measurements using PDMS oligomers as specific molecular probes, for evidencing the strong influence of the degree of silylation on the surface properties of fumed silicas, using IGC at infinite dilution as well as in finite concentration conditions

We observed that the surface properties of the silylated silicas samples do not vary linearly with the TMS coverage ratios of the silica surface and, moreover, that the surface roughness of the fumed silica primary particles influences strongly the capacity of interaction of the silylated silica samples, especially at low silylation ratios.

In the case of a silica exhibiting a relatively smooth and flat surface on the molecular scale, this evolution may be described by calling on a very simple model based on the random distribution of the TMS grafts on a planar silica surface. This model predicts two critical coverage ratios, one at 50 % and one at 25 % coverage.

The former occurs when the TMS groups cover the majority of the silica surface. It corresponds to a sudden decrease in the standard free energies of adsorption of the oligomers and of the free energy of interaction per monomer unit. It is in fact related to the disappearance of the short-range connectivity due to TMS groups acting as barriers that prevent the migration of the adsorbed molecules on the solid surface.

The second critical ratio corresponds to the disappearance of the long-range connectivity when the free non-perturbed silica surface no longer forms the major component of the silylated silica surface. It can only be detected using a population of molecular probes, as in IGC-FC. Indeed, IGC-FC evidences clearly this second transition: a new peak appears, at low energy, in the distribution function of Si2 adsorption energies and, moreover, a break point is observed in the evolution of the BET constant with the TMS coverage ratio.

The wettabilty measurements, according to the Stevens test, and the study of the viscosity of silicon oil /silylated silica mixes, corroborate perfectly the IGC results and therefore confirm the validity of the proposed model.

However, we observed that this model is unable to describe the behavior of a microporous silica submitted to a controlled silylation process. If the jump in surface properties is always evidenced for a TMS coverage ratio of about 50 %, the first transition is no longer detected by either IGC experiments or wettability and rheological measurements. This indicates that the presence of micropores favors the persistence of the long-range connectivity. Looking back to the formation process of the primary silica particles in the flame [15], it is reasonable to admit that the partial coalescence of protoparticles, for silicas having a high specific area and surface roughness, induces both steric hindrance effects and non-regular distribution of the silanol groups on the silica surface, which favor a particular surface topology in which grooves formed by the TMS grafts are separated by channels corresponding to the micropore entrances. This topology allows a persistence of longrange connectivity up to a high TMS coverage ratio.

In short, the study of the surface properties of silicas samples submitted to a controlled silylation process, using IGC-ID and IGC-FC, leads not only to a better understanding of how these surface properties change with increasing degrees of silylation, but also gives access to original information on the surface topology of the unmodified initial silica itself.

References:

[1] H. Balard and E. Papirer, *Prog. Org. Coat.* **1993**, *22*, 1.

[2] H. Balard, A. Saada, J. Hartmann, O. Aouadj, E. Papirer, *Macromol. Chem, Macromol. Symp.* **1996**, *108*, 63.

[3] E. Papirer and H. Balard, in: "*Adsorption and Chemisorption on Inorganic Sorbents*" (Eds.: A. Dabrowski and T. Tertykh), Elsevier, Amsterdam, **1995**, ch. 2.6, pp. 479–502.

[4] A. Khalfi, E. Papirer, H. Balard, H. Barthel, M. Heinemann, *J. Colloid Interface Sci.* **1996**, *184*, 586.

[5] H. Barthel, W. Heinemann, L. Rösch, J. Weiss, *Proc. Eurofillers Mulhouse*, **1995**, pp 157–161

[6] G. M. Dorris, D. G. Gray, *J. Colloid Interface Sci.* **1979**, *71*, 93.

[7] D. W. Sindorf, G. E. Maciel, *J. Phys. Chem.* **1982**, *86*, 5208.

[8] E. A. Nikitina, V. D. Khavryutchenkov, E. F. Sheka, W. H. Barthel, J. Weis, *Surf. Rev. Lett.* **1997**, *4* (5), 879.

[9] J. R. Conder, C. L. Young, *Physicochemical Measurements by Gas Chromatography*, Wiley, New York, **1979**.

[10] W. Rudzinski, D. H. Everett, *Adsorption of Gases on Heterogeneous Surfaces*, Academic Press, London, **1992**.

[11] H. Balard, *Langmuir* **1997**, *13*, 1260.

[12] H. Balard, O. Aouadj and E. Papirer, *Langmuir* **1997**, *13(5)*, 1251.

[13] H. Balard, A. Saada, E. Papirer, B. Siffert, *Langmuir* **1997**, *13(5),* 1256.

[14] P. Stevens, L. Gypen, R. Jennen-Bartholomeussen, *Farma. Tijdschr. Belg.* **1974**, *2*, 150.

[15] G. D. Ulrich, Chem. Eng. News **1984**, *6*, 22.

Modification of Porous Silicon Layers with Silanes

G. Sperveslage, J. Grobe*

Anorganisch-Chemisches Institut, Westfälische Wilhelms-Universität
Wilhelm-Klemm-Str. 8, D-48149 Münster, Germany
Tel.: Int. code + (251)833 6098 — Fax: Int. code + (251)833 6012

Keywords: Porous Silicon / Silanes / Surface Modification / Vibrational Spectroscopy

Summary: The development of materials with well designed surface properties becomes increasingly important in many technological applications. Our studies are particularly concentrated on porous silicon layers (PSLs) as substrates with a great inner surface area, and the modification of PSLs with silanes. On the basis of the results obtained for the modification of PSLs with trimethylsilanes, other functionalized silanes were applied to PSLs as versatile tools for obtaining functionalized surfaces via the SA technique. Vibrational spectroscopic studies of PSLs silanized with different functionalized silanes show that only for 2-chloroethyl-triethoxysilane and benzyltriethoxysilane can the functional group be preserved

Introduction

Porous silicon layers (PSLs) were first obtained by Uhlir and Turner while studying the electropolishing of silicon wafers in dilute hydrofluoric acid solutions [1]. The discovery of visible light emission from PSL in 1990 by Canham has been a source of considerable excitement within the scientific community of chemists, physicists and engineers leading to a boom in research activities with the aim of understanding the fundamental origin of this photoluminescence [2]. Recently, investigation of the reactivity and chemical modification of PSL has gained increasing importance [3]. First experiments concerning the silylation of porous silicon were carried out by Duvault-Herrera, Anderson and Dillon [4–6]. On this basis a thorough investigation was started by applying various silanes to PSLs. It was of special interest whether functionalized silanes can be covalently fixed to the surface with preservation of the functional groups, thus allowing the development of tailor-made surfaces.

Experimental

PSL Formation Method

The Si wafers ((111) p-type with 2–11 Ω cm) were galvanostatically etched in 1:1:2 HF (40 %)/H_2O/EtOH (98 %) at 30 mA/cm^2 for 10 min. After etching the PSLs were washed with deionized water, dried under vacuum and characterized by vibrational spectroscopy (transmission FTIR).

PSL Modification Method

The PSL sample was soaked in neat silane for 120 hours under ambient conditions. After modifying the sample was removed, rinsed with *n*-hexane, placed under vacuum and characterized by vibrational spectroscopy (transmission FTIR).

Surface Modification with Trimethylsilanes

Figure 1 displays an FTIR spectrum of freshly etched PSL containing $\nu(SiH_x)$ stretching modes at 2090 cm^{-1} for SiH- and 2115 cm^{-1} for SiH_2-groups. At 906 cm^{-1} the $\delta(SiH_2)$ scissor mode absorption and at 665 cm^{-1} the torsional mode due to SiH_2 are observed [7, 8].

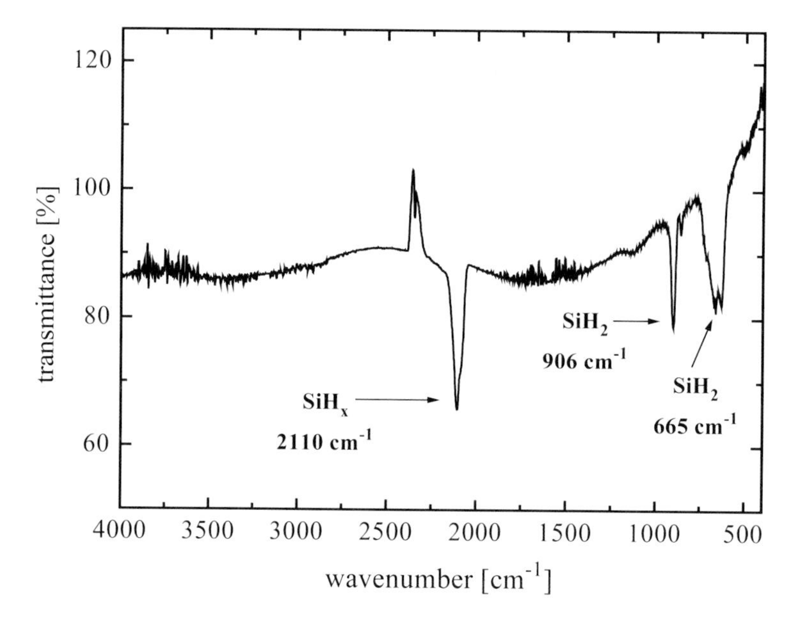

Fig. 1. FTIR spectrum of HF-etched, unmodified PSL.

In a first series, silanes of the type Me_3SiX (X = Cl, OMe, OEt, $OSiMe_3$, $NHSiMe_3$) were applied to PSLs. Tetramethylsilane was used as reference compound. The anchoring of silanes on PSLs can be substantiated easily by means of the very characteristic CH stretching modes located around 2960 cm^{-1}.

For tetramethylsilane the IR spectrum exhibits only absorptions assigned to the oxidation of the porous material. The difference IR spectrum of PSL after treatment with trimethylethoxysilane is dominated by absorption bands corresponding to CH_3 groups (2963 cm^{-1}) and $Si(CH_3)_3$ groups (1254/847 cm^{-1}) (Fig. 2). The anchoring is accompanied by oxidation of the PSL, as indicated by an increase of the band around 1050 cm^{-1} due to SiO or SiOC stretching modes, together with enhanced bands of O_xSiH_y species at 2247 cm^{-1}. The SiH-stretching region is marked by a loss of intensity for the SiH_2 absorption at 2112 cm^{-1}.

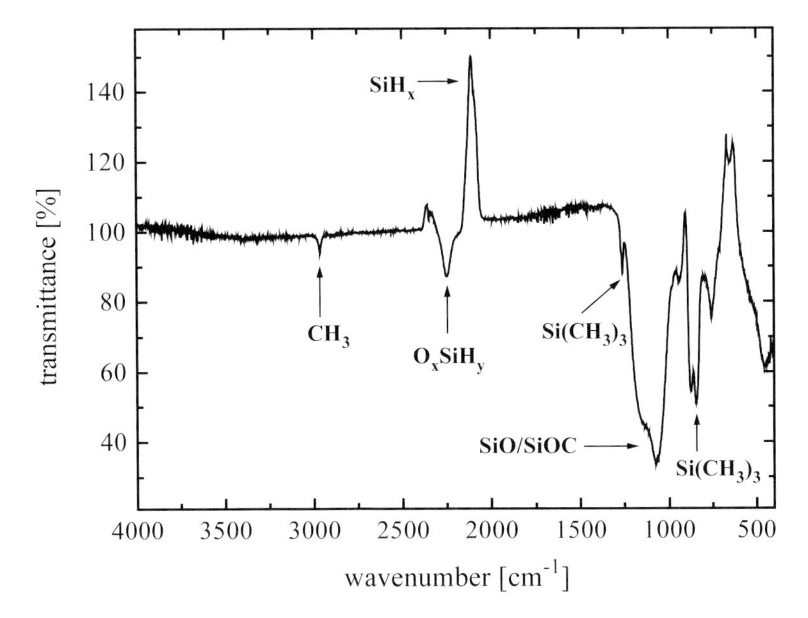

Fig. 2. Difference IR spectrum of PSL modified with trimethylethoxysilane.

The difference IR spectra of PSLs modified with silanes of the type Me_3SiX (with X = Cl, $OSiMe_3$ and $NHSiMe_3$) exhibit the same characteristic absorptions as the spectrum of the PSL modified with trimethylethoxysilane (not shown). For trimethylmethoxysilane the IR spectrum shows an additional absorption band at 2844 cm^{-1} due to the CH stretching mode of methoxy groups.

For the silylation of porous silicon with trimethylsilyl compounds two different models are discussed in the literature. Dubin suggests that moisture transforms the SiH groups of the surface into SiOH groups [9]. Via condensation the silane (hexamethyldisilazane) is attached to the surface and a trimethylsilyloxy-terminated surface is formed. On the contrary, Dillon favors a different reaction mechanism by assuming cleavage of Si–Si bonds and addition of the silane (trimethylchlorosilane) to PSL [5].

Taking into account the sensitivity of Si–X bonds to hydrolysis and the Si–X bond energies of the applied trimethylsilane derivatives we propose the mechanism given in Fig. 3 as a combination of the literature models.

$$Me_3SiX \quad + \quad H_2O \longrightarrow Me_3SiOH \quad + \quad HX$$

Si–H
|
Si–H
+ Me_3SiOH ⟶
Si(H)(H)
Si(H)(O–$SiMe_3$)

Fig. 3. Proposed reaction mechanism for the anchoring of silanes Me_3SiX to PSL.

In the first step the silane is hydrolyzed by water molecules originating from moisture or from a water film on the reaction vessel and the PSL sample, respectively, to give trimethylsilanol, which in the following step reacts with cleavage of Si–Si bonds with PSL to form a trimethylsilyloxy-terminated surface.

Surface Silylation with Functionalized Silanes

On the basis of these results modification of porous silicon layers was also carried out with functionalized silanes of the type $X_{3-n}Me_nSi-(CH_2)_m-Y$ [with X = OMe, OEt; $n = 0, 2$; $m \geq 1$; Y = C=C, Cl, epoxy and benzyl group]. In all cases, the IR spectra indicate an anchoring of the silanes on the substrate. Figure 4 shows the CH stretching region of some difference IR spectra of modified PSLs.

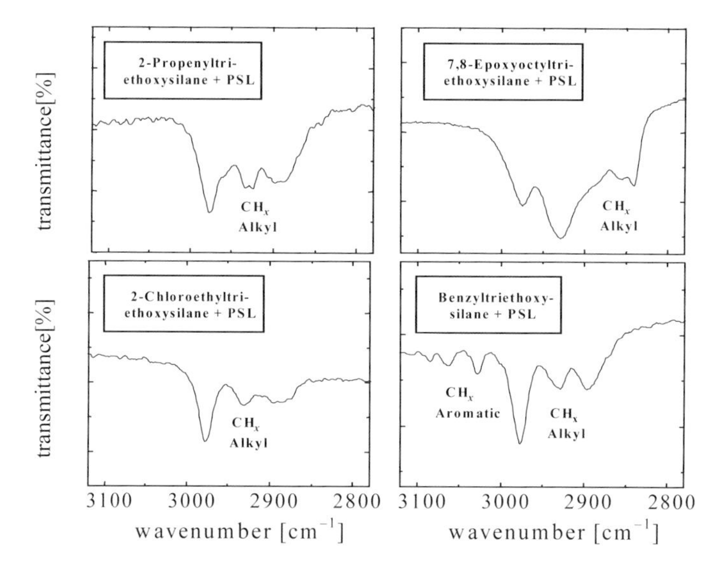

Fig. 4. Difference IR spectra of PSLs modified with functionalized silanes.

However, preservation of the functional group Y was found to be successful only for 2-chloroethyltriethoxysilane and benzyltriethoxysilane. In the corresponding IR spectra of PSLs modified with alkenyl- and epoxyalkyl-functionalized silanes the characteristic absorptions of the olefinic or the oxirane group are not detectable.

Obviously the anchoring process for the alkenyl- and epoxyalkyl-silane derivatives is more complicated than for Me_3SiX because the IR spectra strongly suggest that both functional groups are involved in the surface reaction. Using the same procedure for modifying PSLs with alkenes and epoxides the IR spectra strongly indicate covalent bonding of the molecules to the surface (further details in [10]). These results demonstrate that PSLs show a comparable reactivity towards trimethylsilanes of the type Me_3SiX (X = Cl, OMe, OEt, $OSiMe_3$, $NHSiMe_3$) and alkenes or epoxides, respectively. On the other hand preservation of chloroalkyl groups opens up an alternative preparative route to generate a variety of reactive organic groups after the anchoring process.

Conclusions

PSLs prepared via electrochemical etching of Si wafers in dilute hydrofluoric acid can be easily modified with trimethylsilane derivatives Me_3SiX. The silylation of PSLs with functionalized silanes, however, leads in almost every case to a loss of functionality. Only for 2-chloroethyltriethoxysilane and benzyltriethoxysilane can the functional groups be preserved. Future studies, therefore, will concentrate on the anchoring process and on the characterization of the uppermost layers of modified PSL samples with TOF-SIMS and XPS.

References:

[1] A. Uhlir, *Bell. Syst. Tech. J.* **1956**, *35*, 333.

[2] D. R. Turner, *J. Electrochem. Soc.* **1958**, *105*, 402.

[3] L. T. Canham, *Appl. Phys. Lett.* **1990**, *57*, 1046.

[4] Y. Duvault-Herrera, N. Jaffrezic-Renault, P. Clechet, J. Serpinet, D. Morel, *Colloid. Surf.* **1990**, *50*, 197.

[5] A. C. Dillon, M. B. Robinson, M. Y. Han, S. M. George, *J. Electrochem. Soc.* **1992**, *139*, 537.

[6] R. C. Anderson, R. S. Muller, C. W. Tobias, *J. Electrochem. Soc.* **1993**, *140*, 1393.

[7] Y. J. Chabal, *Phys. Rev. B* **1984**, *29*, 3677.

[8] V. A. Burrows, Y. J. Chabal, G. S. Higashi, K. Raghavachari, S. B. Christman, *Appl. Phys. Lett.* **1988**, *53*, 998.

[9] V. M. Dubin, C. Vieillard, F. Ozanam, J.-N. Chazaviel, *Phys. Stat. Sol. (B)* **1995**, *190*, 47.

[10] G. Sperveslage, *Ph.D. thesis*, University of Münster, **1998**.

In-Situ Controlled Deposition of Thin Silicon Films by Hot-Filament MOCVD with $(C_5Me_5)Si_2H_5$ and $(C_5Me_4H)SiH_3$ as Silicon Precursors

F. Hamelmann*, G. Haindl, J. Hartwich, U. Kleineberg, U. Heinzmann

Fakultät für Physik, Universität Bielefeld
Universitätsstr. 25, D-33615 Bielefeld, Germany
Tel.: Int. code + (521)106 5465 — Fax.: Int. code + (521)106 6001
E-mail: hamelmann@physik.uni-bielefeld.de

A. Klipp, S. H. A. Petri, P. Jutzi

Fakultät für Chemie, Universität Bielefeld
Universitätsstr. 25, D-33615 Bielefeld, Germany
Tel.: Int. code + (521)106 6181 — Fax.: Int. code + (521)106 6026
E-mail: andreas.klipp@uni-bielefeld.de

Keywords: Thin Films / MOCVD / Hot-Filament / Multilayer

Summary: W/Si multilayers with 14 double layers (double layer spacing d = 24 nm) were deposited on Si [100] substrates with hot-filament metal organic chemical vapor deposition (MOCVD). The layer thickness and growth was controlled by an in-situ X-ray reflectivity measurement. Cyclopentadienyl substituted silanes $(C_5Me_5)Si_2H_5$ and $(C_5Me_4H)SiH_3$ were used as silicon precursors, while $W(CO)_6$ was used for the tungsten deposition. The resulting multilayers were characterized by cross-section transmission electron microscopy (XTEM) and sputter auger electron spectroscopy (AES). In addition, the fragmentation of the silicon precursors was studied by mass spectroscopy

Introduction

Metal organic chemical vapor deposition (MOCVD) is a promising method for the preparation of thin films. Metal/silicon multilayers with a single layer thickness of 1-10 nm, as used for soft X-ray mirrors [1, 2], demand very smooth layers (roughness < 0.5 nm).

In our experiments we used $W(CO)_6$ and $Mo(CO)_6$ for the deposition of tungsten and molybdenum. For silicon deposition the cyclopentadienyl silanes $(C_5Me_5)Si_2H_5$ and $(C_5Me_4H)SiH_3$ have recently proven to be easy-to-handle precursors with sufficient vapor pressures and moderate deposition temperatures [3, 4]. During the deposition the thickness of the growing layer is monitored by an in-situ soft X-ray reflectivity measurement by use of 4.4 nm wavelength radiation of the carbon-K line. This method also gives information about the quality of the interfaces.

Since the necessary temperatures for producing silicon films with thermal MOCVD proved to be above 500 °C, interdiffusion of the layers [1] and crystallite formation result in an interface roughness which makes a controlled multilayer deposition impossible. To reduce the deposition temperature an electrically heated tungsten filament is mounted above the substrate. By using this "hot-filament" or "hot-wire" MOCVD (known for the deposition of diamond [5] and a-Si:H [6]) the substrate temperatures can be reduced to below 200 °C, as for PVD deposition of multilayers [7]. To compare this "hot-filament" method with conventional MOCVD, mass spectrometry was performed during the deposition of silicon films. Ex-situ characterization with sputter Auger electron spectroscopy (AES) provides information about the chemical composition of the films, while transmission electron microscopy of sample crosssections (XTEM) gives detailed images of the multilayer structures.

Silicon is a very important material for the development of optical components for soft X-rays (wavelength range between 2 and 100 nm). The use of multilayers with a layer thickness of a few nanometers is the only way to achieve a sufficiently high reflectivity for optical applications in the soft X-ray region [8]. Due to its optical constants, silicon is a good choice for multilayers, in combination with metals such as tungsten [2] or molybdenum [1]. These layers are usually prepared by physical vapor deposition (PVD) methods, such as electron beam evaporation [1]. In this work we want to show that metal organic chemical vapor deposition (MOCVD) has the potential for being an alternative.

It is a challenge to prepare periodic metal–silicon multilayers with a period thickness of just a few nanometers, because the layers have to be extremely smooth (the roughness parameter σ should be smaller than 0.5 nm). The best deposition temperature for molybdenum-silicon multilayers prepared by electron beam evaporation is about 170°C; at lower temperatures the deposited atoms do not have enough mobility to form smooth layers, while at higher temperatures interdiffusion increases and silicide interlayers are formed [1, 7]. In conventional CVD the necessary substrate temperatures are much higher, which makes the use of a non-thermal energy supply, such as plasma or hot-filament, for the fragmentation of the precursors, inevitable.

While useful precursors for the deposition of tungsten and molybdenum are commercially available [i.e. $W(CO)_6$ and $Mo(CO)_6$], the standard precursor for silicon (SiH_4) has some disadvantages (toxic, pyrophoric). Cyclopentadienyl-substituted silanes have proven to be an easy-to-handle alternative [3]. In this work $(C_5Me_5)Si_2H_5$ and $(C_5Me_4H)SiH_3$ are used to prepare multilayers and to compare the abilities of both precursors under hot-filament CVD conditions.

Deposition Procedure

All film depositions are performed in a stainless steel reactor, as shown in Fig. 1. The base pressure is about 10^{-4} Pa; the working pressure was 1 Pa for the samples shown. The precursors are introduced into the system with a carrier gas flow of 2 sccm N_2. The substrate is heated by a resistive ceramic heater, with a thermocouple controlling the temperature. The electrically heated tungsten filament is placed 4 cm above the substrate; its temperature can be monitored by a pyrometer.

The thickness and interface quality of the growing films is controlled by the in-situ soft X-ray reflectivity measurement (Fig. 2). The anode of the X-ray source is graphite coated to emit the carbon K-line at 4.4 nm wavelength. The radiation is reflected by the interfaces of the growing layer, and measured by a proportional counter. For a fixed wavelength λ the period thickness d of the multilayer depends on the incidence angle ϑ, as given by Bragg's law ($\lambda = 2d\sin\vartheta$). A typical in-situ reflectivity measurement is shown in Fig. 3. The starting reflectivity (of the silicon substrate) increases when tungsten is deposited until it reaches a maximum. This is the optimal thickness of the tungsten layer for the chosen d-spacing of the multilayer. In the next step, silicon is deposited, and the reflectivity decreases until it reaches a minimum. The overall decrease in reflectivity from period to period shows that the roughness of the layers is increasing. The use of the in-situ reflectivity measurement limits the operating pressure to 10 Pa: at higher pressures too much of the 4.4 nm radiation would be absorbed. For our purposes this is no problem; the growth rates achieved typically some nanometers per minute are sufficiently high.

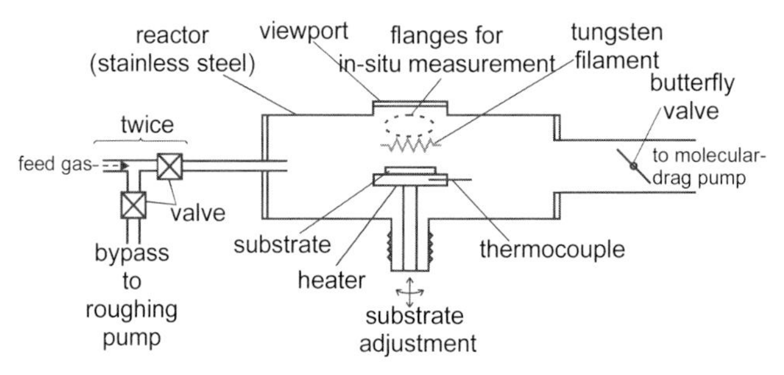

Fig. 1. The hot-filament MOCVD reactor.

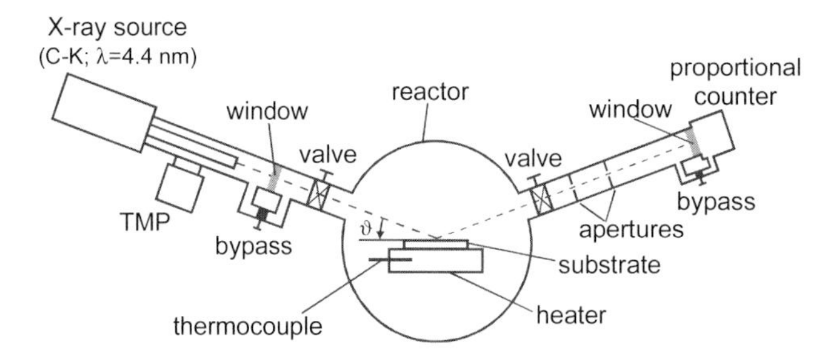

Fig. 2. In-situ reflectivity measurement.

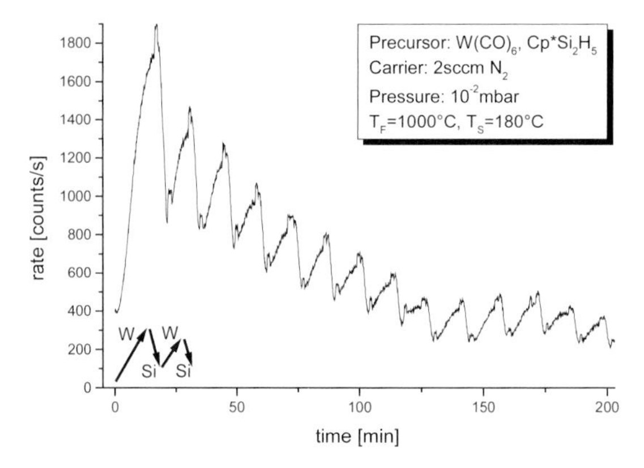

Fig. 3. In-situ reflectivity of a W/Si multilayer.

Characterization of the Multilayers:

To compare the atomic composition of films deposited by hot-filament MOCVD with films deposited by thermal MOCVD, sputter Auger electron spectroscopy (AES) is performed on single layers on silicon substrate. All films were deposited under the same experimental conditions, except for the substrate temperature. The results are listed in Table 1. It is remarkable that the compositions of the silicon layers are almost identical, while the hot-filament deposited tungsten layer contains much more carbon. This can be explained by the fact that the cyclopentadienyl ligands are more stable than the CO ligands of the tungsten hexacarbonyl under hot-filament conditions. It is known that C_5Me_5 is stable in a plasma environment [9]. The relatively high oxygen impurities of the silicon layers cannot originate from the $(C_5Me_4H)SiH_3$ precursor, since it does not contain any oxygen. Hence, the oxygen in the silicon layers might be a result of residual gas in the reactor system.

Table 1. Results of sputter Auger electron spectroscopy.

Layer	W [%]	Si [%]	C [%]	O [%]
Tungsten (thermal)	82.3	—	5.8	11.9
Tungsten (hot-filament)	67.1	—	23.5	9.4
Silicon (thermal)	—	71	8.5	20.5
Silicon (hot-filament)	—	70.4	8.3	21.3

With a quadrupole mass spectrometer the fragmentation patterns of $(C_5Me_5)Si_2H_5$ and $(C_5Me_4H)SiH_3$ under hot-filament conditions are compared. Since the results of mass spectroscopy under thermal conditions and the fragmentation of these precursors is described in [3], only the important fragments are shown in Figs. 4 and 5. All data were acquired with a low ionization energy (20 eV), to keep the fragmentation in the mass spectrometer as low as possible. Nevertheless, the signal of the $(C_5Me_5)Si_2H_5$ precursor is much smaller than the signal of the $(C_5Me_4H)SiH_3$ because of its lower activation energy. In the Figures the values for m/z = 30 (SiH_2) and m/z = 31 (SiH_3) are also plotted; these are, as shown in [3], the relevant species for the formation of the silicon layers. In the case of $(C_5Me_5)Si_2H_5$ formation of SiH_2 and SiH_3 can be observed for filament temperatures higher than 700°C. The signal saturates at about 1700°C, where the process becomes feed-rate limited. As the results from the thermal deposition with these precursors let one expect, the signals of SiH_2 and SiH_3 are about 50 % higher than in the case of $(C_5Me_4H)SiH_3$. This corresponds with the difference of the observed growth rates for the two precursors.

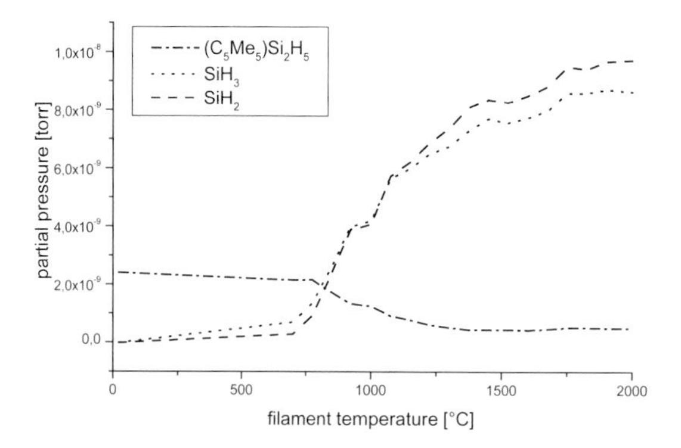

Fig. 4. Mass spectrometry of $(C_5Me_5)Si_2H_5$ at various filament temperatures.

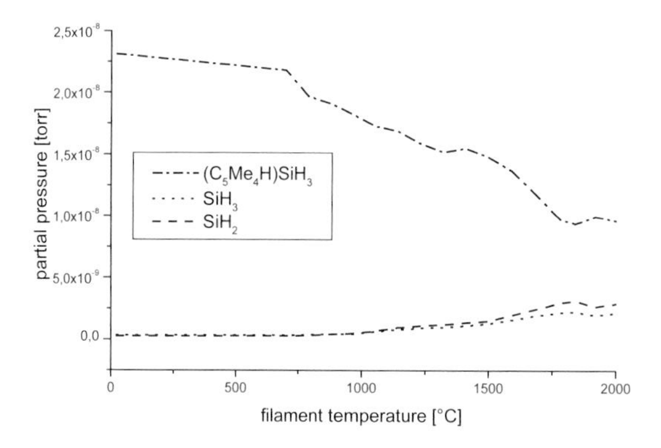

Fig. 5. Mass spectrometry of $(C_5Me_4H)SiH_3$ at various filament temperatures.

To examine whether there are any fragments that are able to contaminate the growing films, the difference of the mass spectra at 2000°C filament temperature and without filament heating is presented in Fig. 6 and 7. In both cases, only the signals of H_2, SiH_2 and SiH_3 increase considerably. This confirms the results of the Auger electron spectroscopy, and shows that the fragments of the cyclopentadienyl unit formed do not affect the growing silicon films, even under the high-temperature conditions of hot-filament MOCVD.

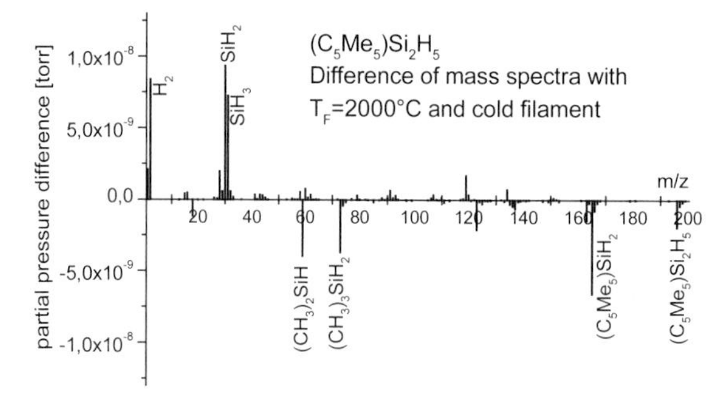

Fig. 6. Difference of $(C_5Me_5)Si_2H_5$ mass spectra.

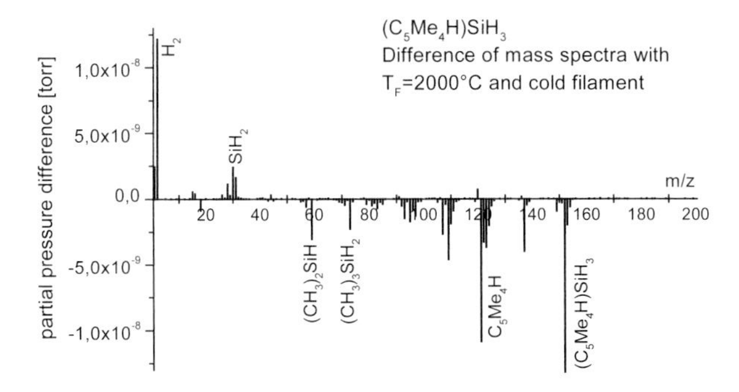

Fig. 7. Difference of $(C_5Me_4H)SiH_3$ mass spectra.

Cross-section transmission electron microscopy (XTEM) images of the samples give views of the layers with an excellent resolution. Figure 8 shows the image of the 14-period W/Si multilayer, whose in-situ reflectivity was shown in Fig. 3. A detail of the first layers is given in Fig. 9.

Fig. 8. XTEM image of a W/Si multilayer.

Fig. 9. Detail of fig. 8.

At the lower right-hand end of the image the silicon substrate can be seen with the regular pattern of a crystal. Next comes the native oxide layer of the substrate, which is amorphous. The first (dark) layer is a tungsten film, also amorphous, followed by the first silicon film. The image of the whole layer shows, that the roughness is indeed increasing from the bottom to the top, small disturbances in lower layers are growing from layer to layer. For comparison Fig. 10 shows the XTEM image of a W/Si bilayer, deposited under thermal conditions at 600°C substrate temperature. In these conditions it was not possible to monitor more than a bilayer with the in-situ measurement; the roughness of the films led to an almost complete loss of reflectivity. The reason is given by the formation of crystallites in the tungsten layer.

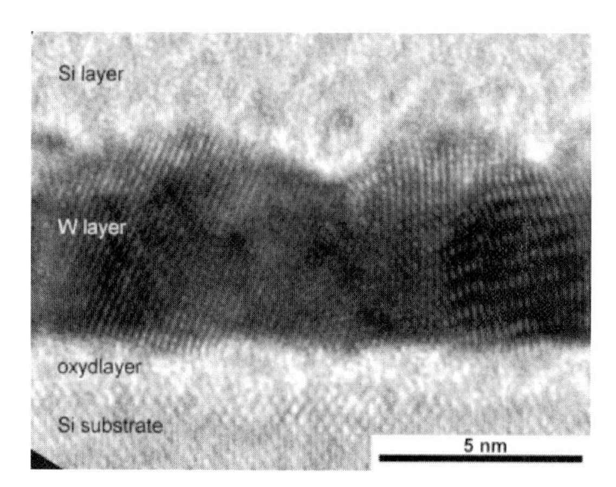

Fig. 10. XTEM image of a thermal deposited W/Si bilayer.

Conclusions

Hot-filament MOCVD was successfully introduced as a new method for the deposition of thin silicon films, using cyclopentadienyl-substituted silanes as precursors. Due to its low fragmentation temperature and the high silicon growth rate $(C_5Me_5)Si_2H_5$ has proven to be a useful material for silicon thin film deposition. The use of a hot filament does not affect the impurities in the film, but leads, due to the lowered substrate temperatures, to a clearly improved interface quality of the silicon–tungsten multilayers.

Acknowledgment: This work is supported by *DFG* as part of the Forschergruppe "Nanometerschichtsysteme".

References:

[1] H.-J. Stock, U. Kleineberg, B. Heidemann, K. Hilgers, A. Kloidt, B. Schmiedeskamp, U. Heinzmann, M. Krumrey, P. Müller, F. Scholze, *Appl. Phys.* **1994**, *371*, A58.

[2] R. Senderak; M. Mergel, S. Luby, E. Majkova, V. Holy, G. Haindl, F. Hamelmann, U. Kleineberg, U. Heinzmann, *J. Appl. Phys.* **1997**, *81*(5), 2229.

[3] A. Klipp, S. H. A. Petri, P. Jutzi, F. Hamelmann, U. Heinzmann, *"The Synthesis of Cyclopentadienyl Silanes And Disilanes and their Fragmentation under Thermal CVD conditions"*, in: *Organosilicon Chemistry IV: From Molecules to Materials* (Eds.: N. Auner, J. Weis), Wiley–VCH, Weinheim, **1999**, 806–811.

[4] F. Hamelmann, S. H. A. Petri, A. Klipp, G. Haindl, J. Hartwich, L. Dreeskornfeld, U. Kleineberg, P. Jutzi, U. Heinzmann, *Thin Solid Films*, in press

[5] C.-P. Klages, *Appl. Phys.* **1993**, *A56*, 513.

[6] P. Broguira, J. P. Conde, S. Arekat, V. Chu, *J. Appl. Phys.* **1995**, *78*(6), 3776.

[7] A. Kloidt, K. Nolting, U. Kleineberg, B. Schmiedeskamp, U. Heinzmann, P. Müller, M. Kühne, *Appl. Phys.Lett.* **1991**, *58*(23), 2601.

[8] E. Spiller, A Segmüller, J. Rife, R.-P. Haelbich, *Appl. Phys. Lett.* **1980**, *37*(11), 1048.

[9] J. Dahlhaus, P. Jutzi, H.-J. Frenck, W. Kulisch, *Adv. Mater.* **1993**, *5*, 377.

The Synthesis of Cyclopentadienyl Silanes and Disilanes and their Fragmentation under Thermal CVD Conditions

A. Klipp, S. H. A. Petri, P. Jutzi*

Fakultät für Chemie, Universität Bielefeld
Universitätsstr. 25, D-33615 Bielefeld, Germany
Tel.: Int. code + (521)106 6181 — Fax.: Int. code + (521)106 6026
E-mail: andreas.klipp@uni-bielefeld.de

F. Hamelmann, U. Heinzmann

Fakultät für Physik, Universität Bielefeld
Universitätsstr. 25, D-33615 Bielefeld, Germany
Tel.: Int. code + (521)106 5465 — Fax.: Int. code + (521)106 6001
E-mail: hamelmann@physik.uni-bielefeld.de

Keywords: Cyclopentadienylsilanes / Cyclopentadienyldisilanes / Thermal CVD / Mass Spectrometry

Summary: The new chloro(cyclopentadienyl)silanes $Cp^xSiH_yCl_{3-y}$ (Cp^x = C_5Me_4Et, $y = 1$ (**1**); Cp^x = C_5Me_4H, $y = 1$ (**2**); $y = 0$ (**3**); Cp^x = $C_5Me_3H_2$, $y = 1$ (**4**)) and pentachloro(cyclopentadienyl)disilanes $Cp^xSi_2Cl_5$ (Cp^x = C_5Me_5 (**5**), C_5Me_4Et (**6**), C_5Me_4H (**7**), $C_5Me_3H_2$ (**8**)) are synthesized via metathesis reactions in good yields. Treatment of **1**–**8** with $LiAlH_4$ leads to the hydridosilyl (Cp^xSiH_3; Cp^x = C_5Me_4Et (**9**), C_5Me_4H (**10**), $C_5Me_3H_2$ (**11**)) and hydridodisilanyl compounds ($Cp^xSi_2H_5$; Cp^x = C_5Me_5 (**12**), C_5Me_4Et (**13**), C_5Me_4H (**14**), $C_5Me_3H_2$ (**15**)). Pyrolysis studies on the volatile cyclopentadienylsilanes **9**–**11** and -disilanes **12**–**15** show their suitability as precursors in the CVD process. The fragmentation pathway of $(C_5Me_5)Si_2H_5$ (**12**) in the thermal CVD process is proved by in-situ mass spectrometry.

Introduction

Thin silicon films are of great interest because of the many applications they have found in fields such as microelectronics, materials science, and optics [1]. The most common silicon precursors used for chemical vapor deposition (CVD) are trichlorosilane ($SiHCl_3$), silane (SiH_4) or disilane (Si_2H_6) [2]. Due to their high reactivity and problematic handling there is a demand for non-hazardous alternative substances. Substituting one hydrogen atom of silane or disilane by an organic group should lead to compounds which meet the desired requirements (e.g. enhanced stability but still sufficient volatility and lower deposition temperatures).

Pentamethylcyclopentadienylsilane $(C_5Me_5)SiH_3$ **16** has been proven to be a very good silicon precursor, e.g. in the low-temperature remote plasma-enhanced CVD process, because a homolytic cleavage of the cyclopentadienyl–silicon bond takes place easily [3]. The resulting pentamethylcyclopentadienyl radical is very stable and reacts preferentially via hydrogen transfer to pentamethylcyclopentadiene and tetramethylfulvene [4]. Modification of the cyclopentadienyl ring or the exchange of the H_3Si group for an H_5Si_2 group should lead to new Si precursors with better deposition parameters.

Synthesis

The chloro(cyclopentadienyl)silanes $Cp^xSiH_yCl_{3-y}$ (y = 0, 1) **1**–**4** and pentachloro(cyclopentadienyl)disilanes $Cp^xSi_2Cl_5$ **5**–**8** are synthesized via metathesis reaction of the cyclopentadienyl lithium salts and trichlorosilane, tetrachlorosilane or hexachlorodisilane in THF or hexane (Scheme 1). Treatment of **1**–**8** with $LiAlH_4$ leads under Cl–H exchange to the corresponding hydridosilyl **9**–**11** and hydridodisilanyl **12**–**15** compounds in good yields.

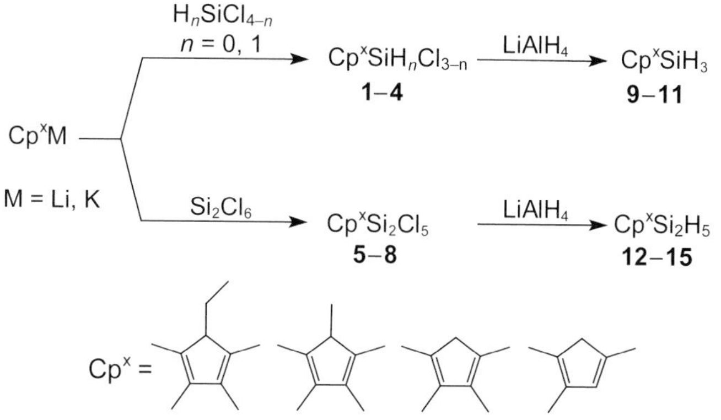

Cp^xM (M = Li, K) $\xrightarrow{H_nSiCl_{4-n},\ n = 0, 1}$ $Cp^xSiH_nCl_{3-n}$ **1–4** $\xrightarrow{LiAlH_4}$ Cp^xSiH_3 **9–11**

Cp^xM (M = Li, K) $\xrightarrow{Si_2Cl_6}$ $Cp^xSi_2Cl_5$ **5–8** $\xrightarrow{LiAlH_4}$ $Cp^xSi_2H_5$ **12–15**

Cp^x =

Scheme 1. Synthesis of the cyclopentadienylsilanes and -disilanes **1**–**15**.

Solid-State Structures

Crystallization of the crude products **6** and **7** from hexane yields large and colorless plates suitable for X-ray structure analysis. ORTEP plots of the molecules are shown in Figs. 1 and 2.

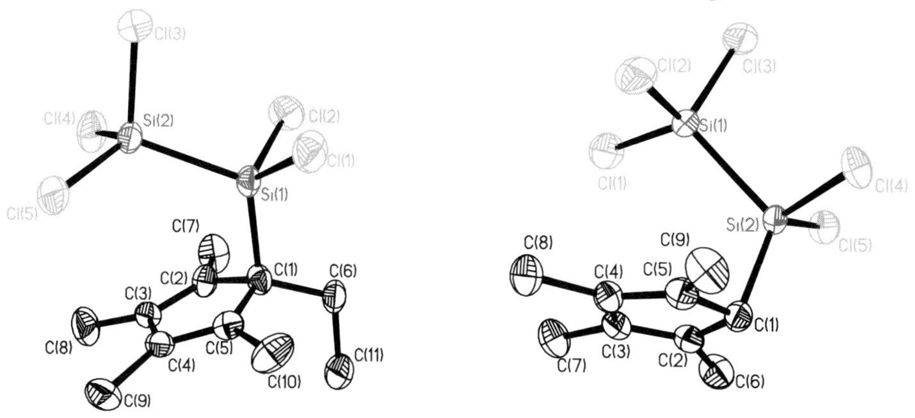

Fig. 1. Molecular structure of $(C_5EtMe_4)Si_2Cl_5$ (**6**).

Fig. 2. Molecular structure of $(C_5Me_4H)Si_2Cl_5$ (**7**).

As expected, all cyclopentadienyl rings are σ-bound with the allylic carbon atom fixed to the pentachlorodisilanyl group. The cyclopentadienyl rings exhibit a diene structure with C–C bond (1.48 Å) and C=C double bond (1.36 Å) lengths typical for σ-bonded species.

Pyrolysis Studies

To evaluate the potential of the silanes **9**–**11** and disilanes **12**–**15** as alternative MOCVD precursors for silicon films and to examine their deposition temperature and chemistry, we carried out first screening experiments with a horizontal hot-wall MOCVD reactor. Deposition conditions are shown in Table 1.

Table 1. Deposition conditions of the CVD experiments.

Compound	Precursor temperature [°C]	Vapor pressure [mbar]	Minimum temperature for film deposition [°C]
$(C_5EtMe_4)SiH_3$ **9**	0	0.15	580
$(C_5Me_4H)SiH_3$ **10**	0	0.1	570
$(C_5Me_3H_2)SiH_3$ **11**	0	1.3	550 (no Si deposition)
$(C_5Me_5)SiH_2SiH_3$ **12**	0	0.05	460
$(C_5EtMe_4)SiH_2SiH_3$ **13**	0–10	0.1	500-520
$(C_5Me_4H)SiH_2SiH_3$ **14**	0	0.3	500-520
$(C_5Me_3H_2)SiH_2SiH_3$ **15**	0	0.5-0.6	490-500
$(C_5Me_5)SiH_3$ **16** [6]	0	0.3	550

Si films were deposited at temperatures above 460°C from the cyclopentadienyldisilanes **12**–**15** and above 550°C from the cyclopentadienylsilanes **9**–**11**, whereas the deposition temperature of SiH_4 is between 630 and 730°C [5].

In order to get information about the fragmentation reactions occurring during CVD, we examined the product mixture. Volatile byproducts were collected in a liquid-nitrogen cold trap, separated and analyzed by ^1H and ^{29}Si NMR spectroscopy. In the case of **12** the reaction products were identified as pentamethylcyclopentadiene, tetramethylfulvene and **16**. The cyclopentadienylsilanes **9**–**11** and disilanes **13**–**15** show complicated fragmentation pathways and higher deposition temperatures compared to **16**.

In order to get more detailed information about the fragmentation pathways of **12** we investigated the thermal CVD process by in-situ mass spectrometry.

For the experimental setup of the MOCVD reactor with in-situ mass spectrometry see ref. 7.

There are three possible initial reactions in the fragmentation pathway of **12** (Scheme 2):

(A) homolytic cleavage of the silicon-silicon bond;

(B) 1,2-hydrogen transfer between the two silicon atoms resulting in pentamethylcyclopentadienylsilylene ($(C_5Me_5)SiH$) and silane (SiH_4);

(C) 1,2-hydrogen transfer between the two silicon atoms resulting in pentamethylcyclopentadienylsilane and silylene (SiH_2).

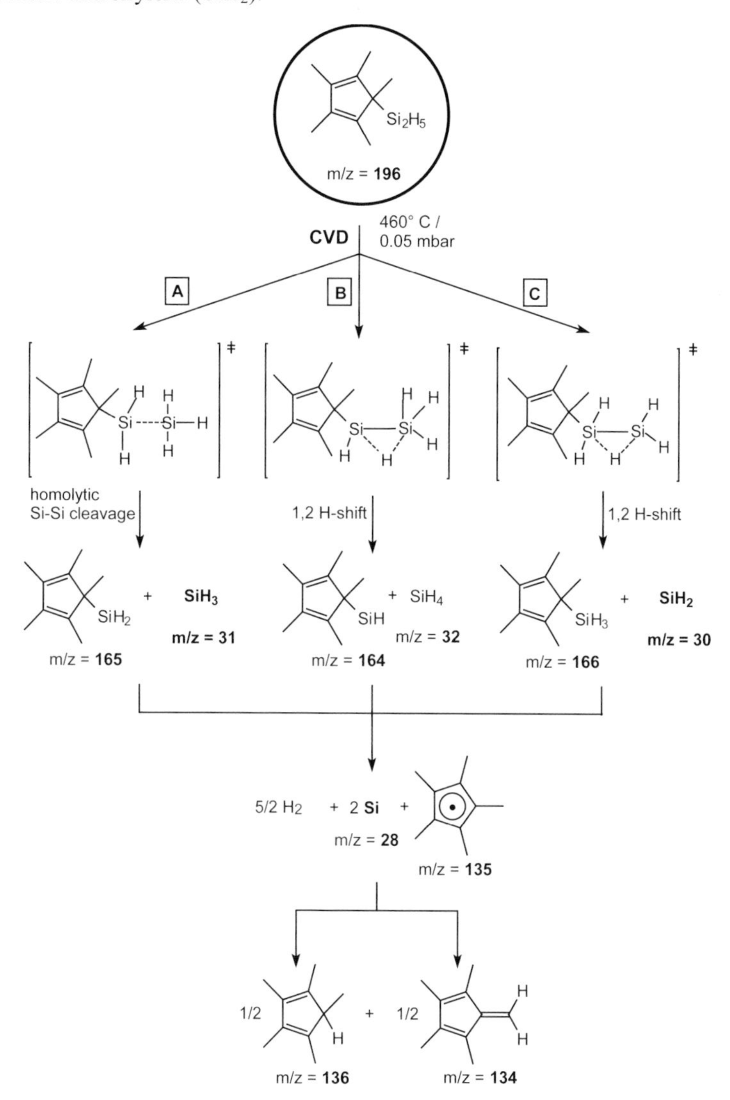

Scheme 2. Possible fragmentation pathways of $(C_5Me_5)Si_2H_5$ (**12**).

The mass spectra of a selected fragments, $(C_5Me_5)Si_2H_5$, and relevant species for the formation of silicon layers (SiH_2 and SiH_3) are shown in Fig. 3. No SiH_4 is detected, so pathway (**B**) is disproved.

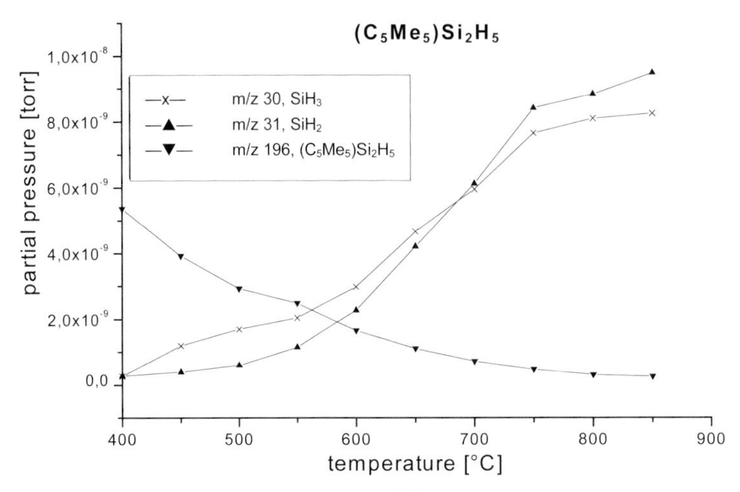

Fig. 3. In-situ mass spectrometry of $(C_5Me_5)Si_2H_5$ **12** at various temperatures.

Conclusions

The best Si precursors seem to be $Me_5C_5Si_2H_5$ **12** and the well-known $Me_5C_5SiH_3$. They behave like disguised Si_2H_6 and SiH_4, but offer high stability and facile handling. Due to their simple fragmentation patterns the only volatile byproducts beside H_2 are pentamethylcyclopentadiene and tetramethylfulvene. It is shown that the deposition temperature of $Me_5C_5Si_2H_5$ **12** is about 100°C lower than that of $Me_5C_5SiH_3$. These precursors should provide a route to silicon films without carbon impurities.

Acknowledgment: This research was generously supported by a grant from the *Deutsche Forschungsgemeinschaft* as part of the Forschergruppe "Nanometerschichtsysteme". *Wacker Chemie*, Burghausen, is acknowledged for a kind gift of Si_2Cl_6.

References:

[1] S. Tsuda, S. Sakai, S. Nakano, *Appl. Surf. Sci.* **1997**, *113/114*, 734–740.

[2] a) J. M. Jasinski, S. M. Gates, *Acc. Chem. Res.* **1991**, *24*, 9–15.
b) H. Karstens, J. Knobloch, A. Winkler, A. Pusel, M. Barth, P. Hess, *Appl. Surf. Sci.* **1995**, *86*, 521–529.
c) I. Herman, *Chem. Rev.* **1989**, *89*, 1323–1357.
d) J. H. Purnell, R. Walsh, *Proc. R. Soc. London, Ser. A* **1966**, *293*, 543–561.
e) W. Ahmed, E. Ahmed, M.L. Hitchman, *J. Mater. Sci.* **1995**, *30*, 4115–4124.
f) H. Habuka, T. Nagoya, M. Mayusumi, M. Katyama, M. Shimada, K. Okuyama, *J. Cryst. Growth* **1996**, *196*, 61–72.

[3] J. Dahlhaus, P. Jutzi, H.-J. Frenck, W. Kulisch, *Adv. Mater.* **1993**, *5*, 377–380.

[4] a) W. R. Roth, F. Hunold, *Liebigs Ann.* **1995**, 1119–1122.
b) P. N. Culshaw, J. C. Walton, L. Hughes, K. U. Ingold, *J. Chem. Soc., Perkin. Trans. 2* **1993**, 879–886.

[5] H. O. Pierson, *Handbook of Chemical Vapor Deposition* , Noyes Publications, Park Ridge, New Jersey, USA, **1992**, pp. 182–185.

[6] P. Jutzi, D. Kanne, M. Hursthouse, A.J. Howes, *Chem. Ber.* **1988**, *121*, 1299.

[7] F. Hamelmann, G. Haindl, J. Hartwich, U. Kleineberg, U. Heinzmann, A. Klipp, S. H. A. Petri, P. Jutzi, "*In-Situ Controlled Deposition of Thin Silicon Films by Hot-Filament MOCVD with (C_5Me_5)Si_2H_5 and (C_5Me_4H)SiH_3 as Silicon Precursors*", in: *Organosilicon Chemistry IV: From Molecules to Materials* (Eds.: N. Auner, J. Weis), Wiley–VCH, Weinheim, **1999**, 798–805.

Novel Silyl-Carbodiimide Gels for the Preparation of Si/C/(N) Ceramics

Edwin Kroke*, Andreas O. Gabriel, Dong Seok Kim*[a]*, Ralf Riedel

Fachgebiet Disperse Feststoffe
Fachbereich Materialwissenschaft
Technische Universität Darmstadt
Petersenstraße 23, D–64287 Darmstadt, Germany
Tel.: Int. code + (6151)166345 — Fax: Int. code + (6151)166346
E-mail: kroke@hrzpub.tu-darmstadt.de

Keywords: Sol–Gel / Ceramics / Carbodiimides / Polymers / Pyrolysis

Summary: The sol–gel process based on alkoxysilanes has been used extensively to prepare oxide glasses and ceramics. Several attempts to apply analogous sol–gel synthesis routes to non-oxide materials were only partially successful. Recently we have found that several di- and trichlorosilanes like $MeRSiCl_2$ (R = –H, $–SiCl_2Me$, $–CH_2CH_2SiCl_2Me$) or $RSiCl_3$ (R = –H, –Cl, –Me, –Ph, $–SiCl_3$, $–CH_2CH_2SiCl_3$), and other silanes as well as silane mixtures react with bis(trimethylsilyl)carbodiimide ($Me_3Si–N=C=N–SiMe_3$) to form polymeric organosilicon gels. Usually no solvent is required and pyridine is used as a catalyst. The highly cross-linked gels are prepared at 20–60°C. In very few cases the sol–gel transition is reversible. Insoluble colorless xerogels are obtained by evaporation of the liquid phase. Heat treatment transforms these polymers into ceramic materials in the ternary Si/C/(N) system.

Introduction

In recent years silicon-based polymers were investigated as precursors for SiC and Si_3N_4 ceramics, as well as for crystalline or amorphous Si/C/N and SiC/Si_3N_4 composite materials [1, 2]. This is due to the very interesting chemical and thermomechanical properties of silicon carbonitrides, such as high hardness, toughness and corrosion resistance. In most of these studies polycarbosilanes, polysilazanes and polycarbosilazanes were applied [3].

Oxygen-containing gels and xerogels prepared from element alkoxides or halides have been used frequently for the preparation of oxidic ceramics [4]. Based on hydrolysis and condensation reactions polymeric or colloidal gels are obtained. Several applications which make use of the sol–gel transition have been developed [4]. Examples are the fabrication of aerogels, fibers and protective coatings. However, only very few reports on the synthesis of non-oxidic materials by a sol–gel process have been published. Seibold and Rüssel [5] found an electrochemical route to aluminum nitride ceramics using a polyaminoalane-sol precursor. Guiton et al. described a sol–gel-like processing for zinc sulfide but usually obtained precipitates instead of gels [6]. Cadmium sufide gels have been synthesized from CdS colloids by Gacoin et al. [7, 8]. Polymeric

---

[a] Now with: Department of Chemistry, Korea Advanced Institute of Science and Technology, Taejon 305-701, Korea

cyanogels were prepared from $K_3Fe(CN)_6$ or K_2PdCl_4 in aqueous solutions, which shows that the non-oxidic sol–gel chemistry is not limited to main group element systems [9]. Narula et al. described a sol–gel process based on oligomerization reactions of substituted borazines with silylamine compounds [10]. Subsequent pyrolysis produced boron nitride ceramics [11]. Attempts have been reported on the synthesis of silicon carbide from organometallic siloxane gels [12], but no non-oxidic sol–gel process for the preparation of silicon carbonitrides is known.

$$\underset{\mathbf{1}}{R_xSiCl_{4-x}} + (4-x)/2\,\underset{\mathbf{2}}{Me_3Si{-}NCN{-}SiMe_3} \xrightarrow{(cat.)} (4-x)\,\underset{\mathbf{3}}{Me_3SiCl} + \underset{\mathbf{4}}{[R_xSi(NCN)_{(4-x)/2}]_n}$$

1a, **4a**: $x = 0$; **1b**, **4b**: $x = 1$, R = -Me
cat. = pyridine

Eq. 1.

Recently, novel carbodiimide polymers have been obtained from reactions of trichloromethylsilane (**1a**) or tetrachlorosilane (**1b**) with bis(trimethylsilyl)carbodiimide (**2**) (Eq. 1) [13-18]. It turned out that in many respects the reactions proceed analogously to the well-known sol–gel process of element halides or alkoxides with water [17, 18]. Phenomenologically, a transformation of a liquid reaction mixture to a solid gel (gelation) occurs, followed by an aging period (syneresis and shrinkage). Chemically, bis(trimethylsilyl)carbodiimide **2** can be considered to replace the water in the classical oxidic sol–gel process, which means that the trimethylsilyl groups adopt the role of the hydrogen atoms, while the oxygen atoms are replaced by the carbodiimide unit (Eq. 2) [20].

$$R_xSiCl_{4-x} + (4-x)/2\ H{-}O{-}H \xrightarrow{(cat.)} (4-x)\,HCl + [R_xSi(O)_{(4-x)/2}]_n$$

Eq. 2.

Many properties of the carbodiimide compounds are surprisingly similar to the corresponding oxygen compounds, which is in accordance to the pseudochalcogenic concept. Examples are the group electronegativity of the NCN unit [19], the chemical shift in ^{29}Si-NMR spectra [20], or the absorption energy in XANES (X-ray absorption near-edge spectroscopy) [20].

Carbodiimide Gels from Chlorosilanes

In the course of our studies of silyl-substituted carbodiimide polymers we first were concerned with materials derived from $MeSiCl_3$ and $SiCl_4$ [13–18]. However, it was found that many other silanes can be used to prepare non-oxidic gels (Table 1).

Syntheses and Characterization

Analogously to the preparation of poly(methylsilsequi-carbodiimide) according to Eq. 1, chlorosilanes were reacted with stoichiometric amounts of bis(trimethylsilyl)carbodiimide at 45°C [17, 18]. No solvent was used and 0.1 equivalent of pyridine was added as a catalyst. The gelation time varied from 0.5 days for Si_2Cl_6 to 181 days for $Cl_2MeSiCH_2CH_2SiMeCl_2$ (Table 1). During an aging period of one month at 45°C the gels derived from chlorosilanes **1c–e** showed a linear shrinkage of approximately 38%. Under the same conditions the gels obtained from **1f–h** showed no shrinkage or phase separation.

Table 1. Comparison of the gelation and aging behavior as well as ceramic yields for different carbodiimide gels prepared in the presence of 0.1 equiv. pyridine.

	Clorosilane used as starting material according to Eq.1	Ideal composition of polymeric gels	Gelation time [days at 45°C]	Shrinkage $\Delta L/L_o$ [%]	Ceramic yield [%]
1b	$MeSiCl_3$	$[MeSi(NCN)_{1.5}]_n$	4.8	44	60
1c	$C_6H_5SiCl_3$	$[C_6H_5Si(NCN)_{1.5}]_n$	1.5	37	58
1d	$Cl_3Si–SiCl_3$	$[Si_2(NCN)_3]_n$	0.5	38	54
1e	$Cl_3Si–(CH_2)_2–SiCl_3$	$[Si(CH_2)_2Si(NCN)_3]_n$	1.5	37	62
1f	$Cl_2MeSi–(CH_2)_2–SiMeCl_2$	$[Me_2Si_2(CH_2)_2(NCN)_2]_n$	60	0	n.d.[a]
1g	$Cl_2MeSi–SiMeCl_2$	$[Me_2Si_2(NCN)_2]_n$	181	0	n.d.[a]
1h	$HSiCl_3$	$[HSi(NCN)_{1.5}]_n$	24	0	n.d.[a]

[a] n.d. = not determined.

After drying in vacuum the polymers were characterized by elemental analyses, FTIR- and solid-state ^{29}Si-NMR spectroscopy. Three examples are shown in Fig. 1. The strong absorption bands around 2150 cm^{-1} in the FTIR spectra indicate that in all cases the carbodiimide unit is present in the polymers. Besides, a significant number of trimethylsilyl endgroups are detected by the signal around 0 ppm in the ^{29}Si-NMR spectra (Fig. 1).

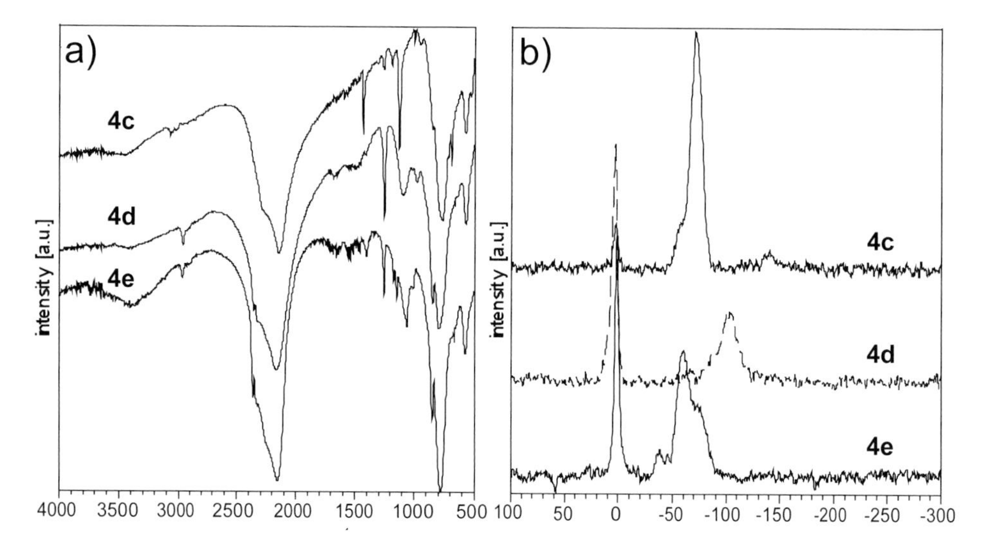

Fig. 1. (a) FTIR spectra of xerogels prepared from $C_6H_5SiCl_3$ (**4c**), Si_2Cl_6 (**4d**), and $Cl_3Si(CH_2)_2SiCl_3$ (**4e**). (b) ^{29}Si-NMR spectra of xerogels prepared from $C_6H_5SiCl_3$ (**4c**), Si_2Cl_6 (**4d**), and $Cl_3Si(CH_2)_2SiCl_3$ (**4e**).

Pyrolysis and Ceramization

The pyrolysis behavior was investigated by TGA/MS (thermal gravimetric analysis with in-situ mass spectrometry of volatile products). The ceramic yield after heating to 1400°C in helium at 2K/min varied between 62 and 54 %. The TGA as well as the mass spectra indicate that up to 500°C crosslinking of remaining endgroups takes place. Between 500 and 650°C substituents of the silicon atoms evolve, e.g. as methane and benzene in the case of **4b** or **4c**. Simultaneously nitriles and HCN are formed, which causes rearrangements in the polymeric framework, since the nitrogen atoms originate from the carbodiimide units. The polymers partially melt and expand due to the evolution of these gases. At temperatures above 600°C elemental hydrogen was detected in all cases. The remaining carbodiimide units decompose to form nitrogen between 800 and 1150°C. At higher temperatures the SiCN ceramics are metastable up to 1440°C, which is in accordance to theoretical calculations [21].

Gels from Silane Mixtures

The novel non-oxidic sol–gel process is useful for the preparation of ceramic coatings, composites and fibers. For these applications it is necessary to control the gelation process and the composition of the polymeric products. The gelation time can be adjusted by the reaction temperature and the pyridine content [17, 18]. For fiber drawing the viscosity is a very important property which has to be set to values around 100 Pa s. In order to control the viscosity increase during the sol–gel transition we tried to use silane mixtures. Depending on the type and ratio of the silanes, a varying gelation and aging behavior was observed.

Table 2. Gelation behavior of Me_2SiCl_2 / $MeSiCl_3$ mixtures.

No.	System: 0.6 equiv. of pyridine/45°C	Mixing ratio	Gelation-time	Remarks
1	$MeSiCl_3$	–	14 h	Rapid viscosity increase at $t = 14$ h
2	$Me_2SiCl_2/MeSiCl_3$	1:10	17,5 h	Viscosity as 1 completely aged after 80 d
3	$Me_2SiCl_2/MeSiCl_3$	1:5	44,5 h	Viscosity comparable to 1 completely aged after 150 d
4	$Me_2SiCl_2/MeSiCl_3$	1:3	93 h	Reversible sol–gel-transition for more than 2 months (70°C/20 min)
5	$Me_2SiCl_2/MeSiCl_3$	1:2	> 50 d	–

With $MeSiCl_3$ a rapid viscosity increase occurs during the sol–gel transition. In less than ten minutes the liquid sol ($\eta < 0.1$ Pa s) solidifies to become a gel ($\eta > 1000$ Pa s). We intended to retard the gelation process by mixing dichlorosilanes like Me_2SiCl_2 or $MeHSiCl_2$ with $MeSiCl_3$. This strategy was partially successful and some interesting observations were made (Tables 2 and 3). As expected, an increasing amount of dichlorosilane causes a longer gelation time. For

$MeHSiCl_2$ this effect is less significant, which is due to its reactivity (Table 3). It is known that Si–H units can contribute to the crosslinking of the polymer via hydrolsilylation reactions, which might also be possible for the carbodiimide group. Besides, disproportionation of $MeHSiCl_2$ in the presence of pyridine can lead to the formation of $MeSiCl_3$ [22, 23].

In the case of $Me_2SiCl_2/MeSiCl_3$ mixtures an increase of gelation and aging time was observed up to ratios of 1:2 (Table 2). Unexpectedly, it was found that for a 1:3 ratio of these silanes the sol–gel transition is reversible. By heating the gel to 70°C for 20 min it becomes liquid again. This is not possible for gel Nos. 1–3 (Table 2). It probably indicates that the carbodiimide gels are not purely formed by the generation of a three-dimensional polymeric network. A colloidal sol might be produced which forms a gel by a large contribution of non-covalent interactions. However, other explanations for the reversibility of the sol–gel transition, such as relatively low energy barriers for redistribution reactions according to Eq. 3, have also to be taken into account.

$$Me_3SiCl + {>}Si{-}NCN{-}Si{<} \xrightarrow{(cat.)} {>}Si{-}Cl + Me_3Si{-}NCN{-}Si{<}$$

Eq. 3.

The same trend as for the gelation time is observed for the aging of the gels. Aging is usually accompanied by syneresis and shrinkage. While $MeSiCl_3$ gels always shrank it was found that the gels obtained from 1:2 and 1:3 $MeSiHCl_2$ / $MeSiCl_3$ mixtures showed no shrinkage upon aging.

Table 3. Gelation behavior of $MeHSiCl_2/MeSiCl_3$ mixtures.

No.	System: 0.6 equiv. of pyridine/45°C	Mixing ratio	Gelation-time	Remarks
1	$MeSiCl_3$	–	14 h	Rapid viscosity increase at $t = 14$ h
2	$MeSiHCl_2/MeSiCl_3$	1:10	41 h	Viscosity as 1 Completely aged after 80 d
3	$MeSiHCl_2/MeSiCl_3$	1:5	49:5 h	Viscosity comparable to 1 Completely aged after 75 d
4	$MeSiHCl_2/MeSiCl_3$	1:3	7 d	reversible sol–gel transition for a few days, no shrinkage
5	$MeSiHCl_2/MeSiCl_3$	1:2	4 d	Reversible sol–gel transition for a few days, no shrinkage

Conclusion

It has been clearly demonstrated that practically all trichlorosilanes and a large number of dichlorosilanes can be used to prepare SiCN gels by reactions with bis(trimethylsilyl)carbodiimide. Gelation and aging time depend on the substituents R in R_xSiCl_{4-x}. Moreover, silane mixtures also form gels which can be used to tailor the gelation behavior as well as the elemental composition of the ceramic products. This is of importance for possible applications of the novel carbodiimide gels like fiber drawing or preparation of coatings and composites.

Acknowledgment: We thank the *Deutsche Forschungsgemeinschaft*, the *Fonds der Chemischen Industrie*, and *Bayer AG* for financial support, as well as Florence Babonneau and Christel Gervais (Paris) for the solid-state NMR investigations.

References:

[1] A. W. Weimer, *Carbide, Nitride and Boride Materials Synthesis and Processing*, Chapman and Hall, London, **1997**.

[2] W. Dreßler, R. Riedel, *Int. J. Refract. Met. Hard Mater.* **1997**, *15*, 13.

[3] M. Birot, J.-P. Pillot, J. Donugues, *Chem. Rev.* **1995**, *95*, 1443.

[4] C. J. Brinker, G. W. Scherer, *Sol–Gel Science*, Academic Press, San Diego, CA, **1990**.

[5] M. Seibold, C. Rüssel, *Mater Res. Soc. Symp. Proc.* **1988**, *121*, 477.

[6] T. A. Guiton, C. L. Czekaj, C. G. Pantano, *J. Non-Cryst. Solids* **1990**, *121*, 7.

[7] T. Gacoin, L. Malier, J.-P. Boilot, *Chem. Mater.* **1997**, *9*, 1502.

[8] T. Gacoin, L. Malier, J.-P. Boilot, *J. Mater. Chem.* **1997**, *7*(6), 859.

[9] B. W. Pfennig, A. B. Bocarsly, R. K. Prud'homme, *J. Am. Chem. Soc.* **1993**, *115*, 2661.

[10] C. K. Narula, R. T. Paine, R. Schaeffer, *Mater. Res. Soc. Symp. Proc.* **1986**, *73*, 383.

[11] C. K. Narula, R. Schaeffer, A. Datye, R. T. Paine, *Inorg. Chem.* **1989**, *28*, 4053.

[12] J. R. Fox, D. A. White; S. M. Oleff, R. D. Boyer, P. A. Budinger, *Mater. Res. Soc. Symp. Proc.* **1986**, *73*, 395.

[13] A. Kienzle, A. Obermeyer, R. Riedel, F. Aldinger, A. Simon, *Chem. Ber.* **1993**, *126*, 2569.

[14] A. Kienzle, K. Wurm, J. Bill, F. Aldinger, R. Riedel, in: *Organosilicon Chemistry II: From Molecules to Materials* (Eds.: N. Auner, J. Weis), VCH, Weinheim, **1996**, p. 725.

[15] A. Greiner, *Ph. D. thesis*, Technische Universität Darmstadt, **1997**.

[16] R. Riedel, A. Greiner, G. Miehe, W. Dreßler, H. Fueß, J. Bill, F. Aldinger, *Angew. Chem.* **1997**, *106*, 657; *Angew. Chem. Int. Ed. Engl.* **1997,** *36*, 603.

[17] A. O. Gabriel, R. Riedel, *Angew. Chem.* **1997**, *109*, 371; *Angew. Chem., Int. Ed. Engl.* **1997**, *36*, 384.

[18] A. O. Gabriel, R. Riedel, S. Storck, W. F. Maier, *Appl. Organomet. Chem.* **1997**, *11*, 833.

[19] H.-D. Schädler, L. Jäger, I. Senf, *Z. Anorg. Allg. Chem.* **1993**, *619*, 1115.

[20] a) R. Riedel, E. Kroke, A. Greiner, A. O. Gabriel, L. Ruwisch, J. Nicolich, P. Kroll, **1998**, *10*, 2954.

b) E. Kroke "*Ceramics: Getting into the 2000's – Part C*" in *9th Cimtec-World Ceramics Congress*, (Ed.: P. Vincenzini), Techna Srl, **1999**, p. 123.

[21] J. Weiss, H. L. Lukas, J. Lorenz, G. Petzow, H. Krieg, *CALPHAD* **1981**, *5(2),* 125.

[22] H. J. Campbell-Ferguson, E. A. V. Ebsworth, *J. Chem. Soc. (A)* **1966**, 1508.

[23] D. S. Kim, E. Kroke, R. Riedel, A. O. Gabriel, *S. C. Shim, Appl. Organomet. Chem.* **1999**, *13*, 495.

Thermodynamics, Kinetics and Catalysis in the System Ni–Si–H–Cl: Thermodynamic Description of Chlorine-Containing Silicides

Jörg Acker*, Klaus Bohmhammel

Institut für Physikalische Chemie
Technische Universität Bergakademie Freiberg
Leipziger Str. 29, D-09596 Freiberg, Germany
Tel.: Int. code + (3731)39 4324 — Fax: Int. code + (3731)39 3588
E-mail: acker@erg.phych.tu-freiberg.de

Gerhard Roewer

Institut für Anorganische Chemie
Technische Universität Bergakademie Freiberg
Leipziger Str. 29, D-09596 Freiberg, Germany
Tel.: Int. code + (3731)39 3174 — Fax: Int. code + (3731)39 4058

Keywords: Chlorine-Containing Silicides / Trichlorosilane / Thermodynamics

Summary: Transition metal silicides, containing small amounts of chlorine, act as catalysts in the hydrodechlorination of silicon tetrachloride into trichlorosilane. Silicides, with comparable properties to the catalytically active phases, can be prepared by the reaction of silicon-rich silicides with the respective metal chloride. For the first time a thermodynamic model of chlorine-containing nickel silicides is given. Chlorine is considered to be dissolved in the silicides and is modeled as a lattice gas. Consequences regarding the bonding state of chlorine in the silicide phases are reported and discussed in relationship with the interactions within the silicide lattice.

Introduction

The catalytic hydrodechlorination of silicon tetrachloride into trichlorosilane [1] is a process of far-reaching importance for the silicon industry (Eq. 1).

$$SiCl_4 + H_2 \rightarrow HSiCl_3 + HCl$$

Eq. 1. Hydrodechlorination of silicon tetrachloride.

It enables the silicon tetrachloride which is produced in high amounts during high-purity silicon production to be converted into trichlorosilane at temperatures between 900 and 1200 K while maintaining the existing high purity. The reaction proceeds only in presence of transition metal silicides. These are formed by the reaction between the transition metals, added during the starting phase of the process, with silicon tetrachloride and hydrogen according to Eq. 2.

In accordance with the thermodynamics of the reaction system, at low temperatures silicon-poor phases are formed, while at high temperatures silicides with a high silicon content are generated. Each of the phases present reaches its catalytic activity after an induction period, in which the thermodynamically stable silicide, corresponding to the temperature of reaction, is formed. The yield of trichlorosilane is independent of the phase used; it corresponds to the equilibrium value calculated for the present temperature. It is concluded that the catalytic reaction proceeds almost without kinetic hindrance and that silicon has a high mobility in the catalyst system [1–3].

$$x\,M + y\,SiCl_4 + 2y\,H_2 \rightarrow M_xSi_y + 4y\,HCl$$

Eq. 2. Formation of transition metal silicides in the starting phase of the catalytic hydrodechlorination.

The basic conclusion of our research is that the presence of chlorine in the silicide phases is one important precondition for catalytic activity.

The subject of this paper is the investigation of chlorine-containing nickel silicide phases. These are obtained by the reaction of silicon-rich phases with nickel(II)chloride in evacuated ampoules during sufficiently long testing periods (Eq. 3). The compounds thus obtained have chlorine contents between 0.02 and 0.6 wt. %. In generally, the chlorine-containing silicides and their route of formation can be used as models to describe the catalytically active phases in the hydrodechlorination process [4].

$$NiSi + x\,NiCl_2 \rightarrow Ni_{1+x}Si_{1-0.5\,x} + 0.5\,x\,SiCl_4$$

Eq. 3. Formation of chlorine-containing nickel silicides.

Characterization of Chlorine-Containing Silicides

The phases described below have been obtained by the reaction of NiSi with nickel(II)chloride according to Eq. 3 in evacuated quartz glass ampoules after a 6-weeks reaction time at temperatures of 973 K and a partial pressure of $SiCl_4$ of 6 bar at this temperature [4].

Table 1. Analysis of the products of the reaction NiSi + x $NiCl_2$ (partial pressure $SiCl_4$ 6 bar, testing period 6 weeks at 973 K).

Initial stoichiometry	Calculated phases	Found phases	Chlorine content (wt. %)	Calculated net formula
NiSi + 0.2 $NiCl_2$	NiSi, Ni_3Si_2	NiSi, Ni_3Si_2	0.018	$NiSi_{0.7519}Cl_{0.0004}$
NiSi + 0.2857 $NiCl_2$	Ni_3Si_2	Ni_3Si_2	0.065	$Ni_3Si_2Cl_{0.0043}$
NiSi + 0.5 $NiCl_2$	Ni_2Si	Ni_2Si	0.118	$Ni_2SiCl_{0.0049}$
NiSi + 0.6071 $NiCl_2$	Ni_2Si, Ni_5Si_2	Ni_2Si, Ni_5Si_2	0.204	$NiSi_{0.4329}Cl_{0.0032}$
NiSi + 0.6652 $NiCl_2$	Ni_5Si_2	Ni_5Si_2	0.401	$Ni_5Si_2Cl_{0.0397}$
NiSi + 0.7142 $NiCl_2$	Ni_5Si_2, Ni_3Si	Ni_3Si, Ni_5Si_2	0.519	$NiSi_{0.3759}Cl_{0.0099}$
NiSi + 0.8071 $NiCl_2$	Ni_3Si, Ni	Ni_3Si (phase range)	0.088	$Ni_3SiCl_{0.0051}$

According to the applied stoichiometry the thermodynamically stable silicide phases are formed. Due to the relatively large inaccuracy of the determination of silicon contents, in contrast to the chlorine contents, the analytically determined amount of chlorine has been related to the theoretically calculated stoichiometric composition of the formed silicides according to Eq. 3 (Table 1). The following uncertainty results: it cannot be decided whether the incorporation of chlorine is accompanied by a slight increase or decrease of silicon comparison with expected composition of the product phases. The consequences for the thermodynamic model are discussed below.

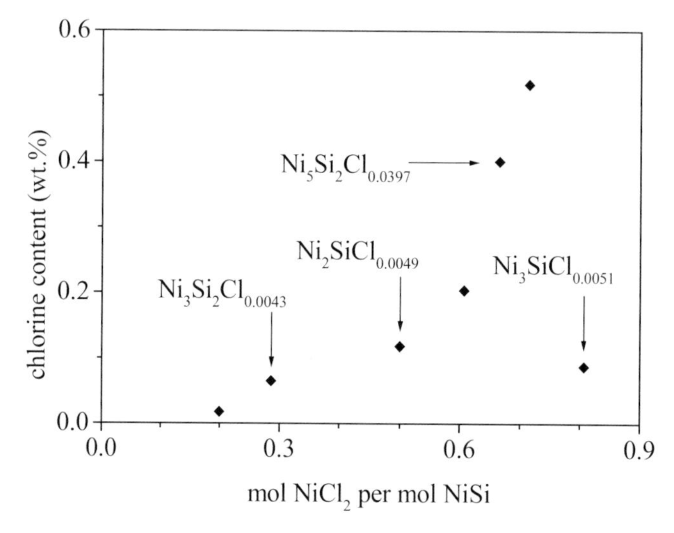

Fig. 1. Stoichiometry of chlorine-containing silicides calculated from analysis data. (Results from Table 1, single silicide phases are labeled, mixture phases unlabeled).

Figure 1 shows the dependence of the chlorine content of product phases on the initial stoichiometry of the educts (mol $NiCl_2$ per mol NiSi). The Ni_5Si_2 phase contains the highest amount of chlorine. Furthermore it is remarkable that the mixture of Ni_5Si_2 and Ni_3Si indicates yet a higher chlorine content. Calculations show that the chlorine contents of the samples in which two silicide phases are present cannot be derived from the chlorine contents of the actual pure silicide phases. Thus increases and decreases in chlorine content as a function of the silicon content of the product phases appear to be steady and not discrete.

Thermodynamic Description of Chlorine-Containing Silicides

By means of the preliminary results it has been possible to thermodynamically model the presence of chlorine in nickel silicides. Chlorine-containing silicides may be taken as diluted solid solutions of chlorine in nickel silicides, which in the thermodynamic sense can be described by a modified Wagner model [5]. The molar free enthalpy of the solution of chlorine in silicide (G_m) is derived from Eq. 4.

$$G_m = \sum_{i=1}^{2} x_i \mu_i^{\circ} + RT \sum_{i=1}^{2} x_i \ln x_i + x_1 RT \ln\gamma_1 + x_2 RT\left(\ln\gamma_1 + \ln\gamma_2^{\circ} + \varepsilon_{12} x_2\right)$$

$\ln\gamma_1 = -\frac{1}{2}\varepsilon_{12} x_2^2$

x_i = molar fraction of the component i

μ_1° = standard potential of silicide

μ_2° = standard potential of chlorine

γ_1 = activity coefficient of silicide

γ_2° = activity coefficient of the dissolved chlorine at infinite dilution

ε_{12} = silicide–chlorine interaction parameter

Eq. 4. The modified Wagner model [5] applied to chlorine-containing nickel silicides.

The parameter of interaction ε_{12} can be divided into an enthalpy fraction and an entropy fraction (per mol of chlorine). The enthalpy contribution corresponds to the solution enthalpy $\Delta_S H_{Cl}$ and the entropy contribution to the solution entropy $\Delta_S S_{Cl}$ of chlorine (Eq. 5).

$$\varepsilon_{12} = \Delta_S H_{Cl} - T \cdot \Delta_S S_{Cl}$$

Eq. 5. Enthalpy and entropy contribution of the parameter of interaction.

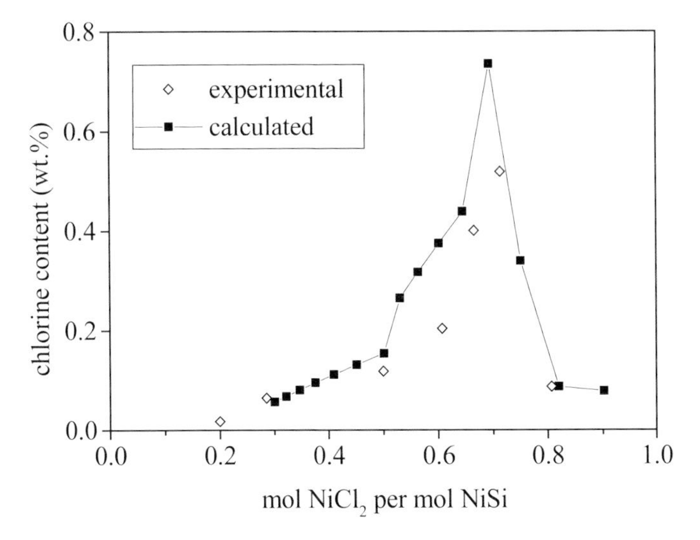

Fig. 2. Comparison of experimental data with the result of modeling.

The standard potential of silicide has been calculated from the data according to [2]. Because the chlorine dissolved in the silicide is considered as lattice gas, the standard potential of gaseous chlorine can be used. Uncertainties in the stoichiometries of the chlorine-containing silicides were taken into account with an estimated error of 5 % as the basis of calculation. Thus the silicide-chlorine interaction parameter, the silicide activity coefficient of and the activity coefficient of the dissolved chlorine at infinite dilution remain as unknown values.

By means of the optimizer [6] implemented by the program ChemSage the unknown values could be determined as the enthalpy and entropy contributions of the interaction parameter. Figure 2 shows that the result of modeling describes the found experimental dependence of chlorine solubility very well. The entropy of solution is almost independent of the stoichiometry and is about $-30\ \mathrm{J\,K^{-1}\,mol^{-1}}$. On the other hand, the enthalpy of solution shows a characteristic dependence on the stoichiometry of the silicides (Fig. 3). The order of magnitude of the found solution enthalpy can be compared with the incremental enthalpy of formation of an Si–Cl bond of about $-166.7\ \mathrm{kJ\,mol^{-1}}$ [7]; however, the conclusion of the presence of Si–Cl bonds is not permitted. Rather, the result of modeling corresponds to the experimental findings, in which neither the characteristic Si–Cl vibrations nor the characteristic Ni–Cl vibrations could be proved spectroscopically. Thus it can be assumed, that chlorine interacts simultaneously with silicon and nickel [8].

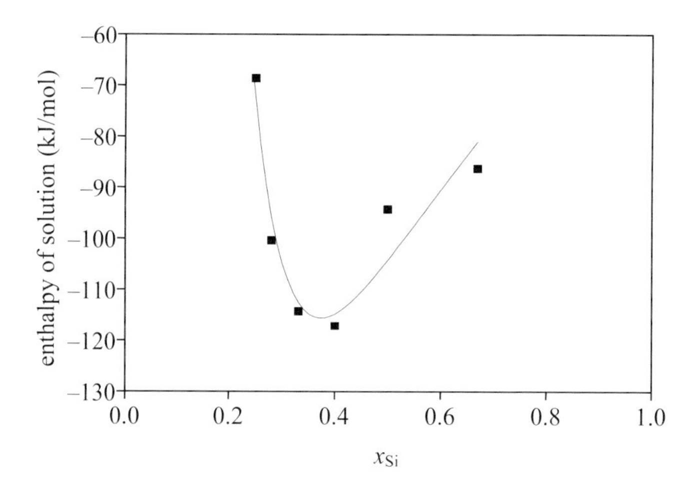

Fig. 3. The enthalpy of a solution of chlorine in nickel silicides as a function of the molar fraction of silicon.

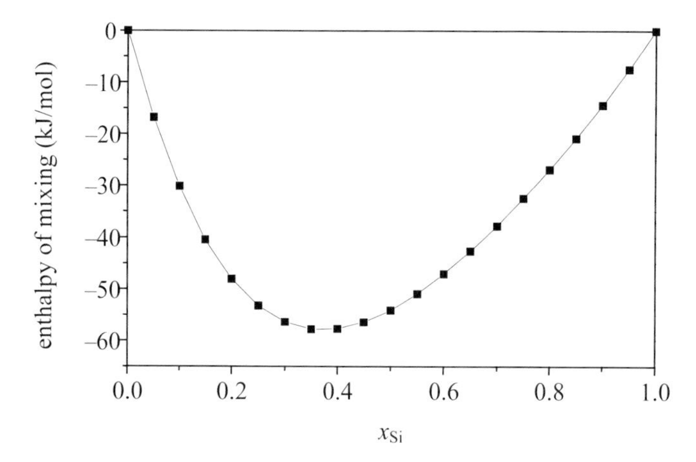

Fig. 4. The molar enthalpy of mixing of a nickel–silicon melt at 1800 K.

The dependence of the calculated enthalpy of solution of chlorine is comparable to the enthalpy of mixing of Ni–Si melts (Fig. 4) [9]. The extreme value of both dependencies is $x_{Si} \cong 0.375$. The negative values of the enthalpy of mixing indicate considerable interactions between nickel and silicon. These interactions should be correlated with chlorine solubility in the nickel silicides. Both findings could point to the already frequently assumed strong metal–silicon interactions in transition metal silicides [10].

Chlorine solubility is essentially restricted by the formation of silicon tetrachloride and, in the presence of hydrogen, of trichlorosilane. The suppression of these reactions, according to our calculations, must lead to the metastable ternary compounds "nickel silicide chlorides". This thesis is supported by earlier theoretical calculations [11], which indicated a possible existence of "calcium silicide bromides" [12], and furthermore by the preparation of "rare earth silicide iodides", "rare earth chlorid silicides" and "rare earth bromids silicides" [13].

References:

[1] H. Walter, *Ph.D. Thesis*, Technische Universität Bergakademie Freiberg, **1995**.

[2] K. Bohmhammel, G. Roewer, H. Walter, *J. Chem. Soc., Faraday Trans.* **1995**, *91*, 3879.

[3] H. Walter, G. Roewer, K. Bohmhammel, *J. Chem. Soc., Faraday Trans.* **1996**, *92*, 4605.

[4] J. Acker, K. Bohmhammel, G. Roewer, "*Thermodynamic and Kinetic Study of the Ni–Si–Cl–H System: The Relevance for the Catalytic Synthesis of Trichlorosilane*", in: *Proceedings of the Conference on Silicon for Chemical Industry IV* (Eds.: H. A. Øye, H. M. Rong, L. Nygaard, G. Schüssler), June 3–5, 1998, Geiranger, Norway, Institute of Inorganic Chemistry, Trondheim, **1998**, p. 133.

[5] A. D. Pelton, C. W. Bale, *Metallurg. Trans.* **1986**, *17A*, 1211.

[6] E. Königsberger, G. Eriksson, *CALPHAD* **1995**, *19*, 207.

[7] M. T. Swihart, R. W. Carr, *J. Phys. Chem. A* **1997**, *101*, 7434.

[8] J. Acker, *Ph.D. Thesis,* Technische Universität Bergakademie Freiberg, **1999.**

[9] K. Schwerdtfeger, J. Engell, *Trans. Met. Soc. AIME* **1965**, *223*, 1327.

[10] R. J. H. Voorhoeve, J. C. Vlugter, *J. Catal.* **1965**, 4, 220, and references cited therein.

[11] M. Jansen, "*Wege zu Festkörpern jenseits der thermodynamischen Stabilität*", in *Vorträge/ Nordrhein-Westfälische Akademie der Wissenschaften: Natur-, Ingenieur- und Wirtschaftswissenschaften; N 420*, Westdeutscher Verlag, Opladen, **1996**.

[12] H. Mattausch, A. Simon, *Angew. Chem.* **1998**, *110*, 498.

[13] H. Mattausch, O. Oeckler, A. Simon, *Z. Anorg. Allg. Chem.* **1999**, *625*, 297.

Author Index

A

B

C

D

E

F

G

H

I

J

K

L

M

N

O

P

R

S

T

U

V

W

Y

Z

Subject Index

A

B

C

N

O

P

Q

R

S